ATLASES IN GEOSCIENCE, 1

CROP PROJECT: Deep Seismic Exploration of the Central Mediterranean and Italy

ATLASES IN GEOSCIENCE, 1

CROP PROJECT: Deep Seismic Exploration of the Central Mediterranean and Italy

Edited by I.R. Finetti

UNIVERSITY OF TRIESTE

TRIESTE – ITALY

ELSEVIER

Amsterdam – Boston – Heidelberg – London – New York – Oxford – Paris –
San Diego – San Francisco – Singapore – Sydney – Tokyo

ELSEVIER B.V.
Radarweg 29
P.O. Box 211, 1000 AE Amsterdam
The Netherlands

ELSEVIER Inc.
525 B Street, Suite 1900
San Diego, CA 92101-4495
USA

ELSEVIER Ltd
The Boulevard, Langford Lane
Kidlington, Oxford OX5 1GB
UK

ELSEVIER Ltd
84 Theobalds Road
London WC1X 8RR
UK

First edition 2005

Library of Congress Cataloging in Publication Data
A catalog record is available from the Library of Congress.

British Library Cataloguing in Publication Data
A catalogue record is available from the British Library.

ISBN: 0-444-50693-4

⊗ The paper used in this publication meets the requirements of ANSI/NISO Z39.48-1992 (Permanence of Paper).

Printed in Italy

About the Editor

Icilio Renato Finetti, born in Parma (Italy), after graduating in Geology at the University of Bologna, entered the world of petroleum geophysics, working as a seismic interpreter in many projects for major international oil companies in various countries of Europe, Africa and the Middle East. Then, attracted by an academic career and scientific research, he joined the University of Trieste as assistant professor of Applied Geophysics and lecturer in Seismic Exploration. At the same time, as supervisor of the seismic division of OGS of Trieste, he promoted the adoption and employment of digital seismic techniques, directing both large-scale offshore seismic profiling projects for scientific purposes and exploration services for the oil industry. In 1974 I.R. Finetti was nominated General Director of OGS and in 1975 Full Professor of Terrestrial Physics at the University of Naples. In 1976 he was appointed to his current position as Chair Professor of Applied Geophysics at the Faculty of Science of the University of Trieste. During his academic career, I.R. Finetti has continued to cooperate with several major oil companies in seismic interpretation and basin studies in Europe, Asia, Africa, Northern and Southern America, the Atlantic margins, the Mediterranean Sea, Black Sea and North Sea, thus gaining world-wide experience. He is author of numerous papers dealing with large-scale seismic exploration studies, including monographs on the Mediterranean Sea, Black Sea and Ross Sea (Antarctica). The last seven years have been devoted full time to the CROP project and the preparation of the present Volume.

I.R. Finetti dedicates this Volume to his children Paola, Claudia and Marco.

Editorial Board

Preface

The Mediterranean Sea has been attracting the attention of numerous geoscientists from various countries and institutions for many years. A complex succession of geodynamic processes interplayed between Africa, Eurasia and the interposed ocean basins has characterized the geological history of the Mediterranean and surrounding regions, with ocean basin openings, plate and crustal block detachments, subductions, collisions, crustal extensions and contractions, arc and back-arc basin formation. In summary, in the Mediterranean area, on a regional scale, the imprint of all those significant plate-tectonics-related events observed on the larger global scale can be recognized.

The occurrence of various multi-staged tectono-dynamic processes in a relatively restricted region, where a deformative phase can be superimposed on previous ones thus masking them, have made it rather difficult to correctly discriminate, and realistically reconstruct the unfolding events.

Geophysical exploration and particularly the *up-to-date multifold seismic profiling* have contributed in a determinant way to understanding the Mediterranean region and its complex geodynamic evolution. Initial local geophysical campaigns were performed in the Mediterranean starting in 1965, before the advent of digital seismic techniques. In the late sixties, various geoscientists and institutions from various countries (Italy, France, UK, USA, Russia) initiated pioneering geophysical investigations of individual provinces of the Mediterranean Sea. In 1968 the Osservatorio Geofisico Sperimentale (OGS) of Trieste, the Italian state group for exploration geophysics, programmed and performed a pluriannual (1969–1980) organic large-scale project of geophysical exploration (seismics, gravity and magnetics) of the whole Mediterranean Sea area, from Gibraltar to the Levantine Sea. The success of this programme was clearly due to the up-to-date multifold seismic techniques employed, unusual at that time in seismic profiling for research. This project provided the geoscience community with data (*MS Seismic Lines*) that contributed significantly to improvement in knowledge of the Mediterranean Sea (see Chapter 1 of this volume). The MS seismic lines were programmed for a penetration ranging from 8 to 12 second (TWT). Such a record length for the basins floored by oceanic basement (Ionian–East Mediterranean, Tyrrhenian and Balearic Seas) was sufficient to obtain complete transcrustal information. In all other sectors, the MS line information was very useful but insufficient to image the deeper crustal intervals where key data are imprinted. Thus the necessity for an Italian deep crustal project became apparent and the CROP programme resulted (see Plate 1). Part of the CROP programme was conducted as an international cooperation: Italy–France (CROP-ECORS, across the Western Alps and Balearic Basin); Italy–Germany–Austria (Transalp, across the Eastern Alps); Italy–Switzerland (CROP-NFP 20, across the Central Alps). The record length of the CROP sections is 20 s in acquisition and 17 to 20 s in processing, with a generally good margin of penetration below the Moho discontinuity. In part of a section of the Southern Tyrrhenian Sea, the acquisition and processing was increased to 40 s in order to follow the Ionian slab. The acquisition of all marine sections and most of the land sections was done by OGS; the remaining land part by Geoitalia.

The data processing was conducted by OGS, AGIP and ENEL (Ismes). The University of Trieste (EGG: Exploration Geophysics Group of the Faculty of Sciences) performed a careful *Interpretative Reprocessing* of a large part of the marine sections, obtaining signal improvements of fundamental importance for understanding the complex features explored.

The interpretation of the CROP seismic network data (almost 10,000 km) and some other integrated sections was done by the Editor with geological contributions from geoscientists of Italian Universities. An extensive cooperation between numerous universities and other institutions produced the results presented in this volume (Universities of Trieste, Bari, Bologna, Cagliari, Catania, Chieti, Ferrara, Firenze, Milano, Napoli, Pavia, Palermo, Perugia, Siena, Granada, Lausanne and ETH Zurich; CNR of Bologna, Firenze and Roma; ENI-AGIP of Milano, and OGS of Trieste).

This volume consists of an organic ensemble of 34 chapters, each one dealing with a specific topic on Mediterranean geodynamics, or with the explorations results obtained. The many new deep CROP seismic imagings made possible both a new understanding of and the reconstruction of well-controlled settings for almost all those complex areas that previously could only be explained on the basis of various schemes, ideas, and suppositions.

The CROP Geoscience Group, coordinated by the Editor, in the preparation of this volume spent over seven years of continuous activity, including: Interpretative seismic reprocessing; 2D lithospheric seismic interpretation; literature analysis and selection; geological validation based on a large amount of available geophysical-geological data; borehole calibration; integration of MS and other seismic lines, and of gravity and magnetic data; refining of 3D lithospheric seismic interpretation; graphical homogenization; and chapter preparation. The 3D refining effort kept the interpretation group engaged for the last two years, but contributed notably to the understanding and reconstruction of various large-scale topics, that were initially difficult to comprehend from a single, local 2D section.

The interpretation of the CROP seismic data was conducted on sections to the scale 1:50,000 and presented in the volume to the reduced scale 1:300,000 with a vertical to horizontal scale ratio true for a velocity of 6000 m/s. To the scale 1:300,000 are also included the plates relative to the uninterpreted CROP seismic sections. This scale is rather small for expressing detailed seismic imagings, but the CD-ROM version of the whole volume and plates allows the reader to plot these fundamental geophysical documents to any desired scale.

Finally, can be said in summary that the results of the CROP project presented in this volume furnish Mediterranean geoscience community with a remarkable increase in information and most of the current hypotheses can be tested against illuminating seismic imagings and basic new data for further studies.

The Editor

Acknowledgements

The Editor is most grateful to the CNR (National Research Council) of Italy, to ENEL and to Eni-AGIP, the CROP partners, for the great honour bestowed by appointing him Scientific Director of the CROP Mare programme. Sincere thanks are due to Dr D. Fenati (Eni-AGIP), Dr S. Martinetti (ENEL) and Prof. A. Praturlon (CNR) of the first CROP Coordinating Committee, who managed the first phase of the CROP operations with greatly appreciated competence and authority. Much gratitude is also expressed to the succeeding Coordinating Committee members Dr G. Bolondi (Eni-AGIP), Dr S. D'Offizi (ENEL), Prof. G. Giglia (CNR), Dr L. Novelli (Eni-AGIP), Dr F. Benelli (Eni-AGIP), Dr A. Bernasconi (Eni-AGIP) and Dr L. Burbi (Eni-AGIP) for their willing cooperation during acquisition, processing and interpretation of the CROP data. Particular thanks are addressed to Dr G. Paulucci, Dr L. Salvador, Dr A. Muzzin, Dr R. Prato, and Dr L. Bertelli of Eni-AGIP for their support and trust during many phases of the CROP project.

Special mention is due to OGS of Trieste who performed the acquisition of the whole CROP Mare programme and did the larger part of the processing. Particular thanks are expressed to all the Seismic Parties of the MV OGS-Explora, which undertook the CROP recording, and to the OGS geophysicists who processed the major part of the CROP data. AGIP and ENEL, who processed part of the CROP sections, are also duly acknowledged.

The seismic Processing Centre of the Faculty of Sciences of the University of Trieste conducted an interpretative reprocessing of over 3,000 km of CROP Mare sections located on complex tectonic features. The improved signal quality obtained was determinant for understanding important key lithospheric features. The laborious graphical work for the preparation of the figures and plates of this Volume benefited from the much appreciated, collaborative efforts of Dr A. Prizzon and Dr E. Forlin of the University of Trieste. Dr P. Guidotti of PGS took care of the English language reviewing. Sincere gratitude is due to the numerous geoscientists who, during the preparation of the Volume, participated in many collegial discussions about the data obtained, thus contributing to improvements in the final results.

The Italian Ministry of University and Research MIUR, the University of Trieste, and the CROP Partners funded the preparation and publication of this Volume.

Icilio Renato Finetti
Editor

Contents

About the Editor v

Editorial Board vi

Preface vii

Acknowledgements ix

SECTION 1
PREVIOUS MAIN BROAD GEOPHYSICAL AND GEOLOGICAL EXPLORATIONS

1. Review of the Main Results of the Pioneering MS Seismic Exploration Programme in the Deep Water of the Mediterranean Sea (OGS, 1969–1980) 1
I.R. Finetti, A. Del Ben, E. Forlin, A. Prizzon

2. Review of Seismic Wide-Angle Reflection–Refraction (WARR) Results in the Italian Region (1956–1987) 31
R. Cassinis, S. Scarascia, A. Lozej

3. Gravity and Magnetic Fields of the Central Mediterranean Region 57
P. Bernardelli, C. Cavalli, R. Longoni, I. Giori

4. Bedrock Geology of the Tyrrhenian Sea Insight on Alpine Paleogeography and Magmatic Evolution of the Basin 69
R. Sartori

SECTION 2
CROP SEISMIC DATA ACQUISITION AND PROCESSING

5. CROP Seismic Data Acquisition, Processing and Interpretative Reprocessing 81
I.R. Finetti, E. Forlin, M. Pipan

SECTION 3
OPENING OF THE IONIAN AND ALPINE NEOTETHYAN OCEANS

6. Ionian and Alpine Neotethyan Oceans Opening 103
I.R. Finetti

SECTION 4
THE ALPS

7. Crustal Tectono-Stratigraphic Sections across the Western and Eastern Alps from ECORS-CROP and Transalp Seismic Data 109
I.R. Finetti

SECTION 5
THE APENNINES-CALABRIAN ARC-SICILIAN/MAGHREBIAN CHAIN

8. Lithospheric Tectono-Stratigraphic Setting of the Ligurian Sea–Northern Apennines–Adriatic Foreland from Integrated CROP Seismic Data 119
I.R. Finetti, M. Boccaletti, M. Bonini, A. Del Ben, M. Pipan, A. Prizzon, F. Sani

9. Crustal Geological Section across Central Italy from the Corsica Basin to the Adriatic Sea Based on Geological and CROP Seismic Data 159
I.R. Finetti, F. Calamita, U. Crescenti, A. Del Ben, E. Forlin, M. Pipan, A. Prizzon, G. Rusciadelli, V. Scisciani

10. Understanding the Ionides and their Geodynamics 197
I.R. Finetti

11. Ionian Tethydes in the Southern Apennines 209
G. Ciarapica, L. Passeri

12. Crustal Tectono-Stratigraphy and Geodynamics of the Southern Apennines from CROP and other Integrated Geophysical-Geological Data 225
I.R. Finetti, F. Lentini, S. Carbone, A. Del Ben, A. Di Stefano, P. Guarnieri, M. Pipan, A. Prizzon

13. Magnetotelluric Profiling along the CROP-04 Section in the Southern Apennines 263
D. Patella, Z. Petrillo, A. Siniscalchi, L. Improta, B. Di Fiore

14. Multiscale Derivative Analysis of the Gravity and Magnetic Fields of the Southern Apennines (Italy) 281
M. Fedi, F. Cella, G. Florio, A. Rapolla

15. Geological Outline of Sicily and Lithospheric Tectono-Dynamics of its Tyrrhenian Margin from New CROP Seismic Data 319
I.R. Finetti, F. Lentini, S. Carbone, A. Del Ben, A. Di Stefano, E. Forlin, P. Guarnieri, M. Pipan, A. Prizzon

16. Comparative NV and WA Seismic Modelling of CROP Sections in the Tyrrhenian Sea 377
A. Del Ben, L. De Luca, I.R. Finetti, E. Forlin, D. Luzio, M. Pipan, A. Prizzon, M. Vitale

17. The Calabrian Arc and Subducting Ionian Slab from New CROP Seismic Data 393
I.R. Finetti

SECTION 6
THE CORSO-SARDINIAN BLOCK

18. Crustal Tectono-Stratigraphic Setting and Geodynamics of the Corso-Sardinian Block from New CROP Seismic Data 413
I.R. Finetti, A. Del Ben, S. Fais, E. Forlin, E. Klingelé, L. Lecca, M. Pipan, A. Prizzon

SECTION 7
DEEP PALEO-NEO-BASINS

19. Crustal Tectono-Stratigraphy of the Ionian Sea from New Integrated CROP Seismic Data 447
I.R. Finetti, A. Del Ben

20. Lithospheric Tectono-Dynamics of the Balearic Basin Opening from CROP-ECORS Seismic Data 471
I.R. Finetti, E. Forlin, M. Pipan

21. Ionian Tethys Lithosphere Roll-Back Sinking and Back-Arc Tyrrhenian Opening from New CROP Seismic Data 483
I.R. Finetti, A. Del Ben

22. Geodynamic Significance of Tyrrhenian Seafloor Morphology 505
M. Marani

SECTION 8
ADRIATIC AND PELAGIAN FORELANDS

23. Crustal Tectono-Stratigraphic Setting of the Adriatic Sea from New CROP Seismic Data 519
I.R. Finetti, A. Del Ben

24. Seismic Modelling Applied to CROP Crustal Section in the Adriatic Sea 549
J.M. Carcione, I.R. Finetti, D. Gei

25. Mechanism and Timing of the Pliocene-Pleistocene Foredeep Migration in the Abruzzo Offshore Area (Central Adriatic) 563
S. Carruba, R. Casnedi, C.R. Perotti

26. Crustal Tectono-Stratigraphic Setting of the Pelagian Foreland from New CROP Seismic Data 581
I.R. Finetti, A. Del Ben

SECTION 9
THE MOHO OF THE CENTRAL MEDITERRANEAN REGION

27. Depth Contour Map of the Moho Discontinuity in the Central Mediterranean Region from New CROP Seismic Data 597
I.R. Finetti

SECTION 10
MAGMATISM

28. Triassic Magmatism and Jurassic Ophiolites at the Margins of the Adria Plate 607
L. Beccaluva, M. Coltorti, E. Saccani, F. Siena, O. Zeda

29. Cenozoic Tectono-Magmatic Evolution of the Central-Western Mediterranean: Migration of an Arc-Interarc Basin System and Variations in the Mode of Subduction 623
L. Beccaluva, G. Bianchini, M. Coltorti, F. Siena, M. Verde

30. Within-Plate Cenozoic Volcanism and Lithospheric Mantle Evolution in the Western-Central Mediterranean Area 641
L. Beccaluva, G. Bianchini, C. Bonadiman, M. Coltorti, G. Macciotta, F. Siena, C. Vaccaro

31. A Reappraisal of the Betic Ophiolitic Association: The Westernmost Relic of the Alpine Tethys 665
E. Puga

SECTION 11
GEODYNAMIC EVOLUTION

32. Evolutionary Reconstruction of the Mediterranean Region: Extrusion Tectonics Driven by the Convergence of Africa Arabia and Eurasia 705
E. Mantovani

33. Plate Tectonics of the Apulia–Adria Microcontinents 747
G.M. Stampfli

34. Geodynamic Evolution of the Mediterranean Region from the Permo-Triassic Ionian Opening to the Present, Constrained by New Lithospheric CROP Seismic Data 767
I.R. Finetti

Subject Index 777

Author Index 779

CROP PROJECT: Deep Seismic Exploration of the Central Mediterranean and Italy
Edited by I.R. Finetti

CHAPTER 1

Review of the Main Results of the Pioneering MS Seismic Exploration Programme in the Deep Water of the Mediterranean Sea (OGS, 1969–1980)

I.R. Finetti[1,*], A. Del Ben[1], E. Forlin[1], A. Prizzon[1]

ABSTRACT

At the end of the sixties, the OGS of Trieste began a quinquennial scientific program of regional seismic exploration in the whole Mediterranean deep water area, using the up-to-date digital multifold techniques which had just appeared in geophysical exploration for hydrocarbons, and which were rather unusual at that time in scientific profiling. The results obtained contributed to revealing most of the unusual tectono-stratigraphic settings of the explored provinces. Basin floored by *old* (*Ionian*) or *younger* (*Balearic, Tyrrhenian*) *oceanic crusts*, complex *thrust belts* (*Apennines, Calabrian and Hellenic arcs, Sicilian Apennines, Maghrebian chain*), *thick Mesozoic and/or Tertiary sedimentary basins, compressive, distensive or shear tectonics were all seismically imaged for the first time, allowing us to understand and reconstruct several features previously unexplored.* The ensemble of results from this pioneering Mediterranean exploration contributed to supply the geoscience community with relevant basic information for further studies. Even after many years this information conserves for the most part its validity and continues to be the support for further investigations.

The MS (Mediterranean Sea) seismic lines were programmed at a record length of 10 s (TWT), and therefore in the deep basins floored by oceanic crust (Ionian, Tyrrhenian, Balearic Seas) provide transcrustal imaging, while on continental or thrust-belt zones, a longer record length is needed to penetrate through the whole crust. In this context the Italian programme of deep seismic exploration (CROP) with 20 s (TWT) penetration was matured. This programme completed the lithospheric seismic exploration of the central Mediterranean region, and in this chapter we summarise the main results of the MS seismic line programme as information base for the new deeper CROP seismic exploration project.

Keywords: MS seismic lines Mediterranean, main pioneering results

1. INTRODUCTION

Until the beginning of the sixties, the Mediterranean basin, except for zones of the continental shelf and some other local shallow water areas, was almost completely unexplored geophysically, and its geological setting practically unknown. The first broad geophysical exploration campaigns for scientific purposes started in 1961 with bathymetric, gravimetric and magnetic surveys by OGS-Saclant, and continued until 1965. Soon after, various geoscientists and institutions of different countries (Italy, France, UK, USA, Russia) performed pioneering geophysical investigations of specific areas or given targets in the Mediterranean region. Supported by CNR (National Research Council of Italy), OGS (Osservatorio Geofisico Sperimentale) of Trieste in 1968 programmed a pluriannual regional exploration project over the whole Mediterranean deep water area, from Gibraltar to the Levantine Sea, started in 1969. From 1969 to 1973 a significant seismic network of over 22,000 km was acquired (Table 1). This wide-scale seismic exploration profiling was conducted using the advanced new digital multifold techniques which had just appeared in hydrocarbon exploration, but were unusual at that time in research geophysics. The seismic

[1]University of Trieste, Department of Geological, Environmental and Marine Sciences, Exploration Geophysics Group, Via Weiss 1, 34127 Trieste, Italy
*Corresponding author: Tel.: +39-40-5582276; e-mail: finetti@units.it (I.R. Finetti)

profiling was integrated with simultaneous bathymetric, gravimetric and magnetic surveys which generated a first, relatively detailed coverage of the entire Mediterranean and Aegean Seas (Figure 2). From 1975 to 1980, OGS continued its regional deep water geophysical program with the exploration of the Black Sea (in cooperation with Russian and Bulgarian Institutions in 1975) and of selected areas of the Mediterranean Sea (1977–1980), producing in total over 32,000 km of seismic lines, of which about 28,000 are located in the Mediterranean basin (Figure 1).

This *MS seismic lines project* provided the Mediterranean geoscience community with basic geophysical data that markedly contributed to disclosing the fundamental characteristics of this basin, to understanding the Mediterranean tectono-stratigraphy and to delineating the different crustal settings involved (Finetti et al., 1970; Finetti and Morelli, 1972, 1973; Finetti, 1976, 1982, 1984; Finetti and Del Ben, 1986). Many other geophysical contributions of relevant importance were made by other authors in this pioneering exploration stage, but the *MS lines project was, and remains today, the only scientific seismic network that regionally covers the whole Mediterranean deep water*. By reprocessing and up-to-date interpretation, the MS seismic lines have continued until present to provide useful information for various investigation targets.

The geophysical information produced by the MS lines and other projects gave rise to a proliferation of studies, proposals, interpretations, and hypotheses presented in a fast growing literature. One of the most impressive and fascinating disclosures was surely that of the young Tyrrhenian basin and its related opening geodynamics, which was subsequently interpreted in various ways: (a) a *back-arc basin related to slab subduction* (Boccaletti et al., 1971; Malinverno and Ryan, 1986; Finetti and Del Ben, 1986; Royden et al., 1987; Wortel and Spakman, 1992; Jolivet and Faccenna, 2000); (b) *lateral extrusion connected to Africa–Europe convergence* (Tapponnier, 1977; Boccaletti et al., 1982; Faccenna et al., 1996; Mantovani et al., 1997); (c) *mantle uprising with gravity-sliding effects* (Van Bemmelen, 1972; Locardi, 1982; Wezel, 1982; and with variations: Lavecchia and Stoppa, 1996; D'Agostino et al., 2001); (d) *late orogenic collapse of overthickened accretionary wedge* (Carmignani and Kligfield, 1990; Carmignani et al., 1994); and (e) *lithospheric delamination* (Roeder, 1980). Various other fundamental topics of the Mediterranean Sea were explained using the subsurface seismic imagings of the MS lines (and other seismic data), such as for example the *crustal type of the Ionian–East Mediterranean and Balearic basins; and the tectono-stratigraphic setting of the Mediterranean Ridge and its geodynamic significance*. The MS lines also strongly contributed to the discovery of many peculiar aspects, such as *the Messinian interval and connected halokinetic tectonics, contractional deformation, extensional and shear tectonics etc*.

In this Chapter 1 of the CROP volume, we summarise some of the important information supplied by the MS lines (10 s TWT penetration), much before the acquisition of the deep seismic lines of the CROP project (20 s TWT penetration).

2. THE MS SEISMIC LINES IN THE MEDITERRANEAN DEEP WATER

When the revolutionary multifold digital seismic reflection techniques entered into industrial application (1966) and for many years more, the seismic reflection profiling for scientific application operated exclusively with analog techniques, mostly single-fold (and single-channel), more rarely with multi-channel stacking techniques. OGS of Trieste, an Italian institution for exploration geophysics, in 1968 programmed a digital seismic profiling of the whole Mediterranean Sea deep water using these advanced multichannel stacking techniques. In most of the Mediterranean basin, affected by complex tectonics, thick sedimentary cover or reverberation effects, the results obtained with the advanced digital techniques were decidedly superior to analog profiling.

In Table 1 are reported the geophysical campaigns of the first five years of activity, for a total of 22,100 km of seismic line. Successively (1976–1980), additional programmes for given Mediterranean areas, where the 1969–1973 seismic network was insufficient to clarify certain topics, were programmed. The MS lines additional to the list in Table 1 (MS-65 to MS-116) amount to about 6,000 km. Thus from 1969 to 1980, in total, 28,000 km of seismic lines were acquired in the Mediterranean deep water.

3. SOME OF THE MAIN INFORMATION OBTAINED FROM THE MS SEISMIC LINES

Here we mention only some of the many new regional settings disclosed by the MS line network, and point out that the transcrustal CROP project matured after a significant previous seismic exploration phase of middle penetration (10 s TWT record length), conducted in the framework of the Italian national programme of marine geo-exploration.

Table 1
List of OGS's seismic reflection lines acquired in the Mediterranean deep water during the period 1969–1973

Year	Coverage (fold)	Area of exploration	Line (MS)	km
1969	12	Tyrrhenian	1–13	2920
1970	12	Straits of Sicily	14–17	3600
		and Ionian Sea	19–22	
1971	6	Southern Adriatic and Ionian Sea	24–33	3770
	12	Ionian Sea	23	230
1972	12	Southern Tyrrhenian and	5 Ext, 6 Ext, 8 Ext	4580
		Western Mediterranean	34–39	
			41–47	
	24	Tyrrhenian	18	500
	40	Western Mediterranean	40	
1973	12	Eastern Mediterranean	49–60	6280
		Ionian Sea	48	
1973		Ligurian Sea and Sardinian Channel	63–64	
	24	Tyrrhenian	61, 62	220
	Total			22,100

A set of selected self-explanatory seismic figures document better than long comments some of the characteristics of the explored Mediterranean basin. A presentation of selected MS seismic examples follows, subdividing the Mediterranean Sea into provinces characterized by different crustal and tectono-stratigraphic characteristics.

3.1. Western Mediterranean Sea

In the western Mediterranean–Ligurian Sea basin, several regional deep-water seismic lines were acquired during the pioneering exploration phase of the MS project (Figure 1). From among these two typical examples have been chosen (Figures 3 and 4): one at the eastern extremity of the Alboran Sea, and the other in the central Balearic basin, between NW Sardinia and the Gulf of Lyon. The first example (Figure 3) exhibits the stratigraphic-tectonic setting of the *Alboran basin* area, where the *Balearic rift stage* (Oligocene–Lower Miocene) stretched the zone, characterized by *thin continental crust*. Syn- and post-rift sequences, jointly with the Messinian–Plio-Quaternary successions are seismically well imaged.

The second example shows the complete cover sequence of the oceanic crust of the Balearic basin, which opened in the homonymous stage (Upper Oligocene–Lower Miocene). In the area of this seismic close-up, the *Messinian evaporite* gave rise to *salt domes* of relatively large dimensions which allow us to see clearly the crustal tectono-stratigraphy even on this unmigrated seismic section. Throughout most of the Balearic basin, where diffused small domes are present, strong diffraction noise severely masks the primary reflections, and only carefully migrated sections allow us to see the geological setting involved.

The oceanic basement (Z) is well evident in this section (Figure 4), while the Moho discontinuity (MO), masked by multiple events, is indicatively picked by assuming that the oceanic basement crust of the Balearic basin, as seismically observed, is rather constantly in the order of 1.1 to 1.3 s TWT thick.

3.2. Tyrrhenian Sea

Since the early sixties, the Tyrrhenian Sea has attracted numerous geoscientists from various countries and institutions. When seismic exploration started in the Mediterranean, the literature on this young basin became very prolific (Moskalenko, 1965, 1967; Finetti et al., 1970; Finetti and Morelli, 1973, 1974; Mulder, 1973; Bijù-Duval et al., 1974, 1977, 1978; Finetti, 1975, 1982; Fabbri and Curzi, 1979). It was soon evident that the Tyrrhenian represents a young basin in which the extreme morphology has relevant tectonic significance.

One of the most important topics concerning the Tyrrhenian in the seventies was the nature of the crust that floors the deepest flat part of the basin. The *seismic exploration* and *calibrating data from the DSDP* (Deep Sea Drilling

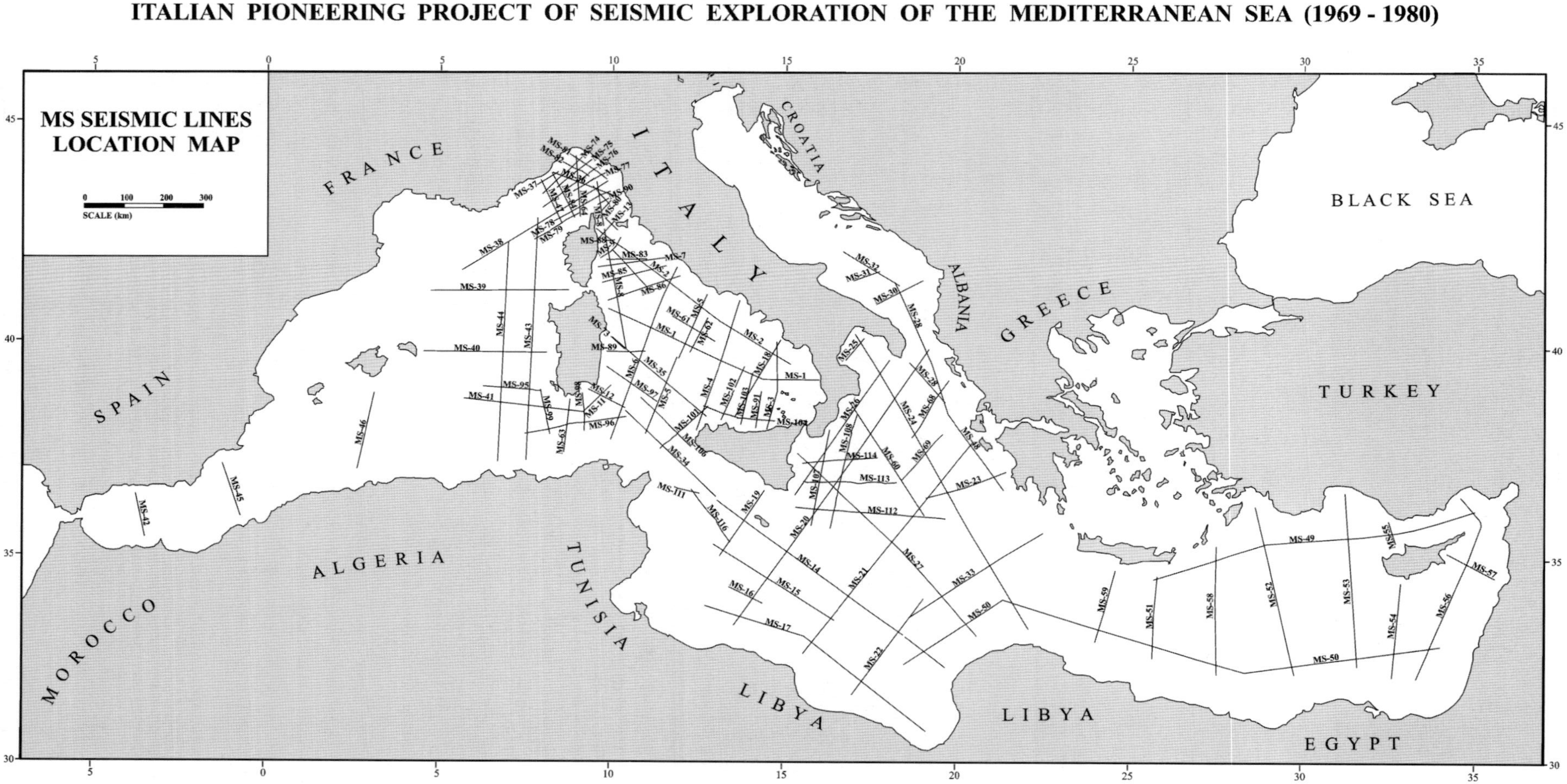

Figure 1: Location map for the MS (Mediterranean Sea) seismic lines, a regional seismic reflection profiling of the whole Mediterranean Sea deep water (1969–1980), supported by the National Research Council (CNR) of Italy, and performed by OGS of Trieste. In the framework of this deep water geophysical programme, with pioneering use of multifold digital seismic techniques for research, also the Black Sea was explored (seismic line positions not reported). A much more detailed bathymetric, gravimetric and magnetic network (see Figure 2) integrates the seismic information.

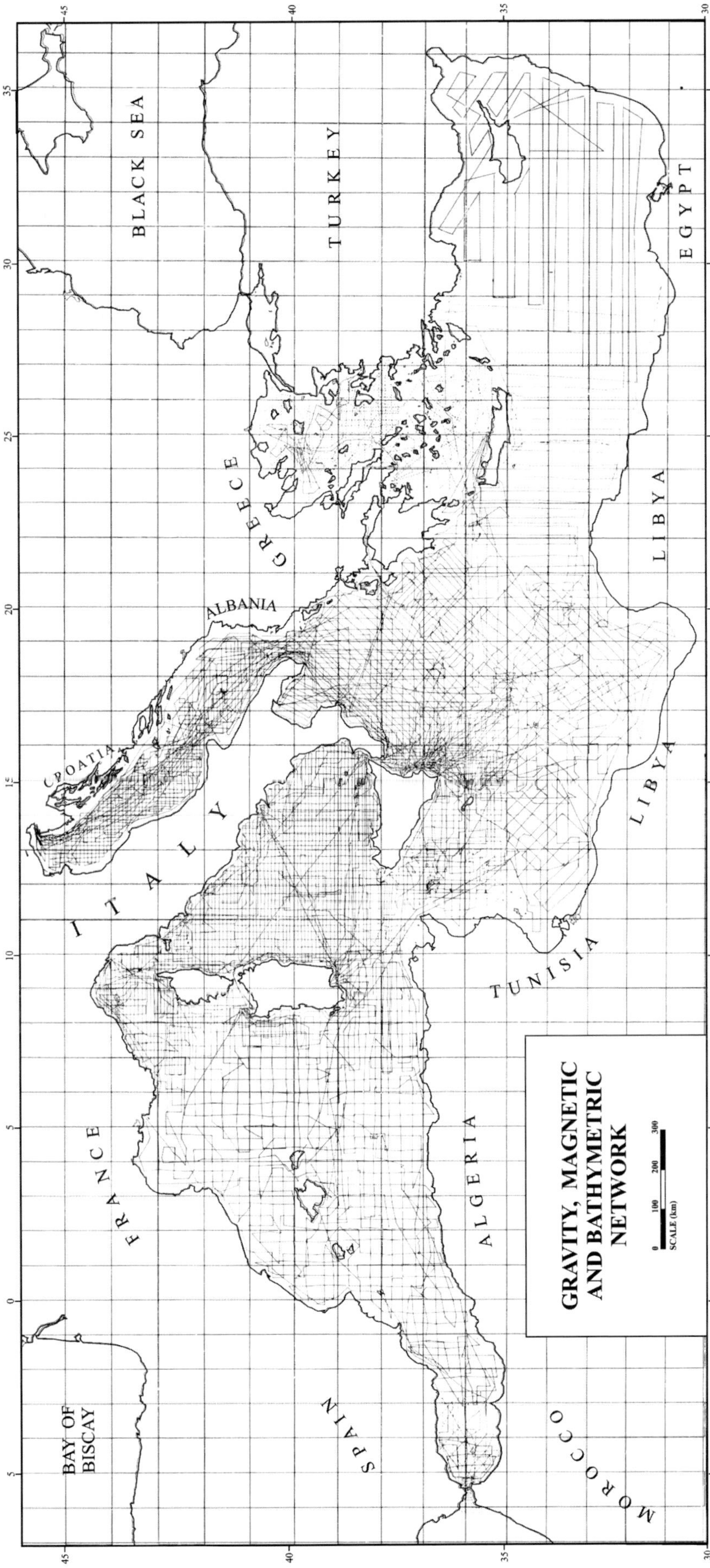

Figure 2: Location map for the gravimetric, magnetic and bathymetric profiling performed by OGS throughout the Mediterranean and Aegean Seas from 1965 to 1973. Part of the network also includes OGS-SACLANT lines (1961–1965) and Cambridge University's profiles in the Levantine Sea (1972).

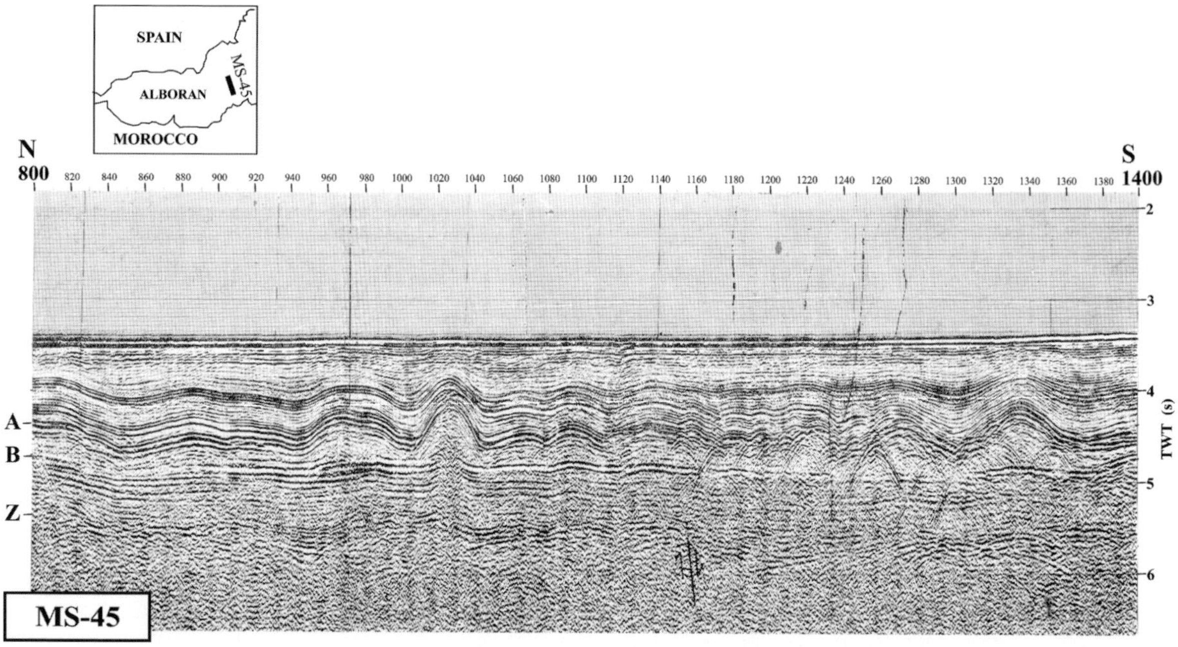

Figure 3: Example of an MS seismic line in the Alboran Sea. The characteristic Messinian evaporite interval "AB" and salt-dome tectonics are clearly evident. Beneath the flat base "B" of the Messinian salt, a pre-Messinian (Miocene–Upper Oligocene) sequence, over a continental substratum (Z), is well imaged seismically.

Programme) *showed that the deep Tyrrhenian is floored by a young oceanic crust.* A digital sonobuoy profiling (Finetti and Del Ben, 1986) exhibits very clearly the oceanic crust of the deep Tyrrhenian Sea (Figures 5 and 6). In Figure 6, in fact, the Moho reflection (RL) and refraction (RR) waves are recorded as good quality arrivals. These data, combined with NV-reflection data in the area allow us to reconstruct the seismically well-controlled crustal settings shown in Figure 7. The Moho is carefully calculated at to be a depth of 9 km. The basaltic basement interval is less than 5 km thick, and the Moho discontinuity is well evident also on NV-seismic sections (Figure 8). The basement cover (Quaternary-Pliocene in this Tyrrhenian zone) indicates the opening time of this back-arc basin sector.

The long line MS-1 passes through the whole Tyrrhenian Sea, from NE Sardinia to the Calabrian coast (Figure 1) and crosses major oceanic-type volcanoes: the *Vavilov volcano* (Figure 9) and the *Marsili volcano* (Figure 10), located within the homonymous Tyrrhenian basins, over thin oceanic crust. There is also a large volcanic body associated with a left-transcurrent fault: the *Palinuro volcano* (Figure 11). In this figure, the transcrustal shear fault separates a northern crust formed by allochthonous crustal fragments of the Corso-Sardinian block (*Calabrides*) from a southern crust of *young oceanic type* (Marsili basin, opened in Quaternary times).

Interpretation of the MS-line network, and the deep water borehole data allowed us to conclude that *the Tyrrhenian Sea is a back-arc basin which opened progressively from the Middle Miocene to Present in connection with outward migration of the Southern Apennines–Calabrian Arc–Sicilian Apennines orogenic belt.* In Figure 25 the tectonic setting of this belt and of its Tyrrhenian back-arc, reconstructed from the MS-line network, is outlined (Finetti and Del Ben, 1986).

On the northern Calabrian margin of the Tyrrhenian Sea, a thick Plio-Quaternary basin (*Paola basin*) was discovered by the MS-1 seismic line (Figure 12). Further investigations showed that this basin grew on a prominent crustal synclinal fold where the substratum is formed by a thrust-contact between Calabrides units (coming from the eastern Sardinian margin) and an allochthonous crustal fragment coming from the NW-Adria plate margin (Finetti, this volume, Chapter 17).

3.3. Ionian Sea

One of the most important topic of the Mediterranean Sea, remained under discussion for a long time, was definition of the Ionian Sea crust type (and East Mediterranean). Until the early eighties, and even several years later, part

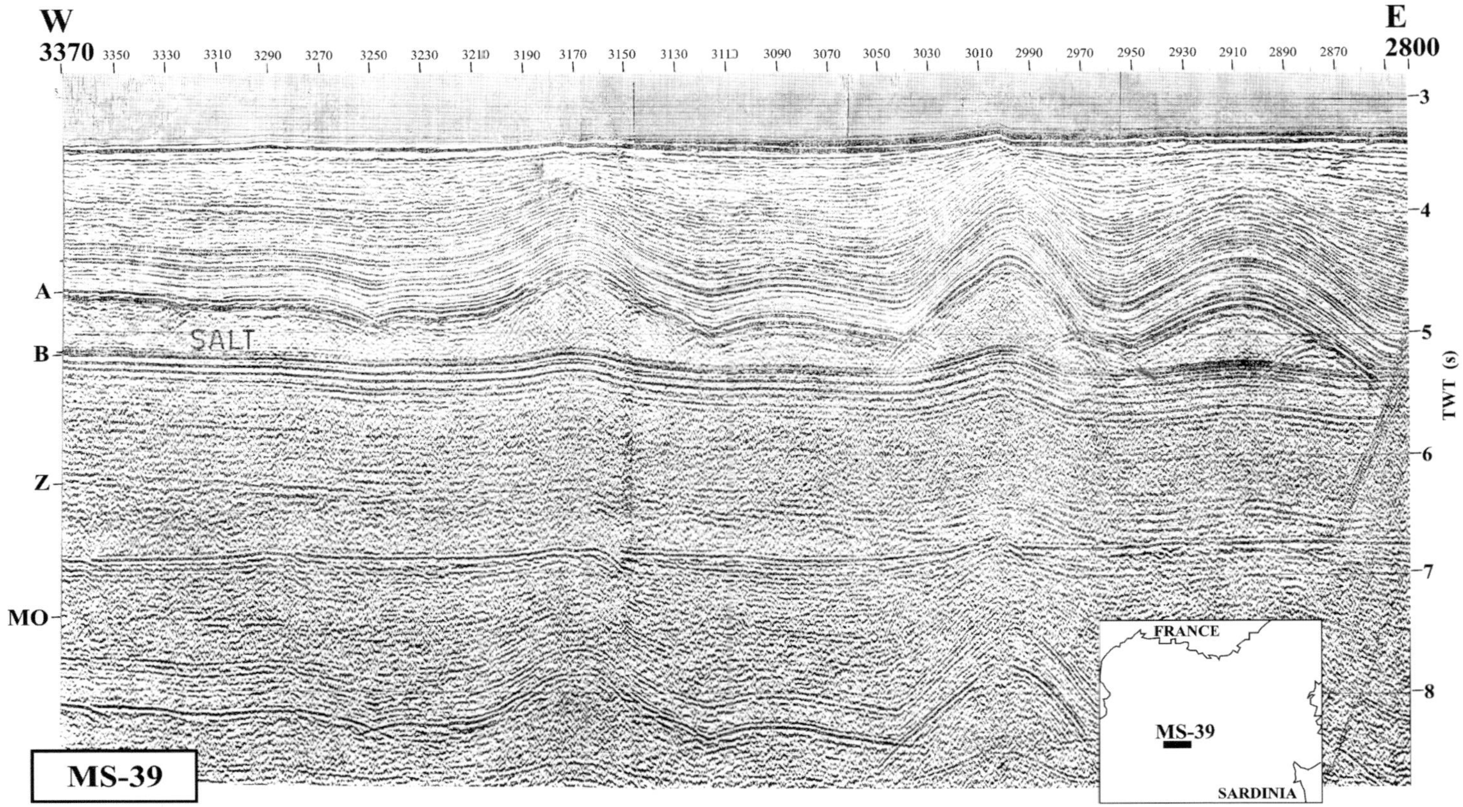

Figure 4: Example of an MS seismic line in the Balearic basin. This section exhibits clear salt dome tectonics, connected to halokinetic movements of the "AB" Messinian interval. The pre-Messinian Miocene–Upper Oligocene sequence (interval "BZ") is also shown with good reflectivity. The Moho discontinuity (MO) of the Balearic oceanic crust (indicatively interpreted around 7.4 s) is completely masked by unremoved multiples.

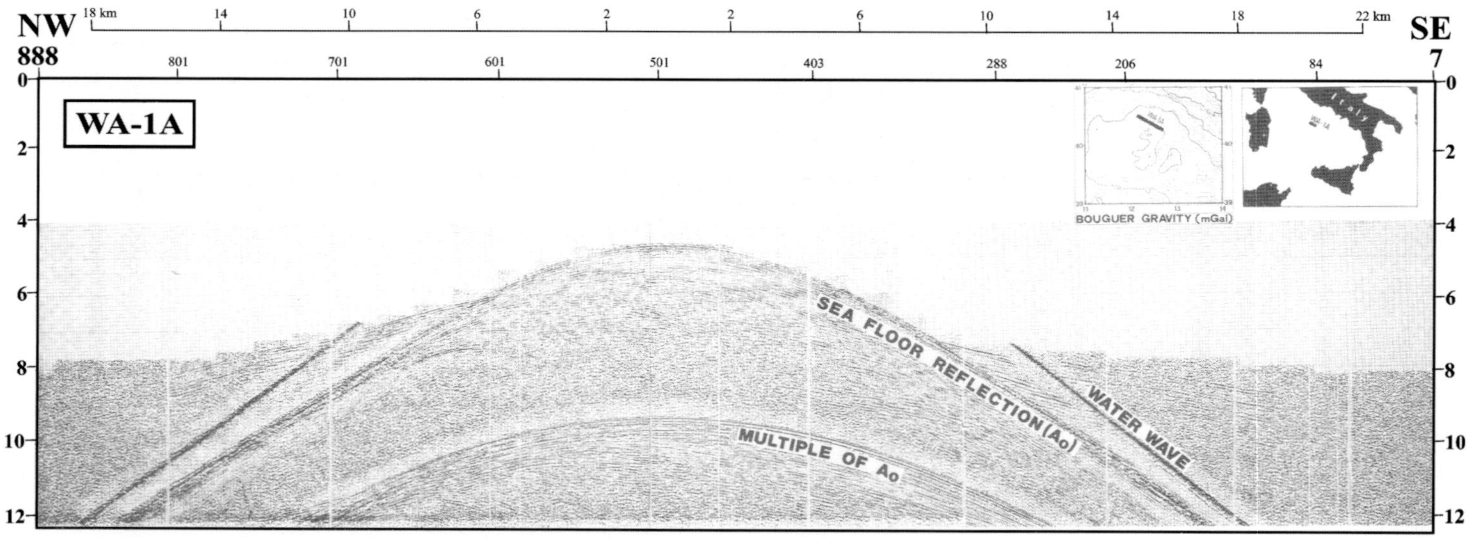

Figure 5: Multichannel-multifold wide-angle (WA) digital sonobuoy profiling in the deep Tyrrhenian Sea (Vavilov basin). To better see the relevant reflected-refracted crustal seismic events for establishing the type of crust in the explored area, it is convenient to examine the more expanded half-profile in Figure 6.

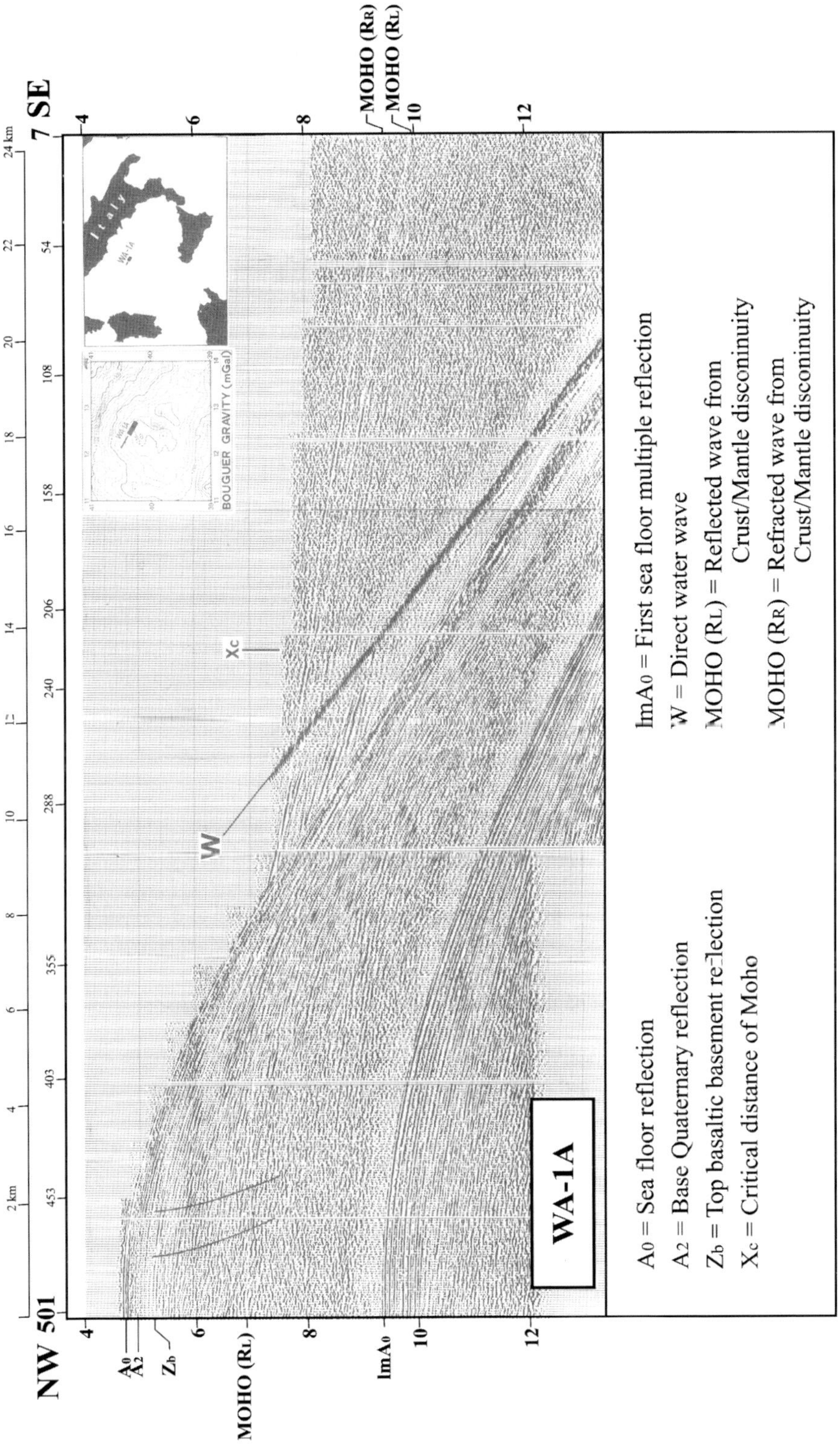

Figure 6: Interpreted half-sonobuoy profile in the Tyrrhenian Sea (Vavilov basin). This sonobuoy section is the southeastern half of Figure 5 (after Finetti and Del Ben, 1986).

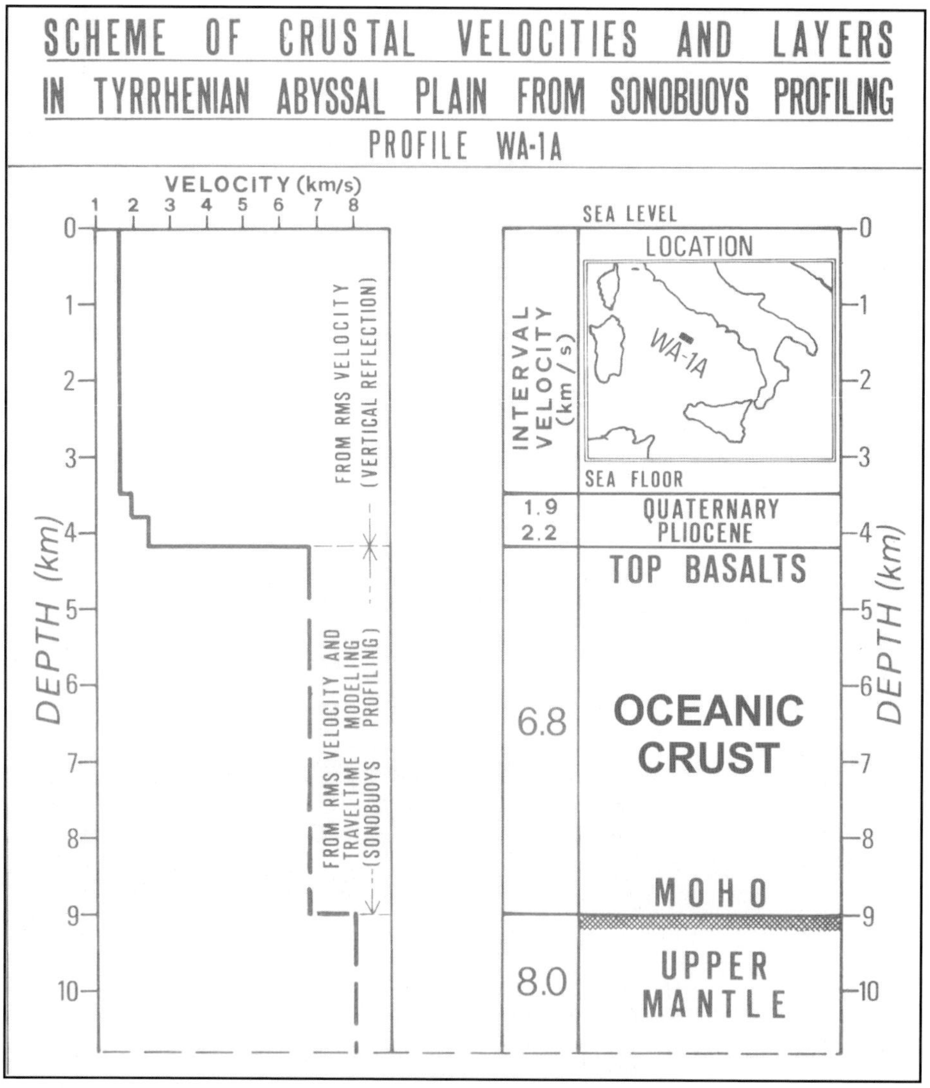

Figure 7: Velocity function and crustal intervals of the Tyrrhenian Sea (Vavilov basin) obtained from NV-reflection and sonobuoy seismic multichannel-multifold profiling.

of the literature considered the Ionian Sea to have a continental crust which deepened in recent times (Scandone, 1979; 1982; Bonardi et al., 1982; Dietrich, 1988; Vai, 1992). But the *MS seismic lines, in the eighties, showed that the Ionian Sea is floored by an old oceanic crust* (Finetti, 1982). The first geophysical document that exhibited, in 1982, the Ionian oceanic crust (interpreted on seismic line MS-26) is reported in Figure 13. At that time, on the basis of both a wide geophysical study and data on relevant magmatic activity in the Triassic and Middle Jurassic in the Hyblean–Maltese escarpment and eastern Pelagian Sea area, the author interpreted the Ionian Sea as being formed by initial rifting in the Early Triassic and spreading in the Middle Jurassic (Finetti, 1982).

Successively, further investigations (Finetti and Del Ben, 1986; Finetti et al., 1996) led to the conclusion that *the Ionian Sea opened in Permo-Triassic times*. The CROP seismic network not only confirmed such an Ionian opening time, but also provided clear seismic images of the Ionian slab subducting under the Calabrian and Hellenic arcs and allowed us to better identify the condensed stratigraphy covering the oceanic crust (Figures 14, 15 and 17). The Ionian Sea is extensively described and illustrated with various CROP sections by Finetti and Del Ben (this volume, Chapter 19).

3.4. Calabrian and Hellenic Arcs

The MS seismic lines explored the Calabrian arc offshore on its Ionian and Tyrrhenian sides, and the Hellenic arc on its Ionian–East Mediterranean wedge, over 20–30 years ago when the geological setting of these areas was

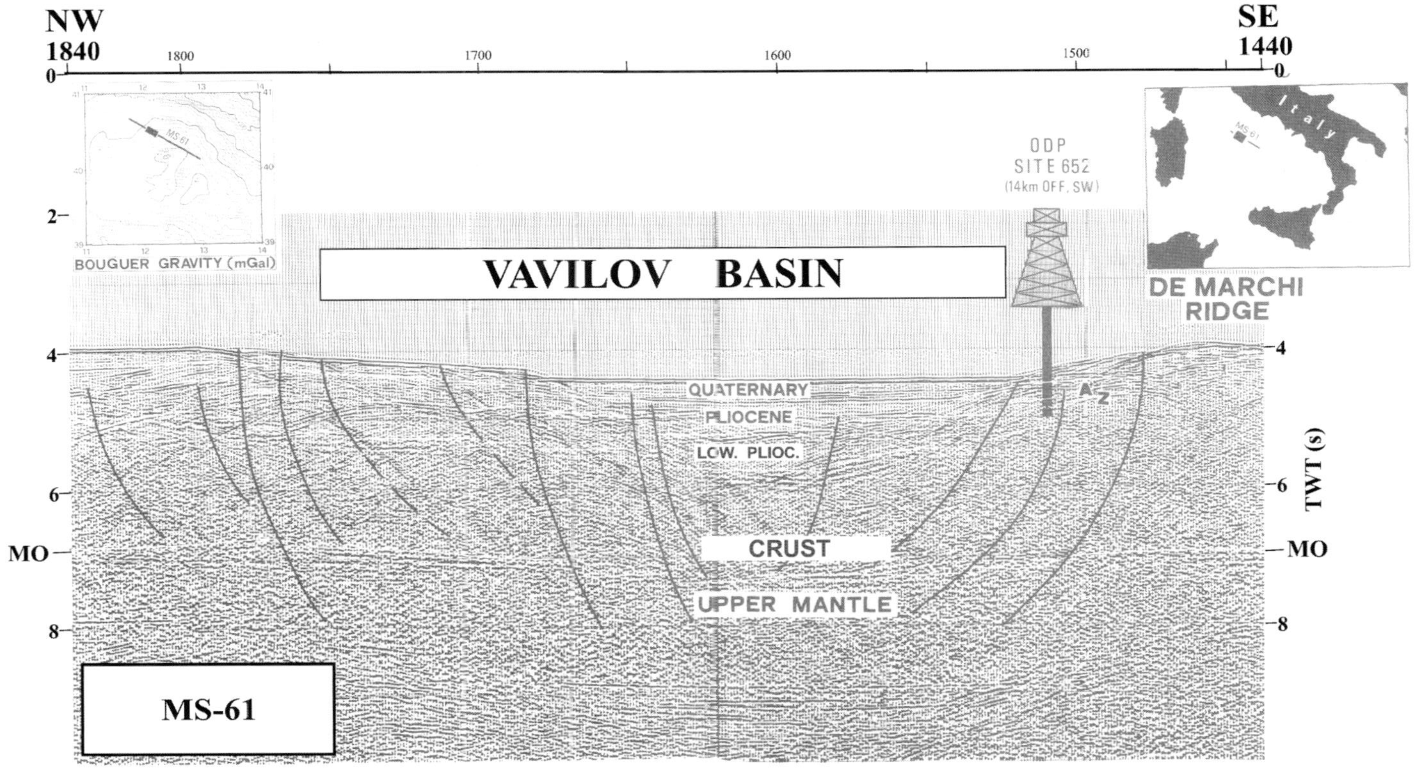

Figure 8: NV-reflection MS seismic section in the deep Tyrrhenian Sea acquired at the same position as the sonobuoy profile in Figure 5. This figure shows a very clear, flat Moho reflector (MO) at about 7 s (TWT).

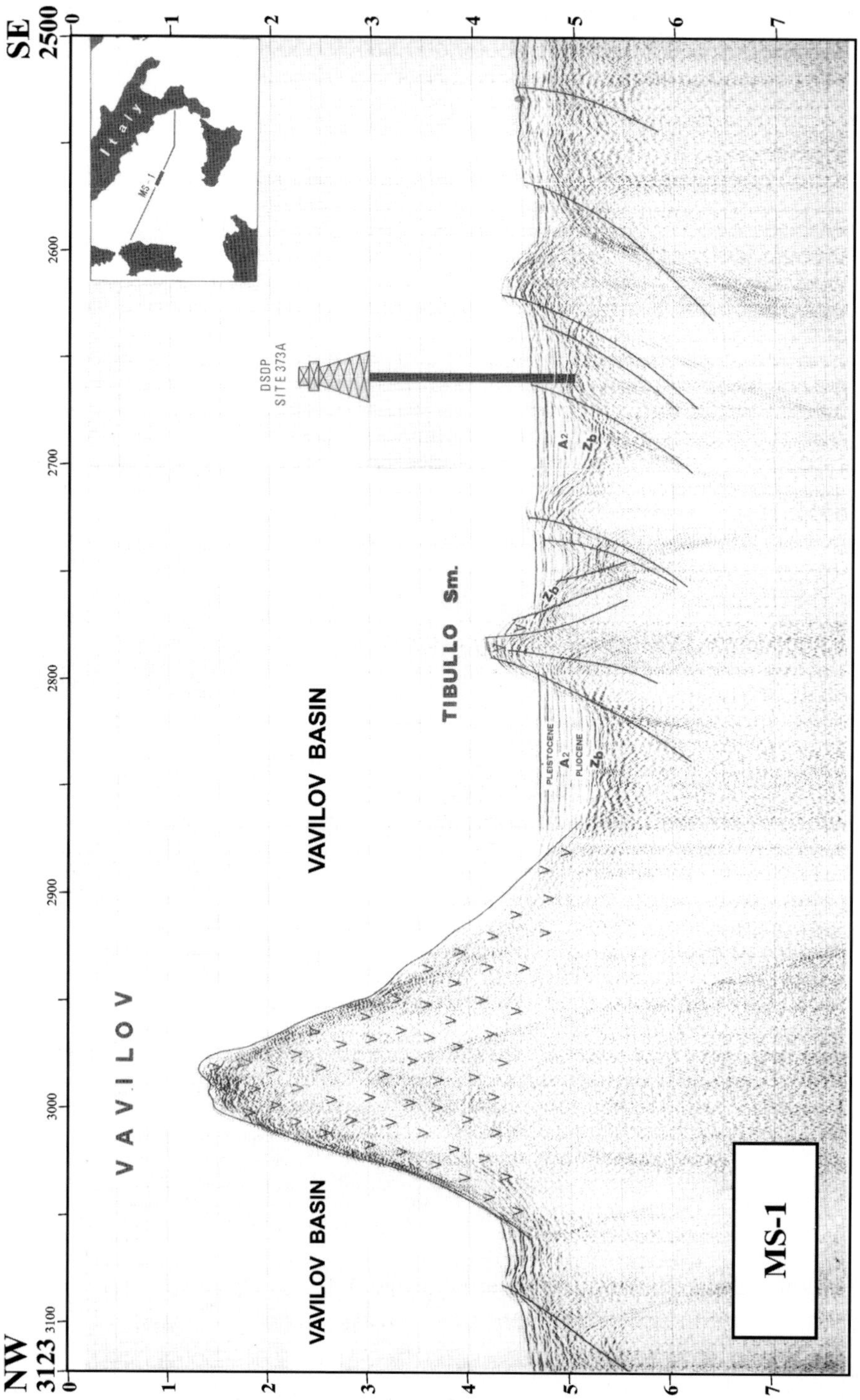

Figure 9: Interpreted seismic close-up of line MS-1, across the *Vavilov volcano*, in the Tyrrhenian Sea (after Finetti and Del Ben, 1986).

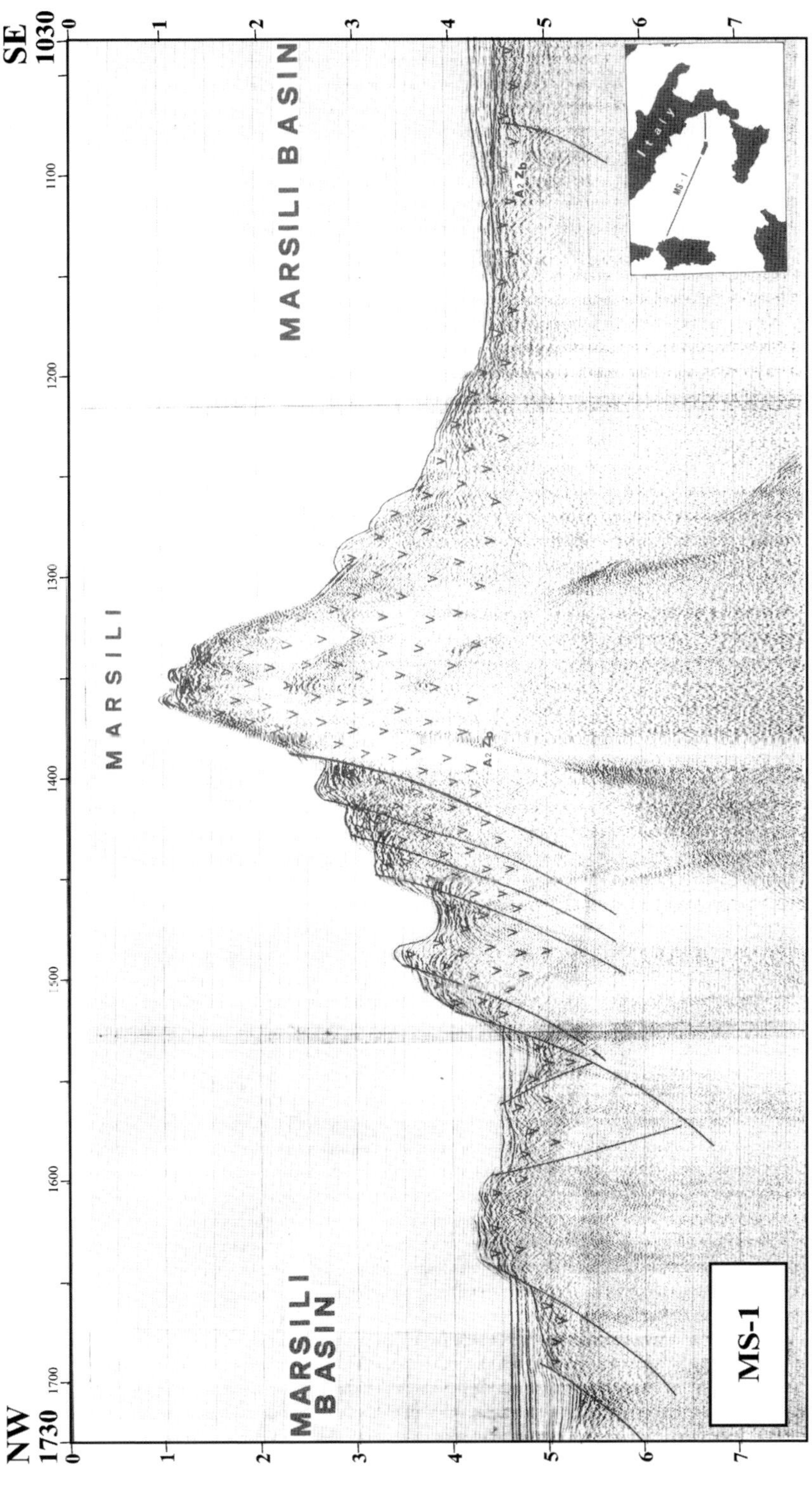

Figure 10: Interpreted seismic close-up of line MS-1 across the *Marsili volcano*, the largest Tyrrhenian volcano of oceanic type, located within the homonymous basin on thin oceanic crust (after Finetti and Del Ben, 1986).

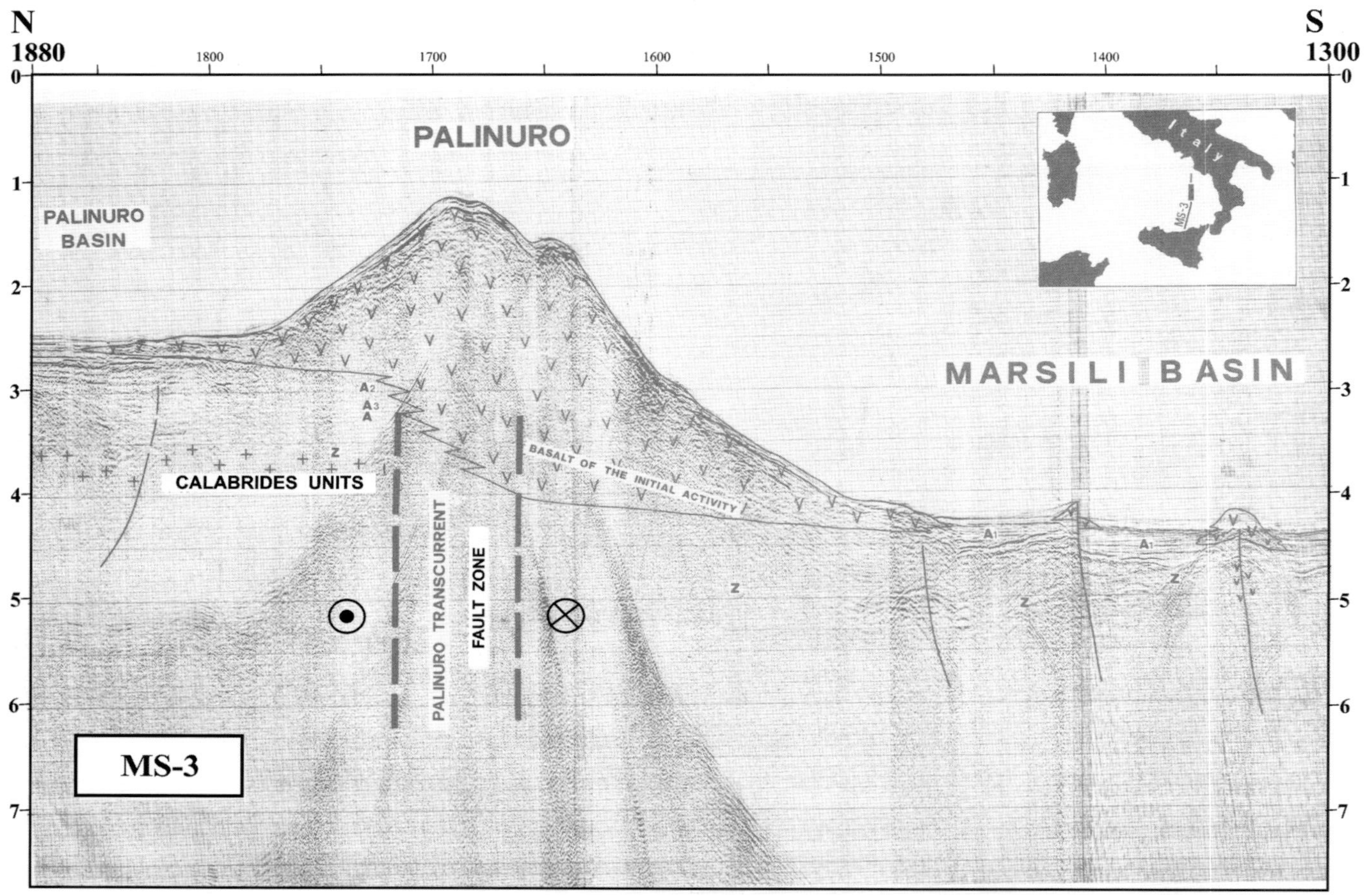

Figure 11: Part of seismic line MS-3 crossing the *Palinuro volcano* connected to the homonymous left-transcurrent lateral fault (fault location given in Figure 25). This fault separates a southern sector of oceanic crust (Marsili basin) from the northern part (Palinuro basin) where the crust is composed of a stack of allochthonous units coming from the Sardinian block (after Finetti and Del Ben, 1986).

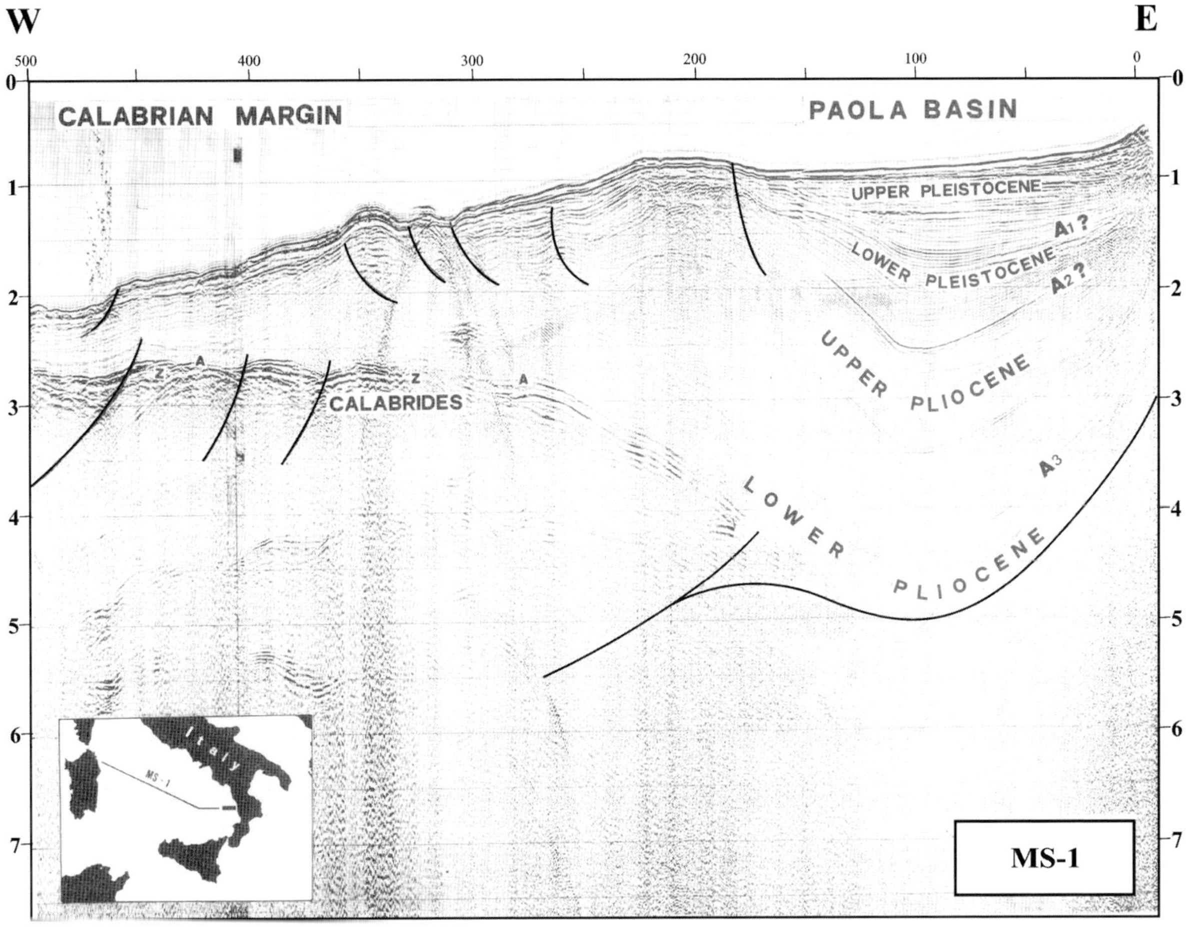

Figure 12: Part of section MS-1 crossing the thick *Paola basin* on the northern Calabrian margin (after Finetti and Del Ben, 1986). This Plio-Quaternary sequence fills a prominent *crustal synclinal fold* generated in the Tyrrhenian geodynamic stage (see Finetti, this volume, Chapter 17).

substantially unknown and highly questioned in the literature. The obtained results were a pioneering exposure of the peculiar characteristics of these orogenic arcs that allowed a progressive elucidation in a correctly controlled setting. It is to be noted that the MS seismic lines, programmed at 10 s (TWT) penetration, in the Tyrrhenian and Ionian oceanic basins cross the whole crust and enter the upper mantle, thus integrating the successive deep CROP sections with comparable information.

In Figure 14 is exhibited a good seismic close-up of the line MS-60 showing the frontal wedge of the Calabrian arc and the Ionian crustal stratigraphy. This outermost Calabrian zone does not deform the condensed Ionian cover of the I–TO interval (Permo-Triassic to Oligocene), but only the Miocene to Quaternary. Going inward the shearing plane deepens so as to detach and imbricate also the I–TO interval. The oceanic basement crustal interval OC deepens progressively below the Calabrian wedge, and in the Tyrrhenian coast zone of the emerged Calabrian arc becomes a subducting denuded slab.

Figure 15 (MS-50, part) reproduces a crustal setting quite similar to that of the previous MS line example at the contact zone between the Hellenic wedge and the Ionian abyssal plain. This transcrustal seismic line shows good reflectivity and provides a clear tectono-stratigraphic setting for the Ionian oceanic crust and its cover. The Messinian interval AB is strongly thickened by the snowplough effect of the outward-migrating Hellenic arc over the horizon B. From the outer sector exhibited in the figure to the Hellenic trench zone, the shearing plane deepens progressively to the basaltic basement TO. The sector of the Hellenic arc internal to the trench is well imaged, on the Ionian side, by the MS seismic example in Figure 16. Clear contractional tectonics and repeated imbrication of thrust-blocks are evident. Another clear seismic imaging of the Ionian Sea–outer Hellenic wedge zone is supplied by the line segment of Figure 17 (MS-33), where the Messinian interval AB, the pre-Messinian Miocene BI, and the condensed Ionian sequence I–TO (Permo-Triassic to Oligocene) are seismically interpreted.

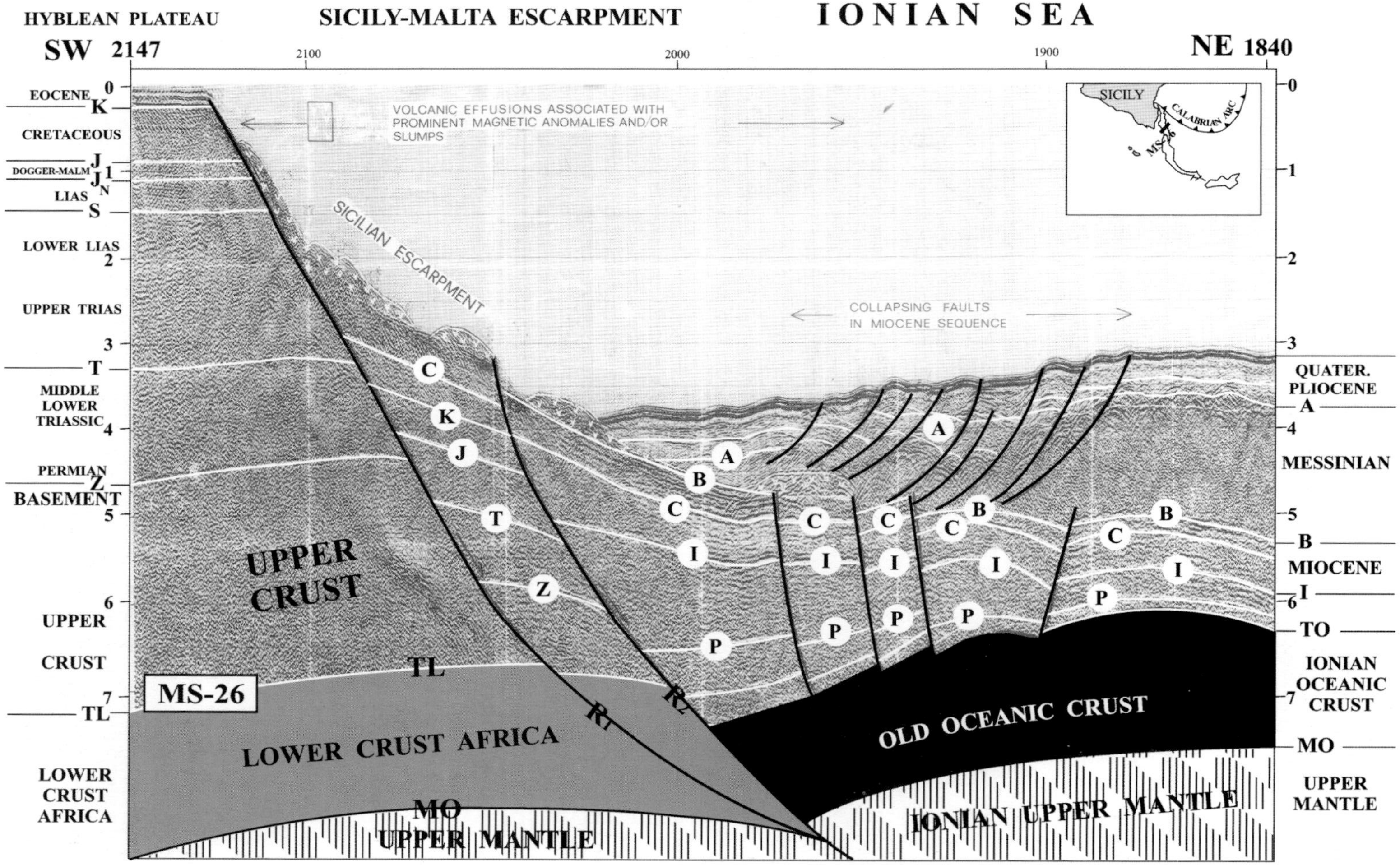

Figure 13: ***First geophysical document in the literature revealing that the Ionian Sea is floored by an old oceanic crust*** **(after Finetti, 1982; revised and implemented). A = base Plio-Quaternary; B = base Messinian; C = within Miocene; I = top condensed Ionian sequence (Permo-Trias to Oligocene); P = top basal Trias-Late Permian sequence; TO = top oceanic crust; MO = Moho. K, J, JN, S, T = horizons within Mesozoic cover sequence of the Hyblean plateau (African crust). TL = top lower crust Africa. R1, R2 = listric rift-faults of Ionian opening, reactivated by loading effects of the Calabrian wedge. Messinian: plastically thickened by snowplough-effect of Calabrian arc propagation.**

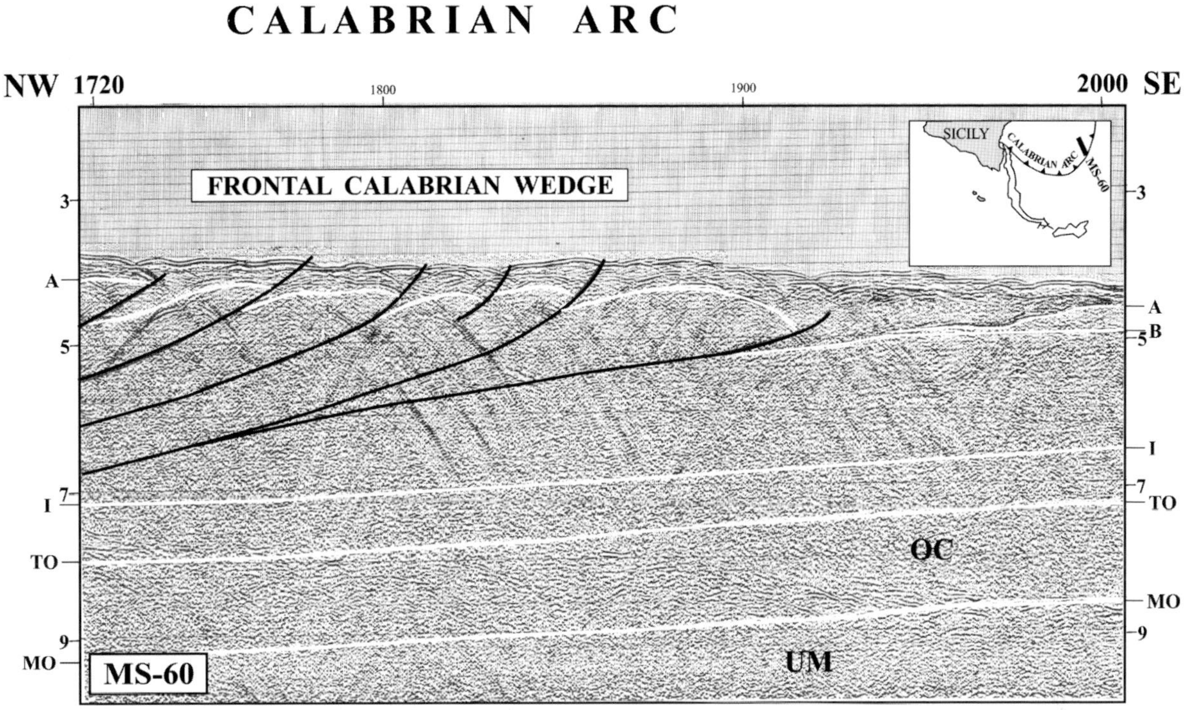

Figure 14: Seismic line example MS-60 (part) on the Ionian Sea floored by old oceanic crust, at the frontal zone of the Calabrian wedge. A = base Plio-Quaternary; B = base Messinian; I = top Oligocene; TO = top oceanic crust; MO = Moho. The interval "I–TO" is a condensed sequence deposited from the Permo-Triassic Ionian opening to the Oligocene. Interval OC is the oceanic basement crust. UM = upper mantle.

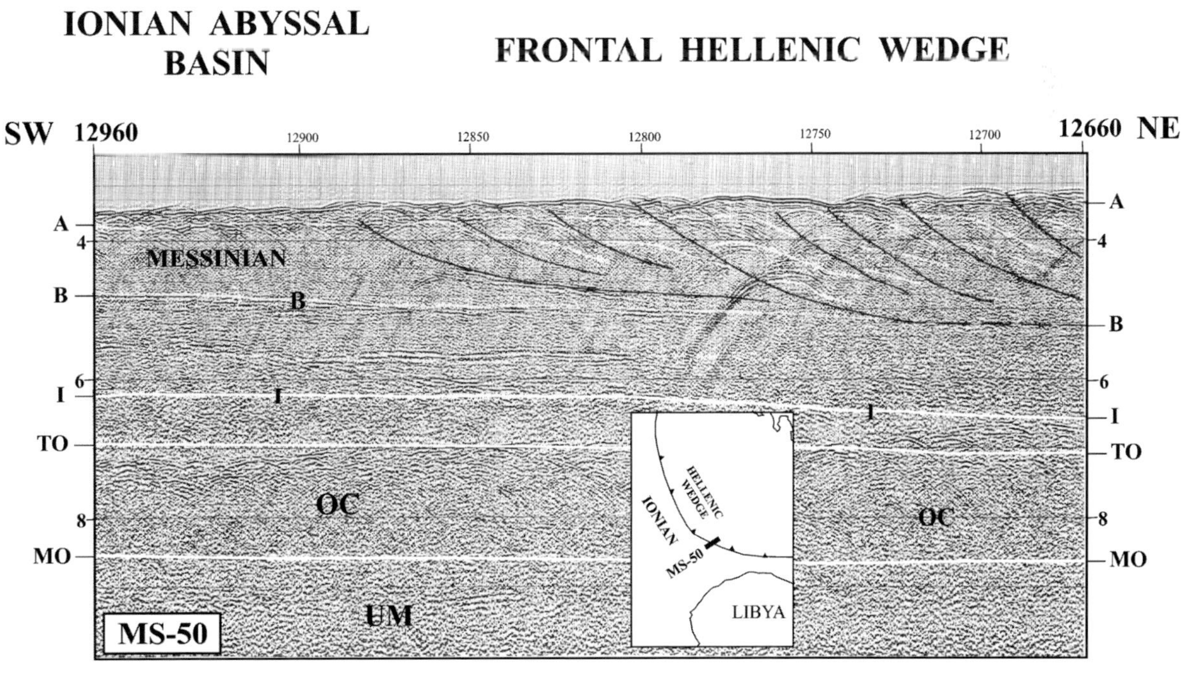

Figure 15: Seismic line example MS-50 (part) on the Ionian Sea at the frontal zone of the Hellenic wedge. A = base Plio-Quaternary; B = base Messinian (thickened by outward arc migration); I = top Oligocene; TO = top oceanic basement; MO = Moho; OC = oceanic crust; UM = upper mantle.

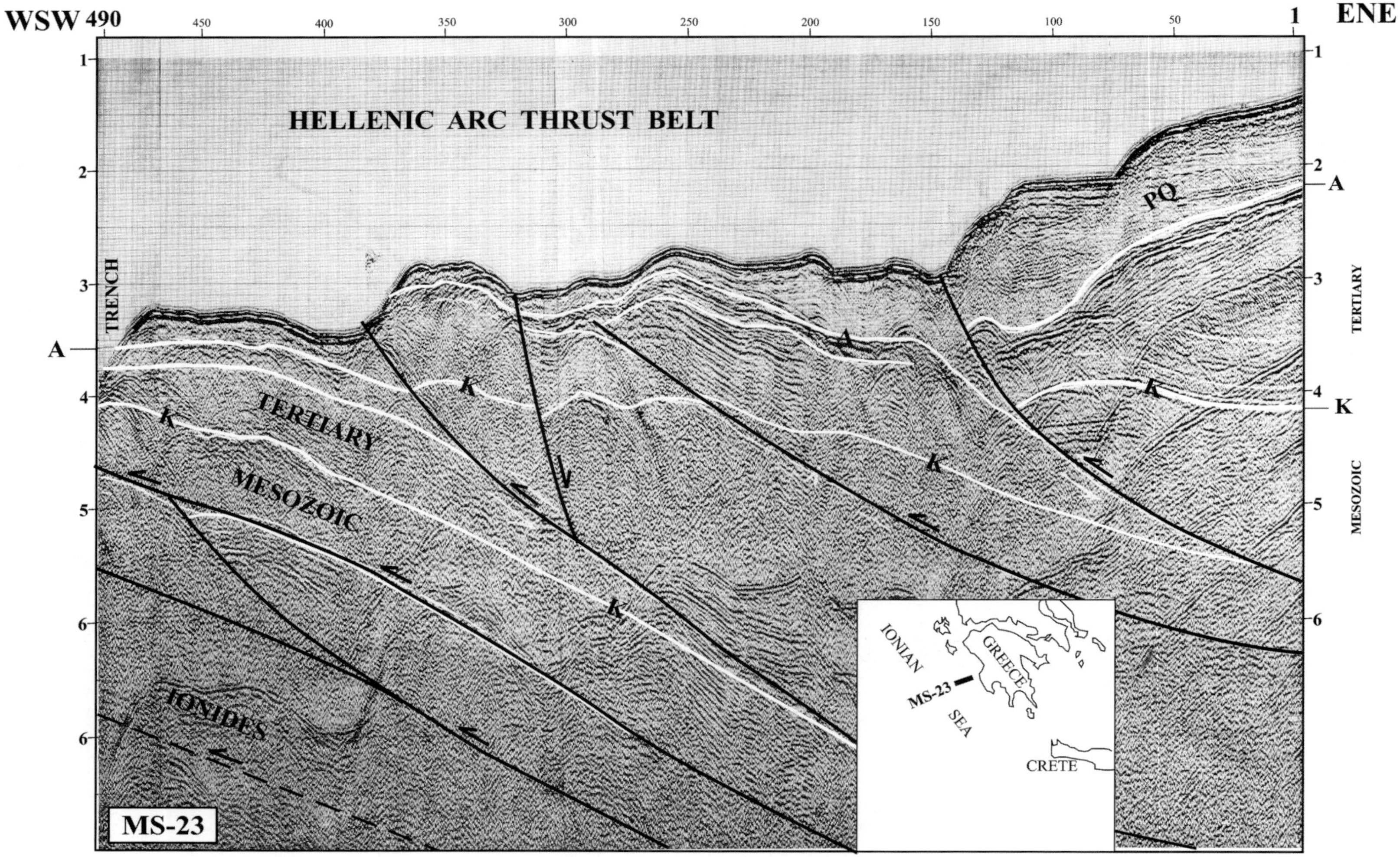

Figure 16: Example of MS seismic line across the Hellenic or Aegean arc. There are evident imbricated thrust-blocks. The trench zone separates an internal orogenic sector (exhibited in this figure) formed by imbrication of continental thrust-blocks of the Hellenic arc, from an external sector (known in the literature as the "*Mediterranean Ridge*") mostly formed by repeated imbrication of *Ionides* (condensed Ionian cover sequence; see Finetti, this volume, Chapter 10) which constitute a significant part of the *wedge of the Hellenic and Calabrian arcs*.

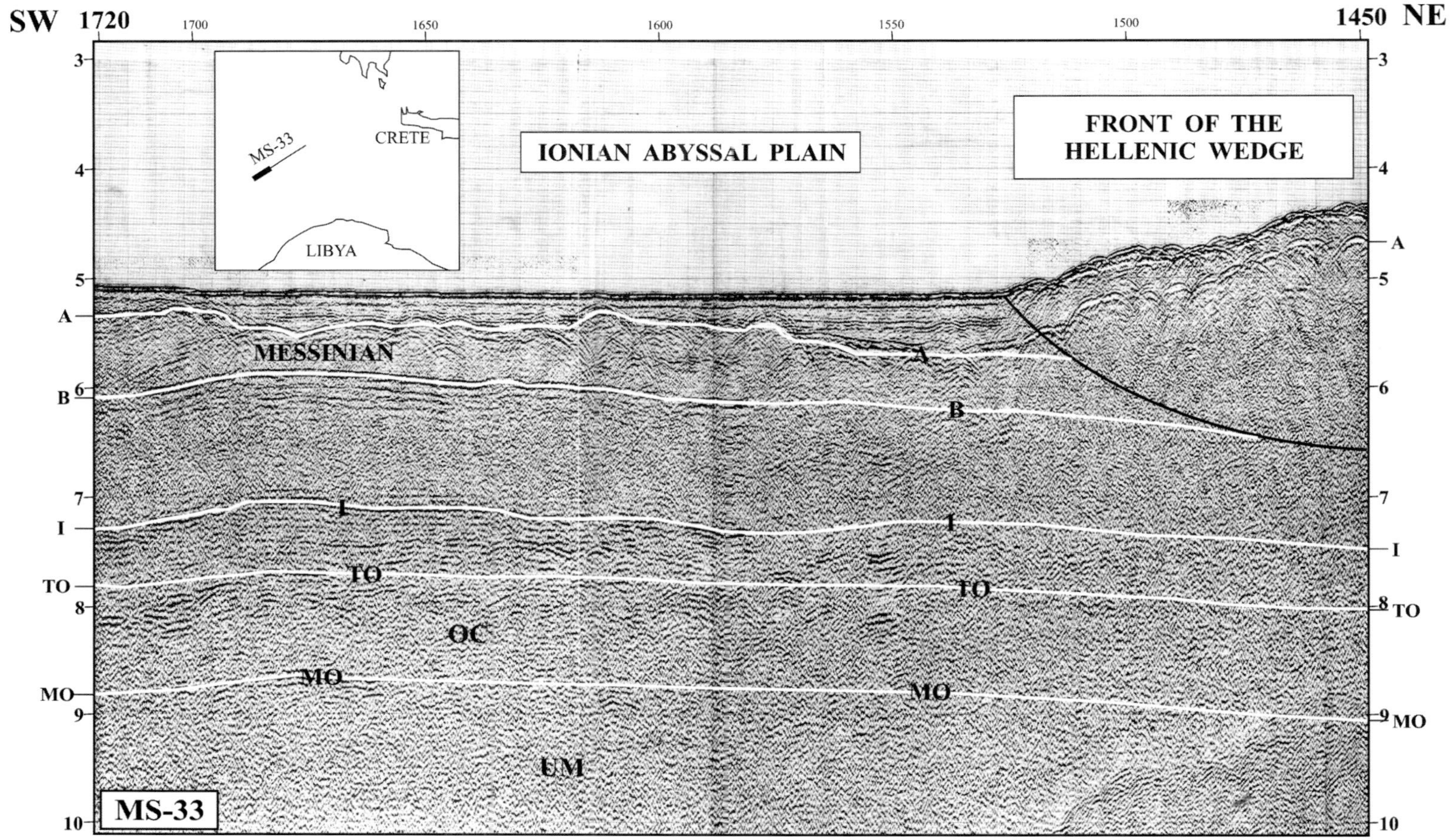

Figure 17: Example of MS seismic line in the Ionian Sea, across the abyssal plain and the frontal accretionary wedge of the Hellenic arc (line MS-33, part). Explanation of picked horizons is as in Figure 15.

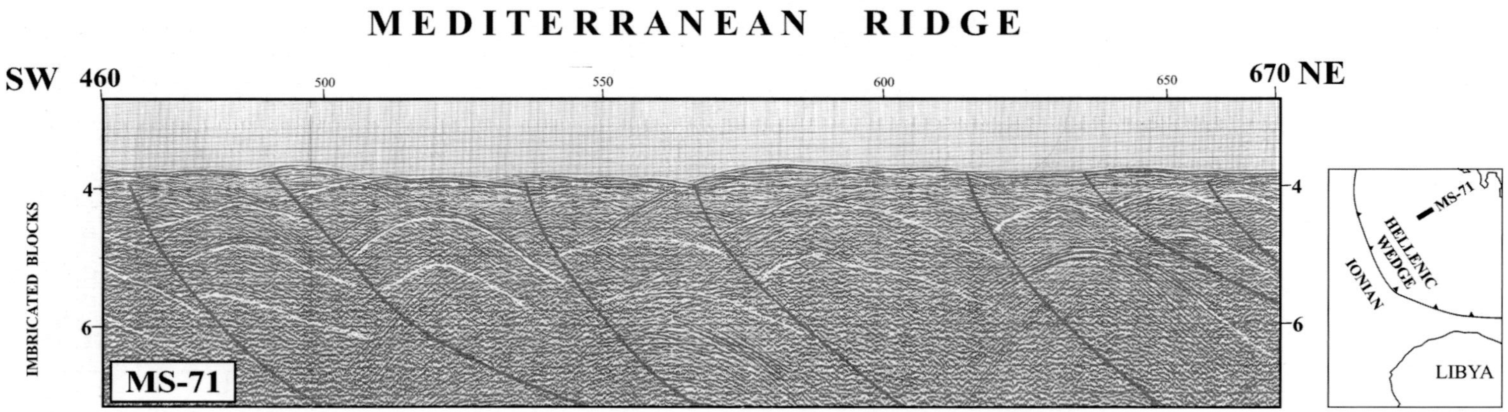

Figure 18: Example of MS seismic line (MS-71, part) on the long-discussed Mediterranean Ridge. ***The MS lines have shown that the Mediterranean Ridge is a wedge feature associated with the Hellenic arc*** **(Finetti, 1976).**

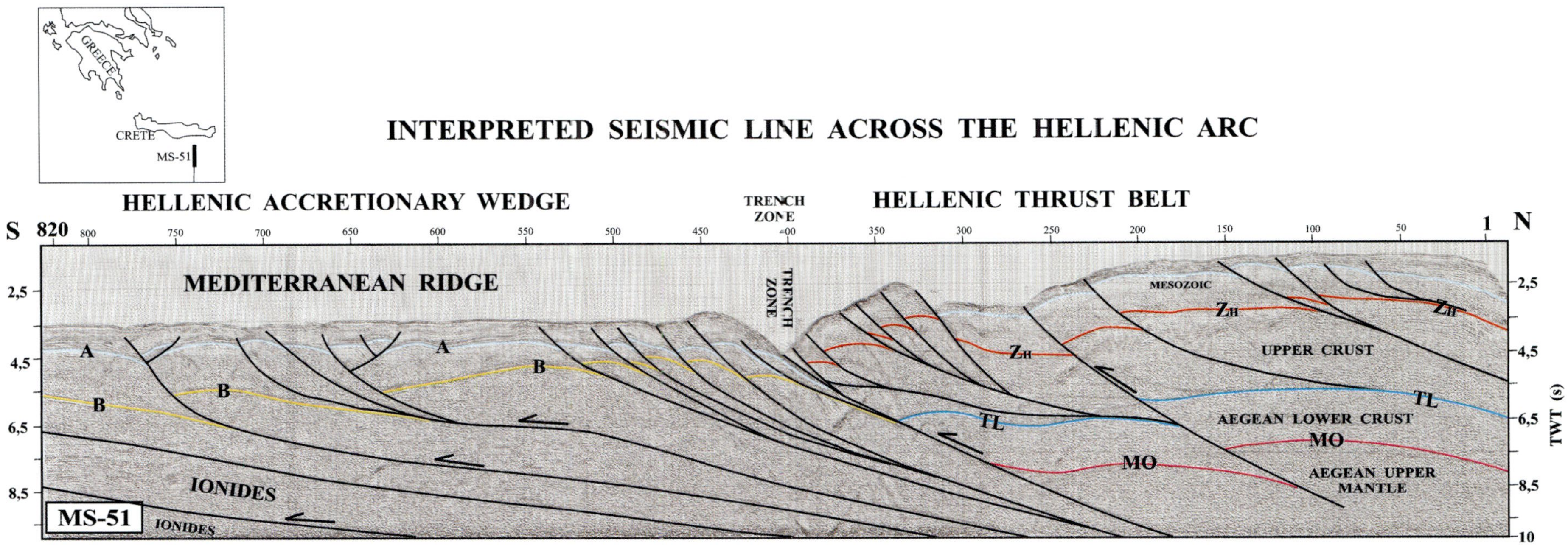

Figure 19: Interpreted seismic line MS-51 on the Hellenic arc, to the south of Crete. The trench zone separates an internal orogenic sector, formed of continental thrust-blocks, from the Hellenic wedge, mostly formed by imbrication of Ionian–East Mediterranean units (Ionides) detached from their subducting oceanic basement (Finetti, this volume, Chapter 10). A = base Plio-Quaternary; B = base Messinian; Z_H = basement of the Hellenic thrust-belt units; TL = top lower crust and MO = Moho of the Hellenic thrust-blocks.

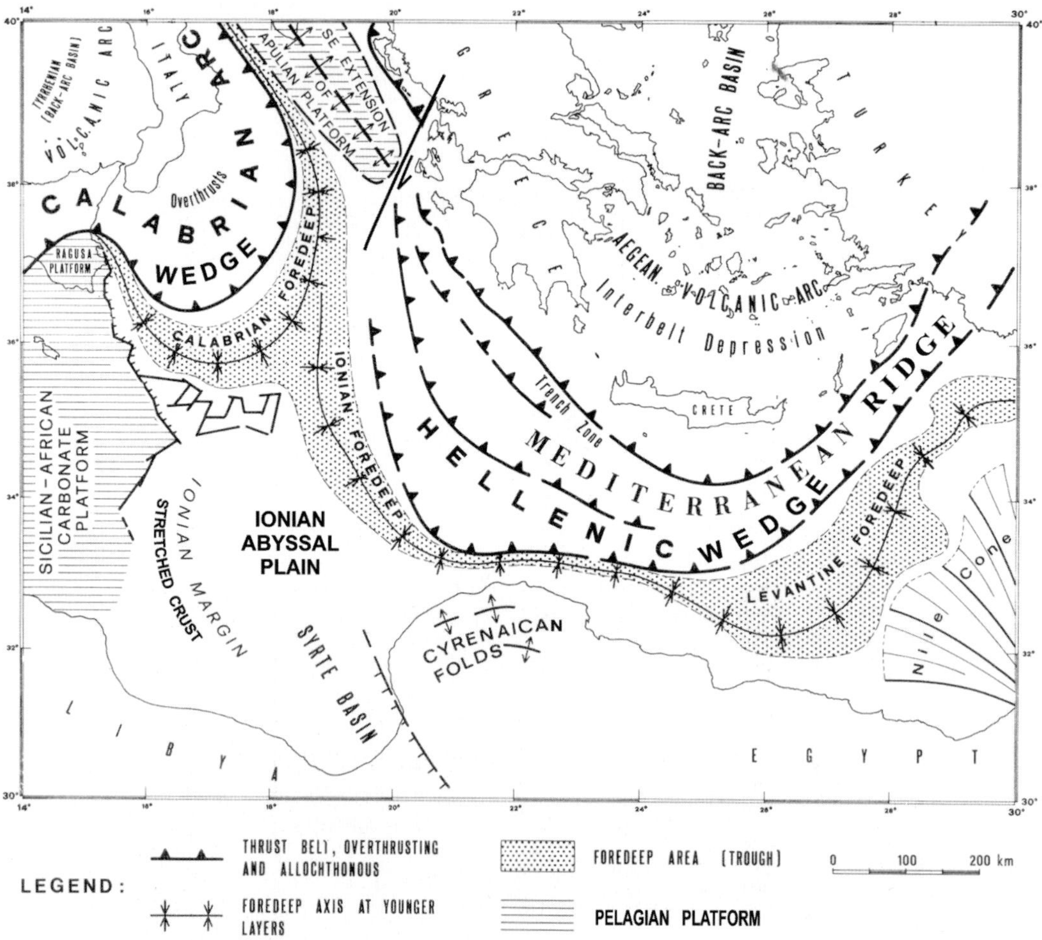

Figure 20: Scheme of the Hellenic arc and its connected wedge (*Mediterranean Ridge*) reconstructed from MS line data a long time ago (Finetti, 1976). At that time, alternative interpretations of the Mediterranean Ridge were under discussion.

It is worth recalling the long-discussed geological significance of the Hellenic wedge feature which extends over a large part of the Ionian–Eastern Mediterranean, known in the literature as the *Mediterranean Ridge*. Various hypotheses were proposed in the seventies, such as arc-associated sedimentary ridge (Rabinowitz and Ryan, 1970); and gravitative sliding of Late Cenozoic masses, including Messinian evaporite (Mulder, 1973; Bijù-Duval, 1974). *On the basis of MS-line evidence it was possible to reveal that the Mediterranean Ridge represents the wedge of the Hellenic arc formed by imbrication of blocks of the Ionian–East Mediterranean sedimentary cover* (Finetti, 1976). In Figure 18, clear imbrication tectonics in the Eastern Mediterranean Ridge, Ionian Sea sector, are exhibited.

A beautiful, almost complete vision of the internal thrust belt and its external wedge of the Hellenic arc can be observed on the interpreted seismogeological section MS-51, to the south of Crete (Figure 19). Between the trench zone, the northern extremity of the section are evident transcrustal thrust-blocks of the Aegean continental realm, translated by a relatively shallow southward-moving Aegean mantle flux. The Mediterranean Ridge, to the south of the trench zone, is formed by an accretionary wedge composed of imbricated cover sheets of the Eastern Mediterranean oceanic crust. A schematic reconstruction of the Hellenic and Calabrian wedge positions, published during the MS exploration phase of the Mediterranean basin, is reported in Figure 20 (after Finetti, 1976).

3.5. Levantine Sea

The Levantine Sea and Eastern Mediterranean have been explored by a coherent regional set of 11 MS seismic lines (Figure 1). From the initial investigation phase, both the complex contractional tectonics of the Hellenic and Cyprus arcs, and the thick, well-reflecting sedimentary cover of the Levantine basin were evident. In particular, a sedimentary deposition of over 5 s (TWT) which in the lower part thins towards the north was imaged on the

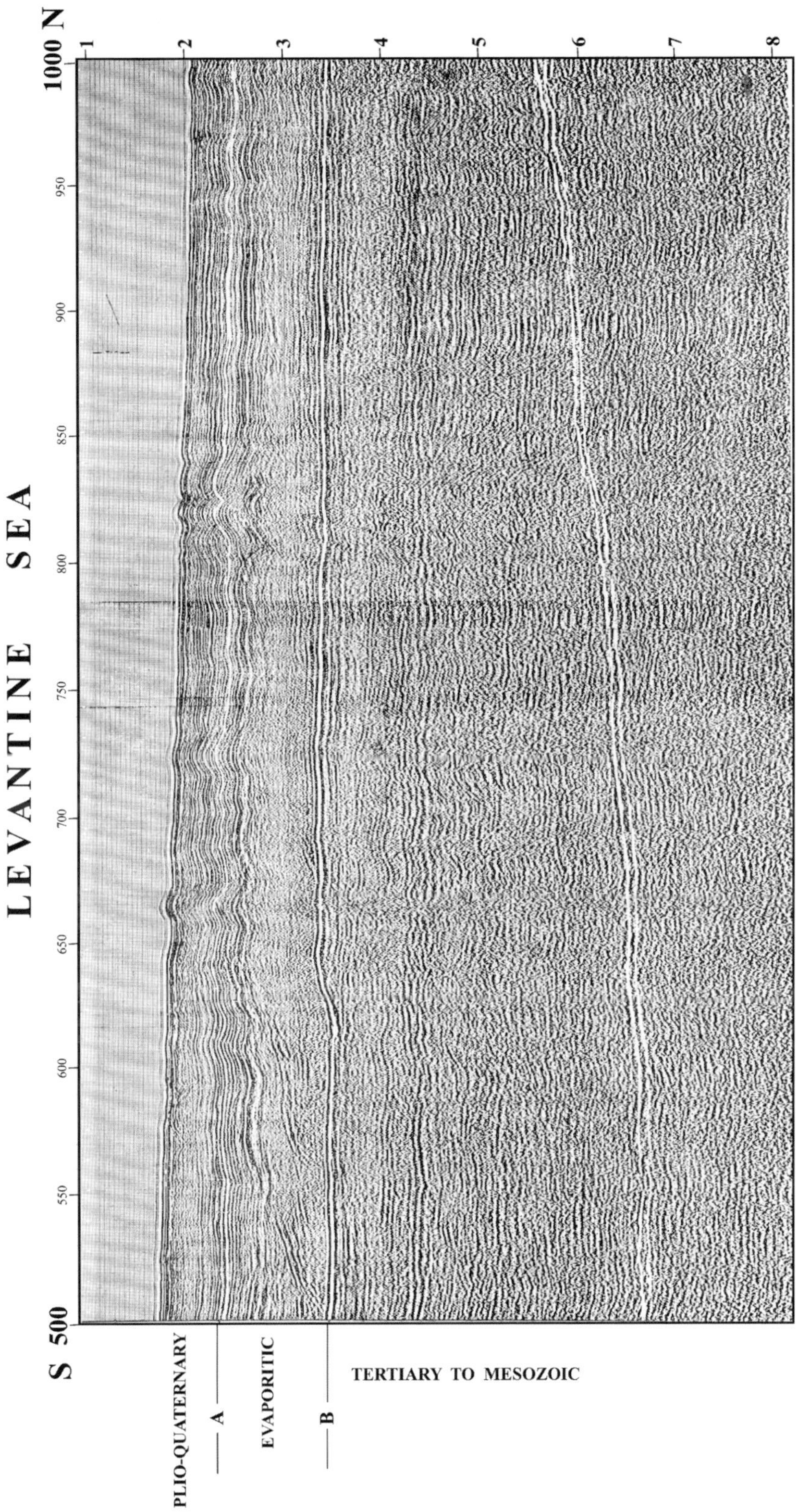

Figure 21: Example of an MS seismic reflection line on the thick sedimentary succession of the Levantine Sea (after Finetti and Morelli, 1973).

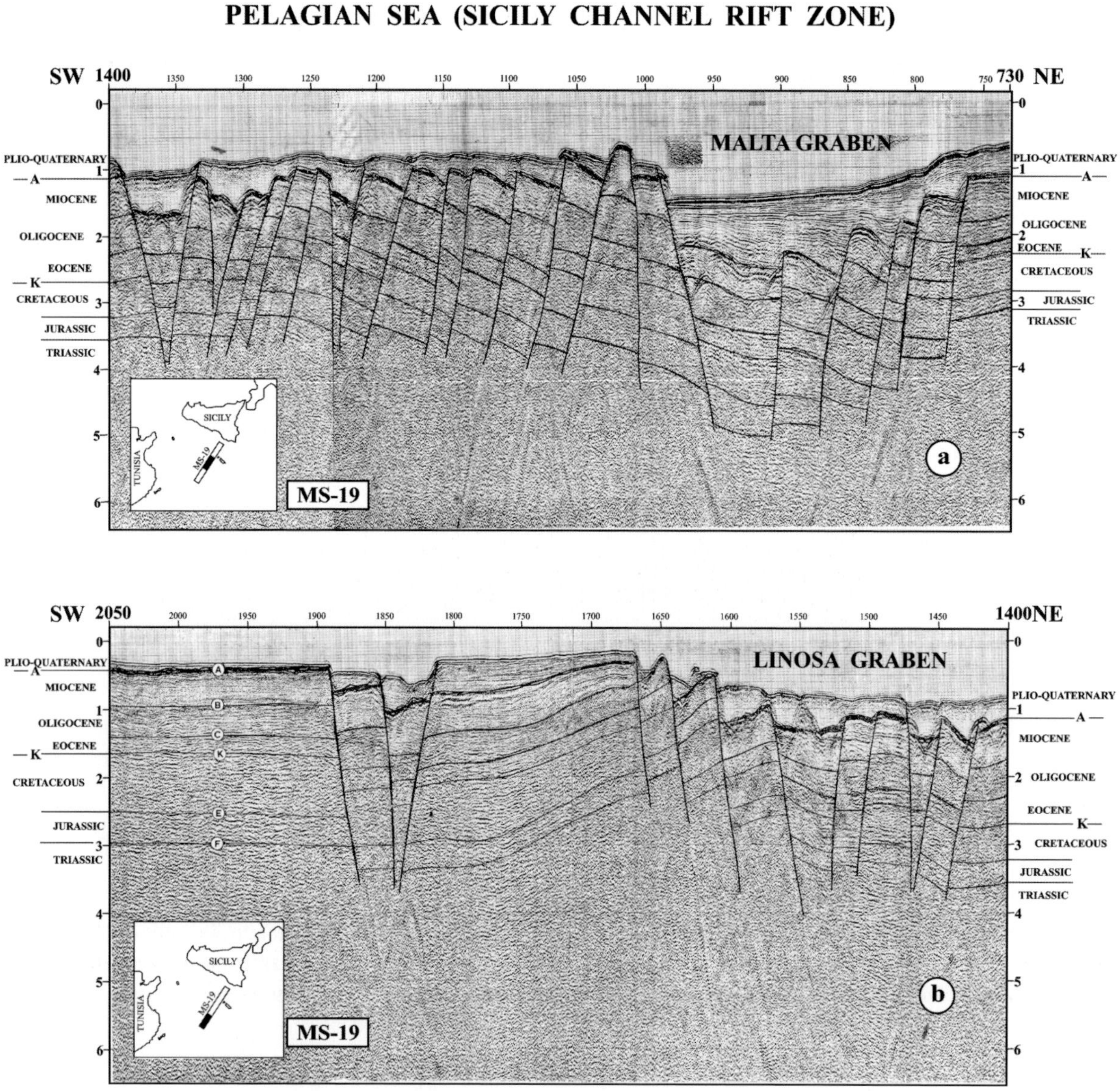

Figure 22: Example of an MS line across *Malta and the Linosa* grabens in the Pelagian Sea. Very impressive rift-tectonics are shown (after Finetti and Morelli, 1973).

Nile cone (Figure 21). Beneath a Plio-Quaternary half a second thick, an undeformed interval AB of Messinian evaporite is evident in this figure. Due to limited Plio-Quaternary lithostatic charge, this Messinian evaporite does not give rise to significant halokinetic doming dynamics.

3.6. Pelagian Sea (Sicily Channel Rift Zone) and Sirte Margin

The Pelagian Sea is morphologically characterized by a flat setting, interrupted by prominent depressions connected with young rift tectonics. The most impressive prominent rift zone is that known as the *Sicily Channel Rift Zone* (SCRZ), located along a NW–SE-trending axial area of the Pelagian Sea (Finetti, 1984). A clear seismic example of the SCRZ can be observed in Figure 22 where the line MS-19 provides a transversal section of the *Malta Graben* (Figure 22a) and *Linosa Graben* (Figure 22b). The deformation of the SCRZ is produced by *transtensive right-slip motion (dynamic pull apart), driven by Africa–Europe convergence kinematics and combined southeastward Ionian slab retreat.* Prominent volcanic activity accompanies the rifting tectonics and huge mantle uplift determines marked crustal thinning. Other elongated grabens of subordinate magnitude have been created in the Pelagian Sea (Figure 25) with the same trend and always by right-transcurrent faults. *The NW–SE transcurrent faults are associated with developed NNE–SSW antithetic left-transcurrent faults* (Finetti and Del Ben, this volume, Chapter 26).

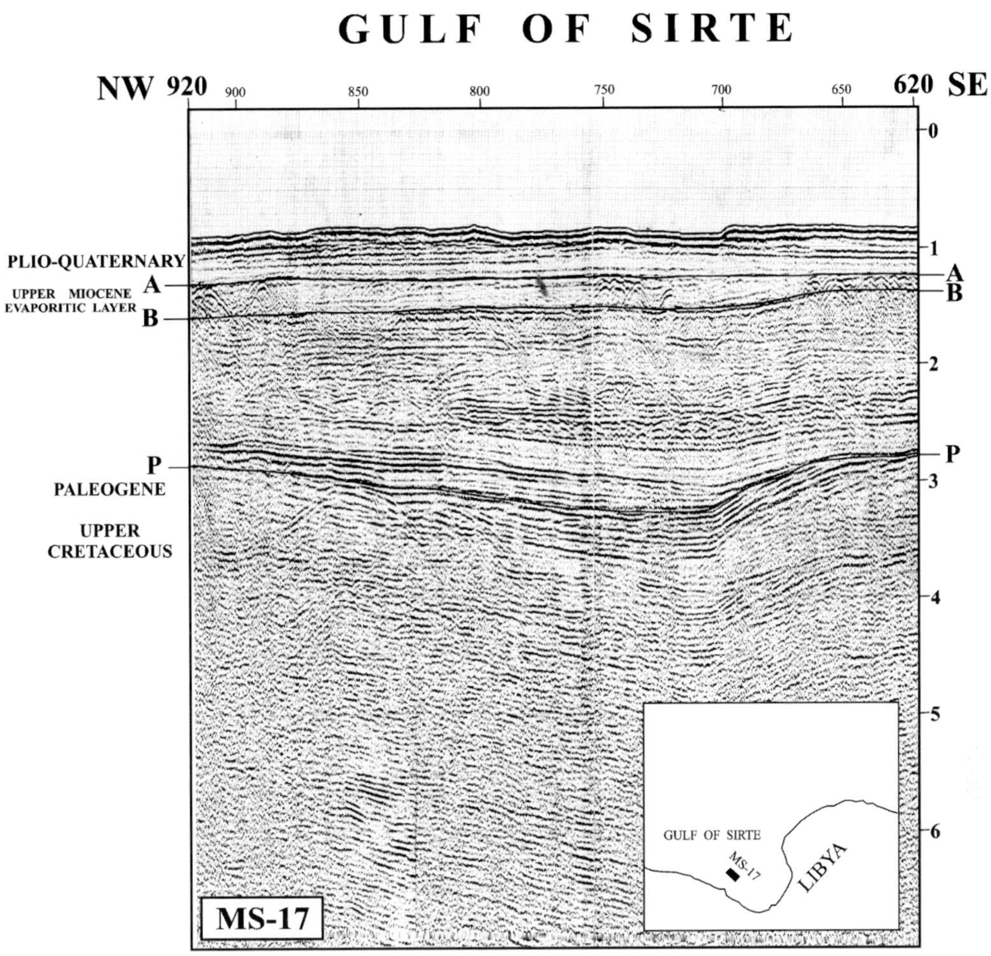

Figure 23: Seismic example (line MS-17; part) in the Gulf of Sirte (after Finetti and Morelli, 1973). It exhibits a well-reflecting sequence with a thick Tertiary–Upper Cretaceous (overlying an unidentified Cambro-Ordovician basement formed by quartzitic sandstone). P = top Paleogenic sequence.

In Figure 23 is exhibited an MS-line close-up (MS-17) which images the upper Sirte margin reflecting sequence.

3.7. Southern Adriatic Basin

For the most part, the Adriatic is a Meso-Cenozoic basin, covered by shallow water, which continues on land in the Po Valley. It represents the foreland of the Apennine orogenic system to the west, of the Dinaric system to the east, and of the Southern Alps to the north. Only the Southern Adriatic is covered by deep water (up to about 1 km). Therefore only this part of the Adriatic was explored by the MS seismic project, whose fixed target was the Mediterranean water over 200 m deep. A well-reflecting transverse seismic line in the Southern Adriatic is shown in Figure 24, where the frontal compressive tectonic deformation of the Southern Dinarides–Albanides of Oligocene–Lower Miocene age is imaged.

It seems evident that Permo-Mesozoic here thins markedly compared to Apulia, very likely because in the Southern Adriatic the sequence is more condensed. This is surely the case also for the Oligocene to upper Lias interval, as indicated by borehole data. The Triassic succession includes thick evaporitic intervals.

3.8. Early Geophysical Reconstruction of the Tyrrhenian Opening-Related Tectonics

Based on the MS seismic lines and a significant amount of other geophysical data for the continental shelf zones (Italian Ministry's public seismic dataset), in the mid eighties, a well-controlled tectonic scheme for the Tyrrhenian Sea and surrounding geological provinces was reconstructed (Finetti and Del Ben, 1986). This tectonic map (Figure 25), provided a framework for the first time in a widely supported, reliable manner of all the most significant

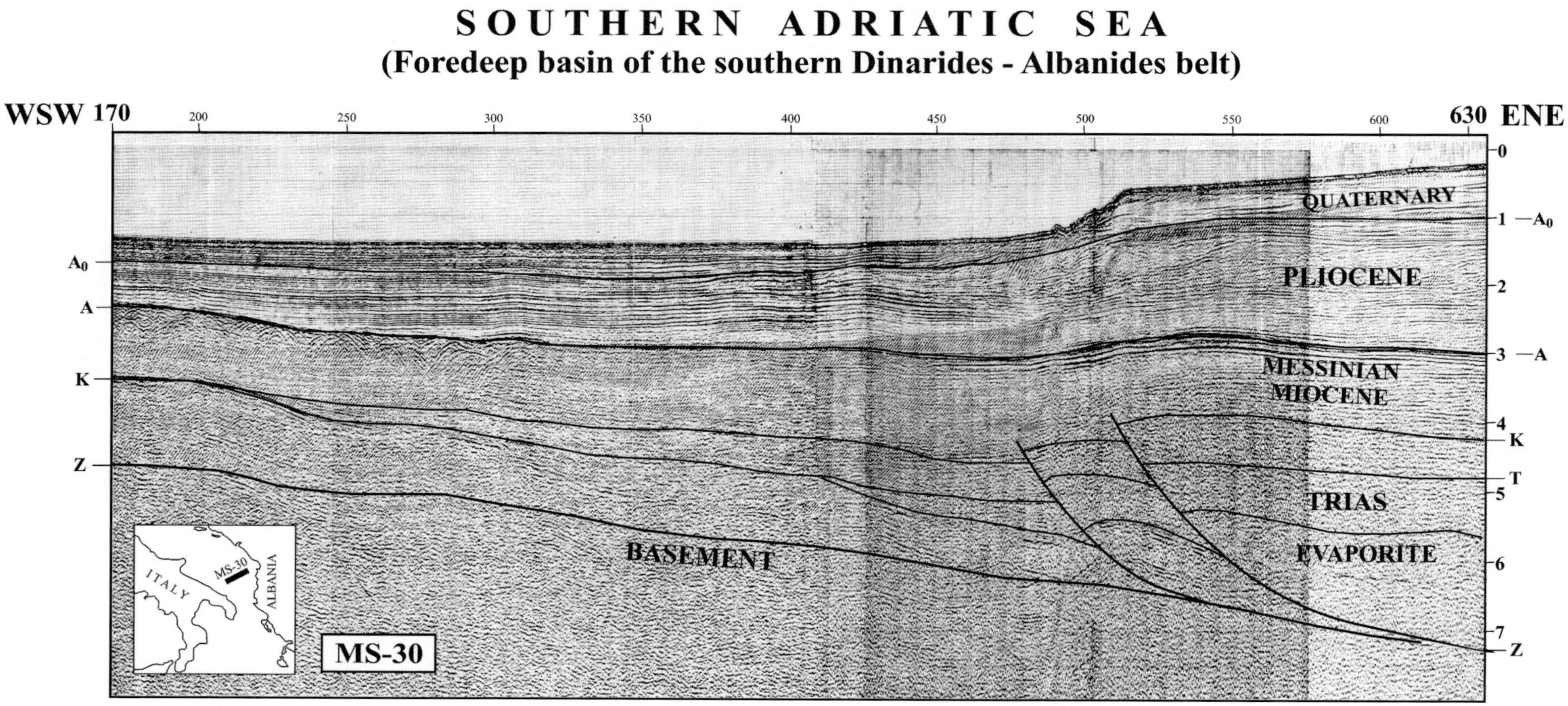

Figure 24: Seismic example (line MS-30; part) in the Southern Adriatic Sea (after Finetti and Morelli, 1973), showing the front of the Southern Dinarides-Albanides covered by a well-reflecting thick Neogenic sedimentary succession. A_0 = base Quaternary; A = base Pliocene; K = top carbonate (Paleogene); T = (top) Triassic; Z = possible basement. A tectonic thickening of the Triassic evaporite seems occurred.

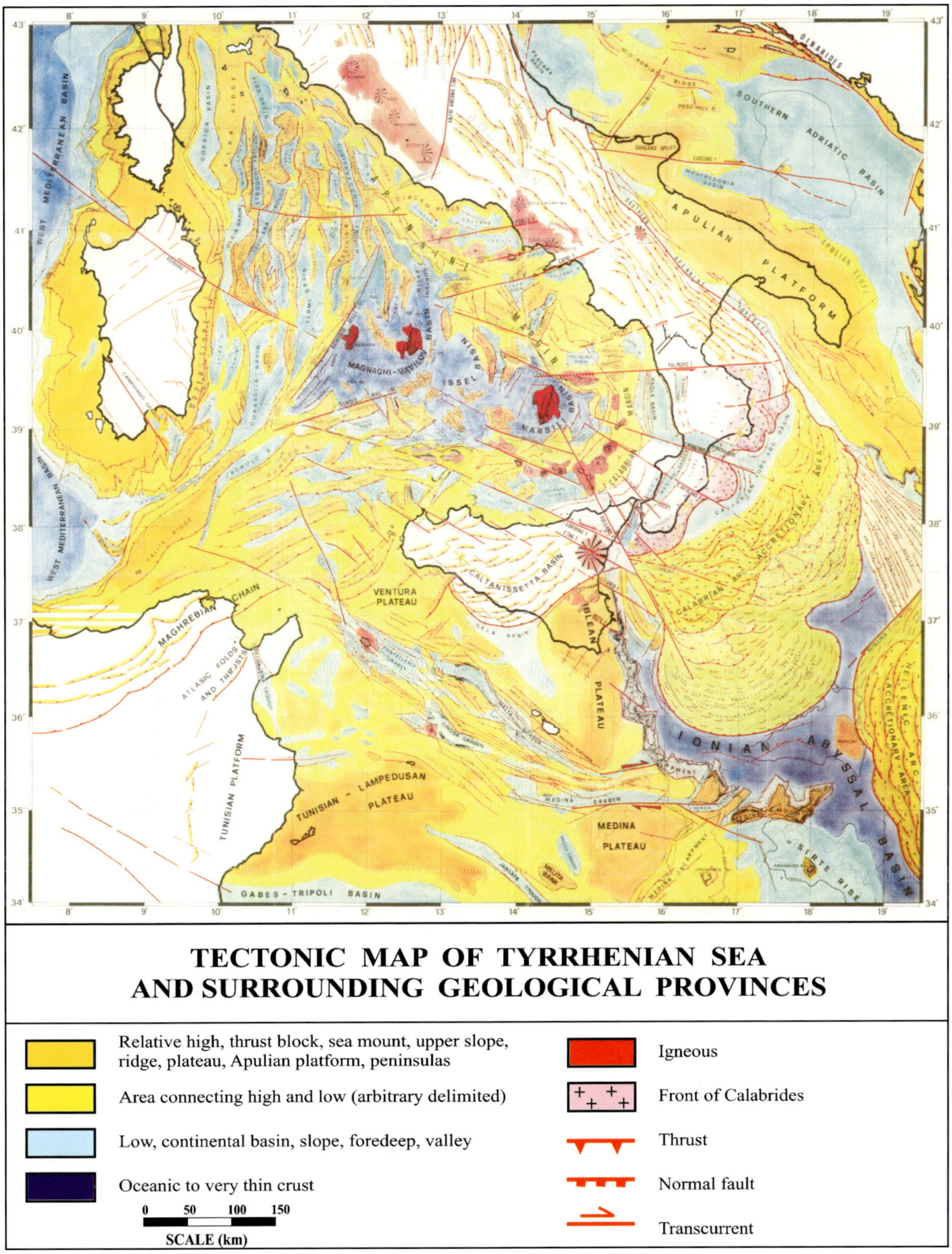

Figure 25: Tectonic sketch map of the Tyrrhenian Sea and surrounding geological provinces, after Finetti and Del Ben (1986) (legend simplified), reconstructed from the MS seismic line network integrated with other seismic data.

features of the complex central Mediterranean region. In summary, the following features were geophysically explored, interpreted, outlined and organically composed in a reliable regional tectonic map:

- Ionian ocean remnant of old opening
- Eastern Balearic ocean of Oligocene–Lower Miocene opening
- Tyrrhenian Sea of Middle Miocene to Present opening
- Corso-Sardinian block and its Tyrrhenian crustal stretching
- Tyrrhenian Margin tectonics
- Maghrebian chain
- Southern Tyrrhenian right-transcurrent system
- Northern Apennines thrust belt outline
- Southern Apennine thrust belt
- Calabrian arc and its prominent wedge
- Hellenic arc in Ionian Sea
- Sicilian Apennines and Gela wedge
- Pelagian Sea and transcurrent tectonics of the Sicily Channel Rift Zone
- Apulian platform and its SE-ward extension seismically defined
- Southern Adriatic and main transcurrent faults (Gargano, Tremiti)

The tectonic map of Figure 25 (Finetti and Del Ben, 1986) has been reconstructed to the scale 1:1,000,000, at which much more detail is exhibited. *It is possible to note that after years and many new investigations, the most fundamental tectono-kinematic settings expressed remain confirmed.* In the deep basins (Balearic, Tyrrhenian, Ionian) the MS lines (10 s TWT penetration) give as full transcrustal information as the new CROP sections.

ACKNOWLEDGEMENTS

Very much gratitude is expressed by I.R. Finetti, former Director, for OGS's geophysicists, seismic data acquisition teams, and data processing operators who all contributed to the pioneering geophysical exploration programme of the MS seismic lines in the Mediterranean deep water.

REFERENCES

Bijù-Duval, B., 1974. Carte géologique et structurale des Bassins Tertiaires du domaine Mediterraneén : Commentaires. Rev. Inst. Fr. Pétrole XXIX (5), 607–639.

Bijù-Duval, B., Letouzey, J., Montadert, L., 1977. Structure and evolution of the Mediterranean Sea basins. Init. Rep. DSDP 42 (I), 951–984.

Bijù-Duval, B., Letouzey, J., Montadert, L., Courrier, P., Mugniot, J.F., Sancho, J., 1974. Geology of the Mediterranean Sea basins. In: Burck, C.A., Drake, C.L. (Eds.), The Geology of Continental Margins. Springer-Verlag, pp. 695–721.

Bijù-Duval, B., Letouzey, J., Montadert, L., 1978. Variety of margins and deep basins in the Mediterranean. AAPG, Mem 29, 293–314.

Boccaletti, M., Elter, P., Guazzone, G., 1971. Plate tectonic models for the development of the Western Alps and Northern Apennines. Nature, Phys. Sci. 234, 108–111.

Boccaletti, M., Conedera, C., Dainelli, P., Gocev, P., 1982. The recent (Neogene-Quaternary) regmatic system of the Western Mediterranean region. J. Petr. Geol. 5, 31–49.

Bonardi, G., Cello, G., Perrone, V., Tortorici, L., Turco, E., Zuppetta, A., 1982. The evolution of the northern sector of the Calabria–Peloritani arc in a semiquantitative palynspastic restoration. Boll. Soc. Geol. It. 101, 259–284.

Carmignani, L., Kligfield, R., 1990. Crustal extension in the Northern Apennines: the transition from compression to extension in the Alpi Apuane core complex. Tectonics 9, 1275–1303.

Carmignani, L., Decandia, F.A., Fantozzi, P.L., Lazzarotto, A., Liotta, D., Meccheri, M., 1994. Tertiary extensional tectonics in Tuscany (Northern Apennines). Tectonophysics 238, 295–315.

D'Agostino, N., Jackson, J.A., Dramis, F., Funiciello, R., 2001. Interactions between mantle upwelling, drainage evolution and active normal faulting: an example from the central Apennines (Italy). Geophys. J. Int. 147, 475–497.

Dietrich, D., 1988. Sense of overthrust shear in the Alpine nappes of Calabria (Southern Italy). J. Struct. Geol. 10 (4), 373–381.

Fabbri, F., Curzi, P., 1979. The Messinian of the Tyrrhenian Sea: seismic evidence and dynamic implications. Giorn. Geol., Bologna 43, 215–248.

Faccenna, C., Davy, P., Brun, J.P., Funiciello, R., Giardini, D., Mattei, M., Nalpas, T., 1996. The dynamics of back-arc extension: an experimental approach to the opening of the Tyrrhenian Sea. Geophys. J. Int. 126, 781–795.

Finetti, I.R., 1975. Isobaths of the base of the Pliocene in the Tyrrhenian Sea. CNR, Quad. Ric. Scientif. 90, 417–426.

Finetti, I.R., 1976. Mediterranean Ridge: a Young Submerged Chain associated with the Hellenic Arc. Boll. Geof. Teor. Appl. 28, 31–62.

Finetti, I.R., 1982. Structure, stratigraphy and evolution of Central Mediterranean. Boll. Geof. Teor. Appl. 24, 247–426.

Finetti, I.R., 1984. Geophysical study of the Sicily Channel rift zone. Boll. Geof. Teor Appl. 12, 263–341.

Finetti, I.R., this volume. Understanding the Ionides and their geodynamics. Chapter 10.

Finetti, I.R., this volume. The Calabrian arc and subducting Ionian slab from new CROP seismic data. Chapter 17.

Finetti, I.R., Morelli, C., 1972. Regional reflection seismic exploration of the Strait of Sicily. In: Saclant Conference, vol. 7. La Spezia.

Finetti, I.R., Morelli, C., 1973. Geophysical exploration of the Mediterranean Sea. Boll. Geof. Teor. Appl. 15, 263–341.

Finetti, I.R., Morelli, C., 1974. Esplorazione geofisica dell'area Mediterranea circostante il Blocco Sardo-Corso. In: Paleogeografia del Terziario Sardo nell'ambito del Mediterraneo Occidentale. Rdc. Sem. Fac. Scienze. Univ. Cagliari.

Finetti, I.R., Del Ben, A., 1986. Geophysical study of the Tyrrhenian opening. Boll. Geof. Teor. Appl. 28, 75–156.

Finetti, I.R., A. Del Ben, this volume. Crustal tectono-stratigraphy of the Ionian Sea from integrated new CROP seismic data. Chapter 19.

Finetti, I.R., A. Del Ben, this volume. Crustal tectono-stratigraphic setting of the Pelagian foreland from new CROP seismic data. Chapter 26.

Finetti, I.R., Morelli, C., Zarudzki, O., 1970. Reflection seismic study of the Tyrrhenian Sea. Boll. Geof. Teor. Appl. 12, 311–345.

Finetti, I., Lentini, F., Carbone, S., Catalano, S., Del Ben, A., 1996. Il sistema Appennino Meridionale–Arco Calabro–Sicilia nel Mediterraneo Centrale: studio geologico-geofisico. Boll. Soc. Geol. It. 115, 529–559.

Jolivet, L., Faccenna, C., 2000. Mediterranean extension and the Africa–Eurasia collision. Tectonics 19, 1095–1106.

Lavecchia, G., Stoppa, F., 1996. The tectonic significance of Italian magmatism: an alternative view to the popular interpretation. Terra Nova 8, 435–446.

Locardi, E., 1982. Individuazione di strutture sismogenetiche dall'esame dell'evoluzione vulcano-tettonica dell'Appennino e del Tirreno. Mem. Soc. Geol. Ital. 24, 569–596.

Malinverno, A., Ryan, W.B.F., 1986. Extension in the Tyrrhenian Sea and shortening in the Apennines as a result of arc migration driven by sinking of the lithosphere. Tectonics 5, 227–245.

Mantovani, E., Albarello, D., Tamburelli, C., Babbucci, M., Viti, D., 1997. Plate convergence, crustal delamination, extrusion tectonics and minimization of shortening work as main controlling factors of the recent Mediterranean deformation pattern. Ann. Geofis. 40, 611–643.

Moskalenko, V.N., 1965. Study of sedimentary series of the Mediterranean Sea by seismic methods. In: Basic Features of the Geological Structure of the Hydrologic Regime and Biology of the Mediterranean Sea. Translation and Interpretation Division, Institute of Modern Languages, pp. 60–72.

Moskalenko, V.N., 1967. Study of the sedimentary series of the Mediterranean Sea by seismic methods. Proc. Acad. Sci. USSR Oc. Comm., Moscow.

Mulder, C.J., 1973. Tectonic framework and distribution of Miocene evaporates in the Mediterranean. In: Messinian Events in Mediterranean. Kon. Acad. Wet., Amsterdam.

Rabinowitz, P.D., Ryan, W.B.F., 1970. Gravity anomalies and crustal shortening in the eastern Mediterranean. Tectonophysics 10, 585–608.

Roeder, D., 1980. Geodynamics of the Alpine–Mediterranean system – a synthesis. Eclog. Geol. Helv. 73, 353–377.

Royden, L.H., Patacca, E., Scandone, P., 1987. Segmentation and configuration of subducted lithosphere in Italy: An important control on thrust-belt and foredeep-basin evolution. Geology 15, 714–717.

Scandone, P., 1979. Origin of the Tyrrhenian Sea and Calabrian Arc. Boll. Soc. Geol. It. 98, 27–34.

Scandone, P., 1982. Structure and evolution of the Calabrian Arc. Earth Evol. Sci. 3, 172–180.

Tapponnier, P., 1977. Evolution tectonique du système alpin en Méditerranée : poinçonnement et écrasement rigide-plastique. Bull. Soc. Géol. Fr. 7, 437–460.

Vai, G.B., 1992. Il segmento calabro-peloritano dell'orogene ercinico. Disaggregazione palinspastica. Boll. Soc. Geol. It. 111, 109–129.

Van Bemmelen, R.W., 1972. Driving forces of Mediterranean orogeny (Tyrrhenian test-case). Geol. Mjinbouw 51, 548–573.
Wezel, F.C., 1982. The Tyrrhenian Sea: a rifted krikogenic-swell basin. Mem. Soc. Geol. Ital. 24, 531–568.
Wortel, M.J.R., Spakman, W., 1992. Structure and dynamics of subducted lithosphere in the Mediterranean region. Proc. K. Ned. Akad. Wet., Biol. Chem. Geol. Phys. Med. Sci. 95, 325–347.

CROP PROJECT: Deep Seismic Exploration of the Central Mediterranean and Italy
Edited by I.R. Finetti

CHAPTER 2

Review of Seismic Wide-Angle Reflection–Refraction (WARR) Results in the Italian Region (1956–1987)

R. Cassinis[1], S. Scarascia[2], A. Lozej[3]

ABSTRACT

The aim of this review is a new synthesis of the results obtained in Italy and in the surrounding areas over more than thirty years (1956–1987) by the seismic exploration method of wide-angle reflection–refraction (WARR), also known as DSS – Deep Seismic Soundings. Italy is among the countries most intensively surveyed by this method. During its long period of application, the technique has undergone remarkable development with regard to instrumentation (from analog to digital), acquisition procedures (density and type of layouts), as well as processing and interpretation (ray-tracing and modelling).

Despite the low resolving power and constraints due to the large distance between the shots and for the low density of recorders, the method is particularly well-suited describing the pattern of the crust-mantle boundary, and is the only way to obtain the distribution of seismic velocities in the lower crust.

A rich literature was been produced both during and since the long period of data acquisition. However, we feel that a further comprehensive survey, compiled using homogeneous criteria, would be useful to discuss and update the geodynamic reconstructions of the central Mediterranean.

On the basis of the most reliable interpretations, we prepared twelve synthetic cross-sections, compiled with homogeneous criteria of presentation. A new map of the Moho boundary is also proposed and compared with the known geodynamic schemes, as well as with the regional Bouguer anomalies. In addition to the depth contour lines of the boundary, which show the complexity of the deep structure, we also describe the crustal types found in the area: besides the two main interacting domains of the Adriatic and European plates, the oceanic–sub-oceanic type is present in the Ligurian and Tyrrhenian Seas. Moreover, a fourth type is defined: the Ligurian–Tuscan–Perityrrhenian strip of thinned, anomalous crust. The thickness and structure of this intermediate crustal type was produced by a multi temporal geodynamic evolution: in the area beneath the Ligurian Sea, Corsica and Tuscany, by the opening of the Provençal Basin and the counterclockwise rotation of the Sardinian-Corsican microplate; in the south, by the opening of the Tyrrhenian Sea.

The structure of the crust-mantle boundary and of the lower crust is outlined in detail beneath the Alpine range, clearly enhancing the contrast between the fairly homogeneous European crust beneath the central Alps and the complex Adriatic-Padan crust along its northern and north western edge; the latter testifies to the effect of the push-and-pull stages responsible for the Alpine evolution.

Keywords: Wide-Angle Reflection Refraction Review Italy

1. INTRODUCTION

Deep seismic exploration of the Italian region began in 1956 as part of the "International Geophysical Year" and was continued in the framework of other international and national projects. The field work and the interpretation are to be credited to the joint effort of an international group (mainly German, Italian, Swiss and French), organised under the scientific sponsorship of the European Seismological Commission. The group operated an average of

[1]Formerly at the University of Milano, Italy. Corresponding author: 30, Viale Lombardia, I-20131; e-mail: roberto.cassinis@infinito.it
[2]Formerly at IRRS-CNR, Milano, Italy
[3]Department of Earth Sciences, Geophysical Division, University of Milano, Italy

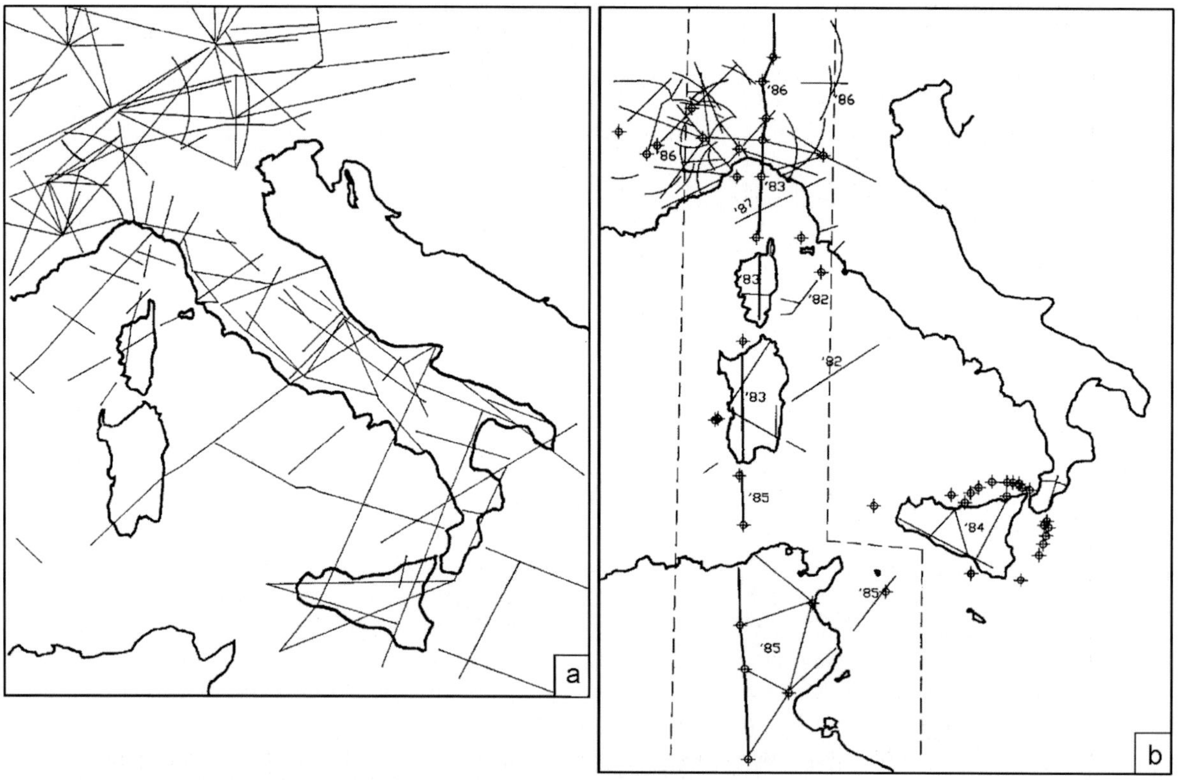

Figure 1: a – WARR profiles and fans 1956–1981 (Morelli and Nicolich, 1990). b – 1982–1987 (Morelli and Nicolich, 1990).

60–80 mobile, homogeneous seismic stations, while the Italian Institutions (mainly the National Research Council-CNR) provided the funds for shooting and drilling and the ships for marine operations.

The WARR profiles and fans recorded from 1956 to 1981 are shown in Figure 1a, while Figure 1b illustrates the activities from 1982 to 1987, mainly in the framework of the EGT (European Geotraverse) Project (Morelli and Nicolich, 1990).

Italy is probably the most intensively surveyed country in the "western world" using the WARR technique aimed at crustal investigations. This can be explained by the exceptionally complex geological and geodynamic setting that was so difficult to explain using plate tectonics theory, then accepted without criticism by the Earth Science Community.

The achievements of the group must be credited to the tireless activity of a number of scientists, among whom P. Giese and C. Prodehl of Germany, S. Mueller of Switzerland and C. Morelli of Italy should be mentioned in particular.

A rich literature was produced during and since the long period of data acquisition (more than 30 years). The reference list at the end of this chapter is far from complete. Plenty of reports were published at the end of each field operation; they contained preliminary interpretations, subsequently revised and modified when new sets of data became available. The same data were reinterpreted several times (sometimes by different authors) following developments in enhancement and processing techniques.

Dedicated volumes or articles summarising the results obtained for particular projects or particular areas were published; e.g. the strip along the EGT, the Alpine range, the seas surrounding the Peninsula (see Blundell et al., 1992; Pfiffner et al., 1997; Roure et al., 1990). Maps of the crust-mantle boundary were proposed for the whole region or just for the Italian Peninsula (Cassinis et al., 1979; Morelli and Nicolich, 1990; Morelli, 2000; Scarascia et al., 1994). However, we believe that a further synthesis of the results, taking into account the last interpretations and revisions done in particular areas, will add to our knowledge of the structure of the deep crust and of its bottom.

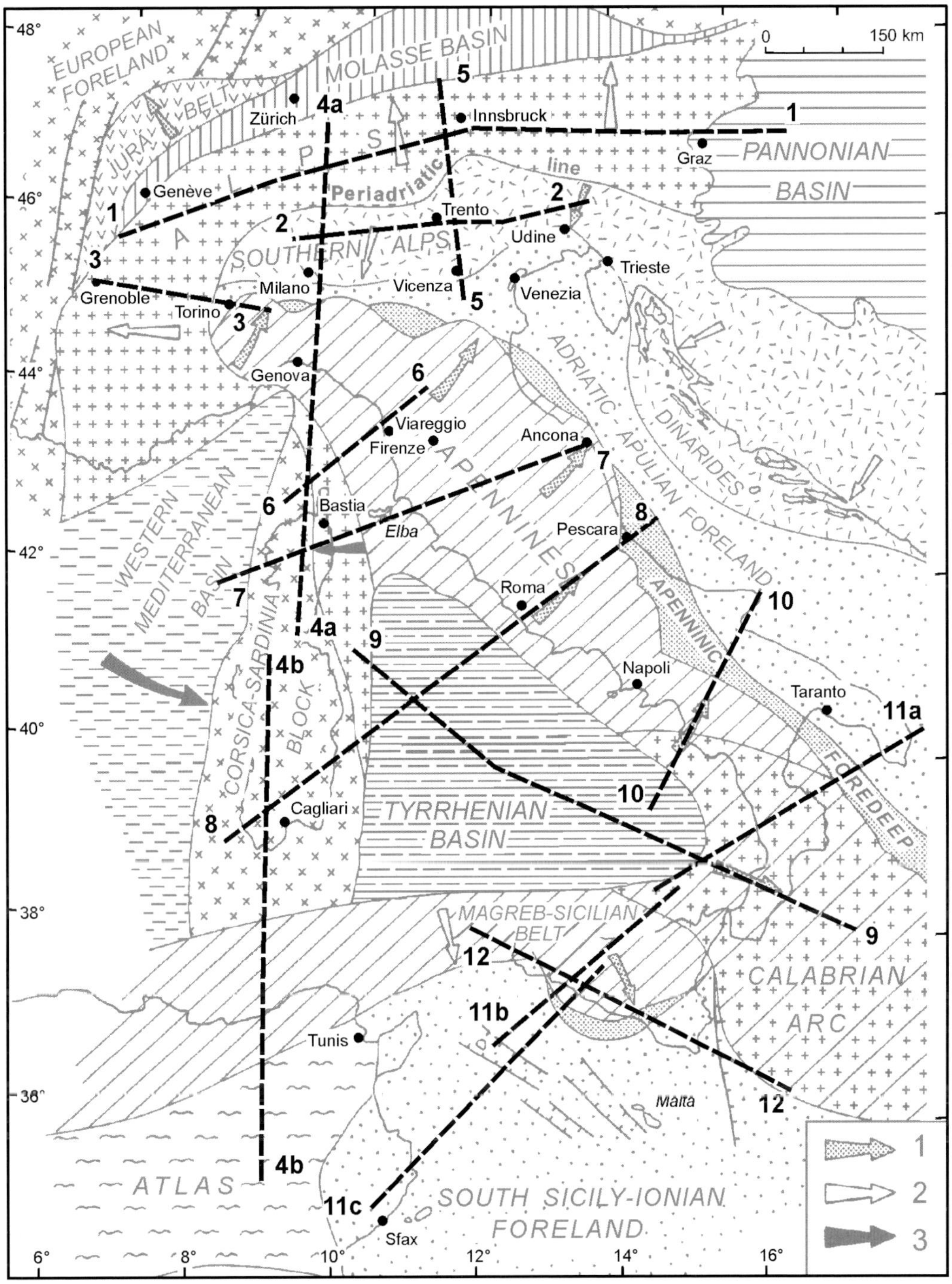

Figure 2: Geodynamic scheme of the Italian Peninsula and surrounding areas (structural model of Italy, CNR, 1983, modified). For Tunisia the scheme follows Boccaletti et al. (1990). Plotted are the traces of the twelve interpretative crustal cross-sections described in this chapter. (See Figures 4.1–4.12.) The arrows represent the direction and ages of drift and last orogenic transport: 1 – Pliocene-Pleistocene, 2 – Middle-Upper Miocene, 3 – Oligocene–Lower Miocene.

To this end, we have prepared twelve interpretative cross-sections drawn using, as much as possible, homogeneous criteria (Figure 4.1). It must be noted that they are based upon data of uneven quantity and quality and, therefore, the reliability of the interpretation is not uniform throughout the area. This is also apparent looking at the density of transepts (Figures 2 and 3) which is higher where networks of profiles based on a larger number of shotpoints are available. The Alpine range and the Tyrrhenian domain have been better investigated than the Adriatic side of the Peninsula.

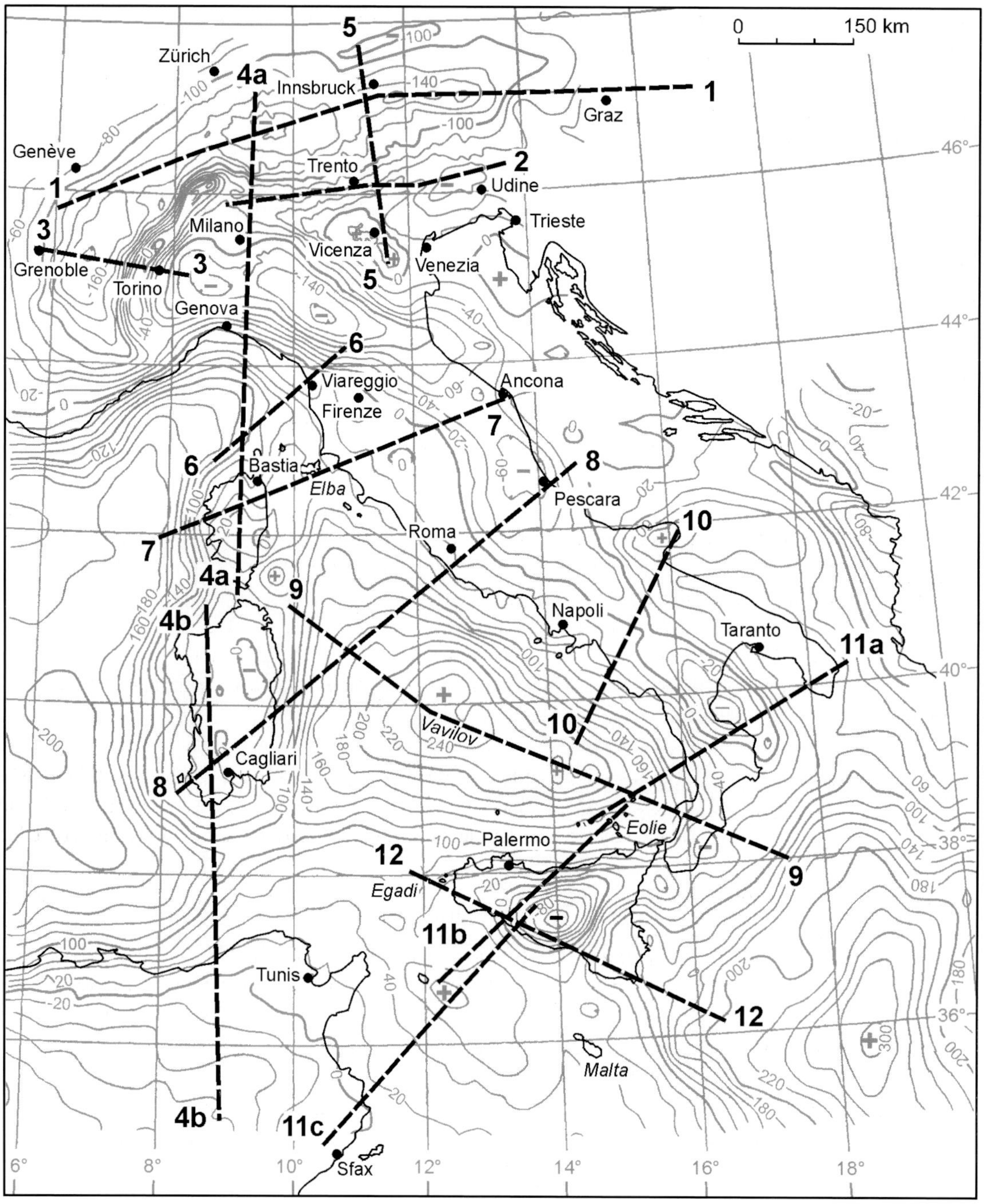

Figure 3: Map of the same region with the traces of the cross-sections and the Bouguer anomalies after Morelli (2000). Contour interval: 20 mGal; compare with Figure 5.

In the attached list we quote the references that have been used to draw the twelve synthetic interpretative cross-sections illustrating the crustal structure of the whole Italian region. When more than one solution was available, we chose the one that seemed better supported by other data in the area.

Up to the late seventies, individual, analog, frequency modulated recorders (mainly four channel, MARS 66) were used for the acquisition; the input consisted of three low-frequency geophones (1 or 2 Hz) recording, as a rule, only the vertical component. A fourth channel recorded the time signal via a radio receiver.

The analog tapes were played back at headquarters at the end of each day of acquisition by groups of demodulators; the traces were photographically reproduced and manually ordered along cross-sections with reference to the time break transmitted from the shot point.

The "digital revolution" reached DSS later than it did industrial geophysical oil prospecting: only in the late eighties were digital instruments adopted (MARS 88).

The coordinates of shots and receivers and their height were read off the available, large scale, maps; only since the early nineties has the GPS positioning system become familiar.

The instruments used always required operator attention. It is apparent that both the type of equipment and the limited resources of the group were a constraint as far the density of recorders and shotpoints is concerned. The average spacing between the recorders was reduced from about 5 km in the early profiles to about 1.5 km in the most recent ones. The density of shotpoints remained very low (one shot every 100 km as a rule).

It is clear that, in general, it was impossible to reliably detect the lateral inhomogeneities in the sedimentary overburden and upper crystalline crust; therefore the information gathered by this technique involves mainly the deep crust and the transition to the upper mantle. Moreover, the low density also negatively affects the reliability of trace correlation for each seismic phase.

As far as processing and interpretation are concerned, developments were especially due to the introduction of digital recording. However, the fundamental elements read on the records remained the same: Pg (penetration waves in the upper crust), PiP wide angle intermediate reflections (if any), PMP reflections from the Moho boundary (in general the overcritical ones) and, for a sufficient length of the profile, the Pn refracted waves from the upper mantle. The maximum shot-receiver distance was, in general, 150–200 km, allowing the recording of overcritical reflections from the base of the crust but, only seldom, the refracted Pn arrivals. These latter arrivals were received especially on the suboceanic and oceanic thin crust. Only where the shot was particularly efficient, were good results obtained at larger distances.

A very clear and exhaustive review of the potential and of the peculiarities of WARR was edited in 1976 by Giese and Prodelh (see also Colombi and Scarascia, 1977). Later, digital recording added, to a certain extent, the dynamic to a purely kinematic approach, also helped by the development of synthetic records and by ray-tracing procedures.

With regard to the final output of the interpretation, up to the late seventies, it consisted mainly of 1D velocity functions connected, whenever possible, by correlation lines to yield a velocity cross-section. This procedure led to the definition of "crustal types". Later, mainly through the application of iterative procedures and of the ray-tracing techniques, 2D models became general. From the pattern of cross-sections, time or depth contour maps were drawn of the main boundary, the Moho. Only seldom were maps of intermediate discontinuities (the "Conrad" and the seismic basement) also plotted in some areas (e.g. the western and central-eastern Alps).

Both the employed techniques and the results obtained for several Italian areas are documented in detail in a group of excellent PhD theses. Also some master theses contain useful information, especially about those experiments that were not fully published in scientific journals or in dedicated volumes.

2. INTERPRETATIVE CROSS-SECTIONS

The tracks of the proposed twelve cross-sections are shown in the geodynamic scheme for the Italian Peninsula and surrounding areas in Figure 2 (Bigi et al., 1990; Boccaletti et al., 1990) as well as on the map of the regional Bouguer anomalies in Figure 3 (Morelli, 2000). The position and interpretation of each transept is taken from already published papers and redrawn with some modifications, chiefly in order to obtain a uniform graphic representation. Cross-sections n. 6, 7, 8, 9, 10, 11a, 11b and 12 (Italian seas and adjacent areas) have been redrawn from Scarascia et al. (1994). In the Alpine region, sections n. 1, 2 and 5 are taken from Alpine Expl. Seism. Group (1976), and from Scarascia and Cassinis (1997), while section 3 was re-elaborated from the interpretation by Giese and Prodehl (1976). For sections 4a and 4b (along the southern sector of the EGT-European Geotraverse), the interpretations by Buness (1992), Egger et al. (1988), and by the Res. Group for Lithospheric Structure in Tunisia (1989, 1992) were used. Lastly, section 11c was redrawn and modified from Scarascia et al. (2000).

The symbols are explained in Figure 4.1; they are common to all transepts (Figures 4.1–4.12).

3. RESULTS IN THE ALPINE REGION

Along-strike profiles:

Cross-section 1 (Figure 4.1) corresponds to the ALP '75 profile (Alpine Expl. Seism. Group, 1976; Yan and Mechie, 1989; Scarascia and Cassinis, 1997); it begins in the northwestern Alps, south of Geneva, and terminates about 100 km east of the city of Graz; starting from the western end, the thickness of the crust increases from about 32 km to about 55 km in the central Alps at the crossing with cross-section 4a (EGT); this change in crustal thickness seems to be solely in the lower crust; then the line exhibits a uniform crustal structure along a stretch of about 400 km. Both the upper and the lower crust are characterized by fairly constant velocities, averaging respectively about 6.1 and 6.3–6.4 km/s, the latter being a rather low value for continental crust.

In the central stretch, the thickness of the two crustal layers is also nearly constant (about 20 km for the upper, and about 25–30 km for the lower crust). The upper and lower crust are separated by a higher velocity layer (6.4–6.5 km/s with a rather strong gradient) about 5 km thick. The profile runs throughout the Tauern Window, but this seems not to affect the seismic homogeneity. West of Graz the crust thins abruptly, clearly marking the transition from the European crust beneath the Alps to the Styrian and Pannonian Basins.

Cross-section 2 of Figure 4.2 (corresponding to the profile SudALP, main references Giese and Prodehl, 1976; Buness, 1992; Valasek, 1992; Ye, 1992; Ye et al., 1995; Scarascia and Cassinis, 1997) lies entirely in the Southern Alps, showing the complicated structure below, thus contrasting with the previous line. The Moho boundary appears to deepen eastwards from 36 to 50 km beneath the Adamello pluton. In the eastern part of the section, Moho depths range from 44 to 38 km, increasing westwards. In the middle, a Moho is seen, rising up to 28 km. This thin crustal block is localised in the Euganei hills area where the gravity field shows a strong positive anomaly. Here the top of the lower crust is 15–20 km deep and deepens to about 30 km eastwards. In the west, its velocity is lower than in the east.

Alpine transepts:

- Western Alps: in section 3 (Figure 4.4a) the interpretation of the WARR data available is shown along a profile approximately corresponding to the CROP-ECORS seismic vertical reflection profile (Grenoble–Torino–Asti). To build the cross-section, we referred mainly to the interpretation by Giese and Prodehl (1976), made prior to the recording of the NVR – Near Vertical Reflection line – CROP-ECORS and, therefore, only after the WARR data. As later confirmed by the fans recorded in preparation for the NVR line (Hirn et al., 1989), the European Moho plunges southeastwards and reaches a depth of about 55 km beneath the western Alpine range; then, before disappearing, its dip seems to reverse: this situation has been variously interpreted. The average velocity in the lower European crust seems higher here than beneath cross-section 1. On the Adriatic (or "Padan") side a fragment (or a sequence of fragments) of Moho-like materiel is observed imbricated into the Alpine crust and reaching its highest point at a depth of about 12 km beneath the Insubric (or Periadriatic) line, only slightly north of the outcropping "Ivrea body" (see also Nicolas et al., 1990). Therefore, in the western Alpine arc, the westwards push of the high density materiel at very shallow depths is very clear.
- Central Alps: the Alpine stretch of EGT (section 4, Figure 4.4a, main references Ye et al., 1995; Holliger and Kissling, 1992; Buness, 1992) is the better controlled WARR profile crossing the Alps, being based on several reversed shots, although they are spaced more than 100 km apart. Moreover it intersects, among others, two longitudinal crustal and lithospheric profiles (SudALP and ALP '75). The interpretation of the Alpine segment of EGT was modified as other data became available, especially after the completion of the Swiss NFP 20 programme of NVR lines. The seismic techniques used in the latter Project (the combination of big shots and vibratory sources, and the use of small and large offsets) gave good data both on the crystalline crust (definition of the "nappes") and on the lower crust. The reflections gathered along the lines recorded in '86 and '88 were

---

Figures 4.1–4.12: Interpretative cross-sections.
Legend (the symbols shown in Figure 4.1 are common to all cross-sections): 1 – Velocities in km/s; 2 – Surface layers (V < 6.0 km/s); 3 – Upper or undifferentiated crust (V = 6.0–6.5 km/s); 4 – Lower crust (6.5–7.5 km/s); 5 – Upper mantle (V > 7.5 km/s); 6 – Low velocity layers; 7 – Coastline; 8 – Intersections between sections. P.A.L.: Periadriatic Lineament. C.S.: cross-section. The major sources assumed for the interpretation of each transept are quoted in the reference list.

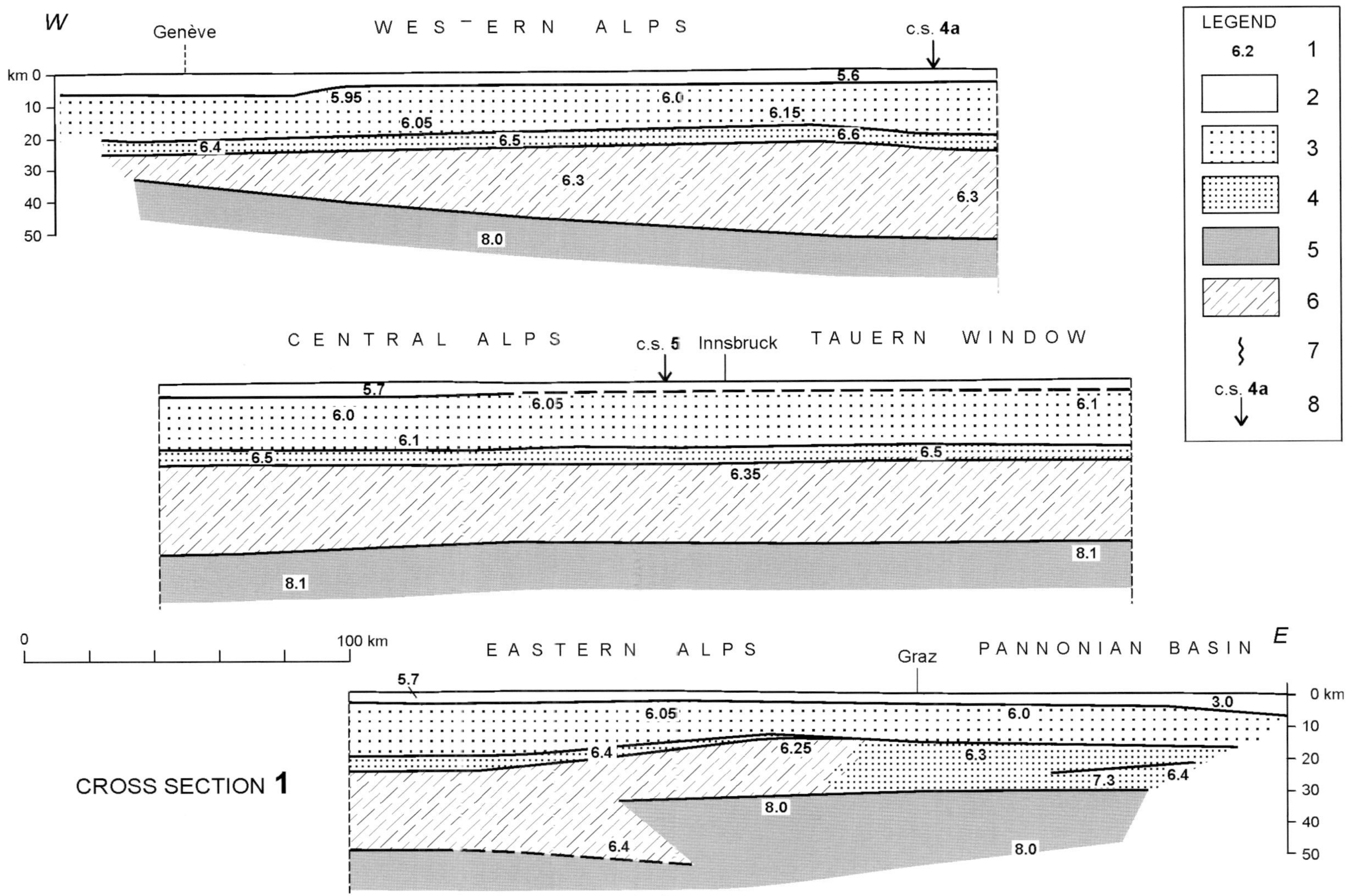

Figure 4.1: Western Alps–Central Alps–Pannonian Basin (ALP '75). The crustal structure appears homogeneous along the central part of the profile, the crustal thickness being around 50 km. The thickness of the upper crust is about 20 km, while that of the lower crust is about 30 km; the seismic velocity of the latter is abnormally low. A gentle rise in the Moho boundary is observed at the western end of the profile up to depth of 30 km beneath the city of Geneva. In the transition zone from the eastern Alps to the Styrian and Pannonian Basin, the Moho suddenly rises from 55 km up to 30 km.

projected onto the EGT velocity model after migration, using the velocity values given by WARR (Valasek, 1992; Ye et al., 1995). The model (mainly derived from Buness, 1992) shows remarkable similarities with the Western Alps transept of Figure 4.4a, as far as the lower crust and the Moho on the European side are concerned. Also here, the deep NVR data become very uncertain south of the Penninic front, while WARR defines clearly enough the uplifting of the Moho beneath the Po Plain, as well as the crustal overriding beneath the Insubric line. The

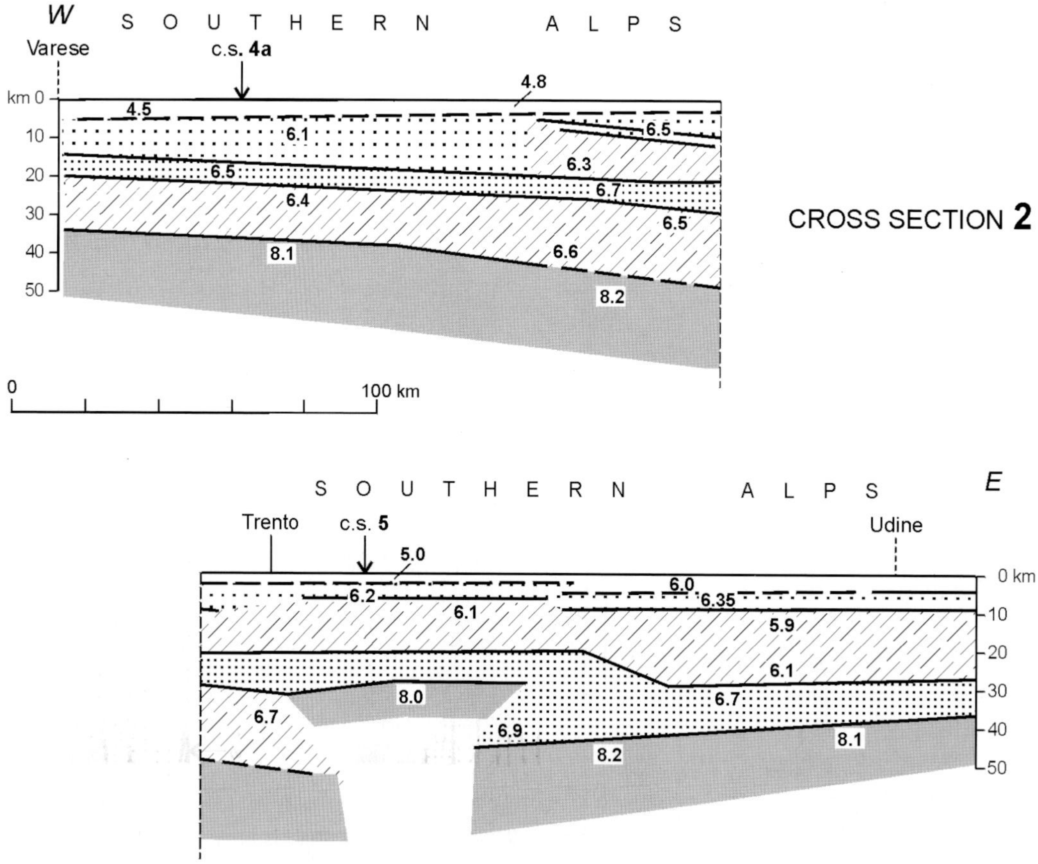

Figure 4.2: Lombardian Southern Alps–Southeastern Alps (SudALP). In the Lombardian domain, the Moho boundary deepens eastwards from 33 to 45 km. Beneath the Euganei hills (crossing with C.S. 5) a sudden uplift of the upper mantle to 28 km is shown, in agreement with the strongly positive Bouguer anomaly (see Figure 3). The thickness of the lower crust is about 20–30 km in the western sector, while it is reduced to about 10 km beneath the Euganei hills and in the eastern sector.

Figure 4.3: Grenoble–Western Alps–Ivrea Zone–Western Po Plain. The overthrusting of the Adriatic plate onto the European one is shown; The European upper mantle deepens from 33 km beneath Grenoble to 52 km under Montcenis, then slowly rises in the ESE direction. The Adriatic Moho boundary is lifted steeply from 30 km beneath the city of Asti to 10 km in the Ivrea zone, corresponding to the strong positive gravity anomaly (see Figure 3). The overriding is observed for nearly 40 km between Turin and the Montcenis pass.

Figure 4.4a: European Geotraverse (EGT): Swiss Alps–Po Plain–Ligurian Sea–Corsica. In this section, four crustal domains can be distinguished: in the Central-Southern Alps, the overthrust of the Adriatic on the European plate is shown, the edge of the Adriatic upper mantle being positioned near the Periadriatic Lineament. The front of the Adriatic lower crust appears shifted about 15 km northwards. The European upper mantle deepens southwards from 38 to 65 km, with a slope therefore of about 22%. The Adriatic Moho is 30–40 km deep beneath the Po Valley; it deepens down to 55 km under the Ligurian Apennines, where it is overridden by a Ligurian thinned crust, about 20–25 km thick. In the Ligurian Sea, a sub-oceanic crustal type is found about 18 km thick. Further southwards the European crustal type, with a thickness of about 30 km, is found again beneath Corsica.

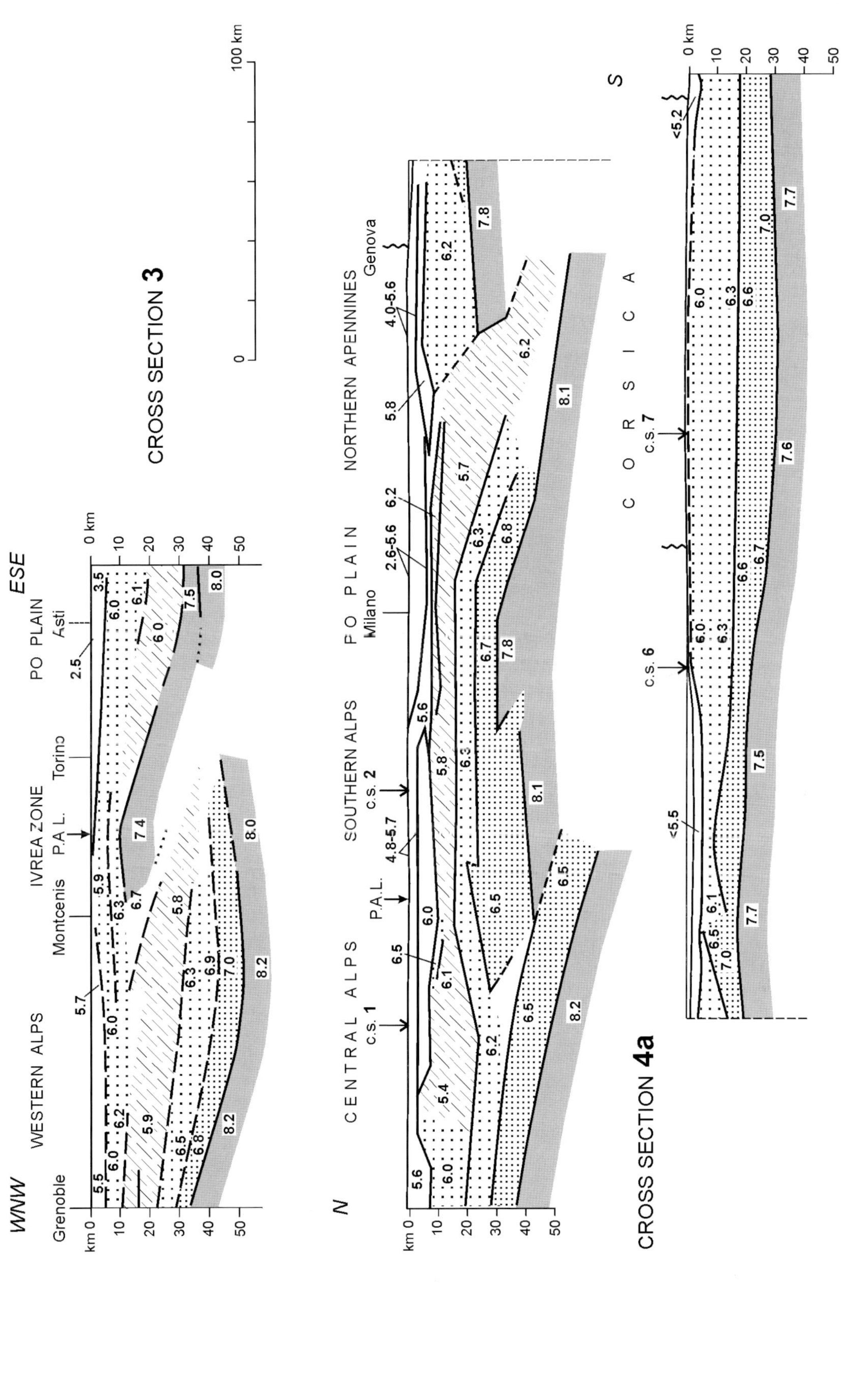
CROSS SECTION 3
WNW
ESE
WESTERN ALPS
IVREA ZONE
PO PLAIN
Grenoble
Montcenis
P.A.L.
Torino
Asti
0
100 km
CROSS SECTION 4a
N
S
CENTRAL ALPS
SOUTHERN ALPS
PO PLAIN
NORTHERN APENNINES
Milano
Genova
c.s. 1
c.s. 2
CORSICA
c.s. 6
c.s. 7

model of overriding, which involves a strong asymmetry of the Alpine structure, in contrast to the previous hypothesis of vertical subduction, is also confirmed by the gravity modelling (Cassinis et al., 1997).

However, even in this transept, the solution to some problems is not unique.

- Eastern Alps: in the Alpine sector east of the Giudicarie line, investigation of crustal structure is less developed than in the western and central sectors, because NVR profiles are not yet available: the first deep reflection seismic line (Project TRANSALP) is still in progress. However, a homogeneous reinterpretation (after digitization of the old seismograms) of several WARR profiles, recorded during the different phases of the early exploration, has shed more light on the structure of the lower crust and on the Moho boundary in the whole area from the central to the Eastern Alps (Scarascia and Cassinis, 1997). Six interpretive cross-sections were proposed as the final output of the seismic modelling, as well as a map of the crust-mantle boundary. We briefly describe the interpretation of cross-section 5 (Figure 4.5), positioned about 170 km east of the EGT, based on several WARR profiles, both across and parallel to the strike. It runs approximately N–S, from Eschenlohe (N of Innsbruck, on the calcareous Alps) to the Euganei hills near Vicenza. The seismic model shows similarities with the EGT transept (the overriding of the Adria Moho on the European Moho, the existence of a wedge of Adria lower crust whose extension northwards seems larger than in EGT). But major distinctions are apparent in the Moho's shape beneath the Southern Alps, where it seems to become flatter. According to the interpretation, the velocity of the European deep crust beneath the Northern Alps is lower than in EGT. The edge of the Adria Moho is well south (about 80 km) of the Insubric line rather then corresponding to this surface expression.

4. LIGURIAN, TUSCAN, NORTH APENNINES REGION, SARDO-CORSICAN BLOCK

– The southern part of section 4a (Figure 4.4a), also coinciding with the European Geotraverse, crosses the Po Plain, the Ligurian Apennines, the Ligurian sea and Corsica, from north to south. To reconstruct the crustal section, the interpretation by Egger et al. (1988), was used. In the northernmost part (Po plain and Apennines), the interpretation by Buness (1992), was also used (see also Giese, 1990). The transition from the Padan-Adriatic crust to the Ligurian transitional type is evidenced beneath the Apennines. The bottom of the Adriatic crust can be observed at a depth of about 30 km in the middle of the Po Plain, then dipping southward and reaching about 55 km beneath the Ligurian coast, thus suggesting a subduction of the Adriatic plate. A shallower Moho-like boundary is shown here at depths of 22–25 km; southwards, it gently rises, reaching a minimum thickness (about 17 km) halfway between Liguria and Corsica, where a sub-oceanic type of crust is found; this "Ligurian" crust seems to override the Adriatic one beneath the Apennines.

Southwards, the crust becomes thicker again, reaching 33 km beneath Corsica, and has a typical continental structure of European crustal type that is repeated in Sardinia (Figure 4.5).

– Cross-section 6 (Figure 4.7), SW–NE trending, crosscuts the Ligurian Sea north of Corsica, and the Apennines between Viareggio and Bologna. The available data were obtained from a line of closely spaced shots fired offshore at a minimum distance of 100 km from the shoreline and recorded by onshore stations; no reversed records are available. For these reasons, only qualitative inferences about crustal structure can be made; however, a refracting

---

Figure 4.4b: European Geotraverse (EGT): Sardinia–Sardinian Channel–Northern Tunisia. This stretch crosses the European type crust of Sardinia (thickness of about 30 km), the sub-oceanic crust in the Sardinia Channel (minimum thickness 20 km), the Magrebides belt and the Atlas range in north Tunisia; here the thickness gradually increases from 25 to 35 km. It should be noted that in northern Tunisia, the upper crust cannot be distinguished from the lower crust, and that the average velocity is abnormally low.

Figure 4.5: Southern Bavaria (Molasse Basin)–Central-Eastern Alps–Euganei Hills. In this transept the European crustal domain appears to extend southwards well beyond the Periadriatic Lineament. The slope of the Moho boundary is 22% in the northern sector (like along C.S. 4a) while it seems flatter in the southern one, where a depth of 55 km is reached. The overthrusting of the Adriatic upper mantle (30 km deep) onto the European (55 km deep) is positioned beneath the city of Trento. The Adriatic lower crustal layers also override the European lower crust, but their front appears to be shifted northwards to a much larger extent than along the transept 4a (about 100 km).

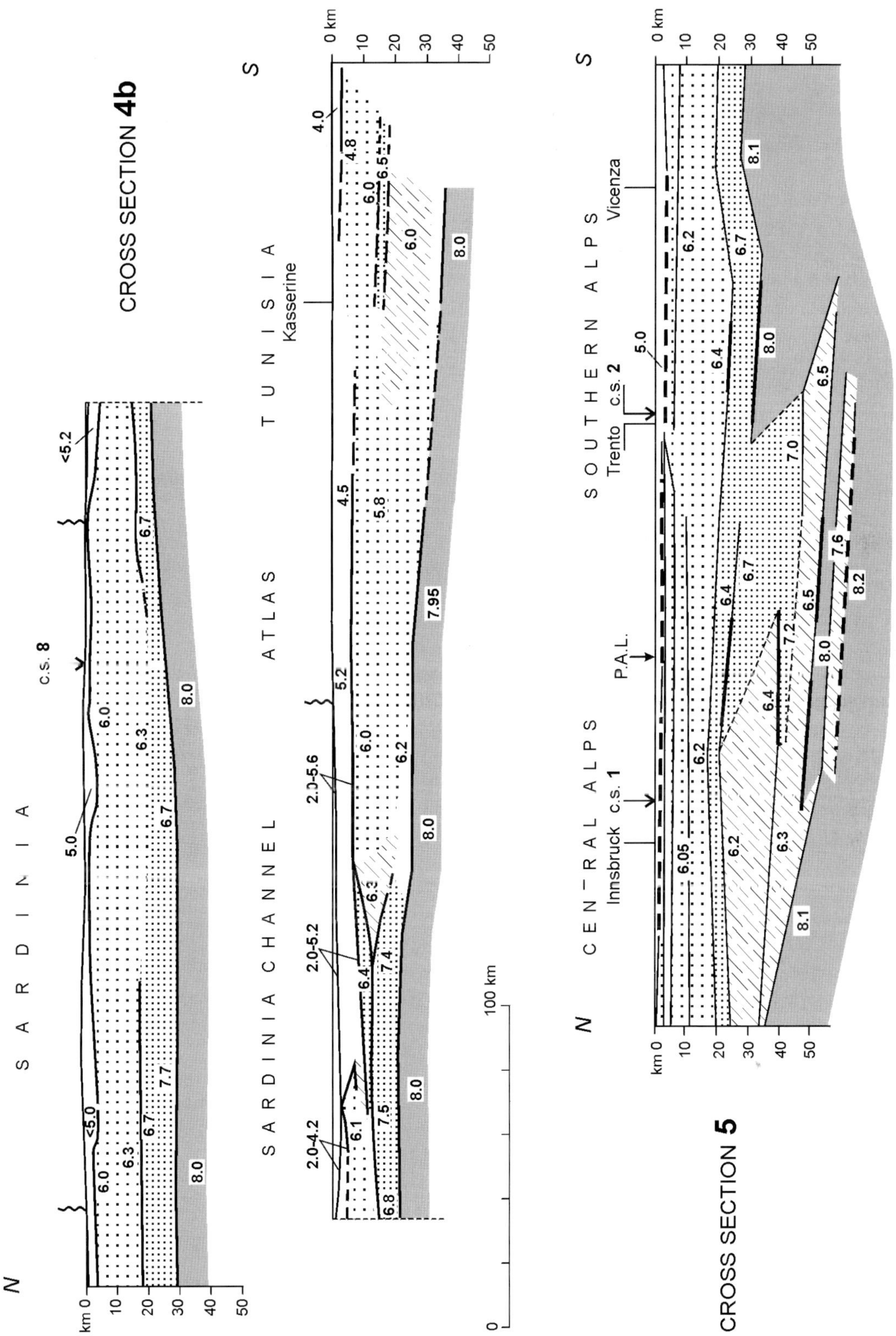
CROSS SECTION 4b
N
S A R D I N I A
c.s. 8
km 0
10
20
30
40
50
S A R D I N I A C H A N N E L
A T L A S
T U N I S I A
Kasserine
S
0 km
100 km
0
CROSS SECTION 5
N
C E N T R A L A L P S
Innsbruck
c.s. 1
P.A.L.
S O U T H E R N A L P S
Trento
c.s. 2
Vicenza
S

horizon (Pn phase) with a velocity of 7.5 km/s can be followed, gently dipping to the NE at a depth of 20 km beneath the Ligurian Sea descending to 25 km under the Apennines. A deeper boundary, with a velocity of 8.0–8.2 km/s, is seen offshore the coast of Tuscany, strongly dipping to the NE at a depth of 45–50 km (Colombi et al., 1977; Scarascia, 1979).

– More information is available on section 7 (Figure 4.7). Data have been obtained both from closely-spaced shots offshore west Corsica and between Corsica and Elba, as well as from land shots on the Tuscan coast near Grosseto, Perugia and Ancona, on the Adriatic coast. Land recording stations were located in Corsica, Elba and in Tuscany, Umbria and the Marches. The section was reconstructed using the interpretations by Hirn and Sapin (1977), Letz et al. (1977) and Wigger (1984) (see also Giese et al., 1978, 1981; Giese, 1990; De Franco et al., 1998).

The cross-section shows that within a distance of about 600 km, at least four major crustal types are met; starting from the west: the sub-oceanic crust of the Western Mediterranean, the continental crust of Corsica (European), the anomalous, transitional, thin crust of western Tuscany and finally the continental crust of the Adriatic microplate margin. Between Corsica and Elba, the Moho boundary seems to sink to about 40 km; further eastwards, a very deep wide-angle reflection is observed at 50–60 km, also dipping ENE. The correlation between the two events is uncertain. The deep reflection could be interpreted as a remnant of the European plate subduction process; the same type of event is seen off the Tuscan coast at Viareggio on cross-section 6. Further eastwards, a sudden crustal thickening is observed near Perugia; even if the data are not sufficient to support the hypothesis of a westwards Adriatic subduction, they clearly show the transition to the Adriatic domain.

5. TYRRHENIAN SEA, CENTRAL-SOUTHERN APENNINES, CALABRIAN ARC

– Section 8 (Figure 4.9) runs from Sardinia (S. Antioco, southwestern coast) to the Adriatic coast at Pescara, crossing the Tyrrhenian Sea and the Apennines. The transept has only been partially explored: data are available only in the coastal belt of Sardinia, in the Tyrrhenian from two O.B.S. (ocean bottom seismometers) and in the area between Latina (Latium coastline) and Pescara. The section is reconstructed on the basis of the interpretations by Scarascia (1980), Nicolich (1981) and the Italian Explosion Seismology Group (1982).

The south western part of the section shows a gradual transition from continental crust under Sardinia to oceanic or sub-oceanic crust towards both the Western Mediterranean and the Tyrrhenian. The smallest crustal thickness in the Tyrrhenian is found at the crossing with section 9 (17 km). Approaching the coast of Italy, the thickness gradually increases to 26 km beneath Latina. Thicknesses of about 30 km can be observed west of Pescara, corresponding to the Adriatic plate. Between Latina and Pescara, a sharp step in the Moho boundary occurs, similar to the one observed beneath Perugia along section 7.

– Section 9 (Figure 4.9) is a long transept (mainly offshore) crossing the Tyrrhenian Sea from NW to SE. It starts near the north eastern coast of Sardinia (Baronie), crossing prominent sea mounts of the Tyrrhenian (Vavilov, Marsili), than Calabria (Cape Vaticano, Punta Stilo), and finally terminating in the Ionian Sea. The Tyrrhenian stretch was explored by O.B.S., recording several closely-spaced shots offshore. In Calabria and in the coastal belts, the shots offshore were recorded by stations inland. Reconstruction of the cross-sections is based on the

Figure 4.6: Ligurian Sea–Viareggio–Northern Apennines. In the Ligurian Sea, a thinned crust (about 20 km) gently deepens east-northeast to 25 km, then disappears beneath the coastline. A deeper Moho-like boundary, plunging from 45 to 55 km in the ENE direction, is also observed, which could be interpreted as a remnant of the subduction of the European plate under the Adriatic.

Figure 4.7: Western Mediterranean Basin–Corsica–Elba–Ancona. Several typologies are revealed in this section: a thinned crust is interpreted in the western Mediterranean Basin (thickness about 20 km), a continental crust beneath Corsica (about 30 km thick) where the upper-mantle layers deepen in the NE direction, then disappear under the Elba Channel with a possible subduction process similar to the one described for line 6. A shallower Moho appears to override the former; east of this boundary, a thinned type of crust is found (Perityrrhenian–Tuscan), characterised by low-velocity layers in the lower crust, that stretches from Elba up to Perugia, (about 140 km). Near Perugia, the crust suddenly thickens and the low-velocity layers disappear, showing that the Adriatic plate is reached.

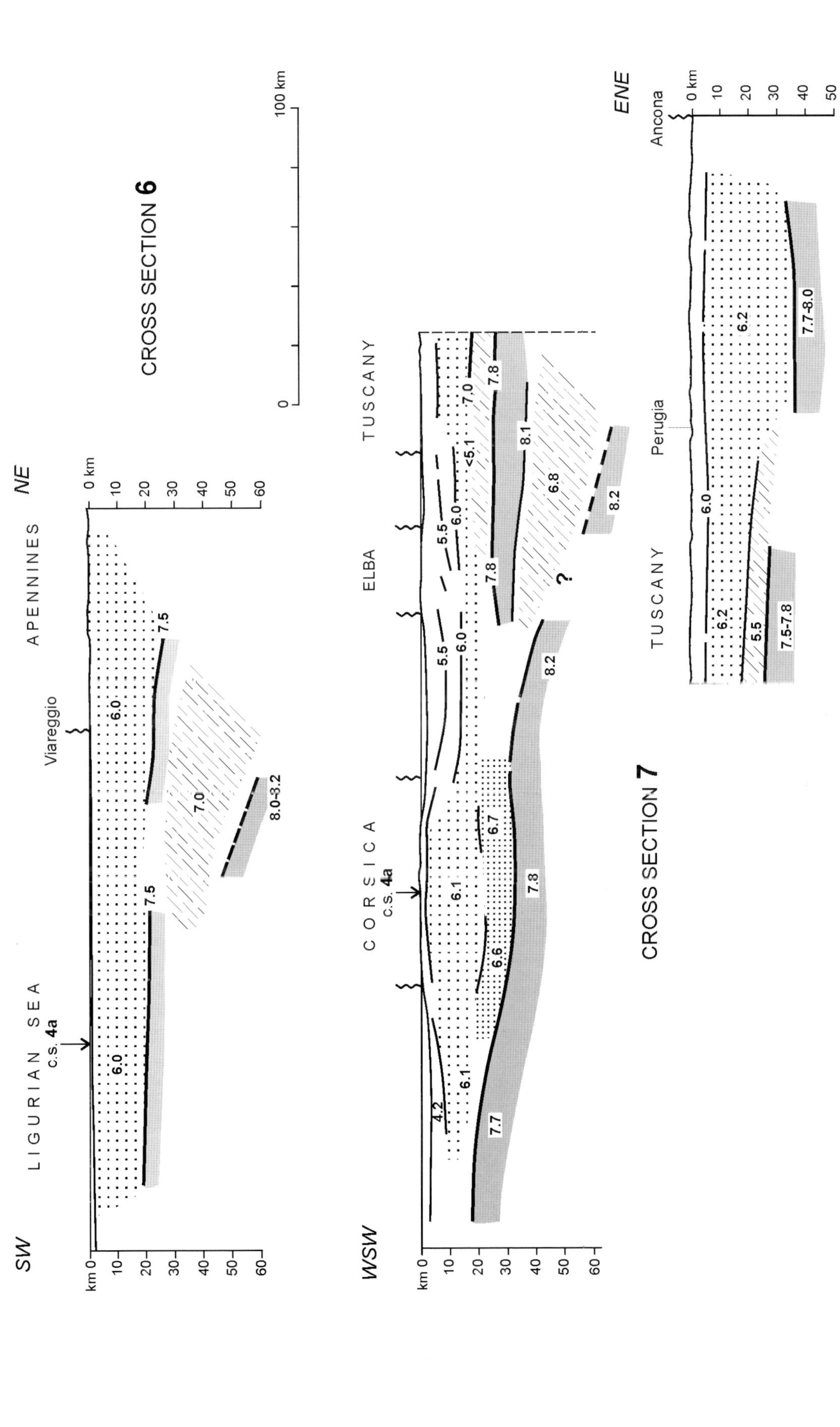
CROSS SECTION 6
SW
NE
LIGURIAN SEA
APENNINES
Viareggio
c.s. 4a
6.0
7.5
7.0
8.0-8.2
0
100 km
CROSS SECTION 7
WSW
ENE
CORSICA
ELBA
TUSCANY
Perugia
Ancona
4.2
6.1
6.6
6.7
7.7
7.8
8.2
5.5
6.0
<5.1
7.0
8.1
6.8
6.2
7.7-8.0
7.5-7.8
km 0
10
20
30
40
50
60

interpretations by Steinmetz et al. (1983) for the Tyrrhenian, by the Italian Explosion Seismology Group (1980), for Calabria and for the coastal belts, and by Linari (1981) for the Ionian Sea.

In the Tyrrhenian, it was possible to determine Moho depth continuously by correlation of the Pn phases and assuming an average crustal velocity of 6.0 km/s. The velocity of the upper mantle ranges between 7.7 and 8.0 km/s. The depth of the crust-mantle boundary in the Tyrrhenian Sea shows two minima (9 and 11 km), corresponding respectively to the Vavilov and Marsili sea-mounts. Between these two areas, there is a broad subsided zone where the Moho is over 20 km deep.

Towards the NW end of the section (Baronie) the crust thickens as the continental structure of the Sardinian-Corsican block is approached. At the opposite end, the crust is relatively thin (between 14 and 20 km), both in the southern Tyrrhenian and beneath Calabria. About 20 km off the Calabrian coast, in the Ionian sea, the shallow boundary (15 km deep) disappears and a deeper discontinuity (over 35 km deep) is observed, gradually rising up to 18 km at about 100 km off the Ionian Calabrian coast (Punta Stilo). This feature looks similar to the one described beneath the Northern Apennines; therefore, obduction of the thin Calabrian and Perityrrhenian crust on the African plate is demonstrated. Off the Calabrian coast, in the Ionian sea, a low crustal velocity is found (about 5.8 km/s).

– Section 10 (Figure 4.10) starts from the Tyrrhenian sea, 50 km off the coast of Cilento, then crosses the Southern Apennines and the Apulian platform, finally reaching the Gargano promontory. The reconstruction of this section is done from data gathered in 1971 (land stations and closely-spaced shots offshore) and 1980 (a series of reversed profiles carried out on land, both parallel and perpendicular to the Apennines), as well as on the interpretations by Colombi et al. (1973), Biella et al. (1994).

The section shows a Perityrrhenian thin crust (22–30 km), extending inland for about 30 km beneath the Cilento, and an Adriatic continental crust, about 25 km thick, beneath the Gargano, dipping towards the SW to a depth of about 50 km below the Tyrrhenian coast. Then an overriding of thinner Perityrrhenian crust on the Adriatic plate is observed; here the evidence is clear of a westwards subduction of the latter.

– Sections 11a and 11b (Figures 4.11a, 4.11b) are dealt with together because the latter is the extension of the former. Section 11a starts at the Aeolian Islands, crosses Calabria and the Gulf of Taranto and ends in the Salentina peninsula. Section 11b starts in the Sicily Channel, north of Pantelleria, crosses Sicily from Sciacca on the south coast to Termini Imerese on the Tyrrhenian coast, ending in the Aeolian archipelago (island of Salina). Seismic data were obtained from surveys carried out in 1971 (Colombi et al., 1973), 1973 (Morelli et al., 1975) and 1985 (Cosentino et al., 1987). The shots were located offshore (Sicily Channel, South Tyrrhenian, Gulf of Taranto and Adriatic Sea) and recorded by land-based stations at Pantelleria, Sicily, the Aeolian islands, Calabria and the Salentina peninsula.

Section 11a shows a crustal thickness of about 18 km under the Aeolian islands. Beneath Calabria the same situation observed in section 9 occurs. In fact, the whole Calabrian stretch of the section is characterised by a crust-mantle boundary at a depth of about 20–25 km, while under the Ionian coastline the evidence for a deeper discontinuity at a depth of 40 km is clear. The latter can be correlated with the bottom of the Salentina peninsula continental crust (about 30 km thick).

Figure 4.8: Sardinia–Tyrrhenian Basin–Pescara. The continental European crust of Sardinia is about 30 km thick, 12 km being the thickness of the lower crust. The crustal structure and type change in the Tyrrhenian Sea, where the minimum thickness is 18 km. In the stretch between Latina and Pescara, a transition between the Perityrrhenian thinned crust and the Adriatic plate is seen, similar to the one observed in cross-section 7, but less clear.

Figure 4.9: Baronie–Tyrrhenian Basin–Calabrian Arc–Ionian Sea. Along the Tyrrhenian Basin the crustal structure can be defined as oceanic or, in some areas, sub-oceanic. The velocity of the crust is assumed constant (6.0 km/s) but that of the upper mantle is accurately determined by the Pn arrivals, and ranges between 7.7 and 8.0 km/s. Two minima crustal thickness were found near the Vavilov and Marsili volcanoes (10 km). The structure becomes more complicated beneath Calabria and the Ionian Sea where a thinned crust (about 15–20 km thick) overthrusts a thicker crustal type, representing the northernmost extension of the African plate.

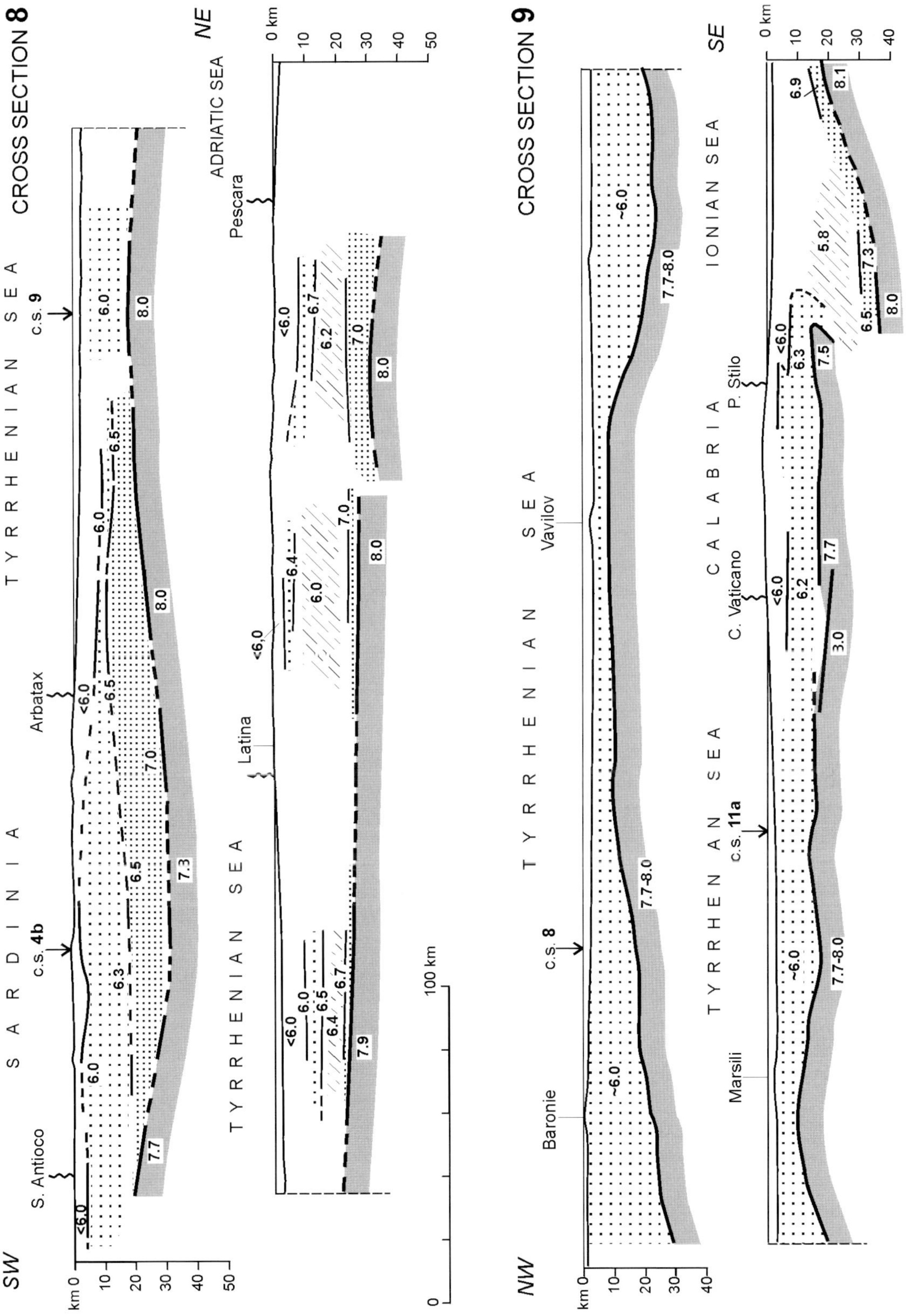
CROSS SECTION 8
SW
NE
S A R D I N I A
T Y R R H E N I A N S E A
ADRIATIC SEA
S. Antioco
c.s. 4b
Arbatax
c.s. 9
Latina
Pescara
100 km
CROSS SECTION 9
NW
SE
T Y R R H E N I A N S E A
Baronie
c.s. 8
Vavilov
Marsili
c.s. 11a
C. Vaticano
P. Stilo
C A L A B R I A
I O N I A N S E A

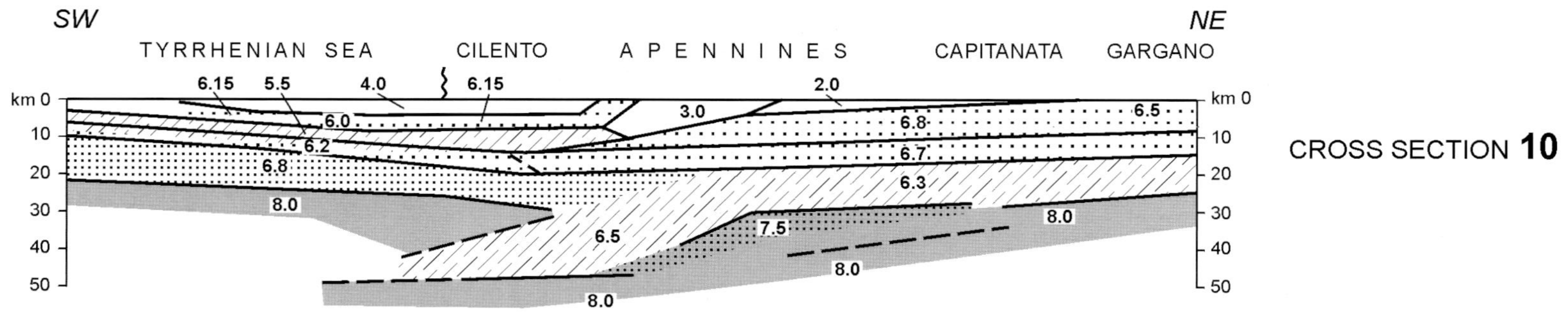

Figure 4.10: South Tyrrhenian Sea–Cilento–Gargano Promontory. In the Perityrrhenian zone, a tinned crust (22–25 km thick) extends from the Tyrrhenian Sea to the Cilento promontory. Here an extended subduction of the Adriatic plate under the Perityrrhenian crust is shown, the Adriatic Moho boundary reaching depth of 50 km. The Adriatic crust thins eastwards approaching the Gargano promontory where the Moho is found at 24 km.

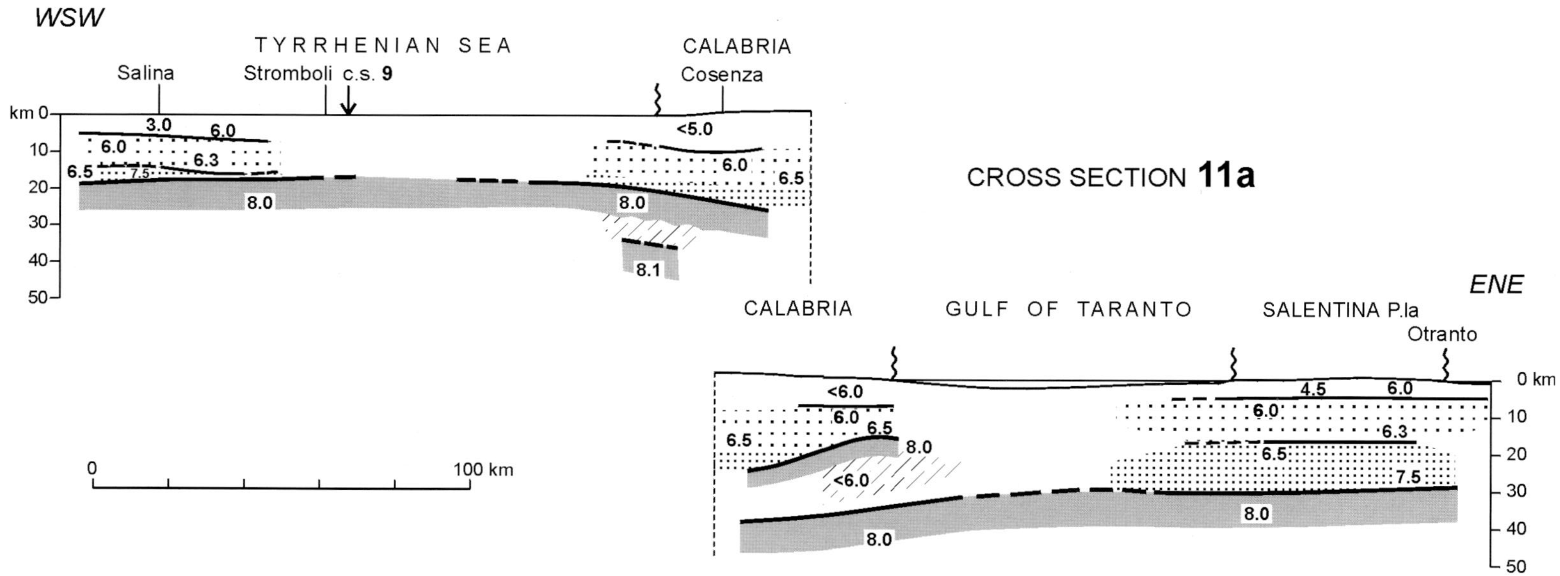

Figure 4.11a: Aeolian Islands–Calabria–Gulf of Taranto–Salentina Peninsula. Below the Tyrrhenian Sea and Calabria, a thinned crust is shown (about 20 km). The Adriatic crust in the Salentina peninsula is 30 km thick and the Moho boundary deepens towards Calabria where an extended subduction zone is interpreted.

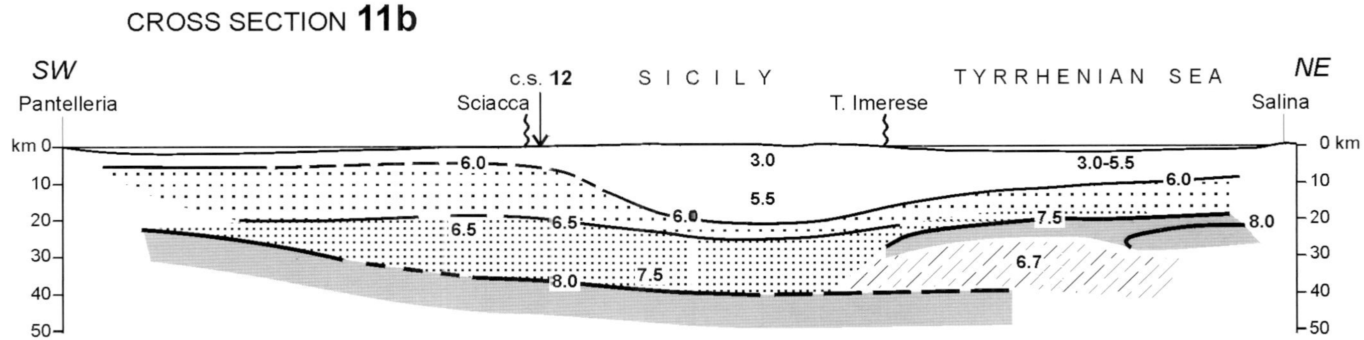

Figure 4.11b: Pantelleria Rift–Western Sicily–Aeolian Islands. In the southwestern section of the transept, the crust, ascribed to the African plate, is 22 km thick in the Sicily Channel and becomes thicker approaching the southern Sicilian coast where it reaches 40 km; this appears to be in agreement with the gravity anomalies (see Figure 3). In the northeastern stretch of the section, the Perityrrhenian thinned crust extends up to the northern coastline of Sicily. In this area, a subduction of the African plate beneath the Perityrrhenian domain is evidenced.

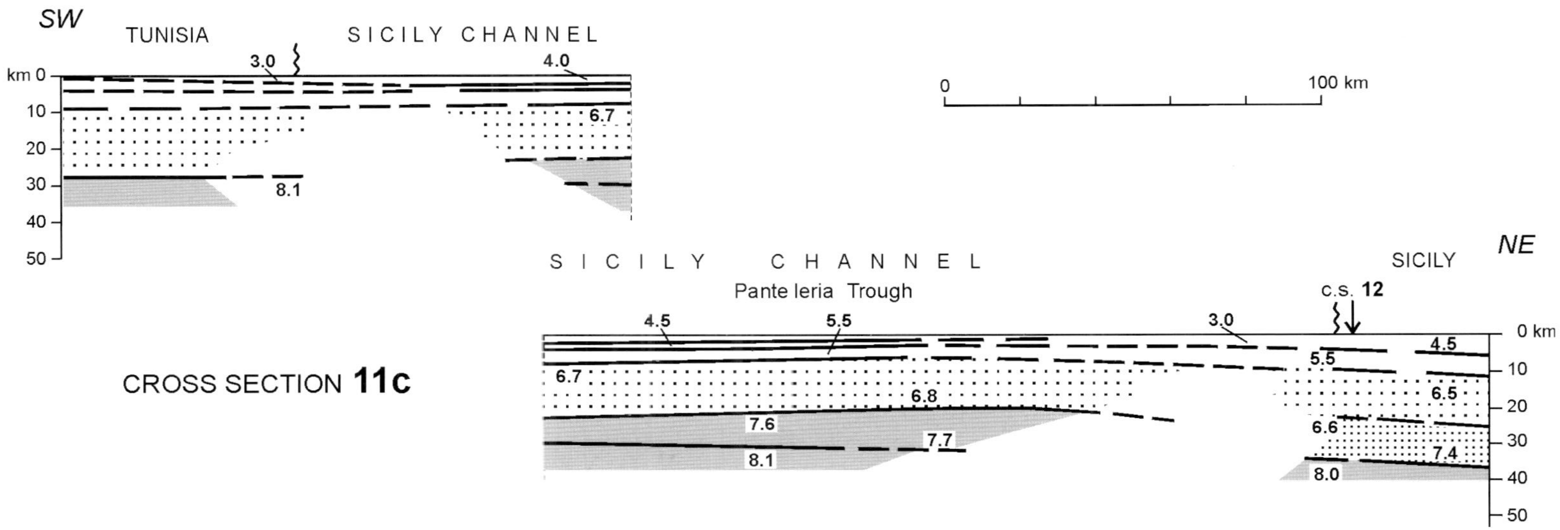

Figure 11.c: Eastern Tunisia–Pantelleria Trough–South Sicily. This section shows a crustal thinning (up to 20 km) in the Sicilian Channel beneath the Pantelleria rift, where the upper mantle reveals relatively low velocity values (7.5–7.7 km/s). The crustal thickness beneath the Tunisia coastline is 28 km, and 38 km beneath that of Sicily.

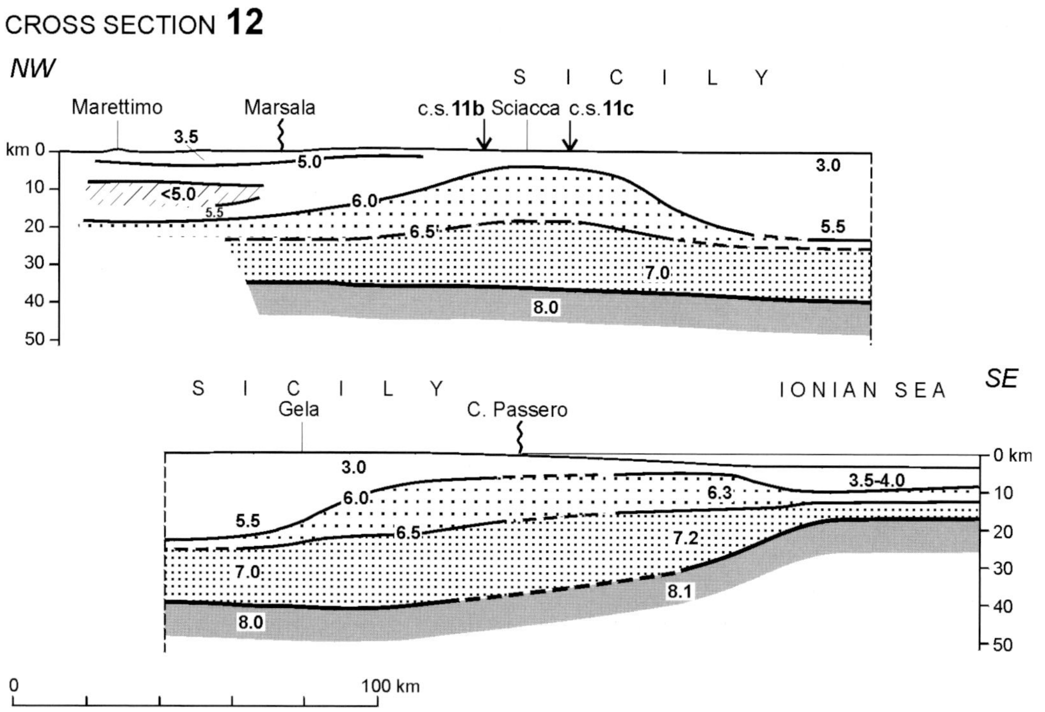

Figure 4.12: Aegadian Islands–Southern Sicilian Coast–Ionian Sea. The Moho depth along the Sicilian coastline ranges between 35 and 40 km and the shallow layers reach a remarkable thickness (20 km). Approaching the southeastern end of the section (towards the Malta escarpment) the Moho boundary suddenly rises to a depth of about 18 km.

Section 11b shows a fast and regular deepening of the bottom of the African crust, from about 20 km beneath Pantelleria to about 40 km beneath central Sicily. In the southern Tyrrhenian Sea, between Termini Imerese and the Aeolian islands, a thin crust of Perityrrhenian type (20–25 km thick) can be recognised, as well as a deeper discontinuity that can be correlated with the bottom of the African crust at a depth of about 40 km (Chironi et al., 2000). This situation is similar to the one observed in section 9 beneath Calabria, indicating obduction of the Perityrrhenian crust on the African plate.

6. SARDINIAN CHANNEL, SICILIAN CHANNEL, NORTHERN TUNISIA

– Section 12 (Figures 4.12) runs along the southern coast of Sicily, between the Aegadian islands and Cape Passero and extends to the SE towards the Ionian abyssal plain. The stretch of the section beneath Sicily was explored using reversed profiles, with shots positioned offshore near the Aegadian islands, in the Gulf of Gela and near the Cape Passero; the shots were recorded by land stations (Cosentino et al., 1987). In the Ionian Sea, O.B.S. and closely-spaced shots offshore were employed (Makris et al., 1986).

Along the Sicilian coast, the thickness of the crust varies between 35 and 40 km. The shallow low-velocity overburden, which around Sciacca (crossings with sections 11b and 11c) is 5–7 km thick, becomes thicker while proceeding to the SE, reaching about 20 km west of Gela (southern edge of the Caltanissetta sedimentary basin of central Sicily). Further SE, the depth of the Moho boundary rapidly decreases, reaching 18 km at the Malta escarpment.

– Section 11c (Figure 11.c): in 1985 a seismic wide-angle profile from eastern Tunisia to the Pantelleria trough was recorded in the framework of EGT, using 15 O.B.S. and explosive sources both onshore and offshore. The O.B.S. records were interpreted only recently (Scarascia et al., 2000) giving a description of the crustal structure along a transept about 100 km long, oriented SW–NE and reaching the southwestern margin of the Pantelleria rift zone.

The Moho boundary rises north-eastwards, its depth ranging between 25 and 20 km. In this sector, the upper mantle is characterized by a low seismic velocity of about 7.6 km/s. A flat, deeper discontinuity in the upper mantle was also interpreted at a depth of about 30 km; below this latter boundary, the velocity reaches the value commonly found in the upper mantle (8.1 km/s).

In order to attempt a geophysical model of the whole transept from Tunisia to Sicily, other published seismic data were used (Maistrello et al., 1990; Res. Group for Litospheric Structure in Tunisia, 1989, 1992) together with the regional Bouguer anomalies.

The Moho is 28 km deep beneath the Tunisian coastline and is found at 34 km under Sicily. In both these areas, the velocity of the upper mantle is about 8.0–8.1 km/s, and therefore normal, in contrast to the structure found across the Sicilian Channel Rift zone (low velocity of about 7.6 km/s and depth of 20 km).

The crustal structure shown by the model is clearly asymmetrical, the northestern flank of the uplifted Moho being steeper than the one towards Tunisia.

A comparison of this model with the published interpretations of transepts across other rifts shows analogies (uplifting of the upper mantle together with low seismic velocity) with the crustal structure of some segments of the Afro-Arabian rift, especially in the Red Sea, where the process of rifting is relatively young and still in progress.

– The southern part of section 4b (Figure 4.5) leaves the south coast of Sardinia, crosses the Sardinia Channel and reaches the coast of northern Tunisia; here, after passing through the coastal belt, it enters into the Atlas Mts. The land portion of the profile in northern Tunisia is about 200 km long. The available data were discussed by the Res. Group for Lithospheric Structure in Tunisia (1989, 1992).

The continental type of crust of the Sardo-Corsican block (see Figures 4.4a and 4.5) south of Sardinia changes gradually, the crustal thickness being reduced from 32 to about 21 km; therefore, in the Sardinian Channel, a transitional type of crust is found. Further to the south, the depth of the crust-mantle boundary increases again, reaching 28 km beneath the Tunisian coast, like that found under the eastern Tunisian coast (section 11c). In the southern sector of the profile, the crustal structure becomes of continental type.

7. A NEW UNIFIED MAP OF THE MOHO BOUNDARY IN THE ITALIAN REGION: CONCLUSIONS

The complex geometry of the Moho revealed by the WARR profiles can be compared with the surface geological features and, therefore, is a basic tool for studying the geodynamic evolution of the region. This is apparent from a comparison of the scheme in Figure 2 (modified from Bigi et al. (1990), Boccaletti et al. (1990)), with the Moho map in Figure 5.

This new map, extended throughout the Italian region, as defined in Figure 2, enables us to better outline the crustal domains and their mutual relationships. On the basis of the crustal thickness and on the distribution of P wave velocities, we distinguish five crustal domains, which are represented by different colours. The yellow indicates the African-Adriatic plate, characterized by a total thickness ranging between about 22 and 35 km (thinned continental crust). This type is found in the Padan-Adriatic side of the Apennines, along the edge of the fragmented Adriatic domain beneath the Southern Alps (where, in some sunken blocks, the Moho depth can reach 40 km), along the Ionian coast of Calabria, in central-southern Sicily, in the Sicily Channel and in northern Tunisia.

The blue areas indicate the European type, found in the northern and western side of the Alpine range and in the Sardo-Corsican block; here the crustal thickness ranges from 35 to 55 km (beneath the Alpine chain) therefore it can be classified as continental or thickened continental crust. A third type (magenta) can be defined as sub-oceanic, its thickness ranging from about 12 km (south Tyrrhenian Sea) to 20–22 km in the Sardinian Channel and in the Provençal Basin. A fourth crustal type (green) is the transitional, Ligurian–Tuscan–Perityrrhenian domain that can be defined as an anomalous, thinned crust. Its thickness ranges between about 20 and 27 km and the seismic velocity is characterised by low average values both in the crust and in the upper mantle, and by inversions in the lower crust, which can be explained by the high thermal regime. Finally, a further type is added (yellow green): it is found only in the northeastern corner of the area, where along cross-section 1 (Figure 4.1) a sudden thinning of the crust is observed east of the city of Graz. This occurrence corresponds to the transition from the Alpine European domain to the Styrian and Pannonian Basins.

The boundaries between these domains are defined with a variable degree of clarity and sharpness. In some instances, they coincide to a satisfactory degree with the present tectonic lines. This is the case of the western-central section of the Periadriatic Lineament corresponding to the front of the Adriatic Moho overriding the European plate, while its eastern section (Pustertal–Gailtal) seems strongly disjointed in respect to the present position of the Moho edge. Along the front of the Padan-Adriatic Moho, the crust appears fragmented in small blocks, alternately subsided and uplifted, and tilted in opposite directions, clearly showing the variable directions of the pull and push stages during the Alpine evolution (Figures 4.2, 4.4a, 4.4a and 4.5).

In contrast to these complex features, the structure of the European crust appears remarkably homogeneous along the whole Alpine range (Figure 4.1), the south-eastward dip of the Moho boundary being of about 10°–12° beneath the Northern Alps and becoming gentler in the central-eastern sector, beneath the southern Alps; here the maximum extension northwards is reached by the wedge of the lower Adriatic crust, overriding the European lower crust (Figure 4.5). This wedge seems to become narrower in the central sector (Figure 4.4a) and almost disappears in the western Alps.

The regular dipping of the European Moho under the Adriatic crust suggests the lithospheric subduction; however, the supposed slab is lost at about 55 km and there is no evidence of a change in the dip angle. Along the western arc the eastward dip of the European Moho boundary (Figure 4.4a) seems similar to that observed in the central sector, although some very questionable data (recorded on fans) could suggest that the boundary rises again before disappearing at about 55 km.

In the western Alps, the material of the Adriatic lower crust and mantle is lifted at the Insubric (Canavese) line, forming the very shallow Ivrea geophysical body. The front of the Adriatic Moho overriding the European plate is traced with fair certainty along the whole Alpine range as far as the Ligurian coast; in the Ligurian Sea the boundary is discontinued while crossing the northeastern end of the Provençal sub-oceanic crust resulting from the opening of this basin. Further southwards it is interpreted as the separation between the Sardo-Corsican block (European domain) and the Ligurian–Tuscan crust; this latter crustal domain is considered as a peripheral part of the Adriatic crust, which was detached, lifted and thinned during the collision with the European plate due to the opening of the Provençal Basin and the counterclockwise rotation of the Sardo-Corsican block. To the south of Elba island, no information is available on the continuation of the boundary.

A second fundamental boundary traced on the map in Figure 5 is the overthrusting front of the thinned Ligurian–Tuscan–Perityrrhenian transitional domain on the Adriatic plate. It actually follows the Apenninic range throughout peninsular Italy and is shaped as a sequence of broken arcs: the Ligurian–Aemilian Apennines, the northern Apennines (Tuscan–Umbrian) and, in the south of the peninsula, the Calabrian Arc. The three bows are separated by discontinuities which can be related only in part to the shallower tectonic features. The overthrusting of the transitional crust on the Adriatic–African foreland is clear both in the northern Apennines and along the Calabrian Arc,

---

Figure 5: Depth contour-lines of the Moho boundary. Contour interval 2.5 km. Crustal types: 1: European plate; 2: Afro-Adriatic plate; 3: Styrian and Pannonian Basin; 4: Ligurian, Tuscan–Perityrrhenian transitional crust; 5: Oceanic–sub-oceanic crust; 6: Over-thrusting fronts of the Moho boundary: of the Adriatic over the European plate (Alpine range); of the Ligurian, Tuscan, Perityrrhenian transitional crust over the Adriatic–African plate (Apennines range); of the Ligurian–Tuscan over the European (Corsica); 7: fragmentation lines in the upper mantle; 8: Moho depth contours (km); 9: Moho depth contours (subducted).

This map distinguishes four main crustal domains (excluding the Stirian and Pannonian Basins) and shows their mutual relationships as well as the depth contours of the Moho boundary. The European plate type extends from the European foreland up to the Periadriatic Lineament and to the Corsica–Sardinia block. The Adriatic–African plate is located in the Padania region, along the northeastern and eastern sector of the Italian peninsula, in southern Sicily and in the Atlas region (the Pantelleria Rift zone was not distinguished in this map). A further crustal type is found along a continuous strip that extends from the Ligurian to the Tyrrhenian coastlines up to the Calabrian Arc and northern Sicily, reaching its maximum breadth in the Tuscany region. Oceanic and sub-oceanic crustal types are found in the Provençal and Tyrrhenian Basins. The interactions between these crustal types are indicated by the overthrusting fronts of the Moho boundaries where they can be clearly defined. The overthrusting front of the Adriatic plate onto the European one appears to coincide with the Periadriatic Lineament in the western and central sectors of the Alpine range, while in the eastern sector it is largely shifted to the south. Evidence of possible remnants of the subducted European crust under the Adriatic are indicated east of Corsica. The front of the Ligurian–Perityrrhenian Moho boundary overthrusting the African–Adriatic plate extends from Liguria to Calabria and Sicily, describing a broad bow in Tuscany and a very prominent arc in Calabria.

as well as in northern Sicily (Figures 4.7, 4.10, 4.9, 4.11a and 4.11b). Especially along the strongly curved bow of the Northern Apennines, where the transitional crust reaches its maximum width, a clear eastward shift of the overthrusting fronts is seen as one proceeds from the Moho to the shallower discontinuities (see also Figures 6a and 6b).

In northern Sicily the overthrusting front is still clear, but is lost offshore the north-western tip of the island (Aegadian islands). The boundary of the African plate is not clearly defined in northern Tunisia also (Figure 4.5). In eastern Tunisia, the structure of the crust is of Adriatic–African type. In the Sicilian Channel, the Pantelleria Rift, separating the western and eastern Mediterranean Basins, is crossed: its structure seems to show that the process of rifting is relatively young and still active (Figure 11.c).

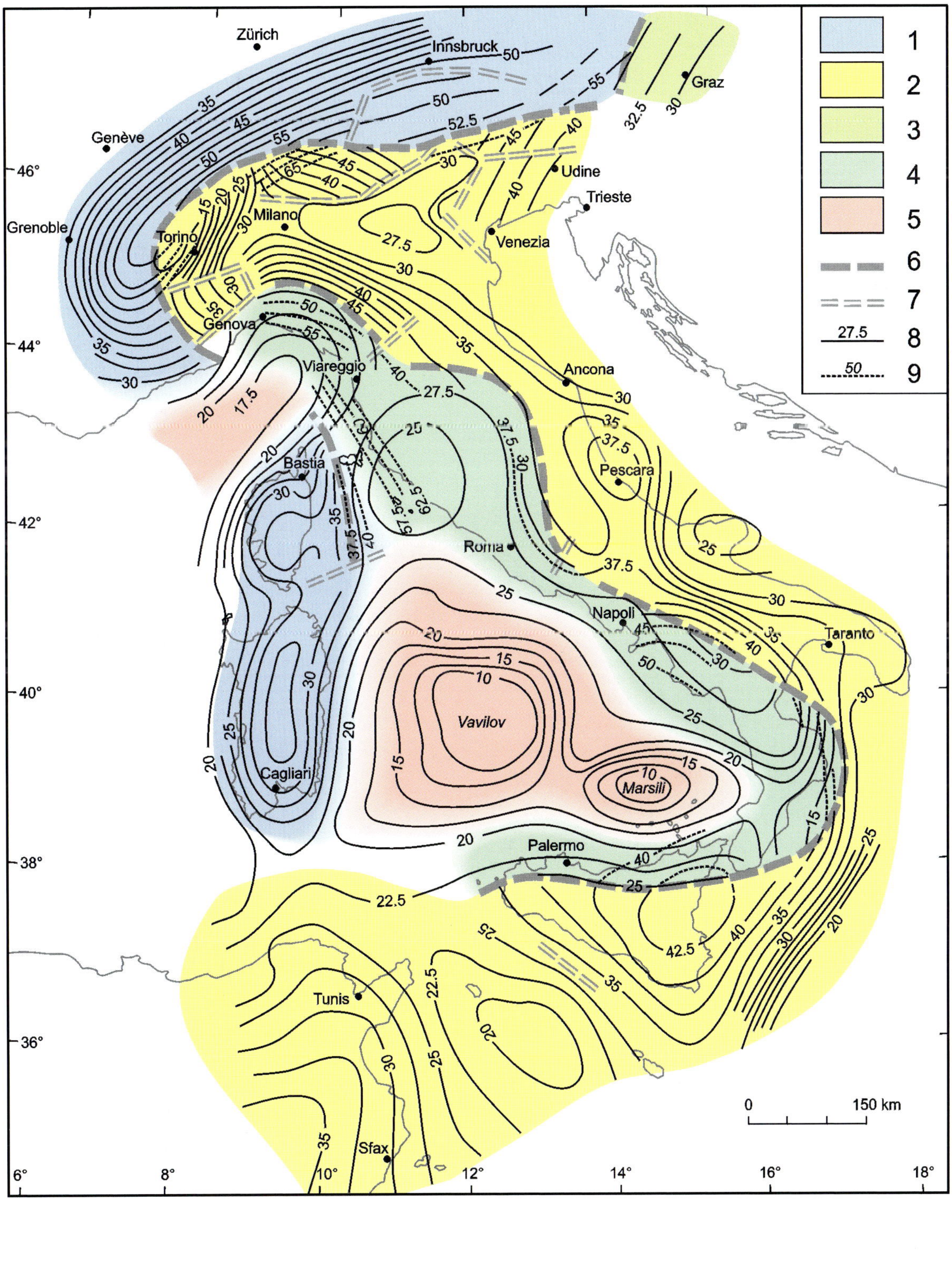

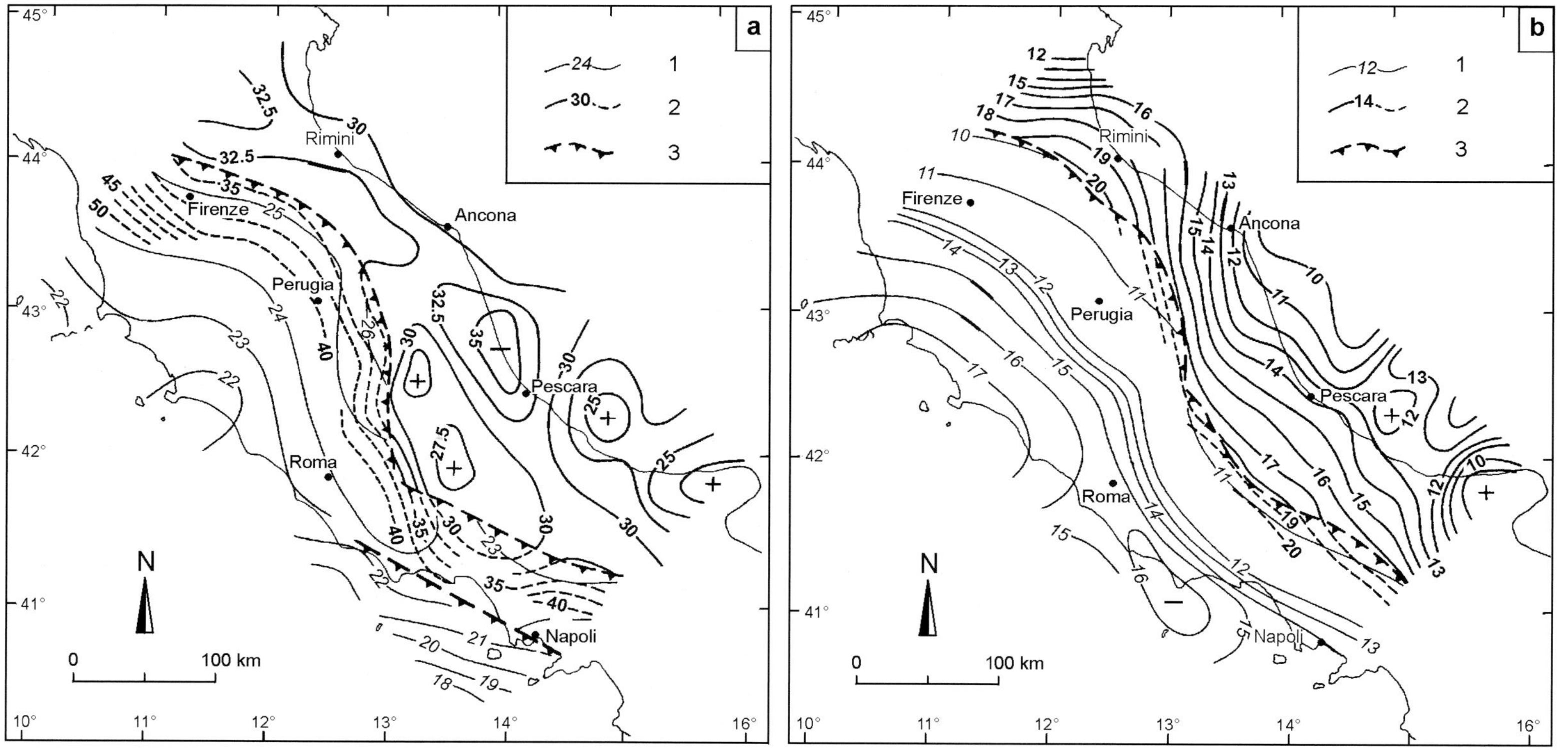

Figure 6: Examples of integration of seismic models using the gravity data (central-northern Apennines). a – Depth contours of the Moho boundary obtained by validating the WARR results with a 3D gravity modelling of the Bouguer anomalies. 1 – depth contours in the Tuscan–Latial–Perityrrhenian domain (contour interval 1.0 km); 2 – depth contours in the Adriatic crust (contour interval 2.5 km); 3 – Overthrusting front of the Moho boundary. b – Depth contours of the intermediate crustal discontinuity. 1 – depth contours in the Tuscan–Latial–Perityrrhenian domain (contour interval 1.0 km); 2 – depth contours in the Adriatic crust (contour interval 1.0 km); 3 – overriding front of the intermediate discontinuity. (Redrawn from Scarascia et al., 1998.)

Comparing the Moho boundary overthrusting front (Figure 6a) with the intermediate discontinuity (Figure 6b), it is noted that the latter is shifted between 20 to 50 km to the NE with respect to the former. The Apenninic foredeep is even more shifted northeastwards (Figure 2).

The map in Figure 5 confirms the general pattern of the geodynamic scheme of Figure 2. However, the picture of the Moho boundary and of the deep crustal structure suggests some observations on the relationships between the deep and shallow structures. Namely:

- the three Alpine sectors (eastern, central and western) behaved differences depending on the time and directions of the tectonic thrusts. While the opening of the Provençal Basin and the rotation of the Sardinian-Corsican block is the mechanism responsible for shaping the western Alpine Arc, the eastern Alpine sector (especially the southeastern one) was influenced by forces coming both from the east–northeast direction (Dinarides) and from the southwest, as consequence of the counter-clockwise rotation of the Italian Peninsula.
- the pattern of the transitional Ligurian–Tuscan–Perityrrhenian strip reflects the effect of the multi-temporal thrusts forming the Apennines chain, enhancing in particular the differences between the northern and southern range. The contrast among the northeastern portions of cross-sections 8 and 10 is quite clear: while in the former only a small vertical step marks the transition from the Moho of the Perityrrhenian to that of the Adriatic domain, in section 10 as well as in other transepts crossing the south Apennines and the Calabrian Arc (11a, 9, 11b) large jumps are observed.

It is clear that two fundamentally different processes are responsible for the evolution of peninsular Italy: the first was the opening of the Provençal Basin and the rotation of the Sardinia–Corsica block; the second was the opening of the Tyrrhenian Sea, from two different centres of expansion (Vavilov and Marsili) and at different times.

Figures 6a and 6b show two maps of the crust-mantle boundary and of an intermediate boundary corresponding approximately to the top of the lower crust for the area of north-central Italy. These maps were obtained by validating the seismic results with 3D gravity modelling (Scarascia et al., 1998). In the northern Apenninic arc, the eastern edge of the overriding Moho front is offset almost 50 km westwards with respect to the shallow formations forming the chain; at the southern end of the bow, where the maximum deposition occurs in the Apenninic foredeep, the fronts are closer.

The regional gravity anomalies are strongly dependent on crustal thickness, as is very clearly shown by a qualitative comparison of Figures 5 and 3. However, it must be noted that the Bouguer anomalies were used to interpolate results of the seismic cross-sections where the density of the latter is low; therefore in these instances the semblance of the two maps is enhanced. Some disagreements are observed where the gravitational effect of the shallow masses of variable density (like the sedimentary basins) is large.

MEMORIAL

After submitted present his last work, Dr Salvatore Scarascia passed away on July 2002. The Editor wants to remember the friend Salvatore with a short profile. He was director of the CNR's Institute for the Geophysics of the Lithosphere in Milano for several years, and is author of many high quality papers, especially in the field of wide-angle seismic profiling for earth crust study. Even after retiring, continued with passion his research activity in the interpretation of seismic and gravity data, giving further remarkable contributions to the understanding of the Alpine–Mediterranean Region. In this last work Salvatore condensed most of his important contributions to the Geophysics. The Geoscience community will hold him in great esteem.

REFERENCES[4]

Alpine Expl. Seism. Group (H. Miller, Reporter), 1976. A lithospheric seismic profile along the axis of the Alps, 1975. Pageoph 114, 1109–1129, **1**.

Biella, G., De Franco, R., Delfrati, L., Corsi, A., 1994. Profilo sismico Gargano-Pantelleria: digitalizzazione e processing dei dati relativi al tratto sud-occidentale (Tirreno-Sicilia). In: Atti del 13° Convegno Gruppo Naz. di Geofisica della Terra Solida, Roma, nov. 1994, pp. 107–112, **10**, **11a**.

Bigi, G., Cosentino, D., Parotto, M., Sartori, R., Scandone, P., 1990. Structural Model of Italy, Scale 1:500000. Edited by CNR, National Research Council of Italy, Progetto Finalizzato per la Geodinamica. Printed by S.E.L.C.A., Florence.

Blundell, D., Freeman, R., Mueller, S. (Eds.), 1992. A continent revealed – The European Geotraverse. Cambridge University Press. 275 p., **4a**, **4b**.

[4] The bold numbers indicate the cross sections described in the text whose interpretation is credited (partly or wholly) to the quoted Authors.

Boccaletti, M., Cello, G., Tortorici, L., 1990. First order kinematic elements in Tunisia and in the Pelagian block. Tectonophysics 176, 215–228.

Buness, H., 1992. Krustale Kollisionstructure an den Randen der northwestlichen Adriaplatte. PhD thesis, Freie Universität Berlin, 221 p., **1**, **2**, **4a**, **5**.

Cassinis, R., Federici, F., Galmozzi, A., Scarascia, S., 1997. A 3D gravity model of crustal structure in the central-eastern Alpine sector. Ann. Geofis. XL (October 1997), 1095–1107, **5**.

Cassinis, R., Franciosi, R., Scarascia, S., 1979. The structure of the Earth's Crust in Italy. A preliminary tipology based on seismic data. Boll. Geof. Teor. Appl. 21 (82), 105–126.

Chironi, C., De Luca, L., Luzio, D., Moretti, A., Vitale, M., Sealand Group, 2000. Crustal structures of the southern Tyrrhenian Sea and the Sicily Channel on the basis of the M25, M26, M28, M39 CROP Mare lines. Boll. Soc. Geol. It. 119, 189–204, **11b**, **11c**, **12**.

Colombi, B., Giese, P., Luongo, G., Morelli, C., Riuscetti, M., Scarascia, S., Schutte, K., Strowald, J., De Visentini, G., 1973. Preliminary report on the seismic refraction profile Gargano-Salerno, Palermo-Pantelleria. Boll. Geof. Teor. Appl. 15, 225–254, **10**, **11b**.

Colomb, B., Guerra, I., Scarascia, S., 1977. Crustal structure along two seismic refraction lines in the northern Apennines (lines 1b and 2). Boll. Geof. Teor. Appl. 19, 214–224, **6**.

Colombi, B., Scarascia, S., 1977. Evoluzione e ruolo della sismica crostale in Italia. Mem. Soc. Geol. It. XIII (suppl.), 97–112.

Cosentino, P., De Franco, R., Luzio, D., Scarascia, S., 1987. Results of DSS experiments in Sicily (EGT '84). In: EUG IV Meeting, Strasbourg, 1987. I.G.L. int. rep., 25 p., **11b**, **12**.

De Franco, R., Ponziani, F., Biella, G., Boniolo, G., Caielli, G., Corsi, A., Maistrello, M., Morrone, A., 1998. DSS-WAR experiment in support of the CROP 03 project. Mem. Soc. Geol. It. LII (1998), 67–90, **7**.

Egger, M., Demartin, A., Ansorge, J., Banda, E., Maistrello, M., 1988. The gross structure of the crust under Corsica and Sardinia. Tectonophysics 150, 363–389, **4a**, **4b**.

Giese, P., 1990. Recent crustal structure and young tectonics along the EGT-segment southern Alps–Ligurian Sea. An attempt of a balancing study. In: The European Geotraverse: integrative studies. Results from the fifth Study Center, pp. 305–310, **4a**.

Giese, P., Morelli. P., Nicolich, R., 1978. Review on the crustal structure of the northern Apennines, the Ligurian Sea and Corsica. In Alps, Apennines, Hellenides. IUGG Scient. Rep., pp. 221–225, **4a**, **6**, **7**.

Giese, P., Prodehl, C., 1976. Main features of crustal structures of the Alps. In: Giese, P., Prodehl, C., Stein, A. (Eds.), Explosion seismology in central Europe. Deutsch. Geophys. Gesell., Springer, pp. 347–375, **1**, **2**, **3**, **4a**, **5**.

Giese, P., Wigger, P., Morelli, C., Nicolich, R., 1981. Sclussbericht zum Forschungsprogram im Bereich der geothermischen Anomalie der Toskana. E.G. Contracts 486 und 487-78-1 (1981), 108 p., **4a**, **6**, **7**, **8**.

Hirn, A., Nadir, S., Thouvenot, F., Nicolich, R., Pellis, G., Scarascia, S., Tabacco, I., Castellano, F., Merlanti, F., 1989. Mapping the Moho in the western Alps by wide angle reflection seismics. Tectonophysics 162, 193–202, **3**.

Hirn, A., Sapin, M., 1977. Crustal structure beneath Corsica. Boll. Geof. Teor. Appl. 19, 233–235, **4a**, **7**.

Holliger, K., Kissling, E., 1992. Gravity interpretation of a unified 2D acoustic image of the central Alpine collision zone. Geophys. J. Int. 111, 213–225, **4a**.

Italian Explosion Seismology Group, 1980. Calabrian arc project: first results of the 1979 seismic experiments. In: EGS-ESC Meeting, Budapest, 1980. I.G.L. int. rep., p. 20, **9**, **11a**.

Italian Explosion Seismology Group, 1982. Crustal structure in the southern Apennines region from DSS data. In: Proceedings of the EGS-ESC Meeting, Leeds, pp. 75–82, **8**.

Letz, H., Reichert, C., Wigger, P., 1977. Results of two seismic refraction lines in the northern Apennines. Boll. Geof. Teor. Appl. 19, 223–225, **6**.

Linari, R., 1981. DSS 1979: interpretazione preliminare del profilo Calabria–Mar Ionio. IMGA, Trieste, int. rep. n. 64, p. 14, **9**.

Maistrello, M., Scarascia, S., Corsi, A., Egger, A., Thouvenot, F., 1990. EGT southern segment. Compilation of data from the seismic refraction experiments in Tunisia and Pelagian Sea, open file report, Part 1, 115 p., Part II, 79 plates, November 1990, CNR-IGL, Milano, **4b**, **11c**.

Makris, J., Nicolich, R., Weigel, C., 1986. A seismic study in the western Ionian Sea. Ann. Geophys. 4B, 665–678, **9**, **12**.

Morelli, C., 2000. The themes of crustal research in Italy and the role of the DSS-WA seismics. Boll. Soc. Geol. It. 119, 141–148.

Morelli, C., Giese, P., Cassinis, R., Colombi, B., Guerra, I., Luongo, G., Scarascia, S., Schutte, G., 1975. Crustal structure of southern Italy. A seismic refraction profile between Puglia–Calabria–Sicily. Boll. Geof. Teor. Appl. 17, 183–210, **11a**, **11b**.

Morelli, C., Nicolich, R., 1990. A cross section of the lithosphere along the European Geotraverse southern segment (from the Alps to Tunisia). Tectonophysics 176, 229–243.

Nicolas, A., Polino, R., Hirn, A., Nicolich, R., ECORS-CROP W.G., 1990. ECORS-CROP traverse and deep structure of the western Alps: a synthesis. In v. 1, p. 15–27, **3**.

Nicolich, R., 1981. Il profilo Latina–Pescara e le registrazioni OBS nel Mar Tirreno. IMGA, Trieste, int. rep n. 64, p. 14, **8**.

Pfiffner, O.A., Lehner, P., Heitzmann, P., Mueller, S., Steck, A. (Eds.), 1997. Deep structure of the Swiss Alps – Results of NFP 20. Birkhäuser, Basel, **1**, **4a**.

Res. Group For Lithospheric Structure In Tunisia, 1989. EGT '85 seismic experiment in Tunisia: a reconnaissance of the deep structures. In: Europ. Geotraverse, EGT, Sixth EGT workshop, Einsiedeld, Europ. Sc. Found., pp. 197–210, **4b**, **11c**.

Res. Group For Lithospheric Structure In Tunisia, 1992. The EGT '85 experiment in Tunisia: a reconnaissance of the deep structures. Tectonophysics 207, 245–252, **4b**, **11c**.

Roure, F., Heitzmann, P., Polino, R. (Eds.), 1990. Deep structure of the Alps. Societé Géologique de France, Societé Géologique Suisse, Società Geologica italiana. 367 p., **3**.

Scarascia, S., 1979. Alcuni risultati dei profili esplorati in Toscana e aree adiacenti. In: Report to the PFG (Progetto Finalizzato per la Geodinamica), Meeting-CNR Rome, pp. 17–19, **6**, **7**.

Scarascia, S., 1980. Struttura crostale della Sardegna dagli esperimenti di sismica attiva condottinel 1979. IGL (Istituto per la Geofisica della Litosfera, CNR), int. rep., 20 p., **4b**, **8**.

Scarascia, S., Cassinis, R., 1997. Crustal structures in the central-eastern Alpine sector: a revision of the available DSS data. Tectonophysics 271, 157–188, **1**, **2**, **5**.

Scarascia, S., Cassinis, R., Federici, F., 1998. Gravity modelling of the deep structure in the northern-central Apennines. Mem. Soc. Geol. It. LII, 231–246, **6**, **7**, **8**.

Scarascia, S., Cassinis, R., Lozej, A., Nebuloni, A., 2000. A seismic and gravimetric model of crustal structures across the Sicily channel Rift zone. Boll. Soc. Geol. It. 119, 213–222, **11c**.

Scarascia, S., Lozej, A., Cassinis, R., 1994. Crustal structures of the Ligurian, Tyrrhenian and Ionian seas and adjacent onshore area interpreted from wide-angle seismic profiles. Boll. Geof. Teor. Appl. XXXVI (141–144), 5–19.

Steinmetz, L., Ferrucci, F., Hirn, A., Morelli, C., Nicolich, R., 1983. A 550 km long Moho traverse in the Tyrrhenian Sea from OBS recorded Pn waves. Geoph. Res. Lett. 10, 428–431, **9**.

Valasek, V., 1992. The tectonic evolution of the Swiss Alpine crust interpreted from a 2D network of deep crustal seismic profiles and an evaluation of 3D effects. PhD thesis, ETH, Zurich, 196 p., **1**, **2**, **4a**.

Wigger, P.J., 1984. Die Krustenstruktur der Nordappennins und angrenzender Gebiete mit besondern Berucksichtigung der geothermichen Anomalie der Toskana. Berliner Geowissenschaftliche Abhandlugen, Reihe B 9, 1–87, **4a**, **6**, **7**, **8**.

Yan, Q.Z., Mechie, J., 1989. A fine structural section through the crust and lower lithosphere along the axial region of the Alps. Geophys. J. 98, 465–488, **1**.

Ye, S., 1992. Crustal structure beneath the central Swiss Alps derived from seismic refraction data. PhD thesis, ETH, Zurich, no. 9631, 125 p., **1**, **2**, **4a**.

Ye, S., Ansorge, J., Kissling, E., Mueller, S., 1995. Crustal structure beneath the eastern Swiss Alps derived from seismic refraction data. Tectonophysics 242, 199–221, **1**, **2**, **4a**.

CROP PROJECT: Deep Seismic Exploration of the Central Mediterranean and Italy
Edited by I.R. Finetti

CHAPTER 3

Gravity and Magnetic Fields of the Central Mediterranean Region

P. Bernardelli[1], C. Cavalli[1], R. Longoni[1], I. Giori[1]

ABSTRACT

Detailed gravity and magnetic maps reveal the regional structural features of the central Mediterranean Sea. These maps incorporate substantial proprietary databases as well as data exchanged with other companies or agencies. They include comprehensive regional Bouguer anomalies, residual magnetic field and its reduction to the pole. In addition to the gravity and magnetic products, a topographic map was created by merging several data sets, including a DTM (digital terrain model) acquired by Getech (Geophysical Exploration Technology of Leeds – UK), together with satellite, land and marine data. This topographic map was implemented in interpretation of the gravity data.

Keywords: Gravity, magnetic, Central Mediterranean region

1. GRAVITY

In order to construct the Bouguer Anomaly maps (Figures 2–5), the following set of data was acquired and utilized:

- over 370,000 onshore stations in Italy;
- over 120,000 measurement points offshore in the adjacent seas;
- almost 9,000 stations acquired using a gravity meter in a bathysphere on the seabed for the areas adjacent to the coast and in the inland lakes of central Italy;
- approximately 53,000 stations obtained as exchange data in the Adriatic Sea, Ionian Sea and Thyrrenian Sea.

Altogether, these amount to approximately 800,000 measurement points.

In addition to the gravity values, each measurement station gives the point's topographic elevation in the event of land surveys, or the water depth in the event of subsea stations. The average density of the measurement points covering Italian territory is approximately one station per sq km, with some areas having greater detail (e.g. in flatter areas such as the Po Valley). Thanks to this coverage, it is possible to calculate the Bouguer Anomaly on a 1×1 km grid cell size (Servizio Geologico d'Italia, 1989).

For the regional purposes of the study, the Bouguer Anomaly was worked out on a 2×2 km grid cell size, with a calculation density of 2.4 g/cc. The Bouguer Anomaly for the Mediterranean Sea not covered by measurement points was calculated on the basis of Free Air satellite data (Sandwell and Smith, 1997), where the topographic effect of the DTM (Getech) data was removed.

The map of Bouguer Anomalies exhibits the following qualitative regional features:

- a wide, triangular-shaped, positive anomaly located in the Tyrrhenian Sea, stretching towards NW in the Ligurian Gulf with reduced intensity on Tuscan archipelago;
- a negative gravity anomaly which follows the axis of the Italian Peninsula from Piedmont to the Gulf of Taranto, then continues along the Crotone coast. From there, it follows the Calabrian ridge and stretches all the way to the Caltanissetta trough;
- another series of positive, less intense anomalies which follow the Alpine mountain trend in the north-western portion and subsequently change their average direction, thus arranging themselves according to a NW–SE (Colli Euganei) and W–E trend in the Gulf of Venice;

[1]ENI – Division Exploration & Production, San Donato Milanese, Italy

TOPOGRAPHY

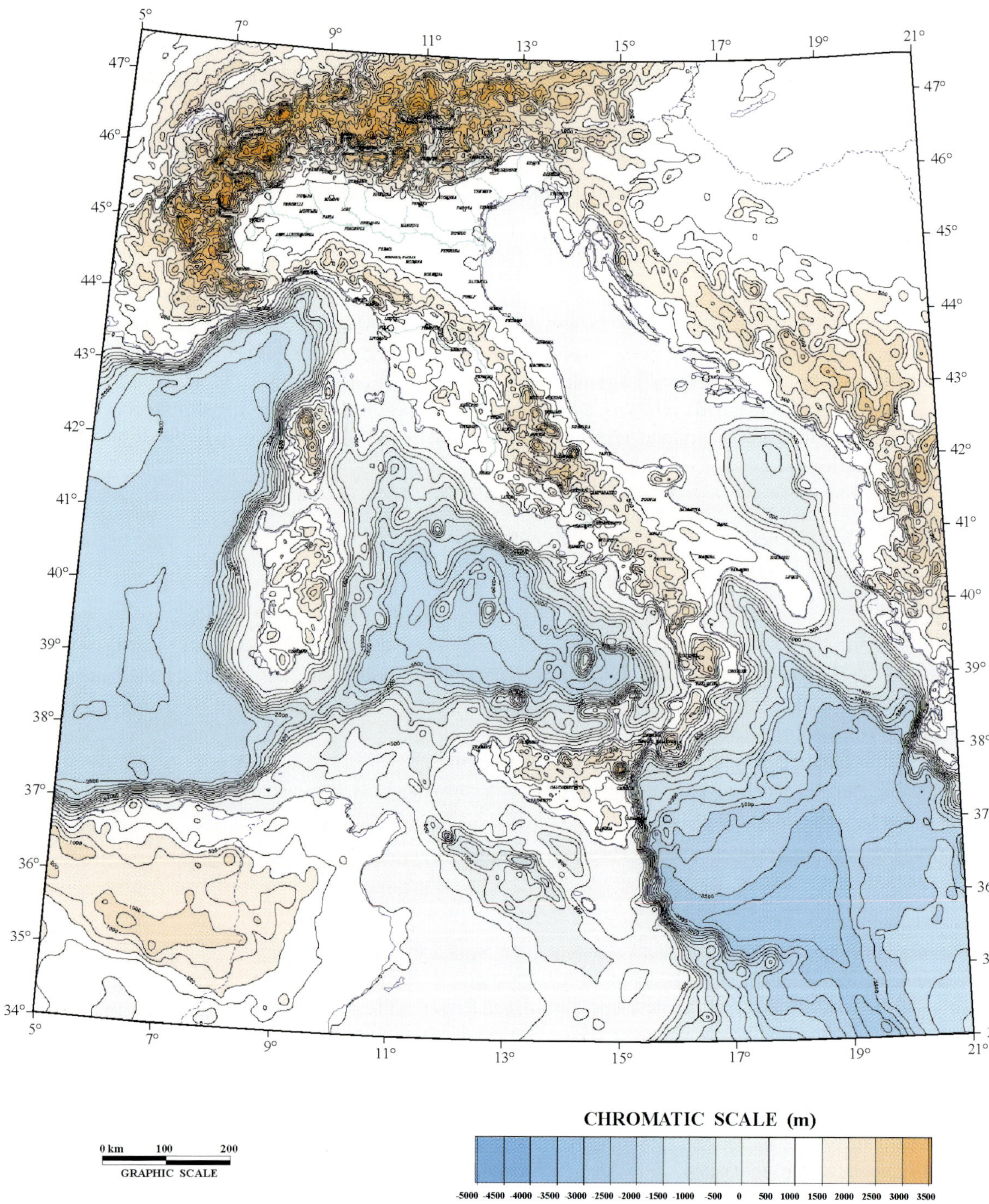

Figure 1: Topographic-Bathymetric map of the Central Mediterranean region.

- in the Adriatic Sea, from the Marches all the way to Apulia, the positive anomalies display NW–SE (Apennine) trends, with the exception of the Gargano area, where there is an intense SW–NE oriented anomaly which subsequently modifies its direction, to an almost W–E direction all the way to the Albanian coast;
- in the Ionian Sea, the positive gravity anomalies are intense again with an average SW–NE direction. In the south-easternmost portion, the trends are modified again since the anomalies are arranged along an average SE–NW direction as far as the north-eastern boundary of the Iblean plateau;
- the map's western portion (to the west of Sardinia) is characterized by an anomaly trending S–N along the Sardinian and Corsican coast. To the N, the anomaly changes its trend, wedging into the Ligurian Gulf;

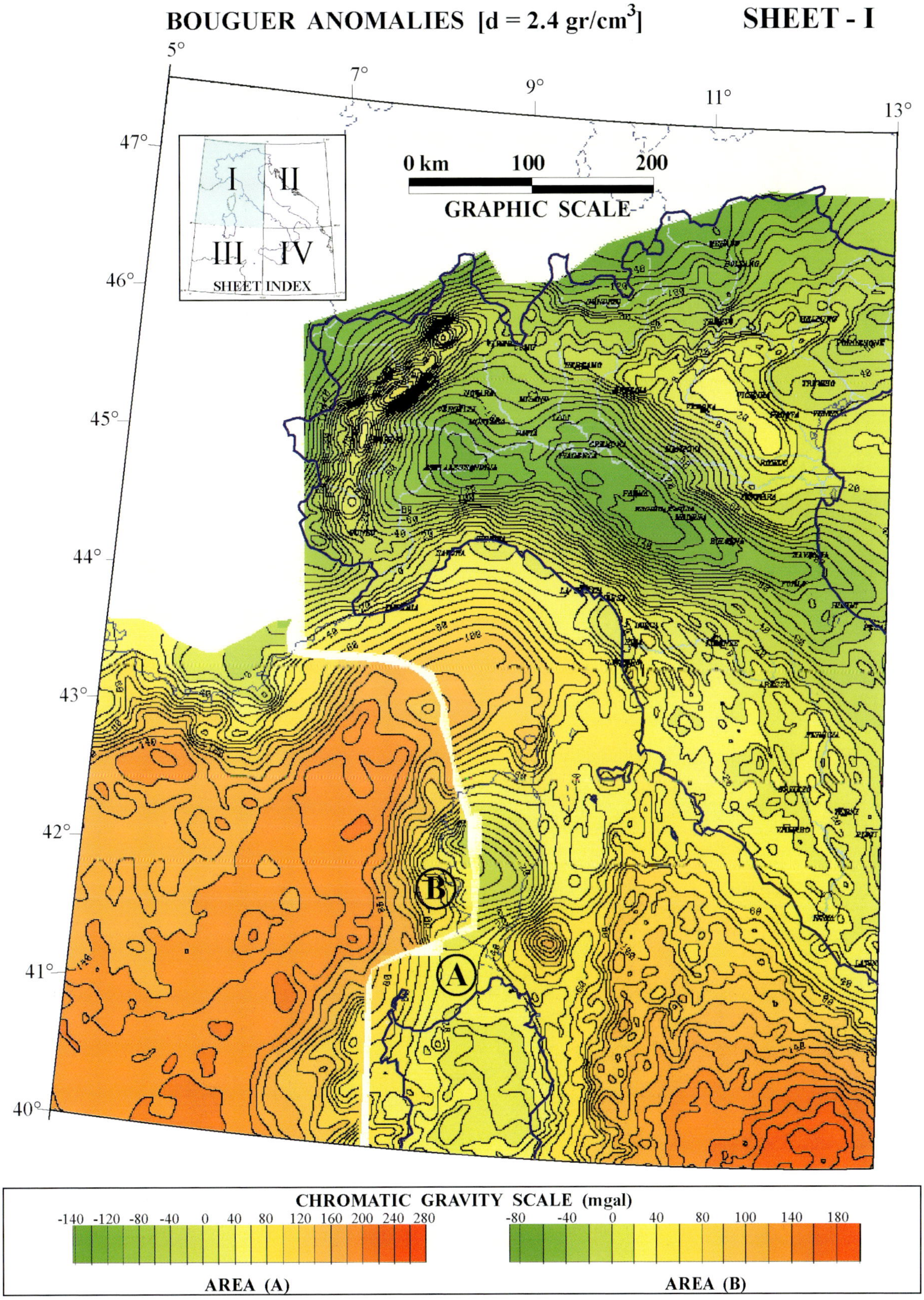

Figure 2: Bouguer gravity map of the Central Mediterranean region: SHEET-I.

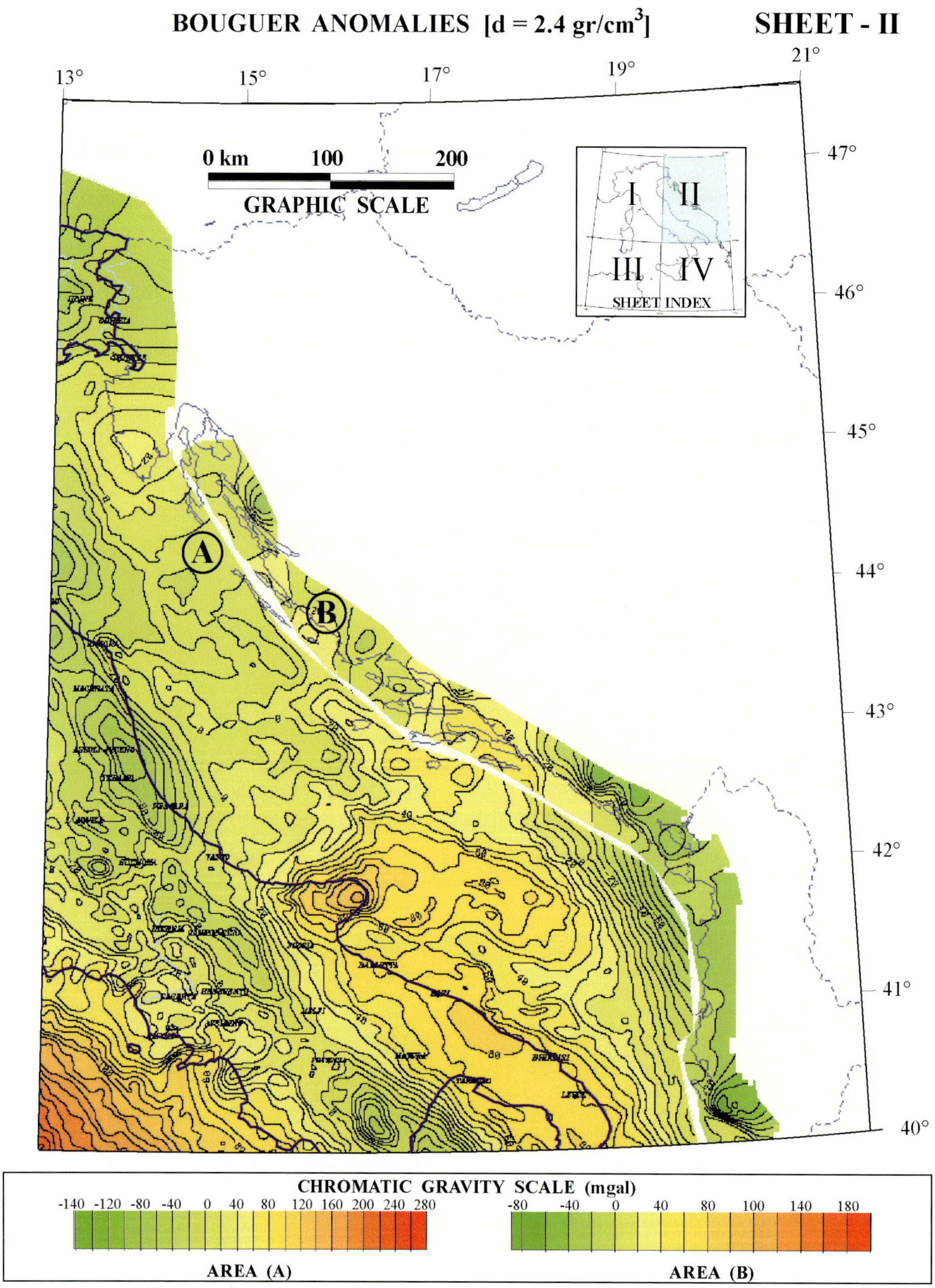

Figure 3: Bouguer gravity map of the Central Mediterranean region: SHEET-II.

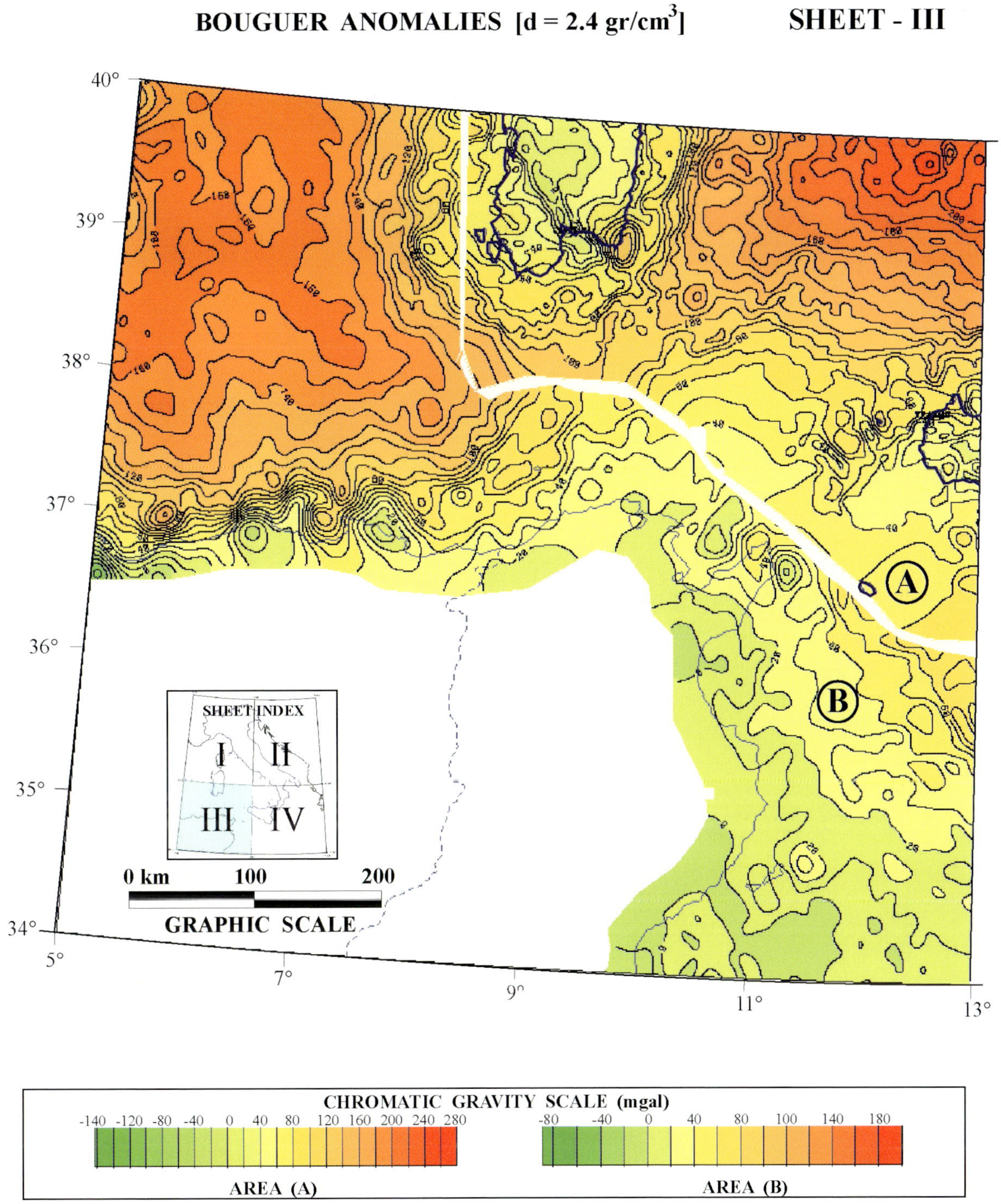

Figure 4: Bouguer gravity map of the Central Mediterranean region: SHEET-III.

- the negative anomaly in the Corsican region, which is not supported by numerous measurement points, is confirmed by the satellite regional anomaly in its western portion.

2. MAGNETOMETRY

The calculation of the Residual Magnetic Field (Figure 6) was carried out using all the aeromagnetic surveys recorded by the Eni – Divisione Exploration & Production not only for potential hydrocarbon detection but also for geothermal purposes in the Italian mainland, islands and seas all the way to the boundary of the territorial waters. In the central Mediterranean Sea, an area under French authority, use was made of data obtained from surveys carried out on behalf of the "Institut de Physique du Globe de Paris" (Cassano, 1991). The magnetic data of the

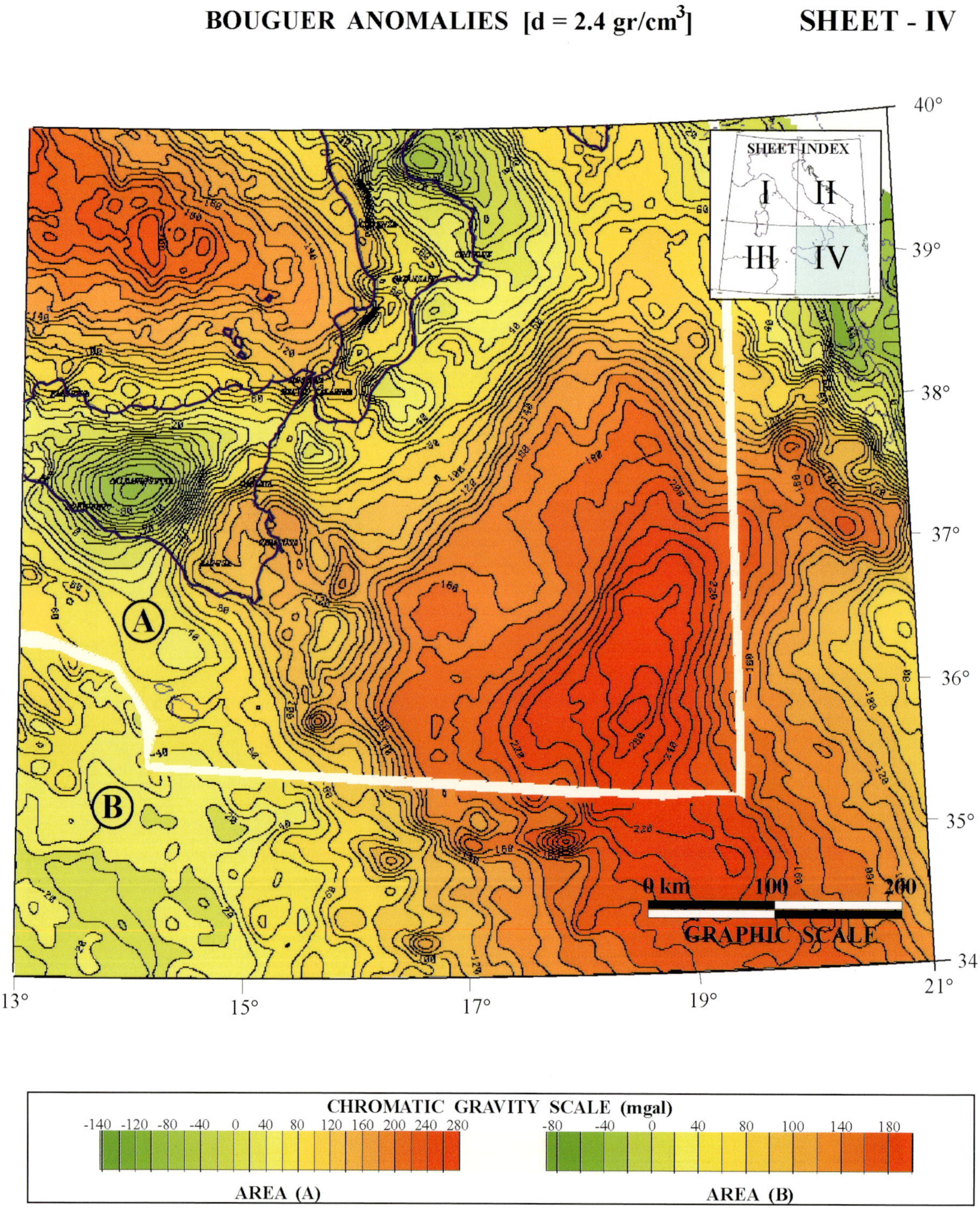

Figure 5: Bouguer gravity map of the Central Mediterranean region: SHEET-IV.

southern portion come from a database acquired by the Eni – Divisione Exploration & Production in the 1990s. Not all the surveys contributing to the calculation of the magnetic anomalies were recorded with the same parameters (flight elevations, acquisition grid, line direction, mesh overlap, etc.). As a result, a significant merging effort was necessary in order to obtain a map of the magnetic anomalies referred to the same barometric elevation of 8500' (2590 m above sea level). Only the portion relevant the Alps range remained on the map at the original elevation of 13100' (approximately 3922 m above sea level) (Cassano, 1984; Servizio Geologico d'Italia, 1994). The magnetic data for the map's southern portion close to the African and Greek coasts were homogenized to an elevation of 1000 m above sea level and, for the sake of uniformity, were processed on the basis of upward prolongation to 2590 m. These data come from a collection of magnetic surveys (AMMP – African Magnetic Mapping Project) acquired by

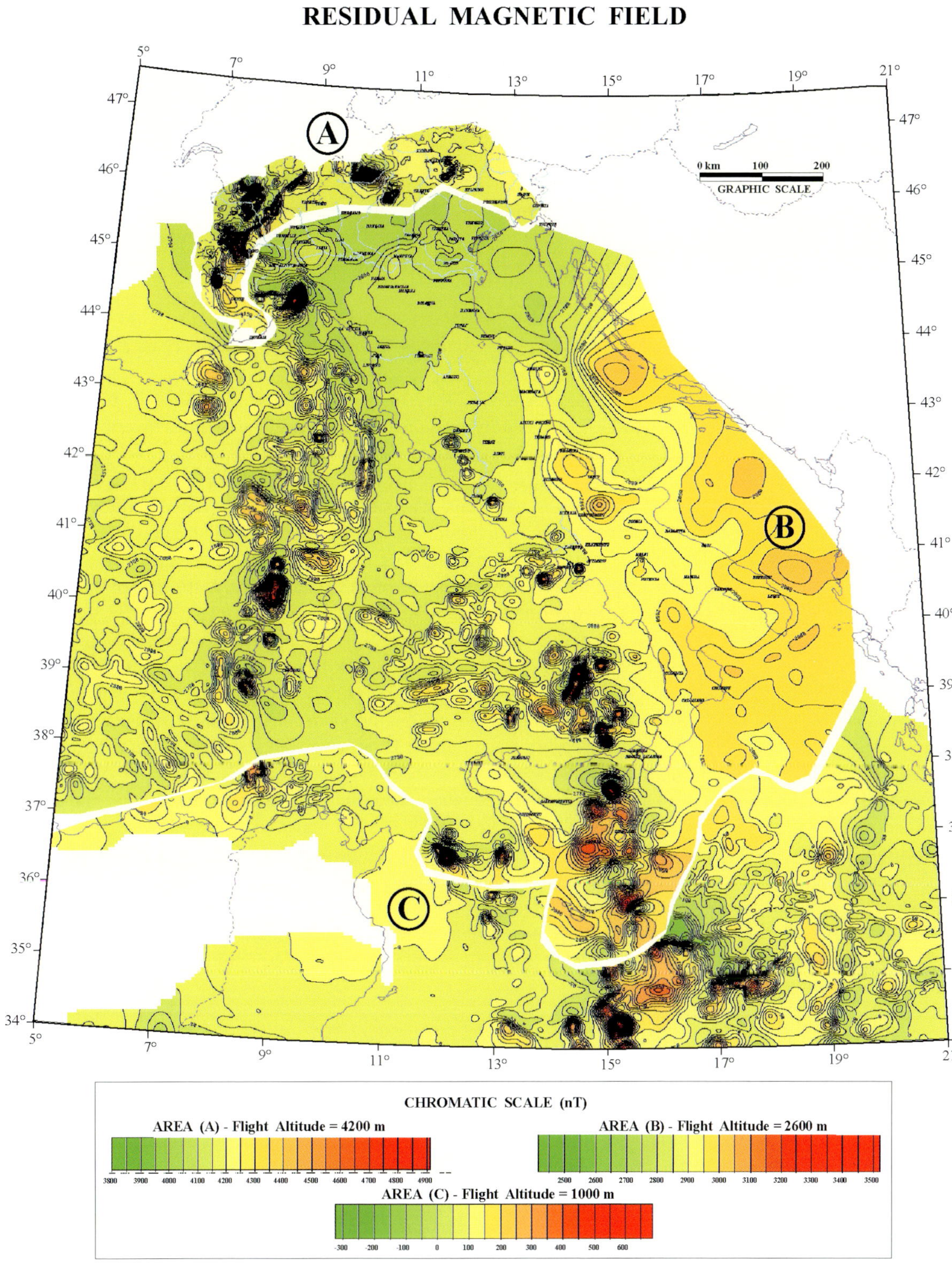

Figure 6: Residual magnetic field of the Central Mediterranean region.

Getech of Leeds, UK (the Eni – Divisione Exploration & Production were an original sponsor, together with other major Oil Companies). Altogether, in order to complete the Magnetic Anomaly map, approximately 430,000 km of original lines and over 490,000 sq km of magnetic data already available on calculated grids were utilized.

The Residual Magnetic Field map is not easy to read since at the average magnetic latitude where the central Mediterranean Sea lies, the anomalies exhibit a marked bipolarity (positive and negative lobes). The median axis of the causative bodies of the magnetic anomalies is located approximately halfway between the positive gradient to the S and the negative one to the N. In order to overcome this interpretative difficulty, the Reduction to the Pole operation is carried out as part of a routine processing procedure. Thanks to this processing, algorithm it is possible, to exactly locate the magnetic anomaly coinciding with the axis of the causative body generating it.

Given the geographic features of the Italian Peninsula and of the surveyed area, which show a wide range in latitude, the parameters used in the calculation of the Reduced to the Pole map have average values. At the N and S ends, there may be boundary effects in the processed anomalies, but these approximations are acceptable. It is useful to bear in mind that the characteristics of the magnetic anomalies (intensity and frequency) are closely related to the lithological nature (percentage of magnetite occurring in the rock) and to the depth of the causative body's top. This makes it possible to discriminate, for instance, shallow intra-sedimentary volcanic bodies from the top of the magnetic basement *latu sensu*.

Basically, the main anomalies that may be observed on the Reduced to the Pole map (Figure 7) are as follows:

- the Western and Ligurian Alps range is characterized by very intense anomalies with SW–NE trends due to the outcrop of the range's most susceptible units, and by medium-high amplitude anomalies determined by the structure of the magnetic basement as observed in the Ivrea area;
- the central portion of the Alpine mountain range exhibits anomalies with a markedly W–E trend to the north of Sondrio and, in the eastern portion, the anomalies are arranged according to an average SW–NE direction;
- the Po Valley shows wide anomalies characterized by medium-low intensity. Their main trend is N 55 °E;
- the northern Apennine range all the way to Tuscany is marked by a magnetic gradient decreasing towards NW without noteworthy anomalies. This magnetic "lull" is interrupted, on the western side, in the area of the volcanic lakes of Lazio, and, on the Adriatic side, by the Ancona magnetic high;
- this distinct behaviour of the magnetic anomalies continues also in the central Apennine portion. The Adriatic side shows large and medium intensity anomalies with a main SE–NW trend (Conero, Pescara, Campobasso and Gulf of Taranto). The Adriatic side exhibits more intense and higher frequency anomalies. From the Gulf of Naples towards the south, the panorama of the anomalies within the Thyrrenian Sea also changes, since there are numerous anomalies caused by volcanic bodies (seamounts). Their main trend is W–E, but in the area to the north of the Eolie Islands, the trend becomes S–N;
- the western Tyrrhenian portion shows intense, high frequency anomalies with a trend that follows the eastern Sardinian and Corsican coasts along a markedly S–N direction in the Ligurian Gulf;
- from Sicily, the south-eastern portion displays intense anomalies (Etna and the entire slope of Malta) which border the panorama of less intense anomalies, whose main trend is N 30° W;
- the map's western portion shows anomalies that resume the regional trend of the southern Tyrrhenian Sea, i.e. W–E, especially in Corsica, Sardinia and in the entire strip to the west of these islands. Only in the portion closer to the Tunisian and in the Sicily Channel, are the magnetic anomalies arranged according to an average trend of N 30° W (area of the Pantelleria graben).

3. GEOLOGICAL & STRUCTURAL CONSIDERATIONS

The identified gravity and magnetic anomalies show a close correlation with the two main Italian orogenic systems: the Alps and the Apennines. This indicates, they are produced by lithological variations reflecting the tectonic setting in which the two systems were formed. Thus conversely, from the observation of the anomalies, it is possible to trace the main geologic and structural elements characterizing these systems and, more generally, the Italian Peninsula. The possible relationships between anomalies and structural elements characterizing the surveyed area are briefly described below. The aim of this description is not to review the models proposed by various authors, but to give some indications about the regional structural features and the bodies that probably cause the gravimetric and magnetic anomalies.

3.1. Gravity

The Bouguer anomaly map appears to be clearly dominated by the set of minima and maxima associated with the Alps and the Apennines system. This system is characterized by a series of negative anomalies aligned in such a way as to draw two arches which affect the entire peninsula both in an E–W direction (Alps) and in a N–S direction (Apennines). The two arches are however characterized by different arrangements of the maxima and minima: the Alps are defined by two large areas of negative anomaly separated by a limited area of relative maxima, whereas

RESIDUAL MAGNETIC FIELD REDUCED TO THE POLE

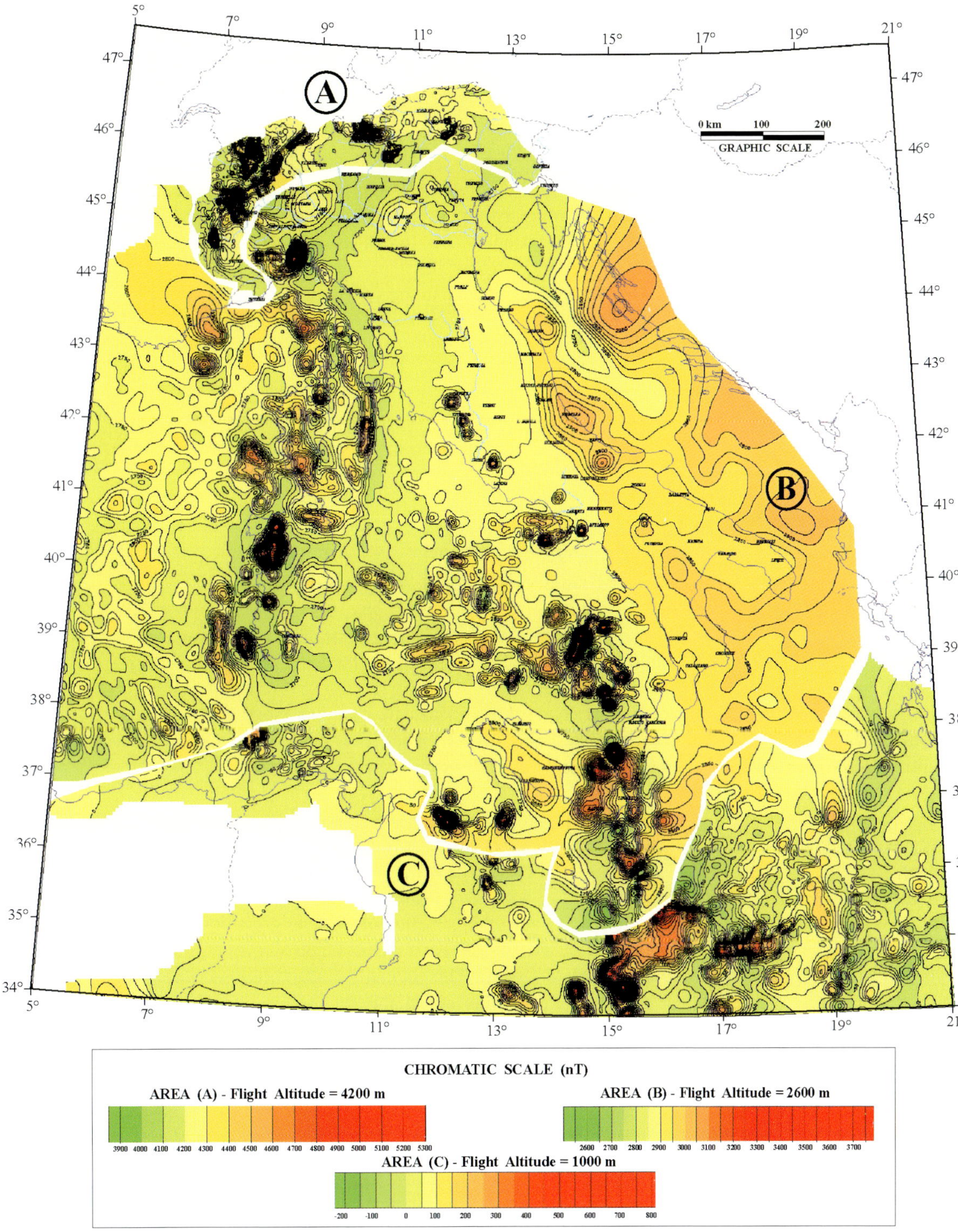

Figure 7: Residual magnetic field of the Central Mediterranean region, reduced to the Pole.

the Apennines are characterized by a limited strip of gravity minima separating wide areas of maxima (Tyrrhenian Sea, Apulian foreland and Ionian Sea).

The configuration of the Alps system is due to the range's double vergence (towards S and towards N) and to the presence of a double foredeep. On the other hand, the gravity maxima are due, at least in the north-western portion, to the outcrop of the root zone of this range. There are, in fact, magmatic rocks – at times outcropping – associated with the Alps basement, while in the eastern sector, the gravity maximum that can be observed appears to be mostly due to the presence of a structural high (Atesino porphyritic plateau of Permian age) on which the so-called Venetian Carbonate Platform settled in the Mesozoic.

The different natures of the rocks coming these maxima may be inferred also from observation of the magnetic map which shows a series of north-western maxima, practically coinciding with the above mentioned gravity ones, while there is no equivalent to the eastern ones.

The negative anomaly related to the foredeep portion of the Alps, visible in the Friuli sector, is well delimited and distinct from the one that may be associated with the Apennines system. By contrast, in the Lombard–Piedmont sector, the two anomalies tend to merge and the anomaly connected to the south-Alpine foredeep is partly masked by the more evident one created by the Apennines system which, in this sector, superimposes the previous.

The evident sequence of negative anomalies in correspondence to the so-called Apennines Foredeep s.l. is attributable to the Apennines system. Within this foredeep, from N to S, a series of basins may be observed:

- Asti – Cuneo Basin
- Emilia – Romagna Basin
- Pescara – Ancona Basin
- Bradanic Foredeep
- Gela – Catania Foredeep (Caltanissetta Basin)

These basins, of mainly Neogenic age, are filled by thick successions of turbiditic sediments where hydrocarbon accumulations may be found. All these basins are structured by a series of average E-verging (from NE-to SE-verging) thrusts which cause the Apennines system to overlap the Apulian foreland.

The sequence of the anomalies has two main interruptions located along the transition between the Pescara–Ancona basin and the Bradanic foredeep and, in Calabria, between the latter and the Gela–Catania foredeep.

The first of these interruptions lies close to a series of lithospheric-scale structural elements (the so-called Campania Lithospheric Discontinuity) that can be associated with the "41° Parallel Fault System" (Serri, 1990). The second interruption develops in Calabria, where the minima are less pronounced.

The negative anomalies of the Apennines system are bounded on the E, SE and W by extensive areas characterized by positive anomalies. The positive anomaly system which develops to the E is mostly an expression of the carbonate lithologies of the Istrian–Dalmatian platform (to the N) and the Apulian platform.

To the SE, an area of positive anomaly associated with the presence of oceanic crust develops off the Calabrian coast (Ionian Sea). The relatively low heat flow ($< 40\ \mathrm{mW/m^2}$) and the presence of a very thick lithospheric mantle (70–90 km) suggest that this crust should be relatively old. This anomaly is also bordered to the W and E by two alignments coinciding with the Malta escarpment (offshore Sicily) and the Salento escarpment (offshore Apulia) which are likely to represent a couple of passive continental margins of Triassic-Jurassic age. With this in mind, the Ionian Sea could represent a full oceanic section containing an oceanic Mesozoic ridge, whose morphological expression has been lost due to the thermal cooling and the subsequent sedimentary cover.

To the W, the large positive anomaly in the Tyrrhenian Sea, characterized by a triangular shape, opens up. The anomaly's maximum values may be due to the presence of oceanic crust which has risen as a consequence of the continental stretching characterizing this sector. Considerations of a geologic, geodynamic and geophysical nature allow us to interpret the Tyrrhenian Sea as the backarc basin of the Apennines system, characterized by a significant subsidence related to the presence of crustal-scale normal faulting. To the E, it is possible to observe how the fault system and the relevant thinning also affect the Italian Peninsula, and how the grabens and the intermontane basins (especially in the Tuscany and Lazio sectors), which characterize the entire central-western portion of the peninsula, are easily identifiable.

These elements are generally of Plio-Quaternary age and have clastic fillings which cause the classic positive anomaly elongated in a NNW–SSE direction.

Finally, close to Sardinia, the negative anomaly produced by the filling of the Campidano graben is clearly visible. However, due to its age, the latter cannot be associated with the grabens of Tuscany and Lazio, but is more likely connected with to the opening phase of the Balearic basin.

3.2. Magnetometry

At first sight, the map of magnetic anomalies shows that Italy can be divided into two sectors: a northern one characterized, on average, by lower anomaly values and a southern sector, where, on the other hand, higher values prevail. The separation between the two sectors occurs close to the 41° Parallel – Campania Fault System.

A more detailed analysis of the map makes it possible to define the distribution of anomalies associated with the presence of rock bodies of a magmatic and/or metamorphic type, especially with a generally basic chemical characteristic.

In the north-western Alps, the anomalies mainly related to the presence of ultrabasic ophiolitic masses, such as in the Voltri Group (maximum close to Genoa) or to a probable mantle intrusion (ascribed to the Permian) in the Ivrea–Verbano–Lanzo system, are highlighted. Other maxima occur in Valtellina (ophiolites), in the Adamello Group (pluton of the same name) and the dolomite area, where they can be referred to Mid-Triassic volcanites and magmatites.

In the Po Valley sector, the relative maxima ascribable to the Eocenic complex buried in the so-called "Mortara Volcano" (E of Pavia) and to Mid-Triassic volcanites, penetrated by exploration wells in the sector between Verona and Mantua (Rodigo 1), are identified.

Proceeding to the peninsular sector, the anomaly maxima related to the Quaternary volcanic systems of Tuscany and Lazio are clearly evident.

As previously mentioned, to the south of the 41° Parallel – Campania Fault System, there is an area characterized, on average, by high anomaly values attributable to a generally less deep basement. Within the sector, clear local anomalies develop on the Campania volcanoes, among which the following ones can be recognized: Roccamonfina, Vesuvio, Ischia and, in Basilicata, Vulture. Further south in Sicily, the anomaly associated with the Etna volcanic complex is highlighted.

The anomalies that can be observed in the Tyrrhenian Sea are ascribable to the volcanic complexes of the Eolie, Palinuro, Marsili, Vavilov and Magnaghi belt, which is genetically related to the backarc basin of the Apennines system. In particular, Marsili, Vavilov and Magnaghi are volcanoes typical of the extensional area (with possible oceanic crust) lying behind the belt, whereas the Eolie Islands and Palinuro belong to the belt system proper: in fact, these latter volcanoes have magmas with a chemical characteristics different from the previous ones (Serri, 1990).

Finally, in Sardinia, the anomaly associated with the alkaline-subalkaline lavas of Monti Ferru of Upper Miocene-Pleistocene age, related to the opening of the Campidano graben, is clearly visible.

REFERENCES

Cassano, E., 1984. Rilievi Magnetici per la Ricerca Mineraria. In: Atti del Convegno di Geomagnetismo, I.N.G., Roma, Ottobre, pp. 117–163.

Cassano, E., 1991. Tyrrenian and Western Mediterranean geomagnetic domain. Terra Nova 2, 638–644.

Sandwell, D.T., Smith, W.H.F., 1997. Marine gravity anomaly from geosat and ERS-1 altimetry. J. Geophys. Res. 102 (B5), 10039–10054.

Serri, G., 1990. Neogene-Quaternary magmatism of the Thyrrenian region: characterization of the magma source and geodynamic implications. Mem. Soc. Geol. It. 41, 219–242.

Servizio Geologico d'Italia, 1989. Carta Gravimetria d'Italia. Istituto Poligrafico e Zecca dello Stato.

Servizio Geologico d'Italia, 1994. Carta Aeromagnetica d'Italia. Istituto Poligrafico e Zecca dello Stato.

CROP PROJECT: Deep Seismic Exploration of the Central Mediterranean and Italy
Edited by I.R. Finetti

CHAPTER 4

Bedrock Geology of the Tyrrhenian Sea Insight on Alpine Paleogeography and Magmatic Evolution of the Basin

Renzo Sartori[1]

ABSTRACT

This chapter is an updated synthesis of the geology of bedrocks recovered from the seafloor and drillings in the Tyrrhenian Sea. Several decades of sampling indicate that the foundation of this back-arc basin consists of late Miocene to Recent igneous rocks plus oceanic crust (accompanying basin extension) and of rocks passively affected by rifting since Tortonian times.

The rifted continental substrate connects Sardinia and Corsica (Alpine chain and its European foreland) to peninsular Italy and Sicily (Apenninic–Maghrebian chain and Calabria terrane). The types and distribution of recovered bedrock samples compared to geology of emerged areas indicate that the Mesozoic Tethyan Ocean was limited to the south by a line connecting the Orosei Canyon (to the west) with northern Calabria (to the east). South of that limit, only Apenninic, Oligocene to Recent compressive deformations with Apulian-African vergence occurred prior to the onset of Tyrrhenian stretching. It turns out that the Cretaceous-Paleogene, Europe-verging Alpine chain did not develop all around Sardinia or further west of it. The chain units deformed until Middle Miocene in the Calabria terrane appear strongly dispersed throughout the extremely thinned southern basin, and were probably the domains most affected by stretching.

In the southern Tyrrhenian area, the igneous rocks coeval to extension form two different groups as regards location and magma character. From Sardinia to the extremely thinned central Vavilov Plain, Pliocene-Quaternary volcanism is largely of ocean island basalt (plus subordinated MORB) type, while Pliocene arc-type magmas are very minor. Ages of volcanic occurrences are rather casually distributed across this wide area. In the southeastern basin, comprising the extremely thinned Marsili Plain and the Eolian Islands (separated from the central Vavilov Plain by a thicker crustal sector), volcanism is instead only of arc-type and is essentially of Pleistocene to Recent age. Taking into account the history of compressive deformation in the southern Apennines, coeval to extension in the southern Tyrrhenian Basin, the age–space distribution of different magma types may be linked to the different characters of the Ionian lithosphere (thinned continental versus oceanic) subducting at different times.

Keywords: Tyrrhenian Sea, Bedrock geology, Igneous rocks, Alpine paleogeography

INTRODUCTION

The Tyrrhenian Sea back-arc basin has been forming since the Late Miocene by subduction of Eastern Mediterranean lithosphere. A large number of papers dealing with the geophysics and geology of this area have been published in the past, so that kinematics of basin extension are well known (e.g. Kastens, et al., 1990, with references).

The portion of the basin located between Corsica and the northern Apennines in Tuscany is a shallow domain rifted since the Miocene, with modest crustal thinning (Figure 1). The southern Tyrrhenian basin, located between Sardinia, Sicily and peninsular Italy, reaches a depth of some 3,600 m and displays an irregular seafloor. Here

[1]Università di Bologna, Dipartimento di Scienze della Terra e Geologico-Ambientali, Via Zamboni 67, 40127 Bologna, Italy. Fax: +39 51 2094522

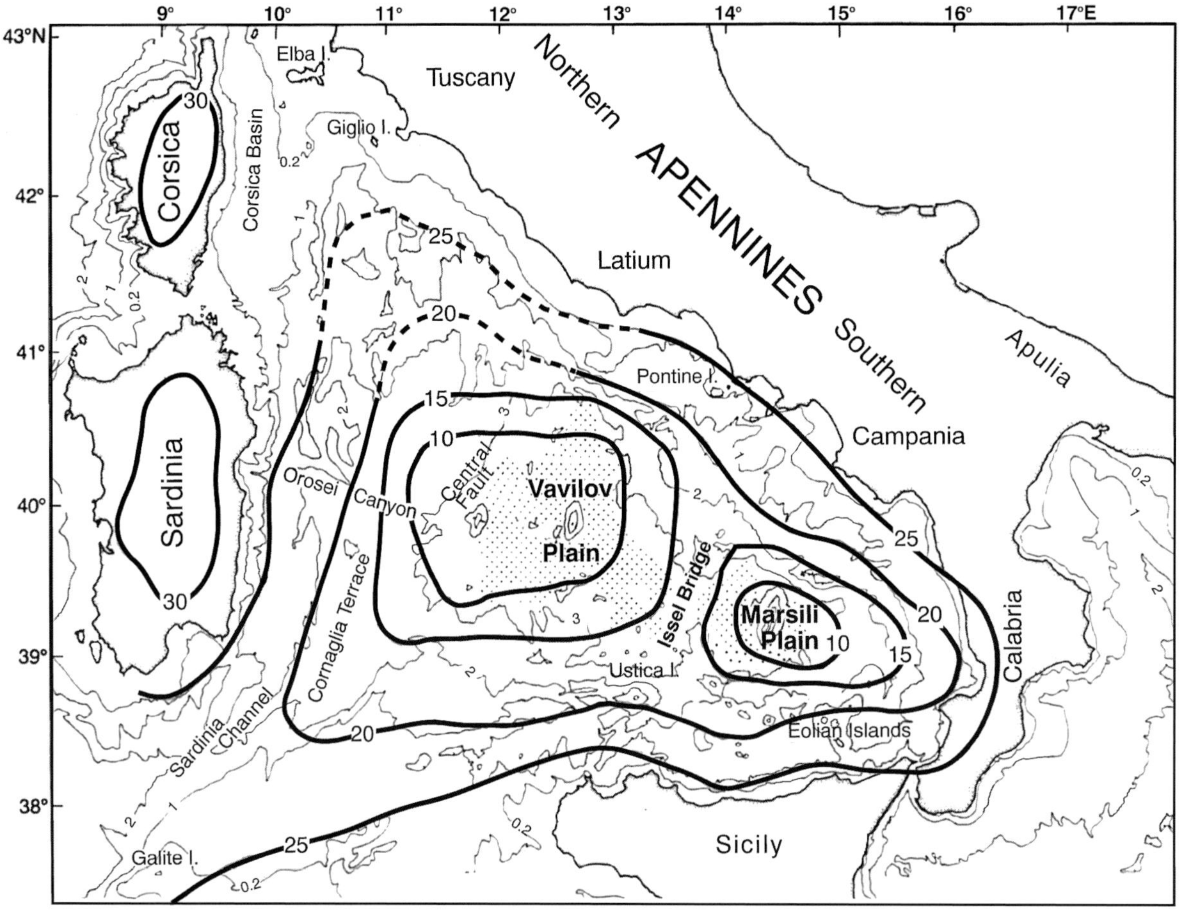

Figure 1: Schematic bathymetry (in km) of the Tyrrhenian Basin with superposed crustal thickness (from Rehault et al., 1987). Indicated also are the main features quoted throughout the text, except for the main seamounts that are indicated as letters in Figures 2 and 3 and referred to in the text.

crustal thickness rapidly decreases from 30 km (Sardinia) to less than 10 km in two discrete areas located in the central (Vavilov Plain) and southeastern portions (Marsili Plain); these areas are separated by a region where the crust is about 15 km thick (Issel Bridge) (Figure 1). Subduction of the Ionian Sea lithosphere is indicated by a Benioff Plane reaching 500 km in depth and plunging NW beneath the southeastern portion of the basin. Rifting started off Sardinia during the Tortonian and migrated, in steps, to the ESE. The two plains with shallow Moho host oceanic crust, emplaced essentially during the Pliocene in the Vavilov Plain and during the Pleistocene in the Marsili Plain. While the Tyrrhenian Basin experienced extension, the orogenic system of the southern Apenninic arc experienced shortening and migration towards the east (Apulia), the southeast (Ionian Sea), and the south (Sicily).

The bedrock foundation of the Tyrrhenian basin, usually making up the acoustic basement in reflection seismic lines, has been sampled from drillings and seafloor outcrops by a large number of researchers during the recent decades. Geological syntheses of these samples have been reported in a few papers. Segre (1958) first distinguished volcanic rocks from other rocks referable to the geology of the encircling land areas. Selli (1974) considered the metamorphic rocks collected in the southern Tyrrhenian as making up a Variscan block also comprising Sardinia and Calabria. Colantoni et al. (1981) published a map with location of samples and information on their ages and lithology. In the maps of Bigi et al. (1990–1992), the bedrock outcrops were delimited and correlated, when possible, with the emerged structural elements. Sartori (1986) tried to synthesize these data at the basin scale. As regards only magmatic rocks, several syntheses have been produced, the last and most updated by Serri et al. (2001, with previous references).

This chapter revises and updates, at the basin scale, a synthesis of the geology of Tyrrhenian bedrocks. It is based both on direct analysis of rock samples, recovered during recent decades, and on revision of recently published data. The chapter aims to discuss the Tyrrhenian geology in comparison with that of encircling emerged areas, focussing on its relevance to the evolution of the basin (igneous rocks) and to the reconstruction of the geology of the area prior to its stretching (pre-rift rocks).

1. PRE-RIFT ROCKS AND ROCK ASSOCIATIONS (FIGURE 2)

The Tyrrhenian area links different geological domains outcropping on land and largely formed during the Alpine orogenic cycle. They are: a fragment of the European continent (Corsica and Sardinia); a segment of the Alpine chain (in north-eastern Corsica); and the Apenninic–Maghrebian chain which includes a large crystalline terrane (Calabria–Peloritani). For an account of the complex geology and evolution of these areas, reference is made to Sartori et al. (1987) and to Serri et al. (2001).

Figure 2 shows type and distribution of the pre-rift bedrocks sampled in the Tyrrhenian area with their possible correspondence to the emerged geological domains. The map is updated to include the results of Bouillin et al. (1999), Mascle et al. (2001), Sartori et al. (2001 and submitted). The numbers in the lower left corner of the figure correspond to those of the geological domains listed here below. Capital letters in the text and in the figure refer to the main sampled features (seamounts, etc.). Black figures are radiometric ages (or age-ranges) obtained at sea.

1.1. Corsica–Sardinia Variscan basement (including pre-Oligocene cover rocks)

This is a segment of the European foreland of the Alpine chain that was displaced from its original location due to the Oligocene-Miocene opening of the Western Mediterranean Basin. It is composed of low to high grade metamorphic rocks making up a Variscan chain, intruded by late orogenic granitoids showing ages of about 280–300 Ma. These bedrocks crop out along the continental slope of Sardinia, and make up the western backbone of the Baronie Smt (C), the Cornaglia Smt (B), and various outcrops around the Ichnusa Smt (A). Towards the Sardinia Channel (Figure 1) radiometric ages tend to decrease to 190–250 Ma. Pre-Oligocene cover rocks of the Hercynian basement have rarely been collected in the Tyrrhenian Basin. Oligocene-Miocene sedimentary rocks occurring on this basement type have been grouped separately (4).

1.2. Metamorphic units of "Alpine Corsica" Auctt. (including pre-Oligocene rocks)

Largely metamorphic, cover and basement nappes making up a segment of the Cretaceous-Paleogene Europe-verging Alpine chain, displaced and rotated during the Oligocene-Miocene with the Corsica–Sardinia block. Metamorphic rocks of various degrees, showing blueschist Eoalpine metamorphism, such as those occurring in NE Corsica, Giglio Island, and along the coast of Tuscany, are not abundant on the Tyrrhenian seafloor. Apart from ophiolite samples, which have been grouped separately (6), the only sure occurrence is at the Cialdi Smt (E) in the northern Tyrrhenian Sea. Oligocene-Miocene sedimentary cover rocks of the Alpine units have been reported separately (4).

1.3. Calabride–Peloritanian–Kabilide units (including pre-Oligocene rocks)

The Calabria terrane is a complex tectonic edifice mostly of Variscan metamorphic rocks and granitoids deformed during the Alpine cycle, but with the addition, only in northern Calabria, of Tethyan ophiolite units and of other units of dubious paleogeographic location. The Peloritanian–Kabilide units have been thrust onto the Maghrebian chain, and the units of northern Calabrian have been thrust onto the southern Apenninic chain.

This section includes bedrock samples made essentially of Variscan basement rocks, which show radiometric age-ranges of Alpine or mixed Hercynian-Alpine times (from 7 to 230 Ma). Rocks are often deformed at the sample scale and sometimes show variable to inverted metamorphic degree along one single submarine slope. They crop out all along the escarpment that runs south of the Sardinia Channel (Bouillin et al., 1999), occur as xenoliths in the Eolian Islands and on Galite Island off Tunisia (for location of this island see Figure 1), and make up the F. Gioia Smt. (Q). Judging from the location of submerged and emerged outcrops, this basement type was strongly disarticulated by stretching of the southern Tyrrhenian Basin, since its fragments occur at least from the Sardinia Channel to Calabria. Oligocene-Miocene rocks covering these units have been grouped separately (4).

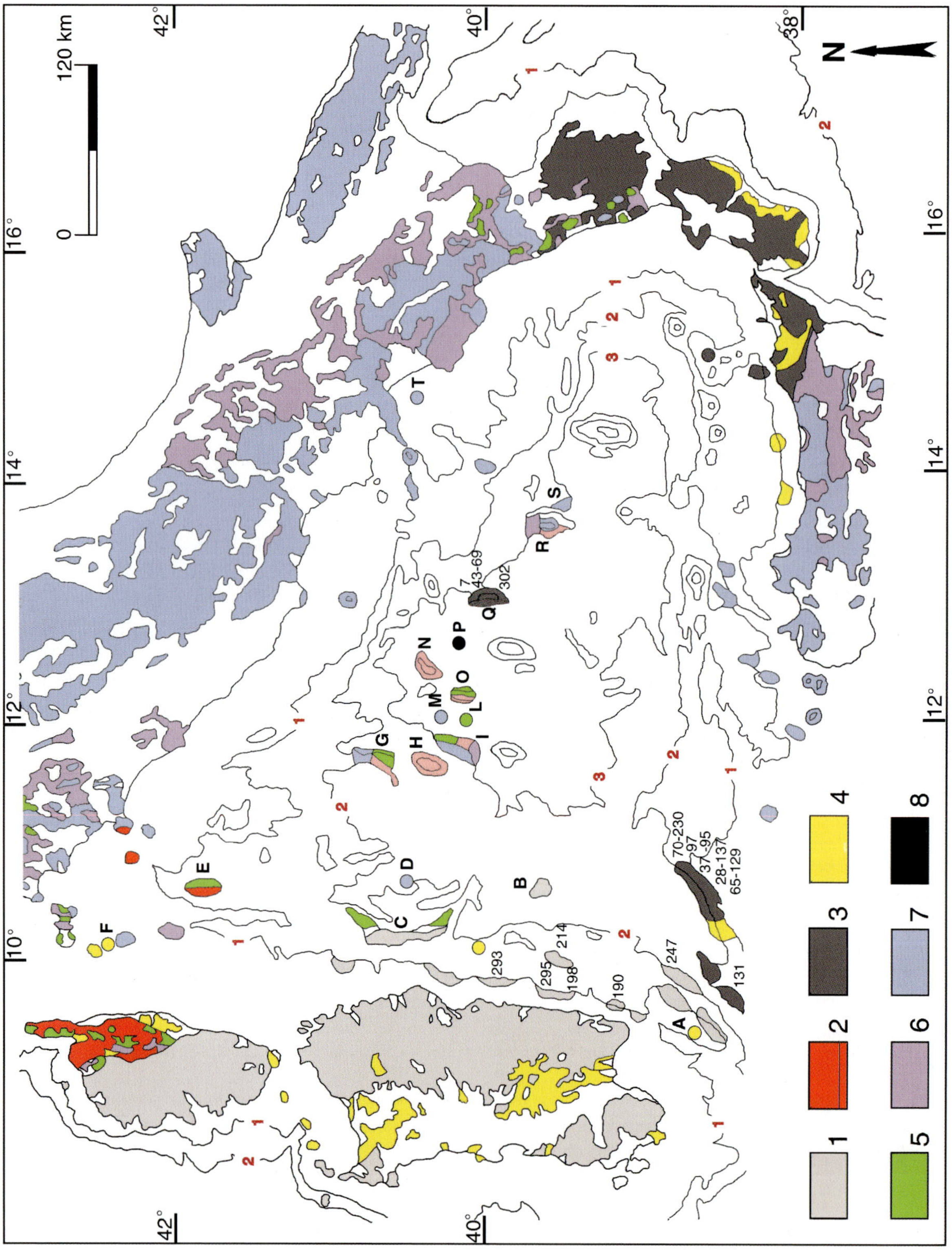

Figure 2: Types and distribution of pre-Tortonian rocks from and around the Tyrrhenian Sea. For explanation of legend and lettering see text. Black figures inside the Tyrrhenian Basin are radiometric ages or age-ranges. Red figures are isobaths in kilometers. Extensions of submarine outcrops are greatly exaggerated.

1.4. Oligocene to Middle-Late Miocene sedimentary rocks in Sardinia, Corsica, and on the Calabride–Peloritanian–Kabilide units

These rocks have been separated from their basements belonging to groups 1–3 because they represent the sedimentary deposition during opening of the Western Mediterranean Basin, and prior to the onset of the Tyrrhenian extension. Clastic, with subordinate carbonate rocks of Oligocene to Middle Miocene age occur in central-western Sardinia, on Alpine Corsica units and on southern Calabria–Peloritanian units. They are notably absent on crystalline-metamorphic units of northern Calabria. At sea, they were recorded from east of the Corsica basin (F), and they occur associated with both Sardinian and Calabrian-type basements in the Cornaglia Terrace, south of the Orosei Canyon, and along the Sardinia Channel. They also occur offshore northern Sicily.

1.5. Ophiolites of the Tethyan domain and their pelagic sedimentary cover

Mesozoic ophiolitic rocks, and their pelagic cover of radiolarites and calpionellid micrites, have been separated from other bedrock types making up the Alpine and Apenninic chain units because of their importance as remains of the subducted Tethys Ocean. When associated to Alpine units they mostly show a blueschist metamorphism (Piemontese Units), while in the Apennines they are mostly unmetamorphosed (except for the oceanic alteration: Liguride units). Ophiolites and their cover are rather common in the Tyrrhenian seafloor, and mostly belong to the Liguride type. In the southern Tyrrhenian Basin, starting from the eastern wings of the Baronie Smt. (C), they crop out at a number of localities, such as the Cialdi Smt. (E), Cassinis Smt. (G), Central Fault (I), ODP Site 656 (L), and De Marchi Smt. (O). They were never found either across the north Sicilian margin or in the Sardinia Channel.

1.6. Apenninic–Maghrebian units: Liguride (except for ophiolites), Sicilide, and associated deep water, often chaotic units (Molise, Sannio, Lucania, etc.)

The units making up the Apenninic (peninsular Italy) and Maghrebid (Sicily) chains are essentially made of sedimentary successions, and they have mostly been deformed since the Oligocene with an African–Ionian–Adriatic vergence. In Figure 2, two different types of units have been distinguished. This section includes sequences originally deposited in deep water in internal domains of the African passive margin, towards the Tethys Ocean. They are usually fine-grained and siliciclastic (often flysch deposits). The clayey sequences frequently display a chaotic texture, assumed to be an indication of their high degree of allochthony inside the chains.

Helmintoides flysch of Liguride type occurs in the northern Tyrrhenian Sea and down to the R. Selli line or Central Fault (I) and to the Issel Smt. (R). The Sicilide and other units of the southern Apenninic arc, mostly made of clayey sequences and often displaying a chaotic texture, are quite uncommon in samplings from the Tyrrhenian seafloor. This could be either a primary feature of their distribution or a sampling bias due to the fact that these fine-grained rocks cannot generate steep slopes and outcrops and/or are washed off the dredges during recovery.

1.7. Apenninic–Maghrebian units: Carbonate sequences of the Adriatic–African margin, including pre-Tortonian clastic or carbonate successions

This section includes all rocks of the Apenninic–Maghrebian chains that were originally deposited in external sectors of the African–Adriatic continental margin along a system of intervening carbonate platforms and basins.

Across the Tyrrhenian Basin, these rocks are mainly shallow-water carbonates, severely altered and often phosphatized, for which an age assignment is rarely possible. Rocks attributed, on the basis of facies similarity, to the carbonate platform and basin systems of the central and southern Apennines have been sampled at Cassinis Smt. (G), Central Fault (I), Issel Smt. (R), Poseidone Smt. (S), and in several localities along the Campanian margin. Recognizable fossils are preserved in some drillings, such as ODP Site 652 (M), and in sample dredges off Sicily, where rocks mostly belong to the Panormide Units of the Maghrebian chain. The westernmost occurrence for this group is at ODP Site 654 (D), east of the Baronie Smt., where low grade metamorphites occur associated with Lower Triassic rocks in Alpine facies, resembling units of the northern Apennines in Tuscany.

1.8. Central Tyrrhenian metamorphic rocks of uncertain paleogeographic attribution

Several dredgings in the central part of the southern Tyrrhenian, such as at Cassinis Smt. (G), Secchi Smt. (H), Central Fault (I), De Marchi Smt. (O), Farfalle E Smt. (N), Issel Smt. (R) only yielded low-grade metamorphic rocks (mostly phyllites) of uncertain paleogeographic attribution. These rocks are however at places associated, on the same outcrop (e.g. G, I, O, R), with shallow water carbonates or with ophiolites. This may suggest that they belong to units of the Apenninic chain, as is for instance the case in Tuscany. Alternatively, they could correspond to the basement of the south Apenninic units, made only of detached sedimentary cover.

1.9. Tyrrhenian mantle serpentinites

Although these are not strictly pre-rift rocks, serpentinized peridotites of the sub-continental mantle have been drilled at ODP Site 651 (P in Figure 2), north of the Vavilov Smt, where they are overlain by pillow lava flows of Pliocene age (I in Figure 3).

2. TERTIARY TO RECENT IGNEOUS ROCKS (FIGURE 3)

Serri et al. (2001) have produced the most updated report on magma characterization of igneous occurrences in and around the Tyrrhenian Basin. The distribution of igneous rocks reported in Figure 3 updates and integrates, with geophysical data, the above report. The numbers in the lower left corner of the figure correspond to the different magma types, modified and simplified from Serri et al. (2001), reported in this paragraph. Localities with only geophysical (magnetic, seismic) evidence of magmatism are contoured by dashed lines. All occurrences are volcanic rocks except where indicated by **s** (sub-volcanic) or by **p** (plutonic). Black figures are rock ages obtained at sea.

2.1. Oligocene to Middle Miocene rocks with island arc basalt type mantle sources (Corsica and Sardinia on land)

At sea, the rocks of the Oligocene-Miocene (30–13 Ma) volcanic arc of Sardinia, connected to the development of the Western Mediterranean back-arc basin, crop out south of Sardinia and between this island and Corsica. A recent sampling (Mascle et al., 2001) has provided calcalkaline volcanics (with lamproitic xenoliths) dated at 12.6 Ma on the Cornacya Smt. (A), east of southern Sardinia. They may represent either the latest stages of Oligocene-Miocene arc volcanism in Sardinia or could be the equivalent of the Sisco lamproite dyke (E) in north-eastern Corsica, assumed by some authors as indication of early stages of rifting in the Tyrrhenian Basin (Mascle et al., 2001).

2.2. Tortonian to Recent rocks either with mantle sources contaminated by crustal materials, or derived from upper crustal sources, or derived by mixing between crustal and mantle sources (Tuscany and Latium on land)

At sea, these are mostly granitoids, sampled at the Vercelli Smt. (D, age 7.3 Ma) and otherwise only inferred from seismic surveys. They occur only north of the 41° North Lineament, which is assumed to separate the northern from the southern Tyrrhenian Basin.

2.3. Pleistocene to recent rocks with ocean island basalt mantle sources enriched by subduction-related sources (Campania)

No igneous rocks of this type, that initiated some 2 Ma ago and is still active on land, have been described so far from the Tyrrhenian Sea.

2.4. Pliocene to Recent rocks with island arc basalt type mantle sources

These rocks appear to constitute the whole southeastern Tyrrhenian area, including the Eolian Island Arc and its submerged edifices, the Marsili volcano (O) and plain (ODP Site 650, N), and the Palinuro volcano (P). Except for the Glauco Smt. (Q), which is probably Late Pliocene in age, this volcanism is Quaternary to Recent and active. More internally, Early to Late Pliocene volcanic rocks of this type occur at the Anchise Smt. (R), west of Ustica island, and at ODP Site 651(I), north of Vavilov Smt.

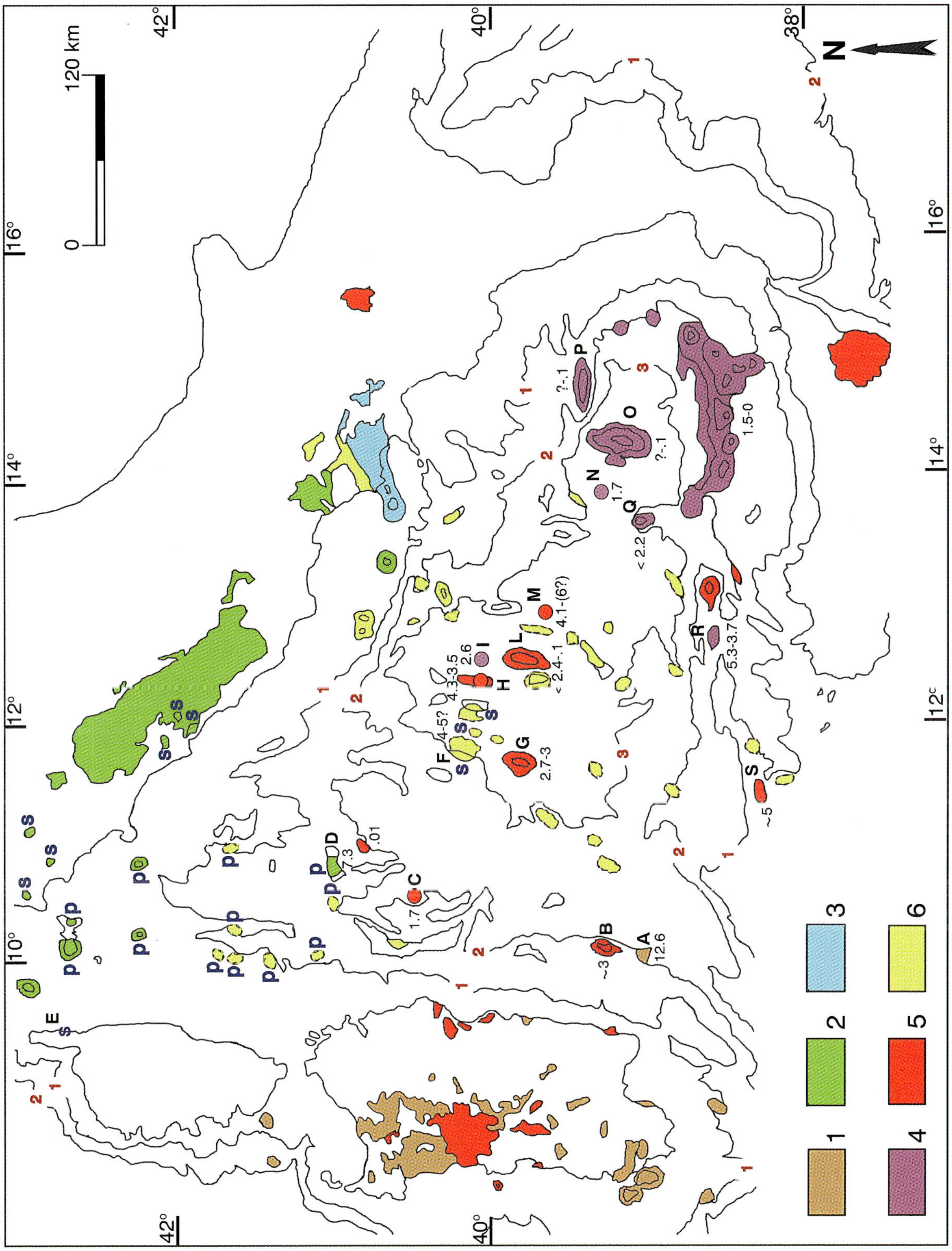

Figure 3: Types and distribution of Oligocene to Recent magmatic rocks from and around the Tyrrhenian Sea. For explanation of legend and symbols see text. Rock ages, determined or inferred, are reported only at sea (black figures). Extensions of seafloor outcrops are greatly exaggerated. Dashed fields are possible occurrences of magmatic rocks suggested by geophysical data.

2.5. Pliocene to Recent rocks derived from ocean island basalt (OIB) type and from mid-oceanic ridge basalt (MORB) mantle sources

MOR basalts have been collected at ODP Site 655 (Gortani Ridge, H) and at DSDP Site 373 (M). Ages are essentially Lower Pliocene (around 4 Ma), although we cannot exclude an older age (around 6 Ma) for part of the basalts drilled at DSDP Site 373 (for a discussion on radiometric ages see Sartori, 1987).

The OIB-type rocks occur from Sardinia as far east as Ustica island. They constitute very large volcanic seamounts such as Magnaghi (G), Vavilov (L), Quirra (B), Drepano (S), and minor occurrences such as the lavas drilled at ODP Site 654 (C). As in Sardinia, this submarine volcanism shows an age range from some 5 Ma (Drepano Smt) to subrecent (0.01 Ma).

2.6. Rocks from undefined magma sources

Several sampled outcrops and all the geophysically inferred igneous occurrences have not been classified in terms of magma sources, and none of them has been directly dated. These are mostly volcanic seamounts and ridges, spread throughout the southern Tyrrhenian Basin. Dykes and laccoliths have also been detected between the Central Fault and the De Marchi Smt. (F). An age of emplacement around 4–5 Ma has been proposed for these sub-volcanic bodies from reflection seismic analysis (Sartori et al., submitted).

3. DISCUSSION

3.1. Pre-Tortonian rocks, their nature and distribution

As already proposed by Sartori (1986), the Tyrrhenian pre-rift rocks include all the geological domains surrounding the basin: the Corsica–Sardinia block, the Alpine chain, the Apenninic–Maghrebian chain and the Calabride–Peloritani–Kabilide units. The latter elements appear as having been those most fragmented by the Tyrrhenian stretching. The distribution of different basement types at sea and on land provides evidence for discussing the paleogeography, age and distribution of deformations in the Tyrrhenian area prior to its Tortonian extension.

Figure 2 seems to indicate that the occurrence of Alpine units (2) and Tethyan ophiolites (5) is limited southwards. To the west, the southern limit of ophiolite occurrences is the Orosei Canyon (OCL Line of Sartori et al., 2001), just south of the Baronie Smt. It is to be noted that Segre (1958) had already questioned the extension of Pennidic units to the south of the Baronie Smt. Moving from the Orosei Canyon to the east, the southern limit is placed along a line reconnecting the Baronie outcrops with those of the southern portion of northern Calabria. South of that limit, only elements of the Sardinia basement (1), of the Calabride–Peloritanian–Kabilide units (3), and of the Maghrebian chain of Sicily (6, 7) occur, all lacking ophiolite units. The Tethyan ocean had then to close in correspondence to the Orosei Canyon line to the west and to the limit between northern and southern Calabria to the east. The Tethys developed during the Mesozoic and was subducted from Cretaceous to Paleogene times to form the Europe-verging Alpine chain. Subsequently, from the Oligocene onwards, the oceanic Liguride units were also included in the east-verging Apenninic chain.

At sea, Oligocene to Miocene clastic sequences (4) were sampled in areas characterized by Calabride- and Sardinia-type bedrocks. On crystalline metamorphic basements on land, Oligocene-Middle Miocene deposits occur in Sardinia and on southern Calabria–Peloritanian units, while they are absent in northern Calabria. In the Sardinia Channel, the Variscan basement of Sardinia is directly thrust above Calabride–Peloritani–Kabilide units, along the so-called Main Sardinia Thrust (Catalano et al., 1987; Sartori et al., 2001). Moving towards this element, radiometric ages of the Sardinia basement appear slightly younger (Figure 2). The Calabride–Peloritanian–Kabilide units were deformed, with African–Ionian vergence, essentially from the Oligocene to Middle Miocene and were thrust above the Maghrebian chain units, as observed in the Sardinia Channel to the west (Drepano Thrust Front of Catalano et al., 1987) and in north-eastern Sicily (Taormina Line, Ogniben, 1960). These basement units show radiometric age-ranges between Hercynian and Miocene (Figure 2). South of the domain in which the Mesozoic Tethys had developed, compressive deformations started in the Oligocene, and until the Lower-Middle Miocene affected the Variscan basement making the Southern Calabria–Peloritani–Kabilide units. Oligocene to Middle Miocene sedimentary sequences, at sea and on land, accompanied these deformations, with different significances at different places: syn- to post-rift in Sardinia, fore-arc basin fill in the SW Tyrrhenian Sea and syn- to post-tectonic in southern Calabria and Peloritani (Sartori et al., 2001). These deformations only had an Adriatic–Ionian–Sicilian vergence prior to the late Miocene extension that occurred in the Tyrrhenian domain. These observations exclude the possibility, assumed in a large number of past paleogeographic reconstructions, of a southward extension of the Europe-verging Alpine chain along the whole of Sardinia and further west.

3.2. Late Miocene to Recent igneous rocks, their nature and distribution

Lacking petrochemical and age characterization for a large number of igneous rocks occurring in the Tyrrhenian Basin (6 in Figure 3), the relationships between magmas and basin evolution is still far from being established in detail. Plutonic rocks are dominant in the poorly thinned northern basin; they represent a portion of the Tuscan magmatic province of the northern Apennines, and are related to a limited subduction of continental Adriatic litosphere (Serri et al., 2001). We will discuss here essentially the igneous rocks of the southern Tyrrhenian Basin, related to subduction of the Ionian lithosphere.

A. Volcanism and oceanic crust in the southern Tyrrhenian Basin

Bedrock samples are only of volcanic nature in a sector limited (compare Figures 1, 2, and 3):

- to the west by the prosecution of the Central Fault south of the Orosei Canyon;
- to the east by the Eolian islands and related submerged edifices;
- to the north by the huge Magnaghi, Vavilov, and Palinuro volcanoes;
- to the south by the Anchise-Ustica and Eolian Islands complexes.

Across this wide area, MOR basalts have been recovered only in the Vavilov Plain region, at ODP Site 655 (Gortani Ridge, H in Figure 3) and at DSDP Site 373 (M in Figure 3). They are of Pliocene age (some 4 Ma), with the possible addition of Messinian basalts at DSDP Site 373.

Gortani ridge is located seaward of the last tilted continental blocks of the Sardinia margin (De Marchi Smt, O in Figure 2), but also to the west (landward) of the sub-continental mantle peridotites drilled at ODP Site 651 (P in Figure 2). By comparison with the Iberia margin, where oceanic crust occurs only seawards of the denudated peridotites of the continental margin, the MORB lavas of ODP Site 655 may not represent a spreading ridge belong instead to a wide volcanic-rich continent-ocean transition (Sartori et al., submitted).

DSDP Site 373 was drilled on a buried circular edifice where volcanic accretion from the Messinian to Lower Pliocene, more than spreading, seems to have occurred (Sartori, 1987).

Although not relatable to spreading ridges, these limited MORB occurrences, in addition to seismic refraction data, indicate that in some sectors of the Vavilov plain oceanic crust was generated, essentially during the Pliocene. The lack of spreading ridges and of elongated magnetic anomalies, as well as the co-occurrence of large volcanic edifices, do not allow us to draw with certainty the continent-ocean boundaries in the area.

B. Pliocene to Recent magma types

It is clear from Figure 3 that the southern Tyrrhenian Basin shares, with other back-arc basins floored with oceanic crust, a petrogenetic complexity (Saunders and Tarney, 1976; Wright et al., 1996). Nevertheless, some rough patterns of distribution of magma types seem to emerge.

Ocean island basalts occur in Sardinia and from its continental slope as far east as the Vavilov Smt. (L) and Ustica island, that is to the west of the Issel crustal bridge (Figure 1). The age of this volcanism, which is spread over a wide portion of the upper plate, ranges from some 5 to 0.01 Ma, with a rather casual distribution of ages in space. In the same wide domain, only the lavas of the Anchise Smt. (R) and of ODP Site 651 (I) have been assigned to the island arc suite, and both are of Pliocene age. These occurrences might call for the existence of a remnant Pliocene volcanic arc submerged in the Tyrrhenian, an argument first advanced by Sartori (1986). More data is needed to solve the problem, but it is worth noting that ODP Site 651 (I, Pliocene island arc volcanism) lies west of DSDP Site 373 (M, Pliocene MORB), which is contrary to the hypothesis that ODP Site 651 belongs to a remnant Pliocene arc.

East of the Issel crustal bridge, across and around the Marsili Plain, Pleistocene to active island arc volcanism appears exclusively, with activity still occurring in the Eolian Islands. Even the huge Marsili volcano (O in Figure 3), which has been interpreted as a super-inflated spreading ridge (Marani and Trua, in press), displays arc-type volcanism. This volcano resembles in shape and elongation the Vavilov and Magnaghi Smts. of the Vavilov Plain (G and L in Figure 3); these are made however of OIB type magmas. Magmatism with subduction influences, though of a type different from that of the southeastern Tyrrhenian, also occurs in Campania, where it started some 2 Ma ago and is still active (Serri et al., 2001).

Summarising the above observations, we can conclude that from Sardinia to the Issel crustal bridge, Pliocene to sub-Recent volcanism of ocean island basalt plus MORB type dominates, while Pliocene arc-type volcanism is very minor. Ages of the volcanic activities are casually distributed in space. East of the Issel crustal bridge, magmatism is instead exclusively of island arc-type and of Pleistocene to Recent age (with only the Glauco Smt., Q, being possibly of Late Pliocene age).

C. Magmatism and history of subduction

Accepting that, notwithstanding the complexities and exceptions, a fundamental bipartition of magma types occurs in the southern Tyrrhenian, an explanation can be found by taking into account the boundary conditions usually assumed to relate magma genesis with dynamic processes.

Volcanic arc-type magmas usually develop where subducting oceanic slabs reach a depth of 100–150 km. The Pleistocene inception of arc volcanism, and its confinement to the southeastern Tyrrhenian (and maybe to Campania), would indicate that an oceanic slab had to have reached 100–150 km depth beneath the area less than some 2 Ma ago (presently the slab is some 250–300 km deep beneath the Eolian Island Arc and even more beneath the Marsili volcano).

However, subduction also must have occurred earlier, from late Tortonian to end Pliocene, when the central Tyrrhenian back-arc basin was generated. The slab presently subducting in the southeastern Tyrrhenian is about 500 km deep and 700 km long, so that the portion presently deeper than 300–350 km (which is some 300 km more in length due to the high dip of the slab) should have experienced subduction prior to the Pliocene. During the initial stages of back-arc rifting (and of subduction), from the Tortonian to Messinian, volcanism in the southern Tyrrhenian was virtually absent. By the Pliocene, extension had eventually produced a wide permeability of the whole upper plate (from the Vavilov Plain to Sardinia). This favoured a rather casual uprising of uncontaminated magmas of deep mantle provenance (ocean island basalt-type) plus minor basalts indicating spreading-type processes. Much less, and randomly located, were Pliocene eruptions, indicating influence from a subducting oceanic lithosphere.

Palinspastic reconstruction in peninsular Italy indicates that from the Tortonian to part of the Pliocene the internal carbonate platform (Campania) of the southern Apennines and the Latium-Abruzzi carbonate platform of the central Apennines, both formerly part of the Mesozoic African margin, became incorporated into the chain units (Patacca et al., 1993). These thick (>5 km) shallow-water sequences of Triassic to Miocene age were not founded on older oceanic crust; their continental basement should, however, have been considerably thinned during the Mesozoic rifting of the African margin. Assuming that these Apennine tectonic units were off-scraped from their subducting lithosphere, we are forced to conclude that the portion of slab, subducted from the Tortonian to the Pliocene and presently deeper than 300 km, has a thinned continental nature. The Ionian type oceanic lithosphere, external to the carbonate platforms, was engaged in subduction only from the Pliocene and reached the critical depth for inception of arc volcanism (100–150 km) at the beginning of the Pleistocene. By the end of the Pliocene, back-arc extension stopped in the central Tyrrhenian area, and moved to the southeastern plain where volcanism became entirely of arc-type. The different nature in time of subducting lithosphere may also have caused the jump in back-arc basin development, with a crustal saddle (Issel Bridge) separating the Vavilov from the Marsili Basin.

CONCLUSIONS

1. The elements of continental crust dissected by the Tyrrhenian rifting consist of the various geological domains occurring in the encircling emerged areas. The distribution of rock types and associations in both submarine and emerged areas depicts the paleogeography produced by the Alpine orogenic cycle prior to basin extension. The Cretaceous-Paleogene Europe-verging Alpine chain, derived from the closure of the Tethyan Ocean, was

limited to the south at a latitude presently corresponding to the Baronie Smt. to the west and to the limit between northern and southern Calabria to the east. These observations exclude the possibility, assumed in several paleogeographic reconstructions, that the Alpine chain had to run from Corsica off the eastern coast of the whole of Sardinia to be reconnected to other Europe-verging chains further west. South of the Tethyan domain, whose portions were also incorporated in the Apenninic chain from the Oligocene onwards, only Adriatic–Africa-verging compressive deformation would have occurred. From the Oligocene to Lower-Middle Miocene (during formation of the western Mediterranean Basin and rotation of Sardinia) they affected essentially the southern Calabria–Peloritanian–Kabilide units. The post-Tortonian stretching of the southern Tyrrhenian Basin strongly disrupted these units and was accompanied by a further migration and shortening of the southern Apenninic arc.

2. The basement of a wide portion of the southern Tyrrhenian Basin is made almost exclusively of Pliocene to Recent igneous rocks plus oceanic crust (whose precise limits cannot be drawn yet). The central-western part of the southern Tyrrhenian, located west of the Issel crustal bridge (and also including Sardinia), essentially host igneous rocks indicating either spreading processes (MORB type) or provenance from deep mantle sources (OIB type), having casually crossed through the extending upper plate. The south-eastern part of the Tyrrhenian (east of the Issel crustal bridge) is instead dominated by Pleistocene to Recent island arc-type products, indicating an influence from subducted oceanic lithosphere. Reconnecting these characters to the palinspastic reconstruction of the southern Apennines, the magmatic differentiation in space and time seems to be related to the nature (thinned continental versus oceanic) of the Ionian lithosphere experiencing subduction with time.

MEMORIAL

After submitting this his last work, Prof. Renzo Sartori passed away at the end of 2002. The Editor remembers him as an author of many high quality papers and dear friend. The Geoscience community will hold him in great esteem as both scientist and gentleman.

REFERENCES

Bigi, G., Cosentino, D., Parotto, M., Sartori, R., Scandone, P., 1990–1992. Structural Model of Italy and gravity maps. Sheets 1–9, Scale 1/500.000. Progetto Finalizzato Geodinamica CNR, SELCA, Firenze.

Bouillin, J.P., Monie, P., Rolfo, F., Tricart, P., Mascle, G., Torelli, L., Compagnoni, R., Mascle, J., Pecher, A., Peis, D., Rekhiss, F., 1999. Données chronologiques ^{40}Ar/^{39}Ar sur les socles sarde et kabilo-peloritain submergés dans le canal de Sardaigne (Méditerranée occidentale). C. R. Acad. Sci. Paris, Science de la Terre et des Planètes 328, 529–534.

Catalano, R., D'Argenio, B., Torelli, L., 1987. From Sardinia Channel to Sicily strait. A geologic section based on seismic and field data. In: Boriani, A., Bonafede, M., Piccardo, G.B., Vai, G.B. (Eds.), The Lithosphere in Italy. Accad. Naz. Lincei Roma Atti Convegni 80, 99–114.

Colantoni, P., Fabbri, A., Gallignani, P., Sartori, R., Rehault, J.P., 1981. Carta litologica e stratigrafica dei mari italiani. Scale 1/1.500.000. Litografia Artistica Cartografica, Firenze.

Kastens, K.A., Mascle, J., Auroux, C., et al. (Eds.), 1990. Proceedings ODP Scientific Results, vol. 107, College Station, TX (Ocean Drilling Program), pp. 1–722.

Marani, M., Trua, T., in press. Thermal constriction and slab tearing at the origin of a superinflated spreading ridge: Marsili volcano (Tyrrhenian Sea). J. Geophysical Research 107.

Mascle, G.H., Tricart, P., Torelli, L., Bouillin, J.P., Rolfo, F., Lapierre, H., Monié, P., Depardon, S., Mascle, J., Peis, D., 2001. Evolution of the Sardinia Channel (Western Mediterranean): new constraints from a diving survey on Cornacya seamount off SE Sardinia. Marine Geology 179, 179–202.

Ogniben, L., 1960. Nota illustrative allo schema geologico della Sicilia nord-orientale. Rivista Mineraria Siciliana 11, 183–212.

Patacca, E., Sartori, R., Scandone, P., 1993. Tyrrhenian Basin and Apennines: kinematic evolution and related dynamic constraints. In: Recent Evolution and Seismicity of the Mediterranean Region. Kluwer Academic, pp. 161–171.

Rehault, J.P., Mascle, J., Fabbri, A., Moussat, E., Thommeret, M., 1987. The Tyrrhenian Sea before Leg 107. In: Kastens, K.A., Mascle, J., et al. (Eds.), Proceedings ODP Initial Reports, vol. 107, College Station, TX (Ocean Drilling Program), pp. 9–35.

Sartori, R., 1986. Notes on the geology of the acoustic basement in the Tyrrhenian Sea. Memorie Società Geologica Italiana 36, 99–108.

Sartori, R., ODP Leg 107 Scientific Staff, 1987. Drillings of ODP Leg 107 in the Tyrrhenian Sea: tentative basin evolution compared to deformations in the surrounding chains. In: Boriani, A., Bonafede, M., Piccardo, G.B., Vai, G.B. (Eds.), The Lithosphere in Italy. Accad. Naz. Lincei Roma Atti Convegni 80, 139–156.

Sartori, R., Mascle, G., Amaudric du Chaffaut, S., 1987. A review of circum-Tyrrhenian regional geology. In: Kastens, K.A., Mascle, J., Auroux, C., et al. (Eds.), Proceedings Ocean Drilling Program Initial Report, vol. 107, College Station, TX, pp. 37–63.

Sartori, R., Carrara, G., Torelli, L., Zitellini, N., 2001. Neogene evolution of the southwestern Tyrrhenian Sea (Sardinia Basin and western Bathyal plain). Marine Geology 175, 47–66.

Sartori, R., Torelli, L., Zitellini, N., Carrara, G., Magaldi, M., Mussoni, P., submitted. Crustal characters of the central Tyrrhenian Sea (Mediterranean).

Saunders, A.D., Tarney, J., 1976. The geochemistry of basalts from a back-arc spreading centre in the East Scotia Sea. Geoch. et Cosmochimica Acta 43, 555–572.

Segre, A.G., 1958. Neue geologische und morphologische Untersuchungen im tyrrhenischen Gebiet. Geol. Rundschau 47, 196–207.

Selli, R., 1974. Appunti sula geologia del Mar Tirreno. In Paleogeografia del Terziario nell'ambito del Mediterraneo occidentale. Seminario Facoltà Scienze Univ. Cagliari 43, 327–351.

Serri, G., Innocenti, F., Manetti, P., 2001. Magmatism from Mesozoic to Present: petrogenesis, time–space distribution and geodynamic implications. In: Vai, G.B., Martini, P. (Eds.), Anatomy of an Orogen: the Apennines and Adjacent Mediterranean Basins. Kluwer Academic, UK, pp. 77–104.

Wright, I.C., Parson, L.M., Gamble, J.A., 1996. Evolution and interaction of migrating cross-arc volcanism and backarc rifting: an example from the southern Havre Trough (35°20′–37°S). J. Geophysical Research 10, 22071–22086.

CROP PROJECT: Deep Seismic Exploration of the Central Mediterranean and Italy
Edited by I.R. Finetti

CHAPTER 5

CROP Seismic Data Acquisition, Processing and Interpretative Reprocessing

I.R. Finetti[1], E. Forlin[1], M. Pipan[1,*]

ABSTRACT

The CROP dataset encompasses nearly 10,000 km of seismic sections. The marine part (8687.15 km) was performed from 1988 to 1995 by the RV OGS-Explora with an 80.4 liter airgun array and a 4500 m long, 180-channel streamer. The basic processing sequence focused on amplitude correction, multiple removal, deconvolution, velocity analysis, stacking, post-stack migration and time-variant filtering. Part of the marine dataset was reprocessed to improve multiple attenuation, deep signal-to-noise ratio and imaging of structurally complex sectors. A proprietary Hough-transform-based algorithm was exploited to attenuate multiple events. Computation of instantaneous attributes by means of wavelet transform improved identification of weak signals in noisy background from deep crustal reflectors. Pre-stack imaging exploited the feedback from the interpretation phase to iteratively refine velocity–depth models and obtain optimum focusing of primary events. The latter interaction between processing and interpretation was the basis for the successful implementation and application of an interpretative strategy for deep crustal seismic data reprocessing.

Keywords: CROP seismic acquisition, processing, interpretative reprocessing

1. INTRODUCTION

Multi-channel stacking reflection seismics is the most effective exploration technique for imaging the deep crust. Several worldwide research programmes have accomplished deep crustal imaging on land and offshore in recent years (Damotte, 1996; Meissner and Bortfeld, 1990; Zhu and Brown, 1986; Steer et al., 1995; Hauck and Brown, 1992). A large number of papers illustrate data characteristics and the processing strategies proposed by the different teams (Stiller and Thomas, 1989; Mazzotti et al., 2000; Damotte, 1996). The processing sequence adopted by the joint Sino-U.S. programme INDEPTH (Hauck and Brown, 1992) is divided up into stages of crooked line geometry assignment, data editing and resampling, filtering to reduce ground roll and random noise, stacking accompanied by velocity and residual statics analysis, and post-stack signal/noise enhancement. The two most important non-routine processing steps are residual statics analysis and FX Deconvolution. Maximum power autostatics correct for source and receiver static shifts, which are not addressed by elevation statics. Frequency–space deconvolution reduces random wind noise. The two latter procedures allow identification of Moho reflections in sectors of the INDEPTH dataset. DEKORP uses a similar generalized processing flow-chart but dedicates a major effort to finding optimum stacking velocity–time functions and muting curves, and to selecting the appropriate velocity models and offset ranges for stacking (Meissner and Bortfeld, 1990; Stiller and Thomas, 1989). Non-routine processing procedures include frequency-wavenumber filtering, zero-phase deconvolution, DMO, slant-stack, true amplitude migration, and coherency filter. Other examples of processing strategies are reported in the references and suggested readings at the end of this chapter (DEKORP Research Group, 1990; Dohr and Fuchs, 1967; Graßl et al., 2004; Pratt et al., 1993; Schilt et al., 1981; Trappe and Bittner, 1989; Wallace, 1980; Yilmaz, 2001).

In short, from the seismic processing standpoint, CROP shares a group of common primary issues with the other deep crust projects:

*Corresponding author: Tel.: +39 40 5582277; e-mail: pipan@units.it (M. Pipan)
[1]Department of Geological, Environmental and Marine Sciences, Exploration Geophysics Group, University of Trieste, Via Weiss 1, 34127 Trieste, Italy

Table 1
List of the marine CROP seismic sections

Section	Length (km)	Acquisition							Processed or reprocessed (R)			Shot		Chapter	Plate
		Year	By	Record length (sec)	Group interv. (m)	Channel	Fold (%)	S. P. Int m	Year	By	Length (sec)	First	Last		
M-2A/I	517.5	1991	OGS	20	25	180	4500	50		R Univ. Trieste	17	100	10450	18	1
M-2A/II	278.5	1991	OGS	20	25	180	4500	50		R Univ. Trieste	17	10451	16022	21	1
M-2A/III	144.3	1991	OGS	40	25	180	3600	62.5		R Univ. Trieste	40	100	2380	17	1
M-2B	310.2	1991	OGS	21	25	180	3600	62.5		R Univ. Trieste	20	100	5027	17	1
M-3	356	1991	OGS	20	25	180	3600	62.5		R Univ. Trieste	20	100	5760	19	1
M-4	266	1991	OGS	20	25	180	3600	62.5		R Univ. Trieste	20	100	4325	17	3
M-5	243	1991	OGS	20	25	180	3600	62.5		R Univ. Trieste	20	100	3956	23	9
M-6A	136.5	1991	OGS	20	25	180	4500	50		R Univ. Trieste	17	100	2785	15	7
M-6B	133.8	1991	OGS	20	25	180	4500	50		R Univ. Trieste	17	100	2730	12	1
M-8	133.3	1995	OGS	20	25	180	4500	50	1997	ENI AGIP	17	100	2721	23	
M-9	178.4	1991	OGS	20	25	180	4500	50		R Univ. Trieste	17	100	3635	21	3
M-10	230.7	1991	OGS	20	25	180	4500	50		R Univ. Trieste	17	100	4686	8	1
M-11	183.3	1991	OGS	20	25	180	4500	50		R Univ. Trieste	17	100	3720	8	3
M-12A	76.8	1991	OGS	20	25	180	4500	50		R Univ. Trieste	17	100	1600	8	7
M-12B	178.9	1991	OGS	20	25	180	4500	50		R Univ. Trieste	17	100	3640	8	5
M-13	168.8	1995	OGS	20	25	180	4500	50	1997	ENI AGIP	17	100	3431	23	7
M-14	101.9	1995	OGS	20	25	180	4500	50	1997	ENI AGIP	17	100	2093	12	
M-15	84.8	1995	OGS	20	25	180	4500	50	1997	ENI AGIP	17	100	1751	9	2
M-16	64.8	1995	OGS	20	25	180	4500	50		Univ. Trieste	17	100	1342	8	7
M-17A	47.7	1995	OGS	20	25	180	4500	50	1997	ENI AGIP	17	100	1009	23	
M-17B	127.9	1995	OGS	20	25	180	4500	50	1997	ENI AGIP	17	100	2613	23	3
M-17C	188.5	1994/95	OGS	20	25	180	4500	50	1997	ENI AGIP	17	100	3825	23	5
M-18	97.3	1995	OGS	20	25	180	4500	50	1997	ENI AGIP	17	100	2001	23	1
M-21	106.3	1994	OGS	20	25	180	4500	50	1996	ENI AGIP	17	100	2195	15	
M-22	181.3	1994	OGS	20	25	180	4500	50	1996	ENI AGIP	17	100	3680	19	3
M-23A1	482.4	1994	OGS	20	25	180	4500	50	1996	ENI AGIP	17	100	9640	19	5
M-23A2	166.8	1994	OGS	20	25	180	4500	50	1996	ENI AGIP	17	100	3392	15	1
M-23B	267.9	1994	OGS	20	25	180	4500	50	1996	ENI AGIP	17	100	5417	15	1
M-24	223	1993	OGS	20	25	120	3000	50	1993	ENI AGIP	17	100	4530	15	8
M-25	156.9	1994	OGS	20	25	180	4500	50	1996	ENI AGIP	17	100	3236	26	3
M-26	127.7	1994	OGS	30	25	90	2300	100	1994	ISMES	17	100	1373	15	10
M-27	143	1994	OGS	30	25	90	2300	100	1994	ISMES	17	100	1507	17	5

Table 1
Continued

Section	Length (km)	Acquisition							Processed or reprocessed (R)			Shot		Chapter	Plate
		Year	By	Record length (sec)	Group interv. (m)	Channel	Fold (%)	S. P. Int m	Year	By	Length (sec)	First	Last		
M-28A	82	1994	OGS	20	25	180	4500	50	1994	ISMES	17	100	1695	15	5
M-28B	267.6	1994	OGS	20	25	180	4500	50	1994	ISMES	17	100	5406	18	3
M-29A	194.8	1994	OGS	20	25	180	4500	50	1994	ISMES	17	100	3953	15	3
M-29B	274.5	1994	OGS	20	25	180	4500	50	1994	ISMES	17	100	1360	21	5
M-30	174.95	1994	OGS	20	25	180	4500	50		Univ. Trieste	12	100	3599	21	7
M-31	105.5	1994	OGS	20	25	180	4500	50	1994	ISMES	17	100	2165	17	
M-32	122.7	1994	OGS	20	25	180	4500	50	1994	ISMES	17	100	2508	18	
M-34	356.2	1994	OGS	20	25	180	4500	50	1996	ENI AGIP	17	100	7178	23	11
M-36	333.3	1994	OGS	20	25	180	4500	50	1994	ISMES	17	100	6721	21	9
M-37	164.1	1994	OGS	20	25	180	4500	50	1994	ISMES	17	100	3386	9	1
M-38	113.8	1994	OGS	20	25	180	4500	50	1996	ENI AGIP	17	100	2331	23	
M-39	124.5	1994	OGS	20	25	180	4500	50	1996	ENI AGIP	17	100	2545	8	5
M-1	205	1988	OGS	20	25	180	4500	50	1988	ENI AGIP	17	100	4090		
Total	8687.15														

Table 2
List of the land CROP seismic sections

Section	Length (km)	Acquisition							Processing			CDP	
		Year	By	Length (sec)	Group interv. (m)	Channel	Fold (%)	Source	Year	By	Length (sec)	First	Last
CROP 03/a	80.7	1994	OGS	25	60	192	3200	Vibroseis	1995	ENI AGIP	25/16	5303	8087
CROP 03/b	69.6	1994	OGS	25	60	192	3200	Vibroseis	1995	ENI AGIP	25/16	2983	5491
CROP 03/c	68.22	1994	OGS	25	60	192	3200	Vibroseis	1995	ENI AGIP	25/16	805	3170
CROP 04	157.96	1990	OGS	25	80	240		Vibroseis	1999	Milan University	10	5	3953
CROP C-ALPS/a	13	1988	OGS	45	80	120	100	Explosive	1990	OGS	21	131	313
CROP C-ALPS/b	38.8	1988	OGS	32	80	120	6000	Vibroseis	1990	OGS	24	134	655
CROP C-ALPS/c	36	1988	OGS	45	80	120	100	Explosive	1990	OGS	24	14	563
CROP C-ALPS/d	14	1988	OGS	45	80	120	100	Explosive	1990	OGS	25	111	306
CROP W-ALPS/a	23.5	1986	OGS	40	80	120	6000	Vibroseis	1986	OGS	24	1	1589
CROP W-ALPS/b	63	1986	OGS	40	80	120	6000	Vibroseis	1986	OGS	24	1	591
TRANSALP/a	159.65	1998	GEOITALIA	18	50	120	6000	Vibroseis	2000	ENI AGIP	18	196	6581
TRANSALP/b	154.05	1999	GEOITALIA	18	50	120	6000	Vibroseis	2002	ENI AGIP	18	6582	12743
CROP 18/a	42.6	1995	DISCOVERY	25	60	192	3200	Explosive	1998	ISMES	12	50	1469
CROP 18/b	73.08	1995	DISCOVERY	25	60	192	3200	Explosive	1998	ISMES	12	112	2546
CROP 11/a	85.38	1996	OGS	25	60	192	3200	Explosive	1999	OGS	15	202	3048
CROP 11/b	89.12	1996/99	OGS/ISMES GEOTEC.	25	40	192	2400	Explosive	2000	OGS	15	3382	7838
CROP 11/c	85.26	1999	CO.GE.PRO	25	40	192	2400	Explosive	2001	OGS	15	7839	12102
Total	1253.92												

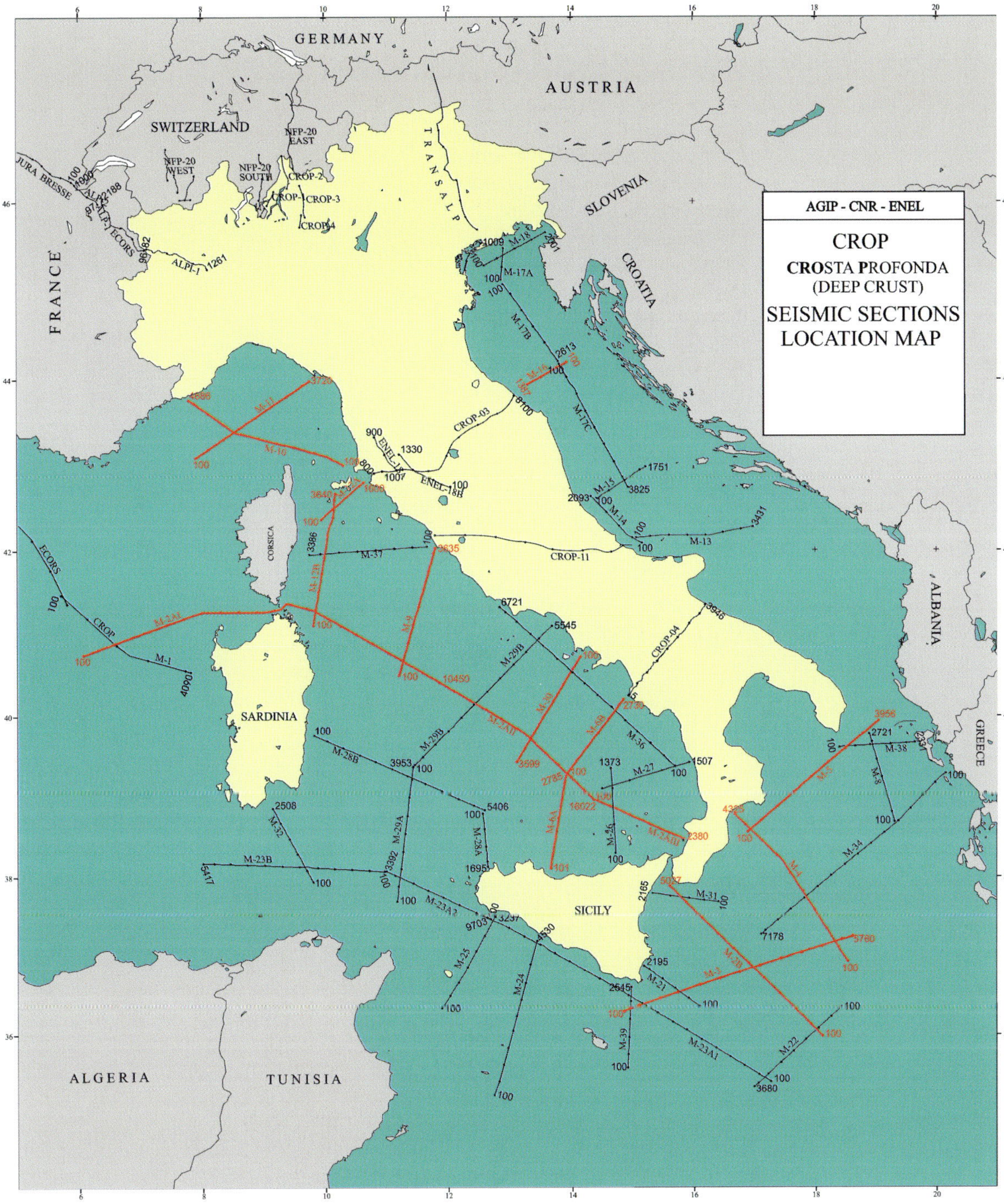

Figure 1: Location map of the CROP seismic section network. Red sections were processed or reprocessed by the University of Trieste.

a. Removal of coherent noise;
b. Enhancement of weak signals in noisy background;
c. Reconstruction of seismic velocity models;
d. Imaging of complex subsurface structures.

All of them are related to the challenging task of identifying deep crustal reflectors from surface seismic observations. Coherent noise is a major problem due to its amplitude and lateral correlation characteristics. Multiple reflections often exhibit energy sufficient to destroy or mask the weak signals from deep crustal discontinuities. The CROP-MARE dataset totals 8687 km of offshore seismic sections obtained with a 5.56 MJ airgun array (see Table 1 for seismic section list). All of the sections exhibit high amplitude multiples and their removal is there-

Table 3
Basic processing sequence for the CROP-MARE dataset

a	Data editing
b	*Preliminary* amplitude correction
c	Multiple reflection velocity analysis
d	Multiple reflection removal
e	*Preliminary* primary reflection velocity analysis
f	Amplitude correction
g	Deconvolution (on NMO corrected CMP gathers)
h	Primary reflection velocity analysis
i	NMO and stack

1	2
Time-variant filtering (TVF)	Migration Time-variant filtering (TVF)

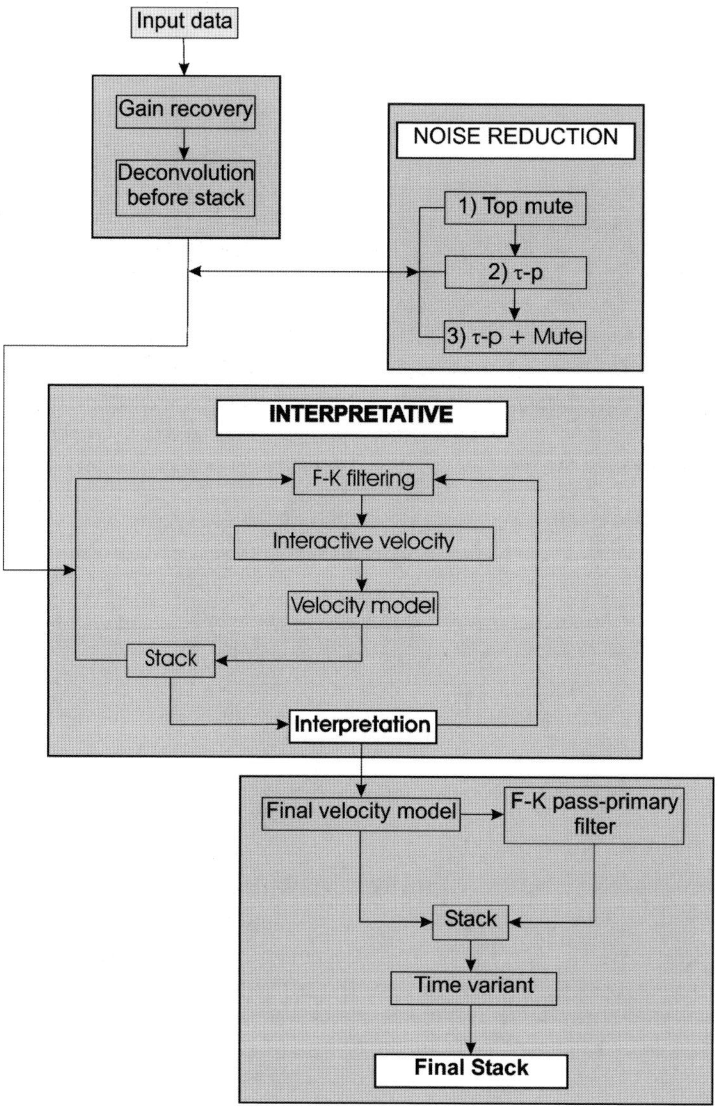

Figure 2: Basic interpretative processing sequence.

Table 4
CROP-MARE processing sequences for vertically mild lateral velocity field variations

<table>
<tr><td colspan="2">Steps “a” to “i” of Table 3</td></tr>
<tr><td>3</td><td>4</td></tr>
<tr><td>Post-Stack Time Migration (TM)</td><td>Post-Stack Depth Migration (DM)</td></tr>
<tr><td>TVF</td><td rowspan="2">Depth-Variant Filtering (DVF)</td></tr>
<tr><td>Time to Depth conversion</td></tr>
</table>

Table 5
CROP-MARK processing sequences for strong lateral velocity field variations

<table>
<tr><td colspan="4">Steps “a” to “i” of Table 3</td></tr>
<tr><td colspan="2">5</td><td colspan="2">6</td></tr>
<tr><td>j</td><td>Dip MoveOut (DMO) processing</td><td>k</td><td>Initial velocity–depth model building</td></tr>
<tr><td colspan="2">Stack</td><td>l</td><td>Kirchhoff pre-stack depth migration (PSDM)</td></tr>
<tr><td colspan="2">Post-Stack Time migration</td><td>m</td><td>Velocity–depth model upgrading</td></tr>
<tr><td colspan="2" rowspan="2">TVF</td><td colspan="2">Final Kirchhoff PSDM</td></tr>
<tr><td colspan="2">DVF</td></tr>
</table>

fore one of the primary targets of the processing sequence. The variable sea bottom characteristics, ranging from shallow water continental shelf in the Adriatic, Pelagian and peri-Tyrrhenian areas to the Ionian and Tyrrhenian abyssal plains, impose different constraints and strategies for their attenuation, which are briefly outlined in this chapter. The second problem is the low signal to noise ratio (SNR) that characterizes the deep part of the seismic section. Lower crust intra-basement reflectors often have reflection coefficients in the range between 0.05–0.1 (e.g. 0.06, for a gabbro/pyroxenite contact with characteristic ρ of 3.0–3.2 g/cm^3 and v_p of 6.8–7.2 km/s). The low seismic energy available to image deeper reflectors combined with such low reflection coefficients results in reflection amplitudes often comparable to the noise level. The proposed strategy to detect low amplitude reflectors in noisy background exploits instantaneous attributes calculated by wavelet transform (WT) to enhance the spatial correlation of deep events. A crucial step in any processing sequence for deep crustal studies is the reconstruction of velocity models. Such models are first exploited in the NMO-stack phase. They are successively refined after DMO, which allows improved focusing in the presence of conflicting dips. Subsurface conditions, characterized by folded, faulted and thrusted blocks are however common in large sectors of the central Mediterranean area and are frequently sampled by the CROP dataset. In such conditions, the appropriate strategy for the velocity model reconstruction and successive imaging phase requires interpreter input. The CROP processing team implemented interpretative processing sequences that allowed optimum stacking velocity–time function reconstructions and VZ model building for pre-stack migration. Sequences and examples of results are outlined in the following sections.

2. DATA ACQUISITION OF THE CROP SEISMIC NETWORK

The CROP dataset amounts to 9941 km of seismic sections, of which 8687 km are offshore (Tables 1 and 2). The location map in Figure 1 shows the seismic section network and the sections re-processed by the University of Trieste. Processing techniques and examples of results presented in this chapter refer to marine data. The seismic source consisted of an airgun array of 32 guns, 80.4 liter volume, maximum output of 103.4 bar-m and rms output of approximately 73 bar-m. A 4500 m long, 180 channel streamer was used for the whole programme with a 143 m minimum offset. The recorded seismic trace length is generally 17–20 sec, with a maximum record length of 40 sec (line M-2A/III). The fold ranges between 2250% and 4500%.

3. BASIC DATA PROCESSING

The basic standard processing sequence is reported in Table 3.

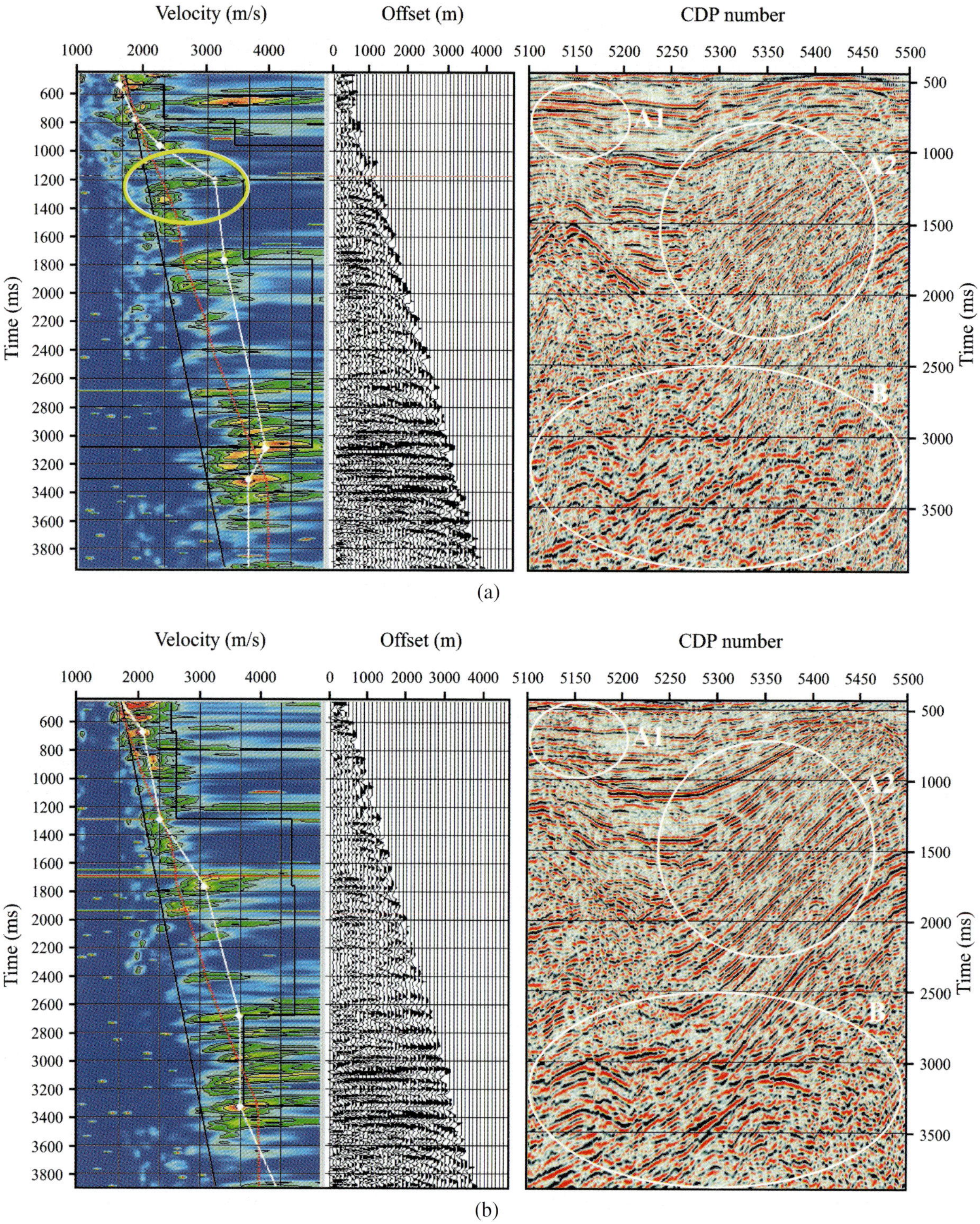

Figure 3: Seismic Line M12B (Tyrrhenian Sea): example of stacking velocity analysis performed (a) before, (b) after DMO correction. DMO enhances velocity analysis results in areas with dipping events and was extensively used in continental shelf, slope and rise sectors.

The procedures defined "*preliminary*" in the flow-chart (b, e) refer to processing steps that require either the detailed reconstruction of velocity–time functions (b, spherical divergence correction) or deconvolved data for improved analysis (e, velocity analysis). They are therefore performed in two steps, through a first approximation and a final refinement.

Sequence 1 produces filtered sections where the position of the imaged targets is not correct and diffraction hyperbola are not collapsed. Such sections are exploited for a preliminary reconstruction of subsurface models and to

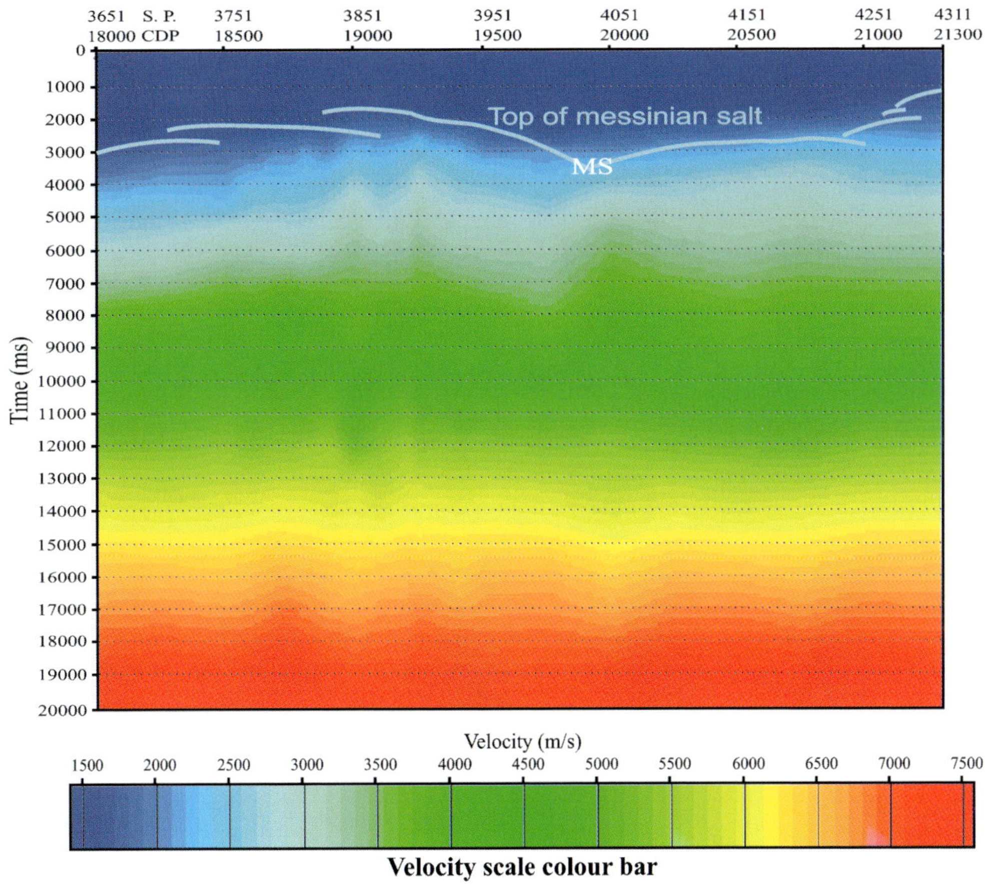

Figure 4: Seismic Line M4 (thick complex wedge of the Calabrian Arc in the Ionian Sea): example of velocity–time model obtained from post-DMO velocity analysis. Line MS represents the top of the Messinian salt.

provide the interpreter input for the successive interpretative processing phase. Sequence 2 produces the migrated and filtered sections, which in several cases, provided improved subsurface imaging.

4. INTERPRETATIVE REPROCESSING

To improve the data quality of complex areas characterized by thick tectonic stacks or deep subducting data of critical importance for understanding lithospheric features, an extensively iterated, careful interpretative reprocessing was done. In several key zones, very evident signal improvement was obtained.

4.1. Velocity Analysis and Multiple Removal

The basic interpretative reprocessing for areas of moderate structural complexity and for multiple removal is reported in the flow-chart of Figure 2. The sequence is based on a first step of coherent noise reduction in the τ–p domain. Noise reduction and primary reflection enhancement is then pursued through iterated applications of NMO correction and FK filtering that help in isolating and detecting the components of the primary field, which are then retained through the application of the final stacking velocity model.

To cope with structural complexity and velocity field characteristics not successfully dealt with in the basic sequence, two alternative processing sequences were designed and applied (Tables 4 and 5).

Velocity–depth model upgrading (step m, Table 5) is performed through an iterative refinement of the velocity model by separately analysing depth gathers and by calculating coherency values for CRP gathers. The analysis of depth gathers and migrated sections allows discrimination of geologically and sedimentologically relevant events and reconstruction of improved velocity models. Where the quality of the post-stack time migrated data is poor and

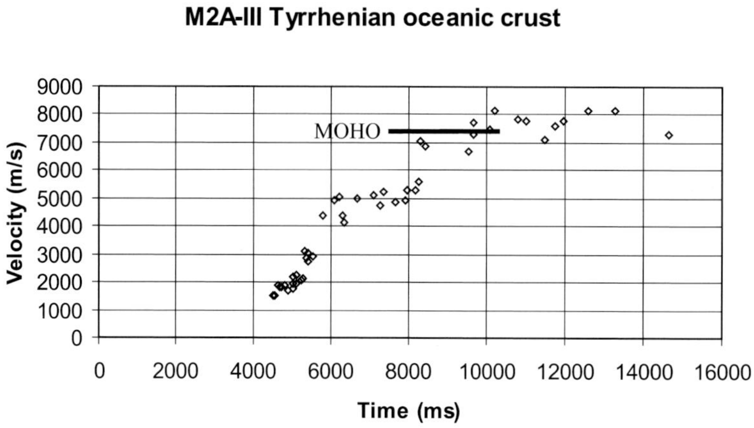

(a)

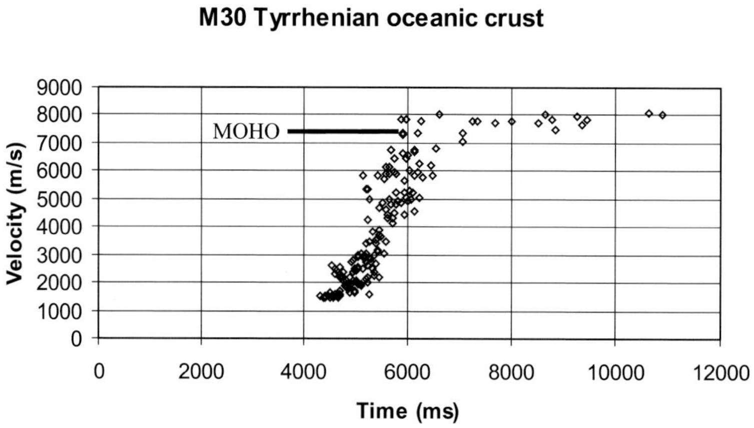

(b)

Figure 5: Interval velocity distribution in the different sectors of the study area: (a) seismic line M-2AIII, Tyrrhenian abyssal plain; (b) seismic Line M-30, Tyrrhenian abyssal plain.

prevents the application of a time–depth conversion based on a set of time horizons obtained from time migration, we chose to reconstruct an initial v–z model by applying a first horizontal smoothing to the RMS-interval velocity conversion. We obtain RMS velocities from accurate stack velocity analysis integrated with interpretation of pre-stack and stack data. Interval velocities were then converted from time to depth by applying a vertical smoothing to avoid major discrepancies between the interval velocity field and reflector geometry. The two-step procedure produce a higher quality initial v–z model. Then an iterative refinement of the v–z model based on residual moveout analysis of CRP gathers was performed. PSDM was performed by means of a Kirchhoff algorithm.

Finally, computation of instantaneous attributes (envelope, phase, frequency) by means of wavelet transform for detection of weak coherent deep crustal signals was performed (Guangyou and Pipan, 2003).

Velocity analysis. An example of stacking velocity analysis from the northern Tyrrhenian Sea is shown in Figure 3. This section is characterized by steeply dipping reflectors and conflicting dips both in the shallow sector (TWT $<$ 2.0 sec), related to a lower Miocene basin, between CDP 5100 and 5400, and in the deeper one, where thrust layers below 3 sec TWT are exhibited. The semblance panel obtained before DMO (Figure 3(a), left) shows smearing of the maximum values of coherency which prevents a direct interpretation of the velocity spectra. Analysis of reflectors in the NMO-corrected CMP gather (Figure 3(a), middle) reveals residual moveout that can't be removed while preserving the proper correction for all of the primary events. The corresponding stack section (Figure 3(a), right) shows poor focusing of most dipping reflectors, with specific reference to shallow events within the Lower Miocene sequence (A1, A2) and deeper events related to thrust structures. The DMO-corrected results show the accurate reconstruction of the proper stacking velocity–time function and the corresponding image enhancement. The stacking velocity–time functions thus obtained can be exploited for a regional characterization of the area of study. Regular sampling of the velocity models obtained after DMO, and interval velocity control and optimization (Figure 4) for the whole reprocessed dataset provided regional interval velocity distributions

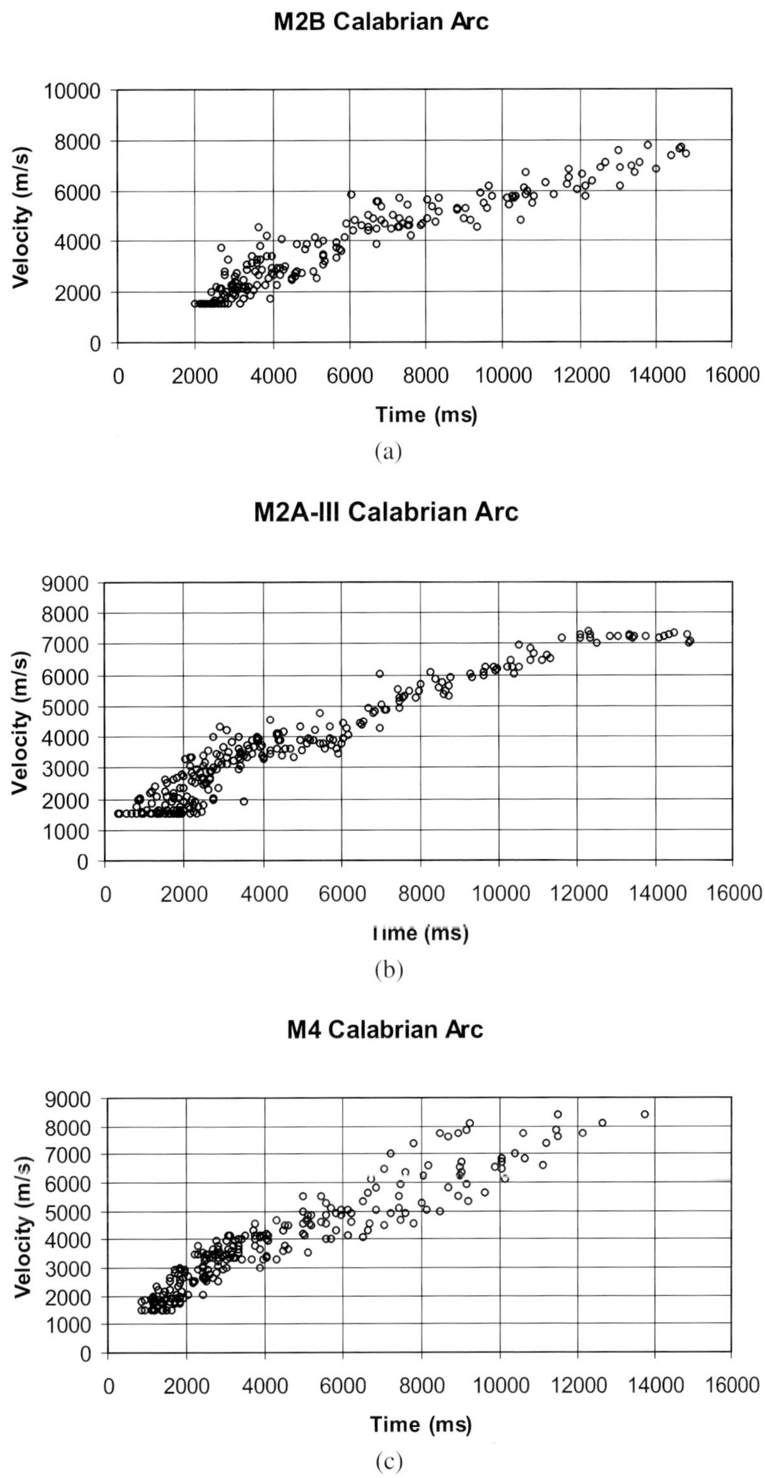

Figure 6: Interval velocity distribution in the different sectors of the study area: (a) seismic line M-2B, Calabrian Arc sector; (b) seismic line M-2AIII, Calabrian Arc sector; (c) seismic line M-4, Calabrian Arc sector.

(Figures 5 and 6). The crust-mantle interface is clearly identified (Figure 5). Interval velocities exhibit an approximately linear increment as a function of two-way-time (TWT) in the Calabrian Arc (Figure 6). Different sectors of the arc exhibit stable and concentrated values of the interval velocity in the deep crust (Figure 6(b)) or pronounced deviations around a generally linear trend (Figure 6(c)).

Multiple removal. Multiple attenuation requires different strategies depending on water depth, water bottom topography and the characteristics of the interference between multiple and primary wavefields. Flat, shallow water

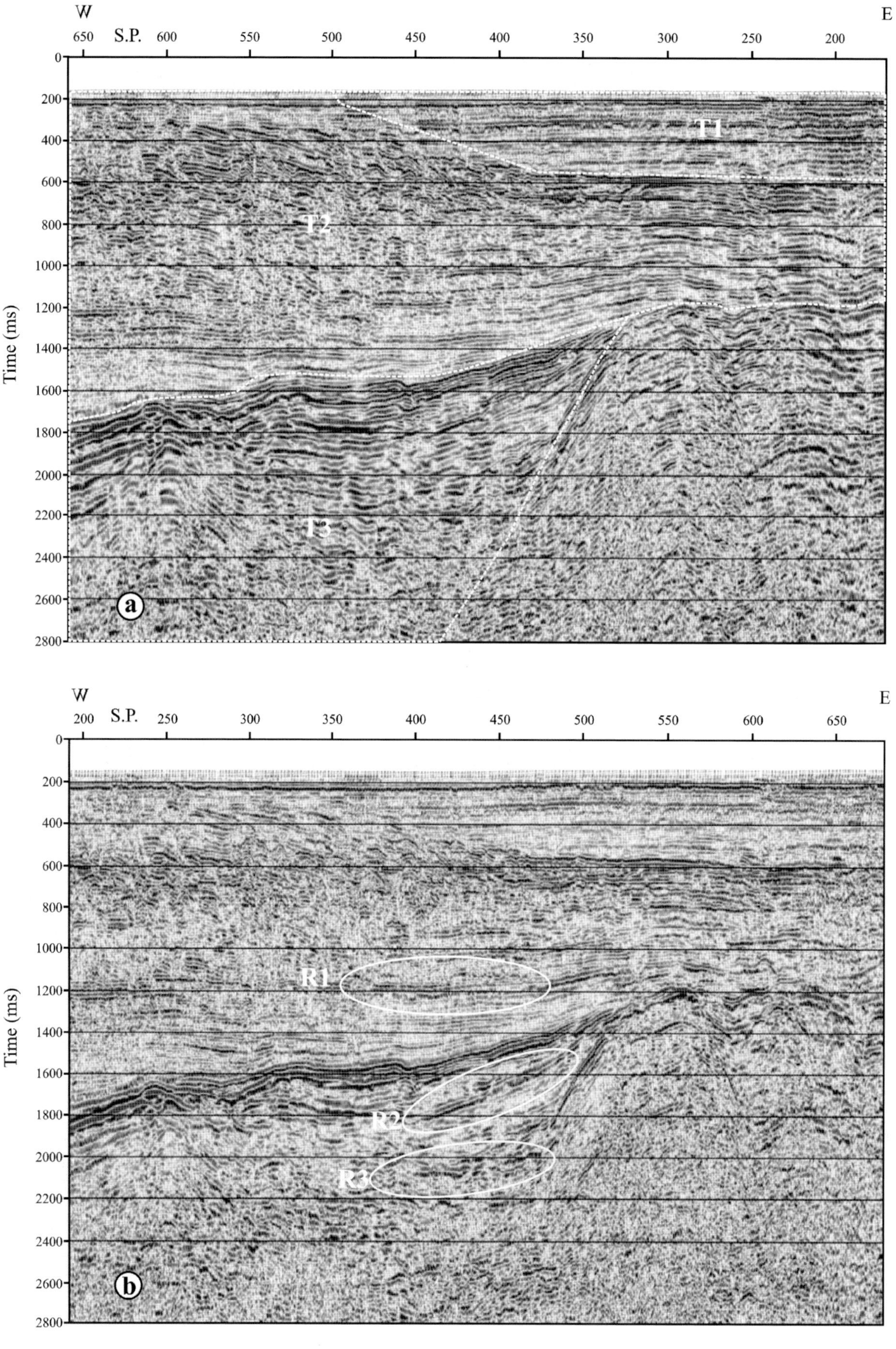

Figure 7: Example of reprocessed CROP section in the Adriatic Sea (M-16): (a) standard stack section; (b) post-stack deconvolution and radon velocity filter (RVF).

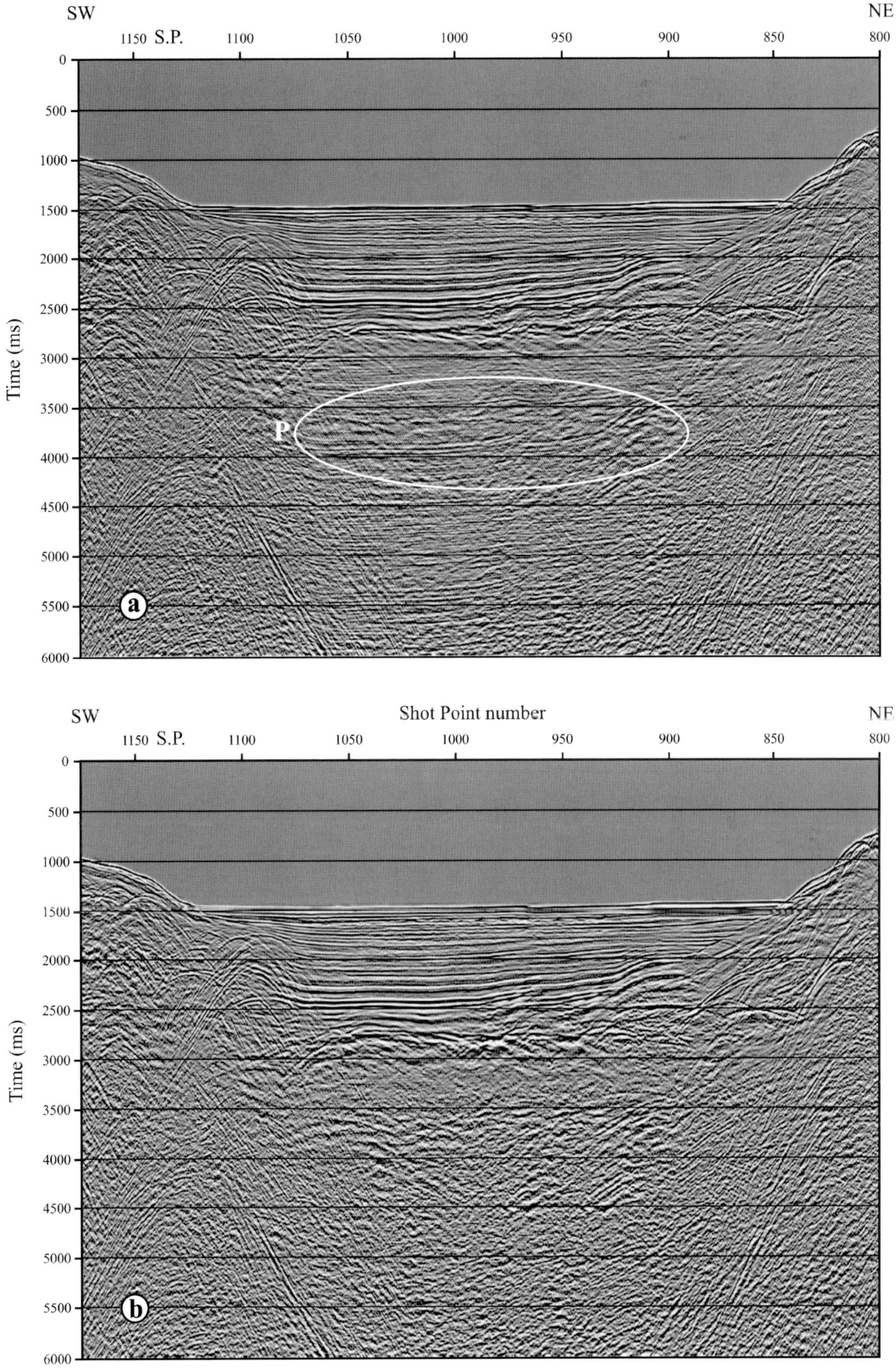

Figure 8: Seismic example of reprocessed CROP section (M-30) in the Tyrrhenian Sea: (a) FK filter for multiple removal; (b) XT filter for multiple removal.

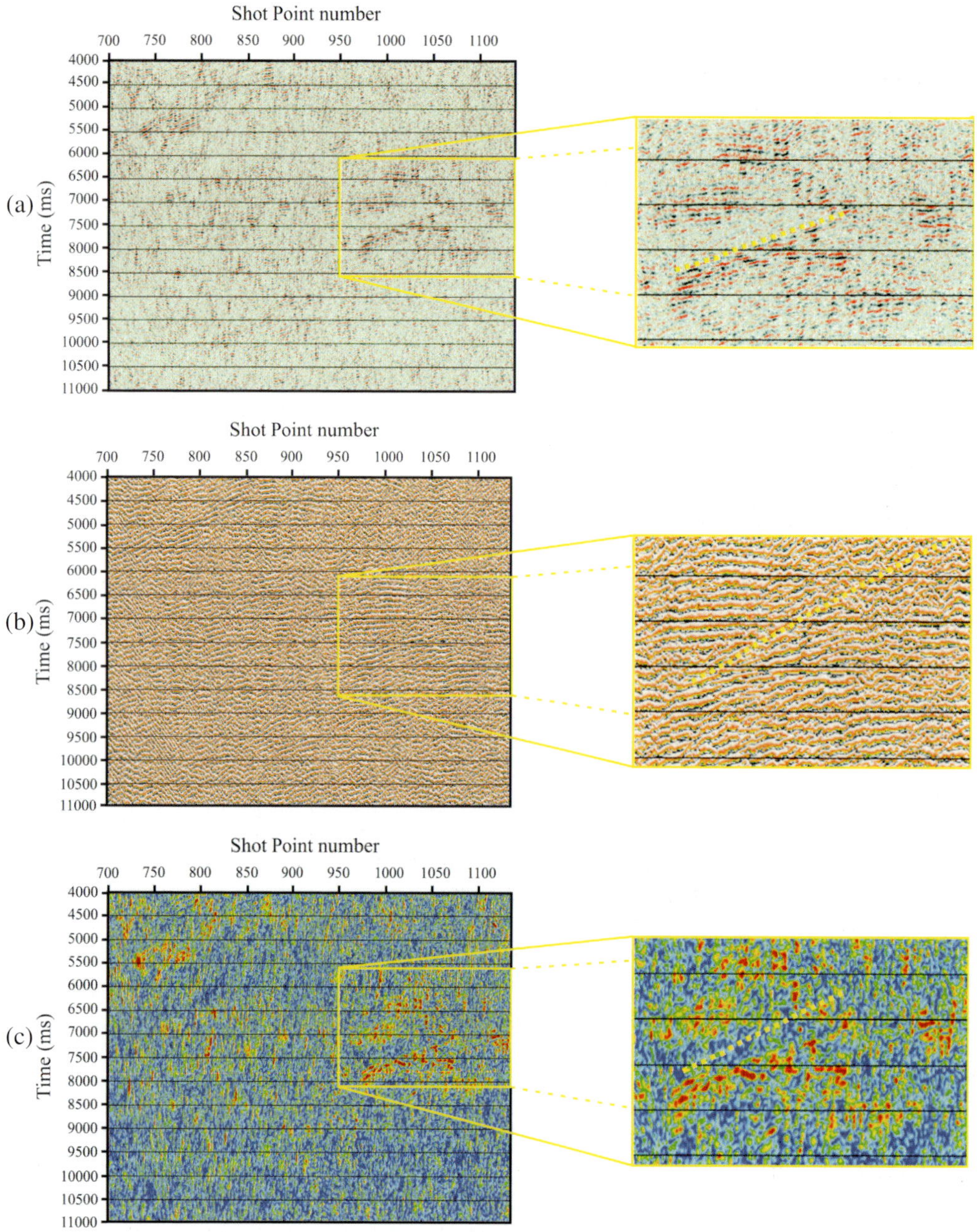

Figure 9: Seismic line M12A (Tuscan archipelago): (a) WEMR-PSPD section; (b) instantaneous phase of WEMR-PSPD section; (c) envelope of WEMR-PSPD section. The obtained results are very helpful for understanding this key area of the Northern Apennines hinterland.

bottom conditions characterize sectors of the Adriatic Sea, and part of the Pelagian Sea continental shelf. In such conditions, stack sections exhibit short period multiples that interfere with primary events at all depths and degrade vertical resolution. An example from the Adriatic Sea is shown in Figure 7(a). The eastern part of the seismic section images an incipient frontal compressive deformation of the Dinaric Belt. The central-western part of the section exhibits the well reflecting Plio-Quaternary sequence of the Apennine foredeep. The water bottom multiples and the multiples of strong primary events prevent the reconstruction of structural features and stratigraphic details, with particular reference to target sectors T1, T2 and T3 (Figure 7(a)). We tested the following multiple attenuation strategies:

a. Wave-equation based multiple removal (WEMR);
b. WEMR and post-stack predictive deconvolution (PSPD);
c. PSPD and radon velocity filter (RVF).

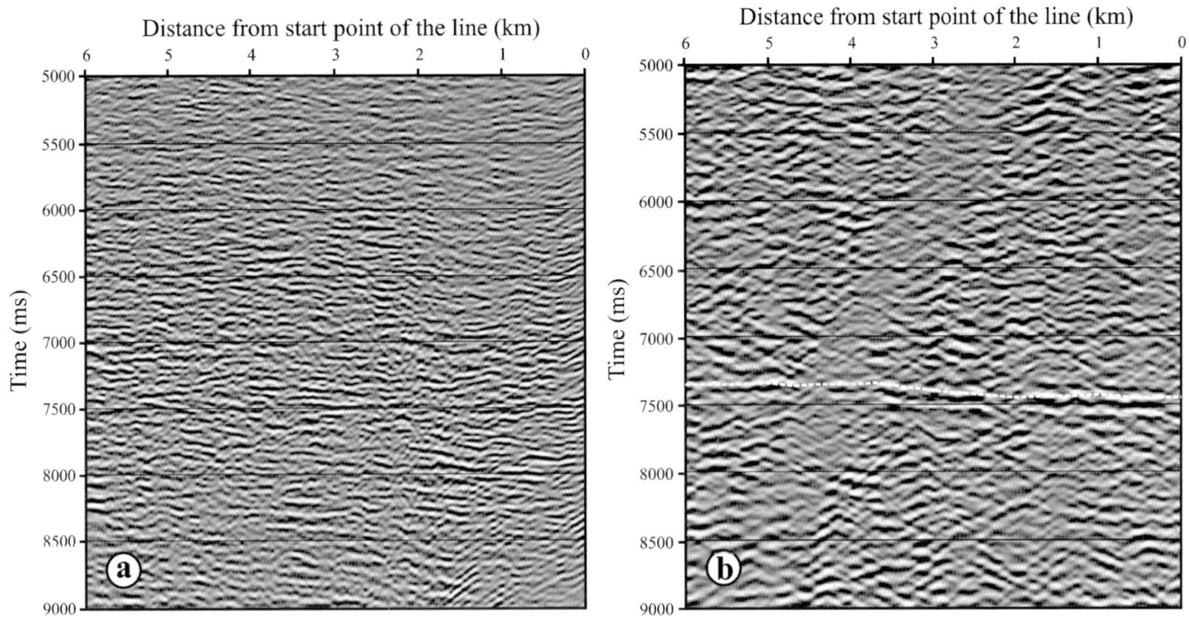

Figure 10: CROP seismic line M30 (Tyrrhenian Sea): (a) standard processing; (b) reprocessed by University of Trieste. White dashed line is the Moho discontinuity, more evident in section b than in the standard processing.

WEMR proves effective in attenuating the multiple wavefield, particularly in the case of a shallow water bottom. An alternative solution is radon velocity filters, while FK multiple removal cannot be successfully applied due to the small moveout difference between primary and multiple events over large parts of the section. A net improvement in the attenuation of multiples is however provided by an integration of WEMR or RVF with post-stack predictive deconvolution (Figure 7(b)). The three considered sectors (T1, T2, T3) exhibit enhanced resolution and discrimination of the primary reflectors, with specific reference to events R1, R2 and R3.

Alternative strategies for multiple attenuation are exploited in the case of deeper seafloor conditions that allow a better NMO separation between multiples and primaries. An example is shown in Figure 8. Here the seismic section samples a Tertiary-Quaternary basin bounded on both sides by extensional faults. The traditional FK filter attenuates the seafloor multiple but, also partly attenuates the primary event between 3.5 and 4.0 s (Figure 8(a), P). A proprietary XT multiple attenuation algorithm (Pipan et al., 1998) successfully removes the multiple wavefield but preserves the primary events well. The XT algorithm was applied in the reprocessing of CROP sections over all the abyssal plain, continental slope and part of the shelf areas.

4.2. Reprocessing of Low Amplitude Signal

The CROP marine dataset was obtained by exploiting a strong source (5.56 MJ) and a long streamer (4500 m), towed at an average 12 m depth, mostly under quiet conditions. These are ideal operating conditions for imaging deep crustal reflectors. It is however still difficult to identify and image low to very low amplitude reflectors in the complex Mediterranean tectonics at depths greater than 6–7 sec TWT. Lossy and attenuating overburden, quite common in the area of study, can reduce deep reflector response to a few μbar, which is around the average streamer noise level. In the processing phase, we calculated instantaneous attributes of the seismic trace by means of wavelet transform (WT). In comparison with methods based on the Hilbert Transform (HT), WT techniques provide enhanced response in the case of weak signals in a noisy background. An example of the application of this reflector-enhancement technique is exhibited in Figures 9 and 10. The WT instantaneous phase (Figure 9(b)) and envelope (Figure 9(c)) enhance reflection continuity and strength respectively. The two attributes are exploited to help in Moho detection (Figure 10).

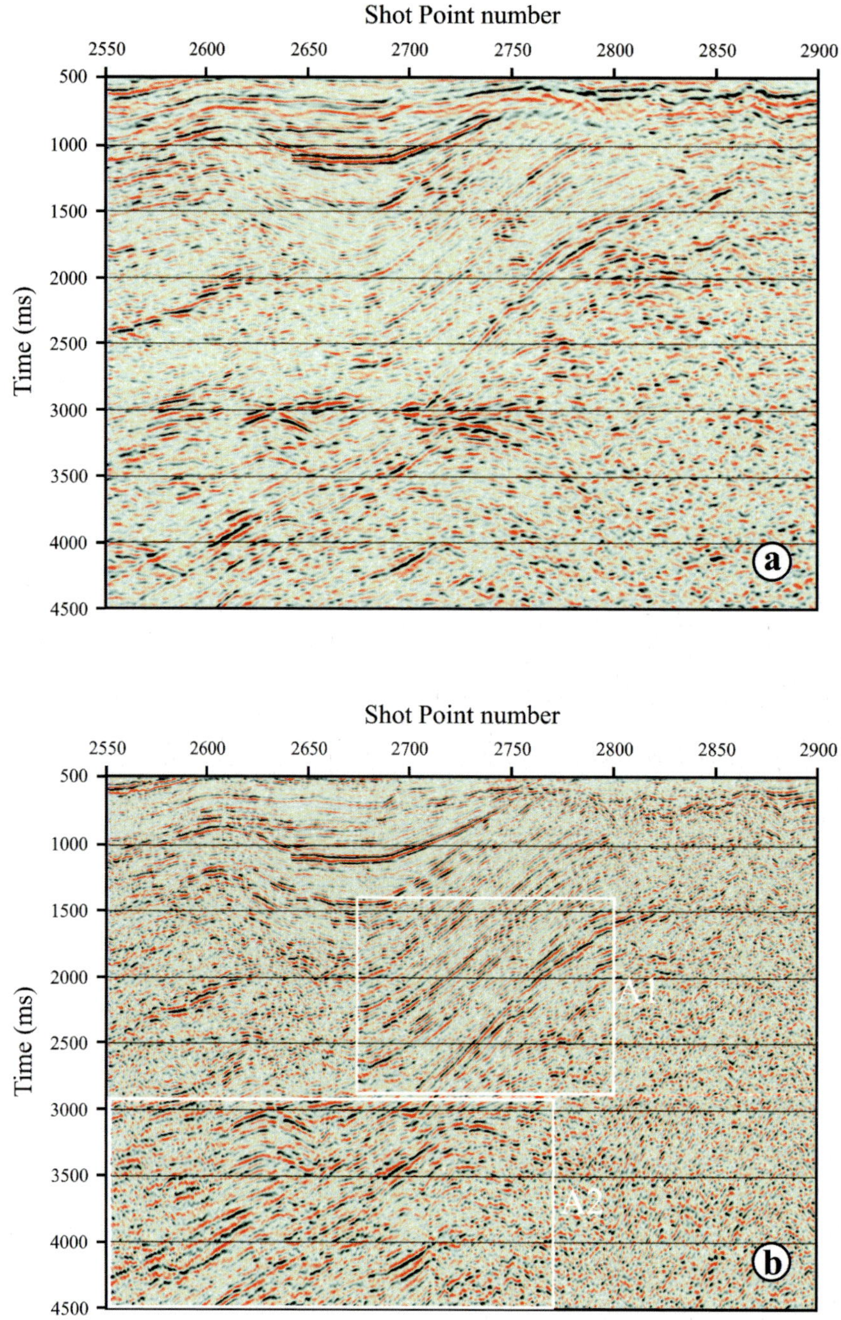

Figure 11: Seismic line M12 B (Tyrrhenian Sea): (a) stack section; (b) DMO section.

4.3. Time and Depth Imaging

Major structural complexities in the marine CROP dataset are located in sectors characterized by compressional and strike-slip deformation, such as the Tyrrhenian margins, the Calabrian Arc, the Corso-Sardinian block, the Tuscan Archipelago, the Sardinia Channel and other areas.

In such conditions, rapid lateral and vertical velocity variations, diffuse scattering and steep, conflicting dips degrade the seismic section and require pre-stack imaging strategies. The structural complexity and the results attainable by the different strategies are illustrated in this paragraph by examples of two seismic sections from the north Tyrrhenian and Ionian Sea. Crustal and mantle duplications, subductions, back-arc basin opening, contractional-extensional and shear tectonics, and igneous intrusion–extrusion are responsible for complex crustal conditions that, in the case of the CROP dataset, occur at most of the margins of the explored basins. The NE Tyrrhenian margin (Figure 11) exemplifies one such case: reflectors in (A1) (Figure 11(b)) bound a shallow depocenter with

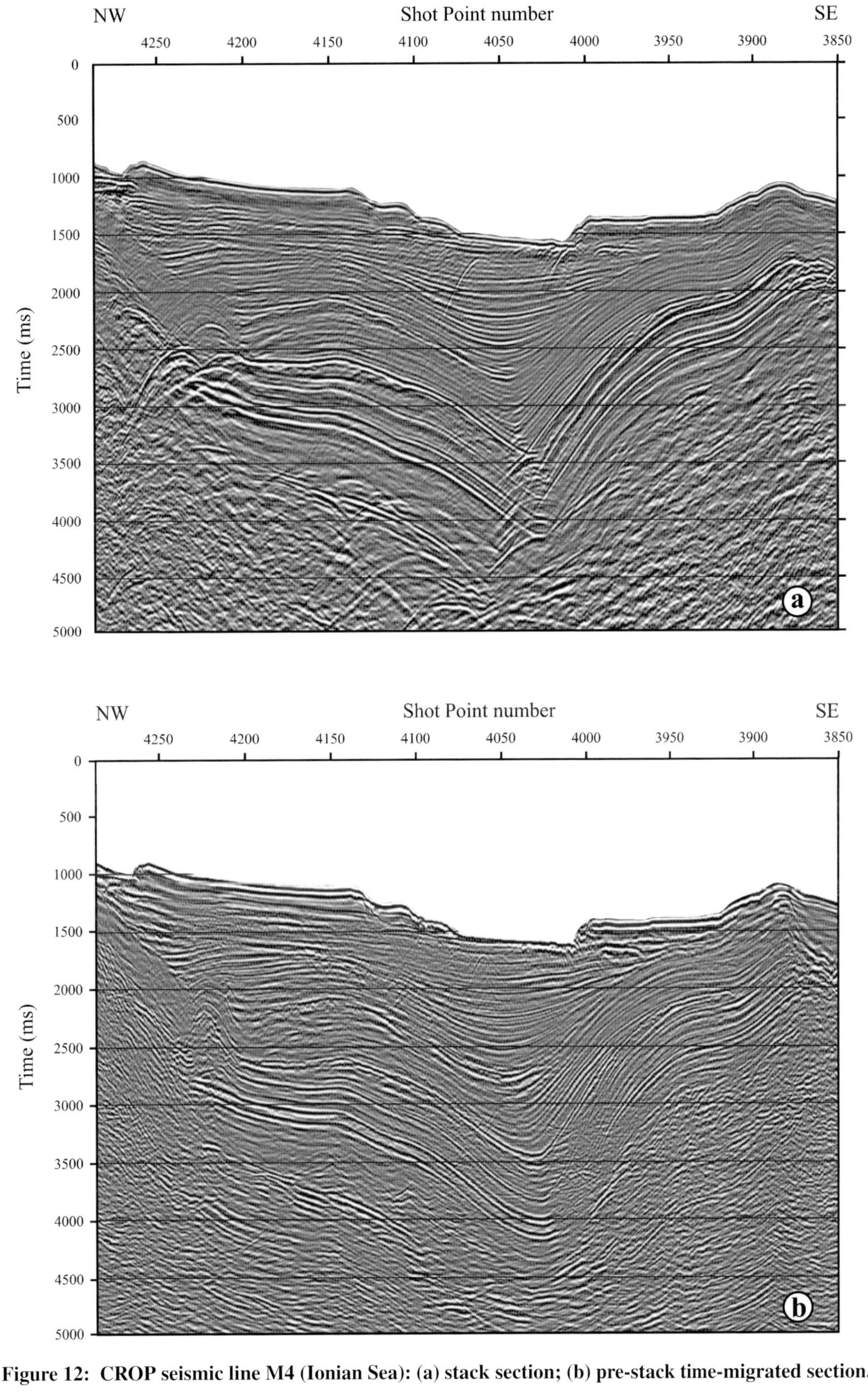

Figure 12: CROP seismic line M4 (Ionian Sea): (a) stack section; (b) pre-stack time-migrated section.

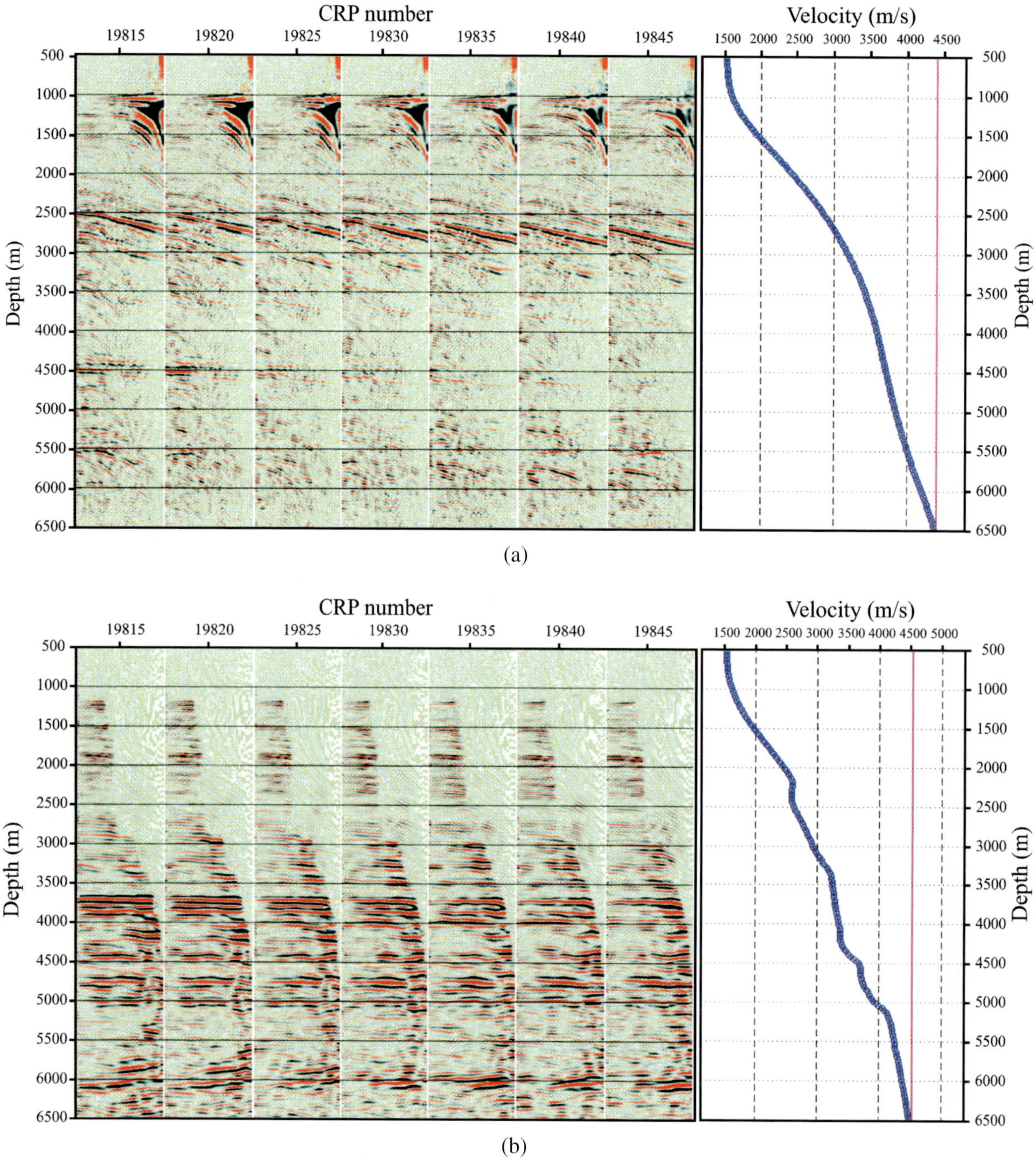

Figure 13: Seismic line M4 (Ionian Sea): (a) CRP gathers before velocity versus depth optimization; (b) CRP gathers after velocity versus depth optimization by integrated Deregowski and focusing techniques.

dips exceeding 20°. The underlying thrust structure (A2) exhibits angular unconformities that, in this case, can be successfully managed by partial pre-stack imaging techniques (DMO). The effectiveness of this procedure can be observed in the example in Figure 11. The stack section (Figure 11(a)) shows the limits of NMO application in the case of dipping events and conflicting dips. Continuity and reflection strength of dipping reflectors are enhanced by DMO. This is particularly the case of the events below 2.0 sec TWT that strongly deviate from the proper conditions for NMO application. Examples from the Ionian Sea (Figures 12(b) to 14) indicate that DMO is not an adequate treatment in the case of larger dips and lateral velocity variations. Comparative analysis of various procedures showed that pre-stack time migration (Figure 12(b)) provides an evident enhancement in the subsurface image. Nonetheless, comprehension of the velocity distribution associated with the observed structural features can provide further image improvements and help in the seismic characterization of the different structural units. To reconstruct the velocity distribution, we performed an iterative refinement of the velocity model by separately analyzing depth gathers, and by calculating coherency values for CRP gathers (Figure 13). An example

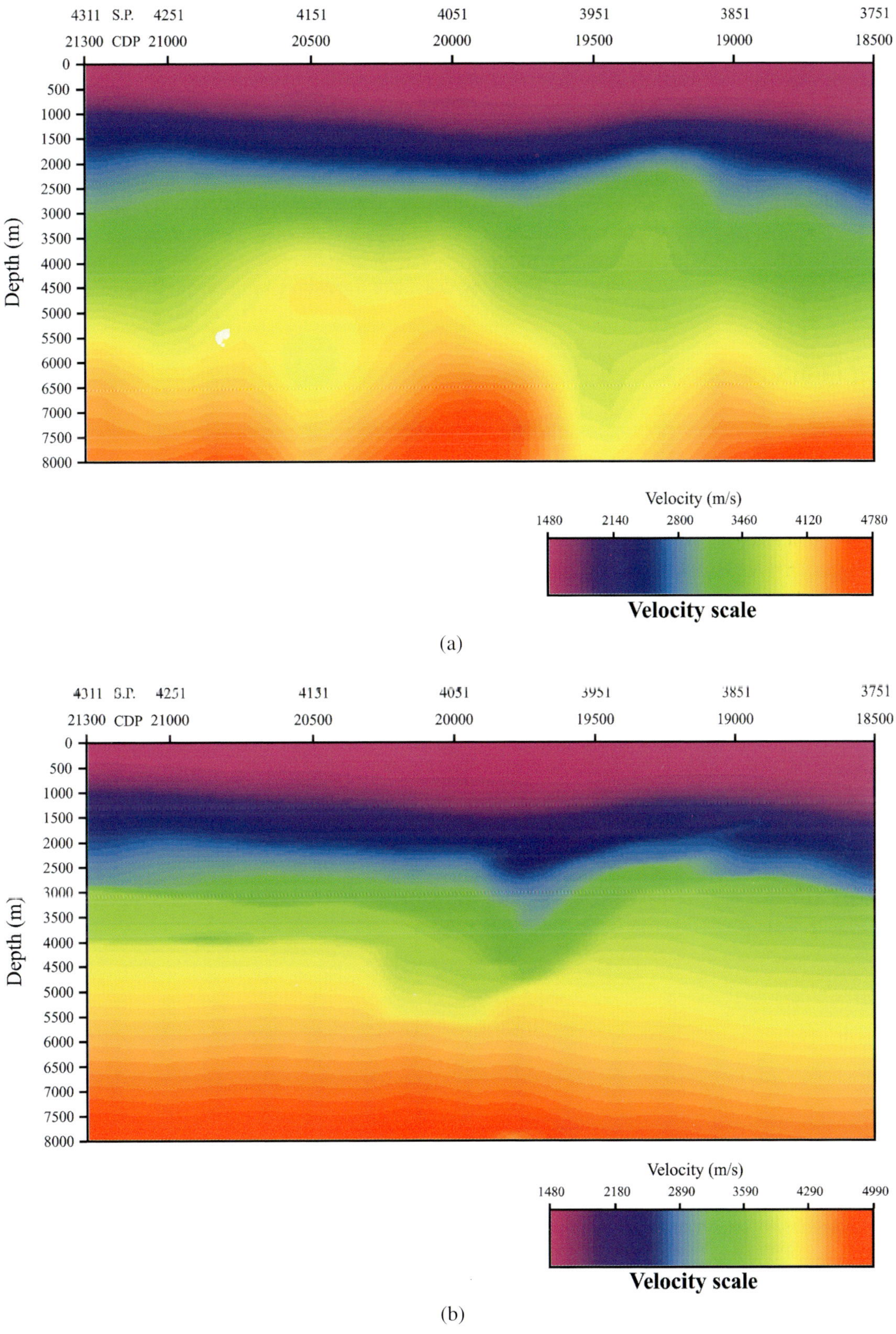

Figure 14: CROP seismic line M4 (Ionian Sea): (a) Initial velocity–depth (VZ) macro-model obtained from stacking velocity analysis by traveltime inversion; (b) optimized VZ macro-model obtained from integrated Deregowski and focusing techniques.

of the initial results of CRP-gather velocity analysis is shown in Figure 13(a). The analysis of depth gathers and migrated sections allows discrimination of geologically and sedimentologically irrelevant events and reconstruction of improved velocity models. Poor quality of post-stack time migrated data often prevents the application of a time–depth conversion based on a set of time horizons obtained from time migration. We thus chose to reconstruct an initial v–z model by applying a first horizontal smoothing to the RMS-interval velocity conversion. We obtained RMS velocities from accurate stack velocity analysis integrated with interpretation of pre-stack and stack data. Interval velocities were then converted from time to depth by applying a vertical smoothing to avoid major discrepancies between the interval velocity field and the geometry of reflectors. This two-step procedure produces a high quality initial v–z model (Figure 14(a)). We then performed an iterative refinement of the v–z model based on the analysis of residual moveout on CRP gathers (Figure 13(b)). An example of the final velocity model is shown in Figure 14(b). The improved correspondence between velocity and geological structure is apparent. Pre-stack depth migration (PSDM) is eventually performed by means of a Kirchhoff algorithm. The comparison between PSDM results before and after velocity model optimization show evident improvement in seismic energy focusing at all depths.

5. CONCLUDING REMARKS

The large amount of seismic data acquired in deep crustal studies, and the wide range of subsurface conditions sampled in the range from shallow sedimentary basins to deep deformed structures impose challenging constraints on data processing. The CROP reprocessing is based on a strong link with the interpretation phase that greatly helps in focusing the processing effort and in optimizing procedures and parameters. A major pre-requisite for a successful implementation of an interpretative processing sequence is the enhancement of the original seismic images through coherent noise removal and deep weak signals extraction. Such operations allow better determination of the velocity field and improved final section quality.

ACKNOWLEDGEMENTS

The editor of the CROP Volume that includes this chapter is very grateful to CNR, ENEL and ENI-AGIP for the confidence shown and honour given by appointing him scientific director of the CROP Mare Programme. This work was funded by MIUR, COFIN 1998, 2000 and 2002 to I.R. Finetti, and by CNR-ENEL-ENI (CROP Partners) with grants from CNR to I.R. Finetti. The University of Trieste continually supported the Seismic Processing Centre and the CROP Seismic Interpretation Team of the Faculty of Science that performed the reprocessing and interpretation of the CROP data.

The authors wish to thank OGS, all Teams of the OGS-Explora Seismic Vessel and the OGS Seismic Division for the CROP data acquisition and processing.

REFERENCES

Damotte, B., 1996. Acquisition and processing of vertical seismic data: the ECORS profile across the Alps, Jura Mountains and Bresse Plain. In: Roure, F., Bergerat, F., Damotte, B., Muignier, J.-L., Polino, R. (Eds.), The ECORS-CROP Alpine seismic traverse. Mem. Société Géologique de France, Nouvelle Série 170.

DEKORP Research Group, 1990. Results of deep-seismic reflection investigations in the Rhenish Massif. Tectonophysics 173, 507–515.

Dohr, G., Fuchs, K., 1967. Statistical evaluation of deep crustal reflections in Germany. Geophysics 32, 951–967.

Graßl , H., Neubauer, F., Millahn, K., Weber, F., 2004. Seismic image of the deep crust at the eastern margin of the Alps (Austria): indications for crustal extension in a convergent orogen. Tectonophysics 380, 105–122.

Guangyou, F., Pipan, M., 2003. Synthetic and Field Examples of GPR profile improvement using two Phase Detection Techniques. Geophysics 68, 554–558.

Hauck, M., Brown, L., 1992. Project INDEPTH Pilot Survey 1992 Technical Report: CDP Survey. Web version at http://www.geo.cornell.edu/geology/indepth/MainMenu/TRep/T1Rep1.html.

Mazzotti, A.P., Stucchi, E., Fradelizio, G.L., Zanzi, L., Scandone, P., 2000. Seismic exploration in complex terrains: a processing experience in the Southern Apennines. Geophysics 65, 1402–1417.

Meissner, R., Bortfeld, R.K. (Eds.), 1990. DEKORP-Atlas. Springer-Verlag, Berlin.

Pipan, M., Geletti, R., Finetti, I., Galuppo, P., 1998. Hough Transform based multiple removal in the XT domain. SEG Annual Meeting, Extended abstracts, S.P.6.5.

Pratt, T.L., Brown, L.D., Mondary, J., 1993. AVO and fluids in the deep crust: a COCORP case study. Leading Edge 12, 186–192.

Schilt, F.S., Kaufman, S., Long, G.H., 1981. A 3-dimensional study of seismic diffraction patterns from deep basement sources. Geophysics 46, 1673–1683.

Steer, D.N., Knapp, J.H., Brown, L.D., Rybalka, S., Sokolov, V.B., 1995. Crustal Structure of the Middle Urals based upon reprocessing of Russian seismic reflection data. Geophys. J. Intern. 123, 673–682.

Stiller, M., Thomas, R., 1989. Processing of reflection-seismic data in the DEKORP Processing Center. Clausthal. In: Emmermann, R., Wohlenberg, J. (Eds.), The German Continental Deep Drilling Program (KTB). Springer, Berlin, pp. 177–232.

Trappe, H., Bittner, R., 1989. Estimation of interval velocities within the Earth's crust. Tectonophysics 157, 59–68.

Yilmaz, O., 2001. Seismic Data Analysis. Society of Exploration Geophysicists, Tulsa, OK, USA.

Wallace, M., 1980. Deep basement reflections in Wind River Line 1. Geophys. Res. Lett. 7, 729–732.

Zhu, T.-F., Brown, L.D., 1986. Consortium for Continental Reflection Profiling Michigan Surveys: reprocessing and results. J. Geophys. Res. 91, 11477–11495.

CROP PROJECT: Deep Seismic Exploration of the Central Mediterranean and Italy
Edited by I.R. Finetti

CHAPTER 6

Ionian and Alpine Neotethyan Oceans Opening

I.R. Finetti[1,*]

ABSTRACT

The CROP seismic network imaged for the first time subducted slabs of the Ionian and Alpine Tethys buried beneath a complex tectonic stack of allochthonous units. These data and the transcrustal tectono-stratigraphic reconstruction of about 10,000 km of seismogeological sections across all the most significant features, allowed the reconstruction of a pre-contractional paleogeographic framework for the central Mediterranean at the end of the Early Cretaceous.
Before the Eo-Alpine geodynamic stage, three continental plates (Europe, Africa, Adria) and two ocean basins (Ionian Tethys opened in the Permo-Triassic and Alpine Tethys opened in the Middle Jurassic–Early Cretaceous) formed the central Mediterranean paleogeography. The Ionian Tethys boundaries are well imaged seismically even where its lithosphere subducted. CROP data show very clearly that the Adria and Africa segmented continental boundaries on the Ionian Tethys, enclosing this basin, can be almost perfectly reconnected. The Alpine Tethys slab is imaged beneath the Alps, NW Adria and the west Panormid Promontory. The Ionian Tethys and Alpine Tethys were openly communicating, and subduction of their lithosphere constituted the dominant driving force of the geodynamics.

Keywords: CROP controlled Ionian, Alpine Tethys opening

1. INTRODUCTION

Generally, geoscientists when investigating a region, collect or collate data before proposing tectonic schemes best fitting their data and/or hypotheses. In the present case, observing that the Mesozoic paleogeographic framework of the Mediterranean, established before the contractional geodynamic stages played a dominant role in its successive evolution, it is more convenient to first reconstruct the paleogeographic setting of the Ionian and Alpine Neotethyan ocean basins, in order to better understand and follow the deformation and driving forces of the Eo-Alpine, Balearic and Tyrrhenian stages. This inversion of the usual investigation mode is in the present case made possible because deep CROP seismic data have allowed us to image and define for the first time the boundaries of subducted Tethyan crusts buried beneath complex tectonic stacks.

As is shown in various successive chapters of this volume, the CROP-controlled paleogeographic schemes reconstructed (Figures 1 and 2) simplify our understanding and give a basic explanation for the successive geodynamic deformation.

2. IONIAN NEOTETHYS OPENING IN TIME AND SPACE

For a while now, on the basis of seismic reflection data, it has been recognized that the *Ionian Sea is formed by an old oceanic crust* (Finetti, 1982). Moreover, the same seismic data have shown that the *Ionian Sea underthrusts the Calabrian Arc forming an active subducting slab* (Finetti and Del Ben, 1986). The new deep seismic data of the CROP programme not only confirmed the Ionian oceanic crust and the remnant Ionian unsubducted area already defined by the MS lines (Finetti, 1982; Finetti et al., this volume, Chapter 1), but also gave clear images of the subducted Ionian slab, buried beneath thick complex allochthonous tectonic stacks, all around the Tyrrhenian

*Corresponding author: Tel.: +39 40 5582276; e-mail: finetti@units.it (I.R. Finetti)
[1]Department of Geological, Environmental and Marine Sciences, Exploration Geophysics Group, University of Trieste, Via Weiss 1, 34127 Trieste, Italy

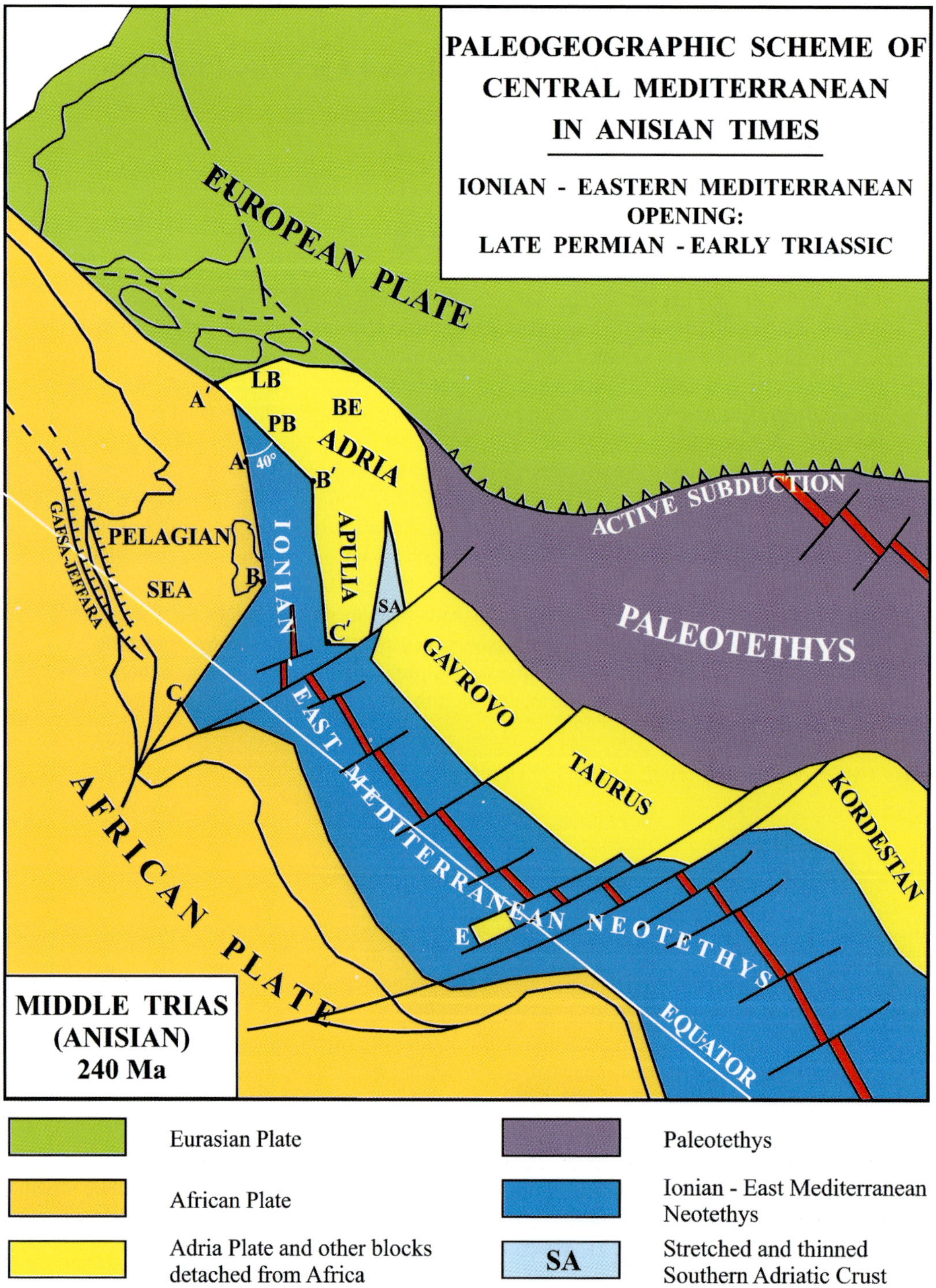

Figure 1: CROP-data-constrained reconstruction of the Ionian opening and Adria plate detachment within a schematic paleogeographic context of the central Mediterranean, during the Anisian (240 Ma). There is very good evidence that the angular points "A", "B" and "C" of the Africa plate–Ionian boundary correspond to the angular points "A′", "B′" and "C′" of the rotated Adria plate–Ionian boundary. The Adria plate is rotated anticlockwise about 40° with respect to Africa. Opening of the Ionian–Eastern Mediterranean took place from the Late Permian to Lower Triassic. Observed Permo-Triassic rift tectonics of Northern Adria (PB = Punta Bianca, LB = Lombard basin, BE = Bellunese basin) were generated by transtensive left-transcurrent faulting during Adria detachment and Ionian opening. "E", in the Eastern Mediterranean, corresponds to the Eratosthene crustal African fragment.

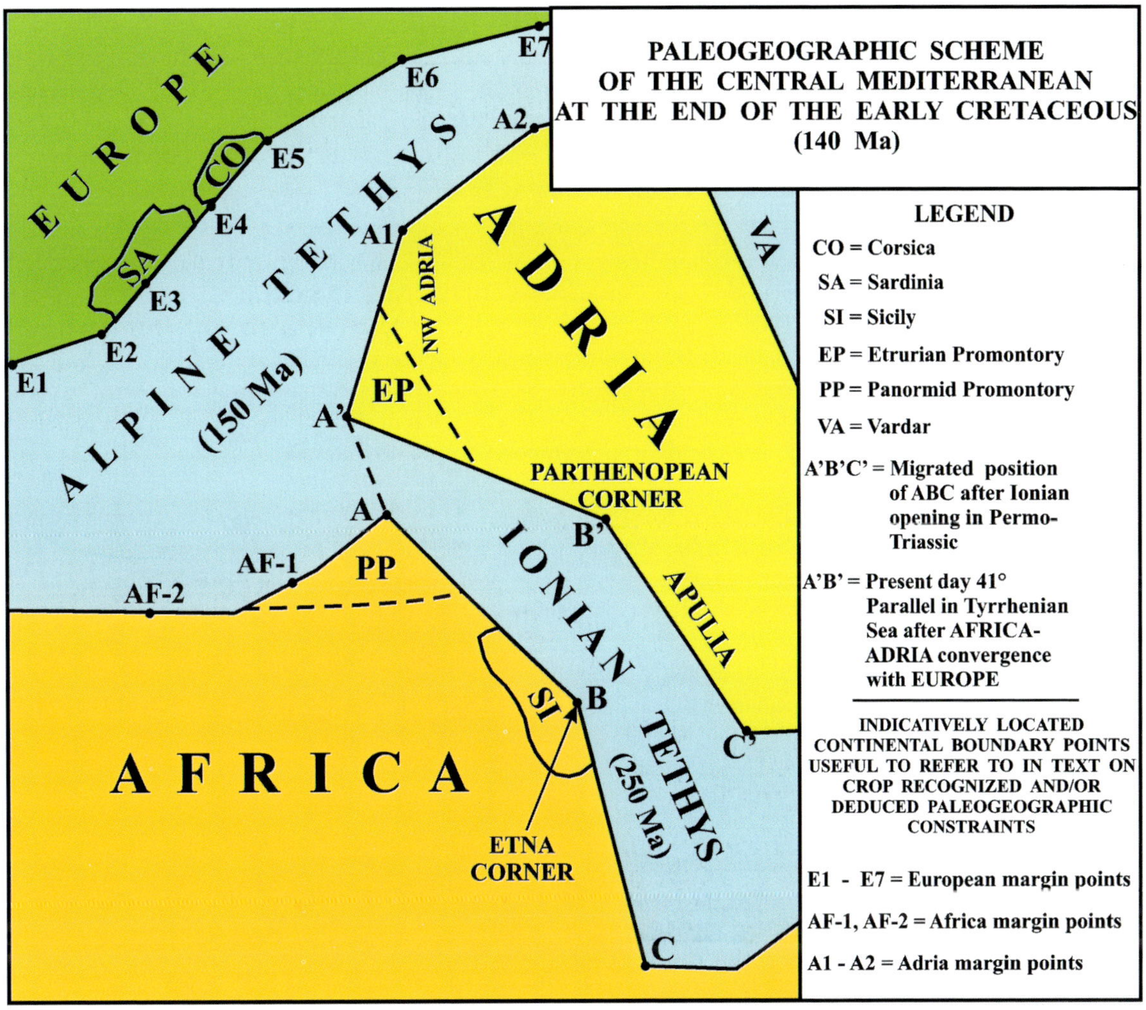

Figure 2: After major Permo-Mesozoic extensional events, at the end of the Lower Cretaceous, in the central Mediterranean region, three continental plates (Europe, Adria, Africa) and two interposed ocean basins (Ionian and Alpine Neotethys) formed the paleogeographic framework. The Ionian ocean basin and the Alpine Tethys, which opened at different times, were fully communicating.

margins of the Southern Apennines–Calabrian Arc–Sicilian Apennines. Moreover, under the Calabrian Arc, the Ionian slab is clearly exhibited as a continuous oceanic crust down to a depth of over 150 km (Finetti, this volume, Chapter 17).

The opening time of the Ionian Tethys can be derived by analysis of units originally forming the Ionian cover that were successively detached from the subducted Ionian slab and tectonically transported on the orogen formed. Such units, as described in other part of this volume (Finetti, this volume, Chapter 10), are formed from condensed deep water deposition layers which were quietly sedimented from the Permo-Triassic Ionian opening to the Oligocene. In the Tyrrhenian stage (Middle Miocene to Present) they were detached and transported on the Southern Apennines–Calabrian Arc–Sicily. In the Southern Apennines, they are known as *Lagonegrese Units* and in Sicily as *Imerese (and Sicanian) Units*. A careful description of the Lagonegrese Units of the Southern Apennines can be found in the Chapter 11 of this volume (Ciarapica and Passeri, this volume), where an initial Late Permian deposition is biostratigraphically identified. Also in Sicily, Permo-Triassic units coming from the Ionian Sea are widely observed (Finetti et al., this volume, Chapter 15). These *Ionian-derived allochthonous units (Lagonegrese, Imerese, Sicanian) have been unified under the indicative name of Ionian Tethydes or Ionides* (Finetti, this volume, Chapter 10; Ciarapica and Passeri, this volume, Chapter 11; Finetti et al., this volume, Chapter 12; Finetti et al., this volume, Chapter 15).

Using all obtained CROP data for the Ionian boundaries, it is possible to reconstruct a well-controlled paleogeography of the Ionian Neotethys (Figure 1). This reconstruction can be schematically extended to the Eastern Mediterranean Sea using data from the literature (Stampfli, 2000). From the CROP-controlled Ionian paleogeography, it is evident that the Adria plate detached from Africa and rotated anticlockwise by about 40°. It is evident also that the African and Adrian Ionian boundaries can be easily reconnected by rotating Adria back. In fact, the segments AB and BC of the Ionian boundary correspond to the other two CROP-controlled Adria segments A′B′ and B′C′. The segment A′B′ now constitutes the northern boundary of the subducted Ionian Tethys, located on the 41°parallel. It represents the northern boundary of the subducted Ionian ocean lithosphere.

3. ALPINE NEOTETHYS OPENING IN TIME AND SPACE

Deep depositional units, coming from the Alpine Tethys (or Ligurian–Piedmont basin) and tectonically transported into a wide range of complex features of the Alps, Apennines, Calabrian Arc and Sicily, are described in numerous papers. There is ample scientific consensus about the age of the Alpine Tethydes (Ligurides) units which started to sediment at the end of the Lias (Toarcian)–Middle Jurassic, contemporary with the initial rifting of the Atlantic ocean opening. Well-argued paleogeographic schemes for the Alpine Tethys ocean basin have been proposed in the literature (Stampfli, 2000, this volume, Chapter 33). Another sketch of the Alpine Tethys, that appears to be basically influenced by its present arcuate chain form, is proposed by Dal Piaz (1999).

In this work, new constraining data obtained from interpretation of various CROP seismic sections and of the Alpine seismic traverses (CROP-ECORS in the Western Alps; CROP-NFP 20 East in the Central Alps; Transalp in the Eastern Alps) is taken into account. Taking also into account the CROP-controlled crustal deformations and subduction processes whose slab imagings were very helpful for reconstruction of the original pre-orogenic frame, the Central Mediterranean paleogeographic setting of Figure 2 was constructed.

In this scheme, the Ionian Tethys boundary ABC and A′B′C′ is fully controlled by numerous CROP sections. From these data it is evident that in the Eo-Alpine, Balearic and Tyrrhenian stages no relevant relative plate movements between Africa and Adria took place. Therefore such Ionian boundaries are directly and fully constrained without any tectonic inversion.

It is more complex to constrain the Alpine Tethys, but the CROP seismic network provides images of the Alpine Tethys slab in several zones, so giving a significant new contribution to the reconstruction of this Tethyan basin paleogeography. In NW Adria and on the Etrurian Promontory (EP), the E-dipping slab of the Alpine Tethys has been seismically imaged, laterally subducted beneath Adria in the Eo-Alpine stage. The Corsica Basin is seen to be floored by a bulge of the Alpine Tethys that exhibits a W-dipping slab beneath Corsica, from the Balearic stage. The Corsica block collided with NW Adria and the Etrurian promontory, and the segment E3–E2–E1 of Europe collided with the African segment A–AF1–AF2, all in the Balearic stage. From the CROP-ECORS and Transalp seismic profiles, it is possible to derive a realistically estimated Adria-Europe convergence and relative thrust displacement. From the Balearic back-arc basin opening, it is possible to estimate the amount of subduction of the Alpine Tethys remnant beneath the Corso-Sardinian block in the Balearic stage. All these cross-controlled constraints led to the reconstruction in Figure 2, where, despite the inevitable use of some imagination, the possible spatial variations were rather restricted.

As described in various chapters of this volume, the paleogeography of the central Mediterranean Region prominently conditioned the evolving geodynamics, and the reconstructed framework, in the author's opinion, explains well the observed tectono-dynamics and their driving forces.

ACKNOWLEDGEMENTS

This work was funded by MIUR, COFIN 1998, 2000 and 2002 to I.R. Finetti. Much gratitude is due to the CNR, ENEL and ENI-AGIP, CROP partners, for the honous given, and faith shown in entrusting me with the task of conducting the management and interpretation of the CROP Mare programme.

REFERENCES

Ciarapica, G., Passeri, L., this volume. Ionian Tethydes in the Southern Apennines. Chapter 11.

Dal Piaz, G.V., 1999. The Austroalpine-Piedmont nappe stack and the puzzle of Alpine Tethys. In: Gosso, G., Jadoul, F., Sella, M., Spalla, M.I. (Eds.), Proceedings of 3rd Workshop on Alpine Geological Studies. Mem. Sci. Geol. 51 (1), 155–176.

Finetti, I.R., 1982. Structure, stratigraphy and evolution of Central Mediterranean. Boll. Geof. Teor. Appl. 24, 247–426.

Finetti, I.R., this volume. Understanding the Ionides and their geodynamics. Chapter 10.
Finetti, I.R., this volume. The Calabrian arc and subducting Ionian slab from new CROP seismic data. Chapter 17.
Finetti, I.R., Del Ben, A., 1986. Geophysical study of the Tyrrhenian opening. Boll. Geof. Teor. Appl. 28, 75–156.
Finetti, I.R., Del Ben, A., Forlin, E., Prizzon, A., this volume. Review of the main results of the pioneering MS seismic exploration programme in the deep water of the Mediterranean Sea (OGS, 1969–1980). Chapter 1.
Finetti, I.R., Lentini, F., Carbone, S., Del Ben, A., Di Stefano, A., Guarnieri, P., Pipan, M., Prizzon, A., this volume. Crustal tectonostratigraphy and geodynamics of the Southern Apennines from CROP and other integrated geophysical-geological data. Chapter 12.
Finetti, I.R., Lentini, F., Carbone, S., Del Ben, A., Di Stefano, A., Forlin, E., Guarnieri, P., Pipan, M., Prizzon, A., this volume. Geological outline of Sicily and lithospheric tectono-dynamics of its Tyrrhenian margin from new CROP seismic data. Chapter 15.
Stampfli, G., 2000. Tethyan oceans. In: Bozkurt, E., Winchester, J.A., Piper, J.D.A. (Eds.), Tectonic and Magmatism in Turkey and the Surrounding Area. Geol. Soc., London, Special Publications 173, 1–23.
Stampfli, G., this volume. Plate tectonics of the Apulia–Adria microcontinents. Chapter 33.

CROP PROJECT: Deep Seismic Exploration of the Central Mediterranean and Italy
Edited by I.R. Finetti

CHAPTER 7

Crustal Tectono-Stratigraphic Sections across the Western and Eastern Alps from ECORS-CROP and Transalp Seismic Data

I.R. Finetti[1,*]

ABSTRACT

Based on published CROP-ECORS and Transalp seismic data, seismogeological trans-crustal sections across the Western and Eastern Alps are presented. *Identification and reconstruction of the seismically evident geometry of the Adria–Europe collisional deformation and closure of the interposed Alpine Tethys is the basic target of interpretation. Obtained results indicate that the Alpine orogenic process did not involve particular mechanisms, but reflects widely recurrent tectonic styles of thrusting, back-thrusting and shear tectonics. The ECORS-CROP profile* across the Western Alps images the following lithospheric sectors: (a) the *European crustal foreland* that is affected eastward by the frontal thrust fault of the Alpine system; (b) a prominent tectonic stack of European upper crustal laminas forming the *Belledonne wedge*, where two interpreted thrust faults also affect the lower crust; (c) the *Penninic wedge* separating the Europe-derived crustal units from those of Adria-plate derivation; (d) *an Adria plate (crust and upper mantle) mega-thrusting over the subducting European plate* (Eo- and Meso-Alpine collisional deformation and relative crustal shortening); (e) the *metamorphic complex of the Gran Paradiso belt*; and (f) the *Insubric transpressive, right-transcurrent zone* forming a *huge positive flower* that exhumes the Adria mantle (*Ivrea–Verbano zone*) and separates northward-verging *Austroalpine thrusts* from *Southalpine backthrusting*.
Interpretation of the *Transalp profile* images the following transcrustal features: (a) the *European foreland* and its thin pre-orogenic cover and Molasse unit; while more internally, Molasse thrust units are overlain by the *Penninic nappe* and the Adria-plate-derived allochthonous units of the *Northern Calcareous Alps*; (b) the *Tauern Window tectonic wedge formed by stacked thrust sheets of European upper crust*, pushed up and exhumed in the late orogenic stage, while *Penninic thrust-sheets* of the Alpine Tethydes, *metamorphic units* and *Oligocene plutons* complete the Tauern stack; (c) the *Periadriatic or Insubric right-transcurrent line separates the Tauern stack and north-verging Austroalpine tectonic units from the Southalpine backthrusts.*
The Western and Eastern Alps exhibit fundamentally similar regional Adria–Europe collisional features, where the main differences are due to late orogenic Insubric-related tectono-dynamics.

Keywords: Western-Eastern Alps, CROP-ECORS-TRANSALP tectono-stratigraphic setting

1. INTRODUCTION

The Alps are generally interpreted as the result of a prominent continental collision between the European and Adria plates with closure of the interposed Alpine Tethys. For years, studies on the Alps have been the dominant investigation focus of various schools of the European Earth Science community. After years of data collection, hypotheses, and proposals, it became evident that, for a complete trans-Alpine understanding, an international scientific cooperation among the countries bordering the Alps to perform advanced deep seismic reflection exploration profiles across selected transversal tracks was necessary. COCORP (USA) and BIRPS (UK) results contributed to

*Corresponding author: Tel.: +39 40 5582276; e-mail: finetti@units.it (I.R. Finetti)
[1]Department of Geological, Environmental and Marine Sciences, Exploration Geophysics Group, University of Trieste, Via Weiss 1, 34127 Trieste, Italy

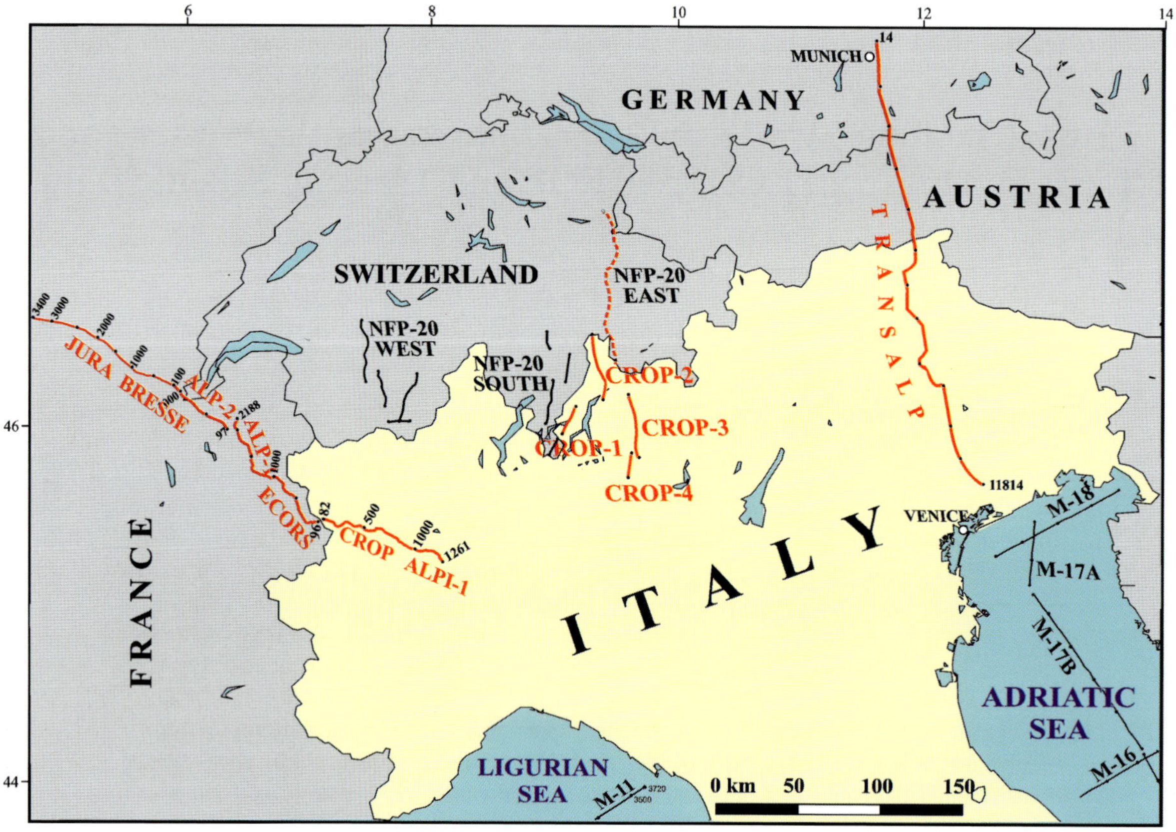

Figure 1: Location map of the crustal seismic profiles across the Alps acquired under international cooperation: France–Italy (Western Alps); Italy–Switzerland (Central Alps); Italy–Austria–Germany (Eastern Alps). In this chapter, an interpretation of the CROP-ECORS and Transalp profiles is presented.

the maturation of such inter-Alpine geophysical cooperations. As concrete results of the European cooperation on deep seismic exploration of the Alps, the following transcrustal profiles were performed (Figure 1):

(a) France (ECORS)–Italy (CROP) cooperation in the Western Alps: profiles Alp-1/Alpi-1 (Figure 1). ECORS profiles continue northwestward up to the Massif Central (Alp-2/Jura–Bresse). The NFP-20 Swiss programme in the Western Alps acquired the profiles NFP-20 WEST.
(b) Italy (CROP)–Switzerland (NFP-20) cooperation in the central Alps: Profiles CROP 1, 2, 3, 4; NFP-20 EAST; NFP-20 SOUTH.
(c) Italy (CROP)–Austria (OEKORP)–Germany (DEKORP) cooperation in the Eastern Alps: the TRANSALP profile.

Descriptions and interpretations of the above listed deep seismic profiles crossing the Alps have been presented in several works. For the high-quality CROP-ECORS seismic profiles across the Western Alps refer to Damotte (1996), Thouvenot et al. (1996), Mugnier et al. (1996), Nicolas et al. (1996). The CROP sections in the central Alps are not discussed in this chapter. NFP-20 East data are extensively described by Pfiffner et al. (1990). The Italian profiles in the central Alps (CROP-1, 2, 3, 4) are interpreted and commented on by Montrasio et al. (1994). One of the most important and successful deep seismic profile in the Alps is certainly *Transalp*, characterized by a relatively good-to-fair crustal reflectivity in large part. Data presented (Lammer et al., 2003; Castellarin et al., 2003) show substantially clear data on the subducting European crust and the overthrusting Adria plate. The above mentioned authors present alternative tectonic interpretations of the *Tauern window feature* (a wedge of European upper crustal sheets).

In this chapter, seismic interpretations of the CROP-ECORS profile across the Western Alps, and the Transalp profile across the Eastern Alps are presented and described. These interpretations are mostly focussed on discrimination of lower and upper basement crustal layers and reconstruction of their tectonic setting. The right-transcurrent transpressive *Insubric zone* and related thrusting, backthrusting and exhumation processes require an understanding of complex features of the Adria plate lithosphere and upper crustal European wedges, which is made difficult by their signal quality, especially where stratigraphic-tectonic interfaces become strongly dipping to sub-vertical, as in the *Ivrea flower zone* and *Tauern window* area.

The basic crustal deformation pattern of the Western Alps presents a fundamental similarity to that of the Eastern Alps.

2. ECORS-CROP PROFILE ACROSS THE WESTERN ALPS

This profile, acquired under a French–Italian cooperation, is well described in the monograph *The ECORS-CROP Alpine Seismic Traverse*, F. Roure et al. (1996).

2.1. Tectonic Framework of the Profile

The tectonic framework of the trans-Alpine area of the profile is presented in an introductory note of the above cited monograph by Nicolas et al. (1996). It is largely accepted that the Alpine orogenesis is strictly due to relative motion of Adria towards Europe, with subduction of the interposed Alpine Tethys lithosphere beneath Adria. This subduction started in Late Cretaceous time, (Barremian) about 130 Ma ago (radiometric age on HP/LT alpine minerals) and continued for about 80 Ma. Indicative petrological evaluations estimate that basement and ophiolite metamorphic units reached a P/T condition compatible with depths of 40–50 km or more. Closure of the Alpine Tethys in the sector of the Alps and initial Adria–Europe continental collision occurred at about 60–40 Ma. From 40 to about 25 Ma, the Europe–Adria crustal shortening continued giving rise to uplift and wedging that involved the upper and lower European crust. More externally, thrusting deformation involved only the upper crust and cover, with detachment of the Jura cover. Around 40 to 35 Ma, the Insubric transpressive right-transcurrent fault system started to be active on the Adria lithosphere, and generated a complex tectonic deformation. This is well imaged by CROP data, and is described later.

2.2. Seismic Interpretation of ECORS-CROP Trans-Western Alps Profile

In Plate 1, the Italian CROP seismic section Alpi-1 is exhibited to the reduced horizontal scale 1:300,000 and vertical scale 1 s $=$ 7.5 mm (scale ratio true for velocity $V_p = 4500$ m/s). In Plate 2, a seismic interpretation of the Alpi-1 CROP section (made to the scale 1:40,000 and plotted to the scale 1:300,000) and of ECORS sections Alp-1/Vanoise, and Alp-2/Bresse–Jura, made using data from the literature published to the scale 1:160,000 (Mém. Soc. Geol. Fr., 1996, no. 170) are plotted. The most significant interpreted seismic horizons and stratigraphic intervals are listed and described directly in the seismogeological section of Plate 2. The Jura–Bresse and Alp-2 ECORS sections are characterized by a very clearly identifiable and well reflecting lower crust interval, which is substantially flat. Only two transcrustal Alpine reverse faults (EU-1 and EU-2) seem to affect the lower crust of the Jura–Bresse and Alp-2 sections. Normal faults, probably related to Late Paleogene shear motion affect the crust of the eastern Bresse–Jura section (Plate 2; Figure 2(a)). The upper basement crust, as is normal, shows generally poor reflectivity but with local zones of good reflections. The interpreted transcrustal thrust fault EU-2 splits upward into various components that form with EU-1 the frontal compressive deformation of the Alpine system.

From thrust fault EU-3 to the Penninic prism, the Alp-2/Alp-1 ECORS sections exhibit a huge wedge mostly formed by thrust blocks of the European upper crust. This wedge outcrops with the *Belledonne external Massif*, where 6 to 7 thrust crustal sheets are tectonically piled up. In the Belledonne wedge, the upper basement crust thickens to over 30 km, while on the undeformed crust, the UC interval is about 16 km. Its relatively good reflectivity allows us to discriminate and reconstruct the tectonic setting of the lower crustal interval (TL-MO) as the most reliable basic lithospheric element on which to set up the overlying tectonic units. This reconstructed European lower crustal setting shows that the thrust-faults EU-3, EU-9 and EU-11 also displace the lower crust interval as well as the upper mantle. From the thrust-fault EU-11 to the SE extremity of the Alpi-1 CROP section (Po plain, Turin city area), the European lower crust interval is formed by a single continuous subducting layer over which the upper crustal interval has been tectonically laminated and transported into the Belledonee thrust wedge. At a depth of about 50 km (15 s, TWT) the European UC has been completely removed (roughly beneath the Insubric zone) and only the LC is subducting.

On the sedimentary cover of the uppermost thrust-block of the Belledonne wedge (Cormet de Roselend) a clear wedge of Penninic units (Alpine Tethydes) is seismically imaged with good reflectivity (Plate 2). Over subducting European crust, complex tectono-dynamic and petrophysical processes, related to the Adria plate crust-mantle

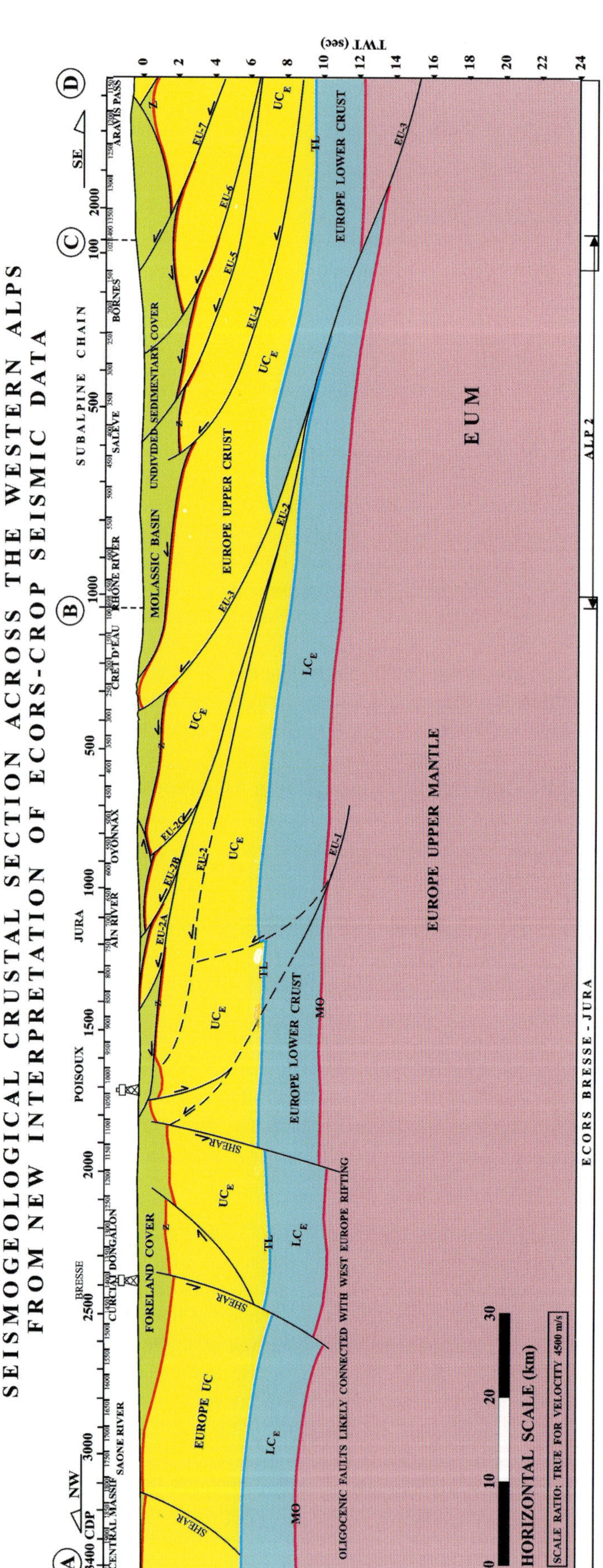

Figure 2: Reduced version of interpreted ECORS-CROP profiles across the Western Alps, from the Central Massif to the Turin city area.

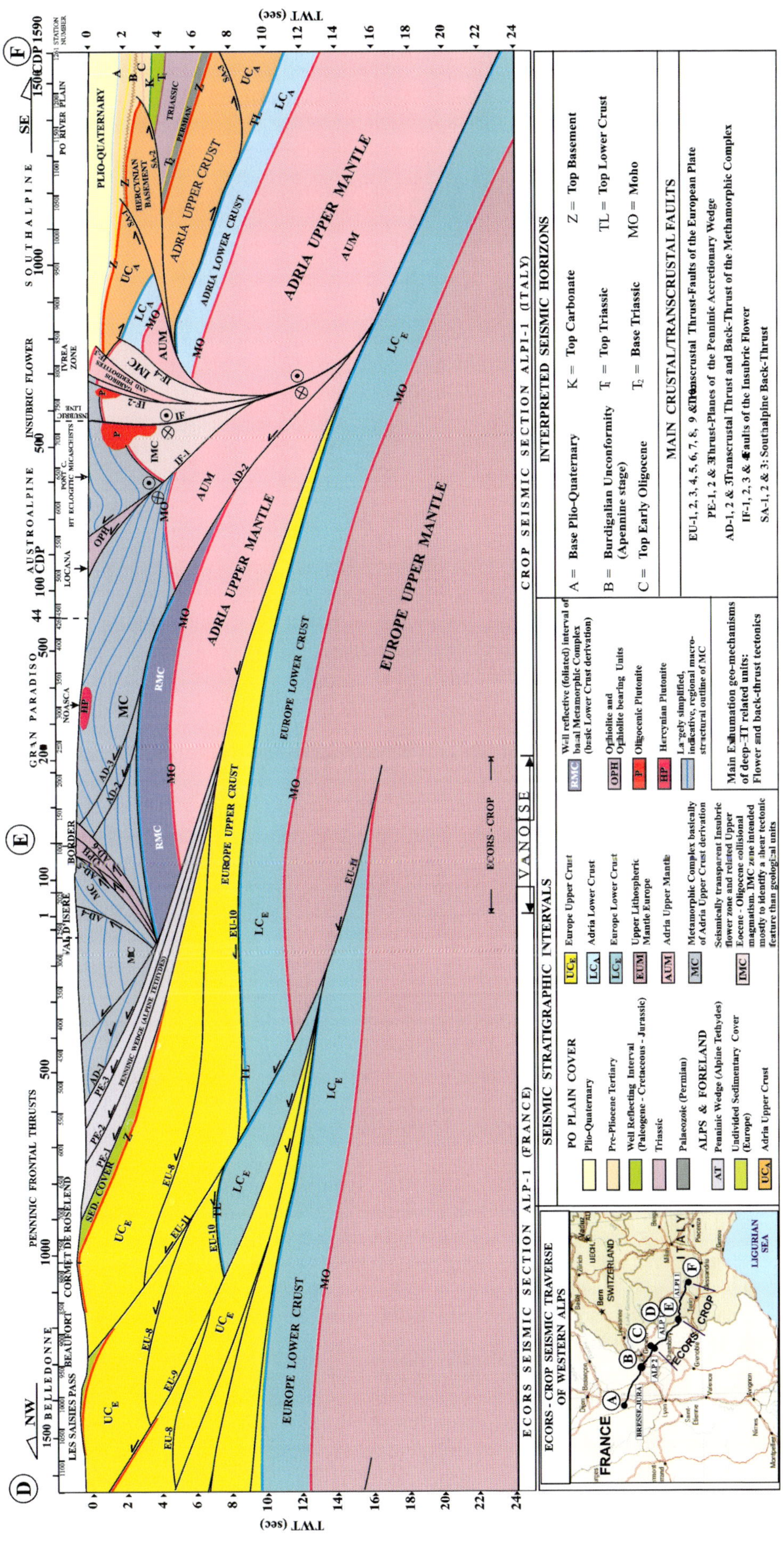

Figure 2: *Continued.*

megathrusting, took place. The CROP-ECORS seismic profile of this doubled crust segment (distance along section 230 to 358 km) exhibits a seismic reflection framework which is rather difficult to identify and set up as a reliably interpreted seismogeological section. Reconstruction started with the identification of the two lower crustal intervals (Europe and Adria) and interpretation of the Po Plain backthrusting (Southalpine). Interpretation of the Austroalpine, Gran Paradiso–Val d'Isere followed, where geological data significantly contributed to outlining the poorly reflective sectors of the Adria-plate-derived thrust units as integrating outcropping calibration data. Finally, it was possible to reconstruct a scheme for the Insubric zone, formed by a prominent transpressive right-transcurrent flower which separates the Europe-verging Austroalpine orogenic deformation from the SE-verging Southalpine backthrusting tectonics.

Backthrusting and Insubric transpressive transcurrency are responsible for the exhumation of deep-originating HT-related metamorphic units. Plutonites (P) of Oligocene age are generated along the Insubric shear zone. *The purpose of this chapter is* not the study of petrophysical events but of *the geometry of the transcrustal setting that with widely recurrent tectonic style, may best fit the seismic reflection field.* The fundamental regional crustal setting obtained, in the author's opinion, seems well supported seismically.

3. TRANSALP PROFILE

Two interpretations of the Transalp profile have been presented in the literature (Lammer et al., 2003; Castellarin et al., 2003). They differ mostly with regard to the Tauern Window and Adria lower crustal setting. Another interpretation of the Transalp profile is now presented and briefly discussed in this chapter (Figure 3). This interpretation highlights two points that, in the author's opinion, characterize and more simply account for the seismic reflection field of the complex central part of the Transalp profile. These topics are the same ones as suggested the two different interpretation models in the literature: *the Tauern Window feature (TW)* and *the tectonic framework of the Adria plate crust.*

The European plate, as clearly proposed by Lammer et al. (2003) and Castellarin et al. (2003), subducts under the Adria plate with no significant lower crustal shortening. Beneath the Molasse basin and the Northern Calcareous Alps, the European crust is gently dipping southward with little bending. In the TW zone, the profile exhibits a very poor- to non-reflective feature which is first sight difficult to understand. Observing the global tectono-dynamic context, which is seismically clearer around the TW block where the good reflectivity is very helpful, and examining carefully the few, deformed reflectors of low amplitude in the TW zone, it seems evident that this *feature is composed of a tectonic stack (wedge) of fragments of the upper European crust thrust over the subducting shear plane, uplifted and exhumed by the transpressive right-transcurrent Insubric (or Periadriatic) line-related tectonics.* Accurate analysis of the few, poorly reflected signals of the TW structure seems to allow the delineation of four stacked thrust-blocks of European upper crust. The two southernmost blocks are almost completely lacking in reflections, and their interpreted vertical position may fully explain this matter. Returning them to their original undeformed position using a cylindrical approach, their horizontal displacement ranges from 35 to 60 km. Beneath the TW the subducting European crust starts to bend more consistently. Between the southernmost stacked block of the TW wedge and the Insubric right-transcurrent line, the Penninic Nappe (PN) made of Alpine Tethydes, and igneous-methamorphic sheet units (IM) are sandwiched. Plutonic bodies (P) of Oligocene age are associated with the Insubric shear zone.

The crustal tectono-stratigraphic setting of the colliding-overthrusting Adria plate is the most difficult part to interpret of the Transalp profile. The Insubric line separates a northern area (TW wedge) of poor- to none-reflectivity from a southern zone characterized by south-dipping reflections of good quality throughout the crust. The explanation of these good reflections evidently cannot be connected to peculiar crustal stratigraphy because they terminate with high amplitude at a same distance from the IS tectonic line throughout the upper and lower crust. We must consequently think of a tectonic-related explanation. Another helpful fact is that the dynamite seismic line of the southern Transalp (Castellarin et al., 2003) indicates rather clearly a duplication of the Adria Lower crust. Considering the interpretation of the contractional tectonics of the sedimentary cover, which is supported by good seismic data (Bertelli et al., 2003), the duplication (thrust) of the lower crust, and the major controlled upper crustal thrust-faults, it becomes evident that *to accommodate the Southalpine crustal context we must introduce a compressive backthrusting lamination of the Adria upper crust* (Figure 3). In the author's opinion, this is the real meaning of the good reflectivity just to the south of the Insubric line. Very probably, the laminating interfaces are water-saturated, which accounts for the high amplitudes of the reflected signals. Immediately to the north of the

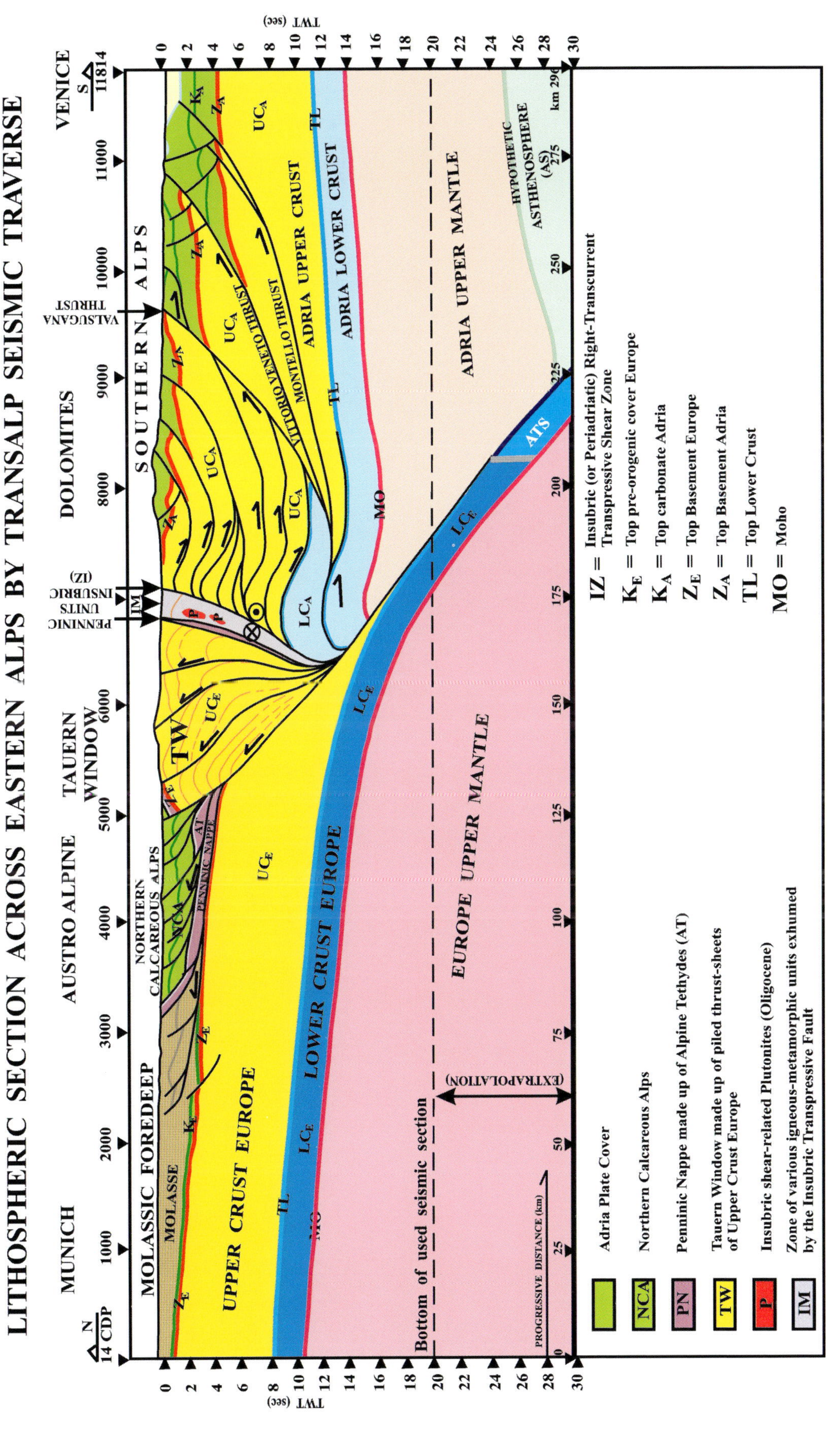

Figure 3: Seismogeological time section across the Eastern Alps, reconstructed by new interpretation of Transalp data from the literature (Castellarin et al., 2003). See Figure 1 for location. Downward extrapolation to 30 sec is purely hypothetical.

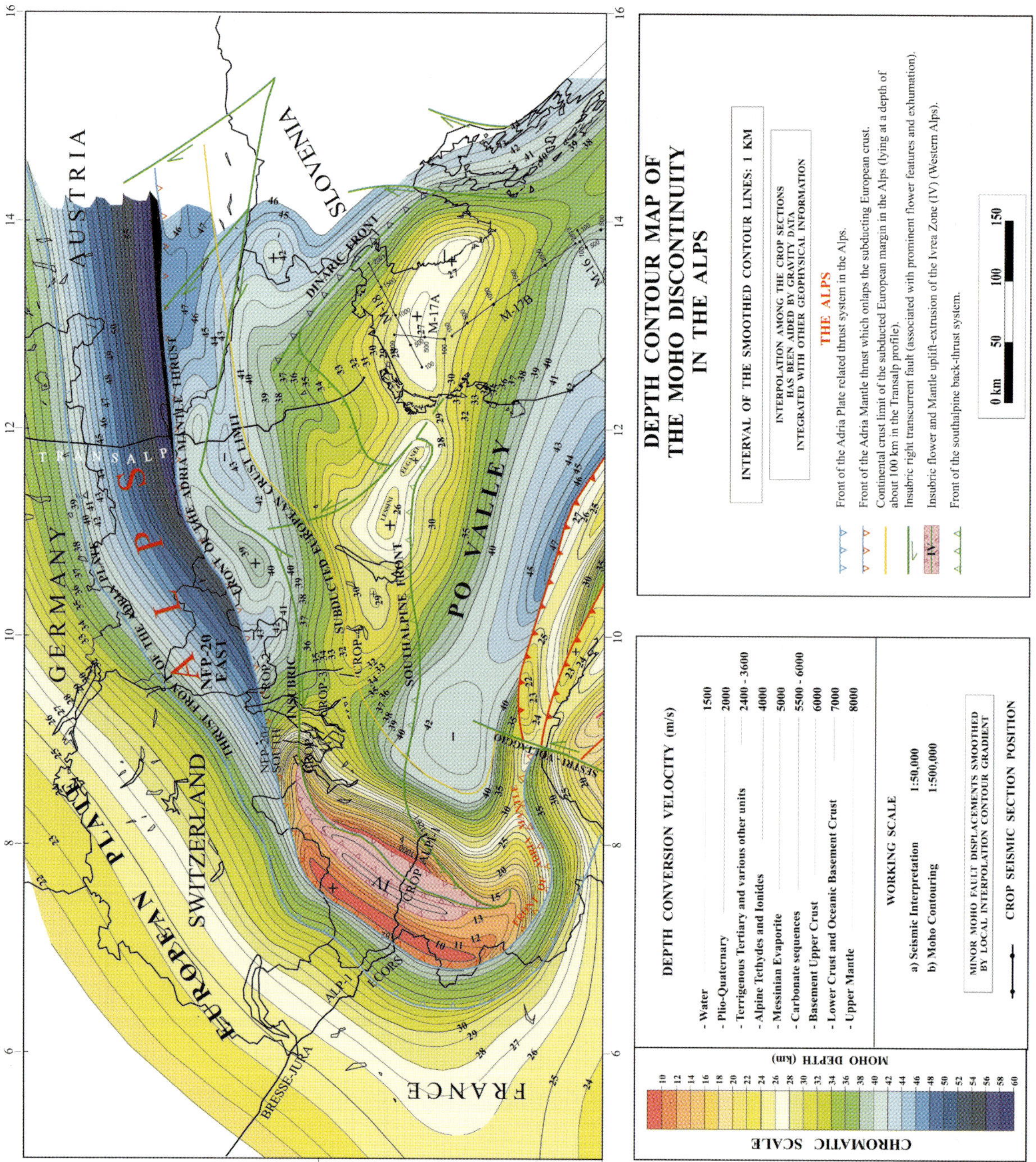

Figure 4: Depth contour map of the Moho discontinuity across the Alps and adjacent Europe–Adria plate areas. Contour interval 1 km. Data controlled by CROP network and trans-Alpine profiles of Figure 1.

Insubric line, the contrasting non-reflective TW zone is very likely due to the fact that its tectono-stratigraphic planes are subvertical.

4. DEPTH CONTOUR MAP OF THE MOHO INTERFACE IN THE ALPINE REGION AND TECTONO-KINEMATIC CONSIDERATIONS

Using CROP, ECORS, NFP-20 and Transalp Alpine seismic sections, interpolated with gravity data, it has been possible to reconstruct a *new depth contour map of the Moho discontinuity* in the Alpine region and surrounding geological provinces (Figure 4). This map very clearly exhibits the subducting European plate and the overthrusting Adria plate crustal setting. Within the Adria plate structure, the Southalpine structural setting is regionally outlined. Very evident prominent mantle uplift and extrusion of the Ivrea–Verbano zone is reconstructed. The subducted edge of the European continental margin (Figure 3; Figure 4: yellow line) has been estimated on the basis of the whole CROP dataset and its plaeogeographic reconstruction (Finetti, this volume, Chapter 34).

Regarding the Adria–Europe convergence, assuming an Alpine Tethys about 250 km wide in the Alps sector, from the interpreted CROP data, it is possible to estimate an indicative relative motion of about 480–500 km. After closure of the interposed Tethys, the Adria plate northern boundary would have thrust about 230–250 km over subducting Europe. Adria crustal shortening due to the Southalpine tectonics, according to interpretation, in the Eastern Alps (Figure 3) amounts to about 40–45 km, and about 30–35 km in the Western Alps (Figure 2, Plate 2).

In the Meso-Alpine stage (Oligocene–Lower Miocene) the Adria–Europe convergence-related deformation encountered higher friction-force along the N–S component (high-angle subducting plane), consequently the shortening was accommodated by moving westward along the Insubric shear zone. The Moho map of Figure 4 clearly evidences this, showing that the NW Adria Moho obliquely overthrusts the subducting European Moho in the sector lying between the Central Alps (NFP-20) and the Ivrea–Verbano zone.

ACKNOWLEDGEMENTS

This work was funded by MIUR, COFIN 1998, 2000 and 2002 to I.R. Finetti, University of Trieste. Many thanks are due to R. Polino for the fruitful discussions on the Western Alps, and to A. Prizzon for the graphical contributions.

REFERENCES

Bertelli, L., Cantelli, L., Castellarin, A., Fantoni, R., Mosconi, A., Sella, M., Selli, L., 2003. Upper crustal style, shortening and deformation age in the Alps along the southern sector of the TRANSALP Profile. Mem. Sci. Geol. 54, 123–126.

Castellarin, A., Dal Piaz., G.V., Fantoni, R., Vai, G.B., Nicolich, R., TRANSALP Working Group, 2003. Lower crustal style and models along the southern sector of the Transalp Profile. Mem. Sci. Geol. 54, 245–248.

Damotte, B., 1996. The ECORS profile across the Alps, Jura Mountains and Bresse Plain: implementation and data processing. Bull. Soc. Géol. France 170, 27–41.

Finetti, I.R., this volume. Geodynamic evolution of the Mediterranean Region, from the Permo-Triassic Ionian opening to the Present, constrained by new lithospheric CROP seismic data. Chapter 34.

Lammer, B., TRANSALP Working Group, 2003. The "crocodile" model and balancing the seismic section. Mem. Sci. Geol. 54, 243–244.

Montrasio, A., Bersezio, R., Forcella, F., Jadoul, F., Sciesa, E., 1994. Geological interpretation of the profile CROP-Central Alps (Passo Spluga – Bergamo). In: Proceedings of Symposium"CROP – Alpi Centrali". Quaderni di Geodinamica Alpina e Quaternaria 2, 171–186.

Mugnier, J.L., Bergerat, F., Damotte, B., Guellec, S., Nicolas, A., Polino, R., Roure, F., Tardy., M., Truffert, C., 1996. Crustal structure of the western Alps and their forelands. Bull. Soc. Géol. France 170, 73–97.

Nicolas, A., Mugnier, J.L., Polino., R., Roure, F., 1996. General conclusions of the ECORS-CROP Alpine survey. Bull. Soc. Géol. France 170, 99–104.

Nicolas, A., Mugnier, J.L., Bergerat, F., Polino., R., Roure, F., Thouvenot, F., 1996. The location of the deep seismic profile and the geological and geophysical framework. Bull. Soc. Géol. France 170, 11–26.

Pfiffner, O.A., Frei, W., Valasek, P., Stäuble, M., Levato, L., Du Bois, L., Schmid, S.M., Smithson, S.B., 1990. Crustal shortening in the Alpine Orogen: results from deep seismic reflection profiling in the eastern swiss Alps, line NFP20-East. Tectonics 9, 1327–1355.

Roure, F., Bergerat, F., Damotte, B., Mugnier, J.L., Polino, R. (Eds.), 1996. The ECORS-CROP Alpine Seismic Traverse. Bull. Soc. Géol. France 170.

Thouvenot, F., Senechal, G., Truffert, C., Guellec, S., 1996. Comparison between two techniques of line-drawing migration (ray tracing and common tangent method). Bull. Soc. Géol. France 170, 53–59.

CHAPTER 8

Lithospheric Tectono-Stratigraphic Setting of the Ligurian Sea–Northern Apennines–Adriatic Foreland from Integrated CROP Seismic Data

I.R. Finetti[1,*], M. Boccaletti[2,3], M. Bonini[3], A. Del Ben[1], M. Pipan[1], A. Prizzon[1], F. Sani[2,3]

ABSTRACT

The structure and geodynamics of the Northern Apennines, and of the Northern Tyrrhenian and Ligurian basin are reconstructed and discussed on the basis of deep seismic reflection data from the Italian lithospheric exploration CROP project in the Central Mediterranean region. Three connected sections cross uninterrupted the Northern Tyrrhenian Sea (CROP M12A), the Northern Apennines (CROP 03) and the Adriatic Sea (CROP M16). The sections CROP M10 and M11 cross the northern Balearic basin (Ligurian Sea) and the northern extension of the Corsica–Sardinia block, while section CROP M12B crosses the Corsica basin sub-longitudinally. Data interpretation indicates very clearly that the Northern Apennines are dominated by a compressional tectonic style expressed by a well developed lithospheric thrust system. In fact, the CROP sections image crustal-lithospheric, E-verging thrusting structures which affect the Adria plate from its western edge (Corsica basin) to the Adriatic Sea. Below the sub-horizontal shear plane AP-2, the boundary between brittle and ductile crust of the western Tuscan Apennines, evident W-verging transcrustal thrusts are preserved. Time–space deformation analysis shows that the Europe-verging buried thrust structures were generated during the Eo-Alpine geodynamic stage (Upper Cretaceous–Eocene) when the segment of the Alpine Tethys interposed between Adria and Europe (Alps) closed with lateral subduction of the oceanic crust beneath NW Adria. In the Balearic stage (Oligocene–Lower Miocene) a generalized Adria-verging thrust propagation took place, with rotation of the Corso-Sardinian block and subduction beneath the European margin of the remaining Alpine Tethys west of the Adria–Europe collisional zone. This process was accompanied by formation of an eastward-migrating Tethyan asthenospheric bulge, Corsica–Adria collision and generation of the first Apenninic chain. South of the 41° parallel, the eastern margin of the Sardinian block entered into the Ionian Neotethys, fully communicating with the Alpine Neotethys ocean. Therefore here the Tyrrhenian stage geodynamics (Middle Miocene to the Present) continued without interruption the thrust propagation southeastward, as far as the Calabrian arc. The prominent compressive asthenospheric bulge of the Tuscan archipelago (Etrurian bulge) continued to act as the main eastward driving geomotor of the Northern Apennine thrust belt, jointly with a slower Adria–Europe convergence velocity.

Opening of the Tyrrhenian back-arc basin occurred over the subducted Ionian Neotethys crustal area, and in the Tuscan archipelago evident extensional deformations accompanied the southeastward Tyrrhenian mantle drag. Crustal extensions hypothesized in some of the literature for the Tuscan Apennine area are second order tectonic deformations, mostly connected to trans-Apenninic transcurrent faults or to the accommodation of the basic Apennine thrust propagation and crustal shortening.

The oceanic crust, the architecture of the sedimentary fill, and the geometry of the margins of the northern Balearic basin are imaged in the western parts of profiles CROP M10 and M11. Taking into account both the presented and other available geophysical data, we discuss the 3D lithospheric/crustal-scale fault pattern and the structures transverse to the

*Corresponding author: Tel.: +39 40 5582276; e-mail: finetti@units.it (I.R. Finetti)
[1]Department of Geological, Environmental and Marine Sciences, University of Trieste, Via Weiss 1, 34127, Trieste, Italy
[2]Department of Earth Sciences, University of Firenze, Via G. La Pira 4, 50121 Firenze, Italy
[3]CNR, Institute of Geosciences and Earth Resources (IGG), Section of Firenze, Via G. La Pira 4, 50121 Firenze, Italy

Northern Apennines, together with their possible implications for geothermal fields. The Corsica basin and the more external foredeep basins filled by siliciclastic deposits formed in response to these large-scale geodynamics. Migration of compression toward the east resulted in major thrusts displacing the Moho beneath the Northern Apennines watershed (Late Miocene–Present; Second Apenninic Stage). The deformation of this lithospheric-scale thrust system propagated far into the foreland and controlled the deposition and sedimentation in the easternmost Pliocene–Quaternary foredeep basin (CROP M16). In the axial zone of the Northern Apennines, this thrust system gives rise to a marked basement culmination, above which blocks of sedimentary cover units are gravitationally collapsing and generating high seismic activity.

Keywords: Deep seismic CROP sections, Northern Apennines, Balearic basin, lithospheric tectono-stratigraphy, geodynamics

1. INTRODUCTION

The Mediterranean Sea, especially the central part, is considered one of the most geodynamically interesting regions for geoscientists, because various plates and intraplate ocean basins and geodynamic stages gave rise to the complex geological framework we observe today. The Ligurian Sea (northeastern sphenochasm of the Balearic basin), the Tyrrhenian basin and the Northern Apennines represent a key area in which a significant part of the long and complex geological history of the Mediterranean region is recorded.

Based on (1) the interpretation of new CROP deep seismic sections (Figure 1: sections M10, M11 and M12B); (2) further refinements in the interpretation of CROP sections M12A-CROP03-M16 with respect to previous a publication (Finetti et al., 2001); and (3) taking into account new geological-geophysical and structural field data, we have reconstructed the lithospheric tectono-stratigraphic setting of this wide complex area and its 3D time–space history.

The Northern Apennines (NA) constitute an arc-shaped NE- to E-verging fold-and-thrust belt, extending from the Monferrato region to the north as far as the roughly N–S trending Olevano-Antrodoco *Auctt.* Line, to the south, where the NA units thrust at high angle over the central Apennines (Figure 2). The NA are composed of two main groups of tectonic units of two distinct plaeogeographic domains: (1) the Tuscan (TU) and Umbria–Marche (UMU) units, constituting the sedimentary cover of the western continental margin of the Adria Plate; and (2) the Ligurian or Alpine Tethys units (AT), representing allochthonous terrains scraped from the original oceanic crust (ophiolites and overlying Jurassic to Eocene sedimentary cover of the Ligurian–Piedmont, or Alpine Tethys ocean) and tectonically transported over the TU and UMU. In the framework of the Mediterranean tectonic region, the Alpine Tethys separated the European and Adria plates to the north-east (Alps sector), and the European and African plates to the south-west (Finetti, this volume, Chapter 34).

The deformation of the AT started in the Late Cretaceous following the onset of the Alpine Tethys ocean closure in the sector of the Adria–Europe collision (Alps). Before the availability of CROP data, two different interpretations of the subduction polarity of the Alpine Tethys had been proposed: (1) a west-dipping subduction below the Corsica–Sardinia block (which was still connected to the European margin) (Scholle, 1970; Wortel and Spakman, 1992; Bortolotti et al., 2001, and reference therein); and (2) an east-dipping oceanic lithospheric subduction below the Adria Plate postulated by Boccaletti and Guazzone (1970, 1974) and Boccaletti et al. (1971). According to this latter hypothesis, the Apenninic orogeny *s.s.* started after the continental collision, i.e. at the end of Late Oligocene (Boccaletti et al., 1980). Following Argand's (1924) original ideas, during the phase of continental collision, which followed the Ligurian–Piedmont ocean closure, the rotation of the Corsica–Sardinia block started together with the opening of the Balearic basin to the west (Dewey et al., 1973; Biju Duval et al., 1977; Burrus, 1984; Finetti and Del Ben, 1986). The time–space migration toward the east of the foredeep system has been commonly associated with this rotation, without precise specification of the driving forces involved. The new CROP data here presented clears up this old question, and the subsurface deep seismic images presented allow us also to better understand which driving forces acted in the area, and in which geodynamic stages.

After rotation of the Corsica–Sardinia block and consequent eastward migration of the NA tectonics, a basin started to develop in the southern Tyrrhenian area in the Middle Miocene, as is well constrained by a large regional seismic dataset (Finetti and Del Ben, 1986). Extensional tectonics associated with the Tyrrhenian basin formation have been

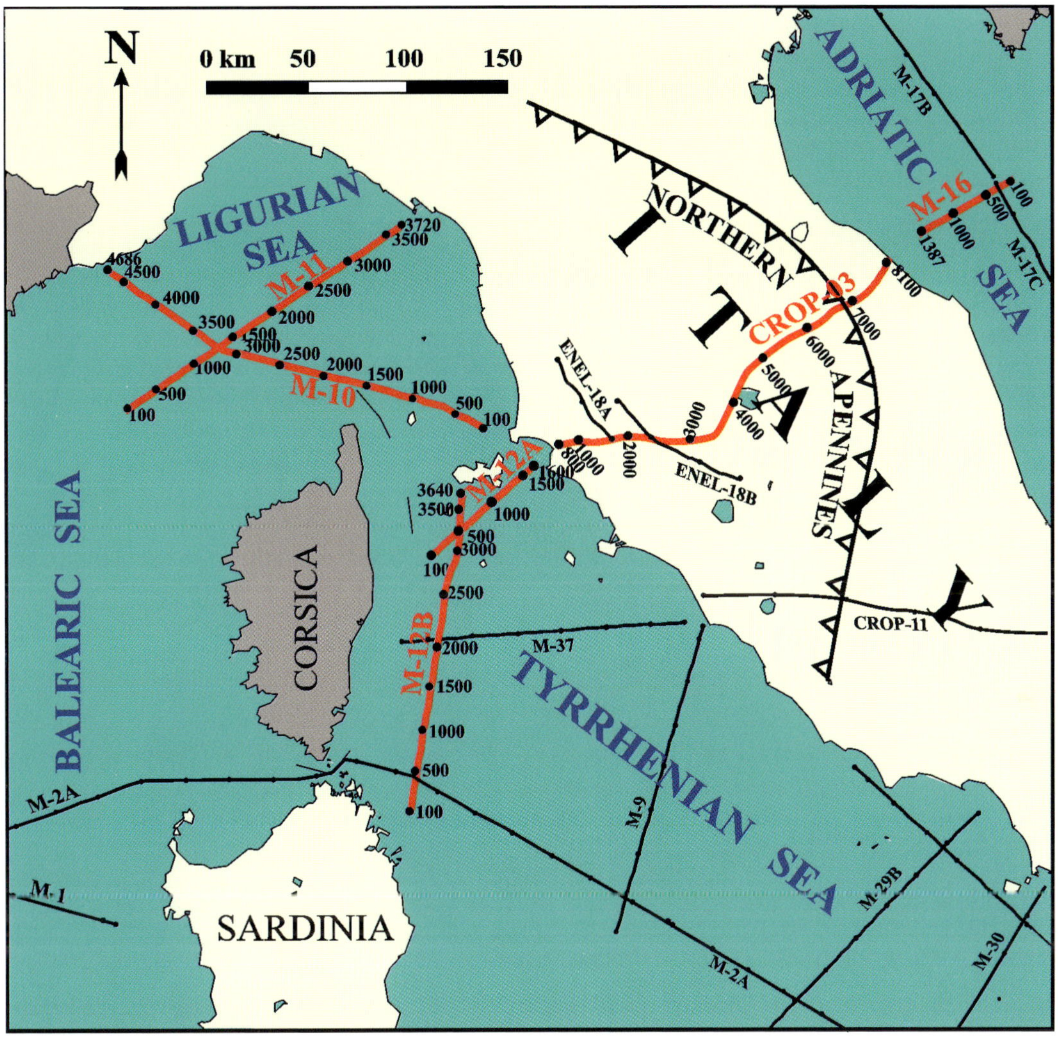

Figure 1: Location Map of the CROP seismic sections used in this chapter. All sections are presented as annexed plates (Scale 1:300,000): Seismic section (SEIS) and interpreted section (INT). CROP M10: Plate 1 (SEIS), Plate 2 (INT); M11: Plate 3 (SEIS), Plate 4 (INT); M12B: Plate 5 (SEIS), Plate 6 (INT); M12A/CROP 03/M16: Plate 7 (SEIS), Plate 8 (INT).[4]

interpreted with various geodynamic models: (a) *back-arc-related extension in subduction models* (Boccaletti et al., 1971; Boccaletti and Guazzone, 1972, 1974; Barberi et al., 1973; Malinverno and Ryan, 1986; Finetti and Del Ben, 1986; Royden et al., 1987; Wortel and Spakman, 1992; Jolivet and Faccenna, 2000); (b) *lithospheric delamination* (Roeder, 1980); (c) *mantle uprise determining gravity-slide phenomena* (Van Bemmelen, 1972; Locardi, 1982; Wezel, 1982), and variations (Lavecchia, 1988; Lavecchia and Stoppa, 1996; D'Agostino et al., 2001); (d) *lateral extrusion of the Apennines following the roughly N–S trending collision between the African and European plates* (Tapponnier, 1977; Boccaletti et al., 1982; Faccenna et al., 1996; Mantovani et al., 1997); and (e) *late orogenic collapse of an overthickened accretionary wedge* (Carmignani and Kligfield, 1990; Carmignani et al., 1994).

More recently, CROP data crossing the Tuscan archipelago (M-12A) and Tuscany (CROP-03), showed that an Alpine Tethys slab subducted laterally beneath the north-western Adria plate margin in the Eo-Alpine stage (Finetti et al., 2001). We present in this chapter a completion of study on this matter, supported by CROP data, that clearly images the presence of two distinct subducted slabs (Plates 2, 4 and 6).

Most of the literature on the NA assumes that the extensional tectonics connected to the Tyrrhenian opening also propagated on-shore. But CROP sections M12A/CROP 03/M16 across the NA show that in the Tuscany sector the extensional tectonics are a second order feature accommodating the first order crustal shortening (Finetti et al., 2001). Moreover, the entire CROP network of the region shows that whereas the extensional tectonics associated

[4] Legend for plate see separate plate section after the references at the end of this chapter.

with the Ionian slab retreat and Tyrrhenian opening are macroscopically evident and impressive south of the 41° parallel, northwards in the southern Tuscan archipelago–NW Adria plate margin of the Etrurian promontory (Finetti and Del Ben, this volume, Chapter 21; Finetti, this volume, Chapter 34) such extension is much more limited and represented by transtensive effects related to left transcurrent faults generated by the southeastward-propagating Tyrrhenian Mantle flux (Plio-Pleistocene).

2. REGIONAL SETTING

Geodynamic regions of great significance in the development and evolution of the Central Mediterranean are located in the wide sector investigated using CROP seismic sections. Starting from the west, there is the Balearic basin which has triangular shape and is delimited by the Provençal-Catalan coasts to the west, the Algerian coast to the south, and the Corsica–Sardinia margin to the east. This basin is floored by oceanic crust and it opened from the Oligocene (30 Ma; Cherchi and Montadert, 1982; Burrus, 1984) to the Langhian (Vigliotti and Langenheim, 1995; Speranza et al., 2002). The counterclockwise rotation of the Corsica–Sardinia block and oceanic spreading occurred between 21 and 16 Ma (for a review see Speranza et al., 2002). From a geodynamical point of view the Balearic basin has been explained as a back-arc basin generated by a roughly NNW-directed subduction of the African plate beneath Europe (Boccaletti and Guazzone, 1974; Rehault et al., 1984; Jolivet and Faccenna, 2000). Makris et al. (1999) suggest that, at the northern corner of the Balearic basin (Ligurian Sea), the boundary between oceanic crust and the continental crust of the Corsica–Sardinia block is characterized by wrench faulting and transtension.

Soon after the rotation and the eastward movement of the Corsica–Sardinia block, another basin formed in the Tyrrhenian area from the Middle-Late Miocene, dissecting the pre-existing NA fold-and-thrust belt (Boccaletti and Guazzone, 1972, 1974, Barberi et al., 1973; Elter et al., 1975; Finetti and Del Ben, 1986; Sartori, 1990).

The NA structural setting consists of a pile of NE-verging units delimited by thrust faults that frequently are accompanied by second-order back-thrusts and local normal faults, mostly due either to transtensive effects produced by transverse transcurrent faults, or to accommodation of the uplift generated by crustal shortening. The Alpine Tethydes are composed of ophiolites and their Jurassic to Eocene sedimentary cover, and tectonically overlie the TU and the UMU. These latter units consist of a lower $\sim$ 2000 m-thick succession of mainly carbonate Mesozoic–Cenozoic rocks resting on about 1000–1500 m of Triassic evaporites (Burano Fm.) that grade laterally (eastwards) into dolomites. Notably, the TU and UMU are detached along this thick basal evaporitic layer. The carbonate sequence is overlain by a Late Oligocene–Miocene succession of 2000–5000 m thick siliciclastic foredeep sediments. Whereas the Mesozoic sequence gives rise to slight gradual variations moving outward, the foredeep-related tectono-stratigraphic units (Oligocene–Miocene) become gradually younger eastward and show consistent variability (Boccaletti et al., 1990). These siliciclastic units, from west to east, are the Macigno, Cervarola-Falterona, and Marnoso-Arenacea units (Figure 2). CROP data (Finetti et al., 2001) suggest that the major foredeep basins are connected to major transcrustal thrust faults with large horizontal and vertical displacements and cumulative crustal shortening of remarkable magnitude.

3. INTERPRETATION OF CROP SEISMIC SECTIONS

This chapter illustrates and describes the novel results obtained from interpretation of the seismic sections CROP M10 and M11 across the Ligurian Sea–Corsica/Adria collisional zone (Plates 1–4), the section M12B across the frontal thrusts of the Corsica–Sardinia block–Corsica basin–NW Adria margin (Plates 5 and 6), and the composite section CROP M12A/CROP 03/M16 from the Corsica Basin to the Adriatic Sea, across the Northern Apennines (Plates 7 and 8). Locations of the CROP sections used is reported in Figures 1 and 2. The results obtained from

---

Figure 2: Geological sketch map of the Northern Apennines, Corsica basin, Balearic basin, Northern Tyrrhenian region. The positions of the six CROP sections presented in this chapter (CROP M10, M11, M12B and M12A/CROP03/M16) are reported. These deep seismic sections cross the whole Northern Apennines orogenic system, from the Corsica basin to the outermost deformation of the Dinarides (sections CROP M12A/CROP03/M16), and the Balearic basin–Corsica block–Tyrrhenian basin (sections CROP M10, M11 and M12B). Off-shore structures (indicated in blue) are from the CROP data integrated with other seismic data (MS Lines).

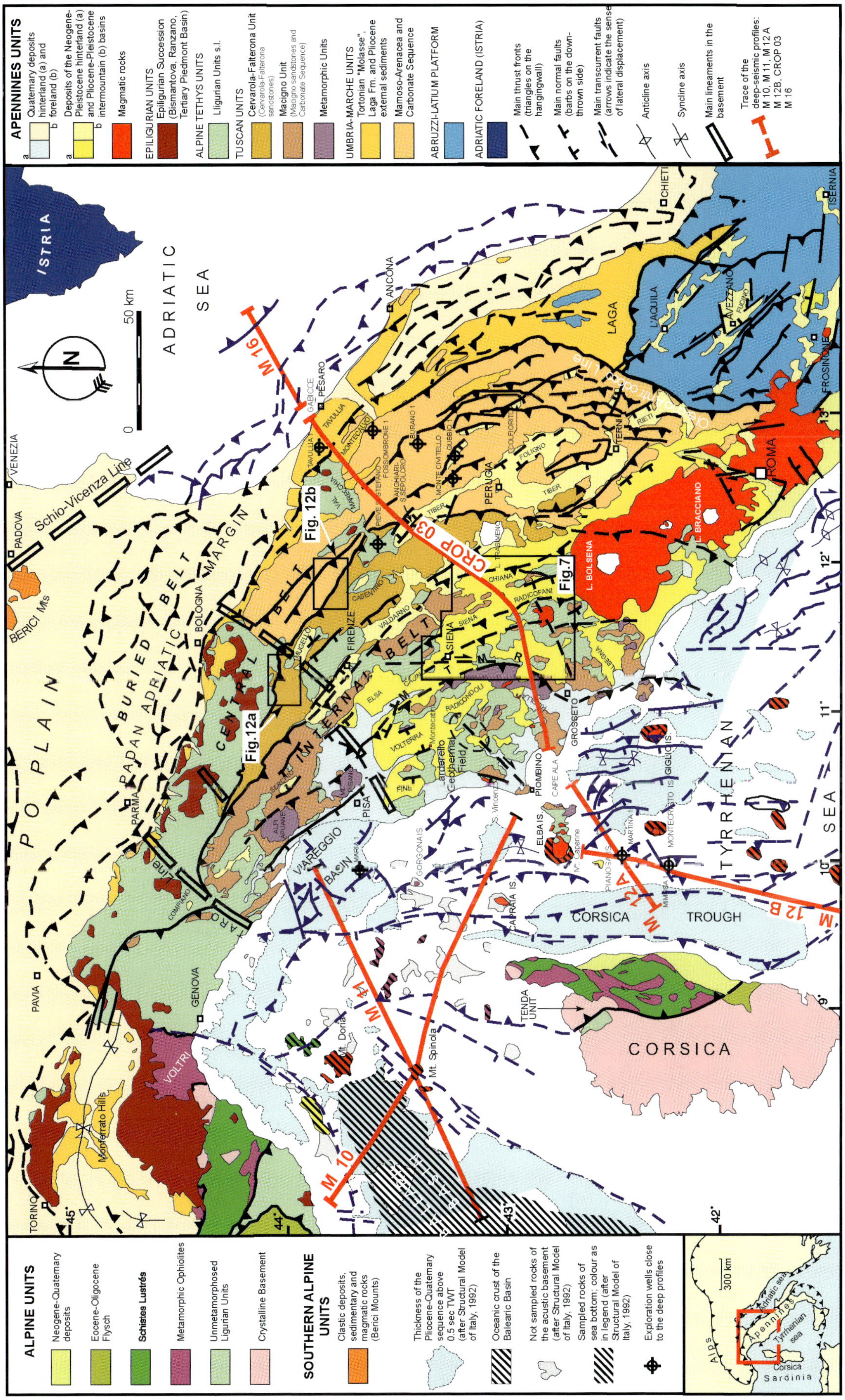
APENNINES UNITS
Quaternary deposits hinterland (a) and foreland (b)
Deposits of the Neogene-Pleistocene hinterland (a) and Pliocene-Pleistocene intermountain (b) basins
Magmatic rocks
EPILIGURIAN UNITS
Epiligurian Succession (Bismantova, Ranzano, Tertiary Piedmont Basin)
ALPINE TETHYS UNITS
Ligurian Units s.l.
TUSCAN UNITS
Cervarola-Falterona Unit
Macigno Unit
Metamorphic Units
UMBRIA-MARCHE UNITS
Tortonian "Molasse", Laga Fm. and Pliocene external sediments
Marnoso-Arenacea and Carbonate Sequence
ABRUZZI-LATIUM PLATFORM
ADRIATIC FORELAND (ISTRIA)
Main thrust fronts (triangles on the hangingwall)
Main normal faults (barbs on the down-thrown side)
Main transcurrent faults (arrows indicate the sense of lateral displacement)
Anticline axis
Syncline axis
Main lineaments in the basement
Trace of the deep-seismic profiles: M 10, M 11, M 12 A, M 12B, CROP 03, M 16
ALPINE UNITS
Neogene-Quaternary deposits
Eocene-Oligocene Flysch
Schistes Lustrés
Metamorphic Ophiolites
Unmetamorphosed Ligurian Units
Crystalline Basement
SOUTHERN ALPINE UNITS
Clastic deposits, sedimentary and magmatic rocks (Berici Mounts)
Thickness of the Pliocene-Quaternary sequence above 0.5 sec TWT (after Structural Model of Italy, 1992)
Oceanic crust of the Balearic Basin
Not sampled rocks of the acoustic basement (after Structural Model of Italy, 1992)
Sampled rocks of sea bottom; colour as in legend (after Structural Model of Italy, 1992)
Exploration wells close to the deep profiles
ISTRIA
ADRIATIC SEA
PO PLAIN
TYRRHENIAN SEA
CORSICA
CORSICA TROUGH
Schio-Vicenza Line
PADAN ADRIATIC BURIED BELT
CENTRAL BELT
INTERNAL BELT
CROP 03
M 16
M 10
M 11
M 12 A
M 12 B
Fig. 12a
Fig. 12b
Fig. 7
ROMA
FIRENZE
BOLOGNA
GENOVA
TORINO
PAVIA
PARMA
PADOVA
VENEZIA
PESARO
ANCONA
PERUGIA
SIENA
PISA
GROSSETO
PIOMBINO
L'AQUILA
CHIETI
FROSINONE
ISERNIA
L. BOLSENA
L. BRACCIANO
ELBA IS.
GIGLIO IS.
CAPRAIA IS.
GORGONA IS.
MONTECRISTO IS.
VIAREGGIO BASIN
LAGA
TENDA UNIT
VOLTRI
Monferrato Hills
BERICI Mts
Mt. Doria
Mt. Spinola
Larderello Geothermal Field
300 km
50 km
Alps
Apennines
Adriatic sea
Tyrrhenian sea
Corsica
Sardinia

interpretation of the long trans-Apenninic section M12A/CROP 03/M16 have been already described in the literature (Finetti et al., 2001). After re-examination and refinement of previous seismogeological sections, all the basic data of this interpretation are now fully confirmed and find further supporting evidence (Plates 7 and 8). New data, obtained from interpretation and 3D integration of the CROP seismic sections of the region, allow us for the first time to extend the geophysical control to most of the key deep features that have been widely debated in literature because they constitute critical elements for understanding the lithospheric setting and its geodynamics.

Careful reprocessing of the sections CROP M10, M11, M12A and M12B performed at the processing centre of the University of Trieste by EGG, the Exploration Geophysics Group of the Faculty of Sciences (Finetti et al., this volume, Chapter 5) consistently improved the seismic signal quality of the previous poor sections obtained in the initial standard processing. Thus the new sections allowed a much more reliable interpretation. The geological position of the many selected seismic horizons for the interpretation are fully specified on each reconstructed seismogeological section, together with their stratigraphic intervals (Plates 2, 4, 6 and 8).

In the interpretation of deep crustal seismic sections, it is not so important to carefully pick reflectors characterizing detailed sedimentary interfaces, but rather to obtain regional control of the ensemble of principal crustal macro-intervals and all major tectonic elements useful for a globally correct macro-reconstruction of the tectono-stratigraphic setting, and for a constrained understanding of the evolving geodynamics. In a complex, noisy seismic context, as is a thrust belt like the Apenninic system, such an interpretation can be considered satisfactory when each delimited tectonic block shows full individual coherence in all its seismic signals, including primary reflections of low to very low amplitude. Particular attention was thus dedicated to this point in order to improve the final result and its reliability, by conducting an extensive program of iterated interpretive reprocessing and progressive refinement. Over 50% of the interpretation time was devoted to assembling the results of each single CROP section into a 3D coherent organic setting. The global improvement derived from corrective refinement by correlation between all the 3D CROP network sections was considerable.

To obtain a more immediate identification and positioning of each tectonic and/or stratigraphic element, all the most significant faults and tectonic blocks of the upper crust (UC) and lower crust (LC) of the trans-Apenninic section (Plate 8) were labelled with a combination of letters and/or numbers. Faults and thrust faults referred to the Eo-Alpine geodynamic stage (Late Cretaceous to Early Eocene) are coded with the letters AL and numbered progressively from SW to NE (AL-1, 2, 3, 4 and 5). AL-X, at the SW-extremity of line M12A, indicates an Alpine fault that was likely reactivated during the Apennine phases. Similarly, the most important crustal blocks identified in the seismic interpretation are progressively numbered from SW to NE. Crustal blocks of the Adria plate LC are numbered within white circles (1 to 14), while those of the upper crust are numbered within white squares (1 to 17).

All the major lithospheric faults, thrust faults and shear planes of the trans-Apenninic CROP section are associated with the Apennine thrust-belt construction (Late Balearic–Tyrrhenian stages), and are coded with letters AP and numbered progressively from SW to NE (AP-1, 2, 3, 4 and 5).

3.1. Ligurian–Balearic Sea and Corsica–Sardinia Block

The new CROP seismic sections M10 (Plates 1 and 2, and Figure 3) and M11 (Plates 3 and 4, and Figure 4) explore the Ligurian sphenochasm, the northeastern extremity of the Balearic basin (Algero-Provençal or Western Mediterranean). Interpretation of these sections shows a marked difference in tectonic style between the Ligurian Sea sector to the west and the Corsica–Sardinia block–northern Tyrrhenian one to the east (Plates 1–4). The western sector of the lines crosses the northern part of the Balearic basin, dominated by extensional structures that propagate from the crust and upper mantle within the cover which controls the synrift deposition. Conversely, the eastern sector is characterized by thrust faults affecting the lithosphere and the sedimentary cover. The subordinate extensional structures occurring at the Corsica–Adria collisional zone of section M10 (Plate 2 and Figure 3) is connected to the Tyrrhenian opening process and driven by the southeastward mantle flux. Both sections (Plates 2 and 4) illustrate very well the crustal tectono-dynamic conditions of the area and the pre-rift, syn-rift and post-rift stratigraphy of the Ligurian Sea, providing new information about lithospheric structures. Crustal opening is very clear on both sections, with section M10 (Plate 2 and Figure 3) favourably oriented to give a complete seismic imaging of the Balearic oceanic crust and of the Provençal and Corsica margins. In particular, the axial zone of the northern Balearic basin is characterized by oceanic crust that is delimited from both the Corsica–Sardinia block and European margins by normal faults systems. The rifted western Corsica margin is characterized by listric faults extending over a wide area (Plates 2 and 4). These faults are controlled by a major west-dipping detachment fault

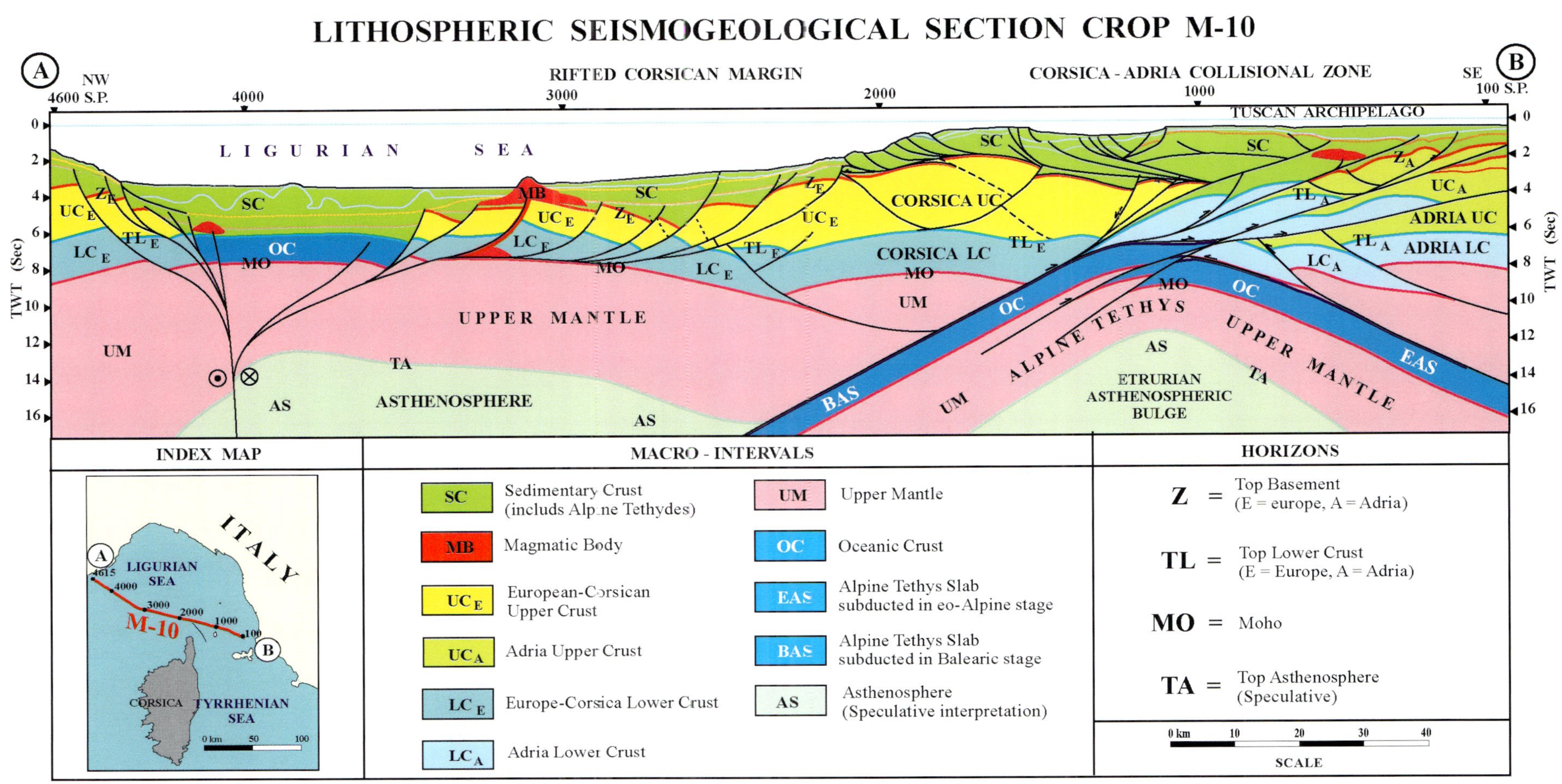

Figure 3: Simplified version of the interpreted seismic section CROP M10 of Plate 2. This shows the Ligurian sphenochasme opening, the rifted Corsican margin, the collisional tectonic deformation between Corsica and the NW Adria margin, and the impressive Etrurian asthenospheric bulge confined by two Alpine Tethys slabs subducted in two different geodynamic stages.

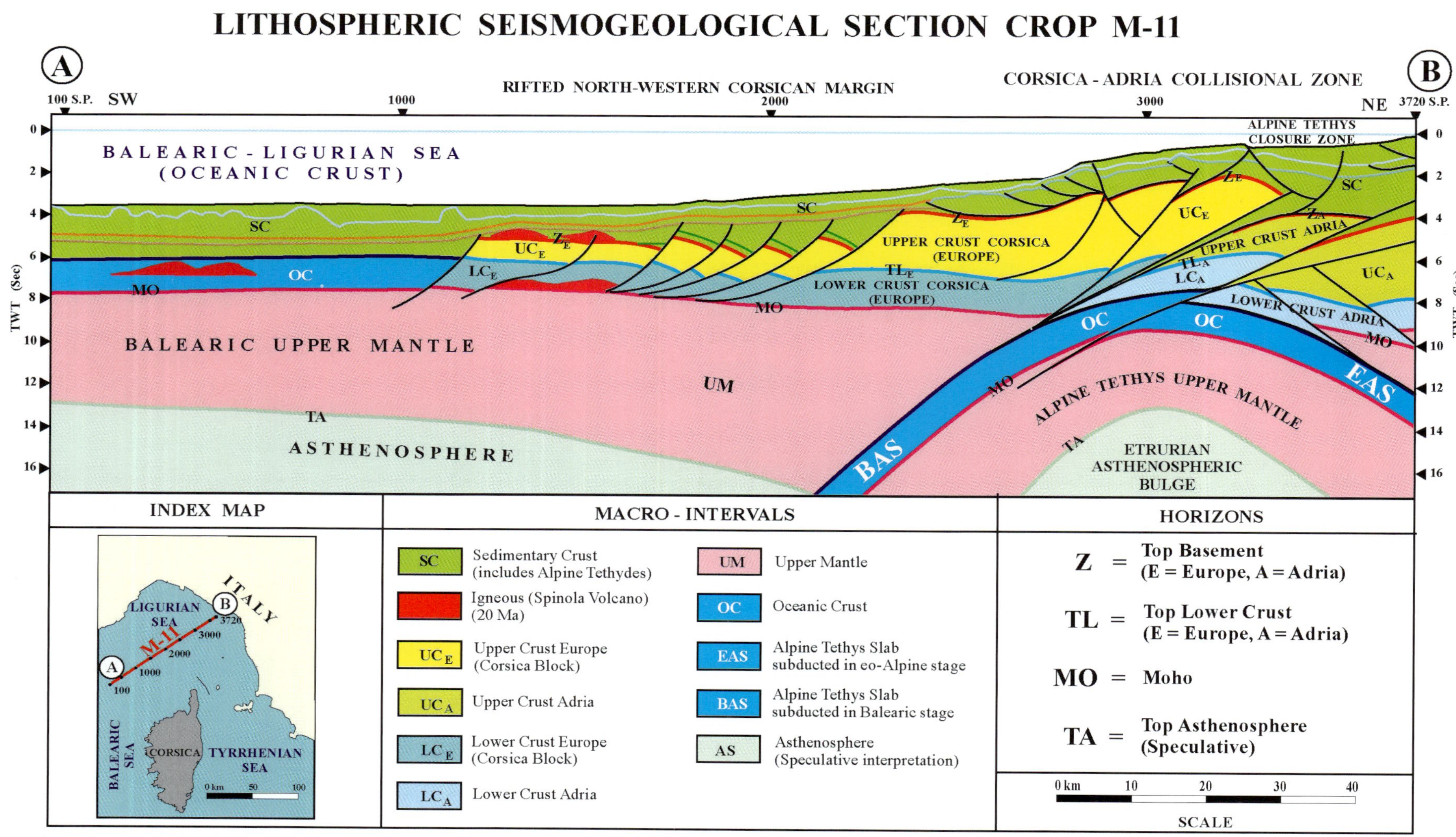

Figure 4: Simplified version of interpreted seismic section CROP M11 of Plate 4. The Corsica–NW Adria collisional tectonics, the rift tectonics of the Corsica margin related to the Balearic back-arc ocean basin opening, and the impressive Etrurian asthenospheric bulge are shown.

cutting into the upper mantle and flattening upsection along the lower crust-upper mantle interface. The European margin (Ligurian escarpment; Plate 2) is instead narrow and characterized by steeper and fewer listric normal faults. The CROP-reconstructed Ligurian opening structures are very clearly associated with a negative flower structure where a notable left transcurrent thrusting took place.

The timing of tectonic activity affecting both margins is well constrained by the architecture of the sediments. Within the sedimentary cover, the pre-rift (PR: Mesozoic–Lower Oligocene), syn-rift (SR: Upper Oligocene–Middle Miocene) and post-rift sequences (Middle Miocene to Present) can be discerned. The pre-rift sequence is rather thin or missing. The syn-rift to syn-spreading succession shows a marked wedging of sediments against fault planes, due to the syndepositional activity of the normal faults. The post-rift succession is characterized by mostly flat layers of Tortonian age overlain by Messinian salt thickening notably on the basin floored by the oceanic crust, and forming well developed salt domes. Nevertheless, some of the listric faults cut the post-rift succession and the Messinian-Quaternary deposits, pointing to a later fault reactivation stage (Plate 2).

On this basis, the northern Balearic basin evolution can be framed into two main stages: (1) a former, mainly extensional stage (Late Oligocene–Middle Miocene) giving rise to an asymmetric fault system, and (2) a successive stage of predominant transtension (since the Tortonian) mainly along the European margin. The occurrence of earthquakes mostly on the western (European) oceanic crust boundary (Bethoux et al., 1988) indicates that ongoing deformation concentrates on this margin, whereas only minor activity is associated with the eastern (Corsica) oceanic crust boundary. In the first extensional stage, magmatism at the Mt. Spinola Volcano (dated at 18 Ma by Rehault et al., 1984) was coeval with the processes that generated the Balearic basin oceanic crust (21–16 Ma, see above). The magma chamber feeding the Mt. Spinola volcano is imaged at the footwall of a major listric normal fault at the margin of the extension zone (Plates 2 and 4). The border basin diapirs of Messinian salt which developed above the Balearic oceanic crust are located over normal faults in the substratum (Plates 2 and 4). The central diapirs seem very likely due to salt, and to the tendency of Lower Pliocene cover to collapse from both border zones toward the central basin during the deepening by thermal cooling process. Halokinetic mobilization started evidently in the terminal Late Pliocene. The Plio-Quaternary sediments not affected by halokinetic deformation terminate with a rather flat sea floor.

Regarding interpretation of the Corsica–Sardinia block–Northern Tyrrhenian sector, the contractional structures superposing the Corsica–Sardinia block onto the Adria Plate margin (along fault AP-1) are still preserved because of the limited effects of the Tyrrhenian basin opening (Plates 2 and 4), which are, however, enhanced southwards. In the seismic profiles, the oceanic Tethyan slab representing the fossil, east dipping subduction of the Alpine Tethys below the Adria Plate (EAS on Figures 3 and 4) is imaged (Finetti et al., 2001), coupled with a west-dipping slab (BAS on Figures 3 and 4) representing the Alpine Tethys slab subducted during the Balearic stage (Plates 2 and 4; Finetti, this volume, Chapter 34). East-dipping thrust faults are likely associated with the former subduction phase and affect the whole Adria lithosphere. These structures were successively truncated by the major fault AP-2, which is rooted below the Corsica–Sardinia block and represents the sole thrust of a major crustal thrust system in the Tyrrhenian–Northern Apennines hinterland area (Plates 2 and 4).

At the scale of the sedimentary cover, the Alpine Tethys units (including the Schistès lustrèes units) appears to be detached westward along the basement-cover interface (horizon Z in Plates 2 and 4) with numerous thrusts splaying from this basal detachment. The geometry of this structure presents great similarities with intercutaneous wedging, termed "passive-roof duplex" by Banks and Warburton (1986). In this interpretation the passive-roof thrust carrying the Ligurian units accommodates the closure of the western Alpine Tethys remnant. The lower limit for the age of this structure is constrained to the post Langhian (post 16 Ma) because the leading edge of the passive-roof thrust involves sediments of this age deposited at the eastern Balearic Basin margin (shot point 2100; Plates 1 and 2). Sections M10 and M11 image an impressive asthenospheric bulge confined between the two reconstructed Alpine Tethys slabs "EAS" and "BAS" (Plates 2 and 4; Figures 3 and 4).

3.2. Corsica Basin

The Corsica basin develops as zone of convergence between NW Adria and the Corsica block (Figures 2 and 17). This basin is sub-longitudinally crossed by the CROP M12B section, whereas its northeastern margin is shown on transverse profile CROP M12A (Figures 1, 2, 5, 6; Plates 5, 6, 7, 8).

Interpretation of these CROP sections shows that the Corsica basin is located above the collision zone between the Corsica block and the north-western margin of the Adria Plate, implying that the Tuscan archipelago represents

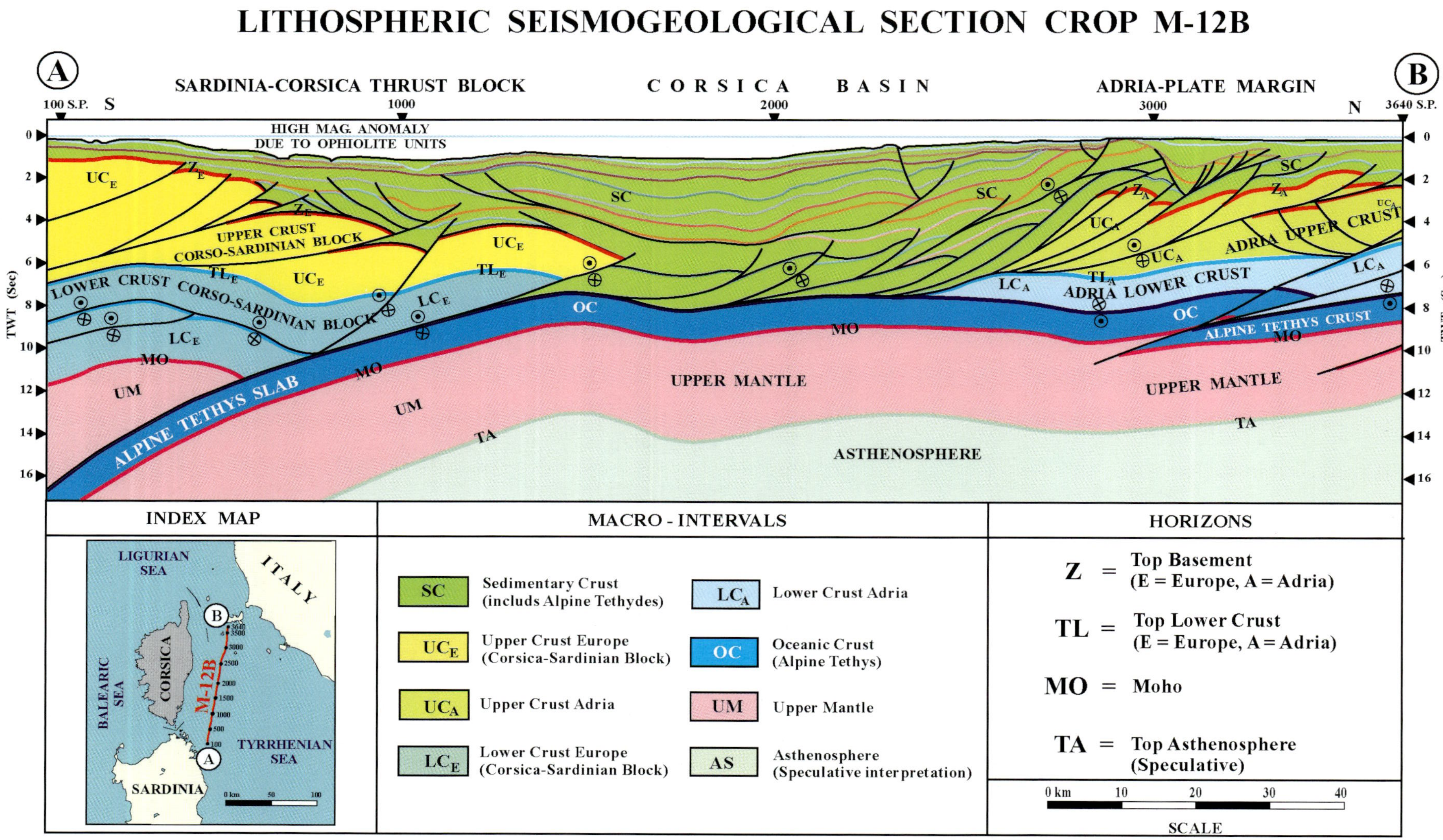

Figure 5: Simplified version of the interpreted seismic section CROP M12B of Plate 6. After the lateral subduction of the Alpine Tethys beneath NW Adria in the Eo-Alpine stage (see section segment from S.P. 2400 to 3640), the remnant oceanic crust lying west of the Alps (see section segment from S.P. 100 to 1600) subducted under the European margin (Corsica block) during the Balearic stage. Since section CROP M12B is subparallel to the Corsica basin–Adria plate boundary, the Eo-Alpine subduction of the Alpine Tethys oceanic crust beneath Adria is visible only at the northern end of the section, where it shows the topmost sub-horizontal structural culmination.

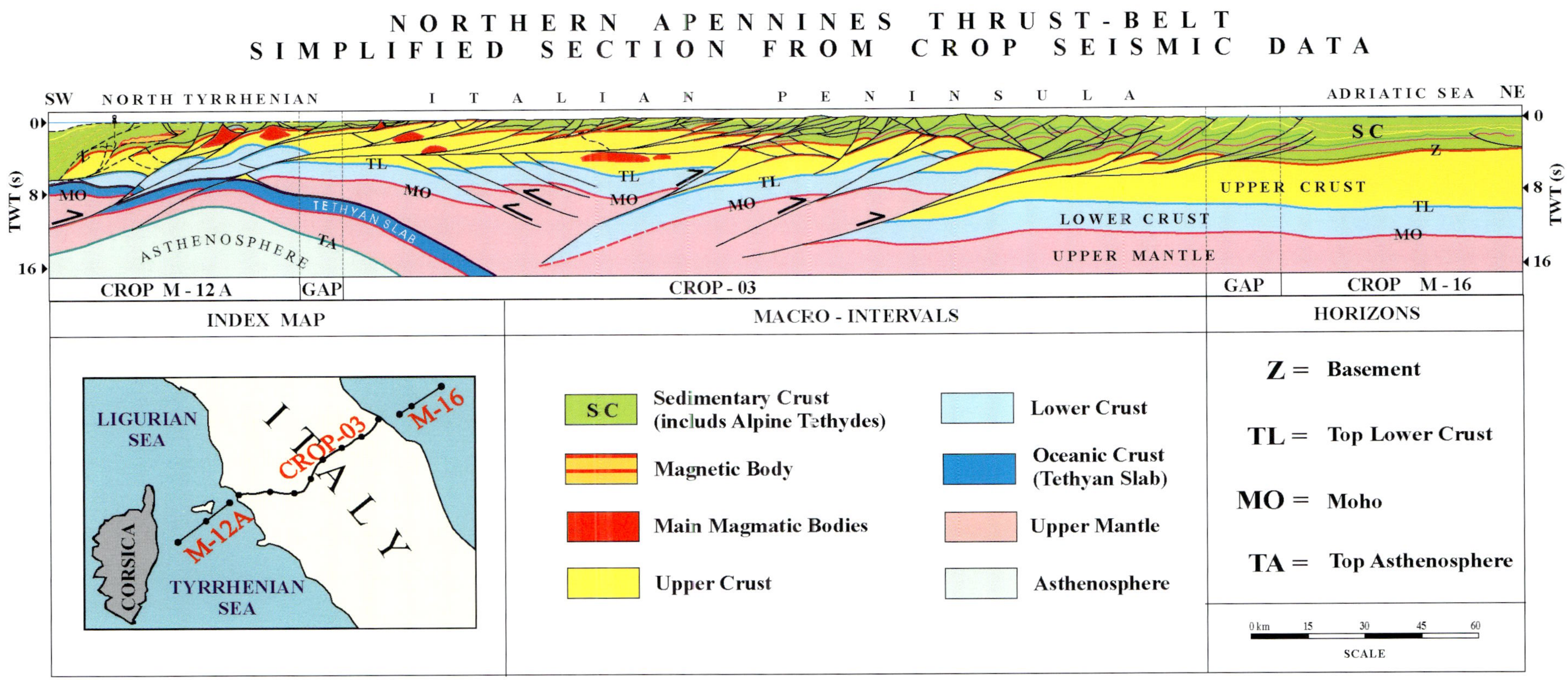

Figure 6: Simplified version of the interpreted trans-Apenninic composite seismic section CROP M12A/CROP 03/M16 of Plate 8. The east-dipping Tethyan slab subducted laterally beneath the north-western Adria margin during the Eo-Alpine stage. In the Balearic (Upper Oligocene–Lower Miocene) and Tyrrhenian stage (Middle Miocene to Present) the lithospheric E-verging NA deformation took place. The Etrurian asthenospheric bulge is well exhibited as a compressive feature confined between the E-dipping Alpine Tethys slab and the W-dipping Alpine Tethys slab of Balearic subduction.

the innermost deformed sector of the Adria Plate (Figures 5 and 6; Plates 6 and 8). North of Corsica Island, the collision zone is marked by overthrusting of the Corsica block continental crust over the continental crust of the Adria Plate (Plates 2 and 4). Interpretation of CROP M12B, which is sub-parallel to the tectonic trends, reveals that the Alpine Tethys units are tectonically interposed between the Corsica block and the Adria continental crust (Figure 5 and Plate 6). Remnants of the oceanic Tethyan crust below the Corsica basin are also imaged. Note that the large extent of the Alpine Tethys units is apparent, as the profile M12B runs almost parallel to the trends of the major structures (Figures 2 and 17; Plate 6).

Along the sub-longitudinal CROP M12B section, the Corsica basin attains a broad synformal structure, reaching a maximum sediment thickness of around 5 sec. TWT in the axial zone (Figure 5; Plate 6). Section M12A reveals instead a marked wedging of sediments that sharply increase their thickness basinward, reaching a maximum value of about 7 sec. TWT (Plate 8). Various seismostratigraphic units have been distinguished within the basin fill, whose age is well constrained by the Martina 1 well located close to the intersection point between sections M12A and M12B (Figure 2). In particular, ages of the deposits filling the Corsica basin range from Oligocene to Late Miocene. Pliocene-Quaternary sediments extending also outside the area of the basin overlie the Corsica basin deposits.

The basin fill represents in effect a closure wedge of the Alpine Tethys (Plates 6 and 8). The northern part of the Corsica basin (between Pianosa and Capraia islands) is well imaged by the transverse deep seismic profiles presented in Mauffret et al. (1999). Here, basin deposits show a clear synformal structure with wedging of sediments.

3.3. Etrurian Asthenospheric Bulge

One of the most geodynamically important features revealed by interpretation of sections CROP M10, M11 and M12A, is the existence of an impressive prominent asthenospheric bulge (Plates 2, 4 and 8, Figures 3, 4 and 6). This asthenospheric high culminates below the Corsica–Adria collisional zone at about 9 sec. (TWT) in section M12A (Figure 6, Plate 8), at about 12 sec. (TWT) in section M10 (Figure 3, Plate 2) and at about 13 sec. (TWT) in section M11 (Figure 4, Plate 4), thus showing a marked northward plunge. This high is coherently accompanied by fairly reliable reflection data. The inferred asthenosphere culmination is in substantial agreement with surface wave dispersion results, although the horizon "TA" must not be regarded as a precise seismic limit but as an approximate mantle – asthenosphere boundary of correct position for the culmination.

This bulge forms an NA-like arc in the Tuscan archipelago and is delimited by two oceanic slabs of the Alpine Tethys subducted during two different geodynamic stages. While CROP M12A images the culmination beneath Elba Island and the slab that subducted the northwestern margin of the Adria Plate in the Eo-Alpine stage, sections CROP M10 and M11 show also the western slab of the Alpine Tethys subducted in the Balearic stage. The origin of this bulge is related to compression of the asthenosphere confined between the two Alpine Tethys slabs (BAS and EAS in Figures 3 and 4) due to the eastward retreat of the Alpine Tethys slab during the Balearic stage (Finetti, this volume, Chapter 34).

This interpretation is supported by the observation that the asthenospheric bulge formed only in the collision zone between the rotated Corsica block and the northwestern boundary of the Adria plate. The E-dipping Eo-Alpine slab does not propagate south of the 41° parallel because here slab retreat continued southeastward (Tyrrhenian stage) in the Ionian Neotethys which was fully communicating with the Alpine Neotethys. It seems therefore appropriate to denominate this primary structure the "Etrurian Asthenospheric Bulge" (Finetti, this volume, Chapter 34).

3.4. Northern Tyrrhenian–Northern Apennines Thrust Belt

The deep seismic data of CROP M12A-CROP03-M16 (Plates 7 and 8) allow investigation of the deep structures of the NA along a transverse lithospheric section extending from the Corsica basin in the hinterland to the Dinaric front in the foreland. The structural characteristics along this section have already been discussed in detail by Finetti et al. (2001). The interpretation of structures imaged on the profiles shows that the basic geometry of the NA is dominated by compressive lithospheric structures, with notable crustal shortening of the deformed Adria plate. The major Apennine thrust-faults AP1-5 nucleate in the Upper Mantle (UM) and offset the Moho boundary at high angle, then propagate at low angle upward into the crust and sedimentary cover, where they splay into numerous thrusts (Plate 8). The analysis of this transect is described here from west to east.

In the western sector, the transect crosses the eastern part of the Corsica basin and the Tuscan archipelago (CROP M12A). The general tectonic picture indicates a substantially similar setting to that imaged in CROP-M10 and CROP-M11, with major west-dipping thrust faults affecting the Adria lithosphere and the recognition of the east-dipping oceanic Tethyan slab (Plate 8). The well evident asthenospheric compressive bulge formed by progressive eastward migration of the Alpine Tethys slab retreat (Balearic stage). In Late Balearic–Early Tyrrhenian stage, bulge-related eastward mantle drag determined the crustal shortening and thrust deformation of the internal Apennines. Successively, when the Southern Apennines–Calabrian Arc dynamics (Ionian slab retreat related) propagated into Ionian Neotethys (south of the 41° parallel), the greater arc of the Balearic stage was dissected along the 41° parallel (boundary Ionian Neotethys–NW Adria), separating the kinematism of the NA from that of the SA (Southern Apennines). While Etrurian bulge-related mantle flux continued to drive the North Apenninic thrust belt deformation eastward also in the Tyrrhenian stage, the southeastward Ionian slab retreat also perturbed the NW Adria lithosphere, north of the 41° parallel (Finetti, this volume, Chapter 34). This perturbation created a NW–SE-trending, right transcurrent and transtensive flower on the triangular offshore Tyrrhenian area extending from Elba Isle, Corsica Basin and the 41° parallel (Finetti and Del Ben, this volume, Chapter 21). In the Corsica basin and bordering Adria plate margin, the Late Tyrrhenian stage partially inverted the displacement of previously formed thrust faults. The extensional effects proposed in some of the literature on the Tuscan Apennines (Liotta and Ranalli, 1999), turn out to be on CROP sections and in surface geology a second order tectonic deformation representing transtensive structures of trans-Apenninic transcurrent faults (i.e. Larderello and Mt. Amiata areas; Figure 13), or local accommodating elements of a fundamental crustal contraction. The western extremity of section M12A is characterized by the very thick and well-reflecting sedimentary sequence of the Corsica basin (see Section 3.2). To the east, two major batholiths outcropping on Elba (the Mt. Capanne and Mt. Calamita batholiths) are imaged in profile CROP M-12A as buried bodies at shallow depth within the upper crust. These batholiths are flanked by normal faults generated by the collapse of the sedimentary sequence over the magmatic bodies.

Further east, in western Tuscany, the CROP03 profile shows the major thrust AP2 running sub-horizontally within the upper crust as far as the Mt. Cetona area, where it corresponds at the surface to the Macigno thrust front (S.P. 4150 in CROP03; Plates 7 and 8; Figure 2). At depth, this important discontinuity locally corresponds to the so-called K-Horizon (Finetti et al., 2001). The hangingwall of AP2 is characteristically deformed by splay thrusts emanated from the major basal detachment and propagating upsection within the sedimentary cover. This deformation gives rise to the thrust anticlines typically bounding the hinterland basins that formed from the Late Miocene (see below Section 4.1).

The west-dipping lithospheric thrust fault AP3 generates a prominent Moho offset below the Mt. Amiata volcano (seen as a buried magmatic body below its outcropping extrusion), as well as thrust faults AP-4 and AP-5 that cause another significant Moho displacement more eastwards, approximately below the Tiber Valley (Plates 7 and 8). Most of the displacement is taken up by the most external lithospheric thrust fault AP-5 that produces a horizontal displacement of as much as 36 km and a vertical throw of over 16 km (Finetti et al., 2001). Forward propagation of deformation of AP-5 into the upper crust and sedimentary cover gave rise to a complex system of thrust faults.

At the main Apennine watershed, from Monterchi village (at about SP 5650 in section CROP03) to the Tiber Valley (about SP 6750), the crustal structure is characterized by a marked thickening due to tectonic stacking by the main lithospheric Apennine faults (AP-3, AP-3B, AP-4A, AP-4B, AP-4C and AP-5A; Plate 8). This geometry produces a marked basement culmination (SP 5650, basement at 1.6 sec TWT), above which the sedimentary cover is affected by large E-dipping normal faults resulting from gravity collapse at the base of Triassic evaporites of the Burano Fm. (GS-1, GS1A, GS2, GS3). The main E-dipping normal fault GS-1, which can be traced down to 5.6 s (about 12 km, SP 6750), affects the eastern part of the crustal culmination displacing earlier thrust faults (AP4 A-C in Plate 8). These normal faults represent a gravitational adjustment of the uplift produced by still-active thrusting processes since, at depth, these structures are seen to die out along the thrust fault AP-5A. Gravitational tectonics generated the basal sliding plane GS-1 and the associated gravity-sliding faults GS-1A, GS-2 and GS-3. These four fault planes truncated part of the whole sedimentary sequence of the thrust blocks associated with faults AP-3, 3B, 4A, 4B and 4C (Plate 8). There is clear seismic evidence that no distension occurs between the two slipping basement blocks of thrust fault AP-5, which is, on the contrary, still active as a compressive fault.

The main lithospheric thrust fault AP-5 propagates toward the foreland into two main fault segments, the fault AP-5-A and thrust AP-5-B (Plates 7 and 8). AP-5-A propagates directly into the overlying sedimentary cover, while AP-5-B propagates toward the foreland within the upper crust, splaying into three crustal thrusts (AP-5-B1, 5-B2, 5-B3), which in turn branch into the sedimentary cover. The resulting complex pattern of thrust faults has also controlled the tectono-sedimentary evolution of some thrust-top basins (the Tavullia and Montecalvo in Foglia basins), as well as the Pliocene-Quaternary foredeep basin delimited to the east by the most external AP-5-related splay thrusts (AP-5B3a-c) forming the Apennine front (Plate 7 and 8). Deformation of the Adriatic Sea bottom around thrust fault AP-5B-3a imaged on CROP M16 (around S.P. 1300 in Plate 8) demonstrates that the current Apennine thrust front is still active. Section CROP M-16 shows, at its eastern extremity, the compressive deformation in Oligocene-Miocene times related to the leading edge of the Dinaric belt, manifested by the gentle thrusting of the platform above the basinal succession.

4. SURFACE DATA

A geological field study along a wide transect extending from the hinterland areas to near the Adriatic margin was undertaken in order to observe directly the surface structures connected with those imaged in the profile CROP03, which is the only on-shore CROP section of the NA. This study allowed us to constrain the age of these structures, and to examine their progression along strike. In reporting the results of these studies we distinguish two main sectors in the NA fold-and-thrust belt: (1) the hinterland and (2) the main thrust belt. These two sectors are described separately below by analysis of some key areas exemplifying their tectonic evolution and structural setting.

4.1. A transect across the hinterland basins

In this section, we describe the structural setting of some important continental or marine basins in the NA hinterland. These basins sedimented on a substratum deformed during the Late Oligocene to Early Miocene phases that built an early Northern Apennines chain. This chain was characterized by a general duplex geometry, in which the Ligurian or Alpine Tethydes units represented the roof thrust sheet overlying the foreland Tuscan units affected by link thrusts connecting the basal décollement to the roof thrust itself. Successively, a number of such link thrusts were reactivated during later deformation phases, often controlling the development of hinterland basins (Boccaletti and Sani, 1998; Boccaletti et al., 1999).

Some of the hinterland basins crossed by profile CROP03 will be discussed in terms of sediments architecture, deformation of basin fill and related structures on the nearby substratum. These basins lie to the east of the mid-Tuscany metamorphic ridge (MTMR), and, from west to east, are the Cinigiano–Baccinello, Velona and Siena–Radicofani basins (Figure 7). The MTMR represents an outcropping crustal thrust exposing the metamorphic units and Palaeozoic basement units. Based on the stratigraphic position of unconformities, the Neogene–Quaternary successions of the basins have been outlined and regionally correlated throughout the whole hinterland area. In the interpretation, the stratigraphy of the basin fills has been re-organized in terms of UBSUs (Unconformity-Bounded Stratigraphic Units; Salvador, 1987), thus highlighting major evolutionary stages in the geological history of this sector.

The lowermost unit (1st UBSU) is represented by the so-called Ponsano Sandstone (Giannini and Tongiorgi, 1959) consisting of shallow-water arenites of Late Serravallian–Early Tortonian age (Foresi et al., 1997). In the study area, these deposits are exposed only in a small outcrop (near Rencine village in Figure 7) but were encountered by the well Radicofani 1 located in the Radicofani basin (location in Figure 7).

The following 2nd UBSU consists of a Late Tortonian–Messinian continental succession composed of lacustrine sandy-clays and interbedded conglomerate levels. The age of this unit is well constrained by radiometric age determinations (K/Ar), yielding a value of 8.07 ± 0.11 Ma, carried out on a tuff layer embedded in an equivalent continental succession to the Radicondoli–Volterra Basin (D'Orazio et al., 1995). This datum suggests an age of about 9 Ma for the base of the 2nd UBSU. In the Velona and Siena basins this unit can be differentiated into two sub-units separated by a minor unconformity (Bonini and Sani, 2002). The 3rd sedimentary unit is composed of a lower part consisting of Late Messinian reddish continental conglomerates (the Montebamboli Conglomerates) and an upper part composed of Early-Middle Pliocene marine clays, silty-clays, sands and conglomerates. The

Figure 7: Schematic geological-structural map of the Cinigiano, Velona and Siena–Radicofani hinterland basins (location also in Figure 2). Locations of CROP03 and other hydrocarbon exploration seismic lines and are plotted in black. Structures indicated with dashes refer to buried or blind faults.

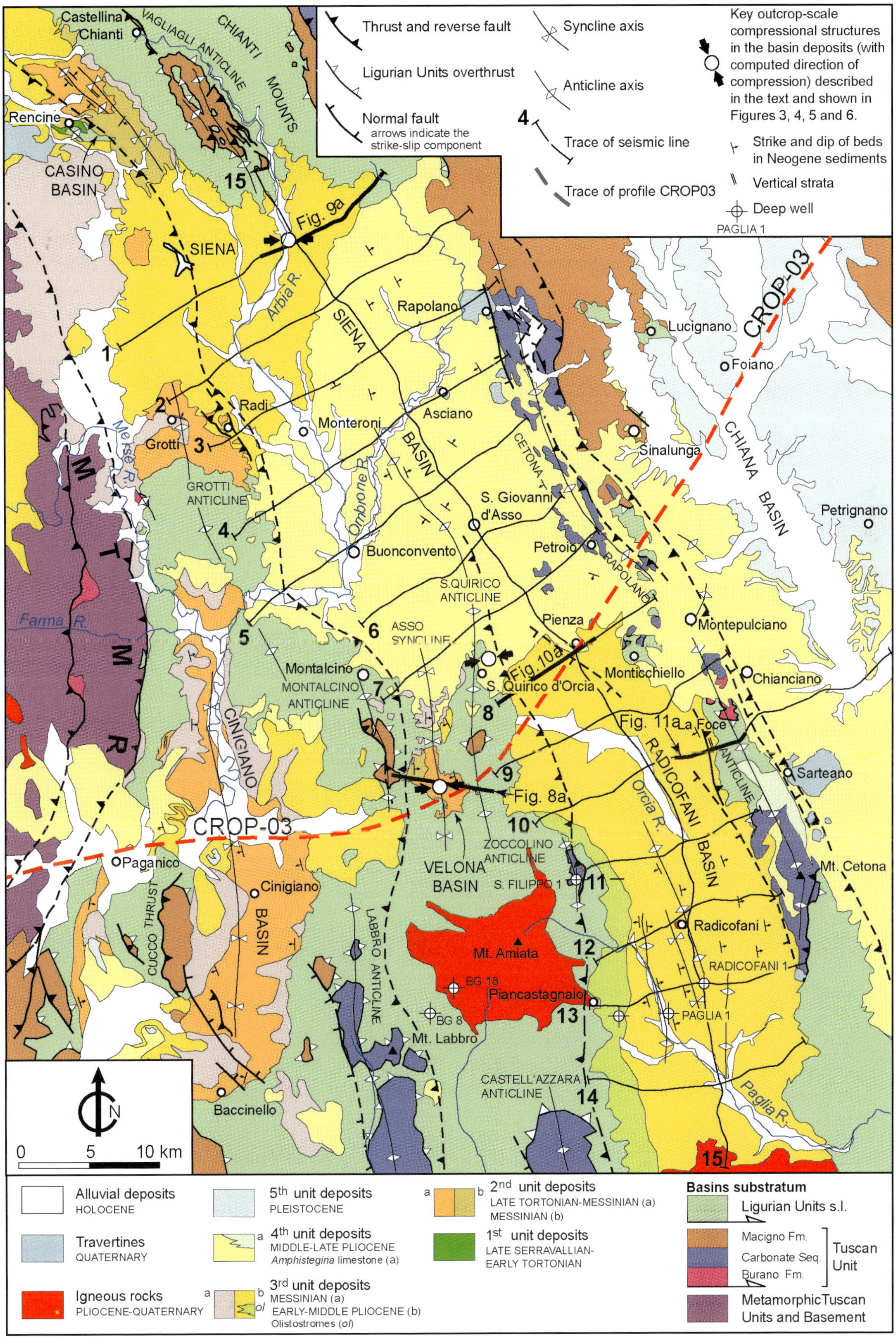
Thrust and reverse fault
Ligurian Units overthrust
Normal fault
arrows indicate the strike-slip component
Syncline axis
Anticline axis
Trace of seismic line
Trace of profile CROP03
Key outcrop-scale compressional structures in the basin deposits (with computed direction of compression) described in the text and shown in Figures 3, 4, 5 and 6.
Strike and dip of beds in Neogene sediments
Vertical strata
Deep well
PAGLIA 1
Castellina Chianti
VAGLIAGLI ANTICLINE
CHIANTI MOUNTS
Rencine
CASINO BASIN
SIENA
Fig. 9a
Arbia R.
SIENA BASIN
Rapolano
Lucignano
Foiano
CROP-03
CHIANA BASIN
Radi
Grotti
Monteroni
Asciano
Sinalunga
Merse R.
GROTTI ANTICLINE
Ombrone R.
CETONA
S. Giovanni d'Asso
Petrignano
Buonconvento
Petroio
RAPOLANO
S.QUIRICO ANTICLINE
Farma R.
Pienza
Montepulciano
ASSO SYNCLINE
Montalcino
MONTALCINO ANTICLINE
Fig. 10a
Monticchiello
S. Quirico d'Orcia
Chianciano
CINIGIANO
Fig. 11a
La Foce
RADICOFANI BASIN
ANTICLINE
Sarteano
Fig. 8a
Orcia R.
CROP-03
ZOCCOLINO ANTICLINE
Paganico
VELONA BASIN
S. FILIPPO 1
Mt. Cetona
Cinigiano
CUCCO THRUST
BASIN
LABBRO ANTICLINE
Mt. Amiata
Radicofani
RADICOFANI 1
BG 18
Piancastagnaio
BG 8
PAGLIA 1
Mt. Labbro
CASTELL'AZZARA ANTICLINE
Baccinello
Paglia R.
N
0 5 10 km
Alluvial deposits
HOLOCENE
Travertines
QUATERNARY
Igneous rocks
PLIOCENE-QUATERNARY
5th unit deposits
PLEISTOCENE
4th unit deposits
MIDDLE-LATE PLIOCENE
Amphistegina limestone (a)
3rd unit deposits
MESSINIAN (a)
EARLY-MIDDLE PLIOCENE (b)
Olistostromes (ol)
2nd unit deposits
LATE TORTONIAN-MESSINIAN (a)
MESSINIAN (b)
1st unit deposits
LATE SERRAVALLIAN-EARLY TORTONIAN
Basins substratum
Ligurian Units s.l.
Macigno Fm.
Carbonate Seq.
Burano Fm.
Tuscan Unit
MetamorphicTuscan Units and Basement

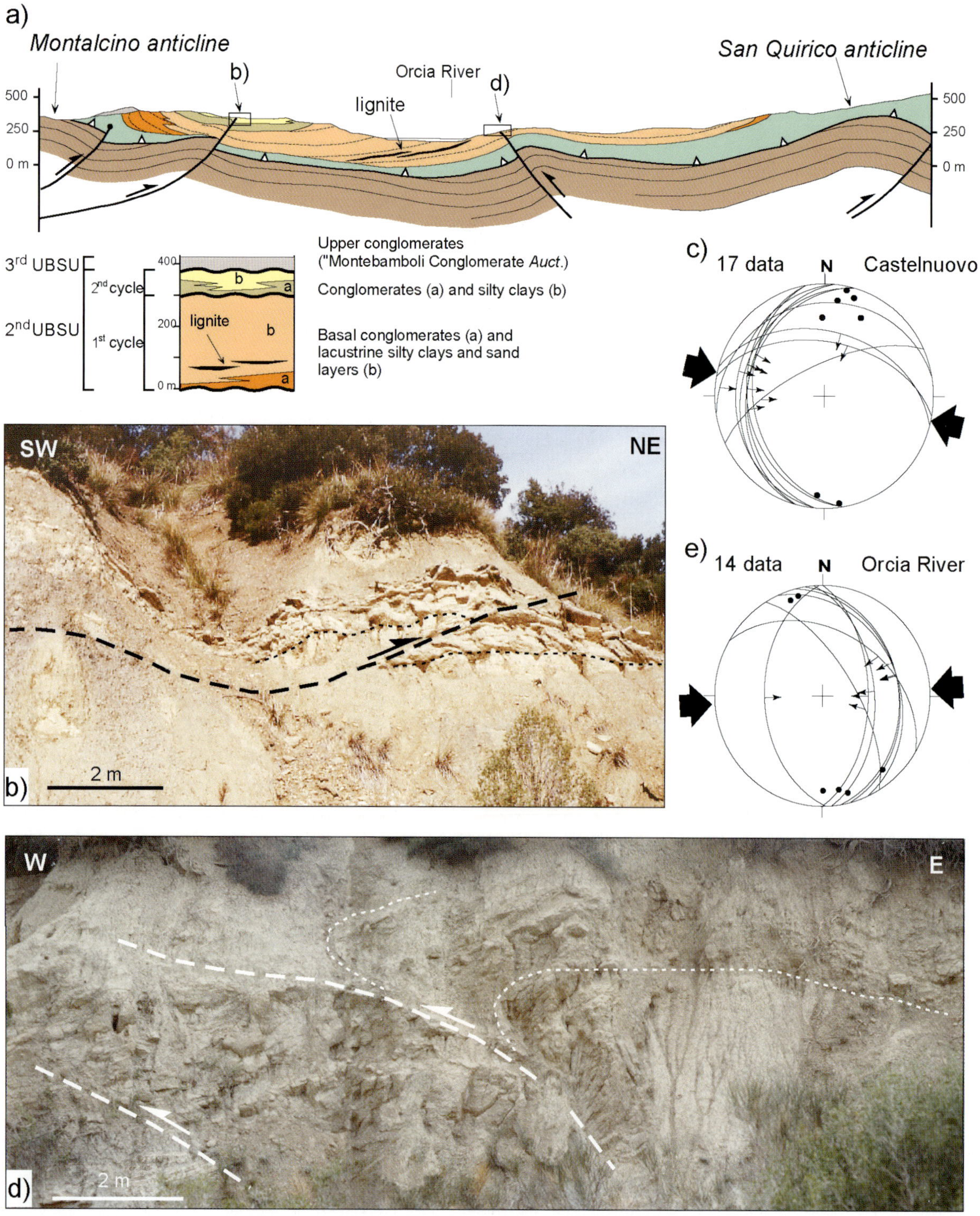

Figure 8: (a) Geological cross-section and stratigraphic column of the Velona basin. (b) Outcrop-scale reverse fault in the Messinian (2nd UBSU, 2nd cycle deposits) and (c) the stereonet of the corresponding mesoscopic structures plotted on a lower hemisphere (Schmidt net projection). (d) Outcrop-scale thrust related folds and reverse faults in the Messinian (2nd UBSU, 1st cycle deposits) and (e) the stereonet of the corresponding mesoscopic structures plotted on a lower hemisphere (Schmidt net projection). Black arrows indicate the direction of compression determined by the inversion methods of Angelier and Mecheler (1977) and Carey (1979). Reverse faults are indicated by cyclographic traces and arrows, folds axes are indicated by dots. The positions of the outcrops (b) and (d) with respect to the structures associated with the major thrust anticlines deforming the basin are reported in (a). Colours of the units are as in Figure 7.

4th sedimentary unit deposits consist mainly of Middle to Late Pliocene marine clays and sands, and outcrop extensively in the Siena basin (Figures 7 and 8). The deposits of the 5th UBSU consist of fluvio-lacustrine sediments of late Villafranchian age. These deposits outcrop extensively in the Chiana basin and occur as small-scattered outcrops in the central-southern part of the Cinigiano basin (Figure 7).

The Cinigiano basin is bordered by the mid-Tuscany metamorphic ridge to the west, and by the Montalcino–Monte Labbro ridge to the east, which corresponds to a thrust anticline possibly involving the basement (Bally et al., 1986; Figure 7). The occurrence of progressive unconformities (*sensu* Riba, 1976) on the eastern basin margin is consistent with syndepositional activity of the Montalcino–Labbro anticline. Field mapping combined with structural analysis carried out in the Cinigiano basin fills allowed a distinction between four different compressive phases characterized by different stress field orientations (Landi et al., 1995). Some reverse faults and folds have been found both at the mesoscale and macroscale, mostly close to the eastern basin margin (where the beds frequently become vertical). These compressive deformations mainly affect the deposits of the 2nd and 3rd UBSUs, while the 4th UBSU deposits are only slightly deformed (Figure 7). Normal faults in the Baccinello area exhibit an oblique direction to the basin trend showing no relation to sedimentation.

The NNW–SSE-oriented small continental Velona basin is located between two main ridges corresponding to E-verging thrust anticlines, the Montalcino anticline to the west and the S. Quirico anticline to the east (Figure 7). These thrust anticlines involve both the Ligurian Units and the underlying Tuscan succession and separate the Velona basin from the Cinigiano basin to the west and from the Siena–Radicofani basin to the east (Figure 7). The Velona basin has been previously interpreted as a half-graben with the master fault controlling the sedimentation located along its western margin (Martini and Sagri, 1993; Rook and Ghetti, 1997). The general setting of the deposits is a gentle syncline, but no large normal faults have been found to affect the basin fill (Bonini et al., 1999, 2001b). The main synformal structure is complicated by the occurrence of folds affecting both the 2nd UBSU sediments and the substratum (Figure 8(a)). Mesoscopic compressive structures mostly concentrate near the basin margins and are well developed both in the first and second cycle deposits distinguished within the 2nd UBSU. An E–W-to ESE–WNW-trending direction of compression coherent with the trend of the major thrust anticlines affecting the basin was found at all sites of structural measurement distributed over the Velona basin (Bonini et al., 1999). These deformations are exemplified by the outcrop-scale contractional structures consisting of E-dipping reverse faults and associated thrust-related folds affecting the 2nd UBSU (1st cycle) deposits at the Orcia riverbed (Figures 8(b)–(e)). The younger 3rd UBSU deposits are not affected by folding as they blanket the older deposits through a nearly planar surface. However, above the western Velona basin margin this erosional surface is clearly dipping toward the west, a setting that may indicate residual activity of the Montalcino thrust anticline during this period (Figure 8(a)).

The Siena–Radicofani basin is perhaps the major hinterland basin; it can be subdivided into two sub-basins, the Siena basin to the north and the Radicofani basin to the south (Figure 7). A roughly NNW-trending thrust fault bounds the western Siena basin. This structure is manifested at surface by the occurrence of aligned thrust anticlines, such as the Grotti, Montalcino and Labbro anticlines. Thus this structure can be traced southward to delimit the western Velona basin margin (Montalcino anticline) and connected further south to the Mt. Labbro structural high (Labbro anticline; Figure 7). The same structure can be reasonably hypothesized to delimit the western Casino basin margin (northwest of Siena). In this hypothesis, the thrust fault would has been responsible for lifting the 1st UBSU deposits (Ponsano Sandstones) to the surface (Rencine area), as well as for inducing the relevant compressional deformations (mostly overturned folds) affecting the Casino basin fill described in Lazzarotto and Sandrelli (1977).

Similarly, the western Radicofani basin margin is delimited by a thrust fault whose continuity is suggested by the outcropping of thrust anticlines parallel to the margin (i.e., the S. Quirico, Zoccolino and Castell'Azzara anticlines; Figure 7). This thrust anticline continues, more or less with the same trend, into the axial part of the Siena basin and can be traced up to the Vagliagli anticline delimiting the Casino basin to the east. Both the Siena and Radicofani basins are delimited to the east by the Cetona–Rapolano thrust anticline that is delineated by aligned cores of pre-Neogene substratum (mainly Tuscan and Ligurian units; Figure 7). At depth, the Cetona–Rapolano anticline corresponds to a thrust involving the upper crust and also transferring deformation further east (Plates 7 and 8).

The tectono-sedimentary evolution and structural style of the Siena–Radicofani basins has been reconstructed by analysis of good quality industrial (hydrocarbon exploration) seismic profiles (location in Figure 7; Bonini and Sani, 2002). In that study, the geometry of the Siena basin was described as a major synformal structure affected by an axial structural high that has been interpreted as the continuation into the basin of both the Vagliagli and S. Quirico thrust anticlines (Figure 7). This structure isolated to the west a mostly buried basin, named the Buonconvento basin, whose sediments are currently exposed only at the Velona and Casino basins. The absence of asymmetry and wedging of sediments toward the basin margins argues against the hypothesis proposed by Costantini et al. (1982) that a normal fault system controlled sedimentation along the eastern Siena basin margin. The normal fault affecting the eastern Siena basin margin in the Rapolano area does not exhibit any relevant seismic expression, merely representing a recent and/or active deformation (see Hancock et al., 1999; Figure 7). The sediment architecture observed both in the seismic profiles and in the field, characterized by pinch-out of sediments toward the basin margins, suggests that the Late Miocene–Pliocene evolution of this basin was controlled rather by the activity of the marginal or internal thrust anticlines (Bonini and Sani, 2002). Syn-depositional uplift of these structures caused the development of angular unconformities, especially close to the basin margins.

Based on available seismic lines, a similar reconstruction has also been proposed for the Radicofani basin. The main difference from the Siena basin is the recognition of a major normal fault along the eastern northern Radicofani basin margin controlling sedimentation during deposition of the 2nd UBSU. Since this fault is visible in lines 11 and 12 only, it should represent a local, not tectonically significant element. Similarly to the Siena basin, the Cetona–Rapolano anticline is limited to the west by an active post-sedimentation normal fault that propagates obliquely into the basin fill from Mt. Cetona to Pienza. The axial thrust anticline in the Siena basin delimits the western margin of the Radicofani basin. In the axial part, gentle folds affecting the Radicofani basin deposits are related to lateral shortening.

The hypothesis of Acocella et al. (2002) relating the development of the axial anticline in the Radicofani basin to the emplacement of a laccolith associated with the Radicofani volcanic centre (age 1.3 Ma; D'Orazio et al., 1991) contrasts with several pieces of evidence: (1) the magmatic body was emplaced after the Messinian-to-Pliocene contractional phases generating the folding of the basin fill; dating of such phases is constrained by the synshortening sedimentation observed in the field and imaged in the seismic profiles crossing the Radicofani basin. (2) The small and local Radicofani volcanic centre is unlikely to have produced the 10–15 km-long axial anticline; in addition, there are two other anticlines in the basin fill parallel to the axial anticline: the claimed existence of a buried magmatic body below the basin axis has been identified neither in the seismic profiles nor in the deep wells Radicofani 1 and Paglia 1, the former very close to the surface termination of the axial anticline, and the second in the middle of the axial line of a 10-km long anticline (Figure 7). (3) The decrease in layer dip toward the bottom taken as evidence for magma inflation can be explained in other ways, for instance (a) by reverse faulting superposing steep strata onto sub-horizontal strata; (b) synshortening intrabasin detachment along ductile layer; (c) lateral (sub-)depocentre migration during basin contraction; and (d) post-sedimentary rotation of angular unconformities. In our opinion, the emplacement of the magmatic body at Radicofani was instead a passive process in the sense that the magma rose at the core of one or more pre-existing (axial) anticlines representing relative low-pressure zones.

In summary, seismic, stratigraphic and structural data obtained in the Siena–Radicofani basin indicate that this basin formed and evolved in a compressional setting resulting from the syn-(or post)-depositional activity of marginal and internal thrust anticlines. Activity of these thrust anticlines can be traced from the deposition of the 1st UBSU sediments in Late Serravallian–Early Tortonian times until (or post) the deposition of the 3rd or 4th UBSU deposits (depending on the local setting) in the Early to Late Pliocene. Analysis of mesoscopic deformation yielded a direction of compression trending around E–W, which is consistent with the trend of thrust anticlines delimiting the Siena–Radicofani basins or affecting the basin fill (Bonini and Sani, 2002). In the case of favourable outcropping conditions, well-developed outcrop-scale compressive structures have usually been detected on the major structures involving the basin fill and, in any case, the substratum. Compressional deformations are also imaged in the interpreted line drawings of the seismic lines, allowing good correlation between deep and surface structures. This correlation is exemplified below:

Seismic line 1 (line drawing) shows a W-dipping reverse fault affecting the Tuscan and Ligurian units in the northern Siena basin, and folding the basin-substratum interface (Figure 9(a)). Propagation of deformation into the basin deposits is supported by the outcrop-scale reverse faults representing back thrusts associated with the deeper faults imaged in the seismic profile (Figures 9(b)–(d)).

Seismic line drawing 8 shows the eastern limb of the E-verging S. Quirico thrust anticline that overthrusts the 2nd and (partially) the 3rd UBSU sediments (Figure 10(a)). This interpretation is supported by the well-developed compressional deformation consisting of thrust-related folds and reverse faults affecting the 4th UBSU sediments

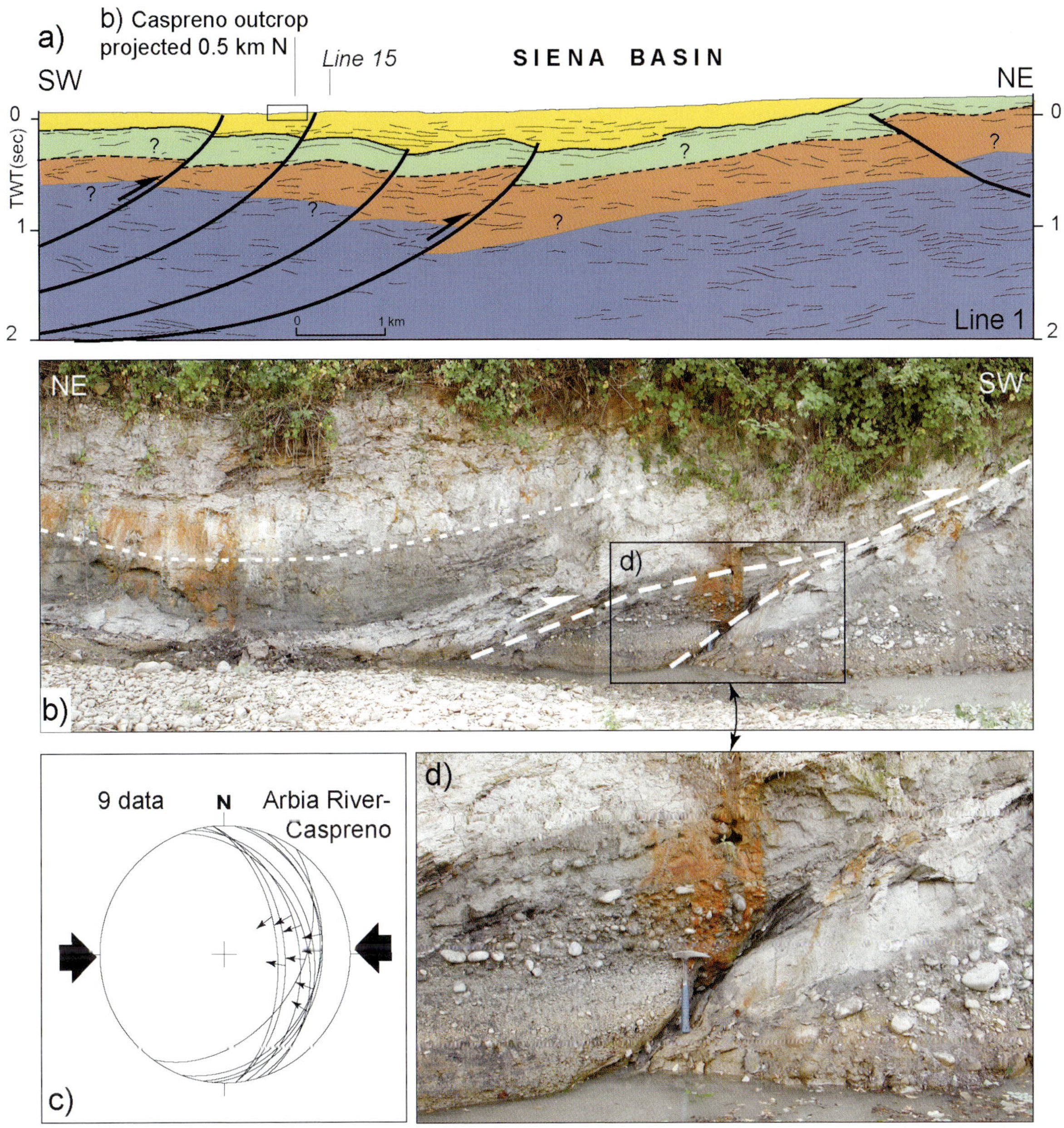

Figure 9: (a) Interpreted line drawing of part of seismic line 1 (location in Figure 7) illustrating E-verging reverse faults in the basin substratum and (b) the W-verging outcrop-scale reverse faults affecting the 3rd UBSU Pliocene (?) continental deposits near Caspreno, along the Arbia riverbed. The position of outcrop (b) with respect to the major structures affecting the substratum-basin contact is reported in (a). The reverse faults in (b) are interpreted as back thrusts to the major faults. (c) Stereoplot of the mesoscopic compressive structures and (d) close-up of (b). The reverse faults propagate along lignite-bearing clay levels and are often characterized by pervasive calcite shear veins. Colours of the units and symbols of the stereoplot are as in Figures 7 and 8.

at the front of the S. Quirico anticline, about 3 km to the north of the track of line 8 (Figures 7 and 10(b)–(c)), which provide independent evidence for shortening.

Seismic line drawing 10 illustrates the deformation style at the eastern Radicofani basin margin, characterized by E-dipping back-thrust faults affecting the Tuscan and Ligurian units (Figure 11(a)). The major W-dipping thrust "injects" the Burano evaporites into the core of the Cetona anticline. Evaporites are indeed exposed near the track of line 10 (Figure 7). The occurrence of E-dipping slightly overturned vertical beds of conglomerates (Figure 11(b)) where the major back thrust is inferred to crop out, supports the seismic line interpretation.

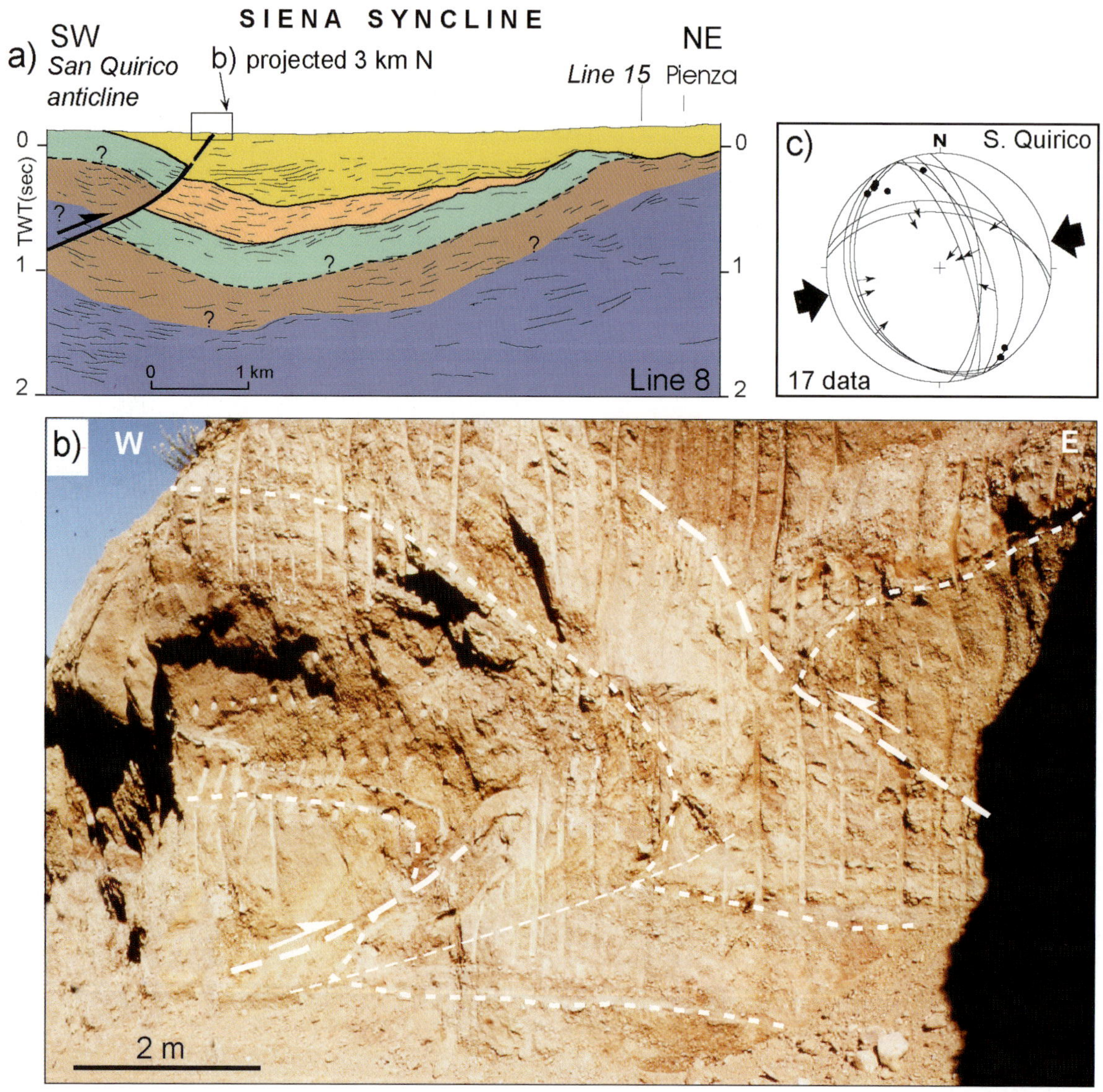

Figure 10: (a) Interpreted line drawing of part of seismic line 8 (location in Figure 7) illustrating the overthrusting of the substratum over the Siena basin deposits along the W-dipping thrust fault generating the S. Quirico anticline. (b) Outcrop-scale E-verging thrust-related anticline associated with a W-dipping reverse fault deforming the Middle-Late Pliocene deposits (4th UBSU), and (c) stereoplot of the corresponding mesoscopic structures (quarry near S. Quirico d'Orcia, location in Figure 3). The outcrop is located at the front of the S. Quirico anticline, and its deformation is interpreted as being related to the emergence at surface of the thrust fault generating this thrust-related fold. The position of outcrop (b) with respect to the major structure affecting the substratum-basin contact is reported in (a). Colours of the units and symbols of the stereoplot are as in Figures 7 and 8.

Notice that seismic line drawings 8 and 10 (Figures 10(a) and 11(a)), together with the geological cross-section illustrated in Figure 8(a), define a nearly complete transverse transect across the Velona and Siena–Radicofani basins (Figure 7). This ideal profile emphasizes the close relation between the marginal thrust anticlines and the contractional deformation in the intervening basin fills.

4.2. Thrust geometry in the axial zone of the Northern Apennines

The Northern Apennines fold-and-thrust belt has previously been described in terms of thin-skinned tectonics, since the Triassic evaporitic Burano formation is the main décollement level resting above the crystalline basement (e.g., Elter, 1960, 1973; Baldacci et al., 1967; Hill and Hayward, 1988; Barchi et al., 1988). Nevertheless, involvement of the basement in thrusting in the internal sector of the NA has been proven for several decades (e.g., Bally et al.,

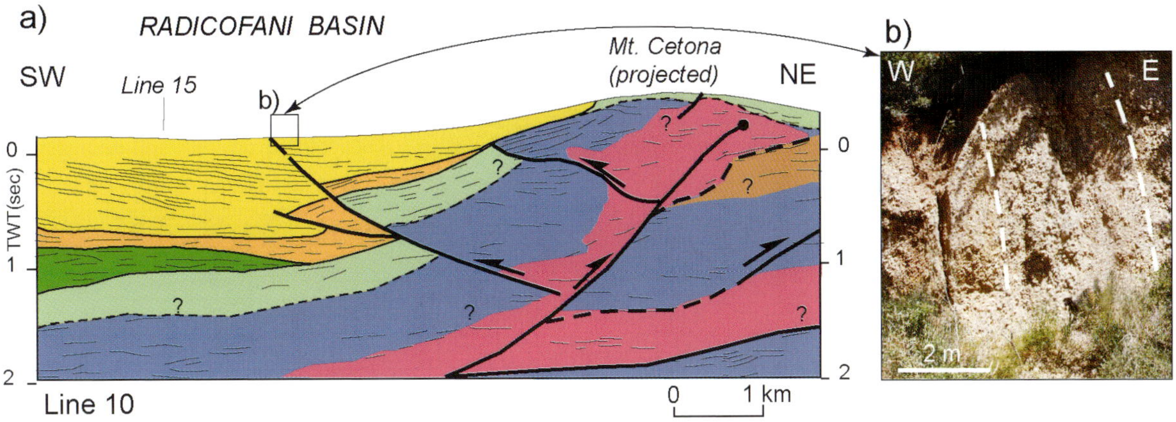

Figure 11: (a) Interpreted line drawing of part of seismic line 10 (location in Figure 7) illustrating the deformation of the Radicofani basin deposits by W-dipping thrust faults and E-dipping back-thrusts. (b) Slightly overturned E-dipping 3rd UBSU conglomerate beds (western Mt. Cetona margin, La Foce area); the attitude of these conglomerates is consistent with a hangingwall anticline to the major back-thrust emerging at surface. The position of outcrop (b) with respect to the major structure is reported in (a). Colours of the units are as in Figure 7.

1986), and the widespread involvement of the same in more external sectors has been recently pointed out from analysis of deep seismic profiles (Finetti et al., 2001, and references therein). The relationships between thrust sheets emplaced along progressively deeper detachment levels into the sedimentary cover and basement produced very complex superficial interference patterns, characterized by recurrent thrust reactivations and out-of-sequence thrusting (e.g., Boccaletti et al., 1980; Roure et al., 1990, 1991; Boccaletti and Sani, 1998).

Reactivation and true out-of-sequence thrusting are well documented in the NA main watershed area, where detailed field mapping integrating stratigraphic revision, and recognition of the main structures delimiting the syn- and/or post-sedimentation tectonic units, as well as mesoscopical structural analysis, has been carried out (Sani, 1990, 1991; Bendkik et al., 1994). At a larger scale, the general structure is a thrust system emplaced in a piggy-back sense of progradation from southeast towards the northeast. Nevertheless, both crosscutting relationships among large structures, and thrust faults cutting fore and/or back limbs of folds at many places account for the presence of out-of-sequence thrusting throughout the area. The available detailed structural analyses confirm this general interpretation. In order to exemplify the complex structural evolution we describe two key areas, whose locations are reported in Figure 7.

The first area is located north of Florence and is characterized by a sequence of thrust sheets involving different units of the Oligocene–Miocene turbiditic succession, articulated in progressive turbidite systems (*sensu* Mutti and Normark, 1987; Mutti, 1992) belonging to the Tuscan foreland units. The formations composing the main thrust sheet units exposed in this area are named (from west to east) the Acquerino Fm. (AQR), Torrente Carigiola Fm. (TCG), Stagno Fm. (STA) and Castiglion dei Pepoli Fm. (CDP) turbidite systems (Bettelli et al., 2002; Figure 12(a)). The exposed thrust sheets may be composed of one or more formations. The upper unit is given by the Ligurian units, normally representing the roof thrust of the NA fold-and-thrust belt, whose emplacement is generally interpreted to be responsible for interruption of sedimentation in the foredeep basins. The thrust faults display a general N-to-NE vergence and a trend varying from E–W to NW–SE. Folding is normally associated with the thrusting. The involvement of the Ligurian units in the thrusting points to the occurrence of significant thrust reactivation phases, that affected the axial NA after the closure of turbiditic sedimentation in the foredeep basins.

In this area, thrust faults are grouped into two different fault sets exhibiting different orientations and referred to superposed deformation phases: an early fault set characterized by NW–SE-trending thrust faults and associated folds is in fact deformed by a later roughly E–W-trending thrust fault set (Figure 12(a)). The tectonic stacking gives rise to a composite thrust system exhibiting complex geometric relations among the units, which are described here from southwest to northeast.

The southern sector is dominated by the overthrusting of the AQR unit (including the AQR Fm. and the tectonically overlying Ligurian Unit remnants) onto the TCG unit (mostly containing the TCG Fm. and in part the overlying STA Fm.) along a thrust fault trending around E–W (Figure 12(a)). This thrust cuts older WNW–ESE-trending

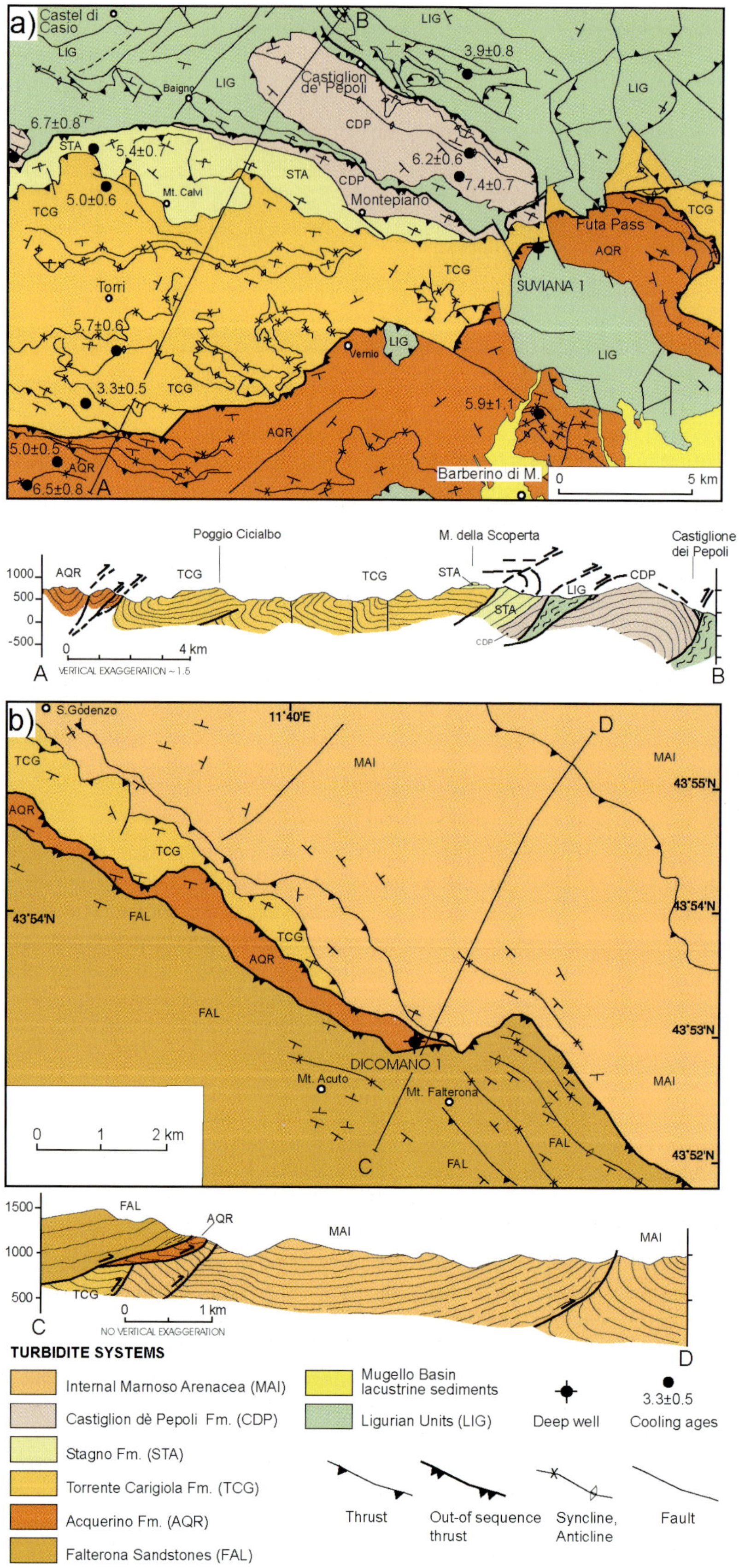

Figure 12: Schematic geological-structural maps of two key areas in the Northern Apennines axial zone illustrating the complex geometric relationships (with the occurrence of well-developed out-of-sequence thrusts) among thrust sheets (location in Figure 2). (a) Castiglion dei Pepoli (modified from Bettelli et al., 2002) and (b) Mt. Falterona areas. Apatite fission track ages in (a) are from Ventura et al. (2001).

thrust faults and associated folds affecting the underlying TCG unit. Moreover, other minor thrust faults affecting the AQR unit have also been transported over the hangingwall of this thrust. To the northeast, a roughly E–W-trending thrust fault superposes the TCG unit onto the STA unit (composed of the STA and CDP Fms.). Another E–W-trending thrust fault superposes the STA and CDP Fms. over the Ligurian units (Figure 12(a)). Finally, the CDP Fm. is involved in the major Castiglion dei Pepoli thrust anticline superposing the CDP Fm. sediments over the Ligurian Units, and cutting former thrust faults in this latter unit (Figure 12(a)).

The above structural observations and geometric relations among thrust faults allow us to interpret a number of thrusts as out-of-sequence structures (e.g., Morley, 1988), such as those at the front of the AQR and STA units, and that at the front of the Castiglion dei Pepoli thrust anticline. Out-of-sequence thrusting developed indeed after the first phases of thrust emplacement closing the turbidite sedimentation, as demonstrated by the different orientations and cross-cutting relationships among thrusts faults, as well as by the occurrence of thrusts cutting the limbs of previous thrust-related folds. This implies that the hangingwall of the out-of-sequence thrusts may often include portions of older thrust sheets. Data from apatite fission tracks collected across a couple of NE–SW-oriented transects transverse to these thrust sheets show an irregular trend of exhumation ages (Ventura et al., 2001) that is consistent with a structural setting dominated by out-of-sequence thrusting. In particular, recent cooling ages ranging around 3.3 to 2.6 Ma have been found interposed between older cooling ages ranging from 9.8 to 5.4 Ma (Ventura et al., 2001). Notably, the data indicating recent apatite fission track ages are clustered immediately at the front of the AQR unit thrust front and aligned with it (Figure 12(a)). These ages are interpreted as being related to the uplifting along the deformation zone associated with the out-of-sequence AQR thrust, whose exhumation is therefore younger than the nearby structures.

Another fascinating structural setting in the NA axial zone is recognizable at the south-eastern termination of the AQR and TCG thrust sheets that are delimited at the front by NW–SE-trending thrust faults (Figures 2 and 12(b)). Like in the former key area, the AQR thrust displays the characteristics of an out-of-sequence thrust, being superimposed obliquely on the TCG thrust sheet. However, both the AQR and TCG thrust sheets terminate abruptly against the thrust carrying the Falterona (FAL) unit, which (in the Mt. Falterona area) completely overthrusts these sheets and reaches the MAI (Internal Marnoso-Arenacea Fm.; Figure 12(b)). In this sector, field mapping was supplemented by a detailed analysis of mesoscopic data, mostly of calcite shear veins in argillaceous sediments commonly marking the emergent thrust zones (Sani, 1990). Although the NW–SE trend of thrust faults could account for a rough NE-trending direction of compression, the slip vectors deduced from this analysis revealed a general N NNE-trend in tectonic transport that is oblique to the mean thrust trend. This fact may be reasonably attributed to reactivation phase(s) successive to the first emplacement of the thrust sheets. The cross-cutting relations among the major structures, together with the results of the mesoscopic structural analysis, lead us to interpret the FAL thrust fault as a major out-of-sequence structure. The fact that the FAL thrust is superimposed obliquely onto the AQR thrust (which also represents a recent out-of-sequence fault, see above) may suggest that the FAL does represent the last reactivated thrust fault.

In summary, the structural data on the hinterland basins and thrust system in the axial NA are consistent with a coherent scenario dominated by thrust reactivations, giving rise to the complex structural geometries, and locally favouring conditions for the establishment of sedimentary basins. It is notable that the structural picture obtained from surface data is in good agreement with the results of the independent interpretation of CROP seismic sections described in Section 3. Definition of the tectonic regime in which the hinterland basins formed and evolved is indeed of crucial importance for the assessment of the Northern Apennines evolution. Our conclusion that this sector was dominated by compressional tectonics is mostly based on analysis of the hinterland basins, which are interpreted as thrust-top basins, since they deformed by pervasive and locally strong contraction. A number of facts support such an interpretation: (1) the compressional deformations are distributed throughout the basins and not limited to transverse zones (as suggested by some authors); (2) the compressional deformation often correlate with thrust anticlines bounding the hinterland basins (see for example Figures 8, 10 and 11) and the sedimentary fill architecture points to synshortening sedimentation (i.e., the development of progressive unconformities at the basin margins; (3) a striking relation between the compressional structures observed in the field and those imaged in the seismic sections is well exemplified in the Siena–Radicofani basin; (4) despite the fact that in the national geological cartography, the basin margins are generally marked by normal faults, these faults have actually been recognized only locally, and in such cases they show local effects or represent post-sedimentary (often active) structures frequently oblique to the basin margins (Sani et al., 2001); (5) the occurrence of strong deformation in the axial part of some basins, like the large extent of vertical to overturned strata in the Radicondoli basin (Moratti and Bonini, 1998), can hardly be explained other than by compression.

5. TRANVERSE STRUCTURES AND GEOTHERMAL FIELDS

Deep seismic profiles of the CROP data set provide the evidence for a reliable interpretation of the deep structure of the Apennines and Tyrrhenian system. As discussed above, the structures mainly consist of thrust faults affecting the upper mantle and/or the continental crust. The continental crust appears deeply involved in thrusting, although the distribution of structures is not homogeneous but varies from place to place. The NA north-western sector is characterized by a single main structure whose trend roughly corresponds to the main trend (NW–SE) of the Ligurian Apennines, whereas southwards the mean trend of structures attains a more N–S direction, such as that corresponding to the well known Olevano-Antrodoco structural lineament (Figure 2). The splaying and breaching exhibited by the crustal thrust faults, frequently portrayed on the seismic lines (Plates 2, 4, 6, and 8) accounts for the structural complexity characterizing the NA hinterland.

The main crustal thrust AP2 can be traced approximately below the western margin of the Siena–Radicofani basin (Figures 2 and 7). In cross-section, this thrust superposes an internal, early deformed portion of the NA fold-and-thrust belt onto another more external one. This tectonic doubling is achieved by detachment along fault AP2, as suggested by the cross-cutting relationship of this structure to early thrusts AL1-AL5 (see also Finetti et al., 2001). The frontal splays of thrust AP2 propagated into the sedimentary cover involving the torbiditic sequences of Tuscan and Umbria foreland units, giving rise to a complex structural geometry that has been partly illustrated above in Section 4.2. The occurrence of further and relevant remobilization along AP2 during crustal doubling is revealed by the sharp cross-cutting relationships related to thrust AP-2G, which truncates earlier thrusts both in the cover and in the upper crust (Plate 8). Thrust AP-2G represents, indeed, the new and more external leading edge of the major crustal thrust sheet responsible for doubling the upper crust in the NA hinterland (Plate 8).

The main structures affecting the crust and lithospheric mantle interpreted from section CROP03 and identified as faults AP-3 and AP-5, continue to the south, well beyond Rome. These two major lithospheric structures appear to be displaced by a major N–S trending structure corresponding to the Olevano-Antrodoco structure at surface (Figure 2). Northward, they are deflected, or possibly interrupted, by a transverse NE–SW, trending lineament roughly corresponding at surface to the "Piombino–Faenza Line" (Boccaletti et al., 1977). This structure is also partly recognizable in Figure 13 and is marked by the presence of scattered magmatic bodies, like the Mt. Capanne granodiorites pluton (at Elba Is.), the S. Vincenzo rhyolites, the Montecatini lamproites, and the large Larderello geothermal field (compare Figure 13 with Figure 2). The two lithospheric thrusts AP-3 and AP-5 can be also traced north-westward. The more internal thrust roughly follows the Tyrrhenian coast and is therefore internal to the Alpi Apuane–Mt. Pisano lineament, where the basement outcrops. The second thrust runs along the main Apennine divide and ends close to Florence, still against the Piombino–Faenza Line. It partially coincides with the basement thrust detected by Argnani et al. (1997). It should also be noted that in this area recent seismicity shows deep earthquakes with a compressive stress field characterized by N10-20E trending σ_1, allowing for a good correlation with deep thrust fault activity. Another major lithospheric thrust displacing the Moho boundary has been identified in profile CROP M12A. This is fault AP2, which is rooted in the Tyrrhenian area and propagates westward into the NA hinterland as a low-angle thrust within the Adria upper crust (Plate 8).

The presence of transverse lineaments has been pointed out in many works (Ghelardoni, 1965; Bortolotti, 1966; Boccaletti et al., 1977; Fazzini and Gelmini, 1982; Bemporad et al., 1986). Based on the fact that these structures extend from the innermost to the externalmost sectors of the NA, their structural significance has been related to lateral thrust ramps in the external sectors characterized by ongoing compression, or to transfer zones in the internal sectors affected by extension (e.g., Liotta, 1991). These structures are also thought to be characterized by significant strike-slip components (Boccaletti et al., 1977; Fazzini and Gelmini, 1982) and possibly to separate along-strike sectors undergoing different stress fields (extension and compression).

The occurrence of magmatic bodies along these structures represents another significant characteristic. Scattered, exposed or hidden magmatic-bodies are in fact aligned along the Piombino–Faenza Line. The recognition of mantle contamination in, or even as origin of, the magmatic bodies located above this transverse lineament, such as the mafic dykes on Elba Island (Rocchi et al., 2002), the mantle signature in the fluid at Larderello Geothermal field (Minissale et al., 2000), and the relatively primitive mantle-derived magmas at Montecatini (Serri et al., 2001), is consistent with a Piombino–Faenza Line propagating downwards through the whole crust. Section CROP03 runs very close to the Larderello and Mt. Amiata geothermal fields. According to Cameli et al. (1993, 1998), the reflective seismic K-horizon represents the brittle-ductile transition in the continental crust of the internal NA sector. Plan-view contour lines of the K-horizon depth reveal that the Larderello area is characterized by an elongate shape (Cameli et al., 1998), with maximum axis slightly oblique to the trend of the Piombino–Faenza

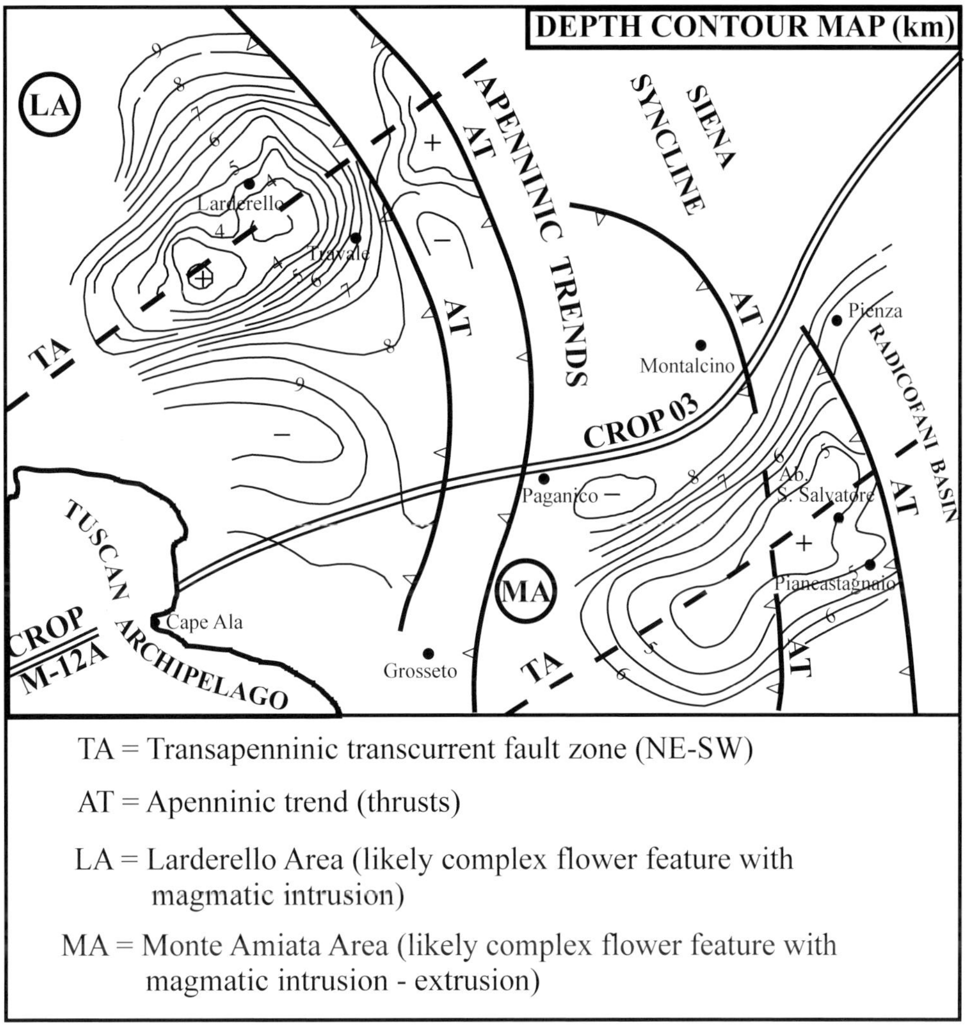

Figure 13: Structural map of the Larderello (LA) and Monte Amiata (MA) geothermal fields. Contour lines of the horizon K (brittle-ductile limit) in km are reconstructed on the basis of a detailed seismic dataset and borehole calibration (Cameli et al., 1998). Both the Siena and Radicofani basins represent asymmetric synclines interposed between two bordering (compressive) E-verging thrust faults of Apenninic trend (NW–SE), locally accompanied by back-thrust. Seismic examples in the literature of extensional faults in the Larderello area (Liotta and Ranalli, 1999) correspond to local, intrusion-related periplutonic faults, limited to the upper crust, not representative of the (seismically controlled) basic regional tectonic style of the Tuscan Apennines.

Line (Figure 13). Apart from the debate about the meaning of the K-horizon and its relation to the magmatic body below Larderello, this contour lines' geometry should also reflect the shape of the Larderello pluton, as deduced by the analysis of teleseismic travel time residuals (Foley et al., 1992). Such a geometry has been also reconstructed for the geothermal field below the Mt. Amiata volcano (Figure 13). The evolution of this volcano is apparently still controlled by a roughly ENE-trending transverse structure (Ferrari et al., 1996), which can be traced further east and west of Mt. Amiata itself (Figure 13).

The emplacement of the causative magmatic bodies of such geothermal fields can be reasonably interpreted as the result of a process of strain localization arising from the interaction between a transfer-like zone (or lateral ramp) and a magma layer at depth, in a similar fashion to the results of the centrifuge analogue models reported in Corti et al. (2002). These analogue models investigated the deformation produced in the overlying crust by a crustal-scale transfer-zone propagating across a magma layer at the base of the continental crust (Figure 14). Model results show that a large quantity of the magma analogue material (glycerol) accumulated above the intersection between the transfer zone and the basal magma layer, giving rise to a large magma body that lifted and folded the brittle-ductile crust interface (Figure 14(b), (c), and (d)). Both in nature and in the model, the shape of the emplaced magma bodies, approximately defined by the contour lines of the brittle-ductile crust interface, show a plan-view

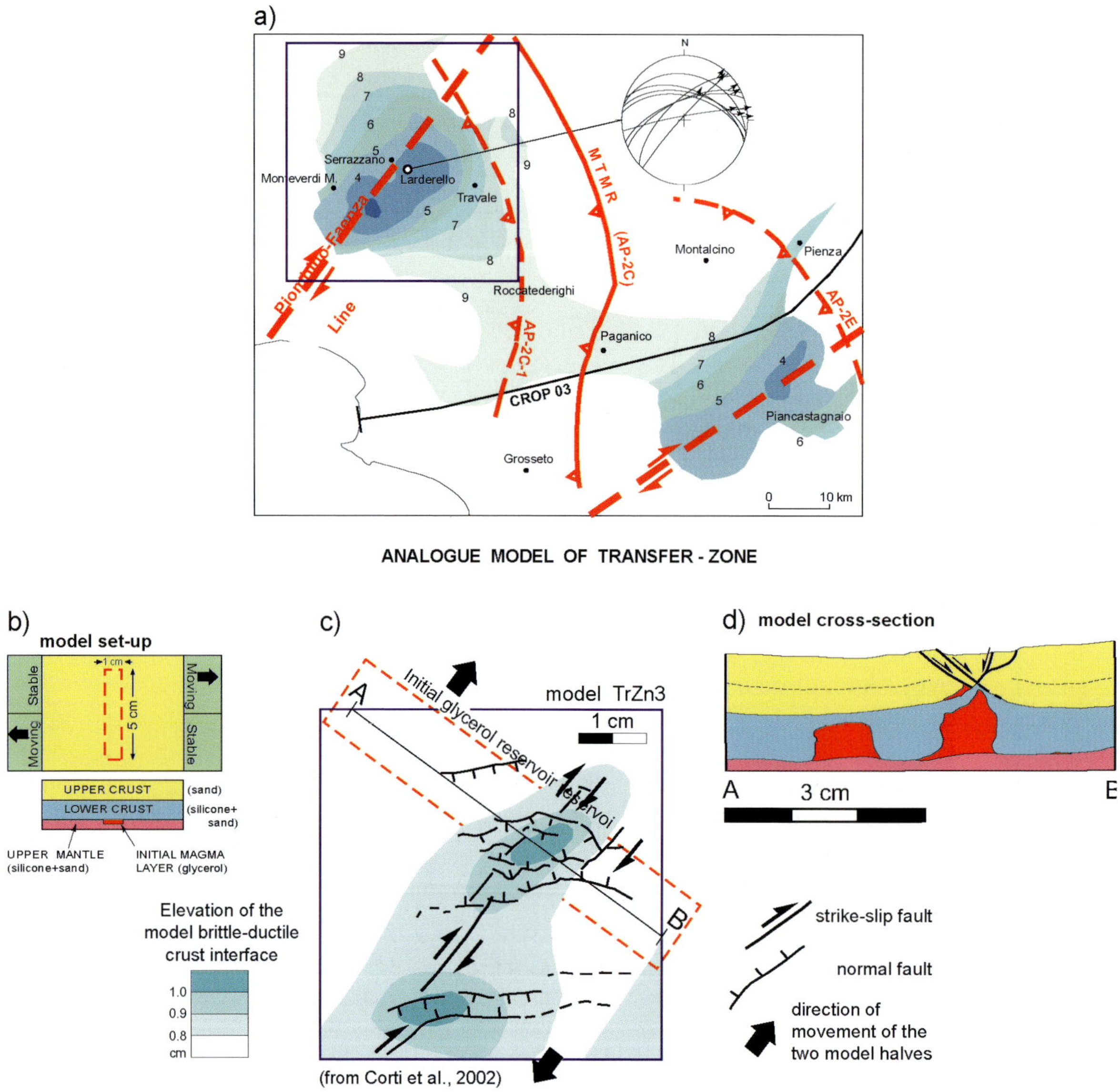

Figure 14: (a) Sketch map of the Larderello and Mt. Amiata geothermal fields with contour lines of the K-horizon depth in km (Cameli et al., 1998). Both geothermal fields are located along transverse tectonic lineaments and show a marked elongation parallel to such anti-Apenninic structures. These transverse lines are mostly characterized by a dextral component of movement (see the stereoplot of fault-slip data collected close to Serrazzano on the Larderello structure; Schmidt projection, lower hemisphere). (b) Model set-up and (c) top-view contour lines of the brittle-ductile crust transition depth in an analogue model simulating the interaction between a dextral transfer (shear)-zone and an initial deep magma layer (indicated by the dashed red box; see Corti et al. (2002) for details of the experimental work). The blue box indicates the Larderello area to which the analogue model is compared. The model has then been rotated in order to place the model transfer zone parallel to the corresponding structure in the Larderello area. (d) Model cross-section AB. Model deformation shows that the ductile crust and magma rose along the model transfer-zone resulting in a dome exhibiting a plan-view elongation slightly oblique to the transfer zone (modified from Corti et al., 2002). Such characteristics are strikingly similar to those of the natural prototype, so that we argue that the Larderello and Mt. Amiata geothermal fields developed by a similar dynamic process. Note also that, both in the model and in nature, the shear zones are dextral, and the contour lines refer to (or, in the Larderello area, at least close to) the brittle-ductile crust transition.

elongation slightly oblique to the trend of the transfer zone (compare Figure 14(a) with Figure 14(c)). The model deformation, characterized by a magma body rising from a deep layer of magma-simulating material, suggests that magmatic bodies in the NA hinterland may be more extensive at depth than suggested by the scattered and small outcrops of magmatic rocks at surface (Figure 14(c), (d)).

The geothermal fields of Larderello and Mt. Amiata, which may also be characteristic of the emplacement of other magmatic rocks, are therefore interpreted as resulting from the uprise and emplacement of crustal and melt materials along hinterland transverse structures. This structural setting is also consistent with the heat flow distribution pattern, which shows peaks markedly localized at the geothermal fields (Mongelli and Zito, 1991; Mongelli et al., 1998; see below). Additionally, the model deformation is characterized by small normal faults oblique to the main transfer-zone and localized above the dome in the lower crust (Figure 14(c)). Such structures accommodated the deformation connected with the domes deformation, representing therefore second-order structures to the transfer (shear) zone itself. Preliminary field work suggests that a similar structural setting can also be hypothesized for the Larderello area. In this study, fault-slip data consistent with a significant dextral component of displacement along the Piombino–Faenza Line have indeed been collected above the Larderello geothermal field (east of Serrazzano; Figure 14(a)). Therefore, our findings and all the relevant CROP and other seismic data discussed above confirm that the claimed first-order E-dipping listric normal faults (Brogi et al., 2003), for the Larderello area also are not representative of its broad fundamental lithospheric tectonic style.

Depending on the operating stress field, the transverse structures in the NA hinterland could have assume the structural significance of transfer zones or lateral ramps. For instance, the Larderello pluton, whose oldest closure age dates around 3.8 Ma (Villa and Puxeddu, 1994), could be coeval with the phases of shortening related to the activity of the thrusts bounding the Siena–Radicofani basin and responsible for the deformation of the 3rd and 4th UBSU sediments. In this view, the Piombino–Faenza Line, at 3.8 Ma, could have acted as a genuine lateral ramp. And in fact, the Larderello and Amiata geothermal fields show clear relationship with basement thrusts which apparently delimit these fields at their base and eastward (Figure 14(a)).

From our interpretation of the CROP sections it is directly evident that the high heat flow is due to the existence of the prominent Etrurian asthenospheric bulge. The pattern of the regional heat flow matches rather well the culmination of this bulge (CROP M10, M11 and M12A) to which correspond the higher heat flow values. It should be also noted that magmas with a significant mantle-derived component are found above the asthenospheric culmination, such as at Capraia Island and the mafic dykes on Elba Island (Serri et al., 2001; Rocchi et al., 2002). Another view assumes that the high heat flow can be correlated to Tyrrhenian opening and crustal extension (Mongelli et al., 1998). CROP data show that thinning of the continental crust is partially inherited from the Jurassic phases of rifting, since the Tyrrhenian area localized at the western *palaeo* passive margin of Adria (see Plate 8). In addition, the heat flow characterizing the internal Tuscany area (Tuscan Domain with $> 100\ \mathrm{mW\,m^{-2}}$; Della Vedova et al., 2001) partly corresponds to the plan-view of thrusts AP2, AP-2G and AP-2C. This raises the possibility that part of the heat flow might be transporting from the asthenospheric culmination area into the hinterland via the basal thrusts AP2-AP-2G (and associated splays), as well as by the emplacement of magmatic bodies along the transverse lines (Larderello and Amiata geothermal fields). In this scenario, the large heat flow triggers the crustal anatectic processes leading to the formation of acidic magmas, which may be emplaced in various structural settings, such as (1) at the footwall of pluton intrusion-related normal faults; (2) along transverse transfer zones; (3) along structures characterized by significant strike-slip components; or perhaps (4) at the core of anticlines representing low-pressure zones. *We therefore assume that the hinterland basins did not develop in an extensional regime because the heat flow is high, but that the deformation mechanisms clearly allow the coexistence of compressional basins and high heat flow conditions.*

6. DEFORMATION ANALYSIS OF THE NORTHERN APENNINES

The presented CROP seismic data integrated with regional and detailed geological-geophysical data enable us to define the deformation processes in the examined Apenninic sector. Below, we focus our attention on the major structures controlling the main tectonic stages in the NA evolution. The E-dipping AL1 to AL5 thrust faults and the remnant of the oceanic slab dipping eastward below the NW Adria margin (Plates 1, 2, 3, 4, 7 and 8) are attributed to early stages of deformation tied to the closure of the western Alpine Tethys ocean (Ligurian–Piedmont) during the Late Cretaceous–Eocene ("Eo-Alpine" stage).

After the closure of the Alpine Tethys sector of the Alps, and restriction of this ocean basin west of the Alps with Europe-verging obduction of Alpine Tethydes units (ophiolites and ophiolite bearing units) on the southern European margin, subduction began. This thrust polarity change is evident in the interpreted CROP sections: E-dipping (W-verging) AL1-AL5 thrust faults are truncated by the E-verging thrust fault AP2 (Plates 7 and 8). During this stage, the Alpine Tethydes (Ligurian Units) were emplaced onto the foreland units (Tuscan and Umbria–Marche units) mainly along the low-angle AP0 thrust fault (Plate 8). This Ligurian Units overthrusting is traditionally considered the last compressive event in the present NA hinterland area, and the age of thrust activity in the NA axial zone is inferred from the age of closure of sedimentation into the underlying foredeep basins progressively involved in the eastward progressing deformation. However, both seismic interpretation and field evidence (Sections 3 and 4) indicate that, after the Alpine Tethydes Unit overthrusting, the Northern Apennines were affected by continuing compressive tectonics (post-Middle Miocene times) that caused the development of extensive out-of-sequence thrusts superposing the foreland units onto the Alpine Tethydes units, thus leading to the local inversion of the original stacking order (e.g. Bendkik et al., 1994; Boccaletti and Sani, 1998; Bonini, 1999; Bettelli et al., 2002). Out-of-sequence thrusting has also been identified in the NA hinterland area, where these thrust faults normally delimit the Neogene basins and locally affect their sedimentary fill (Section 4; Plate 8).

Dating of the AP2 fault activity, representing the main structure in the NA hinterland area, can be deduced from the age of sediments involved in the deformation, and filling these on-shore hinterland basins (Figures 2 and 7). Such basins are indeed located on the AP2 hangingwall and are affected by thrust faults (AP-2C, AP-2E1-3, AP-2, AP-2F) splaying from this basal detachment (Plate 8). In particular, in-sequence crustal thrusts splaying from the basal AP2 detachment gave rise to the out-of-sequence cover thrusts controlling basin development. The AP2 fault itself is however seen to emerge at the surface in front of the Macigno unit (Figure 7 and Plate 8). On this basis, the activity of AP2 is inferred to extend from the Late Serravallian (deposition of the 1st UBSU Ponsano Sandstones in thrust-top basins) until at least the Middle-Late Pliocene, as argued from the compressive deformation affecting the 4th UBSU sediments in front of both the S. Quirico thrust anticline (Figure 10(a)) and the Cetona–Rapolano anticline (Bonini and Sani, 2002). However, the strongest deformation in the basin fills is observed in the Late Tortonian–Messinian sequences.

Further east, in the NA axial zone, the lithospheric thrust faults displacing the Moho at high-angle can be geometrically divided into the more internal fault AP-3 and the more external faults AP-4 and AP-5. Fault AP-3 is hypothesized to have been active since the Late Miocene because in probably controlled (at the surface) the deposition of the external Marnoso Arenacea foredeep basin. Thrusts AP-4 and AP-5 are hypothesized to have developed later. The doubling of the Moho by faults AP-4 and AP-5 is, in fact, attributed to Late Messinian–Early Pliocene times on the basis of the age of the syn-shortening thrust-top-basins associated with these structures (Finetti et al., 2001). Minor thrust faults emanating from the main AP-5 fault caused, in the Pliocene, the development of piggyback basins in the NA eastern sector (Montecalvo and Tavullia syncline basins) and a foredeep clastic wedge at the AP-5 leading edge (Plate 8). The combined prominent lithospheric thrust formed by the AP-4 and AP-5 Apenninic faults determined an E-dipping slope in the basement of the Monterchi to Urbania segment of CROP 03 (Plate 8). From the Lower Pliocene to Present, over this composed basement dipping plane, a gravity sliding of thrust sedimentary cover blocks occurred. Out-of-sequence thrust fault AP-5A2 accommodated this sliding process and the consequent sedimentary block accumulation of the Tiber valley (Plate 8). Deformation at the fault AP5-B3a leading edge in section M-16 supports even a Late Quaternary activity for this first-order structure (Plates 7 and 8; Finetti et al., 2001). The AP-5 related deformation shows that a major advancement of the northern Apennines thrust front took place in the Early Pliocene.

More or less synchronously with the development of fault AP-5, the whole AP2 basal detachment continued its activity during the Late Messinian–Late Pliocene, probably in connection with widespread emplacement of igneous bodies in the hinterland area (Finetti et al., 2001). During this stage, the hinterland basins were passively transported above the AP2 hangingwall, though both minor compressive deformation of the basin fill and high-angle normal faults may have accommodated this Pliocene translation. The upper boundary of AP2 late activity is constrained by the Middle-Late Pliocene sediments (4th UBSU) deformed at its leading edge, such as at the western margins of the Upper Valdarno and Chiana basins (Figure 2). The lower boundary of this late phase of activity could correspond to the boundary of the basal Pliocene marine transgression, whose related sediments onlap the western side of the Chianti Mts.–Mt. Cetona thrust system (Figure 2; Plate 8). This structural interpretation implies that during specific faulting events some of the magmatic bodies (i.e. Mt. Amiata) may be partly allochthonous, carried on the hangingwall of fault AP-2 (Plate 8).

Analysis of Plate 8 also shows that the thrust fault AP-2G (external to AP2) cross-cuts older structures, both at basement and cover levels, a fact that points to an out-of-sequence geometry for the thrust sheets (see Section 4.2). The truncation of the lithospheric high-angle splay AP3-A by the sub-horizontal fault AP-2G is, in particular, very clear. Evidently, this points to a younger (re)activation of AP-2G with respect to AP2. This evolution is again reflected in the age of the hinterland basins that presumably developed in response. Under this hypothesis, activity of AP-2G can be attributed to the Early-Middle (?) Pleistocene on the basis of the contractional deformation affecting the Casentino and Mugello basins developing (further east to the Upper Valdarno and Chiana basins) on the hangingwall of out-of-sequence cover thrusts emanating at depth from the basement thrust AP-2G (Figure 2 and Plate 8). A compressional setting has indeed been postulated for the evolution of the Mugello and Casentino basins (Benvenuti, 1995; Boccaletti et al., 1995).

Analysis and approximate but realistic quantitative estimates of geometric features have been performed along the section across the Tyrrhenian–Northern Apennines given by profiles M12A-CROP03-M16 (Figure 15 and Plate 8). This analysis was made possible by the availability along the onshore profile of field data providing a precise age determination of sediments involved in the deformation, and by the observation of the cross-cutting relationships (chronological and geometrical) among the structures. As a first approach, we evaluate the aspect ratio of individual thrust sheets (i.e. the ratio between the spacing of thrust sheets, L, and their depth of detachment, D (Cello and Nur, 1988; Calamita et al., 1994). The aspect ratio (L/D) along the profile CROP03 was evaluated for three main detachment levels: (1) the base of sedimentary cover (represented by the Triassic evaporitic Burano Fm.); (2) within the upper crust or at the upper/lower continental crust boundary; and (3) the thrust faults displacing the Moho.

It is normal to observe a gradual decrease in structural complexity with depth. This is explained by the rheological variability of the sedimentary cover as well as by the dependence of thrust sheet wavelengths on the detachment depth, since deep thrust sheets (involving the continental crust) are usually very long and less internally deformed than those in the sedimentary cover (Plate 8). The aspect ratios of sedimentary cover thrust sheets show a large dispersion, with depth of detachment ranging from 2 to more than 6 km, and length variable from 5 to 25 km. This may be due to several reasons: (1) the local occurrence of thick synshortening deposits may greatly increase; (2) polyphase thrust reactivations may decrease L (by splitting an older thrust sheet into two or more parts) while out-of-sequence thrusting may increase L (by including more, or part of, previous thrust sheets into a larger thrust sheet); and (3) the presence of décollement layers within the sedimentary cover (like the shales and marls of the Scaglia Fms. in the NA) may increase L and decrease D. As a result, correlation among the various thrust sheet aspect ratios is greatly dispersed (Figure 15(b)). However, a more accurate analysis of the crude data involves elimination of the anomalous cases (red circle data points in Figure 15(c)) represented by thrust sheet data points with large D, due to a great thickness of syntectonic deposits (thrust sheets C21, C22), as well as those with shallow depth of detachment D together with long length L of the thrust sheets (thrust sheets C5, C10, C11, C15 C19; cf. Figures 15(b) and 15(c) with Plate 8). In this case, we obtain a satisfactory correlation between length ($L = y$) of thrust sheets and depth ($D = x$) of detachment, given by a linear regression curve $y = 4.0278x + 0.8721$ (with $R^2 = 0.86$; Figure 15(c)). This analysis suggests that the transverse distribution of aspect ratios across a given fold-and-thrust belt might also be used to evidence the out-of sequence thrusts, such as sheets C10 and C11, which can reasonably be related to out-of-sequence thrusting at the leading edge of the Cervarola thrust (Figure 15(c) and Plate 8). The same reason can be invoked for C19 which, although it shows a great depth of detachment vs. a small length, it is clearly cut by another shallower thrust sheet (C18; Figure 15(a)). It can be noted that the major out-of-sequence thrusts in the sedimentary cover are mostly connected at depth to the major crustal or lithospheric thrusts, like thrust sheets C10-C11 connected to thrust AP-2, and C19 connected to AP-5 (Figure 15(a) and Plate 8).

The aspect ratios for the upper crust, evaluated in the depth-converted and restored profile of CROP03 (reported in Finetti et al., 2001), show two clusters around depths of 7 and 10 km, whereas the lengths of the corresponding thrust sheets vary from 5 to 33 km (Figure 15(d)). The cluster around 7 km is associated with the doubling of the upper crust along the crustal thrust AP-2, while the 10 km deep detachment corresponds instead to the surface of decoupling at the base of the upper crust in the NA internal area (M12A; Plate 8).

Finally, the thrust faults displacing the Moho at high angle show a roughly constant length ranging between 100 and 150 km. Four major lithospheric thrust sheets have been identified throughout the profile CROP03. It is interesting to note that the predictive theoretical values for the lithospheric wavelength and deformation geometry (under compression) expected by Martinod and Davy (1992, 1994) are substantially validated by the real data on thrust wavelengths and tectonic style for AL2-AL5 and AP-3.

The shortening estimates calculated for each of the main lithospheric and crustal thrust faults AP-1 to AP-5 along the depth-converted and restored profile CROP03 are mainly reported in Finetti et al. (2001). For the internal thrust faults AP-1 and AP-2 the evaluation of shortening must also take into account partial displacement inversion related to the Tyrrhenian Basin opening, as well as the out-of-sequence geometry of AP-2. However, the shortening

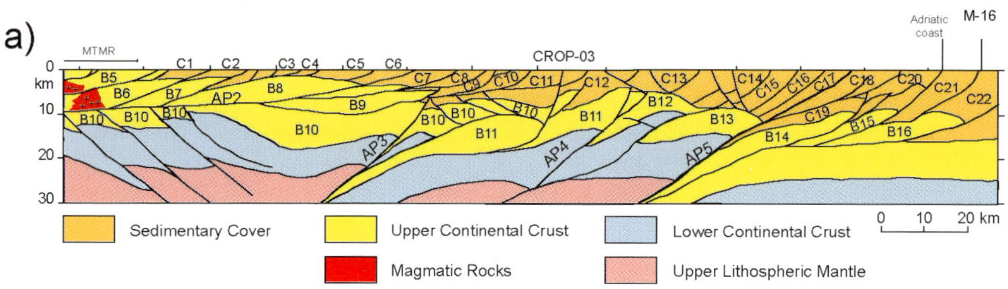

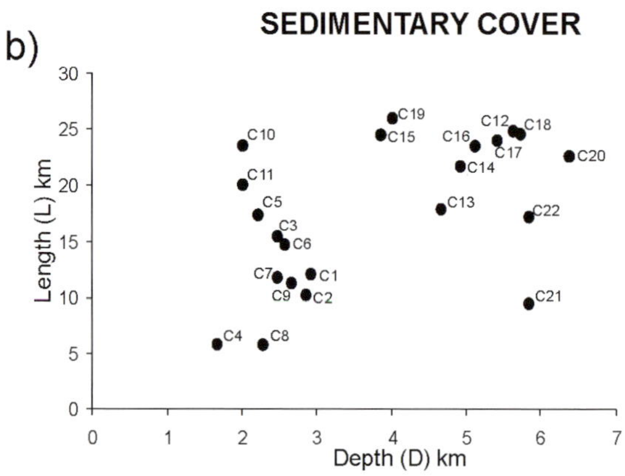

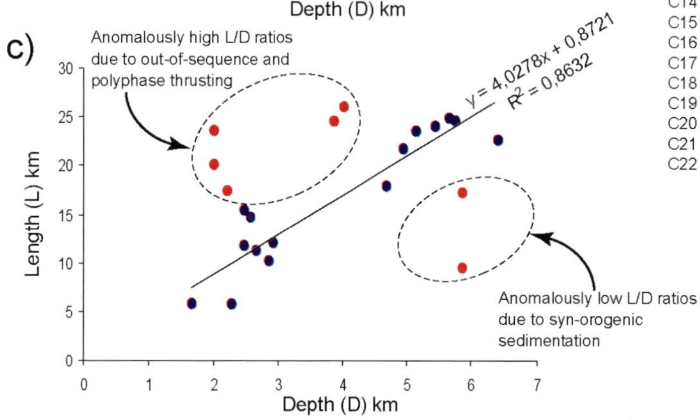

Sedimentary Cover crude data

Thrust sheet number	Depth (D) km	Length (L) km	L/D
C1	2,9	11,9	4,10
C2	2,85	10,1	3,54
C3	2,45	15,3	6,24
C4	1,65	5,6	3,39
C5	2,2	17,3	7,86
C6	2,55	14,6	5,73
C7	2,45	11,7	4,78
C8	2,25	5,6	2,49
C9	2,65	11,2	4,23
C10	2	23,6	11,80
C11	2	20,2	10,10
C12	5,6	24,7	4,41
C13	4,65	17,8	3,83
C14	4,9	21,6	4,41
C15	3,85	24,5	6,36
C16	5,1	23,41	4,59
C17	5,4	23,96	4,44
C18	5,7	24,51	4,30
C19	4	25,88	6,47
C20	6,35	22,5	3,54
C21	5,85	9,4	1,61
C22	5,85	17,1	2,92

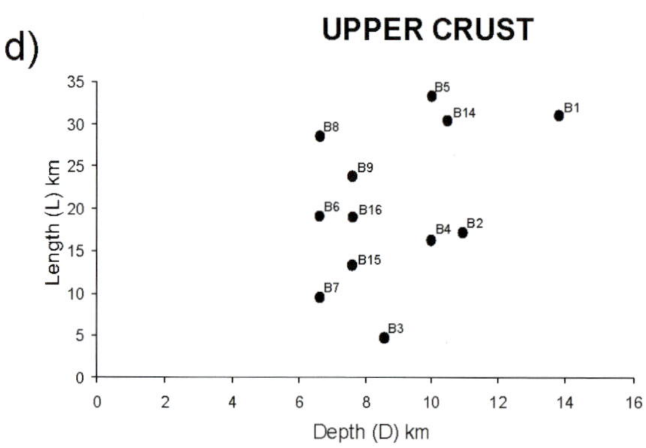

Upper crust crude data

Thrust sheet number	Depth (D) km	Length (L) km	L/D
B1	13,81	30,95	2,24
B2	10,95	17,14	1,57
B3	8,57	4,76	0,56
B4	10	16,19	1,62
B5	10	33,33	3,33
B6	6,66	19,05	2,86
B7	6,66	9,52	1,43
B8	6,66	28,57	4,29
B9	7,62	23,81	3,12
B14	10,48	30,48	2,91
B15	7,62	13,33	1,75
B16	7,62	19,05	2,50

Figure 15: (a) Depth-converted profiles M12A/CROP03/M16 (slightly modified from Finetti et al. (2001) and Plate 8). (b) Length (*L*) and depth (*D*) crude data of sedimentary cover thrust sheets; and (c) linear regression curve of *L*/*D* data points. Data points used to constrain the linear best-fit regression curve are shown in blue, whereas those not considered (either because of large syn-orogenic thickness or because of out-of-sequence and polyphase thrusting) are indicated in red. (d) Aspect ratios (*L*/*D*) of upper crust thrust sheets. See text for details.

accommodated by these structures can be approximately evaluated as 7 km and 64 km for AP-1 and AP-2, respectively (Finetti et al., 2001). The external lithospheric thrusts are more constrained because they were not affected by successive deformations. In particular, AP-3 is hypothesized to have been active since the Late Miocene, while thrusts AP-4 and AP-5 are hypothesized to have been active from the Late Messinian–Early Pliocene up to the Quaternary. Considering that the total amount of shortening accommodated by AP4 and AP-5 can be estimated as about 50 km (Finetti et al., 2001), an average slip rate of ~ 8.3–10 mm/yr is derived by taking the Late Messinian–Quaternary (~ 5–6 Ma) as time span of deformation. This rate is in agreement with that of 5–10 mm/yr estimated for the migration of the NA Late Oligocene–Late Miocene foredeep basins (Ricci Lucchi, 1986), as well as with the deformation rates of 8–15 mm/yr estimated for the Pliocene shortening at the NA leading edge (Calamita et al., 1994).

It should be noted however that shortening estimates cannot be cylindrically extrapolated far from the area of the section because the deformation is neither constant nor cylindrical longitudinally to the chain. As an example of this, we consider the trend of the metamorphic rocks exposed in Tuscany from the Alpi Apuane in the north to the MTMR in the south (Figure 2). The metamorphic units of the chain and the basement rocks outcrop along this ridge because they are lifted by the crustal thrust fault AP-2C (Plate 8). The distance between the AP-2C leading edge in the Alpi Apuane area and the leading edge of the regional Cervarola–Falterona thrust front (Figure 2) is about 20 km, whereas further south (in the MTMR area) it is about 60 km. This variation in distance is apparently inversely proportional to the elevation of the metamorphic ridge, which decreases southwards: the Alpi Apuane core reaches an altitude of about 1950 m, Mt. Pisano 916 m, whereas the MTMR reaches a maximum elevation of about 620 m (Figure 2). This along-strike difference in elevation likely corresponds to dissimilar structural settings, with the Alpi Apuane sector characterized by few and closely-spaced crustal thrusts, whereas many and widely-spaced crustal thrusts (with the hinterland thrust-top basins developing in between) characterize the MTMR (Figures 2 and 13). From this setting, as first approximation, it can be reasonably inferred that the shortening was accommodated by different deformation styles: distribution of shortening along many and widely spaced thrusts resulted in gentle elevations, whereas the same deformation on fewer thrust faults was presumably compensated by increasing the vertical throw. As a consequence, we expect several reactivations of the same thrust faults in the northern sector (i.e., the Alpi Apuane–Mt. Pisano area), whereas more in-sequence thrust sheets should have developed in the southern one (MTMR area).

Normal faulting may accommodate transverse strike-slip motions and gravity collapse above the backlimb and at the forelimb of a composite crustal-scale thrust anticline. Indeed, repeated reactivations of deep thrust faults may cause significant uplifts, so that extensional structures resulting from passive collapse may develop either as synthetic normal faults on the rear of thrust anticlines or as antithetic normal faults at their front.

7. SEISMICITY OF UMBRIA–MARCHE ACTIVE ZONE

The interpretation of the deep seismic reflection data of CROP03 is very useful for understanding the active seismicity of the Umbria–Marche area. From Monterchi to Urbania (Plate 8) CROP 03 is characterized by a prominent composite high thrust of basement, generated by the AP-4 and AP-5 Apenninic faults. This lithospheric high forms an E-dipping basement plane over which, since the Lower Pliocene, a gravity sliding of sedimentary cover thrust sheets has been taking place, with the normal faults GS-1, GS-2 and GS-3 as the most evident expression, as already reported in the literature (Finetti et al., 2001). These extensional structures partly correspond to the so-called Altotiberina Fault that bounds the Anghiari–Sansepolcro basin to the west, and spatially correlates well with to the Umbria–Marche sector, which is characterized by strong and frequent seismicity (Figure 2). In particular, the Umbria–Marche area was damaged during the 1997–1998 seismic sequence, with shocks of higher magnitude ranging from Mw = 4.6 to Mw = 6.0. Fault-plane solutions of the numerous shallow events mainly indicate normal faulting as the most important deformative mechanism (Amato et al., 1998; Ekstrom et al., 1998). Based on these data, active crustal extension was invoked as the main geodynamical model (Boncio and Lavecchia, 2000; Calamita et al., 2000; Collettini et al., 2000), although other authors relate the normal faults to active strike-slip tectonics (Cello et al., 1997). However, some crustal earthquakes presenting clear compressive to strike-slip seismological characteristics were also recorded in the axial zone of the NA (Gasparini et al., 1985; Eva and Solarino, 1992; Tomaselli et al., 1992).

Based on the along-strike (south-eastward) correlation of the axial basement culmination and the Anghiari–Sansepolcro basin (Section 3.4) with the Martani Mounts and the Foligno and Colfiorito basins, respectively, the trans-Apenninic section CROP03 interpretation (Figure 16; Plate 8) shows that a unique geodynamic process generating compression in the lithosphere explains the two different seismicity mechanisms. The shallow seismicity

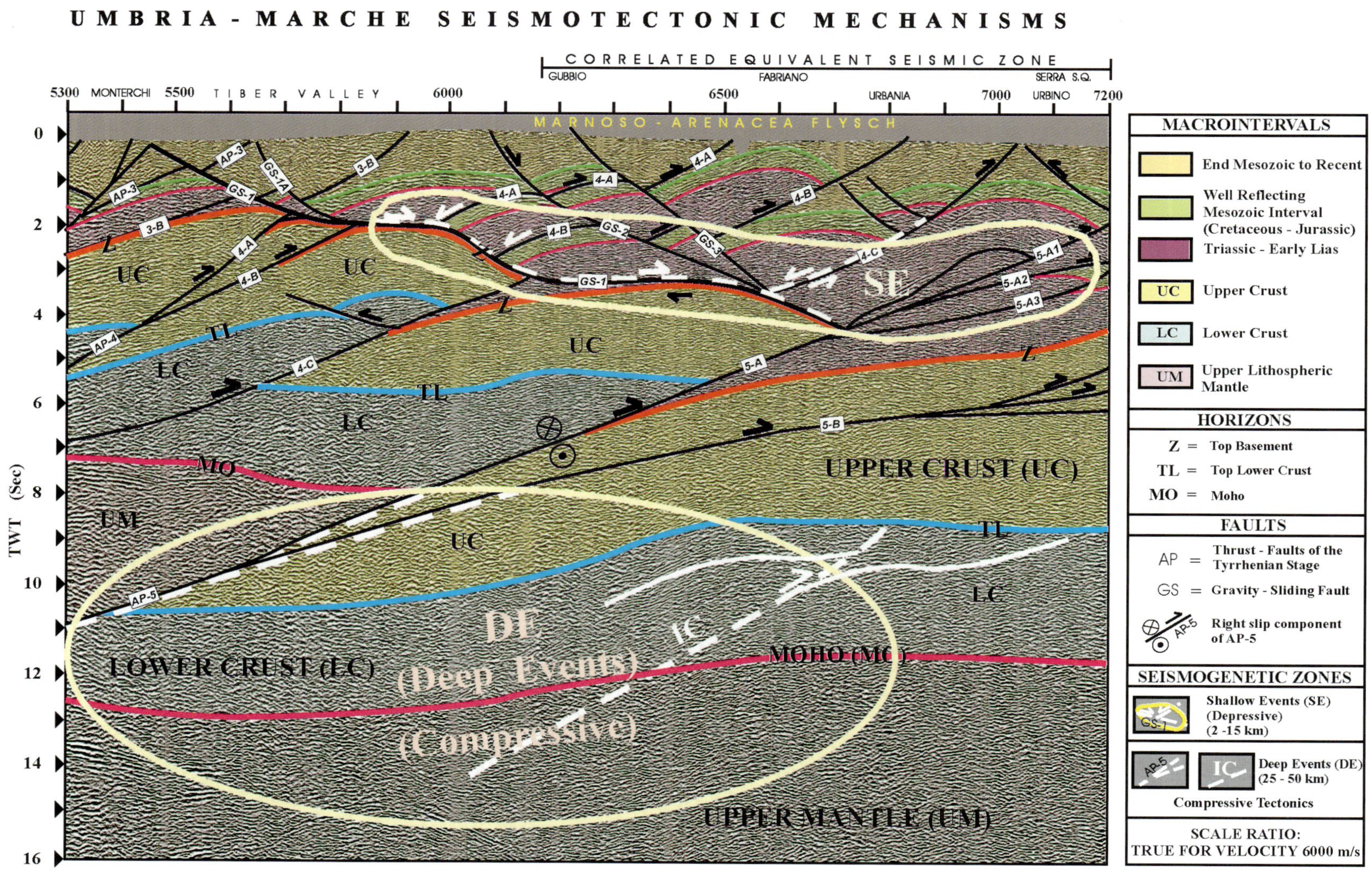

Figure 16: CROP seismic data used for deriving tectonic mechanisms for the Umbria–Marche shallow and deep earthquakes (section CROP-03, part). Shallow seismicity (2–15 km) is generated by a gravity sliding over fault plane GS-1, or by combination of gravity sliding and a right-(anticlockwise) slip component of fault AP-5 with inverted reactivation of previous thrust faults of piled cover sheets as west-dipping normal faults, antithetic to fault GS-1 (i.e. white arrows: faults 4-C, 4-B and 4-A). Deep earthquakes (25–50 km) are generated by compressive tectonics (crustal shortening) along the AP-5 thrust fault, as well as by incipient creeping along the most external nucleating crustal thrust fault (IC).

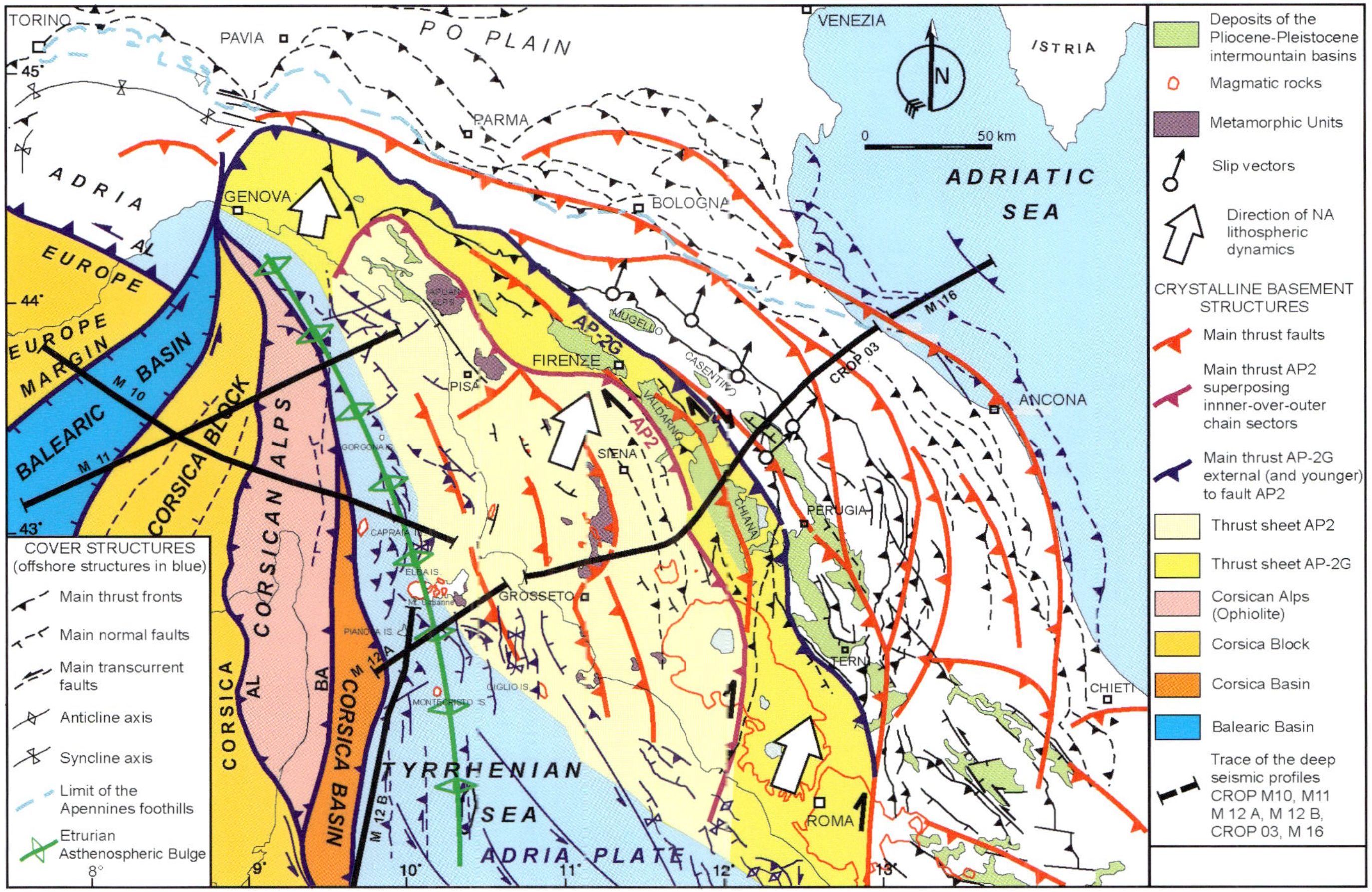

Figure 17: Structural sketch map of the Northern-Central Apennines showing the main lithospheric tectono-dynamic elements. The culmination of the Etrurian asthenospheric bulge is shown. This litho-asthenospheric compressive bulge is confined between an E-dipping Alpine Tethys slab subducted laterally to NW Adria in the Eo-Alpine stage (Upper Cretaceous–Lower Eocene) and a W-dipping Alpine Tethys slab subducted beneath Europe (Corsica) in the Balearic stage (Oligocene–Lower Miocene). The Corsica basin represents the closed foredeep of the Adria–Corsica convergence/collision and Alpine Tethys W-dipping subduction. The basin is floored by Alpine Tethys crust located near the culmination zone of the bulge and is filled by a wedge of flysch and Alpine Tethys units. To the south of the Corsica basin, the collision between the Corso-Sardinian block and W-Adria terminates because here Sardinia faced directly the Ionian Neotethys and its slab-retreat-related dynamics.

(2–15 km) may be induced by gravity sliding of the sedimentary pile along its sloping basement, or by a combination of gravity sliding and a right strike-slip motion (anticlockwise) of the AP-5 megathrust with inverted thrust planes of the pile (Figure 16). The deep events (25–50 km) are directly related to the shortening in the continental crust (Figure 16). In this model, crustal shortening along AP-5 (as well as its associated splay thrusts) produces the basement culmination which has been favouring the E-dipping gravitational sliding along normal faults GS1-3 responsible for the shallow seismicity (Figure 16). Active compressional tectonics along AP5 are also documented by the consistent deformation of the Adriatic sea bottom at its leading edge (fault AP5B-3a, s.p.1270 in profile M16; Plates 7 and 8) and account for the deep seismicity recorded in the Umbria–Marche area. Moreover in seismic signal analysis in the area identifies a deep compressive deformation in the lower crust interval related to an evident embrional reverse fault to which may be attributed initial creep activity (Figure 16: "IC").

8. GEODYNAMICAL INTERPRETATION AND CONCLUSIONS

The interpreted CROP sections, together with other sections in the CROP network provide new insights into the geodynamic evolution of the Balearic basin–Tyrrhenian basin–Northern Apennines system, which are briefly discussed below starting from the Northern Apennines–Northern Tyrrhenian basin system.

Analysis of the CROP sections shows that the NA is evidently dominated by a prominent lithospheric compressional deformation. Extensional structures play a secondary role, being restricted mostly to local and/or accommodating features unrepresentative of the basic tectonic style. In fact the trans-Apenninic CROP M16/CROP 03/M12A combined section through the whole NA orogenic system, including the hinterland sector, exhibits thrusting processes both at lithospheric and cover levels (Plate 8), and involving out-of-sequence cover thrusting, recurrent thrust reactivation and associated thrust-top basins. In the internal chain, recurring reactivations along the already emplaced main thrust fronts determined the detachment of thrust sheets at shallower levels, for example at the base of the main turbiditic foredeep sequences (Macigno, Cervarola and Marnoso-Arenacea units). These kinematics led to the local superposition of Tuscan units onto Ligurian units, especially in the thrust front areas.

Notably, the trans-Apenninic CROP-derived deep seismic data highlight for the first time that the NA are really composed of two superposed major chain portions: (1) an internal portion rooted in the Tyrrhenian area (thrusts AP1-2), and (2) an external one rooted below the axial area of the chain (thrusts AP3-5) and characterized by deep structures affecting the upper mantle. The internal chain portion was transported onto the external along the intracrustal thrust fault AP-2 and carries passively onto its edifice remnants of the Eo-Alpine chain (i.e., the Alpine Tethys Units). Another implication of this interpretation is that, during the Pliocene, two distinct deformation zones with the development of thrust-top basins characterized the Northern Apennines: an internal area controlled by splays of AP-2, and a more external area, at the leading edge of thrusts AP-4 and AP-5 (Plate 8).

This new interpretation of the CROP seismic sections and other geological-geophysical data allows us to provide a condensed tectono-dynamics of the NA thrust belt and regional features examined in the chapter as follows:

(a) Pre-orogenic Paleogeography. Two intraplate Neotethyan oceans separated the Africa, Europe and Adria plates: the *Ionian Neotethys* of Permo-Triassic opening, interposed between Adria and Africa, and the *Alpine Neotethys* Ligurian-Piedmontese basin of MU Jurassic opening, interposed between Adria and Europe in its NE portion, and between Africa and Europe in the SW part (Finetti, this volume, Chapter 34). The two basins floored by oceanic crust were communicating openly. Two fully CROP-seismic-controlled promontories, of great geodynamic importance, delimited the Alpine Neotethys from the Ionian Neotethys: the *Etrurian promontory* to the N and the *Panormid promontory* to the S (see Plate 8, small scale paleogeographic scheme; Finetti, this volume, Chapter 34).

These two ocean-floored paleobasins quietly trapped condensed sedimentary sequence from their opening to the first orogenic deformation, which occurred at different times in different places.

(b) Eo-Alpine Geodynamic stage (Upper Cretaceous–Lower Eocene). The Africa–Ionian Neotethys–Adria plate as one solid block converged with Europe, determining the closure of the Alpine Tethys sector interposed between Adria and Europe, and the continental collision with obduction of the Ophiolite, ophiolite-bearing units and Jurassic–Eocene cover on the southern margin of Europe, from the Betic area to the front of the Alps (Penninic Units).

The Alpine Tethys subducted under the Adria plate, and beneath NW Adria the laterally subducted Alpine Tethys slab was seismically identified by the CROP data and exhibited for the first time (Finetti et al., 2001). CROP sections M10 and M11 confirm now the existence of such a slab of Eo-Alpine subduction (Figures 3 and 4; Plates 2 and 4). Moreover, truncated by the Apennine fault plane AP-2, on section CROP 03, is exhibited the Europe-verging thrust fault AL2-5 of the Eo-Alpine stage (Plate 8). During this stage the western Adria margin located north of the present-day 41° parallel (WE-trending northern boundary of the Ionian Tethys) became an active sinistral transpressive zone along which the Eo-Alpine chain developed, with probable initial E-thrusting of Alpine Tethydes (Ligurides Units) over NW Adria, as Eo-Alpine backthrusting.

(c) South Alpine and First Apenninic–Balearic stage. The CROP network data show that, after the closure of the Alpine Tethys sector interposed between Adria and Europe (Alps segment) and collision of the two plates (Upper Cretaceous–Lower Eocene), *a wide unsubducted Tethys remnant persisted west of the Ligurian Alps* (Finetti, this volume, Chapter 34). This implies that the obduction of the Ophiolite, ophiolite-bearing units and cover of the Alpine Tethydes units (Ligurides) onto the European margin (Betic, Corsica, Coastal Chain of Calabria) may have been related to an intra-ocean, Europe-verging subduction process. Assuming this and observing that on CROP sections M10 (Plate 2; Figure 3) and M11 (Plate 4; Figure 4) a W-dipping Alpine Tethys slab subducted beneath Corsica (Europe) in the Balearic stage (Oligocene–Lower Miocene) is exhibited, it can be concluded that *the Corso-Sardinian block rotation and contemporaneous backarc Balearic basin opening and first Northern Apenninic thrust belt formation are driven by a relatively fast eastward retreat of the Alpine Tethys slab within a persistent context of much slower Africa–Europe convergence. At the end of the Balearic stage, Corsica and West Adria were colliding in the Corsica basin zone and its northward prolongation, and a prominent asthenospheric bulge (Etrurian Bulge) formed at the Adria margin, north of the* 41° *parallel.*

In the Alps sector, during this stage, the formation of the south-Alpine deformation took place. This is represented by a high-angle crustal backthrust emanating from a huge regional right transcurrent motion and consequent generation of an important complex positive flower along the Insubric Line (Finetti, this volume, Chapter 7). In the literature, the cause of rotation of the Corso-Sardinian block, among other hypotheses, has been attributed to indentation of the Adria plate (Schmid et al., 1996).

(d) Tyrrhenian stage. At the end of the Balearic stage (end of the Lower Miocene) the NA-Southern Apennine–Calabrian Arc–Sicilian Apennine–Maghrebian belt constituted a unique continuous arc extending from the Ligurian Sea to the Maghrebian chain. This prominent arc-shaped thrust belt was colliding (continent to continent) at its extremities with the two promontories separating the Alpine Tethys from the Ionian Tethys: the *Etrurian promontory* to the N (collision Corsica–Adria), and the *Panormid promontory* to the S (collision SE Sardinia–Africa). *These collisions stopped the anticlockwise rotation of the Corso-Sardinian block. The two promontories gave rise also to dissection of the large migrating arc mentioned above: the Etrurian promontory dissected the NA from the Southern Apennines, and the Panormid promontory dissected the SW–NE-trending Maghrebian chain from the NW–SE right-transcurrent related system of the Sicilian Apennines* (Finetti, this volume, Chapter 34).

Between the Etrurian and Panormid promontories, where the Ionian Neotethys continued the Alpine Tethys intraplate ocean, the Ionian slab retreat continued previous geodynamics, with southeasternward propagation of the orogenic deformation from the Middle Miocene to the Present, and opening of the Tyrrhenian backarc basin over the former Ionian Neotethys ocean (Finetti, this volume, Chapter 34).

Along the 41° parallel boundary of the Ionian Tethys, a fast eastward-migrating new Tyrrhenian mantle flux contributed to late orogenic anticlockwise motion accompanied by thrust-parallel strike-slip components, as observed in the field.

ACKNOWLEDGEMENTS

This work was supported by MIUR, COFIN 1998, 2000 and 2002, with fundings to I.R. Finetti, and COFIN 99 to the University of Florence. The authors are grateful to FINA Italiana (now TOTAL FINA ELF Italia SPA) for kindly providing the seismic lines on which the interpreted line-drawings of the Siena and Radicofani basins presented in the chapter are based. Many thanks are due to OGS, Seismic Division and to the OGS – Explora's seismic crews for CROP seismic data acquisition, to ENI, AGIP Division for the processing of section CROP 03, and to the EGG (Exploration Geophysics Group) of the University of Trieste, Faculty of Sciences, for their contribution to the iterated interpretive reprocessing of CROP sections M10, M11, M12A, M12B and M16. The signal quality improvement obtained was of fundamental importance for understanding the complex lithospheric setting of the studied area.

REFERENCES

Acocella, V., Pascucci, V., Dominici, G., 2002. Basin deformation due to laccolith emplacement at Radicofani (Southern Tuscany, Italy). Boll. Soc. Geol. Ital. spec. vol. (1), 749–756.

Amato, A., Azzara, R., Chiarabba, C., Cimini, G.B., Cocco, M., Di Buona, M., Margheriti, L., Mazza, S., Mele, F., Selvaggi, G., Basili, A., Boschi, E., 1998. The 1997 Umbria–Marche, Italy, earthquake sequence: a first look at the main shocks and aftershocks. Geophys. Res. Lett. 25, 2861–2864.

Angelier, J., Mecheler, P., 1977. Sur un mèthode graphique de recherche des contraintes principales ègalement utilisable en tectonique et en sismologie: La mèthode des diedres droits. Bull. Soc. Geol. Fr. 19, 1308–1318.

Argand, E., 1924. La tectonique de l'Asie. Comptes Rendus Congrès Géologique International, XIII, Belgique, 1922, v. 1, pp. 171–372.

Argnani, A., Bernini, M., Di Dio, G.M., Papani, G., Rogledi, S., 1997. Stratigraphic record of crustal-scale tectonics in the Quaternari of the Northern Apennines. Il Quaternario 10, 595–602.

Baldacci, F., Elter, P., Giannini, E., Giglia, G., Lazzarotto, A., Nardi, R., Tongiorgi, M., 1967. Nuove osservazioni sul problema della Falda Toscana e sulla interpretazione dei flysch arenacei di tipo "Macigno" dell'Appennino Settentrionale. Memorie della Società Geologica Italiana 6, 218–244.

Bally, A.W., Burbi, L., Cooper, C., Ghelardoni, R., 1986. Balanced sections and seismic reflection profiles across the central Apennines. Mem. Soc. Geol. Ital. 35, 257–310.

Banks, C.J., Warburton, J., 1986. "Passive-roof" duplex geometry in the frontal structures of the Kirthar and Sulaiman mountain belts. Pakistan. J. Struct. Geol. 8, 229–237.

Barberi, F., Gasparini, P., Innocenti, F., Villari, L., 1973. Volcanism of the southern Tyrrhenian Sea and its geodynamic implications. J. Geophys. Rcs. 78, 5221–5232.

Barchi, M., Guzzetti, F., Lavecchia, G., Lolli, O., Bontempo, R., 1988. Sezioni geologiche bilanciate attraverso il sistema a pieghe Umbro-marchigiano: 1. La sezione Trevi – Valle dell'Ambro. Boll. Soc. Geol. Ital. 107, 109–130.

Bemporad, S., Conedera, C., Dainelli, P., Ercoli, A., Facibeni, P., 1986. Landsat imagery: a valuable tool for regional and structural geology. Mem. Soc. Geol. Ital. 31, 287–298.

Bendkik, A., Boccaletti, M., Bonini, M., Poccianti, C., Sani, F., 1994. Structural evolution of the outer Apennine chain (Firenzuola-Città di Castello sector and Montefeltro area, Tuscan-Romagnan and Umbro-Marchean Apennine). Mem. Soc. Geol. Ital. 48, 515–522.

Benvenuti, M., 1995. Il controllo strutturale nei bacini intermontani plio-pleistocenici dell'Appennino Settentrionale: l'esempio della successione fluvio-lacustre del Mugello (Firenze). Il Quaternario 8, 53–60.

Bethoux, N., Cattaneo, M., Delpech, P.Y., Eva, C., Rehault, J.P., 1988. Mècanismes au foyer de seismes en Mer Ligure et dans le sud des Alpes Occidentales: rèsultats et interpretations. C. R. Acad. Sci. Paris, Ser. II 307, 71–77.

Bettelli, G., Boccaletti, M., Cibin, U., Panini, F., Poccianti, C., Rosselli, S., Sani, F., 2002. Note illustrative della carta geologica d'Italia alla scala 1:50000. FOGLIO 252–BARBERINO DI MUGELLO. Servizio Geologico Nazionale. 130 p.

Biju Duval, B., Dercourt, J., Le Pichon, X., 1977. From the Tethys Ocean to the Mediterranean Sea: plate tectonic model of the evolution of the western Alpine system. In: Biju Duval, B., Montadert, L. (Eds.), Structural History of the Mediterranean basins. Technimp, Paris, pp. 143–164.

Boccaletti, M., Guazzone, G., 1970. La migrazione terziaria dei bacini toscani e la rotazione dell'Appennino Settentrionale in una "zona di torsione" per deriva continentale. Mem. Soc. Geol. Ital. 9, 177–195.

Boccaletti, M., Guazzone, G., 1972. Gli archi appenninici, il mar Ligure ed il Tirreno nel quadro della tettonica dei bacini marginali retro-arco. Mem. Soc. Geol. Ital. 11 (2), 201–216.

Boccaletti, M., Guazzone, G., 1974. Remnant arcs and marginal basins in the Cainozoic development of the Mediterranean. Nature 252, 18–21.

Boccaletti, M., Sani, F., 1998. Cover thrust reactivations related to internal basement involvement during Neogene-Quaternary evolution of the northern Apennines. Tectonics 17, 112–130.

Boccaletti, M., Elter, P., Guazzone, G., 1971. Plate tectonic models for the development of the Western Alps and Northern Apennines. Nature, Phys. Sci. 234, 108–111.

Boccaletti, M., Coli, M., Napoleone, G., 1977. Nuovi allineamenti strutturali da immagini landsat e rapporti con l'attività sismica negli Appennini. Boll. Soc. Geol. Ital. 96, 679–694.

Boccaletti, M., Coli, M., Decandia, F.A., Giannini, E., Lazzarotto, A., 1980. Evoluzione dell'Appennino settentrionale secondo un nuovo modello strutturale. Mem. Soc. Geol. Ital. 21, 359–373.

Boccaletti, M., Conedera, C., Dainelli, P., Gocev, P., 1982. The recent (Neogene-Quaternary) regmatic system of the Western Mediterranean region. J. Petr. Geol. 5, 31–49.

Boccaletti, M., Calamita, F., Deiana, G., Gelati, R., Massari, F., Moratti, G., Ricci Lucchi, F., 1990. Migrating foredeep-thrust belt system in the northern Apennines and southern Alps. In: Boccaletti, M., Moratti, G. (Eds.), Neogene Paleogeography of the Perityrrhenian Area. Palaeogeogr. Palaeoclimatol. Palaeoecol. 77, 3–14.

Boccaletti, M., Bonini, M., Moratti, G., Sani, F., 1995. Nuove ipotesi sulla genesi e l'evoluzione dei bacini post-nappe in relazione alle fasi compressive neogenico-quaternarie dell' Appennino Settentrionale. Accad. Naz. Sci., Scritti Doc. 14, 229–262.

Boccaletti, M., Bonini, M., Moratti, G., Sani, F., 1999. Compressive Neogene-Quaternary tectonics in the hinterland area of the Northern Apennines. In: Boccaletti, M., Dahmani, M., Moratti, G. (Eds.), Neogene Sedimentation and Tectonics in the Western Mediterranean. J. Petrol. Geol. 22, 37–60.

Boncio, P., Lavecchia, G., 2000. A structural model for active extension in Central Italy. J. Geodyn. 29, 233–244.

Bonini, M., 1999. Basement-controlled Neogene polyphase cover thrusting and basin development along the Chianti Mountains ridge (Northern Apennines, Italy). Geological Magazine 136, 133–152.

Bonini, M., Sani, F., 2002. Extension and compression in the Northern Apennines (Italy) hinterland: Evidence from the Late Miocene–Pliocene Siena–Radicofani Basin and relations with basement structures. Tectonics 21 (3), 1/1–1/35, 10.1029/2001TC900024.

Bonini, M., Moratti, G., Sani, F., 1999. Evolution and depocentre migration in thrust-top basins: inferences from the Messinian Velona Basin (Northern Apennines, Italy). Tectonophysics 304, 95–108.

Bonini, M., Boccaletti, M., Moratti, G., Sani, F., 2001. Neogene crustal shortening and basin evolution in Tuscany. Ofioliti 26 (2a), 275–286.

Bortolotti, V., 1966. La tettonica trasversale dell'Appennino.I – La linea Livorno–Sillaro. Boll. Soc. Geol. Ital. 85, 529–540.

Bortolotti, V., Fazzuoli, M., Pandeli, E., Principi, G., Babbini, A., Corti, S., 2001. Geology of central and eastern Elba island, Italy. Ofioliti 26 (2a), 97–150.

Brogi, A., Lazzarotto, A., Lotta, D., Ranalli, G., 2003. Extensional shear zones as imaged by reflection seismic lines: the Larderello geothermal field (central Italy). Tectonophysics 363, 127–139.

Burrus, J., 1984. Contribution to a geodynamic synthesis of the provençal basin (north-western Mediterranean). Mar. Geol. 55, 247–269.

Calamita, F., Cello, G., Deiana, G., 1994. Structural styles, chronology rates of deformation and time–space relationships in the Umbria–Marche thrust system (central Apennines, Italy). Tectonics 13, 873–881.

Calamita, F., Coltorti, M., Piccinini, D., Pierantoni, P.P., Pizzi, A., Ripepe, M., Scisciani, V., Turco, E., 2000. Quaternary faults and seismicity in the Umbro-Marchean Apennines (Central Italy): evidence from the 1997 Colfiorito earthquake. J. Geodyn. 29, 245–264.

Cameli, G.M., Dini, I., Liotta, D., 1993. Upper crustal structure of the Larderello geothermal field as a feature of post-collisional extensional tectonics (southern Tuscany, Italy). Tectonophysics 224, 413–423.

Cameli, G.M., Dini, I., Liotta, D., 1998. Brittle/ductile boundary from seismic reflection lines of Southern Tuscany (Northern Apennines, Italy). Mem. Soc. Geol. Ital. 52, 153–162.

Carey, E., 1979. Recherche des directions principales des contraintes associees au jeu d'une population de failles. Rev. Geol. Dynam. Geogr. Phis. 21, 57–66.

Carmignani, L., Kligfield, R., 1990. Crustal extension in the Northern Apennines: The transition from compression to extension in the Alpi Apuane core complex. Tectonics 9, 1275–1303.

Carmignani, L., Decandia, F.A., Fantozzi, P.L., Lazzarotto, A., Liotta, D., Meccheri, M., 1994. Tertiary extensional tectonics in Tuscany (Northern Apennines). Tectonophysics 238, 295–315.

Cello, G., Nur, A., 1988. Emplacement of foreland thrust systems. Tectonics 7, 261–271.

Cello, G., Mazzoli, S., Tondi, E., Turco, E., 1997. Active tectonics in the central Apennines and possible implications for seismic hazard analysis in peninsular Italy. Tectonophysics 272, 43–68.

Cherchi, A., Montadert, L., 1982. Il sistema di rifting oligo-miocenico del Mediterraneo occidentale e sue conseguenze paleogeografiche sul Terziario sardo. Mem. Soc. Geol. Ital. 24, 387–400.

Collettini, C., Barchi, M., Pauselli, C., Federico, C., Pialli, G., 2000. Seismic expression of active extensional faults in northern Umbria (Central Italy). J. Geodyn. 29, 309–321.

Corti, G., Bonini, M., Mazzarini, F., Boccaletti, M., Innocenti, F., Manetti, P., Mulugeta, G., Sokoutis, D., 2002. Magma-induced strain localization in centrifuge models of transfer zones. Tectonophysics 348, 205–218.

Costantini, A., Lazzarotto, A., Sandrelli, F., 1982. Conoscenze geologico-strutturali. In: Il Graben di Siena. C.N.R. Progetto Finalizzato Energetica, RF9, pp. 11–32.

D'Agostino, N., Jackson, J.A., Dramis, F., Funiciello, R., 2001. Interactions between mantle upwelling, drainage evolution and active normal faulting: an example from the central Apennines (Italy). Geophys. J. Int. 147, 475–497.

Della Vedova, B., Bellini, S., Pellis, G., Squarci, P., 2001. Deep temperatures and surface heat flow distribution. In: Vai, G.B., Martini, I.P. (Eds.), Anatomy of an orogen: The Apennines and the Adjacent Mediterranean Basins. Kluwer Academic, pp. 65–76.

Dewey, J.F., Pitman, W.C., Ryan, W.B.F., Bonnin, J., 1973. Plate Tectonics and the evolution of Alpine system. Geol. Soc. Am. Bull. 84, 3137–3180.

D'Orazio, M., Laurenzi, M.A., Villa, I.M., 1991. $^{40}Ar/^{39}Ar$ dating of a shoshonitic lava flow of the Radicofani volcanic center (southern Tuscany). Acta Vulcanologica 1, 63–67.

D'Orazio, M., Foresi, L.M., Laurenzi, M., Sandrelli, F., Testa, G., 1995. Studio petrografico e cronologico di un livello tufitico intercalato nella successione continentale pre-evaporitica dell'area di Sassa (Provincia di Livorno). Studi Geologici Camerti spec. vol. 1995/1, 373–381.

Ekstrom, G., Morelli, A., Boschi, E., Dziewonski, A., 1998. Moment tensor analysis of the central Italy earthquake sequence of September–October 1997. Geophys. Res. Lett. 25, 1971–1974.

Elter, P., 1960. I lineamenti geologici dell'Appennino a nord ovest delle Apuane. Boll. Soc. Geol. Ital. 79, 273–312.

Elter, P., 1973. Lineamenti tettonici ed evolutivi dell'Appennino Settentrionale. In: Moderne vedute della geologia dell'Appennino. Atti Acc. Naz. Lincei 183, 97–109.

Elter, P., Giglia, G., Tongiorgi, M., Trevisan, L., 1975. Tensional and compressional areas in the recent (Tortonian to present) evolution of the northern Apennines. Boll. Geofis. Teor. Appl. 17, 3–18.

Eva, C., Solarino, S., 1992. Alcune considerazioni sulla sismotettonica dell'Appennino nord-occidentale ricavate dall'analisi dei meccanismi focali. Studi Geologici Camerti spec. vol. 1992/2. Appendix, pp. 75–83.

Faccenna, C., Davy, P., Brun, J.P., Funiciello, R., Giardini, D., Mattei, M., Nalpas, T., 1996. The dynamics of back-arc extension: an experimental approach to the opening of the Tyrrhenian Sea. Geophys. J. Int. 126, 781–795.

Fazzini, P., Gelmini, R., 1982. Tettonica trasversale nell'Appennino Settentrionale. Mem. Soc. Geol. Ital. 24, 299–309.

Ferrari, L., Conticelli, S., Burlamacchi, L., Manetti, P., 1996. Volcanological evolution of the Monte Amiata, Southern Tuscany: New geological and Petrochemical data. Acta Vulcanologica 8, 41–56.

Finetti, I.R., this volume. Crustal tectono-stratigraphic sections across the Western and Eastern Alps from ECORS-CROP and TRANSALP seismic data. Chapter 7.

Finetti, I.R., this volume. Geodynamic evolution of the Mediterranean Region, from the Permo-Triassic Ionian opening to the Present, constrained by new lithospheric CROP seismic data. Chapter 34.

Finetti, I.R., Del Ben, A., 1986. Geophysical study of the Tyrrhenian opening. Boll. Geof. Teor. Appl. 28, 75–156.

Finetti, I.R., Del Ben, A., this volume. Ionian Tethys lithosphere roll-back sinking and back-arc Tyrrhenian opening from new CROP seismic data. Chapter 21.

Finetti, I.R., Forlin, E., Pipan, M., this volume. CROP seismic data acqusition, processing and interpretative reprocessing. Chapter 5.

Finetti, I., Boccaletti, M., Bonini, M., Del Ben, A., Geletti, R., Pipan, M., Sani, F., 2001. Crustal section based on CROP seismic data across the North Tyrrhenian–Northern Apennines–Adriatic Sea. Tectonophysics 343, 135–163.

Foley, J.E., Toksoz, M.N., Batini, F., 1992. Inversion of teleseismic travel time residuals fro velocity structure in the Larderello geothermal field, Italy. Geophys. Res. Lett. 19, 5–8.

Foresi, L.M., Pascucci, V., Sandrelli, F., 1997. Sedimentary and ichnofacies analysis of the Epiligurian Ponsano Sandstone (Northern Apennines, Tuscany, Italy). Giorn. Geol. 59, 301–314.

Gasparini, C., Iannaccone, G., Scarpa, R., 1985. Fault-plane solutions and seismicity of the Italian peninsula. Tectonophysics 117, 59–78.

Ghelardoni, R., 1965. Osservazioni sulla tettonica trasversale dell'Appennino Settentrionale. Boll. Soc. Geol. Ital. 84, 277–290.

Giannini, E., Tongiorgi, M., 1959. Stratigrafia neogenica Toscana. 1 – L'Arenaria Elveziana di Ponsano (Volterra). Boll. Soc. Geol. Ital. 78, 83–100.

Hancock, P.L., Chalmers, R.M.L., Altunel, E., Çakir, Z., 1999. Travitonics using travertines in active faults studies. J. Struct. Geol. 21, 903–916.

Hill, K.C., Hayward, A.B., 1988. Structural constraints on the Tertiary plate tectonic evolution of Italy. Mar. Pet. Geol. 5, 2–16.

Jolivet, L., Faccenna, C., 2000. Mediterranean extension and the Africa–Eurasia collision. Tectonics 19, 1095–1106.

Landi, B., Tanini, C., Sani, F., Carobbi, S., 1995. Evoluzione geologico-strutturale del bacino di Cinigiano-Baccinello (Toscana meridionale). Stud. Geol. Camerti spec. vol., 1995/1, 475–488.

Lavecchia, G., 1988. The Tyrrhenian–Apennines system: Structural setting and seismotectogenesis. Tectonophysics 147, 263–296.

Lavecchia, G., Stoppa, F., 1996. The tectonic significance of italian magmatism: An alternative view to the popular interpretation. Terra Nova 8, 435–446.

Lazzarotto, A., Sandrelli, F., 1977. Stratigrafia e assetto tettonico delle formazioni neogeniche nel bacino del Casino (Siena). Boll. Soc. Geol. Ital. 96, 747–761.

Liotta, D., 1991. The Arbia–Val Marecchia Line, Northern Apennines. Eclog. Geol. Helv. 84, 413–430.

Liotta, D., Ranalli, G., 1999. Correlation between seismic reflectivity and rheology in extended lithosphere; southern Tuscany, inner Northern Apennines, Italy. Tectonophysics 315, 1–4, 109–122.

Locardi, E., 1982. Individuazione di strutture sismogenetiche dall'esame dell'evoluzione vulcano- tettonica dell'Appennino e del Tirreno. Mem. Soc. Geol. Ital. 24, 569–596.

Makris, J., Egloff, F., Nicolich, R., Rihm, R., 1999. Crustal structure from the Ligurian Sea to the Northern Apennines – a wide angle seismic transect. Tectonophysics 301, 305–319.

Malinverno, A., Ryan, W.B.F., 1986. Extension in the Tyrrhenian Sea and shortening in the Apennines as a result of arc migration driven by sinking of the lithosphere. Tectonics 5, 227–245.

Mantovani, E., Albarello, D., Tamburelli, C., Babbucci, D., Viti, M., 1997. Plate convergence, crustal delamination, extrusion tectonics and minimization of shortening work as main controlling factors of the recent Mediterranean deformation pattern. Ann. Geofis. 40, 611–643.

Martini, I.P., Sagri, M., 1993. Tectono-sedimentary characteristics of late Miocene–Quaternary extensional basins of the Northern Apennines, Italy. Earth Sci. Rev. 34, 197–233.

Martinod, J., Davy, P., 1992. Periodic instabilities during compression or extension of the lithosphere, 1, Deformation models from an analytical perturbation model. J. Geophys. Res. 97, 1999–2014.

Martinod, J., Davy, P., 1994. Periodic instabilities during compression of the lithosphere, 2, Analogue experiments. J. Geophys. Res. 99, 12057–12069.

Mauffret, A., Contrucci, I., Brunet, C., 1999. Structural evolution of the Northern Tyrrhenian Sea from new seismic data. Marine Petrol. Geol. 16, 381–407.

Minissale, A., Magro, G., Martinelli, G., Vaselli, O., Tassi, G.F., 2000. Fluid geochemical transect in the Northern Apennines (central-northern Italy): fluid genesis and migration and tectonic implications. Tectonophysics 319, 199–222.

Mongelli, F., Zito, G., 1991. Flusso di calore nella regione Toscana. Studi Geologici Camerti spec. vol. 1991/1, 91–98.

Mongelli, F., Palumbo, F., Puxeddu, M.I., Villa, M., Zito, G., 1998. Interpretation of the geothermal anomaly of Larderello, Italy. Mem. Soc. Geol. Ital. 52, 305–318.

Moratti, G., Bonini, M., 1998. Structural development of the Neogene Radicondoli–Volterra and adjoining hinterland basins in Western Tuscany (Northern Apennines, Italy). Geolog. J. 33, 223–241.

Morley, C.J., 1988. Out of sequence thrusts. Tectonics 7, 539–561.

Mutti, E., 1992. Turbidite sandstones. Agip Spa – S. Donato Milanese. p. 165.

Mutti, E., Normark, W.R., 1987. Comparing examples of modern and ancient turbidite systems: problems and concepts. In: Leggett, J.K., Zuffa, G.G. (Eds.), Marine clastic sedimentology. Graham and Trotam, London, pp. 1–38.

Rehault, J.P., Boillot, G., Mauffret, A., 1984. The western Mediterranean basin geological evolution. Mar. Geol. 55, 447–477.

Riba, O., 1976. Syntectonic unconformities of the Alto Cardener, Spanish Pyrenees: a genetic interpretation. Sedim. Geol. 15, 213–233.

Ricci Lucchi, F., 1986. The Oligocene to Recent foreland basins of the northern Apennines. Spec. Publ. Int. Assoc. Sediment. 8, 105–139.

Rocchi, S., Westerman, D.S., Dini, A., Innocenti, F., Tonarini, S., 2002. Two-stage growth of laccoliths at Elba Island, Italy. Geology 30, 983–986.

Rook, L., Ghetti, P., 1997. Il bacino neogenico della Velona (Toscana, Italia): stratigrafia e primi ritrovamenti di vertebrati fossili. Boll. Soc. Geol. Ital. 116, 335–346.

Roure, F., Casero, P., Vially, R., 1991. Growth processes and melange formation in the southern Apennines accretionary wedge. Earth Planet Sc. Lett. 102, 395–412.

Roure, F., Howell, D.G., Guellec, S., Casero, P., 1990. Shallow structures induced by deep-seated thrusting. In: Letouzey, J. (Ed.), Petroleum and Tectonics in Mobile Belts. Technip, Paris, pp. 15–30.

Roeder, D., 1980. Geodynamics of the Alpine–Mediterranean system – a synthesis. Eclog. Geol. Helv. 73, 353–377.

Royden, L.H., Patacca, E., Scandone, P., 1987. Segmentation and configuration of subducted lithosphere in Italy: An important control on thrust-belt and foredeep-basin evolution. Geology 15, 714–717.

Salvador, A., 1987. Unconformity-bounded stratigraphic units. Geol. Soc. Am. Bull. 98, 232–237.
Sani, F., 1990. I thrust appenninici esterni nel settore Firenzuola-Bocca Trabaria. Evoluzione deformativa ed implicazioni strutturali. Studi Geologici Camerti Spec. vol., 87–97.
Sani, F., 1991. Rilevamento strutturale della catena Appenninica nel settore Firenzuola-Bocca Trabaria. Memorie Descrittive Carta Geologica d'Italia 46, 327–337.
Sani, F., Moratti, G., Bonini, M., Landi, B., Tanini, C., Piccardi, L., Menichetti, B., 2001. A transect in southern Tuscany from Baccinello basin to the Cetona ridge. Excursion 25–29 September 1997. In: Proceedings of the meeting "The Elba Island: a key puzzle linking the Corso-Sardinian Massif and Adria". Ofioliti 26 (2a), 281–400.
Sartori, R., 1990. The main results of ODP Leg 107 in the frame of Neogene to recent geology of Perityrrhenian areas. In: Kastens, K.A., Mascle, J. (Eds.), Proceeding of the Ocean Drilling Program, Scientific Results, vol. 107, pp. 715–730.
Schmid, S.M., Pfiffner, O.A., Froitzheim, N., Schönborn, G., Kissling, E., 1996. Geophysical-geological transect and tectonic evolution of the Swiss–Italian Alps. Tectonics 15, 1036–1064.
Scholle, P.A., 1970. The Sestri–Voltaggio Line: a transform fault induced tectonic boundary between the Alps and the Apennines. Am. J. Sci. 269, 343–359.
Serri, G., Innocenti, F., Manetti, P., 2001. Magmatism from Mesozoic to Present: petrogenesis, time–space distribution and geodynamic implication. In: Vai, G.B., Martini, I.P. (Eds.), Anatomy of an orogen: The Apennines and the Adjacent Mediterranean Basins. Kluwer Academic, pp. 77–104.
Speranza, F., Villa, I.M., Sagnotti, L., Florindo, F., Cosentino, D., Cipollari, P., Mattei, M., 2002. Age of Corsica–Sardinia rotation and Liguro–Provençal Basin spreading: new paleomagnetic and Ar/Ar evidence. Tectonophysics 347, 231–251.
Tapponnier, P., 1977. Evolution tectonique du système alpin en Méditerranée : poinçonnement et écrasement rigide-plastique. Bull. Soc. Géol. Fr. 7, 437–460.
Tomaselli, A., Pastore, S., Augliera, P., Eva, C., 1992. Sismicità dell'Appennino nord-occidentale. Studi Geologici Camerti spec. vol. 1992/2, 43–50.
Van Bemmelen, R.W., 1972. Driving forces of Mediterranean orogeny (Tyrrhenian test-case). Geol. Mjinbouw 51, 548–573.
Ventura, B., Pini, G.A., Zuffa, G.G., 2001. Thermal history and exhumation of the Northern Apennines (Italy): evidence from combined apatite fission track and vitrinite reflectance data from foreland basin sediments. Basin Res. 13, 435–448.
Vigliotti, L., Langenheim, V.E., 1995. When did Sardinia stop rotating? New Palaeomagnetic results. Terra Nova 7, 424–435.
Villa, I., Puxeddu, M., 1994. Geochronology of the Larderello geothermal field: new data and the "closure temperature" issue. Contrib. Mineral. Petrol. 315, 415–426.
Wezel, F.C., 1982. The Tyrrhenian Sea: a rifted krikogenic-swell basin. Mem. Soc. Geol. Ital. 24, 531–568.
Wortel, M.J.R., Spakman, W., 1992. Structure and dynamics of subducted lithosphere in the Mediterranean region. Proc. K. Ned. Akad. Wet., Biol. Chem. Geol. Phys. Med. Sci. 95, 325–347.

LIST OF CHAPTER PLATES

Plate 1: CROP seismic sections M10.

Plate 2: Interpreted seismogeological section M10.

Plate 3: CROP seismic sections M11.

Plate 4: Interpreted seismogeological section M11.

Plate 5: CROP seismic sections M12B.

Plate 6: Interpreted seismogeological section M12B.

Plate 7: CROP seismic sections M12A/CROP 03/M16.

Plate 8: Interpreted seismogeological section M12A/CROP 03/M16.

CROP PROJECT: Deep Seismic Exploration of the Central Mediterranean and Italy
Edited by I.R. Finetti

CHAPTER 9

Crustal Geological Section across Central Italy from the Corsica Basin to the Adriatic Sea Based on Geological and CROP Seismic Data

I.R. Finetti[1,*], F. Calamita[2], U. Crescenti[2], A. Del Ben[1], E. Forlin[1], M. Pipan[1], A. Prizzon[1], G. Rusciadelli[2], V. Scisciani[2]

ABSTRACT

Geological, geophysical and new CROP seismic data have been used to reconstruct and discuss a new crustal setting for the Central Apennines. The Apenninic chain of the Italian peninsula is formed by two main arcs (that of the Central-Northern Apennines and that of the Southern Apennines), decoupled along the Sangro-Volturno line.

In the central sector of the Apennine chain, prominent folds and thrusts involve the Triassic-Miocene carbonate succession of the Adria continental margin and of the foreland and foredeep Neogene-Quaternary basins (Apennine Units *s.s.*). To the SE of the Sangro-Volturno line, the Molise Allochthonous Units, related to the Neogene-Quaternary closure of the Ionian Neotethys, overthrust the Apennine Units s.s. In the westernmost zone (Tuscan-Latium area), the Apennine Units remained below the internal allochthonous ones (Alpine Tethydes on Ligurian Units) connected to the closure of the Alpine Neotethys. In the peri Adriatic area, the frontal zone of the chain is buried by the Quaternary succession. The main tectono-stratigraphic units outcropping east of the Quaternary volcanic deposits and the Pliocene-Quaternary deposits of the peri-Tyrrhenian area are formed by the Sibillini Mts.–Antrodoco-Olevano thrust (Sabine Units), the Gran Sasso (Latium-Abruzzi Units) and Mt. Morrone thrusts, and by the Teramo and Maiella thrusts (external units of the Messinian and Pliocene foredeeps). These thrusts have project out, and are locally conditioned by the Mesozoic distribution of the pelagic basins and carbonate platforms established since the Triassic-Liassic rifting processes. The architecture of the paleomargin has also constrained the physiography of the Neogene foredeep basins, whose depocenters are in many cases located in correspondence with the pelagic paleobasins. The Pliocene thrusts located on the Marchean-Abruzzi Apennine mountain front (M.gna dei Fiori-Gran Sasso–Mt. Morrone, Maiella) have produced a structural elevation of the top of the carbonate succession of about 10 km and horizontal displacements of about 45 km. Southward, these thrusts are joined to the NE-SW-oriented Sangro-Volturno dextral transpressive oblique ramp. An analogous kinematic role characterizes the Sibillini Mts.–Antrodoco-Olevano thrust, the southern segment of the Ancona-Anzio (*Auct.*) thrust system.

The Sangro-Volturno oblique ramp constrains to the south the Central-Northern Apennine arc, which is segmented into the Northern and Central Apennine arcs by the Sibillini Mts.–Antrodoco-Olevano oblique ramp. The structural elevation of the Apennine Units s.s. within the Central-Northern Apennines is due to a greater shortening than in to the Southern Apennines, where the allochthonous units outcrop, whereas the latter are eroded in the Marchean-Abruzzi Apennines. The axial culminations of the arcs of the Northern Apennines (the Sibillini Mts. area) and the Central Apennines (the Gran Sasso area) correspond to the apex zones of their arcs where maximum shortening took place.

In the Central-Northern Apennines axial zone, NW–SE trending normal faults with Quaternary activity are present, characterized by a SE-ward increase in the down-throw of up to 1000 m in the Abruzzi area. They are generated by gravity sliding of the sedimentary

*Corresponding author: Tel.: +39 40 5582276; e-mail: finetti@units.it (I.R. Finetti)
[1]Dipartimento DISGAM, University of Trieste, Via Weiss, 1, 34127 Trieste, Italy
[2]Dipartimento di Scienze della Terra, Università degli Studi "G. D'Annunzio" di Chieti e Pescara, Campus Universitario Madonna delle Piane, Via dei Vestini, 30 Chieti Scalo (CH), Italy

cover over the eastward sloping basement of the major composed thrust block of the Northern Apennines. Sliding and combined inversion of previous thrust faults of the piled cover stack constitute the basic mechanism of the high active shallow seismicity in the area. Deep compressive events derive directly from major thrust faults.
New CROP data (lines M-15 and M-37), integrated with other geological and geophysical data, were used to reconstruct a lithospheric section across Central Italy, from the Corsica Basin to the Adriatic Sea (Plate-2). The reconstructed section leads to new observations on the evolution of the Central Apennines. The pre-orogenic paleogeography, as constrained by the new CROP data, consisted of an oceanic domain, opened during Permo-Triassic times, and corresponding to the Ionian Tethys, delimited to the north by the present-day 41° parallel; this ocean was subsequently in communication with the Alpine Tethys, related to Jurassic opening.
The reconstructed lithospheric setting allows us to discriminate between the various geodynamic stages. After the Eo-Alpine subduction of the Alpine Tethys beneath the Adria plate (Upper Cretaceous–Lower Eocene) a wide Alpine Tethys remained unsubducted west of the Alps. In the Balearic stage, the subduction of the Alpine Tethys remnant beneath the rotated Corsica–Sardinia block and the westernmost European margin took place. Between the eastward Eo-Alpine and the westward Balearic subductions, a prominent eastward migrating asthenospheric bulge was created between Corsica and the NW Adria margin. It acted as the basic geomotor for the Northern Apennines deformation during the last 25 Ma (Late Balearic–Tyrrhenian stage). In continuity with the end of the Balearic stage, the Ionian subduction (Middle Miocene to present) started with the formation of the Tyrrhenian back-arc, as a consequence of slab retreat of the Ionian oceanic crust, with formation of wedge skinned nappes (Ionides or Lagonegro), tectonically transported into the Southern Apennines orogenic system, to the south of the 41° parallel (northern boundary of the Ionian Tethys).

Keywords: Central Apennines, Crustal Seismogeological Section, Evolutive Geodynamics

1. INTRODUCTION

The tectonic setting of the Northern Apennines has been recently reconstructed from seismic interpretation of "CROP" seismic data (Barchi et al., 1998; Decandia et al., 1998; Finetti et al., 2001). Thrusts in the crust and the lithospheric mantle have been recognized on the Apennine chain mountain front (Barchi et al., 1998; Finetti et al., 2001). Other authors (Decandia et al., 1998) think that the Umbria-Marche chain developed according to a thin-skinned tectonic style.

Various geodynamic models have been proposed for the Tyrrhenian–Apennines system: (1) back-arc-related extension (Boccaletti et al., 1971; Malinverno and Ryan, 1986; Finetti and Del Ben, 1986; Royden et al., 1987; Wortel and Spakman, 1992; Finetti et al., 1996); (2) delamination (Roeder, 1980); (3) mantle uprise causing gravity-sliding (Van Bemmelen, 1972; Locardi, 1982; Wezel, 1982), or variations on this (Lavecchia, 1985; Lavecchia and Stoppa, 1996); (4) lateral extrusion of the Apennines chain following the N–S trending collision between the African and European plates (Tapponier, 1977; Boccaletti et al., 1982; Faccenna et al., 1996; Mantovani et al., 1997); (5) late orogenic collapse of an overthickened accretionary wedge (Carmignani and Kligfield, 1990; Carmignani et al., 1994).

The geological and CROP datasets presented in this chapter were used to analyse the lithospheric setting of Central Italy (Figure 1; Plates 1, 2 and 3), from the Corsica Basin to the Adriatic Sea, with the aim of reconstructing: (a) the pre-orogenic paleogeographic architecture of the Adria, Africa and European continental margins and interposed oceanic domains (Ionian and Alpine Neotethys); (b) the evolving geodynamic context of the Tyrrhenian-Apennines system.

The Apennine chain is formed from internally derived structural units (Ligurian/Sicilides) overthrust on the carbonate units belonging to the Adria paleomargin (Apennine Units s.s.). In the central-southern sector, the Lagonegro-Molise Units are interposed between the Ligurian-Sicilides and the Apennine Units s.s. The Ligurian-Sicilides and Lagonegro-Molise Units derive, respectively, from the cover of the oceanic crust of the

Figure 1: Position of the reconstructed Lithospheric Seismogeological section ABCDE, based on: ***Segment AB*****: CROP Section M-37.** ***Segment BC*****:** ***Sedimentary Cover*** **from direct and/or extrapolated Geophysical data (Seismics, Gravity and Magnetics) and Geological data;** ***Pre-Sedimentary Crust*** **from extrapolated-interpolated CROP data (mostly CROP-03 and CROP-04) and gravity data.** ***Segment CD*****:** ***Sedimentary Cover and Upper Basement*** **from available seismic data of hydrocarbon exploration;** ***Pre-Sedimentary Cover*** **from interpolation–extrapolation of CROP data (CROP-03 & CROP-04) and gravity data.** ***Segment DE*****: CROP section M-15.**

Alpine and Ionian Neotethys basins, separated by the Panormid and Etrurian promontories, which are the origin of the internal carbonate platform allochthonous units (Finetti and Del Ben, 2000; Lentini et al., 2002; Finetti, this volume, Chapter 34).

The evolution of the Apennine orogeny started in the Tertiary as a consequence of subduction of Alpine Tethys and Ionian oceanic crust, accompanied by the convergence of the European and Adria-African continental blocks (Ben-Avraham et al., 1990; Finetti et al., 1996; Gattacceca and Speranza, 2002; Ciarapica and Passeri, 2002; Lentini et al., 2002; Finetti et al., this volume, Chapter 8; Finnetti et al., this volume, Chapter 12).

The Ligurian-Sicilide units represent the accretionary prism, connected to the Cretaceous-Eocene subduction beneath Adria of the Alpine Neotethys oceanic crust, followed by the Adria–Europe (Alps) collision and lateral (Balearic stage) subduction beneath Europe (Finetti et al., this volume, Chapter 8). They were involved, starting from the Upper Oligocene, in the deformation of the Northern Apennine Adria paleomargin (Boccaletti et al., 1990). Subsequently, the subduction of the Ionian ocean, beginning in the Middle Miocene, produced the accretionary prism formed by the Lagonegro-Molise Units and the external flysch of the Southern Apennines that eastward overthrust the Adria paleomargin Apennine Units s.s., formed by the Apulian carbonate platform (Mostardini and Merlini, 1986; Casero et al., 1988; Finetti et al., 1996; Menardi Noguera and Rea, 2000; Lentini et al., 2002).

The Apennines of the Italian peninsula are organized into two main arcs (Figure 1): the Central-Northern Apennine Arc and the Southern Apennine Arc. The Lazio-Abruzzi sector (Central Apennines) represents the zone of junction between the Central-Northern Apennine Arc and that of the Southern Apennines, respectively with NE and SE convexity. This sector is delimited to the west and east by two important tectonic elements known as the Olevano–Antrodoco line (Parotto and Praturlon, 1975) and the Sangro–Volturno line (Ghisetti and Vezzani, 1997).

The Olevano-Antrodoco line has been interpreted as an oblique ramp, which envelops various thrust planes, probably connected to the presence of pre-existing discontinuities that have conditioned the geometry of the orogenic system (structures of the chain and geometry of the foredeep basins: Cavinato et al. (1986), Bally et al. (1988), Cipollari and Cosentino (1992), Calamita and Deiana (1995)).

In the Umbro-Marche-Sabina and Lazio-Abruzzi sectors of the Apennine chain, the carbonate units related to the deformation of Adria crust (articulated as carbonate platform and pelagic basins, and differentiated in the Upper Triassic rifting process) and the foredeep siliciclastic deposits, connected with the migration of the orogenic system, outcrop. In the Northern Apennines (Umbria-Marche-Romagna sector) the tectonic style of the chain is characterized by an imbricate geometry with involvement of the basement (Barchi et al., 1998; Finetti et al., 2001), while a duplex tectonic style characterizes the Southern Apennines, starting from the Molise sector (Mostardini and Merlini, 1986).

The Umbria-Marche-Sabina and Lazio-Abruzzi Apennines, in particular, show a marked axial culmination of the chain structures, emphasized by the presence of Triassic carbonate units with elevations of up to 3000 m in the Abruzzi sector, and the presence of very different structural trends (Figure 3): N–S (Sabine Units), E–W (Gran Sasso), N–S and NNW–SSE (Gran Sasso, Montagna dei Fiori and Maiella), NW–SE (Mt. Simbruino-Ernici, Mt. Sirente, Mt. Morrone). This structural trend complexity is attributed to constraints by the Adria Mesozoic paleomargin on the geometry of the chain structures and the physiography of the foredeep basins (Koopman, 1983; Calamita and Deiana, 1988; Scisciani et al., 2001, 2000b; Calamita et al., 2003; Tozer et al., 2002).

Other authors connect this structural complexity to the out-of-sequence development of the chain structures and to the strike slip tectonics (Lavecchia et al., 1988; Ghisetti et al., 1991; Mattei et al., 1992). The orogenic system involves the sector of the Apennines analyzed in this chapter beginning from the Lower Miocene, and progressively developing, eastward, separate foredeep basins located in front of the chain and characterized by the presence of thrust top basins (Patacca et al., 1991). Along the axial zone of the Apennine ridge, prominent tectonic depressions are present (Fucino, Sulmona, Campo Imperatore – "Gran Sasso" and L'Aquila intramontane basins), accompanied by normal faults systems with down-throws of even up to 1000 m. These faults are located in an active tectonic contest, as shown by the intense seismicity of the area (Calamita et al., 1999, 2000; Boncio et al., 2000). The Fucino trough is evidently associated with a prominent left-transcurrent fault running on its NW boundary (Figure 21).

The main structural units and relative structural setting of the Central Apennines are herein described, with respect to their structural order, and to the most accredited distribution of pre- and syn-orogenic paleogeographic domains.

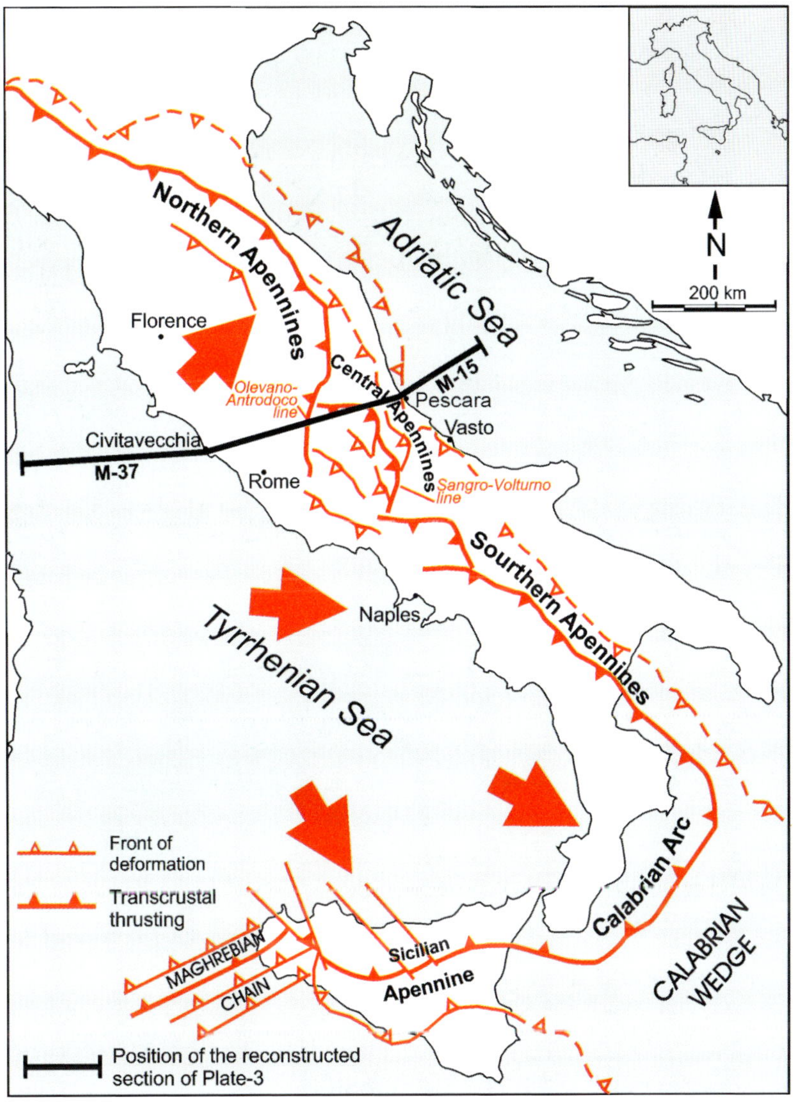

Figure 2: Regional scheme of the Apennines orogenic system: Northern Apennine Arc; Southern Apennine Arc; Calabrian Arc (on Ionian Oceanic crust forming the Ionian slab); Sicilian Apennine Arc connected to the Tyrrhenian stage geodynamics. The Maghrebian Chain is connected to the Balearic stage geodynamics with Alpine Tethys subduction.

The geological, structural and geophysical data were of basic importance for the reconstruction of the sedimentary cover internal to the trans-Apenninic segment (Civitavecchia–Pescara) of the crustal section in Plate 3. Basement, upper crust and lower crust of this section are reconstructed from interpolation-extrapolation of CROP data (CROP 03, 04) and other geophysical data (gravity and seismic lines of medium penetration).

2. REGIONAL STRATIGRAPHIC AND TECTONIC SETTING

The main structural units and relative structural domains of the Central Apennines are here described, with particular respect to their structural order, and referring to the most accredited distribution of pre- and syn-orogenic paleogeographic domains.

2.1. Surface and sub-surface geological-structural and geophysical data of Central Italy

In the Umbria-Marche-Sabine and Lazio-Abruzzi Apennines, various tectonic stratigraphic units are present, related to the deformation of the Adria paleomargin. They are also related to thrust planes that, in general, project out with an eastward convexity, and enveloped by the whole arc of the Central-Northern Apennines (Figures 2

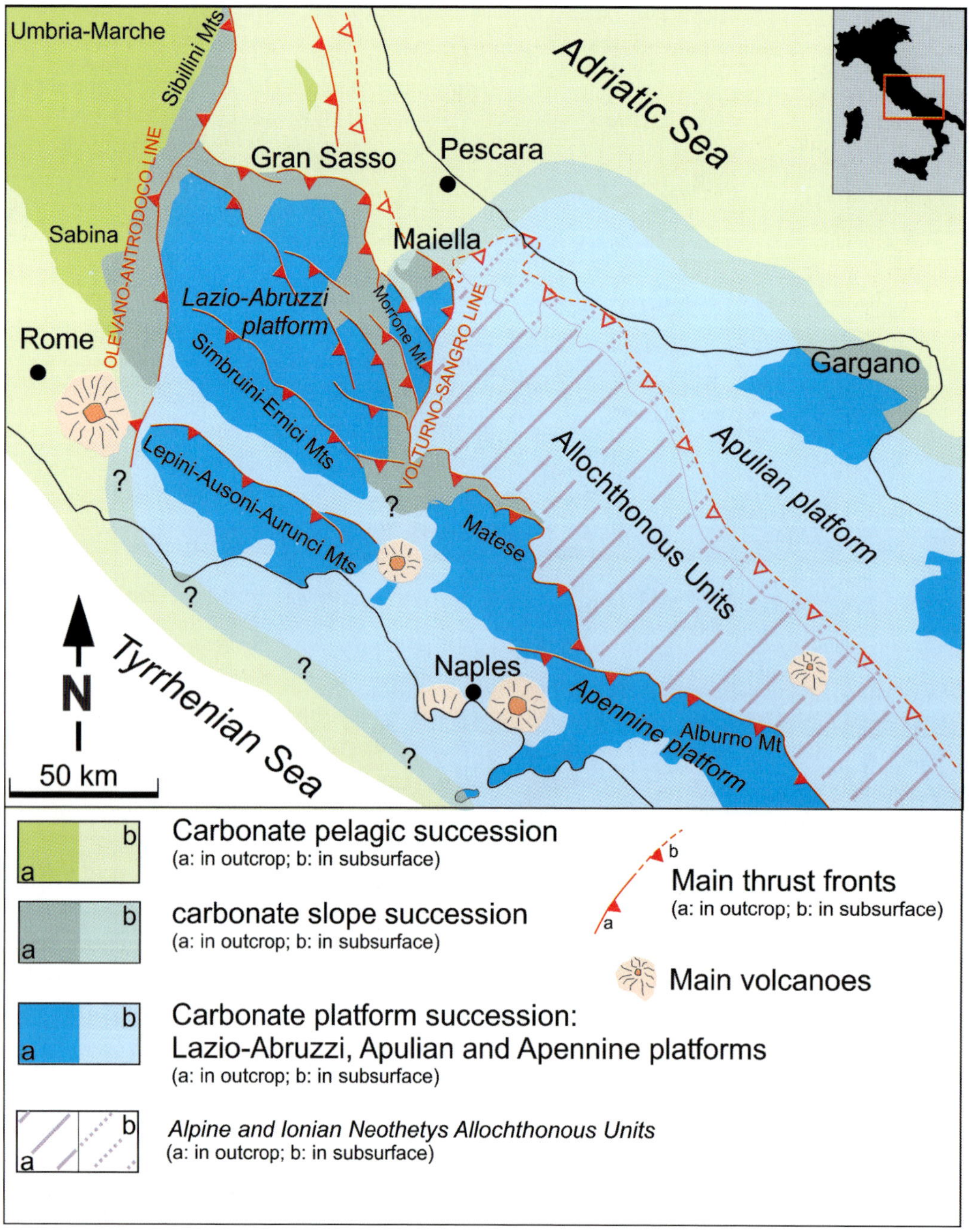

Figure 3: Meso-Cenozoic paleo-domains (carbonate platforms and pelagic basins) involved in the Central-Southern Apennines.

and 4). The thrusts of the Sibillini Mts., Gran Sasso and Mt. Morrone juxtaposed the carbonate sequences onto the siliciclastic deposits belonging to the Messinian-Pliocene foredeep basins of the peri-Adriatic foothills (outer units of the Messinian and Pliocene foredeeps).

In general, these minor arcs coincide with the Mesozoic paleogeographic domains, that trend N–S and E–W (Figure 3). The Sibillini Mts.–Olevano-Antrodoco and the Gran Sasso thrusts are located in correspondence with the areas of transition in the carbonate platforms and pelagic basins. The elements of this paleomargin are preserved where they are nearly perpendicular to the structures of the chain (the Mt. Morrone and Maiella Units).

In the central sector of the Apennine chain (Figure 4), the following tectono-stratigraphic units are present:

(a) *Ligurian-Sicilides Units*: "*Allochthonous Alpine Neotethys*";
(b) *Tuscan Units*;

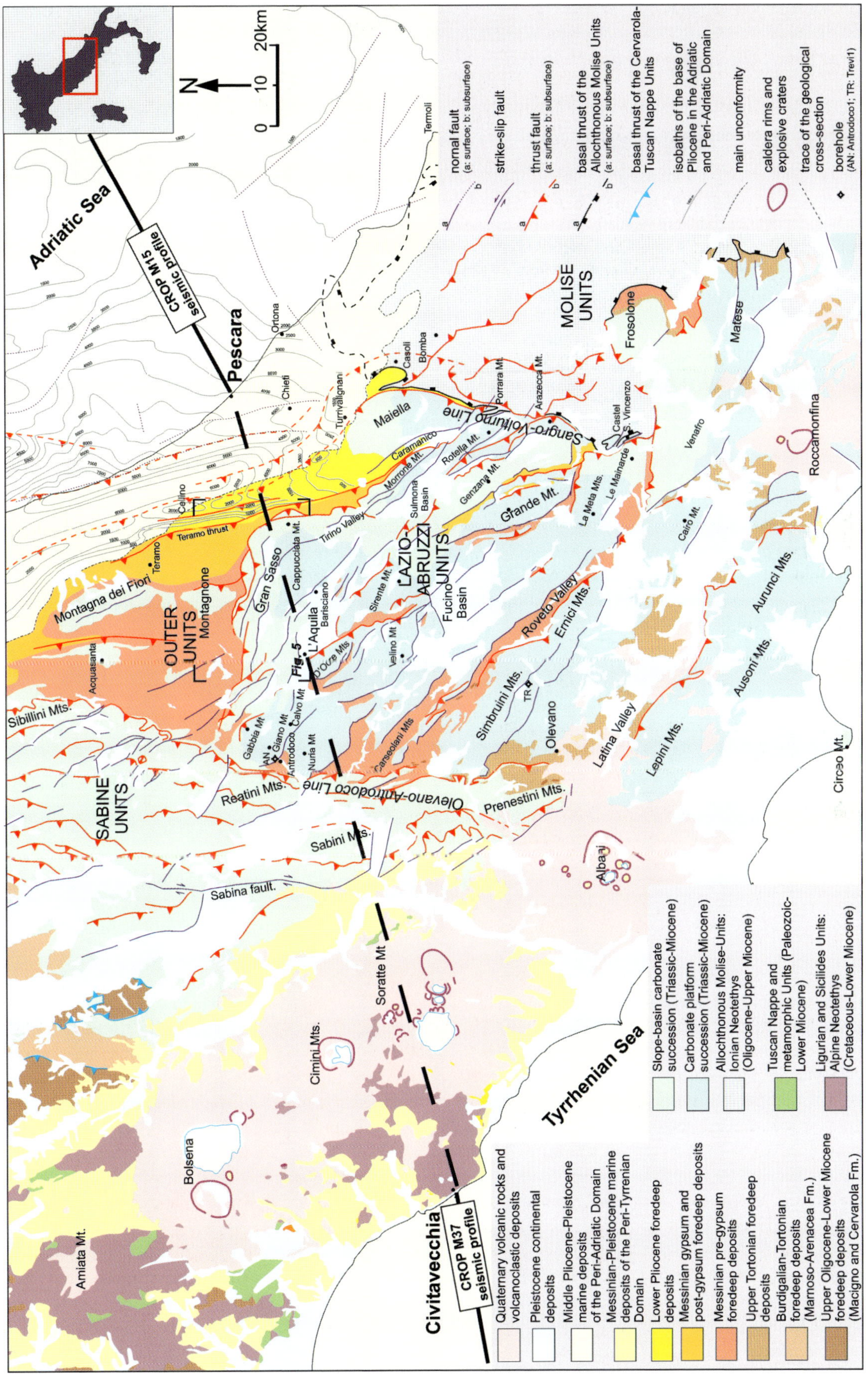

Figure 4: Structural map of the Central Apennines.

(c) *Sabine Units*;
(d) *Lazio-Abruzzi carbonate platform Units*;
(e) *Outer Units of the Messinian and Pliocene foredeeps*;
(f) *Molise Units*: *"Allochthonous Ionian Neotethys"*.

From W to E, and therefore from the upper and more ancient units to the more recent lower ones, and external to the orogenic prism of the Apennine chain, the units are characterized as follows:

(a) The Alpine Neotethys Units – these are formed from shales, marls, limestones and sandstones (Upper Cretaceous–Oligocene) that represent the flysch generated with the construction of the accretionary prism at the closing of the Alpine Tethys. These units together with the Tuscan ones are buried mainly by the Lazio Quaternary volcanoclastic deposits. They crop out in the Mt. Amiata and Civitavecchia sectors, and have been recorded in wells, where they lie on the Tuscan and Sabine Units (Cosentino and Parotto, 1992).

(b) The presence of a "Tuscan succession" in the structure of Mt. Soratte suggests its affiliation with the Tuscan Nappe, as is also confirmed by the timing of chain involvement (Upper Burdigalian: Cipollari et al., 1995).

(c) The Sabine Units – these derive from the deformation of the transitional Mesozoic paleodomain located between the Lazio-Abruzzi carbonate platform and the Umbro-Marche pelagic basin (Cosentino and Parotto, 1986). The succession is formed from Triassic dolomites and Liassic neritic limestones belonging to the carbonate platform, passing upward into calcareous-silico-marly terms (Middle Lias–Middle Miocene), with calcareous-clastic deposits enriched progressively near the Lazio-Abruzzi carbonate platform. The Sabine Units are a N–S trending overturned fold and related thrust system, which developed during the Upper Miocene–Lower Pliocene (Parotto and Praturlon, 1975); the thrust planes become horizontal in the unconsolidated levels of the succession (Cosentino and Parotto, 1992; Corrado, 1995).

A N–S strike slip fault, parallel to the folds and thrusts, with dextral kinematics (Sabina Fault: Alfonsi et al., 1991) characterizes the Sabine sector of the Apennine chain. This structure is about 30 km long and has inverted the Jurassic fault that truncated the Sabine pelagic plateau to the east, with kinematics from dextral transpressive to transcurrent (Pierantoni, 1997). The latter author interpreted this structure as a strain partitioning feature induced by a transpressive regime along the pre-existing discontinuity. Alfonsi et al. (1991) associated the Sabina fault with crustal strike-slip tectonics, which post-date the chain development, responsible for the rotation of the folds and thrusts until they assumed their present-day direction (Mattei et al., 1992).

Westward, the Sabine Units are buried below the Plio-Pleistocene succession of the peri-Tyrrhenian basin, present in the hanging wall of the NNW–SSE trending normal fault. The basal thrust of the Sabine Units is represented by the Olevano-Antrodoco line (Cosentino and Parotto, 1992) that continues northward in the Sibillini Mts. thrust.

In the southern sector, this structural feature overthrust the Sabine Units onto the Lazio-Abruzzi Units, while in the northern, the Umbro-Marche Units lie on the "outer unit of the Messinian foredeep": the Laga Unit (Koopman, 1983; Lavecchia, 1985; Calamita and Deiana, 1988). The Sibillini Mts.–Olevano-Antrodoco thrust is constrained therefore by the Jurassic discontinuity (Figure 3) which had allowed the drowning of the carbonate platform in the Umbro-Marche-Sabine sector that developed the pelagic basin and the relative zone of transition (Castellarin et al., 1978; Calamita and Deiana, 1988; Decandia et al., 2002). The presence of en echelon thrust planes and related folds in the southern sector of the structure, and the axial culmination of the Sabine Units northward (sector of the Sibillini Mts.) point to its dextral transpressive character and the increase in tectonic translation toward the north, compatible with the kinematics of the whole arc of the Central-Northern Apennines. Therefore, the structural feature in question constitutes an oblique ramp, conditioned by the Mesozoic paleomargin architecture, to which minor thrusts are joined.

The different trends among the N–S-oriented Umbro-Marche-Sabine structures and the Lazio-Abruzzi ones (NW–SE) are considered primary and related to the influence of the pre-existing discontinuities on the geometry of the thrust ramps (Koopman, 1983; Calamita and Deiana, 1988; Calamita, 1990). Other Authors have associated these structural complexities with out of sequence or synchronous thrust development (Cipollari and Cosentino, 1992; Corrado, 1995). The development of this thrust is dated to the Upper Messinian–Lower Pliocene (Scisciani et al., 2000b).

(d) The Lazio-Abruzzi Carbonate Platform Units. These structural units are derived from deformation of the Lazio-Abruzzi carbonate platform, both of the transition domain to pelagic basin, and of the siliciclastic foredeep deposits.

(d1) The Simbruini Mts.–Ernici Unit Mts. This unit is represented by an overturned anticline, overthrust onto the siliciclastic deposits belonging to the Messinian foredeep, intensely deformed into metric folds. The emplacement of the unit in question is dated to the Lower Messinian (Cipollari et al., 1995). The tectonic juxtaposition of Triassic dolomites on the Cretaceous-Miocene succession, recorded in the "Trevi 1" well, indicates a thrust displacement of about 15 km (Damiani, 1992).

NW-ward, the Messinian siliciclastic deposits of the Val Roveto are in tectonic contact with the carbonate succession along a fault system oriented NW–SE. These structures have been interpreted as Messinian faults that controlled the physiography of the foredeep basin (Castellarin et al., 1978). Montone and Salvini (1993) recognized sinistral strike-slip kinematics along these faults. Toward the NE, the structures of Mt. Gabbia, D'Ocre Mts. and Mt. Velino outcrop, while more to the east the Gran Sasso Salient constitutes the frontal thrust of the Lazio-Abruzzi Units.

The presence of the Quaternary continental deposits, related to the Fucino and Sulmona tectonic depressions, seals any continuity between these and the structures of the La Meta Mts.–Le Mainarde, Montagna Grande, Mt. Genzana that outcrop SW-ward (Figure 4). Moreover, ENE–WSW trending morpho-tectonic lineament can be recognized in the northern sector of the Fucino Basin, characterized by the correspondence with strike-slip meso-faults. Vezzani and Ghisetti (1995) interpreted this lineament as a dextral strike-slip regional fault that control the Mio-Pliocene counter-clockwise rotation of the Gran Sasso area.

(d2) The Mt. Gabbia, D'Ocre Mts.–Mt. Velino structures. These are three major folds, in the Mesozoic-Tertiary carbonate platform and pelagic sequences, related to NW–SE trending thrust planes. They extend from Antrodoco southward to the Fucino Basin and they are overthrust onto the Messinian siliciclastic deposits.

For the structure of Mt.Gabbia, ca. 10 km of translation was reconstructed (Parotto and Praturlon, 1975), given the juxtaposition of the Triassic dolomites on the Cretaceous limestones, recorded in the Antrodoco 1 well. Capotorti et al. (1992) associated this tectonic contact with the presence of backthrusts, proposing a more conservative interpretation. The D'Ocre Mts.–Mt. Velino folds are an en echelon anticline system, with vertical forelimbs; their backlimbs are characterized by SW-dipping normal faults. Klippen and tectonic windows related to the Velino Mt. thrust planes crop out in the Gole di Celano (to the north of the Fucino Basin) (Angelucci and Praturlon, 1968).

The beginning of chain involvement was dated to the Messinian *"lago-mare"* by Cipollari et al. (1995).

(d3) The La Meta–Le Mainarde Mts., Montagna Grande and Mt. Genzana Structures. These are folds trending NNW–SSE, with similar characteristics to the structures described above, cropping out to the south of the continental Quaternary basins of Sulmona and Fucino. They are bounded by WSW normal-transtensive faults and subordinately by backthrusts, and have sub-vertical forelimbs overthrust onto the Messinian siliciclastic deposits and, in some cases (M.gna Grande), displaced by strike-slip faults (Corrado et al., 1990; Corrado, 1995).

In particular, they involved the border of the Lazio-Abruzzi carbonate platform (Montagna Grande), and the succession of the pelagic throw (Mt. Genzana – Colacicchi, 1967) between the Lazio-Abruzzi and Apulian (Mt. Morrone-Maiella) carbonate platforms.

The thrusts of these structures join toward the SE on the NNE–SSW Sangro-Volturno line (Auct.) that can be interpreted as a transpressive oblique ramp, to which a dextral strike slip motion can be related (Figure 21).

(d4) The Gran Sasso Salient. This represents the northern mountain front of the Abruzzi Apennines, and overthrusts the Triassic-Miocene successions of the Lazio-Abruzzi carbonate platform and related transition domains on the outer units of the Messinian foredeep (Laga Unit: Figures 3–5). This salient reaches a high structural elevation documented by the Triassic succession at about 3000 meters. It trends E–W to the north and N–S in the

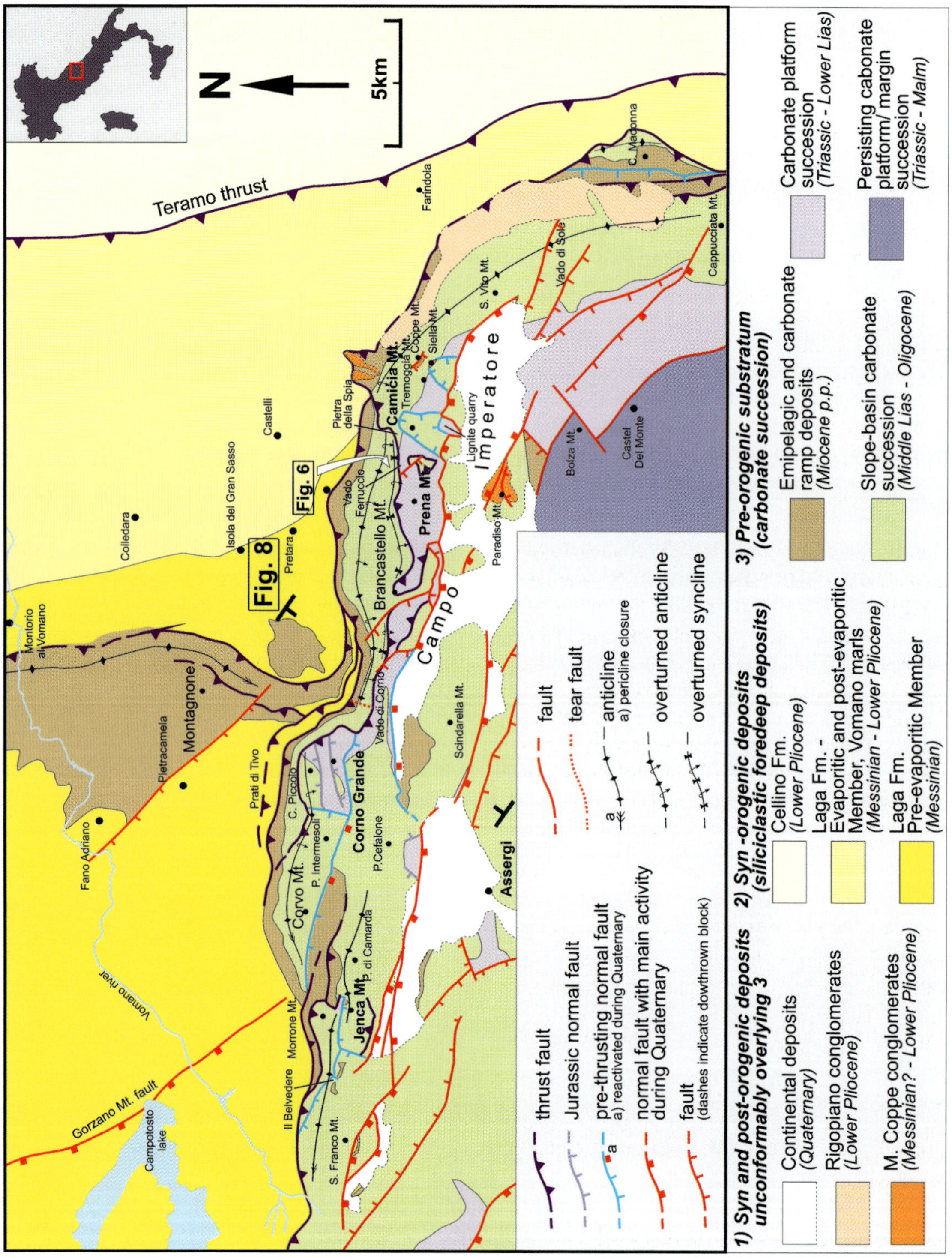

Figure 5: Structural sketch map of the Gran Sasso area; see Figure 4 for location.

southern sector (Mt. Cappucciata), and juxtaposes the carbonate sequence onto the Messinian siliciclastic deposits of the Laga Formation and the N–S oriented structure of the Montagna dei Fiori-Montagnone.

Overturned anticlines and related thrusts, with en echelon pattern, characterize the E–W segment of the salient that shows sinistral transpressive kinematics (Figures 4, 5). The Gran Sasso Salient, characterized by the presence of tectonic lenses, shows an eastward increase in the amount of tectonic transport, as suggested by the juxtaposition of the Triassic Dolomites on the pelagic Jurassic succession (Figure 6), while toward the west it has a pin line at Mt. Corvo (Figure 7). The upper thrust plane is folded, with respect to the development of the lower thrust and related anticline in its footwall, according to a piggy-back sequence of structure propagation during the Upper Messinian–Lower Pliocene. As frequently observed, behind the uppermost Gran Sasso thrust, normal faults, with SW downthrow, are present in the hangingwall (Figure 8). Pre-thrusting faults, rotated during the chain development (Mt. Camicia), or reactivated in the Quaternary (Mt. Corvo) have been recognized (Calamita et al., 2002a). Moreover, faults with mainly Quaternary activity and related intermontane basins (Campo Imperatore) can be observed.

The Gran Sasso Salient reflects the architecture of the Mesozoic paleomargin (Figure 3); it shows an increase in displacement ESE-ward, compatible with the paleomagnetic data that suggests an anticlockwise rotation in the eastern sector. Toward the east, the Gran Sasso Salient assumes a N–S direction, parallel to the axial trend of the Mt. Cappucciata anticline, and joins with the Mt. Morrone thrust. Together, they represent the carbonate mountain front of the Abruzzi Apennines (Figures 3 and 4). The Gran Sasso–Mt. Morrone thrusts, together with the thrusts related to the Laga and Maiella Units, reached a structural elevation of about 10 km, estimated using the top of the carbonate succession at 4000 m elevation in the Gran Sasso Salient and documented at a depth of about 6000 m in the peri-Adriatic area (Pescara); toward the south, they are joined to the Sangro-Volturno dextral transpressive oblique ramp.

(d5) The Mt. Morrone Thrust. Towards the south, the N–S sector of the Gran Sasso Salient is in an en echelon relationship with the Mt. Morrone thrust. The latter corresponds to an evident NNW–SSE anticline with northeast-vergence (Figure 4). It is formed from the Triassic-Miocene carbonate platform in the southern sector (Apulian carbonate platform), while in its central-northern part the transition succession to the pelagic basin outcrops (Figure 3). Traces of the thrust plane can be observed at the base of the fold forelimb; it overthrusts the carbonate succession onto the Messinian siliciclastic deposits of the Laga Fm., involved in the narrow and overturned folds of the Caramanico Valley. The Mt. Morrone thrust continues southward to Mt. Porrara. For this thrust also, which developed during the Lower-Middle Pliocene, kinematic connection with the Sangro-Volturno oblique ramp can be proposed. Behind the head of the thrust, younger normal faults are present in the south-western side of Mt. Morrone (Sulmona tectonic depression)

(e) The Outer Units of the Messinian and Pliocene Foredeeps

(e1) The Laga Unit. The Laga Unit is formed from the sedimentary succession of the Marche-Abruzzi Mesozoic-Tertiary pelagic basin and the Messinian–Lower Pliocene siliciclastic foredeep deposits (Figures 3–5). It makes up the footwalls of the Sibillini Mts. and the Gran Sasso thrusts; toward the east, the Plio-Quaternary succession of the peri-Adriatic basin lies unconformably on the Laga Unit which is overthrust NE-ward, along the Teramo thrust, onto the Maiella Unit (Bigi et al., 1997).

The Teramo thrust has a general N–S direction, and envelops the Gran Sasso Salient and the M.gna dei Fiori fold; it crops out from Teramo to Caramanico Valley to the south, where it is joined to the Mt. Morrone thrust. These relationships suggest a kinematic connection between the thrust planes along the M.gna dei Fiori–Porrara Mt. regional ramp.

The Laga Unit is involved in N–S trending folds (the Acquasanta, Montagna dei Fiori), related to local thrusts, which developed from the Upper Messinian to Lower Pliocene.

(e2) The Maiella Unit and buried units of the peri-Adriatic area. The Maiella anticline constitutes the most impressive external structure of the Abruzzi Apennines. Its axial culmination is the Maiella mountain, while toward the north the fold continues below the Plio-Pleistocene deposits in the adjoining Villadegna-Cellino structure. Toward the south, the fold forms a salient confined by the Sangro-Volturno oblique ramp. It also forms a structural

Figure 6: Uppermost piled unit of the Gran Sasso thrust system, represented by the Triassic platform (Dp: Dolomia Principale Fm.) juxtaposed onto slope deposits (Te: Terratta Fm.; Ma: Maiolica Fm.). The footwall block is involved in an overturned syncline trending approximately E–W. Mt. Prena is one of the highest peaks in the Gran Sasso, which is the highest mountain of the Apennine system.

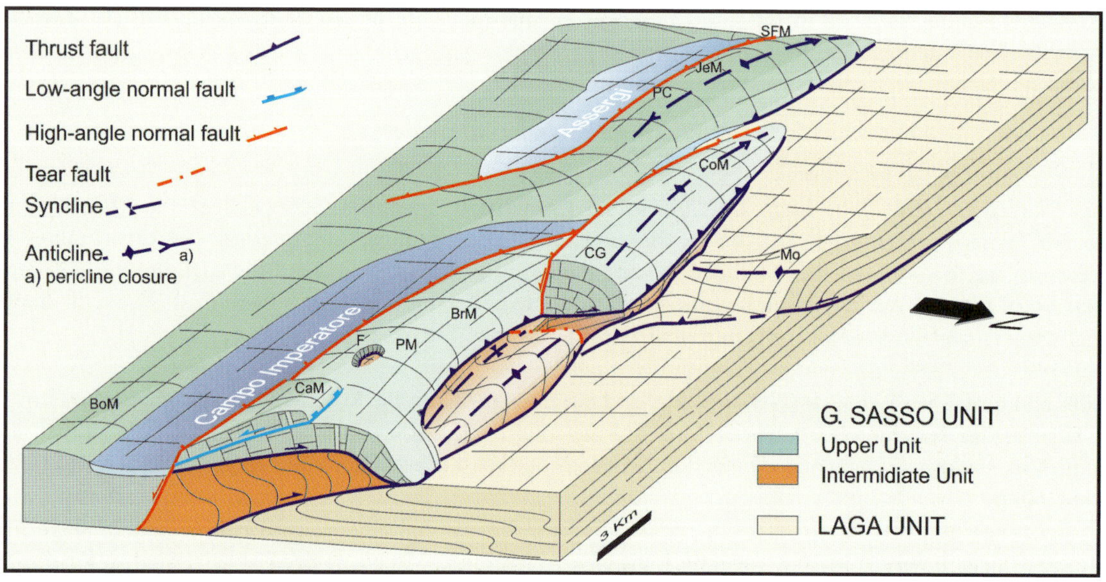

Figure 7: Sketched 3D setting of the E–W trending Gran Sasso thrust front from Mt. Camicia to Mt.S. Franco; CaM: Mt. Camicia; BoM: Mt. Bolza; PM: Mt. Prena; F: Fornaca tectonic window; BrM: Mt. Brancastello; CG: Corno Grande; Mo: Montagnone, CoM: Mt. Corvo; PC: Pizzo di Camarda; JeM: Mt. Jenca; SFM: Mt. S. Franco.

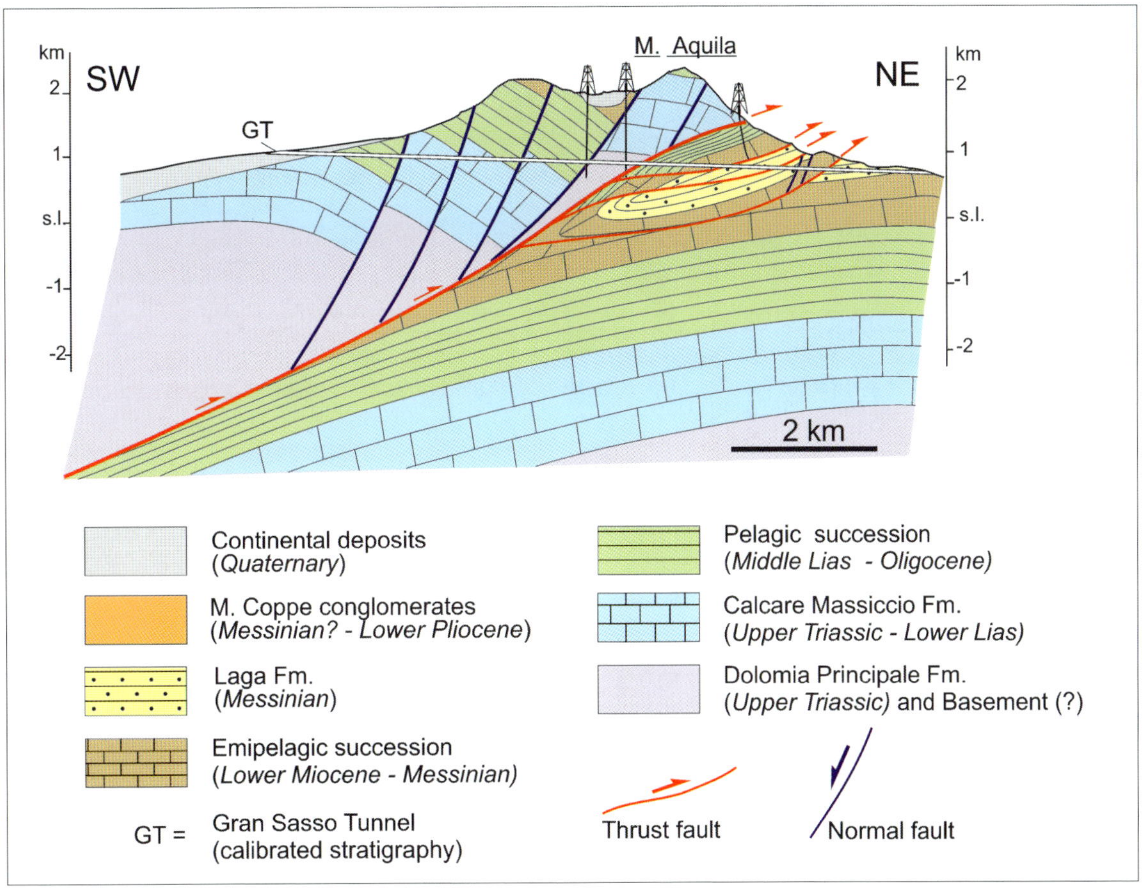

Figure 8: Section across the E–W trending Gran Sasso thrust system; see Figure 5 for location.

depression to the SE of the Apennine Units s.s., present below the Molise Allochthonous Units cropping out in the outer zone of the Southern Apennines (Figure 4).

The Maiella anticline involves the carbonate platform successions (Apulian carbonate platform domain) in the southern sector, the slope-basin in the northern area and the siliciclastic deposits of the Lower Pliocene foredeep.

On the western side of the Maiella thrust, there is a W-dipping normal fault (the Caramanico fault), extending for a length of about 30 km and with maximum down-throw, along the southern sector, of about 3 km, as pointed out by Donzelli (1968), on the basis of data from the Maiella well.

In the north-eastern sector of the Maiella anticline, the Upper Pliocene–Lower Pleistocene succession of the Mutignano Formation (Donzelli, 1968) lies in angular unconformity on the Maiella fold (Calamita et al., 2002b). This relationship allows us to date the main development of the Maiella anticline to the Middle Pliocene. Its axial culmination is constrained by the foreland architecture of the top of the carbonate succession. In fact, the top of the carbonates rises toward the south from a depth of 5000 m in the Pescara zone to about 2000 m at Ortona, in proximity to the northern paleomargin of the Apulian carbonate platform (Scisciani et al., 2000a).

The isobaths of the carbonate platform top point to the influence of the Mesozoic paleomargin architecture on the Lower Pliocene foredeep physiography, which has a depocenter located above the Mesozoic pelagic basin (Figure 9).

The Maiella and Casoli thrusts (identified by seismic profile interpretation: (Calamita et al., 2002b)) are connected to the "Coastal Structure" that forms the buried frontal structure of the Central Apennine chain, to the east of which the Middle-Upper Pliocene to Quaternary foredeep has developed. It also involves, toward the south, the Allochthonous Molise Units, and its main activity is sealed by the Quaternary deposits (Figure 10).

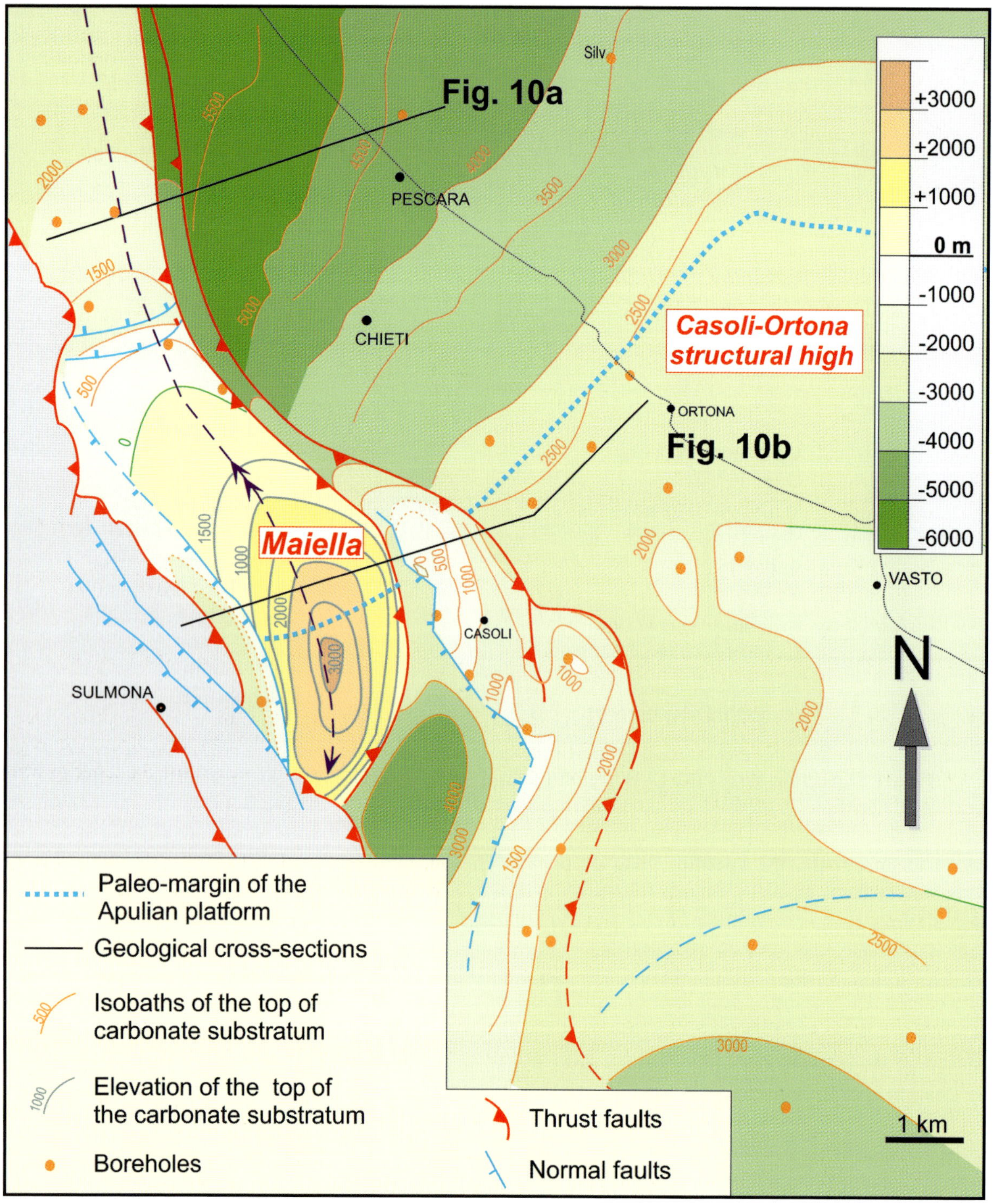

Figure 9: Structural setting of carbonate sequence in the outer Abruzzi area. Isobaths and elevation of the top of carbonate succession are shown.

(f) The Molise Units. To the east of the Sangro-Volturno oblique ramp, the allochthonous Molise Units outcrop, overthrust onto the Apennines Units s.s. involving the carbonate platform sequence of the Apulian domain (Mostardini and Merlini, 1986). They are formed from "Argille Varicolori" clays and marly limestones (Oligocene–Miocene) and Messinian–Lower Pliocene siliciclastic deposits.

These allochthonous units, together with the Lagonegro of the Southern Apennines, are interpreted as an accretionary prism connected to the Ionian Neotethys closure starting in the Middle Miocene (Finetti et al., 1996; Lentini et al., 2002; Finetti, this volume, Chapter 34).

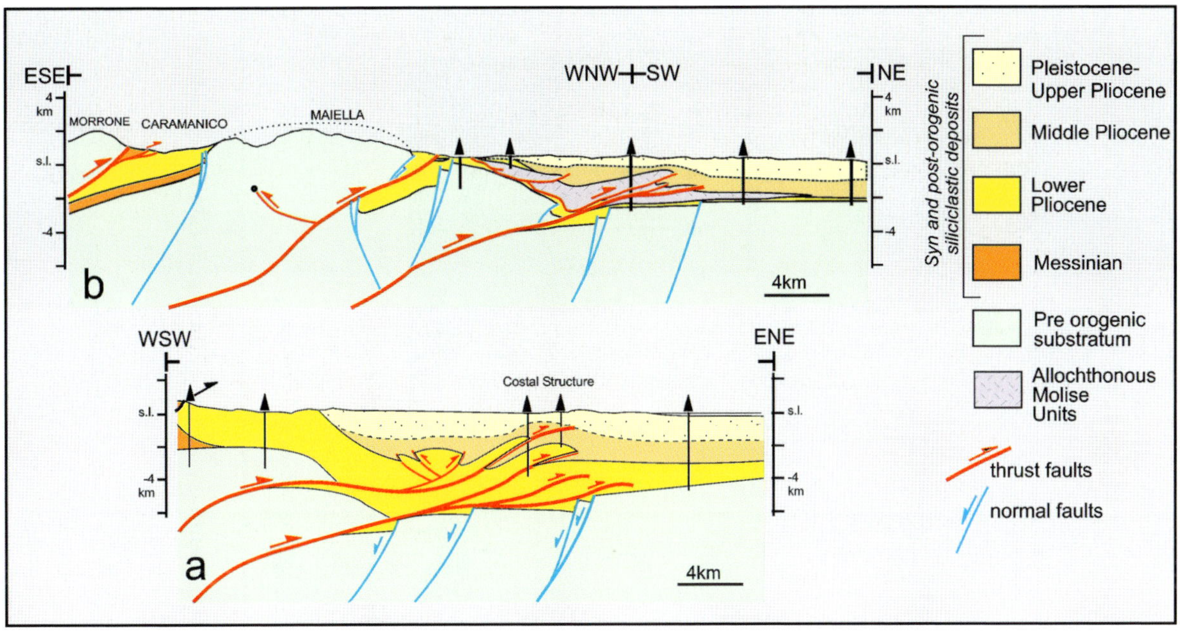

Figure 10: Geological cross-sections along the Apennine thrust front. See Figure 9 for location.

2.2. Meso-Cenozoic stratigraphy of the Apulia Platform and Pliocene evolution of the Abruzzi Basin

The geodynamic evolution of the Lazio-Abruzzi area indicates a first broad subdivision into successions related to the pre-, syn-, and post-rift evolution, and successions connected with the building of the Apenninic chain (foreland and thrust-top basins). The former correspond to Meso-Cenozoic carbonates, while the latter to Mio-Pliocene terrigenous deposits (Figure 12).

2.2.1. *Macrostratigraphy and evolution of the Central Apennines through the Mesozoic.* On the basis of facies associations and paleontological data, four types of succession (platform, margin, slope and basin) characterize the Mesozoic through the rifting and passive margin stages. The spatial distribution of depositional features of these type-successions indicates that their related environments occupy persistent positions, at least throughout the Mesozoic. In spite of this, important changes in the organization of successions are recorded downslope at different stratigraphic intervals, due to regional tectonics, regional and global biological, climatic and eustatic changes. By influencing the growth potential of the platform and the production rates of the resedimented materials, these factors are primarily responsible for the configuration of the platform margin depositional profile, and they influence water circulation patterns and control the amount of material exported down the slope. Tectonic events affected platform areas during the Late Trias to Middle Lias, the latermost Early Cretaceous and from Late Cretaceous throughout the Tertiary. Important biological changes, with ensuing variations in margin benthic communities, produced variations in margin and slope depositional dynamics. These changes characterize the Dogger–Malm interval dominated by hydrozoa, corals and echinoderms; the Cretaceous, dominated by coral and rudist associations; and the Tertiary, dominated by benthic macroforaminifera associations typical of temperate platforms.

(a) Platform successions. The platform type-succession ranges from the Late Trias to Late Cretaceous, for about 3500 meters. Depositional conditions are regulated by the balance between platform growth, sediment production and variations in relative sea level. Generally, they are characterized by cyclothems that record subtidal to supratidal conditions, dominated by algae and benthic foraminifera associations, emersion levels and dissolution breccias.

The *Dolomia Principale* (Norian-Rethian) represents the most ancient platform constituent and is made up of dolostones and cyclothemic dolomitic limestone. The maximum outcropping thickness is about 600 meters, recorded in the Gran Sasso. The transition to the overlying *Calcare Massiccio* is defined by the presence of *Triasina hantkeni* on top of the *Dolomia Principale*, which allows us to indirectly refer the base of the *Calcare Massiccio* Formation to the Hettangian. This unit, about 600 meters thick, organized into massive shallowing-upward cycles culminates

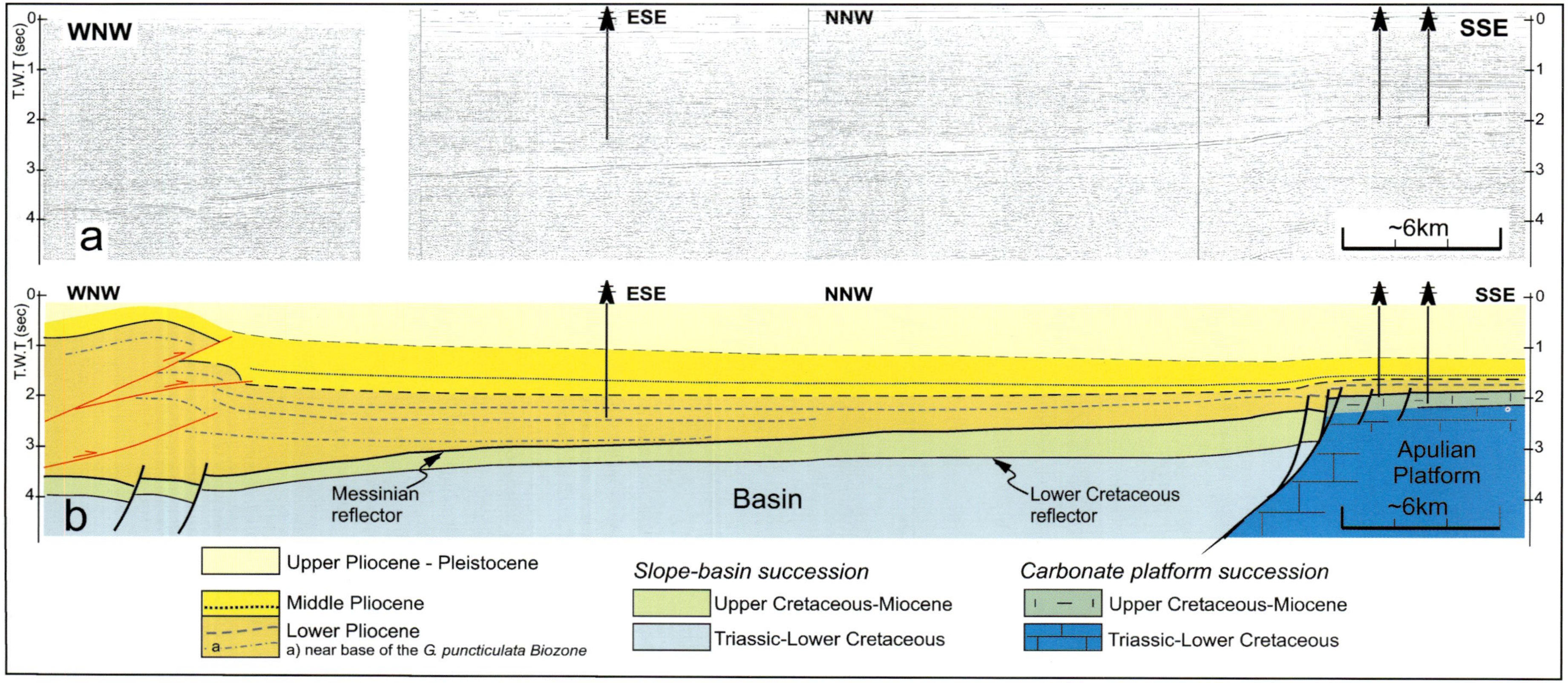

Figure 11: Seismic reflection profile and its interpretation across the Apulian carbonate platform – pelagic basin boundary – front of the Apenninic thrust belt. This section located along the Chieti-Ortona alignment (see Figure 9).

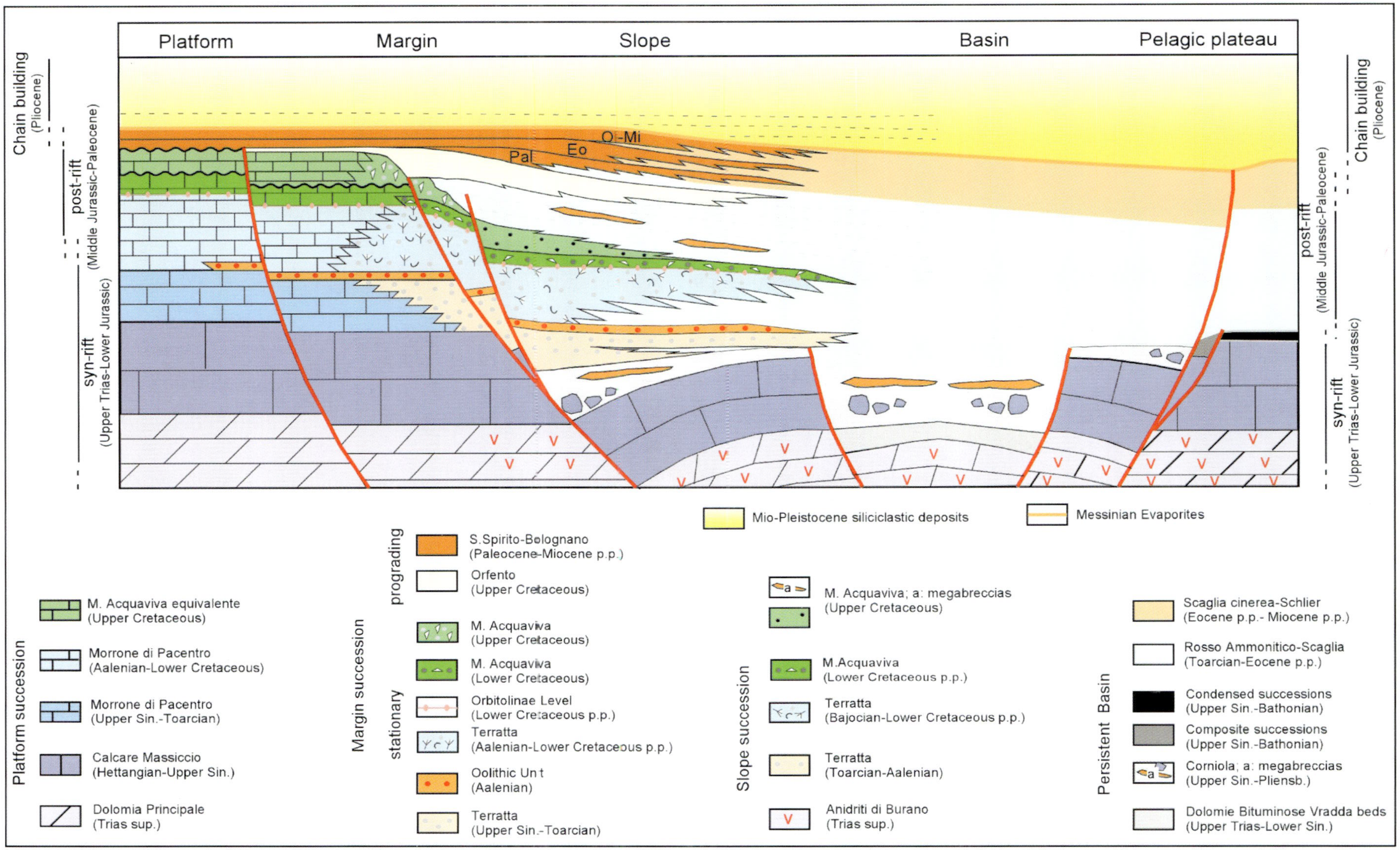

Figure 12: Tectono-sedimentary scheme of the eastern Apulian platform to basin transition zone.

in emersion levels. In the upper part of the formation, oolithic intervals indicate the transition to higher hydrodynamic conditions, testifying to the formation of sunken steps linked to the platform break-up by fault activity during the initial stages of Liassic rifting. Successive to the climax in rifting, the *Morrone di Pacentro* Formation represents the shallow water deposition starting from the Late Sinemurian. Open platform conditions with high hydrodynamic regimes characterize the lower part of this formation up to the Toarcian, with the *Lithiothis* level capped by a widespread emersion event. One other regional key level is the Oolitic unit, with a thickness of up to ten meters and a base of Dogger age.

Starting from the Dogger, and at least until the lower part of the Early Cretaceous, platform areas record the presence of broad uniform depositional conditions forming a 1500 m thick carbonate bank. During the Early Cretaceous, a greater sensibility of the platform to environmental variations is indicated by the frequent alternation of emersion events and high energy facies associations, within successions dominated by lagoon micritic facies. The Lower Cretaceous platform is capped by a regional emersion outlined by the formation of bauxitic soils. The Upper Cretaceous platform is represented by the *Monte Acquaviva Equivalente* Formation, which is a few hundred meters thick. The Upper Cretaceous platform does not extend uniformly over the whole platform area, but it is often truncated by the Paleocene-Eocene regional unconformity. In more subsiding sectors, the *Monte Acquaviva Equivalente* Fm. is overlain by the *Orfento* Fm. or equivalents, which represent the uppermost Cretaceous depositional event and mark the transition between a rimmed or escarpment-bounded platform to a ramp-like depositional profile. During the Tertiary, inner platform facies are almost completely missing and platform tops were dominated by bioclastic ramps.

(b) Margin successions. Margin successions developed in the Lazio-Abruzzi area starting from the Middle Lias, when sunken steps and morphologic gradients formed, following the break up of the *Calcare Massiccio* platform. Older margin facies have been spotted in the massive Upper Trias *Dolomia Principale* facies in the Gran Sasso. They developed until the Late Cretaceous, with a thickness of about 1500 meters. The Tertiary was dominated by low-angle depositional profiles that favoured the ramp configurations. Generally, the facies associations of margin successions are characterized by grainstones, packstones and bio-intraclastic rudstones, by coral boundstones, hydrozoans, rudist biostromes and benthic macroforaminifera.

The Mesozoic margin succession of the Lazio-Abruzzi platform is classically divided into two formations which also include slope facies associations: the *Terratta* Formation (Lias, Middle and Lower Cretaceous p.p.) (Colacicchi, 1967; Crescenti et al., 1969; Crescenti, 1969) and the *Monte Acquaviva* Formation (Lower Cretaceous p.p.–Upper Cretaceous) (Crescenti et al., 1969; Crescenti, 1969).

During the Late Cretaceous, basinal areas were progressively filled due to the decreasing rates of the differential subsidence between platform and basin settings, and to elevated rates of exportation of bioclastic resedimented deposits produced by the benthic factory. As a consequence, morphologic differences between platform and basin settings are minimized and a broad prograding phase of the depositional systems occurs, followed by the formation of ramp-like depositional profiles. The latter developed up to the Messinian p.p. (Figure 12) and were favoured by temperate climate conditions, adverse to the development of frame-building biota, such as corals. Ramp successions are dominated by bioclastic grainstones and foramol-type associations, with benthic macroforaminifera, red algae and siliceous sponges. The formations that characterize this interval are the *S. Spirito* Fm. and the *Bolognano* Fm. Geometric features of the depositional profile favoured frequent intercalations of pelagic facies and large facies displacements.

(c) Slope successions. Carbonate slope and base-of-slope deposits are well represented and widely exposed in kilometre thick successions (2500 m as average) in the Lazio-Abruzzi area. A common feature of slope successions is the presence of reworked skeletal and non-skeletal deposits exported from the platform margin. They are exported from the platform toward the basin as a consequence of the excess of material produced on the platform margin by the carbonate factory. Generally, this type of deposit alternates along the whole stratigraphic record of platform to base-of-slope successions (*Terratta* and *Monte Acquaviva* Fms.), but reaches a very high frequency in the Middle to Upper Liassic interval of the *Terratta* Fm., in its middle part (uppermost Bathonian-Lower Kimmeridgian) and, to a smaller degree, in the upper part of the *Terratta* Fm. (Lower Cretaceous). Large amounts of resedimented deposits exported from the platform characterize the lower (Cenomanian p.p.–Turonian p.p.) and the upper part (Campanian p.p.-Maastrichtian) of the *Monte Acquaviva* Fm. and the slope successions during the Tertiary.

Other important lithologic components of slope resedimented deposits of the Lazio-Abruzzi successions are breccias and megabreccias. They consist of coarse margin-derived and shallow-water debris or blocks transported downslope as a consequence of platform progradation, by erosion or partial dismantling of the platform edge. They become important lithologic components of the slope, at least during the Pliensbachian (*Corniola* Fm.), the Tithonian p.p. (*Terratta* Fm. p.p.) and at various stratigraphic levels of the *Monte Acquaviva* Fm., during the Cretaceous (Albian, Cenomanian, Santonian p.p. and Campanian p.p.).

(d) Basin successions. With few exceptions, basin successions of the Lazio-Abruzzi area present quite similar features of those of the Umbria-Marche Basin. The most ancient facies associations, which indicate the presence of more or less restricted basinal conditions, are those of the Triassic *Dolomie Bituminose*, and they have been documented in basinal areas of Emma (Central Adriatic), Vradda (Gran Sasso) and Filettino (Simbruini). Later on, the great development of basinal conditions is a consequence of the dissection of the *Calcare Massiccio* platform during the Liassic phase of the Tethys rifting. From the Liassic, the particular features of basinal successions can be seen all over the adjacent areas of the Apulia Platform, corresponding to the Umbria, Marche, Sabine, and Adriatic sectors. In these areas, pelagic successions are characterized by fine to very fine white, beige or grey mudstones and wakestones and, more rarely, by medium to fine grained packstones, alternating with black, pink or grey chert levels and shaly-limestone or blackish, greenish or yellowish limey-marls. The main lithofacies are mainly made up of pelagic carbonates where open sea biota, such as radiolarians, pelagic lamellibranches (filaments), sponge spicules, planktonic Foraminifera, Globochaete, calcispheres, Saccocoma, Stomiosphaera moluccana and calpionellids, are dispersed, and by platform-derived shallow-water mud mixed up with skeletal grains (mainly planktonic foraminifera).

In addition to a major separation between platform and basin, the extensional Liassic tectonic created an articulated palaeography in the basin areas as well, by determining the formation of structural highs and lows and pelagic plateaus characterized by condensed and/or incomplete successions, with limestone, marly-limestone and nodular facies developing all over the Dogger–Malm interval.

Evidence for Mesozoic rifting geodynamics and paleogeographic overview. As a consequence of rifting and drifting connected with the opening of the Ionian and Alpine oceans, the Adria microplate is segmented into various crustal sectors. Thick continental crust sectors are dominated by shallow-water carbonate sedimentation, such as the Dinaric Platform, the Apulia Platform and the Etrurian Promontory. Thinned continental or oceanic crust sectors hosted deep marine conditions, such those recorded in the Tuscan-Romagna, Umbria-Marche, Adriatic and Alpine-Ionian Neotethys.

Rifting of the Adria microplate, is well documented from the Lias, but hydrocarbon exploration and CROP seismic data also evidence prominent Permo-Triassic left-slipping transtensive tectonics connected with the counterclockwise rotation of Adria (Africa/Adria–Europe convergence). Variations in thickness and facies associations of pre-Liassic deposits allow us to separate areas of different stratigraphic evolution, and to document the presence of extensional phases in the Lazio-Abruzzi area. One of these major lineaments separates the evaporitic and dolomitic succession (*Anidriti di Burano*) of the Umbria-Marche sector to the north from the carbonate and dolomitic successions (*Dolomia Principale*) of the Lazio-Abruzzi area to the south. At present this separation corresponds to the Olevano-Antrodoco or Ancona-Anzio line. The most important newly proposed crustal boundary is, by far, the one separating the Ionian Neotethys from the Apulia-NW Adria, discovered in the CROP network data (Finetti, this volume, Chapter 34) and reported in Figure 21. This crustal setting played a determinant role in the Tyrrhenian stage geodynamics and simplifies our understanding of the evolution. In the Adriatic, Triassic facies associations (*Anidriti di Foresta Umbra*) are similar to those of the Umbria-Marche basin, and suggest the presence of another important paleogeographic lineament, which probably bounded the eastward extent of the Apulian Platform, during the late Trias. Different depositional settings associated with areas of different subsidence rates, can be observed within the *Dolomia Principale* and the *Anidriti di Burano* or equivalent. These are the buried euxinic Norian-Raethian basins of Emma in the Adriatic offshore, of Pelagruza in the Dinaric offshore, of Vradda in the Gran Sasso, and of Filettino in the Simbruini. The markedly different sedimentation rates of the two Triassic formations (45 m/Ma as average for the *Dolomia Principale* and 200 m/Ma as average for the Triassic evaporites) suggest the existence, during the Trias, of areas with subsidence linked to rifting.

The Triassic palaeographic framework characterized by large carbonate banks separated by deep pelagic basins remains essentially the same during the Liassic stage of the Alpine Tethys rifting. In the Adriatic offshore, the

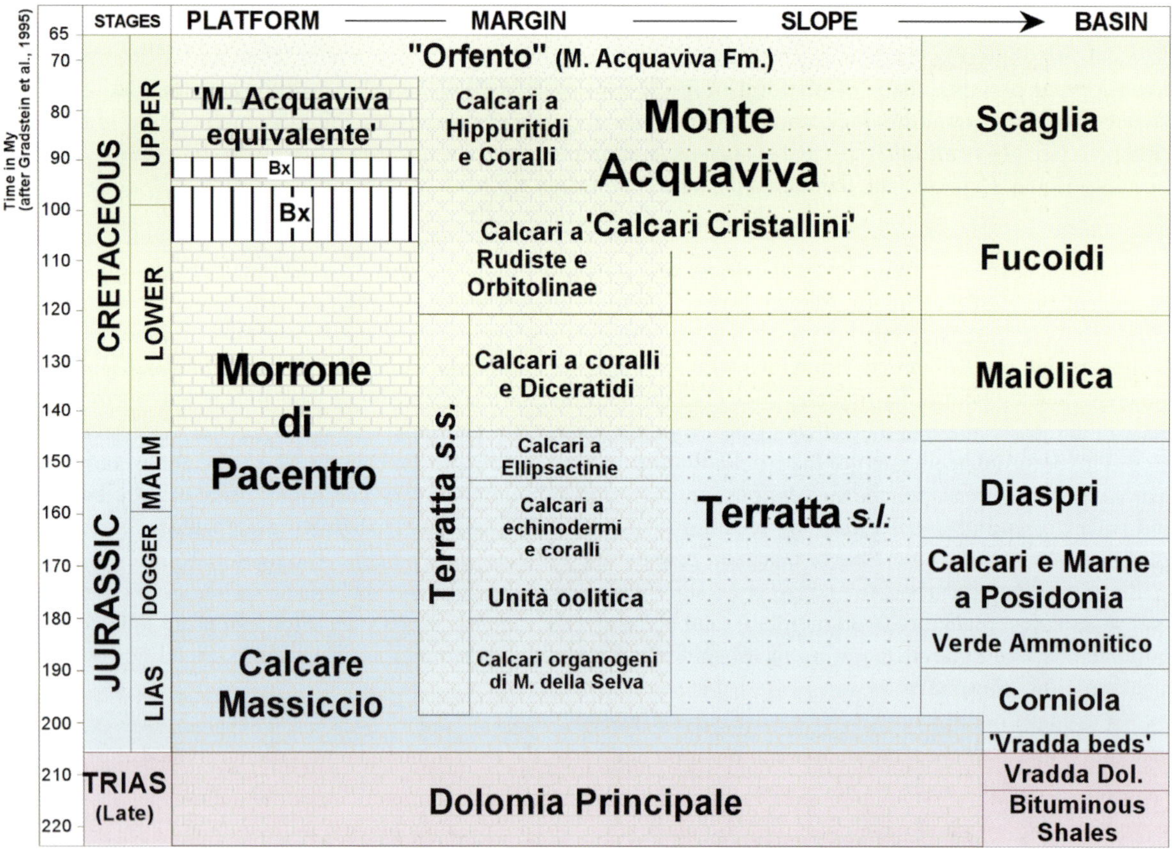

Figure 13: Lithostratigraphic framework of the Lazio-Abruzzi Platform showing formation terminology and the time distribution of the main depositional environments.

transition from the Apulian Platform to the Basin is mostly buried under the Tertiary deposits, but is indicated by seismic exploration data and outcrops on the Gargano Promontory.

Using the outcropping facies distribution (Figures 3, 4) and subsurface data, the Liassic paleogeography of the northern sectors of the Apulia Platform can be outlined. From the Lower Sinemurian, a phase of increasing and differentiated subsidence affects the early Liassic platform (*Calcare Massiccio* Fm. and equivalents), and causes partial dissection of the main Apulia platform bank into a complex system of structural highs and lows. This produces an articulated paleogeography composed of elongated minor carbonate banks and intra-platform basins which persists throughout the Mesozoic evolution. Intra-platform basins are connected to larger basins (e.g. Umbria-Marche, Adriatic and Ionian), but were characterized by shallower bathymetries. The positions, sizes and geometries of these intra-platform basins are generally difficult to reconstruct.

2.2.2. *The Pliocene Abruzzi Basin (Central Apennines).* The Abruzzi Pliocene Basin is bounded by the outcrops of the Laga Fm. westward, just where the structures of Montagna dei Fiori and of the Gran Sasso develop; by the present Adriatic sea eastward, the Maiella thrust southward, and by the Vomano Valley northward (Figure 16). This area is characterized by a system of minor anticlines and depocenters formed as result of the progressive migration of the orogenic wedge toward the Adriatic Foreland. Four main stratigraphic-structural elements can be identified from the innermost to more external areas (Figures 16, 17, 18):

(1) *Villadegna – Cellino Thrust-Anticline*, the innermost structural trend representing the northern prolongation of the Maiella thrust (Figure 16). This structure involves the pelagic carbonate successions and records the Pliocene siliciclastic succession, with outcrops of older Pliocene elements. Eastward, Upper Pliocene deposits onlap the external flank of this feature.
(2) *Atri Basin*, a depocenter located between Villadegna–Cellino Thrust-Anticline and the Coastal Structure. It outcrops only in its Upper Pliocene portion.

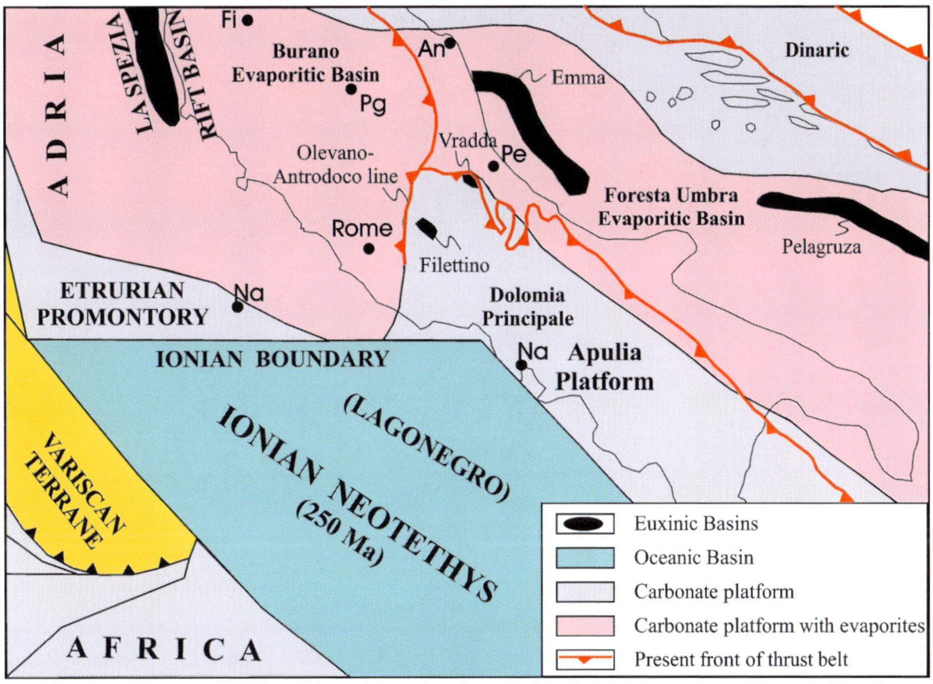

Figure 14: Interpreted distribution of regional environments during the Late Trias (referred to present geographic position). The Ionian Neotethys of Permo-Triassic opening and the Etrurian Promontory as zone of origin of the allochthonous Apenninic Platform units have been reconstructed from CROP sections (Finetti, this volume, Chapter 34).

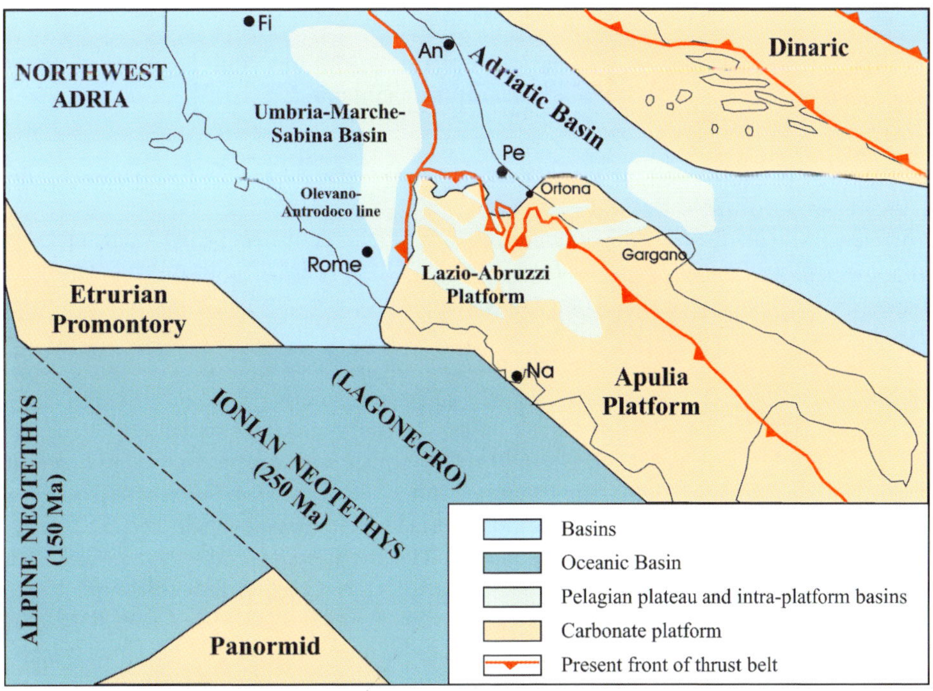

Figure 15: Interpreted distribution of the main depositional systems during Late Lias to Late Malm (referred to present geography). The Ionian-Alpine Neotethys and the Etrurian Promontory as zone of origin of the allochthonous Apennine platform units have been reconstructed from CROP sections (Finetti, this volume, Chapter 34).

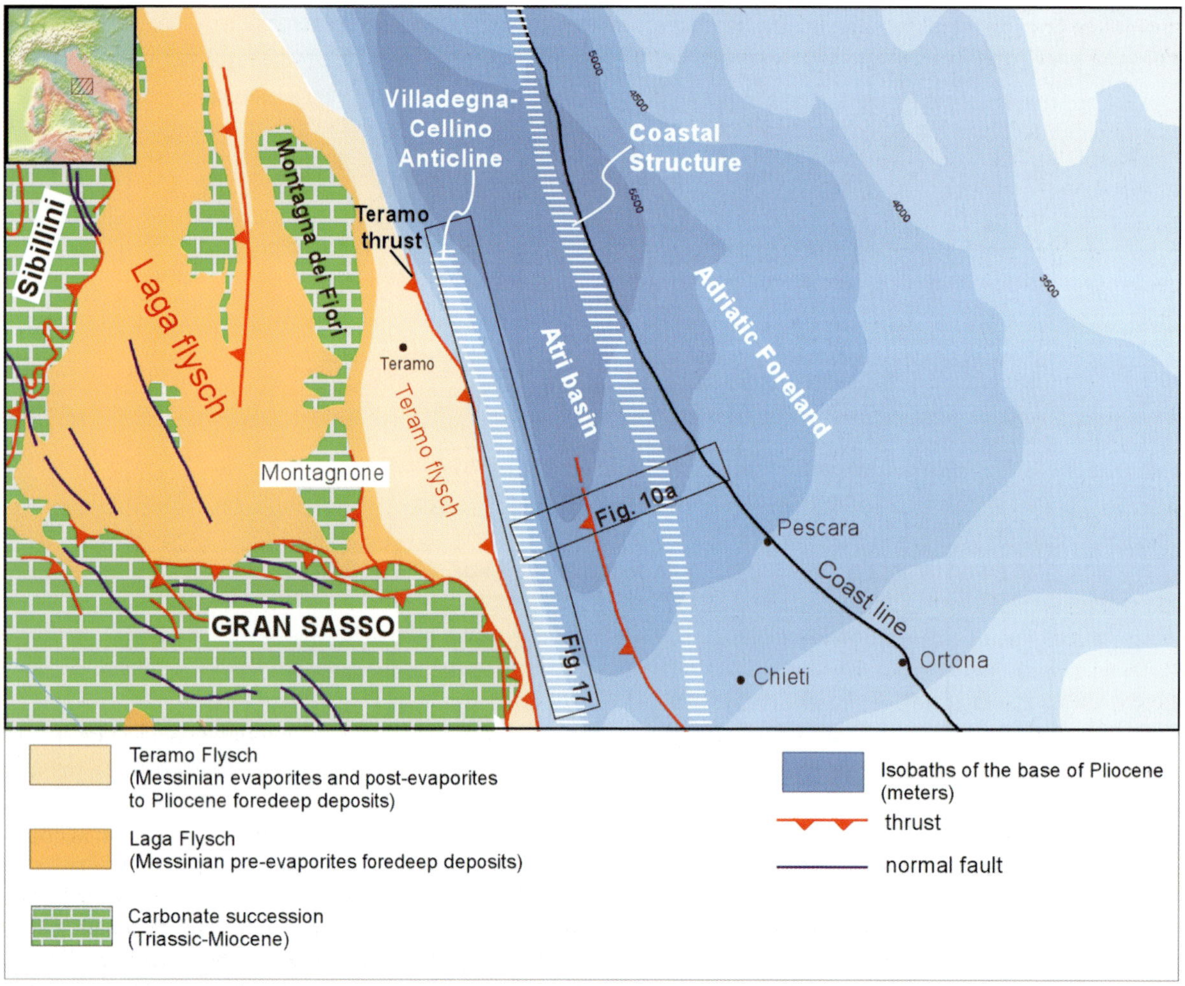

Figure 16: Geological sketch map of the outer Central Apennines and location of the main structural-stratigraphic elements.

(3) *Coastal Structure*, a frontal sector of the Apenninic chain formed by a N–S oriented asymmetric thrust-anticline, located east of the Atri Basin.

(4) *Adriatic Foreland*, east of the Coastal Structure, mainly offshore, entering land south of Ortona city. This area is only marginally affected by the deformation of the last Pliocene-Pleistocene movements. Stratigraphic discontinuities and progressive condensed signature characterize successions, bounded eastward by the Middle Adriatic swell.

The Pliocene succession of the Abruzzi Basin is made up of terrigenous deposits reaching a maximum thickness of 7000 m. It is generally characterized at its base by a polygenic conglomerate (S. Valentino Conglomerate of Donzelli, 1968; Crescenti et al., 1980), and by clayey facies ascribable to the *Sphaeroidinellopsis* biozone which mark a return to normal marine conditions after the Messinian salinity crisis (Crescenti, 1971; Crescenti et al., 2002). Above, the succession is characterized by the stacking of deep-sea siliciclastic sandy deposits. Starting from the Middle Pliocene, successions are dominated by clay-silt deposits, with conglomeratic horizons intercalated at different stratigraphic levels and located close to the Villadegna-Cellino Anticline, Maiella Anticline and Coastal Structure (Crescenti et al., 1980; Ori et al., 1991; Casnedi and Serafini, 1994; Milia, 2002), with a main provenance from the west. Shallow marine water conditions develop progressively during the Upper Pliocene.

Characteristics and organization of the successions, and the relevance of processes associated with the unconformities (erosive truncations and onlap), allowed us to establish a hierarchy of these depositional sequences into five major units representing response of the different stages of an orogenic system migrating toward the foreland, as characterized the Abruzzi basin during the Pliocene. These stages are: Foreland Stage, Incipient Foredeep Stage, Foredeep Stage, Piggy-back Stage and Regional Uplift Stage. Apart from the stages preceding (Foreland Stage) and following (Regional Uplift Stage) the development of the Pliocene Abruzzi Basin, the Incipient Foredeep, Foredeep and Piggy-back stages characterized, in different times, the four main stratigraphic-structural elements.

The Incipient Foredeep Stage corresponds to a transition from the stable undeformed foreland to the highly subsiding foredeep stage. During the incipient foredeep stage, the tectonic subsidence resumes with moderate rates after a period of quiescence, in connection with the involvement of the external zones in the flexure of the lithosphere. It precedes both a rapid increase in the tectonic subsidence and the incoming of turbiditic deposits. A bathymetric deepening, with the deposition of a condensed, mainly clayey succession, occurs. This stage characterizes the sector east of the Coastal Structure from the base of the Pliocene, while the likely presence of pre-Pliocene terrigenous deposits, referable to the post-evaporitic interval (Figure 17), could mean that the internal sectors, between the Villadegna-Cellino Anticline and the Atri Basin, had already been involved in the foredeep stage during the late Messinian. However, the presence of clayey condensed facies (Argille a Sphaeroidinellopsis) above the siliciclastic pre-Pliocene successions (Figure 17) suggests that also the internal sector of the Pliocene Abruzzi Basin is characterized by a pause in migration of stratigraphic-structural elements. Starting from the Pliocene, this stage characterizes all the sectors of the Pliocene Abruzzi Basin. At the Villadegna-Cellino alignment, the Atri Basin and the Coastal Structure it corresponds to the LPl sequence (Figure 17); in the internal areas of the Adriatic Foreland, it extends until the LP2 sequence.

The Foredeep Stage is that of maximum subsidence and sedimentation rates, characterizing the internal and middle zones starting from the LP2 sequence (Figure 17). Sectors east of the Coastal Structure are involved only later, with the sequence LP3, but the highest subsidence rates are reached during the sequence MP1. The external Adriatic Foreland is involved in this stage during the Upper Pliocene, with the sequence UP1, and records the highest subsidence rates between the Upper Pliocene and Quaternary. During this stage, an important migration of the onlap surface is recorded on the foreland ramp.

The Piggy-back Stage divided the main basin into two secondary depocenters separated by growing structures. The innermost depocenter identifies the piggy-back basin *s.s.*, while the conditions defining the foredeep stage (high subsidence and sedimentation rates) migrate toward the foreland. It is characterized by thrust activity, by the formation of thrust-anticlines and is coupled with decrease in subsidence and sedimentation rates and by a bathymetry decrease. This stage is recorded only in internal sectors and corresponds to the onset of the sequence LP5 that marks the deformation of the present Coastal Structure. In the innermost sectors of the Pliocene Abruzzi Basin, the beginning of this stage belongs to the sequence MP2, during which the growth rates of the Villadegna-Cellino anticline are lower than subsidence. This subsidence causes a generalized phase of retrogradation of successions along the outer flank of the structure. This stage is completely lacking in the sectors of the Adriatic Foreland, characterized during its evolution only by foreland to foredeep stages.

3. CRUSTAL GEOLOGICAL SECTION FROM THE CORSICA BASIN TO THE ADRIATIC SEA

The reconstructed crustal geological section (Plate 3) crosses the Corsica Basin, the Southern Tuscan Archipelago, Central Italy (Figures 1–4) and the Adriatic sea, offshore Pescara city. It elucidates various geologic provinces of critical importance for the understanding of questioned areas in the Mediterranean Region, such as the Corsica Basin, the Tuscan Archipelago and the Central Apennines. While the western extremity of the reconstructed section is in the Corsica Basin, which represents (Finetti et al., 2001) the foredeep and filling wedge of the Corsica–Adria convergence, the remaining longer part of the section, all on the Adria Plate, crosses the entire Central Apennines and its hinterland, the Pescara Basin and external Dinaric features with thrust polarity opposite to that of the Apennines. In total, the section has a length of 470 km, of which about 164 km is constrained by CROP section M–37, about 82 km by CROP section M–15 (total direct CROP control km 246), and the remaining km 224 are based on extrapolation of other land CROP sections (CROP-03 and CROP-04, see Figure 1 for locations) and integration of available industrial seismic lines of medium penetration (hydrocarbon exploration), gravity data boreholes data, and geological data, including various specific control data obtained during the two years of collaboration between of the CROP team of this chapter, formed by geophysicists and geologists of the universities of Trieste and Chieti–Pescara.

3.1. Basic data utilized

In summary, the section ABCDE (Figure 1 and Plate 3) is based on the data listed below:

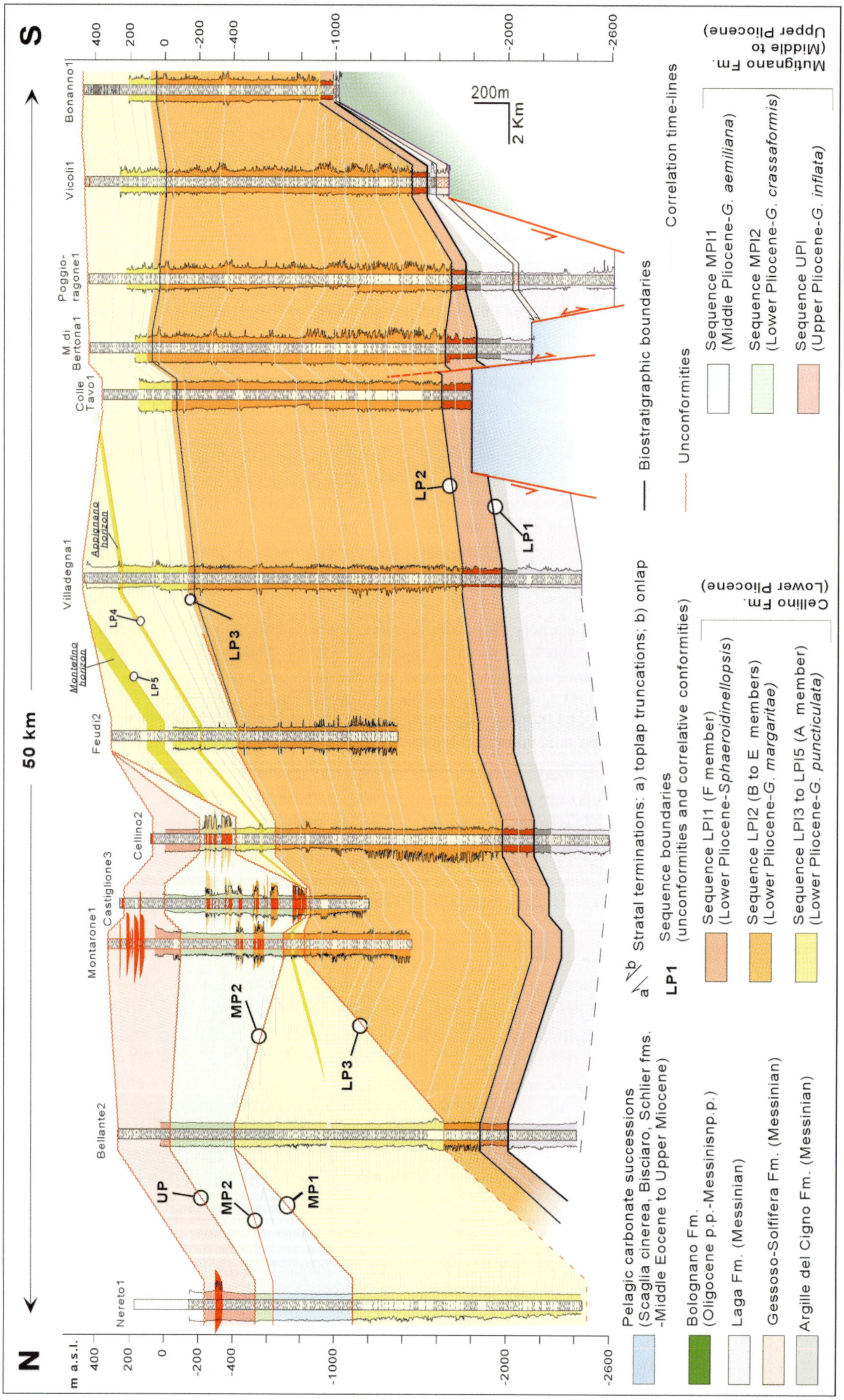

Figure 17: North-south well-log transect showing high resolution correlations and facies distribution of Pliocene successions along the Villadegna-Cellino anticline (see Figure 16 for location).

(a) Segment AB. CROP section M-37 (Plate 1).

(b) Segment BC. *Sedimentary Crust.* Reconstructed directly and/or by extrapolation of various available geological – geophysical data (i.e. scattered seismic lines of medium penetration from hydrocarbon exploration, the published part of the western extremity of section CROP-11 on the Latial volcanoes area (Parotto et al., 2000), gravity data and magnetic data, DSS data, geological data from the literature and/or other specifically collected geological data, and borehole data extrapolated from various distances. The consulted DSS data are those in the chapter by Cassinis et al. (this volume, Chapter 2). The gravity and magnetic data used are reported in the chapter by Bernardelli et al. (this volume, Chapter 3).

Pre-sedimentary Crust. Reconstructed from extrapolation–interpolation of CROP seismic data (CROP M-37, M-15, CROP-03, CROP-04) and integrating gravity magnetic and geological data. The assembled seismic, geological, gravimetric and magnetometric data coherently show that two transcrustal thrust faults must be introduced into the Apenninic sector of the interpreted section in Plate 3 (i.e. the high-angle Olevano-Antrodoco Thrust Fault of nearly N–S trend, and the Gran Sasso Thrust Fault of NW–SE trend).

Obviously, this Apennine segment of the section must be regarded as a schematic crustal reconstruction coherent with other CROP constrained trans-Apenninic sections (CROP-03 in particular: Finetti et al., 2001). However, the major fundamental crustal tectono-stratigraphic elements exhibited are considered of acceptable reliability.

(c) Segment CD. *Sedimentary Crust.* Reliably reconstructed from available seismic lines of medium penetration of hydrocarbon exploration campaigns, and borehole data.

Pre-Sedimentary Crust. Extrapolated from deep CROP seismic data of the section M-15. The westward crustal thinning of the pre-sedimentary Upper Crust from "D" to "C" is equivalent to that seismically observed on section CROP-03 (Finetti et al., 2001) at the corresponding geological position.

(d) Segment DE. Crust fully constrained by CROP seismic section M-15, with partial borehole calibration. The Moho (MO) and Top Lower Crust (TL) reflectors (Plate 3) are seismically well identifiable from the suite of Adriatic CROP lines (Finetti and Del Ben, this volume, Chapter 23).

3.2. Reconstructed Tectono-Stratigraphic Section

The reconstructed section in Plate 3, 470 km long, crosses various important geological provinces of Central Italy and the adjacent Tyrrhenian and Adriatic Seas, where no previous deep multifold seismic reflection data were available. Therefore the new geophysical information given on the two marine sectors of the section, as well as the reconstruction done for the land Apenninic area pass through various critical points, discussed in the geological literature of the region. Indeed, this section, starting from the Corsica Basin, which is interposed between the rotated Corsica block and the deformed slope of the NW Adria plate, crosses structural domains connected to different tectono-dynamic processes on and between the Northern and Southern Apennines. Moreover, in the Adriatic Sea, CROP M-15 explores structures created by the Adria–Europe convergence with compressive deformation of Dinaric polarity. All these tectonic deformations occurred over Adria plate crust with time-space mechanisms discussed in the chapter.

The section in Plate 3 is plotted at horizontal scale 1:300.000 and vertical scale 1 cm = 1 sec two-way reflection time. The scales are equivalent, consequently, for a velocity of 6.000 m/s, which is fairly realistic for a thick crustal interval running from the top Mesozoic carbonate sequence (generally horizon K) to the top of the lower crust (horizon TL). On the lower crust, where the seismic velocity is approximately 7000 m/s, the vertical compression is negligible. However, on the soft young sediments, a notable vertical exaggeration is obviously introduced.

A description of the crustal section is given in the following paragraph, starting from the Corsica basin and going east. Subdivision of the geological sectors is made according to the most evident regional characteristics.

(a) Corsica Basin. As already presented in the literature (Finetti et al., 2001) and in this volume (Finetti et al., this volume, Chapter 8), the Corsica Basin is of great importance for a correct understanding of the complex tectonodynamics of the area. Indeed, this basin represents the closure of the Alpine Tethyan ocean between Corsica

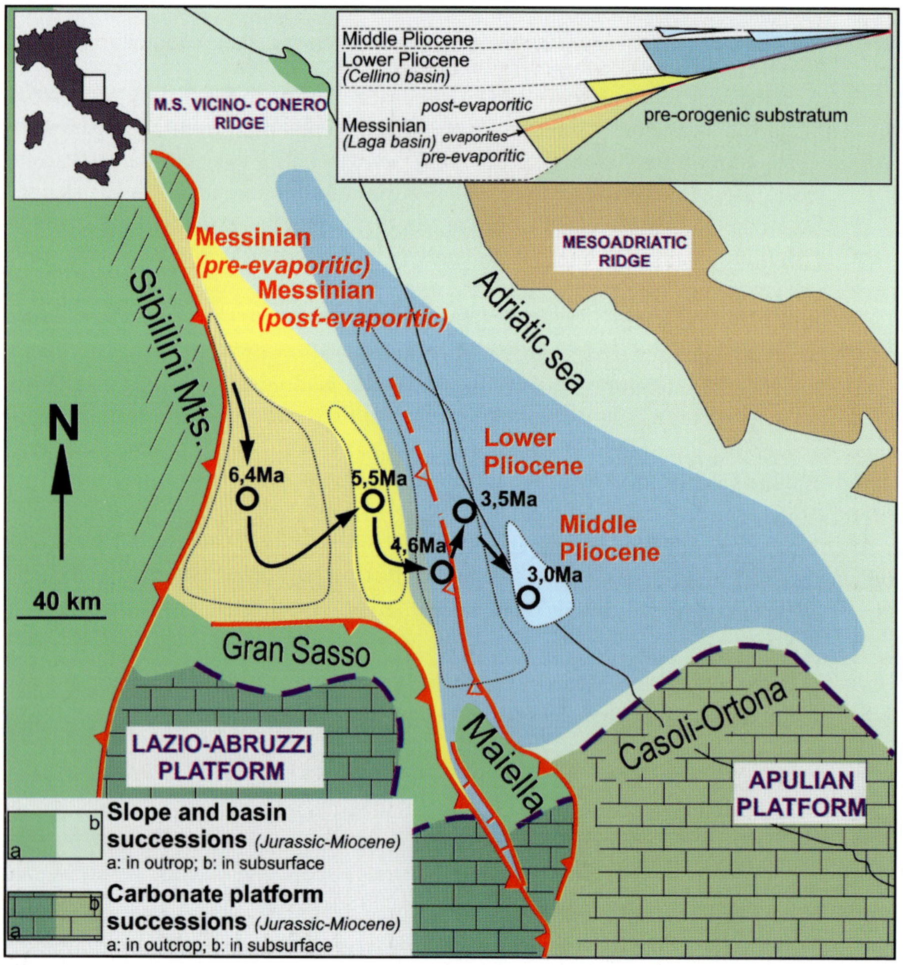

Figure 18: Distribution of main depocentral zones of the Messinian and Lower Pliocene foredeep basins. They are located in correspondence to the pelagic paleo-domains, located, respectively, to the north of the Lazio-Abruzzi and Apulian carbonate platforms.

and Adria with formation of a thick trapped wedge of Apenninic polarity. To simplify the description of the Corsica Basin, it is convenient to give here a short introduction to the evolving geodynamics of the area, starting from the first convergence events. This reconstruction is based on new CROP data described in more detail in other chapters of this volume (Finetti et al., this volume, Chapter 8; Finetti, this volume, Chapter 34).

In the Eo-Alpine stage (Upper Cretaceous–Eocene), when Adria, Africa and the interposed Ionian Sea converged as a single block with Europe, giving rise to formation of the Alps, the northwestern margin of the Adria plate thrust laterally on the Alpine Tethys with no notable sinking of the laterally subducted short slab, leaving a still large unsubducted Alpine Tethys west of the Adria margin, where convergence occurred directly between Africa and Europe, without interposition of Adria (Adria as a Plate detached from Africa).

Lateral subduction of the Alpine Tethys slab beneath the Tuscan Archipelago (Adria) zone is confirmed from other CROP sections (i.e.: CROP M-12A, M-10). Over this, substantially horizontal, laterally subducted slab of Alpine Tethys, we find the deformed western margin of the Adria plate and the Corsica Basin fill. The stratigraphic identification of the Corsica basin fill is in large part based on calibrating borehole Martina-1, on section M-12A (Finetti et al., 2001), located some 60 km to the north and interpolation with full seismic control to section M-37 via section M-12B (Figure 1). Section M-37 shows only the eastern part of the Corsica Basin, while for its setting on the frontal zone of the east-verging thrusts of the Corsica block (about 11–12 km off S.P. 3386) we refer to section M-10 (Finetti et al., 2001; Finetti et al., this volume, Chapter 8).

Over the basement (Alpine Tethys slab and/or the Adria lower and upper crust) lies a thick strongly reflecting sedimentary sequence (over 6 s twt) formed by sub-horizontal to low-angle thrusts of units deposited in the foredeep

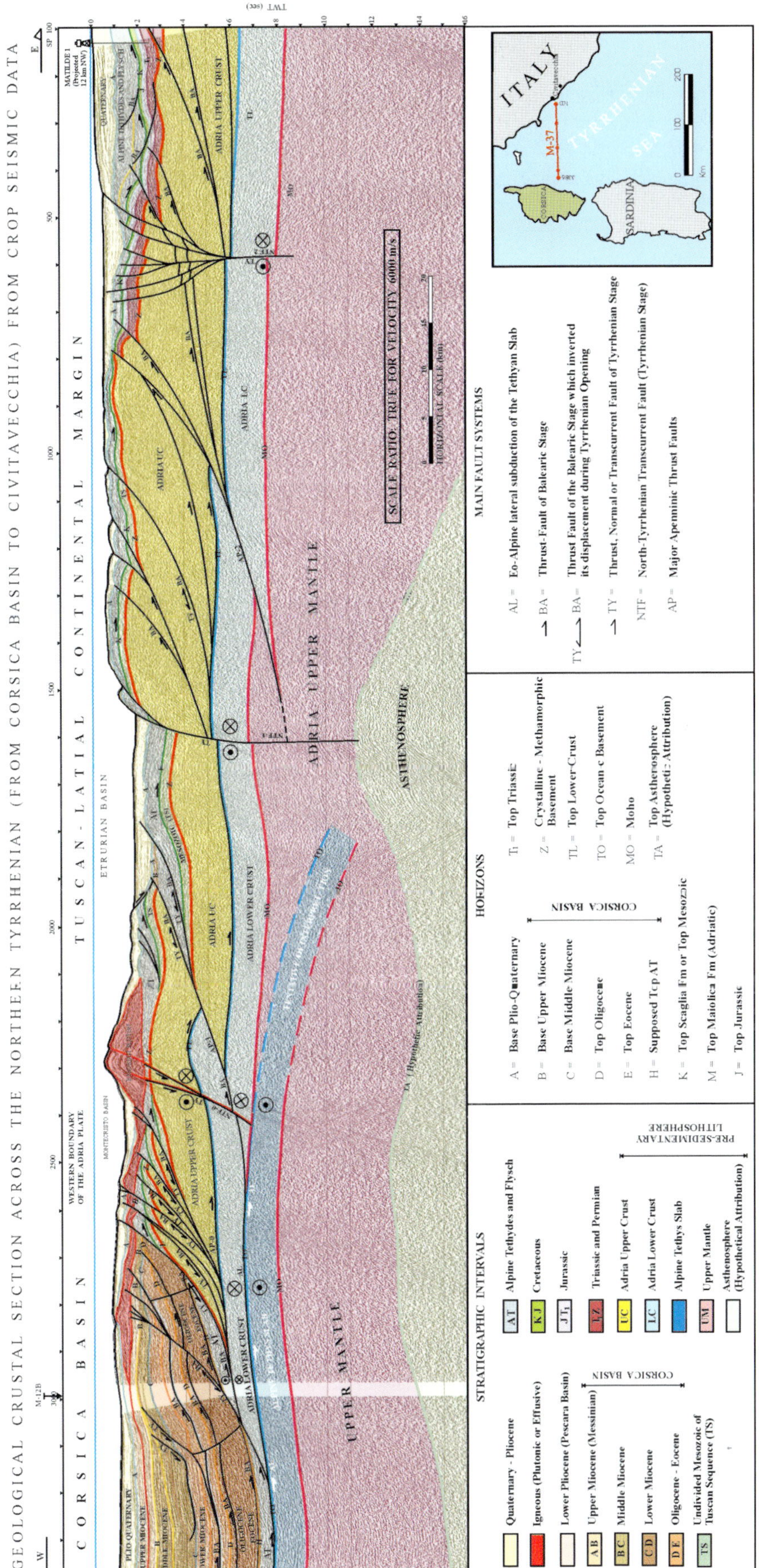

Figure 19: Reduced version of the CROP M-37 seismogeological section across the Tyrrhenian Sea.

and tectonically active during the Corsica rotation (Balearic stage). At the base, layers of Alpine Tethydes (AT) may also have been preserved (Plate 3 and Figure 19).

Very likely, thrust activity took place mostly from the Upper Eocene to Lower Miocene, simultaneously with the westward subduction of the Alpine Tethys beneath Europe and counterclockwise rotation of the Corsica–Sardinia block. In the Middle-Upper Miocene–Pliocene, when the opening of the Tyrrhenian Sea occurred, the Corsica Basin was affected by southeastward stretching effects of the Ionian slab connected with mantle flux, and most thrust faults formed in the Balearic stage inverted their displacement partially (Plate 3). Distensive activity in Upper Miocene–Pliocene times was accompanied by volcanic effusions and very shallow plutonic intrusions in the Corsica Basin and on the marginal area of the Adria plate. The most prominent shallow intrusive body, named the Etrurian Pluton is shown between S.P. 2100 and 2400 of section M-37. It seems to be associated with a sinistral strike-slip fracture driven by the Ionian subduction.

(b) Tuscan – Latial Continental Margin and Hinterland of the Northern Apennines Orogenic System. From the deformed western edge of the Adria plate to the Latial Volcanoes area (Plate 3: from km 30 to about 210) there is a wide internal geological province of relatively thin crust where the Apenninic tectonics did not create significant mountains, and the lower crust, with few exceptions, remains in large part rather flat. But this does not means that tectonic activity was not of significance. Indeed, the CROP seismic section M-37 exhibits the imprint of the strong tectonic effects of two principal geodynamic stages: the Balearic stage (Oligocene–Lower Miocene), and the Tyrrhenian stage (Middle Miocene to Present).

At the western extremity of the Adria plate (Plate 3: from km 30 to nearly 90 of the progressive section length as reported on the bottom of the section) a severe compressive deformation with numerous prominent thrust faults and crustal shortening is clearly visible. Here the crust is very thin: about 2 s twt for the pre-sedimentary upper crust and 1.5 s for the lower crust. Over the crystalline basement is a thin Mesozoic (and Permian) succession of the Tuscan sequence (TS), covered by allochthonous units of the Alpine Tethydes (Ligurides), tectonically transported over allochthonous, or in any case deformed but rooted TS and other Mesozoic-Tertiary units from the Balearic and early Tyrrhenian stages. Compressive deformation is particularly severe in the Montecristo and Etrurian basins with repeated thrust-faults. But also in the more external sector of the continental margin and Apennine hinterland, where the crust is thicker, various thrust faults are evident and contribute to crustal thickening.

From km 130 to the east, on the reconstructed seismogeological section (Plate 3) it is possible to subdivide the Mesozoic sequence by calibration with the Matilde-1 borehole and other available well data. Around km 190 to 220, the section shows the segment containing the Latial volcanic area (i.e. Bracciano and Martignano lakes, enclosed in volcanic craters). The tectono-dynamics of this volcanic system is still questioned. As regard the genesis of the magmatism, from specific petrological analysis, it is observed that we are dealing with sources of upper crust hybridized by mantle (Serri, 1990). From scattered indications on industrial seismic data of medium penetration on the Roman geological province, it is possible that, at least in some cases (i.e. Bracciano), magmatic activity effuses through the intersections of young normal faults of Apenninic trend (about NNW–SSE in the case) with strike-slip faults trending nearly perpendicular to them (almost E–W). This young fault activity is attributable to the Tyrrhenian stage (Middle Miocene to Pleistocene) and is imposed over a thin crust previously deformed by compressive tectonics of the Late Balearic stage (Lower Miocene).

Moving east again, on the crustal section area now examined, the activation of some notable transcurrent faults, all of left polarity took place in the Tyrrhenian stage. These faults (NTF: North Tyrrhenian Transcurrent Faults) belong to an important transcurrent system in the Northern Tyrrhenian sea which is well evident in seismic and magnetic data (Finetti et al., 2001).

In Plate 3 and Figure 19, the left transcurrent fault NTF-0 is associated with pluto-volcanic activity (Etrurian Pluton), while NTF-1 separates different crustal thicknesses, and NTF-2 overprints a previous upper crustal low-angle compressive deformation.

Magnetic data show very clearly southeastward trans-Tyrrhenian lineaments which are recognized as traces of the kinematic evolution of the area and the paths of the allochthonous units of the Southern Apennines and Northern Calabrian Arc (i.e. *Apennine Platform* blocks). Note that global 3-D analysis, and reconstructions based on the CROP seismic data for the whole Tyrrhenian Sea, support the conclusion that from the Adria plate paleogeographic corner denominated the *Etrurian Promontory* (Finetti, this volume, Chapter 34; Finetti, this volume, Chapter 6) a crustal block probably fragmented was geodynamically transported southeastward during the Ionian slab retreat and connected Tyrrhenian opening process (Finetti, this volume, Chapter 17; Finetti, this volume, Chapter 34; Finetti and Del Ben, this volume, Chapter 21). This crustal block constituted a triangular promontory of the NW Adria plate margin with its base nearly on the 41st parallel (Figures 14 and 15) which separated the Alpine Tethys on the west side from the Ionian Tethys south of the 41st parallel. Its sedimentary cover was detached and thrust obliquely onto the Southern Apennines. In the literature these Apenninic allochthonous Meso-Cenozoic cover sheets are considered as coming from an unidentified internal area and given the generic name *Apenninic Platform*. These allochthonous units, in the Southern Apennines and in the Calabrian Arc, are here identified with a new unifying and more significant name which refers to their original starting position: *Etrurian Platform units* (Finetti, this volume, Chapter 17).

On the western edge of the Adria plate, the imbricating thrust faults (AP-0, 1, 2) were generated during the Balearic stage (Upper Oligocene–Lower Miocene) and continued to be active also in the Tyrrhenian stage, with partial inversion of some of them. Such partial inversion probably took place also on the transcrustal thrust fault AP-1. The CROP section M-37 indicates that the thrust faults AP-1 and AP-2 slice through the entire crust.

Deep reflectors forming undulating structures have been attributed to the top asthenosphere. Corresponding time depths are in line with time depths of the interpreted top asthenosphere of section CROP M-12A (Finetti et al., 2001). A prominent asthenospheric high interpreted around S.P. 1500 seems connected to the left transcurrent fault NTF-1.

(c) Northern Apennine Lithospheric Thrusting Arc, Southern Segment (Olevano-Antrodoco). The Northern Apennine orogenic system is characterized by a large arcuate thrust belt which runs from the western Po valley (Monferrato) to Ancona and then turns south, extending as far as the coastal Tyrrhenian zone at Anzio (Figure 21). The southern part is known in the literature as the Ancona-Anzio tectonic line. This wide Northern Apennine orogenic arc is associated with a prominent transcrustal thrust plane which displaces the whole crust and the upper mantle (AP-3). Horizontal and vertical displacement rates are maxima at the apex of the arc, near the CROP-03 section (Finetti et al., 2001) and diminish progressively toward the two extremities of the arc. In the section of Plate 3, AP-3 becomes a transpressive high-angle right transcurrent fault

Gravity data show a major arcuate negative anomaly that follows the front of the whole Northern Apennine arc, including the band located east of the Antrodoco-Olevano line. Such a geophysical framework supports a reconstruction in which the trans-crustal thrust plane clearly checked on the CROP-03 section (Finetti et al., 2001) continues with diminishing displacement as far as at least the Plate 3 zone. This southern tectonic element of the Northern Apennines (Figure 4) composed of basinal terrains (Sabine Units) is macroscopically very distinct from the tectonic setting of the Central Apennine units (i.e. Gran Sasso and Maiella thrusts and innermost features) which involve platform units and form complex structures of clearly different trend.

(d) Gran Sasso and Maiella Lithospheric Thrusts characterizing the Central Apennine Belt. Between the southern segment of the Northern Apennine orogenic arc of N–S trend described in the previous paragraph and the typical dominating NW–SE trends of the Southern Apennines, a deformation process took place that gave rise to the deep-rooted compressive structures of the Central Apennines (Plate 3), where the tectono-dynamics are rather complex and not attributable to simple cylindrical compressive activity. The two impressive major structures of the Central Apennines, Mt. Gran Sasso (the highest Apenninic Mountain) and Mt. Maiella evidently form prominent thrusts. The Gran Sasso thrust moves the Jurassic-Triassic to an elevation of about 3000 m, while in the adjacent Pescara foredeep, the Pliocene base reaches a depth of about 7 km.

Regional correlation with section CROP-03 (Finetti et al., 2001) and with a reconstructed crustal section across the Southern Apennines (Finetti et al., this volume, Chapter 12), together with the gravity data (Bernardelli et al., this volume, Chapter 3), make it very evident that the Gran Sasso and Maiella thrust must be connected to a transcrustal thrust plane which displaces the entire crust and upper mantle. The tectonic scheme reported in Figure 21 exhibits a peculiar Central Apennine arcuate tectonic trend which starts to the north with a W–E direction perpendicular to the Olevano-Antrodoco line that, at Gran Sasso–Maiella, turns southward. It is geologically evident that the basin-platform distribution strongly conditioned the local tectonic setting. Indeed, the northern thrust boundary of the Central Apennines follows closely the basin (Umbro-Marchesan)–platform (Apulian platform) boundary (Figure 21).

But for a more complete understanding of the Central Apennine tectonics, it is necessary to take into account a wider paleogeographic area, in particular the pre-orogenic boundary of the Ionian Neotethys. New CROP seismic

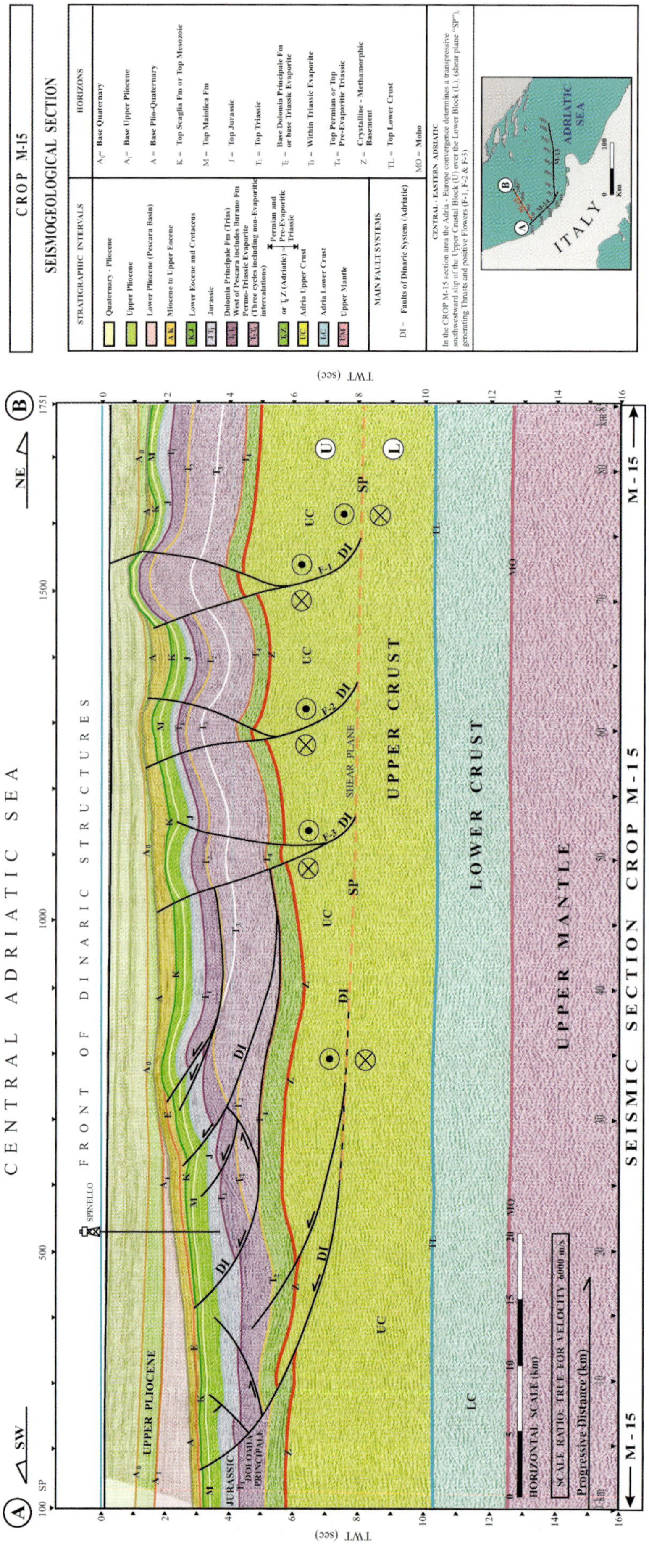

Figure 20: Reduced version of the CROP M-15 seismogeological section across the Adriatic Sea, from the Pescara basin to the frontal Dinaric structures.

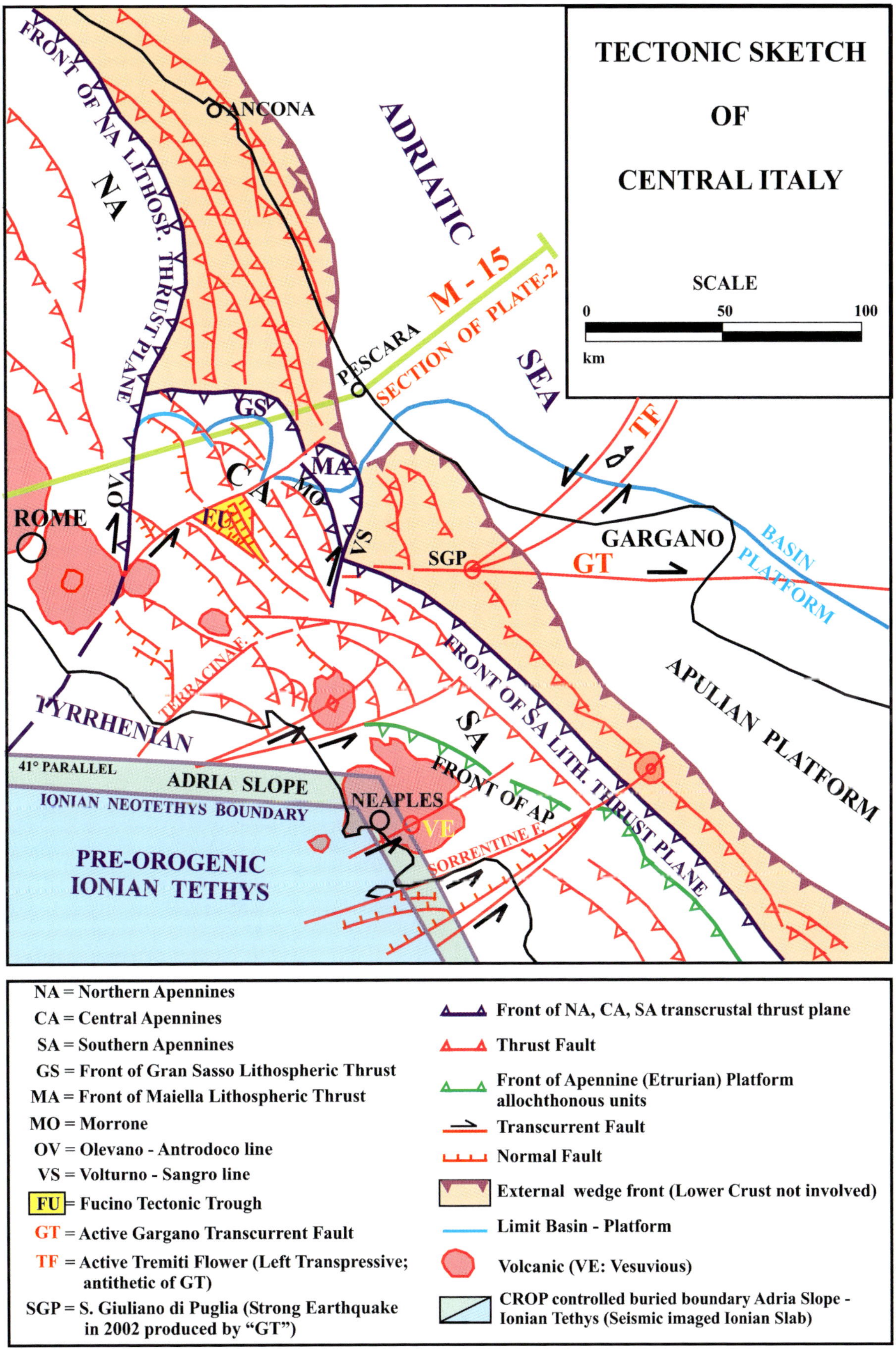

Figure 21: Scheme of the main tectono-dynamic features characterizing the chain sector connecting the Northern, Central and Southern Apennines.

data allow reconstruction with full reliability of the Ionian Ocean buried boundary (Finetti, this volume, Chapter 34; Finetti, this volume, Chapter 6). It is very clear that the northern boundary of the Ionian Ocean was around the present 41° latitude, limited to the east by a NW–SE Apulian slope which forms an obtuse angle with the 41° parallel whose vertex is located several kilometers north of the city of Naples (Figure 21). Assuming this Ionian corner area, and examining the reconstructed main tectonic setting in Figure 21, it is now possible to observe that the Central Apennines is an orogenic area in which a combination of two different geodynamic processes took place over a very differentiated paleogeographic context. Indeed, while in the Northern Apennines region the subduction of the Alpine Tethys determined the closure of this intraplate ocean and the collision Corsica–Adria, with the formation of a first Northern Apennine orogenic arc (Finetti et al., 2001 and Plate 3), south of the 41° parallel, only in the Middle Miocene did the Ionian slab start to subduct and form the Tyrrhenian sea over the Ionian Tethys. Southeastward, Ionian slab retreat and consequent formation of the Southern Apennines–Calabrian Arc–Sicily orogenic system was a relatively fast process, and in the Ionian corner area (*Parthenopean Corner*; Plate 3b and Figure 21) a complex deformation took place in the orogenic area of the Central Apennines as the accommodating zone between the northern and southern Apennines. Northeastward compressive dynamics accompanied by the Fucino left transcurrent and Volturno-Sangro right transpressive fault, torsional movements, together with counterclockwise rotation of the Gran Sasso and Maiella-Morrone structural sector, acted to form the present tectonic setting of the Central Apennines.

An evident young SW–NE directed fault bordering the NW extremity of the Fucino trough explains the extended and collapsed triangular form of this trough.

(e) The Adriatic Crustal Section. The Adriatic segment of the reconstructed crustal section, almost completely constrained by the new CROP data of line M-15 (Figure 20 and Plate 3), is composed of two different geological provinces: the Pescara foredeep, and the frontal structures of the Dinaric system. The quality of the CROP data is very good in the sedimentary sequence and fair to good in the pre-sedimentary crustal part. The Moho discontinuity (MO) and the top of the lower crust are reliably identified. Some boreholes help to calibrate the upper part of the strongly reflecting sedimentary succession. The interpreted basement (Z) can be considered fully acceptable in general and is well evident over part of M-15.

From the outcropping Gran Sasso thrust to the initial part of M-15 (segment CD) the sedimentary crust is interpreted from several seismic lines of various hydrocarbon exploration campaigns, while the upper crust (UC) and lower crust (LC) were derived by extrapolation of CROP data since in this crustal segment sizeable thickness variations are not foreseen.

As well exhibited in Plate 3, the segment CD shows the impressive buried frontal zone of the Central Apennine thrust belt. Beneath the Gran Sasso Mountain, the vertical time displacement of the AP-4B megathrust is of the order of 5.5 s twt (or about 13 km) and the horizontal displacement of the Gran Sasso thrust is of the order of 45 km. Similar vertical and horizontal displacements can be expected for the Morrone-Maiella transcrustal thrust.

The Pescara foredeep with its thick depositional prism is the best time recorder of Central Apennine geodynamic activity. It is fully evident that the major compressional deformation intensity occurred in the Lower-Middle Pliocene.

A part of CROP M-15 which explores the Dinaric structures is also very interesting. Here the tectonic deformation occurs on a thick Triassic evaporite succession (Burano Fm) that favours also plastic tectonic behaviour. In particular the segment DE is of great importance for understanding the tectono-dynamics involving the many Islands of the Yugoslavian Adriatic sea which trend almost E–W and are affected by compressive high-angle south-verging structures.

From a careful study of the CROP M-15 section and of several other seismic lines, it is evident that the compressive structures of Dinaric polarity are all rooted in a common shear plane located in the pre-sedimentary upper crust (Plane SP on Plate 3). The flower structures F-1, F-2 and F-3, as well as the other more external reverse dinaric faults, are deformed by a compressive geodynamic mechanism which is associated with the Adria-Europe plate convergence, that in this case is acting by shortening of the upper crustal block (U) over the SP shear plane, keeping the lower crust block below SP, or part (L) of the crust undeformed. Beneath the shear plane SP the lower crustal part (L) moves northward relative to (L), as indicated in Plate 3.

Combinations of NW–SE trending, currently right-transcurrent faults and antithetic NE–SW left-transcurrent faults are rather common in the Adriatic sea area. There is evidence also that some major NW–SE transcurrent faults inverted their slip polarity after the Eo-Alpine (Upper Cretaceous–Lower Eocene) Adria–Europe collision.

3.3. Evolutive geodynamic scheme

The results obtained from the crustal section in Plate 3, within the global context of new information coming from the whole CROP dataset (Finetti, this volume, Chapter 34), allow us to make some new basic observations on the Central Apennines:

(a) The *Pre-orogenic paleogeography* was composed of an *Ionian Neotethys* of Permo-Triassic opening limited to the north by the present day 41° parallel, in open communication with the *Alpine Neotethys* of Middle Upper Jurassic opening. Two continental promontories (*Panormid* and *Etrurian*; see small scale schemes: Plate 3a and 3b) delimited the passage between the two intraplate oceans.

(b) In the *Eo-Alpine geodynamic stage* (Upper Cretaceous–Lower Eocene), the closure of the segment of the Alpine Tethys interposed between Adria and Europe (Alps) took place, with subduction of the Alpine Tethys beneath Adria (lateral subduction for the NW Adria or Etrurian Promontory). West of the Alps, between Europe and Africa, even if partially restricted, a wide Alpine Tethys remained unsubducted.

(c) Near the *End of the Lower Oligocene* (32–33 Ma), very likely, an intra-ocean, Eo-Alpine backthrust of the westernmost Alpine Tethys remnant, due to Africa–Europe convergence, developed such as to cut the ocean crust, so giving origin to the prominent geodynamics of the *Balearic stage*, with subduction of the Alpine Tethys beneath Europe (Corsica–Sardinia Block and Iberian margin).

(d) In the *Balearic stage* (Upper Oligocene–Lower Miocene) the Alpine Tethys subducted beneath the rotated Corsica–Sardinia block and westernmost European margin (Kabilides). The internal Northern Apennine tectonics started and Corsica–Adria collided along a N–S trending boundary (i.e. Corsica Basin). *A prominent eastward migrating litho-asthenospheric bulge, confined between the eastward Eo-Alpine lateral subduction and the westward Balearic subduction, acted as basic geomotor for the Northern Apennines deformation during the last 28–30 Ma, determining a significant crustal shortening of the Adria plate* (Finetti, this volume, Chapter 34).

(e) In the *Tyrrhenian stage* (Middle Miocene to present), in continuity with the previous stage, the *slab retreat and subduction of the Ionian crust with formation of the Tyrrhenian backarc basin floored by oceanic crust* took place. *The Etrurian and Panormid continental promontories stopped the rotation of the Corsica–Sardinia block at the end of Balearic stage by collision, while, across the communicating Ionian ocean, crustal fragments of the Etrurian, Panormid, Corso–Sardinian blocks and the Alps were transported obliquely on the Southern Apennines Calabrian arc and Sicily.*

(f) Westward subduction of the Alpine Tethys in the Balearic stage, as well as north-westward subduction of the Ionian Tethys, was accompanied by *relevant stripping off of whole ocean cover sequences ("Alpine Tethydes" or "Ligurides" and "Ionides" or "Lagonegro"* respectively) with formation of wedges and skinned nappes covering the orogenic systems involved.

(g) Along the 41° parallel, the northern boundary of the Ionian Tethys, a *fast W–E moving Tyrrhenian mantle drag* contributed, together with the asthenospheric bulge, to an anticlockwise rotation of the Northern Apennines, where crustal shortening is accompanied by transtensional effects on the Tuscan Archipelago driven by the southeastward Tyrrhenian mantle flux. Moreover, this eastward mantle dynamics along and south of the 41° parallel took part also in the Central Apennines deformation between the Northern and Southern Apennines. The Volturno-Sangro lithospheric ramp separates the Northern Apennines from the Southern (Figure 21).

(h) *Ionian slab retreat* dynamics acted at higher speed as a fast geomotor (up to 10 cm/y) superimposed on a slower Africa–Europe convergence (averagely about 0.4 cm/y). Before the Tyrrhenian stage, Africa, Adria and the interposed Ionian Tethys converged as a unique solid block.

CONCLUSIONS

The new CROP data, jointly with geological, structural, and stratigraphic data and integrating other geophysical studies carried out in central Italy, lead us to the following concluding remarks:

(1) The Central Apennines are the result of two combined geodynamic processes, strongly conditioned by the different paleogeographic contexts involved: (a) *North of the 41° parallel, in the Balearic stage, the closure of*

the Alpine Tethys took place with the Corsica–Adria collision, significant deformation of the internal Northern Apennines, and formation of a prominent east-migrating compressive asthenospheric bulge in the Tuscan Archipelago as basic geomotor of the Northern Apennines deformation (Finetti et al., this volume, Chapter 8; Finetti, this volume, Chapter 34); (b) *South of the 41° parallel* (northern boundary of the Ionian Neotethys) *in the Middle Miocene, a fast southeastward Ionian slab retreat started, followed by a Tyrrhenian Mantle flux that created the Southern Apennines–Calabrian Arc–Sicilian Apennines orogenic system.*

(2) *The Sangro-Volturno line constitutes a dextral transpressive oblique ramp that limits southward the whole Central-Northern Apennines arc, decoupling it from that of the Southern Apennines;*

(3) *The Sibillini Mts.-Antrodoco-Olevano oblique ramp, conditioned by the Mesozoic paleomargin, separates the Northern Apennines arc, with greater shortening in the apex sector, from that of the Central Apennines, with greater shortening in the Gran Sasso sector;*

(4) *The Gran Sasso Salient is a smaller arc within the Central Apennines arc, conditioned by the Mesozoic paleomargin architecture;*

(5) *In the frontal zone of the Central Apennines thrust belt there is a vertical crustal displacement of about 10 km, and a horizontal one of about 45 km, with lithospheric mantle involvement. The higher structural elevation of the Apennine Units s.s. within the Northern Apennines with respect to those of the Southern Apennines is due to greater crustal shortening.*

ACKNOWLEDGEMENTS

This work was supported by the CROP partners (CNR, ENEL, ENI) with CNR funding to I.R. Finetti, Grant CTB 97.04525, and by MIUR, COFIN 2000 and COFIN 2002 to I.R. Finetti, and COFIN 2003 to F. Calamita.

REFERENCES

Alfonsi, L., Funiciello, R., Mattei, M., 1991. Strike-slip tectonics in the Sabine area. Boll. Soc. Geol. It. 110, 481–488.

Angelucci, A., Praturlon, A., 1968. Raddoppio tettonico della serie meso-cenozoica nelle gole di Celano a Nord del Fucino (Appennino Centrale). Geol. Romana 7, 431–446.

Bally, A.W., Burbi, L., Cooper, C., Ghelardoni, R., 1988. Balanced sections and seismic reflection profiles across the Central Apennines. Mem. Soc. Geol. It. 35, 257–310.

Barchi, M.R., De Fayter, A., Magnani, M.B., Minelli, G., Pialli, G., Sotera, B.M., 1998. The structural style of the fold and thrust belt. Mem. Soc. Geol. It. 52, 557–578.

Ben-Avraham, Z., Boccaletti, M., Cello, G., Grasso, M., Lentimi, F., Torelli, L., Tortorici, L., 1990. Principali domini strutturali originatesi dalla collisione neogenico-quaternaria nel Mediterraneo centrale. Mem. Soc. Geol. It. 45, 453–462.

Bernardelli, P., Cavalli, C., Longoni, R., Giori, I., this volume. Gravity and magnetic fields of the Central Mediterranean Region. Chapter 3.

Bigi, S., Calamita, F., Centamore, E., 1997. Caratteristiche geologico – strutturali dell'area abruzzese ad oriente del Gran Sasso. Studi Geol. Camerti vol. spec. 1995/2, 67–76.

Boccaletti, M., Elter, P., Guazzone, G., 1971. Plate tectonic models for the development of the Western Alps and Northern Apennines. Nat., Phys. Scr. 234, 108–111.

Boccaletti, M., Conedera, C., Dainelli, P., Gocev, P., 1982. The recent (Neogene – Quaternary) regmatic system of the Western Mediterranean regio. J. Pet. Geol. 5, 31–49.

Boccaletti, M., Ciaranfi, N., Cosentino, D., Deiana, G., Gelati, R., Lentini, F., Massari, F., Moratti, G., Pescatore, T., Ricci Lucchi, F., Tortorici, L., 1990. Palinspatic restoration and paleogeographic reconstruction of the peri-Tyrrhenian area during Neogene. Palaeog. Palaeocl. Palaeoec. 77, 41–50.

Boncio, P., Brozzetti, F., Lavecchia, G., 2000. Architecture and seismotectonics of a regional low-angle normal fault in central Italy. Tectonics 19, 1038–1055.

Calamita, F., 1990. Thrusts and fold-related structures in the Umbria-Marche Apennines (Central Italy). Ann. Tectonicae IV, 83–117.

Calamita, F., Deiana, G., 1988. The arcuate shape of the Umbria-Marche-Sabina Apennines (Central Italy). Tectonophysics 146, 139–147.

Calamita, F., Deiana, G., 1995. Correlazioni tra gli eventi deformativi neogenico-quaternari del settore tosco-umbro-marchigiano. Studi Geol. Camerti vol. spec. 1995/1, 137–152.

Calamita, F., Coltorti, M., Piccinini, D., Pierantoni, P.P., Pizzi, A., Ripete, M., Scisciani, V., Turco, E., 2000. Quaternary faults and seismicity in the Umbro-Marchean Apennines (Central Italy): evidence from the 1997 Colfiorito earthquake. J. Geodynamics 29, 245–264.

Calamita, F., Coltorti, M., Pierantoni, P.P., Pizzi, A., Scisciani, V., Turco, E., 1999. Relazioni tra faglie normali quaternarie e la sismicità nella dorsale appenninica umbro-marchigiana: l'area di Colfiorito. Studi Geol. Camerti XIV, 177–191.

Calamita, F., Paltrinieri, W., Pelorosso, M., Scisciani, V., Tavarnelli, E., 2003. Inherited Mesozoic architecture of the Adria continental Palaeomargin in the Neogene Central Apennines orogenic system, Italy. Boll. Soc. Geol. It. 122, 307–318.

Calamita, F., Scisciani, V., Adamoli, L., Ben M'Barek, M., Pelerosso, M., 2002a. Il sistema a thrust del Gran Sasso D'Italia (Appennino Centrale). Studi Geol. Camerti, nuova serie 1, 1–32.

Calamita, F., Scisciani, V., Montefalcone, R., Paltrinieri, W., Pizzi, A., 2002b. L'ereditarietà del paleomargine dell'Adria nella geometria del sistema orogenico centro-appenninico: l'area abruzzese esterna. Mem. Soc. Geol. It. 57, 355–368.

Capotorti, E., Centamore, E., Chiocchini, U., Civitelli, A., Mancinelli, G., Mariotti, A., Romano, A., Salvucci, R., 1992. Dati preliminari geologico-strutturali sull'unità di Monte Giano e Monte Gabbia. Studi Geol. Camerti vol. spec. 1991/2, 119–124. CROP 11.

Casero, P., Roure, F., Edignoux, L., Moretti, I., Muller, C., Sage, L., Vially, R., 1988. Neogene Goedynamic evolution of the Southern Appennines. Mem. Soc. Geol. It. 41, 109–120.

Casnedi, R., Serafini, G., 1994. Interpretazione geologica della sezione sismica nella Valle del Vomano (Abruzzo). Atti Tic. Sc. Terra, serie speciale 2, 45–49.

Castellarin, A., Colacicchi, R., Praturlon, A., 1978. Fasi distensive, trascorrenze e sovrascorrimenti lungo la linea "Ancona-Anzio" dal Lias medio al Pliocene. Geol. Romana 17, 161–189.

Carmignani, L., Kligfield, R., 1990. Crustal extension in the northern Apennines: the Transiction from comprenssion to extension in the Alpi Apuane core complex. Tectonics 9, 1275–1303.

Carmignani, L., Decandia, F.A., Fantozzi, P.L., Lazzarotto, A., Liotta, D., Meccheri, M., 1994. Tertiary extensional tectonics in Tuscany (Northern Apennines). Tectonophysics 238, 295–315.

Cassinis, R., Scarascia, S., Lozej, A., this volume. Review of the exploration results of deep crust and upper mantle in the italian region by wide-angle seismic reflection–refraction technique (1956–1987). Chapter 2.

Cavinato, G.P., Salvini, F., Tozzi, M., 1986. Evoluzione strutturale del settore centrale della linea Olévano-Antrodoco. Mem. Soc. Geol. It. 35, 591–601.

Ciarapica, G., Passeri, L., 2002. The palaeogeographic duplicity of the Apennines. Boll. Soc. Geol. It. vol. spec. 1, 67–75.

Cipollari, P., Cosentino, D., 1992. La linea Oléveno-Antrodoco: contributo della biostratigrafia alla sua caratterizzazione cinematica. Studi Geol. Camerti vol. spec. 1991/2, 143–149.

Cipollari, P., Cosentino, D., Parotto, M., 1995. Modello cinematico-strutturale dell'Appennino centrale. Studi Geologici Camerti vol. spec. 1995/2, 135–143.

Colacicchi, R., 1967. Geologia della Marsica orientale. Geologica Romana 6, 189–316.

Corrado, S., 1995. Nuovi vincoli geometrico-cinematici all'evoluzione neogenica del tratto meridionale della linea Olevano-Antrodoco. Boll. Soc. Geol. It. 114, 245–276.

Corrado, S., Cosentino, D., Gioia, C., 1990. Sistema di retroscorrimenti nella Marsica orientale (Abruzzo-Appennino centrale). Mem. Soc. Geol. It. 45, 591–603.

Cosentino, D., Parotto, M., 1986. Assetto strutturale dei Monti Lucretili settentrionali (Sabina). Geol. Romana 25, 73–90.

Cosentino, D., Parotto, M., 1992. La struttura a falde della Sabina (Appennino Centrale). Studi Geol. Camerti vol. spec. 1991/2, 381–387. CROP 11.

Crescenti, U., 1969. Stratigrafia della serie calcarea dal Lias al Miocene nella regione Marchigiano-Abruzzese (Parte I – Descrizione delle serie stratigrafiche). Mem. Soc. Geol. It. 8, 155–204.

Crescenti, U., Crostella, A., Donzelli, G., Raffi, G., 1969. Stratigrafia della serie calcarea dal Lias al Miocene nella regione Marchigiano Abruzzese. (Parte II – Litostratigrafia, biostratigrafia, paleogeografia). Mem. Soc. Geol. It. 9, 343–420.

Crescenti, U., 1971. Sul limite Mio-Pliocene in Italia. Geol. Romana 10, 1–21.

Crescenti, U., D'Amato, C., Balduzzi, A., Tonna, M., 1980. Il Plio–Pleistocene del sottosuolo abruzzese – marchigiano tra Ascoli Piceno e Pescara. Geol. Romana 19, 63–84.

Crescenti, U., Biondi, R., Raffi, I., Rusciadelli, G., 2002. The S. Nicolao section (Montagna della Maiella): a reference section for the Miocene-Pliocene boundary in the Abruzzi area. Boll. Soc Geol. It. vol. spec. 1.

Damiani, A.V., 1992. Riflessioni sull'assetto tettonico della dorsale dei Monti Simbruini–Affilani–"pre-Ernici"–Ernici e nuove ipotesi di lavoro. Studi Geologici Camerti vol. spec. 1991/2, 63–66. CROP.

Decandia, F.A., Lazzarotto, A., Liotta, D., Cernobori, L., Nicolich, R., 1998. The CROP 03 traverse. Insights on post-collision evolution of northern Apennines. Mem. Soc. Geol. It. 52, 427–439.

Decandia, F.A., Tavarnelli, E., Alberti, M., 2002. La doppia riattivazione della Linea della Valnerina: implicazioni per l'evoluzione tettonica dell'Appennino Umbro-Marchigiano. Studi Geol. Camerti, nuova serie 1, 77–86.

Donzelli, G., 1968. Studio geologico della Maiella. In: Studio della Maiella, Università G. D'Annunzio, Dipartimento di Scienze della Terra, 1997, 49 p.
Faccenna, C., Davy, P., Brun, J.P., Funicello, R., Giardini, D., Mattei, M., Nalpas, T., 1996. The dynamics of back-arc extension:an experimental approach to the opening of the Tyrrhenian Sea. Geophys. J. Int. 126, 781–795.
Finetti, I.R., this volume. The Calabrian Arc and subducting Ionian slab from new CROP seismic data. Chapter 17.
Finetti, I.R., this volume. Geodynamic evolution of the Mediterranean Region, from the Permo-Triassic Ionian opening to the Present, constrained by new lithospheric CROP seismic data. Chapter 34.
Finetti, I.R., Del Ben, A., 1986. Geophysical study of the Tyrrhenian opening. Boll. Geof. Teor. Appl. 28, 75–156.
Finetti, I.R., Del Ben, A., 2000. Ionian opening, crustal stratigraphy and tectono-dynamics of the Pelagian sea Region from new CROP seismic data. In: Proceedings of the International Conference of the Tripoli. Libya, 2000.
Finetti, I.R., this volume. Ionian and Alpine Neotethyan oceans opening. Chapter 6.
Finetti, I.R., Del Ben, A., this volume. Ionian Tethys lithosphere roll-back sinking and back-arc Tyrrhenian opening from new CROP seismic data. Chapter 21.
Finetti, I.R., Del Ben, A., this volume. Crustal tectono-stratigraphic setting of the Adriatic sea from new CROP seismic data. Chapter 23.
Finetti, I.R., Lentini, F., Carbone, S., Catalano, S., Del Ben, A., 1996. Il sistema Appennino meridionale-Arco Calabro-Sicilia nel Mediterraneo centrale: studio geologico-geofisico. Boll. Soc. Geo. It. 115, 529–559.
Finetti, I.R., Boccaletti, M., Bonini, M., Del Ben, A., Gelati, R., Pipan, M., Sani, F., 2001. Crustal section based on CROP seismic data across the North Tyrrhenian–Northern Apennines–Adriatic Sea. Tectonophysics 343, 135–163.
Finetti, I., Boccaletti, M., Bonini, M., Del Ben, A., Pipan, M., Prizzon, A., Sani, F., this volume. Lithospheric tectono-stratigraphic setting of the Ligurian Sea–Northern Apennines–Adriatic foreland from integrated CROP seismic data. Chapter 8.
Finetti, I.R., Lentini, S., Carbone, S., Del Ben, A., Di Stefano, Guarnieri, P., Pipan, M., Prizzon, A., this volume. Crustal tectono-stratigraphic and geodynamics of the Southern Apennine from CROP and other integrated geophysical-geological data. Chapter 12.
Gattacceca, J., Speranza, F., 2002. Paleomagnetism of Jurassic to Miocene sediments from the Apenninic carbonate platform (southern Apennines, Italy): evidence for a 60° counterclockwise Miocene rotation. Earth and Planetary Sci. Lett. 201, 19–34.
Ghisetti, F., Follador, U., Lanza, R., Vezzani, L., 1991. La zona di Rigopiano-Bussi-Revisondoli: svincolo transpressivo al margine nord-orientale della piattaforma laziale-abruzzese. Studi Geol. Camerti vol. spec. 1991/2, 215–220.
Ghisetti, F., Vezzani, L., 1997. Interfering paths of deformation and development of arcs in the fold-and-thrust belt of the Central Apennines (Italy). Tectonics 16, 523–536.
Koopman, A., 1983. Detachment tectonics in the Central Apennines, Italy. Geol. Ultraiectina 30. 155 p.
Lavecchia, G., 1985. Il sovrascorrimento dei Monti Sibillini: analisi cinematica e strutturale. Boll. Soc. Geol. It. 104, 161–194.
Lavecchia, G., Minelli, G., Pialli, G., 1988. The Umbria-Marche arcuate fold-belt. Tectonophysics 146, 125–137.
Lavecchia, G., Stoppa, F., 1996. The tectonic significance of italian magmatism: an alternative view to the popular interpretation. Terra Nova 8, 435–446.
Lentini, F., Carbone, S., Di Stefano, Guarnieri, A., 2002. Stratigraphical and structural constraints in the Lucanian Apennines (Southern Italy): tools for reconstructing the geological evolution. J. Geodynamics 34, 141–158.
Locardi, E., 1982. Individuazione di strutture sismogenetiche dall'esame dell'evoluzione vulcano-tettonica dell'Appennino e del Tirreno. Mem. Soc. Geol. It. 24, 569–596.
Malinverno, A., Ryan, B.F., 1986. Extension in the Tyrrhenian sea and shortening in the Apennines as a result of arc migration driven sinking of the lithosphere. Tectonics 5, 227–245.
Mantovani, E., Albarello, D., Tamburelli, C., Babbucci, D., Viti, M., 1997. Plate convergence, crustal delaminations, extrusion tectonics and minimization of shortening work as main controlling factors of the recent Mediterranean deformation pattern. Ann. Geofis. 40, 611–643.
Mattei, M., Funicello, F., Kissel, C., Laj, C., 1992. Rotazione di blocchi crostali neogenici nell'Appennino centrale: analisi paleomagnetiche e di anisotropia della suscettività magnetica (AMS). Studi Geol. Camerti vol. spec. 1991/2, 221–230. CROP 11.
Menardi Noguera, A., Rea, G., 2000. Deep structure of the Campanian-Lucanian arc (Southern Apennine, Italy). Tectonophysics 324, 239–265.
Milia, M.L., 2002. Architettura, evoluzione spazio-temporale e formazione del bacino pliocenico abruzzese (Appennino centrale, Italia). PhD Thesys, Università G. d'Annunzio Chieti.
Montone, P., Salvini, F., 1993. Geologia strutturale dei rilievi tra Colli di Monte Bove (Carsoli) e Tagliacozzo, Abruzzo. Geologica Romana 29, 15–29.

Mostardini, F., Merlini, S., 1986. Appennino centro-meridionale: Sezioni geologiche e proposta di modello strutturale. Mem. Soc. Geol. It. 35, 177–202.

Ori, G.G., Serafini, G., Visentin, C., Ricci Lucchi, F., Casnedi, R., Colalongo, M.L., Mosna, S., 1991. The Pliocene-Pleistocene adriatic foredeep (Marche and Abruzzo, Italy): an integrated approach to surface and subsurface geology. In: Adriatic Foredeep Field Trip, Guide Book, 3rd E.A.P.G. Conference, Florence, May 26–30, 1991, 85 p.

Parotto, M., Praturlon, A., 1975. Geological summary of the Central Apennines. Quaderni de "La Ricercerca Scientifica" 90, 257–306.

Parotto, M., Tozzi, M., Carinato, G.P., 2000. Dal Tirreno all'Adriatico, la realizzazione di un vecchio sogno nel sottoprogetto CROP-11. Protecta 12/99-2/00, 46–48.

Patacca, E., Scandone, P., Bellatalla, M., Perilli, N., Santini, U., 1991. La zona di giunzione tra l'arco appenninico settentrionale e l'arco appenninico meridionale nell'Abruzzo e nel Molise. Studi Geol. Camerti vol. spec. 1991/2, 417–441. CROP 11.

Pierantoni, P.P., 1997. Faglie trascorrenti sin-thrusting come ripartizione della deformazione: l'esempio della faglia sabina (Appennino Centrale). Studi Geol. Camerti 14, 279–290.

Roeder, D., 1980. Geodynamics of the Alpine-Mediterranean system – a synthesis. Eclogae Geol. Helv. 73, 353–377.

Royden, L.H., Patacca, E., Scandone, P., 1987. Segmentation and configuration od subducted lithosphere in Italy: an important control on thrust belt and foredeep-basin evolution. Geology 15, 714–717.

Scisciani, V., Calamita, F., Bigi, S., De Girolamo, C., Paltrinieri, W., 2000a. The influence of syn-orogenic normal faults on Pliocene thrust system development: the Maiella structure (Central Apennines, Italy). Mem. Soc. Geol. It. 55, 193–204.

Scisciani, V., Calamita, F., Tavarnelli, E., Rusciadelli, G., Ori, G.G., Paltrinieri, W., 2001. Foreland-dipping normal faults in the inner edges of syn-orogenic basins: a case from the Central Apennines, Italy. Tectonophysics 330, 211–222.

Scisciani, V., Rusciadelli, G., Calamita, F., 2000b. Faglie normali nell'evoluzione Tortoniano-messiniana dei bacini sinorogenici dell'Appennino centrale esterno. Boll. Soc. Geol. It. 119, 715–732.

Serri, G., 1990. Neogene-Quaternary magmatism of the Tyrrhenian region: characterization of the magma sources and geodynamic implication. Mem. Soc. Geol. It. 41, 219–242.

Tapponnier, P., 1977. Evolution tectonique du système alpin en Méditerranée: poinçonnement et écrasement rigide-plastique. Bull. Soc. Géol. Fr. 7, 437–460.

Tozer, R.S.J., Butler, R.W.H., Corrado, S., 2002. Comparing thin- and thich-skinned thrust tectonic models of the Central Apennines, Italy. EGS Stefan Muller Special Publication Series 1, 281–301.

Van Bemmelem, R.W., 1972. Driving forces of Mediterranean orogeny (Tyrrhenian test-case). Geol. Mjinbouw 51, 548–568.

Vezzani, L., Ghisetti, F., 1995. Domini in compressione ed in distensione al retro dei fronti del Gran Sasso–M. Picca e del M. Morrone: Il ruolo della zona di taglio Avezzano–Bussi (Appennino centrale). Studi Geologici Camerti vol. spec. 2, 475–490.

Wezel, F.C., 1982. The Tyrrhenian sea: a rifted krikogenic swell basin. Mem. Soc. Geol. It. 24, 531–568.

Wortel, M.J.R., Spakman, W., 1992. Structure and dynamics of subducted lithosphere in the Mediterranean region. Proc. K. Ned. Akad. Wet., Biol.Chem. Geol. Phys. Med. Sci. 95, 325–347.

CROP PROJECT: Deep Seismic Exploration of the Central Mediterranean and Italy
Edited by I.R. Finetti

CHAPTER 10

Understanding the Ionides and their Geodynamics

I.R. Finetti[1]

ABSTRACT

Among the various allochthonous tectono-stratigraphic units of the Southern Apennines and Sicily thrust belt, a problem of major importance is the correct understanding and definition of the deep water basin where the Lagonegrese and Imerese condensed succession deposited, and which evolutive tectono-dynamic mechanism transported it over the Apulian and Sicilian platforms. The interpretation of the deep seismic data of the CROP project, and of other "MS" and industrial seismic lines, proved determinant for the reconstruction of the original deep deposition basin and of the mechanism of transport in the Southern Apennines–Calabrian Arc–Sicily belt. Starting from outcropping and buried Lagonegrese units from significant boreholes in the Southern Apennines, it is possible to seismically discriminate and follow these tectono-stratigraphic units across the Gulf of Taranto area, and carefully extend reliable interpretation of the correlated thrust-sheets as far as the Calabrian wedge and the undeformed Ionian sequence. *There is good seismic evidence that the Lagonegrese units correspond to the thin characteristic reflecting interval that covers the Ionian oceanic crust. Over this Ionian interval, a deposition of Miocene and Plio-Quaternary continues.*

So, the flat, relatively thin, Pre-Miocene sedimentary cover of the Ionian ocean crust is the original stratigraphic succession which fed the Lagonegrese tectonic units. Piled Lagonegrese sheets similar to those found in boreholes of the Southern Apennines are seismically visible on the wedge of the Calabrian Arc and in Sicily (Caltanissetta wedge). The Imerese units of Sicily are the equivalent of the Lagonegrese Ionian-derived sheets. To characterize these Ionian derived allochthonous units distributed over a wide area of the Southern Apennines–Calabrain Arc–Sicily with a more appropriate and immediately self explanatory name, the new term ***"Ionides" (Ionian Tethydes)*** has been adopted [Finetti, I.R., 2002a. Lithospheric tectono-stratigraphic setting and time–space stacking process of the Calabrian arc from new CROP seismic data. In: Abstract Proceedings of 81st Riunione Estiva of the Società Geologica Italiana, Torino, 10–12 September 2002; Finetti, I.R., 2002b. Understanding Ionides and Calabrian arc from new CROP seismic data. In: Expanded Abstract, Proceedings of the 21° Convegno Nazionale of the GNGTS-EAGE 2nd Italian section convention, Rome, 19–21 November, 2002]. *Therefore, the term "Ionides" represents allochthonous units originally sedimented into the Ionian Tethys and on its bordering slope area from basin opening (Permo-Triassic) to the Oligocene, which, successively, driven by the Ionian roll-back sinking (Middle Miocene to Present), were stripped off their oceanic or bordering slope basement and thrust over the Southern Apennine–Calabrian Arc–Sicily belt.*

Keywords: Ionides: from Ionian to South Apennines–Calabrian wedge–Sicily

1. INTRODUCTION

Widely exposed allochthonous thrust-units of basinal deposition have been recognized in the Southern Apennines for over three decades (Scandone, 1967, 1972), and have continued to attract the attention of geoscientists. One of the most characteristic outcropping zones of these tectono-stratigraphic units is that of the "Lagonegro" (Southern Apennines) which gave its name to the whole complex of such thrust-sheets (*Lagonegrese*). For a long time most of the literature interpreted the Lagonegrese as deposited in a continental basin, located not far from the present thrust

[1]Department of Geological, Environmental and Marine Sciences, Exploration Geophysics Group, University of Trieste, Via Weiss 1, 34127 Trieste, Italy

position, tectonically deformed during the orogenic activity which generated the Southern Apennines. Over the last two decades, geophysical exploration has revealed that the Ionian Sea is floored by an "*old oceanic crust*" which is *underthrusting the Calabrian Arc* (Finetti, 1982). Further seismic exploration studies indicated that the Ionian Sea extended northward, and communicated openly with the Alpine Tethys. Moreover, *seismic studies and geological investigations of the Southern Apennines lead to the conclusion that the Lagonegrese represents a condensed sequence stripped-off the subducted Ionian Sea and tectonically transported into the Southern Apennines* (Finetti et al., 1996). The same investigations pointed out that Ionian-derived allochthonous basin sequences also thrust over Sicily (*Imerese*).

The deep seismic exploration of the CROP programme now provides lithospheric information of major importance for understanding these allochthonous basinal units, for identifying and delimitating their original paleogeographic basin, and for reconstructing the tectono-dynamics driving the migration process and the widespread distribution.

2. TIME–SPACE OPENING OF THE IONIAN TETHYS

The CROP seismic data clearly imaged the boundaries of both the preserved and subducted Ionian Tethys and allowed us to clearly conclude that the Ionian Neotethys, floored by oceanic crust, opened as consequence of detachment of the Adria plate from Africa (Figure 1). The results obtained show that the Africa plate/Ionian boundary segments ABC can be reconnected to the Adria plate/Ionian boundary segments A′B′C′ by rotating Adria back (clockwise) by about 40° (Figure 1).

From a joint study of the Lagonegrese thrust units and of the Ionian cover sequence, and by seismic correlation of the two stratigraphic successions across the Southern Apennines–Gulf of Taranto–Ionian Sea undeformed, it was possible to recognize that the Lagonegrese units of the Southern Apennines (as well as the Imerse units of Sicily) originated in that part of the Ionian Tethys which subducted during the Tyrrhenian stage (Middle Miocene to Present) (Finetti et al., 1996).

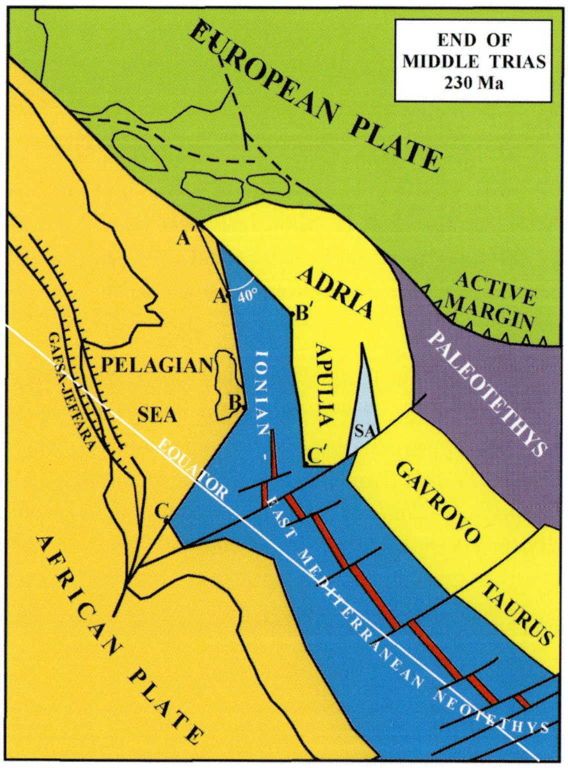

Figure 1: Palaeogeographic scheme of the Ionian–Eastern Mediterranean Neotethys. The AB and BC CROP-imaged boundary segments of the African plate correspond to the A′B′ and B′C′ CROP-imaged boundary segments of the Adria plate, after the Ionian opening. This CROP reconstruction shows that the Ionian Sea opened as a consequence of detachment (and rotation anticlockwise) of Adria from Africa. The Adria segment A′B′ is at present located on the 41° parallel. Geological and seismic stratigraphic studies indicate an Ionian opening in the Permo-Triassic. SA = South Adriatic thinned crust.

CLOSE-UP OF CROP SEISMIC SECTION M3 IN WESTERN IONIAN SEA (SP 1800-2000)

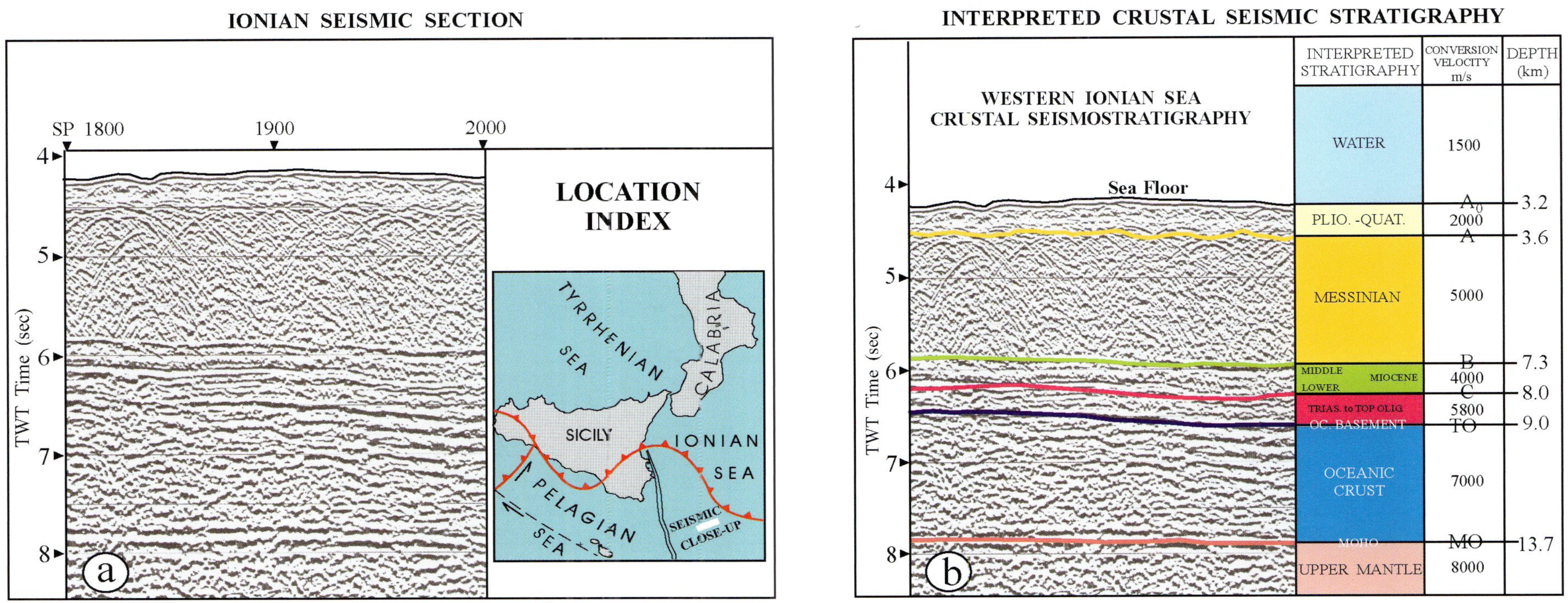

Figure 2: Seismic close-up showing the undeformed crustal stratigraphy of the Ionian ocean remnant area: crustal seismic section (a), and interpreted seismic stratigraphy of the Ionian crust (b). The cover interval "C" (top Oligocene) to "TO" (top oceanic basement) represents the condensed basinal sequence of Figure 3 (Permo-Trias to Oligocene) that was stripped-off the subducting basement and tectonically stacked in the Calabrian wedge for the most part, or transported by oblique thrusting into the Southern Apennines and Sicily.

From careful recent stratigraphic studies of the Lagonegrese Ionian-derived units it was possible to observe that the basal *M. Facito Fm.* is dated *Late Permian* (Ciarapica and Passeri, this volume, Chapter 11). More diffused Permian units of attributed Ionian provenance (*Mufara Fm.*) have been recognized in Sicily (Finetti et al., this volume, Chapter 15). The M. Facito Fm. (Scandone, 1967) consists of quartzarenitic-litharenitic terrigenous deposits, slumped blocks of massive limestone, calcarenites, breccias, with shales, radiolarites and thin bedded micrites to the top. These data indicate that the initial rifting process initiated in the Upper Permian with deposition in the initial rift-trough, while the deep-water condition after basin spreading occurred at the end of the Middle Trias.

It is possible to infer that in the Southern Apennines, the Ionian-derived units left their basal interval dispersed by the oblique thrusting process. The Permo–Triassic Ionian–East Mediterranean Neotethyan opening is confirmed by the basinal units of the Oman region (Stampfli, 2000). It seems unnecessary to introduce a diachrony between the Oman and the Ionian sectors of the Neotethys.

3. DEFINITION AND STRATIGRAPHY OF THE IONIDES

The Lagonegrese allochthonous units of the Southern Apennines and the Imerese thrust-sheets of Sicily have been attributed to a common deep basinal deposition: the subducted Ionian ocean (Finetti et al., 1996). In a careful

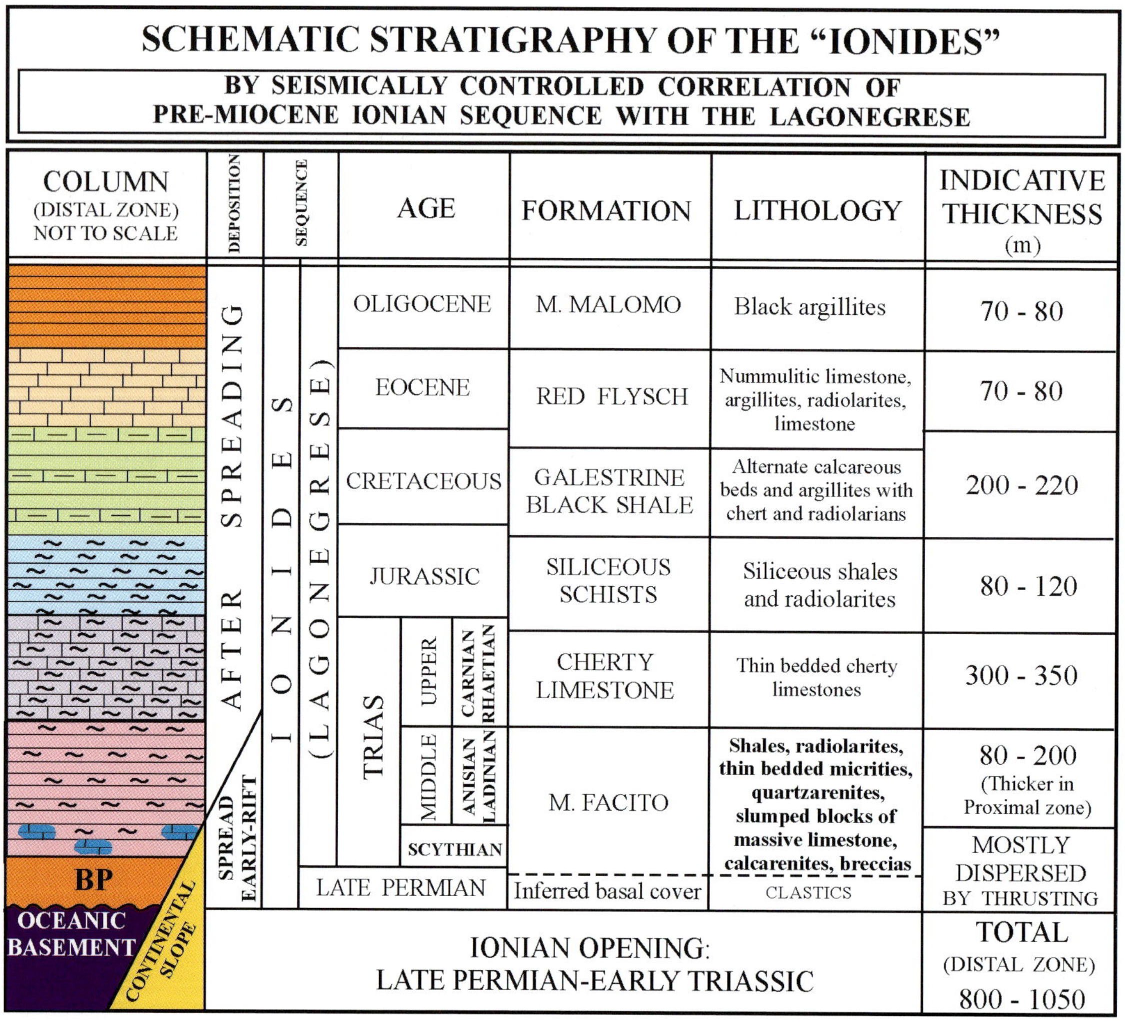

Figure 3: Stratigraphic interpretation of the *Pre-Miocene condensed seismic sequence of the Ionian Tethys cover* by correlation with calibrated deep-water allochthonous seismic units of the *Lagonegrese succession* in the Southern Apennines. Flat and/or deformed (imbricated) seismostratigraphic interval of the Pre-Miocene cover of the Ionian oceanic crust has been reliably identified and correlated across a continuous seismic network over the Ionian Sea–Gulf of Taranto–Southern Apennines to the Lagonegrese Units. The seismic interval from "C" to "TO" of the previous Figure 2 is attributed to the sequence represented in this figure.

dedicated study, it was pointed out that the outcropping Lagonegrese units and those ones found in boreholes of the Southern Apennines can be seismically correlated across the Gulf of Taranto area and the Calabrain wedge with the undeformed cover of the Ionian oceanic crust (Finetti and Del Ben, 2000). Further geophysical investigation revealed that the Calabrian wedge is mostly formed by imbricated thrust-sheets of the lower cover sequence of the Ionian basement crust (Finetti, 2002a). To identify these Ionian-derived allochthonous units, distributed over a variety of tectonic features of the Southern Apennines–Calabrian Arc–Sicilian Apennines, with a common unifying and immediately self-explanatory name, the new term ***"Ionides" (Ionian Tethydes)*** has been proposed (Finetti, 2002a, 2002b).

In the author's definition, the term "Ionides" represents allochthonous units originally sedimented into the Ionian Tethys and on its bordering slope areas, from basin opening (Permo-Triassic) to the Oligocene. Successively, during the Ionian slab retreat (Middle Miocene to Present), they were stripped off their subducting basement and thrust over the Southern Apennines and Sicilian Apennines, and stacked mostly into the Calabrian wedge. In the literature, the Ionides take different names in the different arrival areas (Lagonegrese, Imerese, Sicanian). A schematic reconstruction of the stratigraphy of the Ionides in the Southern Apennines is exhibited in Figure 3. The basal deposition is the lower *M. Facito Fm.* of Late Permian age (Ciarapica and Passeri, this volume, Chapter 11), and is very likely connected to the initial rift-opening of the Ionian Sea. More diffused Late Permian deposits are found in the Sicilian Ionides (Mufara Fm.; equivalent of the M. Facito). The persistent deep basin condition is recognized in the Ionides of the Southern Apennines only at the end of the Middle Trias (radiolarites of the Upper M. Facito Fm.). The *Cherty limestone Fm.* (Upper Trias) marks the passage to the post-spreading condition. The stratigraphic reconstruction of Figure 3 should be self-explanatory for the whole post-spreading Ionian deposition.

It is to be observed that while the cherty limestone and successive depositions (siliceous schists, Galestrine black shale and Paleogenic Fms) in the thrust process generally conserved their laminar structure in the imbricated thrust-sheets, for the basal Ionides of the M. Facito, the thrust units are less regular. The inhomogeneous petrophysical nature of the basal M. Facito Fm. leads us to think that during the stripping-off and extensive migration, this formation was in part tectonically dispersed, going to feed reworked flysch units.

4. TECTONO-DYNAMICS AND DISTRIBUTION OF THE IONIDES IN THE SOUTHERN APENNINES–CALABRIAN ARC–SICILIAN APENNINES

In the Balearic stage (Upper Oligocene–Lower Miocene) the Mediterranean geodynamics were characterized by the W-dipping subduction of the Alpine Tethys (a remnant preserved to the west of the Alps after the Adria–Europe collision in the Eo-Alpine stage) beneath the anticlockwise-rotating Corso-Sardinian block. At the end of the Balearic stage (end of the Aquitanian), the subduction of the Alpine Tethys was completed with collision between the Corsica block and the Etrurian promontory (NW Adria) and between the SE-Sardinian margin with the Panormid promontory (Africa). Such collisions stopped the rotation of the Corso-Sardinian block, but the W-dipping subduction process of the Alpine Tethys continued without stop in the Ionian Tethys which communicated openly with the Alpine Tethys (Finetti et al., 1996; Finetti, this volume, Chapter 34). From this point (Burdigalian), driven by the Ionian lithosphere roll-back sinking, the Tyrrhenian geodynamical stage (Burdigalian to Present) starts. The Ionides detach from their subducting oceanic basement and move outward, forming an accretionary wedge of imbricated thrust-sheets. The subduction process and slab-retreat move at high speed (over 10 cm/a).

Most of the imbricated Ionides go to form the wedge of the SE-migrating Calabrian Arc over the flexured oceanic crust (Figure 4). No significant amount of Ionian cover sediments follows the subducting slab. The Ionides (or Ionian Tethydes, IT) are covered by a pre-Messinian Miocene MC deposition (Aquitanian to Tortonian) of higher sedimentation rate, followed by the Messinian evaporite MS and a thin Plio-Quaternary. Flexural faults affect the underthrusting Ionian crust. The imbrication of the Ionides in the Calabrian wedge is a mostly regular sequential deformation of substantially cylindrical inversion.

The Ionides of the Southern Apennines and Sicily, on the contrary, arrived into these orogenic belts through a more complex inhomogeneous oblique thrusting process, which occurred on the two sides of the migrating arc, where the stacked Ionian features suffered irregular fragmentation. In these orogenic belts, the resulting tectonic deformation is rather far from cylindrical. A reconstructed seismogeological section example from the Ionides (IT) in the Southern Apennines is shown in Figure 5. An extensive description of the complex tectono-dynamics involved in the Southern Apennines is reported in Finetti (this volume, Chapter 12), and in Finetti (this volume, Chapter 17) for the section of Figure 5.

In Figure 6 a reconstructed seismogeological section across Eastern Sicily and its Tyrrhenian margin is plotted. It is very interesting to see the upper slab of the subducted Ionian crust and the thrust position of the Ionides, as revealed by the CROP data. A wedge of four stacked sheets of Ionides fills the erroneously called Caltanissetta basin, which represented a foredeep at the end of the Lower Pliocene when the Ionides initiated their thrusting migration over this part of the Sicilian platform. The figure is fairly self-explanatory but a full description is given in Chapter 15.

On the basis of the new interpreted CROP data, integrated with several MS lines (Finetti et al., this volume, Chapter 1) it is possible to define with satisfactory reliability the following features (Figure 7):

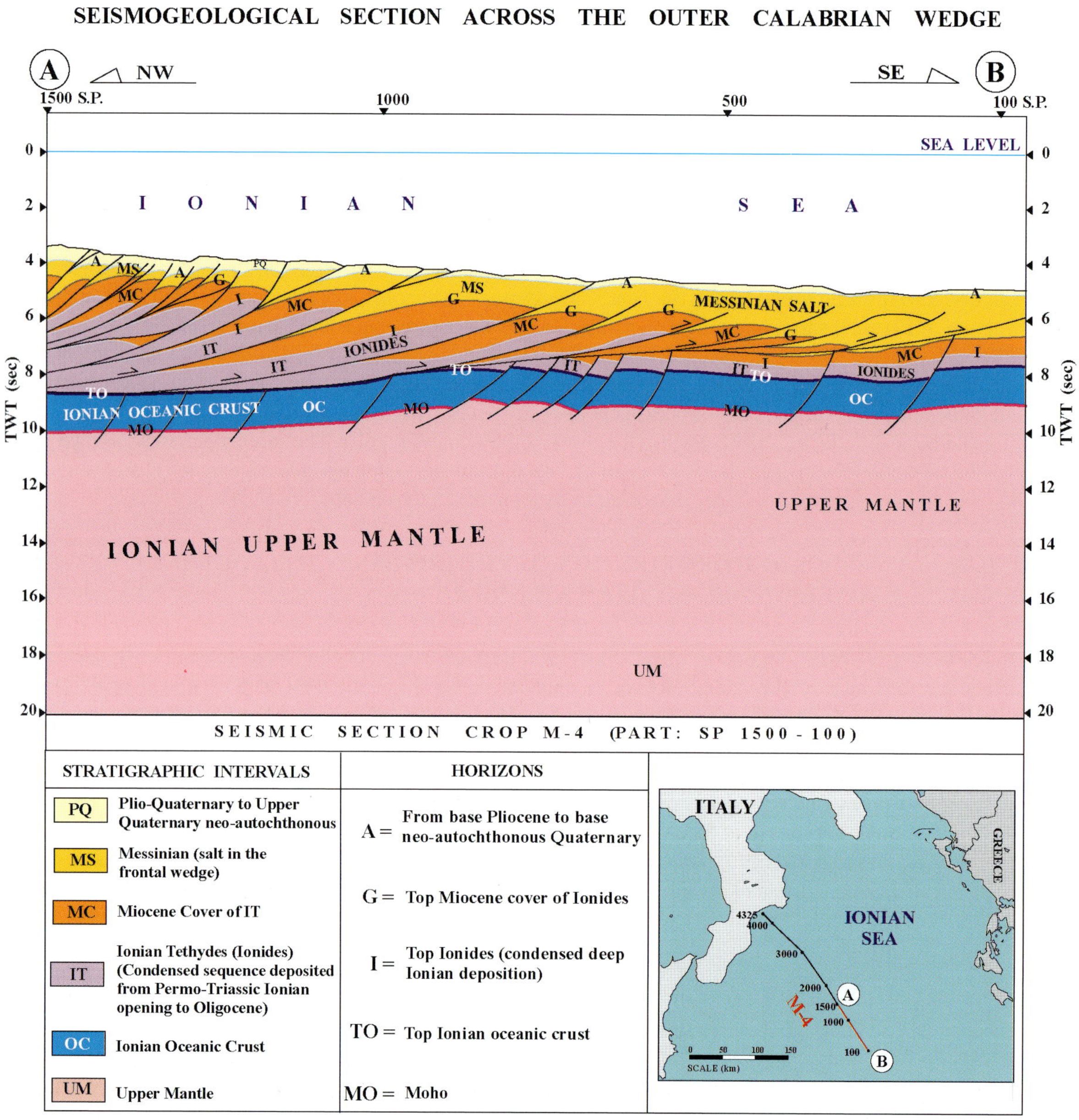

Figure 4: The Ionides in the externalmost part of the Calabrian wedge.

Figure 5: Reconstructed seismic tectono-stratigraphic example of the allochthonous Ionides in the Gulf of Taranto. These Ionian-derived units have been correlated with the undeformed cover sequence of the Ionian oceanic crust by using the Italian Ministry's public seismic lines and MS lines across the Gulf of Taranto. Repeated imbrications of Ionides outcrop and have been penetrated by boreholes in the Southern Apennines. In Figure 5(a): FSA = front of the Southern Apennines; BIA = buried boundary between the Apulian and the subducted Ionian ocean crust.

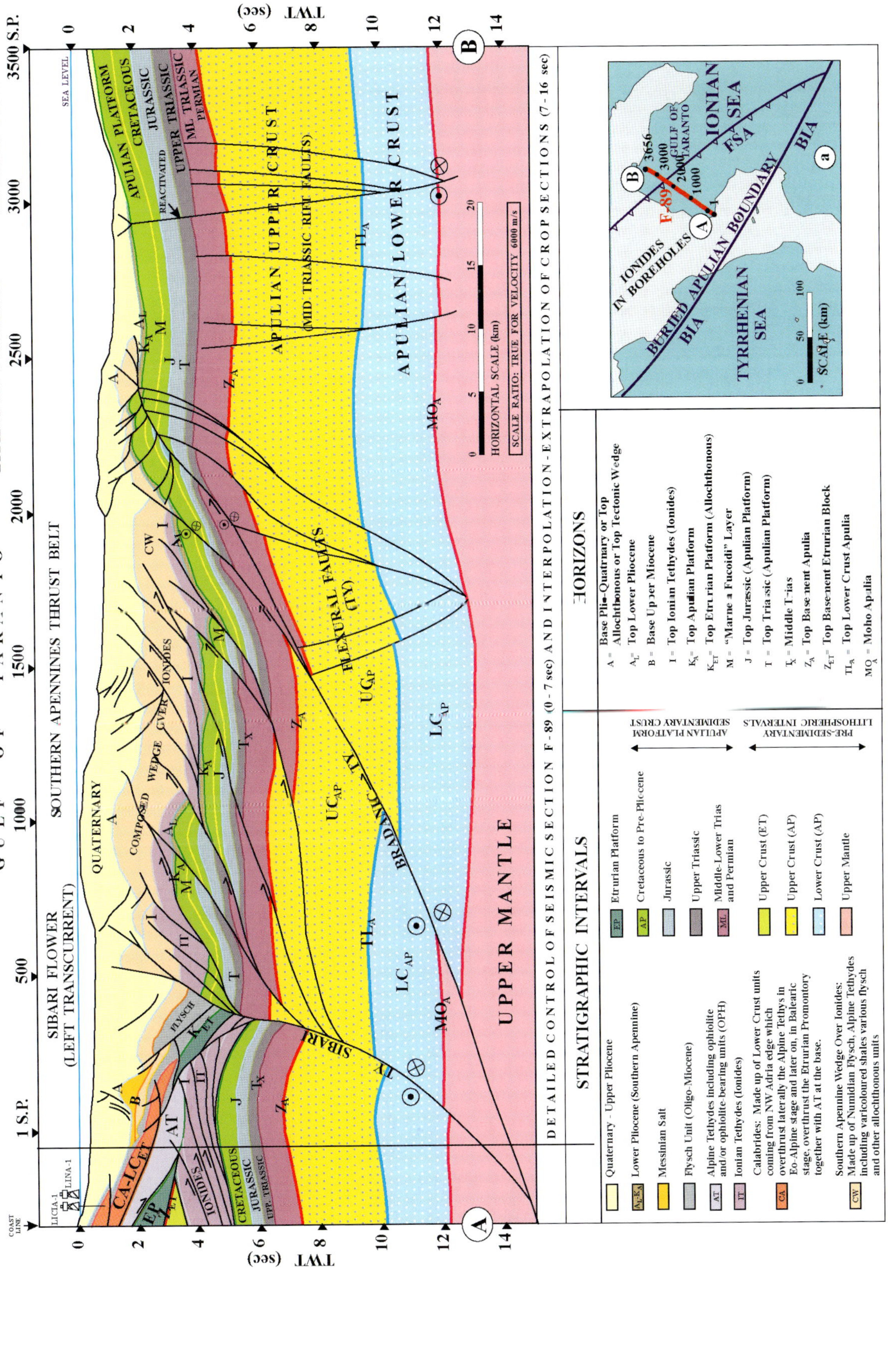

GULF OF TARANTO
APULIAN PLATFORM
BRADANIC FOREDEEP
SOUTHERN APENNINES THRUST BELT
SIBARI FLOWER (LEFT TRANSCURRENT)
UPPER MANTLE
APULIAN UPPER CRUST (MID TRIASSIC RIFT FAULTS)
APULIAN LOWER CRUST
FLEXURAL FAULTS (TY)
TWT (sec)
HORIZONTAL SCALE (km)
SCALE RATIO: TRUE FOR VELOCITY 6000 m/s
DETAILED CONTROL OF SEISMIC SECTION F-89 (0 - 7 sec) AND INTERPOLATION-EXTRAPOLATION OF CROP SECTIONS (7-16 sec)
STRATIGRAPHIC INTERVALS
Quaternary - Upper Pliocene
Lower Pliocene (Southern Apennine)
Messinian Salt
Flysch Unit (Oligo-Miocene)
Alpine Tethydes including ophiolite and/or ophiolite-bearing units (OPH)
Ionian Tethydes (Ionides)
Calabrides: Made up of Lower Crust units coming from NW Adria edge which overthrust laterally the Alpine Tethys in Eo-Alpine stage and later on, in Balearic stage, overthrust the Etrurian Promontory together with AT at the base.
Southern Apennine Wedge Over Ionides: Made up of Numidian Flysch, Alpine Tethydes including varicoloured shales various flysch and other allochthonous units
Etrurian Platform
Cretaceous to Pre-Pliocene
Jurassic
Upper Triassic
Middle-Lower Trias and Permian
Upper Crust (ET)
Upper Crust (AP)
Lower Crust (AP)
Upper Mantle
APULIAN PLATFORM SEDIMENTARY CRUST
PRE-SEDIMENTARY LITHOSPHERIC INTERVALS
HORIZONS
A = Base Plio-Quaternary or Top Allochthonous or Top Tectonic Wedge
Top Lower Pliocene
B = Base Upper Miocene
I = Top Ionian Tethydes (Ionides)
Top Apulian Platform
Top Etrurian Platform (Allochthonous)
M = "Marne a Fucoidi" Layer
J = Top Jurassic (Apulian Platform)
T = Top Triassic (Apulian Platform)
Middle Trias
Top Basement Apulia
Top Basement Etrurian Block
Top Lower Crust Apulia
Moho Apulia
IONIDES IN BOREHOLES
BURIED APULIAN BOUNDARY
TYRRHENIAN SEA
IONIAN SEA
GULF OF TARANTO
SCALE (km)

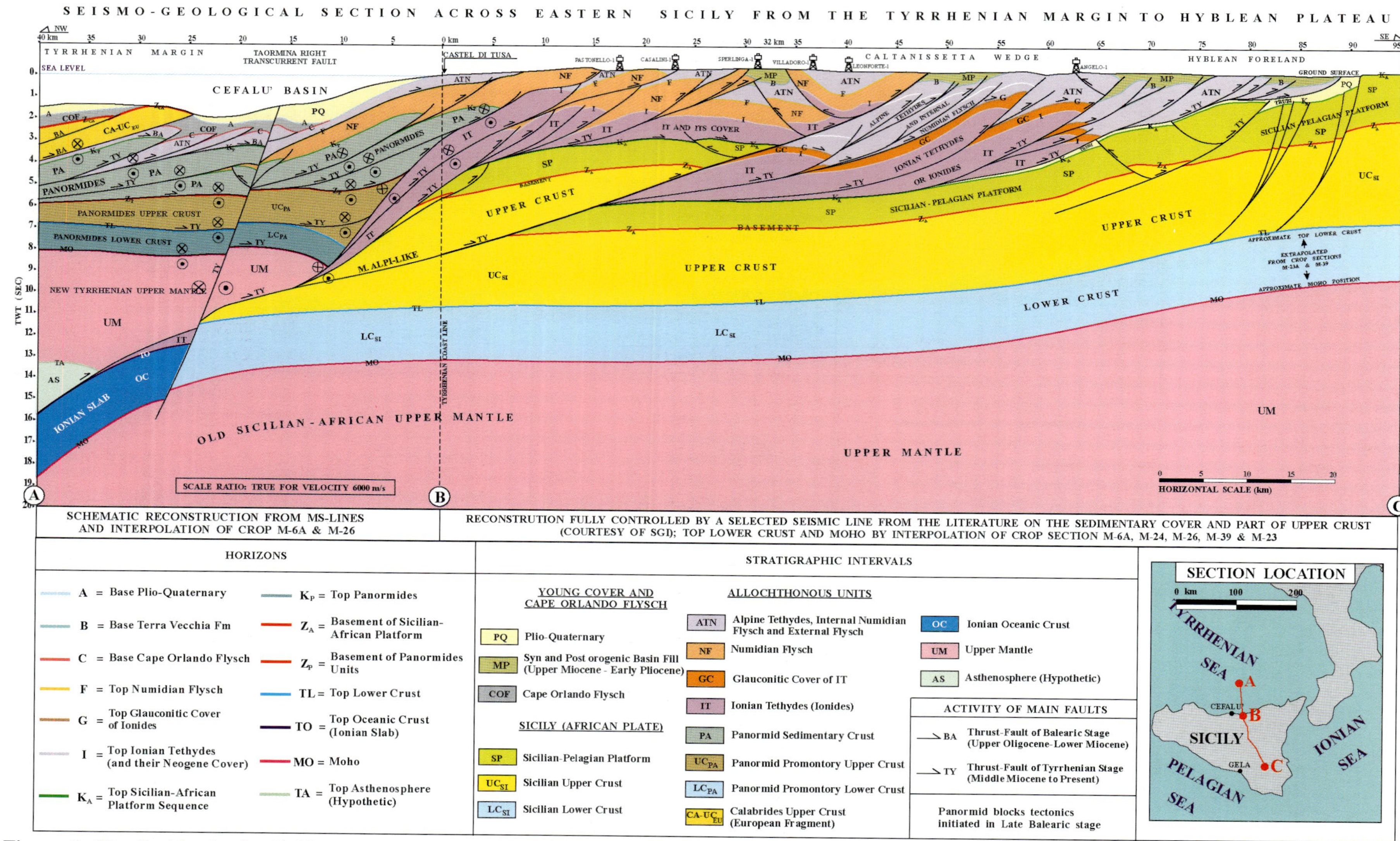

Figure 6: The Ionides in the Sicilian Apennines. These Ionian-derived units thrust over the Sicilian platform in Pliocene times, and more precisely, the thick Caltanissetta wedge arrived over the Trubi Fm. in the Middle Pliocene. The M. Alpi-like crustal thrust-fault is a late orogenic event generated at the closure of the Ionian Tethys when the stacked crustal units of the Tyrrhenian margin collided with the Sicilian slope.

QUANTITATIVE COMPUTATION OF IONIDES (IONIAN TETHYDES, "IT") STRIPPED OFF THE IONIAN SLAB AND SCHEME OF THEIR TECTONIC DISTRIBUTION

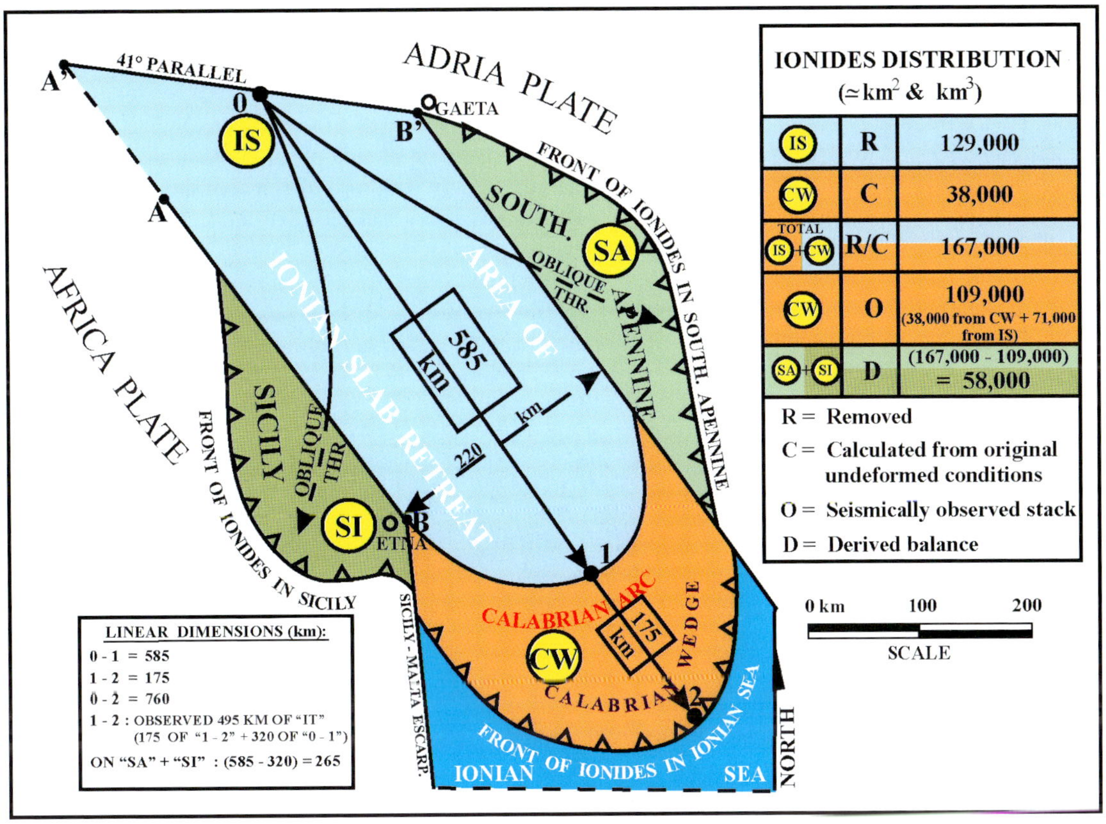

IS AREA OF STRIPPING OFF OF IONIDES FROM IONIAN SLAB (km 585 x 220) ≃ 128,700 km²

CW CALABRIAN WEDGE AREA OVER IONIAN OCEANIC CRUST: Seismically observed 495 km of imbricated Ionides with substantially regular sequence along the segment "1 - 2" (km 175 originated in "CW" and 320 come from "IS" area)

SA SOUTHERN APENNINE AREA WHERE PART OF IONIDES OF "IS" AREA THRUST OBLIQUELY

SI SICILY AREA WHERE PART OF IONIDES OF "IS" SUBDUCTED IONIAN AREA THRUST OBLIQUELY

<u>LINEAR BALANCE</u> OF "0 - 2" IONIDES DISTRIBUTION: (760 km)
"CW" : ≃ 495 km
"SA" + "SI" : ≃ (760 - 495) ≃ 265

CW = 65 %
SA + SI = 35 %

<u>AREAL AND VOLUMETRIC ESTIMATION</u> (km² & km³)
(Assumptions: Average "IT" thickness 1 km; Average Ionian width 220 km; No significant "IT" subduction; Cylindrical approch for CW stack; Regular dynamic progression from NW to SE).

- Original "IS" Source : (585 x 220) ≃ 128,700
- "CW" : (495 x 220) ≃ 108,900 (70,400 from "IS" + 38,500 from "CW")
- "SA" + "SI" : [(585 - 320) x 220] ≃ 58,300

Figure 7: Schematic distribution of the Ionides in the Southern Apennines, Calabrian Arc and Sicilian Apennines thrust-belt system. About 65% of the Ionides are stacked in the Calabrian wedge and the remaining 35% thrust into the Southern Apennines and Sicily.

(a) The paleogeographic area of the Ionian Tethys;
(b) The area of the Ionian Tethys subducted in the Tyrrhenian stage;
(c) The area of the Calabrian wedge where the Ionides are forming an imbricated stack;
(d) The total length of the Ionides stacked in the Calabrian wedge, measured from a CROP section along the axial line of the Ionian Tethys.

We obtain approximate, but realistic, data on the distribution of the Ionides. Figure 7 reproduces in scale the areal setting of the subducted Ionian crust, and of the Ionides in the Calabrian wedge. The area of the subducted Ionian Tethys (area "IS"), along the axial line, has a length of 585 km (Figure 7: segment 0 to 1). In the Calabrian wedge "CW" (segment 1 to 2 of the axial zone = 175 km) are stacked *Ionides* for a total observed length of 495 km. So, along the axial zone, running from point 0 to 2, Ionides have been removed from their basement for a total length of 760 km, of which 495 km (65%) are in the Calabrian wedge and, assuming no subduction-dispersion occurred, km 265 (760–495) (35%) are in the Southern Apennines and Sicily. Because of the complex tectonics involved, it is difficult to attempt a subdivision between the Southern Apennines and Sicily of the linear balance of Ionides obtained from the total originated minus that observed in the Calabrian wedge.

Assuming a width of the subducted Ionian of 220 km, and an average thickness of the Ionides of 1 km, it is also possible to make an indicative areal and volumetric estimate of the distribution of the Ionides, as reported in Figure 7.

The repeated imbrications of the Ionides observed in the Calabrian wedge have also been found in boreholes for hydrocarbon exploration in the Southern Apennines.

ACKNOWLEDGEMENTS

This work was supported by MIUR, COFIN 1998, 2000 and 2002 with fundings to I.R. Finetti, University of Trieste. The University of Trieste helped support the interpretative seismic reprocessing and final interpretation. Many thanks are due to the EGG (Exploration Geophysics Group) of the University of Trieste, Faculty of Sciences, for the much appreciated contribution given by the iterated interpretative reprocessing of the CROP data. The obtained signal quality improvement was determinant for understanding the Ionides and their geodynamics.

REFERENCES

Ciarapica, G., Passeri, L., this volume. Ionian Tethydes in the Southern Apennines. Chapter 11.

Finetti, I.R., 1982. Structure, stratigraphy and evolution of Central Mediterranean. Boll. Geofis. Teor. Appl. 24, 247–426.

Finetti, I.R., 2002a. Lithospheric tectono-stratigraphic setting and time–space stacking process of the Calabrian arc from new CROP seismic data. In: Abstract Proceedings of 81st Riunione Estiva of the Società Geologica Italiana, Torino, 10–12 September 2002.

Finetti, I.R., 2002b. Understanding Ionides and Calabrian arc from new CROP seismic data. In: Expanded Abstract, Proceedings of the 21° Convegno Nazionale of the GNGTS-EAGE 2nd Italian section convention, Rome, 19–21 November, 2002.

Finetti, I.R., this volume. The Calabrian arc and subducting Ionian slab from new CROP seismic data. Chapter 17.

Finetti, I.R., this volume. Geodynamic evolution of the Mediterranean Region from the Permo-Triassic Ionian opening to the Present, constrained by new lithospheric CROP seismic data. Chapter 34.

Finetti, I.R., Del Ben, A., 2000. Ionian opening, crustal stratigraphy and tectono-dynamics of the Pelagian Sea region from new CROP seismic data. In: Proceedings of the International Symposium on Geology of NW Libya, Tripoli, 2000.

Finetti, I., Lentini, F., Carbone, S., Catalano, S., Del Ben, A., 1996. Il sistema Appennino Meridionale–Arco Calabro–Sicilia nel Mediterraneo Centrale: studio geologico-geofisico. Boll. Soc. Geol. It. 115, 529–559.

Finetti, I.R., Del Ben, A., Forlin, E., Prizzon, A., this volume. Review of the main results of the pioneering MS seismic exploration programme in the deep water of the Mediterranean Sea (OGS, 1969–1980). Chapter 1.

Finetti, I.R., Lentini, F., Carbone, S., Del Ben, A., Di Stefano, A., Guarnieri, P., Pipan, M., Prizzon, A., this volume. Crustal tectono-stratigraphy and geodynamics of the Southern Apennines from CROP and other integrating geophysical-geological data. Chapter 12.

Finetti, I.R., Lentini, F., Carbone, S., Del Ben, A., Di Stefano, A., Forlin, E., Guarnieri, P., Pipan, M., Prizzon, A., this volume. Geological outline of Sicily and lithospheric tectono-dynamics of its Tyrrhenian margin from new CROP seismic data. Chapter 15.

Scandone, P., 1967. Studi di geologia lucana: la serie calcareo-silico-marnosa e i suoi rapporti con l'appennino calcareo. Boll. Soc. Natur. Napoli 76, 301–469.
Scandone, P., 1972. Studi di geologia lucana: Carta dei terreni della serie calcareo-silico-marnosa e note illustrative. Boll. Soc. Naturalistica in Napoli 81, 255–300.
Stampfli, G., 2000. Tethyan oceans. In: Bozkurt, E., Winchester, J.A., Piper, J.D.A. (Eds.), Tectonic and Magmatism in Turkey and the Surrounding Area. Geol. Soc., London, Special Publications 173, 1–23.

CROP PROJECT: Deep Seismic Exploration of the Central Mediterranean and Italy
Edited by I.R. Finetti

CHAPTER 11

Ionian Tethydes in the Southern Apennines

G. Ciarapica[1], L. Passeri[1]

ABSTRACT

The Triassic-Jurassic stratigraphic succession of the Lagonegro domain is discussed in order to point out that it belongs to the Ionian Ocean.
In the paleogeographic panorama of the Mesozoic, the Ionian Ocean is the southwestern branch of the Neotethys. The steps in the sedimentary evolution of the so-called "Lagonegro Basin" (Southern Apennines) mark the first phases of the Ionian rifting, characterized by tectonic instability of its continental margins. Radiolaritic sedimentation, since the Ladinian, was related to the ocean spreading.
In the same western Tethyan realm, the Liguria-Piedmont Ocean (Alpine Tethys) developed later. This oceanic trough is considered to be the northeastern prolongation of the Atlantic Ocean. The geodynamic and the sedimentary history of the Northern Apennines and of the Alps reflect its evolution.
According to this complex scenario of the west Mediterranean area, the Apenninic thrust-belt involves two different continental margins: in the northern part, that belonging to the Liguria-Piedmont Ocean, in the southern part, that belonging to the Ionian Ocean.
A comparison between the Lagonegro sequence (Southern Apennines) and the La Spezia sequence (Northern Apennines) clearly shows their connection with the evolution of two distinct oceans: the continental margin uplift that preceded the oceanic spreading occurred during the Carnian in the Northern Apennines, where it corresponds to the beginning of the Alpine sedimentary cycle; on the other hand the Ionian cycle begun in the Lower Anisian, as documented in the Lagonegro sequence. The first radiolaritic deposits are referred to the Middle Triassic in the Lagonegro domain, while they appeared only in the Middle Jurassic (Callovian) in the Ligurian domain.

Keywords: Apennines, stratigraphy, paleogeography, Triassic-Jurassic, Ionian Tethys, Lagonegro

1. INTRODUCTION

According to the traditional view, the Mesozoic sedimentary history of the Alps and Apennines was linked to the rifting of the Liguria-Piedmont Ocean. Over the last two decades, studies in the Mediterranean area have led to the belief that a more complex panorama could have existed in the western Tethys realm (Abbate et al., 1986; Argnani et al., 1995; Catalano et al., 1991, 2000; Ciarapica and Passeri, 1998, 2002; Dercourt et al., 1993; Finetti, 1982, 1985; Finetti et al., 1996, 2001; Serri, 1995; Stampfli et al., 1991, 2003; Stampfli and Marchant, 1995).

The evolution of the Mediterranean area now seems related to two main geodynamic processes: (a) consecutive opening and closure of various oceanic branches in the Tethyan gulf (Vardar, Pindos and Ionian ocean; (b) opening of the central Atlantic Ocean with its Alpine prolongation (Liguria-Piedmont Ocean or Alpine Tethys) to the west of the Tethyan gulf. In this scenario the sedimentary history of the Alps and Northern Apennines appears to be related to the evolution of the Liguria-Piedmont Ocean, while that of the Southern Apennines is related to the Ionian Ocean (east Mediterranean Ocean or Mesogea), which is the southwestern branch of Neotethys.

The Liguria-Piedmont rifting began in the Late Triassic and intersected the older Middle Triassic seaways connected with the Tethyan Gulf. A crustal uplift preceded the Liguria-Piedmont rifting so that two main sedimentary cycles are well evident in many places (Alps and La Spezia area in the Northern Apennines): the lower one, from the Permian to Carnian, and the upper one, from the Carnian to Cenozoic (Passeri, 1985). In other places, as in

[1] Dipartimento di Scienze della Terra, Università di Perugia, Perugia, Italy

the main part of the Northern Apennines, only the upper cycle is present, since until the Carnian these areas were continental environments.

The basinal facies cropping out in the Southern Apennines (the so-called Lagonegro facies) show a different evolution. They contain: Permian shallow marine carbonates probably reworked in the Permian; Early Triassic deep water facies which pass to shallow water facies in the lower part of the Middle Triassic; deep water facies, with Ammonitico Rosso and radiolarites, that begun to develop again from the Late Anisian (?)–Early Ladinian. If a crustal uplift occurred, it most likely took place before the Late Anisian.

Sedimentological and stratigraphic studies carried out over many years in the various tectonic units that form the Northern, Central and Southern Apennines have provided large amounts of field data for analyzing the main steps of the rifting phases.

The outcropping units of the Northern Apennines show that a pelagic subsiding basin developed during the Early Jurassic from eastern Liguria (La Spezia area) to the Adriatic Sea. This basin is known as the Tuscany–Umbria–Marche basin (TUM).

In the Central Apennines, which is made of up many tectonic units, the northeastern part of the Apenninic carbonate platform developed during the Mesozoic.

In the tectonic units piled to form the Southern Apennines, different paleogeographic Mesozoic domains are documented: the Apenninic platform, the Apulian platform and the Lagonegro-Molise basin (Figure 1).

To better understand the relationships among all these paleogeographic domains and the causes of their existence, it is necessary to look at the very first stage of their development, when the basins were only seaways inside a continental plate. The sedimentary evolution of the Northern Apennines basin (Tuscany–Umbria–Marche basin) will shortly be described to point out the main differences from the Southern Apennines basin (Lagonegro-Molise basin); this latter will be described in more detail due to its close relationship with the Ionian Ocean.

2. TRIASSIC AND EARLY JURASSIC FACIES IN THE NORTHERN APENNINES

Most of the sequences outcropping in the Northern Apennines are characterized by the presence of continental deposits until the Early Carnian (lower part of the Verrucano Group), when the first shallow marine siliciclastic facies appeared. The Norian-Raethian is represented by dolostones (Grezzoni fm.), dolostones and evaporites (Burano Anhydrites fm.) and then by limestone and shales (*Rhaetavicula contorta* facies of the Mt. Cetona fm.). The Early Jurassic is made of a carbonate platform (Calcare Massiccio fm.) which drowned between the end of the Hettangian and the end of the Sinemurian. Then it was followed by pelagic carbonate facies starting in the Sinemurian, and by radiolarites in the Late Jurassic.

Only in the La Spezia area, is it possible to recognize the existence of two main sedimentary cycles: a "Middle Triassic cycle" and an "Alpine cycle".

(a) The "Middle Triassic cycle" is well exposed at the southern end of the eastern promontory of La Spezia Gulf, at Punta Bianca (Elter and Federici, 1964; Martini et al., 1986; Passeri, 1985) (Figure 2). The transgressive part of the cycle is represented by a polymict conglomerate unconformably overlying the Paleozoic basement, followed by quartzitic sandstones and thin-bedded limestones with Middle Triassic dasycladales. The middle part of the cycle contains carbonate lenses, alkali basaltic lenses, a thin horizon with Late Anisian ammonites (Venturi and Tenneroni, 1981) and paraconglomerates with carbonate clasts in a shaly matrix. The regressive part of the cycle is represented by a coarsening-upward sequence of a prograding delta with shales, quartzarenites and quartzitic conglomerates. It marks the regression which affected the La Spezia area between the Late Ladinian and the Early Carnian (Figure 3).

(b) The "Alpine cycle" began with quartzitic sandstones and evaporites, followed by shallow water carbonates with Carnian forams (Coregna Dolostones; Ciarapica and Passeri, 1980; Ciarapica and Zaninetti, 1984). The Coregna Dolostones are characterized by the presence of abundant quartz grains, indicating the close presence of exposed lands under erosion. The drowning of this carbonate platform occurred at the end of the Carnian, with the deposition of shales and limestones in a shallow marine basin (La Spezia fm.; Norian-Raethian). In the western promontory of La Spezia Gulf, the whole Jurassic is represented by pelagic facies from the Early Hettangian; radiolarites sedimented in the Late Jurassic, as in Tuscany.

PRESENT DISTRIBUTION OF THE MAIN PALEOGEOGRAPHIC UNITS
IN THE APENNINIC THRUST BELT

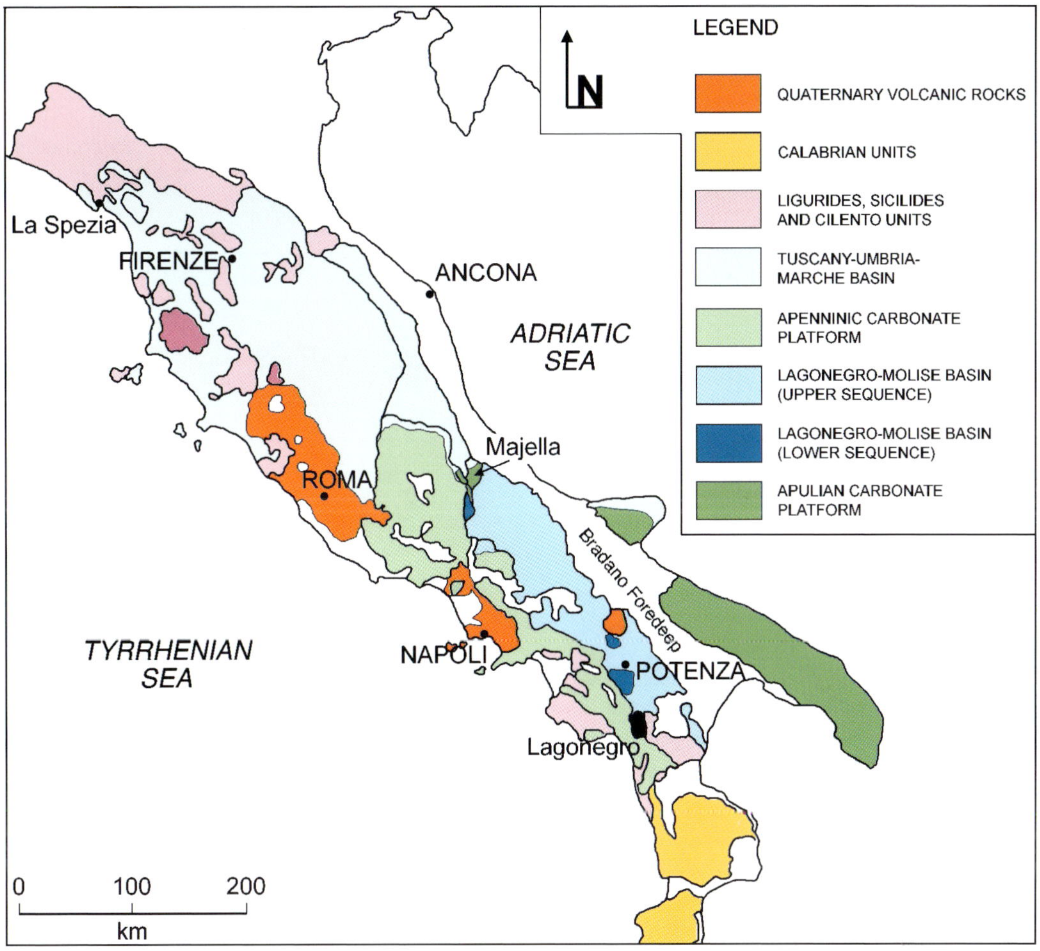

Figure 1: The paleogeographic units illustrated in this schematic map of the Apennine thrust belt are actually split up in many tectonic units, variously piled up from the Oligocene to the Pliocene with a main counterclockwise rotation. In the southern Apennines, the Lagonegro basinal deposits and the Apenninic carbonate Platform units are completely detached from their basement and overlay the Apulian carbonate Platform.

3. SOUTHERN APENNINES

Stratigraphic studies of the various tectonic units piled up to form the Southern Apennines allow us to restore a complex Mesozoic paleogeographic setting, characterized by the existence of carbonate platform/basin systems. Many paleogeographic restorations have provided models which differ in the relative position and number of carbonate platforms and basins (D'Argenio and Scandone, 1970; D'Argenio et al., 1973; Sgrosso, 1988; Mostardini and Merlini, 1986; Marsella et al., 1995; Ciarapica and Passeri, 1998, 2002). Geophysical data and field analyses support the hypothesis sketched by Mostardini and Merlini (1986) who suggested the presence of two carbonate platforms (Apenninic platform and Apulian platform) separated by a deep basin (Lagonegro-Molise basin).

Despite the persistence of the platform environment from the Late Triassic up to the Miocene, the Apenninic platform, that crops out from Pollino (Calabria) to Marsica (Abruzzi), had a complex sedimentary history marked by phases of local or general emersions and periods of local temporary drowning. For this reason, during some short periods it became a dissected complex of more or less isolated carbonate platforms. Other basins developed mainly in the Jurassic to the west (Liguria-Piedmont basin-to-ocean) and to the east (east Gargano basin or south Adriatic basin). Euxinic facies sedimenting in restricted basins, developed inside the platform complex during the Late Triassic (Emma Basin, Mt. Camicia, Filettino, Picentini Mts. and Verbicaro; Ciarapica et al., 1987; Adamoli et al., 1990; Zappaterra, 1990; Cirilli, 1993; Iannace and Zamparelli, 1999). Some of these restricted basins persisted during the Mesozoic, becoming parts of larger basins (Verbicaro, Mt. Camicia and Emma) while others where filled

AREAL DISTRIBUTION OF THE MAIN TECTONO-STRATIGRAPHIC UNITS IN THE NORTHWESTERN PART OF THE NORTHERN APENNINES

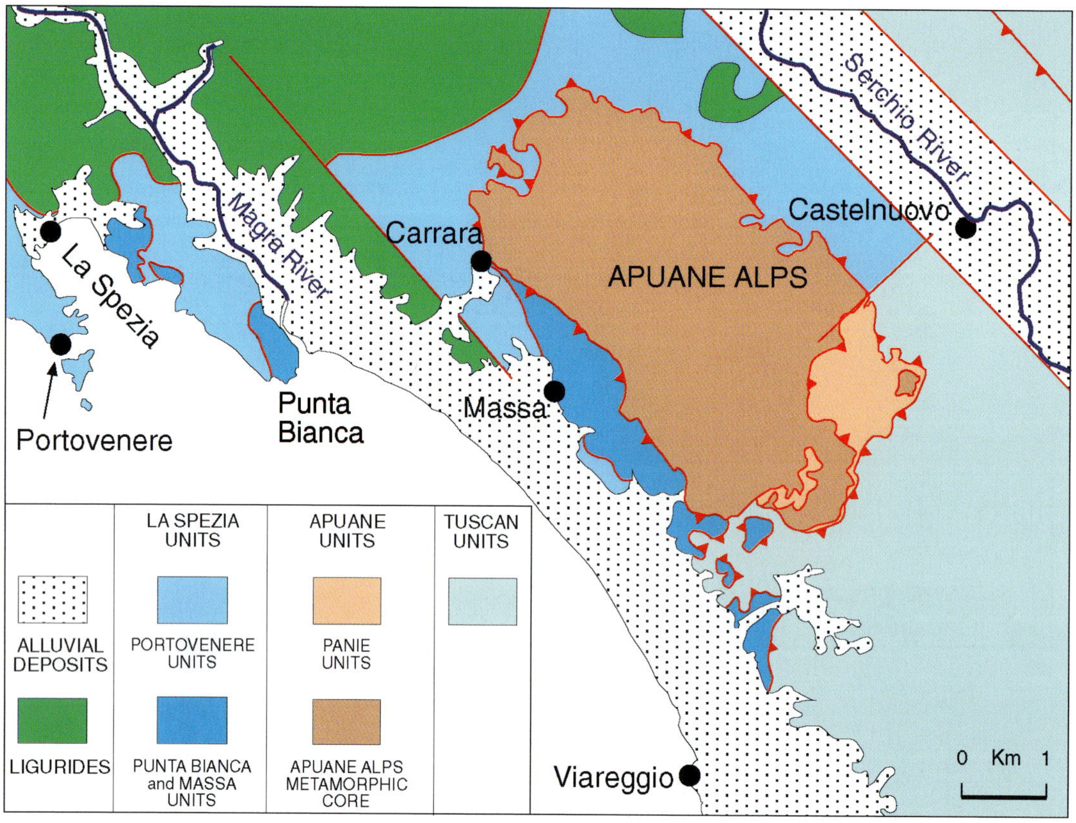

Figure 2: The La Spezia, Apuane and Tuscany units belong to the Meso-Cenozoic Tuscany–Umbria–Marche Basin. The La Spezia sequence differs from the others: it contains Middle Triassic marine facies while the Apuane and Tuscany are characterized by continental deposits. During the late Triassic and Early Jurassic, pelagic facies appeared in the La Spezia domain while in Apuane and Tuscany carbonate and evaporitic platforms developed (for details see Ciarapica and Passeri, 1998).

by sediments, and carbonate platform conditions were restored (Picentini Mts. and Filettino). On the contrary, the Lagonegro basin is characterized by persistent pelagic facies in a deepening-upwards trend, starting in the Middle Triassic at least.

4. LAGONEGRO SEQUENCE

Among the various Lagonegro-Molise successions it is possible to distinguish between proximal and distal situations: in the former, huge breccias and calcarenitic bodies coming from the adjacent platforms are contained in the pelagic facies; whereas the distal areas are characterized by condensed successions of open marine deep-sea deposits, similar to those of the modern oceans (radiolarites and shales) (Figure 4).

Mostardini and Merlini (1986) described a "Lagonegro Lower Sequence" and a "Lagonegro Upper Sequence". The first, from base to top, consists of the Monte Facito fm., Cherty Limestone fm., Siliceous Schists fm., Flysch Galestrino, Flysch Rosso (Figure 5); the second is formed from the younger part of the succession which is always detached from its basement: Flysch Rosso, "Argille Varicolori", Numidian Flysch and Irpinian units p.p.

Figure 3: The Anisian-Ladinian basin (Punta Bianca gr.) is considered as a southwestern prolongation of Middle Triassic northern branches of the Tethys; the Carnian uplift and the successive transgression are instead related to the first phases of the Liguria-Piedmont rifting.

LA SPEZIA STRATIGRAPHIC SUCCESSION

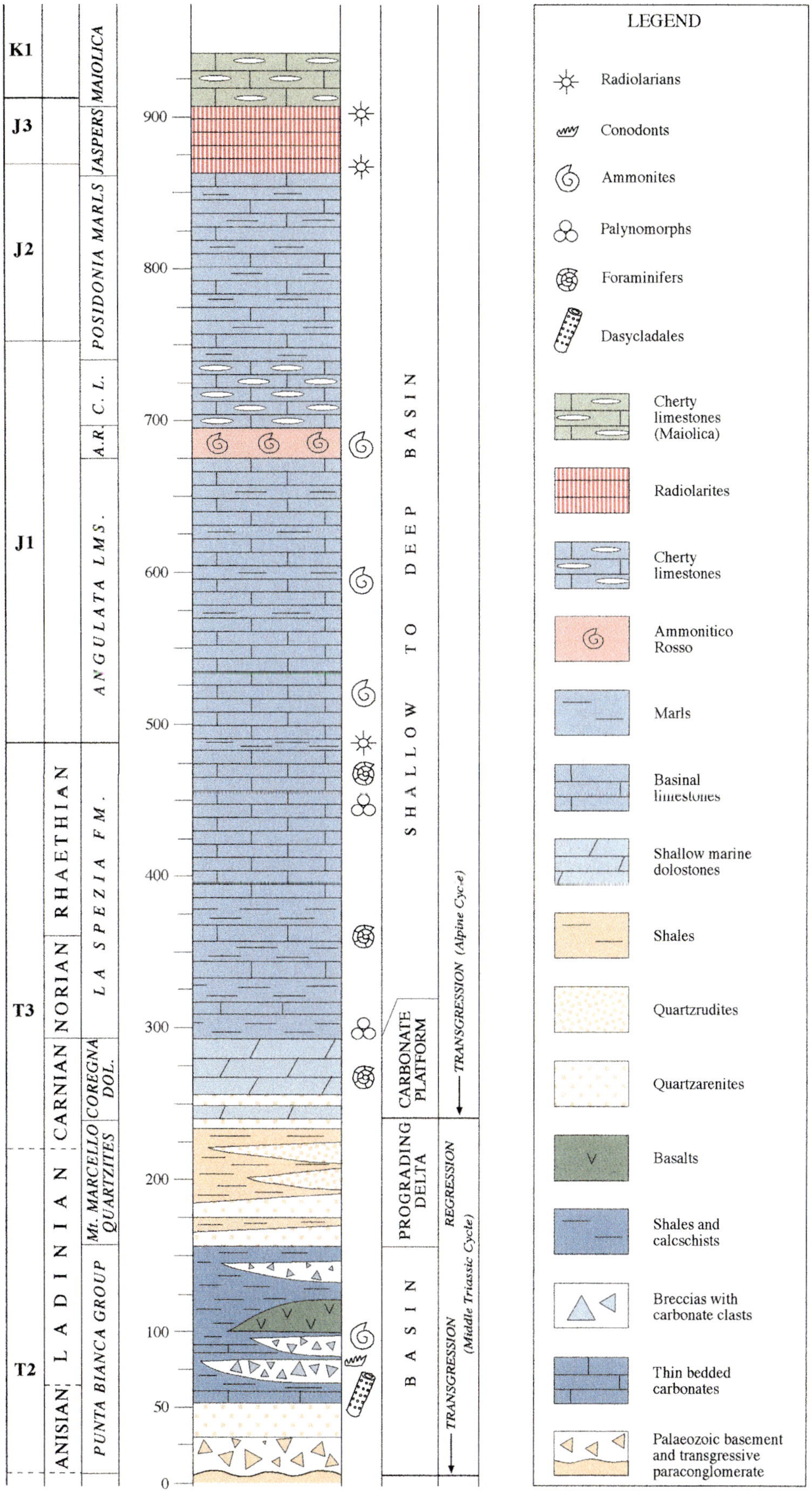

AREAL DISTRIBUTION OF THE MAIN STRATIGRAPHIC UNITS
IN THE SOUTHERN APENNINES

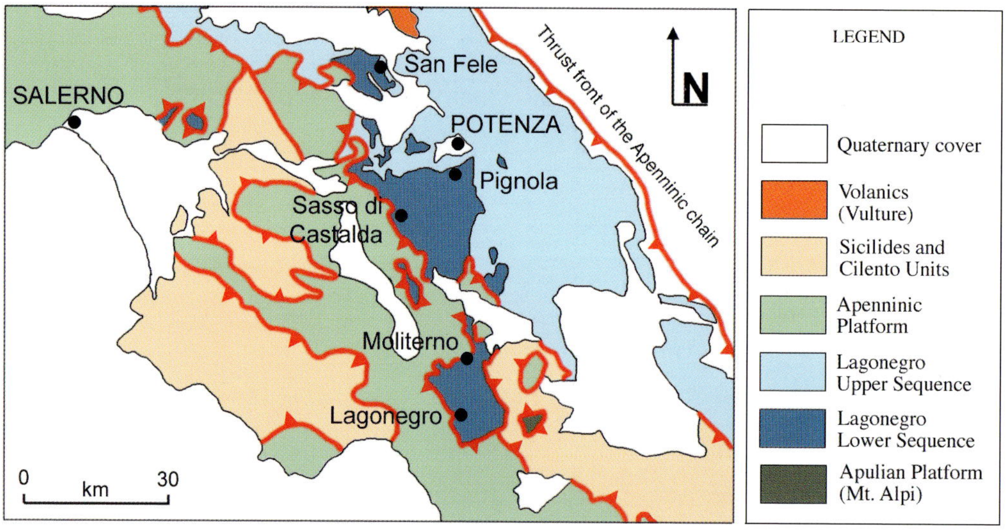

Figure 4: Main outcropping areas of the Lagonegro Lower sequence in Southern Apennines. Only the main thrusts are illustrated in this schematic map. Actually the Lagonegro domain was split up in many tectonic elements all detached from their basement and piled the one over the others; few of them contain the Monte Facito fm. as illustrated in Figure 5, the lower part of the others is completely unknown.

4.1. Lagonegro Lower Sequence

The oldest deposits of Southern Apennines belong to the Monte Facito formation (Scandone, 1967). This lower part of the Lagonegro succession crops out in Lucania, and it was reached by wells in many areas of the Southern Apennines.

4.1.1. *Monte Facito fm.* The Monte Facito fm. was recently revised by Ciarapica and Passeri (2000) into two main outcropping areas: Sasso di Castalda and Moliterno. It is made up of many different lithologic units, from terrigenous sandstones, shales and quartzarenites to limestones and radiolarites. The Apenninic orogenesis affected this "formation", which was one of the preferential levels of detachment. Besides this situation, the large variety of included lithofacies and their age compel us to consider the Monte Facito as a group of broken formations and not a single formation s.s. (as defined in the stratigraphic nomenclature international code). Many authors consider the Monte Facito formation as a huge olistostrome of Middle Triassic age (Miconnet, 1988; Martini et al., 1989; Marsella et al., 1993). Ciarapica and Passeri (2000) pointed out that only part of the Monte Facito fm. could be considered as a mass flow deposit, while the major part is a tectonic melange. This statement is founded on the presence of clasts derived from rocks which suffered deep burial diagenesis before being embedded in the melange.

The first attempts at dating lithofacies of the Monte Facito formation were made at the beginning of the sixties: foraminifera considered of Permian age (*Pachyfloia* and *Abriolina*) were found in the so-called "Abriola Limestone" (Azzaroli, 1962; Luperto, 1965a, 1965b); later studies demonstrated that the "Abriola Limestone" contains a large amount of Middle Triassic fauna (algae, foraminifera, brachiopods and conodonts) and its age is Anisian and Ladinian (Scandone, 1967; Taddei Ruggiero, 1968; Ciarapica et al., 1990a, 1990c; Panzanelli Fratoni, 1991; Ciarapica and Passeri, 2000). Fusulinids from the Monte Facito area were illustrated by Donzelli and Crescenti (1970) in a different facies.

Figure 5: The Anisian uplift and the successive transgression, documented in the Monte Facito fm., are related to the initial phases of the Ionian rifting. Deep water facies begun in the Middle Triassic and persisted all through the Mesozoic. The change from radiolaritic to carbonate sedimentation at the end of the Ladinian (Cherty Limestones) is not linked to a regressive phase, but it indicates a major carbonate production in the adjacent carbonate platforms.

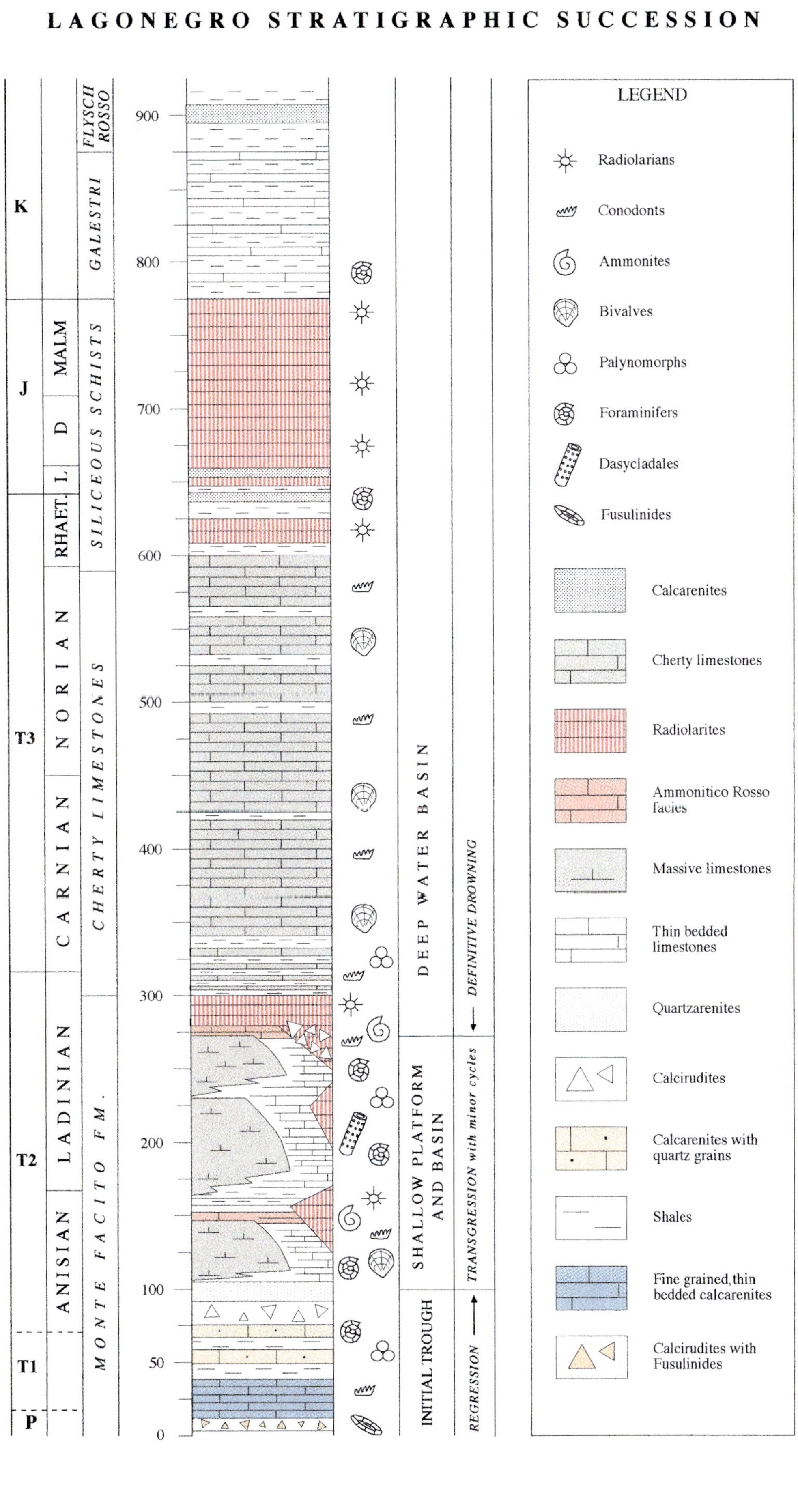
LAGONEGRO STRATIGRAPHIC SUCCESSION
K
J
T3
T2
T1
P
MALM
D
L
RHAET.
NORIAN
CARNIAN
LADINIAN
ANISIAN
FLYSCH ROSSO
GALESTRI
SILICEOUS SCHISTS
CHERTY LIMESTONES
MONTE FACITO FM.
900
800
700
600
500
400
300
200
100
50
0
DEEP WATER BASIN
SHALLOW PLATFORM AND BASIN
INITIAL TROUGH
DEFINITIVE DROWNING
TRANSGRESSION with minor cycles
REGRESSION
LEGEND
Radiolarians
Conodonts
Ammonites
Bivalves
Palynomorphs
Foraminifers
Dasycladales
Fusulinides
Calcarenites
Cherty limestones
Radiolarites
Ammonitico Rosso facies
Massive limestones
Thin bedded limestones
Quartzarenites
Calcirudites
Calcarenites with quartz grains
Shales
Fine grained, thin bedded calcarenites
Calcirudites with Fusulinides

A long and careful study of the paleontological content (foraminifera, algae, brachiopods, conodonts, ammonites, radiolarians) together with detailed sedimentological and stratigraphic analyses and new facies mapping, provided a tentative interpretation of the early sedimentary history of the Lagonegro basin (Ciarapica and Passeri, 2000).

In the Mt. Facito and Sasso di Castalda areas, fine-grained conglomerates with Permian fossils were found in boulders and pebbles associated with cataclasites (Donzelli and Crescenti, 1970; Ciarapica et al., 1986; Panzanelli Fratoni et al., 1987; Ciarapica et al., 1990b). The Permian conglomerates occur as boulders in a tectonic melange with a shaly matrix; they could represent small tectonic slices of the oldest rocks of the Lagonegro basin. The hypothesis of an Early Triassic or Middle Triassic reworking of these conglomerates is not substantiated by any stratigraphic evidence, because Permian materials were never found in Early Triassic (Capelluzzo beds) or Middle Triassic facies.

In the Monte Facito fm. Ciarapica and Passeri (2000) found eight lithofacies differing in age, petrographic and sedimentological features and sedimentary environment.

Lithofacies A (Shales and calcarenites lithofacies) was determined to be the oldest one. It is made up of a shallowing-upwards sequence with gray and pink shales, black mudstones in the lower part, cross-bedded calcarenites with variable amounts of quartz grains, conglomerates with ooids, quartz and glauconitic grains in the matrix in the upper part. It sedimented in an off-shore to coastal environment during the Spathian-Aegean at least (Mietto et al., 1991).

Lithofacies B includes shales and sandstones with variable amounts of feldspars, mica and quartz up to quartzarenites.

Lithofacies C consists of shales and carbonate build-ups containing encrusting organisms, calcareous algae, sponges and scattered madreporarians (Late Anisian and Ladinian). The boundstone facies laterally pass either to massive or thin-bedded pelagic micrites with radiolarians.

Lithofacies E is pink nodular limestone or pink calcareous breccias which often overlies the organic build-ups, and can be Early Fassanian, Late Fassanian or Early Longobardian in age according to content in conodonts and ammonites (Mietto and Panzanelli Fratoni, 1990; Panzanelli Fratoni, 1991); these facies represent the drowning of the carbonate build ups.

Lithofacies D consists of bedded cherts and radiolarites with calcarenites and calcirudites. This lithofacies is referred to a deep water environment. The carbonate bodies are considered to be mass flow deposits. The age spans from the Illirian (?)–Early Fassanian to Late Longobardian.

Lithofacies F is red shales with high Al_2O_3/Na_2O ratio which indicates their origin as residual clays.

Lithofacies G includes shales, marls and micritic limestones and is considered transitional to the Cherty Limestones (Calcari con Selce Auctt. or Sirino Formation, Selli, 1962) which follow the Monte Facito formation in the Lagonegro succession.

Lithofacies H is a tectonic melange, containing shales, calcarenites, arenites and also sedimentary conglomerates.

The scheme in Figure 6 illustrates the stratigraphic relationships among the different lithofacies. Some observations follow on this setting:

(1) The oldest deposits known in the Southern Apennines sedimented in a marine environment from the Permian. They provide evidence of the existence of a basinal area with terrigenous, siliceous and carbonate sedimentation before the Mesozoic.

(2) This basin was affected by extensional tectonics, so that areas with different subsidences created structural highs and lows. During the Middle Triassic, in the shallower areas, carbonate build-ups developed, while in the deeper zones thin successions of pelagic mudstones and shales settled. After a first phase of drowning marked by Ammonitico Rosso facies followed by radiolarites (Fassanian), other build-ups grew in the shallower parts of the basin, but they drowned definitively in the Late Longobardian when the radiolaritic facies became widespread all over the basin.

(3) At the end of the Ladinian, sedimentation inside the basin changed from radiolarites to carbonate wackestones and packstones with radiolarians; this variation suggests that large carbonate platforms, able to export fine-grained carbonates toward the basin, began to develop in adjacent areas. The carbonate sedimentation persisted throughout the Norian.

(4) Stratigraphic studies suggest the following sequence starting from a Permian–Early Triassic basin:
   1. Early Anisian (Aegean) uplift evidenced by the Early Triassic regressive cycle;

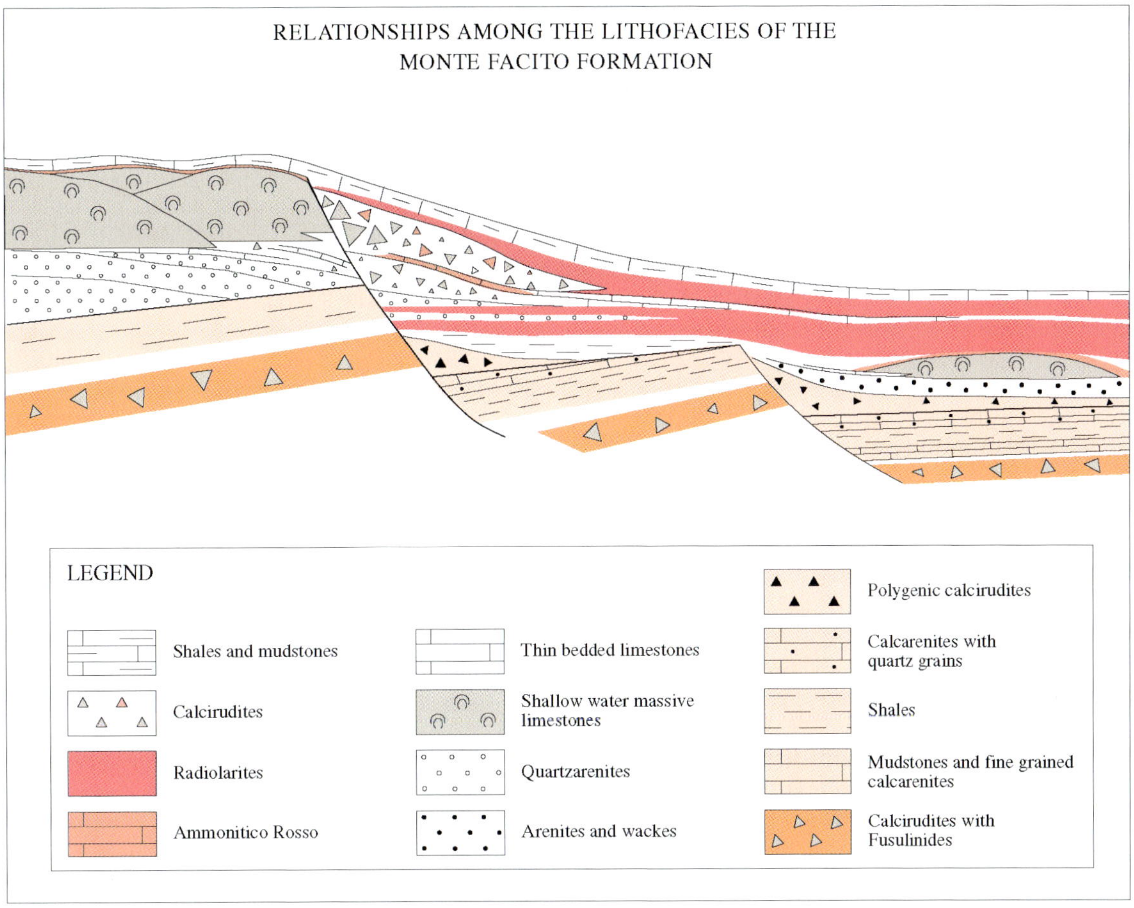

Figure 6: The Monte Facito is a group of broken formations that includes lithologies variable for facies and age. This schematic restoration of the original stratigraphic relationships among them arose from studies in the outcrops of Moliterno area (for the left side) and in Sasso di Castalda area (right side) (after Ciarapica and Passeri, 2000, modified).

2. Anisian-Ladinian deepening connected with extensional tectonics, shown by widespread radiolarites and by the occurrence of repeated gravity flow deposits.
3. Definitive drowning of the carbonate build-ups at the end of the Late Ladinian (Early Longobardian).

In this tentative interpretation, derived from the study of some of the widest outcrops of the Monte Facito fm., there is no evidence of a Carnian crisis similar to the one that separates the two sedimentary cycles in the Northern Apennines.

4.1.2. *Cherty Limestones (Calcari con selce).* The Cherty Limestones (Scandone, 1967) are also known as the Sirino fm. (Selli, 1962). In the northern outcrops (Abriola, Mt. Pierno, S. Fele, Frosolone well, Mt. Marrone) they are dolomitized. This formation crops out in many localities of Lucania, but only scattered outcrops show the transition to the underlying Monte Facito formation. In two of these sites Mietto and Panzanelli Fratoni (1990) found Ladinian conodonts in the lower part of the Cherty Limestones. Norian conodonts of the Lacian to Sevatian characterize the upper part of the calcareous pelagic succession (Miconnet, 1982).

This formation, 300 to 500 m thick, is made up of thin-bedded micritic limestones with cherty nodules and layers. Shales are often present as interbeds; when they are absent, stylolithes give a massive appearance. Horizons laminated with *Halobia* are frequent, so that the Cherty Limestones are also indicated as "*Halobia* Limestones" (De Capoa Bonardi, 1970; Scandone, 1975). In some instances slumpings and paraconglomerates made of pelagic material are found.

The Cherty Limestones pass upwards to the Siliceous Schists fm. through a "transitional interval" containing cherty limestones and red shales (Amodeo and Baumgartner, 1994; Amodeo, 1996, 1999). Scandone (1967) did not evidence this transitional interval and considered the second important red shale intercalation as the base of the Siliceous Schists, when the cherty limestones actually disappear.

4.1.3. *Siliceous Schists (Scisti Silicei).* In the Italian geological panorama, this lithologic unit is peculiar to the Lagonegro basin for its persistence from the Late Triassic up to the Late Jurassic. It is made of red, green and black bedded cherts and siliceous shales with radiolarians and manganese. Its thickness varies from 50 m to more than 200 m according to the position inside the basin: the thickest sequences contain calcirudites and calcarenites of platform origin (San Fele facies) documenting a proximal location; the more distal sequences are less than 100 m thick and contain only few calcarenitic beds (Lagonegro, Sasso di Castalda). This reduction is not only due to the amount of reworked deposits but could be related both to very low sedimentation rates in the more distal portions of the basin, and to phases of erosion of the siliceous oozes caused by bottom currents or gravitative processes. Detailed biostratigraphic studies are now in progress to test these hypotheses. The age of the Siliceous Schists ranges from Late Triassic (Rhaetian) to Late Jurassic (Early-Middle? Tithonian) (Miconnet, 1982; Miconnet et al., 1983; Amodeo, 1996, 1999). This formation pertains to a sedimentary environment ranging from the lower bathyal to the abyssal zone.

4.1.4. *Galestri (Flysch Galestrino).* These outcrop widely in Lucania on top of the Siliceous Schists. This formation is made up by alternation of siliceous shales, mudstones and calcarenites with a large amount of manganese. In the proximal facies the calcarenites show a normal gradation in the lower part and sometimes flute casts on the bottom. The shales are often brown or black in colour due to the high manganese and organic matter content. Brown radiolarites and jaspers are present subordinately. The sedimentary environment is referred to a deep, distal basin.

Fossils are scarce but allow us to assign the age of this formation to the Early Cretaceous (De Stasio, 1971; Brönnimann et al., 1971).

4.1.5. *The Lagonegro Lower sequence in Molise.* The prolongation of the Lagonegro basin toward the Molise tectonic units was suggested by Mostardini (1986), Mostardini and Merlini (1986), Manfredini (1986), Pescatore (1988) mainly on the basis of the transitions between the Lagonegro upper sequence and the Molise uppermost successions. The lower part of the Molise facies was known only near Isernia, in the Frosolone 1 well, which reached a succession made of cherty limestones and dolostones followed by radiolarites. Unfortunately it did not provide any fossils.

Over the last few years, new data have been collected on the southeastern Mainarde range (Mt. Marrone) where a thick sequence of cherty dolostones crops out (Bertinelli, 1999). This succession had been previously considered as platform dolostones belonging to the lower portion of the "Abuzzi" succession and the base of a by-pass margin of Late Jurassic–Cretaceous age (D'Andrea and Urgera, 1986; D'Andrea et al., 1991). Radiolarian fauna in the cherty nodules defined a Late Triassic age for this part of the sequence (Bertinelli et al., 2002). The Mt. Marrone dolostones appear very similar to the Lagonegro proximal facies of San Fele and they confirm the prolongation of the Lagonegro basin toward the Molise basin, so that the original width of this basin was greater than previously hypothesized.

4.2. The Lagonegro-Molise Upper Sequence

Mostardini and Merlini (1986) defined the "Lagonegro-Molise Upper Sequence" as a complex of tectonic units, sedimented from the Late Cretaceous to Miocene, which are detached from their substratum. These include mélanges, shales, radiolarites and turbiditic sandstones. The most complete succession is given by: (a) the so-called "Flysch Rosso" with shales, radiolarites and calcarenites (Late Cretaceous–Oligocene); (b) mature quartzarenites of the Numidian Flysch (Aquitanian-Langhian), coming from the African cratonic area; (c) immature sandstones and conglomerates, grouped as Irpinian flysch (Late Langhian-Tortonian), coming from the orogenic belt. Mélanges ("Argille varicolori") are a chaotic mixture of siliciclastic and carbonate elements in shaly matrix. The original Mesozoic substratum of these units is completely unknown. The lower part of these units was considered as belonging to the Ligurian domain (a part of the "Sicilide complex"), but in recent studies it was described as the upper part of the Lagonegro sequence, according to Mostardini and Merlini (1986) (Carbone et al., 1988, 1991; Ciarapica and Passeri, 1998, 2002; Pescatore et al., 1999).

5. THE PALEOGEOGRAPHIC MEANING OF THE LAGONEGRO-MOLISE SEQUENCES

Since the Miocene, the tectonic units containing the Lagonegro-Molise sequences have been incorporated in the Southern Apennine thrust belt, and now lie between the Apulian platform units to the footwall and the Apenninic platform units to the hanging wall. The original width of this basin cannot be defined with precision but could be similar to that of the modern Tyrrhenian and Balearic Seas together.

Presumed lithologic size and paleontological content of the Lagonegro-Molise facies show that this basin was not isolated but that it was fully connected with other deep marine or oceanic basins. Three different possible connections have been suggested: (a) a northeastward connection with the Budva basin (Scandone, 1975); (b) a northwestward connection with the La Spezia basin and Liguria-Piedmont Ocean (D'Argenio and Pialli, 1975; Marsella et al., 1995); and (c) a southward connection with the Ionian Ocean (Catalano et al., 1991; Ciarapica and Passeri, 1998, 2002).

The connection with the Budva-Pindos basin across the modern Adriatic Sea was suggested by various lithologic and paleontological similarities but mainly for the presence of *Halobia*-bearing beds in the Late Triassic cherty limestones of the Budva-Pindos and Lagonegro basins. This hypothesis cannot be substantiated because the Lagonegro-Molise basin in the Late Triassic was bounded to the north and to the east by the Abruzzi and Apulia carbonate platforms, and no evidence of its existence was found in the Adriatic Sea.

The connection toward the northwest with the La Spezia basin and the Liguria-Piedmont Ocean was mainly suggested by the presence of shales (Galestri) in the Early Cretaceous of Lagonegro and of the Ligurian units. This hypothesis presumes that the Lagonegro sequence was sedimented to the west of the Apenninic platform but is in conflict with field and geophysical data. Moreover the stratigraphic evolution of the Lagonegro basin and that of the La Spezia basin or of the Liguria-Piedmont Ocean are very different: deep-marine facies appeared in the Lagonegro basin well before those in the La Spezia basin. The La Spezia basin suffered an emersion in the Carnian, while the Lagonegro was deepening. Argillitic facies were diffused in the Norian-Rhaethian of La Spezia and Western Alps while radiolarites spread out in the Lagonegro Basin (Figure 7).

The postulated connection with the Ionian Ocean is the simpler and more realistic way to understand the paleogeographic and geodynamic meaning of the Lagonegro-Molise facies.

6. CONCLUSIONS

(a) *Northern and Southern Apennines* – Analyses of the Apenninic stratigraphic successions point out the different evolutions of the Northern and Southern Apennines: the former was related to the rifting of the central Atlantic and of its Liguria-Piedmont prolongation with oceanic spreading in the Late Jurassic; the latter was linked to the rifting of the southwestern branch of the western Tethys which caused the opening of the Ionian Ocean. At present this ocean is almost completely subducted with the exception of the two remnant basins known as the Levantine and Ionian basins.

(b) *Ionian Ocean* – Evidence for the existence of the Ionian Ocean is manifold. The most important data come from geophysical investigations on the Mediterranean floor (Finetti, this volume, Chapter 10; Finetti, this volume, Chapter 34; Finetti and Del Ben, this volume, Chapter 6; Finetti, 1982, 1985; Finetti et al., 1996; Le Pichon et al., 1982; Cita and Camerlenghi, 1990; Chamot-Rooke et al., 1990) and from petrological studies of the Eolian and Hellenic volcanic arcs. These studies point out the existence of a west-dipping subduction of oceanic crust below the Calabrian arc (Malinverno and Ryan, 1986; Serri, 1990, 1995; Catalano et al., 2000) and a north-dipping subduction below the Hellenic arc (Fytikas et al., 1984).

Other data come from sedimentological and stratigraphic studies on the sedimentary sequences outcropping in the circum-Mediterranean area, and from paleogeographic restorations (Stampfli et al., 1991; Robertson et al., 1996; Ciarapica and Passeri, 1998, 2000, 2002).

In our opinion the Lagonegro Lower Sequence, outcropping in the alloctonous units of the Southern Apennines with Early-Middle Triassic terrigenous and carbonate shallow water facies, would have been deposited on the northern continental margin of the Ionian Ocean. On other hand we cannot be sure of the presence of Early and Middle Triassic shallow water facies throughout the Lagonegro basin. During the orogenesis, the Lagonegro basin

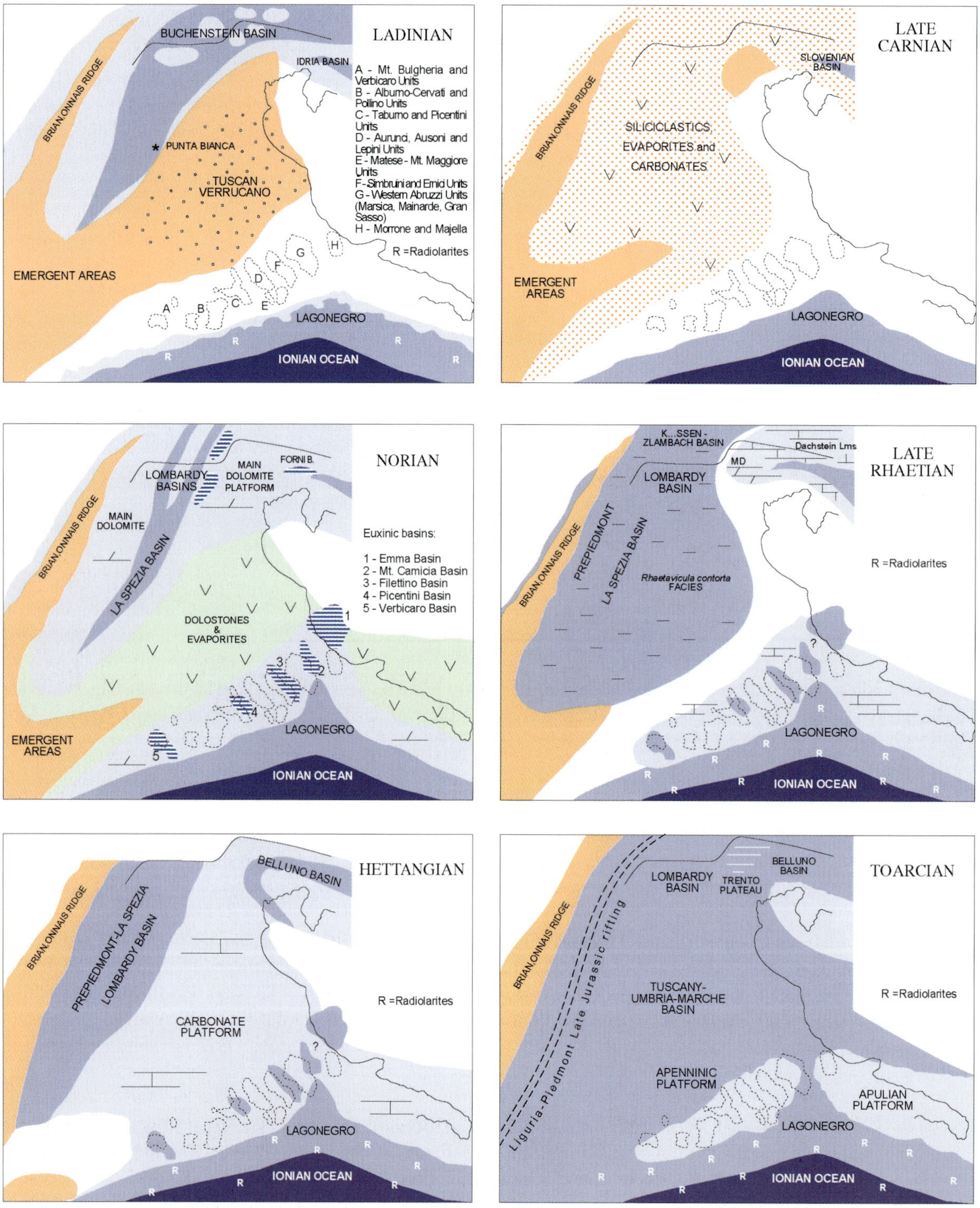

Figure 7: Paleogeographic maps of the westmediterranean area before the Liguria-Piedmont oceanic spreading. During the orogenesis (since the Miocene) the western part of the Apenninic carbonate Platform was detached from its basement and split up in many tectonic units that piled over the Lagonegro tectonic units; then, all of them overrode the Apulian Platform.

underwent a strong shortening: in some wells (San Fele–Mt. Foi) the succession is repeated many times and this suggests that only a minor part is outcropping (10–15%). The same considerations arise from the Crop 4 seismic profiles (Finetti, this volume). As a consequence, the possibility that the more distal facies of the Lagonegro succession were deposited on oceanic floor cannot be excluded. The "Lagonegro Upper Sequence" itself, completely detached from its substrate, demonstrates that the assumption of a Lagonegro basin completely settled on continental crust is arbitrary.

A part of the Late Cretaceous–Miocene sequences of Sannio and Irpinia, detached from their substrate and called the "Lagonegro-Molise Upper Sequence" by Mostardini and Merlini (1986), could be considered as parts of the oceanic cover.

(c) *Evolution of the Ionian basin through the analysis of the Lagonegro sequences* – Basin analysis of the "Lagonegro Lowermost Sequence" is made uncertain by the strong tectonization of the Monte Facito formation. Field data suggest the existence of a shallowing-upward sequence from Permian (?)–Early Triassic to the Early Anisian, followed by local emersions and a hiatus in sedimentation. A successive deepening sequence started with the Late Anisian: the drowning of the algal buildups (Pelsonian, Illirian and Fassanian in age) occurred from the Early Fassanian to the Early Longobardian; deep water pelagic facies (Ammonitico Rosso and radiolarites) developed from the Early Fassanian in the distal parts of the basin and from the Early Longobardian in the proximal ones. They contain large bodies of calcirudites and calcarenites. The sudden deepening of the depositional environment during the Ladinian (from the Early Fassanian to the Early Longobardian) suggests that the Lagonegro area was involved in an active rifting phase at that time.

In our opinion, the change from the Late Ladinian radiolaritic sedimentation of the M. Facito (Longobardian) to the calcareous one of the Cherty Limestones (Late Ladinian to Norian) is not related to a shallowing trend but to an increase in fine-grained carbonate input inside the basin, due to the development of large carbonate platforms in adjacent areas (i.e., Apenninic and Apulian platforms).

The radiolaritic sedimentation of the Siliceous Schists from the Late Triassic (Raethian) to Late Jurassic, when the adjacent carbonate platforms were fully developed, can be referred to an abyssal plain environment. This interpretation is supported also by the presence of a high content of manganese in the siliceous beds.

During the Early Cretaceous, radiolarites with calcarenites and calcirudites were sedimented in the basinal units now outcropping in Molise, while the other units (Lagonegro) were characterized by the presence of abundant shales and radiolarites with scattered beds of calcarenites. The origin of the shales in this part of the Lagonegro-Molise basin is unknown. Just as a hypothesis of a relationship with the eo-Hellenic orogenesis and with the closure of the Vardar and Pindos oceans is possible. Anyway the Ionian basin, from the Cretaceous to the Miocene, acted as a trough where sediments coming from the cratonic (African) areas and from the Apenninic orogenic belt accumulated.

Briefly, the field data do not allow us to define the kind of marine basin existing during the Late Permian in the Southern Apennines, but conodont-bearing mudstones testify to a relatively deep water environment in the Middle Spathian (Mietto et al., 1991). This sedimentary cycle ended with a relative sea-level fall (uplift); then deep water facies of the successive cycle begun to develop from the Early Fassanian, and radiolarites of the Siliceous Schists spread out from the Raethian to the end of the Jurassic.

These data suggest that the rifting of the Ionian basin began in the Permian, and that the oceanic spreading started during the Middle Triassic.

The sedimentary evolution of the Lagonegro-Molise basin presented in this chapter is very similar to the one described by Robertson et al. (1996) for the Taurides (Antalya) and Cyprus, belonging to the eastern prolongation of the Ionian basin. The diffusion of radiolarites and megabreccias during the Ladinian, and the presence of alkali and MORB transitional basalts in Antalya suggest that the main phase of rifting occurred at that time for the whole southern branch of the Neotethys.

REFERENCES

Abbate, E., Bortolotti, V., Conti, M., Marcucci, M., Principi, G., Passerini, P., Treves, B., 1986. Apennines and Alps ophiolites and the evolution of the Western Tethys. Mem. Soc. Geol. It. 31, 23–44.

Adamoli, L., Bigozzi, A., Ciarapica, G., Cirilli, S., Passeri, L., Romano, A., Duranti, F., Venturi, F., 1990. Upper Triassic bituminous facies and Hettangian pelagic facies in the Gran Sasso Range. Boll. Soc. Geol. It. 109 (1), 219–230.

Amodeo, F., 1996. Il Triassico terminale – Giurassico del Bacino Lagonegrese. Tesi di Dottorato in Geologia del Sedimentario. Università degli Studi di Napoli "Federico II" e Université de Lausanne, Napoli, 223 p.

Amodeo, F., 1999. Il Triassico terminale – Giurassico del Bacino Lagonegrese. Studi stratigrafici sugli Scisti Silicei della Basilicata (Italia meridionale). Mémoires de Géologie (Lausanne) 33, 1–121.

Amodeo, F., Baumgartner, P.O., 1994. Stratigraphic revision of the "Scisti Silicei" Formation (Upper Triassic–Upper Jurassic), Lagonegro Basin, southern Italy. Preliminary Report. Paleopelagos 4, 35–46.
Argnani, A., Marani, M., Savelli, C., Galassi, B., 1995. Migrazione del vulcanismo di arco cenozoico nel contesto geodinamico intraorogenico del Mar Tirreno meridionale: un riesame. In: Polino, R., Sacchi, R. (Eds.), Atti del Convegno Rapporti Alpi–Appennino. Acc. Naz. delle Scienze, Scritti e Doc. XIV, 377–396.
Azzaroli, A., 1962. Affioramento di calcare permiano presso Abriola. Boll. Soc. Geol. It. 81 (1), 85–86.
Bertinelli, A., 1999. La successione bacinale di M. Marrone e confronti con le facies mesozoiche molisane. Tesi in Sedimentologia. Univesità di Perugia; a.a. 1997/98, Perugia, 176 p.
Bertinelli, A., Ciarapica, G., Passeri, L., 2002. The stratigraphic succession of Mt. Marrone (Mainarde range) as a northern part of the Jurassic Molise Basin. Boll. Soc. Geol. It. vol. spec. 1 (1), 481–487.
Brönnimann, P., Duran-Delga, M., Grandjaquet, C., 1971. Présence simultanée de Protopleneroplis striata Weinschenk et de Calpionelles néocomiennes dans le "Flysch Galestrino" de Lucanie (Italie méridionale). Rev. Micropal, Spec. Publ., Paris 14, 96–101.
Carbone, S., Catalano, S., Lentini, F., Monaco, C., 1988. Le unità stratigrafico-strutturali dell'alta Val d'Agri (Appennino lucano) nel quadro dell'evoluzione del sistema catena-avanfossa. Mem. Soc. Geol. It. 41, 331–341.
Carbone, S., Catalano, S., Lazzari, S., Lentini, F., Monaco, C., 1991. Presentazione della carta geologica del Bacino del Fiume Agri (Basilicata). Mem. Soc. Geol. It. 47, 129–143.
Catalano, R., Di Stefano, P., Kozur, H., 1991. Permian circumpacific deep-water faunas from the western Tethys (Sicily, Italy) – new evidences for the position of the Permian Tethys. Palaeogeography, Palaeoclimatology, Palaeoecology 87 (1–4), 75–108.
Catalano, R., Doglioni, C., Merlini, S., 2000. On the Mesozoic Ionian Basin. Geophys. J. Int. 143, 1–24.
Chamot-Rooke, N., Truffert, C., Devoogt, B., Huchon, P., Lallemat, S., Le Pichon, X., 1990. Crustal structure of the Eastern Mediterranean: Results of the Pasiphae Cruise. EOS 71 (43), 1634.
Ciarapica, G., Cirilli, S., D'Argenio, B., Marsella, E., Passeri, L., Zaninetti, L., 1987. Late Triassic open and euxinic basins in Italy. Rend. Soc. Geol. It. 9, 157–166.
Ciarapica, G., Cirilli, S., Martini, R., Rettori, R., Salvini Bonnard, G., Zaninetti, L., 1990a. Carbonate buildups and associated facies in the M. Facito Formation (Southern Apennines). Boll. Soc. Geol. It. 109 (1), 151–164.
Ciarapica, G., Cirilli, S., Martini, R., Panzanelli Fratoni, R., Zaninetti, L., Salvini Bonnard, G., 1990b. Reworked foraminifera in the Triassic M. Facito Formation Auctt., Lagonegro Basin (Southern Apennines, Italy). Boll. Soc. Geol. It. 109 (1), 143–149.
Ciarapica, G., Cirilli, S., Martini, R., Zaninetti, L., 1986. Une microfauna à petits Foraminifères d'age permien remaniée dans le Trias moyen de l'Apennin méridional (Formation du M. Facito, Lucanie occidentale). Description de Crescentia verebralis n. gen. n. sp. Rev. Paléobiol. 5, 207–215.
Ciarapica, G., Cirilli, S., Panzanelli Fratoni, R., Passeri, L., Zaninetti, L., 1990c. The M. Facito Formation (Southern Apennines). Boll. Soc. Geol. It. 109 (1), 135–142.
Ciarapica, G., Passeri, L., 1980. La litostratigrafia della serie triassica del promontorio occidentale del Golfo di La Spezia. Mem. Soc. Geol. It. 21, 51–61.
Ciarapica, G., Passeri, L., 1998. Evoluzione paleogeografica degli Appennini. Atti Tic. Sc. Terra 40, 233–290.
Ciarapica, G., Passeri, L., 2000. Le facies del Triassico inferiore e medio (Fm. di Monte Facito Auctt.) nelle aree di Sasso di Castalda e di Moliterno (Basilicata). Boll. Soc. Geol. It. 119 (2), 339–378.
Ciarapica, G., Passeri, L., 2002. The paleogeographic duplicity of the Apennines. Boll. Soc. Geol. It. vol. spec. 1, 67–75.
Ciarapica, G., Zaninetti, L., 1984. Foraminifères et biostratigraphie dans le Trias supérieur de la série de La Spezia (Dolomie de Coregna et Formation de La Spezia, nouvelle formation), Apennin Septentrional. Rev. Paléobiol. 3, 117–134.
Cirilli, S., 1993. Il Trias di Filettino e Vallepietra (Monti Simbruini, Appennino Centrale). Boll. Soc. Geol. It. 112 (1), 371–394.
Cita, M.B., Camerlenghi, A., 1990. The Mediterranean ridge as an accretionary prism in collisional context. Mem. Soc. Geol. It. 45, 463–480.
D'Andrea, M., Miccadei, E., Praturlon, A., 1991. Rapporti tra il margine orientale della piattaforma laziale-abruzzese ed il margine occidentale della piattaforma Morrone-Pizzalto-Rotella. Studi Geol. Camerti vol. spec. 1991/2, 389–395.
D'Andrea, M., Urgera, A., 1986. Correlazioni stratigrafiche ed alcune considerazioni paleogeografiche sul mesozoico dei Monti della Meta (Appennino centro-meridionale). Mem. Soc. Geol. It. 35, 27–32.
D'Argenio, B., Scandone, P., 1970. Jurassic facies pattern in the Southern (Campania-Lucania) Apennines. In: Colloquium on the Mediterranean Jurassic, Budapest, September 1969. Ann. Hung. Geol. Inst. 54, 383–396.
D'Argenio, B., Pescatore, T., Scandone, P., 1973. Schema geologico dell'Appennino meridionale (Campania e Lucania). In: Accad. Naz. Lincei, Convegno sul tema: Moderne vedute sulla geologia dell'Appennino. Quad., vol. 183, pp. 49–82.

D'Argenio, B., Pialli, G., 1975. Tentativo di correlazione tra unità paleogeografiche mesozoiche dell'Appennino meridionale e settentrionale. Atti Acc. Pont., n.s. 23, 29.

De Capoa Bonardi, P., 1970. Le Daonelle e le Halobie della serie calcareo-silico-marnosa della Lucania (Appennino meridionale). Studio paleontologico e biostratigrafico. Mem. Soc. Nat. in Napoli 78 (1969), 1–127.

Dercourt, J., Ricou, L.E., Vrielynck, B. (Eds.), 1993. Atlas Tethys Paleoenvironmental Maps and Explanatory Notes. Gauthier-Villars, Paris. 307 p., 14 maps.

De Stasio, L.M., 1971. Su di alcune microfaune rinvenute nel flysch galestrino della Lucania (serie calcareo-silico-marnosa). Mem. Soc. Nat. in Napoli 78 (1969), 409–419.

Donzelli, G., Crescenti, U., 1970. Segnalazione di una microbiofacies permiana, probabilmente rimaneggiata, nella Formazione di M. Facito (Lucania Occidentale). Boll. Soc. Natur. in Napoli 79, 13–19.

Elter, P., Federici, P.R., 1964. Sulla presenza di terreni attribuibili al Trias medio nel promontorio orientale del Golfo di La Spezia. Boll. Soc. Geol. It. 83 (2), 395–402.

Finetti, I.R., 1982. Structure, stratigraphy and evolution of Central Mediterranean. Boll. Geof. Teor. Appl. 24 (96), 247–312.

Finetti, I.R., 1985. Structure and evolution of the Central Mediterranean (Pelagian and Ionian Sea). In: Stanley, D.J., Wezel, F.C. (Eds.), Geological Evolution of the Mediterranean Basin. Springer-Verlag, New York, pp. 215–230.

Finetti, I.R., this volume. Understanding the Ionides and their geodynamics. Chapter 10.

Finetti, I.R., this volume. Geodynamic evolution of the Mediterranean region, from the Permo-Triassic Ionian opening to the Present, constrained by new lithospheric CROP seismic data. Chapter 34.

Finetti, I.R., Del Ben, A., this volume. Ionian and Alpine Neotethyan oceans opening. Chapter 6.

Finetti, I.R., Lentini, F., Carbone, S., Catalano, S., Del Ben, A., 1996. Il sistema Appennino Meridionale–Arco Calabro–Sicilia nel Mediterraneo Centrale: studio geologico-geofisico. Boll. Soc. Geol. It. 115 (3), 529–559.

Finetti, I.R., Boccaletti, M., Bonini, M., Del Ben, A., Geletti, R., Pipan, M., Sani, F., 2001. Crustal section based on CROP seismic data across the North Tyrrhenian–Northern Apennines–Adriatic Sea. Tectonophysics 343, 135–163.

Fytikas, M., Innocenti, P., Manetti, P., Mazzuoli, R., Peccerillo, A., Villari, L., 1984. Tertiary to Quaternary evolution of volcanism in the Aegean region. In: Dixon, J.E., Robertson, A.H.F. (Eds.), The Geological Evolution of the Eastern Mediterranean. Geol. Soc. London Spec. Publ. 17, 687–700.

Iannace, A., Zamparelli, V., 1999. Lattari and Picentini Mountains. In: Zamparelli, V., Cirilli, S., Iannace, A., Jadoul, F. (Coord.), Palaeotectonic and palaeoceanographic controls on microbial-serpulid communities in the Norian-Rhaetian carbonates of Italy: a synthesis. Palaeopelagos, spec. publ., vol. 3, pp. 32–35.

Le Pichon, X., Augustithis, S.S., Mascle, J. (Eds.), 1982. Geodynamics of the Hellenic Arc and Trench. Tectonophysics, spec. issue 86, 304 p.

Luperto, E., 1965a. Foraminiferi del "Calcare di Abriola" (Potenza). Boll. Soc. Pal. It. 4 (2), 161–207.

Luperto, E., 1965b. Sezioni di forme nodosaroidi permiane del "Calcare di Abriola" (prov. Potenza, Appennino meridionale). Boll. Soc. Pal. It. 4 (2), 208–215.

Malinverno, A., Ryan, W.B.F., 1986. Extension in the Tyrrhenian Sea and shortening in the Apennines as result of arc migration driven by sinking of the lithosphere. Tectonics 5, 227–245.

Manfredini, M., 1986. Explanation notes for the geological map of southern Italy between parallels 40° and 41°20 In: Int. Symp. Engineering geology problems in seismic areas. Bari.

Marsella, E., Kozur, H., D'Argenio, B., 1993. Monte Facito Formation (Schitian–Middle Carnian). A deposit of the ancestral Lagonegro Basin in the Southern Apennines. Boll. Serv. Geol. It. CX (1991), 225–248.

Marsella, E., Bally, A.W., Cippitelli, G., D'Argenio, B., Pappone, G., 1995. Tectonic history of the Lagonegro Domain and Southern Apennne thrust belt evolution. Tectonophysics 252 (1–4), 307–330.

Martini, R., De Wever, P., Zaninetti, L., Denelian, T., Kito, N., 1989. Les radiolarites triasique de la Formation du M. Facito Auctt. (Bassin de Lagonegro, Italie méridionale). Rev. Paléobiol. 8 (1), 143–161.

Martini, I.P., Rau, A., Tongiorgi, M., 1986. Syntectonic sedimentation in a Middle Triassic Rift Northern Apennines, Italy. Sed. Geol. 47, 191–219.

Miconnet, P., 1982. Précisions stratigraphiques et tectoniques dans un secteur du Lagonegro (Italie méridionale). Ann. Soc. Geol. Nord 102, 17–24. Lille.

Miconnet, P., 1988. Evolution mesozoique du secteur de Lagonegro. Mem. Soc. Geol. It. 41, 321–330.

Miconnet, P., Ciarapica, G., Zaninetti, L., 1983. Faune a foraminiferes du Trias supérieur d'affinité sud-tethysienne dans l'Apennin Méridionale (Bassin de Lagonegro, province de Potenza, Italie); comparaison avec l'Apennin septentrionale. Rev. Paléobiol., Genève 2 (2), 131–147.

Mietto, P., Panzanelli Fratoni, R., 1990. Conodonts from the Monte Facito Formation and from the base of the Monte Sirino Formation (Lagonegro sequence). Boll. Soc. Geol. It. 109 (1), 165–169.

Mietto, P., Panzanelli Fratoni, R., Perri, M.C., 1991. Spathian and Aegean conodonts from the Capelluzzo Calcarenites of the M. Facito Group (Lagonegro Sequence–Southern Apennines). Mem. Sc. Geol. 43, 305–317. Padova.

Mostardini, F., 1986. Southern Apennines: structural model supported by subsurface and geophysical data. A geological cross section through the irpinian sector. In: Int. Symp. Engineering Geology Problems in Seismic Areas. Bari.

Mostardini, F., Merlini, S., 1986. Appennino centro-meridionale. Sezioni geologiche e proposta di modello strutturale. Mem. Soc. Geol. It. 35, 177–202.

Panzanelli Fratoni, R., 1991. Analisi stratigrafica della "Formazione del M. Facito" Auctt. (serie di Lagonegro-Appennino Meridionale). Tesi di Dottorato in Scienze della Terra, Università degli Studi di Perugia, Perugia, 215 p.

Panzanelli Fratoni, R., Limongi, P., Ciarapica, G., Cirilli, S., Martini, R., Salvini Bonnard, G., Zaninetti, L., 1987. Les foraminifères du Permien superieur remaniés dans le "Complex terrigène" de la Formation triasique du M. Facito, Apennin méridional. Rev. Paléobiol. 6 (2), 293–319.

Passeri, L., 1985. Il Trias dell'Unità di Punta Bianca. Mem. Soc. Geol. It. 30, 105–114.

Pescatore, T., 1988. La sedimentazione miocenica nell'Appennino campano-lucano. Mem. Soc. Geol. It. 41, 37–46.

Pescatore, T., Renda, P., Schiattarella, M., Tramutoli, M., 1999. Stratigraphic and structural relationships between Meso-Cenozoic Lagonegro basin and coheval carbonate platforms in southern Apennines, Italy. Tectonophysics 315, 269–286.

Robertson, A.H.F., Dixon, J.E., Brown, S., Collins, A., Morris, A., Pickett, E., Sharp, I., Ustaömer, T., 1996. Alternative tectonic models for the Late Palaeozoic–Early Tertiary development of Tethys in the eastern Mediterranean region. In: Morris, A., Tarling, D.H. (Eds.), Palaeomagnetism and Tectonics of the Mediterranean Region. Geol. Soc. London Spec. Publ. 105, 239–263.

Scandone, P., 1967. Studi di geologia lucana: la serie calcareo-silico-marnosa e i suoi rapporti con l'Appennino calcareo. Boll. Soc. Nat. in Napoli 76, 1–175.

Scandone, P., 1975. Triassic seaways and the Jurassic Tethys Ocean in the central mediterranean area. Nature 256 (5513), 117–119.

Selli, R., 1962. Il Paleogene nel quadro della geologia dell'Italia meridionale. Mem. Soc. Geol. It. 3, 737–790.

Serri, G., 1990. Neogene-Quaternary magmatism of the Tyrrhenian region: characterization of the magma sources and geodynamic implications. Mem. Soc. Geol. It. 44, 219–242.

Serri, G., 1995. Magmatism and Plate Tectonics. In: Ranalli, G. (Ed.), Plate Tectonics: The First Twenty-Five Years. Proc. of the VIII Summer School Earth and Planetary Sciences, Siena, pp. 105–133.

Sgrosso, I., 1988. Nuovi elementi per un più articolato modello paleogeografico nell'Appennino centro-meridionale. Mem. Soc. Geol. It. 41, 225–242.

Stampfli, G.M., Marchant, R.H., 1995. Plate configuration and kynematics in the alpine region. In: Polino, R., Sacchi, R. (Eds.), Atti del Convegno "Rapporti Alpi–Appennino". Acc. Naz. delle Scienze, Scritti e Doc. XIV, 147–166.

Stampfli, G., Marcoux, J., Baud, A., 1991. Tethyan margins in space and time. Palaeogeography, Palaeoclimatology, Palaeoecology 87 (1–4), 373–409.

Stampfli, G., Vavassis, I., De Bono, A., Rosselet, F., Matti, B., Bellini, M., 2003. Remnants of the Paleotethyis oceanic suture-zone in the western Tethyan area. Boll. Soc. Geol. It. vol. spec. 2, 1–23.

Taddei Ruggiero, E., 1968. Brachiopodi triassici della Pietra Maura (Lucania), studio paleontologico statistico. Boll. Soc. Natur. in Napoli 77, 349–392.

Venturi, F., Tenneroni, S., 1981. Ammoniti di presunta età anisica nella sezione di Punta Bianca. Rend. Soc. Geol. It. 4, 327–328.

Zappaterra, E., 1990. Carbonate paleogeographic sequences of the Periadriatic region. Boll. Soc. Geol. It. 109 (1), 5–20.

CROP PROJECT: Deep Seismic Exploration of the Central Mediterranean and Italy
Edited by I.R. Finetti

CHAPTER 12

Crustal Tectono-Stratigraphy and Geodynamics of the Southern Apennines from CROP and other Integrated Geophysical-Geological Data

I.R. Finetti[1,*], F. Lentini[2], S. Carbone[2], A. Del Ben[1], A. Di Stefano[2], P. Guarnieri[2], M. Pipan[1], A. Prizzon[1]

ABSTRACT

Using CROP geophysical data integrated with other information, a new lithospheric seismogeological section across the Southern Apennines, from the Marsili Basin in the Tyrrhenian Sea to the Adriatic coast of Apulia, has been reconstructed. This section provides a clear regional deep subsurface imaging of the tectono-stratigraphy, and interpretation of the CROP network allows us to propose a new, much more constrained, geodynamic model.

Across the Southern Apennines thrust belt, from the young (Quaternary) back-arc basin of Marsili to the Apulian Permo-Mesozoic foreland, various tectono-stratigraphic units with different histories are imaged and described.

The first part of the paper comments on the geological data obtained by geological fieldwork carried out mainly in the Lucania sector, integrated with available subsurface data (seismic, borehole); the second presents the new interpreted crustal-lithospheric seismogeological section.

From the Tyrrhenian ocean basin to the outerward undeformed Apulian foreland, the Southern Apennine thrust belt is formed by: (a) a deformed European crustal fragment, detached from the Corso-Sardinian block in the Tyrrhenian stage; (b) a deformed crustal block of the Etrurian Promontory (NW Adria corner), which collided with the Apulian slope, whose Mesozoic cover gave rise to the Apennine platform thrust-sheets. Between (a) and (b), Alpine Tethydes Units (Ligurides), detached from subducted Alpine Tethys slab, were closed in the Balearic stage; (c) A wedge of Ionian Tethydes or Ionides (Lagonegrese) stripped off the subducted Ionian slab and thrust obliquely onto Apulia (over Lower Pliocene cover) in the Tyrrhenian stage; (d) two main families of flysch: the Albidona group and the Gorgoglione more external flysch group, plus the Numidian flysch derived from displaced Paleozoic basal quartzites of denuded Adria–Africa zones (Etrurian and Panormid promontories); and e) rooted or severely displaced buried thrust structures of the Apulian platform.

A well-constrained description of the tectono-stratigraphy of the units which compose the allochthonous nappe is given, and some geological sections are illustrated. The field geology with careful biostratigraphic control shows that the units of the Southern Alpine chain (SAC), deriving from sedimentary sequences originating in the Alpine Tethys realm, started to deform in the Late Oligocene and continued later on involving the external domains.

The reconstructed lithospheric seismogeological section exhibits clearly that two main tectonostratigraphically distinct complexes are present on the onshore Southern Apennines thrust belt: (1) a buried Apulian platform, deformed by Pliocene compressive tectonics, which produced folds and reverse faults, mostly of relatively moderate horizontal displacements, where a major part of the thrust features remain substantially rooted; (2) a strongly tectonized composite nappe over Apulia, made up of allochthonous units and internal-external-reworked flysch. A stack of allochthonous crustal blocks colliding with the Apulian slope on the Tyrrhenian side characterizes the Tyrrhenian margin.

*Corresponding author. Tel.: +39 40 5582276; e-mail: finetti@units.it (I.R. Finetti)
[1]Department DISGAM, University of Trieste, Via Weiss, 1, 34127 Trieste, Italy. Fax: +39 40 5582290
[2]Department of Geological Science, University of Catania, Corso Italia, 55, 95129 Catania, Italy

Careful interpretative seismic reprocessing allowed us to obtain for the first time a clear seismic imaging of the subducted Ionian slab, the buried slope of Apulia on the Tyrrhenian side, and the stacked allochthonous crustal blocks. From the Marsili basin to the Apulian foreland, the major crustal features have been identified and their tectono-stratigraphy outlined. On the Apulian slope, a previously formed tectonic stack of two crustal fragments, which took part in the closure of the interposed Alpine and Ionian Neotethyan oceans, collided. The Corso-Sardinian block in the Balearic stage (Late Oligocene–Early Miocene) collided with the northwestern Adria margin with thrusting of the Alpine Tethydes over Adria (Etrurian Promontory). In the Tyrrhenian stage (Middle Miocene to Present) the stacked Corso-Sardinian and Etrurian fragments followed the Ionian slab retreat, closing the interposed basin with tectonic transport of the Ionian ocean cover ("Ionides", "Lagonegrese" in the literature) over the Apulian platform.
The Bradanic foredeep and the Mt. Alpi Unit are connected to the two major transcrustal thrust faults in the examined area. The total crustal shortening of the Apulian platform during the Southern Apennines tectonics amounts to about 30 km, very much less than the observed shortening of the Adria plate during the Northern Apennines tectonics (about 150 km).

Keywords: Southern Apennines, thrust propagation, CROP, seismogeological section, geodynamics

1. INTRODUCTION

The first part of this paper describes the tectono-stratigraphy of the Southern Apennines orogen, mainly studied on the Lucanian Apennines where geological field data are integrated with abundant subsurface data from hydrocarbon exploration activity. The second part of the paper, based essentially on CROP data, presents a *New Crustal-lithospheric Seismogeological Section across the Southern Apennines, from the Marsili Basin to the Adriatic coast of Apulia* (Figure 1).

The land area discussed in this work corresponds to the southern sector of the Apennines, confined to the north by the ENE–WSW-alignment of the Sorrento peninsula and the Ofanto river (Bari), and to the south by the Sibari plain. To the southwest and northeast the area is limited by the Tyrrhenian and Adriatic coasts, respectively. Data used to compile the map of Figure 2 are partly from the literature and partly newly acquired. The detailed geological profiles and stratigraphic sections constructed allowed us to correlate different formations and so obtain the new regional structural map of Figure 2. A description of the stratigraphic successions of the main outcropping tectonic units of the Apenninic Chain is given. The tectonic units forming the backbone of the Apenninic Chain are the result of deformation of palaeogeographic domains which differ in terms of both crustal characteristics and their position relative to the orogenic propagation. The progressive consumption of intra-plate ocean areas is marked by the development of foredeep basins.

Several authors have proposed reconstructions and evolutionary models for the Southern Apennines thrust belt (Ogniben, 1969; Scandone, 1972; D'Argenio et al., 1973; Pescatore, 1978; Ben-Avraham et al., 1990; Marsella et al., 1992; Finetti et al., 1996; Monaco et al., 1998; Patacca and Scandone, 2001; Lentini et al., 2002).

In the central Mediterranean region the Southern Apenninic orogen started during the Balearic geodynamic stage (Late Oligocene–Early Miocene) when the Alpine Tethys crust subducted under the European margin and the Corso-Sardinian block rotated anticlockwise. The recognizable structural domains are the foreland domain, the orogenic domain and the hinterland domain (Ben-Avraham et al., 1990; Lentini et al., 1994, 1995; Finetti et al., 1996). The foreland domain includes the still-undeformed continental areas of the Adria microplate (Apulian Platform), which had been separated since the Late Paleozoic–Early Triassic from the main African plate by the oceanic crust of the Ionian Neotethys. The orogenic domain of the Southern Apennines is composed of three stacked tectonic belts: the external, almost completely buried, thrust system (ET), the Apenninic chain (AC) and the Calabrides chain (CC). The ET is formed by thrust structures of the Apulian platform slightly displaced and still rooted, and by a late orogenic, out of sequence, "Mt. Alpi-like" highly tectonized and very prominently displaced thrust block. The AC is formed by imbrication of the sedimentary sequences belonging both to the oceanic-crust-type sectors (Alpine and Ionian Tethys) and to continental-crust-type sectors (inner carbonate platform: Apennine platform, coming from the Etrurian Promontory) and by detached-drifted fragments of the European margin.

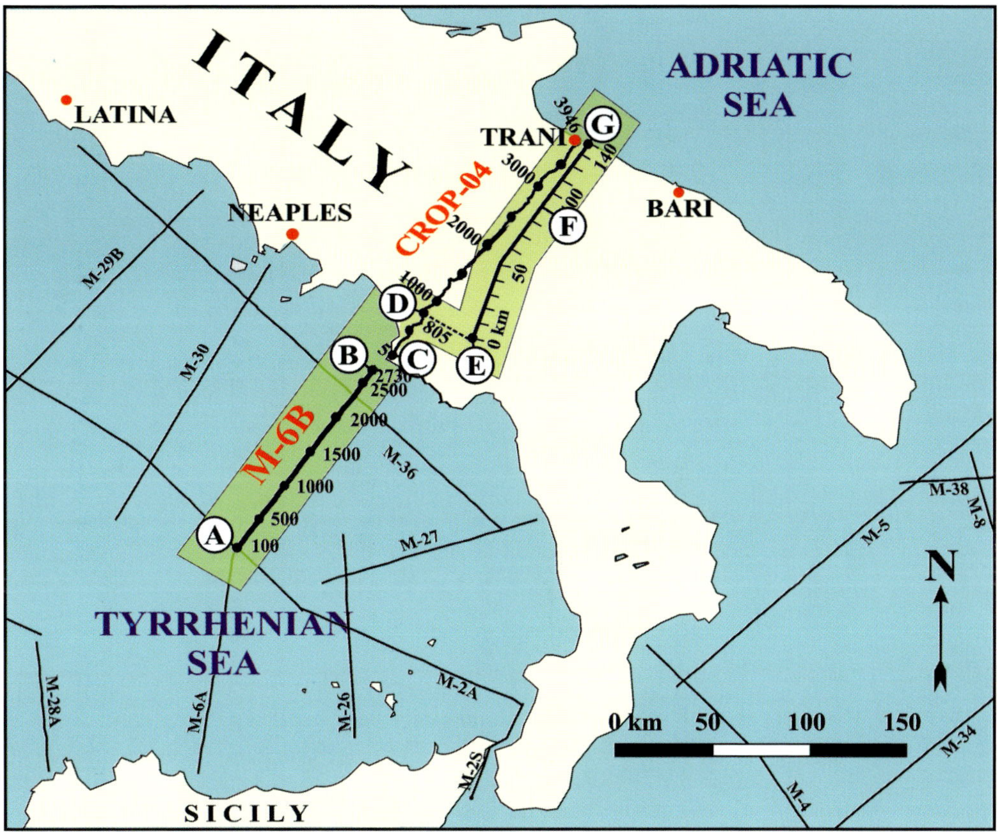

Figure 1: Location map for the reconstructed *Crustal Seismogeological Section* across the Southern Apennines, from the Tyrrhenian (Marsili basin) to the Adriatic Sea. *Section segments*: AB = CROP M-6B; BC = Gap between CROP M-6B and CROP-04; CD = Southwestern extremity of CROP-04 (CMP 5 to 805) from the literature; DE = connecting zone between CROP-04 (805 CMP) and the data of EF (courtesy of ENI-AGIP); FG = combined CROP data (correlation with CROP-04 and M-5). Point C is the initial zone of CROP-04, near Agropoli. Along AB are shown the shot point numbers; along CD the CMP numbers; along EG the progressive distance in km from E. Note that the connecting zone DE, being substantially parallel to the major structural features, in practice does not involve large problems of correlation between D and E.

The hinterland domain is represented by the Sardinian-Etrurian blocks and the Tyrrhenian basin. The later, in its abyssal plane zone, is floored by an oceanic crust, where the opening process started in the Middle Miocene.

The most evident tectonic characteristics of the Southern Apennines lie in its general duplex geometry. The hanging wall, several thousand meters thick, is made up of allochthonous units of the Apenninic chain, while the footwall is represented by thrusts of the Adria Plate edge. Only at Mt. Alpi does the Apulian thrust system outcrop.

2. REGIONAL TECTONO-STRATIGRAPHIC SETTING OF THE SOUTHERN APENNINES

A general description of the stratigraphy of the outcropping tectonic units of the Apenninic orogen is here given (Figure 2). Units of the Lucania sector, where the field data are well integrated with subsurface data, are carefully examined. The Apenninic Chain is exposed at numerous outcrops in the Southern Apennines. They consist of allochthonous east-verging thin-skinned thrusts including Mesozoic sedimentary sequences, detached from oceanic and continental crusts, and of Cenozoic flysch-type cover. A general structural characteristics of this allochthonous chain is the decoupling between the Mesozoic sequences and the Tertiary flysch-type terrigenous cover. In the external areas this thrust system overlies the preserved pre-orogenic Pliocene cover of the Apulian Platform (Figure 3). This regional setting is rather complicated in detail, but has been carefully studied and reconstructed with satisfactory reliability thanks to specific new geologic data, and by integrating seismic and borehole data.

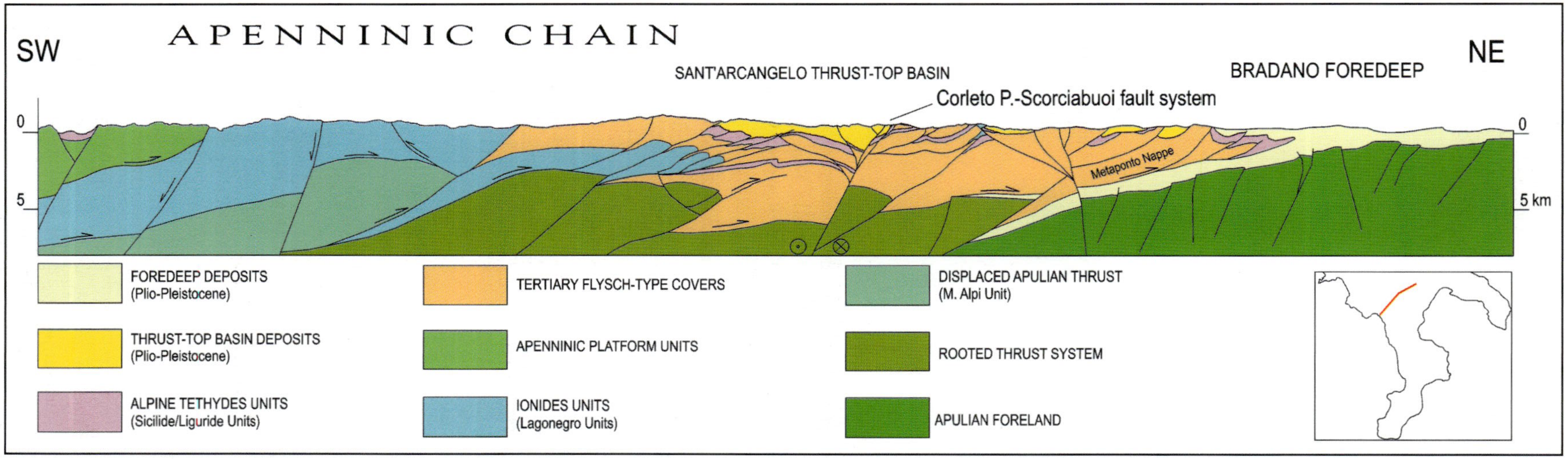

Figure 3: Schematic structural section across the Southern Apennines reconstructed from surface geology and subsurface oil exploration data.

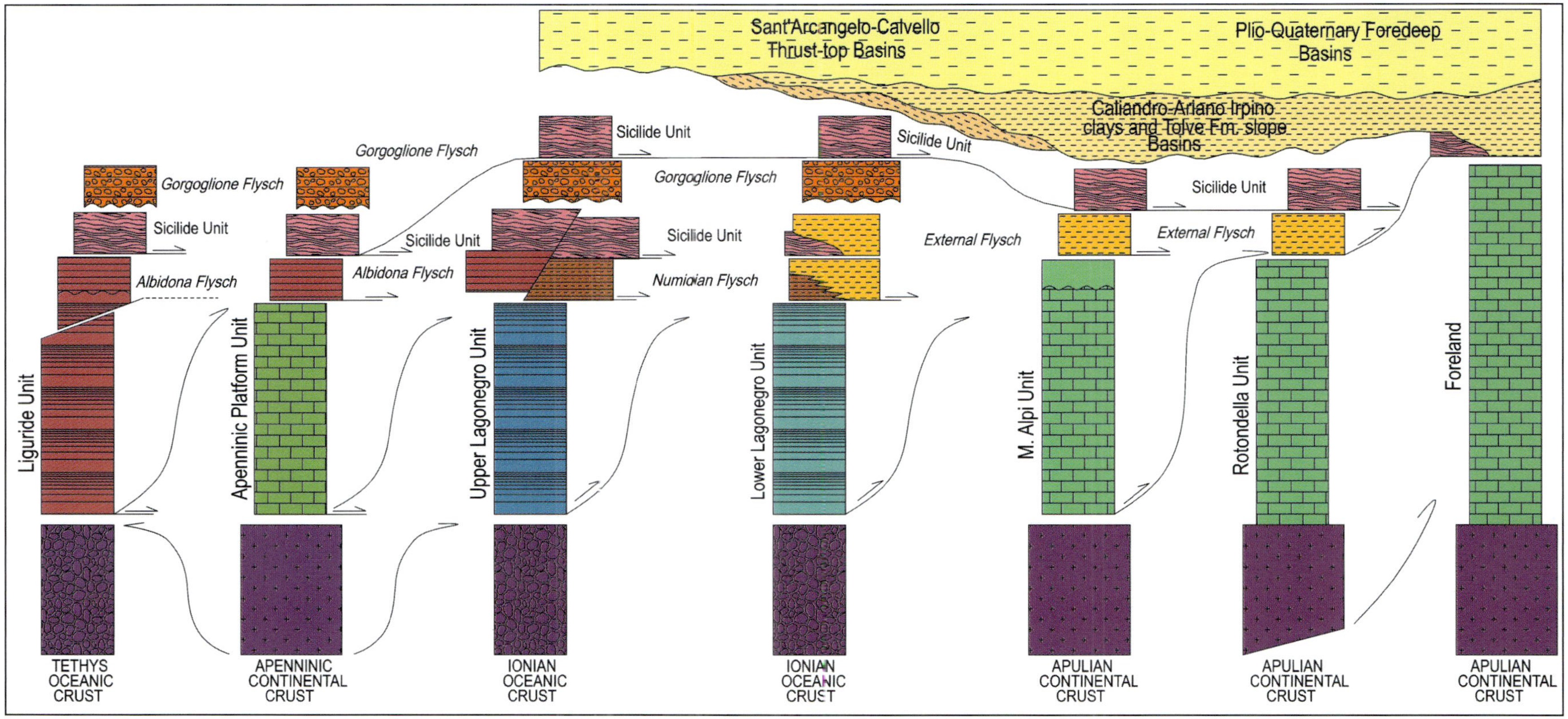

Figure 4: Schematic stratigraphic columns of the Southern Apennines Units located in their palaeogeographic positions. Major detachment planes and thrust polarity indicated by arrows. Two main basinal sequences can be distinguished: the Alpine Tethydes (Liguride and Sicilide Units) and the Ionides (Lagonegro Units). The Foreland is represented by the Apulian platform, whose inner margin is involved in thrusting giving rise to the Mt. Alpi and Rotondella Units (External thrust system). The Flysch-type cover of the Early Miocene, coeval with the Balearic stage, is represented by the thrust-top basin of the Albidona Flysch, and by the foreland deposits of the Numidian and External Flysch. The Flysch-type cover of Middle-Late Miocene is represented by the thrust-top basin deposits of the Gorgoglione Flysch, and is partially coeval with the External Flysch. These tectonic units travelled outward with the Sicilide Units and constitute the frontal wedge of the Apenninic chain.

The Apenninic Chain shows an architecture originating from polyphase tectonics which affected the sedimentary sequences involved in different times and different places. The assumption that the uppermost structural units should be the innermost is not always correct. Nevertheless, knowledge on the geology of this Apenninic sector has much improved in recent years, and it is now well known that the tectonic units forming the backbone of the Lucanian sector of the SAC are derived from the deformation of paleogeographic domains which differ both in crustal characteristics and in geometry and relationships with adjacent units and with respect to the orogenic front. The large amount of data used, in particular the new CROP data, allow us to outline the original palaeogeographic derivation of the sedimentary units (Plate 2 and Figure 4).

The geophysical, biostratigraphical, petrological and structural constraints are essential for defining the architecture of the orogen and for inverting its time–space evolution. In the Apenninic thrust belt it is possible to recognize two oceanic sequences, of which the more external one, the Ionian Meso-Cenozoic deep water succession comes from the subducted Ionian slab (Lagonegro Units) (Finetti et al., 1996; Finetti and Del Ben, 2000; Lentini et al., 2002; Finetti, this volume, Chapter 10). The innermost basinal deposits, the Ligurides-Sicilides units, or more appropriately the "*Alpine Tethydes*", derive from subduction of the Alpine Tethys oceanic crust, where the sedimentary cover was stripped off from its oceanic basement during the Balearic stage (Late Oligocene–Early Miocene). The Ionides were detached from the Ionian slab in the Tyrrhenian stage (Middle Miocene to Quaternary).

2.1. Apulian Platform

2.1.1. *Apulian Foreland.* The Apulian Foreland is formed from a very thick Permo-Meso-Cenozoic carbonate sequence and thin Plio-Pleistocene cover. A 1000 m basal Permo-Triassic sequence displaying alluvial and deltaic facies has been encountered by the Puglia 1 well (from 6100 to 7070 m depth) (Figure 5). These sediments overlie a pre-Cambrian crystalline basement, postulated from geophysical data (Agip Mineraria, 1968; Colombi et al., 1973; Cassinis et al., 1979; Morelli et al., 1979). In the uppermost portion of the sequence, shallow-water carbonate facies, bounded by an erosional surface, has been recognized. The whole succession presents a remarkable analogy with the "Verrucano facies" interpreted as a syn-rift sequence (Ricchetti, 1980) of the northern African Plate.

The Late Triassic anhydritic-dolomitic succession (Anidriti di Burano Auct.), which rests unconformably on the "Verrucano facies" is the result of sedimentation in a carbonate-evaporitic platform environment, under a syn-rift tectonic regime. Deep wells in the Apulian area have revealed varying thicknesses for this unit, oscillating between 1000 m in the Puglia 1 well (northeastern Murge) and 2500 m in the Foresta Umbra well (eastern Gargano), where young tectonic activity of the Gargano, and its antithetic Tremiti transcurrent faults, created a diapiric thickening.

The succession is followed in stratigraphic continuity by a Jurassic-Cretaceous carbonate sequence, more than 3000 m thick, with typical subsiding carbonate platform facies. The Jurassic sequence is characterized by calcareous dolomites of internal platform facies, while in the Gargano area the outcropping late Jurassic to Cretaceous sequences belong to external platform or basin facies. The interval from the Early Cretaceous to the Cenomanian is represented by a succession about 2000 m thick, which also shows an internal platform facies with intertidal micrites and episodic intercalations of rudist limestones. A stratigraphic gap, characterized in the surrounding areas by bauxitic levels and various continental deposits, marks a regional angular unconformity (Ciaranfi et al., 1988). Thicknesses of the successions above the stratigraphic gap oscillate between 1500 and 650 m for the platform and basin lithofacies, respectively.

In summary, the open sea "Upper Jurassic to Cretaceous system" shows an evolution to a platform-basin context connected with rift and passive continental margin dynamics; the "internal platform Cretaceous" is an expression of sedimentation that was constantly controlled by subsidence and eustatic factors induced by tectonics related to a passive margin regime (Masse and Borgomano, 1987; Ricchetti et al., 1988).

The Paleogene deposits display the typical features of coastal marine and scarp environments, with transgressive clastic carbonate facies along the Adriatic margin and toward the Bradanic foredeep, and show a distinct palaeogeographic modification corresponding to the emergence of the Mesozoic platform. The thicknesses of these successions vary from 350 m in the Gargano to about 100 m in the Salento. During the Eocene, modest intraplate

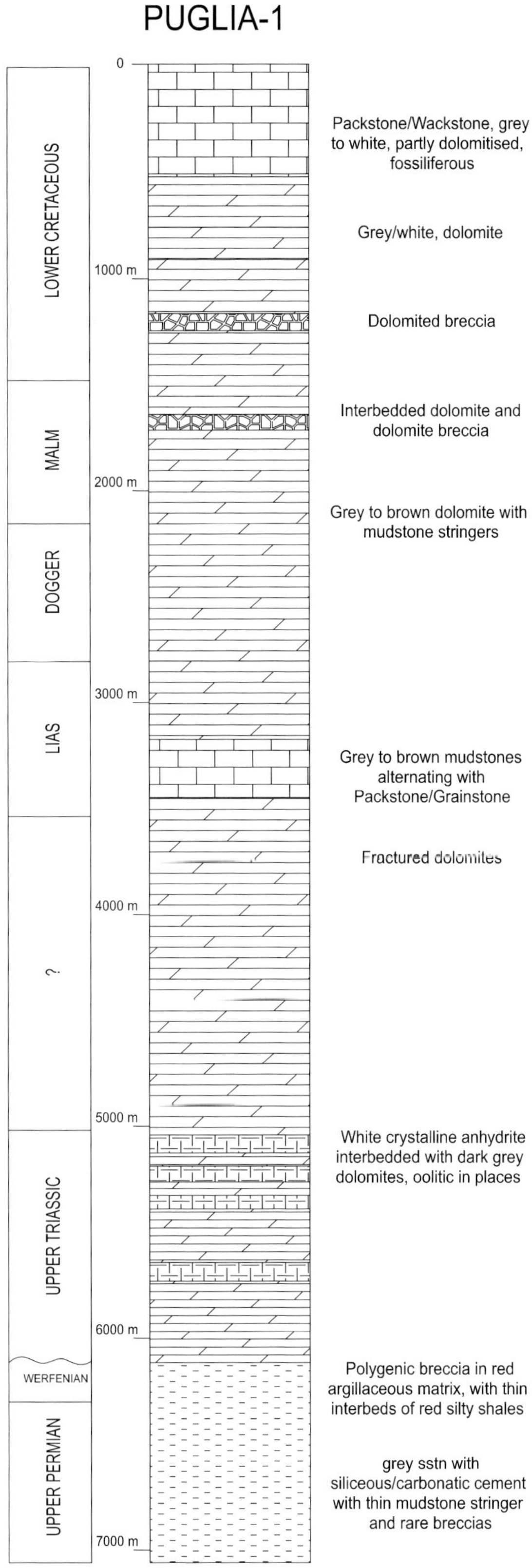

Figure 5: Stratigraphy of the Puglia-1 well located in the Apulia foreland domain.

volcanism occurred in the area (Carella, 1963; Dell'Anna and Pizzirani, 1964; De Fino et al., 1981), whose products constitute dikes aligned along fault systems, that lie on the western margin of the Gargano, and occur in the subsurface over vast areas of the Mesozoic carbonate body.

The Neogene-Pleistocene cover corresponds to different and distinct sedimentary events and is characterized by Serravallian-Tortonian organogenic and clastic carbonate facies, red vertebrate-bearing bauxitic clays (soils?),

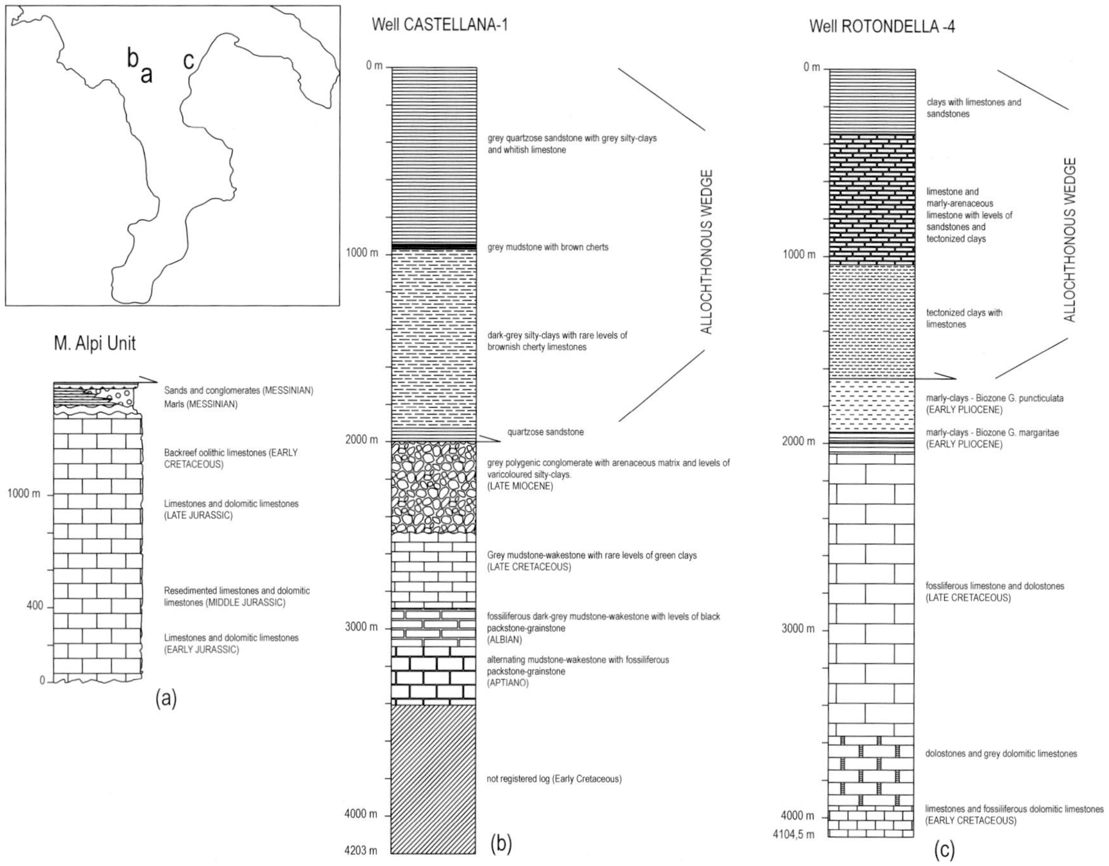

Figure 6: Stratigraphy of the External Thrust System Units. (a) Mt. Alpi Units, derived from surface data; (b), (c) Castellana-1 and Rotondella-4 boreholes.

oolithic limestones, marls and calcareous sands with thin diatomitic levels, that can be referred to the pre-evaporitic Messinian. The Plio-Pleistocene sequences consist of clastic glauconite-bearing carbonates typical of a neritic environment, of varying depth, and with a number of intervals of regressive sequences and stratigraphic gaps (Bossio et al., 1987; D'Alessandro et al., 1987). In contrast, the thick Upper Pliocene to Pleistocene calcarenitic-clayey successions (Calcareniti di Gravina e Argille subappennine Auctt.) correspond to sedimentation in a shallow neritic environment under a subsidence regime. They represent the most significant depositional event in the Apenninic foredeep.

Successions of Miocene to Middle Pleistocene age outcrop in the northern Gargano and in the southern Salento; those of Late Pliocene to Early Pleistocene age are widely exposed in Apulia.

The facies and geometry of the Neogene-Quaternary sequences reflect the tectono-sedimentary evolution of the Apulian foredeep-foreland system, where compressive bilateral effects of the Apenninic and Dinaric dynamics are accompanied by the Murge uplift and consequent active mantle bulge.

2.1.2. *External Thrust System.* The External Thrust System is almost everywhere buried, but well known thanks to hydrocarbon exploration. Numerous boreholes, below the allochthonous wedge of the Apenninic Chain, crossed horizons of variable age, ranging from the Late Miocene to Early Pliocene, which stratigraphically rest upon Cretaceous limestones deposited in a neritic environment. These sequences represent a portion of the Apulian cover overridden by the allochthonous wedge of the chain that followed the deformation of the External Thrust System of Apulia.

Mt. Alpi Unit. This unit (Ortolani and Torre, 1971; D'Argenio et al., 1973) is formed from a Mesozoic succession about 1000 m thick, with limestones, calcirudites alternating with dolomitic limestones that pass to oolithic limestones of Bajocian to Late Tithonian age (Sartoni and Crescenti, 1962) and Lower Cretaceous limestones of back reef facies (Figure 6(a)). They are stratigraphically followed by Miocene sediments that can be subdivided into two cycles (Ortolani and Torre, 1971). Those of the lower cycle paraconformably overlie the units of the substratum and have thicknesses varying between 15 and 30 m; they are formed from organogenic *Lithothamnia-* and *Pecten*-bearing calcarenites and marly limestones and marls, sometimes containing gravels, attributed in the literature to the Late Langhian to Serravallian. The second unconformable cycle has been dated to the Tortonian, and is characterized by conglomerates, calcarenites, sands and siltites of proximal marine and deltaic facies. Sgrosso (1988) attributes the first depositional cycle to the Messinian, due to the presence, from the base of the clastic sequence upward, of *Globorotalia acostaensis* (d) and *Globigerina multiloba* among the planktonic foraminifers, and *Amaurolithus primus* among the nannoplankton. The sediments of the upper cycle, 100–170 m thick, on the basis of the new chronological attribution by Sgrosso (1988) should not be older than Late Messinian in age. The Castellana 1 well penetrated about 400 m of Upper Miocene conglomerates below the allochthonous beds, lying on top of the carbonate sequence referred to the Mt. Alpi Unit (Figure 6(b)).

The Mt. Alpi sequence tectonically underlies the Liguride Units and, in part, remnants of the Lagonegro Units transported since the Late Messinian.

Rotondella Unit. This represents the sector of the deformed Apulian margin which is confined between the Mt. Alpi Unit to the west and the flexured foreland to the east. It is buried and can be recognized only on seismic lines and in hydrocarbon exploration boreholes (Rotondella 1-4). The top levels of the carbonate sequence are represented by Upper Pliocene marly clays. Probably it corresponds to the Apulian Units, which in the upper Val d'Agri constitute the Monte Alpi Oil Field (Figure 6(c)).

2.2. Apenninic Chain Units

The Apennine chain is formed by a nappe of complex stacked allochthonous units coming from two different Neotethyan basins, an interposed promontory of the NW Adria plate, the eastern margin of the Corso-Sardinian block, and accompanied prominent flysch sequences. This nappe covers the deformed structures of the Apulian platform defined as the "External Thrust System". The externalmost Neotethyan basin, from which derive the deep depositional units of the Lagonegrese succession (Ionides), is the Ionian Tethys, floored by an old oceanic crust of Late Permian–Early Triassic opening. The other more internal basin of derivation from the Alpine Tethydes units (AT) is the Alpine Tethys, floored by an oceanic crust which opened in Middle Jurassic times.

In between these two openly communicating oceanic basins there were two promontories, one of the Adria plate and the other of the Africa plate, which delimited the two basins to the north and south respectively (see paleogeographic sketch of Plate 2 and Finetti, this volume, Chapter 34). The NW Adria promontory, named here the "Etrurian Promontory" is the place of origin of the Apennine platform units tectonically transported in the Southern Apennines.

2.2.1. *Ionides.* The deepest tectonic units of the **AC** are made up of the Lagonegro sequences. They show basinal facies belonging to an original Ionian palaeobasin (Finetti, 1982; Finetti et al., 1996; Finetti and Del Ben, 2000; Lentini et al., 2002; Finetti, this volume, Chapter 10), and are widely exposed in tectonic windows (from north to south, respectively: Frigento-Monte Forcuso, Campagna, S. Fele, Agri Valley and Lagonegro-Monte Sirino) along the NW–SE culmination of the chain and tectonically underlie the Apenninic Platform Units (Figures 3 and 4). Starting from outcropping and buried Lagonegrese units of oil exploration boreholes in the southern Apennines it is possible seismically to discriminate and follow these tectono-stratigraphic units across the Gulf of Taranto area, extending as far as the Calabrian wedge and the undeformed Ionian sequence. *There is good seismically-controlled evidence that the Lagonegrese units correspond to the thin Pre-Miocene characteristic reflecting interval that covers the Ionian oceanic crust.* Taking into account that this interval fed a large quantity of tectonic units of the Southern Apennines–Calabrian arc–Sicily thrust belt, the more appropriate and immediately self-explanatory name of "***Ionides***" has been proposed (Finetti, 2002a, 2002b; Finetti, this volume, Chapter 10).

They show tectonic repetitions and form at least two major units, and are characterized by decoupling of the Oligo-Miocene cover, which, deprived of its Mesozoic substratum, forms sub-units that have advanced further outward. Classically interpreted as deposits of an original external basin, the Lagonegro sequences are in fact often exposed lacking their original Tertiary cover. This may be found, in an attempt to obtain a balance on a regional scale, within the enormous volumes of detached terrigenous formations, that were involved in a series of superimposed tectonic slices, which outcrop mostly in the external areas of the chain. In this sense, many of the flysch sequences that lack their Mesozoic substratum have been considered the original cover of the Lagonegro Units (Lentini et al., 1990; Carbone et al., 1988, 1991).

On the whole, the Mesozoic sequences constitute a thrust system extensively superimposed onto flysch intervals, which in turn override Late Miocene horizons at the top of the buried Apulian units. Numerous hydrocarbon exploration wells have penetrated an allochthonous layer that lies between the Lagonegro successions and the underlying Apulian units; this consists of flysch successions of Middle Miocene age. Such geometric relationships were attained during the final phases of deformation, when the entire Lagonegro thrust system overrode the frontal wedge of the chain.

Field mapping, carried out in the entire area between Lagonegro and the upper Val d'Agri (Figure 2), confirms that the Lagonegro successions exposed within the study area are characterized by extensive overthrusting, and can be largely referred to the units identified by Scandone (1972). The deepest of these can be ascribed to the Lagonegro I Unit (corresponding to the Monte Sirino Unit *sensu* Carta geologica del Bacino del Fiume Agri, 1991), while the upper one can be attributed to the Lagonegro II Unit, which is equivalent to the Mt. Torrette–Mt. Malomo Unit (Carbone et al., 1991). According to Scandone (1972) the different geometric positions correspond to different vertical evolutions of the successions, which show the characteristics of proximity to the platform domain in the upper unit, and more distal characteristics in the lower unit. During the Plio-Pleistocene tectonic phases the Apulian deep-seated duplex system influenced the roof thrust system, producing remarkable detachments.

The lowermost formation of the *Lower Lagonegro Tectonic Unit* is represented by the "*Monte Facito*" Fm. (Scandone, 1967). It consists of quartzarenitic-litharenitic terrigenous deposits of resedimented carbonate rock occurring both in a fine-grained and coarse-grained form, the latter associated with blocks, sometimes of notable dimensions. Furthermore, there are nodular limestones, tuffitic sandstones and locally terrigenous resediments and fine-grained carbonate breccias containing *Fusulina* and small Permian foraminifers. It is difficult to reconstruct the original stratigraphic succession, but the recognized ages range from the Scythian to the Middle Carnian (Donzelli and Crescenti, 1970; Wood, 1981; Miconnet, 1983, 1988; Ciarapica et al., 1986, 1990; Martini et al., 1989; Mietto and Panzanelli Fratoni, 1990; Marsella et al., 1991) (Figure 7(b)). Recently (Ciarapica and Passeri, this volume, Chapter 11), a re-examination attributes the basal deposition of the Monte Facito to the Late Permian.

The Monte Facito deposits are followed by the "*Calcari con Selce*" (*cherty limestone*) Fm., which is formed from gray, nodular siliciferous calcilutites. At various stratigraphic elevations there occur dolomitic limestones and dolomites with bands and nodules of chert, whose lateral and vertical passages are blurred. Locally there are banks of intra-formation dolomitic breccias and intramicritic calcirudites, which at times are oolithic, slightly dolomitizied or silicified, and contain radiolarians and sponge spicules. These correspond to the Pignola-Abriola facies of the Calcari con Selce Fm. of Scandone (1967, 1972), whose Carnian-Norian age has also been confirmed by Miconnet (1983, 1988) on the basis of conodont and radiolarian associations. The thickness varies from a few tens of meters to about 400 m.

Passing upwards, is the "*Scisti Silicei*" Fm. and an interval of siliceous claystones. The Scisti Silicei Fm. consists of an alternation of red siliciferous claystones occurring in thin beds, marls, marly limestones, and red, medium to fine grained calcarenites, that partly correspond to the Pignola-Abriola facies of the same named Jurassic formation of Scandone (1972) (Figure 7(b)). The thickness of this formation ranges from a few to some hundreds of meters, and the age of the formation is generally Jurassic (Crescenti, 1966; Scandone, 1963a, 1963b, 1967; Scandone and De Capoa, 1966).

In its upper part, the succession passes to blackish-brown claystones and siliciferous calcilutites with thin intercalated jasperine beds, that constitute the "*Flysch Galestrino*" Fm. of Late Jurassic to Early Cretaceous age, containing microfaunas marked by *Protopeneroplis* sp., *Trocholina* spp., rare *Calpionella*, Radiolaria and sponge spicules (Scandone, 1967). Brönnimann et al. (1971) note the presence of *Calpionella alpina*, *C. elliptica* and *Remaniella cadischiana*, indicating a Berriasian age. The thickness of the Flysch Galestrino Fm., generally estimated at about 300–400 m, due to detachment and intense folding, shows notable variations.

The siliceous claystones are exposed on the eastern slope of Mt. Tangia, at Mt. La Ricciola and at Mt. Malomo, where the Lagonegro succession is characterized by Jurassic–Paleogene intervals with siliceous claystones and radiolarites passing upward to the *Mt. Malomo* Fm. The microfaunas recognized in this portion (radiolarians and small Globotruncanidae) extend the age to at least the Turonian. This formation is therefore partially coeval with the Galestri Fm., and where the latter pinches out, the upward passage leads directly to the Mt. Malomo Fm. The

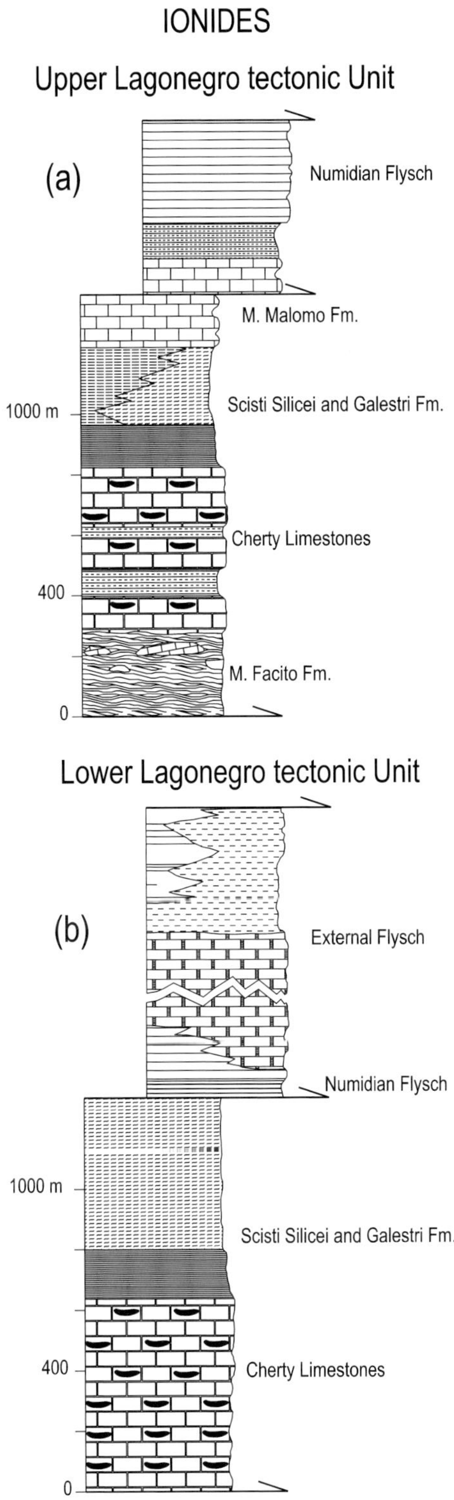

Figure 7: Stratigraphy of the Ionides (Lagonegro Units). The Cenozoic cover is tectonically decoupled from that of the Mesozoic, in particular (a) the Numidian Flysch Units are detached from the Mt. Malomo Fm. and (b) from the Galestri Fm.

siliceous claystones seem, in part, to correspond to a lower interval of the "*Flysch Rosso*" Fm. and to the Scisti di Pecorone of Scandone (1967). The comprehensive thickness of the described sequence can hardly be evaluated but it does not seem to exceed 200 m.

In the area of Pergola, at the contact with the underlying Scisti Silicei Fm., some recognized associations in the Galestrine Flysch allow them to be attributed to the Jurassic–Early Cretaceous, confirming the Late Jurassic as the oldest age for this formation (Marsella et al., 1995). About 100 m above the base, in the rare horizons of graded, fine-grained arenitic breccias, a significant microfaunal association has been observed, which is characterized by *Coscinoconus alpinus*, *Protopeneroplis striata*, *Nautiloculina oolitica*, *Trocholina* sp., *Ammobaculites* sp., *Robulus* sp., rare *Calpionelle* and nannofloras characterized by *Cruciellipsis cuviellieri*, *Watzanaueria barnesae*, *Nannoconus steinmanni*, and *Stradneria crenulata* of the Early Cretaceous. Miconnet (1988) recognizes, from the lowest levels of the succession onward, associations of *Calpionella* of Berriasian age. At Brindisi di Montagna, nannofloras of *Micula* spp. and *M. stauropora* extend the age of this formation to the Turonian. The thickness varies from 0 to 200 m, and it is likely that the formation is laterally substituted by the siliceous claystones that overlie the Scisti Silicei Fm.

To the northwest of Abriola the gradual transition to the "*Monte Malomo*" Fm. is visible. This is characterized by increase in the marly and carbonate fraction. At Mt. Malomo itself the type succession is exposed. It is represented by a dense alternation of clays and red, sometimes also green and gray, marls, of blackish siliciferous claystones, whitish calcilutites, calcarenites and calcirudites with rudist fragments and macroforaminifers. The formation is well exposed along the monocline of the Mt. Malomo ridge. The calcarenitic fraction, which is characterized by recrystallized grainstones, is seen to be unfossiliferous, while the most fine-grained component is a mudstone within which a calcareous nannofossil association shows this unit to be Late Cretaceous. The thickness of the formation varies from area to area due to tectonic repetitions or elisions, oscillating between 0 and 250 m.

Further to the south, near Nova Siri, an alternation of calcarenites, gray calcirudites and red, gray and green claystones and marls is exposed. It can be interpreted as a more carbonatic lithofacies of the Mt. Malomo Fm. (Nova Siri Limestones) of the Late Oligocene. In continuity with the Mt. Malomo Fm. there is a passage toward brown claystones, marly clays and red and green marls, which occur in thin strata and contain intercalations of fine quartzose sandstones, dark gray bioclastic calcarenites, which at times, due to detachment, rest directly on top of the Scisti Silicei Fm. (Mt. Tangia). In the calcarenites, foraminifera like *Amphistegina* sp., *Miogypsinoides* sp., *Lepidocyclina* sp., *Heterostegina* sp. and Globigerinidae as well as Algae (*Lithothamnium*), Bryozoa and echinoids, show this outcrop to be Late Oligocene. This formation represents the basal interval of the Numidian Flysch decoupled from the Lagonegro successions and outcropping in more distal areas. Its age is Late Oligocene, and its thickness can be estimated at around 100 m. The Numidian Flysch (Ogniben, 1969) (Arenarie di Stigliano by Selli, 1962) is represented by a clayey-quartzarenitic alternation several hundred meters thick (Figure 7).

This formation extends from Sicily, where it is subdivided into repeated tectonic units with thicknesses of some thousands of meters (Bianchi et al., 1987), to the southern Apennines, where it occupies an "external" belt from Lucania to Molise, even though occurring only in limited and discontinuous units (Ortolani, 1974; Narciso et al., 1975; Ciaranfi and Loiacono, 1983; Patacca et al., 1992a, 1992b).

The Numidian Flysch has been interpreted as the Lower Miocene terrigenous cover stratigraphically overlying the so-called Late Cretaceous-Paleogene "Flysch Rosso" at the top of the Lagonegro Units (Scandone, 1967, 1972; Ogniben, 1969; D'Argenio et al., 1973; Cocco et al., 1974; Carbone et al.,1987, 1988; Servizio Geologico d'Italia, 2002). The original sequence is preserved along the outer border of the chain, from Potenza toward the north, and in some restricted areas like Pergola (Marsico Nuovo village), but the Numidian Flysch is mostly exposed completely detached from its original substratum and is involved in a series of imbricated slices. In the Agri Valley between Corleto Perticara and Stigliano, at least two structural slices have been recognized with the help of seismic lines and boreholes. At the base of the uppermost slice, extremely tectonized levels of Mt. Malomo Fm. are observed, and this indicates to recognize the original stratigraphic relationship before detachment.

The entire Cretaceous-Miocene sequence is similar to that which characterizes the thinner Sannitic Units (Selli, 1962), outcropping in the Sannio, Molise, Lucania, and in the Maddalena Mountains. To the arenites of Numidian type of the Sannitic units Patacca et al. (1992a, 1992b) attribute a Langhian age, while they date the overlying arkosic sandstones to the Serravallian.

As a whole, the outcropping Numidian sequences display a duplex geometry formed by a series of imbricated slices, that developed below the tectonic contact with the Sicilide mélange, resting on top of the Burdigalian horizons of the more internal Numidian successions, and unconformably overlain by the Middle Miocene levels of the Gorgoglione Flysch. This structural horizon rests tectonically on top of further buried flysch-type sequences, which from well logs are generally attributed to the Langhian.

In the Lucanian sector, the Numidian Flysch appears widely along the ridges of Campomaggiore-Stigliano, where it is truncated at the Burdigalian level by the Sicilide nappe of Cirigliano (Lentini, 1979), and further south-east on the Valsinni ridge, where it passes upward to a marly-clayey-sandy alternation (Marne arenacee di Serra Cortina, Ogniben, 1969) of Late Burdigalian to Early Tortonian age (Servizio Geologico d'Italia, F. Pisticci, in press). The Numidian Flysch Units represent part of the frontal wedge in the crustal profiles (Plate 2).

The micropaleontological content is generally scarce, the quartzarenitic intervals are observed to be sterile, while microfaunal associations have been recognized in the pelitic fraction which contain *Paragloborotalia mayeri, P. opima continuosa, Globigerinoides trilobus*, and *Globoquadrina dehiscens* of the Early Miocene. Nannofossils are represented by *Helicosphaera carteri, H. ampliaperta, H. euphratis, Sphenolitus moriformis, S. conicus* and *Discoaster deflandrei* of the MNN2b Biozone of Fornaciari and Rio (1996), indicating an early Burdigalian age. The Numidian Flysch evolves upwards into lithologic successions characterized by arkose alternating with calciclastic sandstones (Serra Palazzo Fm., attributed to the Serravallian; Palmentola, 1967, 1970; Patacca et al., 1990), or a dense alternation of marls and clays with thin arkosic levels (Marne arenacee di Serra Cortina, Ogniben, 1969, attributed to the Burdigalian-Helvetian by Lentini, 1979, 1980, or to the Langhian-Serravallian by Carbone et al., 1991).

Further to the east, the Numidian Flysch is exposed in tectonic slices oriented more or less N–S from Rotondella to Murge S. Caterina (Figure 2); it is tectonically interposed between Sicilide units. All outcrops along this tectonic alignment form a regular upward transition to the Rotondella Fm., confirming the stratigraphic passage between the Numidian Flysch and the Serra Palazzo Fm. of Boenzi et al. (1968) and Palmentola (1967), and thus the original relationships with the Burdigalian horizons of the External Flysch (Figure 6(a)).

2.2.2. *Ionian Flysch-type sequences (Irpinian Basin).* These sequences include the entire Middle-Upper Miocene turbiditic sequence (Irpinian Units of Cocco et al., 1972, or "Flysch Esterni" after Carbone et al., 1991), which outcrops in the external sectors of the Apenninic thrust system, and are tectonically superimposed onto the Pliocene horizons of the Bradano foredeep. To these sequences were also ascribed the successions resting unconformably on top of the allochthonous cover of internal origin such as the Gorgoglione Flysch. The entirely of these successions has been interpreted by Cocco et al. (1972) as the original deposits of the foredeep, which became distinct from the Langhian onward. The Middle Miocene successions exposed in the frontal areas are grouped in the present paper as the tectonic units of the External Flysch. The Gorgoglione Flysch would apply to the Irpinian basin, which had already been deformed during the early Langhian.

At the current state of knowledge, the original relationships between the *External Flysch* and the Lagonegro sequences seem to be supported by ties on a regional scale, which should include an evaluation of the effective volumes lacking in the balance of the "denuded" Lagonegro successions, and the compatibility of the age of the earliest deformation of the detached terrigenous series with their presumed position in external areas. The External Flysch in effect seems to be sufficiently linked, by the Tortonian age of the deformation and by its position on the front of the allochthonous wedge, directly covering the buried Apulian successions, to an original position on top of the Lagonegro successions. However, continuity between the External Flysch and the Apulian platform is suggested by the mixed siliciclastic-carbonate compositions (Ogniben, 1969; Palmentola, 1970), which indicate a gradual invasion of the siliciclastic fraction of internal origin into the more external areas of the Irpinian basin.

To the *External Flysch* are here ascribed those sequences characterized by a mixed carbonate-siliciclastic composition, which correspond mainly to the Serra Palazzo Formation according to Selli (1962) and presently form an imbricated system in the frontal wedge of the **AC** (Figure 2).

When describing these sequences, one must not ignore the position they occupy within the Middle Miocene foredeep, which can be now recognized in the various tectonic units outcropping in the Lucanian and Foggia sectors of the Southern Apennines.

Ogniben (1969) subdivided the Serra Palazzo Fm. into three lithostratigraphic units: a basal, calcareous one (Masseria Palazzo Fm.), a middle, prevalently marly one (Serra Palazzo Fm. s.s.), and an upper, arenaceous-clayey one (Masseria Luci Flysch), all of Langhian to Tortonian age. Lentini and Vezzani (1974) considered the External Flysch a continuous succession, extending from the Aquitanian to the Early Tortonian, and represented by the "Marne arenacee di Serra Cortina", the "Masseria Palazzo Fm.", the "Serra Palazzo Fm." and the "Masseria Luci Flysch". They believed that these formations were the lateral equivalents to the Gorgoglione Flysch, deposited on a deformed substratum.

Recently (Servizio Geologico d'Italia, 2002) two tectonic subunits have been distinguished within the External Flysch. The upper of these is characterized by terrigenous successions still dominated by siliciclastic components and is indicated as the Masseria Luci Subunit, while the lower one, formed from limey-marly sequences has been named the Serra Palazzo Subunit. These two subunits, considering their actual position, can be respectively referred to the internal margin formed from the chain units, and to the external margin represented by the boundary of the Apulian platform of the original foredeep basin.

The *Serra Palazzo Subunit*. A basal member of calcareous turbidites, about 500 m thick, and an upper marly-clayey member, that contains intercalations of micaceous and arkosic sands, about 300 m thick, can be recognized. To the lower member has been attributed part of the successions previously believed to belong to the Masseria Palazzo Fm.

In the calcareous member, microfaunas characterized by *Globigerinoides bisphericus*, *Praeorbulina glomerosa*, *Orbulina universa* and nannofossils of the MNN5b-MNN6 Biozone of Fornaciari et al. (1996) indicate an age ranging from the Langhian to the Serravallian. In the clayey top levels, nannofloras referable to the MNN8 Biozone of Fornaciari et al. (1996), extend the age of the formation to the early Tortonian.

The uppermost structural level of the External Flysch Unit is represented by the Masseria Luci Flysch (*Masseria Luci Subunit*), made up from quartz sandstones of Numidian type, evolving into arkoses alternating with marls. This flysch has been previously interpreted as the apical formation of a unique succession of the External Flysch, resting stratigraphically on top of the marls of the Serra Palazzo Formation. Recently (Servizio Geologico d'Italia, 2002; Lentini et al., 2002) the tectonic nature of the contact has been recognized at the base of the formation, which is furthermore evident from attributing a Burdigalian-Langhian age to the units ascribed to the Masseria Luci Flysch, and an early Serravallian age to those of the underlying Serra Palazzo Fm.

In the context of the geodynamic evolution of the Apenninic Chain, these successions seem to represent those levels onto which the detached sequences of the more internal portion of the Numidian Flysch were transported, with the overlying Varicoloured Clays of the "Cirigliano Unit" and the Gorgoglione Flysch, before these were also involved, from the early Tortonian, in the thrust system.

The nannofossil assemblages belong to the MNN3a Biozone of Fornaciari and Rio (1996) and show them to be Middle Burdigalian. The formation grades up to an alternation of whitish arkosic sandstones and macroforaminifera-bearing calcareous turbidites, with thin intercalations of marls and marly calcilutites, in which the nannoflora, referable to the MNN4a Biozone of Fornaciari et al. (1996), indicates a Late Burdigalian age. The top levels of the formation are characterized by nannofossil associations, referable to the MNN5a Biozone of Fornaciari et al. (1996) of Langhian age. The overall age range of the formation thus extends from the Middle Burdigalian to the Langhian.

In the easternmost portion of the Apenninic Chain (Foggian Apennines) the Irpinian deposits consist of turbiditic sequences resting in continuity on top of the Numidian Flysch. From the internal to the external areas these deposits are represented by the S. Bartolomeo Flysch, the Serra Palazzo Fm. and the Faeto Flysch, all of Late Langhian to Serravallian age (from the *Orbulina suturalis–Globorotalia peripheroronda* Zone to the *G. menardi* s.l. Zone), which are followed in continuity by Lower Tortonian sediments (Marne argillose del Toppo Capuana of Crostella and Vezzani, 1964).

The *S. Bartolomeo Flysch* (Crostella and Vezzani, 1964) is a piggy-back sequence (Dazzaro and Rapisardi, 1996) about 500 m thick. In the foredeep, the Numidian Flysch is followed by the arenaceous-calcareous-marly sedimentation of the *Serra Palazzo Fm.*, which in these areas reaches a thickness of about 400 m. The *Faeto Flysch* (Crostella and Vezzani, 1964) represents a foreland basin deposit, formed nearly exclusively by about 400-m-thick calcareous-marly turbiditic or hemipelagic sediments, where siliciclastic supply was lacking (Santo and Senatore, 1988).

The *Gorgoglione Flysch* (Ciaranfi, 1972) is a terrigenous deposit resting unconformably on various units of the chain, which Boenzi and Ciaranfi (1970) attribute to the Langhian–Early Tortonian. This flysch as a whole represents the only terrigenous cover of the Lucanian Apennines which, after the final tectonic transport, maintained its original relationship with the units onto which it was deposited. For this reason, this formation assumes a fundamental importance for the analysis and reconstruction of the deformational events on a regional scale (Figure 4). It is dismembered into two tectonic units (Lentini, 1979). The external of these remained solid with the units of the substratum consisting of the Sicilide Units (Cirigliano Unit) and form a nappe on top of the Numidian Flysch.

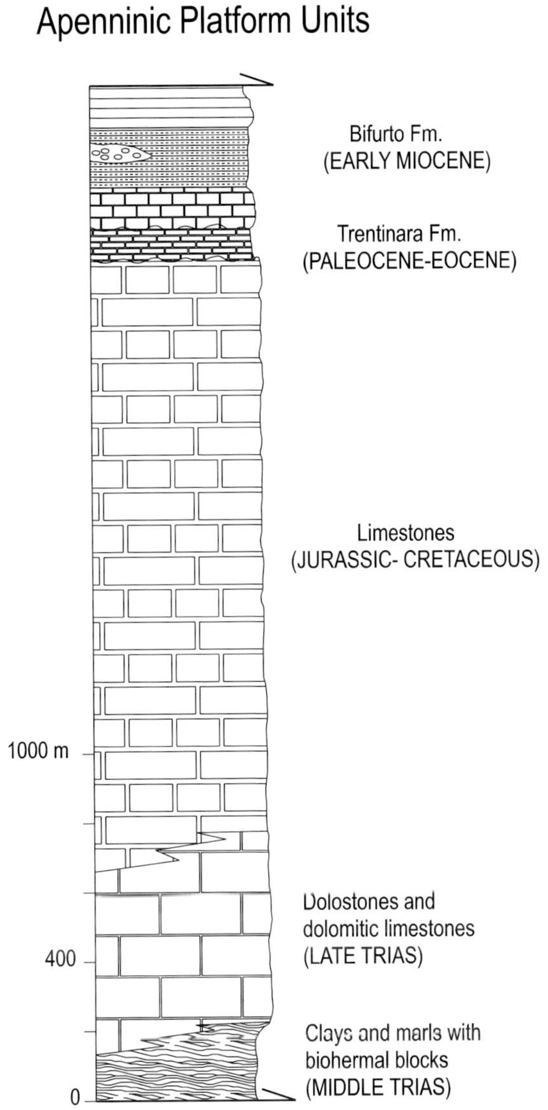

Figure 8: Schematic stratigraphy representative of the Apenninic Platform Units, reconstructed from surface data.

The lower unit forms the widespread flysch cover of Pietrapertosa-Gorgoglione. This formation, up to 1200 m in thickness in the type area (Carta geologica del Bacino del Fiume Agri, 1991) outcrops widely in the section of the Southern Apenninic orogen from the Cilento (cf. Monte Sacro Fm.) to the Taranto gulf (cf. Oriolo Fm.).

There is general agreement regarding the age of the formation as covering the interval from the Langhian to the Early Tortonian (Boenzi and Ciaranfi, 1970; Pescatore, 1978; Lentini, 1979; Lentini et al., 1987; Carbone et al., 1991), although Patacca et al. (1990) assign the entire formation from the base upward a Tortonian age. Evidence of reworking is widespread.

The *Castelvetere Flysch* is a succession with a thickness of several hundred meters, formed from polygenetic conglomerates with sandy matrix, generally evolving at the base into turbiditic sands and marly sands, sometimes with calcareous olistoliths associated with flows of Varicoloured Clays.

2.2.3. *Apenninic Platform Units.* The tectonic units relative to this structural domain are characterized by variable thicknesses of mostly Mesozoic carbonate platform facies, outcropping with discontinuity in the Southern Apennines. These units were detached from their original basement in the Lower Miocene, migrated southeastward in the Tyrrhenian stage across the subducting Ionian sea, and were thrust obliquely onto the Southern Apennines. With regard to their provenance, various schematic proposals have been made in the past. Ogniben (1969) attributed these units to the Panormides, but we know now that the Tyrrhenian magnetic anomalies, which recorded the drift routes of migrated tectonic units, exclude such a hypothesis. Finetti et al. (1996) proposed a paleogeographic scheme where the Apenninic platform units originated from a crustal block located to the north of the Ionian neotethys, close to, but detached from the NW Adria plate.

From the CROP seismic data and the clear kinematic information on the Tyrrhenian Magnetic field, it is now possible to produce a new, much more tightly constrained paleogeography which indicates that *the Apennine platform tectonic units come from the NW corner of the Adria plate called the "Etrurian Promontory"* (Finetti, this volume, Chapter 34). Detached Apennine platform units are called "Etrurian Platform units".

In the following, these "Apennine" (or "Etrurian") platform units will be described.

Marzano Mt.–Maddalena Mts. Unit. This is derived from deformation of the probably more external sectors of the original Triassic-Miocene Etrurian platform in transition to the Lagonegro basin (Scandone, 1964a, 1964b, 1972; Scandone and Bonardi, 1967; D'Argenio et al., 1973) and outcrops in the Lagonegro area at the boundary between Campania and Basilicata. The Maddalena succession is formed from 1000–1500 m of shallow-water dolomites and dolomitic limestones of the Late Triassic to Early Lias, followed in discontinuity by several hundred meters of organogenic limestones of the Late Jurassic and clastic platform margin deposits of Liassic–Early Cretaceous age, and Upper Jurassic to Eocene rudist-bearing limestones and carbonate resediments. The topmost levels of the unit, paraconformably resting on the underlying levels, are Burdigalian glauconitic calcarenites (Calcareniti di Laviano, Selli, 1957; Patacca et al., 1992a), Lower Tortonian turbiditic siliciclastic sediments (Formazione di Monte Sierio, Sgrosso, 1988) and thin arenaceous-conglomeratic levels attributed to the Castelvetere Flysch (Pescatore et al., 1970) of the late Tortonian (Patacca et al., 1990). Sgrosso (1992) attributes the siliciclastic deposits that rest at the top of this succession to the Late Tortonian. This age gives a significant clue to the palaeogeographic collocation of this tectonic unit, which would thus have been affected by the first tectonogenetic phase much more recently and more directly than those which are observed in external areas (Apulian platform).

In the upper Val d'Agri, the Mesozoic sequence of this carbonate unit is characterized by extremely tectonized dolomites, oolithic limestones, and biogenic rudist limestones.

Picentini Mts. Unit. This unit outcrops from the Sorrento peninsula to the ridge of the Picentini Mountains (Bonardi et al., 1988a). It is formed from several hundred meters of dolostones, marls and calcarenites, and bituminous shales (Late Triassic–Early Lias). It is followed by a few tens of meters of calcarenites and breccias containing *Ellipsactinia* and nodular limestones (Liassic–Early Cretaceous). The Mesozoic sequence terminates with rudistid limestones (Late Cretaceous). The Tertiary portion of the succession is represented by several tens of meters of Paleocene–Lower Miocene calcarenites and calcirudites, that contain macroforaminifers; several hundred meters of clays and marls with intercalations of graded Langhian calcirudites, about 400 m of turbiditic sands with calcilutites and *Orbulina* – bearing marls and olistoliths of lithothamnia-bearing limestones and Numidian-type sands (Serravallian).

Foraporta Mt. Unit. Originally described as "Monte Foraporta-Monti della Maddalena Unit" (Scandone, 1972; D'Argenio et al., 1973), this results from the deformation of the southern margin of the Etrurian carbonate platform, characterized by deposits in proximal basin facies. The horizons of this unit outcrop in proximity to the tectonic windows of Lagonegro, along the valley of the Noce river, and to the upper Val d'Agri, between Moliterno and Casalbuono, in several tectonic slices interposed at the base within the Triassic white dolomites of the units of the Maddalena Mountains or directly within the Lagonegro units and the Alburno-Cervati units at the top.

The various tectonic elements of this unit consist of a succession of basinal to restricted circulation facies with a thickness of about 400 m, which extends from the Late Triassic to the Early Lias, with dolomites and gray and black limestones, and to the Dogger–Malm?, with dark limestones and carbonate resediments, at times with gravel (Boni et al., 1974; De Alfieri et al., 1987). The basal levels are represented by greenish shales with biohermal blocks, sands, radiolarites of Middle Trias age (cf. "Moliterno succession" in Carta geologica del Bacino del Fiume Agri, 1991).

Alburno-Cervati-Pollino Unit. The Alburno-Cervati Unit consists of numerous tectonic elements, which can be referred to the central portion of the Etrurian carbonate platform (Scandone, 1972; D'Argenio et al., 1973; Bonardi et al., 1988b). It is exposed, with notable facies variations (shallow water, slope to basin), from the Pollino massif and from the Calabrian coastal chain to the south to as far north as the Lattari Mountains of the Sorrento peninsula, extending over a length of about 300 km. The type succession consists of about 4500 m of dolomites and dolomitic limestones, that can be dated from the Late Triassic to the Jurassic, and of platform limestones of Middle Liassic to Late Cretaceous age. The Paleogene interval is represented by about 100 m of limestones and marls (Trentinara Fm., Selli, 1962). The succession is closed unconformably by calcarenites (Cerchiara Fm.) of Early Miocene age, which rapidly pass upward to siliciclastic deposits.

Various works (De Blasio et al., 1981; Sgrosso, 1981; Santo and Sgrosso, 1987; Patacca et al., 1992a) extend the age of the siliciclastic deposits which constitute the sedimentary cover of this carbonate succession, to the Tortonian.

Capri-Bulgheria-Verbicaro Unit. This unit outcrops in discontinuous manner from the Tyrrhenian coast to northern Calabria and is considered to be derived from the dismembering of the western margin of the Etrurian carbonate platform. It consists of more than 2000 m of platform carbonates in slope and base-of-slope facies, characterized by wide stratigraphic gaps, which extend from the Late Triassic to the Cretaceous, with dolomites and dolomitic limestones (Late Triassic–Early Lias), "Calcari ad Ellipsactinie", carbonate resediments and limestones with gravel of the Middle Lias to Early Cretaceous. These are followed by Upper Cretaceous limestones with gravel and calcareous resediments and marls in "scaglia" facies, that grade upward into Tertiary calcareous-dolomitic sequences in cliff and platform or basin facies. In stratigraphic continuity on top of the condensed deposits in "scaglia" facies there follow calcarenites and terrigenous clastic deposits of Early Miocene age (Brancaccio and Cinque, 1992).

S. Donato Unit. The strata of the S. Donato Unit (Amodio-Morelli et al., 1976; Ietto and Barilardo, 1993) outcrop from the coastal northern Calabria chain to the southwest of Monte Pollino and have been affected by green-schist Barrovian-type metamorphism during the Burdigalian tectonic phase. The sequence consists of Middle Triassic terrigenous deposits (metarenites and metapelites) with lenticular intercalations of massive algal metalimestones and metabasites overlain by Carnian-Paleogene (?) metalimestones and crystalline dolomites with rare intercalations of alkaline metabasites in the lower part. The thickness is about 2000 m. The top levels are represented by a few meters of Neogene siliciclastics (graded metarenites and metapelites, Aquitanian-Burdigalian in age) (Patacca et al., 1992a).

Ietto and Barilardo (1993), on the basis of evident affinities of these Triassic sequences to the less metamorphized ones, hypothesize that they are coeval with the Lagonegro sequences, and belong to the same palaeogeographic domain. Marsella et al. (1995) subscribe to this palaeogeographic reconstruction. We may assume that this unit belonged to the Etrurian promontory–Ionian ocean boundary zone.

2.2.4. *Alpine Tethydes.* These distant allochthonous units are characterized by sedimentary sequences which deposited in the Alpine Tethys basin, located between the European plate and the NW Adria plate (Etrurian promontory). They have been tectonically transported on the Etrurian Platform units and on the Ionides, reaching the frontal wedge into the Bradanic Foredeep.

It is convenient to divide the ophiolitiferous and semimetamorphic units (Frido, Cilento Units) from those sequences characterized by cover, that can be distinguished by the occurrence of tuffitic sands derived from the erosion of an Oligocene volcanic arc (Sicilide Units). This subdivision is facilitated by the presence of terrigenous cover interposed between the earlier two and can be grouped under the name of "Albidona Flysch".

Liguride Units. These units outcrop widely from the Calabria–Lucania boundary to the Cilento; they occupy an elevated structural position in the Apenninic chain and are unconformably overlain by Lower Miocene flysch-type deposits of the Albidona "family".

Originally defined by Selli (1962) as nappes of the Cilento and northern Calabria, they represent the southernmost ophiolitiferous nappe of the Apennines, covered tectonically by the crystalline Calabride Units (Figure 9).

Since the early 1970s, a vast number of publications have proposed stratigraphic, structural and interpretative reorganizations of the Liguride units. Firstly, the tectonic superimposition of the metamorphic units and the overlying sedimentary successions was recognized. The stratigraphic continuity between the metamorphic units of the "Frido" and overlying sedimentary successions has been discussed by various authors (Scandone, 1972; Bousquet, 1973; D'Argenio et al., 1973; Amodio-Morelli et al., 1976; Lanzafame et al., 1978), who considered

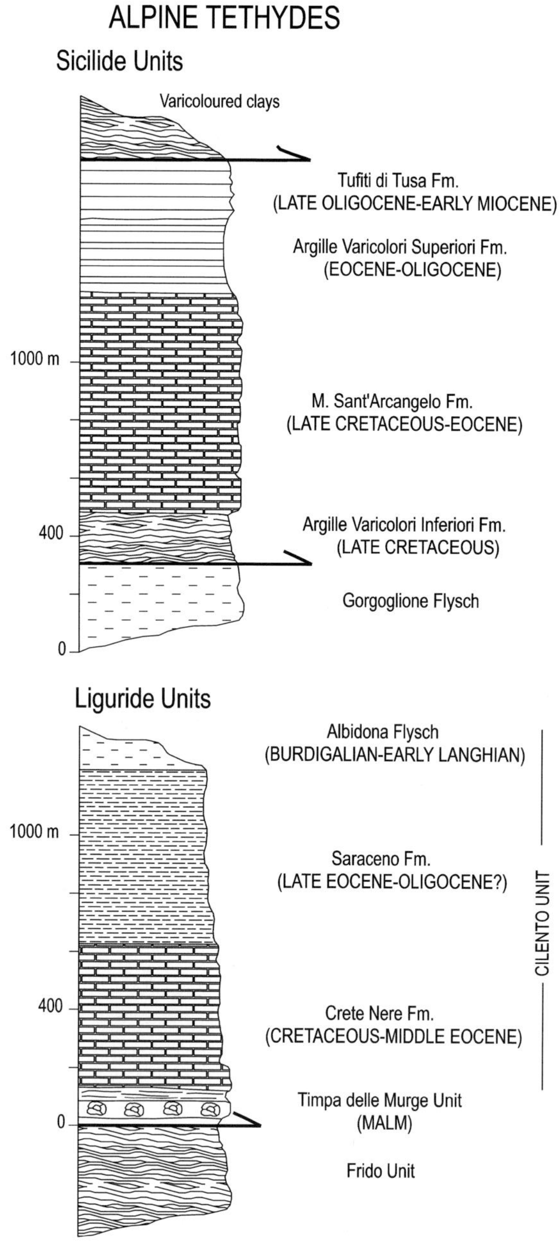

Figure 9: Stratigraphy of the Alpine Tethydes Units reconstructed from surface data.

the superposition contact between the two terrains a tectonic one, and subdivided the Liguride Complex into two different tectonic units: the geometrically lower Frido Unit with low-grade metamorphic rocks and metaophiolites, and the Cilento Unit, tectonically superimposed on the earlier, and formed from sediments deposited on top of the Tethyan accretionary wedge during the Late Oligocene and Early Miocene (Amodio-Morelli et al., 1976).

As stated by Bonardi et al. (1988b), the Liguride group is represented by two tectonic units, from base to top: the Frido Unit and the Northern-Calabrian Unit. Sedimentary rocks represented by the Pollica, S. Mauro, Torrente Bruca and Albidona Fms. rest in angular unconformity on top of the aforementioned units; they range in age from the late Burdigalian to the Langhian.

Ogniben (1969, 1973, 1985), Boullin (1984), Knott (1987), Monaco et al. (1991) considered the Liguride complex an Apulia-vergent element representing a basin interposed between the "Calabrian" and African domains.

According to Haccard et al. (1972), Alvarez (1976), Amodio-Morelli et al. (1976), Grandjacquet and Mascle (1978) and Bonardi et al. (1982, 1988b) the metamorphic rocks of the "Frido" constitute elements of the Eo-Alpine, Europe-verging chain, which were subsequently involved in the construction of the Apenninic chain during the

Neogene. The pertinent original segment of Alpine Tethys ocean was located between the European margin and the NW Adria palaeomargin, from which the Austroalpine crystalline cover, outcropping in the Calabrian arc would be derived, and on top of which the sediments of the Cilento Unit were deposited in unconformity.

Monaco et al. (1991, 1995) subdivide the Liguride Complex, into three main groups of tectonic units. The first of these, named Calabro-Lucanian Flysch Unit, occupying a basal geometrical position, consists of non-metamorphized ophiolithiferous units; the second group includes the metamorphic ophiolitic rocks of the Frido (Amodio-Morelli et al., 1976), and the third is represented by the syn-orogenic turbiditic successions (Saraceno Fm. and Albidona Fm.) of the Late Oligocene to Langhian, and by the Tortonian clayey-marly-arenaceous alternation.

In the present work, on the basis of recent fieldwork conducted at the Calabro-Lucanian boundary, it is believed that the Liguride Units can be subdivided into semimetamorphic units, that are essentially referable to the Frido Unit, and in Cilento units, including the thick siliciclastic and carbonate sequences outcropping in the Cilento area (Ascea Fm. and Pollica Fm.) and at the Calabro-Lucanian boundary (Crete Nere Fm. and Saraceno Fm.). The S. Mauro in Cilento Fm. and the equivalent Albidona Flysch at the Calabro-Lucanian boundary will be treated separately following the original sense of Selli (1962) of "mesoautochthonous" deposits (Figure 10).

The Epimetamorphic units (Frido) and high-grade crystalline units are exposed in the southern extremity of the Campanian Apennines and at the Calabria–Lucania boundary between the village of Pisciotta to the west, at the Catanzaro isthmus to the south, and occur in the Castrovillari outcrops and in the Calabrian coastal chain. The Frido Unit (Amodio-Morelli et al., 1976) rests tectonically upon the internal carbonate units and is composed of a succession of polymetamorphic rocks. With this succession, are associated basic and ultrabasic rocks, which can be referred to an ophiolitic sequence, and blocks of rocks derived from continental crust (garnet gneiss, amphibolites and granitoids), which are characterized by an Alpine high-P/high-T metamorphic imprint (Spadea, 1982) and are interpreted as nappes of the Europe-vergent Alpine chain of African provenance (Haccard et al., 1972; Alvarez, 1976; Amodio-Morelli et al., 1976; Grandjacquet and Mascle, 1978; Bonardi et al., 1982; Scandone, 1982) or as the deepest portion of the Liguride accretionary wedge (Monaco et al., 1991).

The Frido Unit corresponds to the Frido Fm. of Vezzani (1969), which is characterized by five members subdivided into basal, lower metapelitic, quartzarenitic, calc-schisty, and upper metapelitic, over a total thickness of about 1300 m. Excluding the upper metapelitic member, already referred to the Crete Nere Fm. (Bousquet, 1973; Lanzafame et al., 1978; Bonardi et al., 1985) and of the basal portion characterized by a mélange with arenaceous-clayey matrix, that surrounds both the metamorphic and crystalline rocks, graded fine-grained *Nummulites* and *Lepidocyclina* breccias, turbiditic sandstones and quartzarenites, Bonardi et al. (1988b) place these metasediments at the base of the "Liguride Units" and confirm the Early Cretaceous age already established in the literature.

The ophiolitic rocks are formed from lense-shaped bodies of prevalently cataclastic serpentinized peridotites and metabasite blocks (Vezzani, 1966c; Spadea, 1979) to which locally the original metamorphized sedimentary cover is associated (Lanzafame et al., 1979), and which are interpreted as tectonic slices in basal and apical position within the "Frido Fm." (Ogniben, 1969) or as blocks of ophiolitic units in a mélange, corresponding to the entire Frido Unit (1972; Amodio-Morelli et al., 1976).

The crystalline lithotypes of continental origin have been considered klippen of the "Calabride Complex" (Dioritic-kinzigitic Fm. *auct.* of Calabria, Ogniben, 1969), which rest on the "Frido-Crete Nere Fm.", or patches of the Polia-Copanello Unit interposed between the Frido Unit and the Cilento Unit (Amodio-Morelli et al., 1976), although in the area of the Calabria–Lucania boundary (between S. Severino Lucano and Episcopia) these rocks occupy the uppermost geometrical position, and their relationship with the adjacent formations remains poorly constrained.

These rocks, as well as the major part of the serpentinite blocks, have also been interpreted as elements of major dimensions of a mélange (Spadea, 1982) or a shear zone (Knott, 1987), a schisty matrix of serpentinite and metapelites, sandwiched between the Frido Unit and the North Calabrian Unit *sensu* Bonardi et al. (1988b).

The petrogenetic evolution of these rocks is characterized by an event in the amphibole-granulitic facies, followed by a strong retrocessive high-P/low-T event with the more or less complete transformation of the mineral phases and development of aragonite and lawsonite and subsequent retromorphism into greenschist facies (Spadea 1976, 1979). The age of metamorphism by radiometric datations (28 Ma) is Late Oligocene (Spadea, 1982).

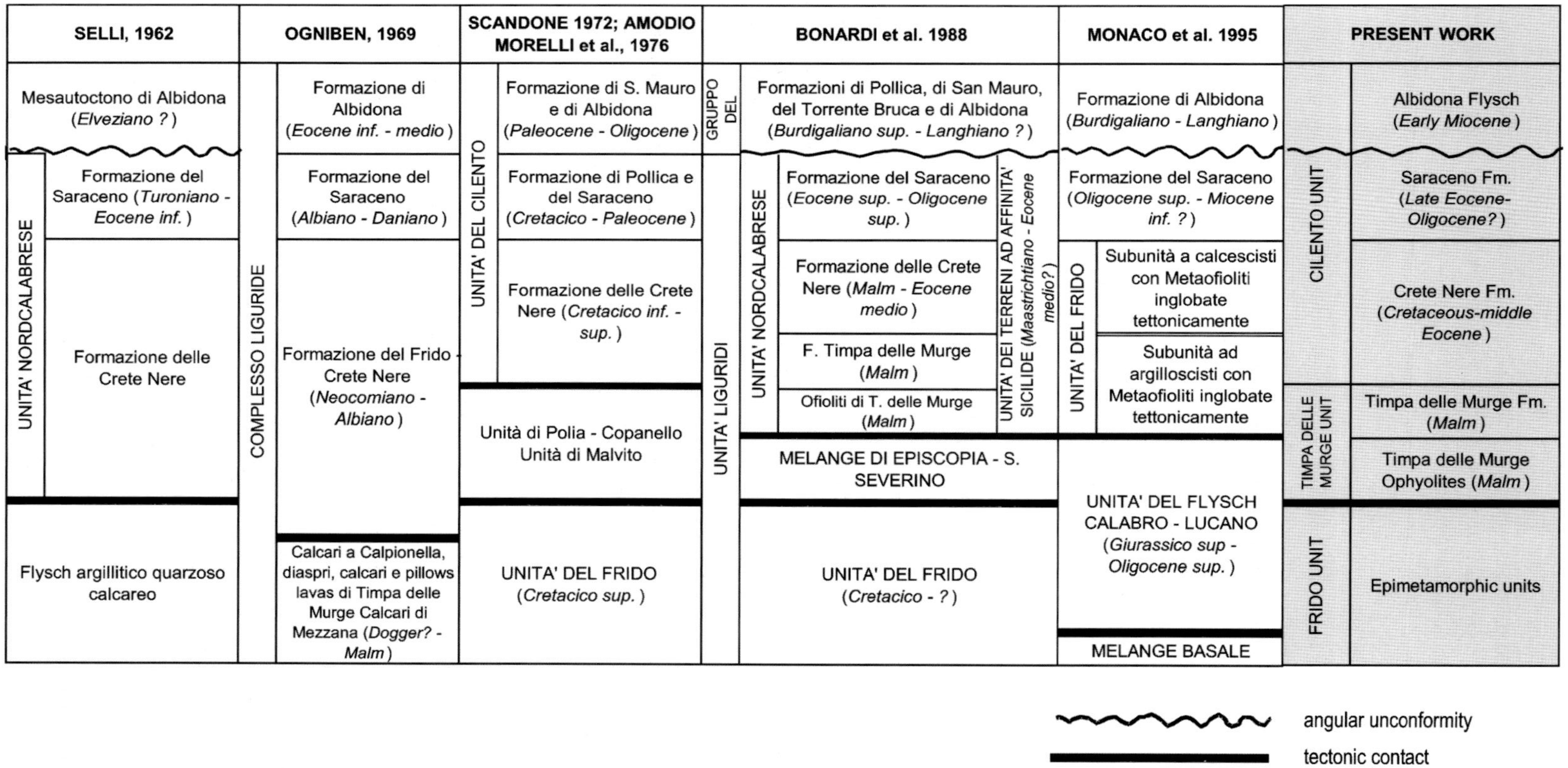

Figure 10: Comparative table of interpreted Eo-Alpine Tethydes by various authors (modified by Bonardi et al. 1988a, 1988b).

Cilento Unit. The *Crete Nere Formation* (Selli, 1962) consists of distal turbidites characterized by greenish-gray or bluish claystones alternating with fine-grained, gray to green quartzarenites grading upward into 500 m of black shales with litharenitic intercalations and, in their upper part, fine to medium grained calcareous *Nummulites*-bearing turbidites.

To the Crete Nere Fm. different ages have been assigned: Neocomian-Albian by Ogniben (1969) for the "Frido-Crete Nere" couple; Early to Late Cretaceous by Scandone (1972) and Amodio-Morelli et al. (1976), and Malm to Middle Eocene by Bonardi et al. (1988a). The lattermost authors attribute a Middle Eocene age only to the upper portion of the succession, which is characterized by nannofossil associations of the NP17 Biozone of Martini (1971).

The *Saraceno Formation*, as named by Selli (1962) and formalized by Vezzani (1968), is a 500 m thick sequence characterized by a basal portion with more or less siliciferous calciclastic and marly strata, and by an upper portion of turbiditic character, with siliciclastic arenitic-pelitic strata (Critelli, 1991). The sandstones vary in composition from litharenites and arkoses and are generally subordinate to the calcarenites, but they become more common up-section, where they constitute a clayey-arenaceous alternation which forms the "prelude" to the overlying Albidona Flysch.

The earliest datings gave a Late Albian-Danian age (Vezzani, 1966b; Ogniben, 1969; Servizio Geologico d'Italia, 1969), but subsequently the formation has been referred to the Middle-Late Eocene by De Blasio et al. (1978), and more recently to the Late Eocene–Late Oligocene by Bonardi et al. (1988a). The formation underwent yet another "rejuvenation", even though indirectly, by Monaco (1993), who attributed a Late Oligocene age to the basal portions of the succession, with a probable extension, of the topmost levels, to the Early Miocene, as already stated by Ghezzi and Bayliss (1963).

Sicilide Units. In the Southern Apennines, there are widely exposed varicoloured shales, known as the "Argille scagliose" or "Argille varicolori", belonging to original internal domains. Formerly distinguished as "coltri lucane" by Selli (1962), and containing "inglobed clods" from the Mt. Sant'Arcangelo and Corleto Perticara Fms., they were subsequently included by Ogniben (1969) in the Sicilide complex due to its analogy with allochthonous successions outcropping in Sicily. The term "Sicilide Units" as used by D'Argenio et al. (1973, 1975) indicates the same successions of the Campano-Lucanian Apennines.

The Sicilide Formations are distributed in various nappes inserted within different structural levels, separated by Middle-Upper Miocene flysch-type deposits and involved in the later tectonic phases. They form the frontal wedge of the Apenninic chain and are involved in nappes differing as regard tectono-stratigraphic composition and time–space stacking position (Cirigliano Nappe, Rocca Imperiale Nappe).

The *Cirigliano Nappe* is composed of chaotic Varicoloured Clays, which form a tectonic mélange, inglobing exotic blocks of Cretaceous rudist-bearing limestones, Numidian quartzarenites, and fairly widespread patches of Tusa Tuffites. The exotic components within the mélange can be referred to a Late Oligocene–Early Miocene age, in analogy with the ages ascertained for the formations from which they were derived (Numidian Flysch and Tusa Tuffites). On the whole, the mélange formation therefore can be assigned to the transition from the Early to Middle Miocene.

The *Rocca Imperiale Nappe* lies in tectonic superposition upon Upper Serravallian–Lower Tortonian flysch-type deposits. In this nappe the distantly derived Gorgoglione Flysch has also been included.

Two deformational phases can be distinguished during the emplacement of this cover, after which the front of the Sicilide Units extended as far as the most external areas. The first one, attributed to the Burdigalian, was responsible for the overriding on top of the Numidian Flysch and for various contacts between the Sicilide formations on top of the Lagonegro sequences; the second, which took place during the Early Tortonian, led to a general overthrusting of the "External Flysch". Further deformational evolution created the relationships between Sicilide Units and Plio-Pleistocene cover. In the area of the Calabria–Lucania boundary the Sicilide Units tectonically overlie the Albidona Flysch.

2.2.5. *Alpine Tethys Flysch (Albidona Flysch).* The Albidona Flysch was first established by Selli (1962) under the name Albidona Fm., of Late Langhian to Early Helvetian (Auct.) age.

A Miocene age has been confirmed by Bonardi et al. (1985), who indicated an age ranging from the Middle Burdigalian to the Langhian-Serravallian. Lentini et al. (1987) believed that the contact of the Albidona Flysch with various underlying units, in particular, the carbonate deposits of the Apenninic platform (Carbone et al., 1991; Carta geologica del Bacino del Fiume Agri, 1991), is of true stratigraphic nature. Recently Lentini et al. (2002) conclude that the Albidona Flysch is extensively detached and was mechanically transported into a more external position in a sort of "cover substitution" of the Numidian sequences in turn displaced toward the frontal wedge of the **AC**. In fact the flysch, predominantly of calcareous-pelitic composition, is characterized by two members: the lower one is chaotic, and composed of polygenic elements (crystalline, volcanic and sedimentary), and the upper member, which is more typically composed of turbiditic arenaceous-clayey material with thick banks of calcareous marls.

The new CROP data show that at the end of the Balearic stage, with the closure of the Alpine Tethys, this flysch unit thrust over the Etrurian Promontory, and its cover (Apenninic Platform Units) together with the Albidona Flysch (possibly both in stratigraphic and tectonic contact with Apenninic Platform Units) moved southeastward in connection with the Ionian slab retreat (Finetti, this volume, Chapter 34).

2.2.6. *Upper Miocene–Lower Pliocene Deposits.* The sequences of this chronological interval (not distinguished in Figure 2) are present in southern Irpinia and in the Salerno area, constituting the deposits of the Altavilla Unit (D'Argenio et al., 1975). These units include two sedimentary cycles: the lower one is formed by clayey sands, diatomites and evaporitic limestones of lagunar facies (Formazione Gessoso-Solfifera *auct.*), and reaches thicknesses of several hundred meters. The second cycle, often with thicknesses of several hundred meters, is characterized at its base by clayey olistostromes derived from rocks of the internal units, which are followed by clays and evaporites, conglomeratic sands, and sands typical of a shallow marine to deltaic environment.

In the external areas of the southern Apenninic chain the sequences of Perosa, which outcrop along the Sinni river between Latronico and Episcopia, and thus on the western margin of the Sant'Arcangelo basin, rest unconformably on the units of the chain (Vezzani, 1966a). They consist of basal conglomerates, marly clays passing into a clayey-arenaceous alternation, and then into clayey sands over an interval of about 400 m.

The Messinian evaporitic horizons have been penetrated by wells within the Metaponto nappe (Recoleta 1, Cavone 1, S. Basilio 1 wells, etc.), where they rest upon the carbonate substratum of the Bradanic foredeep and of the "Apulian chain" (Tursi 1 and Montegiordano 1).

2.2.7. *Pliocene-Quaternary Deposits.* These deposits record numerous sedimentary events, which document on the one hand the palaeogeographic evolution of the Apulian foreland during the last phases of the Apenninic-Dinaric tectogenesis, and on the other those recorded in the Apenninic chain in the post-collisional phase. In the Southern Apennines, the Plio-Pleistocene cover was deposited in a chain-foredeep-foreland system, already outlined, which followed the Late Miocene deformation of the most external allochthonous successions.

During the Pliocene, the successions deposited on top of the Apenninic chain units participated, along the entire allochthonous wedge, in the overthrusting of the Apulian carbonate sequences. In this framework, the cover, that rests unconformably upon the Apenninic units transported along the allochthonous substratum, assumes a different structural role than to the coeval deposits emplaced directly upon the Apulian substratum, which nowadays are only encountered in drillholes at depth. The position at the base of the allochthonous edifice of horizons that become more recent toward the external areas documents the progressive time–space outward migration of the nappes onto the Apulian domain.

From numerous oil exploration boreholes, it is clear that much of the allochthonous chain has been thrust over Apulia since the Middle Pliocene. It is only in the frontal areas, at the base of the Apenninic chain, that units of the Middle Pliocene to Quaternary have been involved, up to the deposits containing *Hyalinea baltica*, within which the front of the so-called Metaponto nappe is located. These latest deposits only extend as far as to suture the allochthonous front, which is thus buried, and they extend without any break in continuity to the relatively undisturbed Plio-Pleistocene foredeep deposits.

Plio-Pleistocene successions were sedimented under continuous local tectonic and eustatic control, and in various basins distributed on different positions of the thrust belt (Sant'Arcangelo, Calvello, Potenza, Alto Ofanto basins, Eboli and Tanagro basins).

Quaternary sediments were deposited in tectonic depressions, prevalently elongated in an Apenninic direction, and constitute the intramontane basins of the Vallo di Diano, the Val d'Agri and the Mercure valley. They are characterized by lacustrine deposits (clays with minor sands) and alluvial fan deposits, hundreds of meters in thickness. Along the Ionian belt, these deposits are represented by a marine succession with Middle Pleistocene sands and conglomerates.

2.2.8. *Bradanic Foredeep Plio-Quaternary Deposits.* The Bradanic foredeep, the southern segment of the Apennine foredeep, is formed by an outward migrating depocenter delimited by an eastern margin with carbonate sedimentation, and an internal western margin characterized by siliciclastic sedimentation fed by the Southern Apennines thrust front.

As in the chain, the successions correspond to three distinct sedimentary events. One dated Early Pliocene is characterized by modest thicknesses of *Globigerina*-bearing glauconitic biomicrites, which rest transgressively on Cretaceous or Miocene limestones and are exposed near the coastal areas or in the hinterland of the Salento. The second event includes a sedimentary succession of Middle Pliocene to Early Pleistocene age, composed of distinct transgressive and regressive units (Ricchetti, 1980), among which those related to the culminating phase of transgression have affected vast areas of the foreland. The transgressive portions of this cycle, known in the literature as the "Plio-Pleistocene Cycle of the Bradanic Foredeep", are formed by massive or cross-stratified biocalcarenites. The Early Pleistocene interval of this cycle is characterized everywhere by macrofossil associations containing *Arctica islandica*. In the depocentral area and along the western margin of the Bradanic Foredeep, the calcarenites pass laterally into clayey-marly deposits of gray-blue color (Argille subappennine *auct.*), which are exposed with thicknesses of hundreds of meters in the foredeep areas. This is the most extensive and thick lithofacies and it is characterized by hemipelagites represented by silts, clays and marls, within which there are intercalations, at various stratigraphic elevations, of medium-to-fine-grained sands or coarse-grained detritic deposits. The presence of slides or slumps testifies to a sedimentation on mobile margins with steep gradients, which were affected by syndepositional tectonics. The "Sub-Apenninic Clay Unit" constitutes the sequence upon which the infilling siliciclastic regressive deposits that close the succession of the Bradanic Foredeep are resting.

3. RECONSTRUCTION OF A LITHOSPHERIC SEISMOGEOLOGICAL SECTION ACROSS THE SOUTHERN APENNINES FROM THE TYRRHENIAN OCEAN BASIN TO THE ADRIATIC APULIAN COAST

The Southern Apennines constitute a complex thrust belt where the surface geology, mainly represented by allochthonous units of debated provenance, does not reflect the structural style and stratigraphy of the rooted tectonostratigraphic setting of the underlying Apulian platform. For understanding the tectono-stratigraphy of the Southern Apennines it is consequently essential to have at one's disposal an adequate seismic dataset for subsurface control.

Hydrocarboon exploration activity over the last decade, especially after the recent discovery of important oilfields in the Lucanian Apennines, has clearly shown that in the Southern Apennines there is a complicated, more-or-less-thick, uppermost tectono-stratigraphic setting represented by the allochthonous edifice of the Apenninic chain, underlain by a compressive rooted structural setting of the inner buried Apulian platform, formed by reverse faults of relatively modest horizontal displacements. Only the thrust fault of Monte Alpi (Mt. Alpi Unit) and the regional transcrustal thrust fault connected to the Bradanic foredeep have relevant displacements. The Mt. Alpi thrust fault displaces the whole upper crust eastward by about 16 km, and south of the section in Plate 2 shows a marked backthrusting in its upper part. The high-angle Bradanic thrust fault produces a transcrustal displacement of about 14 km (Plate 2). Both the Mt. Alpi and the Bradanic thrust faults, due to the obliquity of thrusting, are affected by a late orogenic left-transcurrent component.

The contribution from hydrocarbon exploration, together with surface and borehole geological data, was particularly relevant for progress in regional knowledge of this orogenic system. The contribution by the CROP seismic program, extending exploration of the onshore area downward through the entire crust (CROP-04), and of the offshore Apenninic margin through the whole lithosphere (CROP M6B) allowed for the first time the reconstruction for a crustal seismogeological section of the entire Southern Apennines belt, completely controlled from the young Tyrrhenian ocean basin to the Adriatic coast (Plate 2).

3.1. Basic Data used for reconstruction of the Seismogeological Section

The seismogeological section of Plate 2 represents the result of a specific integrated multidisciplinary programmed investigation conducted in cooperation between the University of Trieste and the University of Catania in recent years. The final results illustrated in Plate 2 are supported by the following data (Figure 1):

(a) *CROP Mare Seismic Section M-6B (Plate 1).*
(b) *Land Seismic Section CROP-04 (from CMP 5 to 805).*
(c) *Available (courtesy ENI-AGIP) oil exploration data for the sedimentary cover along the segment "EF", close and subparallel to CROP-04.*
(c) *Basement upper crust and lower crust data from the nearby CROP-04 for the corresponding segment "EF" of the section.*
(d) *Extrapolated data from the nearby line CROP-04 and line M-5 for the segment "EF" of the section.*
(f) *Simplified stratigraphic data from borehole Puglia-1 (from the literature) for the Permo-Mesozoic cover of the Apulian platform swell.*
(g) *Borehole data available along the section (and not all indicated in the section).*
(h) *Various geological and geophysical data from the literature of regional or local significance.*
(i) *New field geology data specifically produced for the control and/or validation of the seismic interpretation.*
(j) *Gravity and magnetic data.*

It is important to note that a seismic interpretation of the southwestern extremity of the CROP-04 section, without connection to the northeastern part of CROP Mare M-6B, and vice versa, would be very difficult, and probably questionable because these two pieces of section together cover a key area for understanding the most complex internal part of the Southern Apennines.

3.2. Seismic Interpretation

In reconstructing the crustal seismogeological section across the Southern Apennines, from the Tyrrhenian to the Adriatic Sea (Figures 1, 12–16 and Plate 2), the best available exploration data were used. From "A" to "D" the reconstruction is based on full direct control by CROP data; from "E" to "F", which is a large part of the complex Southern Apennines thrust belt, the section on the sedimentary interval of the crust takes advantage of seismic data obtained by hydrocarbon exploration (courtesy of ENI-AGIP) which, in the first 5–7 seconds of reflection time, are obviously of higher resolution than the deep penetrating seismic sections of the CROP Project.

The seismic section CROP M-6B in its initial standard processing was of very poor data quality. Therefore a careful iterated interpretive reprocessing was conducted by the EGG (Exploration Geophysics Group) of the University of Trieste. At the end, a clearly improved section was obtained (Plate 1).

The interpreted key seismic horizons are listed and specified in the legend for Plate 2, and so are not given here in the text.

3.2.1. *Stacked Allochthonous Blocks on the Tyrrhenian Margin, and Ionian Slab.* From the new M-6B version, the seismic imaging of the subducted upper Ionian slab was immediately impressive (Figures 12, 13 and Plate 1, 2). Also evident was a collisional contact, but for a clearer understanding of the complicated tectono-dynamics involved, as explained above, it was advisable to perform an interpretation connecting the NE-extremity of CROP M-6B with the SW-extremity of land section CROP-04, available in the literature (Mazzotti et al., 2000). So, section M-6B was extended up to CMP 805 of CROP-04. This connection, broken by only a small gap (Plate 2, segment "BC") allowed the collisional boundary between the Apulian margin on the Tyrrhenian Sea to be reconstructed and allochthonous crustal blocks to be defined.

Sedimentary cover sheets of the colliding block continue eastward and outcrop in the Southern Apennines (i.e. Plate 2: from Mt. Soprano to Vallo di Diano) as allochthonous Mesozoic platform carbonates, known in the literature as "Apennine Platform" of unprecise, questionable internal provenance. A preliminary hypothesis that the paleo-Ionian ocean, extended northward beneath the present Tyrrhenian Sea, and that the Apennine platform arrived in the Southern Apennines from the northwest, traveling across the subducted Ionian Sea, was proposed some years ago (Finetti et al., 1996).

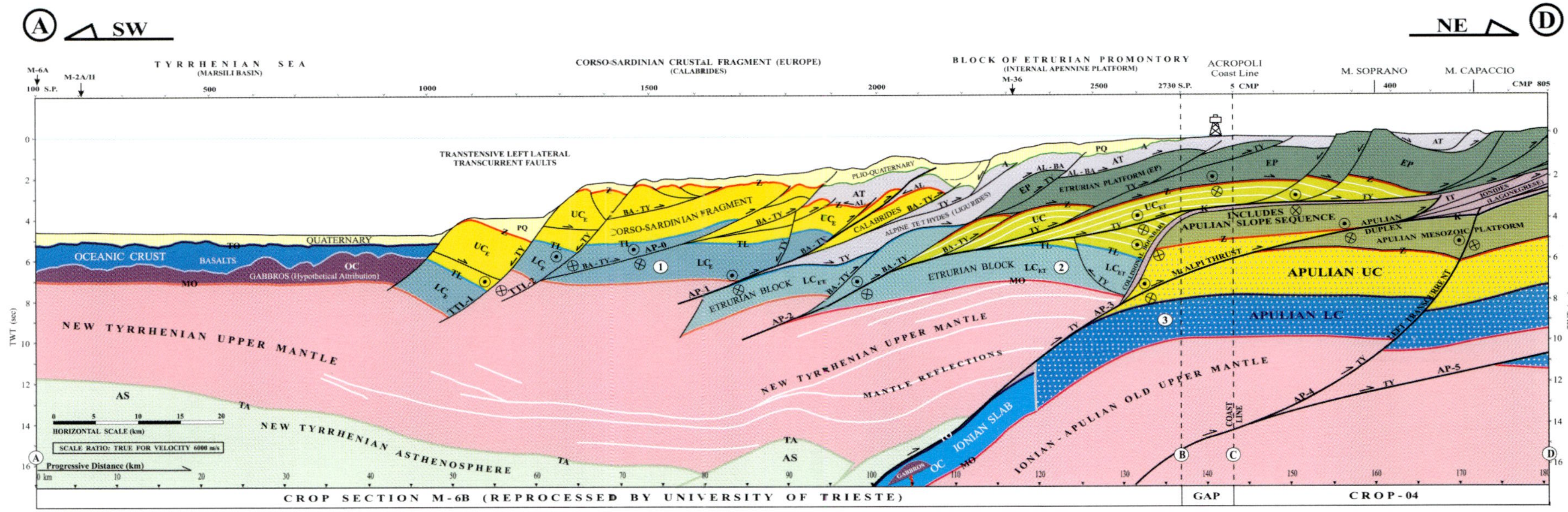

Figure 12: Lithospheric section reconstruction across the Tyrrhenian Sea–Southern Apennine margin (segment AD of Plate 2, see Figure 1 for location). Legend is reported in Plate 2. The reconstruction is obtained from seismic interpretation of CROP section M-6B and the SW part of CROP-04 (CMP 5-805).

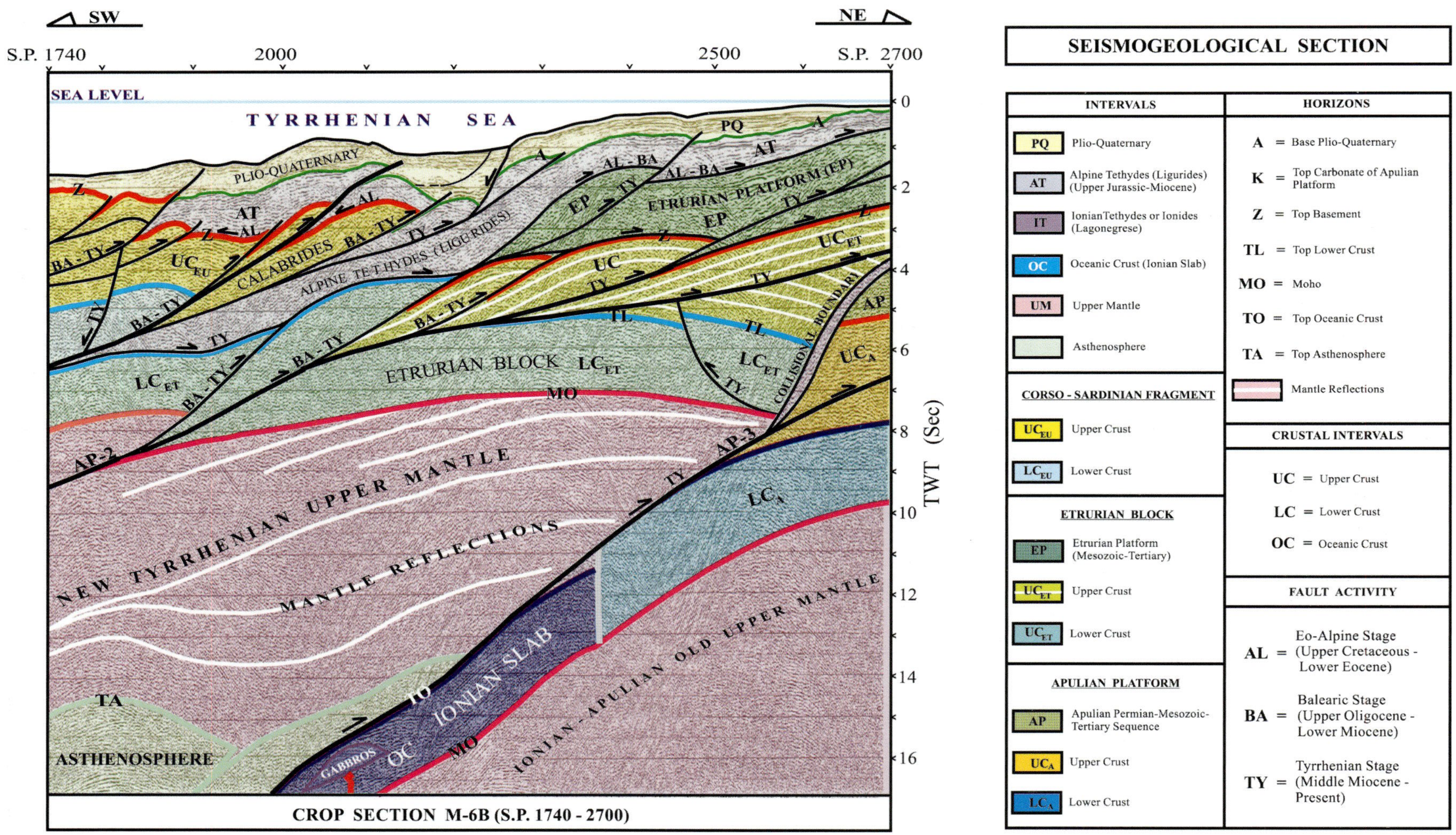

Figure 13: Interpreted key segment of the CROP seismic section M-6B (S.P. 1740–2700), showing the collisional contact between the allochthonous Etrurian crustal block and the steep Adria plate slope. This collision determined a high-angle "Mt. Alpi-like" thrust fault (AP-3) which displaces the upper crust and the Apulian platform cover about 20–22 km eastward. Suture zone of the Alpine Tethys closure, the Calabrides (Europe) thrust fragments and the Ionian slab are also well imaged seismically. A careful interpretive reprocessing by the University of Trieste was determinant in obtaining data quality interpretable with satisfactory reliability.

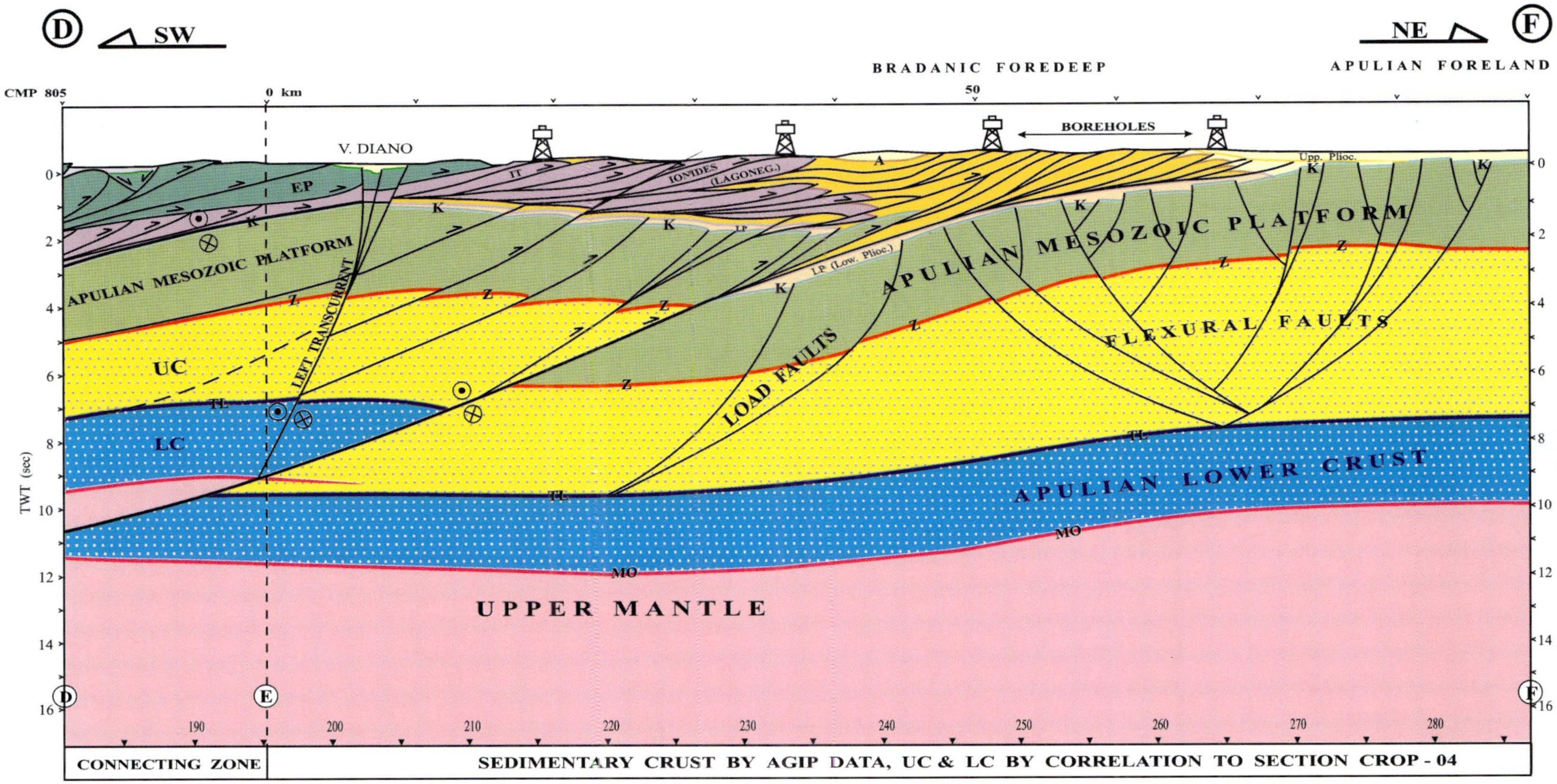

Figure 14: Combine crustal seismogeological section across the Southern Apennines. The *sedimentary crustal data* are courtesy of ENI-AGIP, assembled with minor re-arrangements. *Upper (UC) and lower (LC) crustal data* were obtained by correlation with seismic section CROP-04. For location see Figure 1; for legend see Plate 2.

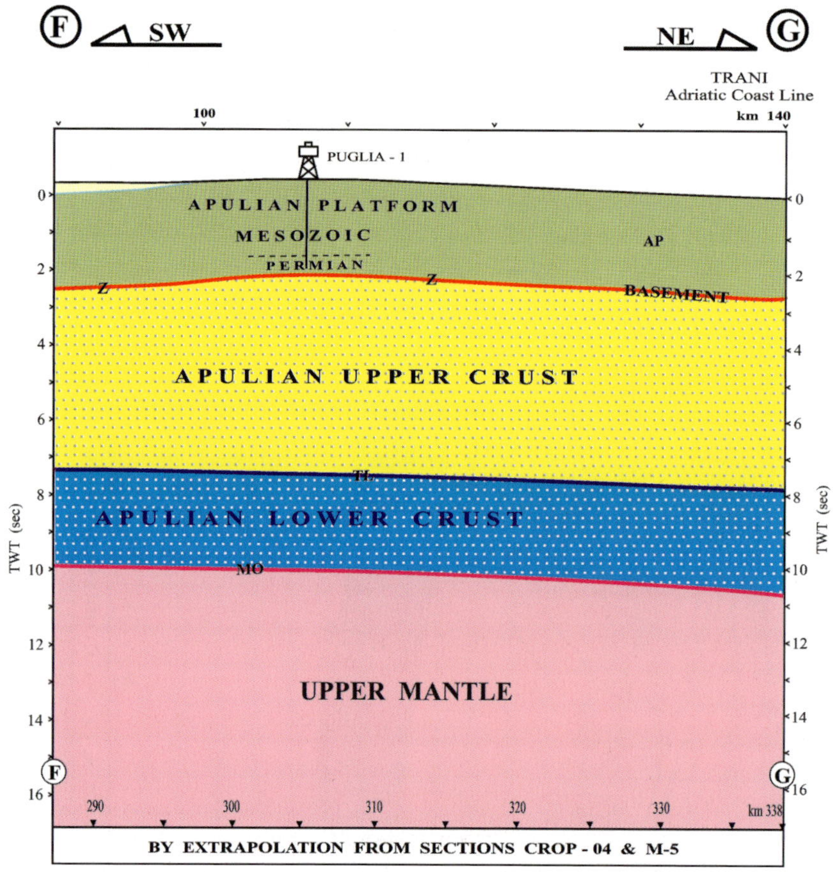

Figure 15: Crustal seismogeological section across the Apulian platform swell. For simplicity, the Permo-Mesozoic sequence is reported as a unique undivided interval. For location see Figure 1.

From a global 3D interpretation of the CROP section network and newer more realistic and precise paleogeographic constraints, and recognizing the magnetic lineaments in the Tyrrhenian Sea as imprinted indicators of geodynamic migration travel-paths (Finetti and Del Ben, this volume, Chapter 21), it was evident that the crustal block colliding with the Apulian slope (Plate 2: block 2 within the white circle) was coming from a promontory area denominated the "Etrurian Promontory" (Finetti, this volume, Chapter 34), located on the northwestern extremity of the Adria plate. In the Upper Oligocene–Lower Miocene, the eastern margin of Corsica collided with the NW Adria promontory, and in the Burdigalian, when the Ionian subduction started, the above mentioned crustal block moved southeastward, until in the Upper Miocene–Early Pliocene, it collided with the Apulian boundary. Since various allochthonous crustal blocks of the Etrurian Promontory arrived on different zones of the Southern Apennines and northern Calabrian arc, it is more appropriate and immediately self-explanatory to describe these blocks with the unifying provenance name of "Etrurian Blocks".

Looking at the whole hinterland of the Southern Apennines with its Tyrrhenian margin, it is seen that this sector is substantially made up of two different allochthonous crustal blocks: the Etrurian block 2 (Adria), as above, and block 1, detached from the Corso-Sardinian block (Europe) (Figures 12, 13, 16 and Plate 2). These blocks arrived in their present position, coming from the northwest, after a long migration. At the end of the Balearic stage (Early Miocene) the Etrurian Promontory collided with the rotating Corsica block and gave origin to the detachment of the Etrurian fragment in Plate 2.

The reprocessed CROP M-6B section shows very clearly that this Etrurian block, the corner of the Adria Plate, in the end, collided with a steep Apulian slope, generating a trans-upper crust regional thrust fault, which is the same regional thrust fault "Mt. Alpi" that, south of the section in Plate 2, exhumes the Apulian platform with a well developed back-thrusting structure in its uppermost part. It is to observe that the "Mt. Alpi-like" regional thrust fault is accompanied by a marked component of left-strike-slip motion consequent to the oblique tectonics.

"Mt. Alpi-like" transcrustal faults, similar to this one, are also observed on CROP sections on the northern Sicilian margin, where the duplexed upper crust is that of the northern African margin in the zone close to the subducted Ionian Tethys slab.

In all the examined CROP sections of the Southern Apennines and Sicily Tyrrhenian margins, it is evident that only a small segment of the lower crust follows, with a marked flexure, the subducting Ionian slab, while the upper crust and its sedimentary cover are detached and thrust outward. Thus, in all the examined cases of the CROP network it is shown clearly that subduction of the Ionian Tethys oceanic crust took place with no relevant underplating effects (Finetti et al., this volume, Chapter 15; Finetti and Del Ben, this volume, Chapter 21). Only the lower crust subducts by some 15–20 km below its original depth, as in Plate 2. Seismic imaging of this oceanic slab is very clear (Plate 1 and Figure 13). In the Alps, subduction of the Alpine Tethys causes a deeper downgoing of the European lower continental crust (Finetti, this volume, Chapter 7).

Starting from the Tyrrhenian sea (Marsili basin), where it is more evident, an hypothetical top Asthenosphere (TA) has been tentatively interpreted. The resulting depth of less than 12 sec reflection time (or about 26 km) is in agreement with the S-wave dispersion data. In the upper mantle fair to good reflections are evident, as drawn in the reconstructed seismogeological section (white lines). These mantle reflections, as well as the Asthenospheric setting, indicate a southeastward mantle flux following the Ionian slab retreat.

In the southwestern extremity of the interpreted CROP-M6B section, the flat and thin oceanic crust of the young (Quaternary) Tyrrhenian Sea (Marsili basin) is seismically well evident (S.P. 100–1000). The undulating reflector within the oceanic crust, as frequently occur in zones close to the continental margin (initial crustal opening), has been interpreted as probable gabbroic plutons, and is very distinct from the mostly flat overlying basaltic floodings.

In the collisional contact between the *Etrurian promontory fragment* and the Apulian platform, the Ionian slab subduction was evidently accompanied by a Tyrrhenian mantle-asthenosphere flux that generated an eastward thrusting of the upper Apulian crust by the lower crust of the Etrurian allochthonous block. The upper crustal basement of this block thrust in part over the "Mt. Alpi-like" Apulian block, overlain by the Etrurian platform cover (Apennine platform in the literature). The Etrurian crustal block detached from the Etrurian promontory and traveled southeastward some 400–450 km across the subducting Ionian Sea area, before entering in oblique collision with the steep Apulian slope. After the collisional contact, the platform cover (Apennine platform of Mt. Soprano; Figure 12) and other EP units, already detached from their basement since the Late Balearic stage (Lower Miocene), thrust obliquely east and southeastward, leaving more internally parts of their lower crust (LCET).

Nearby and around the Etrurian platform units the Numidian Flysch is distributed. This leads us to believe that the source areas of the Numidian Flysch could be represented by quartzitic layers lying between the platform cover and its basement, where platform detachment started in the late Balearic stage, when the Corsica and northern Sardinian block collided with the NW Adria margin (Etrurian Promontory).

3.2.2. *The Alpine Tethydes (Ligurides/Sicilide) and the Ionides (Lagonegrese).* Using the new CROP data, we show that two different deep basins fed the allochthonous deep water units that were tectonically transported in the Southern Apennines nappe: the *Alpine Tethys* from which the Alpine tethydes units (Ligurides) derive, and the *Ionian Tethys* where the Ionian Tethydes or Ionides (Lagonegrese) originated.

Alpine Tethydes (Ligurides/Sicilide Units). Between the Etrurian block (block 2) and the internalmost Corso-Sardinian fragment (block 1), within the tectonic stack, are trapped remnant units of the Alpine Tethydes of which a major part thrust eastward over the Etrurian platform (Ligurides) and other more external allochthonous units (Sicilide Units) which form a significant part of the frontal wedge (Figure 12 and Plate 2). The closure of the Alpine Tethys took place in the Balearic stage (Upper Oligocene–Lower Miocene).The stacked blocks 1 and 2 moved together toward the Apulian margin in the Late Balearic–Early Tyrrhenian stage (Lower-Middle Miocene) and started to collide at the end of the Miocene.

Part of the Alpine Tethydes (AT), during the Eo-Alpine stage (Upper Cretaceous–Lower Eocene) thrust over the European margin (future Corso-Sardinian block) and was successively piggyback transported by crustal block 1 onto the present tectonic stack, where it welded with the AT part that had thrust over the Etrurian block in the Balearic stage.

Ionides (Lagonegrese Units). The allochthonous units formed from a sequence of deep water deposition in the Southern Apennines have been recognized in the literature for many years. They were discussed in Section 2.1.1. The palaeogeography, basin extent and tectonodynamic setting of the Lagonegrese of the Southern Apennines are frequently discussed topics in the international literature for over three decades.

The CROP seismic data together with MS seismic lines and available seismic data from hydrocarbon exploration allow a good continuous correlation of the undeformed deep deposition in the Ionian Sea with the imbricated units of the Ionides wedge of the Southern Apennines, passing along the Gulf of Taranto. This correlation shows *unequivocably that the Lagonegrese Units are nothing less than the pre-orogenic cover (Permo-Triassic to Lower Oligocene) of the Ionian oceanic crust* (Finetti, this volume, Chapter 10). Consequently, we assume for these units the more self-evident name of "Ionides". This new term represents thus, in the Southern Apennines, *allochthonous units, originally sedimented in the Ionian Tethys, from basin opening to the Oligocene. Successively, during the Ionian slab subduction (Middle Miocene to Present), they were stripped off their oceanic or bordering slope basement and tectonically transported into the Southern Apennines over the Apulian platform* (Figures 12–16 and Plate 2). Repeated imbrications of several thrust units of the Ionides have been found in deep boreholes, which basically confirms the image of their wedge in Figure 16 (Ionides wedge roughly located from 10 to 40 km east from point "E" in the section). The relationships between the Lagonegrese Units and other units, the geometry, and the stratigraphy of the Lagonegrese Units has been described in the previous paragraph. A careful stratigraphic study of the Lagonegrese units is also presented by Ciarapica and Passeri (this volume, Chapter 11). It is possible to establish that the basal deposition of the Ionides of the Southern Apennines ("Mt. Facito" Fm.) includes the Upper Permian and, very probably, represents an early rifting stage with deposition on the Apulian slope, when Adria started to detach from the Africa plate.

Lower Pliocene sequences rooted in the Apulian platform beneath the allochthonous edifice of the Apenninic chain have been found in many boreholes. This indicates that the (oblique) thrusting process of the Ionides over central-eastern Apulia took place mostly from the upper part of the Early Pliocene to the Middle Pliocene.

The External Flysch with remains of Sicilide Units and Numidian Flysch constitute a relevant part of the tectonic wedge of the Southern Apennines thrust belt. In the crustal profile they have not been distinguished in the frontal wedge, but details on their geometry can be seen in the geological profiles of Figure 11.

3.2.3. *Apulian Platform Crustal Setting.* The Apulian platform from the Adriatic coast to its collisional Tyrrhenian slope is well defined seismically in the crust. The following main regional seismic intervals of Apulia have been identified and reconstructed:

- *Mesozoic platform and basal Triassic-Permian*
- *Lower Pliocene cover beneath the allochthonous units*
- *Allochthonous units overlying Apulian cover*
- *Basement upper crust*
- *Lower crust*
- *Upper mantle*
- *Upper Ionian slab (Apulia lower crust)*

On the swell area of the Apulian platform, the deep borehole Apulia-1 penetrated the entire existing Mesozoic sequence and entered into the Permian succession (Figures 5, 15, 16 and Plate 2). Here the sedimentary crust is about 2.60 sec thick (about 7.8 km). The well stopped at over 7000 m depth, not far from the basement. The basement upper crust is over 5.2 sec thick (over 15 km) and the lower crust is 2.65 sec thick (over 9 km). The Moho, in the swell (Adriatic coastal area) is at a depth of about 32–33 km. Along the segment FG of the section, there are substantially flat crustal tectonics. Beneath the eastern Bradanic foredeep the Apulian platform is affected by a marked flexure, accompanied by classic flexural faults (Figures 14, 16 and Plate 2) of clear Middle-Upper Pliocene age.

The Bradanic foredeep, as other major foredeeps in the Apennines (i.e., "Northern Apennines foredeep") is associated with a prominent transcrustal high-angle regional thrust-fault (AP-5) that displaces TL (top Lower Crust) and MO (Moho) horizontally by about 14 km and vertically by about 7 km. The horizon Z (basement of the platform) absorbes part of the horizontal displacement with a series of reverse faults confined to the sedimentary cover and upper crust. Two main load faults are present in the underthrusting block of AP-5.

From the Apulian swell area, the crustal thickness decreases, with a sharp thinning below the Bradanic foredeep. Major thinning occurs for the basement upper crust, which passes from 5 sec (15 km) in the swell area to 3.1 sec in the Bradanic zone (9 km). It is interesting to note that also in the Northern Apennines (Finetti et al., 2001) the major equivalent of transcrustal thrust fault AP-5 connected to the main foredeep occurs on a similar marked thinning of the basement upper crust.

The Apulian sedimentary crust thins from 2.6 sec (7.80 km) to 1.2 sec (3.5 km) from east to west. The lower crust diminishes it thickness from 2.65 sec (over 9 km) to 1.9 sec (6.9 km). The other major thrust fault is the Mt. Alpi thrust, already mentioned above (F. AP-3) which displaces the upper crust and the Apulian platform cover about 16 km eastward, duplexing the Permo-Mesozoic sequence. The Mt. Alpi equivalent overthrust block of Apulia probably includes slope sequences, and its well developed reflectivity is probably connected to foliation generated by the compressive tectonics and subsequent water saturation. The Mt. Alpi thrust block is much more intrinsically micro-tectonized than is exhibited in the reconstructed seismogeological section.

Other relevant tectonic elements of the section are the AP-4 left transcurrent thrust and other transcurrent thrusts of the Vallo di Diano. These thrusts as well as the Mt. Alpi fault, were generated in the late orogenic phase (Late Pliocene–Pleistocene), when the general dynamics of the area, on the physical principle of minimum work, acquired a general Ionian-direction tendency (southeastward in the place of previous early orogenic eastward) as the easier escape. The dangerously high seismicity of the Southern Apennines is mostly produced by transtensive-transpressive left crustal strike-slips affecting the Apulian platform, subparallel to the chain.

In summary, the Southern Apennines are formed from a tectonically complicate upper allochthonous edifice, as described in the first part of this paper, and by an underlying Apulian platform deformed by compressive young tectonics, that left a major part of the structures substantially rooted with relatively smaller horizontal displacements than occurred in the Northern Apennines. The total crustal shortening of the Apulian platform in the Southern Apennines amounts to about 30 km.

4. GEODYNAMIC EVOLUTION

All the new deep geophysical and geological data and the re-examination of numerous data from the literature converge toward the reconstruction of an historical geodynamics which is now much better constrained and innovative on several key points. First of all it is evident that the Southern Apennines tectonic evolution is strictly connected to the Ionian slab subduction, which had inherited previous tectonic deformations (Eo-Alpine and Balearic stages). It is also clear that the Apennine platform allochthonous units came from an internal sector, which needs to be better defined. The deep depositional basin of the Lagonegrese Units and its tectonodynamics, together with that of the internal-external flysch units, must be reconstructed in a better 3D time–space historical setting which has to fit the newly obtained CROP constraints.

Now, with the new CROP data, with regard to an historical geodynamic reconstruction we are in a privileged position, because in the place of previous inevitably uncontrolled time–space inversions of the complex allochthonous tectonic stacks observed in the thrust belt outcrops, we may now introduce objective seismic constraints. Using the CROP network data, it is possible for the first time to define the boundary of the Ionian Neotethys with clear seismic imagings of the upper slab, buried beneath the allochthonous blocks of the Tyrrhenian margins (Plate 2, paleogeographic sketch "a"). These seismic imagings show that the Adria plate (detached from Africa in the Permo-Triassic) formed the Ionian Tethys after an anticlockwise rotation of about 40° with respect to Africa, and the A, B, C angular limits of the African plate can be still superimposed on the corresponding A′, B′, C′ angular points of the detached Adria plate. Also the Alpine Tethys slab has been imaged on CROP sections M10, M11, M12A, M12B, M37, parts of M2A/I, and the southwestern part of CROP 03, allowing the reconstruction of the communicating Alpine and Ionian Neotethys of Plate 2, paleogeographic sketch "a".

This new well-constrained pre-orogenic paleogeography is a very important starting element for the understanding and reconstruction of the evolution, in which the initial setting played a major role.

The first contractional geodynamics of the *Eo-Alpine stage* (Late Cretaceous–Early Eocene) determined the closure of the Alpine Tethys sector interposed between Adria and Europe, forming the Alps (this Eo-Alpine deformation continued later on). This geodynamic stage took place with a convergence between Africa and Europe, where Africa–Ionian–Adria moved jointly as a unique block. West of the Adria–Europe collisional sector (Alps), the subduction (with obduction of ophiolites over the European margin) took place as an intra-ocean process and, after the Eo-Alpine geodynamic stage, a wide Alpine Tethys remained unsubducted (Finetti, this volume, Chapter 34). Within this vast remnant ocean, the Balearic geodynamic stage (Late Oligocene–Early Miocene) developed with subduction of the Tethyan crust beneath Europe, rotation of the Corso-Sardinian block, and collision of Corsica with the north-west Adria corner called here the *Etrurian promontory* (Plate 2, paleogeography sketch "a"). In the late Balearic stage, the closure of the Alpine Tethys was accompanied by detachment of crustal fragments of the Etrurian promontory (and Panormid promontory to the south), decoupling of sedimentary cover of such fragments, and formation of Numidian flysch, very likely at the expense of presumed quartzitic sandstone interposed between the sedimentary cover and the basement of removed fragments of the Etrurian and Panormid promontories. In fact, widespread distributions of Cambro-Ordovician quartzitic sandstone are well known as prominent outcrops or in numerous boreholes in Northern Africa, and these two promontories, before the Ionian opening, were part of Northern Africa.

While the Corsica block collided with the NW Adria margin, of which the Etrurian promontory is a part, the SE Sardinian block collided with the Panormid promontory, but most of the eastern Sardinian margin, in oceanic continuity with the subducted Alpine Tethys, entered into the Ionian Tethys (Plate 2, paleogeographic sketch c). Then, the Tyrrhenian geodynamic stage (Burdigalian to Present) started, driven by the Ionian slab retreat, with formation of the Southern Apennine–Calabrian arc–Sicily Apennine thrust belt and backarc Tyrrhenian basin opening.

The condensed cover sequence of the Ionian Tethys was stripped off from its sinking oceanic basement and obliquely transported onto the Apulian platform, where it constitutes the tectonic units known as the *Lagonegrese. Since these Ionian cover tectonic units were transported also onto Sicily, and constitute a thick wedge over the subducting oceanic crust of the Calabrian arc, it is more appropriate and immediately self explanatory to call them Ionides* (Finetti, this volume, Chapter 10).

The sedimentary cover units of the crustal fragments detached from the Etrurian promontory form, in the Southern Apennines, the allochthonous Meso-Cenozoic units known as the *Apennine platform* units (Plate 2). The magnetic anomalies of the Tyrrhenian sea represent recordings of the routes of migrating allochthonous blocks driven by the Tyrrhenian mantle flux.

Crustal blocks of the Etrurian promontory and of the Corso-Sardinian blocks with in between trapped *Alpine Tethydes (or Ligurides)*, detached from their basement (Alpine Tethys) in the Balearic stage, were stacked on the Tyrrhenian margin of Apulia.

A detailed study of the various tectonic stacks, associated flysch dynamics, and deformation history of the Southern Apennine thrust belt structures was given in Section 2.

CONCLUDING REMARKS

Connecting the CROP M-6B and CROP-04 sections, and integrating this basic data with other exploration results for the sedimentary cover (ENI-AGIP courtesy) and with new geological data, it has been possible to reconstruct a long, innovative and continuous seismogeological crustal section across the Southern Apennines, from the Marsili backarc basin to the Adriatic coast of the Apulian platform (Plate 2).

From the new information obtained it is possible to derive the following conclusions:

(1) *The Southern Apennines thrust belt is composed of two distinct tectono-stratigraphic complexes: (a) an overlying, strongly tectonized, allochthonous nappe; (b) an underlying, almost everywhere buried, Apulian platform, deformed by compressive tectonics with relatively little crustal shortening, and where most features are substantially rooted.*
(2) *The allochthonous Lagonegrese deep basin units were detached from the subducted Ionian slab and obliquely thrusted progressively from west to east over the Pliocene top levels of deformed Apulian carbonate sequences. These units are called here with the more appropriate self-explanatory name of Ionides.*
(3) *Allochthonous Apennine Platform Units of the Southern Apennines are made up of Mesozoic cover sheets coming from a reconstructed northwest Adria corner zone named here the Etrurian promontory. The detached basement upper crust and lower crust of these Etrurian promontory Units collided with the steep upper crustal slope of Apulia with generation of the Mt. Alpi regional thrust fault.*

(4) *Crustal fragments of the Corso-Sardinian block, previously thrust (Balearic stage) over the Etrurian promontory margin, subsequently, in the Tyrrhenian stage, driven by mantle flux dynamics connected to the Ionian slab retreat, detached and traveled jointly to the stacked Etrurian block as far as their present position.*
(5) *The subducted Ionian slab has been clearly seismically imaged.*
(6) *The backarc oceanic basin of Marsili, new Tyrrhenian mantle and the hypothetical underlying asthenosphere are seismically well defined.*
(7) *The Bradanic foredeep and the Mt. Alpi unit are connected to the major transcrustal thrust fault of the Southern Apennines.*
(8) *Structural and biostratigraphic analyses allow us to reconstruct times and modalities of thrust propagation of the allochthonous stacked nappe.*
(9) *The total crustal shortening of the Apulian sector of the Adria plate in the Southern Apennines amounts to only about 30 km, much less than what is observed in CROP data for the Northern Apennines, where the cumulative Adria plate shortening due to the Balearic and Tyrrhenian geodynamic stages* (Finetti et al., 2001) *is about 150 km.*

ACKNOWLEDGEMENTS

The Authors are grateful to ENI-AGIP Division for providing the exploration data which was very useful in the reconstruction of the sedimentary cover setting of segment EF in the section of Plate 2. We wish to thank Prof. W.B. Ryan of the Lamont-Doherty Earth Observatory of Columbia University for careful reviewing and fruitful suggestions. Funding for seismic reprocessing, interpretation and field geological research was provided by MURST, Grants COFIN 2000 and 2002 to I. Finetti (University of Trieste) and F. Lentini (University of Catania).

REFERENCES

Agip Mineraria, 1968. Interpretation report reconnaissance marine seismic survey Adriatic Sea: zone D. Geophys. Serv. Intern. Limited, London.

Alvarez, W., 1976. A former continuation of the Alps. Bull. Geol. Soc. Am. 87, 891–896.

Amodio-Morelli, L., Bonardi, G., Colonna, V., Dietrich, D., Giunta, G., Ippolito, F., Liguori, V., Lorenzoni, S., Paglionico, A., Perrone, V., Piccarreta, G., Russo, M., Scandone, P., Zanettin-Lorenzoni, E., Zuppetta, A., 1976. L'Arco Calabro-peloritano nell'orogene appenninico-maghrebide. Mem. Soc. Geol. It. 17, 1–60.

Ben-Avraham, Z., Boccaletti, M., Cello, G., Grasso, M., Lentini, F., Torelli, L., Tortorici, L., 1990. Principali domini strutturali originatisi dalla collisione nogenico-quaternaria nel Mediterraneo centrale. Mem. Soc. Geol. It. 45, 453–462.

Bianchi, F., Carbone, S., Grasso, M., Invernizzi, G., Lentini, F., Longaretti, G., Merlini, S., Mostardini, F., 1987. Sicilia orientale: profilo geologico Nebrodi-Iblei. Mem. Soc. Geol. It. 38, 429–458.

Boenzi, F., Ciaranfi, N., 1970. Stratigrafia di dettaglio del Flysch di Gorgoglione. Mem. Soc. Geol. It. 9, 65–79.

Boenzi, F., Ciaranfi, N., Pieri, P., 1968. Osservazioni geologiche nei dintorni di Accettura e di Oliveto Lucano. Mem. Soc. Geol. It. 7 (3), 379–392.

Bonardi, G., Amore, F.O., Ciampo, G., De Capoa, P., Miconnet, P., Perrone, V., 1988b. Il Complesso Liguride Auct.: stato delle conoscenze e problemi aperti sulla sua evoluzione pre-appenninica ed i suoi rapporti con l'Arco Calabro. Mem. Soc. Geol. It. 41, 17–35.

Bonardi, G., Cello, G., Perrone, V., Tortorici, L., Turco, E., Zuppetta, A., 1982. The evolution of the northern sector of the Calabria-peloritani arc in a semiquantitative palynspastic restoration. Boll. Soc. Geol. It. 101, 259–284.

Bonardi, G., Ciampo, G., Perrone, V., 1985. La formazione di Albidona nell'Appennino calabro-lucano: ulteriori dati biostratigrafici e relazioni con le unità esterne appenniniche. Boll. Soc. Geol. It. 104, 539–549.

Bonardi, G., D'Argenio, B., Perrone, V., 1988a. Carta geologica dell'Appennino meridionale, scala 1:250 000. S.EL.CA., Firenze.

Boni, M., Ippolito, F., Scandone, P., Zamparelli, E., Torre, V., 1974. L'Unità M. Foraporta nel lagonegrese (Appennino Meridionale). Boll. Soc. Geol. It. 93, 469–512.

Bossio, A., Mazzei, R., Monteforti, B., Salvatorini, G., 1987. Studi sul Neogene e Quaternario della Penisola Salentina. II. Evoluzione paleogeografica dell'area di Leuca nel contesto della dinamica mediterranea. In: Atti Convegno: Le conoscenze geologiche del territorio salentino. Quaderni Centro Studi Geotecnici e Ingegneria 11, Lecce.

Boullin, J.P., 1984. Nouvelle interprétation de la liason Appennin-Maghrebides en Calabre : consequences sur la palaéogéographie téthysienne entre Gibraltar et les Alpes. Rev. Géol. Dyn. Géog. Phys. 25, 321–338.

Bousquet, J.C., 1973. La tectonique recente de l'Apennin calabro-lucanien dans son cadre géologique et géophysique. Geol. Romana 12, 1–104.

Brancaccio, L., Cinque, A., 1992. Escursione nell'Appennino campano-lucano. In: Convegno su Evoluzione Geomorfologica e tettonica quaternaria dell'Appennino centro-meridionale, Napoli.

Brönnimann, P., Durand Delga, M., Grandjaquet, C., 1971. Présence simultanée de Protopeneroplis striata WEINSCHENK e de Calpionelles néocomiennes dans le "flysch galestrino" de Lucanie (Italie méridionale). Rev. Microp. 14, 96–101.

Carbone, S., Lentini, F., Sonnino, M., De Rosa, R., 1987. Il Flysch Numidico di Valsinni (Appennino Lucano). Boll. Soc. Geol. It. 106, 331–345.

Carbone, S., Catalano, S., Lentini, F., Monaco, C., 1988. Le unità stratigrafico-strutturali dell'Alta Val d'Agri (Appennino Lucano) nel quadro dell'evoluzione del sistema Catena-Avanfossa. Mem. Soc. Geol. It. 41, 331–341.

Carbone, S., Catalano, S., Lazzari, S., Lentini, F., Monaco, C., 1991. Presentazione della Carta Geologica del Bacino del Fiume Agri (Basilicata). Mem. Soc. Geol. It. 47, 129–143.

Carella, R., 1963. Eruttivi di S. Giovanni in Pane e della Punta delle Pietre Nere (Gargano). Boll. Soc. Geol. It. 82.

Carta geologica del Bacino del Fiume Agri, 1991. Scala 1:50 000 (2 fogli), Diret. rilev. F. Lentini. S. EL.CA., Firenze.

Cassinis, R., Franciosi, R., Scarascia, S., 1979. The structure of the Earth's Crust in Italy. A preliminary typology based on seismic data. Boll. Geof. Teor. Appl. 21.

Ciaranfi, N., 1972. Il Flysch di Gorgoglione. Boll. Serv. Geol. It. 92, 101–114.

Ciaranfi, N., Loiacono, F., 1983. Il Flysch Numidico nel quadro dell'evoluzione tettonico-sedimentaria inframiocenica dell'Appennino Meridionale. Studi Geol. Geof. sulle Regioni Pugliese e Lucania 16, 3–43.

Ciaranfi, N., Pieri, P., Ricchetti, G., 1988. Note alla carta geologica delle Murge e del Salento (Puglia centro meridionale). Mem. Soc. Geol. It. 41, 449–460.

iarapica, G., Cirilli, S., Martini, R., Zaninetti, L., 1986. Une microfaune à petits Foraminifères d'age permien remaniée dans le Trias moyen de l'Apennin méridional (Formation du Monte Facito, Lucanie occidentale). Description de Crescentia Vertebralis, n. gen. n. sp. Rev. Paléobiol. 5, 207–215.

Ciarapica, G., Cirilli, S., Panzanelli Frantoni, R., Salvini-Bonnard, G., Zaninetti, L., 1990. Reworked Foraminifera in the Monte Facito Formation, Lagonegro Basin, Southern Apennines, Italy. Boll. Soc. Geol. It. 109, 151.

Ciarapica, G., Passeri, L., this volume. Ionian Tethydes in Southern Apennines. Chapter 11.

Cocco, E., Cravero, E., Ortolani, F., Pescatore, T., Russo, M., Sgrosso, I., La Torre, M., 1972. Les facies sedimentaires du Bassin Irpinien (Italie meridionale). Atti Acc. Pontaniana, n.s. 21, 13 p., Napoli.

Cocco, E., Cravero, E., Ortolani, F., Pescatore, T., Russo, M., Torre, M., Coppola, L., 1974. Le Unità Irpine nell'area a nord di Monte Marzano, Appennino meridionale. Mem. Soc. Geol. It. 13, 607–654.

Colombi, B., Giese, P., Luongo, G., et al., 1973. Preliminary report on the seismic refraction profile Gargano-Palermo-Pantelleria. Boll. Geofis. Teor. Appl. 15, 225–254.

Crescenti, U., 1966. Osservazioni sulla stratigrafia dell'Appennino meridionale alla luce delle recenti ricerche micropaleontologiche. Boll. Soc. Geol. It. 85 (2), 541–579.

Critelli, S., 1991. Evoluzione delle mode detritiche delle successioni arenitiche terziarie dell'Appennino meridionale. Mem. Soc. Geol. It. 47, 55–93.

Crostella, A., Vezzani, L., 1964. La geologia dell'Appennino foggiano. Boll. Soc. Geol. It. 83, 121–141.

D'Alessandro, A., Loiacono, F., Ricchetti, G., 1987. Note illustrative della Carta geomorfica del Salento meridionale F. 525 Gallipoli, 526 Nardò, 527 Otranto. Quad. Ric. Centro Studi Geot. Ing. 11, 207–222.

D'Argenio, B., Pescatore, T., Scandone, P., 1973. Schema geologico dell'Appennino meridionale (Campania, Lucania). In: Atti Convegno: Moderne vedute sulla geologia dell'Appennino. Acc. Naz. Lincei, Quaderno 183, 49–72, Roma.

D'Argenio, B., Pescatore, T., Scandone, P., 1975. Structural pattern of the Campania–Lucania Apennines. In: Ogniben, L., Parotto, M., Praturlon, A. (Eds.), Structural Model of Italy. Quad. Ric. Scient. 90, 313–327, 2 tavv., Roma.

Dazzaro, L., Rapisardi, L., 1996. Schema geologico del margine appenninico tra il F. Fortore e il F. Ofanto. Mem. Soc. Geol. It. 51, 143–147.

De Alfieri, A., Guzzi, R., Sacchi, M., D'Argenio, B., Perrone, V., Zamparelli, V., 1987. Monte Foraporta Unit: a minor element of Southern Apennine nappe pile. Stratigraphic and tectonic study. Rend. Soc. Geol. It. 9, 171–178.

De Blasio, I., Lima, A., Perrone, V., Russo, M., 1978. Studio petrografico e biostratigrafico di una sezione della Formazione del Saraceno nell'area-tipo (Calabria nord-orientale). Riv. It. Paleont. Strat. 84 (4), 947–972.

De Blasio, I., Lima, A., Perrone, V., Russo, M., 1981. Nuove vedute sui depositi miocenici della Penisola Sorrentina. Boll. Soc. Geol. It. 100, 57–70.

De Fino, M., La Volpe, L., Piccarreta, G., 1981. Geochemistry and petrogenesis of the Paleocene platform magmatism at Punta delle Pietre Nere (Southeastern Italy). N. Jb. Miner. Abh. 142, 161–177.

Dell'Anna, L., Pizzirani, L., 1964. I minerali delle rocce ignee di Punta delle Pietre Nere presso il Lago di Lesina (Foggia). Period. Mineral. 34.

Donzelli, G., Crescenti, U., 1970. Segnalazione di una microbiofacies permiana, probabilmente rimaneggiata, nella Formazione di M. Facito (Lucania occidentale). Boll. Soc. Natur. in Napoli 79, 59–75.

Finetti, I.R., 1982. Structure, stratigraphy and evolution of central Mediterranean. Boll. Geof. Teor. Appl. 24, 247–426.

Finetti, I.R., 2002a. Lithospheric tectono-stratigraphic setting and time–space stacking process of the Calabrian arc from new CROP seismic data. In: Abstract Proceedings of 81st Riunione Estiva of the Società Geologica Italiana, Torino, 10–12 September, 2002.

Finetti, I.R., 2002b. Understanding Ionides and Calabrian arc from new CROP seismic data. Expanded Abstract. In Proceedings of the 21° Convegno Nazionale of the GNGTS–EAGE 2nd Italian Section Convention, Rome, 19–21 November, 2002.

Finetti, I.R., this volume. Crustal tectono-stratigraphic sections across the Western and Eastern Alps from ECORS-CROP and TRANSALP seismic data. Chapter 7.

Finetti, I.R., this volume. Understanding the Ionides and their geodynamics. Chapter 10.

Finetti, I.R., this volume. Geodynamic evolution of the Mediterranean Region, from the Permo-Triassic Ionian opening to the Present, constrained by new lithospheric CROP seismic data. Chapter 34.

Finetti, I.R., Del Ben, A., 2000. Ionian opening, crustal stratigraphy and tectono-dynamics of the Pelagian Sea region from new CROP seismic data. In: Proceedings of the International Symposium on Geology of NW Libya, Tripoli, 2000.

Finetti, I.R., Del Ben, A., this volume. Ionian Tethys lithosphere roll-back sinking and back-arc Tyrrhenian opening from new CROP seismic data. Chapter 21.

Finetti, I.R., Lentini, F., Carbone, S., Catalano, S., Del Ben, A., 1996. Il sistema Appennino meridionale-Arco Calabro-Sicilia nel Mediterraneo centrale: studio geologico-geofisico. Boll. Soc. Geol. It. 115, 529–559.

Finetti, I., Boccaletti, M., Bonini, M., Del Ben, A., Geletti, R., Pipan, M., Sani, F., 2001. New crustal section based on CROP seismic data across the North Tyrrhenian–Northern Apennines–Adriatic Sea. Tectonophysics 343, 135–163.

Finetti, I.R., Lentini, S., Carbone, S., Del Ben, A., Di Stefano, A., Forlin, E., Guarnieri, P., Pipan, M., this volume. Geological outline of Sicily and lithospheric tectono-dynamics of its Tyrrhenian margin from new CROP seismic data. Chapter 15.

Fornaciari, E., Di Stefano, A., Rio, D., Negri, A., 1996. Middle Miocene quantitative calcareous nannofossil biostratigraphy in the Mediterranean region. Micropaleontology 42 (1), 37–63.

Fornaciari, E., Rio, D., 1996. Latest Oligocene to Early Middle Miocene quantitative calcareous nannofossils biostratigraphy in the Mediterranean region. Micropaleontology 42 (1), 1–37.

Ghezzi, G., Bayliss, D.D., 1963. Uno studio del Flysch nella regione calabro-lucana. Stratigrafia, tettonica e nuove idee sul Miocene dell'Appennino meridionale. Boll. Serv. Geol. It. 84, 3–64.

Grandjacquet, C., Mascle, G., 1978. The structure of the Ionian Sea, Sicily and Calabria–Lucania. In: Nairn, Kanes, Stheli (Eds.), In: The Ocean Basin and Margins, vol. 4B. Plenum, New York, pp. 257–329.

Haccard, D., Lorenz, C., Grandjaquet, C., 1972. Essai sur l'evolution tectonogénètique de la liasison Alpes-Apennines (de la Ligurie à la Calabre). Mem. Soc. Geol. It. 11, 309–381.

Ietto, A., Barilardo, A.M., 1993. L'unità di San Donato quale margine deformato cretacico-paleogene del bacino di Lagonegro (Appennino meridionale-Arco Calabro). Boll. Soc. Geol. It. 112, 477–496.

Knott, S., 1987. The Liguride Complex of Southern Italy: a Cretaceous to Paleogene accretionary wedges. Tectonophysics 142, 217–226.

Lanzafame, G., Spadea, P., Tortorici, L., 1978. Provenienza ed evoluzione dei Flysch cretacico-eocenici della regione calbro-lucana. II: Relazioni tra ofioliti e Flysch cabro-lucano. Ofioliti 3 (2–3), 189–210.

Lanzafame, G., Spadea, P., Tortorici, L., 1979. Mesozoic ophiolites of Northern Calabria and Lucanian Apennine (Southern Italy). Ofioliti 4, 173–182.

Lentini, F., 1979. Le Unità Sicilidi della Val d'Agri (Appennino Lucano). Geol. Romana 18, 215–224.

Lentini, F., 1980. Carta geologica del Bacino dell'Agri F. 507, Pisticci. Scala 1:50 000. S.EL.CA., Firenze.

Lentini, F., Carbone, S., Catalano, S., 1994. Main structural domains of the central Mediterranean region and their Neogene tectonic evolution. Boll. Geof. Teor. Appl. 36, 103–125.

Lentini, F., Carbone, S., Catalano, S., Grasso, M., 1995. Principali lineamenti strutturali della Sicilia nord-orientale. Studi Geol. Camerti vol. spec. 2, 319–329.

Lentini, F., Carbone, S., Catalano, S., Monaco, C., 1987. Confronti sedimentologio-petrografici e posizione strutturale dei Flysch di Albidona e di Gorgoglione nella media val d'Agri (Appennino Lucano). Mem. Soc. Geol. It. 38, 259–273.

Lentini, F., Carbone, S., Catalano, S., Monaco, C., 1990. Tettonica a thrust neogenica nella Catena appenninico-maghrebide: esempi dalla Lucania e dalla Sicilia. Studi Geologici Camerti vol. spec., 19–26.

Lentini, F., Carbone, S., Di Stefano, A., Guarnieri, P., 2002. Stratigraphical and structural constraints in the Lucanian Apennines (southern Italy): tools for reconstructing the geological evolution. J. Geodynamics 34, 141–158.

Lentini, F., Vezzani, L., 1974. Note illustrative del foglio 506 S. Arcangelo. I.R.P.I. Cosenza, 46 p.

Marsella, E., Bally, A.W., Cippitelli, G., D'Argenio, B., Pappone, G., 1995. Tectonic history of the Lagonegro Domain and Southern Apennine thrust belt evolution. Tectonophysics 252, 307–330.

Marsella, E., Kozur, H., D'Argenio, B., 1991. Monte Facito Formation (Schitian-middle Carnian). A deposit of the ancestral Lagonegro Basin in the Southern Apennines. Boll. Serv. Geol. It. 110, 225–248.

Marsella, E., Pappone, G., D'Argenio, B., Cippitelli, G., Bally, A.W., 1992. L'origine interna dei terreni lagonegresi e l'assetto tettonico dell'Appennino meridionale. Rend. Acc. Sci. della Soc. Naz. Sc. Let. ed Arti Napoli 59, 73–101.

Martini, E., 1971. Standard Tertiary and Quaternary calcareous nannoplankton zonation. In: Farinacci, A. (Ed.), Proceedings of the II Planktonic Conference, Roma, 1970, vol. 2. Edizioni Tecnoscienza, pp. 738–785.

Martini, R., De Wever, P., Zaninetti, L., Denelian, T., Kito, N., 1989. Les radiolarites triassique de la Formation du M. Facito Auctt. (Bassin de Lagonegro, Italie méridionale). Rev. Paleobiol. 8 (1), 143–161.

Masse, J.P., Borgomano, J., 1987. Un modèle de le transition plate-forme-basin carbonaté controlé par des phénomenes tectoniques: le Crétacé du Gargano (Italie méridionale). C. R. Acad. Sci. Paris 304.

Mazzotti, A.P., Stucchi, E., Fradelizio, G.L., Zanzi, L., Scandone, P., 2000. Seismic exploration in complex terrains: A processing experience in the Southern Apennines. Geophysics 65, 1402–1417.

Miconnet, P., 1983. La region de Lagonegro (Italie méridionale) : évolution géologique d'un bassin dans son cadre alpin. Thèse, Lille, 185 p.

Miconnet, P., 1988. Evolution mesozoique du secteur de Lagonegro. Mem. Soc. Geol. It. 41, 321–330.

Mietto, P., Panzanelli Fratoni, R., 1990. Conodonts from the Monte Facito Formation and from the base of the Monte Sirino Formation (Lagonegro sequence). Boll. Soc. Geol. It. 109, 165–169.

Monaco, C., 1993. Le unità liguridi nel confine calabro-lucano (Appennino Meridionale): revisione dei dati esistenti, nuovi dati ed interpretazione. Boll. Soc. Geol. It. 112, 751–769.

Monaco, C., Tansi, C., Tortorici, L., De Francesco, A., Morten, L., 1991. Analisi geologico-strutturale dell'Unità del Frido nel Confine calabro-lucano (Appennino Meridionale). Mem. Soc. Geol. It. 47, 341–353.

Monaco, C., Tortorici, L., Morten, L., Critelli, S., Tansi, C., 1995. Geologia del versante nord-orientale del Massiccio del Pollino (confine Calabro-Lucano): nota illustrativa sintetica della carta geologica alla scala 1:50 000. Boll. Soc. Geol. It. 114, 277–291.

Monaco, C., Tortorici, L., Palrtinieri, W., 1998. Structural evolution of the Lucanian Appennines, Southern Italy. J. Structural Geology 20, 617–638.

Morelli, C., Carrozzo, T., Ceccherini, P., Finetti, I., Gantar, C., Pisani, M., Schmidt Di Friedberg, P., 1979. Regional geophysical study of the Adriatic Sea. Boll. Geof. Teor. Appl. 11 (41–42), Trieste.

Narciso, G., Ortolani, F., Sanzo', A., 1975. Prime considerazioni sulla presenza del Flysch Numidico nell'Appennino Sannita. Boll. Soc. Natur. in Napoli 86, 1–16.

Ogniben, L., 1969. Schema introduttivo alla geologia del confine calabro-lucano. Mem. Soc. Geol. It. 8, 435–763.

Ogniben, L., 1973. Schema geologico della Calabria in base ai dati odierni. Geol. Romana 12, 243–585.

Ogniben, L., 1985. Relazione sul modello geodinamico "conservativo" della regione italiana. Commissione ENEA-ENEL per lo studio dei problemi sismici connessi con la realizzazione di impianti nucleari, ENEA, 357 p.

Ortolani, F., 1974. Faglia trascorrente pliocenica nell'Appennino Campano. Boll. Soc. Geol. It. 93, 609–622.

Ortolani, F., Torre, M., 1971. Il Monte Alpi (Lucania) nella paleogeografia dell'Appennino meridionale. Boll. Soc. Geol. It. 90, 213–248.

Palmentola, G., 1967. Sui rapporti tra la "Formazione di Stigliano" e la "Formazione di Serra Palazzo" nei dintorni di Tolve (Potenza). Boll. Soc. Natur. in Napoli 76, 291–297.

Palmentola, G., 1970. Nuovi dati e considerazioni sulla Formazione di Serra Palazzo in Lucania. Mem. Soc. Geol. It. 9, 81–90.

Patacca, E., Sartori, R., Scandone, P., 1990. Tyrrhenian basin and apenninic arcs: kinematic relations since late Tortonian times. Boll. Soc. Geol. It. 45, 425–451.

Patacca, E., Scandone, P., 2001. Late thrust propagation and sedimentary response in the thrust belt-foredeep system of the Southern Apennines Pliocene-Pleistocene). In: Vai, G.B., Martini, I.P. (Eds.), Anatomy of an Orogen: The Apennines and Adjacent Mediterranean Basins. Kluwer Academic, Dordrecht, pp. 401–440.

Patacca, E., Scandone, P., Bellatalla, M., Perilli, N., Santini, U., 1992a. The Numidian sand event in the Southern Apennines. Mem. Sci. Geol. Padova 43, 297–337.

Patacca, E., Scandone, P., Bellatalla, M., Perilli, N., Santini, U., 1992b. La zona di giunzione tra l'arco appenninico settentrionale e l'arco appenninico meridionale nell'Abruzzo e nel Molise. Studi Geol. Camerti spec. vol. 11, 417–441, CROP.

Pescatore, T., 1978. Evoluzione tettonica del Bacino Irpino (Italia meridionale) durante il Miocene. Boll. Soc. Geol. It. 97 (5–6), 783–805.

Pescatore, T., Sgrosso, I., Torre, M., 1970. Lineamenti di tettonica e sedimentazione nel Miocene dell'Appennino campano-lucano. Boll. Soc. Natur. in Napoli 78, 337–408.

Ricchetti, G., 1980. Contributo alla conoscenza strutturale della Fossa bradanica e delle Murge. Boll. Soc. Geol. It. 49, 421–430.

Ricchetti, G., Ciaranfi, N., Luperto Sinni, E., Mongelli, F., Pieri, P., 1988. Geodinamica ed evoluzione sedimentaria e tettonica dell'Avampaese apulo. Mem. Soc. Geol. It. 41, 57–82.

Santo, A., Senatore, M.R., 1988. La successione stratigrafica dell'Unità Dauna a Monte Sidone (Castelluccio Valmaggiore-Foggia). Mem. Soc. Geol. It. 41, 431–438.

Santo, A., 1987. Sgrosso, I., Alcune precisazioni sulle trasgressioni mioceniche nell'Appennino meridionale. Mem. Soc. Geol. It. 38, 225–240.

Sartoni, S., Crescenti, U., 1962. Ricerche biostratigrafiche nel Mesozoico dell'Appennino meridionale. Giorn. Geologia 29 (s.2), 159–302.

Scandone, P., 1963a. Stratigrafia degli scisti silicei della Lucania. Mem. Soc. Geol. It. 4 (2), 9 p.

Scandone, P., 1963b. Trasgressioni mesozoiche e terziarie nell'alta Val d'Agri tra Paterno e Marsico Nuovo (Potenza). Boll. Soc. Natur. in Napoli 72, 125–131.

Scandone, P., 1964a. Marnoscisti ad Halobia in Lucania. Boll. Soc. Natur. in Napoli 72 (1963), 207–212.

Scandone, P., 1964b. Nota preliminare sui foraminiferi delle scogliere triassiche della Lucania. Boll. Soc. Natur. in Napoli 72 (1963), 267–269.

Scandone, P., 1967. Studi di geologia lucana: 1a serie calcareo-silico-marnosa e i suoi rapporti con l'appennino calcareo. Boll. Soc. Natur. in Napoli 76, 301–469.

Scandone, P., 1972. Studi di geologia lucana: Carta dei terreni della serie calcareo-silico-marnosa e note illustrative. Boll. Soc. Natur. in Napoli 81, 255–300.

Scandone, P., 1982. Structure and evolution of the Calabrian Arc. Earth Evol. Sci. 3, 172–180.

Scandone, P., De Capoa, P., 1966. Sulla posizione stratigrafica e l'età dei livelli a Daonella e Halobia in Lucania. Boll. Soc. Natur. in Napoli 75, 30–39.

Scandone, P., Bonardi, G., 1967. Synsedimentary tectonics controlling deposition of Mesozoic and Tertiary carbonatic sequences of areas surrounding Vallo di Diano (Southern Apennines). Mem. Soc. Geol. It. 7, 1–10.

Selli, R., 1957. Sulla trasgressione del Miocene nell'Italia meridionale. Giorn. Geologia 26 (s.2), 1–54.

Selli, R., 1962. Il Paleogene nel quadro della geologia dell'Italia meridionale. Mem. Soc. Geol. It. 3, 737–789.

Servizio Geologico d'Italia, 1969. Foglio 211 (S. Arcangelo) della Carta Geologica d'Italia, 1:100 000, direttore L. Ogniben. Libreria dello Stato, Roma.

Servizio Geologico d'Italia, 2002. Carta geologica d'Italia, F. 506 Sant'Arcangelo, scala 1:50 000, coordinatore F. Lentini. Poligrafico e Zecca dello Stato, Roma, in stampa.

Sgrosso, I., 1981. Il significato delle calciruditi di Piaggine nell'ambito degli eventi del Miocene inferiore nell'Appennino Campano-Lucano. Mem. Soc. Geol. It. 100, 129–137.

Sgrosso, I., 1988. Possibile evoluzione cinematica miocenica nell'Orogene centro-sud-appenninico. Boll. Soc. Geol. It. 117, 679–724.

Sgrosso, I., 1992. Nuovi dati biostratigrafici sul Miocene di M. Alpi (Lucania) e conseguenti ipotesi paleogeografiche. Mem. Soc. Geol. It. 41 (1988), 343–351.

Spadea, P., 1976. I carbonati delle rocce metacalcaree della Formazione del Frido della Lucania. Ofioliti 1, 431–465.

Spadea, P., 1979. Contributo alla conoscenza dei metabasalti ofiolitici della Calabria settentrionale e centrale e dell'Appennino Lucano. Rend. Soc. It. Min. e Petr. 35, 251–276.

Spadea, P., 1982. Continental rocks associated with ophiolites in Lucania Apennines. Ofioliti 7, 501–522.

Vezzani, L., 1966a. La formazione trasgressiva supraeocenica di Castelluccio Superiore (Potenza). Boll. Soc. Geol. It. 85 (2), 287–297.

Vezzani, L., 1966b. Nota preliminare sulla stratigrafia della Formazione di Albidona. Boll. Soc. Geol. It. 85, 767–776.

Vezzani, L., 1966c. La sezione stratigrafica di Caliandro nel Pliocene medio-inferiore della Val d'Agri (Lucania). Riv. Ital. Paleont. e Strat. 72, 191–229.

Vezzani, L., 1968. Distribuzione, facies e stratigrafia della Formazione del Saraceno nell'area compresa tra il Mare Jonio ed il Torrente Frido. Geol. Romana 7, 229–276.

Vezzani, L., 1969. La Formazione del Frido (Neocomoiano-Aptiano) tra il Pollino ed il Sinni (Lucania). Geol. Romana 8, 129–176.

Wood, A.W., 1981. Extensional tectonics and the birth of the Lagonegro Basin (southern Italian Apennines). N. Jb. Geol. Palaont. Abh. 161, 93–131.

CROP PROJECT: Deep Seismic Exploration of the Central Mediterranean and Italy
Edited by I.R. Finetti

CHAPTER 13

Magnetotelluric Profiling along the CROP-04 Section in the Southern Apennines

D. Patella[1], Z. Petrillo[2], A. Siniscalchi[3], L. Improta[4], B. Di Fiore[1,2]

ABSTRACT

We show the results of a magnetotelluric (MT) survey along a SW–NE profile from the Tyrrhenian to the Adriatic sea close to the CROP-04 transect, integrated with a NW–SE profile placed on the Apulia foreland. The NW–SE profile displays a nearly 1D apparent resistivity pattern, which is used to constrain the 2D modelling along the profile that crosses the Southern Apennine range. To improve the 2D model across the main SW–NE profile, a combined MT-gravity analysis is carried out until mutual consistency is reached. Starting from the southwestern side of the SW–NE profile, the MT cross-section allows the shallow Tertiary basinal units, with resistivity less than 100 Ωm, to be differentiated from the rootless nappes of the Apennine belt, with resistivity less than 1000 Ωm and thickness up to 7 km, which mainly include the Western carbonate platform and Lagonegro basin Mesozoic sequence. Proceeding downward in the southwestern portion of the MT SW–NE cross-section, a layer, about 1 km thick and with resistivity in the range 3–30 Ωm, is found between the Apennine thrust sheets and a lower stratum with resistivity less than 10 000 Ωm, extending down to the bottom of the model placed at 15 km of depth, which is assumed to correspond to the Apulia carbonate platform. As the conductivity of the thin layer is consistent with the presence of high fluid concentrations, the thin layer is assumed to indicate a major decollement level. The whole SW–NE MT cross-section delineates the large-scale trend of the top of the Apulia carbonate platform. Its SW-bending beneath the Bradano Trough and marked deepening down to 5 km beneath the Ofanto basin and San Fele antiform conforms to the results from the CROP-04 seismic profile. A body with resistivity greater than 10000 Ωm, found beneath the Apulia platform at a depth ranging from about 6 km beneath the Apulia foreland and likely down to more than 15 km beneath the Ofanto basin, is interpreted as the crystalline basement. It appears to be separated from the overlying Apulia platform by a conductive layer ascribed to Permo-Triassic clastic sequences. A noticeable feature of the resistivity model across the SW–NE profile is the presence of two laterally-bounded vertical bodies with resistivity less than 3 Ωm, cutting the whole Apulia platform and part of the Apennine nappes, beneath the Sele graben and Ofanto basin, respectively. Accounting for geochemical, geothermal and active and passive seismological data, such low resistivity regions are interpreted as highly fractured zones characterized by melt intrusions and/or fluid circulation. Referring to the MT model across the NW–SE secondary profile, a quite simple structure is observed crossing the foreland-foredeep areas. A slow bending of the Apulia platform and a thickening of the much less resistive deposits of the Bradano Trough are observed proceeding northwestwards. A remarkable result is that the Apulia platform does not show any important variation of thickness along the entire NW–SE section.

Keywords: Magnetotellurics along CROP-04, Southern Apennines

[1]Department of Physical Sciences, University Federico II, Naples, Italy
[2]National Institute of Geophysics and Volcanology, Vesuvian Observatory, Naples, Italy
[3]Department of Geology and Geophysics, University of Bari, Bari, Italy
[4]National Institute of Geophysics and Volcanology, Rome, Italy

1. INTRODUCTION

The Southern Apennines are a Neogene E-verging thrust-and-fold belt generated above the western margin of the subducting Apulian and Ionian lithosphere. The belt accreted eastwards by incorporating Meso-Cenozoic basinal and platform paleogeographic domains within an accretionary prism (D'Argenio et al., 1973). Compression along the outer fronts of the Southern Apennines was synchronous with extension in the Tyrrhenian back-arc basin and along the internal zone of the belt, and with the flexure of the Apulian foreland lithosphere, which was responsible for the development of a deep foredeep (Figure 1). According to Malinverno and Ryan (1986) and Patacca and Scandone (1989), from the Upper Miocene to Lower Pleistocene, this basin-belt-foredeep system migrated progressively eastwards due to the roll-back of the passively subducting foreland lithosphere. A major change in the tectonic regime occurred about 700 ky ago, when the belt and foredeep areas underwent a NE-oriented extension (Patacca and Scandone, 2001). The intense seismic activity of the Southern Apennines, characterized by earthquakes of up to 7.1 magnitude, together with borehole breakout data, demonstrate that this stress regime is still active (Amato and Montone, 1997).

Despite intense hydrocarbon exploration, the internal geometry in the axial and internal zone of the thrust-and-fold belt is not yet well understood, due to the extreme structural complexity of the Southern Apennines. All structural models so far constructed by integrating surface geology, well data and reflection seismic profiles (Mostardini and Merlini, 1986; Casero et al., 1988; Patacca and Scandone, 1989; Roure et al., 1991; Menardi and Rea, 2000) show a tectonic stack of rootless nappes overlying the Apulia Carbonate Platform (ACP), which belongs to the Apulian continental lithosphere.

The ACP outcrops widely in the foreland (Figure 1). As documented by deep subsurface data (Puglia 1 well; Ricchetti, 1994) the succession is up to 7 km thick and consists of Miocene-Jurassic limestones and Triassic anhydrites, overlying Permo-Triassic clastic deposits (Verrucano fm.). Moving westwards, the top of the ACP, which corresponds to evident reflectors in the commercial seismic profiles (Roure et al., 1991), follows the westward bending of the Apulian lithosphere and is covered by thick terrigenous sequences of the Pliocene foredeep. The ACP progressively dips westwards below the outer thrust system, reaching a depth of 5–6 km. In the axial zone of the belt, the ACP is instead deeply involved in the thrust-and-fold system and often structured as wide antiforms.

Since the '70s, many models have been formulated for the tectonic evolution and paleogeographic position of the units incorporated in the stack of rootless nappes. Most models assume the following arrangement from east to west (Casero et al., 1988; Patacca and Scandone, 1989; Roure et al., 1991; Menardi and Rea, 2000): (a) units of the Lagonegro basin (LB), originally located west of the Apulia domain, which were thrust onto the ACP; (b) carbonate thrust sheets of the Western carbonate platform (WCP), with its cover of Upper Miocene flysch sequences belonging to the foredeep phase; (c) basinal units deposited in more internal oceanic paleogeographic domains as the Liguride and Sicilide units, which were incorporated in the accretionary wedge before the opening of the Tyrrhenian basin and now form the highest structural units of the Southern Apennines. All these units are in turn covered by Late Miocene–Pliocene thrust-sheet-top deposits filling satellite basins (Patacca and Scandone, 2001). For alternative models of tectonic evolution and palinspastic reconstruction see D'Argenio et al. (1973), Sgrosso (1986), Mostardini and Merlini (1986), Pescatore et al. (1988) and Marsella et al. (1995).

In this chapter, we show the results of a magnetotelluric (MT) survey consisting of two profiles, along which 21 reliable sounding stations (Figure 1) with a spacing between adjacent sites varying from 3 to 15 km were selected from a total of 25. The MT primary profile trends SW–NE, parallel to the CROP-04 transect about 15 km to north, and extends from the Tyrrhenian coast to the Adriatic margin, crossing the very complex Apennine range and Bradano Trough. The central part of the profile crosses the epicentral region of the large (Ms 6.9) 1980 Campania-Lucania normal faulting earthquake (Bernard and Zollo, 1989). The MT secondary profile trends NW–SE and extends from the Gargano promontory to the Murge area, where the ACP crops out, crossing the Apulia foreland. The latter profile was designed mainly to calibrate the resistivity sequence across a foreland crustal section where a much less complex structure was to be expected. Both profiles are tied to several deep oil exploration wells.

The purpose of the primary MT profile was to yield useful information on the large-scale structural setting of the upper crust in the Southern Apennines. In particular, we aimed at investigating whether MT might give new insights on:

- the upper crust in the inner axial zone of the belt, where reflection seismic imaging is often hampered by the extreme inhomogeneity of the shallow structure and by significant static problems (Mazzotti et al., 2000; Dell'Aversana et al., 2000);

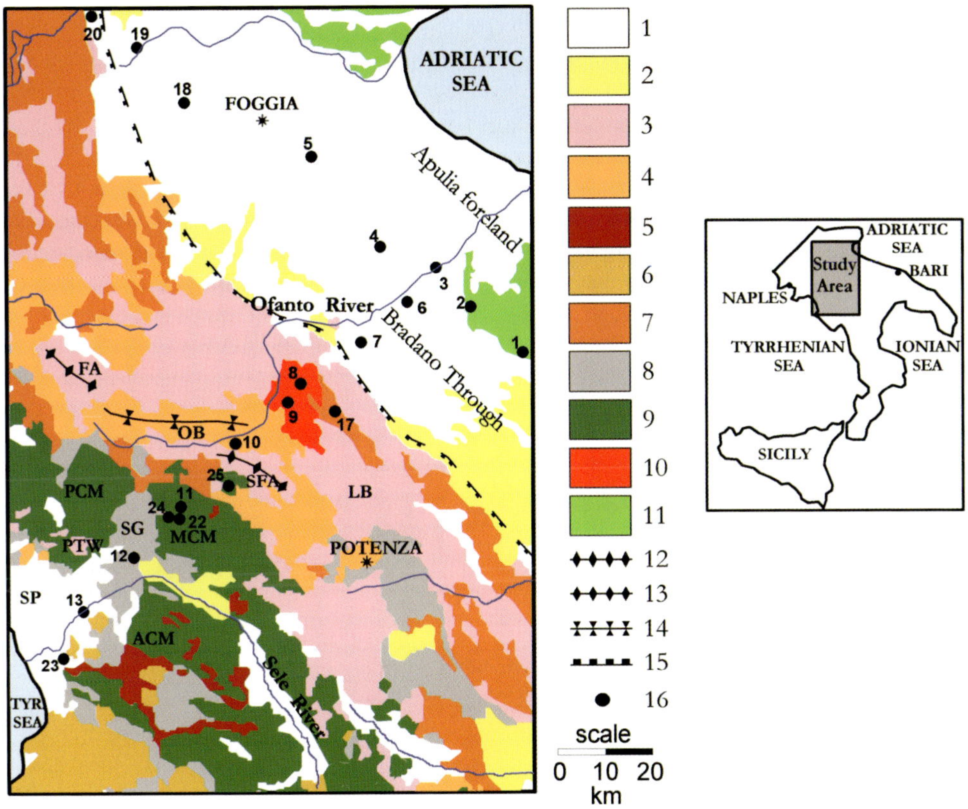

Figure 1: Geological outline of the study area (modified from Bonardi et al., 1988). [1 = Quaternary; 2 = Bradano sedimentary cycle (Pleistocene); 3 = Lagonegro basin sequences (Trias-Miocene); 4 = Thrust-top sheet deposits (Miocene-Pliocene); 5 = Piaggine-Raganello formation; 6 = Pollica-S. Mauro-Albindona formation; 7 = Flysch deposits (Upper Miocene); 8 = Internal units; 9 = Western carbonate platform; 10 = Vulture volcano; 11 = Apulia carbonate platform; 12 = San Fele antiform; 13 = Frigento antiform; 14 = Ofanto basin; 15 = Apennines buried frontal thrust; 16 = MT sounding station. LB, Lagonegro basin; OB, Ofanto basin; SFA, San Fele antiform; FA, Frigento antiform; PCM, Picentini carbonate massif; MCM, Marzano carbonate massif; ACM, Alburni carbonate massif; PTW, Picentini tectonic window; SP, Sele plain; SG, Sele graben].

- the upper crust beneath the ACP in the Apulia foreland, whose nature is still uncertain, as the only two deep wells (Puglia 1 and Gargano 1; Ricchetti, 1994) penetrating the bottom of the ACP do not reach the crystalline basement, and accurate seismic velocity models are not yet available due to the lack of good quality refraction/wide-angle reflection seismic profiles;
- the electrical properties of the carbonate rocks of the ACP buried beneath the axial zone of the Apennine range, where recent geochemical investigations (Italiano et al., 2000) suggest the presence of fluid intrusions along deep lithospheric faults.

2. GEOLOGICAL OUTLINE ALONG THE MT PROFILES

The SW–NE MT primary profile runs nearly orthogonally to the main fronts and shallow tectonic features of the Southern Apennines (Figure 1). Close to the Tyrrhenian margin, the profile crosses the Sele Plain (SP), which is a Plio-Pleistocene tectonic depression located in the inner zone of the Apennines, subject to extensional tectonics since Pliocene times (Cinque et al., 1993). The substratum is formed by deposits of the WCP, which crops out in the surrounding Picentini (PCM), Marzano (MCM) and Alburni (ACM) carbonate massifs. Large normal faults bound the plain to the north and east and lower the WCP, which rapidly deepens beneath the Plio-Pleistocene marine and continental sediments of the plain (Mostardini and Merlini, 1986).

Proceeding northeastwards, the SW–NE MT profile crosses the southern termination of the Sele graben (SG), located between the PCM and MCM horsts. The PCM and MCM consist of highly fractured Mesozoic limestones of the WCP, exceeding a thickness of 4,000 m. Along the north-eastern margin of the MCM horst, the Mesozoic carbonate sequences, with discontinuous cover of Upper Miocene flysch deposits, are thrust onto the units of the Lagonegro basin (LB) and are in turn tectonically overlain by mainly clayey deposits referable to the Sicilide nappe (Patacca and Scandone, 1989). This tectonic stacking, documented by the Contursi 1 well (Mostardini and Merlini, 1986), is also exposed in the Picentini tectonic window (PTW).

A few km northeast of the MCM, the profile crosses the WNW-trending San Fele antiform (SFA), where Mesozoic basinal rocks of the LB outcrop. As documented by the S. Fele well (Improta et al., 2003), the SFA is formed by a tectonic stacking of Mesozoic slices of the LB (mainly cherty-limestones and cherts), more than 5 km thick, alternating with thin Tertiary, mainly clayey decollement levels. On the eastern flank of the SFA, which to the west is tectonically covered by the WCP, thrust-sheet-top Pliocene sequences representing the infill of the Ofanto basin (OB) are folded (Hippolyte et al., 1994). To the north, Pliocene deposits of the OB disconformably overlay Tertiary sequences of Lagonegro and Sicilide units.

In the external zone of the belt, Tertiary basinal sequences, with their cover of thrust-sheet-top Pliocene sequences, are folded and thrust onto Pliocene terrigenous deposits of the Bradano foredeep (Casero et al., 1988). The outer thrusts are in turn covered by Pleistocene volcanites of the Vulture volcano and by post-orogenic shelf-to-continental clastic deposits of the Bradano cycle (Patacca and Scandone, 2001). East of the buried frontal ramp of the Apennines, the ACP rapidly rises toward the foreland, where the Apulia carbonates outcrop a few km east of the profile.

Crossing the foreland and foredeep areas, the NW–SE MT secondary profile is characterized by a simple geological setting (Figure 1). It must be noted that the distance between the profile and the frontal Apennine thrusts progressively decreases moving northwards from the Murge to Gargano area, with the northernmost part of the profile definitely crossing the buried frontal ramp of the Apennines. This is an effect of the local configuration of the Southern Apennines, where the NNW-striking Molise-Sannio arc to the north contrasts with the NW-trending Campania-Lucania arc to the south (Cinque et al., 1993).

3. MT DATA ACQUISITION AND PROCESSING

The MT acquisition system we used consisted of two sets of equipment, each with three induction coils and two electrical channels linked to a 16 bit A/D converter. Almost all soundings were carried out in single site mode, except two pairs of soundings which were acquired in remote reference mode.

MT data were sampled in the frequency bands 10^{-5}–0.25, 10^{-2}–2.5, 0.2–50 and 1–250, all expressed in Hz. After the usual pre-processing (visual inspection of time series, despiking and removal of corrupted data segments), the impedance tensor elements were estimated in the frequency range 0.0013–100 Hz.

To check the influence of noise, the two impedance estimations $Z^H = [EH][HH]^{-1}$ and $Z^E = [EE][HE]^{-1}$ with the autopowers of the magnetic (H) and electrical (E) field components, respectively, were made (Sims et al., 1971; Vozoff, 1972; Gamble et al., 1979). Most MT curves did not show significant mis-tie effects, thus at least proving that the error distribution due to the uncorrelated noise on the electrical and magnetic channels was of low magnitude compared to that of the coherent part. Then, the coherence weighting and robust regression estimate method (Egbert and Livelybrooks, 1996) was applied. No significant impedance estimation differences were observed between remote referenced and single site soundings.

To assess MT data dimensionality and to control whether the MT regional strike were consistent with the main geological trend, tensor analysis and the decomposition method (Swift, 1967; LaTorraca et al., 1986) were applied to both profiles. The two datasets appeared quite heterogeneous since they involved areas of different geological complexity. This feature was very evident in the behaviour of the skew parameter, which reaches the highest values in the sites located across the Apennines along the SW–NE primary profile. In particular, along this profile skew increases versus period, reaching values greater than 0.5 for periods longer than 10 s, thus revealing the influence of some 3D effects. Again along the SW–NE profile, an electric strike direction was recognized and estimated around 50°W, which is consistent with the Southern Apennines trend, tending to slightly rotate westwards for periods greater than 10 s. As an example, Figure 2 shows the behaviour of the tensor polarization for MT stations 24, 22 and 11, located in a sector of the SW–NE profile geologically highly disturbed (see Figure 1). The three MT soundings show a general agreement at periods greater than 1 s, where the tensor polarization appears to align along the strike direction of the Southern Apennines. In conclusion, while taking into account the low density

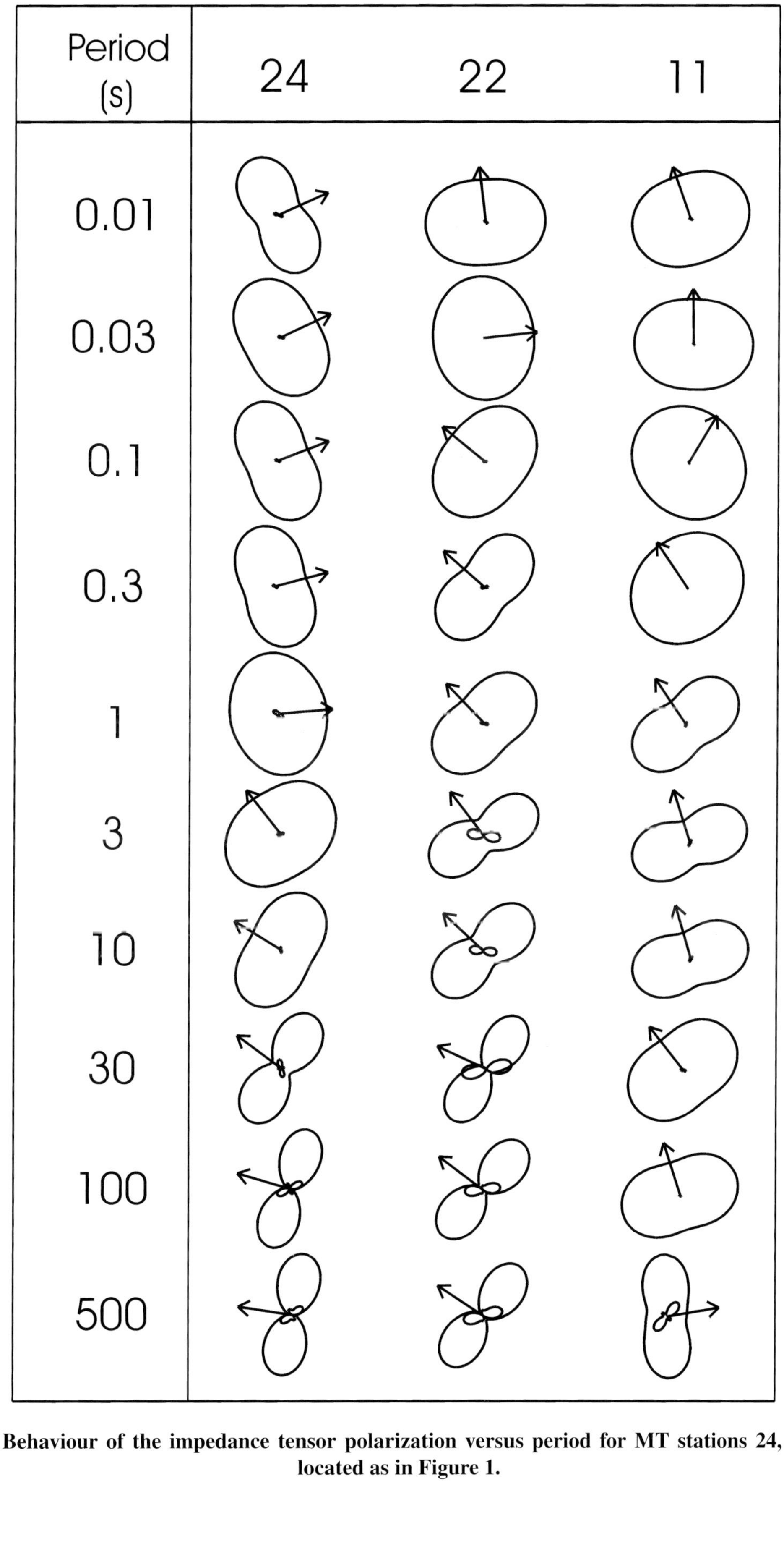

Figure 2: Behaviour of the impedance tensor polarization versus period for MT stations 24, 22 and 11, located as in Figure 1.

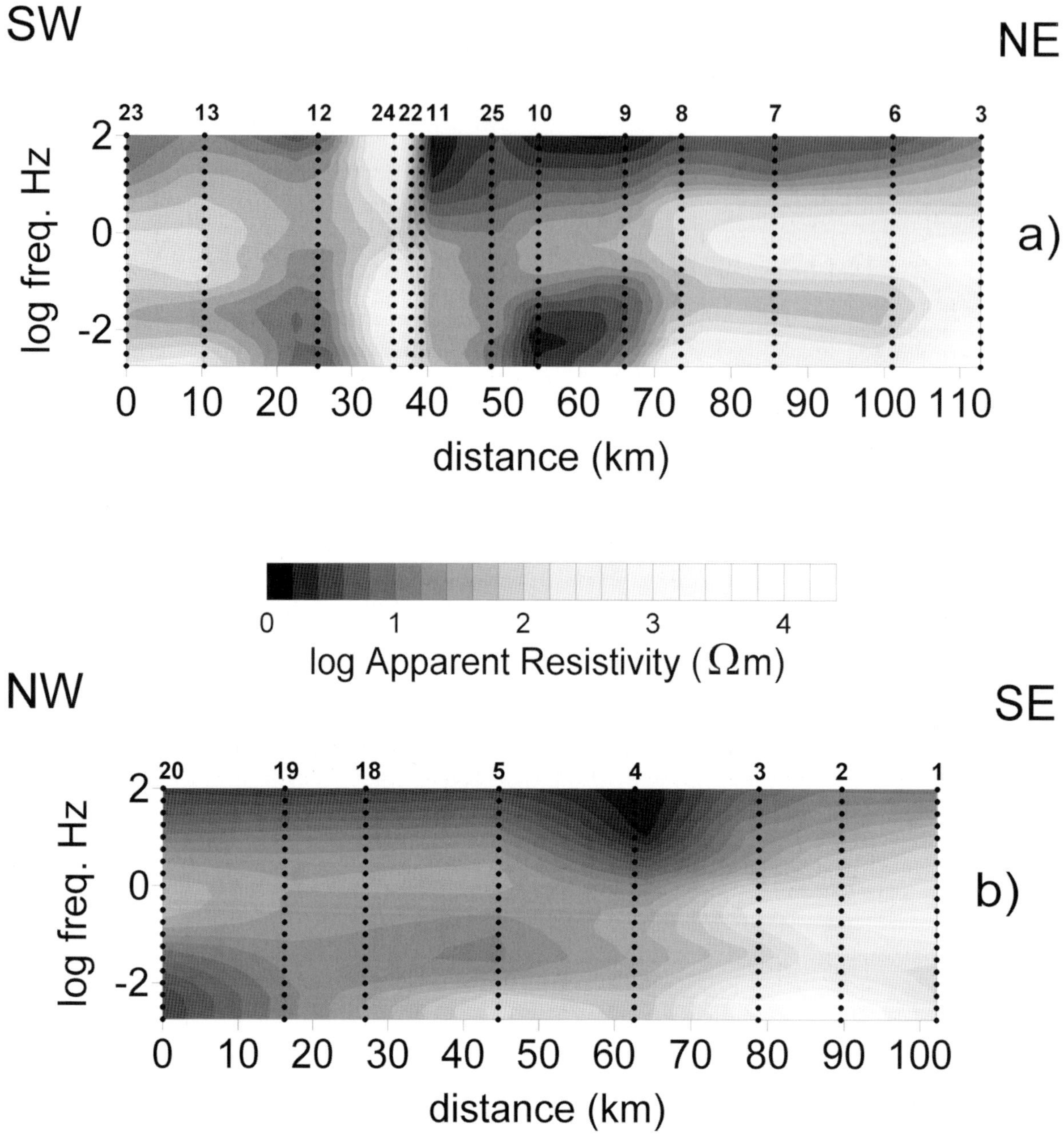

Figure 3: Magnetotelluric field pseudosections along (a) the SW–NE primary profile and (b) the NW–SE secondary profile. Numbers at the top of the sections refer to the MT stations located as in Figure 1. Horizontal distances are counted from the first MT station on the left.

distribution of MT stations, a 1D MT response appears to characterize the NW–SE secondary profile, while a 2D MT response appears to dominate along the SW–NE primary profile.

As is known, MT sounding curves derived from rotationally invariant parameters reduce the risk of misinterpretation when approaching a 2D modelling along profiles with low MT station density (Ingham, 1988). Accordingly, in the study of the Southern Apennines, the MT curves were computed using the determinant of the impedance tensor matrix (Ranganayaki, 1984). Among all known invariants, the determinant is considered an average impedance, which also has the advantage of attenuating the influence of shallow, highly conductive lateral bodies, such as the sea (Park and Livelybrooks, 1989; Ingham, 1992).

A rough analysis of the resistivity pattern across the two profiles can easily be made by inspecting the pseudosections of the determinant of the apparent resistivity tensor, depicted in Figure 3 (a) and (b). While the SW–NE primary profile (Figure 3(a)) is disturbed by evident irregularities, as long as it cuts the complex structure of the

Apennine chain, the NW–SE secondary profile (Figure 3(b)), instead, appears characterized by a notable lateral uniformity, particularly in the foreland area. It is worth observing that the northeastern terminal portion of the SW–NE profile also shows a conformal, nearly 1D pattern.

4. MT DATA MODELLING

The modelling strategy was to start drawing preliminary sections along both profiles as envelopes of the resistivity depth sequences, mostly obtained by 1D Bostick inversion of the determinant of the apparent resistivity curves. Only for MT soundings close to the sea, might the 1D approach have generated wrong sequences. Along the SW–NE primary profile, the influence of the sea was quite evident only in the three westernmost soundings, with a perturbation propagating at lower periods as the MT site approached the Tyrrhenian coastline. The sea effect was thus included in the preliminary inversion of the westernmost group of MT soundings. On the other side, the easternmost sounding of the SW–NE primary profile was located about 20 km from the Adriatic coastline. Hence, accounting also for the relatively small depth of the Adriatic sea, the coast effect was considered negligible. Similarly for all MT soundings of the NW–SE secondary profile, located about 20–30 km from the Adriatic coastline, the perturbation on the determinant was considered negligible.

The two preliminary sections were then discretized to get rough 2D structures, including topography, which were refined by a trial-and-error procedure, using a 2D finite-difference code, and assuming as constraints the information from nearby resistivity logs. To avoid false effects, we used a variable horizontal grid size calibrated to the MT station irregular spacing, and introduced only laterally-confined surficial bodies constrained by other datasets.

Figure 4 (a) and (b) shows the 2D forward models across the two profiles. The models confirm in a quantitative way the electrical pattern which had already emerged from the preliminary analysis of the corresponding pseudosections of Figure 3 (a) and (b). In particular, the MT section across the Southern Apennines (Figure 4(a)) shows significant sharp lateral changes of resistivity, very likely indicative of the important lithological variations expected there due to the presence of the fold-and-thrust system involving carbonate platform and basinal sequences.

In order to extract complementary information, and likely improve the MT cross-section along the most important SW–NE profile, we deemed it useful to model also an available gravity dataset, using the 2D section of Figure 4(a) as input to a new combined MT-gravity modelling as follows.

5. COMBINED MT-GRAVITY MODELLING

We extracted the gravity dataset from the database compiled by Carrozzo et al. (1981), which integrates the results of several surveys made in Southern Italy by public research institutions and oil companies. In this database, which refers to the IGSN71 network (Morelli et al., 1974), the Bouguer anomaly data are computed with respect to the 1980 International Ellipsoid using a reduction density of 2.4 g/cm^3. This value is an average between the 2.2–2.3 g/cm^3 range attributed to Plio-Pleistocene sediments, widely exposed in the Bradano Trough and Apulia foreland, and the 2.5–2.6 g/cm^3 interval assigned to stiff Mesozoic carbonate platform and basinal rocks, widespread in the axial zone of the belt. The distribution of gravity data in the study area is quite irregular. The average station density is very high in the Bradano Trough and axial zone of the belt which is the focus of intense hydrocarbon exploration. On the contrary, the average station density is low on the carbonate massifs of the inner zone of the range due to the very rough topography. Nevertheless, an average station density of about 2 per km^2 allows short wavelength Bouguer anomalies produced by density variations in the upper crust to be detected.

The Bouguer anomaly map in Figure 5 extends from the Tyrrhenian coast to the Apulia foreland and includes the southern end of the Molise-Sannio arc and the Campania-Lucania Apennines. The Bouguer anomaly isolines are characterized by a regular trend in the Apulia foreland and Bradano Trough, where NW-trending contours decrease from +80 mGal down to about +30 mGal moving toward the external zone of the Apennine range. Such a pattern is consistent with the SW-flexure of the Apulia lithosphere and the relatively simple upper crustal structure, which is characterized by the progressive deepening of the ACP beneath the Bradano Trough, as shown in the northeastern portion of the 2D resistivity model across the SW–NE profile shown in Figure 4(a). Conversely, across the Apennine range the Bouguer map is very complex, showing irregular anomalies correlated to surface tectonic structures. Pronounced gravity highs match quite well the horsts of the WCP and the nappe anticlines deforming Mesozoic basinal sequences and underlying carbonates of the ACP, while deep lows correspond to synforms filled with Cenozoic clayey sequences and Plio-Pleistocene clastic deposits.

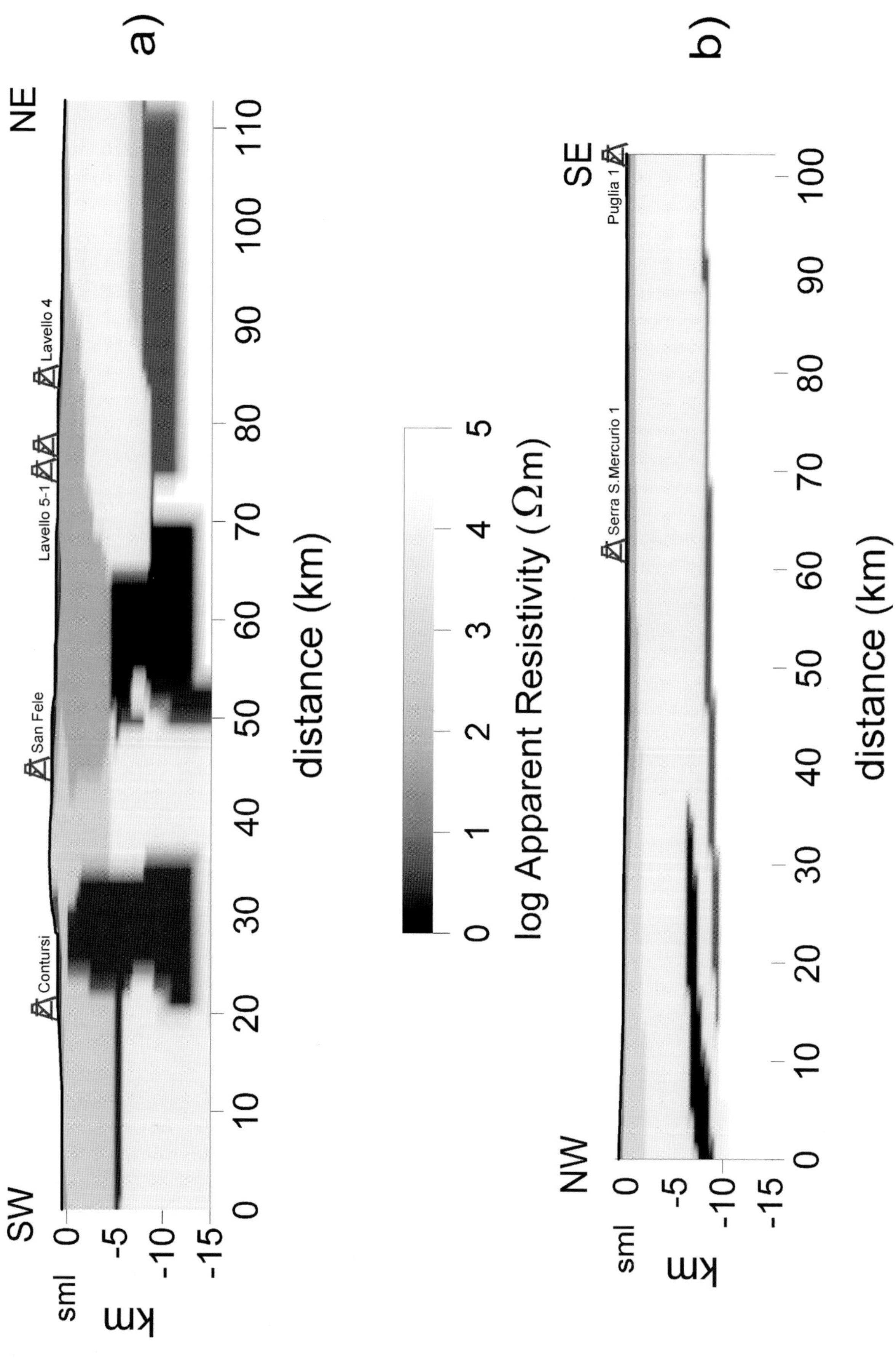
a)
SW
NE
Contursi
San Fele
Lavello 5-1
Lavello 4
km
sml
distance (km)
log Apparent Resistivity (Ωm)
b)
NW
SE
Serra S.Mercurio 1
Puglia 1
km
sml
distance (km)

Figure 4: Magnetotelluric interpreted sections. (a) Preliminary model across the SW–NE primary profile; (b) final model across the NW–SE secondary profile. Oil wells located about the profiles are shown. Horizontal distances are counted from the first MT station on the left, as in Figure 3.

With the aim of enhancing the gravity effects produced by upper crustal heterogeneities, we applied a 2D low-pass filter ($\lambda > 60$ km) in the wavenumber domain. The filtered map (Figure 6) shows regular NW-trending contour lines. The isoanomaly values decrease from the Apulia foreland towards the Bradano Trough, reaching minimum values in the axial zone of the Apennines. Moving southwestwards the filtered anomalies increase, reaching maximum values in the Tyrrhenian basin (outside the map). This pattern is in agreement with the Moho geometry as inferred from deep seismic refraction data (Scarascia et al., 1994) and the interpretation of deep reflection seismic profiles (Finetti et al., this volume, Chapter 12).

The residual Bouguer map (Figure 7) confirms that a close correlation exists between short wavelength anomalies and known upper crustal structures, such as anticlines and synclines. In particular, proceeding northeastwards along the SW–NE profile, and referring to the structural features depicted in Figure 1, we observe the following evident Bouguer anomalies:

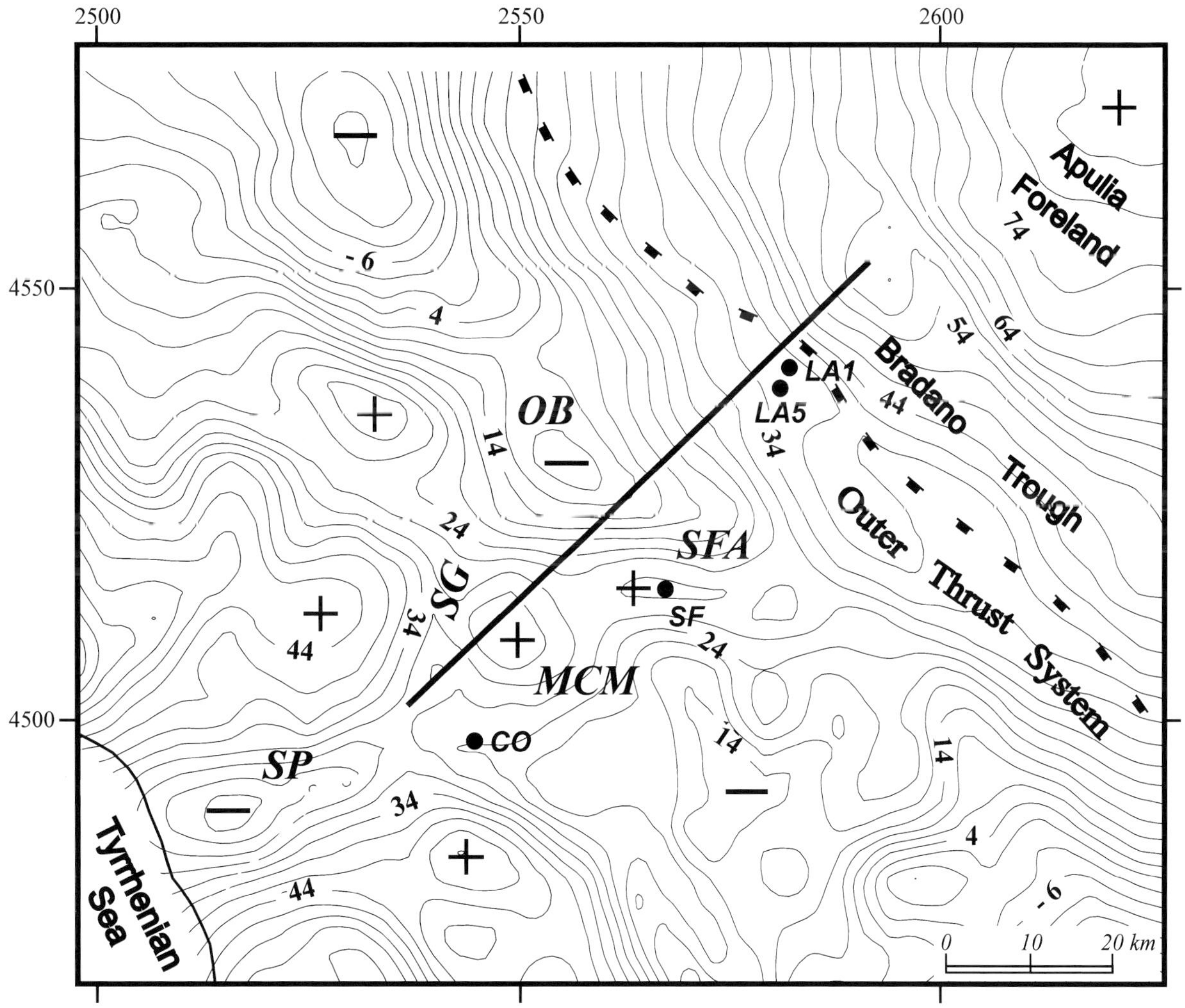

Figure 5: Bouguer map of the Campania-Lucania Apennines and surrounding areas. The reduction density is 2.4 g/cm^3. Bouguer anomaly values are gridded at 2.5 km square cell size and contoured at 5 mGal intervals; + and – denote gravity highs and lows. The interpreted gravity profile (thick continuous line) and trace of the magnetotelluric primary profile (dashed line) are shown. Wells are indicated by solid circles. OB, Ofanto basin; SFA, San Fele antiform; MCM, Marzano carbonate massif; SP, Sele plain; SG, Sele Graben; CO, Contursi 1 well; SF, San Fele well; LA1-5, Lavello wells.

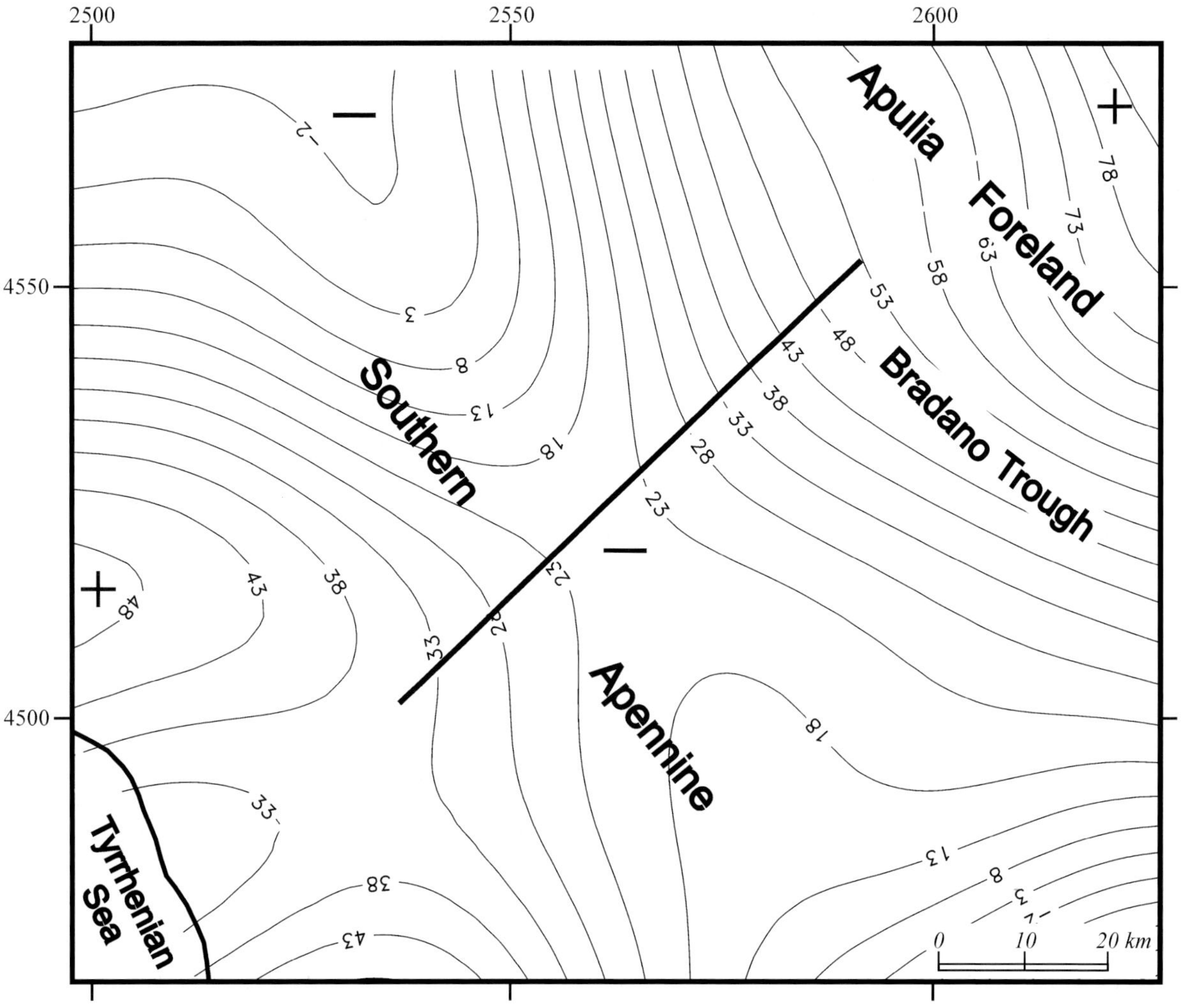

Figure 6: Map of the low-pass filtered Bouguer anomalies (λ > 60 km). Contour interval is 5 mGal; + and − denote regional gravity highs and lows.

- close to the Tyrrhenian coastline a W–E-oriented deep gravity low (−12 mGal) corresponds to the SP tectonic depression (Figure 1). This anomaly merges northwards into a secondary low that strikes N–S following the SG;
- a pronounced circular gravity high (+8 mGal) is associated with the MCM horst, which also seems responsible for the shallow high resistivity region in the MT model of Figure 4(a). The MCM high merges eastwards into a further positive anomaly elongated in the W–E direction. This second positive anomaly matches the SFA structure, which is essentially developed within stiff Mesozoic rocks of the LB (cherty dolomites and cherts);
- both MCM and SFA gravity highs are bounded northwards by a very deep gravity low (−14 mGal) corresponding to the OB synform;
- in the Bradano Trough, wide positive and negative NW-trending anomalies may be related to the outer fold-and-thrust system, which involves Tertiary basinal sequences;
- finally, in the Apulia foreland, a broad positive anomaly (+8 mGal) clearly relates to the outcropping ACP, as also documented by the MT section of Figure 4(b).

In order to combine the MT section along the SW–NE primary profile with gravity data, a $2\frac{1}{2}$D forward technique (Won and Bevis, 1987) was used to model the Bouguer residuals across the same MT profile, which strikes almost orthogonally to the Bouguer regional trend. Since the modelling was focused on detecting upper crustal inhomogeneities down to about 15 km of depth, we reduced the length of the gravity line with respect to the MT profile. The gravity profile is in fact 75 km long and extends from the Apennine inner zone close to MT station 12 to the outer thrust system close to MT station 7, where resistivity shows marked lateral changes.

The gravity profile (Figure 8) exhibits a large-wavelength trend characterized by a slow decrease from the south-western side as far as the OB, followed by a strong rise toward the Bradano Trough. Short-wavelength anomalies

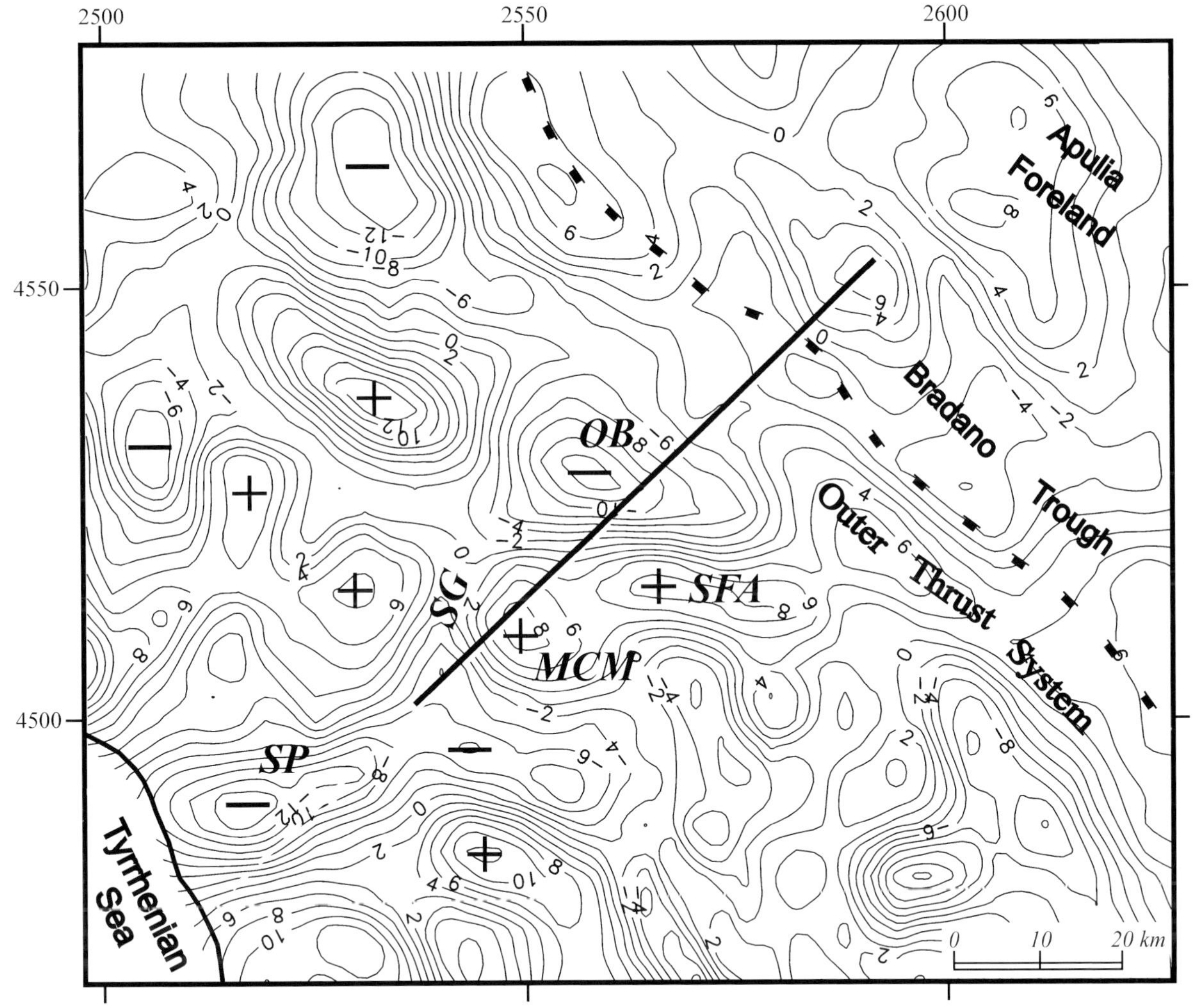

Figure 7: Map of the residual Bouguer anomalies. Contour interval is 2 mGal; + and − denote residual gravity highs and lows. The interpreted gravity profile (thick continuous line) and trace of the magnetotelluric primary profile (dashed line) are shown. OB, Ofanto basin; SFA, San Fele antiform; MCM, Marzano carbonate massif; SP, Sele plain; SG, Sele Graben.

appear in close correspondence with the SG and OB, at about 5 and 35 km from the origin, and with the MCM at about 15 km from the origin.

Our strategy for gravity data modelling was based on drawing a preliminary density model using a simplified version of the resistivity model in Figure 4(a). After correlating each resistivity body to a tectono-stratigraphic unit, according to both surface and subsurface information, we assigned to each unit the reference density range proposed by Improta et al. (2003) after a calibration procedure based on the integration of reflection seismic, well log and gravity data. The following minor changes were also made (Figure 8):

- on the western side, the high resistivity ($< 1000\ \Omega$m) region was differentiated into an upper (2.65 g/cm^3) and a lower (2.50–2.55 g/cm^3) body, corresponding to WCP and LB units, respectively. Such differentiation was based on the Contursi 1 well (Mostardini and Merlini, 1986) and on reflection seismic data (e.g. Casero et al., 1988; Finetti et al., this volume, Chapter 12), which document overthrusting of MCM above LB sequences;
- the two lowest resistivity ($< 3\ \Omega$m) regions, embedded in the high resistivity ($<10\,000\ \Omega$m) layer interpreted as ACP (2.7 g/cm^3), were not included in the starting density model. Indeed, we assumed that these lowest resistivity values are not representative of such large-scale lithological changes as to be responsible for large density variations;
- the thin low resistivity ($< 100\ \Omega$m) layers at the top and bottom of the ACP were not included owing to their negligible gravitational effect;
- some minor near-surface changes were based on surface geological data.

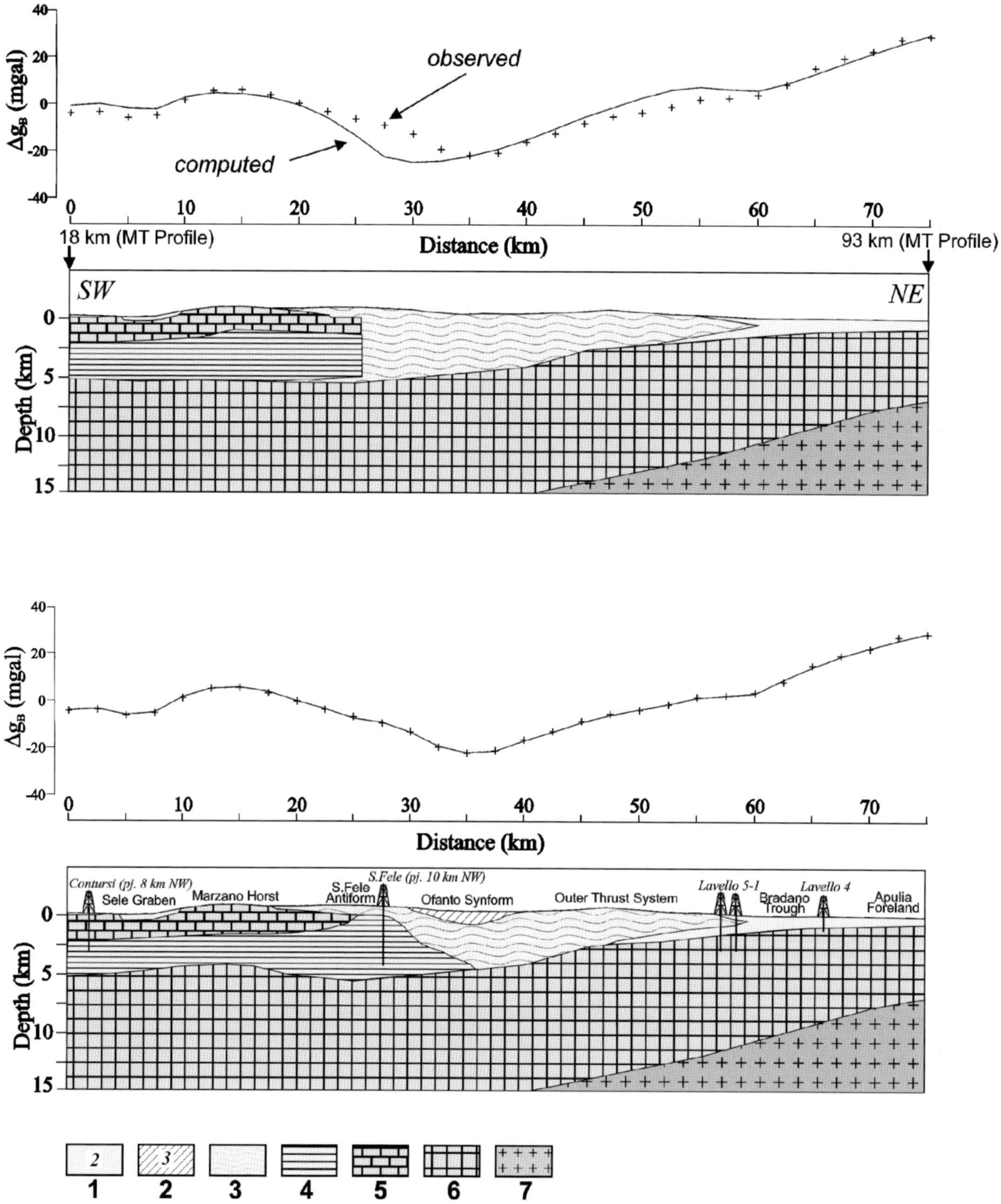

Figure 8: Gravity data modelling across the SW–NE primary profile: (upper panel) preliminary density model featured upon the resistivity section of Figure 4(a); (lower panel) final density model. The observed gravity profile (crosses) and the calculated profile (continuous line) are shown at the top of each model. The wells used to constrain the model are also shown. Horizontal distances are counted from the first gravity value on the left [1 = Pleistocene clastic deposits, ρ = 2.1; 2 = Pliocene thrust-sheet-top clastic deposits, ρ = 2.3; 3 = Cenozoic basinal successions, ρ = 2.4; 4 = Lagonegro basin lower sequence, ρ = 2.50–2.55; 5 = Western carbonate platform, ρ = 2.65; 6 = Apulia carbonate platform, ρ = 2.7; 7 = Paleozoic crystalline basement, ρ = 2.8. Density ρ is in g/cm^3. Numbers inside shallow body screens 1 and 2 denote half-length (in km) used for $2\frac{1}{2}$D forward modelling].

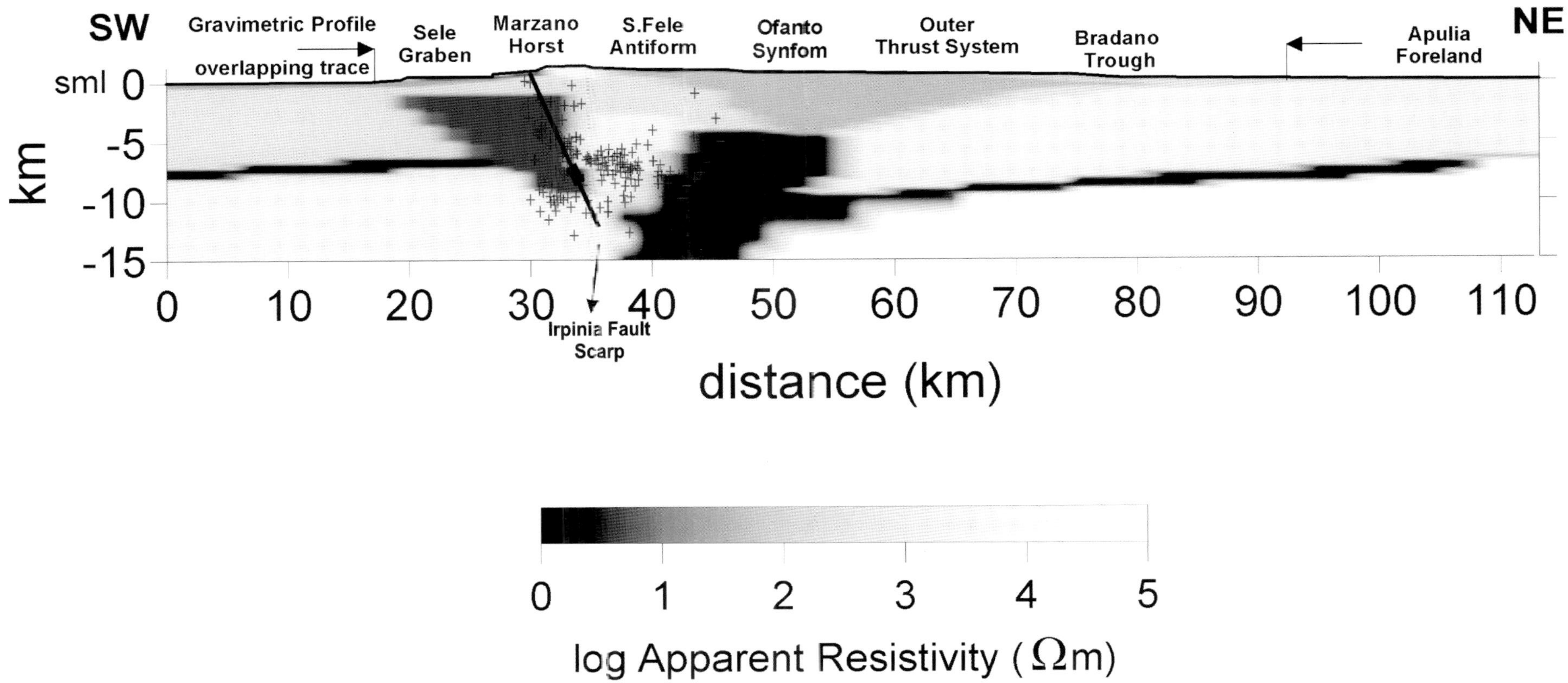

Figure 9: **Magnetotelluric final model across the SW–NE primary profile and location of the 1980 Irpinia earthquake aftershock hypocentres. Horizontal distances are counted from the first MT station on the left, as in Figure 3.**

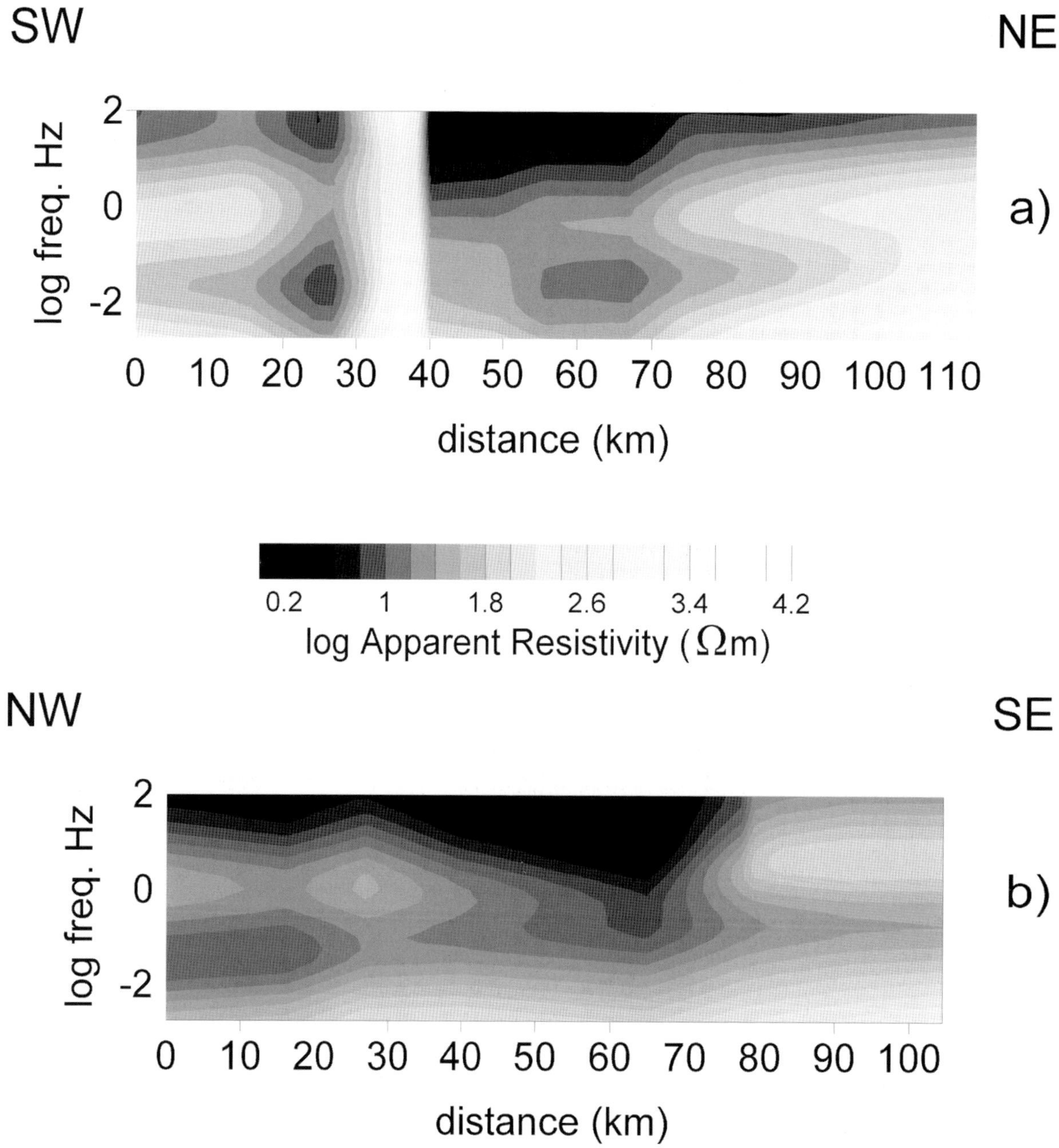

Figure 10: Magnetotelluric synthetic pseudosections along (a) the SW–NE primary profile and (b) the NW–SE secondary profile. Horizontal distances are counted from the first MT station on the left, as in Figure 3.

The densities used for the remaining bodies are 2.4 g/cm^3 for the shallow low resistivity ($< 100\ \Omega m$) region, which can be related to Tertiary basinal units (mainly flysch and clayey deposits), and 2.1 g/cm^3 for shallow Pleistocene layers in the Bradano Trough and SG.

The synthetic gravity profile computed for this preliminary density model matches quite well the observed large-wavelength trend (Figure 8(a)). On the contrary, in correspondence to the SFA and OB, as well as above the outer thrust system, the preliminary density model is not consistent with the observed short-wavelength anomalies. In order to fit them, we modified the upper part of the density model (Figure 8(b)), using as constraints surface, reflection seismic profiles (Hippolyte et al., 1994) and well data (San Fele well, Improta et al., 2003). We modelled the deep gravity low at about 35 km along the profile by introducing a low density (2.3 g/cm^3) shallow body corresponding to the Pliocene filling of the OB. In addition, introducing a bump in the 2.55 g/cm^3 body allowed the secondary gravity high located at the SFA to be fitted. This feature agrees with the presence of a nappe anticline developed within stiff Mesozoic rocks of the LB. At about 15 km along the profile, the rise in the 2.7 g/cm^3 layer,

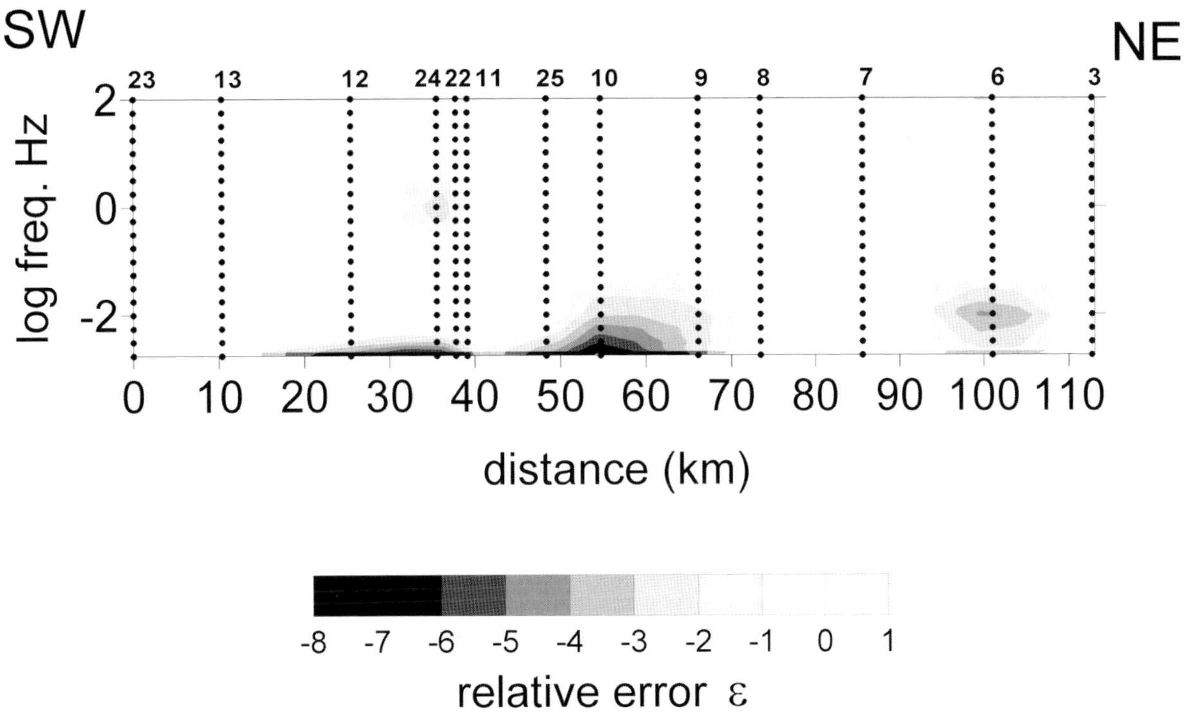

Figure 11: Relative error of the model response with respect to the experimental response for the model in Figure 9.

which corresponds to the ACP, is based on proprietary well data (Improta et al., 2003). On the western side of the model, a surface low density (2.1 g/cm^3) body allowed the SG low to be reproduced. Finally, in the 50–60 km distance range from the origin, we obtained a better fit by thickening the low density (2.1 g/cm^3) layer correlated with the Pliocene terrigeneous deposits of the Bradano foredeep, overthrust by the Apennine outer nappes.

This analysis shows that the starting resistivity model along the SW–NE primary profile is fairly consistent with gravity data. Moreover, the gravity model constrained by surface and subsurface data yields further information on the shallow crust structure. The final model in Figure 8(b) clearly shows the SFA and OB structures, which are, poorly imaged in the resistivity section of Figure 4(a) because of the large spacing (9 km) between MT stations 25 and 10. In addition, on the western side of the gravity model, the WCP can be discriminated from the LB sequence, whereas both are combined into a single high resistivity ($< 1000\ \Omega$m) layer in Figure 4(a). This may be explained by the lack of important variation in electric properties between the highly fractured limestones of the WCP and the cherty-limestones of the LB.

At greater depths, the gravity model confirms the flexure of the very high resistivity ($> 10\,000\ \Omega$m) body interpreted as crystalline basement, beneath the Apennine external zone. Finally, both extremely low resistivity ($< 3\ \Omega$m) zones in the central part of the model are not associated with evident Bouguer anomalies, as the gravity data modelling demonstrates that the gravity lows at about 5 and 35 km are entirely produced by shallow sources (Plio-Pleistocene basins). The lack of strong density anomalies makes the presence of large-scale lithological changes beneath the chain between 5 and 15 km of depth unlikely.

The better definition of the OB in the gravity model allowed a further refinement of the resistivity model across the SW–NE profile to be attempted. The final version of the MT model across the primary profile is presented in Figure 9 and discussed in the following section.

To assess the goodness-of-fit for the final resistivity models along the MT SW–NE (Figure 9) and NW–SE (Figure 4(b)) profiles, the corresponding synthetic pseudosections are shown in Figure 10 (a) and (b) for comparison with the experimental ones reported in Figure 3 (a) and (b). Finally, in order to quantify the model reliability across the SW–NE primary profile, we computed the model relative error as $\varepsilon = (\rho_a^{\text{field}} - \rho_a^{\text{model}})/\rho_a^{\text{field}}$, where ρ_a^{field} and ρ_a^{model} are the experimental and synthetic apparent resistivity determinant, respectively. The distribution of ε is shown in Figure 11, where two zones with a model frequency response significantly higher than the field response appear in the 30–60 km distance range at the lowest frequencies, exactly where the major structural complexity was observed.

6. GEOLOGICAL DISCUSSION AND CONCLUSION

The following geological interpretation of the MT sections is made with the help of well data, published reflection seismic profiles (e.g. Casero et al., 1988; Finetti et al., this volume, Chapter 12) and the results of the gravity data modelling (Figure 8(b)).

Referring to the SW–NE primary profile (Figure 9) and starting from the top, resistivity data allow the Tertiary basinal units to be differentiated from the Mesozoic successions within the rootless nappes. The top low-resistivity ($< 100\ \Omega$m) bodies can be related to Tertiary basinal units, which consist mainly of clayey and clastic deposits. This correlation is supported by a few lines of evidence:

- the thickening found in the 40–60 km distance range is consistent with the structure of the OB synform, where the maximum thickness of the Pliocene basin (Figure 8(b)) and underlying Tertiary clayey units is thought to exceed 5 km (Hippolyte et al., 1994; Patacca and Scandone, 2001);
- the low resistivities found in the SG area correlate quite well with outcrops of Tertiary clayey and Plio-Pleistocene clastic sequences.

The intermediate-resistivity ($< 1000\ \Omega$m) body on the left-hand side of the section can be related to the tectonic stack of nappes referable to the WCP and LB units. Due to the intense fracturation of the WCP, and to the prevalence of Mesozoic cherty limestones and cherts within the LB nappes (Contursi 1 well; Mostardini and Merlini, 1986), resistivity (Figure 9), in contrast to gravity (Figure 8(b)), does not allow the WCP to be discriminated from the LB. Proceeding downwards, a noticeable feature appearing in the left-hand portion of the MT SW–NE cross-section is the thin low-resistivity (3–30 Ωm) layer (about 1 km thick) sandwiched between the LB Mesozoic rocks of the upper intermediate-resistivity ($< 1000\ \Omega$m) body and the high-resistivity ($< 10\,000\ \Omega$m) lower stratum, which is assumed to correspond to the ACP. This thin layer, which likely represents a marker for the ACP geometry, may be ascribed to Tertiary basin deposits. Beneath the SG, this interpretation is locally supported by the Contursi 1 well, which penetrates Miocene, mainly clayey and flysch deposits beneath the LB Mesozoic units (Mostardini and Merlini, 1986). In addition, due to the very low resistivities, we do not exclude the presence of an abnormally high concentration of fluids within this thin layer, which may correspond to a major decollement level in the stack of rootless nappes.

At greater depths, the whole SW–NE MT section of Figure 9 delineates the large-scale trend of the top of the ACP. Its SW-bending beneath the Bradano Trough, and marked deepening down to 5 km beneath the OB synform, conform to reflection seismic evidence (e.g. Casero et al., 1988). It is worth noting the consistency of the SW-deepening of the ACP with the results from the CROP-04 seismic profile which delineate a thickening of the rootless nappes proceeding from the axial to the inner zone of the Southern Apennines (Finetti et al., this volume, Chapter 12).

The very high-resistivity ($> 10\,000\ \Omega$m) body located beneath the Apulia foreland, on the right-hand half of the MT cross-section of Figure 9, can be interpreted as the crystalline basement, whereas the low-resistivity (3–100 Ωm) thin layer sandwiched between the ACP and the basement can be related to the Permo-Triassic clastic sequences (Verrucano Fm.) encountered in the Puglia 1 well (Richetti et al., 1994). The bending of this layer documents well the flexure of the ACP in agreement with deep reflection seismic data (Roure et al., 1991; Finetti et al., this volume, Chapter 12). Furthermore, the MT cross-section clearly shows the continuity of the Permian deposits westwards, and the lack of important lateral variation in thickness (about 2 km) as far as the axial zone of the belt. This may suggest that the thrust system in the Apennine outer zone is confined within the sedimentary cover of the shallow crust.

A noticeable feature of the resistivity model of Figure 9 is the presence of two extremely low-resistivity ($< 3\ \Omega$m) bodies cutting the ACP from about 3–5 km depth down to the bottom of the model. Such low-resistivity regions could be interpreted as highly fractured zones characterized by melt intrusions and/or fluid circulation. This interpretation is supported by the following evidence:

- geochemical investigations in the axial zone of the Campania-Lucania Apennines by Italiano et al. (2000) document a diffuse and intense CO_2-dominated gas emission, which is characterized by a significant mantle-helium

contribution. According to the authors, the phenomenon is too intense to be explained by direct degassing from the mantle, but requires a crustal source that can be sought in mantle melts intruded into the lower-mid crust along lithospheric faults. It is noteworthy that the two anomalous low-resistivity regions match quite well two overlying sites of strong degassing located in the SG and in the OB (Italiano et al., 2000);
- the low-resistivity region beneath the OB synform is located along the southern termination of a broad area where anomalous high heat flow values are reported by Mongelli et al. (1996);
- reflection seismic sections (Finetti et al., this volume, Chapter 12) exhibit deep lithospheric faults, which may explain the presence of highly fractured zones favouring uprise of fluids. In particular, the low-resistivity body starting beneath the SG and dipping northeastwards below the SFA is located in correspondence to the western termination of the WNW-trending left-transcurrent Vallo di Diano-Tanagro fault zone (Cinque et al., 1993). The low-resistivity body beneath the OB can, instead, be related to a major thrust developed within the ACP;
- aftershocks of the 1980 Irpinia earthquake (Amato and Selvaggi, 1993) cluster in the high-resistivity region laterally bounded by the two anomalous bodies. It is worth noting that this high-resistivity region matches well a pronounced P-wave high-velocity anomaly identified by local earthquake tomography from the 1980 event aftershocks (Amato and Selvaggi, 1993; Chiarabba and Amato, 1994).

Referring to the MT model across the NW–SE secondary profile (Figure 4(b)), we observe a quite simple structure crossing the foreland-foredeep areas. We notice a slow bending of the ACP and a thickening of the Plio-Pleistocene clastic deposits ($< 100\ \Omega$m) of the Bradano Trough proceeding northwestwards, as the profile runs transversally to the strike of the Apulia lithosphere flexure. It is worthwhile noting that the ACP does not show any important variation of thickness along the entire section. A noticeable result, which requires investigation with additional MT surveys, is the low-resistivity region located beneath the ACP on the northern side of the profile. It may tentatively be associated with the W-trending Gargano shear-zone, which is also responsible for a locally high heat-flow.

REFERENCES

Amato, A., Selvaggi, G., 1993. Aftershock location and P-wave velocity structure in the epicentral region of the 1980 Irpinia earthquake. Ann. Geofis. 36, 3–15.

Amato, A., Montone, P., 1997. Present day stress field and active tectonics in southern peninsular Italy. Geophys. J. Int. 130, 519–534.

Bernard, P., Zollo, A., 1989. The Irpinia (Italy) 1990 earthquake: detailed analysis of a complex normal fault. J. Geophys. Res. 94, 1631–1648.

Bonardi, G., D'Argenio, B., Di Nocera, S., Marsella, E., Pappone, G., Perrone, V., Pescatore, T.S., Senatore, M.R., Sgrosso, I., Ciaranfi, N., Pieri, P., Ricchetti, G., 1988. Geological Map Southern Apennines. 74th Meeting of the Italian Geological Society.

Carrozzo, M.T., Chirenti, A., Luzio, D., Margiotta, C., Quarta, T., 1981. Carta gravimetrica d'Italia: tecniche automatiche per la sua realizzazione. In: Proc. of the 1st Annual Meeting of the Italian National Geophys. Group, pp. 132–139.

Casero, P., Roure, F., Endignoux, L., Moretti, L., Muller, C., Sage, L., Vially, R., 1988. Neogene geodynamic evolution of the Southern Apennines. Mem. Soc. Geol. It. 41, 109–120.

Chiarabba, C., Amato, A., 1994. From tomographic images to fault heterogeneities. Ann. Geofis. 37, 1481–1494.

Cinque, A., Patacca, E., Scandone, P., Tozzi, M., 1993. Quaternary kinematic evolution of the Southern Apennines. Ann. Geofis. 36, 249–260.

D'Argenio, B., Pescatore, T., Scandone, P., 1973. Schema geologico dell'Appennino Meridionale (Campania e Lucania). In: Atti del Convegno: moderne vedute sulla geologia dell'Appennino Meridionale. Acc. Naz. Lincei 183, 49–72.

Dell'Aversana, P., Ceragioli, E., Morandi, S., Zollo, A., 2000. A simultaneous acquisition test of high density "global offset" seismic in complex geological settings. First Break 18, 87–96.

Egbert, G.D., Livelybrooks, D.W., 1996. Single station magnetotelluric impedance estimate: Coherence weighting and regression M-estimate. Geophysics 61, 964–970.

Finetti, I.R., Lentini, F., Carbone, S., Del Ben, A., Di Stefano, A., Guarnieri, P., Pipan, M., Prizzon, A., this volume. Crustal Tectonostratigraphy and Geodynamics of the Southern Apennines from CROP and other integrated geophysical-geological data. Chapter 12.

Gamble, T.D., Goubau, W.M., Clark, J., 1979. Magnetotellurics with a remote reference. Geophysics 44, 53–68.

Hippolyte, J.C., Angelier, J., Roure, F., Casero, P., 1994. Piggyback basin development and thrust evolution: structural and paleostress analyses of Plio-Quaternary basins in the Southern Apennines. J. Structural Geol. 16, 159–173.

Improta, L., Bonagura, M.T., Iannaccone, G., Capuano, P., 2003. An integrated geophysical investigation of the upper crust in the epicentral area of the 1980, Ms = 6.9, Irpinia earthquake (Southern Italy). Tectonophysics 317, 273–297.

Ingham, M.R., 1988. The use of invariant impedances in magnetotelluric interpretation. Geophys. J. Int. 92, 165–169.

Ingham, M.R., 1992. Audiomagnetotelluric soundings on White Island volcano. J. Volcanol. Geotherm. Res. 50, 301–306.

Italiano, F., Martelli, M., Martinelli, G., Nuccio, P.M., 2000. Geochemical evidence of melt intrusions along lithospheric faults of the Southern Apennines, Italy: geodynamic and seismogenic implications. J. Geophys. Res. 105 (B6), 13569–13578.

LaTorraca, G.A., Madden, T.R., Korringa, J., 1986. An analysis of the magnetotelluric impedance for three-dimensional conductivity structures. Geophysics 51, 1819–1829.

Malinverno, A., Ryan, W.B.F., 1986. Extension in the Tyrrhenian Sea and shortening in the Apennines as a result of arc migration driven by sinking of the lithosphere. Tectonics 5, 227–245.

Marsella, E., Bally, A.W., Cippitelli, G., D'Argenio, B., Pappone, G., 1995. Tectonic history of the Lagonegro domain and Southern Apennines thrust belt evolution. Tectonophysics 252, 307–330.

Mazzotti, A.P., Stucchi, E., Fradelizio, G.L., Zanzi, L., Scandone, P., 2000. Seismic exploration in complex terrains: a processing experience in the Southern Apennines. Geophysics 65, 1402–1417.

Menardi, A., Rea, G., 2000. Deep structure of the Campania-Lucanian arc (Southern Apennines, Italy). Tectonophysics 324, 239–265.

Mongelli, F., Harabaglia, P., Martinelli, G., Squarci, P., Zito, G., 1996. Nuove misure di flusso geotermico in Italia meridionale: possibili implicazioni sismotettoniche. In: Proc. of the 14th Annual Meeting of the Italian National Geophys. Group, pp. 929–939.

Morelli, C., Gantar, T., Honkasalo, T., Mc Connell, P.K., Tanner, J.B., Szabo, B., Uotila, U., Whalen, C.T., 1974. In: The International Gravity Standardization Net 1971 (IGSN71). IUGG, AIG, Publ. spec., no. 4, Paris, pp. 165–179.

Mostardini, F., Merlini, S., 1986. Appennino centro-meridionale. Sezioni geologiche e proposta di modello strutturale. Mem. Soc. Geol. It. 35, 177–202.

Park, S.K., Livelybrooks, D.W., 1989. Quantitative interpretation of rotationally invariant parameters in magnetotellurics. Geophysics 54, 1483–1490.

Patacca, E., Scandone, P., 1989. Post-Tortonian mountain building in the Apennines. The role of the passive sinking of a relic lithospheric slab. In: Boriani, A., Bonafede, M., Piccardo, G.B., Vai, G.B. (Eds.), The Lithosphere in Italy. Acc. Naz. Lincei 80, 157–176.

Patacca, E., Scandone, P., 2001. Late thrust propagation and sedimentary response in the thrust belt-foredeep system of the Southern Apennines (Pliocene-Pleistocene). In: Vai, G.B., Martini, I.P. (Eds.), Anatomy of a Mountain: the Apennines and Adjacent Mediterranean Basins. Kluwer Academic, pp. 401–440.

Pescatore, T.S., Renda, P., Tramutoli, M., 1988. I rapporti tra le Unità Lagonegresi e le Unità Sicilidi nella media valle del Basento, Lucania, Appennino Meridionale. Mem. Soc. Geol. It. 41, 353–361.

Ranganayaki, R.P., 1984. An interpretative analysis of magnetotelluric data. Geophysics 49, 1730–1748.

Ricchetti, G., 1994. Geologia delle aree di avampaese. In: Soc. Geol. It., 77° Congresso Nazionale, pp. 1–86.

Roure, F., Casero, P., Vially, R., 1991. Growth processes and melange formation in the Southern Apennines accretionary wedge. Earth Planet. Sci. Lett. 102, 395–412.

Scarascia, S., Lozej, A., Cassinis, R., 1994. Crustal structures of the Ligurian and Tyrrhenian and Ionian sea and adjacent onshore area interpreted from wide-angle seismic profiles. Boll. Geofis. Teor. Appl. 36, 141–144.

Sgrosso, I., 1986. Criteri ed elementi per la ricostruzione paleogeografica delle zone esterne dell'Appennino Centro-Meridionale. Mem. Soc. Geol. It. 35, 203–219.

Sims, W.E., Bostick Jr., F.X., Smith, H.W., 1971. The estimation of magnetotelluric impedance tensor elements from measured data. Geophysics 36, 938–942.

Swift Jr., C.M., 1967. A magnetotelluric investigation of an electrical conductivity anomaly in the southwestern United States. Ph.D. thesis, MIT, Boston.

Won, I.L., Bevis, M., 1987. Computing the gravitational and magnetic anomalies due to a polygon. Algorithms and FORTRAN subroutines. Geophysics 53, 232–238.

Vozoff, K., 1972. The magnetotelluric method in the exploration of sedimentary basins. Geophysics 37, 98–141.

CROP PROJECT: Deep Seismic Exploration of the Central Mediterranean and Italy
Edited by I.R. Finetti

CHAPTER 14

Multiscale Derivative Analysis of the Gravity and Magnetic Fields of the Southern Apennines (Italy)

M. Fedi[1], F. Cella[2], G. Florio[1], A. Rapolla[1]

ABSTRACT

A multiscale analysis of the gravity and magnetic fields of the Southern Apennines (Italy) is performed. The method, called Multiscale Derivative Analysis (MDA), is based on the good resolution properties of the Enhanced Horizontal Derivative (EHD) and allows the potential field to be explored at different scales. The task of separating the different scale-related effects is traditionally accomplished by classical filtering methods, based on Fourier or wavelet expansion, but some distortion is likely to occur. MDA, instead, does not apply sharp component separations, but simply enhances the contributions at different resolution which are really present in the field.

From a morphological point of view, Southern Italy is characterized by the presence of a main structural element represented by the southern sector of the Apennine chain. This consists in a thrust and fold belt system primarily, with a rather complex geological structure, and represents the geodynamic evolution, developed from Lower Cretaceous up to the Quaternary, resulting from the convergence between the African and European plates.

MDA was performed at three resolutions to highlight regional, intermediate and local trends, for interpreting the gravity and aeromagnetic fields in the area of the Apennine belt and surrounding regions, just in terms of geological structures.

MDA of gravity data provided an almost complete representation of the structural framework of Southern Italy at three different scales. Most of the known geological elements of the Apennine system are clearly shown at intermediate and short scales, together with several trends indicating structures that do not have any reference at the surface. This is, for instance, the case of some regional linear trends in the foredeep area, which in some cases may be related to hidden thrust systems at depth.

MDA of magnetic data allowed localization of several linear and closed trends, identifying anomaly sources the presence of which was not previously detected, even by MDA of gravity data. In addition, interpretation of some magnetic structures may support the hypothesis that within the sedimentary cover of the Apulian foreland, at a first sight considered as relatively non-magnetic, several sources of detectable magnetization are present.

Keywords: Southern Apennines, multiscale derivative gravity – magnetic analysis

1. INTRODUCTION

The Apennine Chain (Figure 1) can be described as a complex thrust and fold belt system, built between the Lower Cretaceous and the Quaternary, as consequence of the convergence between the African and European plates (Finetti and Del Ben, 1986; Dewey et al., 1989; Finetti et al., 1996).

The first phases of this convergence are represented by an oceanic subduction, aged between Upper Cretaceous and Oligocene, with formation of an accretionary prism (Upper Oligocene–Tortonian) and, consequently, the closure of the Tethys margin. The palaeogeographic context was formerly represented by broad carbonate shelfs alternating with deep basinal domains. Finally, the continental collisional and post-collisional phases (Upper Miocene–Pleistocene) followed.

[1]Dipartimento di Scienze della Terra, Università Federico II di Napoli, Largo san Marcellino 10, 80138 Napoli, Italia
[2]Dipartimento di Scienze della Terra, Università degli Studi della Calabria, Arcavacata di Rende, Cosenza, Italia

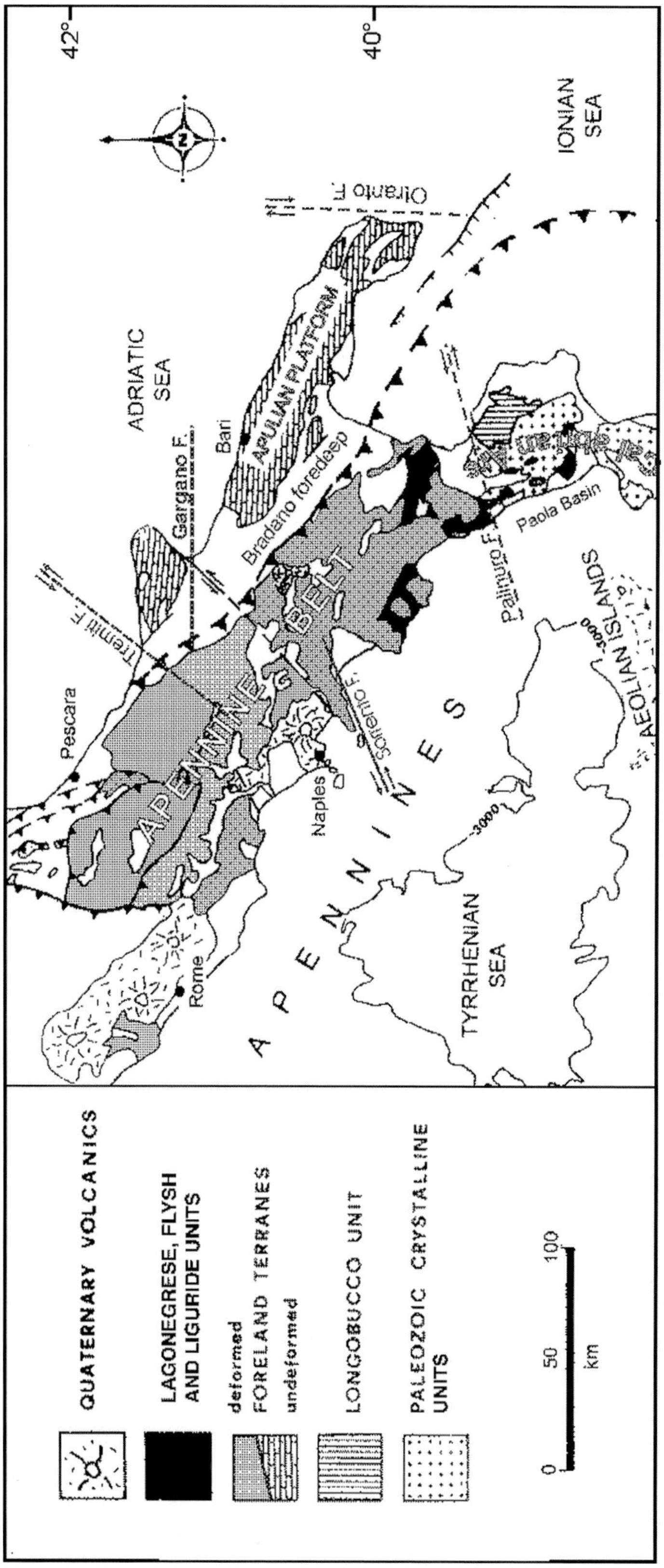

Figure 1: Regional sketch map of the main geological and structural features in Southern Italy (Critelli, 1999, modified).

The space–time migration of the chain-foredeep-foreland system, the formation of a thrust and fold belt along the Apennines and the extensional regime in the external back-arc sector, can be all explained as the final result of a complex geodynamic evolution strictly related to the opening of the Tyrrhenian basin and driven by the passive subduction of the Adria plate (Patacca et al., 1990).

The structural geometry of the tectonic units constituting the Southern Apennine Chain still provides many, strongly debated, open questions. Several works and various, sometimes very different, models (Mostardini and Merlini, 1986; Marsella et al., 1995) have recently been suggested. The main differences concern the palaeogeographic meaning of various units, the palinspastic restoration of the chain, the location at depth of many discontinuities and the hypothetic involvement of the crystalline basement in the tectonic deformation (Casero et al., 1988).

In this chapter, the structural setting of the Southern Apennine region will be studied by means of a multiscale analysis of a specific quantity, named the Enhanced Horizontal Derivative, based on the gravity and magnetic fields.

Due to the complexity of the real geology, the spatial density and magnetization distributions are not homogeneous within the Earth, and many sources related to different depths and extents are normally expected to contribute to the measured anomaly fields. A classical problem of potential field interpretation arises just from the fact that these single effects are merged together, and their identification and isolation is often difficult.

To this end, a number of methods were developed suitable for a mathematical treatment of the measured field as a superposition of effects from sources located at different depths, each corresponding to a different spatial extent (or scale). Among them, we note those based on the Fourier transform (Baranov, 1975), which is function of frequencies, and on the wavelet transform, which depends on scales (i.e. the reciprocal of frequencies) and positions (Fedi and Quarta, 1998). The important thing is that the expansion of the field by any mathematical series of functions may allow filtering of unwanted effects. The coefficients related to these components are first identified, e.g. by power spectra in the Fourier case, and then muted, thus generating a filtered and a residual couple of signal components.

The choice among the many digital filtering methods is made according to the specific requirements of the specific separation problem. For instance, a low-degree polynomial may be thought appropriate to approximate a long period trend, such as that generated in the gravity field by very deep sources (like the Moho). Filtering by the wavelet transform may efficiently isolate the effects from sharp and localized sources, while to filter out a diffuse high frequency disturbance, the better solution may result to be the use of the Fourier transform.

Anyway, none of the above techniques can guarantee perfect separation of contributions, because these effects may overlap with respect to their frequency (or scale) content. In this case, an anomalous effect from a single source can be split into two anomalies, one in the regional map and the other in the residual, both obviously meaningless. Thus, the physical consistency of the anomaly separation by digital filtering has always to be carefully evaluated and whenever possible such mathematical filtering of gravity or magnetic fields should be avoided.

There are however other means to separate regional from local fields that have a more definite physical meaning, such as analytical upward/downward continuation or n-order spatial derivations.

The latter may help to better evidence the effects related to shallow sources with respect to the deeper ones. In fact, the horizontal or vertical derivatives have the property of increasing the high wavenumber spectral components, thus enhancing the contributions of shallow sources. As in the continuation process, there is no suppression of the regional scale effects, but they are strongly reduced with respect to the local ones. Depending on the degree of the derivatives, this relative attenuation follows different rates. The important aspect is the physical consistency of such transformations, which is preserved at any order.

In this chapter, the structural setting of the Southern Apennine region will be studied by analyzing its gravity and magnetic fields using a method that exploits these derivative filtering properties. The method, called Multiscale Derivative Analysis (*MDA*) is based on the resolution properties of the Enhanced Horizontal Derivative (*EHD*, Fedi and Florio, 2001), and will be very useful for enhancing the different contributions of the field relative to various scales involved in the measured field. It has already applied to the Tyrrhenian gravity field (Fedi, 2002). More specifically, three different resolutions will be selected to perform the Multiscale Derivative Analysis of the gravity and the aeromagnetic fields of Southern Italy. This will make it possible to highlight regional, intermediate and local structural trends, thus obtaining a valuable tool for the interpretation of geologic structures.

2. GEOLOGICAL FRAMEWORK OF THE SOUTHERN APENNINES

The central-Southern Apennines can be represented by structural domains displaced in geographic bands, NW–SE oriented and some tens of km wide. These are (from E to W):

- The Apulian foreland, extending for 300 km from the Gargano peninsula to the Salento. It consists of a 6–8 km thick series of Meso-Cainozoic carbonates and is representative of slope or shelf depositional facies (D'Argenio et al., 1973).
- The Bradano foredeep, represented by terrigenous sequences with thickness increasing toward the mountain chain (up to 2.5 km) and aged Plio-Pleistocene. They cover the allochthonous nappes of the internal provenance, with unconformable contact, and suture the tectonic contact between the most external thrust front and the underplating Apulia Units.
- The eastward (external) mountain chain sector, represented by extended outcrops of Meso-Cainozoic basinal units (Lagonegro and Molise Units) and of terrigenous sequences of Mio-Pleistocene age.
- The western (internal) mountain chain sector, represented by several carbonate massifs consisting of sequences of Meso-Cainozoic age with thickness reaching 4000 m. They started to be deformed during the Miocene and occupy different tectonic positions: (a) tectonically underplaced with respect to western allochthonous nappes represented by transitional or basinal units of oceanic or transitional provenance; (b) overlapping the basinal units of the eastern chain sector.

Since the Lower Pliocene to the Lower Pleistocene, the internal carbonate platform units and the basinal units, variously deformed, have drifted eastward and overthrusted on the Apulian carbonate platform, thus causing the passive deformation of the overlapping nappes and the consequent formation of piggy-back basins. The undeformed units of the Apulian platform represent the foreland area (Mostardini and Merlini, 1986; Balduzzi et al., 1982).

Above these units, tectonic nappes are present (Liguridi, Calabridi and Sicilidi) occupying the higher geometric location in the chain. These are relics of older basinal (oceanic or transitional) domains of internal position that were deformed, prior to the Tyrrhenian opening, in a pre-Tortonian accretionary prism during the closure of the neo-Tethyean basin caused by the Europe–Africa collision.

From the structural point of view the Apennine chain is dissected by several transcurrent fault lines of different origins strictly related to the tectonic and palaegeographic evolution of the area. Between them, the Sangineto Line, southward delimiting the edge of the Southern Apennines, and the "Ortona-Roccamonfina" right transcurrent line, divides the central-Southern Apennines into two main arcs.

The northern arc is composed of thrust systems structured as embricated fans in piggy-back sequences. The southern one, characterized by a greater shortening than the northern one, is built as duplex systems with overthrusting of the internal units onto the eastern edge of the Apulian platform (Mostardini and Merlini, 1986; Bally et al., 1986).

3. MULTISCALE DERIVATIVE ANALYSIS OF POTENTIAL FIELDS

Multiscale Derivative Analysis is based on the very specific resolution property of the Enhanced Horizontal Derivative (EHD): according to Fedi and Florio (2001) we may define EHD as:

$$\mathrm{EHD}(x, y) = \sqrt{\left(\frac{\partial \phi}{\partial x}\right)^2 + \left(\frac{\partial \phi}{\partial y}\right)^2}, \tag{1}$$

where

$$\phi(x, y) = w_0 f(x, y) + w_1 f^{(1)}(x, y) + w_2 f^{(2)}(x, y) + \cdots + w_m f^{(m)}(x, y), \tag{2}$$

and $f^{(1)}, \ldots, f^{(m)}$ are the m-order derivatives of the field f. Finally, $w_0, \ldots, w_m$ is a set of weights.

Since EHD may involve numerical computation of high-order vertical derivatives, a stable procedure to perform them has to be followed. The Integrated Second Vertical Derivative method (ISVD; Fedi and Florio, 2001) has proved to be much more stable than standard Fourier procedures. It consists of two steps: (a) computing a (stable) vertical integral of the field in the frequency domain, and (b) computing the second vertical derivative of the transformed field from the Laplace equation. Rather stable horizontal derivatives may be computed in space domain by finite-differences or by means of splines.

Fedi and Florio (2001) showed that the sum of the several terms of Equation (2) reinforces the signal over the source boundaries, and tends to cancel the other spurious maxima that appear within the single derivatives of different order. Source boundaries will therefore be well outlined by considering the location of maxima of the EHD function. Let us assume now the more general case of sources of different depth/extent, so generating effects at various scales. In this case, our choice of weights, starting term and last term is decisive for obtaining different images of the source boundaries. The several derivative terms of Equation (2) may in fact be combined in such a way as to enhance the potential field effects at a given specific scale.

MDA consists in practice of computing EHD according to Equations (1) and (2), for different orders of derivatives m. It was in fact shown (Fedi, 2002) that when a good choice of weights and the starting term of the summation is made, analysis of EHD for different values of m corresponds in practice to performing a multiscale boundary analysis. *MDA* may be fully exploited in the study of gravity and magnetic anomalies over wide areas such as the Southern Apennines, where the superposition of deep and shallow effects (i.e. at large and short scale) is likely to occur.

To demonstrate the utility of this kind of analysis, let us first consider the following synthetic test. In Figure 2(a) the gravity field of a set of three prisms simulating a complex geologic structure is presented, with sources placed at different depth, along the same horizontal location. Two small shallow prisms superimposed on a deeper one, having the highest density, form the structure. Due to the superposition of effects, it is hard to single out the contributions of the shallow structures by a simple visual inspection of the field (Figure 2(a)). However MDA will allow the recognition of the boundaries for both the shallow and deep sources. After some attempts, we decided not to apply weighting to the derivatives, i.e. to use, in Equation (2), $w_m = 1$ for each order m; the gravity field was chosen as the starting term of the summation. When the summation is composed of just a few terms, the only boundaries clearly defined are those related to the deep source (Figure 2(b) and (c)). Increasing the order of derivatives involved in the summation enhances the boundaries of the shallow sources (Figure 2(d)–(f)). The separation and selective enhancement of MDA demonstrated with the previous example is similar to a filtering process. Note however that MDA does not involve true separation of field components, but is instead determined by a selective combination of (derivative-based) components each one having a different inherent resolution.

4. MULTISCALE DERIVATIVE ANALYSIS OF SOUTHERN APENNINES: GRAVITY DATA

The investigated Bouguer anomaly field covers an area from 13.44°E to 17.46°E, and from 39.65°N to 42.91°N. Gravity data are windowed from the Bouguer Gravity Anomaly Map of Italy published by the CNR (Carrozzo et al., 1986). The input data set has the following features:

Density for the Bouguer and terrain corrections: 2.4 g/cm^3.
Normal Gravity: U.G.G.I., (1984).
Reference system: IGSN71.
Grid sampling step: 1 km.

The gravity anomaly field of the studied area (Figure 3) shows the presence of many negative and positive short-wavelength anomalies, together with regional anomalies. The latter are represented by an elongated low running along the main axis of the Apennine belt and by an extended high, whose maximum is in the central area of the Tyrrhenian basin.

In several cases these anomalies roughly coincide with carbonate units or basin structures.

There is a limited correlation with topography, which typically occurs in a Bouguer anomaly map. This is mostly caused by the fact that morphologic elements often correspond to structural discontinuities extended in depth.

A multiscale derivative analysis will be applied to the gravity field of the Southern Apennines corresponding to three different resolutions. Since the concept of resolution is a relative one, we will refer in the following to large, intermediate and short scale maps.

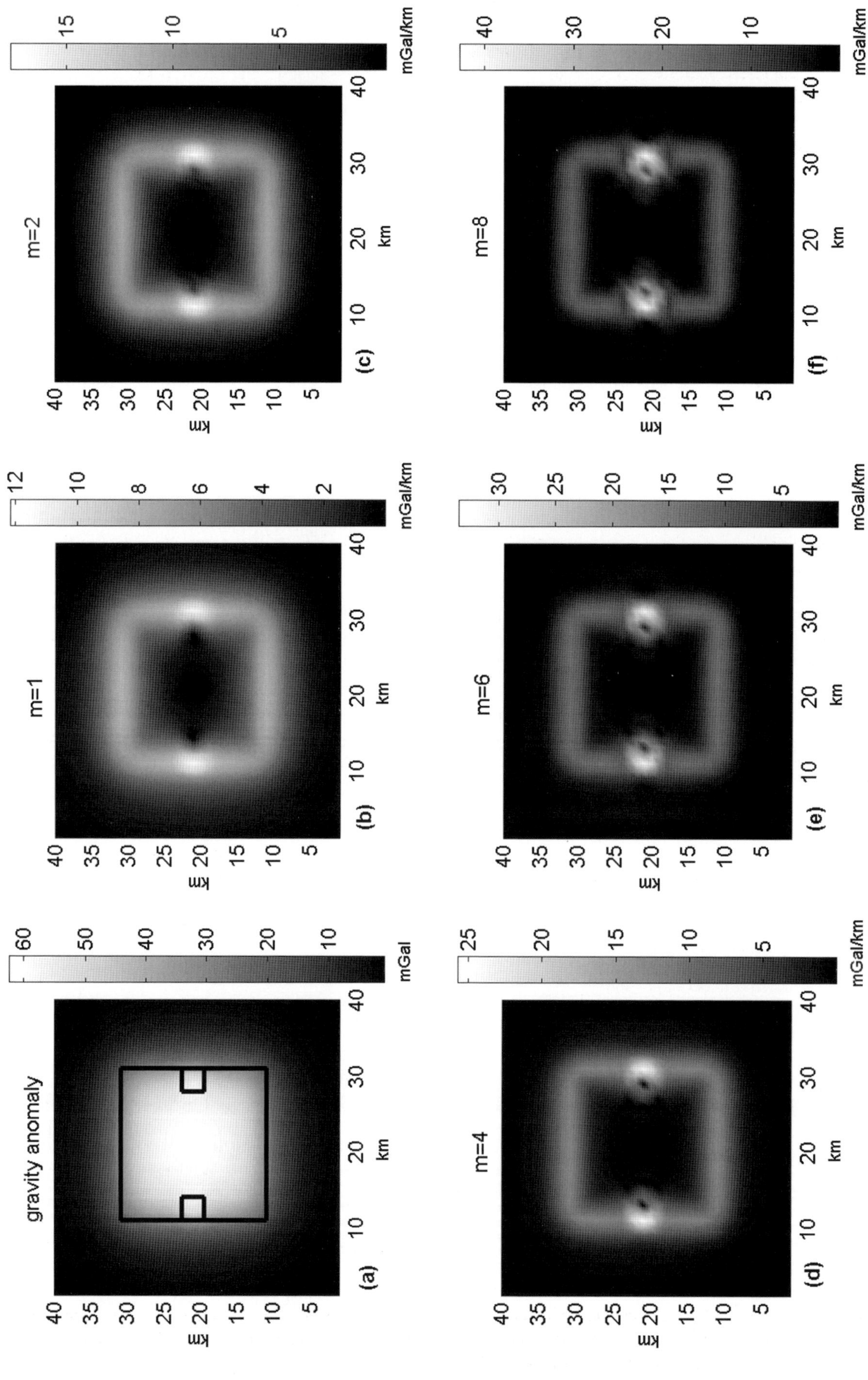
gravity anomaly
m=1
m=2
m=4
m=6
m=8
(a)
(b)
(c)
(d)
(e)
(f)
km
mGal
mGal/km

Figure 2: Example of Multiscale Derivative Analysis. Gravity field generated by a three prismatic sources ensemble (a). The two small and shallower sources have a density contrast of 0.7 g/cm^3, while the deepest has a density contrast of 1 g/cm^3. Solid lines outline the horizontal positions of the prisms. MDA consists in computing EHD (Equation (1)) using vertical derivatives of increasing maximum order m (b, c, d, e, f). The boundary of only the deep source is clearly visible for $m = 1$ and $m = 2$ (b, c). When also higher terms are added on ($m = 4$ and $m = 6$) an image containing boundaries of all sources is obtained (d, e). The boundaries of the shallow sources are better defined by adding further higher-order vertical derivatives ($m = 8$; f).

Correlations between the MDA map and the geological setting of the investigated area were founded on two different sources:

1. *Structural Model of Italy* (Scale 1:500.000), published by the CNR (Bigi et al., 1990) from: "Progetto Finalizzato Geodinamica" – Sottoprogetto: "Modello Strutturale Tridimensionale".
2. *Geological Map of Southern Apennines* (Scale 1:250.000) from: Proceedings of the 74th Congress of the Italian Geological Society (1988).

4.1. Large scale

The MDA map showing gravity source patterns at large scale (Figure 4) was obtained by computing EHD starting from the gravity scalar potential as first term of the summation, and considering derivatives up to $m = 2$ (Equation (2)).

The map singles out two broad trends of maxima. The first is located on the eastern edge of the chain, roughly coinciding with the northern portion of the Gulf of Taranto. Northward, its direction becomes normal to the Apennines axis. The other one is less visible and is located on the opposite side of the Apennines belt. Their meaning is presumably related to the density contrast due the lateral increase in crustal thickness. Consequently, the linear maxima trends at regional scale seem correlated to the regional change in depth of the Moho discontinuity.

The flatness of the signal in the Apulian platform area suggests a crustal thickness without significant variations. The map shows other trends of maxima with a probable intra-crustal origin, but sufficiently extended and deep to be visible at the large scale.

A similar interpretation can be given to the linear trends of maxima bordering westward the Apulian plate. These presumably indicate the presence of regional normal faults along which the Apulia platform sinks toward the foredeep.

Some comments can be made:

- The direction of the regional trend of EHD maxima, on the eastern side of the Apenninic belt, changes from NW to NE. The existence of a structural feature has to be expected at the regional scale, coinciding with the Gargano platform and, presumably, separating two different domains.
- On the eastern side of the map, an EHD pattern marks the boundary between the foreland area (represented by the Apulian platform, relatively undeformed and outcropping near the Murge) and, westward, the foredeep (represented by the Bradano trough).
- The abrupt change in EHD patterns evidenced on the eastern side of the central trend of mimima (corresponding to the main belt) does not coincide with the chain front but runs westward, parallel to it, and highlights the complex post-Tortonian thrust front system that represents the more external sector of the Apenninic allochthonous nappes.
- The areas of EHD maxima located in the Gulf of Naples and in the southern sector of the Gulf of Gaeta are notable. These areas are characterized by poor seismic penetration, probably caused by the presence of large volcano-clastic bodies within the Pliocene-Pleistocene sedimentary sequence.
- Finally, the low EHD of the northern side of the Sele plain broadens toward the Tyrrhenian Sea. In practice, it breaks up the continuity in the maxima running west of the axial belt, and separates the Cilento area from the Apenninic system surrounding the Campanian plain.

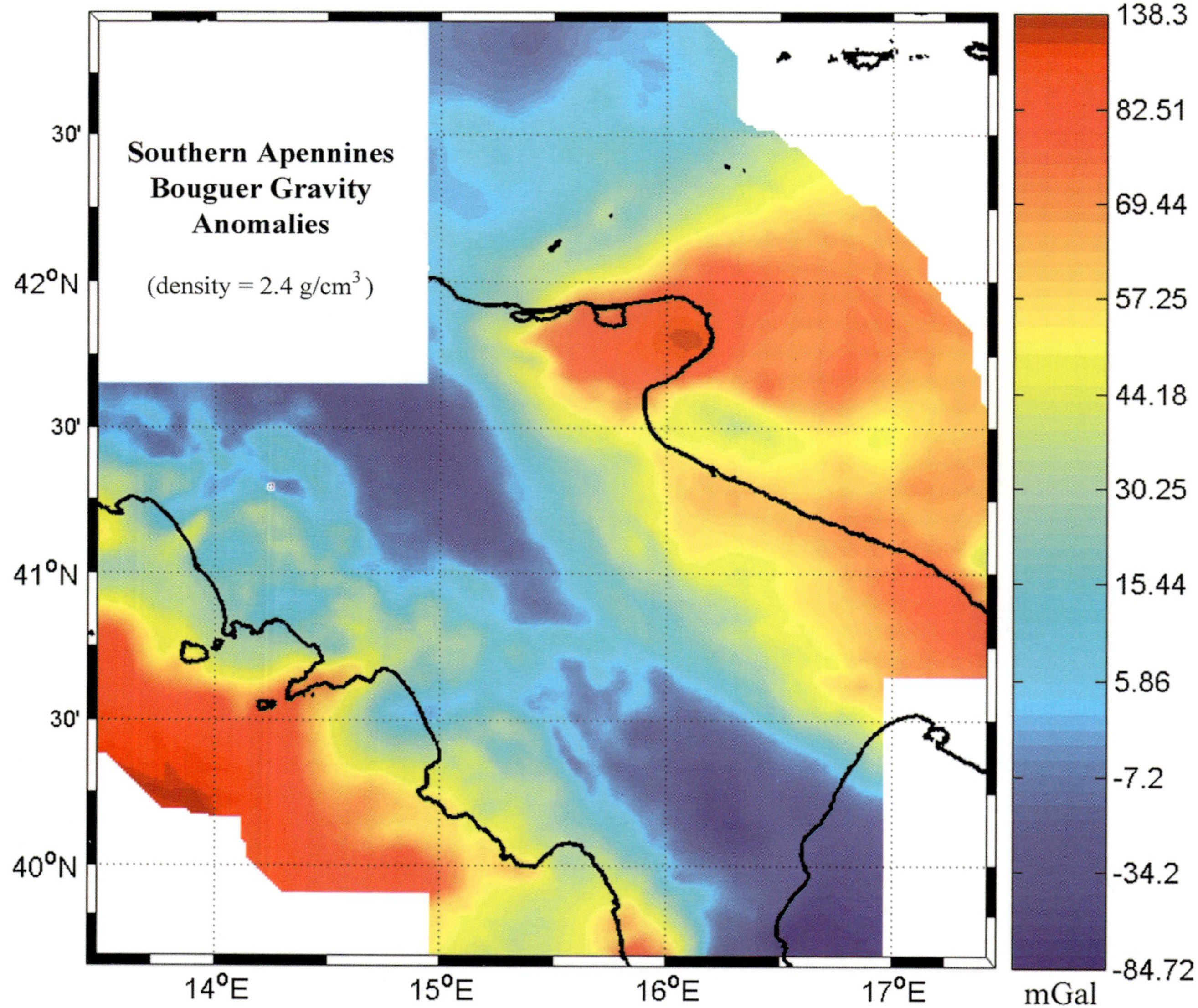

Figure 3: Bouguer anomaly gravity field of Southern Italy. Density for the Bouguer and terrain corrections: 2.4 g/cm^3. Normal Gravity: U.G.G.I. (1984). Reference system: IGSN71. The grid spacing is 1 km.

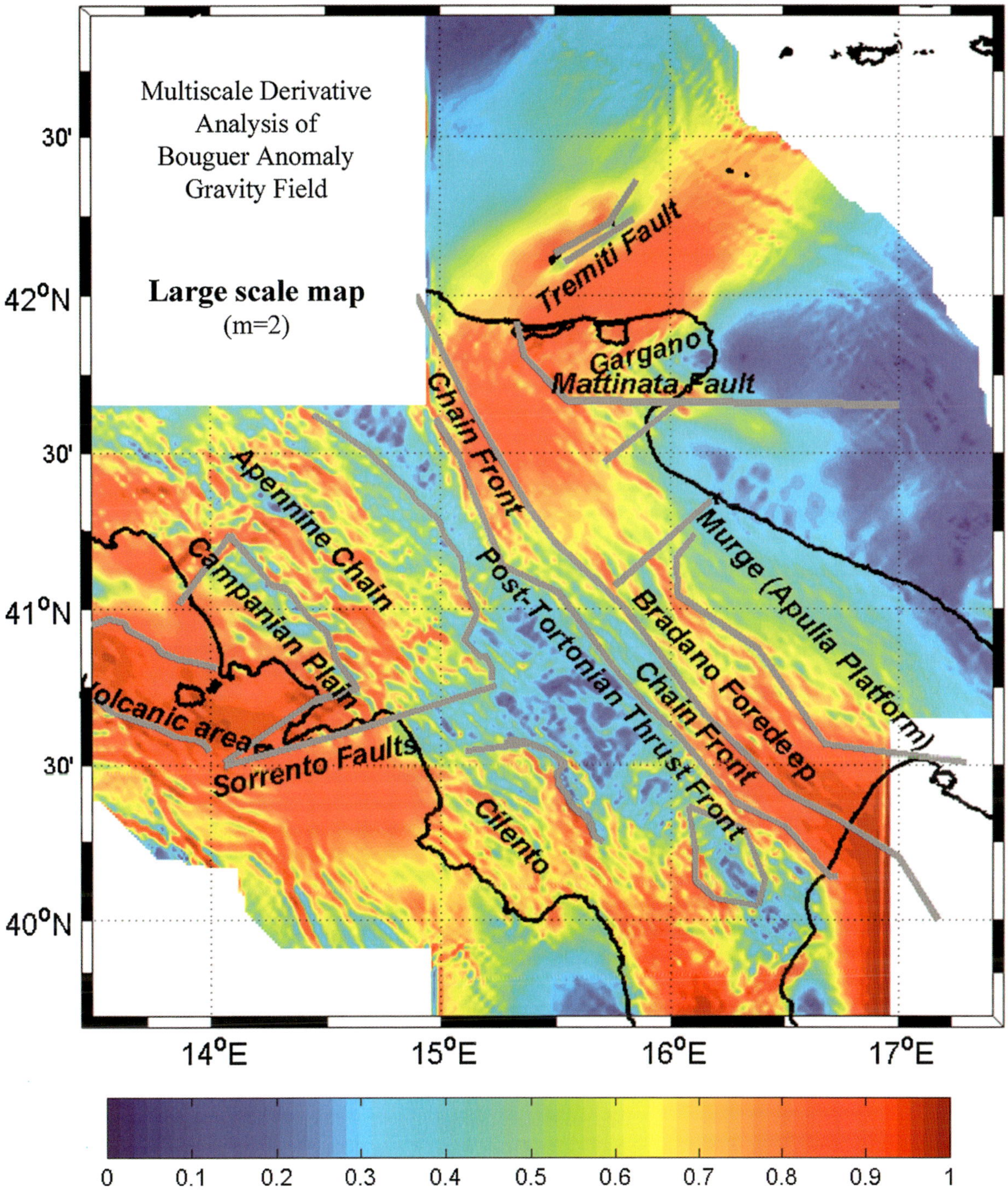

Figure 4: Large scale MDA map of Bouguer anomaly gravity field in Southern Italy. Lineaments are identified by trends of local maxima, which are graphically determined by the color bar shown in normalized units.

- A regional linear trend seems well related to the main normal fault pattern bordering the whole Campanian plain, from the Sorrento Peninsula to the Mt. Massico carbonatic horst. This may be due to the presence of deep crustal tectonic discontinuities beneath the above mentioned normal faults, which are characterized by a very high throw.

In conclusion, analysis of the MDA map at large scale shows the presence of regional trends that allow identification of the main regional morpho-structural domain forming the whole of Southern Italy.

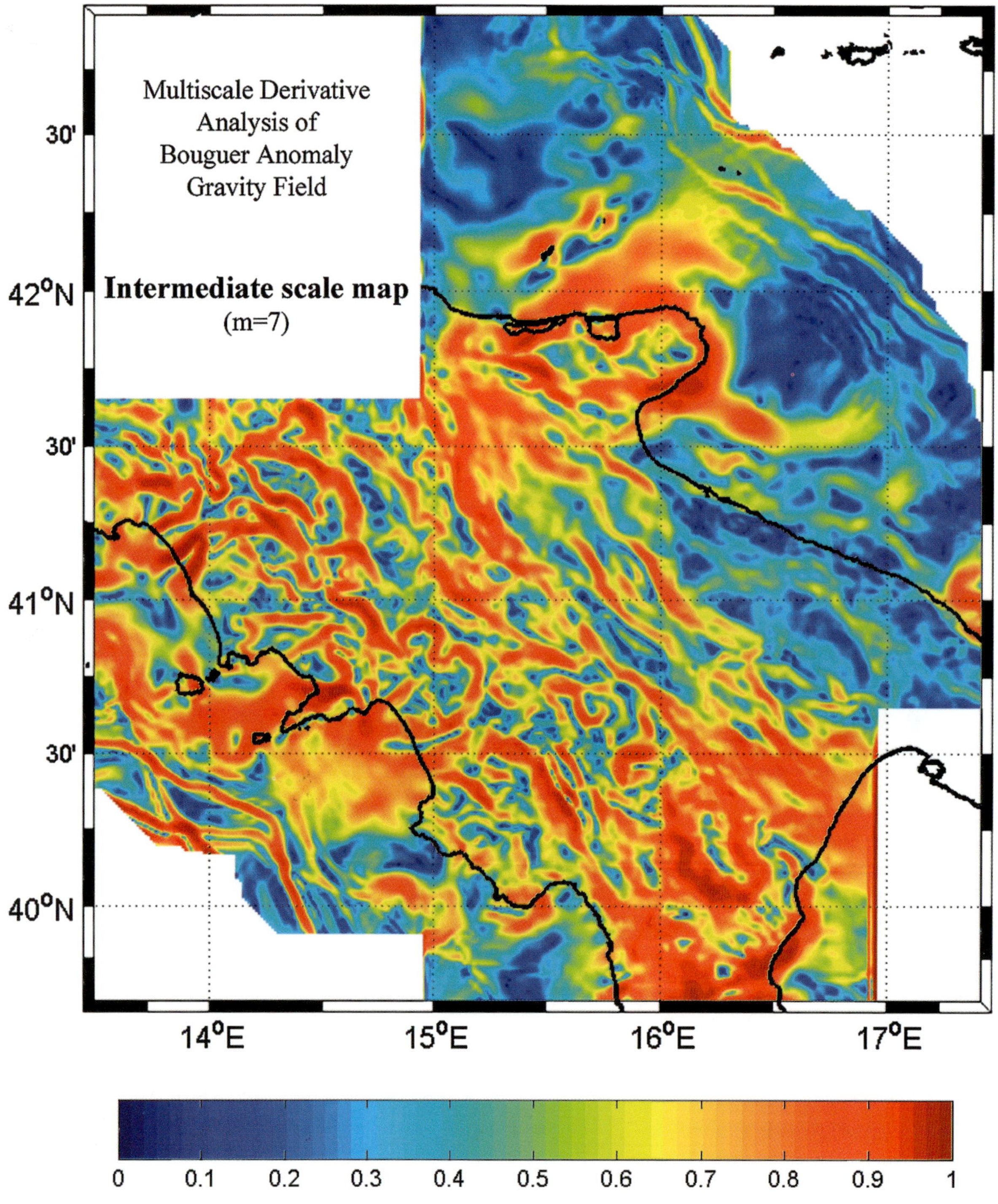

Figure 5: Intermediate scale MDA map of Bouguer anomaly gravity field in Southern Italy. Lineaments are identified by trends of local maxima, which are graphically determined by the color bar shown.

4.2. Intermediate scale

The MDA map showing the gravity source patterns at intermediate scale (Figure 5) was obtained by computing EHD starting from the gravity field as first term of the summation (2), and considering derivatives up to $m = 7$.

As far as the structural setting, the tectonic style and the shallow geological features are concerned, the regional pattern of MDA outlines reveals with extreme comprehensibility the radical change in tectonic style observed along the eastern side of the central and Southern Apennines.

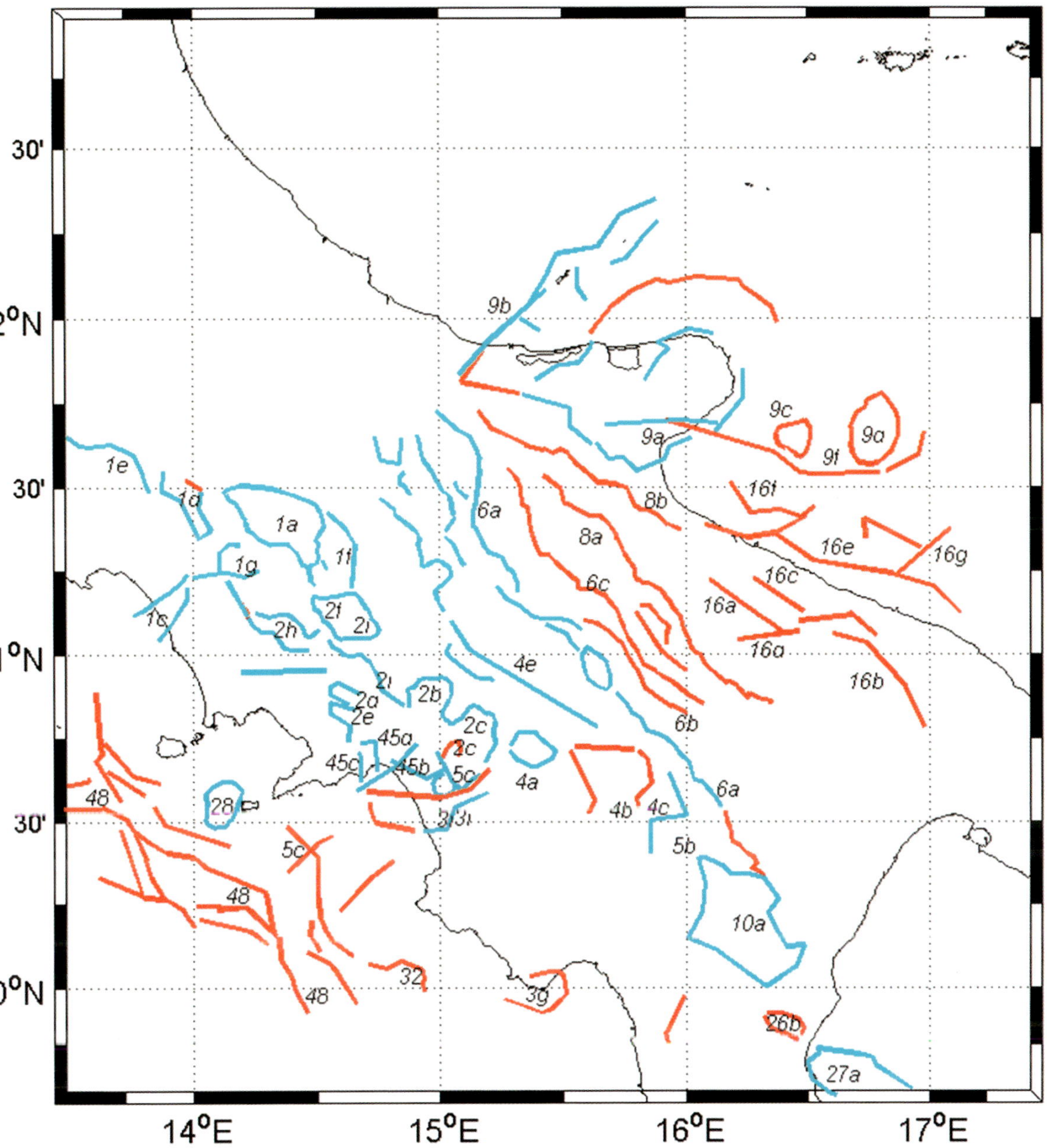

Figure 6: Major lineaments identified by EHD maxima of the intermediate scale MDA map of Bouguer anomaly gravity field in Southern Italy. Cyan: correlated lineaments; red: uncorrelated or partially correlated lineaments. Reference to lineament numbers in Legend I.

Westward of the front chain, the MDA signal shows the prevalence of short, arc-shaped and variously oriented trends, together with some linear trends with anti-Apennine direction.

Most of MDA trends within the chain seem related to its structural elements, and generally coincide with normal faults systems and the major overthrusts fronts. This is clearly related to the geodynamic evolution of the area, since the tectonic history of the structural units within the chain is much more complex than external to the chain front. In fact, since the Tortonian, an intense crustal shortening has occurred, due to a complex, multiphase compressive regime, and to the consequent mountain building forming the present Apenninic chain.

Along the western side of the Apennines, several lineaments can be described as related to the structural units resulting from the deformation of the external Apennine domains. In the following, many of these lineaments will be described, within the framework of structural geological units. For the sake of simplicity, the discussion take the various units separately.

4.2.a. *The Apenninic chain: tectonic units resulting from the deformation of the external Apenninic domains*

The "Matese–M. Maggiore" Units. One of the most clear MDA lineament bounds the morpho-structural limits of the Matese massif (1a; Figure 6) whereas, eastward, the outcropping series of the "Flysh Rosso", ascribed to the Lagonegro II units are delimited by another lineament (1f; Figure 6).

A couple of EHD maxima (1c; Figure 6) NE–SW-oriented, bound the M. Massico massif on both its southern and northern sides, indicating the presence of two noticeable normal fault lines. These trends are prolonged offshore. Eastward, a structural connection (although partially hidden by the volcanic series of the Roccamonfina volcano) is clearly observed between them and other lineaments (1g; Figures 6 and 8). These reveal the complex system of normal faults bounding the carbonate series of Mt. Maggiore on its western and eastern side and are clearly visible at both intermediate and short scales.

East of the Liri valley, a rather visible linear trend (1d; Figure 6) clearly locates the normal fault system dislocating the carbonate complex outcropping east of Venafro on its western, southern and eastern side.

North of the Liri valley, another linear trend marks a fault line dislocating the northern side of the Mt. Cairo massif (1e; Figure 6).

The "Monti Picentini–Taburno–Aurunci" Unit. A marked MDA trend (2f; Figure 6) indicates two carbonate massifs, geologically ascribed to different litho-stratigraphic units: Camposauro massif (Matese–Mt. Maggiore Unit) and Taburno massif (Monti Picentini–Taburno Unit). This trend is located immediately W of another one (2i; Figure 6), marking the eastern side of geological complexes of Miocene-Pliocene age (Altavilla Unit) outcropping W of Benevento.

The Picentini–Taburno Unit is also represented by the carbonate series of the Terminio (2b; Figure 6) and Cervialto (2c; Figure 6) massifs, the structural limits of which are nicely singled out by an MDA trend, except on its southern side.

Two trends bound the western and eastern structural limits of the carbonate series outcropping near Caserta (2h; Figure 6).

Southward, the relations between the MDA signal and the morpho-structural boundaries of the outcropping carbonate massifs are represented by a clear trend marking the western structural limit of the Sarno Mts (2e; Figure 6) and Avella Mts (2d; Figure 6), whereas the eastern one is marked 2i.

The Sorrento peninsula appears laterally displaced by a strike-slip fault system with Apenninic orientation strongly controlling the morpho-structural setting of the ridge, and whose existence is clearly evidenced by secondary MDA linear trends (45c and 45d; Figure 6).

Another trend (5c; Figure 6) occurs between those signed as 45e and 45f. It marks with good precision the boundaries of faulted alluvial conglomerates ("Eboli Conglomerates"), bordering the southern tectonic limits of sedimentary sequences outcropping W of Campagna and ascribed to both the Picentini–Taburno and Lagonegro II Units.

North of the Terminio and Cervialto massifs, a linear trend (4e; Figure 6) is clearly visible at intermediate scale and can be related to the axis of a regional anticline involving the Neogenic lithostratigraphic units from pre- to late orogenic (Irpinian Units and the Ariano Unit). The correlation should presumably involve also deeper structural features.

The presence of a closed trend (28; Figure 6), west of Capri, supports the existence of an uprisen carbonatic block partially buried by younger marine sediments.

The "Mt. Marzano–M. della Maddalena" Unit. Other morpho-structural complexes ascribed to the Campania-Lucania Carbonate Units (inner platform) are precisely identified by means of the MDA signal. Among them, the M. Marzano should be mentioned (4a; Figure 6). Around Potenza, several small MDA trends seem uncorrelated with the outcropping structural setting. This is the case of line 4b (Figure 6), probably locating a structural discontinuity beneath the covering Sicilide Units outcropping SE of Potenza and, eastward, of the trend indicated by 4c (Figure 6) that only partially coincides with the tectonic contact of the Sicilide Units above the Lagonegro II Units.

4.2.b. *The Sele plain.* Some attention should be given to the MDA trends in the Sele plain. Here the existence of two trends of maxima on the northern and southern (3i; Figure 6) sides of the plain, and of a central axis of MDA minimum, provides evidence of a strong structural control on the morphological evolution of the basin. The existence of two main normal fault systems, buried beneath a thick cover of alluvia and, westward, prolonged offshore, can be hypothesized. They seem not to immediately border the laterally outcropping structures, but are slightly displaced toward the intermediate axis of the plain.

The central minimum zone shows evidence of an intermediate sector of basement displaced at some depth. In this regard, MDA seems to indicate that the lower segment of the Sele river has not been formed along the central axis, but along the southern fault.

4.2.c. *The Tyrrhenian region.* Offshore, at great distance from the Tyrrhenian shoreline, both the intermediate and short scale MDA maps describe linear trends with regional extension. These lineaments (48; Figure 6) run with WNW or NWN directions. Their meaning could be reasonably associated with the presence of several normal faults bordering the Tyrrhenian depression and strictly related to its formation. It must be pointed out that correlation between these trends and the known structural setting exists only locally (i.e. in correspondence to Sirene Mt., 13°50'E–40°20′N).

Offshore the Cilento promontory, the MDA at intermediate scale reveals trends of difficult correlation with the structural features traditionally mapped in the area. At the southern end of Cilento, an MDA closed trend occurs (3g; Figure 6) which cannot easily be correlated with the outcropping carbonate series of the Bulgheria–Verbicaro Units.

4.2.d. *The Calabrian-Lucanian Apennines.* The existence of a closed trend (26b; Figure 6) located E of the Pollino massif appears only partially correlated with the outcropping structures represented by the Cilento Flysh (Albidona, Pollica and S. Mauro Formations).

The MDA signal shows a discrete efficiency in locating coastal basins filled by thick sedimentary series: a trend running NW–SE (27a; Figure 6) may be likely correlated with the structural boundaries of the Sibari basin, located in the Taranto Gulf, offshore the Sibari plain.

4.2.e. *The Bradano foredeep and the Apulia-Gargano platform.* Within this region, external to the chain front, MDA clearly shows an abrupt change in tectonic style and structural setting.

A rather complex system of linear MDA trends runs across the Murge platform, thus suggesting that the tectonic setting in this sector is more complex than that inferred from surface evidence. Most of the trends run NW–SE (16a, 16b and 16c; Figure 6), whereas others have a NNE–SSW direction (16d; Figure 6), evidencing a tectonic setting represented by a normal fault system with main stresses along Apenninic and anti-Apenninic directions.

The MDA maxima along these trends are not intense. This may be explained by small density contrasts along normal faults displacing the carbonate series of the Apulia-Gargano platform.

The MDA map shows that the tectonic setting above described is not limited to the Murge platform but continues, with the same directions, offshore into the Adriatic Sea (16e, 16f and 16g; Figure 6).

As far as the regional geologic features are concerned, the MDA of the north-eastern sector of the Apulia-Gargano domains appears very meaningful. The local uprising of the Apulia-Gargano platform, represented by the Gargano promontory, is well marked by intense MDA trends along the whole perimeter of its bordering fault lines. A strong linear trend (9a; Figure 6), running E–W, is easily correlated with the Mattinata fault. The existence of this important structural element is verified also eastward, in the Adriatic Sea, to a longitude of 17°. The trend has a slight change in offshore direction (toward ESE) where it marks a linear zone (9f; Figure 6) coinciding with a local slight deepening of the top of the Mesozoic carbonate series (Del Ben et al., 1994). Eastward it links to a pattern of Adriatic outlines better visible on the short scale map.

North of the '9f' MDA line, two closed trends (9c and 9d; Figure 6) are not correlated to surface evidence and, presumably, are related to two small structural highs of the Mesozoic carbonate basement.

Just north of the Gargano promontory, a linear trend (9b; Figure 6) running SW–NE is visible from the shoreline up to Tremiti Islands. It indicates significant regional discontinuities presumably consisting of a normal anti-Apennine fault system, causing the local rise in the Mesozoic carbonate basement.

4.3. Short scale

The short scale MDA map (Figure 7) describes the finest gravity source patterns. It was obtained starting from the gravity field as first term and computing EHD up to $m = 9$. The great improvement in describing structural-geologic patterns with small extension and local significance is immediately apparent. This makes it arduous to trace here a complete description of the whole map. Therefore, the map has been divided into three different sectors: north-western sector (Figure 8), north-eastern sector (Figure 9) and southern sector (Figure 10).

4.3.a. *The Apenninic chain: tectonic units resulting from the deformation of the external Apenninic domains*

The "Monti Picentini–Taburno–Aurunci" Unit. Several trends WNW-oriented, and roughly parallel, describe the main morpho-structural features between the Gulf of Gaeta and the Liri valley. The first trend delineates the western edge (2a; Figure 8) of the carbonate series of the Aurunci Mts. This series is northward overthrust on the Sicilidi Units whereas, eastward and southward, is in contact with volcanic and alluvial deposits along normal fault systems (2g; Figure 8). The other two trends (2h; Figure 8), WNW-oriented, indicate the presence of two faults controlling the formation of the Liri valley tectonic depression.

MDA shows also a line of maxima (2m; Figure 8) running NNE from the Gulf of Gaeta toward the Liri valley. This marks precisely the tectonic contact between the Picentini–Taburno Units and the Altavilla Unit (Neogenic litho-stratigraphic units of the Apenninic chain, from pre- to late orogenic).

East of the trend denoted by 4a (Figure 6), a trend (23a; Figure 9) is clearly visible around Potenza and should be partially correlated with normal or undetermined faults dislocating the Lagonegro Units with respect to the Campania-Lucania Carbonate Units. These contacts are partially hidden by terrigenous outcropping covers aged Middle Pliocene.

Between trends 6a (Figures 6 and 8) and 23a (Figure 9), the short scale map evidences some lines (23b; Figure 9) with no apparent correlation with the shallow features represented by the extended outcrops ascribed to the 'Flysh Rosso' (Lagonegro II Units).

The short scale MDA shows however a clear correlation with the ridge represented by the Mesozoic carbonate series of the Picentini–Taburno Units and constituting the Sorrento peninsula. In fact, both the northern boundary (45a; Figure 8) and the southern (45b; Figures 6 and 8) are clearly evidenced by two strong linear trends running ENE–SWS and coinciding with two regional normal fault lines. From the eastern end of the southern one, two different linear trends seem to run onshore. The first one (45f; Figure 8) deviates NE, roughly coinciding with the valley of the Picentino river; the second one (45e; Figure 8) continues eastward and marks the southern boundary of the Picentini Group. In any case these trends should be correlated to normal fault systems strictly related to the uplift of the Sorrento peninsula.

East of the Agri and Melandro valleys, a strong MDA trend (5b; Figures 6 and 10) E–W and N–S oriented indicates a deep-seated density contrast, hidden by the sedimentary cover of Plio-Pleistocenic age (Calvello Unit), and maybe related to the tectonic contacts between Sicilide and Lagonegro I Units.

4.3.b. *The Campanian volcanic district.* In the Campanian volcanic district, both the medium and short scale MDA map provide major information about the shallow crustal structures. Some of them are better defined by the short scale MDA, whereas others result are more readable at the MDA intermediate scale.

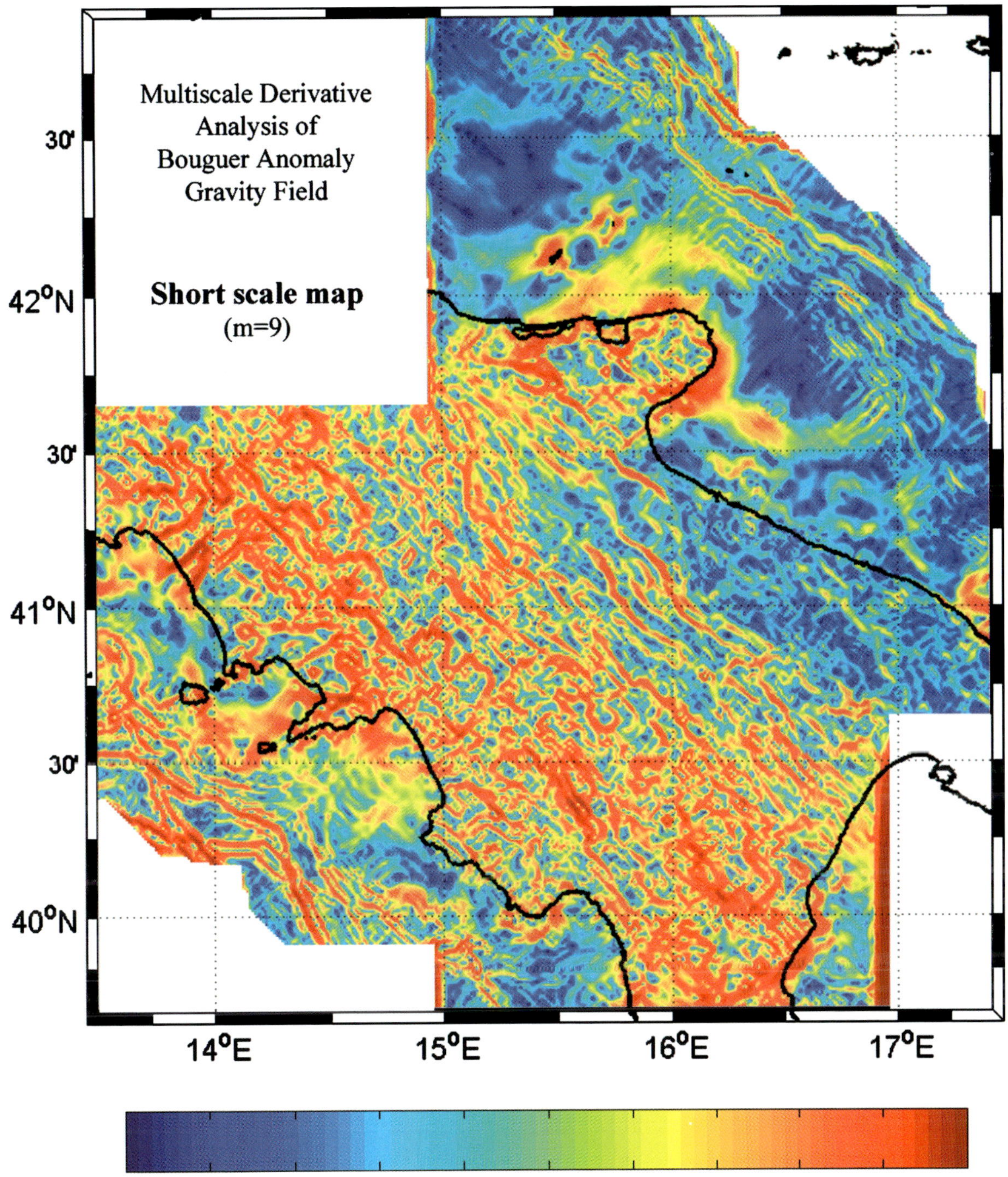

Figure 7: Short scale MDA map of Bouguer anomaly gravity field in Southern Italy. Lineaments are identified by trends of local maxima, which are graphically determined by the color bar shown in normalized units.

Many MDA trends are clearly correlated with the geological evidence: example the closed trends related to the Vesuvius crater (18a; Figure 8) and the Phlegrean Caldera (17a; Figure 8) (Florio, 1993). Other structures, on the contrary, do not seem to have any correspondence with features at the surface. This is the case of a linear trend (18b; Figure 8) running ENE–WSW immediately north of the Somma–Vesuvius volcanic complex (Florio, 1993). The trend continues offshore and probably relates to a deep regional normal fault, due to an extensional tectonic regime responsible for the volcanic evolution of the Somma–Vesuvius complex. Another noticeable linear trend (17b; Figure 8) in the Campanian plain north of Naples runs E–W from longitudes 14°10′ to 14°30′. It agrees with the existence of a small linear carbonate ridge buried below the volcano-sedimentary sequences filling the

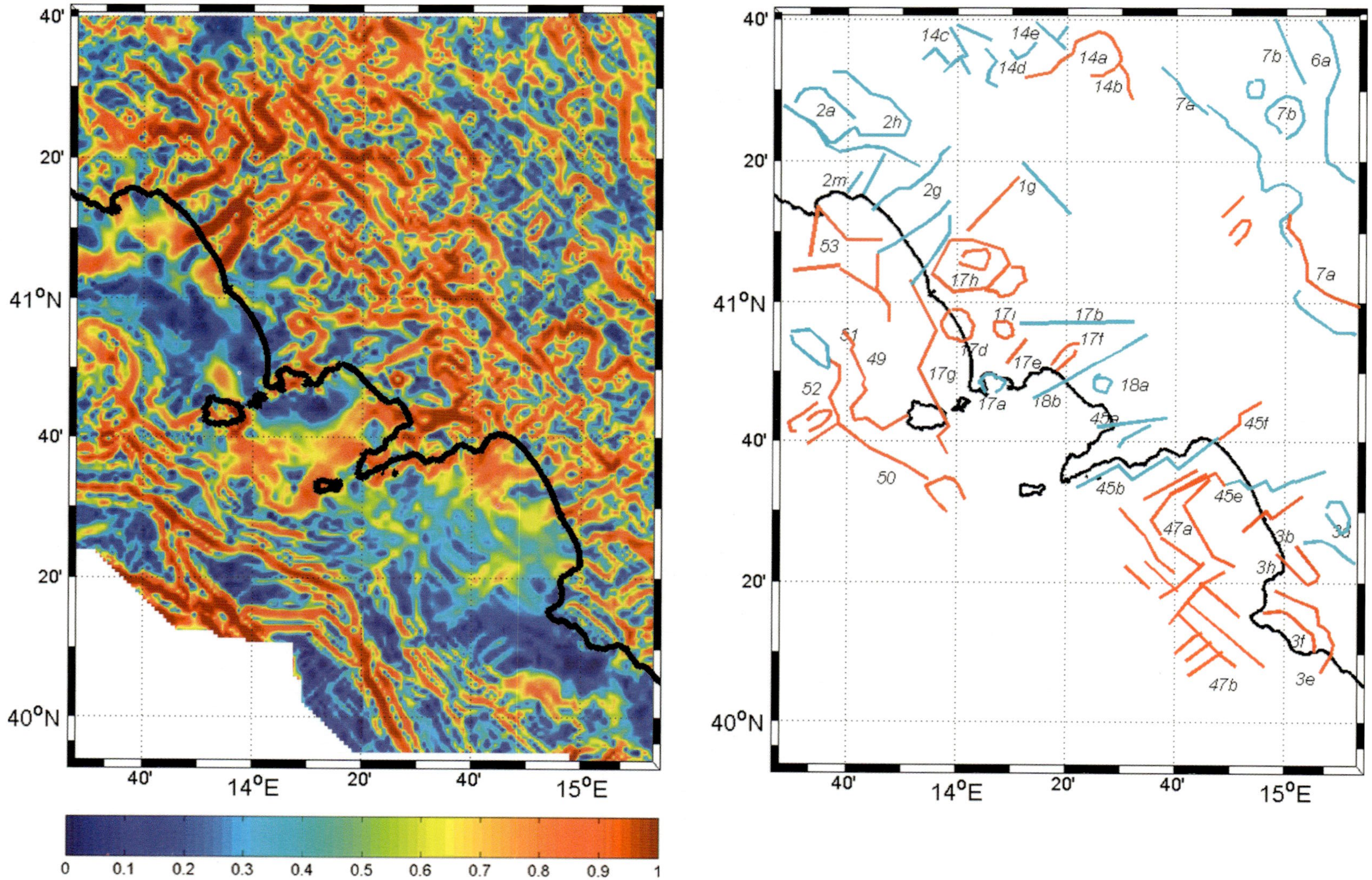

Figure 8: **Short MDA map of gravity data in Southern Italy: *north-western sector*. On the right panel, major lineaments identified by EHD maxima are shown. Cyan: correlated lineaments; red: uncorrelated or partially correlated lineaments. Reference to lineament numbers in Legend II. Lineaments are identified by trends of local maxima, which are graphically determined by the color bar shown in normalized units.**

Campanian plain, whose top deepens westward. This small ridge presumably implies a westward continuation, beneath the volcano-sedimentary cover, for the E–W fault system controlling the structure of the Avella carbonate massif.

The existence of a ring-shaped trend (17i; Figure 8), just to the western end of the previously described element, could be related to an old volcanic structure (Parete) buried below the surface, probably resulting from the extensional regime associated with a local fault system. From the Parete structure, a small linear trend (17e; Figure 8) runs southward. Another MDA trend is visible around the eastern side of Naples (17f; Figure 8) and roughly coinciding with a structure related to the morphologic features of the Sebeto valley.

In correspondence to the Lake Patria (north of the Phlegrean Fields) another ring-shaped trend (17d; Figure 8) has no apparent correlation with shallow structures.

In the northern Campanian plain we also note a broad closed trend (17h; Figure 8) located in the area of the lower course of Volturno valley. This trend marks the edge of a low gravity area, probably related to a thickening of light sediments in that area.

4.3.c. *Neogenic Apenninic lithostratigraphic units from pre- to late orogenic: the Molise units.* In the area between Isernia and the Matese massif, the thick basinal series attributed to the Molise Units are tectonic complexes resulting from the deformation of the external Apennine domains. The mapped limits separate lithologies with presumably small density differences. Nevertheless, trends 14a and 14b (Figure 8) are intense and appear shifted eastwards with respect to the limits mapped at the surface, thus indicating a deep-seated density contrast related to a structural contact running differently from that visible at the surface. A deep tectonic contact with the carbonate series of the Matese Units can be inferred.

East of Isernia, the tectonic contacts between carbonate series of both the Matese and Frosolone Units are deduced from other N–S trends. Some of them are clearly at the surface (14c; Figure 8), whereas others (14d and 14e; Figure 8) are, presumably, deeper.

4.3.d. *The Sele plain.* In the Sele Plain, the presence of two trends of maxima on the northern and southern (3i; Figure 6) sides of the plain, and of a central axis of MDA minimum, provides evidence of strong structural control on the morphological evolution of the basin. The presence of two main normal fault systems, buried beneath a thick cover of alluvia and, westward, prolonged offshore, can be hypothesized. They do not seem to immediately border the laterally outcropping structures, but are slightly displaced toward the intermediate axis of the plain.

The central minimum zone shows evidence of an intermediate sector of basement displaced at some depth. In this regard, MDA seems to indicate that the lower segment of the Sele river has not been formed along the central axis, but along the southern fault.

4.3.e. *The Cilento area: tectonic units resulting from the deformation of the external Apenninic domains (Alburno-Cervati units) and neogenic lithostratigraphic units from pre- to late orogenic (Cilento units).* Valuable MDA results are obtained for the Cilento region. Here the correlation between MDA signal and the morpho-structural features is not easy. Many trends do not coincide with tectonic, stratigraphic or morphologic limits. This implies the presence in the shallow crust of lateral density contrasts not evidenced by the structural setting at the surface. Only the structurally well evidenced Alburni carbonate massif is described by the MDA signal with some precision (3a; Figures 9 and 10).

At the north-western end of the area, there is a trend (3h; Figures 8, 9 and 10) roughly coinciding with the outcrop of the Liguride Units tectonically overlapping the Cilento and Alburno-Cervati Units.

Two small and weak trends mark two different structures. The first one coincides with the Mt. Cervati massif (3b; Figures 8 and 10) whereas the second one, westward located, is not apparently correlated with shallow structures (3c; Figure 10). A partial correspondence is evidenced between a linear trend (3e; Figure 10) and the lower Alento valley; it could be partially due to the presence of a tectonic discontinuity at depth. A linear trend (3f; Figure 10), located just north of P. Licosa, on the contrary, is totally uncorrelated.

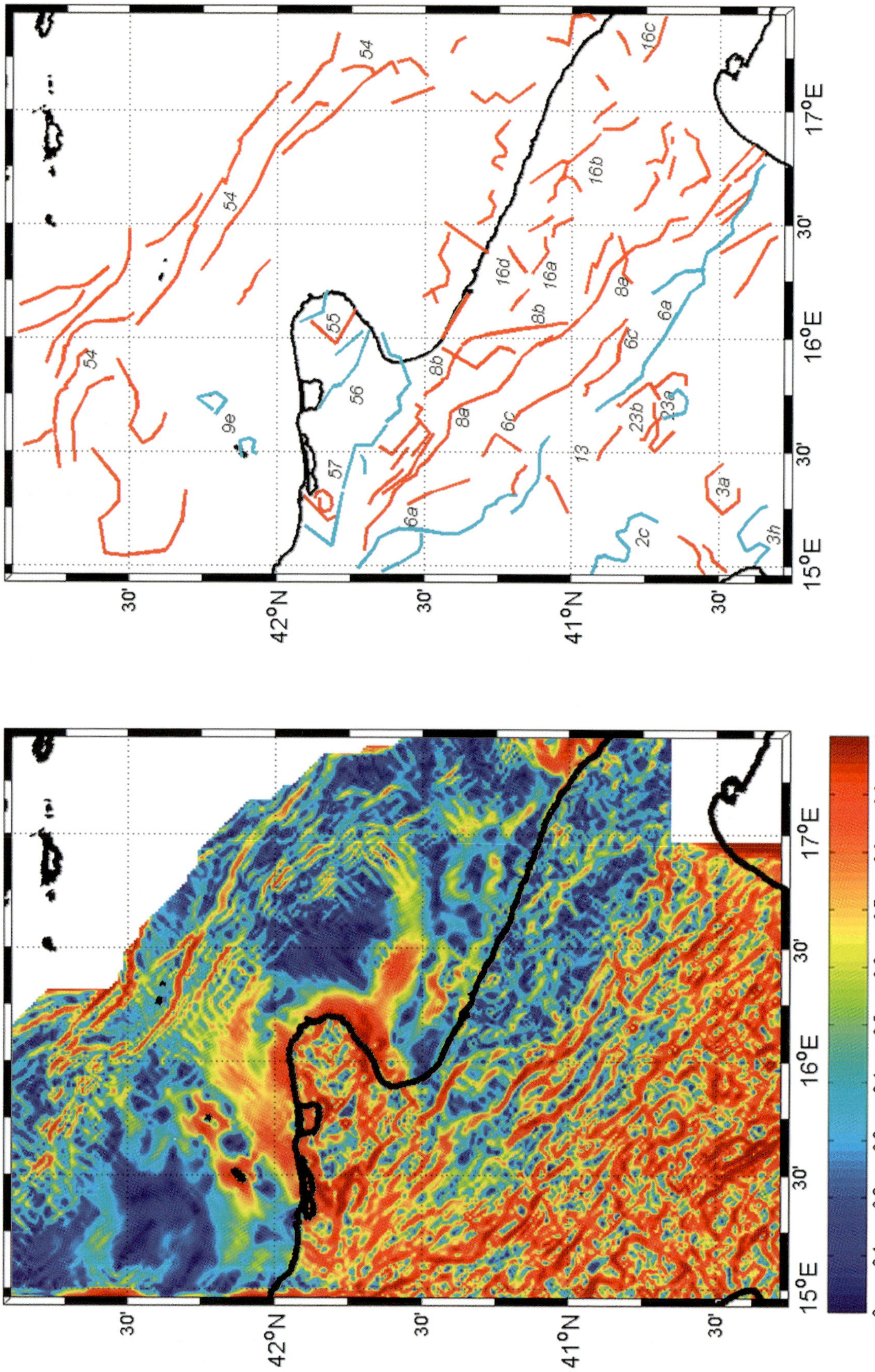

15°E
30'
16°E
30'
17°E
42°N
30'
41°N
30'
54
54
54
9e
55
56
57
6a
8b
8a
6c
16d
16a
8b
8a
6c
6a
16b
16c
13
23b
23a
3a
3h
2c
0
0.1
0.2
0.3
0.4
0.5
0.6
0.7
0.8
0.9
1

Figure 9: Short MDA map of gravity data in Southern Italy: *north-eastern sector*. On the right panel, major lineaments identified by EHD maxima are shown. Cyan: correlated lineaments; red: uncorrelated or partially correlated lineaments. Reference to lineament numbers in Legend III. Lineaments are identified by trends of local maxima, which are graphically determined by the color bar shown in normalized units.

The linear trends roughly evidenced in the south-western side of the Sele plain by the MDA at intermediate scale show many more particulars at a shorter scale (3i; Figure 10). Thus a deep structural complex can be inferred, completely hidden by the recent sedimentary cover, and more complex than expected.

The inner part of the Cilento area presents several small MDA trends only partially correlated with geological features at the surface. One example is represented by the thrust front maintaining contact with the Sicilide Units (Albidona and Saraceno formations) on the silico-clastic and carbonatic units (from pre- to late orogenic) of Neogenic age of the Apenninic chain (3d; Figure 10).

4.3.f. *The Calabrian-Lucanian Apennines.* Just E of Cilento, four long and well marked MDA trends are aligned in a NW–SE direction. They clearly mark a set of morpho-structural boundaries delimiting the Maddalena Mts. They represent the morphological belt separating, westward, the Diano valley (11a; Figure 6) and, eastward, the Melandro valley (24a; Figure 6). The trend 24a continues south-eastward and coincides with the Agri valley (24b; Figure 6) but stops at the western end of the S. Arcangelo basin. The sharpness of this signal and the abrupt interruption of the Agri trend, demonstrate that the discontinuities surrounding the S. Arcangelo basin are noticeably deeper than the structural features associated with the genesis of the Agri valley.

It can be observed that the linear feature marking the eastern edge of the Diano valley seems to develop southward, over the southern morphologic limit of the trough, by means of three linear NW-running trends (11b, 11c and 11d; Figure 10), up to the Noce valley. This could imply the existence at depth of a tectonic contact, represented at the surface by the overthrusting of the carbonate platform series of the Alburno Cervati Units (M. Cocuzzo, M. Coccovello) above the Lagonegro I Units (M. Sirino).

The S. Arcangelo basin, along the Southern Apennines, has one of the more impressive MDA outlines. The boundaries of this significant geological and morphological element of the Southern Apennines can be roughly observed also at the MDA intermediate scale, but their structural details can be visible only at the short scale (10a; Figures 6 and 10).

Also the area located immediately west of the S. Arcangelo basin shows trends (10b, 10c, and 10d; Figure 10) roughly visible at the intermediate scale. Nevertheless, the short scale map shows that these features present many more local features, thus attesting to the really complex structural setting involving the tectonic contacts between the S. Arcangelo basin and the inner domains (Lagonegro I Units, Alburno Cervati Units, Frido Units, and Liguridi Units).

A significant correlation between MDA signal (10c; Figure 10) and the boundaries of the M. Alpi carbonate structure should also be mentioned. Southward, just eastward of the Tyrrhenian shoreline of Calabria, a closed trend (26d; Figure 10) marks the tectonic contact of the "Frido" metasedimentary Formation (Calabride Units) overthrusting the surrounding carbonate series of the Bulgheria–Verbicaro Units.

Southward, the transition towards the North Calabrian Units is characterized by a rather confused MDA signal, but a clear trend (25a and 25b; Figure 10), belonging to the Sangineto tectonic Line, follows the extreme southern limit of the Apenninic Chain. This trend, together with two other short lines, coinciding with the southern edge of the Pollino massif and the eastern edge of the Caramolo massif, bounds the small outcrop of Castrovillari (26e; Figure 6).

Immediately west of 26e, the short scale MDA map clearly shows another trend (26c; Figure 10), NW- and NE-running, that precisely marks: (a) the normal faults delimiting eastward the Verbicaro Carbonate Units and the meta-sedimentary series of the S. Donato Unit; (b) the main normal fault displacing the above mentioned units with respect to the Alburno–Cervati Carbonate Unit, outcropping with the Pollino massif.

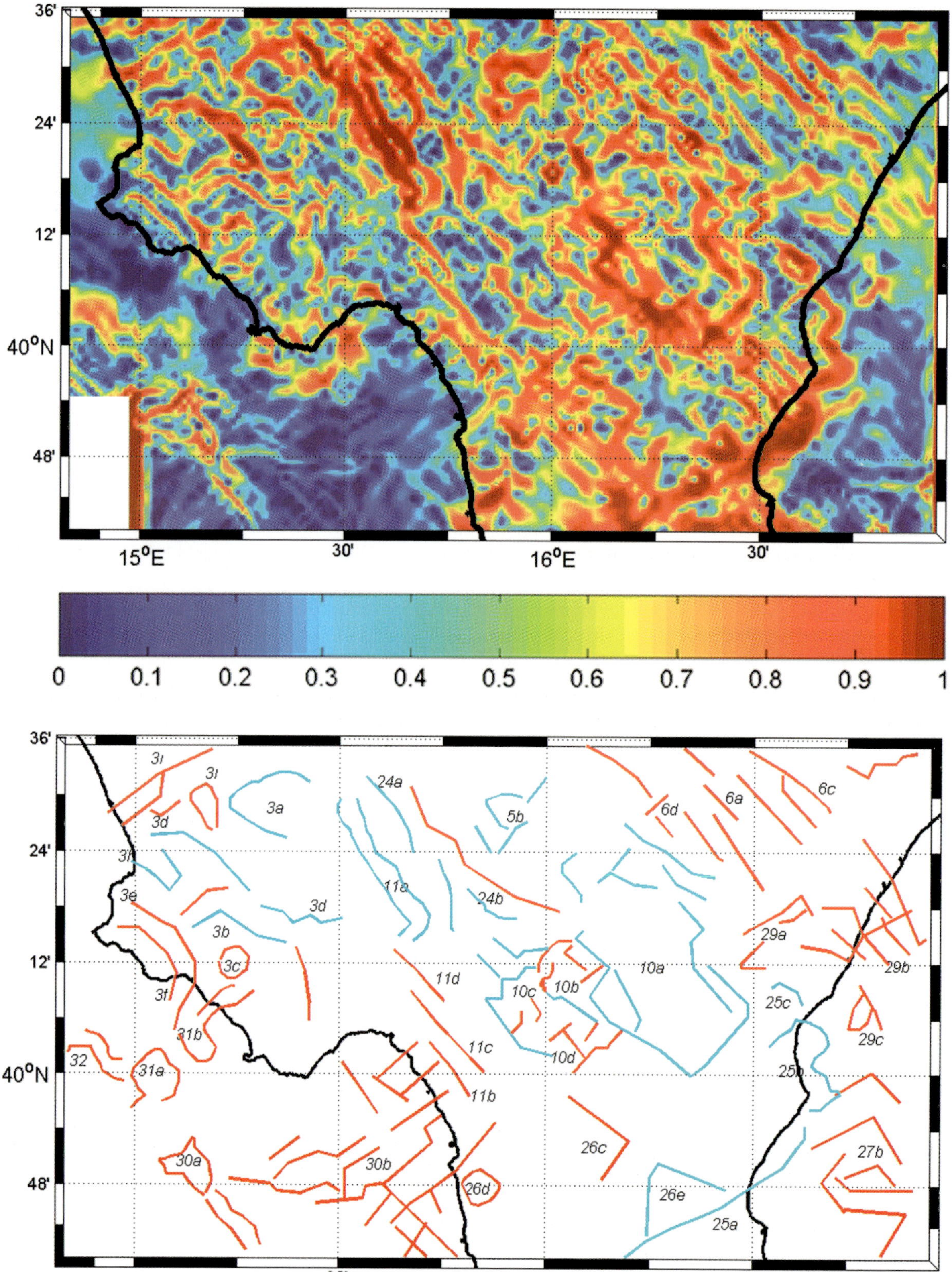

Figure 10: Short MDA map of gravity data in Southern Italy: *southern sector*. On the lower panel, major lineaments identified by EHD maxima are shown. Cyan: correlated lineaments; red: uncorrelated or partially correlated lineaments. Reference to lineament numbers in Legend IV. Lineaments are identified by trends of local maxima, which are graphically determined by the color bar shown in normalized units.

4.3.g. *The Tyrrhenian basin offshore the Salerno, Naples and Gaeta gulfs.* Another area where a lot of information can be obtained from the short scale MDA is the Salerno Gulf, Tyrrhenian Sea. Here a complex system of linear trends occurs (47a; Figure 8), running with NW–SE and NE–SW directions. It seems to extend southward, to the northern Cilento shoreline (47b; Figure 8) and could be correlated with a normal fault system, associated with the formation of the tectonic depression of the Gulf of Salerno and of the thick sedimentary sequence filling it offshore the Sele plain.

Offshore northern Campania, a long trend (17g; Figure 8) runs from the Mt. Massico carbonate complex toward Procida and Ischia Islands. It is not well defined and could be tentatively associated with a normal fault system displacing the carbonate basement and connecting two distinct volcanic districts (M. Roccamonfina and the Ischia-Phlegran Field complex).

Westward, far from the shoreline, two other, roughly parallel, trends (49 and 50; Figure 8) are clearly revealed on the MDA short scale map. They consist in alternate segments running NW and NE and do not seem to have a clear correlation with known structures. Since these features are located in an area with a complex pattern of MDA signals, they could probably be due to the coexistence of structural features associated with the volcanic activity and the normal fault-system bordering the continental shelf along NNW and WNW directions, south of the Ponziane Islands. On the western end of the area, two closed trends are visible. The southern one (52; Figure 8) seems uncorrelated with the geological setting, whereas the second one (51; Figure 8) partially coincides with a small trough between Ventotene island and the Gaeta Gulf.

Northward, some trends (53; Figure 8), located on land, continue offshore. Cross lineaments prevail offshore, with a W–E direction, implying a rather complex structural setting.

4.3.h. *The Tyrrhenian basin offshore of the Cilento area.* In this area the MDA short scale clearly describes structural features located offshore.

The MDA trend marked 30a (Figure 6) shows a structure represented by two normal faults with the same direction. Northward, other two small structures occur (31a and 32; Figure 6), and a structural connection seems to exist between one of them (31a; Figure 6) and the trends 3f and 3e (Figures 8 and 10).

A set of linear trends (30b; Figure 10) with Apenninic and anti-Apenninic direction is visible within the Gulf of Policastro. These lineaments are probably related to a normal fault system controlling the subsidence of the shallow crustal structures and, consequently, the formation of the Sapri basin. This consists of a thick sequence of undeformed (autochthonous) sediments. This system of linear trends develops southward to the Tyrrhenian offshore area respect of the northern Calabrian coast.

Some MDA trends are evidenced by the short scale map in the Cilento Area also. This is the case of a curved lineament (31b; Figure 10) located across the shoreline where Alpine Calabride Units (Frido Formation) overthrust the Cilento Units.

4.3.i. *The Ionian basin offshore Calabria and Lucania.* The area along the shoreline of the Bradano trough, in the Gulf of Taranto, is marked by linear trends (29a and 29b; Figure 10) crossing with NW–SE and SW–NE directions. Some of the lineaments with Apenninic direction seem to be the offshore prolongation of regional trends present northward, along the Bradano trough and described previously. Therefore, these lineaments represent the southern extension of the structural patterns present at regional scale, along the whole foredeep and the chain front.

It should be noted that trends 29a, 29b, 29c and 25c (Figure 10) surround a small basin characterized by a thick sequence of autochthonous sediments.

Southward, E of the Sibari plain, a system of linear trends (27b; Figure 10) has no simple correlation with the elements forming the Sibari basin.

4.3.j. *The Bradano foredeep and the Apulia-Gargano platform.* The presence of several regional linear trends with Apenninic direction, along the eastern side of the belt, suggests the existence of deep tectonic discontinuities located near the external end of the chain. Their structural meaning and role within the geodynamic framework of the central-Southern Apennine is obscure. Nevertheless they may be regional inverse fault systems displacing the sedimentary sequences ascribed to the main platform and basinal palaeo-geographic internal domains (Molise basin, Apennine platform). These overthrust the carbonate series of the external domains, like the Apulia-Gargano foreland.

These significant features of regional relevance are more or less partially hidden beneath tectonic nappes coming from more internal domains, but the MDA maps allows these buried structures to be identified along their whole length.

In the eastern sector of the studied area, the short scale MDA map singles out a large number of linear patterns and, consequently, provides a better representation of many structures already evidenced at the intermediate scale. In some places, instead, it shows some small trends that are completely invisible at larger scales.

More specifically, one of the stronger MDA outlines in this area is represented by an uncorrelated, long regional trend (6a; Figures 8, 9 and 10), extending from Abruzzo to the Gulf of Taranto, and marking with high precision the position of the eastern boundary of the allochthonous chain front. This has formed since the Pliocene and was buried beneath younger sedimentary sequences of Middle Pliocene to the Lower Pleistocene age. These series are ascribed to the foredeep basins (Bradano Units) and outcrop along a band extending from W of Gargano to the Taranto Gulf.

The presence of a well-defined, closed trend (13; Figure 9), located just west of the 6a (Figure 9) outline should be mentioned. It corresponds rather precisely to the boundaries of the Vulture volcanic complex.

Some other inner outlines, with a similar orientation, are located where Neogenic litho-stratigraphic units, from pre to late orogenic, outcrop just east of Campobasso and in a position immediately internal respect to 6a trend. These trends run with Apenninic direction and correspond with good precision to several overthrusts along which Internal and External Irpinian Units and the Lagonegro II Unit ("Flysh Rosso") are in tectonic contact. These strong linear trends (7a and 7b; Figure 8) could probably also indicate older surfaces (and, more significantly, buried subsurfaces) revealing the existence of unit fronts generated during the early Pliocene or Messinian (7a; Figure 8).

Immediately eastward of trend 6a, other two significant MDA lineaments (16b and 16c; Figures 6 and 9) trending NW–SE, are completely uncorrelated with surface evidence, but suggest the existence of buried discontinuities characterized by strong density contrasts.

The regional nature of trends 6a, 6b, 6c (Figures 6 and 9) is evidenced by their clear presence even on the intermediate scale map. Other trends, uncorrelated with shallow features, are oriented with the same direction in the southern sector of the Bradano Foredeep (6d; Figure 10).

The same lack of correlation with the outcropping geological setting characterizes another noticeable linear trend (8a; Figure 9) located immediately eastward of the previous ones and running from the Gargano to the Matera area. It could be related to the presence of an inner margin of the Apulia-Gargano Platform sunk along a regional normal fault system and buried by the Bradano Units. This trend follows a direction roughly parallel to the surface limit of the carbonate series of the Apulia-Gargano platform. The southern end is partially related to the structural features at the surface, where the deep carbonate block, identified to the north, outcrops near Matera. Northward, the trend is divided into two different branches. The first continues in the same direction and is clearly visible at the short scale. The second (8b; Figures 6 and 9) turns northward and probably indicates the internal margin of a buried carbonate sector connecting the outcropping Murge and Gargano carbonate blocks. The northern segment of this branch is clearly seen also at the intermediate scale.

The MDA short scale map shows several NW–SE running trends (16b and 16c; Figure 9), locally more clearly than at the intermediate scale. A larger number of small features can be observed, thus agreeing with the existence of a complex normal fault setting. The related density contrasts are so weak that they can only be seen at the short scale.

As far as the structure of the Gargano promontory is concerned, several short linear trends (55 and 56; Figure 9) are visible, running in the NW or NE direction. They correspond to the normal faults dislocating the foreland carbonate series outcropping in the Gargano promontory. Well defined MDA short scale trends (57; Figure 9) are identified also along the southern boundaries of the Gargano promontory and presumably coincide with the normal faults controlling the uplift movements of this carbonate complex. West of the Gargano promontory, in the Bradano foredeep, several small trend lines are evidenced by the MDA short scale.

Two small strong trends (9e; Figure 9) surround the Tremiti Islands and presumably describe the morphologic features related to the outcropping of the carbonate blocks at these islands.

One of the most noticeable patterns of MDA trends, seen both at short and intermediate scales, is represented by the system of regional MDA lineaments located in the Adriatic Sea, far from the Apulia coast. Most of these trends (54; Figure 9) are oriented with Apenninic direction (NW–SE) and extend for many tens of km. It would be reasonable to infer that they represent a structural feature of the eastern side of the Adria microplate and are correlated with the more external sectors of the deformed plate, overthrusting with SW-vergence. Southward, these elements seem to change direction, joining the eastern end of the Mattinata fault and, therefore, following a complex structural pattern.

5. MULTISCALE DERIVATIVE ANALYSIS OF SOUTHERN APENNINES: AEROMAGNETIC DATA

We analyze now the Southern Italy aeromagnetic data (Figure 11), in the area from 14°21′E to 17°00′E, and from 39°00′N to 41°30′N. The original flight lines followed different patterns and altitudes due to the peculiar topography over the object area, and were acquired along profiles spaced 5 km apart (in a small area the spacing was 10 km). Tie lines were flown perpendicularly to the survey lines, spaced on average 10 km apart. The grid we used was prepared by AGIP (1981), merging flight line data to a unique altitude of around 2600 m a.s.l. (8500 feet), with a 2 km sampling interval. The field in the onshore part of the map is characterized by smooth, long-period anomalies. A belt of dipolar anomalies with wavelength of the order of some tens of km coincide with the position of the front of the Apennines chain. Anomalies due to shallow sources are also present, as are those due to the volcanic structures of Mt. Vulture volcano (15°35′E, about 41°N) and the Somma–Vesuvius complex (14°25′E, 40°45′N). The smooth magnetic anomaly pattern (apart from these volcanic structures) is generally consistent with the outcropping geology. In fact, a thick sedimentary rock sequence forms the Apennines chain, and it may be considered as relatively non-magnetic, at least at the resolution of this survey. On the contrary, in the Tyrrhenian Sea, south of 40°N latitude, the aeromagnetic field is characterized by very intense and local magnetic anomalies, generated by the volcanic structures of the Aeolian Islands as well as by other submerged volcanoes and deeper intrusions.

To perform a multiscale derivative analysis of the above described aeromagnetic field, we computed three different resolution EHD signals. We applied to the EHD expression (Equation (2)) the set of weights defined by $w_m = K^m$, where $K = 4$ and m is the derivative order. The large, intermediate and small scale maps were obtained starting from the magnetic scalar potential as first term of the summation. This was obtained from the magnetic field by a pseudo-gravity transformation (Baranov, 1975). The directions of the field used and magnetization vectors were consistent with average IGRF data for Southern Italy (declination: 0°, inclination: 56°). The ISVD technique was used to compute all the subsequent vertical derivatives up to the 8th order. The three orders of EHD, corresponding to the three MDA maps at different resolution, were chosen to be $m = 3$ ('Large Scale' map), $m = 5$ ('Intermediate Scale' map) and $m = 8$ ('Short Scale' map).

In order to obtain a more balanced representation of the EHD signal, which attains very high values in the southern Tyrrhenian Sea and is much weaker onshore, the logarithm of the EHD signal was imaged using a histogram equalization algorithm. The representation obtained in this way allows the investigation of subtle, but meaningful features that in a standard map would have been totally obscured. The histogram equalization procedure tends in fact to equalize the number of pixels assigned to each color band. This results in a nonlinear relation between data values and colours, allowing a better visualization of subtle trends. However, the colours in a histogram equalized map will be saturated in the data values near the extremes, and in our case this means some lack of graphical resolution in the Tyrrhenian region and other volcanic areas.

5.1. Large scale

Large-scale boundaries are illustrated in Figure 12. In this map, representative of regional scale structures, two main trends of magnetized structures are present. The first (1) follows the Apennines chain, with lower EHD values going toward NW. The second (2) is considerably weaker, and NE–SW oriented. While the presence of

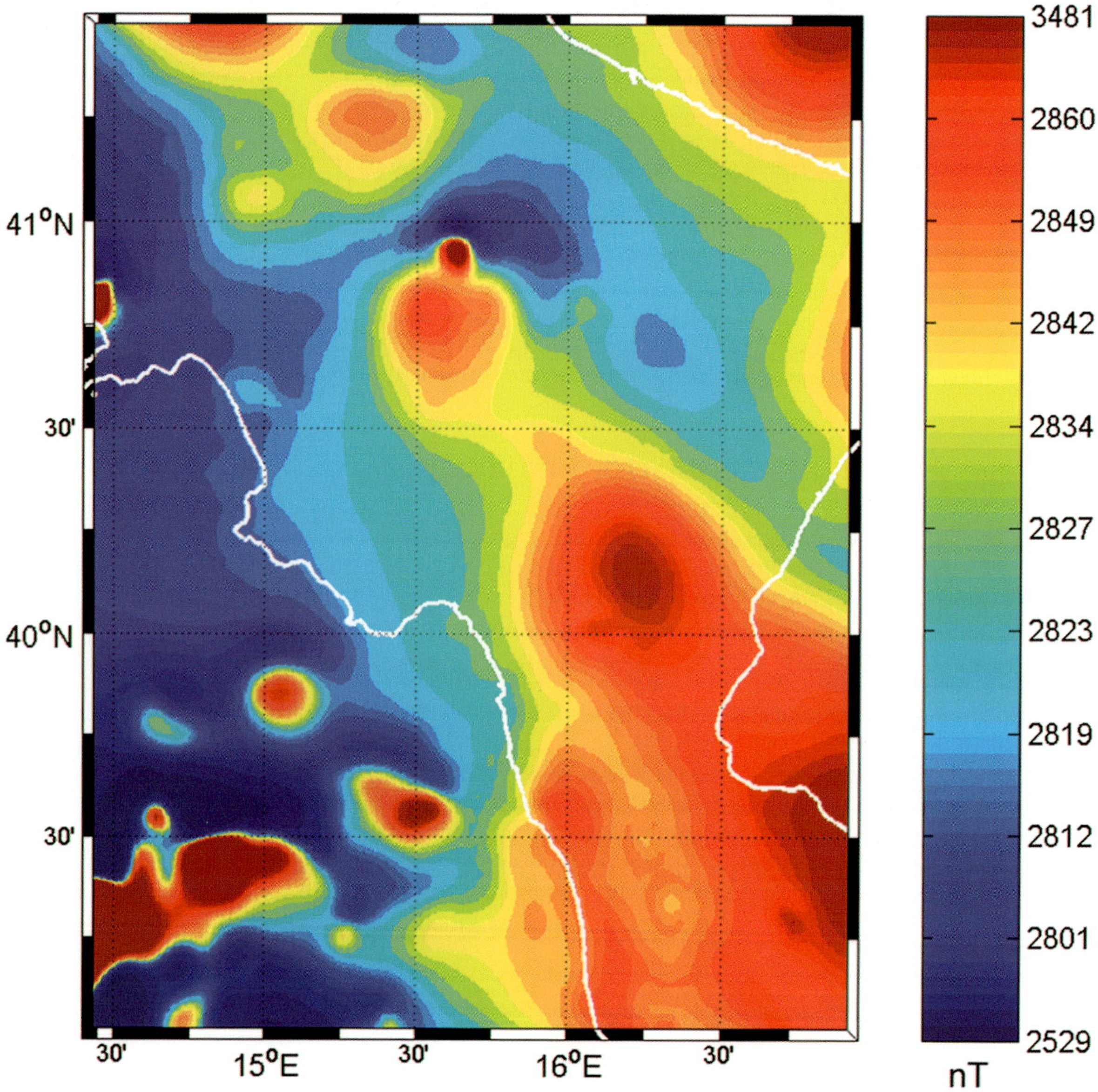

Figure 11: Aeromagnetic anomaly field of Southern Italy (AGIP, 1981). The grid spacing is 2 km.

the Apennines trend is clear from analysis of the aeromagnetic anomaly map, it also indicates the presence of magnetized structures never referred to before. Other maxima in the large scale EHD map are those relative to the volcanic areas onshore and offshore (3).

5.2. Intermediate scale

Structures are delineated with greater detail in the intermediate scale EHD map (Figure 13). Apart from the very intense signal in the volcanic areas of Vesuvius (1), Vulture (2) and the Tyrrhenian Sea (3), there are a number of other lineaments that are not always correlated with known geological structures. The long period trends seen in the previous map are now better evidenced: the Apenninic one (4) appears split into two parts south of the Pollino area. Here, the zone with EHD minima, corresponding to the Crati valley sedimentary basin, separates two EHD maxima (coastal chain to the W (5) and the Sila Mts. to the E (6)) with outcropping of granites and metamorphic rocks (gneiss, kinzigites, granulites). The NE–SW regional trend is now better imaged and appears to be composed by many subtle trends with different directions (7).

At the Crati valley, some sub-circular maxima are also present, apparently not correlated with the surface geology of Plio-Quaternary sedimentary deposits (8). Another magnetic source is located in the Sila area (9). Also here

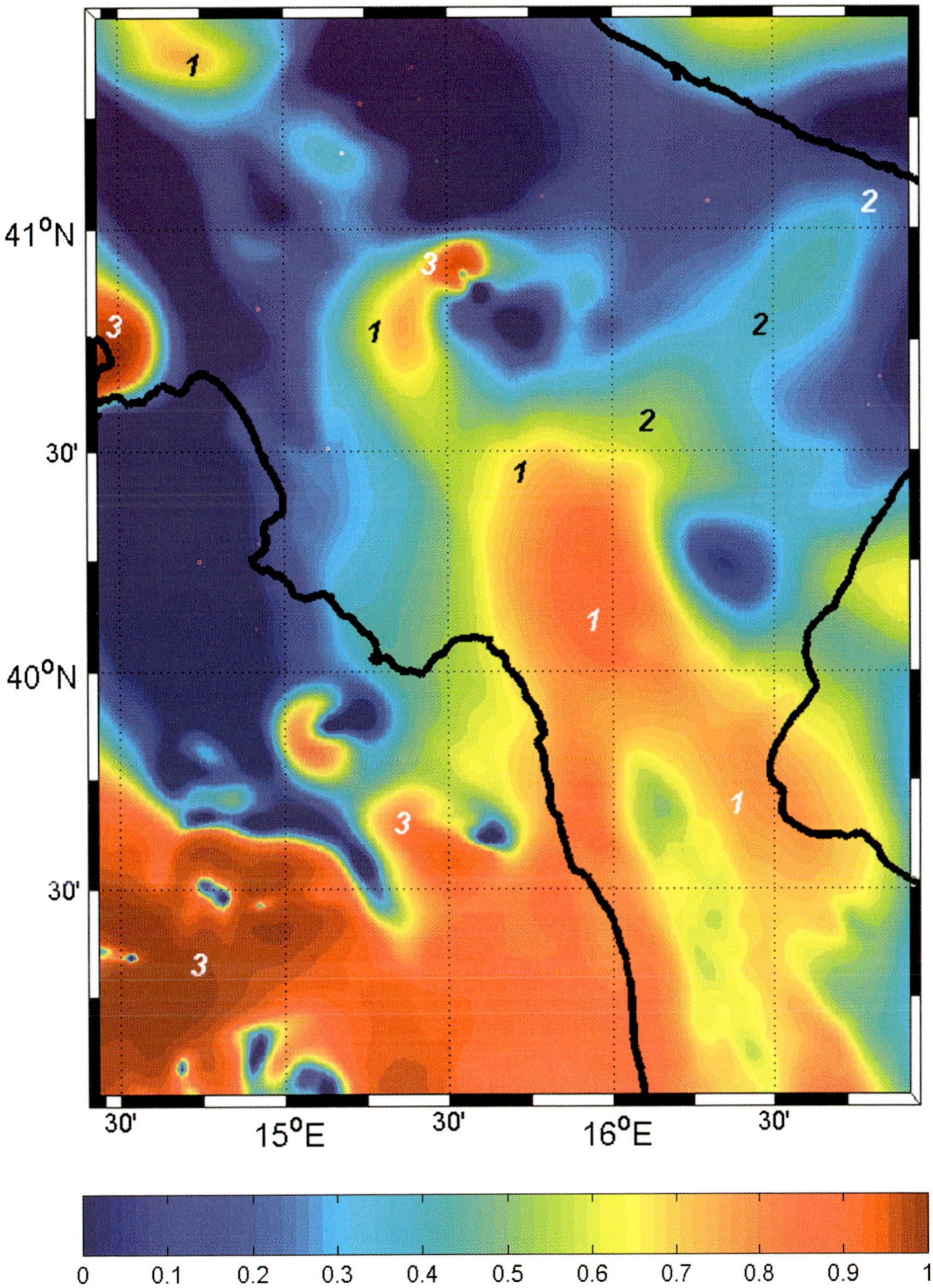

Figure 12: Large scale MDA map of aeromagnetic data in Southern Italy. Reference to lineament numbers in Legend V. Lineaments are identified by trends of local maxima, which are graphically determined by the color bar shown in normalized units.

there is no clear correspondence with surface lithologies to justify a significant magnetization contrast. Offshore, north of the Aeolian Islands, three circular trends are clearly identified, indicating the presence of isolated magnetic bodies. The first could be associated with the known Enotrio and Diamante seamounts (10). In the structural map of Italy (Bigi et al., 1990) these seamounts are interpreted as outcrops of acoustic basement. Due to the strong values of the EHD signal, suggesting high magnetization values for the anomaly sources, this basement may be hypothesized as built by volcanites or intrusive basic rocks in a low susceptible crust. The same reasoning may

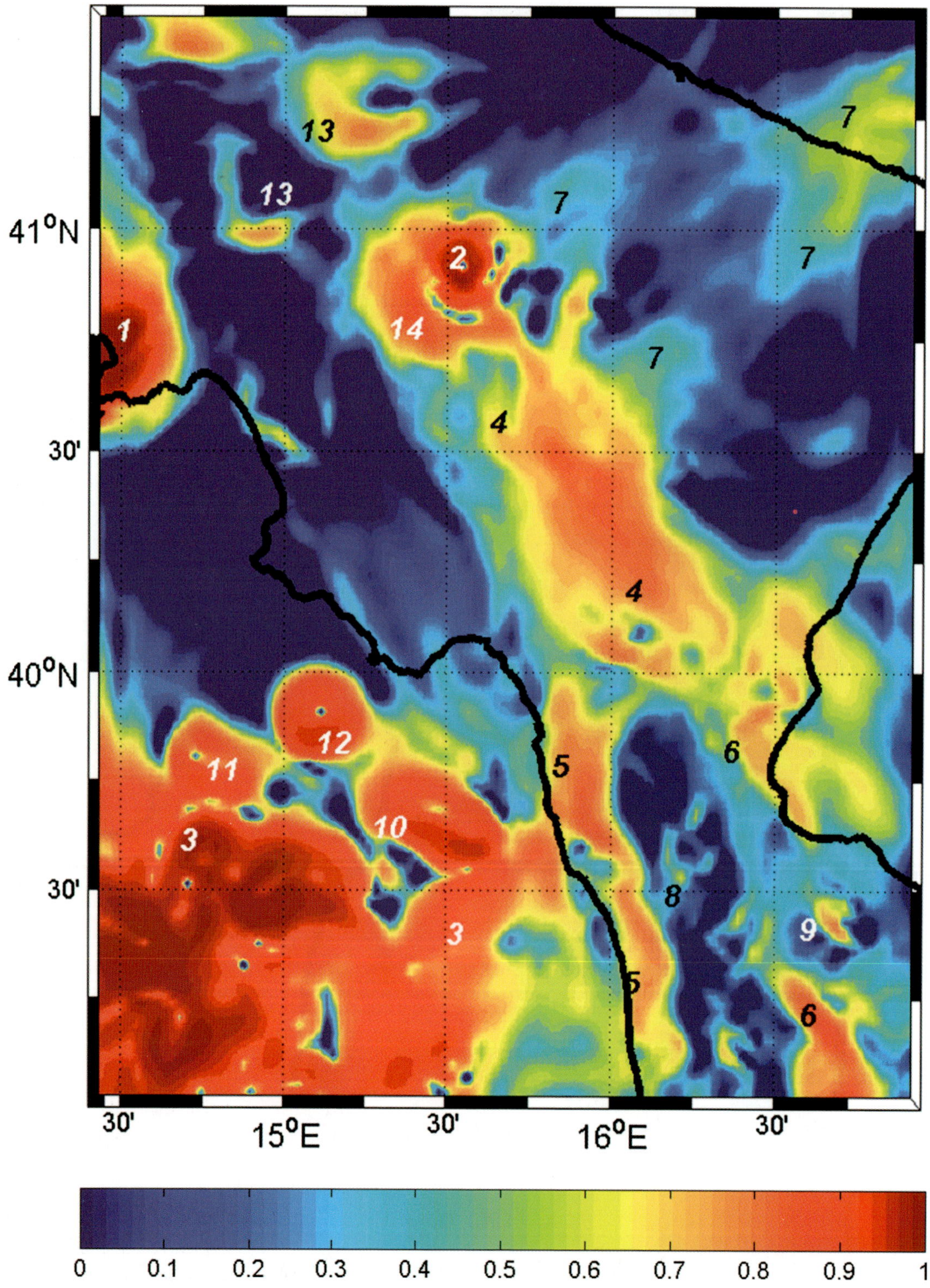

Figure 13: Intermediate scale MDA map of aeromagnetic data in Southern Italy. Reference to lineament numbers in Legend VI. Lineaments are identified by trends of local maxima, which are graphically determined by the color bar shown in normalized units.

apply to other two magnetic anomaly sources mapped as circular trends with high values offshore the Cilento coast (11 and 12), where no igneous or volcanic body is presently known. Further circular structures at a larger scale are those corresponding to the 'Foggia anomaly' (13) and to another long-period anomaly in the Vulture area (14). The sources of these magnetic anomalies were interpreted as intrusions about 15 km deep from the mantle at crustal discontinuities (Florio, 1993).

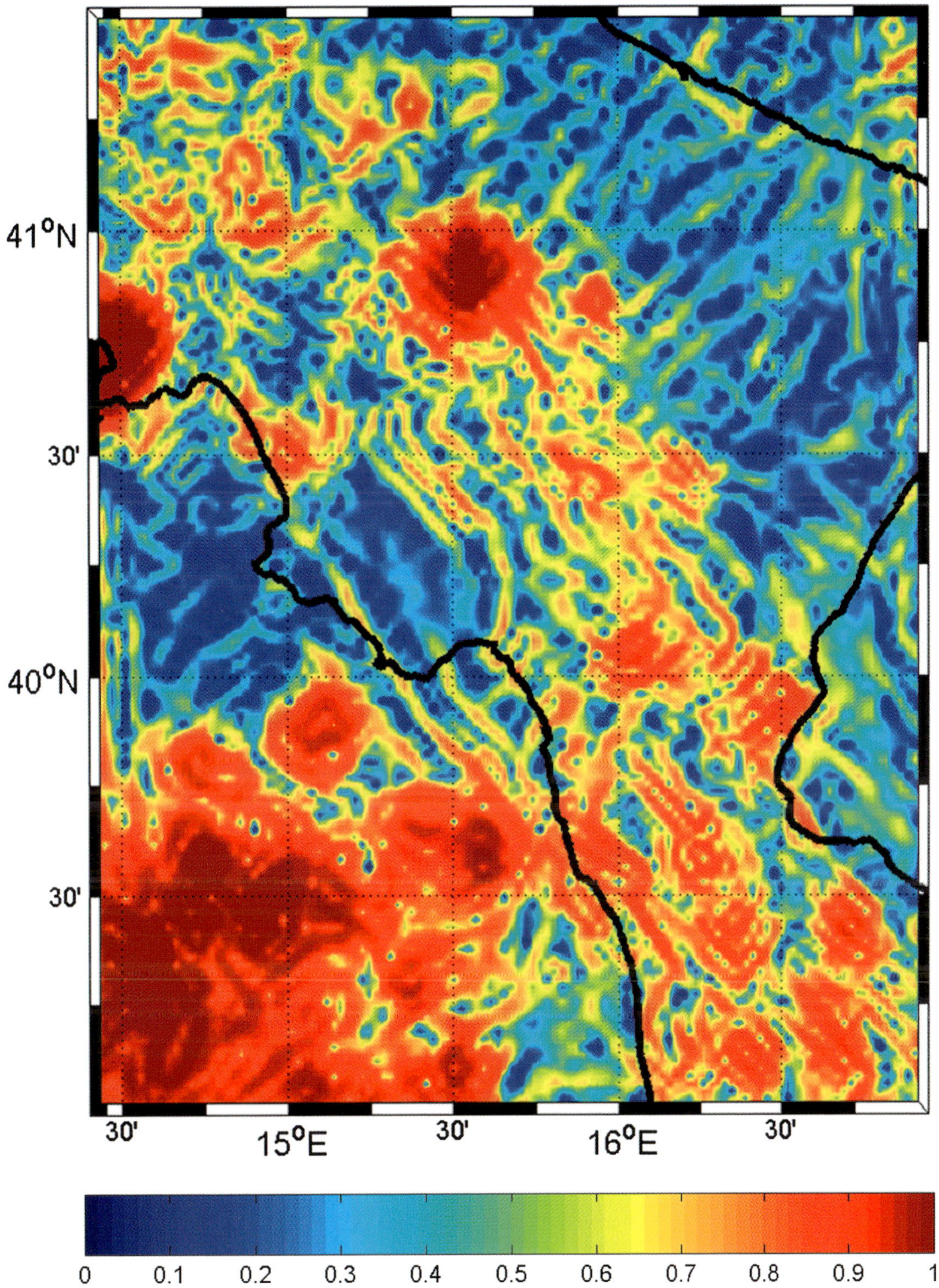

Figure 14: Short scale MDA map of aeromagnetic data in Southern Italy. Lineaments are identified by trends of local maxima, which are graphically determined by the color bar shown in normalized units.

5.3. Short scale

The short scale map (Figure 14) displays at high resolution a number of magnetic trends. Most of them are difficult to recognize in the aeromagnetic anomaly map. Thus, it yields valuable insights for understanding the structural significance of the region. Due to our choice of weights in Equation (2) and to the limitations inherent in application of the reduction to the pole, some noise is evident and doubled lineaments appear, due to an imperfect cancellation of the spurious maxima present in the high-order derivatives. Nevertheless, many subtle trends indicating the positions of local structures are clearly visible.

An analysis of the EHD maxima trends (Figure 14) allows us to clearly single out three main structural domains: the first is located NE of the front of the Apennines chain thrusts (foreland area), in which subtle NE–SW lineaments are displayed; the second is represented by the chain area, and is mainly characterized by NW–SE linear trends; the third is located in the Tyrrhenian area, where a strong EHD signal is correlated to isolated magmatic/volcanic bodies.

5.3.a. *Foreland area.* In Figure 15, a detailed image of the EHD analysis for the foreland area is shown. Here the main lineaments corresponding to EHD maxima are evidenced. They are mostly NE–SW oriented (anti-Apenninic), but some of them trend NW–SE. The lineaments that roughly follow a NE–SW trend tend to disappear just W of the front of the Apennines chain thrusts, and display an anastomosing and intersecting pattern. These lineaments were not evidenced in the gravity multiscale analysis. One of these NE–SW structures exhibits a good correlation with the known Ofanto valley anti-Apenninic normal fault, and a continuation of the same structure offshore is also clearly visible (1; Figure 15). Another NE–SW magnetic lineament is in its central segment correlated with a known structure, an anti-Apenninic normal fault (2; Figure 15) located between the front of the Apenninic thrusts and the outcrops of Murge hills limestones (Bigi et al., 1990). The absence of such structures in the gravity EHD analysis (not even the Ofanto valley fault has a gravity expression) may be explained by the small or zero density contrast between the two faulted blocks, suggesting a limited depth for these faults. If this interpretation of the lack of gravity anomalies is correct, the hypothesis of a source of magnetic anomalies in the sedimentary cover may be favoured. In fact, recent papers deal with the study of aeromagnetic anomalies generated inside the sedimentary cover (e.g. Millegan, 1998; Peirce et al., 1998; Grauch et al., 2001). Such sources of magnetization inside the overall non-magnetic sedimentary section are still debated and can be explained by (a) geochemical processes destroying or creating magnetic minerals along a fault zone, (b) differences in magnetization between lithologies juxtaposed along a fault plane and (c) detrital magnetic material filling paleochannels.

However, if the faulting were related to an old tectonic phase it would only have affected rocks that at present lie at a very deep level. This may imply an involvement of just crystalline basement, with changes in magnetic properties (see above), but without significant changes in the density across the fault. A tentative date for such a tectonic phase, able to produce the whole set of NE–SW trending faults evidenced by MDA in the foreland area, may be Mesozoic, in correspondence to the Ionian opening (see Finetti, 2005, Chapter 6). Under this hypothesis, these lineaments may be interpretable as micro-strike-slip faults. If this is the case, the genesis of magnetized minerals along the fault surface might have been caused by geochemical processes (see point (a) above).

In the foredeep area, there are two NW–SE lineaments tens of kilometers long (3; Figure 15) with very similar positions to some Apenninic trends evidenced by the multiscale analysis of gravity data (previous section) and with similar position to the buried front of the allochthonous units. Another NW–SE lineament is visible near the mouth of the Ofanto river (4; Figure 15) and could be related to a known Apenninic normal fault (Bigi et al., 1990). Finally, a sub-circular structure is evidenced at 4°50′N, 15°55′E, just north of Bradano river near the village of Acerenza (5; Figure 15). It may be ascribed to a small sub-volcanic structure, perhaps tied to Mt. Vulture activity, and to our knowledge never evidenced in surface geologic field work.

5.3.b. *Chain area.* In the chain area, north of about 40°45′ (Figure 16), the structural trends evidenced by MDA analysis show no simple pattern. They reflect the superposition of long period dipolar anomalies and more local effects, resulting in circular features as well as NE–SW-oriented lineaments.

To the south of about 40°45′ many linear features are seen, mainly with NW–SE (Apenninic) direction. The more interesting structures are shown in Figure 16. This NW–SE trend is very likely connected to the Apennine thrusting. However, these lineaments may also be related to the presence of NW–SE-oriented sinistral strike-slip faults evidenced by analysis of seismic data (see Finetti et al., 2005, Chapter 12).

The Cilento area is characterized by low gradients in the aeromagnetic field; nevertheless two NW–SE lineaments cut the region and continue offshore for some tens of km (1a, 1b; Figure 16). Independent geologic structural mapping indicates the presence of structures following this trend (A. Ascione, personal communication). Nevertheless, considering also the low correlation with gravity lineaments, it is probable that these lineaments are related to deep-seated structures.

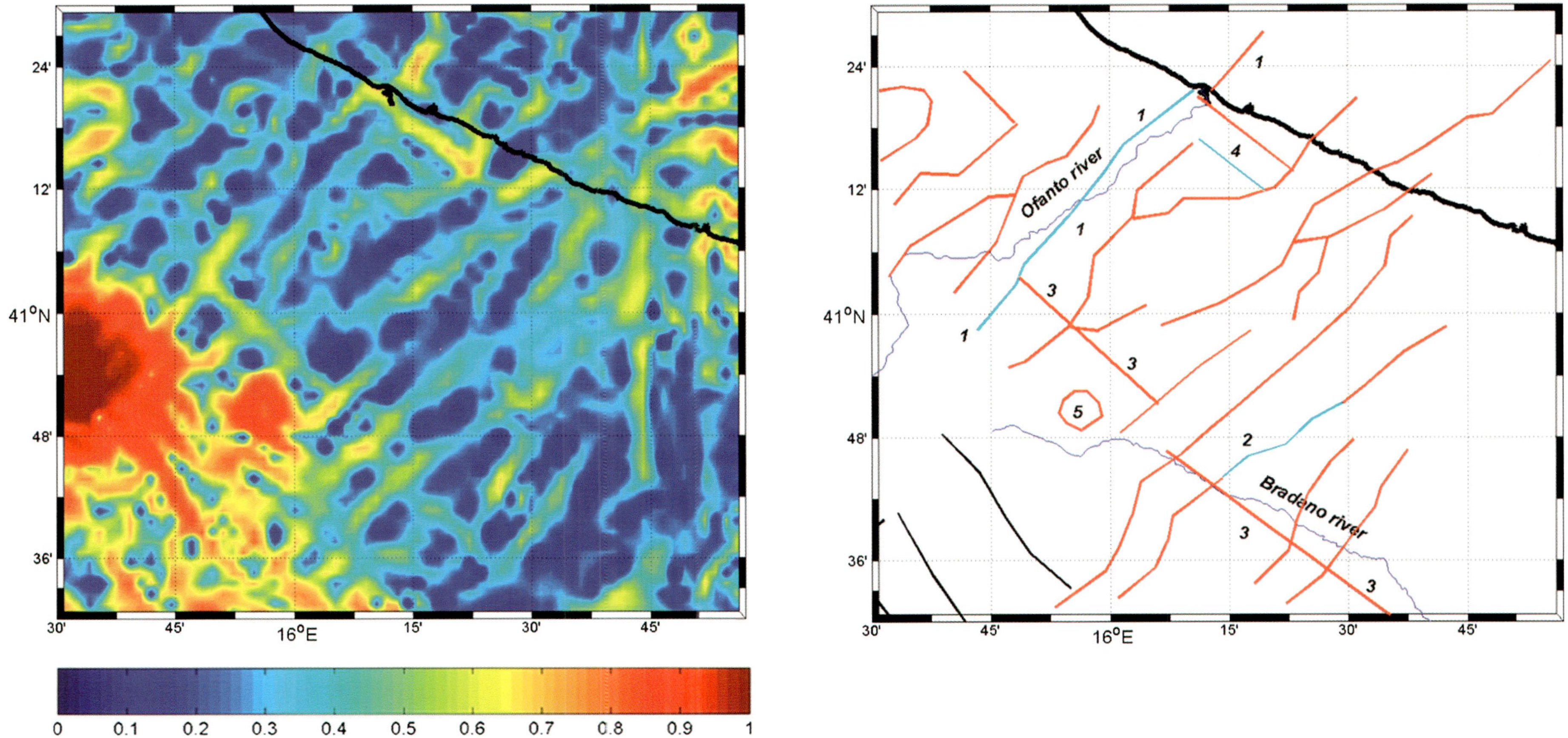

Figure 15: Short scale MDA map of magnetic data in Southern Italy: *foreland area*. On the right panel, major lineaments identified by EHD maxima are shown. Cyan: correlated lineaments; red: uncorrelated or partially correlated lineaments. Reference to lineament numbers in Legend VII. Lineaments are identified by trends of local maxima, which are graphically determined by the color bar shown in normalized units.

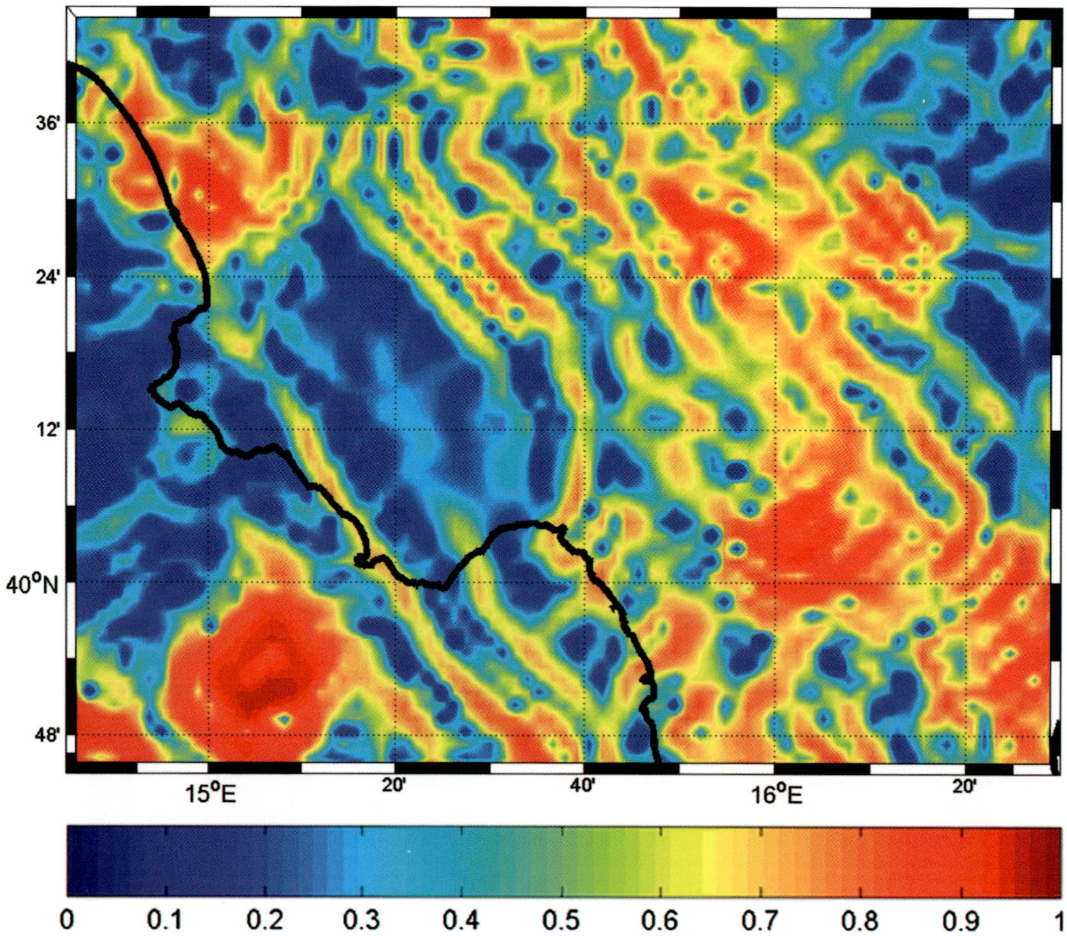
36'
24'
12'
40°N
48'
15°E
20'
40'
16°E
20'
0
0.1
0.2
0.3
0.4
0.5
0.6
0.7
0.8
0.9
1

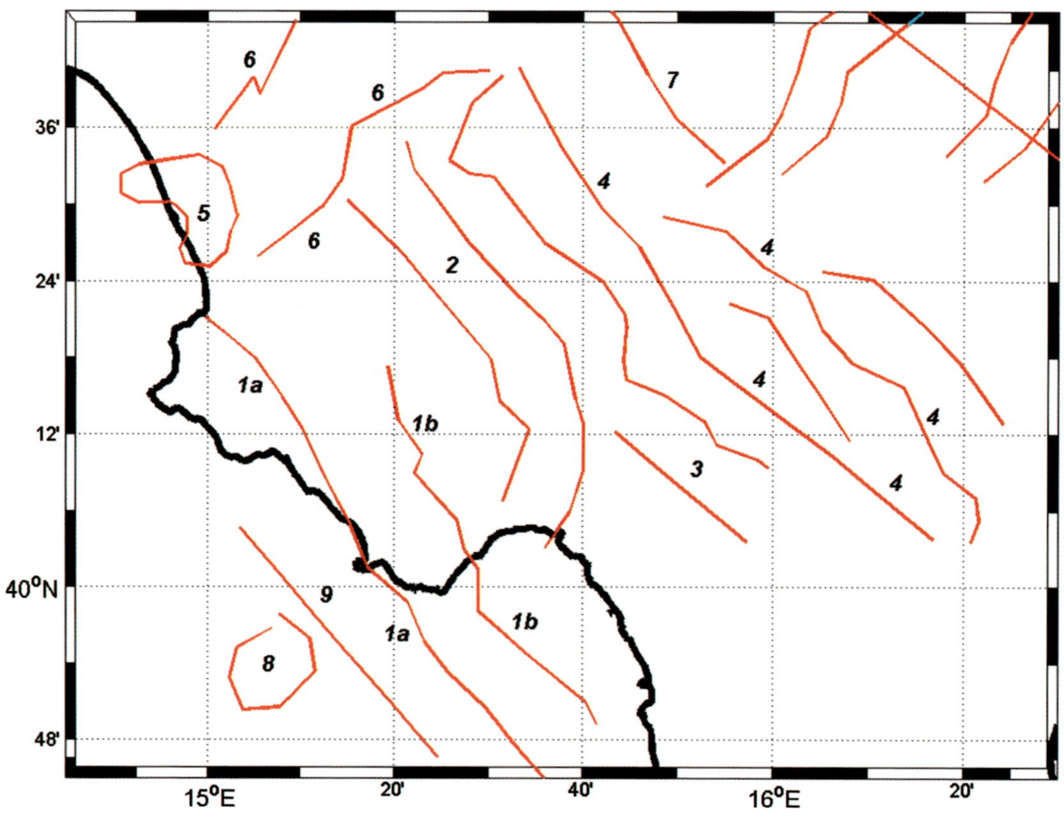
36'
24'
12'
40°N
48'
15°E
20'
40'
16°E
20'
1a
1b
2
3
4
5
6
7
8
9

Figure 16: Short scale MDA map of magnetic data in Southern Italy: *chain area*. On the lower panel, major lineaments identified by EHD maxima are shown. Cyan: correlated lineaments; red: uncorrelated or partially correlated lineaments. Reference to lineament numbers in Legend VIII. Lineaments are identified by trends of local maxima, which are graphically determined by the color bar shown in normalized units.

The same considerations may apply to the Apenninic lineaments of the Diano valley (2; Figure 16) and Agri valley (3; Figure 16) depressions. In those areas, an imperfect correlation with structural and morphologic elements, as well as with gravity lineaments, is observed.

The Sant'Arcangelo basin is crossed by a couple of lineaments from NW to SE (4; Figure 16). The geological significance of these lineaments may be related to the presence of the transtensive strike-slip fault responsible of the evolution of this basin (see Finetti, 2005, Chapter 17: Sibari-like fault). In particular, the rising of magmatic material along this fault may explain the contemporaneous presence of a magnetic anomaly due to a positive magnetization contrast and a negative gravity anomaly.

At the Sele river mouth, there is another structure related to an isolated source (5; Figure 16). This structure and other anti-Apenninic lineaments (6; Figure 16) might be correlated with the presence of an important strike-slip fault trending NE–SW (Incoronato et al., 1985).

Other NW–SE trending lineaments are present in the chain area south of Vulture volcano area (7; Figure 16), following the trend previously seen in the lower resolution maps (Figures 12 and 13). South of 40°N, the main structures evidenced by the small scale MDA analysis in the Calabrian area are rather difficult to interpret and are in general better seen in the lower resolution maps.

In the Tyrrhenian area, the small scale MDA analysis gives very high amplitude EHD signals in correspondence to a number of volcanic structures (Aeolian Islands and other seamounts). In this map, the Cilento offshore ring structure (8, but numbered as 12 in the intermediate scale map; Figure 16) appears bounded by a lineament to the NE (9; Figure 16), maybe as a result of superposition of different source effects. In the Tyrrhenian offshore, the areas characterized by low values of EHD are, to the north, the Salerno Gulf (Sele basin) and the Cilento northern offshore, and the Calabrian offshore in correspondence to the Paola basin. In the Salerno Gulf, some circular as well as linear trends are evidenced in correspondence to very low aeromagnetic field gradients (Figure 14).

6. CONCLUSIONS

Since features of a potential field map are the result of interference between anomalies of different extents and depths, methods for their analysis at different scales are needed. In the attempt to separate the field into its main components, the so-called filtering methods are usually used, but may however introduce some distortion. MDA was defined in such a way as to yield valuable results without resorting to any sharp separation of the effects related to different scales. This is due to the specific enhancement provided by different order derivatives, which may be variously combined to form the Enhanced Horizontal Derivative. MDA at different scales gave very interesting results for the source boundaries related to gravity and magnetic data of the Southern Apennines.

For the MDA of gravity data, most of the structural and geological patterns outcropping along the Apennine chain and surrounding areas are clearly shown by the method. The most significant result is represented by several structural lineaments well evidenced by MDA, but poorly correlated or completely uncorrelated to any outcropping feature. This is the case especially for many trends located within the Cilento area and surrounding zones, the Lucania region, the Benevento province, along the Bradano trough and the Apulia foreland (onshore and offshore). Other similar cases involve the Molise Units outcropping north of the Matese massif, the Phlegrean Fields and Vesuvius.

When MDA is performed by enhancing high order derivatives, the individuation of geological structures is carried out with high detail and allows recognition of a very large number of elements at local scale. The structural meaning of each may not be immediately deduced from the outcropping features and, therefore, may need more careful geological interpretation.

For the MDA of magnetic data, analogous considerations can be made. In fact a large amount of new information can be obtained at different scales by the method.

It was possible to establish that some closed trends, located offshore in the Tyrrhenian Basin, are related to small volcanic bodies partially hidden by sedimentary cover, and previously unknown or interpreted as local outcrops of crystalline basement (Enotrio and Diamante seamounts). Another significant conclusion is support for the hypothesis that sedimentary cover is only apparently non-magnetic. In fact, many trends visible on the MDA short scale map demonstrate the presence of several detectable magnetized sources within the sequence of sedimentary rocks. This could be due to various causes, like the destruction/creation of magnetic minerals along a fault zone by means of geochemical processes, or the simple variations in magnetization of the faulted rock complexes or, alternatively, the presence of palaeochannels filled in a second phase with differently magnetized sediments.

Several trends recognized on the MDA magnetic maps do not have any correspondence with the available geological information. Some examples are the small trends located along the Crati valley (Calabria) and on the Sila massif. Another case is represented by the many lineaments located along the Bradano foredeep and Apulia foreland, and NE–SW oriented.

In conclusion, our results allow us to recognize significant structures, which cannot easily be described by simple analysis of the potential field, or by other traditional methods of signal enhancement. The lineaments are described selectively, depending on their spatial extent. However the information contents is often so rich as to need a much more careful geological interpretation than that we have tried to provide in this chapter.

Legend I – Description of major lineaments identified by EHD maxima. Intermediate scale MDA of gravity data in Southern Italy

1e Linear trend marking a fault line dislocating the northern side of the M. Cairo massif (north of the Liri valley).
1d Linear trend marking the normal fault system dislocating the carbonate complex outcropping east of Venafro (East of the Liri valley).
1a Trend bounding the morpho-structural limits of the Matese massif.
2i Trend marking the eastern side of the Altavilla Unit (Miocene-Pliocene) outcropping W of Benevento.
1f Trend delimiting the outcropping series of the "Flysh Rosso", ascribed to the Lagonegro II Units.
2f Trend outlining Camposauro massif (Matese–M. Maggiore Unit) and Taburno massif (Monti Picentini-Taburno Unit).
2h Two trends WNW-oriented, indicating the presence of two faults controlling the formation of the Liri valley tectonic depression.
1g Trend connection, partially hidden by the volcanic series of the Roccamonfina volcano, revealing the complex system of normal faults bounding the carbonate series of M. Maggiore.
1c Trends bounding the M. Massico massif southward and northward, related to the presence of two normal fault lines.
2d Trend marking the western structural limit of the Avella mountains.
2e Trend marking the western structural limit of the Sarno mountains.
2i Trend marking the eastern structural limit of the Sarno Mts and Avella Mts.
45c/45d Secondary linear trends evidencing a strike-slip fault system displacing the normal faults controlling the structural setting of the Sorrento ridge.
2b/2c Trends marking the eastern structural limits of the carbonate series of the Terminio (2b) and Cervialto (2c) massifs (Picentini–Taburno Units).
5c Trend marking the boundaries of faulted alluvial conglomerates ("Eboli Conglomerates") outcropping W of Campagna.
28 Closed trend west of Capri revealing the existence of an uprisen carbonatic block partially buried by younger marine sediments.
48 Linear trends with regional extension and presumably associated with the presence of several normal faults bordering the Tyrrhenian depression.
3i Trends on the northern and southern sides of the Sele plain, providing evidences of strong structural control on the morphological evolution of the basin.

45b Trend correlated with the fault system bounding the Mesozoic carbonate series (Picentini–Taburno Units) outcropping along the Sorrento ridge.
9b Linear trend indicating the presence of significant regional discontinuities presumably consisting of a normal anti-Apennine fault system causing the outcropping of a carbonate block at the Tremiti Islands.
9a Strong linear trend easily correlateable with the Mattinata fault.
16f/16e/16g Pattern of linear trends indicating the presence of extensive tectonic lineaments immediately offshore the Murge area.
16d Trends evidencing a tectonic setting represented by a fault system with anti-Apenninic direction.
16a /16b/16c Trends evidencing a tectonic setting represented by a normal fault system with Apenninic direction.
2c Trend marking the eastern side of the structural limits of the carbonate series of the Cervialto massif (Picentini–Taburno Units).
4a Trend bounding the M. Marzano carbonate series.
11a Trends correlated to the structural boundaries of the Maddalena Mts. carbonate belt, eastward delimiting the Diano valley.
24a/24b Trends identifying, the morpho-structural boundaries northward of the Melandro valleys and, southward, of the Agri valley.
4b Uncorrelated trend line showing a structural discontinuity beneath the covering Sicilide Units outcropping SE of Potenza.
4c Uncorrelated trend line partially coinciding with the tectonic contact of the Sicilide Units above the Lagonegro II Units.
5b Trend showing a deep discontinuity hidden by the Plio-Pleistocene cover and presumably related to the tectonic contacts between Liguride Units.
6a Intermediate section of a regional trend extending from Abruzzo to the Gulf of Taranto and visible at both short and intermediate scale.
9c/9d Two closed trends uncorrelated with evidence at the surface and, presumably, outlining two small structural highs of the Mesozoic carbonate basement.
8b Northern branch of trend 8a. It probably indicates the internal margin of a buried carbonate sector connecting the outcropping Murge and Gargano carbonate blocks. It is partially evidenced also at the intermediate scale.
4e Linear trend related to the axis of a regional anticline affecting the Neogenic lithostratigraphic units from pre- to late orogenic (Irpinian Units and the Ariano Unit). The correlation should presumably also affect deeper structural features.
9f Eastward prosecution of trend 9a (Mattinata fault). It shows a slight change in direction (toward ESE) and coincides with a local slight deepening of the top of the Mesozoic carbonate series (Del Ben et al., 1994).
6b/6c MDA lineaments uncorrelated with evidence at the surface, but suggesting the existence of buried discontinuities characterized by strong density contrasts.
8a Uncorrelated linear trend running from the Gargano to the Matera zone and probably related to an inner margin of a sector of Apulia-Gargano platform sunk along a regional normal fault system and buried by the Bradano Units.
3i Linear trends on the northern and southern sides of the Sele plain.
10a Regional trend marking the S. Arcangelo basin.
3g Closed trend located southward of the outcropping carbonate series of the Bulgheria-Verbicaro Units.
26b Closed trend partially correlated with the outcropping structures represented by the Cilento Flysh.
27a Trend system correlated to the structural boundaries of the Sibari basin located in the Taranto Gulf.

Legend II – Description of major lineaments identified from EHD maxima. Short scale MDA of gravity data in Southern Italy: north-western sector

53 Trend suggesting a rather complex structural setting between Ventotene island and the Gaeta Gulf.
17h Closed and uncorrelated trend located near the lower course of the Volturno valley.
17b Small ridge presumably indicating the westward extension, beneath the volcano-sedimentary cover, of the E–W fault system controlling the structure of the Avella carbonate massif.
17d Ring-shaped trends at the Patria lake (north of the Flegrean Fields).
52 Two circular trends south of the Ponziane islands; the southern one (52) seems to be uncorrelated with the geological setting, whereas the second one (51) partially coincides with a small trough between Ventotene island and the Gaeta Gulf.
49/50 Two trends located westward of the Gulf of Gaeta, without clear correlation with known structures.
17g Long trend, running from the M. Massico carbonate complex toward Procida and Ischia islands.

17e Small linear trend running from the ‘Parete structure’ toward the western side of Naples.

17a Trend bordering the Phlegrean caldera.

14a/14b Trends mapping tectonic limits between the basinal series of the Molise Units resulting from the deformation of the external Apennine domains (near Isernia and the Matese massif).

18a Trend bordering the Vesuvio crater.

18b Linear trend probably indicating the presence at depth of a regional fault correlated to the volcanic evolution of the Somma–Vesuvius complex.

45a/45b Linear trends coinciding with two regional normal fault lines.

45f Linear trend running onshore and roughly coinciding with the valley of the Picentino river.

45e Linear trend continuing eastward and marking the southern boundary of the Picentini Group.

47a/47b Linear trends correlated to a normal fault system associated with the formation of the tectonic depression of the Gulf of Salerno and of the thick sedimentary sequence filling it offshore the Sele plain.

3h Trend roughly coinciding with the outcrop of the Liguride Units overhrusting on the Cilento and Alburno-Cervati Units.

3e Linear trend partially corresponding with the lower Alento Valley, presumably due to the presence of a tectonic discontinuity at depth.

3f Uncorrelated linear trend, located just north of P. Licosa.

3b Small trends surrounding the Cervati massif.

7a Linear trends correlated with tectonic limits between the Internal/External Irpinian Units and the Lagonegro II Unit (“Flysh Rosso”).

7b Linear trend presumably indicating old tectonic interfaces (and, more significantly, buried subsurfaces) like the chain front aged early Pliocene or Messinian.

6a Regional trend, extending from Abruzzo to the Gulf of Taranto.

17h Closed and uncorrelated trend located near the lower course of the Volturno valley.

17f Ring-shaped trend visible around the eastern side of Naples.

17i Ring-shaped trend related to an old volcanic structure (Parete) buried below the surface.

1g Trend connection, partially hidden by the volcanic series of the Roccamonfina volcano, revealing the complex system of normal faults bounding the carbonate series of M. Maggiore.

2m Linear trend precisely marking the tectonic contact between the Picentini–Taburno Units and the Neogenic litho-stratigraphic units of the Apenninic chain, from pre-to late orogenic (Altavilla Unit).

2h Trends WNW-oriented, indicating the presence of two faults controlling the formation of the Liri valley tectonic depression.

2g Trend delineating the western edge of the carbonate series of the Aurunci Mts. (Picentini–Taburno Units).

14c/14d/14e Trends N–S running, indicating tectonic contacts visible at the surface (14c) and, presumably, at depth (14d/14e), between series of the Matese Unit and of the Frosolone Units (east of Isernia).

51 Closed trends partially coinciding with a small trough between Ventotene island and the Gaeta Gulf.

3d Trend visible near the M. Soprano massif and suggesting the existence of a strong density contrast concerning the contact between the Alburno-Cervati Units, the Piaggine-Raganello Formations and the Liguridi Units.

2a Trend delineating the western edge of the carbonate series of the Aurunci mountains (ascribed to the Picentini–Taburno Units).

Legend III – Description of major lineaments identified from EHD maxima. Short scale MDA of gravity data in Southern Italy: north-eastern sector

57 Trends identified along the southern boundaries of the Gargano promontory and marking the normal faults controlling the uplift movements of this structure.

56 Uncorrelated trend running from the south-eastern coast of the Gargano to the western side of the Varano lake.

55 Short linear trends individuating the normal faults dislocating the foreland carbonate series outcropping in the Gargano promontory.

6a Intermediate section of a regional trend extending from Abruzzo to the Gulf of Taranto and visible at both short and intermediate scale.

8a Uncorrelated linear trend running from the Gargano to the Matera zone and probably related to an inner margin of a sector of the Apulia-Gargano platform sunk along a regional normal fault system and buried by the Bradano Units.

54 Noticeable pattern seen both at short and intermediate scales represented by the system of regional MDA lineaments located in the Adriatic Sea. They presumably represent the external sectors of the eastern margins of the Adria microplate deformed with SW vergence.

13 Circular trend marking the boundaries of the Vulture volcanic complex.
3h Trend roughly coinciding with the Liguride Units outcropping with tectonic overlap on the Cilento and Alburno-Cervati Units.
23a/23b Circular trend partially correlated with normal or undetermined faults dislocating the Lagonegro Units with respect to the Campania-Lucania Carbonate Units.
8b Northern branch of trend 8a. It probably indicates the internal margin of a buried carbonate sector connecting the outcropping Murge and Gargano carbonate blocks. It is partially evidenced also at the intermediate scale.
3a Trend surrounding the structural boundaries of the Alburni carbonate massif.
9e Small trends surrounding the Tremiti Islands and presumably outlining the morphologic features related to the outcropping of the carbonate blocks of these islands.
16a/16b/16c Trends evidencing a tectonic setting represented by a normal fault system with Apenninic direction.
23a Trend clearly visible around Potenza and partially correlated with normal or undetermined faults dislocating the Lagonegro Units with respect to the Campania-Lucania Carbonate Units. These contacts are partially hidden by terrigenous outcropping cover aged Middle Pliocene.
23b Trends apparently uncorrelated with the shallow features represented by extended outcrops ascribed to the 'Flysh Rosso' (Lagonegro II Units).
6b MDA lineament uncorrelated with the evidence at surface, but suggesting the existence of buried discontinuities characterized by strong density contrasts.

Legend IV – Description of major lineaments identified from EHD maxima. Short scale MDA of gravity data in southern Italy: southern sector

3h Trend roughly coinciding with outcrop of the Liguride Units overthrusting the Cilento and Alburno-Cervati Units.
3e Linear trend partially corresponding to the lower Alento valley.
3d Small trend partially correlated with the thrust front of Sicilide Units on the silico-clastic and carbonatic units of Neogenic age of the Apenninic chain.
3b Weak trend surrounding the M. Cervati massif.
3c Small trend apparently uncorrelated with shallow structures.
3f Linear, uncorrelated trend, located just north of P. Licosa.
31b Trend located across the Cilento shoreline, where Alpine Calabride Units (Frido Formation) overthrust above the Cilento Units.
30b Linear trends within the Gulf of Policastro probably related to a normal fault system controlling the subsidence of the shallow crustal structures and the formation of the Sapri basin.
29b Linear trends crossing the Bradano trough and the Gulf of Taranto.
27b Linear trends partially uncorrelated with the structural elements of the Sibari basin.
26d Ring-shaped trend, evidenced along the Tyrrhenian coast of Calabria, roughly corresponding to the tectonic limits represented by the Frido Units overthrusting the Verbicaro Units.
3i Linear trends on the northern and southern sides of the Sele plain.
5b Strong MDA trend indicating a deep seated density contrast, hidden by the sedimentary cover of Plio-Pleistocenic age (Calvello Unit) and, presumably, related to the tectonic contacts between Sicilide and Lagonegro I Units.
3a Trend surrounding the structural boundaries of the Alburni carbonate massif.
25a/25b Trend following the extreme southern limit of the Apenninic chain, represented in outcrop by the Sangineto tectonic line.
11b/11c/11d Linear feature marking the eastern edge of the Diano valley. It seems to develop southward, over the southern morphologic limit of the trough, by means of three linear trends NW-running (11b, 11c and 11d), up to the Noce valley.
6a Intermediate section of a regional trend extending from Abruzzo to the Gulf of Taranto and visible at both short and intermediate scale.
6a Long regional trend (6a), extending from Abruzzo to the Gulf of Taranto, marking with high precision the position of the eastern boundary of the allochthonous chain front.
6c/6d Lineaments uncorrelated with evidence at surface, but suggesting the existence of buried discontinuities characterized by strong density contrasts.
26c Trend, NW and NE-running, that precisely marks (a) the normal faults eastward delimiting the Verbicaro Carbonate Units and the meta-sedimentary series of the S. Donato Unit and (b) the main normal fault displacing the above mentioned units with respect to the Alburno-Cervati Carbonate Unit outcropping with the Pollino massif.

10a S.Arcangelo basin MDA trend.

10c/10b/10d Trends not correlated with the boundary of the complex contact area between the S. Arcangelo basin and the inner domains. Trend 10c outlines the M. Alpi complex but indicates for this carbonatic structure a lateral extension much larger than inferred at the surface.

29a/29b/29c/25c Trends surrounding a small basin characterized by a thick sequence of autochthonous sediments reaching the isochronopache of 1.5 seconds.

24a/24b Trends identifying the morpho-structural boundaries northward of the Melandro valley and, southward, of the Agri valley.

11a Linear trends marking the structural boundaries of the morphological belt delimiting eastward the Diano valley and constituted by the Maddalena Mts.

26e Trend coinciding with the southern edge of the Pollino massif and the eastern edge of the Caramolo massif.

31a/32 Two uncorrelated trends revealing small structures delimited by normal faults.

30a Trend marking a fault system running E–W.

Legend V – Description of major lineaments identified from EHD maxima. Large scale MDA of magnetic data in Southern Italy

1 Trend following the Apennines chain.
2 Broad trend in direction NE–SW.
3 EHD maxima relative to the volcanic areas onshore and offshore.

Legend VI – Description of major lineaments identified from EHD maxima. Intermediate scale MDA of magnetic data in Southern Italy

1 Somma–Vesuvius volcano.
2 Vulture volcano.
3 Volcanic or magmatic sources in the Tyrrhenian Sea area.
4 Trend following the Apennines chain.
5 Calabrian coastal chain.
6 Sila Mts.
7 NE–SW-trending array of lineaments.
8 Sub-circular EHD maxima in the Crati valley.
9 Isolated magnetic boundary in the Sila Mts. area.
10 Enotrio and Diamante seamounts.
11 Sub-circular EHD maxima offshore the Cilento coast.
12 Sub-circular EHD maxima offshore the Cilento coast.
13 Boundary of the Foggia aeromagnetic anomaly source.
14 Boundary of the long-wavelength aeromagnetic anomaly source in the Vulture area.

Legend VII – Description of major lineaments identified from EHD maxima. Short scale MDA of magnetic data in Southern Italy: foreland area

1 Magnetic lineament correlated to the Ofanto valley anti-Apenninic (NE–SW) normal fault.
2 Magnetic lineament correlated to a known anti-Apenninic direct fault.
3 NW–SE lineaments spatially correlated with the buried front of the allochthonous units.
4 NW–SE lineament near the mouth of the Ofanto river related to a known Apenninic fault.
5 Sub-circular structure probably ascribed to a small (sub)volcanic structure.

Legend VIII – Description of major lineaments identified from EHD maxima. Short scale MDA of magnetic data in Southern Italy: chain area

1a/1b NW–SE lineaments in the Cilento area.
2 NW–SE lineaments in the Diano valley area.
3 NW–SE lineaments in the Agri valley.
4 NW–SE lineaments in the Sant'Arcangelo basin.
5 Boundary of an isolated magnetic source near the Sele river mouth.

6 Anti-Apenninic lineament bordering to SE the Sele valley.
7 NW–SE trending lineaments south of the Vulture volcano area.
8 Sub-circular EHD maxima offshore the Cilento coast.
9 NW–SE lineament NE of structure 8.

REFERENCES

AGIP, 1981. Carta aeromagnetica d'Italia (scala 1:500.000). Att. Min., Direz, Espl. Idrocarburi. S. Donato Milanese.

Balduzzi, A., Casnedi, R., Crescenti, U., Mostardini, F., Tonna, M., 1982. Il Plio-Pleistocene del sottosuolo del Bacino Lucano (avanfossa appenninica). Geologica Romana 21, 89–111.

Bally, A.W., Burby, L., Cooper, C., Ghelardoni, R., 1986. Balanced sections and seismic reflection profiles across the central Apennines. Mem. Soc. Geol. It. XXXV, 257–310.

Baranov, W., 1975. Potential Fields and their Transformations in Applied Geophysics. Gebruder Borntraeger, Berlin–Stuttgart.

Bigi, G., Casentino, D., Parotto, M., Sartori, R., Scandone, P., 1990. Structural model of Italy. Consiglio Nazionale delle Ricerche (CNR) – Progetto finalizzato geodinamica/Sottoprogetto "Modello strutturale tridimensionale" – 1:500 000. Quaderni de "La Ricerca scientifica", 114/3.

Carrozzo, M.T., Luzio, D., Margiotta, C., Quarta, T., 1986. Gravity Anomaly Map of Italy. CNR: "Progetto Finalizzato Geodinamica" – Sub-project: "Modello Strutturale Tridimensionale".

Casero, P., Roure, F., Endignoux, L., Moretti, I., Muller, C., Sage, L., Vially, R., 1988. Neogene geodynamic evolution of the southern Apennines. Mem. Soc. Geol. It. 41, 109–120.

Critelli, S., 1999. The interplay of lithospheric flexure and thrust accomodation in forming stratigraphic sequences in the Southern Apennines foreland basin system, Italy. In: Rend. Fis. Acc. Lincei, Roma, pp. 257–326.

D'Argenio, B., Pescatore, T., Scandone, P., 1973. Schema geologico dell'Appennino Meridionale (Campania e Lucania). In: Moderne vedute sulla geologia dell'Appennino. Problemi attuali di Scienza e di cultura, Quaderno 183. Accad. Naz. Lincei, Roma, pp. 49–72.

Del Ben, A., Finetti, I., Mongelli, F., Zito, G., 1994. Seismic and heat flow study of the Southern Adriatic Basin. Boll. Geof. Teor. Appl. 36 (141–144), 29–44.

Dewey, J.F., Helman, M.L., Turco, E., Hutton, D.H.W., Knott, S.D., 1989. Kinematics of the Western Mediterranean. In: Coward, M.P., Dietrich, D., Park, R.G. (Eds.), Alpine Tectonics. Geological Society Spec. Publ. 45, 265–283.

Fedi, M., 2002. Multiscale Derivative Analysis: a new tool to enhance gravity source boundaries at various scales. Geophys. Res. Lett. 29, 16-1–16-4.

Fedi, M., Florio, G., 2001. Potential fields source boundaries detection by an Enhanced Horizontal Derivative. Geophys. Prospecting 49, 13–25.

Fedi, M., Quarta, T., 1998. Wavelet analysis for the regional-residual and local separation of potential field anomalies. Geophys. Prospecting 46, 507–525.

Finetti, I., Del Ben, A., 1986. Geophysical study of the Tyrrhenian opening. Boll. Geof. Teor. Appl. 28 (110), 75–156.

Finetti, I., Lentini, F., Carbone, S., Catalano, S., Del Ben, A., 1996. Il sistema Appenninico Meridionale–Arco Calabro–Sicilia nel Mediterraneo Centrale: studio geofisico-geologico. Boll. Soc. Geol. It. 115, 529–559.

Finetti, I.R., 2005. Ionian and Alpine Neotethyan Oceans opening. In: Finetti, I.R. (Ed.), CROP Deep Seismic Exploration of the Mediterranean Region. Elsevier, pp. 109–117. Chapter 6.

Finetti, I.R., Lentini, F., Carbone, S., Del Ben, A., Di Stefano, A., Guarnieri, P., Pipan, M., Prizzon, A., 2005. Crustal tectono-stratigraphy and geodynamics of the Southern Apennines from CROP and other integrated geophysical-geological data. In: Finetti, I.R. (Ed.), CROP Deep Seismic Exploration of the Mediterranean Region. Elsevier, pp. 225–262. Chapter 12.

Finetti, I.R., 2005. The Calabrian Arc and subducting Ionian slab from new CROP seismic data. In: Finetti, I.R. (Ed.), CROP Deep Seismic Exploration of the Mediterranean Region. Elsevier, pp. 393–412. Chapter 17.

Florio, G., 1993. An interpretation of abnormal shape magnetic anomalies in South-Eastern Italy. Boll. Geof. Teor. Appl. 35 (140), 447–461.

Grauch, V.J.S., Hudson, M.R., Minor, S.A., 2001. Aeromagnetic expression of faults that offset basin fill, Albuquerque basin, New Mexico. Geophysics 66, 707–720.

Incoronato, A., Nardi, G., Ortolani, F., Pagliuca, S., 1985. The Plio-Quaternary Bagnoli Irpino-Torrente Calaggio strike slip fault, Campania-Lucania (Southern Apennines). Mem. Soc. Geol. It. 104, 399–404.

Marsella, E., Bally, A.W., Cippitelli, C., D'Argeni, B., Pappone, G., 1995. Tectonic history of the Lagonegro domain and southern Apennine thrust belt evolution. Tectonophysics 252, 307–330.

Millegan, P.S., 1998. High-resolution aeromagnetic surveying. In: Gibson, R.I., Millegan, P.S. (Eds.), Geologic Applications of Gravity and Magnetics: Case Histories: Soc. Expl. Geophys. and Am. Ass.-Petr. Geol., pp. 18–19.

Mostardini, F., Merlini, S., 1986. Appennino Centro-Meridionale. Sezioni geologiche e proposta di modello strutturale. Mem. Soc. Geol. It. 35, 177–202.

Patacca, E., Sartori, R., Scandone, P., 1990. Tyrrhenian Basin and Appenninic arcs: kinematics relations since Late Tortonian times. Mem. Soc. Geol. It. 45, 425–451.

Peirce, J.W., Goussev, S.A., Charters, R.A., Abercrombie, H.J., De Paoli, G.R., 1998. Intrasedimentary magnetization by vertical fluid flow and exotic geochemistry. The Leading Edge 17 (1), 89–92.

CROP PROJECT: Deep Seismic Exploration of the Central Mediterranean and Italy
Edited by I.R. Finetti

CHAPTER 15

Geological Outline of Sicily and Lithospheric Tectono-Dynamics of its Tyrrhenian Margin from New CROP Seismic Data

I.R. Finetti[1,*], F. Lentini[2], S. Carbone[2], A. Del Ben[1], A. Di Stefano[2], E. Forlin[1], P. Guarnieri[2], M. Pipan[1], A. Prizzon[1]

ABSTRACT

Based on new CROP seismic data, integrated with other geological-geophysical information, an evolutive lithospheric tectono-dynamic reconstruction of the Sicilian region and its Tyrrhenian margin has been done with innovative results. The data and conclusions presented here are supported not only by the here annexed and described sections, but also by relevant results from the organic ensemble of the whole interpreted CROP dataset. The pre-orogenic (Lower Cretaceous) paleogeographic framework of the central Mediterranean, now much better constrained by direct CROP data, consisted of two inter-communicating ocean basins: the *Alpine and Ionian Neotethys*. These basins were interposed between Europe and Africa–Adria, and between the Adria and the Africa the Ionian one. Downward extension of the geophysical imaging to the upper mantle allowed a better understanding of the complex evolution of this wide Mediterranean area. One of the most significant and helpful new results obtained is represented by clear *seismic imaging of the upper Ionian slab buried beneath a stack of allochthonous crustal units colliding with the paleo-Sicily (African) slope*. Such a situation, previously postulated in uncontrolled, extrapolated models, is now well defined.

The stratigraphic and structural data collected onland, and integrated with subsurface geophysical and borehole data allow us to reconstruct the geological setting of Sicily. The *Foreland Domain* is represented by the undeformed portion of the Pelagian block, exposed in the Hyblean plateau and the Sciacca platform. From the Late Miocene, the detachment of inner edge units gave rise to Triassic-through-Miocene carbonate platform imbricates: the *Pelagian–Sicilian Thrust Belt*. This external thrust system is exposed only in western Sicily. The lowermost carbonate platform imbrication reaches the surface in the area north of Sciacca as a large ramp anticline involving Early Pliocene deposits. Geoseismic data show that the wedge of Meso-Cenozoic carbonate platforms is underlying a tectonic melange, mainly composed of Oligo-Miocene Numidian nappes. The thrust belt outcrops in the structural highs of Rocca Busambra and M. Kumeta ("M. Alpi-like thrust"). The deep-seated thrust belt is tectonically overlain by "roof thrust systems", composed of the Apenninic fold-and-thrust belt, originating from the deformation of both basinal and carbonate platform Mesozoic-Tertiary deposits overlain by crystalline units (*Calabrides*).

The lowermost tectonic units of the chain are the *Ionides* (the Sicanian, Imerese, M. Judica basinal carbonate and siliceous thrust sheets). They deposited in the Ionian Tethys basin and are overridden by the Panormides.

A large nappe, composed of Permo-Triassic rocks grading up into a Mesozoic-Paleogene carbonate sequence, the Lercara unit, lies widely over the Middle-Late Miocene top levels of the Sicanian and Imerese units. Therefore it is thought to pertain to a palaeogeographical area originally located along the margin of the Ionian Tethys, and provides useful information about the opening of this Neotethyan Sea. The Ionides are composed of complex basinal thrust sheets, overriding the Late Miocene–Pliocene top levels of the Sicilian–Pelagian Thrust System (Tyrrhenian stage). Sicily is characterized by a huge volume of Late Oligocene–Early Miocene Numidian Flysch, deriving from the substratum of the

*Corresponding author. Tel.: +39 40 5582276; e-mail: finetti@units.it (I.R. Finetti)
[1]Department of Geological, Environmental and Marine Sciences, University of Trieste, Via Weiss 1, 34127 Trieste, Italy
[2]Department of Geological Science, University of Catania, Corso Italia, 55, 95129 Catania, Italy

Panormides. The Numidian Flysch is overlain by the Alpine Tethydes, widely displaced as far as the frontal wedge of the chain.
All reconstructed seismogeological sections show coherent results and thus allow us to invert in a regionally well-constrained manner the tectonic processes which occurred. In the *Balearic stage* (Oligocene–Early Miocene) a NW-dipping subduction of the Alpine Tethys (remnant of a wider ocean closed in the Alps sector by the Adria–Europe collision) gave rise to the formation of a wide orogenic arc which included stack components of the Northern Apennines, the Southern Apennines, the Calabrian arc and the Sicilian Apennines–Maghrebian chain. *The rotation of the Corso-Sardinian block stopped with the collision of Corsica against the NW Adria margin, and SE Sardinia against the Panormid promontory.* In the late Balearic–early Tyrrhenian stage (Burdigalian) *the Panormid promontory dissected the Sicilian Apennines from the Maghrebian chain to the south, and the Etrurian promontory (NW Adria) dissected the Northern Apennines from the Southern Apennines to the north, giving origin to the impressive tectono-dynamics of the SE-directed Ionian slab retreat and connected Tyrrhenian mantle dynamics (Middle Miocene to present).*
The Ionian slab subduction was accompanied by detachment and thrusting of Panormid units over the Ionian basin, detachment of the condensed sedimentary cover of the Ionian ocean and of its marginal zone sequence, with oblique thrusting over the Sicilian platform. Prominent amounts of Ionian Tethydes (Ionides = Lagonegrese, Imerese, Sicanian) made up the Sicilian Apennine wedge filling the Caltanissetta basin.
In the Late Tyrrhenian stage, the collision of the migrating crustal block, and mantle flux pressure against the steep Sicilian slope generated a transcrustal thrust that duplexed the Sicilian platform. New ophiolite-bearing units obducted in the Eo-Alpine stage over the southern European margin are now discovered on the Calabrian units of the SE-stretched Sardinian margin. The innermost allochthonous units facing the back-arc Tyrrhenian basin are all formed by delaminated fragments of Eastern Sardinia within which the new Tyrrhenian ocean started to open.

Keywords: Sicily, geological outline, CROP lithospheric tectono-dynamics

1. INTRODUCTION

Sicily and its Tyrrhenian margin constitute a geological area where various units coming from different paleogeographic domains are stacked, and where the complex tectonic features of major importance for collecting those basic elements that allow a correct understanding of the complex geodynamics of the Mediterranean region are imprinted. Several geological and geophysical studies of the rich literature for the area revealed much stratigraphic and tectonic data at various scales, and proposals of different tectono-dynamic schemes. Sicily, located in the central Mediterranean Sea, is an orogenic segment linking the African Maghrebides with the Calabrian arc-Southern Apennines. The earliest studies, mostly stratigraphical and palaeontological, were carried out from the end of the nineteenth century in a general framework of autochthony. The first work on the regional architecture of the island is that published by Ogniben (1960), which presented a new structural picture of northeastern Sicily, based on the geosyncline model. Further papers illustrating the geological features of the Calabrian–Peloritanian arc followed (Amodio Morelli et al., 1976, and bibliography therein). In eastern Sicily, intense field studies produced detailed geological maps, and helped to illustrate the modern views about the general architecture of the Apenninic–Maghrebian orogen (Carbone et al., 1984, 1986, 1990; CNR, 1991; Lentini, 2000). A tectonic wedge formed by the stacking of several thrust nappes emplaced against the Hyblean foreland was identified (Lentini and Vezzani, 1975; Lentini, 1982). The extension of the Hyblean carbonates beneath the thrust system of the chain, reaching the deep substratum of Mount Etna volcano, was supported by Cristofolini et al. (1979) and by Lentini (1982). The structure of eastern Sicily has been analyzed by Grasso and Lentini (1982), Ghisetti and Vezzani (1984), Lentini et al. (1996a), Lickorish et al. (1999). Bianchi et al. (1987) illustrated the Sicilian geology from the Nebrodi Mountains to the Hyblean foreland with a north to south deep cross-section. Later this geologic cross-section was reinterpreted by Roure et al. (1990), by Lentini et al. (1996b) and recently by Bello et al. (2000).

The expression of the continental collision that, from Tortonian times, involved the African margin (Lentini et al., 1994, 1996a, 1996b) is represented by the structures of the South Tyrrhenian System (Finetti and Del Ben, 1986; Finetti et al., 1996). These structures represent the kinematic junction to the progressive southeastward advance of the Calabrian–Peloritanian arc, related to the northwestward subduction of the Ionian crust (Finetti and Del Ben, 1986; Malinverno and Ryan, 1986; Kastens et al., 1988; Patacca and Scandone, 1989).

Innovative facies analysis of the Mesozoic carbonate and Neogene terrigenous deposits was begun in western Sicily only in the 70s (Giunta and Liguori, 1973). Catalano and D'Argenio (1978; 1982) performed palinspastic restorations of the tectonostratigraphic assemblages. Palaeomagnetic data were also used for tectonic reconstructions (Catalano et al., 1978; Oldow et al., 1990; Channel et al., 1990). Tectonic phases and the development of foreland basins and coeval satellite deposits within the chain were illustrated by Catalano and D'Argenio (1982) and by Vitale (1990). Recent papers (Lentini et al., 1990a, 1996b, 1994; Finetti et al., 1996; Catalano et al., 2000; Del Ben and Guarnieri, 2000, Guarnieri et al., 2002; Catalano et al., 2002), based on seismic lines and geological field data, helped to update previous concepts on the geology of western Sicily.

It is evident that the complex geological problems of Sicily and especially of its Tyrrhenian margin cannot be solved without extension of the investigation to the surrounding marine areas with seismic exploration data. Wide-scale geophysical exploration of the Mediterranean basin, including the Sicilian seas, started years ago with the systematic *pioneering campaigns that contributed greatly to providing the basic setting of the Mediterranean region* (Finetti et al., 1970; Finetti and Morelli, 1972, 1973, 1974; Finetti, 1976, 1982, 1984; Finetti and Del Ben, 1986). Many other local or specifically projected seismic investigations were also conducted by geoscientists of various countries and institutions. *All these geophysical campaigns helped provide the international scientific community with data for understanding the Mediterranean basin.* But no campaign explored the Sicilian seas at crustal depth. On the Sicilian mainland, the major geophysical contribution to exploration was the hydrocarbon exploration activity.

With the CROP project, we now have available for the first time transcrustal seismic data that allow controlled deep reconstructions previously done mostly by more or less hypothetical downward extrapolations. The scope of this work is to present the lithospheric interpretation of seven CROP sections integrated with onland seismic data (Figure 1). The original seismic sections used and seismogeological sections obtained are shown in Plates 1 to 11 and Figures 19 to 29. Seismic lines from hydrocarbon exploration, taken from the literature (Catalano et al., 2000; Bello et al., 2000), after graphic reprocessing were interpreted and extended with other data to produce the schematic crustal sections in Figures 26–28 and 29 on western and eastern Sicily (Figure 1).

The results obtained, very innovative with regard to several important topics, are validated by a critical selection of data from the literature and by new geological data specifically collected and duly described in the following.

2. GEOLOGICAL SETTING AND TECTONOSTRATIGRAPHIC UNITS OF SICILY

In Figure 2 the central Mediterranean's main regional geologic domains are outlined: the *Foreland Domain*, the *Orogenic Domain* and the *Hinterland Domain*.

The Foreland Domain includes the still undeformed continental areas of Africa, represented by the Pelagian block and the Ionian basin. The Orogenic Domain is composed of three main tectonic belts, the *External Thrust System*, the *Apenninic chain* and the *Calabride chain* generated, respectively, by the detachment of the internal sedimentary cover of Africa, by the imbrication of the sedimentary cover of the Alpine Tethys and Ionian Tethys, and by delaminated European units. The Hinterland Domain is represented by the *Sardinia block* and the *Tyrrhenian basin*.

The relationships between the regional structural units can be observed in the geological cross-section of Figure 3. In the next chapters the tectonostratigraphic units, which appear in a schematic map of Sicily (Figure 4) will be described. Moreover, Figure 5 shows a scheme for the stratigraphy of the main structural units and their palaeogeographic distribution.

2.1. The Foreland Domain

In Sicily the Foreland Domain is represented by undeformed area of the Pelagian block, which extends from Tunisia to Sicily (Finetti, 1982; Reuther and Eisbacher, 1985; Burollet et al., 1987). The crust is composed of a cover 6–7 km thick of Mesozoic-Cenozoic shallow-water to basin carbonatic sedimentary succession, with

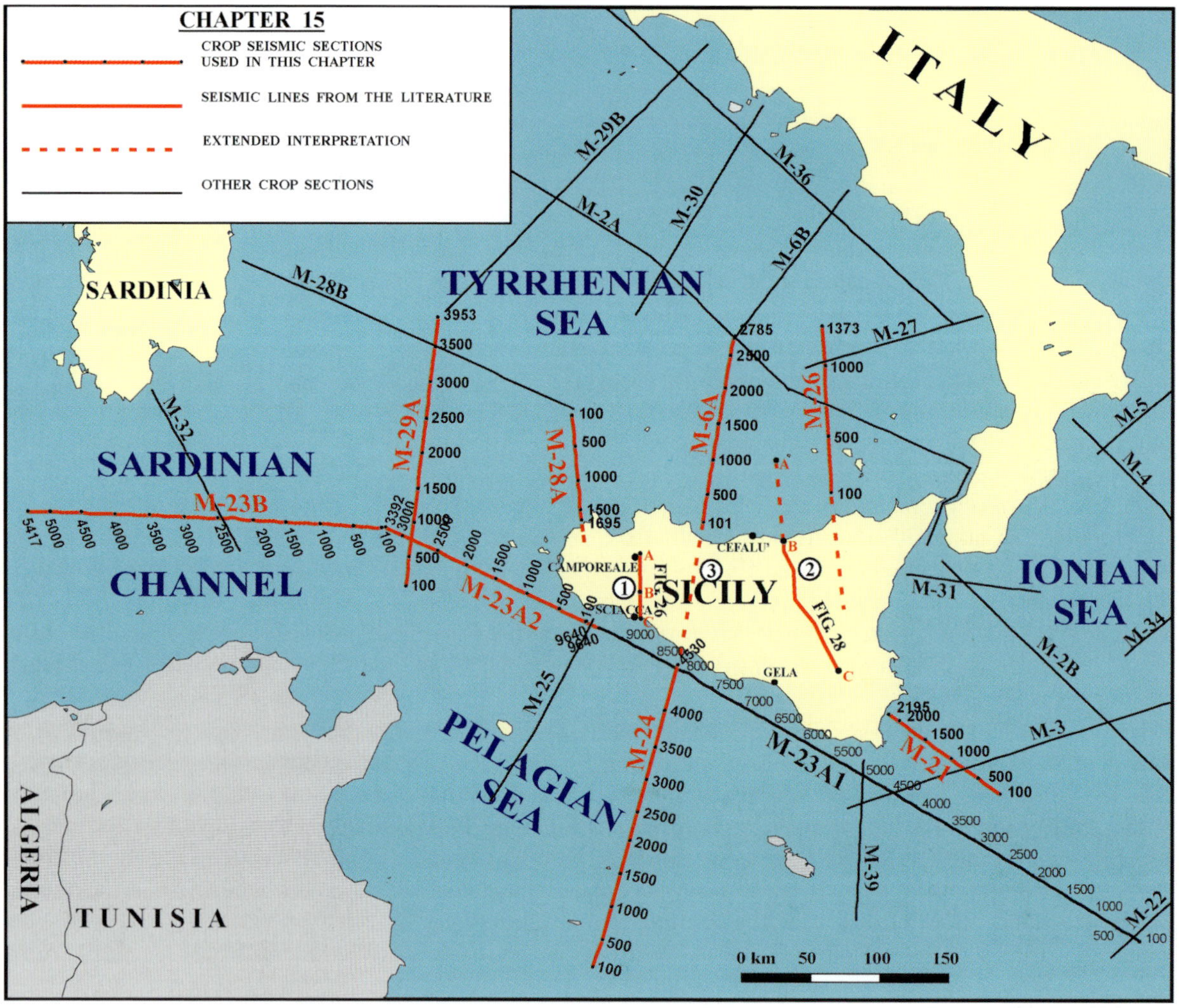

Figure 1: Location map of the interpreted CROP seismic sections described in the chapter (red lines). The positions of selected, graphically reprocessed, and interpreted seismic lines from the literature (Figures 26 and 28) on Sicily (courtesy of SGI) are also traced (sections 1 and 2 within the white circle). Dashed red segments of section indicate extension of crustal interpretation based on extrapolated-interpolated CROP and other seismic data. CROP sections M6A (Tyrrhenian Sea) and M24 (Pelagian Sea), together with the connecting reconstructed indicative "Section 3" (white circle) form the long joint section of Plate 9 and Figure 22, extending continuously from the deep Tyrrhenian Sea to the Lampedusa high in the Pelagian Sea.

intercalated volcanic rocks in various horizons. The Pelagian block represents a sector of the Africa continental margin. The undeformed sequences of the Pelagian block crop out in the Sahel region of Tunisia and extend offshore into the Sicily Channel, emerging in Lampedusa and Malta islands, in the Sciacca area of western Sicily and in the Hyblean Plateau in eastern Sicily. Onshore, to the west, the Pelagian block is delimited by the folded Atlas Mts. (Figure 2). The Hyblean–Maltese escarpment, bounds the Pelagian block to the east in the Ionian basin. The Pelagian block is represented by a shallow water epicontinental sea with an irregular bathymetry. This area is affected by intense rifting, that since the latest Miocene led to several NW–SE-oriented deep throughs (Pantelleria, Malta and Linosa Graben), separated by normal, sub-parallel faults and filled with 1000–2000 m thick Pliocene-Quaternary deposits of turbiditic and hemipelagic sediments (Maldonato and Stanley, 1977; Winnock, 1981). Rift-related alkaline volcanism is typical of intraplate rifting. In correspondence with the axis of these tectonic depressions, the so-called "*Sicily Channel Rift Zone*", crustal thinning is evident, with the moho lying at less than 20 km depth (Colombi et al., 1973; Finetti and Morelli, 1973; Finetti, 1984; Buness et al., 1990; Finetti and Del Ben, this volume, Chapter 26). Bouguer gravimetric anomalies (from +40 to +90 mGal) (Morelli et al., 1975; Scarascia et al., 1994, 2000) and relatively elevated heat flow (Della Vedova and Pellis, 1979; Zolotarev and Sochelnikov, 1980; Della Vedova et al., 1988) characterize this Pelagian zone. The CROP seismic data and MS lines (Finetti et al., this volume, Chapter 1) show that the Pelagian Sea tectonics are dominated by a

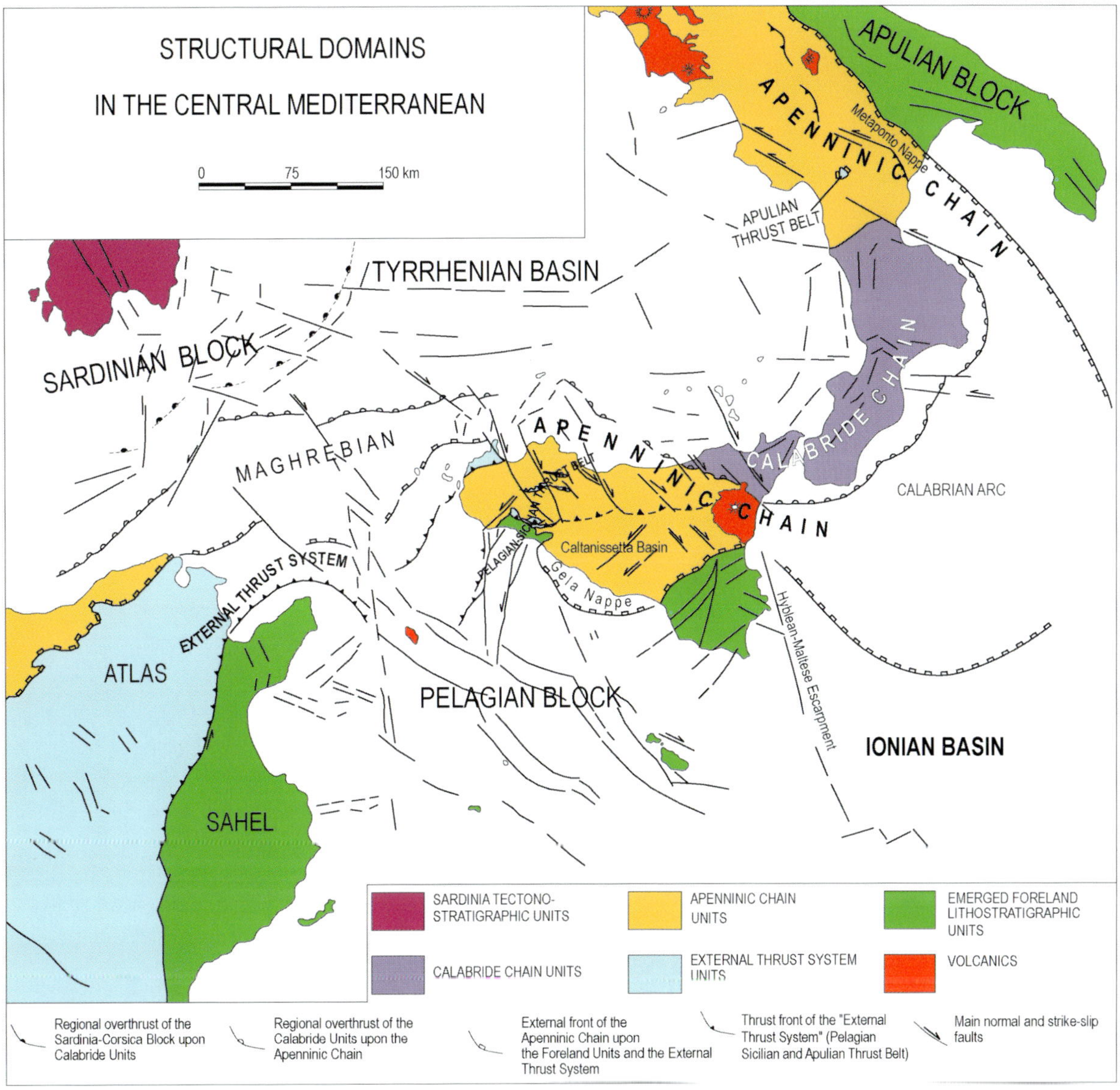

Figure 2: Regional scheme of the distribution of the structural domains in the central Mediterranean (after Lentini et al., 1996a, 1996b). The Foreland Domain consists of two blocks: the Apulian block (Adria continental crust) and the Pelagian block (Africa continental crust), separated by the oceanic crust of the Ionian basin. The lowermost structural level of the orogen is an external thrust system: Atlas in North Africa, Pelagian–Sicilian thrust belt in Sicily and Apulian thrust belt in the Southern Apennines.

NW–SE-trending system of right-shear faults (i.e. Sicily Channel rift zone) and associated antithetic NNE–SSW left-strike-slip faults (i.e. Mazara, Belice, Sciacca, Ragusa, Rosolini, P. Castellazzo and Pachino in Figure 30).

2.1.1. *Hyblean Plateau.* The Pelagian block is flexured below the front of the orogenic domain, and the *Hyblean Plateau* represents an uplifted element, separated from the flexured areas on its northwestern margin by a system of collapse faults oriented NE–SW (Figures 28 and 29). These faults delimit the Gela foredeep, which is fully occupied by the allochthonous units of the front of the chain (Figures 28 and 29). Roughly N–S structures constitute the western margin of the plateau, separating it from a depressed sector of the foreland underplating allochthonous units. To the west of this lineament, the Miocene Hyblean successions have subsided to a depth of about 3000 m, and thus are about 4000 m lower than their counterparts outcropping on the plateau (Cogan et al., 1989). This major collapse of the Hyblean successions is accompanied by a significant southwestward areal extension of the Gela foredeep, within which a distinct advance of the allochthonous units has occurred. A wide axial depression within the Caltanissetta wedge (Figure 28) is known in the literature under the inappropriate name of "Caltanissetta Basin", where thick deposits of Messinian salt are associated, emplaced below the fronts of the thrusts prograding

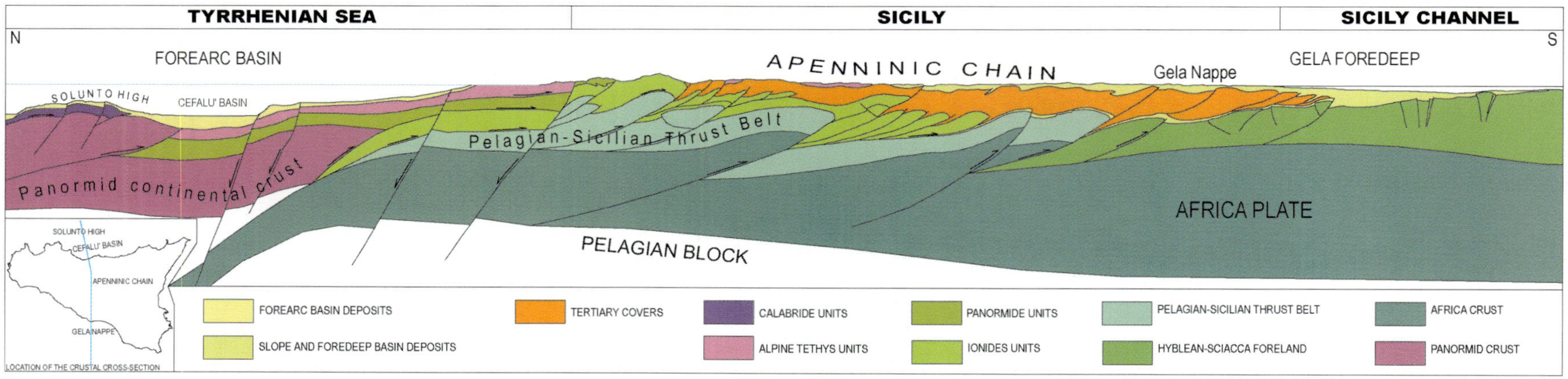

Figure 3: Schematic N–S cross-section showing the regional architecture of Sicily. The African continental crust extends northward beneath the Apenninic chain; this is a thin-skinned thrust system in which a general decoupling of Tertiary terrigenous cover occurs.

towards external areas. The northward extent of the foreland below the main thrust wedge in Sicily is known, thanks to geophysical data and indirect geological reconstructions. Carbonate bodies, connected to the successions of the Hyblean plateau, have been detected below the allochthonous of the orogenic belt, to as far as the northern slope of Mt. Etna (Cristofolini et al., 1979; Lentini, 1982; Bianchi et al., 1987). In western Sicily, Catalano et al. (1994) extend the submerged foreland to the northern coast of Sicily with gradual thinning and facies changes, where it underplates accreted allochthons (Roure et al., 1990). Lentini et al. (1996a, 1996b) emphasize the presence of an External Thrust System, named Sicanian Chain, and originating from the Late Miocene–Pliocene compressive phases affecting the margin of the Pelagian block. The stratigraphy of the Foreland Domain is illustrated by the columns in Figure 6.

The autochthonous sedimentary cover, about 10 km thick in the central area of the plateau, has been detected down to a depth of about 6 km, where Middle Triassic layers have been found, while there is no information about the Permo-Triassic interval. The age of the underlying crystalline basement is poorly known, but on mainland Tunisia, Precambrian granites and metamorphic rocks have been drilled (Burollet, 1991). The sedimentary succession is coupled with its crystalline basement and consists of thick Triassic-Liassic platform carbonates, overlain by Jurassic-Eocene pelagic carbonates and Tertiary open-shelf clastic deposits (Patacca et al., 1979; Lentini, 1982; Bianchi et al., 1987; Montanari, 1987; Catalano and D'Argenio, 1982; Lentini et al., 1987). Limited to the Norian-Rhaetian, the Hyblean sector is characterized by tidal-flat deposits, constituted by dolomites and evaporites (Gela Formation), with intercalations of mafic volcanics of varying thickness revealed in a number of wells. By late Triassic times, rifting split the Gela platform and formed the Streppenosa basin. During the early Jurassic, the shallow marine carbonate sedimentation continued on the highs, forming the Siracusa platform. On the ramp flanking the subsiding margins of Streppenosa basin, characterized by high rates of sedimentation, nodular carbonates and marls were deposited (Streppenosa Fm., up to 3000 m thick). In the Sinemurian-Pliensbachian interval the basin began to subside more slowly and fine grained pelagic and hemipelagic limestones and marls of the Modica Fm. were deposited above the black shales of the Streppenosa Formation. During the late Pliensbachian, with the drowning of the Siracusa platform, the two palaeodomains were homogeneously covered by pelagic carbonate sediments (Buccheri Fm.). During this time, volcanic activity resumed with the formation of isolated seamounts, which served as physiographic highs and locally conditioned the facies distribution ("Rosso Ammonitico" of the Scicli seamounts). From the Tithonian to Middle Eocene, pelagic carbonates and marls (Chiaramonte Fm., Tithonian-Hauterivian; Hybla Fm., Hauterivian-Albian; Amerillo Fm., Cenomanian to Middle Eocene) with scattered and limited mafic lavas, were deposited in the Hyblean region. On the Hyblean plateau, exposed sedimentary rocks are mostly of Tertiary age. The succession has been divided into a western and an eastern facies association (Grasso and Lentini, 1982). In the western part of the Hyblean plateau, well-exposed Late Oligocene–Miocene limestones and marly limestones, known as the Ragusa Formation, grading up to the marls of the Tellaro Formation, were deposited on a carbonate ramp under neritic to pelagic conditions. Pliocene and Quaternary pelagic carbonates passing basinward into clays are present, at places intercalated with mafic volcanics (Behncke, 2001).

The eastern Tertiary succession consists of Miocene carbonates, belonging to the Climiti and Palazzolo Formations and overlying reefal to lagoonal limestones with intercalated pyroclastic rocks (Carlentini and M. Carrubba Formations). These rest upon rudist-bearing carbonates of Upper Cretaceous age overling volcanic seamounts.

The Tertiary palaeogeographic picture, unchanged until Messinian times, shows a totally different arrangement from the Pliocene, when the northeastern sector of the Hyblean plateau ceased to play the role of a structural high, and progressively collapsed in connection with the reactivation of the Hyblean–Maltese escarpment as consequence of southward Calabrian wedge propagation.

2.1.2. *Sciacca Platform.* Seismic and borehole data indicate a lateral facies transition from the Hyblean domain towards the Sciacca domain in western Sicily (Catalano, 1987; Antonelli et al., 1988). The interpretation of CROP section M23A shows a continuity between the Sciacca area and the Hyblean plateau, and Sciacca is associated with a lithospheric left-transcurrent fault (Finetti and Del Ben, this volume, Chapter 26).

The Sciacca carbonate platform indicates a transition from a neritic to pelagic carbonate platform with local seamounts. The sequence consists of Norian-middle Liassic stromatolitic limestones with *Megalodon* sp., followed by crinoid limestones, late Liassic-Malm "Ammonitico Rosso" with basalts, *Calpionella*-bearing limestones (Late Jurassic–Early Cretaceous), reworked nummulitic calcarenites and calcilutites (Eocene–Oligocene), *Lepidocyclina* limestones (Late Oligocene), Serravallian-Tortonian pelagic marls, late Tortonian-Messinian clays, sandstones and evaporites, and Early Pliocene whitish marls in "Trubi" facies.

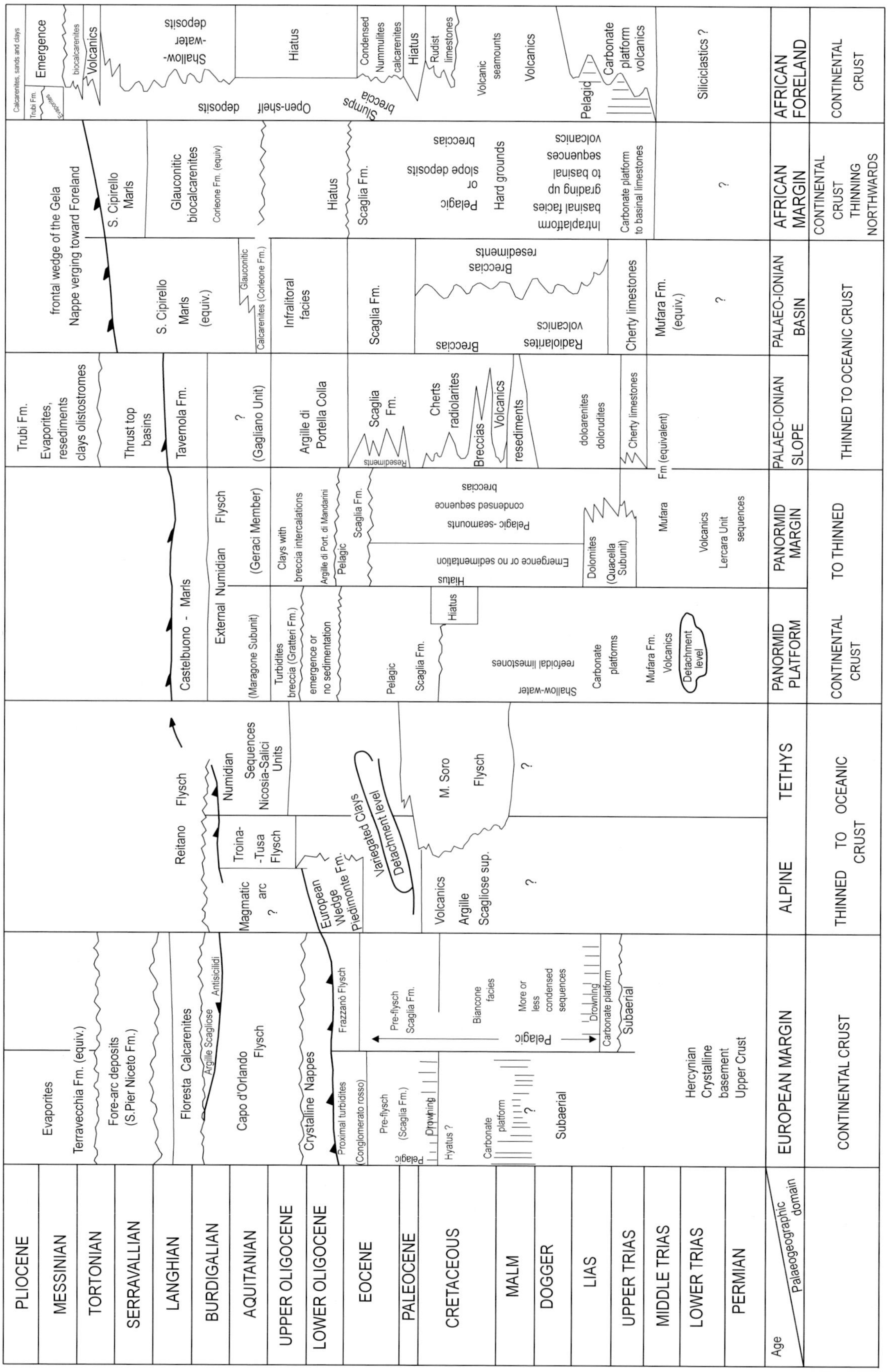

Age
Palaeogeographic domain
PLIOCENE
MESSINIAN
TORTONIAN
SERRAVALLIAN
LANGHIAN
BURDIGALIAN
AQUITANIAN
UPPER OLIGOCENE
LOWER OLIGOCENE
EOCENE
PALEOCENE
CRETACEOUS
MALM
DOGGER
LIAS
UPPER TRIAS
MIDDLE TRIAS
LOWER TRIAS
PERMIAN
EUROPEAN MARGIN
CONTINENTAL CRUST
Evaporites
Terravecchia Fm. (equiv.)
Fore-arc deposits (S.Pier Niceto Fm.)
Floresta Calcarenites
Argille Scagliose
Antisicilidi
Capo d'Orlando Flysch
Crystalline Nappes
Proximal turbidites (Conglomerato rosso)
Frazzanò Flysch
Pre-flysch (Scaglia Fm.)
Pre-flysch Scaglia Fm.
Pelagic
Drowning
Hyatus ?
Carbonate platform
Biancone facies
More or less condensed sequences
Carbonate platform
Subaerial
Hercynian Crystalline basement Upper Crust
ALPINE TETHYS
THINNED TO OCEANIC CRUST
Reitano Flysch
Magmatic arc ?
Troina--Tusa Flysch
Numidian Sequences Nicosia-Salici Units
European Wedge Piedimonte Fm.
Variegated Clays Detachment level
Volcanics Argille Scagliose sup.
M. Soro Flysch
PANORMID PLATFORM
CONTINENTAL CRUST
Castelbuono - Marls
External Numidian Flysch
(Maragone Subunit)
Turbidites breccia (Gratteri Fm.)
emergence or no sedimentation
Pelagic
Scaglia Fm.
Hiatus
Shallow-water reefoidal limestones
Carbonate platforms
Mufara Fm.
Volcanics
Detachment level
PANORMID MARGIN
TO THINNED
(Geraci Member)
Clays with breccia intercalations
Argille di Port. di Mandarini
Pelagic
Scaglia Fm.
Pelagic -seamounts condensed sequence breccias
Hiatus
Emergence or no sedimentation
Dolomites (Quacella Subunit)
Mufara
Volcanics Lercara Unit sequences
PALAEO-IONIAN SLOPE
Trubi Fm.
Evaporites, resediments clays olistostromes
Thrust top basins
Tavernola Fm.
? (Gagliano Unit)
Argille di Portella Colla
Scaglia Fm.
Resediments
Cherts radiolarites
Breccias
Volcanics
resediments
doloarenites dolorudites
Cherty limestones
Fm (equivalent)
PALAEO-IONIAN BASIN
frontal wedge of the Gela Nappe verging toward Foreland
S. Cipirello Marls (equiv.)
Glauconitic Calcarenites (Corleone Fm.)
Infralitoral facies
Scaglia Fm.
Breccias Radiolarites volcanics
Breccias resediments
Cherty limestones
Mufara Fm. (equiv.)
?
AFRICAN MARGIN
CONTINENTAL CRUST THINNING NORTHWARDS
S. Cipirello Marls
Glauconitic biocalcarenites
Corleone Fm. (equiv)
Hiatus
Scaglia Fm.
Pelagic or slope deposits breccias
Hard grounds
Intraplatform basinal facies grading up to basinal sequences volcanics
Carbonate platform to basinal limestones
?
AFRICAN FORELAND
CONTINENTAL CRUST
Trubi Fm.
Calcarenites, sands and clays
Emergence
biocalcarenites
Volcanics
Shallow- -water deposits
Open-shelf deposits
Hiatus
Condensed Nummulites calcarenites
Hiatus
Slumps breccia
Rudist limestones
Volcanic seamounts
Volcanics
Pelagic
Carbonate platform volcanics
Siliciclastics ?

Figure 5: Stratigraphic scheme of the main structural units with their palaeogeographic distributions from the innermost (at left) to the outermost areas (the modern foreland). The development in time and space of the thrusting and related sedimentation is shown.

2.2. The orogenic domains

At the end of the Balearic stage, a single large, continuous arc included the Northern Apennines, the Calabrian arc and the Sicilian Apennines–Maghrebian chain. In Burdigalian times, the Panormid promontory (Figure 30) began to dissect the Sicilian Apennines from the Maghrebian chain by a right-transcurrent system. The studied orogenic domains include the eastern Maghrebian chain and the Sicilian Apennines. The Sicilian orogenic belt is actually located between the foreland and a newly opened oceanic area (Tyrrhenian Basin), which originated in the Tyrrhenian stage. The architecture of the Apenninic edifice consists of three distinct tectonically superimposed chains, deriving from deformation of crustal palaeosectors, which mostly originated during Mesozoic times. It is possible to recognize a regional duplex structure composed of an allochthonous edifice, extensively overlying the External Thrust System, which forms a deep-seated thrust system, mostly buried.

2.2.1. *External Thrust System (ETS): Pelagian–Sicilian Thrust Belt.* This is the name used here to indicate the Sicilian ETS, originating from the detachment of the inner margin of the Pelagian block and mainly buried below the unrooted nappes of the Apenninic AC (Figure 3). The Pelagian–Sicilian thrust belt is exposed in western Sicily, while in the eastern sector of the Island, it has been identified only on seismic lines.

Western Sicily. In western Sicily, the ETS consists of Triassic-Liassic shallow-water carbonates, Middle Jurassic to Early Oligocene pelagic carbonates, and of Late Oligocene to early Tortonian continental shelf to slope, syntectonic terrigenous and biocalcarenitic deposits. The stratigraphic successions are more or less those described in the geological literature as Trapanese or Saccense stratigraphic-structural units (Catalano et al., 1994). The geoseismic and geological cross-sections show that the architecture of the westernmost part of Sicily is that of a thick wedge of Meso-Cenozoic carbonate platforms, beneath a tectonic stack of Late Mesozoic–Miocene nappes ("Pre-panormide" units). The carbonate thrust system consists of northward-dipping imbricates arranged in large antiforms with NW-verging back-thrust faults. This thrust system crops out in Montagna Grande, but it extends beneath the overriding "Pre-panormide" nappes dipping north against the thrust wedge of the S. Vito Peninsula. Lentini et al. (1994), Finetti et al. (1996) suggest that the extent of the "foreland carbonates" below the main thrust wedge continues at depth to the northern coast of Sicily with gradual thinning and facies changes. This is confirmed by the new interpretation of graphically reprocessed seismic lines (Figures 26 and 27).

The *M. Magaggiaro–Pizzo Telegrafo Subunit* is characterized by an 800 m-thick exposed sequence starting from the base with early-middle Liassic peritidal platforms, that are paraconformably overlain by a condensed Dogger to Tithonian succession (Figure 8). Late Tithonian–Albian cherty calcilutites (Lattimusa) follow upwards, and then in unconformity there are marly limestones (Scaglia facies), with thick intercalations of calcareous megabreccias. The Oligocene–Miocene interval is made up by *Nummulites*-bearing and glauconitic biocalcarenites, followed by *Orbulina* marls and discontinuous silicoclastics (Terravecchia Fm.) and Messinian evaporites. The Lower Pliocene chalks (Trubi) directly overlie the different horizons of the sequence without the Late Miocene evaporites. This helps to distinguish the stratigraphic succession of the Pizzo Telegrafo unit from the innermost units. This unit represents the most external unit of the ETS in western Sicily.

Rocca Busambra Subunit. Deep seismicity data seem to indicate, with good confidence, that the ETS extends northward to connect with the structural high of Rocca Busambra (Figure 7), which in this interpretation is not attributed to a Trapanese platform of intrabasinal character (Giunta and Liguori, 1973; Catalano and D'Argenio, 1978; Catalano et al., 1996). The Rocca Busambra Subunit crops out to the north of Corleone as an effect of a complicated thrust system, cut by dextral transcurrent NW–SE-oriented faults. This structural setting shows remarkable analogy with the M. Alpi in the Southern Apennines (Finetti et al., this volume, Chapter 12). Westwards the Rocca Busambra ridge plunges beneath the Miocene clastics and is exposed in the scattered outcrops of the Roccamena area, which corresponds to the Maranfusa thrust fault (see Figures 26 and 27).

The stratigraphic sequence outcropping at R. Busambra is represented from the bottom to the top by Late Triassic–Sinemurian Bahamas-type peritidal carbonates, topped by a typical sharp discontinuity characterized by a crust

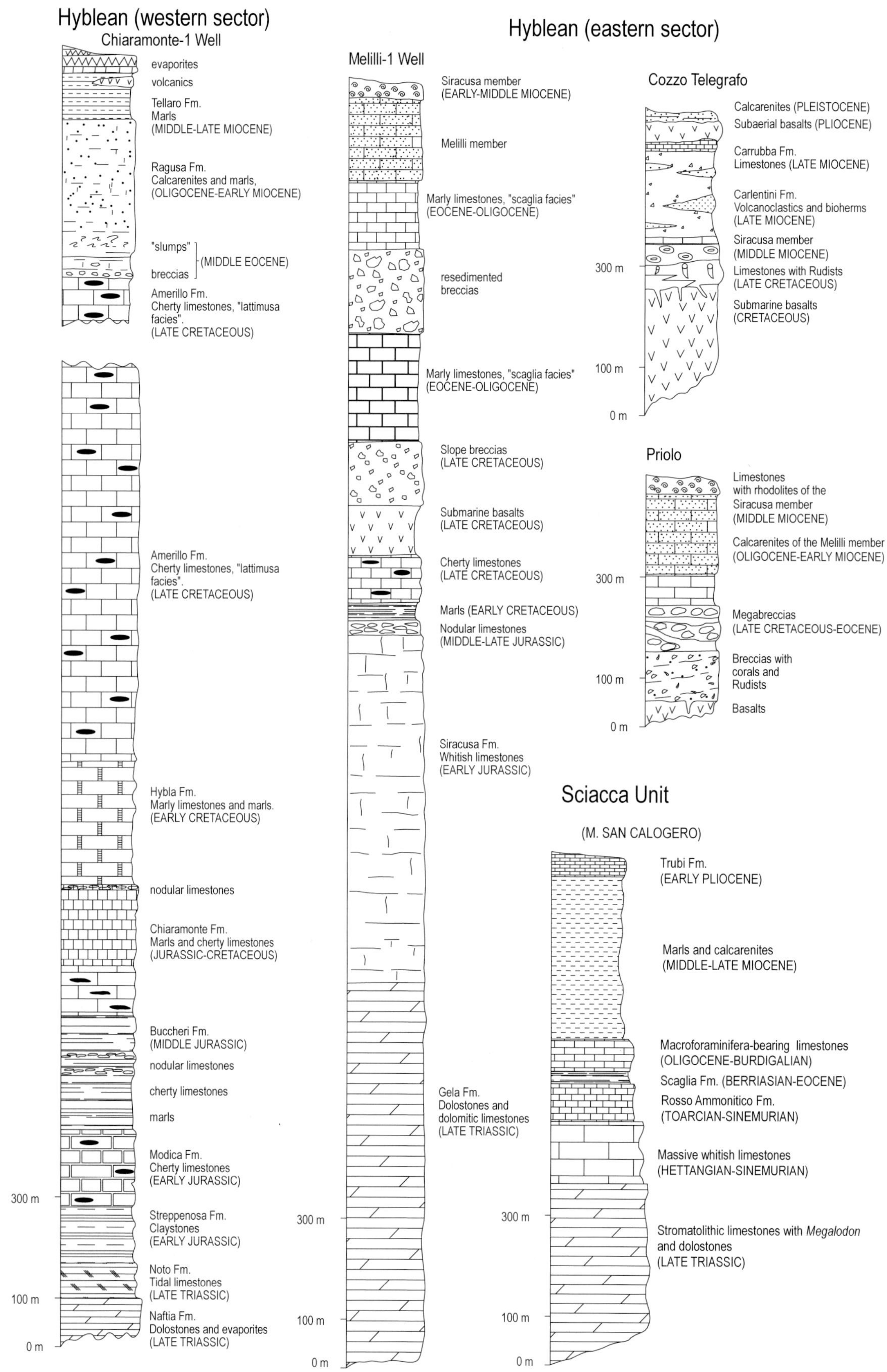

Figure 6: Stratigraphic logs of the Hyblean and Sciacca foreland sequences.

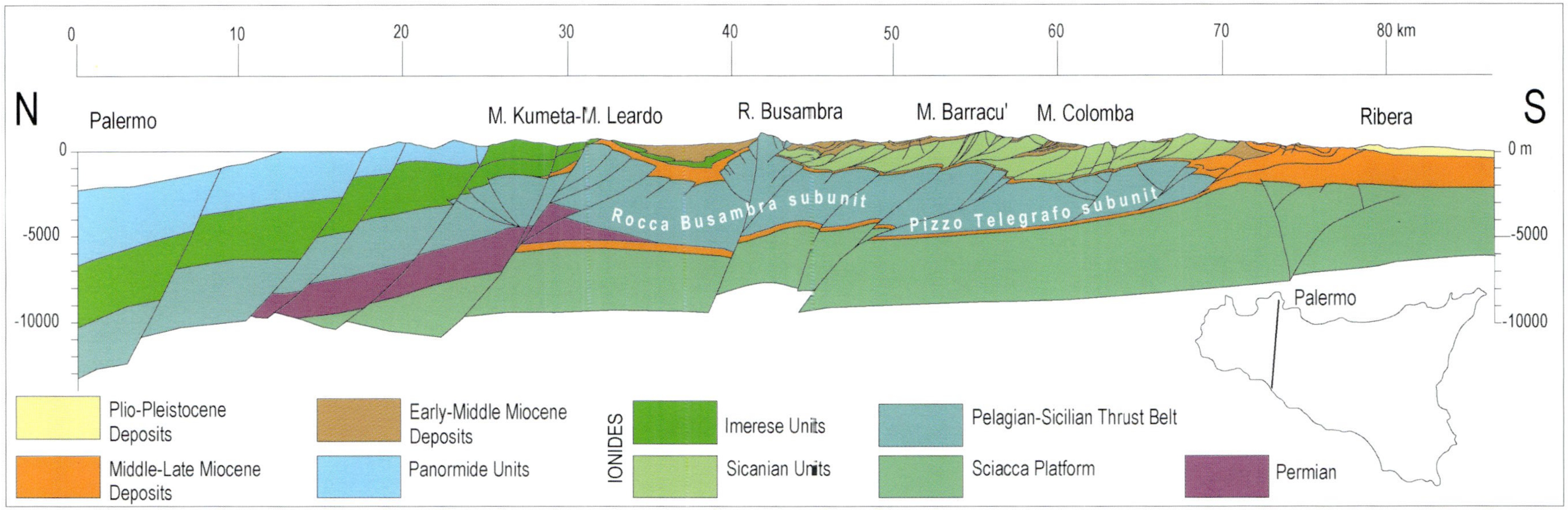

Figure 7: Geological cross-section showing the structural setting of western Sicily. The foreland is represented here by the Sciacca platform. The Pelagian-Sicilian thrust belt is a thick tectonic wedge of Mesozoic carbonate platforms, underlying the Ionides. Mainly buried, this thrust belt crops out in the R. Busambra and M. Kumeta ridges due to transcurrent faults (M. Alpi-like thrust fault).

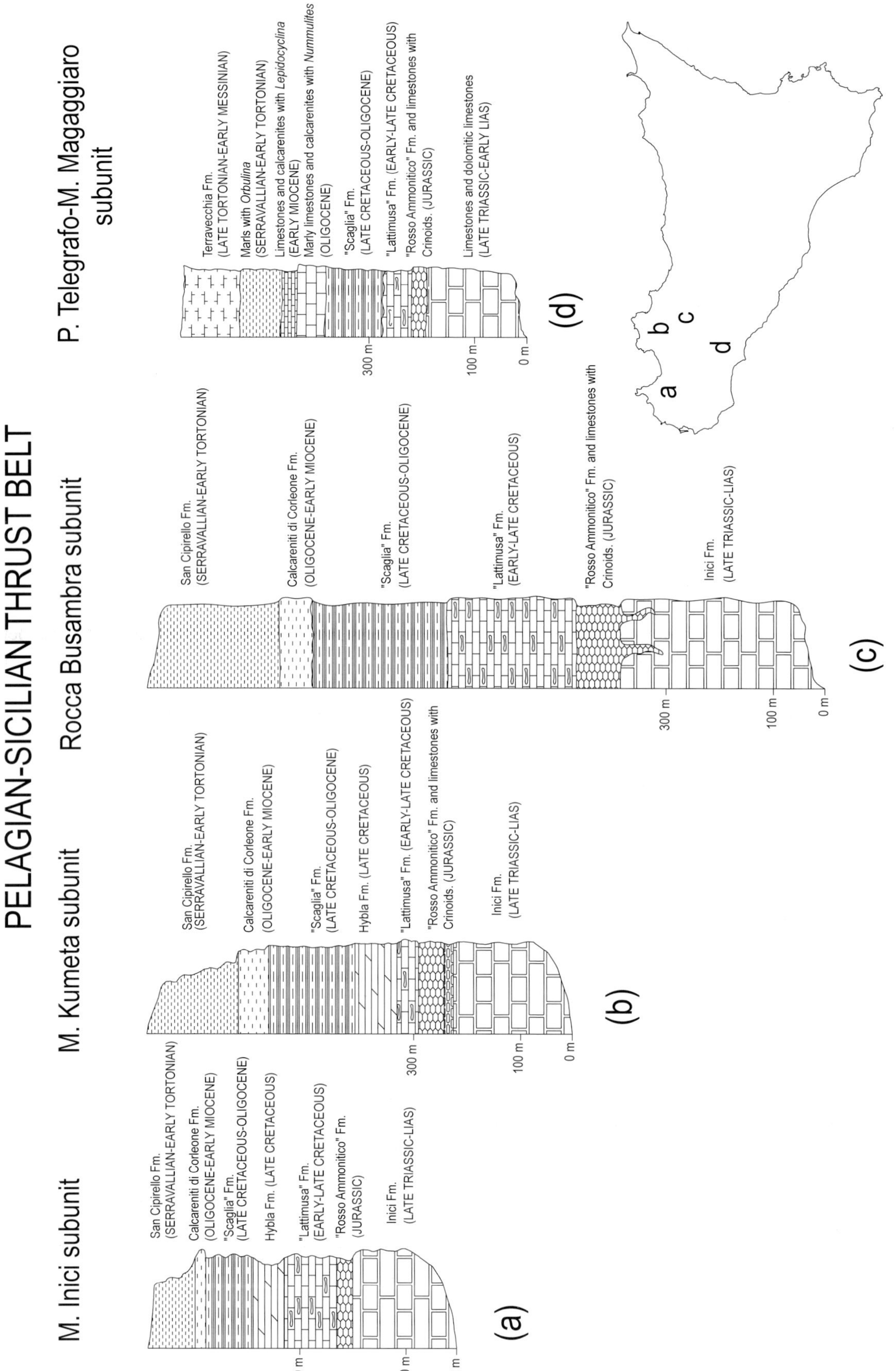

Figure 8: Stratigraphic logs of the units of the Pelagian-Sicilian thrust belt.

of Fe–Mn oxides and by pinnacled morphology. This is overlain by a Middle-Late Jurassic condensed nodular ammonite-bearing limestone (Rosso Ammonitico), on which the Late Cretaceous–Paleogene Scaglia Fm. rests unconformably. A final important discontinuity separates the Scaglia from Early Miocene glauconitic biocalcarenites equivalent to the "Calcareniti di Corleone", belonging to the allochthonous Sicani Units.

M. Kumeta Subunit. North of the Rocca Busambra culmination, the ETS crops out again along the Monte Kumeta–M. Leardo–Marineo ridge, and with the help of seismic lines, Catalano et al. (2000) suggest a probable link between both culminations. The sequence exposed along the E–W-trending M. Kumeta–Marineo ridge (Figure 8(b)) consists of an unexposed substratum of Late Triassic dolostones and of more than 300 m of lower Liassic peritidal to open-shelf carbonates (Inici Fm). This latter is unconformably overlain by white to pink encrinites of Pliensbachian age. A late Liassic–early Tithonian condensed pelagic sequence, consisting of "Rosso Ammonitico inferiore", varicoloured cherts and radiolarian marls, and nodular limestones ("Rosso Ammonitico superiore") follows upwards. The late Tithonian–Neocomian interval is represented by calpionellid-bearing cherty calcilutites (Lattimusa Fm.), grading up to *Aptychus* marls (Hybla Fm.) and to the Scaglia Fm., spanning from Early Cretaceous to Eocene. A deep disconformity at the top of the Scaglia is sealed by Burdigalian-Langhian glauconitic biocalcarenites (equivalent to the Corleone Calcarenites), that are followed in turn by the Serravallian–early Tortonian S. Cipirello Marls (Di Stefano and Mindszenty, 2000).

M. Inici Subunit. The carbonate bodies of "Trapanese facies" outcrop also in the sector lying between the Trapani Mts. to the west, the Capo S. Vito peninsula to the north and Alcamo to the east, including M. Inici, Montagna Grande, Segesta and M. Bonifato.

In the Trapanese area, the ETS is represented by a south-verging thrust system. The sequences are characterized by carbonate platform lithofacies, reaching at least 1000 m in thickness, grading to early-middle Liassic white cyclothemic limestones similar to the underlying sequences and upwards to open-shelf deposits. The middle-late Liassic facies indicate the opening of intraplatform basins preceding the final drowning of the carbonate platforms. The Dogger–Malm intervals are represented by condensed carbonates grading up to nodular limestones. Furthermore, hard grounds have been observed, and locally cherty limestones containing ammonites, belemnites and pelecypods (*Bositra buchi*). Upwards, the succession shows a transition to the Lattimusa Fm. (Tithonian-Neocomian) and to an interval of black shales (Lower Cretaceous). The succession grades up to the Scaglia facies, and to marls (Upper Cretaceous–Lower Oligocene). Basaltic pillow lavas are present as intercalations within the Malm and Lower Cretaceous levels. The succession continues upwards into glauconitic calcarenites of Lower Miocene age and the Middle-Late Miocene S. Cipirello Marls.

The sequences outcropping along the Capo S. Vito peninsula have been variously ascribed to Imerese, Panormide or outermost units in the geological literature. From new CROP data, it is evident that this tectonic feature is an intermediate element of passage from rooted thrust-structures of the Sicilian-Panormid plate margin of Maghrebian trend, lying west of the Trapani right-transcurrent fault, and the disrooted real Panormid thrust units occurring east of the S. Vito cape right-transcurrent fault (Figure 21; Plate 6).

Eastern Sicily. In eastern Sicily, the Pelagian-Sicilian thrust belt is not exposed, but it has been clearly identified on many hydrocarbon exploration seismic lines (i.e. Figure 28) and crossed by boreholes (es. Cerola-2; Vicari-1).

2.2.2. *Apenninic Chain (AC).* The Apenninic chain began in the Late Oligocene, first at the expense of the Alpine Tethys basinal sequences, which floored the oceanic crust (Alpine Tethydes, or Sicilide units). Successively, in the Middle Miocene, there was a tectonic denudation of continental crust sectors, with orogenic transport of the allochthonous carbonatic covers (Panormide units) onto the Ionian basinal successions (Imerese Sicani units), which in turn have been overthrusting the ETS since the Late Miocene (Lentini et al., 1994, 2000; Finetti et al., 1997). The AC consists of a roof thrust system characterized by unrooted nappes (Figures 3, 5, 7) and widely overthrusts the Pelagian–Sicilian thrust belt (PSTB), and in some cases directly the foreland areas with thin frontal nappes, where the ETS is represented by the Apulian Thrust System (Finetti et al., this volume, Chapter 12). The AC on the whole was completely outlined in the Middle-Late Miocene and since that time has suffered passive transport onto outer areas, contemporaneously with the growth of the PSTB and the opening of the Tyrrhenian basin.

In the AC it is possible to distinguish three structural layers: (a) involved terranes referred to the original Alpine Tethys (Tethydes), inherited by an original Oligocene-Miocene tectonic wedge; (b) sequences referred to the shallow water platforms (Panormide unit); and (c) deep water deposits (Ionides). These units, mainly formed by Mesozoic-Eocene sequences, grade upwards into Oligocene–Middle Miocene flysch-type successions.

2.2.2.1. *The Ionides.* In the roof thrust system, the deepest Meso-Cenozoic sequences are the Ionides (Figure 9). They display a complete basinal facies and come from the subducted Ionian Tethys basin (Finetti et al., 1996; Lentini et al., 2002; Finetti, this volume, Chapter 10). Their tectonic removal and transport occurred from the Middle Miocene to the Pliocene. In Sicily, the Ionides are widely exposed in the Sicani Mts., the mountains around Palermo, and the Madonie Mts., while in eastern Sicily, they are mostly buried, except in the M. Judica area. The correspondence between Sicanian sequence and the M. Judica unit is marked by their palaeoenvironmental significance, from an oceanic to deep marine succession, and, most of all, by the presence of Oligocene-Miocene neritic deposits characterized by abundant glauconite. Comparison between these two successions also shows an analogy in the structural conditions: on the Serravallian-Tortonian horizons of both units rests, as a "nappe", the Numidian Flysch, which later was involved, along with the successions of its substratum, in the Mio-Pliocene deformative phases.

Between the Imerese and Sicanian wedges there are neither stratigraphic controls nor physical continuity. It is suspected that the original relationships are buried at depth within the structural depression lying between the ridges of M. Kumeta and Rocca Busambra (Catalano et al., 1998). Their Oligocene–Early Miocene sedimentary cover is represented by clays and glauconitic sandstones in the Sicani and Judica Mts. areas, and by the Numidian Flysch, mostly exposed in central and eastern Sicily. Here a general decoupling of the Numidian cover occurs, giving rise to subunits further advanced toward the foredeep. In Sicily, the Permo-Triassic sequences include siliciclastic, clastic-carbonate rocks and volcanics, and are exposed in a wide area between Palermo and Sciacca and in minor outcrops in central-eastern Sicily. They generally represent the common base of the Ionides as well as the Panormide unit, and played an important role as a preferential detachment horizon during the orogenic transport.

The fossil content was the subject of detailed biostratigraphic studies throughout the 20th century, and allowed palaeogeographic reconstructions (Gemmellaro, 1887–1899; Fabiani and Trevisan, 1937; Scandone, 1975; Catalano and D'Argenio, 1978; Catalano et al., 1991). The recent findings appear significant for the palaeogeography of the Ionian palaeobasin: the studies on Triassic levels in western Sicily indicate a pelagic development since Ladinian times. Unfortunately, the careful biostratigraphical analyses have not been accompanied in the literature by adequate observations on structure and geometries, and the descriptions usually grouped outcrops belonging to different tectonic units. In fact, the very thick clayey and marly Permo-Triassic sequences widely outcropping in the Lercara-Roccapalumba and Cerda areas cannot represent the basal horizon of the Sicanian units. Detailed field work reveals that these sequences tectonically overlie the Late Miocene top levels of the Sicanian units. Furthermore, they are located at the footwall of the recent thrust front of the Pelagian-Sicilian thrust belt in the M. Kumeta and Rocca Busambra ridges. The original basin must be located in inner areas, but still in the Ionian Tethys. These sequences of Ionides will be described under the name of the Lercara unit, which includes more known Permian sequences (Sosio, Lercara, Roccapalumba and Cerda), while the basal horizons of the Sicani, Imerese, M. Judica and Panormide units are represented by a formation, which correlates with the Triassic Mufara Formation, first defined in the Madonie Mounts at the base of the Panormide platform.

Sicanian Units. This name indicates the stratigraphical-structural units outcropping in western Sicily and originating from the deformation of deposits of the "Sicanian basin" (Catalano and D'Argenio, 1978, 1982), ascribed to a Late Miocene–Early Pliocene tectonic phase. The Sicanian thrust sheets originated from the deformation of a sedimentary sequence, consisting of up to 2000 m of Triassic to Late Miocene basinal deposits, that accumulated in the westernmost prolongation of the Ionian Tethys. The sequence ranges from Carnian marls and limestones to Oligo-Miocene marls and biocalcarenites grading into Serravallian-Tortonian marls.

Figure 9: Stratigraphic logs of the Ionides. The Imerese unit was deposited at the inner margin of the Ionian palaeobasin adjacent to the Panormide platform. The Gagliano and M. Judica sequences indicate an oceanic environment with a persistent low sedimentation rate below the CCD limit during the whole Jurassic-Cretaceous interval. The Sicanian units were located to the west and south, where the coalescence of the Africa and panormide platforms occurred, and they rimmed the western extremity of the Ionian palaeobasin.

IONIDES
IMERESE UNIT
Madonie Mts
PANORMIDE UNIT
Portella Colla Fm. (Late Oligocene)
Caltavuturo Fm. (Eocene - Oligocene)
Spongolites (Late Cretaceous)
Calcarenites with Ellipsactinia (Titonian)
Radiolarites (Jurassic)
Fanusi Fm. (Late Triassic)
Scillato Fm. (Late Triassic)
(a)
GAGLIANO UNIT
GAGLIANO 101 well
Portella Colla Fm. (Late Oligocene)
Caltavuturo Fm. (Eocene - Oligocene)
Crisanti Fm. (Jurassic-Cretaceous)
Scillato Fm. (Late Triassic)
F. MUFARA (Carnico)
(b)
M. Scalpello
Sicilide Unit
Numidian Flysch
Glauconitic marly clays (Late Oligocene - Middle Miocene)
Scaglia Fm. (Eocene-Oligocene)
Scillato Fm. (Late Triassic)
Mufara Fm. (Middle-Late Triassic)
(c)
M. JUDICA UNIT
M. Judica
Glauconitic marly clays (Late Oligocene - Middle Miocene)
Scaglia Fm. (Eocene-Oligocene)
Scillato Fm. (Late Triassic)
(d)
M. Gambanera
Glauconitic marly clays (Late Oligocene - Middle Miocene)
Scaglia Fm. (Eocene-Oligocene)
Scillato Fm. (Late Triassic)
Mufara Fm. (Middle-Late Triassic)
(e)
Cozzo Cucca
Scaglia Fm. (Eocene-Oligocene)
Volcanites (Jurassic)
Scillato Fm. (Late Triassic)
(f)
SICANIAN UNITS
M. Barracù
Glauconitic marly clays (Late Oligocene - Middle Miocene)
Scaglia Fm. (Eocene-Oligocene)
Radiolarites
Scillato Fm. (Late Triassic)
(g)
M. Genuardo
S. Cipirello Fm. (Serravallian-Early Tortonian)
Glauconitic sandstones (Burdigalian-Langhian)
Sandy-marls and quartz-glauconitic sandstones (Oligocene)
White calcilutites (Scaglia Fm.) (Eocene)
Megabreccias (Maastrichtian)
White-red calcilutites (Scaglia Fm.) (Campanian)
Marls with Aptici (Aptian-Albian)
Calcilutites with cherts (Titonian-Berriasian)
Radiolarites (Oxfordian-Titonian)
S. Maria del Bosco limestones (Toarcian-Oxfordian)
Oolithic limestones of M. Genuardo (Early-Middle Lias)
Dolomitic limestone and dolostones (Norian-Raethian)
(h)
300 m
100 m
0 m
a
b
c
d
e
f
g
h

The Mesozoic facies, first transitional (M. Genuardo subunit) and then truly basinal (Pizzo Mondello and M. Barracù–M. Colomba subunits) form a thrust system widely overriding the M. Magaggiaro-Pizzo Telegrafo subunit, and involving Mio-Pliocene deposits, as demonstrated by the remnants of the Trubi Fm., which are distributed against the M. Genuardo thrust front (Figure 7).

M. Genuardo Subunit. This subunit overthrust the carbonate platform units of the ETS, and extends westwards below Neogene deposits of the Belice basin and eastwards beneath the Sicanian thrust stack of the M. Barracù–M. Colomba. Seismic data show that the M. Genuardo subunit thrust over the Maranfusa-Roccamena structure belonging to the ETS (see Figure 26). In the crustal section of Figure 27 this subunit appears unrooted and probably completely detached. It is characterized by an Upper Triassic to Neogene thick sequence of carbonates and siliciclastic deposits, whose thickness in outcrop is about 1500 m (Di Stefano and Gullo, 1986). The basal levels consist of shallow water limestones and platform dolostones of the Late Triassic, which are overlain through an angular unconformity by late Lias–Dogger oolithic calcarenites (Figure 9). The "drowning" is testified by slope to basin deposits of Jurassic–Early Cretaceous age represented by late Lias–Dogger nodular calcilutites and by thin levels of radiolarites and cherty calcilutites referred to the Malm. Moreover, a thick basalt layer occurs, as well as the widespread Jurassic magmatism known in western Sicily. On top of Late Jurassic–Early Cretaceous *Calpionella*-bearing calcilutites ("Lattimusa" facies) there are Cretaceous-Eocene marly limestones ("Scaglia" facies) with interbedded carbonate breccias. The sequence ends with Oligocene sandy marls, grading up to quartzose sandstones and glauconite-bearing biocalcarenites ascribed to the Burdigalian-Langhian (*Calcareniti di Corleone* Fm. of Ruggeri, 1966) and finally Middle Miocene marls, ascribed to the S. Cipirello Marls.

Pizzo Mondello Subunit. This is a 1000 m thick basinal sequence. The basal level of Pizzo Mondello is represented by marls and calcilutites of Julian age grading upwards into 450 m thick cherty limestones ascribed by Di Stefano et al. (1998) to the Carnian-Norian, and then into Rhaetian white calcilutites with radiolarians, and to Liassic calcarenites with crinoids and brachiopods. Upwards the sequence continues with Middle-Late Jurassic cherty calcilutites and with Cretaceous-Eocene Scaglia Fm. The uppermost level is made of Oligocene-Miocene open-shelf deposits: marls and glauconitic sandstones and biocalcarenites.

M. Barracù–M. Colomba Subunit. The Barracù stratigraphic sequence strongly resembles that of Pizzo Mondello, but differs from it, along with M. Colomba and M. Triona, by its structural layering. Also in this case, the Oligo-Miocene interval is characteristically represented by open shelf glauconitic sediments (*Calcareniti di Corleone*) (Figure 9).

M. Judica Unit. The M. Judica unit is characterized by a mostly pelagic Meso-Cenozoic succession (Figure 9), starting with claystones, calcilutites, calcarenites and microbreccias with halobids, ammonoids and conodonts, known as Mufara Fm. or "Flysch Carnico", and ascribed by Lentini (1974) to the Aonoides Zone (Carnian). The "Mufara Fm." commonly occurs at the base of the different tectonic units grouped as Ionides, as well as the Panormide units. The type section lies on the southern slope of M. Mufara in the Madonie Mountains, and represents the basal level of the Panormide carbonate platform. In the literature this name has been used to indicate without distinction the Triassic basal level, even where these deposits have lost any stratigraphic relationships with the adjacent successions and are imbricated within the Numidian Flysch or within Permo-Triassic nappes. In this chapter the term of "Mufara Fm.-equivalent" will be used, because this name is correct only for the basal levels of the Panormide platform. In the M. Judica unit, the "Mufara Fm.-equivalent" grades upwards into 200–300 m thick cherty limestones with halobids ascribed to the Norian-Raethian. Nodular facies are common. The Liassic-Cretaceous interval is represented by 10 to 100 m of siliceous claystones and radiolarites with lenses of mafic submarine volcanics. This sequence represents quite a long chronological interval during which the environmental conditions did not change, and indicates sedimentation in a deep basin. The upper part is characterized by slumps and breccias, indicating instability probably due to movements along transform faults. These deposits are overlain by 10–100 m thick marly limestones (Scaglia Fm.) referred to the Middle Eocene–Early Oligocene, passing upwards into 300–400 m thick marly clays with glauconitic sandstones of the Upper Oligocene–Serravallian. Southwards, on the ridges of M. Gambanera and M. Matteo, the stratigraphic sequences are distinct for their reduced thickness and absence of the Jurassic-Eocene interval. The reduction in thickness is most notable in the Cozzo Cucca section (Figure 9), to the west of M. Judica, where an oceanic succession, encompassing Triassic limestones, mafic volcanics with thin levels of Jurassic radiolarites, and Eocene red marls, is exposed within a 15 m interval.

In conclusion, the entire stratigraphic succession shows a basinal character for the Late Trias to Eocene interval, very similar to that of the Imerese Unit outcropping in the Madonie area (Ogniben, 1960; Grasso et al., 1978),

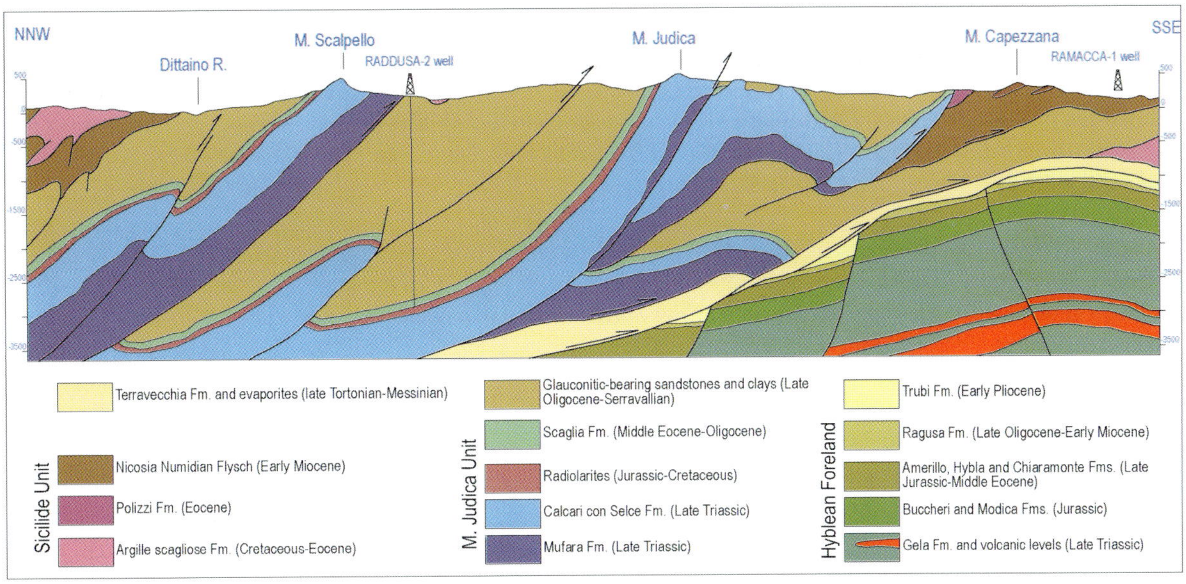

Figure 10: Geological cross-section of the M. Judica area, showing the frontal wedge of the Apenninic chain, composed of Ionides embricates (M. Judica unit), with remains of Alpine Tethydes nappes (Sicilides and inner Numidian Flysch). The south-verging stack overrides an interval of Late Miocene siliciclastics (Terravecchia Fm.) and evaporites and in turn the whole edifice tectonically overlies the Early Pliocene "Trubi" at the top of the Hyblean margin. The interpretation of this structural setting, well constrained by subsurface data, permits a complete reconstruction of times and modalities of the tectonic evolution of the frontal wedge.

but with the latter being different by the intercalations of carbonatic resediments originating from the Panormide platform; from this the succession is distinguished further by the epicontinental-type terrigenous cover comparable to the "Sicani" cover.

The M. Judica unit, tectonically overlain by the most advanced portions of the Alpine Tethydes, has been deformed by a system of imbricated slices, completely unrooted and overriding onto the successions of the ETS and onto the Messinian to Lower Pleistocene horizons of the Gela foredeep (Figure 10). The structural style of the Ionides belonging to the M. Judica unit is well expressed in the crustal section crossing eastern Sicily (Figures 28 and 29).

Gagliano Unit. This unit does not appear in outcrop, but has been crossed by boreholes in the Gagliano Castelferrato area (Enna) and seismically identified (Bianchi et al., 1987). The Gagliano 101 well shows that the stratigraphic sequence of this unit is characterized by a Carnian basal carbonate and clayey interval ("Mufara Fm.-equivalent"), grading upwards into 800 m thick Late Triassic cherty limestones (Scillato Fm.). These latter show a transition to several hundreds meters of cherty marls, shales and radiolarites, Jurassic-Cretaceous in age. The top level is represented by the Eocene Caltavuturo Fm., grading upwards into an 800 m thick Oligocene-Miocene interval of brown clays followed by an alternation of clays and quartzarenites referred to the Numidian Flysch. This terrigenous sequence represents the original Tertiary cover, even if it is more or less detached.

The aforementioned succession belongs to a tectonic unit overriding the Pelagian-Sicilian thrust belt and underlying a 2000 m thick interval, composed of a wedge of Sicilide nappes and inner Numidian Flysch (Figures 28 and 29).

Imerese Unit. This unit is well exposed in the western Madonie Mts, in the Trabia and Termini Imerese area and in the Palermo Mts, and represents the sequence deposited at the original internal margin of the Ionian Tethys (Figure 9). Lower Triassic records are rare: pebbles referred to late Scythian-early Anisian age have been found by Di Stefano and Gullo (1997) in Carnian debrites in the Palermo mountains. In the Termini Imerese area and the western side of the Madonie Mts., the succession commonly starts with Ladinian-Carnian black marls with intercalations of fine-graded breccias, calcilutites associated with pillow basalts ("Mùfara Fm.-equiv."), followed by 500 m thick cherty limestones (Scillato Fm.) of Carnian-Norian age. These are the well known *Halobia* limestones, widespread from the western Madonie Mts to Termini Imerese and the Palermo Mts. They are overlain by the turbiditic dolostones of the Fanusi Fm. (200 m). Slump-breccia intercalations 150 m thick are found in upper Norian sequences in the Palermo area. The Fanusi Fm. is unconformably overlain by a Jurassic-Cretaceous radiolaritic succession with mafic volcanics (Crisanti Fm.) and more or less pronounced hiatuses, containing calcareous breccia intercalations (Grasso et al., 1978) giving way gradually to 160 m thick late Cretaceous-Eocene "Scaglia"-type calcilutites (Caltavuturo Fm.). On it rests in conformity the Numidian Flysch (Late Oligocene–Early Miocene). In western Sicily the "Imerese terrains" occur as imbrications of thrust sheets, subdivided into main stratigraphic-structural units by Catalano and D'Argenio (1982).

In the Palermo Mts the Imerese sequence consists of Carnian marls and calcilutites ("Mufara Fm.-equiv.") passing upwards into Carnian–late Norian cherty limestones (Mirabella Fm.), with the uppermost levels characterized by the dolorudites and doloarenites of the Fanusi Fm., probably Rhaetian-Liassic in age, well exposed in the Valle Cuba area, and radiolarites and cherty limestones of the Crisanti Fm. and marls and calcarenites of the Caltavuturo Fm. Upsection there follow Oligocene brown shales with *Nummulites*-bearing resedimented calcirudites and quartzarenites belonging to the Numidian Flysch. This sedimentary cover as well as the uppermost Langhian-Tortonian marls are often more or less detached.

Lercara Unit. The Lercara Unit consists of repeated Permian and Triassic slices up to 3000 m thick, with remains of a Mesozoic cover alternating with thin slices of Miocene clays, in which remains of Tethyan Numidian Flysch are recognizable. This unit overrides southwards the Middle-Late Miocene top levels of the Sicanian sequence, which indicates that the first tectonic displacement of this allochthon is not older than Tortonian. In the Lercara-Roccapalumba area, this unit is composed of south-verging imbrications, well documented by seismic and borehole data, and this structure is inflected northwards below the thrust front of the Ionian Numidian Flysch or the Rocca Busambra carbonate platform. Catalano et al. (1993, 1996) recognize three different duplex stacks consisting of Paleozoic-Carnian, Late Triassic–Paleogene and Neogene rocks, respectively; but the general opinion is that these sediments were the basal horizon of the Sicanian sequence. However, field work has revealed that they are overthrust by the External Numidian Flysch or Rocca Busambra carbonates out-of-sequence structures. Moreover the highest structural location of these nappes is also well defined by scattered limbs of the upper Tortonian Terravecchia Fm., unconformably resting upon the Permo-Mesozoic mélange.

In the Roccapalumba, Lercara and Vicari areas the Permian siliciclastic deposits are widespread and were formerly studied by Fabiani and Trevisan (1937) and subsequently by Castany (1956), Schmidt di Friedberg (1964–1965), Broquet (1968), Mascle (1979), Catalano et al. (1991). There has been disagreement among some of the authors on the age of these deposits; the Permian age of these rocks was not supported by convincing biostratigraphic data: Ruggeri and De Vita (1972) ascribed the whole succession to a Miocene mélange; Montanari (1968) and Catalano and Montanari (1979) considered the Permian fossils as resedimented in younger Carnian deposits.

Field work and stratigraphic analysis have allowed Catalano et al. (1988, 1991) to differentiate the Lercara type Fm. into three lithostratigraphic units: (1) a Flysch unit (Kungurian, Early Permian); (2) an Olistostrome unit (Early-Middle Permian); (3) a Red Clay unit (Late Permian) (Figure 11).

The Kungurian Flysch consists of micaceous quartz wackes alternating with silty shales. An increase in the carbonate content occurs upwards. These beds contain *Mesogondolella idahoensis* and *M. intermedia*, which indicates a Kungurian age. Other sections show graded resedimented microbreccias and calcarenites with parallel and cross-laminations interbedded with micaceous quartz wackes and silty shales. The microbreccias contain fusulinids, crinoids, bryozoans and abundant reef-derived clasts. Near the Roccapalumba station, an alternation of micaceous quartz wackes, silty shales and rare sandy calcarenites containing the richest fossil association found in this flysch is exposed in a 10 m thick section. Several outcrops of calcareous sandstones contain conodonts indicating latest Artinskian or earliest Kungurian age. At Cozzo Intronata (Roccapalumba area) chaotic lithologies of the Kungurian Flysch crop out, overlain by middle and late Carnian marls and limestones containing pelagian conodonts and radiolarians. Some blocks with brachiopods bearing dark grey limestones can be ascribed to early Chihsian age (Catalano et al., 1991).

Diabasic sills have been found in the flysch, but no absolute age determinations of these magmatic rocks are available. Censi et al. (2000) have studied magmatic rocks included in deep-water turbidites and their geological setting suggests a mafic magma intruding into unlithified sediments. The tholeiitic composition led them to ascribe this phenomenon to continental rifting during Permian times. Younger Permian rocks are exposed to the south in the Sosio Valley. They have been described by Catalano et al. (1991) under the names of Olistostrome unit and Red Clay unit, followed by the well-known Permian megablocks of the Palazzo Adriano area. The oldest

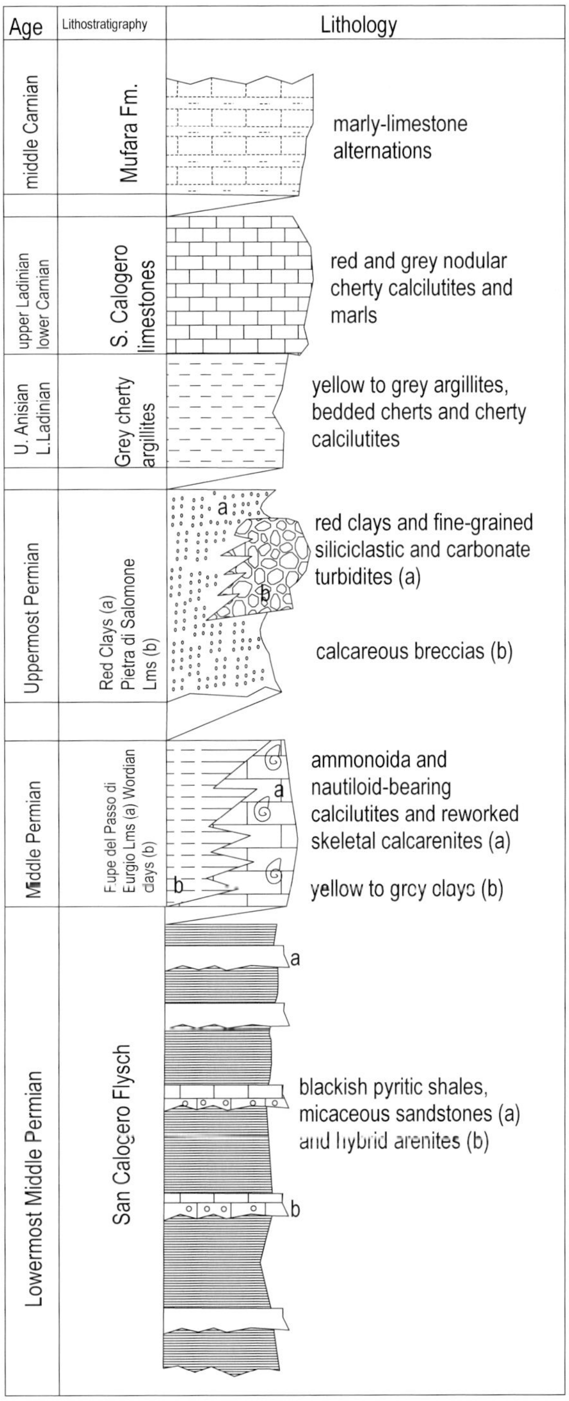

Figure 11: Stratigraphic log of the Permo-Triassic Lercara unit (after Di Stefano and Gullo, 1997).

lithostratigraphic unit recognized in this area is the S. Calogero Flysch, consisting of grey to blackish pyritic shales and siltstones with interbedded micaceous sandstones (Di Stefano and Gullo, 1997). The age of this interval, based on conodonts, is lowermost Middle Permian (Gullo, 1993). In the same Torrente S. Calogero area, similar deposits appear as a chaotic clayey mass containing sandstones, and the conodonts indicate the same age as the flysch, suggesting that these deposits have been affected by synsedimentary sliding and reworking.

The Olistostrome unit is overlain by middle-late Permian red and light-grey clays, containing sponge spiculae, deep-water ostracods, and abundant radiolarians. Resedimented bioclastic calcarenites are very rich in pelagic conodonts. The fossil contents indicate that this horizon spans from the latest Middle to the earliest Late Permian. Moreover the radiolarian faunas from the red clays indicate deep-water environments similar to the Middle-Late Permian radiolarite sequences in the circum-Pacific and Oman areas. The megablocks of the Sosio Valley consist of limestones, reef, back-reef and reef-slope deposits and partly element supported megabreccias (Di Stefano, 1990; Catalano et al., 1991). These blocks contain a rich fauna, studied by Gemmellaro (1887–1899), Montanaro-Gallitelli (1956), Kozur and Mosler (1989), and assumed to derive from shelf areas flanking deep-water basin, as confirmed by ammonoids and conodonts.

The best known outcrop is the Pietra di Salomone block, an up to 70 m thick fossil-rich limestone, which was deposited in a base-of-slope position. A Late Permian age for the matrix is suggested by fusulinids and conodonts occurring with a large mass of reworked Middle Permian fossils (Fluegel et al., 1991; Di Stefano and Gullo, 1997). Other blocks, such as the Rupe del Passo di Burgio Limestones, referred to Wordian-age hemipelagic carbonates, deposited in a distal slope area (Di Stefano and Gullo, 1997).

The Triassic sequences of the Lercara Unit display radiolarians, pelagic bivalves and conodonts dated as old as early Ladinian. The records of Scythian age are very rare. Middle Triassic pelagic deposits have been documented in the Sosio Valley. Younger deposits are late Ladinian–early Carnian varicoloured clays and nodular cherty calcilutites. In the Lercara, Vicari and Roccapalumba areas, the Triassic sequences rest upon the Permian "Lercara Formation" and are composed of siliceous limestones, radiolarites, red and grey nodular limestones and marls (Ladinian-Cordevolian), followed upwards by marls and calcilutites (middle Carnian), referred to the Mufara Fm. A thick sequence of micaceous sandstones, marls and calcirudites, exposed to the north in the area of Cerda-Cozzo Rasolocollo and containing halobids, has been referred to the Mufara Fm. by many authors and considered an Imerese deposit.

Field evidence shows that this thick sequence, crossed by the Cerda 1 borehole, is located in a structural depression, surrounded by Varicoloured Clays of the Alpine Tethys unit and directly underlain the late Tortonian Terravecchia Fm. Therefore it seems to belong to an unrooted nappe, derived from a more internal basin compared to at least the Imerese domain, as well as to the Lercara unit. The Permo-Triassic rocks of the Lercara unit pass upwards into a condensed carbonate sequence with volcanics, ascribed to a Dogger–Malm and Cretaceous age, and fragmented into blocks scattered in the Vicari and Roccapalumba areas (Trevisan, 1937; Caflish and Schmidt di Friedberg, 1967). Late Triassic–early Liassic calcareous-dolomitic olistolithes, locally with inclusions of pillow lavas, have been recognized within the Caltavuturo Fm. (Imerese) near Termini Imerese village (Palermo). Pillow lavas occur within carbonate olistolithes, outcropping at the Rocca di Papa locality. Petrographic, geochemical and radiometric data suggest for these exotic volcanics (i) an alkaline character, (ii) a significant affinity with tectonic settings of within-plate basalts, and (iii) a K/Ar isochron age of about 196 My (Guarnieri et al., 2000).

Outcrops of the shales and marls, as well as remains of Mesozoic calcareous cover widely occur westwards, between Cefalà Diana and Marineo (M.S. Angelo), on the southern slope of Rocca Busambra and in many small blocks. These rocks belong to a structural depression between the Numidian Flysch of Godrano-Mezzojuso and the M. Kumeta-Marineo ridge, and are uncomformably overlain by the Terravecchia Fm. Remains of the original Meso-Cenozoic carbonate cover occur southwards on the eastern flank of M. Cammarata (La Montagnola). They rest tectonically upon the Middle-Late Miocene top levels of the Sicanian sequence of Mt. Cammarata (Broquet, 1970).

Evidence of a widespread original distribution of this tectonic unit is represented by the Triassic fragment of a sill with a thin skin of its wall rocks, consisting of a well bedded pelagic sequence, outcropping in central Sicily. The sequence has been attributed to the Lercara and Mufara Fms. The dominant igneous rock consists of fine-grained albite-augite diabase; this is consistent with magmatic activity known elsewhere in Sicily and attributed to the early stages of rifting, which during the Ladinian evolved to deep basin.

Ionian Terrigenous Sequences. The Tertiary cover of the basinal sequences belonging to the Ionides is represented by foreland/foredeep deposits, whose relationships with the substratum are sometimes preserved, although large detachments occurred with further forward transport of the cover, which generated repeated slices with an apparent increase in the thickness penetrated by the boreholes.

The Oligo-Miocene deposits consists of mostly Numidian Flysch, which represents the terrigenous cover of the Imerese and Gagliano units, whereas the M. Judica and Sicanian units support an Oligo-Miocene glauconite-bearing sequence, with calcarenites, sandstones and clays passing upwards into Serravallian–early Tortonian marls (San Cipirello Fm.).

Numidian Flysch. The Numidian Flysch is composed of an alternation of brown clays and yellowish quartzarenites. The lower portion of the sequence is mainly clayey and is ascribed to the Late Oligocene, whereas the arenaceous strata become thicker and more frequent in the Aquitanian-Burdigalian interval. The restored thickness of this formation reaches up to 1500 m. The volumes of this formation are much bigger than those calculated on the basis of surface data. The analysis of the boreholes and seismic lines in eastern Sicily, in fact, revealed considerable thicknesses (some thousands of meters) completely buried below the outcropping units. These values result from tectonic repetitions within the stratigraphic succession. Actually, the formation consists of several superimposed structural levels corresponding to distinct stratigraphic-structural units that can be attributed to different complexes, and constitutes the Oligo-Miocene cover of various Mesozoic to Tertiary successions. Moreover this formation suffered large detachments, as well as a forward tectonic transport, giving rise to imbricates forming part of the frontal wedge of the AC. Due to its wide distribution, this formation was already described (in the Nebrodi Mts) in the geological literature of the late 19th century (Baldacci, 1886). Ogniben (1960) recognized the structural continuity between the north African "Numidien" and the Sicilian formation, proposing the name "Numidian Flysch". Caire and Mattaeur (1960) confirmed the lithologic identity of the Sicilian outcrops of the Numidian Flysch with the "Numidien" of Morocco, Algeria and Tunisia. The structural setting of the Numidian Flysch in the framework of the orogen has been the subject of intense controversy in the geological literature. This controversy has always concerned the provenance of the quartzose detritus that constitutes the formation, especially regarding the original source area of the sediments – whether these were derived from orogenic zones or cratonic areas in Africa (Ogniben, 1960, 1964; Dueé, 1969, 1970; Broquet, 1970; Caire, 1970; Wezel, 1973, 1974; Grasso et al., 1978; Giunta, 1985; Lentini and Tortorici, 1986; Bianchi et al., 1987).

On the basis of surface and subsurface analysis, Bianchi et al. (1987) proposed a revision of the units of the Numidian Flysch, observing that this formation consists of repeated slices detached from their original Mesozoic substratum, and distinguished the following tectonic units (going from base to top and, presumably, from the more external to the more internal): Gagliano, Serra del Bosco, Maragone, M. Salici, and Nicosia. A vast lower structural horizon of the Numidian units constitutes the *External Numidian Flysch* (Figure 5), formed by imbricated slices, originally the cover of the Imerese and Gagliano units, and more or less solidly with the Panormide carbonate platforms, of the Maragone unit. The overlying allochthonous horizon originally was the cover of the Tethydes, and belongs to the Nicosia and the M. Salici units, which form the nappes of the "far travelled" Numidian Flysch, filling the wide depression of the so-called Caltanissetta basin. In the crustal section of Figure 29, it is included in the Alpine Tethydes (AT), and indicated as NF. The Numidian Flysch grades up to marls with thin arenitic levels, characterized by a litharenitic or arkosic composition, that indicates the end of the quartzarenitic sedimentation. These marls are Langhian in age and will be discussed later.

2.2.2.2. *Panormide Platform Units.* The Panormide sequence crops out in the northern sector of the Palermo Mts and in the Madonie Mts, but they are detected eastwards connected in the subsurface to the S. Agata di Militello area (Maragone 1 well in Bianchi et al., 1987). The Panormide sequence starts with Late Triassic euxinic marls, interbedded with graded biocalcarenites (Mùfara Formation) overlain by reefal carbonates, ranging in age from Norian to Middle Cretaceous, with a Dogger–Malm hiatus. Upper Cretaceous-Eocene wackestones and red marls (Scaglia facies), Oligocene fine-grained marls, quartzarenites and calcarenites follow upsection. This sequence is stratigraphically overlain by the arenaceous-pelitic deposits of the Numidian Flysch (Late Oligocene–Early Miocene).

The Panormide Units tectonically overlie the Ionides, and in particular override the Imerese sequences. This is clearly shown in the Madonie Mountains and in the Palermo area, where this nappe has been recognized for a long time (Trevisan, 1960; Ogniben, 1960). After recent studies carried out on the Palermo Mountains, Catalano and Di Maggio (1996) proposed a new interpretation of the structural relationships between the Panormide and Imerese units. In their opinion the Imerese Units overthrust the Panormide. But the evidence for the original emplacement of the Panormide platforms onto the Imerese basinal sequences is supported by several field observations and represent a constraint for inverting the tectono-dynamic evolution of the orogen. A local inversion of the relationships between Imerese and Panormide units could have been caused by later tectonic contacts, such as the M. Cuccio back-thrust located south of Palermo town. In the Egadi Islands and adjacent offshore, there are extensive Triassic carbonate platform facies, evolving to basinal conditions during the Jurassic. This panormide succession outcrops only in westernmost Sicily and is made up of Triassic–early Liassic carbonate platforms, grading into Jurassic slope or pelagic deposits, to Cretaceous-Eocene cherty, turbiditic limestones, unconformably underlain by Oligocene-Miocene marls and calcarenites, and followed by Numidian quartzarenites. Analysis of the distribution of the platform facies suggests that in a regional context, with the closure of the Ionides westwards, there was probably a sealing of the Panormide or Pre-Panormide-type carbonate platforms with the external African-type ones.

In the Madonie Mts, two different stratigraphic sequences of the Panormide Unit can be distinguished. From the Middle Pliocene they underwent a NW–SE shortening, as indicated by the Early Pliocene Trubi Fm. (Abate et al., 1993), sandwiched between two tectonic subunits. Further investigations have shown a good correlation between the Madonie and the Palermo Mts. The Panormide sequences forming nappes on top of the Imerese units belong essentially to two subunits: the lower one is prevalently constituted by an incomplete dolomitic sequence that can be ascribed to the M. Quacella unit, outcropping in the M. Gibilmesi–M. Cuccio range in the Palermo area; the upper one, of Pizzo Carbonara, is equivalent to the M. Palmeto–M. Gallo subunits in the neighbourhood of Palermo. The *M. Quacella subunit* starts with the Triassic Mùfara Fm., a Carnian terrigenous-carbonate horizon, grading upwards into dolomites or dolomitic breccias with *Megalodon* (M. Quacella Fm., Figure 12). The Rhaetian top of the platform is truncated by an erosional surface and unconformably covered by the Eocene-Oligocene Scaglia Fm. (Grasso et al., 1978), underlying the Numidian Flysch. In the M. Quacella subunit, a widespread subaerial exposure occurred after the Rhaetian-Hettangian boundary. The Jurassic tectonics affected the hinge zones between platform and basin, causing the disintegration of the platform and producing huge aprons of resedimented dolomites in the adjacent Imerese basin.

The *Pizzo Carbonara subunit* displays a thicker sequence, consisting of dolomitic limestones and stromatolitic and lopheritic dolostones in areas of lagoons, and coral and sponge bioliths in cliff areas. The Dogger–Malm interval is represented by nodular calcilutites containing ammonites and gastropods, calcarenites with crinoids and lamellibranchia (Rosso Ammonitico), which are followed by bioliths of sponges and corals, *Ellipsactinia*, ammonoids, gastropods, echinoids and rudists, *Nerinea* sp. and algae, whose ages range from the Tithonian to the Cretaceous (Figure 12). From the Late Cretaceous onward, the sedimentation became pelagic and was characterized by Scaglia-type calcilutites with foraminifers, which frequently fill sedimentary dykes. The Oligocene follows upwards. In the Palermo Mts., this subunit corresponds to the M. Palmeto and Cozzo di Lupo units, constituted by inner carbonate platform sequences characterized by tidal flats and back-reef lagoons as well as margin facies.

Panormide Terrigenous Covers. The basal horizon of the terrigenous cover of the *Pizzo Carbonara subunit* is represented by the *Gratteri* Fm., unconformably overlying the Pizzo Carbonara carbonate sequence, and made up of an alternation of marls and calcarenites with macroforaminifers. At its base, the formation rests, with carbonatic breccias, upon various levels of the Panormide sequence. The age is Middle-Early Oligocene. Interpreted already by Ogniben (1960) as a Tertiary cover of the Panormide units and incorporated with these into a nappe resting on the Imerese domain, the formation passes upwards into the Numidian Flysch of Maragone (Figure 12).

To the Numidian Flysch of the *Maragone subunit* have been ascribed siliciclastics lying directly upon the carbonates of the Pizzo Carbonara unit or upon the Gratteri Fm., and forming the backbone of the Nebrodi Mts. They crop out as imbricates more or less rooted at the top of the Mesozoic carbonates or partially detached. For this link with the platform domain, the unit was formerly assigned to the Panormide Units. The Numidian succession of Maragone shows characteristic levels of brown claystones with nodular algal inclusions and coral fragments, which testify to contiguity with the carbonate series. Upwards, the Numidian sequence grades into Castelbuono Marls, late Burdigalian-Langhian in age. The siliciclastic cover of this subunit begins with the "Argille di Portella di Mandarini", and unconformably overlies the Mesozoic limestones with some intercalations of red marls of Scaglia facies, through a clayey level with megabreccias. The Late Oligocene sandy shales contain amygdaloid or lens-shaped intercalations of carbonate breccias. These breccias consist of reefoidal or shallow-water carbonates and were derived from the progressive dismantling of the Panormide platform; they originated from the combined effects of the synsedimentary tectonics and eustatic lowering. The Argille di Portella di Mandarini grade upwards into the Numidian Flysch (*Geraci subunit*), a thick alternation of silty clays and quartzarenites (Figure 12). On topsection there is a gradual passage towards an interval of brown silty clays with thicknesses varying from 500 to 1000 m (*Castelbuono Marls*). These deposits can be more or less detached from their own substrate.

2.2.2.3. *Alpine Tethydes.* The widely distributed allochthonous units coming from the Alpine Tethys ocean basin, and globally called *Alpine Tethydes* in the Sicily literature, are known with the local name of *Sicilide units* (Ogniben, 1960). They are characterized by Upper Jurassic–Oligocene basinal carbonates and sandy mudstones, evolving into Upper Oligocene–Lower Miocene terrigenous turbiditic successions, that are mainly detached from their substratum. Ogniben (1960) described the "Complesso Sicilide" as composed of two stratigraphic-structural units emplaced one upon the other: the Cesarò nappe and the underlying Troina nappe.

According to Lentini et al. (1987, 1990a, 1990b), the Sicilide units are composed of a substratum of variegated clays with distinct terrigenous cover, successively dismembered and forming several nappes grouped in the Nicosia and

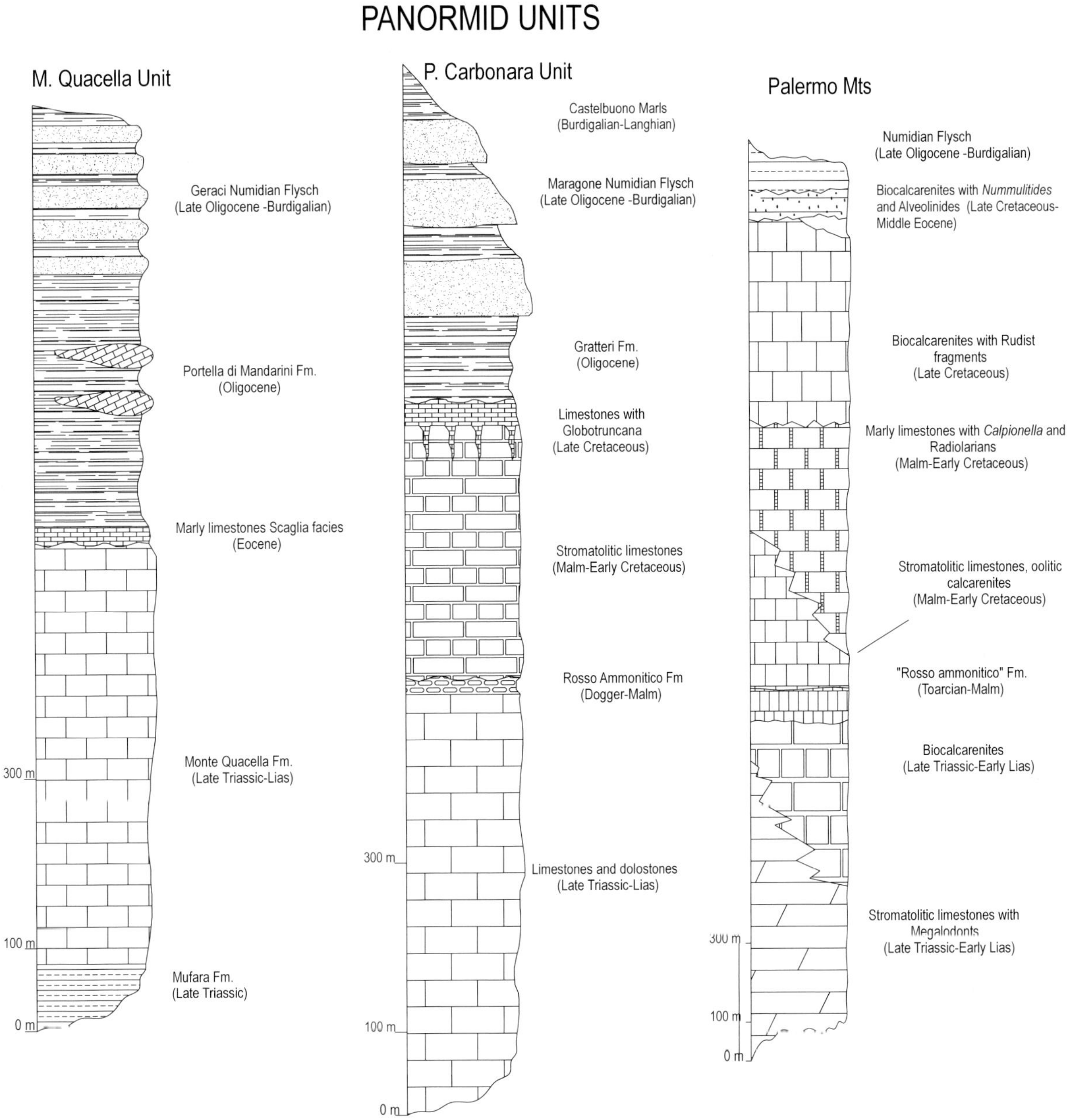

Figure 12: Stratigraphic logs of the Panormide units. The succession of the M. Quacella unit is characterized by a gap in the Jurassic-Cretaceous interval and probably suffered deep erosion, as demonstrated by the clasts resedimented in the Imerese sequence. This is why this unit is thought to have been originally located on the edge of the platform. The Mesozoic platforms are topped by siliciclastics grading upward into the Numidian Flysch.

the Troina units. The Nicosia unit is composed of the innermost Numidian Flysch. The Troina unit is characterized by an Upper Oligocene volcanoclastic sequence (*Tufiti di Tusa*) evolving upsection into a Lower Miocene flysch-type. Lentini et al. (1996a, 1990b, 2000) considered that the architecture of the Sicilides is that of a tectonic wedge, progressively thinning, until disappearing, towards the internal sector, below the Kabilo-Calabride chain. Inside this wedge at least four tectonic units have been distinguished: two made up of Mesozoic sequences, the M. Soro unit and the "Argille Scagliose superiori" unit, and two Paleogene to Early Miocene successions, the Nicosia and Troina subunits. The latter might represent the Cenozoic cover of the Argille Scagliose. The Sicilide units reach their maximum thickness in eastern Sicily, where they are preserved within a wide depression in the chain, now well documented in the seismogeological cross-section illustrated in Figures 28 and 29. In central-western Sicily, the Sicilides generally overthrust the Numidian Flysch slices and are unconformably covered by discontinuous Serravallian clayey deposits (Castellana Fm.) and by late Tortonian terrigenous sediments (Terravecchia Fm.).

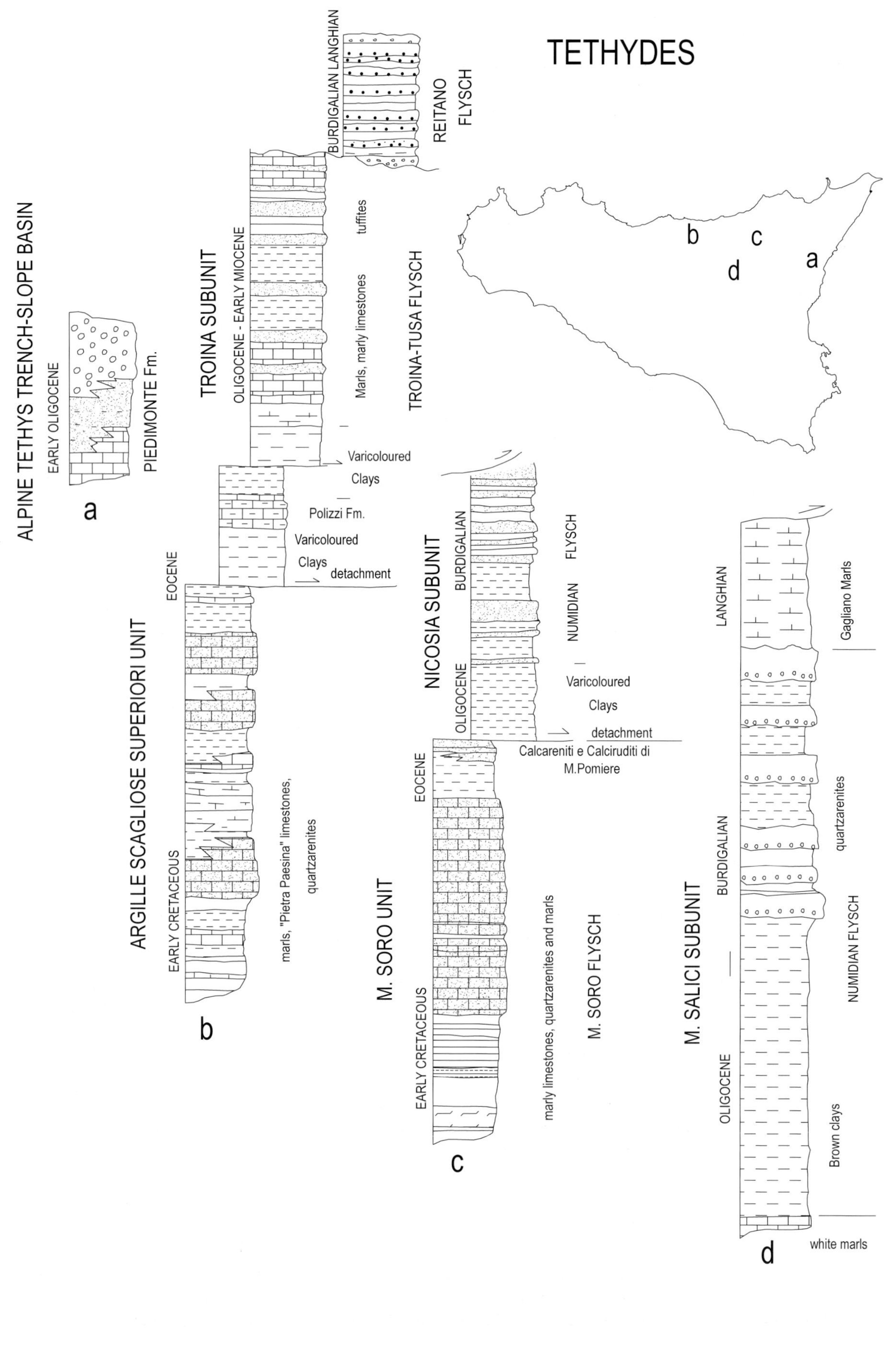
TETHYDES
ALPINE TETHYS TRENCH-SLOPE BASIN
EARLY OLIGOCENE
PIEDIMONTE Fm.
a
TROINA SUBUNIT
OLIGOCENE - EARLY MIOCENE
Marls, marly limestones
tuffites
TROINA-TUSA FLYSCH
BURDIGALIAN LANGHIAN
REITANO FLYSCH
Varicoloured Clays
Polizzi Fm.
Varicoloured Clays
detachment
ARGILLE SCAGLIOSE SUPERIORI UNIT
EOCENE
EARLY CRETACEOUS
marls, "Pietra Paesina" limestones, quartzarenites
b
NICOSIA SUBUNIT
OLIGOCENE
BURDIGALIAN
NUMIDIAN FLYSCH
Varicoloured Clays
detachment
Calcareniti e Calciruditi di M.Pomiere
M. SORO UNIT
EOCENE
EARLY CRETACEOUS
marly limestones, quartzarenites and marls
M. SORO FLYSCH
c
M. SALICI SUBUNIT
LANGHIAN
BURDIGALIAN
OLIGOCENE
Gagliano Marls
quartzarenites
NUMIDIAN FLYSCH
Brown clays
white marls
d
b
c
d
a

Figure 13: Stratigraphic logs of the Alpine Tethydes. The successions were sedimented in the Alpine Tethys realm and suffered repeated tectonic transport during the Balearic stage and thereafter, passively, during the Tyrrhenian stage. At present they form a tectonic wedge sandwiched between the Calabride nappes and the external Numidian Flysch, and have travelled far, reaching the extreme front of the thrust system, the Gela nappe. The geometric relationships within the wedge are the result of a progressive accretionary process accompanied by a large detachment of the Tertiary cover, followed by the breaching of the Cretaceous-Eocene interval. The sequences of the Nicosia and M. Salici subunits belong to the "far travelled" Numidian Flysch.

The *Argille Scagliose superiori unit* was the innermost deposit, because it actually lies everywhere upon the M. Soro Unit (Figure 13). It is represented by a tectonic mélange, within which there are exotic blocks, formed by clayey quartzarenitic alternations very similar to those attributed to the M. Soro unit, from which they can be distinguished only by geometrical position. They form a continuous horizon at the base of the overlying Calabride nappes. An Early Cretaceous age to Eocene age has been recognized.

The *Troina subunit* has been interpreted as the detached original cover of the previously described unit, presently forming the Sicilide accretionary wedge. The sequence starts from the bottom with varicoloured clays, silts and calcarenites, containing *Tubotomaculum* and microfauna with *Globigerinoides primordius* and *Globoquadrina praedehiscens* mixed with Eocene resedimented faunas. The nannofossils indicate an Oligocene age. Upwards, the varicoloured clays evolve to the Troina-Tusa Flysch. The *M. Soro unit* consists of a thrust system lying between the Numidian Flysch at the base and the Argille Scagliose Superiori unit; it overrides out-of-sequence the Nicosia and Troina units, the contact being exposed along the S. Fratello-Cesarò alignment. The *Nicosia subunit* is a Numidian-type sequence, with a basal interval made up of varicoloured clays and with micaceous sandstones interbedded within the quartzarenites.

On top of the Calabride units in northeastern Sicily, there are Sicilide nappes. The *Argille Scagliose Antisicilidi* rest in tectonic cover on the Capo d'Orlando Flysch and, locally, on the basement. Sometimes these are isolated occurrences, but their distribution over the whole of the Peloritani Mts. as well as Calabria suggests that they previously formed a continuous sheet before they were affected by erosion. The Argille Scagliose Antisicilidi have the aspect of polychromic clays with sliced texture, often with a chaotic habit, with intercalations of bituminous clayey schists alternating with variegated radiolarites, thin clayey-arenaceous strata, and enclosing yellowish quartzarenites.

2.2.3. *Calabride Chain.* The Calabride chain is formed from a thrust system composed of Hercynian basement units with Meso-Cenozoic cover coming from the European plate domain. Its deformation occurred in the Paleogene, and saturation in the late orogenic phase starting from the Oligocene. The chain constitutes a thrust edifice overriding the Apenninic units along a thrust surface, exposed along the S. Fratello-Alcantara alignment. This orogenic feature originated in the Late Eocene–Early Oligocene when the first Balearic rifting processes took place, and the SE margin of the European plate started to move southeastward, driven by NW-dipping subduction of the Alpine Tethys remnant. Thrusting and migration continued throughout the Balearic and Tyrrhenian stages. Stacked Calabride units are exposed in the Calabro-Peloritani chain. The present structural position shows that the uppermost units are those that contain basement with metamorphic rocks of the highest grade. The deepest tectonic units are composed of Hercynian epimetamorphites overlain by Meso-Cenozoic sedimentary cover (Capo S. Andrea, Longi-Taormina and S. Marco units). The *Capo S. Andrea Unit* is formed by epimetamorphites (m_1) and a predominantly carbonatic sedimentary succession, with reduced thicknesses and repeated gaps, extending from the early Lias to the Cretaceous-Eocene. The *Longi-Taormina unit*, in contrast, is characterized by epimetamorphites (m_2) and a continuous sedimentary succession with passages, in the Liassic levels, from carbonate platform deposits to pelagites (Figure 14). The *S. Marco d'Alunzio* unit is represented by a thick epimetamorphic (m_3) basement and an entirely carbonatic and condensed sedimentary succession rich in gaps.

The units described above are tectonically covered by moderately low- and high-grade metamorphics and, subordinately, by remaining portions of Mesozoic-Tertiary sedimentary successions. These rocks are represented by the *Mandanici unit*, which is formed from a polyphase Hercynian basement, developed at variable low-P and T from a greenschist facies over the chlorite zone to an incipient amphibolite facies. Its sedimentary cover is referred to the Rocca Novara unit (Lentini and Vezzani, 1975; Atzori et al., 1977), given by a sequence beginning with a platform facies of the Malm, then passing into Aptian pelagic facies and Scaglia

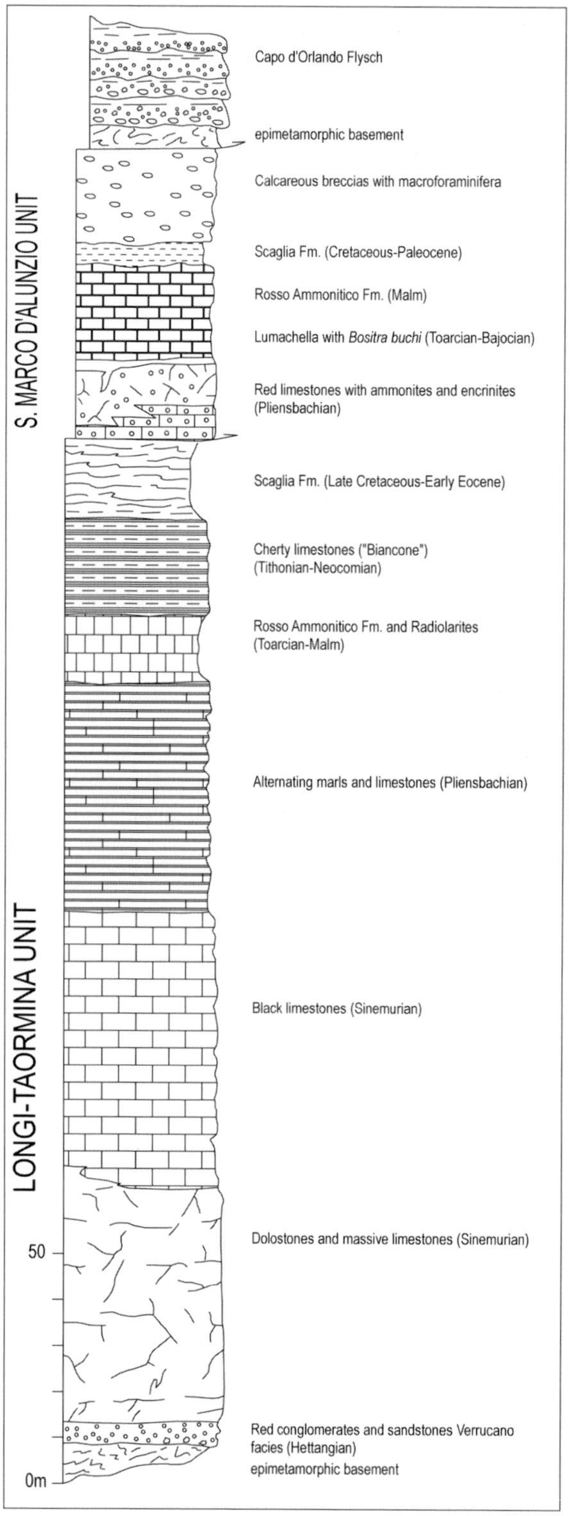

Figure 14: Stratigraphic sequence of the Longi-Taormina and S. Marco units, analyzed in the Longi area (Peloritani Mts.). These successions represent an example of original cover of the epimetamorphic nappes, and its relationships (after Lentini, 1975).

Fm. The proximal character of this sequence is confirmed by the extension of the carbonate platform facies into the Malm, which had been "drowned" elsewhere already during the middle Lias. The Rocca Novara unit, detached from the phyllitic basement (Mandanici unit), lies presently next to the boundary of the extreme front of the overriding phyllites. In the more internal portions of the edifice, at the top of the Mandanici unit, traces of an original sedimentary cover are preserved, which are characterized by a terrigenous Permo-Triassic interval and another, carbonatic one, ranging in age from early Lias to Late Cretaceous, and affected by a very low grade alpine metamorphism (*Unità di Alì* of Atzori, 1968; Truillet, 1968). This cover marks the tectonic contact between the Mandanici unit and the overlying *Aspromonte unit* (Ogniben, 1960; Lentini and Vezzani, 1975). This latter consists of a basement of Hercynian, H–T amphibolite facies metamorphites, with pre-Hercynian granulitic relicts and intrusions of late Hercynian plutons, and affected by a polyphase alpine overprint varying from medium-T and medium-P amphibolite facies to low-P and low-T greenschist facies; the Mesozoic sedimentary cover is completely absent.

2.2.4. *Tertiary Terrigenous Covers.* The Sicilian orogenic belt originated in Oligocene times and progressively migrated from the innermost domains toward the outer ones. The Miocene-Pliocene terrigenous cover thus well records the entire tectonic evolution of the studied area.

The following different sedimentary settings can be distinguished:

(1) Migrating forearc and intra-arc basins of the Balearic and Tyrrhenian stages.
(2) Migrating foredeep basins (present Gela-Catania foredeep; previous Caltanissetta foredeep now substituted by a thick wedge.
(3) Thrust-top basins.
(4) Pelagian foreland (Hyblean-Sciacca platform).

With the progression of deformation there is generally an evolution from foreland deposits to foredeep deposits. Figure 15 summarizes the evolution of the main Mio-Pliocene deposits, from the internal to the external areas. Two main orogenetic stages are distinguishable: a Balearic stage, from the Oligocene to the Burdigalian, and a Tyrrhenian stage starting in the Langhian.

2.2.4.1. *Hinterland Areas.* In the northeastern sector of Sicily, terrigenous cover outcrops, ranging in age from Late Eocene to Miocene. This are sometimes present at the top of the Meso-Cenozoic sequences and thus seems to precede the tectonic phases that led to the emplacement of the various crystalline units, and elsewhere it sutures the thrust contacts and is thus to be considered posterior to the construction of the Calabride edifice. The composition of this cover suggests a provenance of the detritus from the areas of the Corso-Sardinian block (Cherchi and Montadert, 1982; Rehault et al., 1984; Carmignani et al., 1986; Lentini et al., 1995a). The Oligo-Miocene terrigenous formations outcropping at the top of the Calabride units also extend constantly onto the more internal Apenninic units. The oldest deposits have been partially involved in late movements along the thrust of the Calabride chain onto the Apenninic units. The upper Burdigalian horizons mark the definitive sealing between the two structural units.

The *Frazzanò Flysch* (Ogniben, 1960) represents the interval of terrigenous sedimentation that follows the Eocene "Scaglia", and is significant in that it was deposited immediately preceding the emplacement of the Calabride crystalline units. However, the tectonics have frequently overwhelmed the original relationships and reduced the sequence into repeated slices pseudoparallel to the stratification. The sequence consists of an alternation of turbiditic sandstones and silty clays with minor conglomerates. The sandstones show a compositional evolution from basal feldspathic litharenites and lithic arkoses, very rich in carbonatic and epimetamorphic fragments derived from the dismantling of the original basement and its sedimentary cover, to feldspathic arkoses at the top, which contain moderately high-grade crystalline fragments (Carmisciano and Puglisi, 1979). The latter show marked analogies with the sandstones of the overlying Capo d'Orlando Flysch. The northernmost cycle is composed of the *Conglomerato Rosso Fm.*, deposited within the mountain range directly on the epi-metamorphic units.

Capo d'Orlando Flysch. This flysch can be interpreted as a *thrust top basin* deposit, posterior to the main emplacement of crystalline nappes, but involved in further tectonic phases (Figure 15). The Late Oligocene–Early Miocene Capo d'Orlando Flysch (Ogniben, 1960) consists of molasse-type sediments, evolving into flysch-type deposits. The early Chattian conglomeratic interval grades upward into an arenaceous-argillaceous facies, ascribed

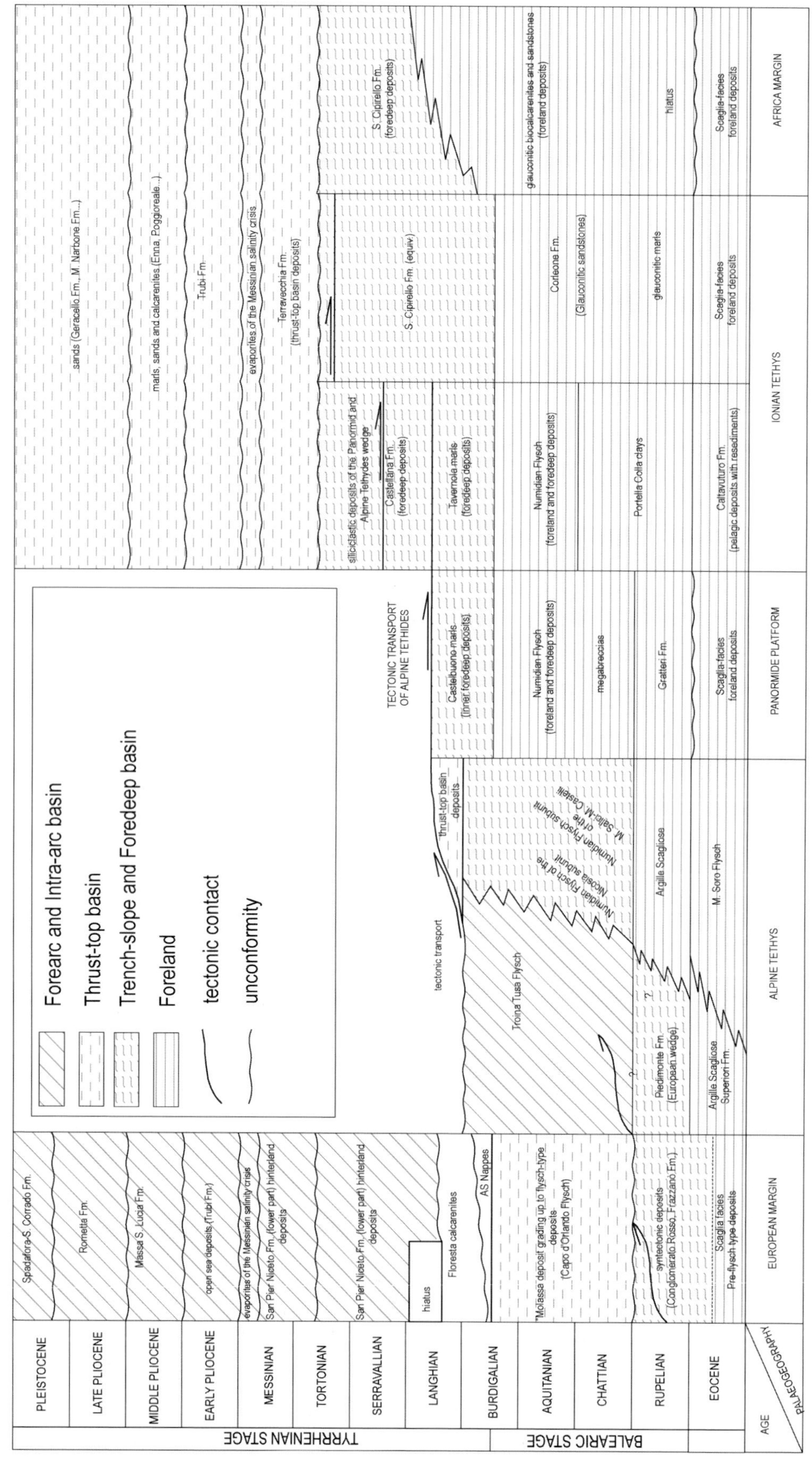

Figure 15: Stratigraphic scheme of the Tertiary terrigenous cover outcropping in Sicily. It shows the location in the structural setting and the geodynamic evolution of the sedimentation areas.

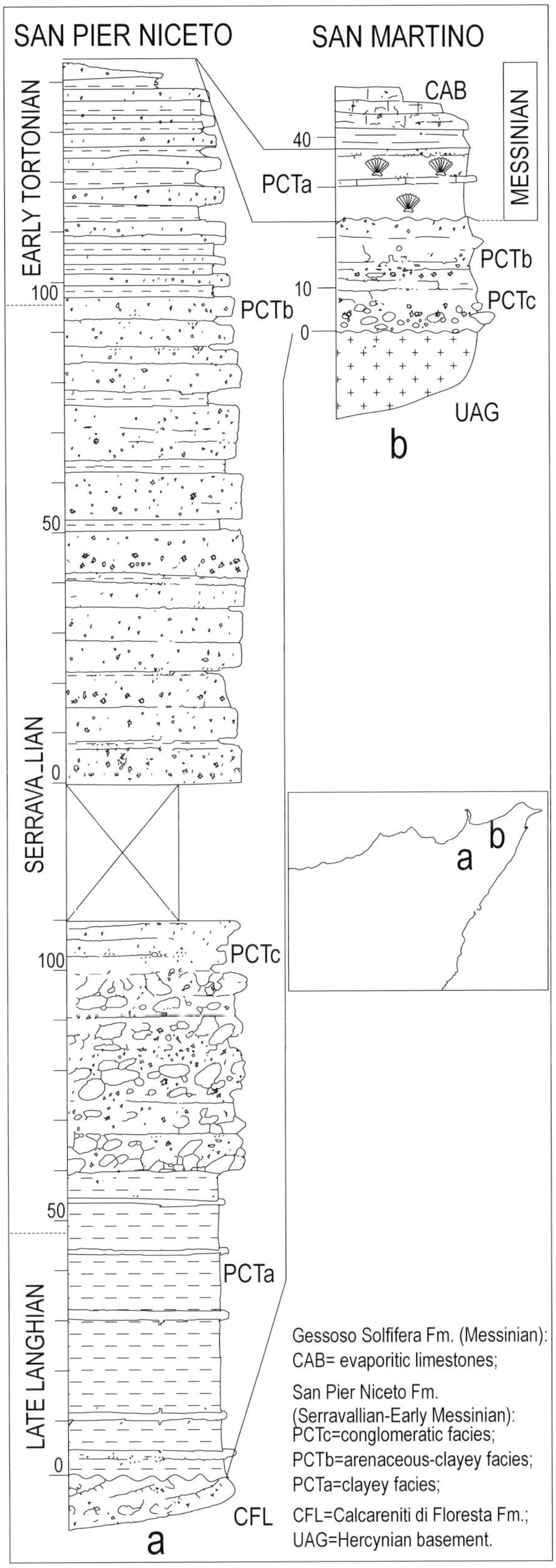

Figure 16: Miocene stratigraphic log of the forearc deposits exposed in the Tyrrhenian slope of M. Peloritani ridge (NE Sicily) (after Lentini et al., 1997).

to the Chattian–lower Burdigalian. This deposit, characterized by variable thickness and facies distribution, rests unconformably on the different tectonic units and post-dates the relative tectonic events.

Facies distribution and sedimentological character suggest that the Capo d'Orlando Flysch was deposited in perched basins floored by imbricated slices of the Calabride units and characterized by several depocentres, according to the distribution of structural depressions within the thrust and fold belt (Lentini et al., 1995a, 1995b; Catalano and Di Stefano, 1996; Catalano et al., 1996). Its present distribution is a result of the activation, from the Tortonian onward, of impressive systems of transcurrent dextral faults of the South Tyrrhenian system (Finetti et al., 1996), that have severely deformed the geometry of the Early Miocene thrust system. Bonardi et al. (1980) unified the Stilo Flysch (Bonardi et al., 1971), outcropping in Calabria, with the Capo d'Orlando Flysch and thus created the Stilo-Capo d'Orlando Fm., Aquitanian-Langhian in age. Lentini (2000) assumes a Late Oligocene–early Burdigalian age for this flysch.

Floresta Calcarenites. The *Argille Scagliose Antisicilidi* are overlain by late Burdigalian–Early Langhian deposits, consisting of mostly shallow-water carbonates and bioclastic calcarenites (*Floresta Calcarenites*; Ogniben, 1960; Carmisciano et al., 1981b) associated with marls, sandstones and conglomerates. They are scattered over the central and northern areas of the Peloritani Mountain belt, but they were originally widely distributed before undergoing deep erosion, as their distribution in some structural depressions suggests. The Floresta Calcarenites unconformably overlie the Antisicilide nappe and sometimes seal the tectonic contact between the nappe and the Capo d'Orlando Flysch. In the easternmost areas they directly rest upon the crystalline nappes.

Post-Burdigalian Migrating Forearc Basin. Along the Tyrrhenian side of the island, marls and marly clays with thin lenses of sandstones overlie, in a top-lapping arrangement, the south-dipping foresets of the Floresta Calcarenites or unconformably rest on a Langhian erosional surface. They are mostly pelagic fine-grained sediments and indicate a sudden change in the tectono-sedimentary regime, that lead to the demise of source areas that fed the clastic horizons of the Floresta Calcarenites, followed by an overall marine transgression over the previously emerged areas (Figures 15 and 16).

Along the north-facing slopes of the Nebrodi-Peloritani ridge, from S. Stefano di Camastra through the largest outcrop on the Milazzo Peninsula to Rocca Valdina and Rometta, there are a number of exposures of Late Miocene reefal carbonates (Grasso and Pedley, 1997). The reefs are part of a Late Miocene circum–Mediterranean reef development, which is well represented in the central Mediterranean region. The Pliocene and Pleistocene deposits are preserved in downfaulted areas, adjacent to the modern coastline. They were deposited within structural depressions and then successively modified by Recent tectonics. The early Pliocene sediments are composed of white marls (Trubi), deposited during rising sea-level, following the Messinian salinity event. The Plio-Pleistocene deposits consist of shallow-water calcarenites, sands and clays, and form distinct cycles, unconformably covering the substratum and filling depocentres formed during the Messinian (Figure 17).

2.2.4.2. *Alpine Tethys Terrigenous Cover.* The analysis of Tertiary cover of the Tethydes allows us to recognize the role of the terrigenous deposits and provides useful information on the basin evolution (Figure 15).

The *Piedimonte Formation* consists of completely detached sequences forming several tectonic slices located upon the Apenninic chain in front of the leading edge of the Calabride chain. It consists of conglomerates passing laterally into flysch-type facies (Carmisciano et al., 1981a, 1981b; Lentini et al., 2000). The distribution of the lithofacies was interpreted as a vertical sequence from marly clays at the bottom evolving into sandstones and progressively into conglomerates (Carmisciano et al., 1981b, 1981b). New field data and microfauna analysis showed that the different facies are coeval, Lower Oligocene in age, and that the conglomerates represent the proximal facies, and the marly clays the distal facies (Lentini et al., 2000). The present geometry is that of an accretionary wedge, located between the Calabride chain and the Apenninic chain. It allows us to date the tectonic events, the rotation of the Sardo-Corsica block and the geodynamic evolution of these structural domains.

Figure 17: Stratigraphic logs of the Late Miocene to Quaternary deposits of NE Sicily (after Di Stefano and Lentini, 1995).

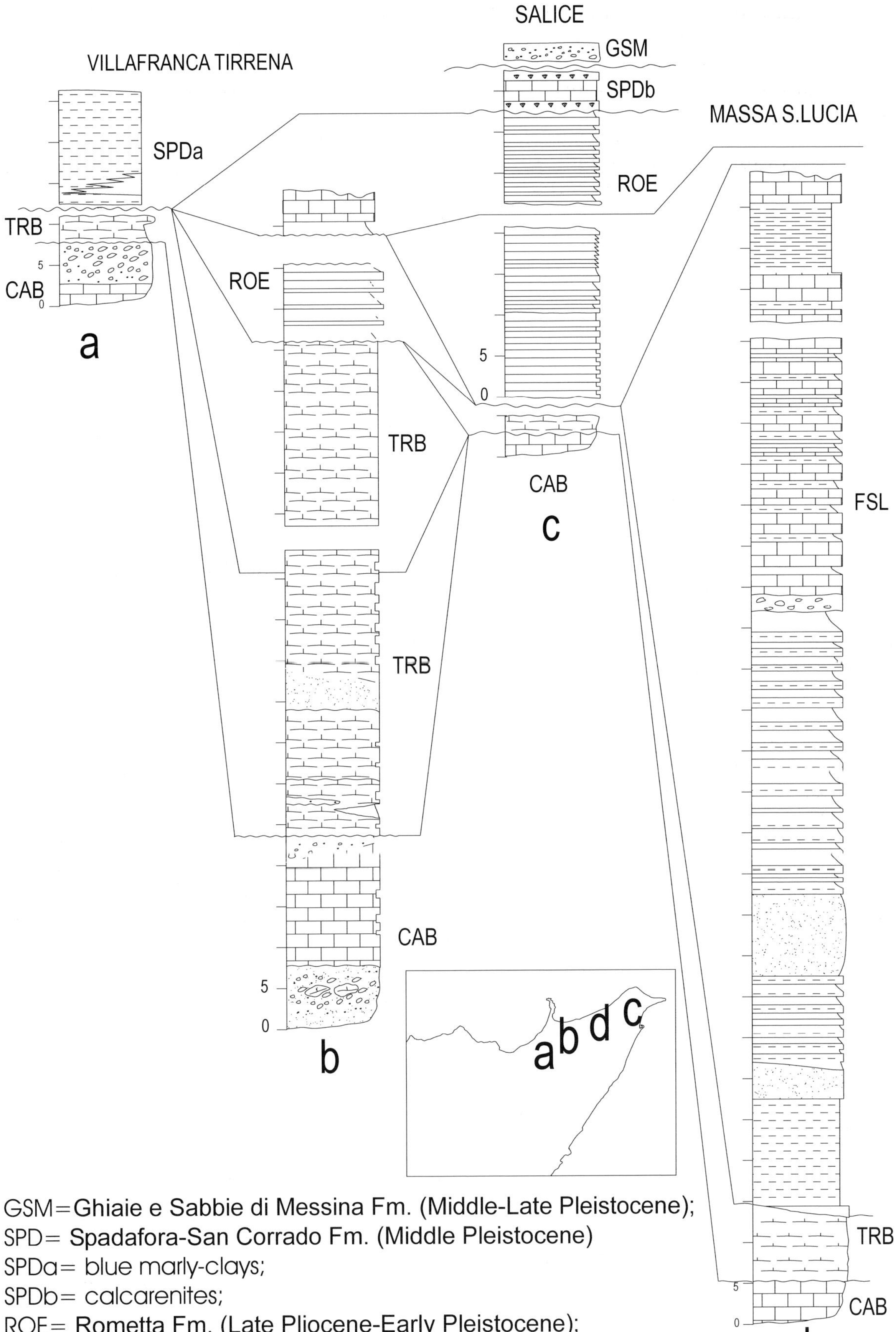

GSM= Ghiaie e Sabbie di Messina Fm. (Middle-Late Pleistocene);
SPD= Spadafora-San Corrado Fm. (Middle Pleistocene)
SPDa= blue marly-clays;
SPDb= calcarenites;
ROE= Rometta Fm. (Late Pliocene-Early Pleistocene);
FSL= Massa S. Lucia Fm. (Middle Pliocene);
TRB= Trubi Fm. (Early Pliocene);
CAB= evaporitic limestones (Messinian)

Troina-Tusa Flysch. This flysch has been described above as the original cover of the "Argille Scagliose Superiori", later detached and underthrusting the Mesozoic substratum. The composition of the sandstones, which are rich in tuffites, induced Ogniben (1964) to postulate a volcanic arc as their source. It is difficult to define the original palaeogeographic location of this flysch. It seems to be a subduction-related deposit, linked to the late Balearic stage (Figure 15).

"Far Travelled" Numidian Flysch. This term is intended to indicate the group of Numidian sequences that migrated from a distant original deposition. They were originally sedimented in the outermost part of the Alpine Tethys, more or less in continuity with the Alpine Tethydes, and were reached by the deformation only during the post-Burdigalian compressive phases. The lateral transition into the Troina-Tusa Flysch is supported by the composition of the arenites of the top-levels (Grasso et al., 1987). The *Nicosia Subunit* has been described above as the original cover of the M. Soro Flysch. The horizons, probably marking a transition, are constituted by blackish clays containing Paleocene nannofossils and by the calcarenites and calcirudites of M. Pomiere. In a structurally deeper position, the *M. Salici Subunit* is mainly represented by the outermost Numidian sequence of the Alpine Tethys accretionary wedge. It includes part of the successions that were formerly believed to be rooted at depth, since they form fairly regular monoclines characterized by Lower Miocene quartzarenitic banks alternating with brown clays that rest upon a substratum of grayish and sometimes varicoloured clays of Late Oligocene age. To this subunit has also been assigned an apical interval of marly clays (*Marne di Gagliano*) with a thickness of about 200 m (Figure 15). The overall age of the Numidian succession extends from Late Oligocene to Burdigalian. The Gagliano Marls are Langhian in age (Amore, 1969; Fornaciari et al., 1996) and play the role of a syntectonic deposit.

Reitano Flysch. This flysch is composed of polygenic conglomerates with rounded metamorphic, plutonic or calcareous clasts, set in a micaceous arenaceous calcareous or micaceous quartzose feldspathic matrix (Puglisi, 1979, 1987, 1992; La Manna et al., 1995) and by coarse-grained, graded and cemented micaceous sandstones, alternating with marly clays. This formation is exposed in Northern Sicily from Cefalù to Pettineo, Reitano, and Caronìa, in the area of Capizzi, and occurs occasionally at Cerami and Troina.

The sandstones vary in composition from volcanic sandstones to arkose, which testifies to a setting in a basin formed on already deformed Alpine Tethydes units, and delimited toward the interior by the Calabride crystalline units. The age ranges from late Burdigalian to Langhian.

This formation postdates the overriding of the Alpine Tethydes onto the Numidian Flysch of the Nicosia Subunit, while along with these it seems to have participated in the transport onto the Langhian horizons of the more external sectors.

2.2.4.3. *Neogene-Quaternary Foredeep and Thrust Top Basin Deposits.* As the deformation proceeded, the sedimentation areas were involved in the orogenesis and tectonically placed upon the Messinian to Early Pliocene deposits of the Hyblean-Sciacca foreland. The external front of the allochthonous units constitutes the Gela wedge, whose most advanced portion is in part sutured by the Quaternary foredeep deposits (Gela-Catania).

During the Balearic Stage the Panormide platform played the role of foreland with the complete closure of the Alpine Tethys, after which its detached blocks overrode the Ionian domain starting in the late Langhian (beginning of the Tyrrhenian Stage). In this geodynamic context, the aforementioned Numidian covers occupied a wide area of the foreland, which evolved into an extensive foredeep (Figure 15). During the Oligocene, the Gratteri Fm. and the Argille di Portella di Mandarini Fm., whose megabreccias can be interpreted as ramp deposits originating from the carbonatic platforms, had been deposited on the Panormide platform before but not during the tectonic transport of the Panormide units. Upward in the sequence, the Aquitanian-Burdigalian interval with its thick quartzarenitic banks represents the characteristic lithotype of the Numidian Flysch, as previously described. This sequence is terminated by a prevalently marly interval, which marks the end of the exclusively quartzose sedimentation. The formations are of late Burdigalian to Langhian age and are named Castelbuono Marls where resting on the Panormide Units, while they are called Tavernola Fm. where they rest upon the Imerese units (Ogniben, 1960). These horizons were deposited on wide foredeeps, where they were subsequently covered tectonically by the Alpine Tethydes.

Towards the external areas, the Numidian cover is coeval with the glauconitic sequences of the Sicanian units and of M. Judica, which thus represent epicontinental deposits affected by the tectogenesis only from the upper

Tortonian onward. They consist of sandy clays with a calcarenitic intercalation at the Oligocene-Miocene passage (Corleone Calcarenites) and are overlain by the S. Cipirello Marls. These latter underthrust an allochthonous melange composed of the Lercara unit, the Alpine Tethydes and the "far travelled" Numidian Flysch, overlain unconformably by the Terravecchia Fm. On the outermost areas the sedimentation was more or less continuous since the Langhian (e.g., the Licata Fm., in southeast Sicily), testifying to a late involvement of these areas. The carbonatic sequences of the Pelagian-Sicilian thrust belt outcropping at M. Kumeta, at Rocca Busambra and near Alcamo, as well as in the Trapani mountains and S. Vito lo Capo, are considered external. Mesozoic deposits are unconformably overlain by Lower Miocene glauconitic biocalcarenites that pass upward into marls, named *S. Cipirello Marls*, indistinguishable from those mentioned above for the Sicanian Units. The Terra Vecchia Fm. is a transgressive sequence that rests on a substratum made up of the allochthonous deposits of the Alpine Tethydes, and passes upward into the Messinian evaporites and the Early Pliocene chalks of the Trubi Fm. The entire sedimentary succession is strongly deformed with short amplitude folds, and represents one unit of the allochthonous wedge of Messinian–Early Pliocene age.

2.3. Crustal Tectonic setting of Central Sicily

The crustal structural style of central Sicily, reconstructed by new interpretation of seismic data from the literature and CROP data interpolation, can be well observed in the seismogeological sections of Figures 28 and 29. Field studies have revealed the structural character of the Nicosia syncline, formed from the Lower Pliocene. Its axis, oriented east-west, shows a clockwise rotation, connected to the NW–SE-oriented dextral fault system. This system is known as the South Tyrrhenian Thrust-Transcurrent System (Finetti et al., 1996), and it has been clearly recognized on-land, first in NE Sicily and then, by extensive specific field work, over the whole island.

It is now possible to distinguish two vast, structurally depressed areas, for which in the literature the term "basins" has been inappropriately used: the "Castelvetrano Basin" and the "Caltanissetta Basin". *They are in effect inner foredeeps, filled by the thrust-wedge of the Sicilian orogen.* The Castelvetrano Foredeep is confined to the north by the ridges belonging to the Pelagian-Sicilian thrust belt (the San Vito lo Capo Mountains, M. Inici and M. Bonifato) and by the Palermo Mountains, and to the east by the Monti Sicani. This sector is characterized by a deep-seated carbonate thrust system within which the Mesozoic successions are unconformably overlain by glauconitic biocalcarenites (equivalent to the *Calcareniti di Corleone*) which pass upward into the *Marne di S. Cipirello*). The S. Cipirello Marls are tectonically overlain by the allochthonous Numidian Flysch, which in turn is overlain unconformably by the Terravecchia Fm. Toward the south, the allochthonous units pinch out and the Mio-Pliocene sedimentation assumes an external domain character, although repeated stratigraphic gaps are present.

The Caltanissetta foredeep is bounded to the north by the Trabia-Termini Imerese Mountains, and by the Madonie–Nebrodi Ridges; to the west by the Sicanian Mountains, from which it extends eastward to cover the entire south-central sector of the island, across the Enna-Caltanissetta area to that of Centuripe–M. Judica, and southward to the Agrigento-Licata area, including the frontal Gela wedge of the Sicilian Apennines. The Serravallian–early Tortonian siliciclastics cropping out in the northern sector (Castellana Fm.) form a tectonic wedge with the Alpine Tethydes overriding the Langhian Tavernola Fm. at the Numidian top section. They are unconformably overlain by the clastic deposits of the Terravecchia Fm. In central Sicily, two major Messinian evaporitic cycles have been recognized, separated by a regional angular unconformity. The lower cycle starts with pre-evaporitic sediments (Tripoli Fm.) and carbonates deposited on pre-existing highs, while gypsum and chlorides were formed within lows. The lower evaporitic sequence is folded and truncated by an intra-Messinian unconformity, and underlies deposits characterized by resediments originating from the lower units interbedded by selenitic gypsum. The Trubi Fm. was deformed together with the evaporites during the Early Pliocene tectonic phase. A second Pliocene cycle is represented by marls and blue clays grading upward into sands and calcarenites ("*Marne di Enna*" and "*Calcareniti di Capodarso*"). A third regressive sequence, unconformably lying upon different horizons, has been ascribed to a Late Pliocene–Early Pleistocene age ("Geracello Cycle"). Borehole and seismic data provide evidence that the allochthonous front of the Apenninic chain (the "Gela wedge") rests tectonically upon the Pliocene deposits, and in the outermost zone, its tectonic propagation seems to stop almost completely in the Upper Quaternary. Furthermore, the Early Pleistocene horizons, sometimes slightly deformed, lap upon the frontal thrusts (Di Geronimo et al., 1978). Thus, since the Upper Pliocene the allochthonous front has tended to slow its advance, coming to a complete halt in the Lower Pleistocene (Figure 18).

The Gela nappe winds southwestward in the direction of the Gela offshore, from which it extends to the Adventure Shelf (Argnani et al., 1987). The Gela foredeep breaks off in western Sicily where the present foreland-chain boundary consists of a complicated Plio-Pleistocene transcurrent belt (Argnani et al., 1987).

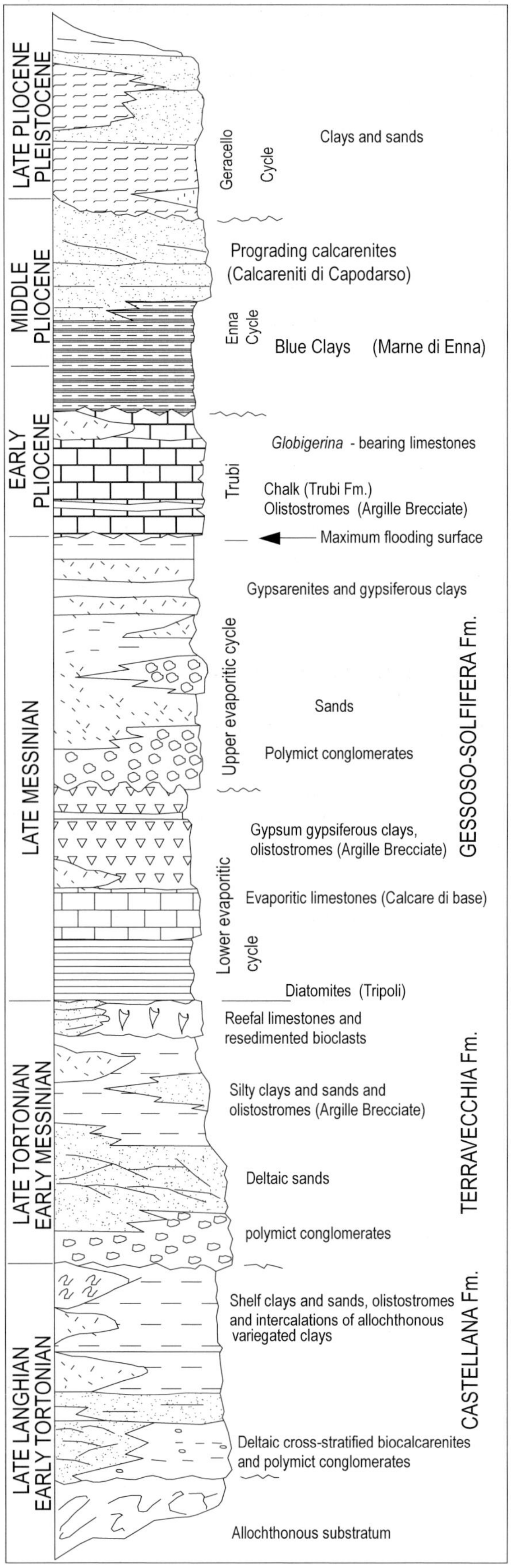

Figure 18: Stratigraphic column of the Middle Miocene to Pleistocene deposits of Central-Southern Sicily.

3. SEISMIC INTERPRETATION OF CROP SECTIONS AND SELECTED OTHER LINES

The interpretation of the CROP seismic sections described in this work is not an independent local study of the investigated area of Sicily and surrounding seas, but a coordinated part of an organic project of deep seismic exploration of the Mediterranean region and Italian mainland. The area of study contains several items of great scientific importance for best understanding the complex geodynamics involved. In particular, interpretation of the origin of the allochthonous units, their space–time tectono-dynamics and the related driving forces have been discussed in the literature for a long time.

Now, the CROP project furnishes for the first time deep seismic exploration data that allow us to study these complex lithospheric topics with support from real data from direct subsurface imagings. Using the whole CROP section network, it is now possible to reconstruct with more severe constraints a pre-orogenic paleogeographic setting of higher reliability than previously. To describe in better order the many topics clarified by the interpretation of the deep CROP seismic sections and selected lines shown in Figure 1, it is convenient to give a general outline of the original paleogeographic framework in the Lower Cretaceous, before the Eo-Alpine geodynamic stage. *This paleogeography played a major role in the following evolution. In summary, the central-western Mediterranean region was composed of two fully communicating deep basins, floored by oceanic-type crust, interposed between continental plates: the Ionian Neotethys, open in Permo-Triassic times, separating Africa from Adria, and the Alpine Neotethys, open in Middle Jurassic–Lower Cretaceous times, separating Africa–Adria from Europe* (Finetti, this volume, Chapter 34; Plate 2a). The area explored here includes several contentious questions about the tectono-dynamics connected to the Eo-Alpine, Balearic and Tyrrhenian stages, and related effects on the European and African margins (Plates 2, 4, 6, 9 and 11; Figures 19–23, 26–28 and 29).

3.1. CROP Section M23B/M23A2/M23A1 (Part) in the Sicily Channel–NW Pelagian Sea

This long (437 km) interpreted CROP section (Plates 1 and 2) explores for the first time the crustal-upper mantle conditions of the complex Sardinian Channel–NW Pelagian geological province, where geological features of major importance for the understanding of the complex geodynamic events in the region are stacked. Describing our interpretation of this and all other Tyrrhenian CROP sections of this chapter (M29A, M28A, M6A and M26) we will proceed from the internal sector outward.

Table 1
CROP and other integrated seismic sections exhibited and commented on in this work (Chapter 15)

<table>
<tr><th rowspan="3">CROP section</th><th rowspan="3">SP</th><th rowspan="3" colspan="2">Length (km)</th><th colspan="4">Chapter 15</th></tr>
<tr><th colspan="2">Figure N°</th><th colspan="2">Plate N°</th></tr>
<tr><th>Seismic</th><th>Int.</th><th>Seismic</th><th>Int.</th></tr>
<tr><td>M23B
M23A2
M23A1(P)</td><td>100–5413
100–3340
9400–9640</td><td colspan="2">437</td><td>–</td><td>19</td><td>1</td><td>2</td></tr>
<tr><td>M29A</td><td>100–3953</td><td colspan="2">195</td><td>–</td><td>20</td><td>3</td><td>4</td></tr>
<tr><td>M28A</td><td>100–1695</td><td colspan="2">85</td><td>–</td><td>21</td><td>5</td><td>6</td></tr>
<tr><td>M6A</td><td>100–2785</td><td>139</td><td rowspan="2">460*</td><td rowspan="2">–</td><td rowspan="2">22</td><td>7</td><td rowspan="2">9</td></tr>
<tr><td>M24</td><td>100–4530</td><td>225</td><td>8</td></tr>
<tr><td>M26</td><td>100–1360</td><td colspan="2">128</td><td>–</td><td>23</td><td>10</td><td>11</td></tr>
<tr><td>M21</td><td>100–2195</td><td colspan="2">107</td><td>24</td><td>25</td><td>–</td><td>–</td></tr>
<tr><td colspan="8">Integrated seismic sections**</td></tr>
<tr><td>Western
Sicily</td><td>–</td><td colspan="2">32
54*</td><td>26
27</td><td>–</td><td>–</td><td></td></tr>
<tr><td>Eastern
Sicily</td><td>–</td><td colspan="2">104
149*</td><td>28
29</td><td>–</td><td>–</td><td></td></tr>
</table>

*Including interpolated-extrapolated data on Sicily and the Tyrrhenian margin.

**Courtesy Società Geologica Italiana.

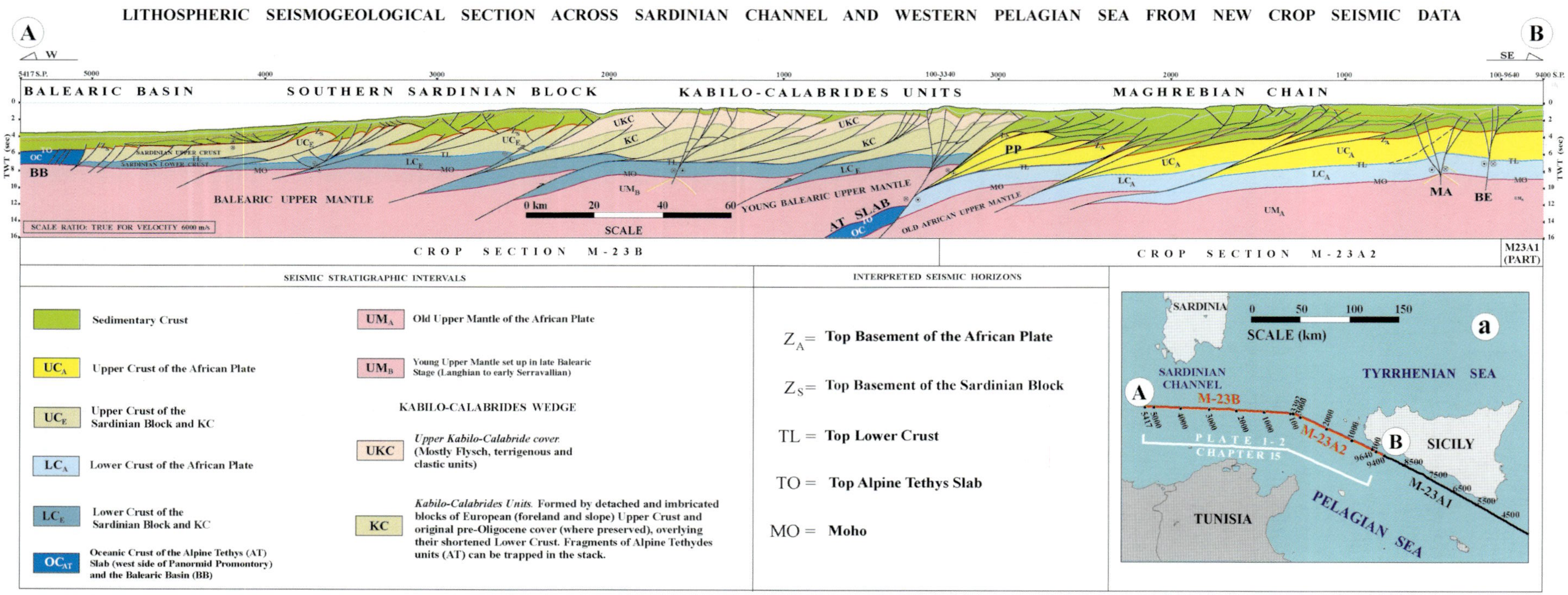

Figure 19: Reduced and simplified version of the seismogeological section of Plate 2 (CROP M23B/M23A2/M23A1 Part) in the Sardinian Channel and western Pelagian Sea, across the SE edge of the Balearic basin, the southern extremity of the rotated Sardinian block, the Kabilo-Calabride wedge and the Maghrebian chain. The section ends on the African plate foreland zone, where two positive flower structures are present ("MA" = Mazara flower, and "BE" = Belice flower) generated by NNE–SSW left-transcurrents, antithetic to the main NW–SE right-transcurrent rift system of the Pelagian Sea. The Panormid promontory units (PP) derive from tectonization of the NW-side of the Panormid promontory in the late Balearic stage.

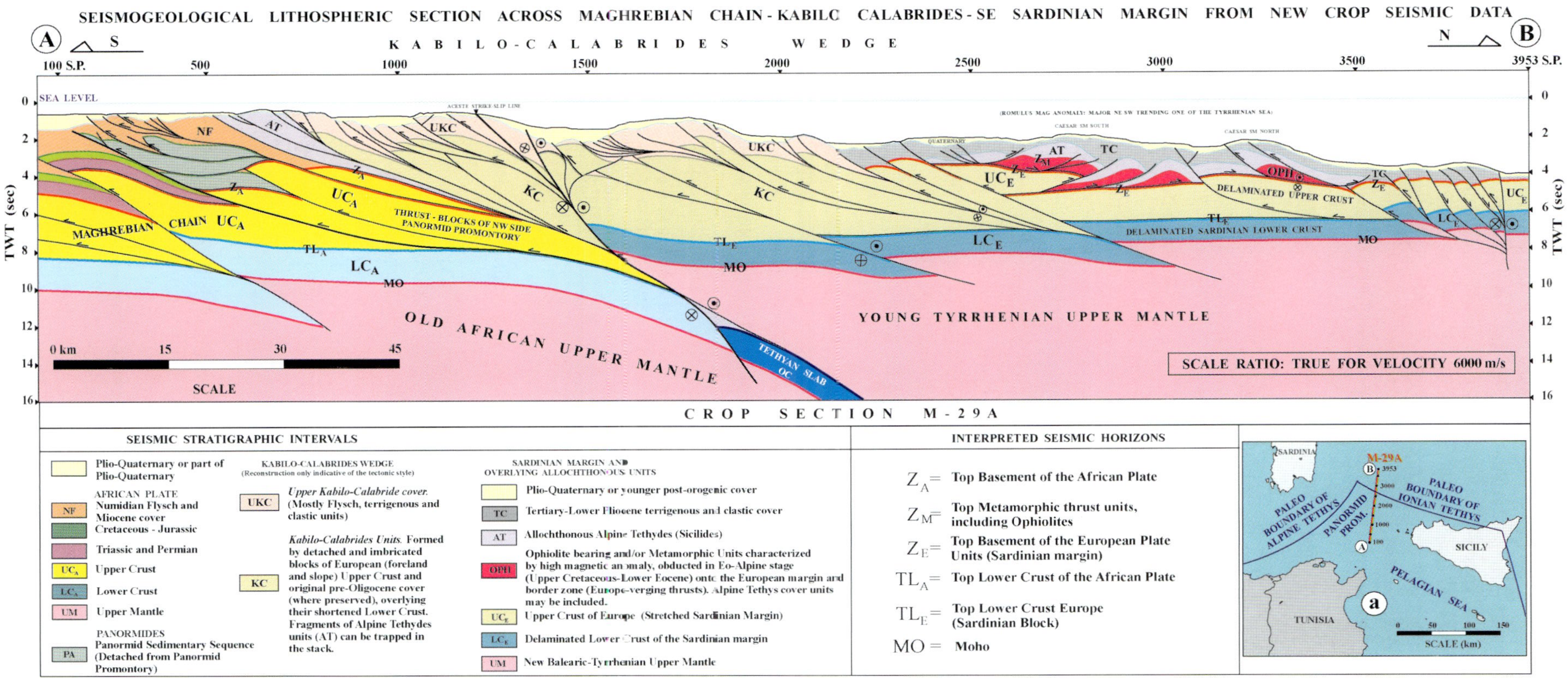

Figure 20: Reduced version of the seismogeological section of Plate 4 (CROP M29A) across the complex area of the SW Tyrrhenian Sea–Kabilo-Calabride wedge–Panormid thrust units–Maghrebian chain. Among the various important crustal elements exhibited, the section shows newly discovered ophiolite units of major importance for the reconstruction of the geodynamics, successive to their Eo-Alpine obduction on the European margin.

At its left extremity, the section CROP M23B shows the SE edge of the Balearic basin with oceanic crust and its Tertiary-Quaternary cover, delimited to the east by the rifted-thinned (delaminated) margin of the rotated Sardinian block. The Sardinian block can be identified with good reliability (Plate 2) from about 18 to 150–160 km (progressive section distance from left) where east-verging thrust blocks preserve their upper crustal intervals substantially rooted to the lower crust. From the Sardinian block to the Elimi flower structure (from about 120 to 280 km) there is a wedge of imbricated units formed by upper crustal blocks of eastern Sardinia, delaminated during the Balearic stage (Oligocene–Lower Miocene) and stacked by compression in the late Balearic–early Tyrrhenian stage when the Corso-Sardinian block rotation gave rise to collision of the SE-Sardinian margin with the NW side of the Panormid promontory ("PP" in Plate 2c). This wedge constitutes the Kabilo-Calabrides "KC" units, tectonically transported and stacked on Kabilian and Calabrian orogenic systems. The upper crustal KC units and any preserved pre-Oligocene cover are overlain by a generically defined "Upper Kabilo-Calabride Cover" (UKC) formed by mostly flysch of Oligo-Miocene age. Strong reverberations within the UKC interval produce a severe deterioration in the underlying stratigraphic intervals which are generally completely masked. The interpreted horizon HKC is thus a tentative interpretation using the interval velocity discrimination > 4500 m/s (KC) and < 4500 m/s for UKC. Details of the interpretation in Plate 2 are sufficiently self-explanatory not to require comments.

The CROP discovery of the Alpine Tethys slab (Plate 2, km 230–255) is very important and explains and delimits the convergency zone between the European crustal block of Sardinia and the African Panormid promontory, the closure of the Alpine Tethys and the compressive deformation and cover detachment of the Panormides. Compressive deformation also affects the NW margin of the Africa plate (km 290 to about 380) forming a NE–SW-directed thrust belt which is the easternmost continuation of the Maghrebian chain (Plate 2). Two positive flower structures are present on right extremity of the section: the "MA" Mazara and "BE" Belice flowers. They were generated by NNE–SSW left-transcurrent faults acting as antitethic fractures of the main WNW–ESE right-transcurrent system occurring prominently in the Pelagian Sea (Finetti, 1984; Finetti and Del Ben, this volume, Chapter 26).

3.2. CROP Section M29A

This NS-directed deep seismic section (Plates 3 and 4, Figure 20) crosses a complex tectonic area in the central Mediterranean not yet sufficiently explored, whose crust still remains poorly defined. The interpretation in Plate 4 shows some key new subsurface imagings that contribute to improving the understanding of the tectono-dynamics in the area. First of all, the seismic imaging of the upper Tethyan slab in a zone which represents the passage from the Alpine Tethys to the Ionian Tethys, is to be noted. In Plate 4a and Figure 20(a), the paleo-boundary of two ocean basins (Alpine and Ionian Tethys) with respect to their northern Africa slope, as clearly derived from various CROP Sections (M23B, M29A, M28A, M6A, M26, M21, M3, M23A1 and several other MS lines) is illustrated. The CROP seismic network integrated with other MS seismic lines allow us to reconstruct the boundary of the Panormid promontory (Plate 4a) which delimits the Alpine Tethys from the Ionian to the south.

This sections, like all other CROP sections in the Southern Tyrrhenian (M28A, M6A and M26), exhibits oblique compressive collisional crustal tectonics between SE-migrating crustal blocks detached in the Tyrrhenian stage (Ionian slab subduction) from the Sardinian Margin. In the Balearic stage (Oligocene–Early Miocene) a large continuous arc including the Northern Apennines–Southern Apennines–Calabrian arc–Sicily–Maghrebian chain was formed. At the beginning of the Tyrrhenian stage (Burdigalian–Serravallian), this arc was dissected to the south by the Panormid promontory, and while the stacked units of the segment facing the NW side of the Promontory remained trapped against this obstacle, to the north of this segment (Plate 2b) the arc, driven by the Tyrrhenian mantle flux connected to the Ionian slab retreat, started to stretch, delaminate and transport the Kabilo-Calabride crustal fragments southeastward, opening progressively the back-arc Tyrrhenian basin. Since the Panormid promontory

---

Figure 21: Reduced and simplified version of the seismogeological section of Plate 6 (CROP M28A with small extension on land). It shows the Calabrides units, firstly delaminated (late Balearic–early Tyrrhenian stage) and successively stacked by collision of their crustal fragment with the margin of the Panormid promontory. To the top of the tectonized European (Sardinian) upper crust units are exhibited the Europe-verging thrust units of the "Drepano high", associated with a prominent magnetic anomaly. Tectonic style and Mag anomaly support the interpretation that these features represent ophiolite and ophiolite-bearing units. The Panormid thrust block and deformed Africa margin are shown on the right extremity of the figure.

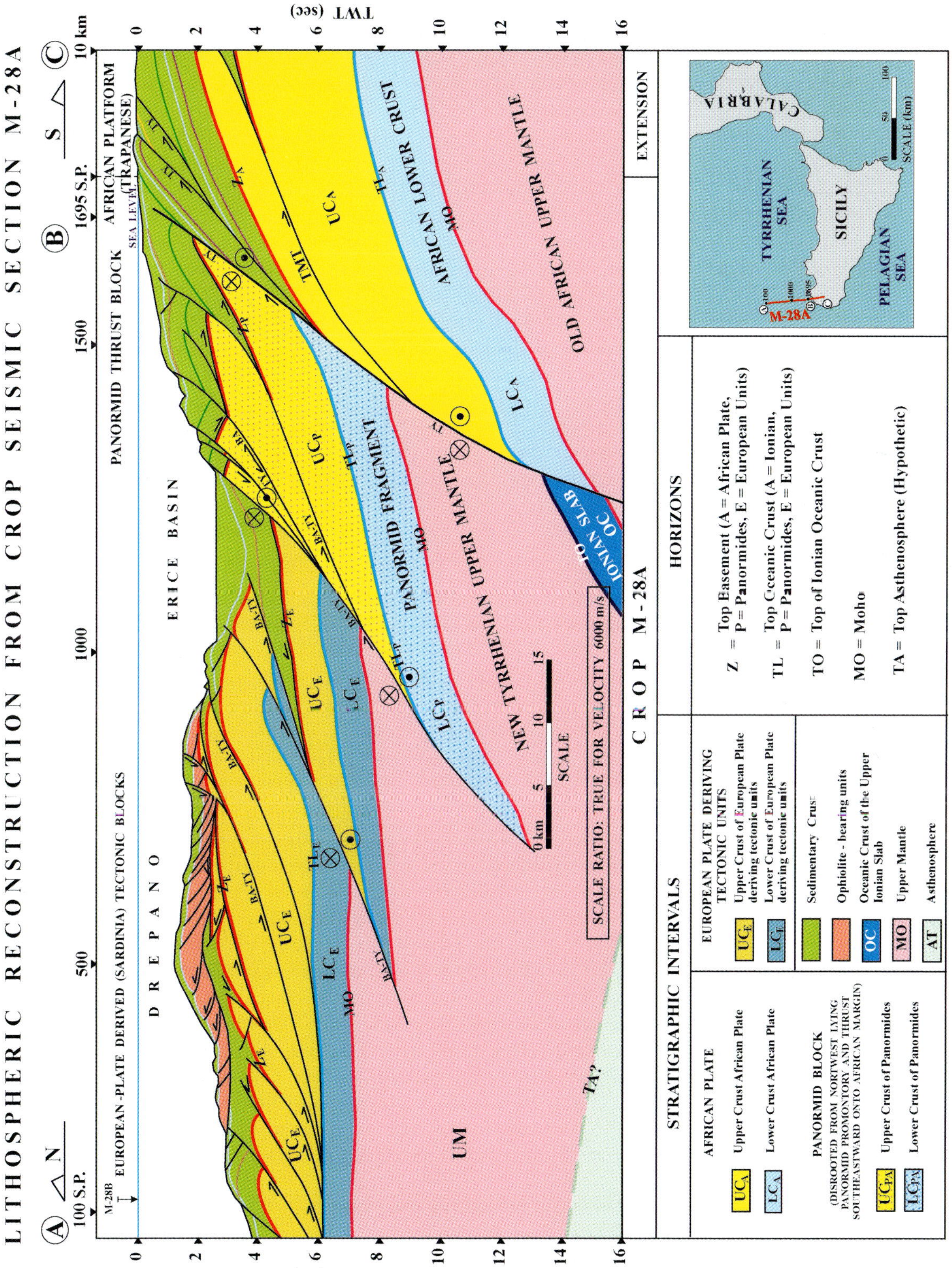
LITHOSPHERIC RECONSTRUCTION FROM CROP SEISMIC SECTION M-28A
A N
100 S.P.
B
1695 S.P.
S C
500
1000
1500
10 km
EUROPEAN-PLATE DERIVED (SARDINIA) TECTONIC BLOCKS
PANORMID THRUST BLOCK
AFRICAN PLATFORM (TRAPANESE)
SEA LEVEL
DREPANO
ERICE BASIN
TWT (sec)
UM
TA?
MO
NEW TYRRHENIAN UPPER MANTLE
PANORMID FRAGMENT
AFRICAN LOWER CRUST
OLD AFRICAN UPPER MANTLE
IONIAN SLAB
SCALE
SCALE RATIO: TRUE FOR VELOCITY 6000 m/s
CROP M-28A
EXTENSION
STRATIGRAPHIC INTERVALS
AFRICAN PLATE
Upper Crust African Plate
Lower Crust African Plate
PANORMID BLOCK
(DISROOTED FROM NORTWEST LYING PANORMID PROMONTORY AND THRUST SOUTHEASTWARD ONTO AFRICAN MARGIN)
Upper Crust of Panormides
Lower Crust of Panormides
EUROPEAN PLATE DERIVING TECTONIC UNITS
Upper Crust of European Plate deriving tectonic units
Lower Crust of European Plate deriving tectonic units
Sedimentary Crust
Ophiolite - bearing units
Oceanic Crust of the Upper Ionian Slab
Upper Mantle
Asthenosphere
HORIZONS
Z = Top Basement (A = African Plate, P = Panormides, E = European Units)
TL = Top Oceanic Crust (A = Ionian, P = Panormides, E = European Units)
TO = Top of Ionian Oceanic Crust
MO = Moho
TA = Top Asthenosphere (Hypothetic)
CALABRIA
TYRRHENIAN SEA
SICILY
PELAGIAN SEA
M-28A
SCALE (km)

dissected the southern leg of the arc, stopping the Corso-Sardinian rotation and separating the SW–NE-trending Maghrebian chain and overstacked Kabilides from the Sicilian Apennine, dominated by the NW–SE-directed Tyrrhenian right-transcurrent system (Finetti, this volume, Chapter 34), going eastward, it seems more correct to call these Sardinian allochthonous units simply *Calabrides*, keeping the term *Kabilides* to the Europe-derived units stacked west of the panormid apex, and *Kabilo-Calabrides* only to this transition zone between the tectonics of the Balearic stage and that of the Tyrrhenian. Here the Kabilo-Calabrides wedge was stretched and delaminated in the Balearic stage by mantle dynamics connected to the eastward slab-retreat of the Alpine Tethys. In the late Balearic–early Tyrrhenian stage, the collision with the Panormid promontory determined a compressive tectonic setting for these delaminated units. Between the Kabilo-Calabrides wedge and the collided-deformed Panormid zone, Alpine Tethys units remained trapped as evidence of the ocean basin closure. The Tyrrhenian–Balearic mantle dynamics produced a remarkable thrust-deformation of the Panormid and Maghrebian chain zone of the African plate. The detachment of the Panormid cover units delivered their basal Cambro–Ordovician quartzitic sandstone that fed the Numidian flysch units. On the northernmost sector of the section, delaminated crustal units belonging to the same Sardinian block as those of the Kabilo-Calabride wedge are present (Plate 4: *Delaminated Sardinian Margin*). We keep separate this part of the section because it is seismically much clearer and not involved in the compressive wedge stacking of the Balearic–early Tyrrhenian stage. Also the cover and allochthonous units are seismically well defined, while in the KC wedge the interpretation must be regarded as purely indicative. In this sector of the section, it was possible indeed to clearly recognize important ophiolite and ophiolite-bearing units of Eo-Alpine obduction associated with the major Tyrrhenian magnetic anomaly ("Romulus" anomaly, Finetti and Del Ben, 1986). The ophiolite units of the *Caesar Seamount South* (km 120–150) still preserve the original Europe-verging thrust position (N-verging component also after rotation by about 40°; opposite to the S-verging polarity of the KC units of the Balearic–Tyrrhenian stage tectonization). The ophiolite units of the *Caesar Seamount North* (km 150–170) have evidently been involved in the successive younger Tyrrhenian stage tectonization.

On extremity "B" of the section (km 190) the Trapani right-transcurrent fault is visible, and is the first prominent westernmost one of the NW–SE South-Tyrrhenian transcurrent system connected to the Tyrrhenian back-arc opening (Finetti and Del Ben, this volume, Chapter 21).

3.3. CROP Section M28A

This section (Plates 5 and 6, Figure 21) substantially repeats most of the stratigraphic and tectonic items seen on the previous section M29A, with the main difference that crop M28A enters fully into the domain of the south Tyrrhenian dynamics and into the Sicilian Apennine tectonics, completely dissected by that of the Maghrebian chain segment which collided with the NW side of the Panormid promontory (Maghrebian chain SW–NE directed). The CROP-section interpretation has been extended some 7–8 km on land on the basis of industrial seismic lines, geology and CROP-data extrapolation. Simplifying the descriptive scheme of the section, it is possible to subdivide it into three main crustal blocks: (a) a wedge of Sardinian upper crustal units, delaminated in the Balearic stage, and stacked in the late Balearic–early Tyrrhenian stage by compression over their shortened lower crust; (b) a crustal fragment interpreted as of Panormid promontory derivation; (c) the northern edge of the African (Sicily) plate. Also in this section, as is seen on CROP M29A, the stacked upper crust units of the Sardinian margin (Europe) overlie a lower Sardinian crust which in the stretching phase (Balearic stage) was delaminated together with its upper crust. In the compressive phase of the late Balearic–early Tyrrhenian stage, the upper crust units assumed the imbricated position we observe in the section, while the lower crust remained mostly subhorizontal and absorbed the cumulative shortening by means of a few transcrustal thrust faults (Plate 6 and Figure 21). The Europe-deriving units (km 0–55) and any preserved pre-Oligocene cover are overlain by Alpine Tethydes (AT) of Eo-Alpine (AL) overthrusting. To the top of the *Drepano Seamount*, a very clear allochthonous element characterized by multilayered thrust-sheets of N-vergence (Europe), associated with a high magnetic anomaly, has been seismically defined. *All clear geophysical characteristics, and the tectono-stratigraphic and paleogeographic position make it evident*

Figure 22: Reduced and simplified version of the seismogeological section of Plate 9 (CROP M6A in the Tyrrhenian Sea, M24 on the Pelagian foreland, and interposed Western Sicily). This reconstruction exhibits a continuous crustal setting from the Tyrrhenian Sea floored by oceanic crust to Lampedusa island in the Pelagian Sea, giving a useful global overview of the various tectono-stratigraphic provinces crossed. The Sicilian land portion of the section must be considered as an approximate but regionally realistic scheme supported by geophysical-geological data from the literature and CROP extrapolations–interpolations.

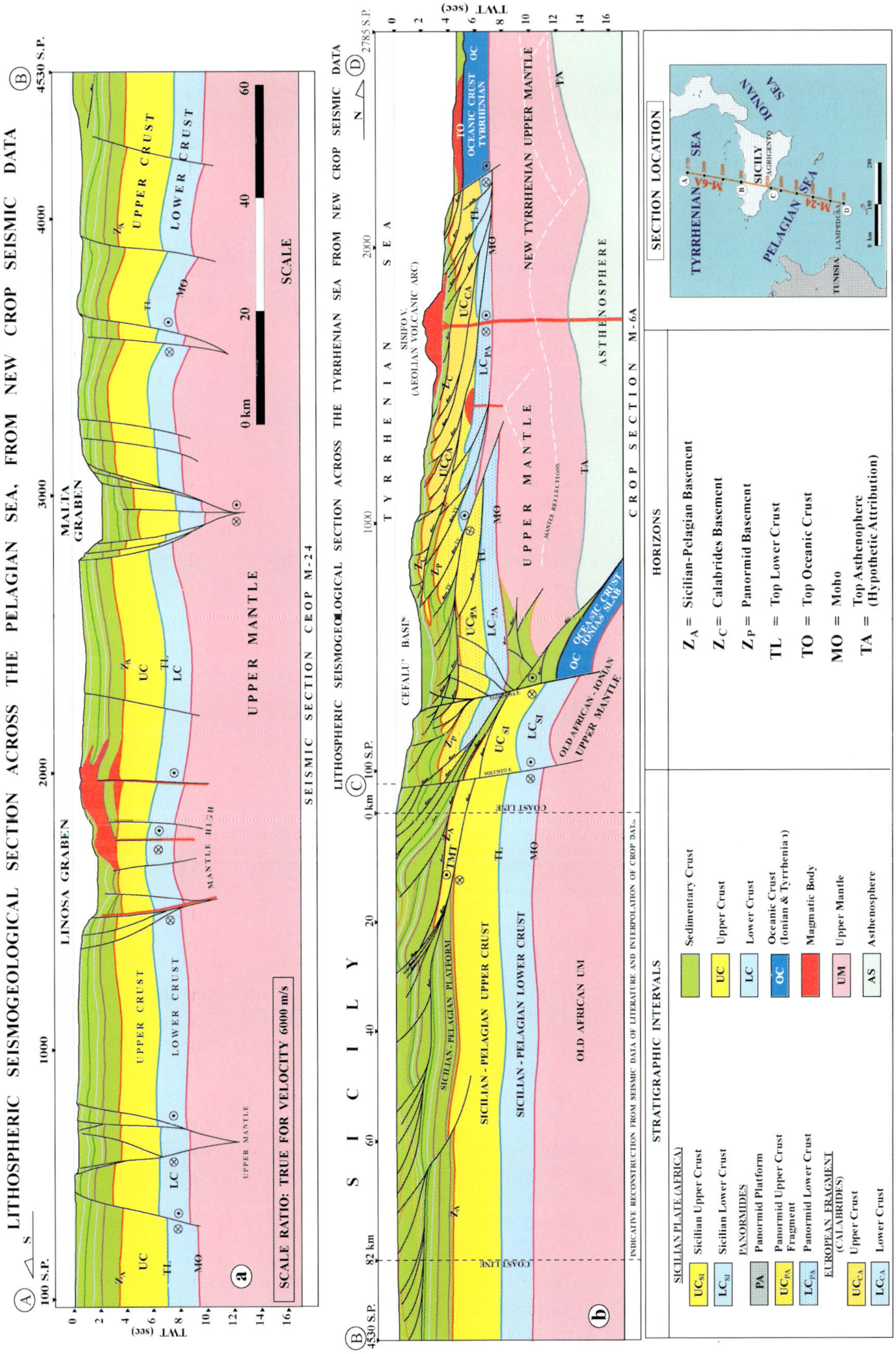
LITHOSPHERIC SEISMOGEOLOGICAL SECTION ACROSS THE PELAGIAN SEA, FROM NEW CROP SEISMIC DATA
SEISMIC SECTION CROP M-24
SCALE RATIO: TRUE FOR VELOCITY 6000 m/s
LITHOSPHERIC SEISMOGEOLOGICAL SECTION ACROSS THE TYRRHENIAN SEA FROM NEW CROP SEISMIC DATA
CROP SECTION M-6A
INDICATIVE RECONSTRUCTION FROM SEISMIC DATA OF LITERATURE AND INTERPOLATION OF CROP DATA
STRATIGRAPHIC INTERVALS
HORIZONS
SECTION LOCATION

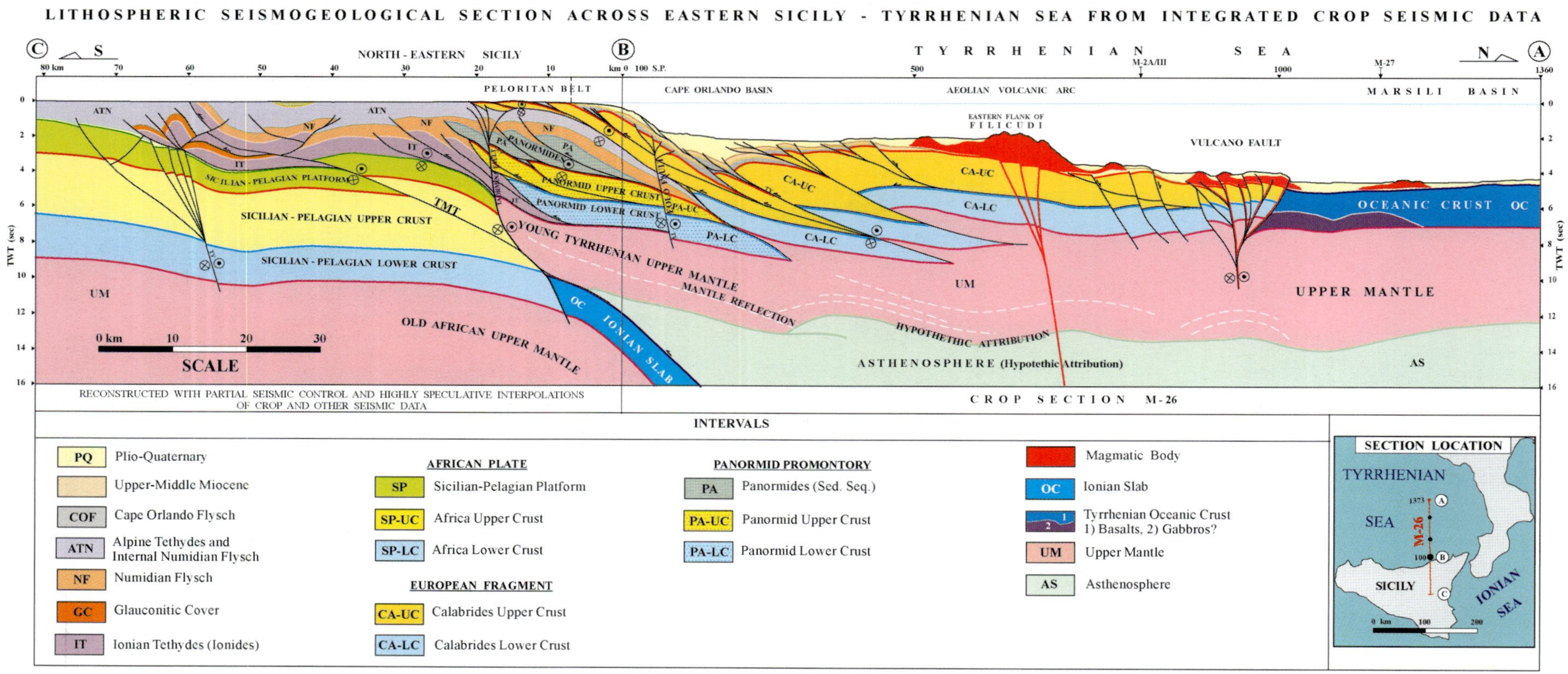

Figure 23: Reduced version of the seismogeological section in Plate 11 (CROP M26, with extension on the land side).

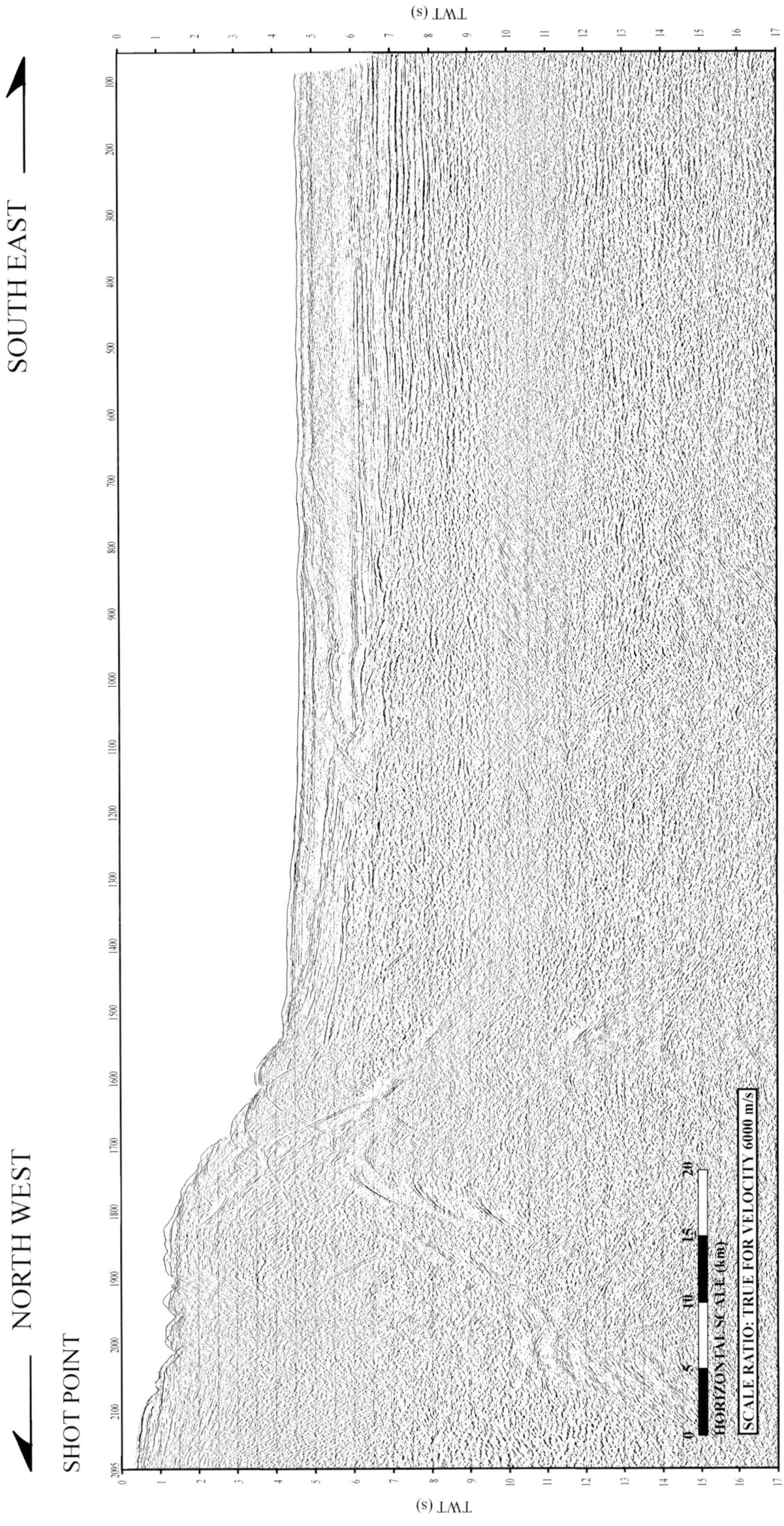

Figure 24: Seismic section CROP M21 across the Hyblean offshore–western Ionian Sea (see Figures 1 and 25 for location).

that the Drepano Sm. represents a segment of ophiolite/ophiolite-bearing units obducted over the European margin in the Eo-Alpine stage.

This, together with the ophiolite units seen on section CROP M29A (Plate 4, Figure 20) constitutes important new information both in its role as a previously missing segment of the Eo-Alpine ophiolite belt, and as a geodynamic marker for the Mediterranean evolution (Finetti, this volume, Chapter 34).

The Panormid crustal block migrated to its present position, coming from the NW, driven by the mantle dynamics of the late Balearic–early Tyrrhenian stage and kinematically conducted by the right-transcurrent NW–SE South Tyrrhenian fault system. The African plate edge is lithospherically delimited by a major right-transcurrent that also slices the Ionian crust/continent contact. This slab imaging is the southeasternmost one in the Ionian Tethys explored by the CROP seismic network. The Africa plate edge zone is severely bent and its sedimentary cover and upper crust are affected by high-angle thrust faults.

3.4. CROP section M6A/M24 and interposed Western Sicily Terrain

To reconstruct a transcrustal section representing the regional tectono-stratigraphic setting of the Sicilian Apennines thrust belt and of its African (Pelagian) foreland without gaps, not having CROP data on the Sicily mainland, the best available information has been combined from a CROP section through the Sicilian margin of the Tyrrhenian Sea (M6A), a CROP section in the Pelagian Sea (M24), and interpolation of integrated geophysical-geological data (Plates 7, 8 and 9; Figure 22). The basement upper crust and the lower crust of Sicily is the simple interpolation of flanking CROP data from sections M24 and M6A. The sedimentary cover of the Sicily land segment is based on an integration of: (a) various available local geophysical seismic sections from hydrocarbon exploration; (b) extrapolated re-interpretation of seismic data in the literature (see for example Figures 26, 27 and 28); (c) new geological data specifically collected; and (d) borehole data. Obviously the land segment "BC" of the reconstructed seismogeological section in Plate 9 must be regarded as a simplified, indicative regional crustal setting of the area. But the fundamental features exhibited are inserted in a general context of the realistic new CROP reconstruction.

The Tyrrhenian part of the section, fully CROP-controlled (M6A), shows the basic tectono-stratigraphic characteristics dominating the whole Tyrrhenian margin of Sicily. At the north extremity the back-arc Tyrrhenian basin is visible, floored by oceanic crust, and delimited by the right-lateral transcurrent "Glauco fault". From the ocean basin to the coast zone, two different types of imbricated thrust units are present: the Calabrides units, formed by fragments of Europe-derived terranes, and the Panormid units coming from the Panormid promontory (Plate 9, c).

In the Balearic stage (Oligocene–Lower Miocene) the European crustal fragments followed the closure of the Alpine Tethys, and the deep basin sequences (Alpine Tethydes) were stripped off their subducting basement and transported outward over the Panormides and Sicilian Apennines. In the Tyrrhenian stage (Middle Miocene to Present) the Ionian slab retreat was accompanied by an impressive southeastward flux of new Tyrrhenian mantle. These mantle dynamics constitute the basic driving force responsible for the generation of the young active orogenic arc that includes the Southern Apennines, the Calabrian arc and the Sicilian Apennines. On the Tyrrhenian margin of the Sicilian (African) platform, various allochthonous units transported by the Tyrrhenian mantle are stacked. Going inward we find: Ionian Tethydes "IT" or Ionides, Panormid units and Panormid crustal fragments, Numidian Flysch, Alpine Tethydes and Calabrides units or European-derived crustal blocks. On these last units, the Tyrrhenian back-arc basin floored by oceanic crust opened. Various major right-transcurrent faults of the South Tyrrhenian system are crossed by the CROP sections (i.e. Trapani, Solunto, Taormina, Eolo, Alicudi, Glauco, and Vulcano).

The CROP section M6A exhibits very clearly the upper Ionian slab, the Solunto fault and the moho of the colliding Panormid fragment. Between the Ionian slab and the Panormid moho there are reflecting bodies that have been interpreted as Ionian Tethydes trapped during the oblique collisional dynamics. The crust of the Pelagian Sea (section M24) is affected by a principal right-transcurrent system of NW–SE direction (i.e. Malta graben, Linosa

---

Figure 25: Interpreted CROP seismic section M21 on the Hyblean offshore (Ionian Sea). This figure shows the quite different crustal characteristics existing between the Hyblean-Maltese continental margin of the Africa plate and the Ionian ocean basin of Permo-Triassic opening.

SEISMOGEOLOGIC SECTION ACROSS HYBLEAN MARGIN - IONIAN SEA FROM NEW SECTION CROP M-21

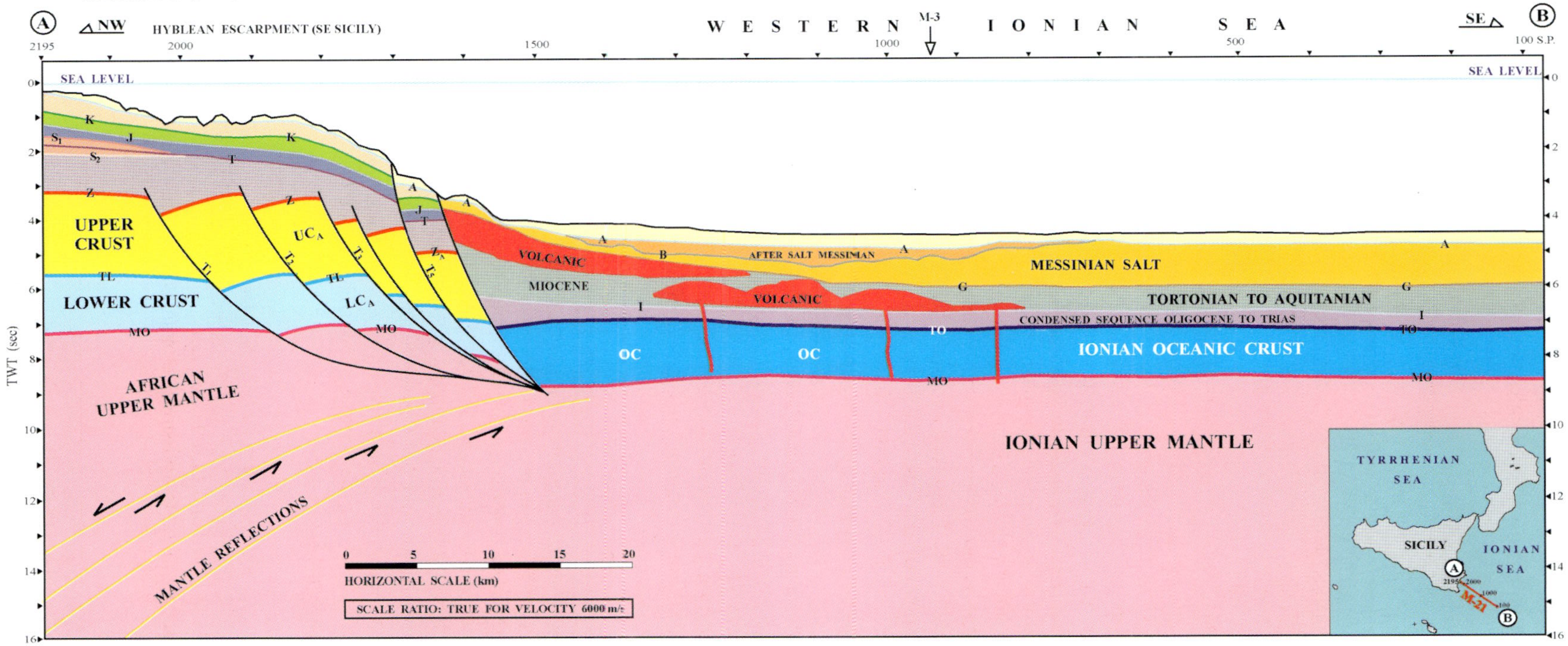

SEISMIC STRATIGRAPHIC INTERVALS

HYBLEAN PLATEAU (ESCARPMENT)

- Plio-Quaternary
- Miocene (or Other Tertiary)
- Cretaceous
- Jurassic
- Streppenosa Fm
- Trias (and Upper Permian)
- UC_A Upper Crust Africa
- LC_A Lower Crust Africa
- UM Upper Mantle

REMARK:
Prominent multilayered effusive magmatic activity (mostly basaltic) in the Hyblean and whole Pelagian eastern margin area during Triassic to Middle Jurassic, not reported in this section.

IONIAN SEA

- Plio-Quaternary
- Non-Salt Messinian
- Messinian Salt
- Tortonian to Aquitanian
- Condensed Sequence (Oligocene to Trias-Permian)
- OC Oceanic Crust
- UM Upper Mantle
- Volcanic

INTERPRETED SEISMIC HORIZONS

HYBLEAN PLATEAU (ESCARPMENT)

- A = Base Plio-Quaternary
- K = Top Cretaceous or Top Carbonate
- J = Top Jurassic
- S_1 = Top Streppenosa (Upper Hettangian)
- T = Top Trias
- S_2 = Base Streppenosa (Late Norian-Rhaetian)
- Z_A = Top Basement
- TL = Top Lower Crust
- MO = Moho

IONIAN SEA

- A = Base Plio-Quaternary
- B = Base post-Salt Messinian
- G = Base Messinian Salt
- I = Top Condensed Sequence
- TO = Top Oceanic Basement
- MO = Moho

graben and other transtensive deformations) which form the so-called *Sicily channel rift zone* (Finetti, 1984). This system, connected to the Africa–Europe convergence and to Tyrrhenian mantle flux dynamics, is accompanied by antithetic left-transcurrents of NNE–SSW trend. All these left-transcurrent faults produce flower structures (i.e. *Marzara, Belice, Sciacca, Ragusa, Rosolini, P. Castellazzo and Pachino flower*, Figure 30). In the Linosa graben (negative flower of NW–SE direction) prominent igneous activity (Plate 9, d) and mantle uplifting took place.

3.5. CROP Section M26, Southeastern Tyrrhenian Margin

This CROP section, as the previous ones in the Tyrrhenian margin of Sicily, crosses an ensemble of stacked allochthonous tectonic units coming from different paleogeographic domains (Plate 11). The interpretation has been extended 80 km southward on the margin of Sicily in order to delineate the regional crustal relationships between the colliding allochthonous blocks and the Sicilian platform units. The cover of zone “d” (Plate 11), part of the extended section, is controlled by detailed direct seismic data taken from the literature (Bello et al., 2000; courtesy of SGI, Società Geologica Italiana), which have been graphically reprocessed and interpreted. The remaining cover and the whole basement upper crust and lower crust of the “BC” segment of the section were reconstructed by extrapolation–interpolation of CROP data and by examination of various other geophysical and geological data.

The results of the interpretation are sufficiently self-explanatory. The oceanic Tyrrhenian basin is here delimited by the very impressive “Vulcano Fault” (Figure 23) which is the most important lateral right-transcurrent of the Calabrian arc, because it dissects the thrust system, setting it free to expand more easily southward within the Ionian Sea, along the Hyblean-Maltese escarpment (Figure 30). It should be pointed out that in the Vulcano fault the transcurrent motion occurs in the thrust propagating stack. Outward, in the Ionian Sea, it acts as lateral transfer fault of the Calabrian wedge formed by repeated imbrication of the Ionian cover (Ionides) stripped off their subducting oceanic basement (slab). In the Ionian basement, the Vulcano transcurrent generated an active overload-related normal fault where the displacement increases inward proportionally to the thickness of the migrating stack system.

From the *Vulcano fault* to the *Taormina right-transcurrent fault*, the section (Plate 11) is composed of a complex stack of crustal allochthonous tectonic units. Starting from the inner zone we find fragments of European (Sardinian) margin stretched and delaminated during the Balearic-Tyrrhenian stage and re-compressed during the oblique thrusting over the Sicilian margin. In the *Balearic stage*, the Alpine Tethys closed and the European (Sardinian) crustal units thrust over the Panormid promontory (Plate 11, c). In the *Tyrrhenian stage*, part of the Ionian cover (*Ionian Tethydes* or *Ionides*) detached from their subducting oceanic basement and thrust obliquely over Sicily, while most of the composed stack of different Sardinian and Panormid units remained on the deformed Sicilian margin.

The Ionian slab is seismically well imaged (SP 100–200). A typical recurrent compressive fault of the Tyrrhenian margin zone is represented by a *late orogenic (mostly out-of-sequence) thrust fault that displaces upper crustal blocks and overlying units outward, as consequence of pressure from the Tyrrhenian mantle flux against the steep palaeoslope* (i.e. Tyrrhenian margin thrust “TMT” of Plates 6, 9 and 11, and Figures 28 and 29). *This type of fault is equivalent to the M. Alpi fault in the Southern Apennines* (Finetti et al., this volume, Chapter 12). Similar thrust faults are present on the NW side of the Panormid promontory, where the pressure against the slope was produced by the Balearic mantle dynamics (Plates 2 and 4). In Figure 27 (NW Sicily) the corresponding TMT fault can be considered the *Maranfusa thrust fault* (Tf-2).

3.6. Crustal sections across Western and Eastern Sicily

On the basis of seismic data of medium penetration from hydrocarbon exploration and integrated CROP and geological data, two important transcrustal seismogeological sections have been reconstructed: one across the western part of the island (Figures 26 and 27) and the other across the eastern sector (Figures 28 and 29). The section in Figure 26 derives from a seismic line from the literature (Catalano et al., 2000; kind permission of SGI, Società

Figure 26: Seismogeological section across Western Sicily, reconstructed by new interpretation of a graphically reprocessed seismic line from the literature (Catalano et al., 2000; kind permission of SGI, Società Geologica Italiana). Top lower crust (TL) reconstructed by interpolation of CROP seismic data. A regional scheme of major tectonic features is inserted (26 “a”). For legend see next Figure 27.

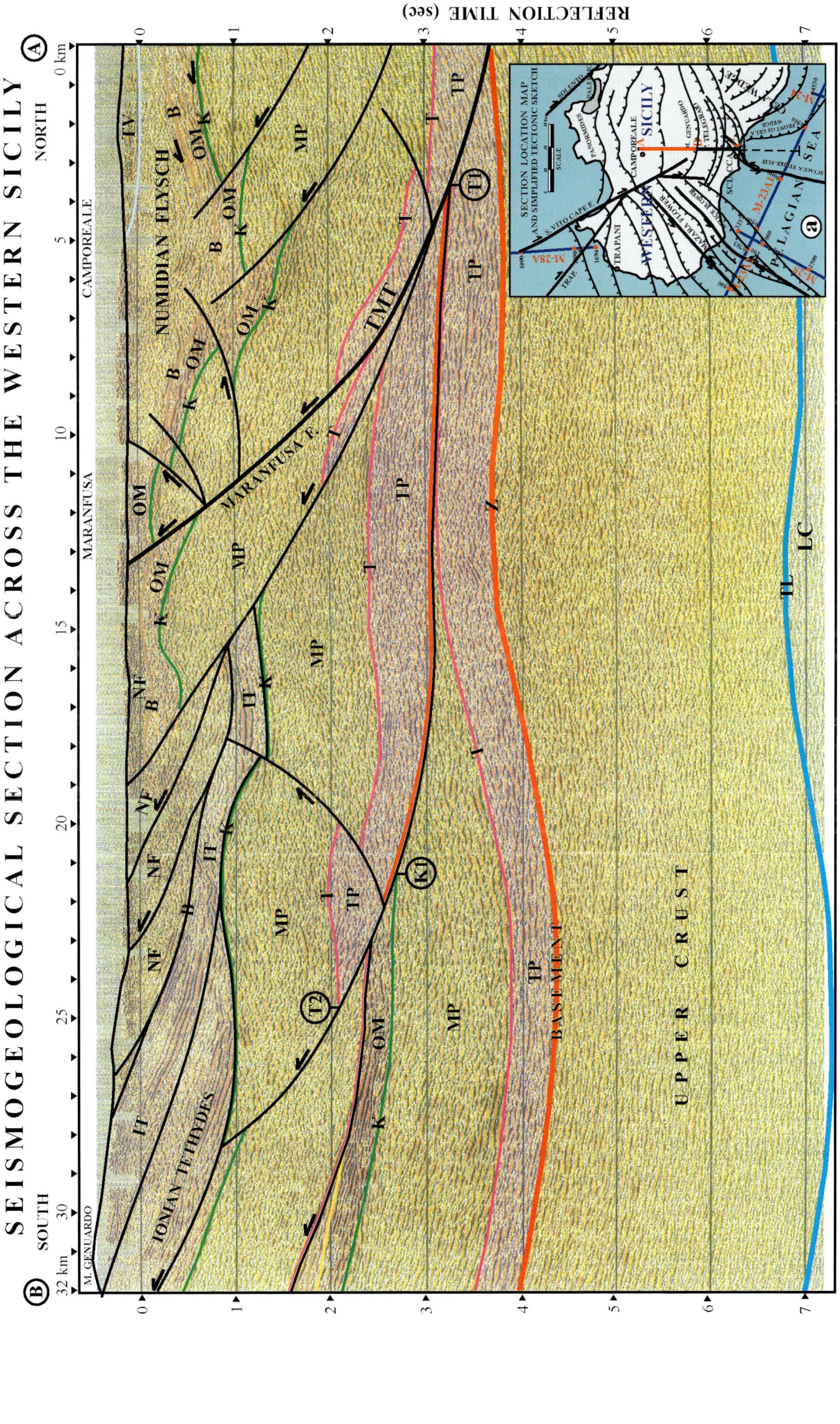
SEISMOGEOLOGICAL SECTION ACROSS THE WESTERN SICILY
SOUTH
NORTH
REFLECTION TIME (sec)
M. GENUARDO
MARANFUSA
CAMPOREALE
NUMIDIAN FLYSCH
MARANFUSA F.
IONIAN TETHYDES
BASEMENT
UPPER CRUST
SECTION LOCATION MAP AND SIMPLIFIED TECTONIC SKETCH
PELAGIAN SEA

Geologica Italiana) graphically reprocessed and re-interpreted. Integrating this line with geological, borehole, and other seismic data, and CROP extrapolated-interpolated data, the interpretation was extended from M. Genuardo to the coast, and downward into the lower crust (Figure 27). The combined section allows us to outline the regional stratigraphic and structural crustal setting of Western Sicily, from Camporeale (some 12 km from the northern coast of the Gulf of Castellammare) to the southern coast of Sciacca.

The seismic data in Figure 26 are of good quality, and therefore allow a reliable reconstruction. The sedimentary cover of the Sicilian platform is characterized by a thick, poorly reflective Mesozoic succession (MP) overlying a basal Lower Trias–Permian (TP) of distinct and good reflectivity. This seismostratigraphic characteristic is very helpful for the interpretation. The Maranfusa thrust fault (TF-2) displaces the entire sedimentary crust southeastward as exhibited. But CROP data for the Tyrrhenian margin of Sicily (M28A and M6A, faults "TMT", or the "M. Alpi-like" thrust fault in the Southern Apennines; see Finetti et al., this volume, Chapter 12) consistently show that this fault continues across the basement upper crust and terminates over the slope bent by the Ionian slab subduction. These data (Figures 26 and 27) also allow us to clearly measure the duplication (shortening) of the Mesozoic platform cover produced by the thrust fault "TF-1" during the Tyrrhenian opening geodynamics. From "K1" to "K2" for the "MP" sequence, and from "T1" to "T2" for the reflective "TP" sequence, a thrust displacement of respectively 27 and 22 km is found.

It is interesting to observe the *new seismic evidence for the Ionian Tethydes (IT) or Ionides being thrust obliquely over Sicily*. Here, the Ionides (IT) thrust over the Maranfusa structure as far as M. Genuardo (km 32 of the section). At the southern extremity of the section, the *Sciacca left-transcurrent flower* is exhibited, which, as above mentioned, represents an antithetic fault of the main NW–SE right-transcurrent system.

In Figure 28, the new interpretation of a graphically reprocessed very good and important seismic line from the literature is shown (Bello et al., 2000, kind permission of SGI Società Geologica Italiana). This line is important because it crosses Eastern Sicily from the Tyrrhenian coast (Castel di Tusa) to the Hyblean plateau, controlling almost the whole cover and a large part of the basement upper crust. But, considering that the pre-sedimentary upper crust and the lower crust from the Tyrrhenian (M6A and M26) and Pelagian (M39 and M23A1) CROP sections can be reliably interpolated downward to reconstruct the entire crust, in practice, the seismogeological section of Figure 28 can be seen substantially as an additional CROP section across Eastern Sicily. This section was extended northward to the Tyrrhenian margin for a schematic CROP-correlated view of the allochthonous crustal thrust blocks transported by the mantle-asthenosphere dynamics.

The late orogenic trans-upper-crustal thrust fault TMT (Figures 28 and 29) of the Sicilian (African) margin which duplicates the Mesozoic platform sequence and part of pre-sedimentary upper crust is drown also on this section. The roof-thrusting of the Ionian Tethydes (IT) and of their cover is evident. It is very interesting to observe that the Caltanisetta foredeep is filled by a regular, geometrically well-defined *wedge of Ionides* (IT) overlain by their *glauconitic cover* (GC). Over the Ionides, *Panormid units* to the north (PA), *Numidian flysch units* (NF) and *Alpine Tethydes* with *internal Numidian flysch* units, followed by syn and post-orogenic basin fills, are present.

3.7. Section CROP M21 across the Hyblean–Maltese Escarpement–Ionian Sea

The reduced version of the CROP seismic section M21 is shown in Figure 24. In Figure 25 is exhibited the interpreted seismogeological section which clearly evidences the sharp geological passage from the Sicilian (African) platform margin and the Ionian Sea floored by an old crust of oceanic-type (Finetti, 1982, 1985). The Hyblean margin shows a rifted zone (listric faults T1, T2, T3 and T4) connected to the Permo-Triassic opening of the Ionian sea, when the Adria plate (and the Cimmerian terranes) detached from Africa, creating an interposed ocean basin.

Figure 27: Seismogeological section across Western Sicily, from Camporeale to Sciacca, reconstructed by new interpretation of the graphically reprocessed seismic line of the previous Figure 26 and CROP seismic data. Platform crustal shortening, measured on the outer thrust fault "TF-1" amounts to 22 km at the base of sedimentary sequence (T1 to T2), and 27 km to the top platform "K" of the sequence "MP" (K1 to K2). The high-angle late orogenic Maranfusa thrust-fault "TF-2" cuts and displaces the entire upper crust, terminating on the lower crust bending zone connected to the Ionian slab, in the Tyrrhenian Sea. Scale ratio: true for velocity 6000 m/s.

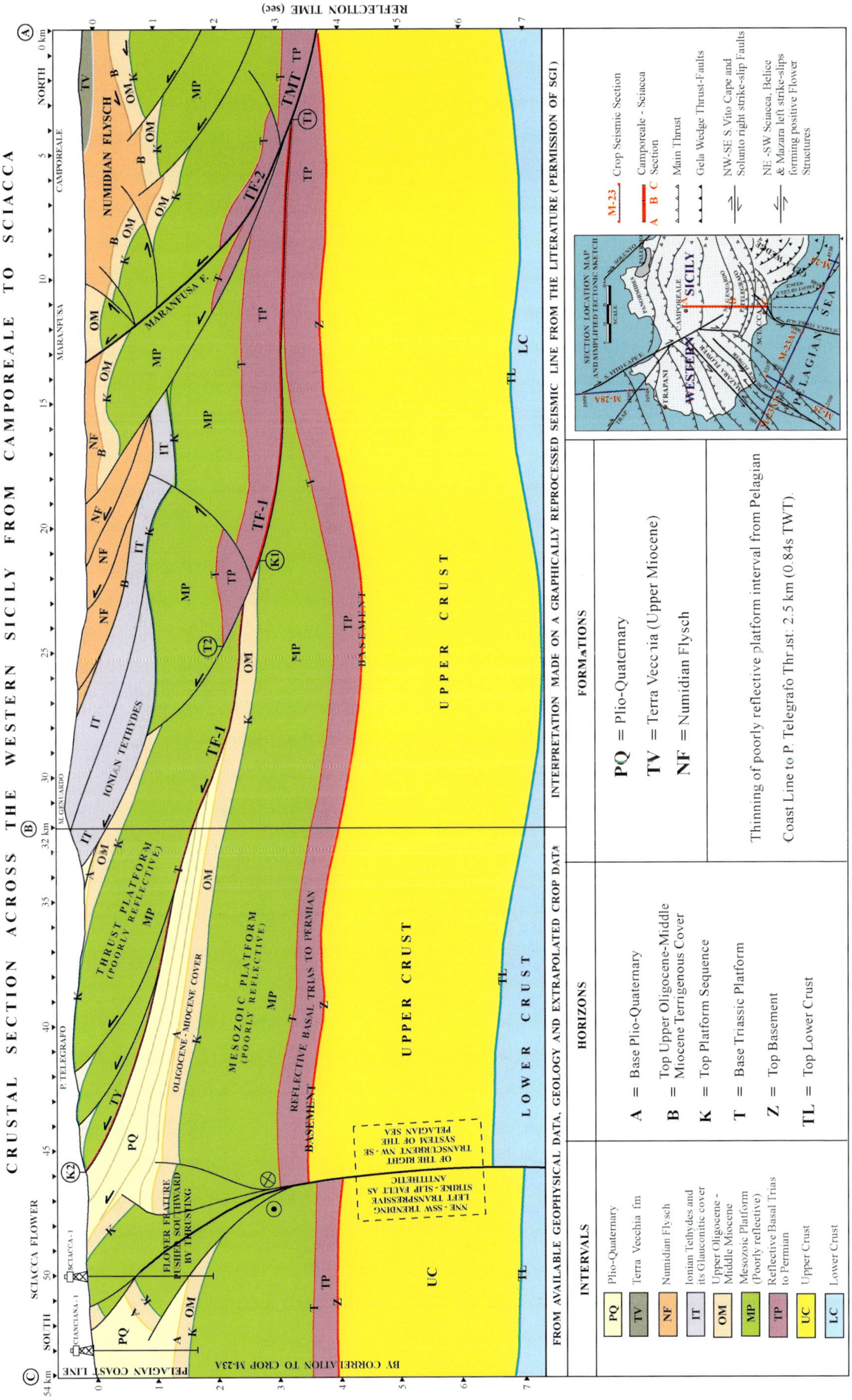
CRUSTAL SECTION ACROSS THE WESTERN SICILY FROM CAMPOREALE TO SCIACCA
REFLECTION TIME (sec)
BY CORRELATION TO CROP M-23A
PELAGIAN COAST LINE
SOUTH
SCIACCA FLOWER
P. TELEGRAFO
M. GENUARDO
MARANFUSA
CAMPOREALE
NORTH
FLOWER FEATURE PUSHED SOUTHWARD BY THRUSTING
NNE - SSW TRENDING LEFT TRANSPRESSIVE STRIKE - SLIP FAULT AS ANTITHETIC OF THE RIGHT TRANSCURRENT NW - SE SYSTEM OF THE PELAGIAN SEA
THRUST PLATFORM (POORLY REFLECTIVE)
MESOZOIC PLATFORM (POORLY REFLECTIVE)
OLIGOCENE - MIOCENE COVER
REFLECTIVE BASAL TRIAS TO PERMIAN
BASEMENT
UPPER CRUST
LOWER CRUST
IONIAN TETHYDES
NUMIDIAN FLYSCH
MARANFUSA F.
FROM AVAILABLE GEOPHYSICAL DATA, GEOLOGY AND EXTRAPOLATED CROP DATA
INTERPRETATION MADE ON A GRAPHICALLY REPROCESSED SEISMIC LINE FROM THE LITERATURE (PERMISSION OF SGI)
INTERVALS
PQ Plio-Quaternary
TV Terra Vecchia fm
NF Numidian Flysch
IT Ionian Tethydes and its Glauconitic cover
OM Upper Oligocene - Middle Miocene
MP Mesozoic Platform (Poorly reflective)
TP Reflective Basal Trias to Permian
UC Upper Crust
LC Lower Crust
HORIZONS
A = Base Plio-Quaternary
B = Top Upper Oligocene-Middle Miocene Terrigenous Cover
K = Top Platform Sequence
T = Base Triassic Platform
Z = Top Basement
TL = Top Lower Crust
FORMATIONS
PQ = Plio-Quaternary
TV = Terra Vecchia (Upper Miocene)
NF = Numidian Flysch
Thinning of poorly reflective platform interval from Pelagian Coast Line to P. Telegrafo Thrust: 2.5 km (0.84s TWT).
SECTION LOCATION MAP AND SIMPLIFIED TECTONIC SKETCH
WESTERN SICILY
PELAGIAN SEA
M-23 Crop Seismic Section
A B C Camporeale - Sciacca Section
Main Thrust
Gela Wedge Thrust-Faults
NW-SE S.Vito Cape and Solunto right strike-slip Faults
NE -SW Sciacca, Belice & Mazara left strike-slips forming positive Flower Structures

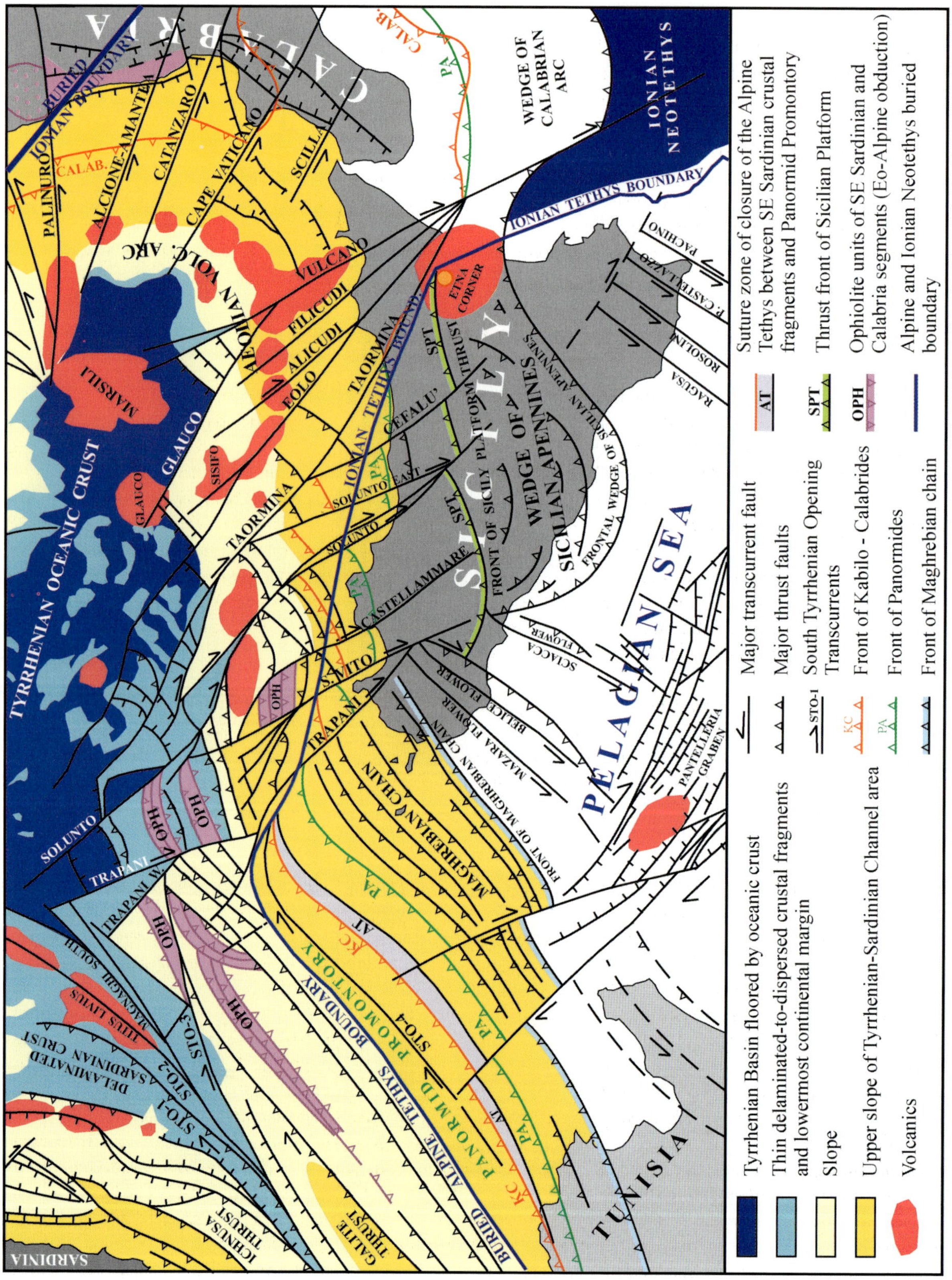
Tyrrhenian Basin floored by oceanic crust
Thin delaminated-to-dispersed crustal fragments and lowermost continental margin
Slope
Upper slope of Tyrrhenian-Sardinian Channel area
Volcanics
Major transcurrent fault
Major thrust faults
STO-1 South Tyrrhenian Opening Transcurrents
KC Front of Kabilo - Calabrides
PA Front of Panormides
Front of Maghrebian chain
AT Suture zone of closure of the Alpine Tethys between SE Sardinian crustal fragments and Panormid Promontory
SPT Thrust front of Sicilian Platform
OPH Ophiolite units of SE Sardinian and Calabria segments (Eo-Alpine obduction)
Alpine and Ionian Neotethys buried boundary
SARDINIA
TUNISIA
PELAGIAN SEA
SICILY
CALABRIA
TYRRHENIAN OCEANIC CRUST
IONIAN NEOTETHYS
WEDGE OF CALABRIAN ARC
IONIAN TETHYS BOUNDARY
BURIED ALPINE TETHYS BOUNDARY
PANORMID PROMONTORY
MAGHREBIAN CHAIN
WEDGE OF SICILIAN APENNINES
FRONT OF SICILY PLATFORM THRUST
ETNA CORNER
AEOLIAN VOLC. ARC
MARSILI
PANTELLERIA GRABEN
FRONT OF MAGHREBIAN CHAIN
DELAMINATED SARDINIAN CRUST
ICHNUSA THRUST
GALITE THRUST

Figure 30: Tectonic sketch map of Sicily and surrounding marine areas showing major transcurrent and thrust faults, related to Ionian slab retreat and consequent Tyrrhenian mantle dynamics.

The Mesozoic sequence of Middle-Upper Triassic to Cretaceous and covering Tertiary is not affected by faults, and is gently east-dipping, probably due both to thermal cooling and, successively, to sedimentary and tectonic load of the innermost Ionian basin by the Calabrian arc. Due to the reduced scale of the figure, it is impossible to draw the numerous basaltic layers found in the Triassic-Jurassic sequence by boreholes. *The Ionian sequence is characterized by condensed deep-water deposition from the Ionian opening (Permo-Triassic) to the Oligocene (interval "I–TO"). This sequence constitutes the Ionian Tethydes (Ionides) thrust onto Sicily and the Southern Apennines (Lagonegrese units). Over the condensed sequence, a Tortonian to Aquitanian succession of higher sedimentation rate is deposited, overlain by thick Messinian salt pushed outward by the Calabrian arc.* Prominent igneous (volcanic) intercalations are observed. West-dipping mantle reflections seem evident at the NW extremity of the section.

4. CROP-CONSTRAINED GEODYNAMIC EVOLUTION

Using deep seismic exploration data from the CROP project, new geological investigations specifically programmed, and selected data from the literature, it is now possible to better understand and constrain than in the past the tectono-dynamics that generated the various stacked complexes formed by tectono-sedimentary units coming from different paleogeographic domains. This is because key data are deeply buried beneath the Tyrrhenian margins and cannot be realistically outlined without a wide-scale deep seismic control. Such is surely the case also of the paleogeographic framework existing before the orogenic deformation processes. *From the whole CROP dataset, integrated with other geophysical-geological data, it is possible to reconstruct the buried boundaries of the Ionian Tethys even where this oceanic crust is subducted, because its upper slab portion has been recognized everywhere on the CROP sections* (see Plates 6, 9 and 11, and Figures 21, 22 and 23). *Also, the Alpine Tethys boundary has been reliably identified and reconstructed* (see for example Plate 2 and Figure 19: the small Alpine Tethys slab-zone present in the study area of this chapter).

It is well CROP-data-constrained that at the end of the Lower Cretaceous, the Ionian Sea of Permo-Triassic opening and the Alpine Tethys (Neotethys) were fully communicating (Finetti and Del Ben, this volume, Chapter 6; Finetti, this volume, Chapter 34). It is also clear that *during the Eo-Alpine geodynamic stage (Late Cretaceous–Early Eocene) Africa, Adria and the interposed Ionian Tethys (Neotethys), as a single block, converged with Europe, leaving the Ionian totally undeformed. This behaviour persisted also through most of the Balearic stage from the Oligocene to the end of the Aquitanian.*

It is possible to support with CROP data the observation that in the Eo-Alpine stage the continental collision occurred only in the Alpine Tethys segment of the Alps, while west of the Alps the south-dipping subduction was of intra-ocean type, and a wide remnant of Alpine Tethys, at the end of the Eo-Alpine geodynamics, remained unsubducted (Finetti, this volume, Chapter 34). *Within this Alpine Tethys remnant, at least from the Early Oligocene, a NW-dipping inverted subduction which gave rise to the Balearic geodynamic stage started, with rotation of the Corso-Sardinian block and formation of a large unique arc including the Northern Apennines, Southern Apennines, Calabrian arc and the Sicilian–Maghrebian chain. Corsica collided with the NW-Adria margin, closing the Alpine Tethys (Corsica basin–Etrurian block). To the south, SE Sardinia collided with the Panormid promontory which separated the Alpine Tethys from the Ionian Tethys* (see small figure "c" of the Burdigalian paleogeographic scheme inserted into Plates 9 and 11). *The Panormid promontory dissected the Maghrebian chain from the Sicilian Apennines to the south, and the Etrurian promontory dissected the Northern Apennines from the Southern Apennines. From the Burdigalian, the Corso-Sardinian block rotation substantially stopped and the W-dipping subduction continued in the Ionian Tethys during the Tyrrhenian stage with a fast slab retreat (average: 45–50 mm/a; during the Lower Pliocene: over 100 mm/a) and SE-propagating new Tyrrhenian mantle flux. These mantle dynamics are the main driving force which generated the Southern Apennines–Calabrian arc–Sicilian Apennine systems, stretched and delaminated the eastern Sardinian margin, opening the Tyrrhenian back-arc basin over the Ionian Tethys, and transported cover units and crustal fragments of the Panormid promontory firstly over the Ionian sequence and successively, together with detached Ionian cover (Ionides), over Sicily in oblique thrusting.* The progressive stacking process of the different paleogeographic units and flysch-generated units is sufficiently self-explanatory on the reconstructed seismogeological sections but is described in detail in Paragraph 4. It is notable that on the steep buried slope of Sicily, crustal blocks of the Panormides, followed by crustal blocks of Sardinia (Calabrides) collided. In the same way on the Apulian (Adria) slope, crustal blocks and tectonic units of the Etrurian promontory (NW Adria; Finetti et al., this volume, Chapter 12), followed by Calabrides, collided. In the arc sector floored by Ionian oceanic crust (Calabrian arc) the Ionides constitute a large wedge of repeated, almost sequential imbrications, internally followed by a block of Adria (Etrurian promontory) in northern Calabria and by blocks of Panormides in the southern Calabrian arc. Topmost basement thrust-sheets are formed by Calabrides (Europe) in Peloritani and the south Calabrian sector.

A simplified tectonic scheme of the studied area is shown in Figure 30, where the south Tyrrhenian right-transcurrent system that characterizes the Sicilian Apennines is evident.

ACKNOWLEDGEMENTS

This work was partially funded by CROP Partners (CNR, ENEL, ENI) with grants from CNR to the Scientific Director of the CROP Mare Programme (I.R. Finetti): grants CTR 96.00797; and CTB 97.04525.CT05. Major financial support was given by MIUR with funding from the national coordinated Italian programmes: grants COFIN 98, COFIN 2000 and COFIN 2002 to I.R. Finetti (University of Trieste) and to F. Lentini (University of Catania). Relevant financial support was granted also by the University of Trieste. The authors express their gratitude to all supporters. Many thanks are due to OGS, Seismic Division, to all teams of the OGS-Explora Seismic Vessel for the CROP seismic data acquisition, to ENEL and ENI for the processing of the CROP seismic sections used, and to the EGG (Exploration Geophysics Group) of the Faculty of Sciences of the University of Trieste for iterated interpretative reprocessings of part of the CROP data, where signal improvement was determinant for a better understanding of the complex features involved. We wish to thank SGI (Società Geologica Italiana) for granting permission to use seismic figures from its journals for the Sicily onshore area.

REFERENCES

Abate, B., Di Stefano, E., Ferruzza, G., Incandela, A., Renda, P., 1993. Fase tettonica pliocenica nelle Madonie (Sicilia centro-settentrionale). Riv. Min. Siciliana 6 (168), 37–46.

Amodio Morelli, L., Bonardi, G., Colonna, V., Dietrich, D., Giunta, G., Ippolito, F., Liguori, V., Lorenzoni, S., Paglionico, A., Perrone, V., Piccarreta, G., Russo, M., Scandone, P., Zanettin Lorenzoni, E., Zuppetta, A., 1976. L'arco Calabro-Peloritano nell'orogene Appenninico-Maghrebide. Mem. Soc. Geol. It. 17, 1–60.

Amore, T., 1969. La sezione stratigrafica di Gagliano nelle marne azzurre mesoautoctone del Miocene inferiore-medio della Sicilia. Riv. Min. Siciliana 20, 129–172.

Antonelli, M., Franciosi, R., Querci, A., Ronco, G.P., Vezzani, F., 1988. Paleogeographic evolution and structural setting of the Northern side of the Sicily Channel. In: Soc. Geol. It. 74° Congr. Naz., Relazioni, pp. 79–86.

Argnani, A., Cornini, S., Torelli, L., Zitellini, N., 1987. Neogene-Quaternary foredeep system in the strait of Sicily. Mem. Soc. Geol. It. 36, 123–130.

Atzori, P., 1968. Studio geo-petrografico dell'affioramento mesozoico di Alì Terme (Messina). Atti Acc. Gioenia Sc. Nat. Catania 20, 134–172.

Atzori, P., Carveni, P., Lentini, F., Pezzino, A., Vezzani, L., 1977. Posizione strutturale dei lembi mesocenozoici dell'Unità di Rocca Novara nei Monti Peloritani (Sicilia nord-orientale). Boll. Soc. Geol. It. 96, 331–338.

Baldacci, L., 1886. Descrizione geologica dell'Isola di Sicilia. Mem. Descr. della Carta Geol. d'It., vol. 1, 408 p.

Behncke, B., 2001. Volcanism in the Southern Apennines and Sicily. In: Vai, G.B., Martini, I.P. (Eds.), Anatomy of an Orogen: the Apennines and Adjacent Mediterranean Basins. Kluwer Academic, pp. 105–120.

Bello, M., Franchino, A., Merlini, S., 2000. Structural model of eastern Sicily. Mem. Soc. Geol. It. 55, 61–70.

Bianchi, F., Carbone, S., Grasso, M., Invernizzi, G., Lentini, F., Longaretti, G., Merlini, S., Mostardini, F., 1987. Sicilia orientale: profilo geologico Nebrodi-Iblei. Mem. Soc. Geol. It. 38, 429–458.

Bonardi, G., Pescatore, T., Scandone, P., Torre, M., 1971. Problemi paleogeografici connessi con la successione mesozoico-terziaria di Stilo (Calabria meridionale). Boll. Soc. Natur. in Napoli 80, 1–14.

Bonardi, G., Giunta, G., Perrone, V., Russo, M., Zuppetta, A., Ciampo, G., 1980. Osservazioni sull'evoluzione dell'Arco Calabro-Peloritano nel Miocene inferiore: la Formazione di Stilo–Capo d'Orlando. Boll. Soc. Geol. It. 99, 365–393.

Broquet, P., 1968. Etude géologique de la région des Madonies (Sicilie). Thèses Univ. Lille AO3037, 797 p.

Broquet, P., 1970. The geology of the Madonie Mountains of Sicily. In: Alvarez, W., Gohrbandt, K.H.A. (Eds.), Geology and History of Sicily. Petroleum Expl. Soc. Libya, pp. 201–230.

Buness, H., Giese, P., Eva, C., et al., 1990. EGT'85 seismic experiment in Tunisia: a reconnaissance of deep structures. In: Freeman, R., Muller, S. (Eds.), European Geotraverse Project – Data Compilations and Synoptic Interpretations, Proc. 6th Workshop, ESF, Strasbourg, pp. 197–210.

Burollet, P.F., 1991. Structures and tectonics of Tunisia. Tectonophysics 195, 359–369.

Burollet, P.F., Mugniot, G.M., Sweeney, P., 1987. The geology of the Pelagian Block: the margins and basins of Southern Tunisia and Tripolitania. In: Nairn, A., Kanes, W., Stelhi, F.G. (Eds.), The Ocean Basins and Margins, vol. 4B. Plenum Press, pp. 331–339.

Caflish, L., Schmidt di Friedberg, P., 1967. L'evoluzione della Sicilia e sue relazioni con la tettonica e la naftogenesi. Mem. Soc. Geol. It. 6, 449–474.

Caire, A., 1970. Sicily in its Mediterranean setting. In: Alvarez, W., Gohrbandt, K.H.A. (Eds.), Geology and History of Sicily. Petroleum Expl. Soc. Libya, pp. 145–170.

Caire, A., Mattaeur, M., 1960. Comparaison entre la Berbérie et le territoire Siculo-calabrais. C. R. Ac. Sc. 251, 1804–1806.

Carbone, S., Grasso, M., Lentini, F., 1984. Carta Geologica della Sicilia Sud-Orientale, scala 1:100.000, S. El.Ca., Firenze.

Carbone, S., Grasso, M., Lentini, F., 1986. Carta Geologica del settore Nord-Orientale Ibleo, scala 1:50.000, S.El.Ca., Firenze.

Carbone, S., Catalano, S., Grasso, M., Lentini, F., Monaco, C., 1990. Carta geologica della Sicilia centro-orientale. Scala 1:50.000. S.El.Ca., Firenze.

Carmignani, L., Cocozza, T., Ghezzo, C., Pertusati, P.C., Ricci, C.A., 1986. Guidebook to the excursion on the Paleozoic basement of Sardinia, IGCP Proj., 5.

Carmisciano, R., Gallo, L., Lanzafame, G., Puglisi, D., 1981a. Le Calcareniti di Floresa nella costruzione dell'Appennino Calabro-Peloritano (Calabria e Sicilia). Geol. Romana 20, 12 p.

Carmisciano, R., Lentini, F., Puglisi, D., 1981b. Caratteri petrografici ed evoluzione tettonico-sedimentaria della Formazione di Piedimonte (Sicilia nord-orientale). Rend. Soc. It. Min. Petr. 37, 91–104.

Carmisciano, R., Puglisi, D., 1979. Caratteri petrografici delle arenarie del Flysch di Capo d'Orlando (Monti Peloritani, Sicilia nord-orientale). Rend. Soc. It. Min. Petr. 34, 403–424.

Castany, G., 1956. Essai de synthèse géologique du territoire Tunisie–Sicile. Ann. Min. Géol., Tunis 16, 101.

Catalano, R., 1987. Northeastern Sicily Straits. Stratigraphy and structures from seismic reflection profiles. Rend. Soc. Geol. It. 9, 103–112.

Catalano, R., D'Argenio, B., 1978. An essay of palinspastic restoration across the western Sicily. Geol. Romana 17, 145–159.

Catalano, R., D'Argenio, B., 1982. Schema geologico della Sicilia. In: Catalano, R., D'Argenio, B. (Eds.), Guida alla geologia della Sicilia Occidentale. Soc. Geol. It., Palermo.

Catalano, R., Di Maggio, C., 1996. Sovrapposizione tettonica delle Unità Imeresi sulle Panormidi nei Monti di Palermo (Sicilia). Naturalista Siciliano 3–4, 147–166.

Catalano, R., Montanari, L., 1979. Geologia dei Monti di Trabia-Termini Imerese e dei Sicani orientali. Rend. Acc. Sc. Fis. Mat., Soc. Naz. Sc. Lett. e Arti Napoli IV 46, 1–27.

Catalano, R., Di Stefano, P., Nigro, F., Vitale, F.P., 1993. Sicily Mainland and its offshore: a structural comparison. In: Max, M.D., Colantoni, P. (Eds.), Geological Development of the Sicilian-Tunisian Platform. UNESCO Report in Marine Science, vol. 58, pp. 19–24.

Catalano, R., Di Stefano, P., Sulli, A., Vitale, F.P., 1996. Paleogeography and structure of the central Mediterranean: Sicily and its offshore area. Tectonophysics 260, 291–323.

Catalano, R., Franchino, A., Merlini, S., Sulli, A., Agate, M., Basilone, L., 1998. Materiali per la comprensione dell'assetto profondo della Sicilia Centro-Occidentale. In: Catalano, R., Lo Cicero, G. (Eds.), Guida escursioni 79° Congr. Soc. Geol. It., vol. 1, pp. 175–185.

Catalano, R., Infuso, S., Sulli, A., Vitale, F.P., Agate, M., 1994. Plio-Pleistocene of Sicily and surroundings marine areas. Tectonosedimentary evolution. In: Riassunti 77° Congr. Naz. Soc. Geol. It., Bari, pp. 235–236.

Catalano, R., D'Argenio, B., Montanari, L., Renda, P., Abate, B., Monteleone, S., Macaluso, T., Pipitone, G., Di Stefano, E., Lo Cicero, G., Di Stefano, P., Anesi, V., 1978. Contributo alla conoscenza della struttura della Sicilia Occidentale: il profilo Palermo-Sciacca. Mem. Soc. Geol. It. 19, 485–493.

Catalano, R., Di Stefano, P., Kozur, H., 1988. New results in the Permian and Triassic stratigraphy of Western Sicily with special reference to the section at Torrente San Calogero SW of the Pietra di Salamone (Sosio Valley). In: Atti 74° Congr. Soc. Geol. It., vol. A, Sorrento, pp. 126–135.

Catalano, R., Di Stefano, P., Kozur, H., 1991. Permian circum Pacific deep water fauna from the Western Tethys (Sicily–Italy) – New evidences for the position of the Permian Tethys. Palaeog. Palaeocl. Palaeoecol. 87, 75–108.

Catalano, R., Franchino, A., Merlini, S., Sulli, A., 2000. Central western Sicily structural setting interpreted from seismic reflection profiles. Mem. Soc. Geol. It. 55, 5–16.

Catalano, R., Merlini, S., Sulli, A., 2002. The structure of western Sicily, central Mediterranean. Petrol. Geosci. 8, 7–18.
Catalano, S., Di Stefano, A., 1996. Nuovi dati geoloogici e stratigrafici sul Flysch di Capo d'Orlando nei Peloritani orientali (Sicilia NE). Mem. Soc. Geol. It. 51, 149–164.
Censi, P., Chiavetta, S., Ferla, P., Speziale, S., Di Stefano, P., 2000. Tholeiitic magmatites in Lower Permian Turbidites from Western Sicily. Mem. Soc. Geol. It. 55, 307–313.
Channel, J.E.T., Oldow, J., Catalano, R., D'Argenio, B., 1990. Paleomagnetically determined rotations in the Western Sicilian Fold and Thrust Belt. Tectonics 9, 641–660.
Cherchi, A., Montadert, L., 1982. Il sistema di rifting oligo-miocenico del Mediterraneo occidentale e sue conseguenze paleogeografiche sul terziario sardo. Mem. Soc. Geol. It. 24, 387–400.
Cogan, J., Rigo, L., Grasso, M., Lerche, I., 1989. Flexural tectonics of SE Sicily. J. Geodynamics 11, 189–241.
Colombi, B., Giese, P., Luongo, G., Morelli, C., Riuscetti, M., Scarascia, S., Shutte, K.G., Strowald, G., De Visintini, G., 1973. Preliminary report on the seismic refraction profile Gargano–Salerno–Pantelleria. Boll. Geofis. Teor. Appl. 15, 225–254.
C.N.R., 1991. Structural model of Italy, sheet n. 6, scale 1:500.000. Publ. P.F. Geodinamica, Dir. F. Barberi, S.El.Ca., Firenze.
Cristofolini, R., Lentini, F., Patane, G., Rasa, R., 1979. Integrazione di dati geologici, geofisici e petrologici per la stesura di un profilo crostale in corrispondenza dell'Etna. Boll. Soc. Geol. It. 98, 239–247.
Del Ben, A., Guarnieri, P., 2000. Neogene Transpression in the Cefalù Basin (Southern Tyrrhenian): comparison between land and marine data. Mem. Soc. Geol. It. 55, 27–33.
Della Vedova, B., Pellis, G., 1979. Risultati delle misure di flusso di calore eseguite nel Tirreno sud-orientale. In: Proc. Conv. Sc. Naz. P.F. Oceanografia fondi marini, pp. 693–712.
Della Vedova, B., Pellis, G., Corubolo, P., 1988. Evidenze termiche del rifting continentale nel Canale di Sicilia. Atti 6° Conv. Gruppo Naz. Geof. Terra Solida, pp. 687–698.
Di Geronimo, I., Ghisetti, F., Lentini, F., Vezzani, L., 1978. Lineamenti neotettonici della Sicilia orientale. Mem. Soc. Geol. It. 19, 543–549.
Di Stefano, P., 1990. The Triassic of Sicily and the Southern Apennines. Boll. Soc. Geol. It. 109, 21–37.
Di Stefano, P., Gullo, M., 1986. Late Triassic-Early Jurassic sedimentation and tectonics in the Monte Genuardo Unit (Saccence domain–Western Sicily). Rend. Soc. Geol. It. 9, 179–188.
Di Stefano, P., Gullo, M., 1997. Permian deposits of Sicily: a rewiew. Geodiversitas 19, 85–101.
Di Stefano, P., Gullo, M., Muttoni, G., 1998. Stratigrafia della successione carnico-retica di Pizzo Mondello. In: Guida all'escursioni 79° Conr. Naz. Soc. Geol. It., La Sicilia, un laboratorio naturale nel Mediterraneo, vol. 1, Palermo, pp. 121–128.
Di Stefano, A., Lentini, R., 1995. Ricostruzione stratigrafica e significato paleotettonico dei depositi plio-pleistocenici del margine tirrenico tra Villafranca Tirrena e Faro (Sicilia nord-orientale). Studi Geologici Camerti vol. spec. 1995/2, 219–237.
Di Stefano, P., Mindszenty, A., 2000. Fe–Mn encrusted "kamenitza" and associated features in the Jurassic of Monte Kumeta (Sicily). Subaerial and/or submarine dissolution? Sedimentary Geology 132, 37–68.
Dueé, G., 1969. Etude géologique des Monts Nebrodi (Sicile). Thèse Fac. Sci., 2 voll., 221+169 p., Paris.
Dueé, G., 1970. The geology of the Nebrodi Mountains of Sicily. In: Geology and History of Sicily. Petrol. Expl. Soc. Libya, pp. 187–200.
Fabiani, R., Trevisan, L., 1937. Di alcune novità geologiche nel territorio del foglio Termini Imerese (Palermo). Boll. Soc. Sc. Nat. Econ. 19, Palermo, p. 1–8.
Finetti, I.R., 1976. Mediterranean Ridge: A Young Submerged Chain associated with the Hellenic Arc. Boll. Geofis. Teor. Appl. 28, 31–62.
Finetti, I.R., 1982. Structure, stratigraphy and evolution of Central Mediterranean. Boll. Geofis. Teor. Appl. 24, 247–312.
Finetti, I.R., 1984. Geophysical study of the Sicily Channel rift zone. Boll. Geofis. Teor Appl. 12, 263–341.
Finetti, I.R., 1985. Structure and evolution of Central Mediterranean (Pelagian and Ionian Seas). In: Stanley, D.J., Wezel, F.C. (Eds.), Geological Evolution of the Mediterranean Basin. Springer, pp. 215–230.
Finetti, I.R., this volume. Understanding the Ionides and their geodynamics. Chapter 10.
Finetti, I.R., this volume. Geodynamic evolution of the Mediterranean Region, from the Permo-Triassic Ionian opening to the Present, constrained by new lithospheric CROP seismic data. Chapter 34.
Finetti, I.R., Del Ben, A., 1986. Geophysical study of the Tyrrhenian opening. Boll. Geofis. Teor. Appl. 28, 75–156.
Finetti, I.R., this volume. Ionian and Alpine Neotethyan oceans opening. Chapter 6.
Finetti, I.R., Del Ben, A., this volume. Ionian Tethys lithosphere roll-back sinking and back-arc Tyrrhenian opening from new CROP seismic data. Chapter 21.
Finetti, I.R., Del Ben, A., this volume. Crustal tectono-stratigrapic setting of the Pelagian foreland from integrated new CROP seismic data. Chapter 26.

Finetti, I.R., Morelli, C., 1972. Regional reflection seismic exploration of the Strait of Sicily. Saclant Conference, vol. 7, La Spezia.
Finetti, I.R., Morelli, C., 1973. Geophysical exploration of the Mediterranean Sea. Boll. Geofis. Teor. Appl. 12, 263–341.
Finetti, I.R., Morelli, C., 1974. Esplorazione geofisica dell'Area Mediterranea circostante il Blocco Sardo-Corso. In: Paleogeografia del Terziario Sardo nell'ambito del Mediterraneo Occidentale. Rdc. Sem. Fac. Scienze, Università di Cagliari.
Finetti, I.R., Del Ben, A., Forlin, E., Prizzon, A., this volume. Review of the main results of the pioneering MS seismic exploration programme in the deep water of the Mediterranean Sea (OGS, 1969–1980). Chapter 1.
Finetti, I.R., Lentini, F., Carbone, S., Catalano, S., Del Ben, A., 1996. Il Sitema Appennino Meridionale–Arco Calabro–Sicilia nel Mediterraneo centrale: studio geologico-geofisico. Boll. Soc. Geol. It. 115, 529–559.
Finetti, I.R., Lentini, F., Carbone, S., Catalano, S., Del Ben, A., Geletti, R., Pipan, M., 1997. Domini geologico-strutturali del segmento orogenico Appennino meridionale–Arco Calabro–Sicilia. In: Riass. Conv. Naz. Prog. CROP, Trieste, giugno 1997.
Finetti, I.R., Morelli, C., Zarudzki, E.F.K., 1970. Reflection seismic study of the Tyrrhenian Sea. Boll. Geofis. Teor. Appl. 12, 311–345.
Finetti, I.R., Lentini, F., Carbone, S., Del Ben, A., Di Stefano, A., Guarnieri, P., Pipan, M., Prizzon, A., this volume. Crustal tectono-stratigraphy and geodynamics of the Southern Apennines from CROP and other integrated geophysical-geological data. Chapter 12.
Fluegel, E., Di Stefano, P., Gullo, G., 1991. Microfacies and depositional structure of allochthonous carbonate base-of-slope deposits: the Late Permian Pietra di Salomone Megablock, Sosio Valley (Western Sicily). Facies 25, 147–186.
Fornaciari, E., Di Stefano, A., Rio, D., Negri, A., 1996. Middle Miocene quantitative calcareous nannofossil biostratigraphy in the Mediterranean region. Micropaleontology 42 (1), 37–63.
Gemmellaro, G.G., 1887–1899. La fauna dei calcari con fusuline della valle del Fiume Sosio, nella provincia di Palermo. G. Sci. Nat. Econ. 1988, 20, 26 p; Atti Acc. R. Lincei, Rend. Cl. Sci. Fis. Mat. Nat. 1989, s. 5, 7, 3 p.
Ghisetti, F., Vezzani, L., 1984. Thin-skinned deformations of the western Sicily thrust belt and relationships with crustal shortening: Mesostructural data on the Mt. Kumeta–Alcantara Fault Zone and related structures. Boll. Soc. Geol. It. 103, 129–157.
Giunta, G., 1985. Problematiche ed ipotesi sul Bacino Numidico nelle Maghrebidi siciliane. Boll. Soc. Geol. It. 104, 239–256.
Giunta, G., Liguori, V., 1973. Evoluzione paleotettonica della Sicilia. Boll. Soc. Geol. It. 92, 903–924.
Grasso, M., Lentini, F., 1982. Sedimentary and tectonic evolution of the Eastern Hyblean Plateau (South-eastern Sicily) during Late Cretaceous to Quaternary time. Palaeog. Palaeocl. Palaeoecol. 39, 261–280.
Grasso, M., Pedley, H.M., 1997. Upper Miocene peri-Tyrrhenian reefs of the Calabrian Arc: overview. Guidebook Interim Colloquium R.C.M.N.S., Catania, 4–9 novembre 1997.
Grasso, M., Lentini, F., Vezzani, L., 1978. Lineamenti stratigrafico-strutturali delle Madonie (Sicilia centro-settentrionale). Geol. Romana 17, 45–69, 1 carta geol.
Grasso, M., Guerrera, F., Loiacono, F., Puglisi, D., Romeo, M., Balenzano, F., Carmisciano, R., Di Pierro, M., Gonzalez-Donoso, J.M., Martin-Algarra, A., 1987. Caratterizzazione sedimentologica, biostratigrafica e mineralogico-petrografica di "successioni miste" infra-mioceniche affioranti in Spagna (Catena Betica) e in Italia meridionale (Monti Nebrodi e Appennino Lucano). Boll. Soc. Geol. It. 106, 475–516.
Guarnieri, P., Carbone, S., Di Stefano, A., 2002. The Sicilian orogenic belt: a critical tapered wedge? Boll. Soc. Geol. It. 121, 221–230.
Guarnieri, P., Puglisi, D., Balogh, K., 2000. Age and provenance of the exotics within the Eocene Imerese succession (NW Sicily). Mem. Soc. Geol. It. 55, 235–241.
Gullo, G., 1993. Studi stratigrafici sul Permiano ed il Trias pelagico della Sicilia occidentale. Tesi di Dottorato, Palermo, pp. 1–203.
Kastens, K., et al., 1988. ODP leg 107 in the Tyrrhenian Sea: Insights into passive margin and back-arc basin evolution. Geol. Soc. Am. Bull. 100 (7), 1140–1156.
Kozur, H., Mosler, M., 1989. Echinoderm remains from the middle Permian (Wordian) from Sosio Valley (Western Sicily). Neues Jahrb. Geol. B–A 132 (4), 677–685.
La Manna, F., Grasso, M., Romeo, M., Maniscalco, R., Di Stefano, A., 1995. Evoluzione tettonico-sedimentaria neogenica del bordo tirrenico dei Monti Nebrodi (Sicilia settentrionale). Studi Geol. Camerti vol. spec. 1995/2, 293–305.
Lentini, F., 1974. Caratteri stratigrafici e strutturali della zona di M. Judica (Sicilia orientale). Riv. Mineraria Siciliana a. XXV 145–147, 3–22.
Lentini, F., 1975. Le successioni mesozoico-terziarie dell'Unità di Longi (Complesso Calabride) nei Peloritani occidentali (Sicilia). Boll. Soc. Geol. It. 94, 1477–1503.

Lentini, F., 1982. The geology of the Mt. Etna basement. Mem. Soc. Geol. It. 23, 7–25.

Lentini, F., 2000. Carta Geologica della Provincia di Messina, scala 1:50.000, 3 fogli. S.El.Ca., Firenze.

Lentini, F., Tortorici, L., 1986. Tentative correlation and paleogeographic evolution of the Sicilian-Tunisian orogen. Giorn. Geol. ser. 3° 48 (1–2), 93–98.

Lentini, F., Vezzani, L., 1975. Le unità meso-cenozoiche della copertura sedimentaria del basamento cristallino peloritano (Sicilia nord-orientale). Boll. Soc. Geol. It. 94, 537–554.

Lentini, F., Di Stefano, A., Strazzulla, S., 1997. The Middle-Late Miocene terrigenous deposits of the Peloritani Mounts. In: Interim Colloquium R.C.M.N.S.: Neogene Basins of the Mediterranean Region – Controls and Correlation in Space and Time, Catania, pp. 14–18.

Lentini, F., Grasso, M., Carbone, S., 1987. Introduzione alla geologia della Sicilia e guida all'escursione. In: Conv. Soc. Geol. It. "Sistemi Avanfossa-Avampaese lungo la Catena Appenninico-Maghrebide", Naxos-Pergusa, 22–25 aprile 1987.

Lentini, F., Carbone, S., Catalano, S., Grasso, M., Monaco, C., 1990a. Principali elementi strutturali del Thrust Belt Appenninico-Maghrebide in Sicilia centro-orientale. Mem. Soc. Geol. It. 45, 495–502.

Lentini, F., Carbone, S., Catalano, S., Monaco, C., 1990b. Tettonica a thrust neogenica nella Catena Appenninico-Maghrebide: esempi dalla Lucania e dalla Sicilia. Studi Geol. Camerti vol. spec., 19–26.

Lentini, F., Carbone, S., Catalano, S., 1994. Main structural domains of the central mediterranean region and their tectonic evolution. Boll. Geofis. Teor. Appl. 36, 103–125.

Lentini, F., Carbone, S., Catalano, S., Di Stefano, A., Gargano, C., Romeo, M., Strazzulla, S., Vinci, G., 1995a. Sedimentary evolution of basins in mobile belts: examples from tertiary terrigenous sequences of the Peloritani Mts. (NE Sicily). Terra Nova 7, 161–170.

Lentini, F., Carbone, S., Catalano, S., Grasso, M., 1995b. Principali lineamenti strutturali della Sicilia nord-orientale. Studi Geol. Camerti vol. spec. 1995/2, 319–329.

Lentini, F., Carbone, S., Catalano, S., Grasso, M., 1996a. Elementi per la ricostruzione del quadro strutturale della Sicilia Orientale. Mem. Soc. Geol. It. 51, 179–195, 1 carta geol.

Lentini, F., Catalano, S., Carbone, S., 1996b. The External Thrust System in Southern Italy: a target for petroleum exploration. Petrol. Geosci. 2, 333–342.

Lentini, F., Catalano, S., Carbone, S., 2000. Note illustrative della Carta geologica della Provincia di Messina, scala 1:50.000. S.El.Ca., Firenze.

Lentini, F., Carbone, S., Di Stefano, A., Guarnieri, P., 2002. Stratigraphical and structural constraints in the Lucanian Apennines (Southern Italy): tools for reconstructing the geological evolution. J. Geodynamics 34, 141–158.

Lickorish, H., Grasso, M., Butler, R.W.H., Argnani, A., Maniscalco, R., 1999. Structural styles and regional tectonic setting of the "Gela Nappe" and frontal part of the Maghrebian thrust belt in Sicily. Tectonics 18, 655–668.

Maldonato, A., Stanley, D.I., 1977. Lithofacies as a function of depth in the Starit of Sicily. Geology 2, 11–117.

Malinverno, A., Ryan, W.B.F., 1986. Extension in the Tyrrhenian Sea and shortening in the Apennines as a result of arc migration driven by sinking of the lithosphere. Tectonics 5, 227–245.

Mascle, J., 1979. Etude géologique des Monts Sicani. Mem. Riv. It. Pal. Strat. 16, 1–431.

Montanari, L., 1968. Materiale per la geologia dell'Alto Lercarese (Sicilia). Boll. Soc. Geol. It. 87, 133–142.

Montanari, L., 1987. Lineamenti stratigrafico-paleogeografici della Sicilia durante il ciclo Alpino. Mem. Soc. Geol. It. 38, 361–406.

Montanaro-Gallitelli, E., 1956. Il Permiano del Sosio e i suoi coralli. Paleont. Ital. 49, 1–98.

Morelli, C., Gantar, G., Pisani, M., 1975. Batymetry, gravity and magnetism in the Strait of Sicily and in the Ionian Sea. Boll. Geofis. Teor. Appl. 17 (65), 39–58.

Ogniben, L., 1960. Nota illustrativa dello Schema geologico della Sicilia nord-orientale. Riv. Min. Siciliana 64–65, 183–212.

Ogniben, L., 1964. Arenarie tipo Taveyannaz in Sicilia. Geol. Romana 3, 125–170.

Oldow, J.S., Channell, J.E.T., Catalano, R., D'Argenio, B., 1990. Contemporaneous thrusting and large scale rotations in the western Sicilian fold and thrust belt. Tectonics 9, 661–681.

Patacca, E., Scandone, P., 1989. Post Tortonian mountain building in the Apennines. The role of the passive sinking of a relic of lithospheric slab. In: Boriani, A., Bonafede, M., Piccardo, G.B., Vai, G.B. (Eds.), The Lithosphere in Italy, Advances in Earth Science. Res. Atti Conv. Lincei 80, 157–176.

Patacca, E., Scandone, P., Giunta, G., Liguori, V., 1979. Mesozoic paleotectonic evolution of the Ragusa zone (southern Sicily). Geol. Romana 18, 331–369.

Puglisi, D., 1979. Variazioni composizionali nelle arenarie del Flysch di Reitano (M. Nebrodi, Sicilia centro-settentrionale). Miner. Petrogr. Acta 23, 13–46.

Puglisi, D., 1987. Le successioni torbiditico cretacico-terziarie della Sicilia nord-orientale nel quadro dell'evoluzione del settore meridionale dell'Arco Calabro-Peloritano e della Catena Maghrebide siciliana. Giorn. Geologia 49, 167–185.

Puglisi, D., 1992. Le successioni torbiditiche "tardorogene" della Sicilia orientale. Giorn. Geologia, s. 3 54 (1), 181–194.

Rehault, J.P., Boillot, G., Mauffret, A., 1984. The Western Mediterranean Basin geological evolution. Marine Geol. 55, 447–477.

Reuther, C.D., Eisbacher, G.H., 1985. Pantelleria Rift-crustal extension in a convergent intraplate setting. Geolog. Rundschau 74, 585–597.

Roure, F., Howell, D.G., Muller, C., Moretti, I., 1990. Late Cenozoic subduction complex of Sicily. J. Struct. Geology 12 (2), 259–266.

Ruggeri, G., 1966. Primi risultati di ricerche sulla tettonica della Sicilia occidentale. Geol. Romana 5, 453–456.

Ruggeri, G., De Vita, G., 1972. L'età della Formazione Lercara. Boll. Soc. Geol. It. 91 (4), 631–637.

Scandone, P., 1975. Triassic seaways and the Jurassic Tethys Ocean in the central Mediterranean area. Nature 256/5513, 117–119.

Scarascia, S., Lozej, A., Cassinis, R., 1994. Crustal structure of the Ligurian, Tyrrhenian and Ionian seas and adjacent onshore areas interpreted from wide-angle seismic profiles. Boll. Geofis. Teor. Appl. 36 (141–144), 1–19.

Scarascia, S., Cassinis, R., Lozej, A., Nebuloni, A., 2000. A seismic and gravimetric model of crustal structures across the Sicily Channel Rift Zone. Boll. Soc. Geol. It. 19, 213–222.

Schmidt di Friedberg, P., 1964–1965. Litostratigrafia petrolifera della Sicilia. Riv. Min. Siciliana 88/90, 188–217; Riv. Min. Siciliana 91/93 (1965) 73–140.

Trevisan, L., 1937. Scoperta di basalti accompagnati di tufi fossiliferi del Bajociano a Vicari (Palermo). Boll. Soc. Sc. Nat. Econ. Palermo 19, 3–15.

Trevisan, L., 1960. Elementi di Geologia. 4[a] Ediz., Pisa, 406 p.

Truillet, R., 1968. Etude géologique des Péloritains orientaux (Sicile). Thèse Fac. Sci., Paris. 2 voll., 547 p.

Vitale, F.P., 1990. Studi sulla Valle del medio Belice (Sicilia centro-occidentale). L'avanfossa Plio-Pleistocenica nel quadro dell'evoluzione paleotettonica dell'area. Tesi di Dottorato, Palermo, p. 1–202.

Wezel, F.C., 1973. Nuovi dati sull'età e posizione strutturale del Flysch di Tusa in Sicilia. Boll. Soc. Geol. It. 92, 193–211.

Wezel, F.C., 1974. Flysch successions and the tectonic evolution of Sicily during the Oligocene and Early Miocene. In: Geology and History of Sicily. Petrol. Expl. Soc. Libya, pp. 1–23.

Winnock, E., 1981. Structure du Bloc Pelagien. In: Wezel, F.C. (Ed.), Sedimentary Basins of Mediterranean Margins. Technoprint Bologna, pp. 445–464.

Zolotarev, V.G., Sochelnikov, V.V., 1980. Geothermal conditions of the Africa–Sicilian rise. Izv. Earth Phys. 16, 202–206.

CROP PROJECT: Deep Seismic Exploration of the Central Mediterranean and Italy
Edited by I.R. Finetti

CHAPTER 16

Comparative NV and WA Seismic Modelling of CROP Sections in the Tyrrhenian Sea

A. Del Ben[1], L. De Luca[2], I.R. Finetti[1], E. Forlin[1], D. Luzio[2,*], M. Pipan[1], A. Prizzon[1], M. Vitale[2]

ABSTRACT

Comparative NV and WA seismic modelling of CROP sections from the Southern Tyrrhenian Sea (M26 and M28A) was performed. After an opportune processing (static correction, amplitude balancing, frequency filtering, stacking and trace binning, polarization-analysis-based noise attenuation) of WA data, modelling reconstruction started from depth conversion of NV time-interpretation data. Picking of observed WA traveltimes, optimization by ray-tracing of velocity models, and search for the one in agreement with the NV-based lithospheric structures were the subsequent operations. Velocity-to-density conversion and iterated adjustments followed. Satisfactory seismic and gravity modelling were at the end obtained.

Keywords: CROP, vertical/wide-angle seismics, comparative modelling

1. INTRODUCTION

Wide-angle reflection/refraction (WARR) data can be collected simultaneously with near-vertical reflection (NVR) exploration, and the two integrated. Advantages of combined WARR and NVR studies have been described in Braile and Chiang (1986) and Mooney and Brocher (1987), among others. Processing and interpretation of NVR and WARR data should interact in many of the steps. For example, the inversion of WARR data can be constrained by velocity analysis and boundary geometries coming from the interpretation of NVR profiles. This interpretation, in its turn, can be improved using the velocity functions deduced by WARR modelling.

The advantage of WA/NV integration was demonstrated by results obtained from various experiments which highlighted the well-known theoretical statement that reflectors characterized by a low acoustic impedance may be nearly invisible if the incident rays are nearly perpendicular, but more clearly visible when the incident rays have angles near the critical value, or near to 90° when the incident ray has a greater velocity in the upper medium than in the lower one. This has increased interest in wide-angle data migration techniques, with the aim of integrating the two techniques for imaging buried structures. On the basis of such issues and of experimented results (BABEL Working Group, 1991), an informal Sea Land Group was constituted in 1994 in Italy, having the objective of integrating the CROP Mare project with the acquisition of high-density, large-offset, three-component data.

In this chapter, the results from the WA data of the CROP M26 and M28-A sections (Figure 1), constrained by the interpretation of the coincident NV sections (Finetti et al., this volume, Chapter 15), are described. For this purpose, the smoothed shapes of the geological bodies observed in NVR sections were mapped with an initial distance-depth model on the basis of the interval velocities. The boundary depths and the velocity and density distributions of the different geological structures were corrected to account for the numerous phases, picked out on the wide-angle sections, and the Bouguer anomalies.

[1]Department of Geological, Environmental and Marine Sciences, University of Trieste, Via Weiss 1, 34127 Trieste, Italy
[2]Department of Chemics and Physics of the Earth, University of Palermo, Via Archirafi 26, 90123 Palermo, Italy
[*]Corresponding author. Tel.: +39 091 6169260; e-mail: luzio@unipa.it (D. Luzio)

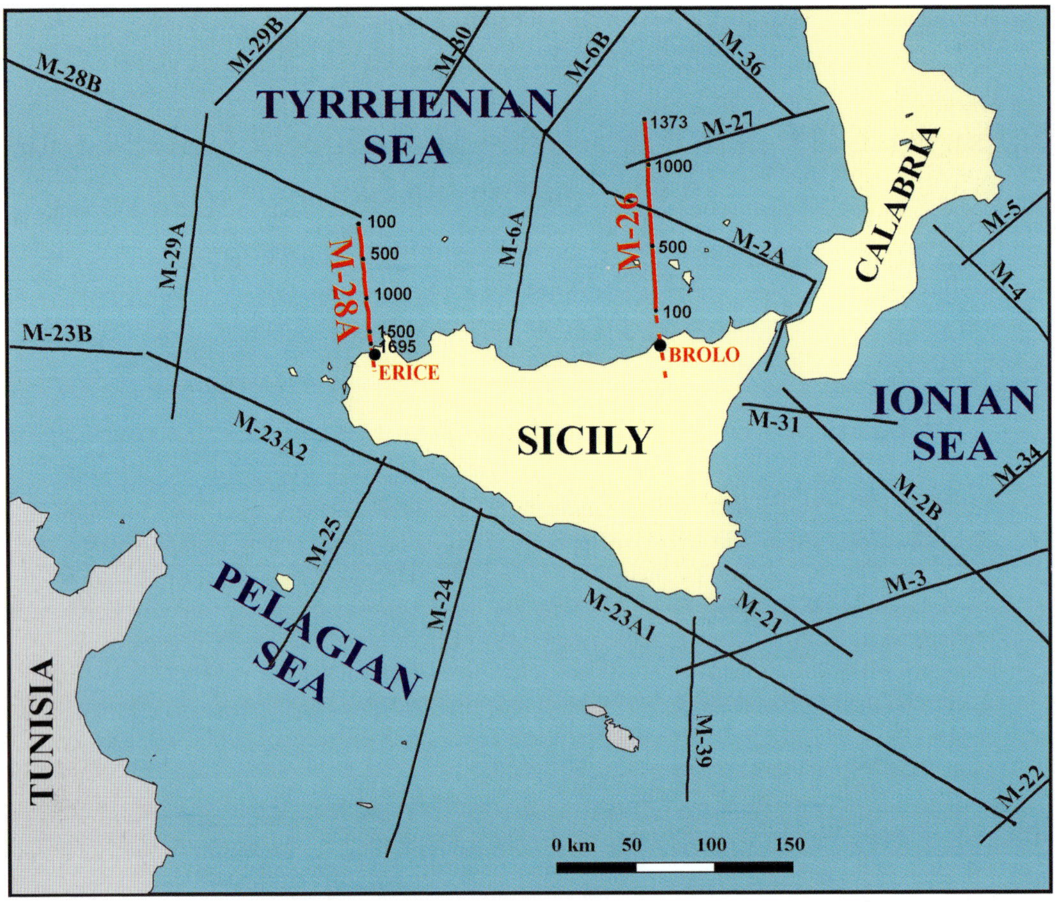

Figure 1: Location map of the CROP sections around Sicily. Red-traced sections are those used in the WA–NV seismic reflection modelling of this work.

2. LITHOSPHERIC OUTLINE OF THE SOUTHERN TYRRHENIAN-SICILY MARGIN

On the Northern Sicily margin are imprinted lithospheric tectono-dynamic data of major importance for understanding the Tyrrhenian stage geodynamics. In fact, beneath a complex tectonic stack of geological units coming from different realms, the subducted slab of the Ionian Tethys of Permo-Triassic opening is geophysically identified. In the Balearic stage (Upper Oligocene–Lower Miocene) the Corso-Sardinian block rotated and the SE Sardinian margin collided with the Panormid Promontory of NW Sicily. Crustal Panormid units were detached and transported over Ionian Tethys. In the Tyrrhenian stage, driven by Ionian slab retreat, the Ionian cover units (Ionides thrust over Sicily) are followed by Panormides, Calabrides Alpine Tethydes and related flysch units.

All these allochthonous units are stacked on the deformed Sicilian margin (Figures 5 and 6).

3. WIDE-ANGLE DATA ACQUISITION AND PROCESSING

The Sea Land Group data for the M26 and M28-A CROP sections were recorded by stations located in low-seismic noise-sites near the onshore extensions of the shot lines (Figure 1). The M-26 WA profile was recorded at Brolo and the M28-A near the town of Erice. Energization was carried out using 32 airgun of 90 l by the M/N OGS EXPLORA. The shot intervals adopted in the M26 and M28-A seismic profiles were 40 s and 20 s respectively, corresponding to a trace spacing of about 100 m and 50 m in the seismic sections. Since most of the energy of airgun signals lies in the frequency range 4–20 Hz, the data were recorded with a sampling rate of 125 Hz (M26) and 62.5 Hz (M28-A); anti-aliasing filters with cut-off frequencies 50 Hz and 25 Hz respectively were applied. In Figure 2 (a) and (b) the seismic sections M26 (recorded at Brolo) and M28-A are shown with a simple band pass filter in the range 2–18 Hz. The shot interval of 20 s produced wraparound noise in the wide-angle data. This effect occurs at the offsets for which slow phases, typically T-waves, arrive at the recording station with travel times greater than the interval between consecutive shots. These T-waves interfere with the first-refracted arrivals

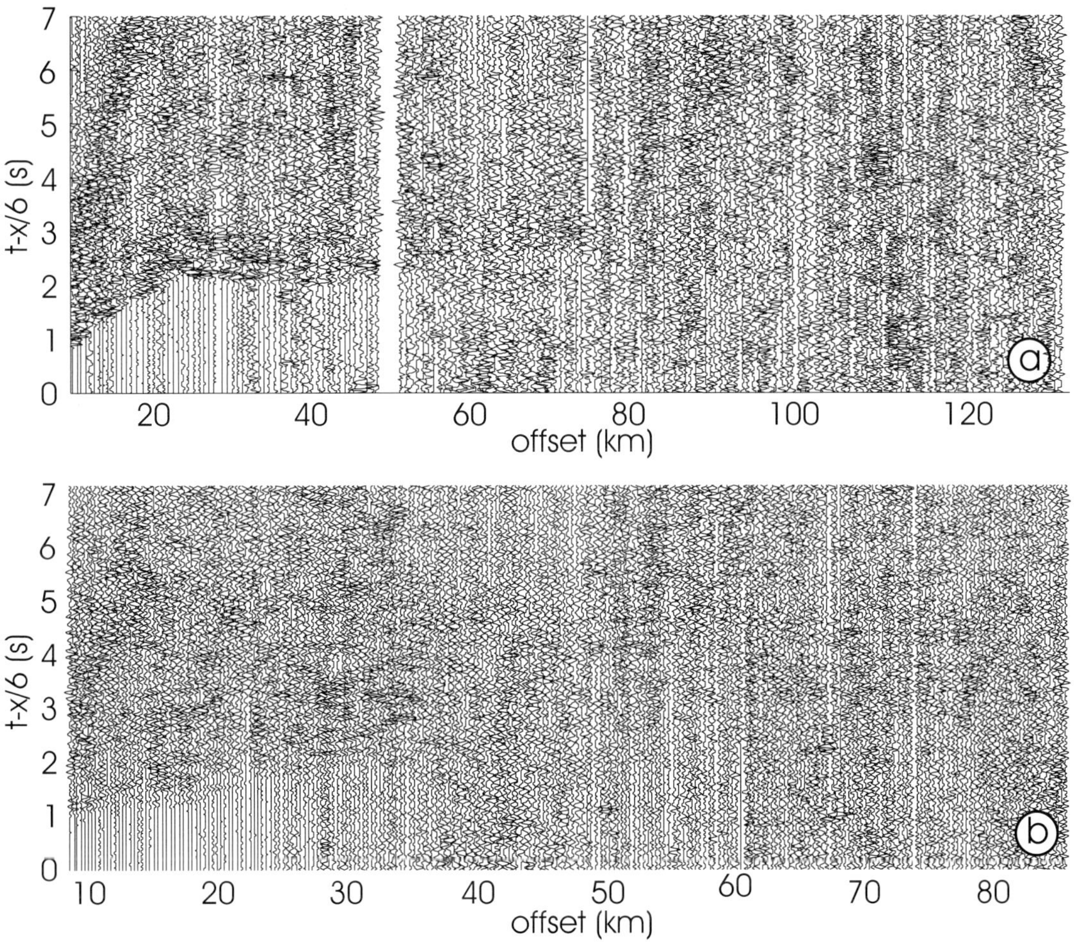

Figure 2: CROP wide-angle seismic sections (vertical component) after the application of a frequency filter (2–18 Hz): (a) M26 (recorded at Brolo); (b) M28-A.

Pg of the next shot, at an offset of about 40 km. An example of this effect is shown in Figure 2(b), in which a significant decrease in the S/N ratio can be observed in the offset range 35–50 km produced by the interference. The simultaneous determination of the boundary settings and velocity functions from a set of travel time and AVO curves is a very unstable inverse problem. In fact, even small perturbations of the boundary shape can lead to large modifications in the raypaths.

Many techniques have been designed to decrease this instability and to constrain the final models to the starting assumptions. Possible strategies consist of inverting sequentially only for one kind of parameter at each step or representing the boundary and velocity features by smoothed functions. Nevertheless, such strategies are not sufficient to solve the instability problem, which is also connected with the number of unknown parameters as related to that of the experimental data. A decrease in the variance of parameter estimates can be achieved by applying effective denoising techniques.

Even though the traces of the CROP WA sections are often very noisy, their high spatial density and their vector nature allow the application of denoising techniques capable of separating signal from noise using the coherence and polarization properties of seismic phases. The result of an effective processing is the identification of numerous phases on the sections which have different physical meanings, a necessary step for constructing detailed crustal models.

Since a number of tests performed using commonly available software for the analysis of NVR sections yielded unsatisfactory results when applied to WA data, new algorithms for data processing were designed and implemented on the Matlab platform. These allow optimization of the processing sequence and algorithm input parameters for each experimental data set.

The main algorithms of the processing sequence are described in the following.

Construction of data matrix. In the first stage, the data recorded blockwise in a format typical of the acquisition instrument are converted into a format readable by Matlab or by widely used processing softwares, and then stored in a single time sequence. In the second stage, the time sequence is arranged into an n-sample by m-trace (shot) matrix.

Static corrections. These include corrections for the effects of source and receiver height above a datum level, as well as for the variable thickness of water and/or the shallowest sediment layer. The corrections carried out in the first processing stage were aimed at smoothing the phase correlation lines traced on the seismic sections, in order to increase the effectiveness of the denoising algorithms based on the correlation analysis of traces. In order to carry out the bathymetric correction along each profile, the seafloor depth h was measured by sonar. The bathymetric function $h(x)$ and the function $v_a(x)$, the latter describing the mean value of the P-wave apparent velocity vs. offset for an average crustal structure, are used to calculate the static corrections in time and offset for each trace using the formula

$$s^c_{x'}(t') = s_x(t), \tag{1}$$

in which

$$t' = t - \frac{h(x)v_a(x)}{v_m\sqrt{v_a^2(x) - v_s^2}}\left(\frac{1}{v_m} - \frac{1}{v_s}\right), \qquad x' = x + \frac{h(x)v_s}{\sqrt{v_a^2(x) - v_s^2}} - \frac{h(x)v_m}{\sqrt{v_a^2(x) - v_m^2}}. \tag{2}$$

Equations (1), (2) simulate the substitution of the uppermost layer with velocity v_m (e.g. sea water) with another one whose velocity v_s approaches that of the underlying layer, under the assumption that the ray emersion angle depends on v_a, as in a 1D medium.

An inverse correction with respect to (1), (2) is applied after the denoising, so as to solve the direct kinematical and/or dynamic seismic problem and attribute the real velocities to the structures. The static correction for local effects, generally produced by small sedimentary basins, can be applied interpretatively. Such corrections should be made by considering the information coming from NVR data and the available geological evidence.

Amplitude balancing function. An amplitude balancing function whose temporal trend is offset dependent is often applied to recover the amplitude decay effects. It is defined as

$$A_i^c = A_i\big(t_i + t_0(x)\big)^{\alpha}, \tag{3}$$

where A_i and t_i are the amplitude and the time of the ith sample respectively, α is the positive proportionality constant between the instantaneous logarithmic gain and the logarithm of a properly corrected time; and $t_0(x)$ is a correction function.

1D and 2D frequency filters. In many steps of the adopted processing sequences, Butterworth band-pass frequency filters were applied, setting the low and the high cut-off frequencies respectively in the ranges 4–8 Hz and 12–18 Hz. The Matlab algorithms were optimized to minimize the phase distortion effects.

Since the spectral characteristics of noise and signal can change noticeably within a seismic section, a more effective filtering is obtained by offset- and/or time-variant filters. This preserves the high-frequency content of shallower phases, and so increases the resolution in the model's shallower portion. They prove particularly useful in the processing of long traces (over 20 s) for the S-phase study.

Stacking and trace binning. The stacking of high-density WA seismic data was performed between signals coming from the same seismic source and recorded by small-extension arrays (array stacking), or coming from several similar close sources and recorded by a single station (horizontal binning).

The analysis of the effect of the horizontal binning on seismic sections having a trace interval of about 50 m and reduction velocity 6 km/s showed that the S/N ratio regularly grows as the window width increases up to about 450 m. The S/N ratio tends to a saturation value for larger intervals because of the destructive interference in the signal. In some applications, clear artefacts appear for window widths larger than 300 m. The optimal window width ranges between 200 m and 400 m.

The MCW algorithm (Chironi et al., 1997) allows the use of much wider horizontal binning windows, obtaining a larger increase in the S/N ratio. It operates by replacing the central trace of a space–time moving window with that resulting from a weighted stacking between the traces inside the window. Before stacking, the traces are re-phased with respect to the central one through a cross-correlation analysis; the weighting factors assigned to each trace depend on the position of the trace and on its correlation with the central one. It has been observed that MCW stacking can be applied to CROP WA sections using windows up to 800 m wide without producing visible artefacts.

A second algorithm was also implemented (Chironi et al., 1997) to increase the efficiency of the array stacking. It performs a modulation of the stacked section by a weight matrix obtained by smoothing a coherency attributes matrix. A further denoising algorithm, called PCCSTACK (Vitale, 2003), performs a weighted array stacking between all the components of the recording stations located on the same site. The weighting factors depend on instantaneous coherence attributes computed from the signal's Hilbert transforms. Such attributes are based on a sample-by-sample phase correlation between corresponding signals recorded with different stations and horizontal phase coherence within either section.

Coherency filters. Some denoising algorithms for WA sections discriminate signal from noise through the much larger lateral coherency of the first than the latter (Carrozzo et al., 2002). A filter that was designed specifically for the analysis of CROP data and extensively applied to them is CDF, aimed at the enhancement of coherency between traces. It is based on the assumption that the difference in lateral coherence extent between signal and noise also holds for each multi-resolution detail resulting from the discrete wavelet transform. After the multi-resolution analysis, a characterization of the space coherency in each detail section is performed. The individual details are then attenuated in the time intervals where a lower lateral coherence indicates a lower S/N, and finally added up to obtain the denoised trace.

Comparison between the results obtained by CDF and other coherency filters based on different seismic-trace decomposition techniques (Ulrych et al., 1999) showed the greater efficiency of the CDF algorithm.

Polarization analysis. The acquisition of three-component seismic data makes it possible to study the polarization properties of wave fields. The determination of these properties may help in noise reduction, in the discrimination of interfering seismic phases from waves with different polarization features, and principally in the identification of P or S phases.

An algorithm (De Luca, 2000) provides, through the analysis of a cross-energy matrix, estimates of the polarization vector that prove to be unbiased by the presence of random noise in the case of linearly polarized wave trains.

In the processing of wide-angle sections, some of the algorithms described in this paragraph can be applied with different control parameters and in different steps of the processing. The M26 (recorded at Brolo) and M28-A sections were processed following the scheme represented in Figure 4. As regards the M26 section recorded at Capo d'Orlando, since only the vertical component section was available, a partial version of this scheme was applied.

4. GEOPHYSICAL MODELLING

The seismic model's resolution and reliability can be significantly increased if the inversion of the travel times of wide-angle reflected, diving or head waves is carried out jointly with those of near-vertical reflected waves. WA and NVR travel times were not used in a simultaneous inversion for the modelling along the M26 and M28-A profiles, but the interpretation of this latter identifies the structures which must be included in the interpretative model and their relative position in the starting model of the WA data interpretation.

The approach used to design the velocity and corresponding density models for the M26 and M28-A profiles consisted of the following steps:

1. sketching out, and rough conversion to depth models of the TWT lithospheric seismogeological sections obtained from NVR data interpretation (Finetti et al., this volume, Chapter 15);
2. synthesis and representation of all other constraints;
3. parameterization of the starting P-wave velocity models, compatible with the interpretation algorithm used;
4. tracing of the observed travel times on the wide-angle sections;
5. a priori physical interpretation of each experimental travel-time curve;

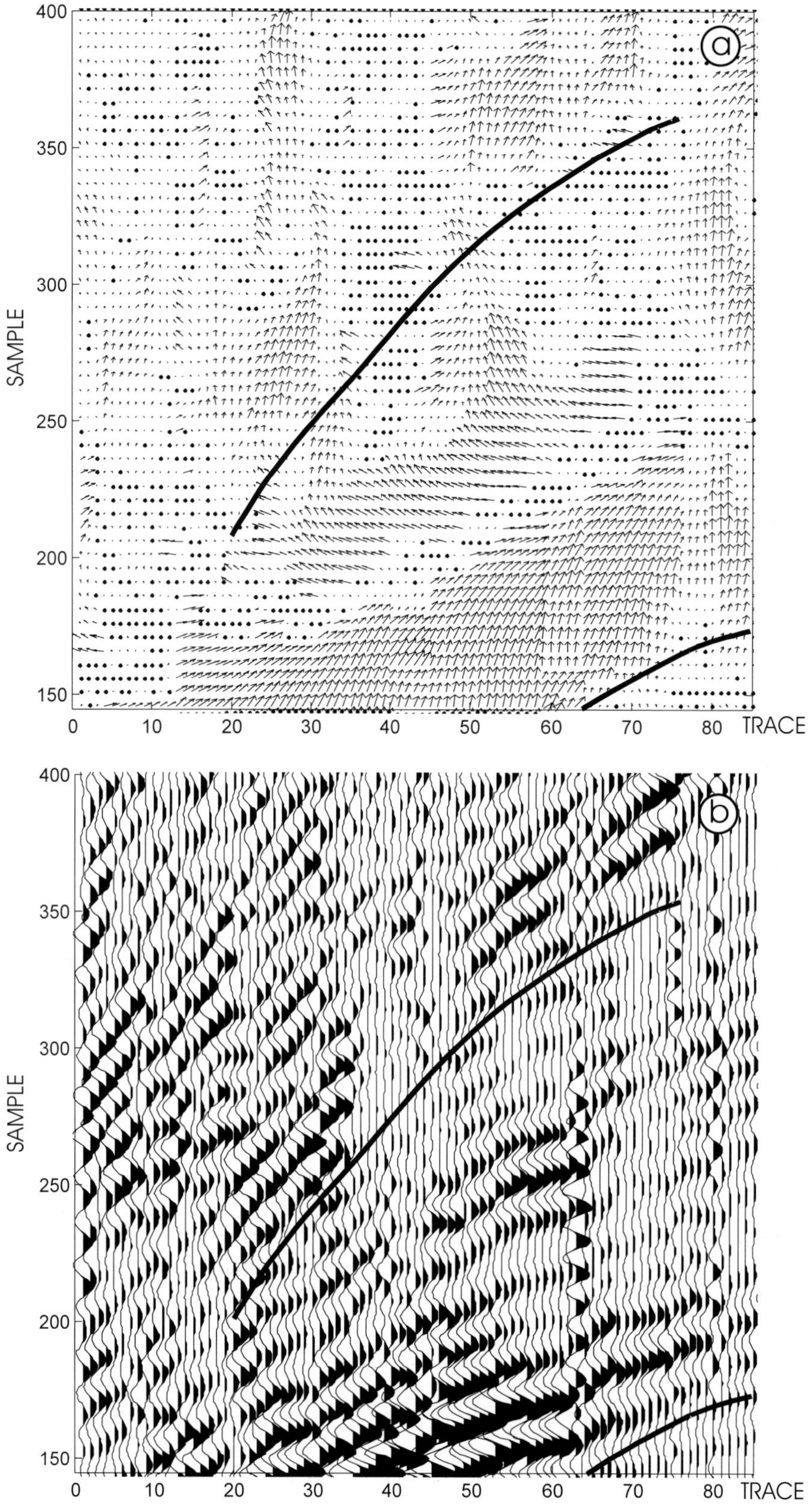

Figure 3: (a) Polarization vectors divided by their respective space variance. (b) A portion of the CROP wide-angle seismic section M27 with the interpreted phases evidenced by lines.

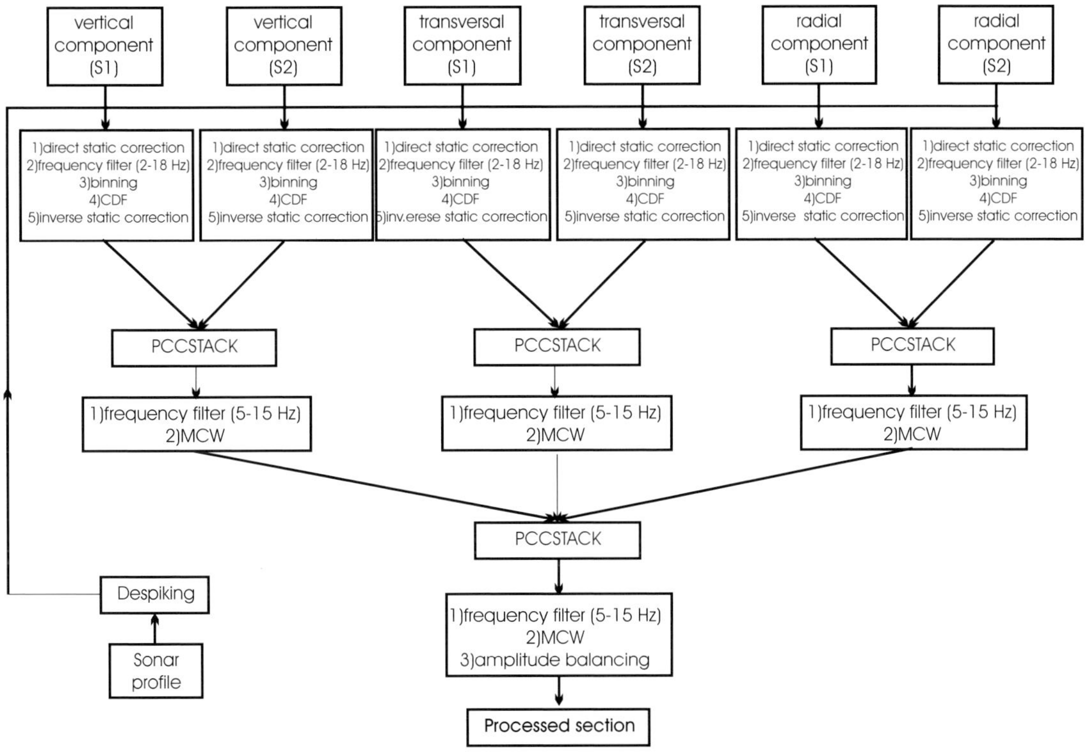

Figure 4: Block diagram of the processing sequence applied to the M26 and M28-A CROP sections. For details about each algorithm see Paragraph 3. S1 and S2 indicate the recording stations.

6. optimization of the velocity models with the 2D ray-tracing and inversion algorithm Rayinvr by Zelt and Smith (1992), using a quasi-inversion approach;
7. search, within the class of equivalent models, for the velocity models most in agreement with the main lithospheric structures deduced from NVR sections;
8. conversion from velocity to density models by means of the Nafe and Drake correlation law (Nafe and Drake, 1963);
9. adjustment of density models by the 2.5D forward-modelling and gravity inversion program Gravmag (Pedley et al., 1993);
10. iteration of the process (back to step 5) when the required corrections to density models exceed the uncertainty of the Nafe and Drake correlation law.

4.1. Seismic Modelling

The 2D P-wave velocity models along the M26 and M28-A profiles were constructed using the ray-tracing and inversion algorithm by Zelt and Smith (1992), in which the model parameterization is well suited to the inversion of reflection/refraction data. The velocity model is defined in terms of layer boundaries and top and bottom-layer velocities, which are specified by an arbitrary number and spacing of boundary and velocity nodes. Each boundary is defined by a linear interpolation between consecutive nodes; the velocity field in each layer is determined by bilinear interpolation between four adjacent nodes, two of which are relative to the top-layer velocity and the remaining two relative to the bottom-layer velocity.

In the upper portion of the crust, velocity data from NV is mostly required. In Figures 5 and 6 are shown the lithospheric seismogeological sections M26 and M28-A obtained from the NVR data (Finetti et al., this volume, Chapter 15). These TWT sections were initially simplified and converted to depth sections.

Starting from such models, the travel-time curves correlated on the M26 and M28-A wide-angle sections (Figures 7 and 8(a)) were interpreted by a quasi-inversion approach consisting in optimization steps from the surface layers down to deeper levels. A reasonable fit between observed and calculated travel times was obtained by the application of the inversion algorithm, together with manual adjustments to model parameters when the solution of the inverse problem failed.

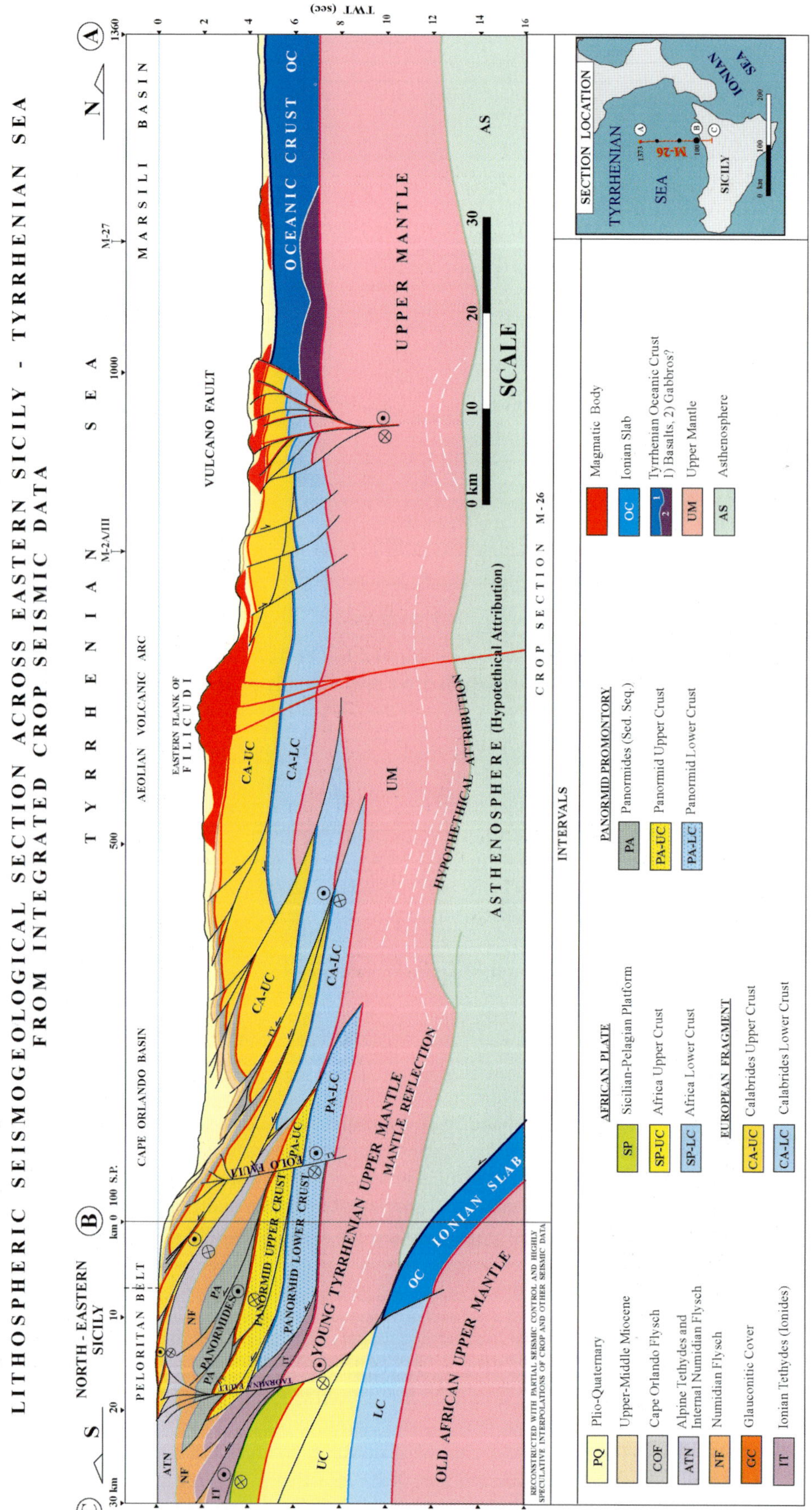

Figure 5: Lithospheric seismogeological section CROP M26 (after Finetti et al., this volume, Chapter 15).

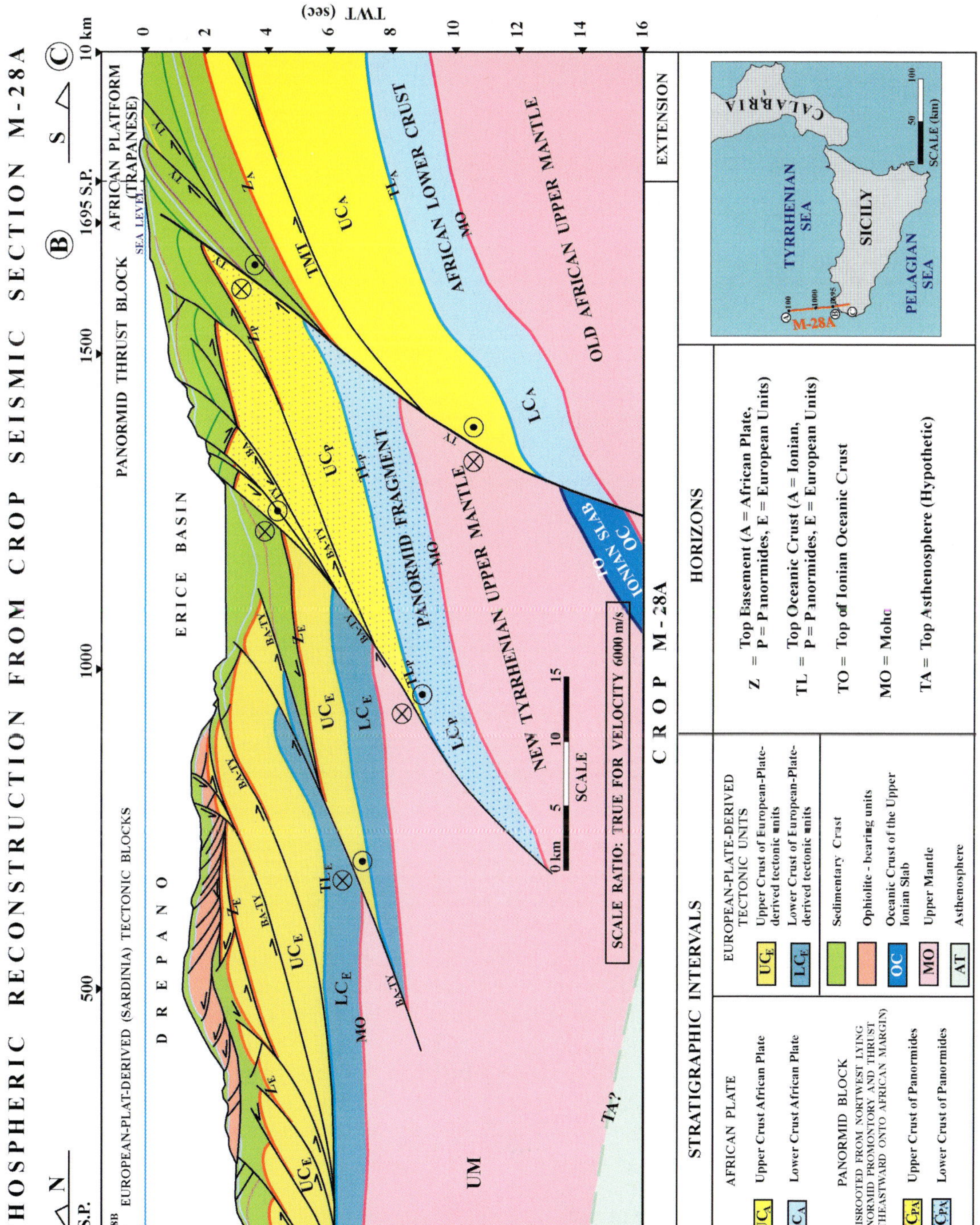

Figure 6: Lithospheric seismogeological section CROP M28-A (after Finetti et al., this volume, Chapter 15).

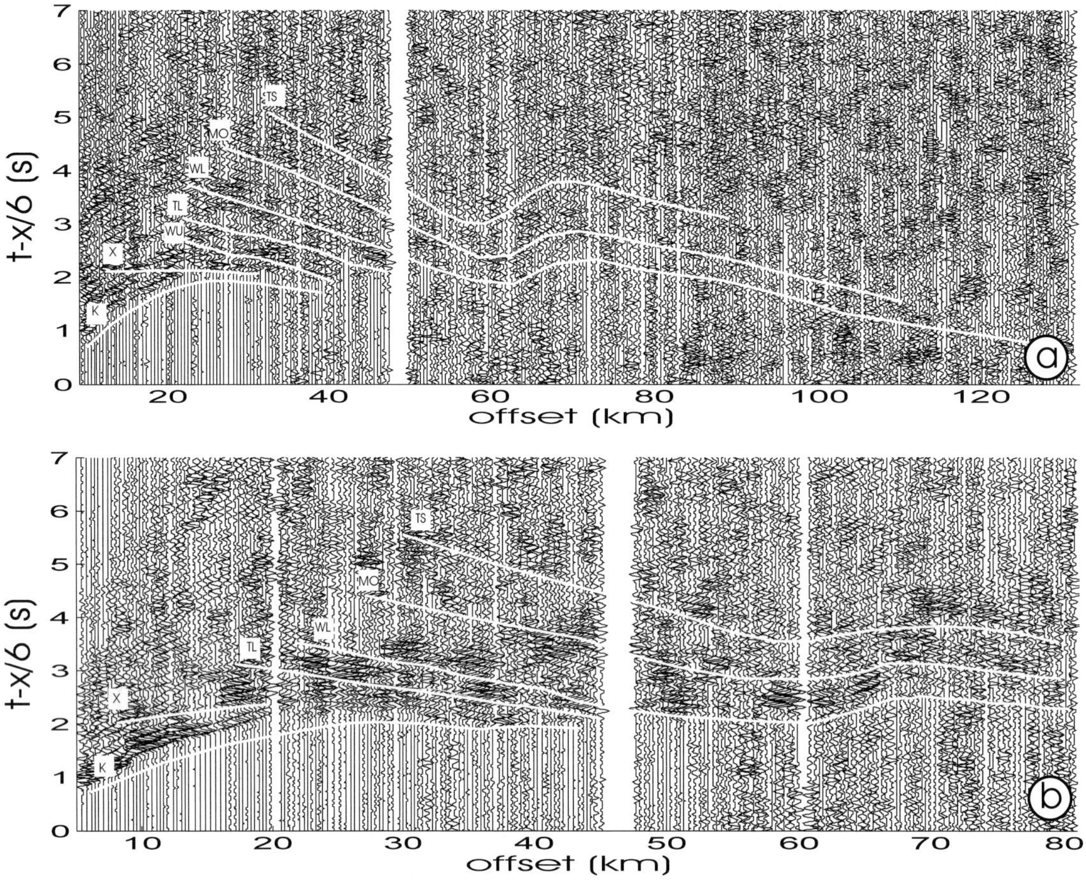

Figure 7: Processed wide-angle seismic sections CROP M26 recorded at: (a) Brolo (processing carried out according to the scheme in Figure 4); (b) Capo d'Orlando (processing carried out only on the vertical component section). The travel-time curves used in the interpretation are: K = first arrivals from sedimentary cover to basement; X = buried basement beneath thrust; WU = within upper crust; TL = top lower crust; WL = within lower crust; MO = Moho; TS = top subducted Ionian slab.

CROP M26 Section. The final velocity model and the ray tracings along the M26 profile, relative to both the Brolo and Capo d'Orlando stations, are shown in Figure 9. The model consists of 83 different spatial positions and 141 velocity parameters. The RMS of the residuals between calculated and observed travel times is about 0.15 s. The refracted K phase from the upper crust allows us to determine the velocity field of the crust's shallowest portion in the first 45 km of the profile. The K travel-time curve displays an evident concavity, indicating a relatively steep vertical velocity gradient in the sediments and underlying basement. The average velocity is 5.6 km s^{-1} beneath the surface sedimentary cover, and 6.0 km s^{-1} at about 9 km depth. The upper crust thins in the central part of the profile, beyond the Aeolian Arc, which has an average velocity of about 4.0 km s^{-1}, and before the transition to the oceanic crust at about 90 km along the profile, reaching a thickness of about 4 km. The reflected phase X constrains the bottom of a low-velocity sedimentary body belonging to the Panormide block, which in proximity to the coast line dips under fragments of the European upper crust (UC_E). This body has an average velocity of 4.8 km s^{-1} and is located in the first 25 km of the model at a depth between 4 km and about 7 km.

The velocities of the lower crust and the presence of a layered structure within it are constrained by the TL, WL and MO phases. The TL phase is from P-waves which reflect from the top of the Panormide continental lower crust (LC_P); this boundary lies at 11.5 km in the first 55 km along the profile. An intermediate interface within the lower crust, whose trend is parallel to the lower crust top, was modelled from the WL arrivals, which are from waves reflected from the portion of this boundary lying within the Panormide continental lower crust and diving in the upper layer, which is a fragment of the European-lower crust (LC_E). Using this phase we derived a possible velocity distribution in the shallower part of the continental lower crust and confirmed the transition to the oceanic crust (OC), observed on the NVR sections, at about 85 km from the coast line. Even though the oceanic crust is scarcely constrained by the WL arrivals and reflected waves from the Moho, the travel times interpretation assigned 6.4 km s^{-1} and 7.2 km s^{-1} to the top and bottom velocity of the oceanic lower crust, respectively.

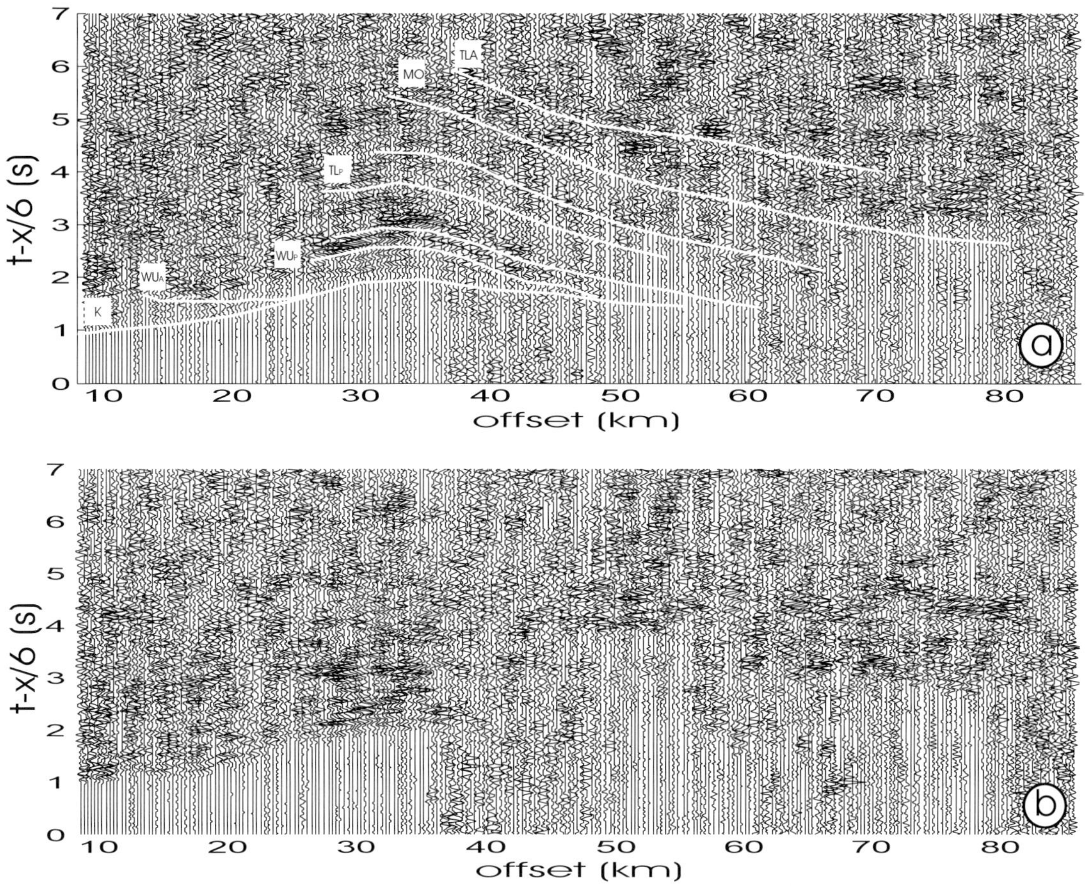

Figure 8: Processed wide-angle seismic sections CROP M28-A recorded near Erice. (a) The processing of this seismic section was carried out according to the scheme in Figure 4. The travel-time curves used in the interpretation are: K = first arrivals from sedimentary cover to basement; WU_A = within African upper crust; WU_P = within Panormide upper crust; TL_P = top Panormide lower crust; MO = Moho; TL_A = top African lower crust. (b) The processing of this seismic section was done only on the vertical component.

The Moho is modelled by the MO phase from sub-critical and post-critical reflections, its depth ranges from 21 km to 24 km in the first 55 km of profile and remains constant at 21 km in the following sector as far as the transition to the oceanic crust. On average, the continental lower crust along the M26 profile proved to be characterized by a high velocity; in particular, the average velocity ranges between 6.9 km s^{-1} and 7.4 km s^{-1}.

The infra-mantle TS phase was observed on both seismic sections and was attributed to reflected arrivals from a very steep interface, which Finetti et al. (this volume, Chapter 15) interpreted as the top of the Ionian slab (IS).

CROP M28A Section. The final P-wave velocity model and the ray tracing along the M28-A profile are shown in Figure 10. The model uses 97 different spatial positions and 228 velocity parameters. The RMS of the residuals between calculated and observed travel times is about 0.20 s. The K phase was correlated on the seismic section for about 40 km and is for waves diving beneath the shallow sediments in a very heterogeneous upper crust, which belongs to the African and European plates.

In the first part of the model, the K-phase arrivals sample the basement of the African upper crust, whose average velocity is 4.9 km s^{-1} beneath the surface sedimentary cover, and 5.9 km s^{-1} at the boundary with the upper crust shallower portion. Close beneath the sediments of the Erice basin, the upper crust of the Panormide block (UC_P) is very shallow and, as indicated by the interpretation of the phase K, its velocity at top and bottom is on average 6.2 km s^{-1} and 6.4 km s^{-1}, respectively.

The reflected phases WU_A and WU_P constrain a discontinuity within the African and Panormide upper crust, whose depth increases from about 6 km to about 10 km in the first 14 km of profile, and becomes very small, less

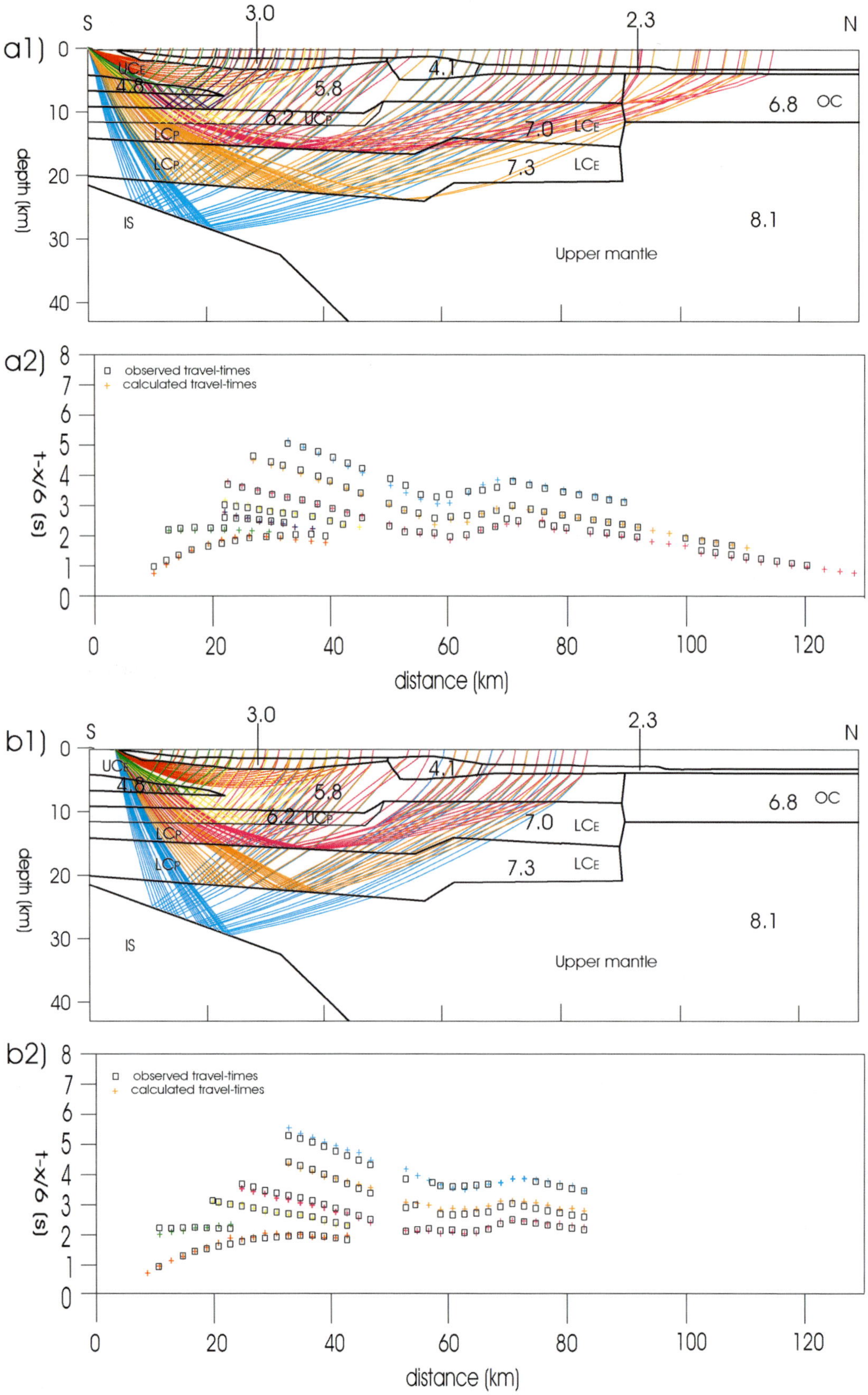

Figure 9: (a1), (a2) Velocity model and ray tracing. (b1), (b2) Comparison between observed and calculated travel times, along the seismic profile CROP M26 (a1/b1 Brolo station; a2/b2 Capo d'Orlando station).

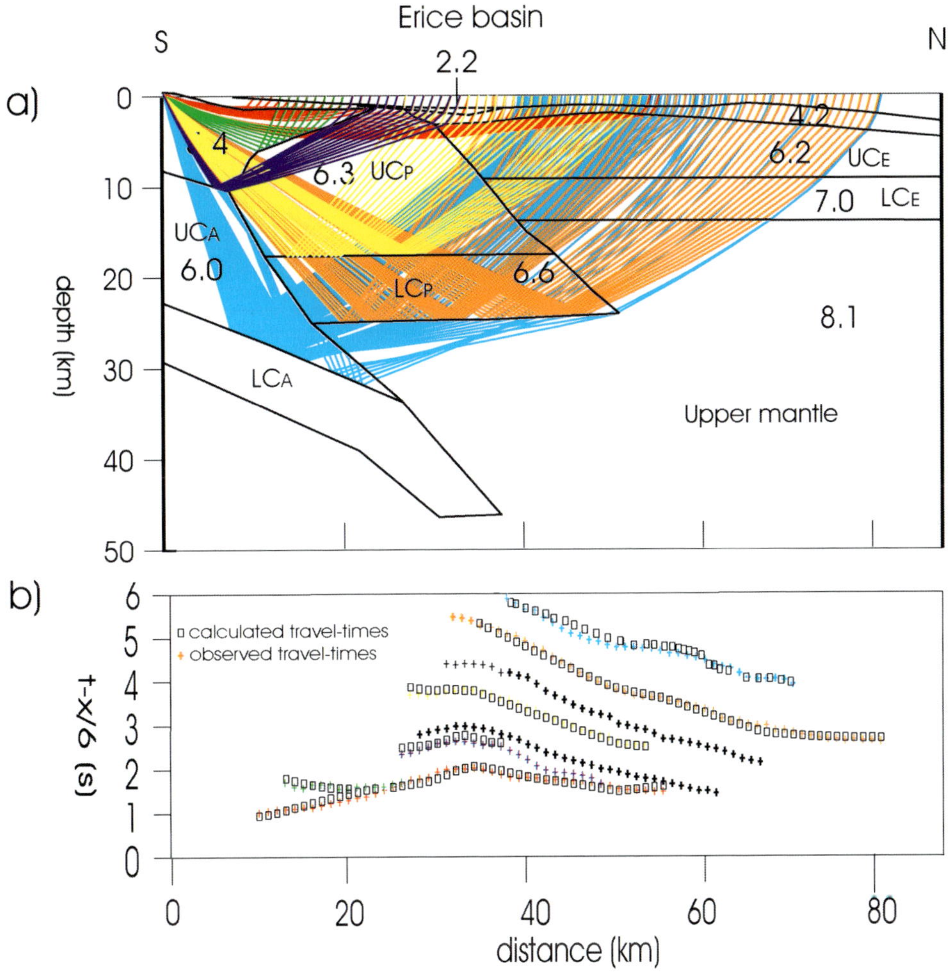

Figure 10: (a) Velocity model and ray tracing. (b) Comparison between observed and calculated travel times along the seismic profile CROP M28-A.

than 2 km, nearly below the Erice basin. In the central northern part of the profile, phases reflected from the upper crust top and bottom were not correlated, therefore the geometrical setting of this layer results from the analysis of NVR data, and is confirmed by the interpretation of the travel times observed on the WA section, for waves reflected from deeper boundaries.

The TL_P and TL_A phases are for P waves which reflect from the top of the Panormide and African lower crust (LC_P and LC_A). The first lies in the central part of the model, between about 10 km and 45 km from the origin of the profile, at a nearly constant depth of 17 km; the second interface is characterized by a depth which decreases from about 20 km to about 30 km in the first 30 km of the profile, and is connected with the African plate underthrusting the Panormide fragment. Even though no reflected phase was correlated from the top of the European lower crust (LC_E), such an interface was represented as sub-horizontal at a depth of about 9.5 km, in agreement with the apparent structural geometries observed on NVR sections, and with the travel time curves for deeper phases.

It turned out that the Panormide lower crust has on average a lower velocity than the European one. In particular, the velocities range from $\sim$ 6.4 km s^{-1} to $\sim$ 6.7 km s^{-1}, and from $\sim$ 6.8 km s^{-1} to $\sim$ 7.2 km s^{-1} respectively in the Panormide and European lower crusts.

The depth of the Moho of the Panormide block is 25 km and its trend is parallel to the top of the lower crust. The Moho is shallower in the European plate, being at a depth of about 14 km.

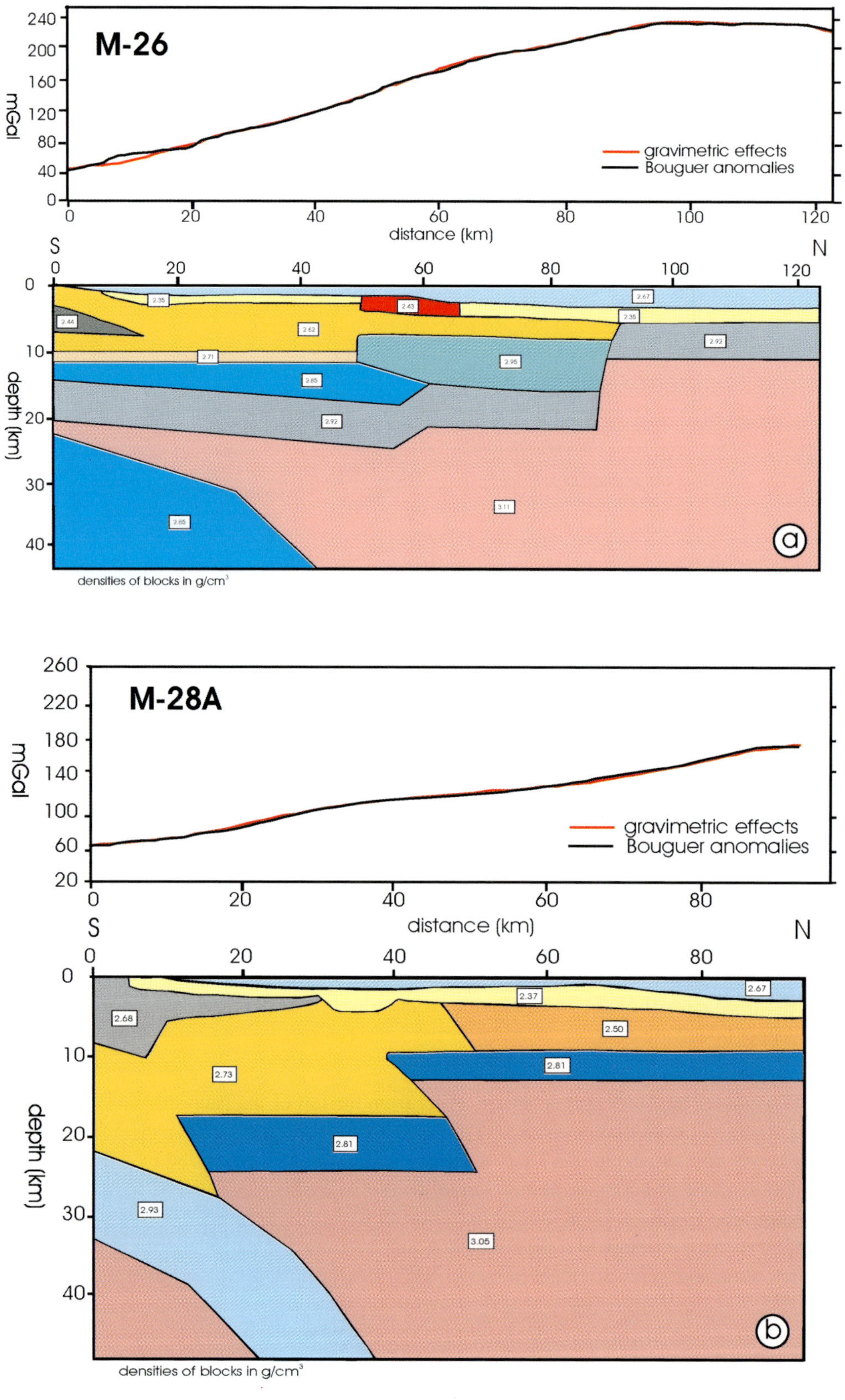

Figure 11: Comparison between Bouguer anomalies and gravimetric effect calculated for the density model for profiles: (a) M26; (b) M28-A.

4.2. Gravity Modelling

The compatibility of the obtained velocity models with the Bouguer anomalies along the same profiles was verified by assuming Nafe and Drake's (1963) velocity–density correlation law. The Bouguer anomalies available for the marine areas around Sicily are by Morelli (1975), those inland by Carrozzo et al. (1992).

The density models corresponding to the velocity distributions were adjusted by the 2.5D gravity forward-modelling and inversion program Gravmag (Pedley et al., 1993). When the corrections to density values turned out higher than the errors predicted by the correlation law, and when significant adjustments to the geometrical parameters were necessary to fit the Bouguer anomalies, the seismic data interpretation was repeated until the final crustal model agreed with both seismic and gravity data.

In Figure 11(a) are shown the final density model and the comparison between Bouguer anomalies and gravimetric effects along the CROP profile M26.

The shallow sediments are characterized by an average density of 2.35 $g\,cm^{-3}$, while a density of 2.43 $g\,cm^{-3}$ was attributed to the igneous rocks forming the Aeolian Arc. The basement, which has a density of 2.62 $g\,cm^{-3}$, overlies a thin layer with density 2.71 $g\,cm^{-3}$ present in the first 50 km of the profile, which represents the deeper portion of the upper crust. The transition from continental to oceanic crust, the latter having densities of 2.35 and 2.92 $g\,cm^{-3}$ in its upper and lower parts, respectively, was modelled at about 90 km from the profile origin. As regards the continental lower crust, the travel-time curve interpretation evidences a boundary within it at about 14 km depth below the coastline; it separates the Panormide lower crust (in the first 55 km of profile) into two sectors characterized by average densities of 2.85 $g\,cm^{-3}$ and 2.92 $g\,cm^{-3}$, and the European one into two sectors characterized by average densities of 2.95 $g\,cm^{-3}$ and 2.92 $g\,cm^{-3}$.

The upper mantle was assigned a constant density of 3.11 cm^{-3}. A density of 2.95 $g\,cm^{-3}$ was assigned to the portion of the Ionian slab observed from evident arrivals of waves reflected from its top on seismic sections.

In Figure 11(b) are shown the final density model and the comparison between Bouguer anomalies and gravimetric effects along the CROP profile M28-A. The basement, whose density was set to 2.68 $g\,cm^{-3}$, overlies the African upper crust which has density 2.73 $g\,cm^{-3}$ in the northern part of the profile behind the Erice basin. Beyond the Erice basin, the upper crust of the European plate has a slightly lower average density (2.50 $g\,cm^{-3}$). A single density value (2.81 $g\,cm^{-3}$) was attributed to the Panormide and European lower crusts, being compatible with their velocities according to the correlation law. The upper mantle was modelled with a homogeneous distribution of density set to 3.05 cm^{-3}.

ACKNOWLEDGEMENTS

This work was funded by MIUR, COFIN 1998, 2000 and 2002 with grants to I.R. Finetti. The authors are grateful to OGS for CROP NVR data acquisition and to ENEL-ISMES for data processing.

REFERENCES

BABEL Working Group, 1991. Recording marine airgun shots at offsets between 300 and 700 km. Geophys. Res. Lett. 17 (4), 645–648.

Braile, L.W., Chiang, C.S., 1986. The continental Mohorovicic Discontinuity: results from near-vertical and wide-angle seismic reflection studies. In: Barazangi, M., Brown, L. (Eds.), Reflection Seismology: A Global Perspective. American Geophysical Union, pp. 257–272.

Carrozzo, M.T., de Franco, R., De Luca, L., Luzio, D., Primiceri, R., Quarta, T., Vitale, M., 2002. Wavelet correlation filter for wide-angle seismic data. Geophys. Prospecting 50, 547–564.

Carrozzo, M.T., Luzio, D., Margiotta, C., Quarta, T., 1992. Gravity map of Italy. In: P. Scandone (resp.), Structural Model of Italy. Quad. Ric. Scient., CNR, Rome, vol. 114, no. 3.

Chironi, C., De Luca, L., Luzio, D., Vitale, M., Gruppo Sea Land, 1997. Algoritmi di stacking per il processing di sezioni sismiche litosferiche a grande angolo. In: CD-ROM degli Atti del 16° Convegno G.N.G.T.S., Rome, n. 03.16, 15 p.

De Luca, L., 2000. Nuove tecniche di elaborazione ed interpretazione di profili sismici crostali a grande angolo. PhD thesis, Palermo University, 296 p.

Finetti, I.R., Lentini, F., Carbone, S., Del Ben, A., Di Stefano, A., Forlin, E., Guarnieri, P., Pipan, M., Prizzon, A., this volume. Geological outline of Sicily and lithospheric tectono-dynamics of its Tyrrhenian margin from new CROP seismic data. Chapter 15.

Mooney, W.D., Brocher, T.M., 1987. Coincident seismic reflection/refraction studies of the continental lithosphere: a global review. Rev. Geophys. 25 (4), 723–742.

Morelli, C., 1975. The gravity map of Italy. In: Ogniben, L., Parotto, M., Praturlon, A. (Eds.), Structural Model of Italy. Quad. Ric. Sci., CNR 90, 427–447.

Nafe, J.E., Drake, C.L., 1963. Physical properties of marine sediments. In: Hill, M.N. (Ed.), The Sea, vol. 3. Interscience, pp. 794–815.

Pedley, R.C., Busby, J.P., Dabek, Z.K., 1993. Gravmag User Manual – Interactive 2.5D Gravity & Magnetic Modelling. British Geological Survey.

Ulrych, T.J., Sacchi, M.D., Graul, J.M., 1999. Signal and noise separation: art and science. Geophys. Prospecting 64 (5), 1648–1656.

Vitale, M., 2003. Seismic tomography techniques for crustal modelling. Theory and applications. PhD thesis, Palermo University, 221 p.

Zelt, C.H., Smith, R.B., 1992. Seismic traveltime inversion for 2D, crustal velocity structure. Geophys. J. International 108, 16–34.

CROP PROJECT: Deep Seismic Exploration of the Central Mediterranean and Italy
Edited by I.R. Finetti

CHAPTER 17

The Calabrian Arc and Subducting Ionian Slab from New CROP Seismic Data

I.R. Finetti[1,*]

ABSTRACT

On the basis of significant new deep seismic exploration data from the CROP project, an innovative lithospheric reconstruction of the tectono-stratigraphy and geodynamics of the Calabrian arc has been done. For a long time, the detection of deep seismic activity has suggested the existence of a Calabro-Tyrrhenian slab, and seismic exploration data showed that the Ionian Sea was floored by old oceanic crust subducting beneath the Calabrian arc. CROP data allow us now to reconstruct a well-controlled pre-orogenic paleogeographic framework and the evolving tectono-dynamics, imaging for the first time the Ionian slab down to a depth of over 150 km.

Two communicating ocean basins (*Ionian Tethys* of Permo-Triassic opening, interposed between Africa and Adria, and the *Alpine Tethys* of Middle Jurassic–Lower Cretaceous opening, interposed between Africa and Europe to the west, and Adria and Europe to the east), three plates (*Africa*, *Eurasia* and *Adria*) and two promontories (*Etrurian Promontory* on NW Adria and *Panormid Promontory* to the NW of Sicily) played leading roles in the evolution of the Mediterranean geodynamics. In the Eo-Alpine stage, the northeastern Alpine Tethys closed in the Alps sector with formation of the Eo-Alps, leaving unsubducted a wide western Alpine Tethys remnant. In this stage lower crust units of NW Adria (Etrurian Promontory) thrust over the western Tethys remnant; we find at the end these basement sheets at top stack of the Northern Calabria.

Driven by W-dipping subduction of the western Alpine Tethys remnant, in the Balearic stage (Oligocene–Lower Miocene) the Corso-Sardinian block (CSB) rotated, causing collision between Corsica and the Etrurian Promontory to the north, and between SE Sardinia and the Panormid promontory to the south. CSB rotation stopped, but subduction continued in the Ionian Sea in the Tyrrhenian stage (Middle Miocene to Present). Driven by Ionian lithospheric roll-back sinking, the tectono-dynamics of the Southern Apennines, Calabrian arc, Sicilian Apennines and the Tyrrhenian back-arc basin opening took place. The Etrurian Promontory dissected the Northern Apennines from the Southern along the 41° parallel, northern Ionian boundary, and the Panormid Promontory dissected the Maghrebian chain from the Sicilian Apennines. Crustal blocks of the Etrurian Promontory corner (overlain by Europe-verging austro-Alpine Etrurian lower crust units and ophiolite-Alpine Tethydes units, thrust over in the Balearic stage) were detached and transported southeastward. Part thrust obliquely over the Apulian slope, leaving detached cover units free to move outward forming the Apennine platform topmost sheets; part went to form the reset stack-pile of Northern Calabria (*Sila–Northern Serre Mts*), followed by Corso-Sardinian crustal fragments.

Crustal blocks of the Panormid Promontory, stacked with the SE-Sardinian margin in the Balearic stage, were detached and tectonically transported eastward over the Ionides. Part thrust over the northern Sicilian slope and part were stacked in the Southern Calabrian arc, overlain by Alpine Tethydes and other units and, topmost, by basement units of the Sardinian margin (*Southern Serre*, *Aspromonte* and *Peloritani Mts*). Deep deposition cover of the Ionian Tethys slab was stripped off from its subducting basement, going to form allochthonous units (*Ionides*) in the Southern Apennines (*Lagonegrese*) and Sicily (*Imerese–Sicanian units*) and a prominent wedge of imbricated units of the Calabrian arc in the Ionian basin.

[1]Department of Geological, Environmental and Marine Sciences, Exploration Geophysics Group, University of Trieste, Via Weiss 1, 34127 Trieste, Italy
*Corresponding author. Tel.: +39 40 5582276; e-mail: finetti@units.it (I.R. Finetti)

Keywords: Calabrian Arc, Ionian slab, CROP Exploration

1. INTRODUCTION

The Calabrian orogenic system is formed from a very complex assemblage of tectono-stratigraphic units coming from different geological domains, which evolved under different geodynamic stages, and were stacked in a very active migrating arc, of which only a minor part is emerged land. For several years, the Calabrian arc has attracted geoscientists of various countries and institutions: the fact that outcropping crystalline basement units interrupt the continuity of the Southern Apennines and Sicilian chain formed from sedimentary rocks, stimulated the search for a geological explanation.

A prolific literature accompanied the many studies performed on the matter. Nevertheless, even if remarkable progress in our knowledge of various aspects has been achieved, there is no substantial agreement as regard provenance (paleogeography) of the stacked units; the relationship between the Southern Apennines and Sicilian Apennines and between the northern and southern Calabrian arc; the definition and meaning of a tendentially increasing number of tectono-stratigraphic units; or geological history of the ophiolite and ophiolite-bearing units. The coastal chain of Northern Calabria, the Sila and Northern Serre still remain the most contentious Calabrian zones in the literature.

The initial studies on Calabria (Cortese, 1895; Lugeon and Argand, 1906; Limanowski, 1913) were followed by various others accurately reported and analyzed by Ogniben (1969, 1973). The latter (Ogniben, 1973), after examining the data of previous authors and his new data on northeastern Sicily and the Calabrian-Lucanian boundary, identified in Calabria three allochthonous complexes: (1) *Panormid* (carbonate platform); (2) *Ligurid* (metamorphic and heavily tectonized, including ophiolites and radiolarites); (3) *Calabride* (correlated to Sardinia, and characterized by metamorphic rocks of high degree over others of low degree: the piled thrust units of *Longi*, *Galati*, *Castagna* and *Aspromonte*.

Haccard et al. (1972) proposed the Calabrian thrust units as belonging to a previously continuous chain running from the western Alps to the Corsican Alps, Southern Italy, Sicily and Northern Africa. Alvarez (1976) also considered a unique orogenic belt fragment in Miocene-Pliocene times. According to Amodio-Morelli et al. (1976) the Calabro-Peloritan arc is a fragment of the Alpine chain formed by Europe-verging *austro-Alpine* units and *Ligurides ophiolites*, thrust later (Lower Miocene) on the Africa-verging Apennino-Maghrebian chain. Scandone (1979) introduced Africa-verging insubric elements between the Europe-verging Alpine thrust units and the Apenninic Africa-verging ones of Miocene age. Since 1980, various authors (Bonardi et al., 1980, 1982, 1992, 1994; Boccaletti et al., 1984; Scandone, 1982) have subdivided the Calabro-Peloritan arc into two distinct sectors: (1) a *northern sector* formed by the *Coastal Chain and Sila*, where the interpretation of Amodio-Morelli et al. (1976) and Scandone (1979) is substantially accepted; and (2) a *southern sector* formed by the *Aspromonte and Peloritani Mts.* of still discussed interpretation but of different evolution. *In the Coastal Chain–Sila–Northern Serre sector of the Calabrian arc are recognized* (from bottom upward): (a) *Apenninic units composed of dolomite, limestone and evaporite observed in tectonic windows of the Coastal Chain* (*Lower Apennine complex* of Ogniben, 1969, 1973; *S. Donato* and *Verbicaro units* of Amodio-Morelli et al., 1976); (b) *Ophiolitic Units* (*Liguride complex* of Ogniben, 1969, 1973; *Frido-Diamante Terranova-Gimigliano* and *Malvito units* of Amodio-Morelli et al., 1976); and (c) *Crystalline Units*, Hercynian metamorphites intruded by Late Hercynian magmatites partly affected by Alpine metamorphism with and/or without Mesozoic cover, representing middle-lower crust (*Calabride complex* of Ogniben, 1973; *Longobucco–Longi Taormina units*, *Bagni–Castagna units*, *Polia–Copanello* and *Mt. Garglione units* of Amodio-Morelli et al., 1976). These crystalline sheets are delimited to the north by the *Palinuro left-transcurrent fault* (*Sangineto Line* in the literature).

The major distinctive characteristics of the southern Calabrian sector are considered: (a) *the absence of ophiolite units;* (b) *the absence of Apenninic platform units;* (c) *the absence of Europe-verging Eo-Alpine tectonics;* (d) *the absence of Alpine metamorphism HP/LT;* (e) *the age of main deformation ranging from Upper Eocene–Lower Miocene in the southern sector and Upper Cretaceous in the northern sector;* (f) *no correlations possible between the units of the two sectors with the exception of a possible correlation between the Longobucco and Longi-Taormina units;* (g) *the presence of the Stilo-Cape Orlando fm. deposited contemporaneously to thrust activity over the Apennine domains in the northern sector; and* (h) *the presence of a nappe of varicoloured shales.*

Van Dijk et al. (2000) perform a very careful and extensive analysis of the various units of the Calabrian arc already documented in the literature, and proposed new ones. They recognize a left-transcurrent tectonic system, which is prominently imaged by the CROP dataset. Moreover they reconstruct a crustal section of Northern Calabria by interpretation of seismic and magnetotelluric data. For a complete list of the bibliographic references to the complex ensemble of Calabrain units the reader is addressed to this exhauriant recent work.

On the various proposed tectono-stratigraphic units of the Calabrian arc, there is a wide range of interpretations of their original paleogeographic domains and tectono-dynamic histories. A large-scale 3D transcrustal seismic interpretation of the entire CROP dataset, on the one hand does not permit detailed local tectono-stratigraphic analyses, but on the other is the best basis for a controlled reconstruction of the pre-orogenic paleogeographic setting and lithospheric macro-evolution of the Central Mediterranean region. Here I present the results obtained from interpretation of the CROP data, describing those specific sections that more directly control the Calabrian arc. Having reconstructed a pre-orogenic paleogeography, controlled for the first time by reliable lithospheric seismic imagings, which shows that this paleogeography strongly conditioned the tectono-dynamics of the region, it can be said that the CROP project has provided a better understanding of Mediterranean and Calabrain arc geodynamics in general. In fact, it is possible to outline the original positions, stacking-migrating stages and mechanisms of questioned allochthonous units, such as the basement units of northern and southern Calabria, the Alpine Tethydes (including the ophiolite and ophiolite-bearing units), the Apennines platform outcropping in tectonic windows of northern Calabria, the Panormid platform units buried in the southern Calabria–Peloritani thrust stack area, and the Lagonegrese–Imerese deep deposition units.

The new CROP deep seismic exploration data have contributed to a remarkable progress in our knowledge of the Mediterranean lithosphere. The interpretation here presented, with regard to fundamental innovative regional features, can be considered of acceptable to satisfactory reliability, although on a smaller scale, improvements, refinements and amendments can be introduced everywhere.

2. CROP SEISMIC DATA DESCRIBED

The thrust stack of the Calabrian arc is growing over the subducting slab of the Ionian ocean crust and, laterally, over the slopes and margins of the buried Ethnean corner of the NE-Sicily platform boundary at one side, and on the slope and margin of the buried Apulian platform at the other side (northern Calabrian sector). Knowing that northern Calabria differs from southern, and that the two chain sectors are geologically disputed, two CROP sections were acquired across the orogenic arc: one crossing the southern sector (CROP M2A/III–M2B) and one on the Tyrrhenian margin of northern Calabria (CROP M27). The original program included also the prosecution of line M27 into the Gulf of Taranto, but for temporary state reasons this prolongation was not allowed (Figure 1). So, in the Gulf of Taranto, CROP section M27 is connected to the hydrocarbon exploration line F-89 (Italian Ministry's public seismic dataset).

The sections M2B and M4 give transcrustal information on the thick Calabrian wedge, where a substantially sequential imbrication of *Ionides* (*deep depositional condensed cover of the Ionian Tethys*) is seismically well evidenced. *This unit* (*Ionides*) *according to new views* (Finetti et al., 1996; Finetti, this volume, Chapter 10) *corresponds to the questioned allochthonous units that in the literature are known as the Lagonegrese* (*Southern Apennines*) *and Imerese–Sicanian units* (*Sicily*). The exhibited CROP section M5 (part: SP 100–2300) transversely crosses the eastern side of the Calabrian wedge and the margin zone and steep slope of the Apulian platform. The CROP section M31, along a transverse direction to the wedge propagation, images the western side of the Calabrian wedge pile.

CROP seismic data were integrated with various other MS seismic lines of 10 s penetration (Finetti et al., this volume, Chapter 1; Finetti, 1976, 1982; Finetti and Del Ben, 1986). Those MS lines, in the basins floored by oceanic basement (Ionian and Tyrrhenian Seas), explore the entire crust, providing a very useful supplement to the CROP data. On the Ionian and Tyrrhenian shelf and upper Calabrian slope area, a very useful detailed geophysical control was provided also by the Italian Ministry's public seismic dataset, and by stratigraphic calibration of numerous boreholes on the Ionian shelf.

The Ionian crust has been recognized to be of oceanic type for several years (Finetti, 1982). One of the key features of the Calabrian arc is the *subducting Ionian slab* whose presence is supported by deep seismicity data and seismogeological studies that analyzed possible detachments. To obtain a direct seismic imaging of the Ionian slab,

Table 1
CROP seismic sections exhibited and described in this work

CROP section	SP	Length (km)	Figure N°		Chapter 17 Plate N°	
			Seismic	Int.	Seismic	Int.
M2A/III M2B	100–2380 1300–5027	144 232		2	1	2
M4	100–4325	266		3	3	4
M31	100–2165	106	4	5		
M27 (F-87 + BC Interpolation)	100–1607	145 160 (Tot. 305)		6	5	6
M5	100–2300	135		7		
Deep Ionian Slab CROP Section M2A/III–40s	1870–2380	Penetration Depth (km) 150 (40 s TWT)	8			

a specific CROP programme was performed: a section of doubled recorded length (40 s in place of the 20 s for all normal CROP sections). The location was selected on a segment of CROP M2A/III, SP 1830–2380 (Figures 1 and 8). Recording was done in two parts: a first run with record length 21 s, and a second run with record start delayed 20 s with respect to the airgun shot (recording interval 20–40 s; 1 s overlap). Processing by the University of Trieste (Exploration Geophysics Group of the Faculty of Sciences) provided for the first time a seismic slab imaging, extending to a depth of 40 s (Figure 8).

3. CROP SEISMIC SECTIONS INTERPRETATION

Combining the CROP section M2A/III (144 km, Tyrrhenian margin of the Calabrian arc) and M2B (231 km, wedge of the Calabrian arc in the Ionian Sea) gives seismic control of the whole submerged area of the southern Calabrian arc, which is by far the major part of this orogenic system (Plates 1 and 2; Figure 2). But the section segment "BC" (57 km) which is almost entirely formed from the emerged part of the arc, remains uncontrolled. Looking at the location of the section in Plate 2 (Figure 1; Plate 2c) it is seen that the perpendicular distance between "B" and "C" is really less than measured on the oblique BC interval. The reconstructed BC section segment is based on: (a) correlation between the controlled B and C areas; (b) the observed CROP-controlled stacks of the Southern Apennines and Sicilian Apennines (contiguous Apenninic zones of the Calabrian arc) where, respectively, Etrurian and Panormid promontory-derived crustal units left their basement units collided on Apulian and Sicilian platform slopes, followed by Sardinian basement units (Figure 11); (c) the objectively CROP-controlled assumption that along the Palinuro left-transcurrent faults to the north, and along the Taormina-Vulcano right-transcurrents to the south, different identified basement units moved outward and thrust over piles of units identified by CROP network data on contiguous Apenninic zones; (d) the assumption that the Tyrrhenian mantle between B and C onlaps the subducting Ionian slab with no significant underthrusting of Ionian cover units, as assessed by CROP sections on the Tyrrhenian margins of the Southern Apennines and Sicily; and (e) geological data for Calabria and borehole data. Obviously, the interpolated segment BC of the reconstructed section contains a significant number of features conceptualized from a general 3D viewpoint that the author matured at the end of extensive work on the CROP project. But, starting from fixed situations at points "B" and "C", the number of alternative solutions is in effect much less than imaginable at first sight. In any case, the CROP network, in its totality, indicates that, if the individual geometries and thicknesses of some thrust-sheets in section "BC" are rather schematic, the tectonic position and identification of the piled thrust units are deductively correct. A careful iterated interpretive reprocessing conducted by the University of Trieste on sections CROP M2A/III and M2B was determinant in obtaining a marked signal improvement over the original poor quality standard processing.

The Northern Calabrian arc is reconstructed from CROP section M27 (Plates 5 and 6; Figure 6) on the Tyrrhenian margin, and from the Italian Ministry's public dataset line F-89, with interpolation on the land segment "BC" and

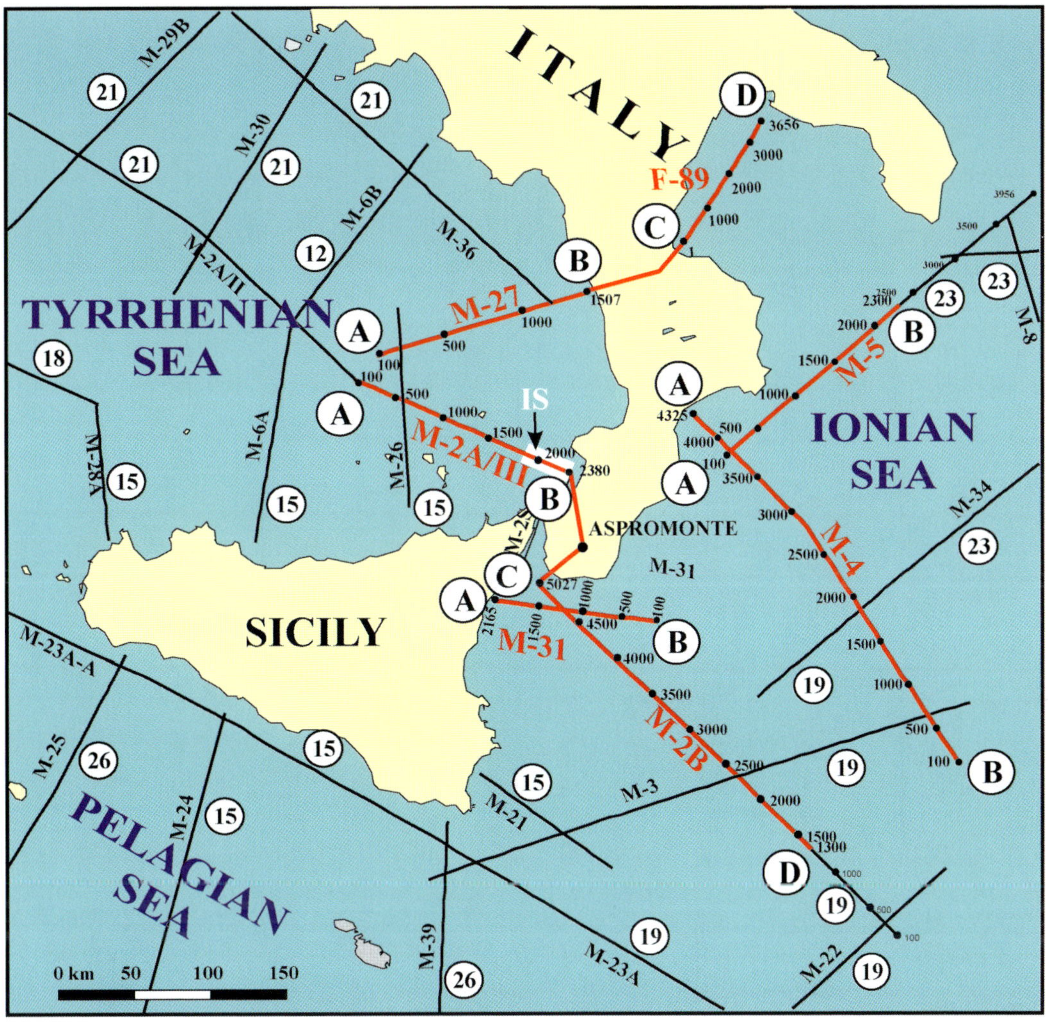

Figure 1: Location map of the CROP sections interpreted and presented in this work (red-traced lines). Other CROP sections reported in the figure (black lines) are exhibited and described in different chapters of this volume (chapter number indicated within white circles). *IS* is the segment of CROP section M2A/III where a 40 s deep seismic penetrating section to explore the *Ionian slab* was acquired, as shown in Figure 8.

on the deeper part (7 to 16 s) of the segment "CD". In this latter case, the reconstruction is on the whole much better outlined than in the section across Southern Calabria, since the limits and macro-features of the Apulian crust are better defined by the CROP programme.

3.1. Southern Calabrian Arc

The *Southern Calabrian arc* has been interpreted along a composite lithospheric section, which from the Tyrrhenian to the Ionian Sea, crosses the internal margin, the emerged chain sector, the forearc basin and a prominent active wedge (Plate 1, 2; Figure 2). The whole section runs through an area which belongs to the Ionian Tethys of Permo-Triassic opening, partially subducted or under subduction and in part preserved as a restricted ocean remnant. The internal sector (Tyrrhenian) is fully controlled by CROP section M2A/III which, in the reprocessed version, exhibits fair to good reflectivity, and so the reconstruction can be considered of high reliability.

The Tyrrhenian oceanic crust (of Quaternary age in the Marsili basin) is very clearly imaged, followed by delaminated crustal blocks affected by listric faults. The entire Tyrrhenian margin of the arc explored by section M2A/III is composed of crustal fragments detached from the Sardinian margin during the Tyrrhenian opening, except for a basal crustal interval at the SE extremity attributed to delaminated basement units from the Panormid promontory (Plate 2; 2a and 2b). This attribution seems well supported by the global 3D CROP-reconstructed distribution of the Panormid units and their calibrating data from Sicily. Attribution of the Calabrides crustal blocks to the Sardinian

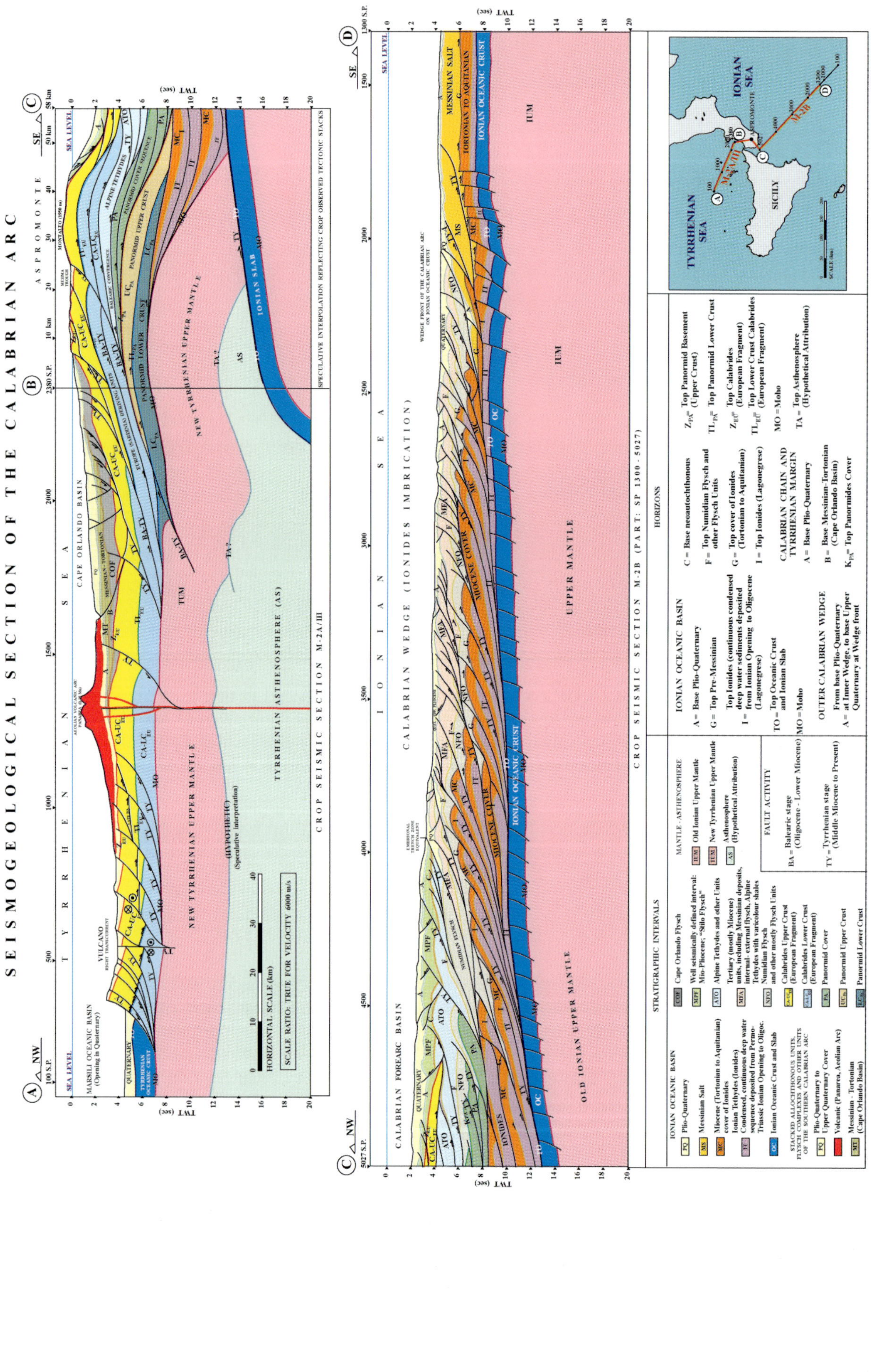
SEISMOGEOLOGICAL SECTION OF THE CALABRIAN ARC
CROP SEISMIC SECTION M-2A/III
CROP SEISMIC SECTION M-2B (PART: SP 1300 - 5027)
STRATIGRAPHIC INTERVALS
HORIZONS
TYRRHENIAN SEA
IONIAN SEA
SICILY

Figure 2: Lithospheric seismogeological section crossing the *Southern Calabrian Arc* from the Tyrrhenian Sea to the frontal orogenic wedge in the Ionian Sea. Reduced version of Plate 2.

margin is supported without doubt by the whole 3D control of the CROP seismic network in the Tyrrhenian Sea. Such Sardinia (European)-derived crustal units are distributed all around the Tyrrhenian margins (Figure 11). The sedimentary succession of the Cape Orlando basin, as well as the Volcano Panarea (Aeolian Volcanic arc), are well imaged by the seismic data. It can be observed that, while the lower crust has a substantially normal thickness, the upper crust is comparatively thin. On this point, it is evident from the CROP data that the crustal delamination process of the Sardinian units took place predominantly in the upper and lower crust on that part of the Sardinian margin which remained connected to Sardinia (i.e. to the west of the Selli line: Cornaglia terrace).

The Sardinian (European) crustal units imaged on section CROP M2A/III were delaminated in the Balearic stage (Upper Oligocene–Lower Miocene), when the Corso-Sardinian block rotated anticlockwise, and in Late Balearic stage, when SE-Sardinia collided with the Panormid Promontory, they were recompressed and stacked over Panromid units, closing the Alpine Tethys, and moving its cover units eastward. In the Late Balearic–Early Tyrrhenian stage (Burdigalian), the subduction of the Alpine Tethys terminated, and the Ionian Tethys started to subduct. Driven by roll-back sinking of the Ionian lithosphere, stacked crustal fragments of Panormides started to migrate over Ionides detached from their subducting oceanic basement.

Consideration of all data and processes above outlined led to the reconstruction of section segment "BC", with the thrust pile of southern Calabria (Aspromonte) schematically composed from top to bottom as follows (Plate 2):

- Delaminated units of Calabrides (deriving from European Sardinia) upper crust (CA-UC$_{EU}$), checked against surface geology and two projected boreholes in the Ionian shelf area; identified and interpreted on northern CROP section M2B.
  Delaminated thrust sheets of Calabrides (deriving from European Sardinia) lower crust (CA-LC$_{EU}$), extrapolating from CROP section M2A/III (SP 2000–2380) and terminating the extrapolation southeastward before the northern extremity of section M2B (SP 5027) where no evidence of such basement sheets has been found.
- Alpine Tethydes and other (flysch units) closed between Calabrides and Panormid sheets in the Balearic stage. These units have been identified and interpreted on northern section M2B and very speculatively extrapolated over segment BC.
- Panormid cover units (PA) interpreted on northern section M2B (mainly by velocity interval amplitude) and speculatively extrapolated on segment BC.
- Panormid upper crust (UC$_{PA}$) and lower crust (LC$_{PA}$) sheets interpreted on section M2A/III by lateral geophysical correlation to similar stacks on the Tyrrhenian margin of Sicily where PA outcrops.
- Ionides (IT) units and Miocene cover (MC) clearly reconstructed on CROP section M2B and arbitrary extrapolated within a questionable area interposed between the Panormid and top Tyrrhenian mantle.

With regard to the top of the Tyrrhenian mantle MO, at the point "B" of the section, this horizon is very clearly defined at a reflection time of 6.6 s. It is also evident that at point C, the southeastward-moving Tyrrhenian mantle flux has not yet arrived. Having established these two important constraints, isolation of the Tyrrhenian mantle can be done by onlapping of horizon MO on the subducting Ionian slab, as in the section of Plate 2, or leaving a part of the imbricated Ionides to subduct. The solution given in Plate 2 reflects what is observed on almost all CROP sections crossing the Tyrrhenian margins.

Due to vertical velocity variation effects, the Ionian slab on section segment BC shows an apparent time-slope markedly smaller than in the effective depth-slope (Plate 2: see comparison). Using deep low-amplitude primary reflectors (around 12–16 s) and considering S-wave dispersion indications, horizon TA has been speculatively picked, attributing it hypothetically to the top asthenosphere. This horizon is substantially coherent with a 3D TA setting resulting from interpretation and correlation of similar deep intermittently distributed reflectors of the other Tyrrhenian CROP sections. The southern part of Plate 2 (CROP section M2B) explores the wedge zone of the Calabrian arc which is imaged by four CROP sections and globally described in Section 3.3.

The section segment CD reconstructed by interpretation of CROP M2B images the thick wedge of the Calabrian arc, over 200 km long, made up of a rather regular sequence of imbricated units of Ionian cover. This cover is composed of a condensed deep-water succession from the Ionian opening (Permo-Triassic) to the Oligocene (IT),

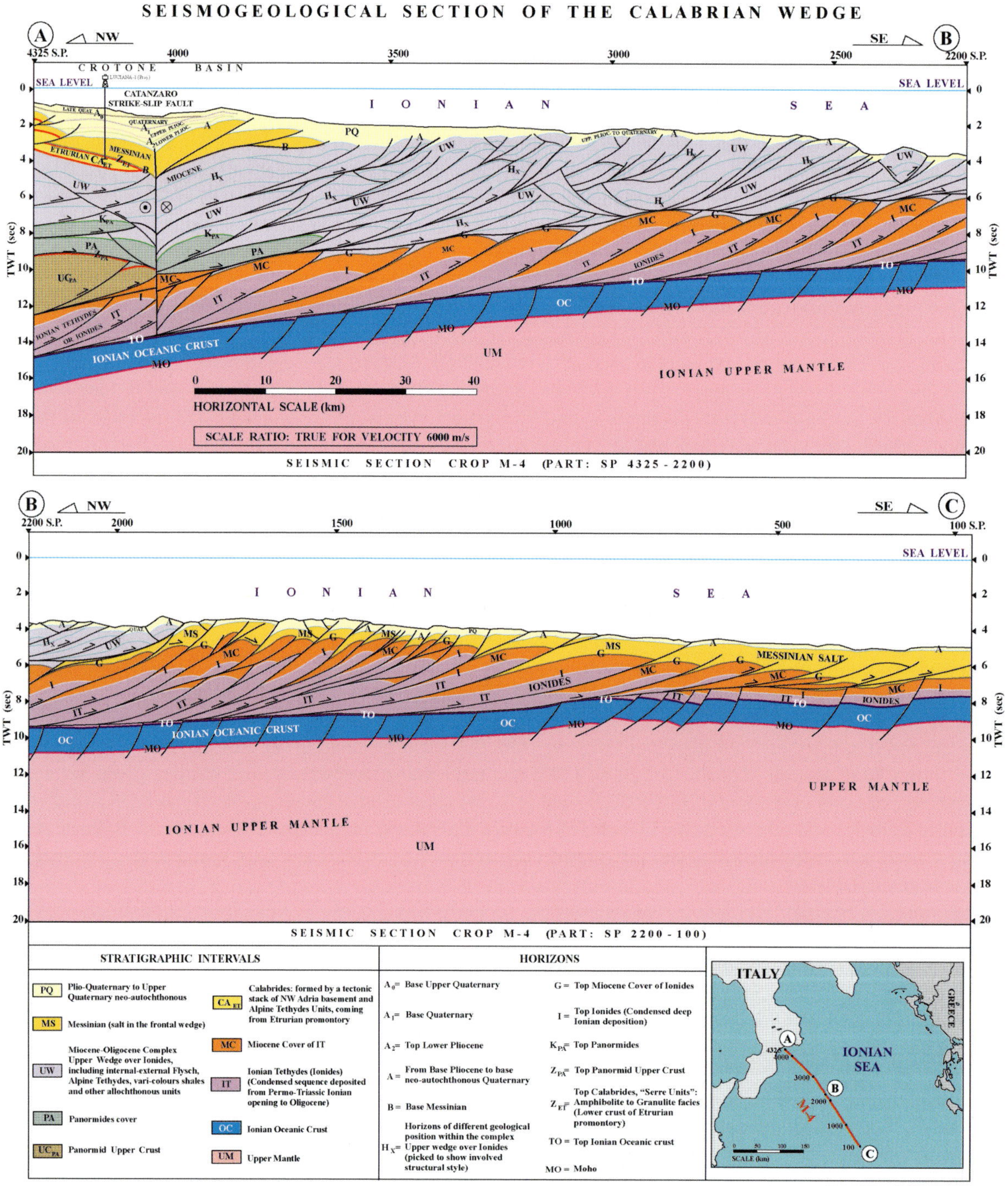

Figure 3: Lithospheric seismogeological section of the *Calabrian wedge* growing over the subducting oceanic crust of the Ionian Sea, reconstructed from new CROP seismic data (section CROP M4). Reduced version of Plate 4.

followed by a Miocene cover (MC) interval of Aquitanian to Tortonian age of higher sedimentation rate. This condensed deep-water Ionian sequence was stripped off its subducting Ionian basement (Ionian slab) and went to form the *Lagonegrese allochthonous units* in the Southern Apennines, the *Imerese–Sicanian units* in the *Sicilian Apennines* (oblique thrusting in both cases), and the *Calabrian wedge* seen in CROP sections M2B (Plate 2; Figure 2); M4 (Plate 4; Figure 3); M31 (Figure 5); M5 (Figure 7). *As a unifying name for these common basin-derived Southern Apennines and Sicilian allochthonous units, the author proposes to call them Ionian Tethydes* or *Ionides* (Finetti, this volume, Chapter 10). *Such a name gives an immediate provenance and simplifies the nomenclature of units that, until now, their common origin being unknown, were characterized by various names related to the different provinces where the Ionides migrated. Understanding the common basinal origin of the Ionides thrust on*

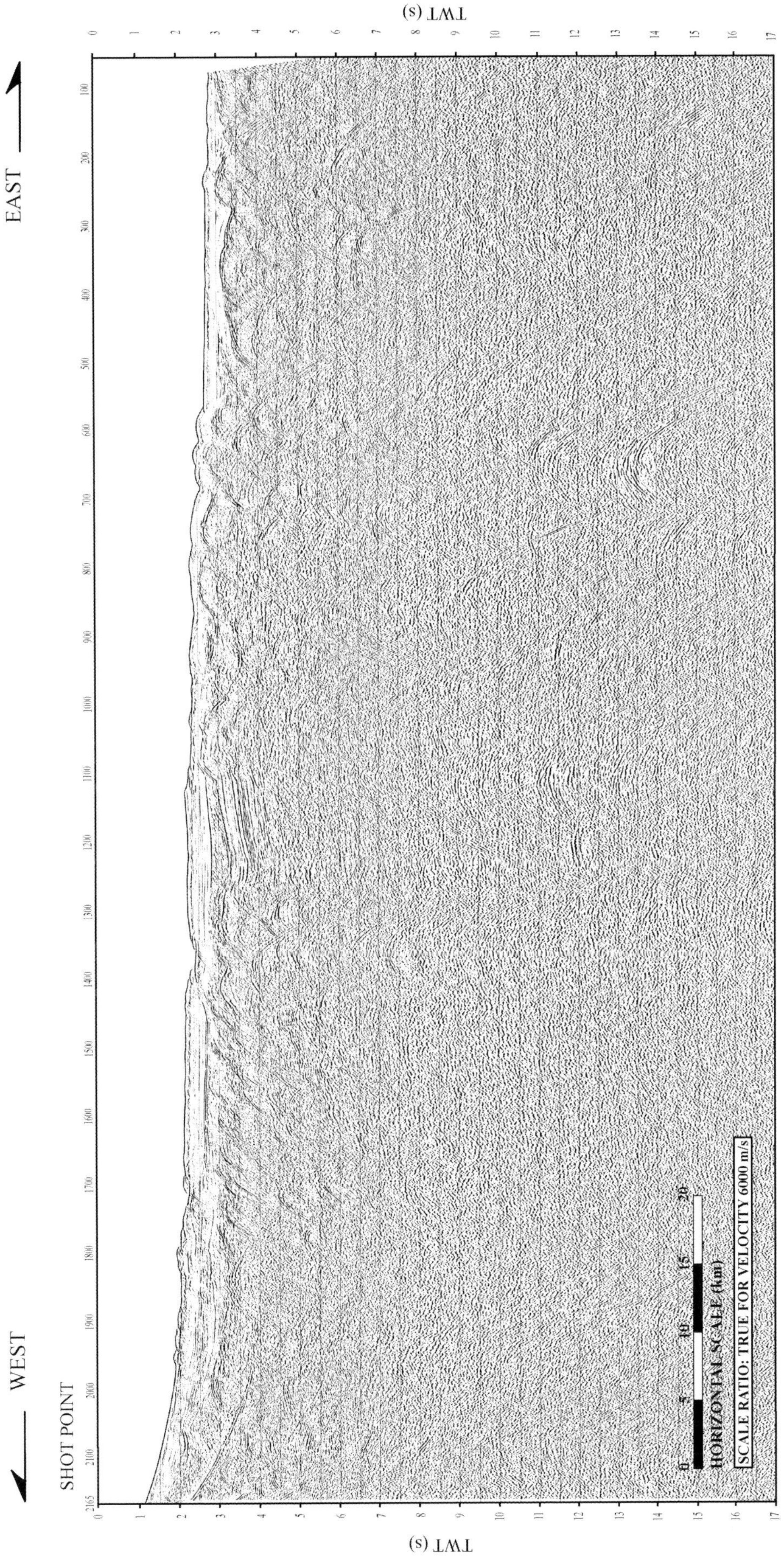

Figure 4: CROP Seismic Section M31 on the Calabrian Arc. See Figures 1 and 5 for location.

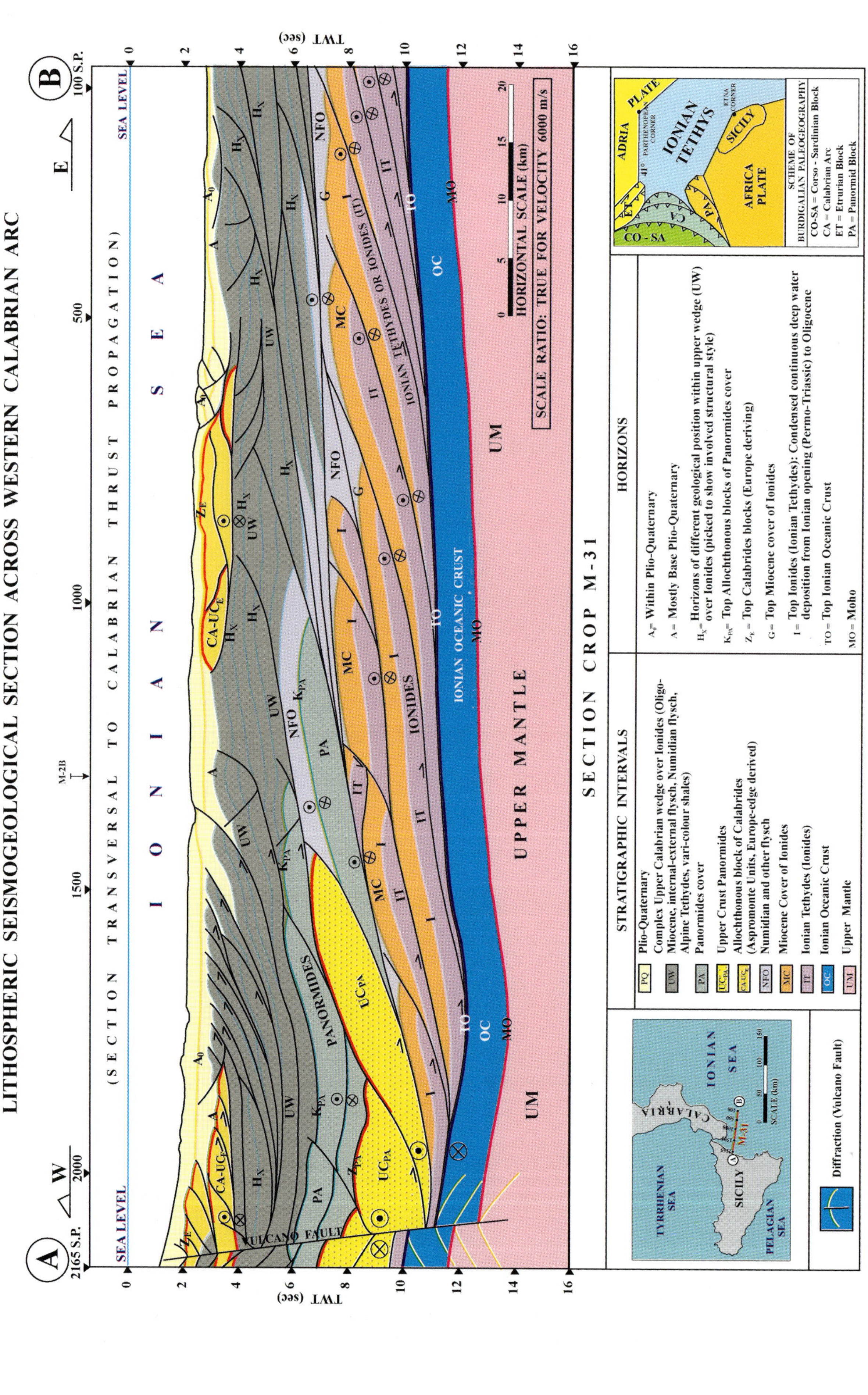

LITHOSPHERIC SEISMOGEOLOGICAL SECTION ACROSS WESTERN CALABRIAN ARC
(SECTION TRANSVERSAL TO CALABRIAN THRUST PROPAGATION)
IONIAN SEA
SECTION CROP M-31
SEA LEVEL
TWT (sec)
VULCANO FAULT
PANORMIDES
IONIAN OCEANIC CRUST
UPPER MANTLE
IONIDES
IONIAN TETHYDES OR IONIDES (IT)
HORIZONTAL SCALE (km)
SCALE RATIO: TRUE FOR VELOCITY 6000 m/s
STRATIGRAPHIC INTERVALS
PQ Plio-Quaternary
UW Complex Upper Calabrian wedge over Ionides (Oligo-Miocene, internal-external flysch, Numidian flysch, Alpine Tethydes, vari-colour shales)
PA Panormides cover
UCPA Upper Crust Panormides
CA-UCE Allochthonous block of Calabrides (Aspromonte Units, Europe-edge derived)
NFO Numidian and other flysch
MC Miocene Cover of Ionides
IT Ionian Tethydes (Ionides)
OC Ionian Oceanic Crust
UM Upper Mantle
HORIZONS
A0 = Within Plio-Quaternary
A = Mostly Base Plio-Quaternary
HX = Horizons of different geological position within upper wedge (UW) over Ionides (picked to show involved structural style)
KPA = Top Allochthonous blocks of Panormides cover
ZE = Top Calabrides blocks (Europe deriving)
G = Top Miocene cover of Ionides
I = Top Ionides (Ionian Tethydes): Condensed continuous deep water deposition from Ionian opening (Permo-Triassic) to Oligocene
TO = Top Ionian Oceanic Crust
MO = Moho
TYRRHENIAN SEA
PELAGIAN SEA
SICILY
CALABRIA
IONIAN SEA
SCALE (km)
Diffraction (Vulcano Fault)
ADRIA PLATE
IONIAN TETHYS
AFRICA PLATE
SICILY
CO-SA
SCHEME OF BURDIGALIAN PALEOGEOGRAPHY
CO-SA = Corso - Sardinian Block
CA = Calabrian Arc
ET = Etrurian Block
PA = Panormid Block

Figure 5: Interpreted CROP section M31 on the NW-wedge of the Calabrian arc. The section crosses the Calabrian wedge transversally, and therefore we see the thick thrust pile sub-perpendicular to the orogenic propagation.

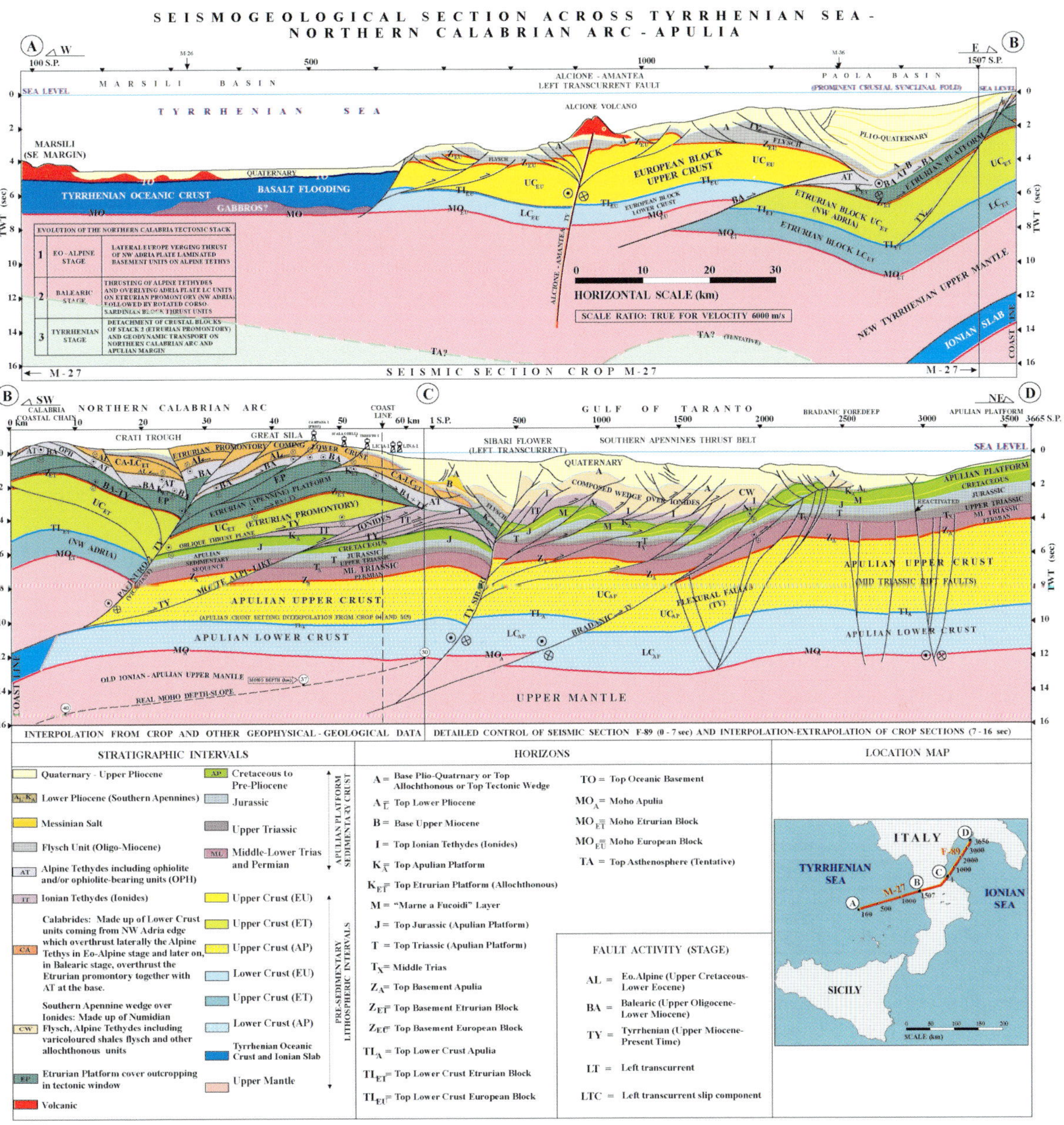

Figure 6: Lithospheric seismogeological section across the *Northern Calabrian Arc*, reconstructed from the new CROP section M27 in the Tyrrhenian Sea (segment "AB"), and section F-89 (0–7 sec record length; from Italian Ministry's public dataset) in the Gulf of Taranto (segment "CD"), and from interpolation of CROP and other geophysical-geological data for the segment "BC". Below 7 sec penetration in section F-89 the reconstruction has been done by interpolation–extrapolation of CROP data with no particular difficulty, and with acceptable reliability. Reduced version of Plate 6.

the Southern Apennines–Calabrian arc–Sicilian Apennines also very much simplifies the correct understanding of the evolving geodynamics of the Tyrrhenian stage in the central Mediterranean region and clearly shows that the Ionian slab roll-back is its basic driving force.

3.2. Northern Calabrian Arc: evolving tectono-dynamics of staked units

As previously noted, various authors have proposed interpretations on the Northern Calabrian arc and remarked on the differences with respect to the southern sector of the arc. A reconstruction of the tectono-stratigraphic setting of Northern Calabria, based on the new CROP seismic data and integrating other information, is exhibited in Plate 6. The internal part of this transcrustal section (segment AB) is controlled by CROP M27 which crosses the Tyrrhenian margin. The segment CD is interpreted from a good seismic line of the Ministry's public dataset (Line F-89) in the time interval 0–7 s and by interpolation–extrapolation of CROP sections in the interval 7–20 s. The segment BC, for the Apulian crust, is reconstructed from highly reliable interpolation of CROP data and from a more complex derivation of various geological-geophysical sources for the Calabrian thrust-stack.

The Apulian platform area is reconstructed from line F-89, a good quality section, with calibration from projected borehole data. Basement upper crust and lower crust are interpolated from CROP data and are rather uniform with regard to their thicknesses. Therefore the author is of the opinion that, not only the cover interval but, the whole crustal setting in section segment CD is quite realistic and tectonically self-explanatory. The transcrustal *Bradanic fault* is well correlated with a corresponding fault seen in the Southern Apennines (Finetti et al., this volume, Chapter 12). The *Sibari flower* is generated by a left-transcurrent motion, which with other shear faults, fits a marked tendency of the Southern Apennines–eastern Calabrian wedge to propagate along a trend subparallel to the Apulian slope.

The Apulian platform reconstruction in section segment BC is the interpolation resulting from crustal correlation of sections CROP-04 (Southern Apennines) and CROP M5 (Ionian Sea), with the important constraint of the seismically imaged Ionian slab at the eastern extremity of CROP M27. The platform crustal settings obtained from correlated regional CROP sections and the directly imaged Ionian slab constraint fit very well (Plate 6). A prominent compressive mantle bulge beneath the coastal chain of northern Calabria is clearly CROP supported.

The interpretation of CROP section M27 in Plate 6 images the Tyrrhenian margin of the northern Calabrian arc formed by two different crustal blocks: (1) an internal block formed from a *fragment of the Corso-Sardinian (Europe) block*, and (2) a block representing a *fragment of the Etrurian Promontory* (NW Adria). The Corso-Sardinian crustal fragment, thinned and delaminated, is overlain by flysch units and Plio-Quaternary cover. The promontory coming from a crustal block is overlain by Alpine Tethydes, flysch units and Plio-Quaternary sediments that thicken prominently in the Paola basin, which is underlain by a crustal synclinal fold (Plate 6).

The outcropping northern Calabria of section of Plate 6 (progressive section distance 150 to 205 km) is reconstructed by taking into account all the relevant geological data widely discussed in the literature. It is geologically clear and largely accepted that from top to bottom in northern Calabria the following units are stacked:

(a) Basement sheets of lower crust (provenance questioned);
(b) Ionian Tethydes (including ophiolite and ophiolite-bearing units;
(c) Platform units outcropping in tectonic windows.

By reconstructing from CROP network data the pre-orogenic paleogeography, and inverting the observed crustal deformation, *it is evident that the paleogeography strongly conditioned the orogenic processes*. One important result obtained from the interpretation and study of the whole CROP network is the conclusion that the *discussed "Apennine platform" thrust units of the Southern Apennines, as well as the crustal block exhibited on eastern CROP M27 section and the allochthonous platform sheets of northern Calabria, come from the Etrurian Promontory corner area (NW Adria)*. They where detached when Corsica–northern Sardinia collided with the Etrurian corner at the end of the Balearic stage (Figure 10). In the Tyrrhenian stage, these Etrurian crustal fragments migrated and were thrust obliquely over the Apulian slope; their cover was detached, laminated, and thrust outward over the stacked pile (Plate 6).

Top stack basement sheets of northern Calabria are geodynamically explained as follows (Figures 10, 11): in the Eo-Alpine stage the Alpine Tethys closed to the north in the Alps sector with lateral subduction beneath the NW

Figure 7: Lithospheric seismogeological section across the Calabrian wedge-western Apulian margin in the Ionian Sea; reconstructed from CROP M5 (part: SP 100–2300). The section is located transversally to the wedge propagation.

SEISMOGEOLOGICAL SECTION ACROSS THE CALABRIAN ARC-APULIAN PLATFORM FROM NEW CROP SEISMIC DATA

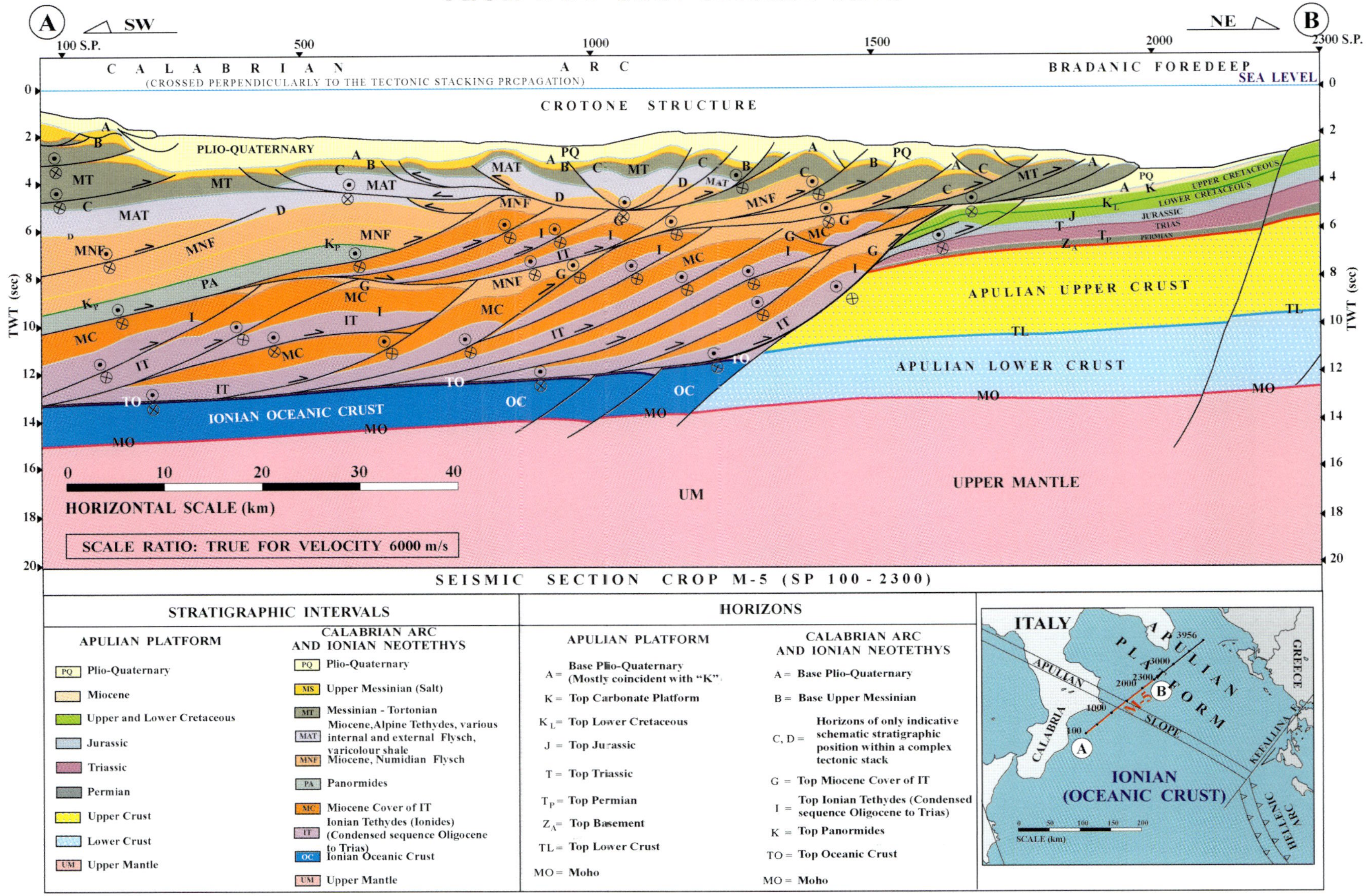

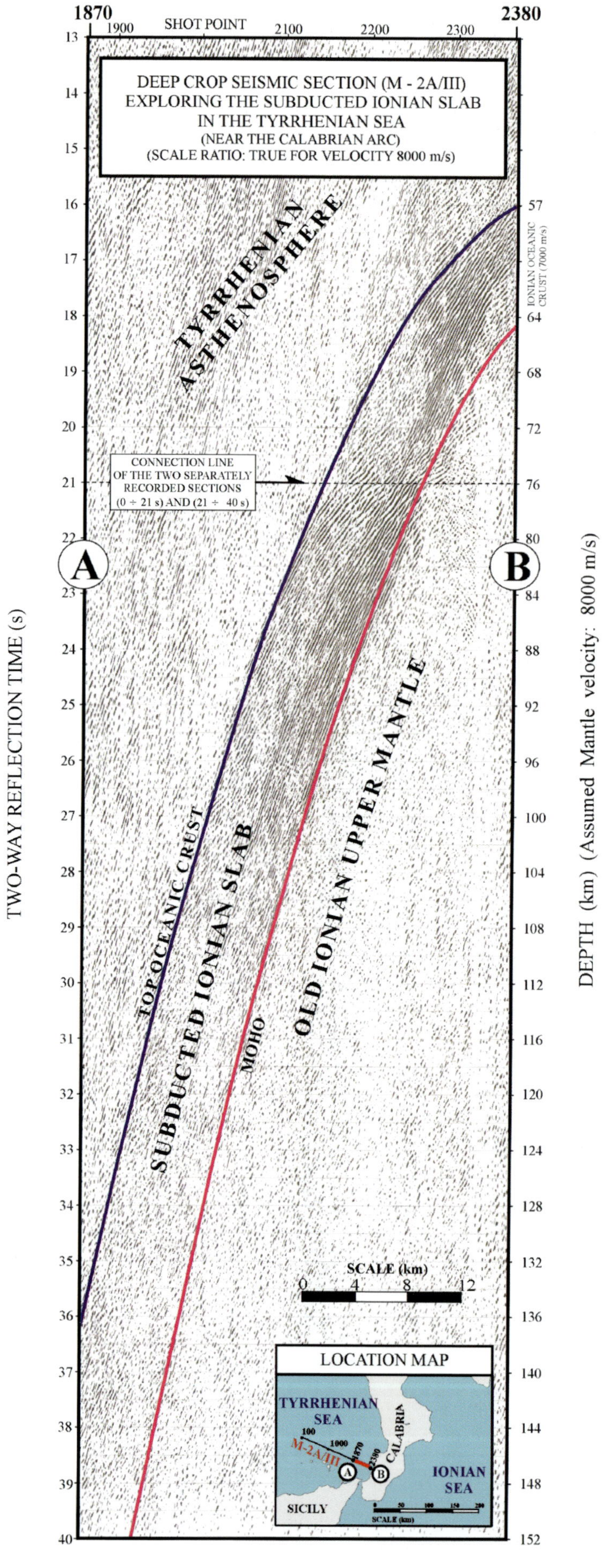
1870
SHOT POINT
2380
DEEP CROP SEISMIC SECTION (M - 2A/III)
EXPLORING THE SUBDUCTED IONIAN SLAB
IN THE TYRRHENIAN SEA
(NEAR THE CALABRIAN ARC)
(SCALE RATIO: TRUE FOR VELOCITY 8000 m/s)
TYRRHENIAN ASTHENOSPHERE
IONIAN OCEANIC CRUST (7000 m/s)
CONNECTION LINE OF THE TWO SEPARATELY RECORDED SECTIONS (0 ÷ 21 s) AND (21 ÷ 40 s)
A
B
TWO-WAY REFLECTION TIME (s)
DEPTH (km) (Assumed Mantle velocity: 8000 m/s)
TOP OCEANIC CRUST
SUBDUCTED IONIAN SLAB
MOHO
OLD IONIAN UPPER MANTLE
SCALE (km)
LOCATION MAP
TYRRHENIAN SEA
CALABRIA
M-2A/III
IONIAN SEA
SICILY

Figure 8: Deep litho-asthenospheric CROP section showing the *subducted Ionian slab*. The exhibited results are the combination of two processed CROP seismic sections recorded at two different times on the same segment (SP 1870–2380) of the CROP M2A/III section: from 0 to 21 s on the first run and, successively, with 20 s delayed record start, from 20 to 40 s on the second run (1 s overlap). The Ionian slab can be well identified and seismically controlled down to a depth of over 150 km. Even though recorded separately, the splicing of the two sections at 21 s reflection time is satisfactory.

Adria margin zone (Etrurian Promontory), leaving a wide western Alpine Tethys remnant unsubducted. This lateral subduction beneath NW Adria determined an Austro-Alpine-like thrusting of Etrurian Promontory lower crust units over the Alpine Tethys. In the Balearic–initial Tyrrhenian stage, driven by W-dipping subduction of the Alpine Tethys remnant, rotation of the Corso-Sardinian block took place, with collision with the Etrurian corner and detachment of Etrurian crustal blocks overlapped by W-verging lower crust units of the Etrurian area previously thrust over the Alpine Tethys, together with ophiolite and Alpine Tethydes at the base. The Tyrrhenian stage continued with southeastward migration, resetting and thrust-stacking on northern Calabria of these multistage composed stack units (Figures 10, 11).

3.3. Wedge of the Calabrian Arc

Two CROP sections cross the entire wedge of the Calabrian arc in the Ionian Sea lengthwise: M2B (Plates 1, 2; Figure 2) and M4 (Plates 3, 4; Figure 3). The other two CROP sections, exhibited in this chapter, cross transversely the western (M31; Figures 4 and 5) and eastern (M5; Figure 7) Calabrian wedge. It has been realistically calculated (Finetti, this volume, Chapter 10) that the major part of the *Ionides* (about 65%) removed from their Ionian basement is imbricated in the Calabrian wedge that is growing over the subducting oceanic crust. The remaining 35% thrust obliquely onto the Southern Apennines and Sicilian Apennines. While it is possible to conduct a geophysically well-controlled evaluation of the total Ionian surface where Ionides were stripped off, and a realistic computation of the Ionides stacked in the Calabrian wedge, it is difficult to estimate individually the Ionides of the Southern Apennines and Sicily. The global amount of Ionides of the two orogenic areas (35%) is obtained from the difference between the total Ionian area of detachment and the observed stacked amount of Ionides in the Calabrian wedge, assuming that no significant amount subducted, as suggested by CROP sections. An estimate of about equal distributions of Ionides between the Southern Apennines and Sicily is probably not far from reality, and would indicate some 17–18% of the total for each area.

It is immediately evident seismically that the Ionides form a prominent part of the Calabrian wedge, where the tectonic stacking process has conserved substantially good sequentiality (Plate 2, 4, Figures 2, 3). *The Ionian Tethydes (IT) or Ionides are formed from a deep depositional sequence from Permo-Triassic Ionian opening to the Oligocene, and are imbricated solidly with their pre-Messinian Miocene cover (MC)*. Over the Ionides, tectono-sedimentary intervals formed from various internal-external flysch units, Alpine Tethydes units, including varicoloured shales, Messinian units, and other terrigenous tertiary units have been seismically distinguished. The lowermost seismic interval distinguished (NFO) is mostly made up of Numidian flysch, which is distributed where cover units of the Northern Africa plate were detached (in this case Panormid and Etrurian units), allowing that the Paleozoic quartzitic sandstone, interposed between crystalline basement and cover, might have moved to feed these flysch units. The Messinian salt, during the wedge-growing process, was tendentially pushed outward with a snowplough-like action.

On the interpreted CROP section M4 (Plate 4; Figure 3) the upper complex wedge interval overlying the imbricated Ionides has not been stratigraphically subdivided, as on CROP M2B, and the interpretation here has the scope only of outlining the structural style. To the northern extremity of the section is visible the Catanzaro left-transcurrent fault which dissects the Sila Mts. from the Serre Calabrian Mts.

The Calabrian wedge is also explored transversally by CROP sections M31 (Figure 5) and M5 (Figure 7). In Figure 5 the western side of the wedge to the east of the important and very active *right-transcurrent Vulcano fault* is imaged. *This fault dissects the Calabrian arc from the Sicilian Apennines* and transfers the Calabrian wedge southeastward, along a direction subparallel to the Hyblean-Maltese escarpment. In Figure 5 the Vulcano fault cuts also the Ionian oceanic crust, but it acts as a strike-slip fault only for the thrust wedge-pile, and is a dip-slip fault for the Ionian crust. The section M5 (part) transversely images the eastern Calabrian wedge, accreting on the Ionian oceanic crust, and the Apulian platform margin.

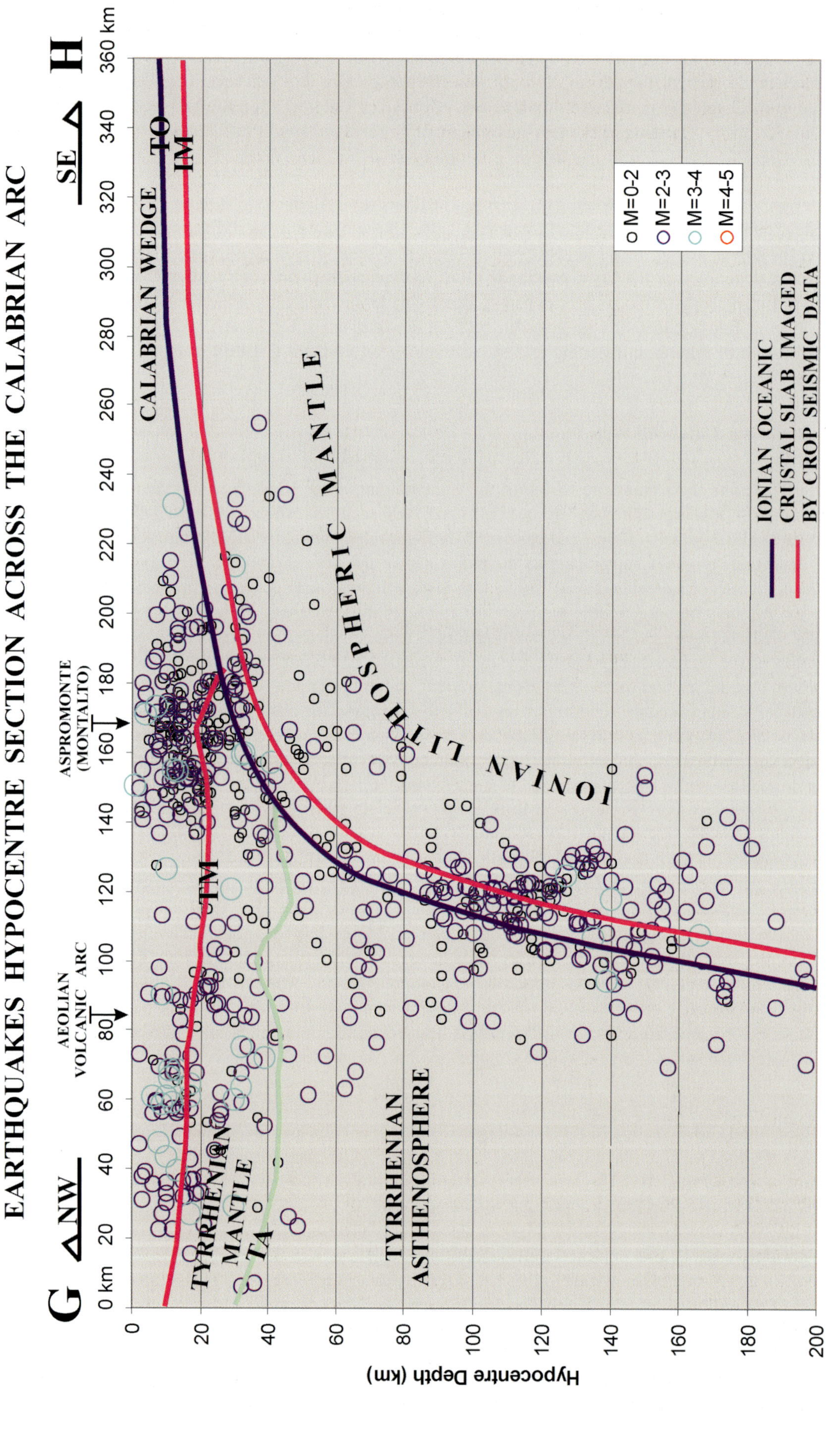
EARTHQUAKES HYPOCENTRE SECTION ACROSS THE CALABRIAN ARC
G
NW
SE
H
AEOLIAN VOLCANIC ARC
ASPROMONTE (MONTALTO)
0 km
20
40
60
80
100
120
140
160
180
200
220
240
260
280
300
320
340
360 km
CALABRIAN WEDGE
TO
IM
TM
TYRRHENIAN MANTLE
TA
TYRRHENIAN ASTHENOSPHERE
IONIAN LITHOSPHERIC MANTLE
M=0-2
M=2-3
M=3-4
M=4-5
IONIAN OCEANIC CRUSTAL SLAB IMAGED BY CROP SEISMIC DATA
Hypocentre Depth (km)

Figure 9: Litho-asthenospheric section showing the *seismicity of the Calabrian arc*. Hypocenters from GNDT – Working Group, 2001. *Instrumental Catalog of Italian Earthquakes (1981–1996), version 1.0* CD-ROM. Clueb, Bologna. Events plotted orthogonally on section centred on a band of data 40 km wide. Position of the section is that of Plate 2 along the sector 0–360 km (progressive distance). TM = Tyrrhenian Moho. TO = Top oceanic crust (Ionian Sea). IM = Ionian Moho. TA = top Tyrrhenian asthenosphere. Data dispersion is mostly due to assumed velocity models, and partially to orthogonal projection onto section plane of lateral events on non-planar slab geometry.

It is interesting to observe that the Ionian basement crust is intensely faulted (loading faults) by the propagating frontal increase in lithostatic charge of the wedge. Then, going inward, the vertical block displacements of crustal fracturation are recomposed. This means that the old Ionian oceanic crust-lithospheric mantle has a buoyancy such that a small loading change is sufficient to produce a relevant isostatic rebound.

3.4. The Ionian slab

Deep earthquake activity of the Calabrian arc has for a long time shown the existence of a slab beneath the south-eastern Tyrrhenian Sea. Peterschmitt (1956) recognized a Benioff plane beneath the Calabrian and SE-Tyrrhenian Sea. Years later, Caputo et al. (1970), Gasparini et al. (1982), Finetti and Del Ben (1986), and various other authors published investigations, based on progressively increasing basic seismogeological data and refined shock locations obtained from seismic networks. If seismology provided significant data on deep seismicity connected to this slab, the advanced multifold seismic exploration allowed an early understanding that the Ionian Sea was floored by old oceanic crust over which the Calabrian arc was thrusting (Finetti, 1982). But no direct seismic imaging of this slab has never been produced. In the meantime continued seismic reflection investigations arrived at the conclusion that the Ionian ocean, before the Tyrrhenian opening, extended considerably northward, up to the area where the deep Tyrrhenian basin is today located (Finetti et al., 1996); and it was also shown that the slab in question represents the subduction of the Ionian lithosphere. It is therefore correct to call it the *Ionian slab*.

With the precise purpose of obtaining a deep seismic reflection imaging of the Ionian slab, within the CROP Mare programme, a 40-second two-way-time seismic section was acquired on the Tyrrhenian margin of the Calabrian arc (Figure 1, segment IS). The section, recorded in two seismic ship passages (0–21 s the first, and 20–40 the second) was combined in processing, obtaining the successful imaging in Figure 8. The Ionian oceanic crust of the slab is clearly imaged. In the larger scale section, the slab is better visible to the bottom (40 s) of the section (over 150 km). It is evident that no slab truncation or detachment occurs in the litho-asthenospheric interval exhibited. Previous seismogeological investigations (Mele, 1998) arrived at the correct conclusion that "the Tyrrhenian slab is not detached down to 80 km depth".

In Figure 9 the seismicities of the Calabrian arc and Ionian slab are plotted. The deep hypocentre distribution follows the subducting slab with dispersion due to assumed velocity models and lateral events. The Tyrrhenian seismicity (0–40 km) is confined within its lithospheric interval. Maximum shallow event concentrations occurs in the stacked chain sector (Aspromonte in the section), while the outer Calabrian wedge is growing with a substantially aseismic deformation process.

ACKNOWLEDGEMENTS

This work was partially funded by CROP Partners (CNR, ENEL, ENI) with CNR grants to I.R. Finetti: CTR 96.00797; CTB 97.04525.CT05. Major financial support was given by MIUR: Italian programmes; grants COFIN 98, COFIN 2000 and COFIN 2002 to I.R. Finetti (University of Trieste). Relevant financial support was granted also by the University of Trieste for reprocessing and interpretation. The author express gratitude to all supporters. Many thanks are due to OGS, Seismic Division, to all teams of the OGS-Explora seismic vessel for the CROP seismic data acquisition, to the EGG (Exploration Geophysics Group) of the Faculty of Sciences of the University of Trieste for iterated interpretive reprocessings of a major part of the CROP data used, where signal improvement was determinant for a better understanding of the complex features involved. Particular thanks are due to Alessia Taboga for a careful work of literature research. We wish to thank William B.F. Ryan of the Lamont-Doherty Earth Observatory, Columbia University, USA, for reviewing the manuscript and for much appreciated comments.

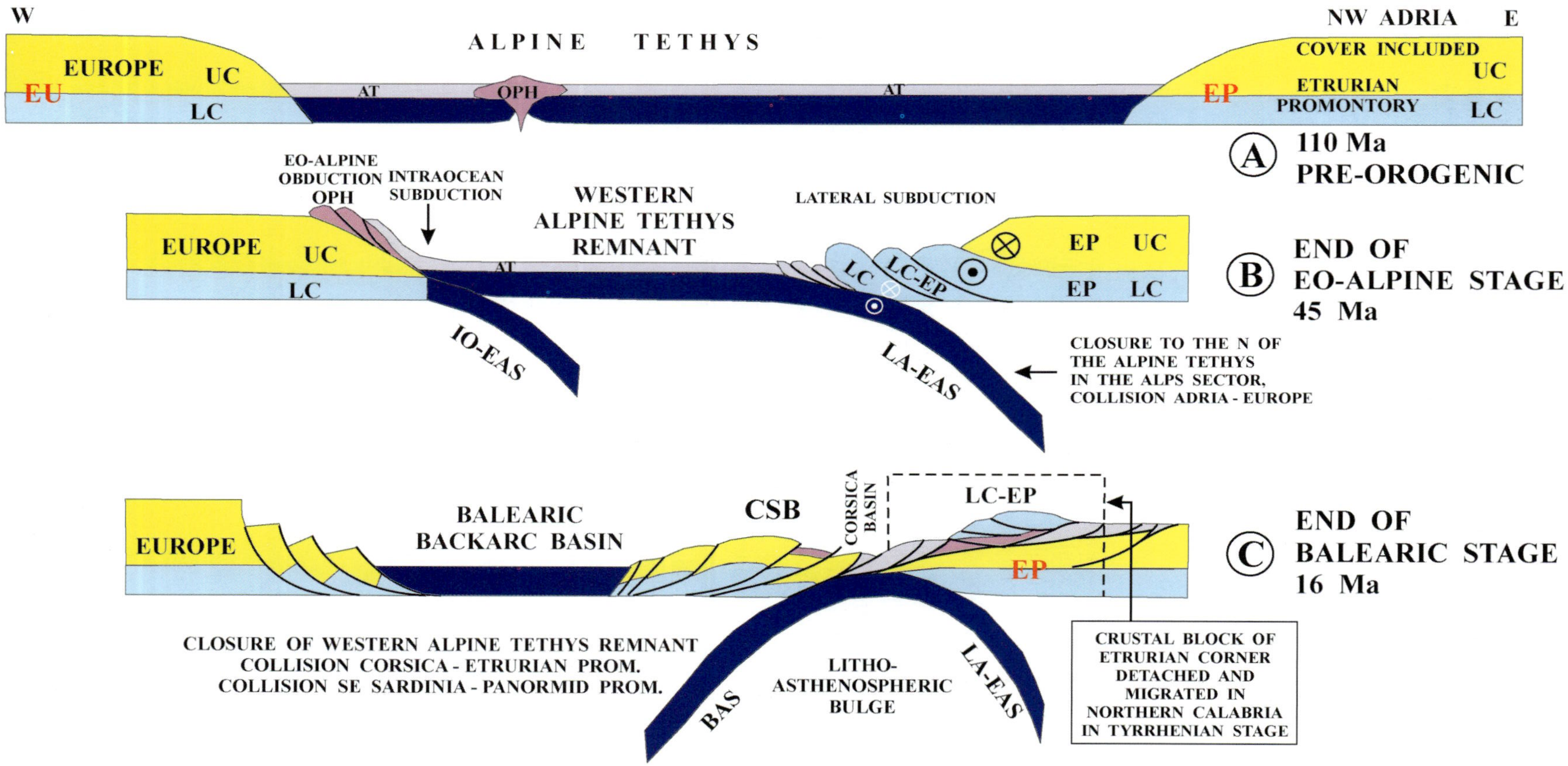

Figure 10: *Scheme of the pre-Tyrrhenian opening evolution*, constructed for a better understanding of the highly questioned *northern Calabria tectonic stack* and of its difference with respect to southern Calabria and Peloritani Mts. AT = Alpine Tethydes, including ophiolite (OPH). IO-EAS = Intra-ocean SE-dipping Eo-Alpine subduction (obduction of ophiolite over southern European margin). LA-EAS = Lateral E-dipping subduction of NW-Alpine Tethys beneath Adria, driven by dominant converging forces between Europe and joint Adria-Ionian-Africa block. BAS = W-dipping slab of subducted western Alpine Tethys remnant beneath Europe (CSB Corso-Sardinian block) in the Balearic stage. Between the BAS and LA-EAS a prominent litho-asthenospheric bulge originated. LC-EP = lower crustal fragment thrust over Alpine Tethys in Eo-Alpine stage and transported jointly to basal AT and OPH units over the Etrurian promontory (EP) in the Balearic stage. The whole LC-EP stacked crustal corner detached, migrated and thrust onto the northern Calabria orogenic belt in the Tyrrhenian stage.

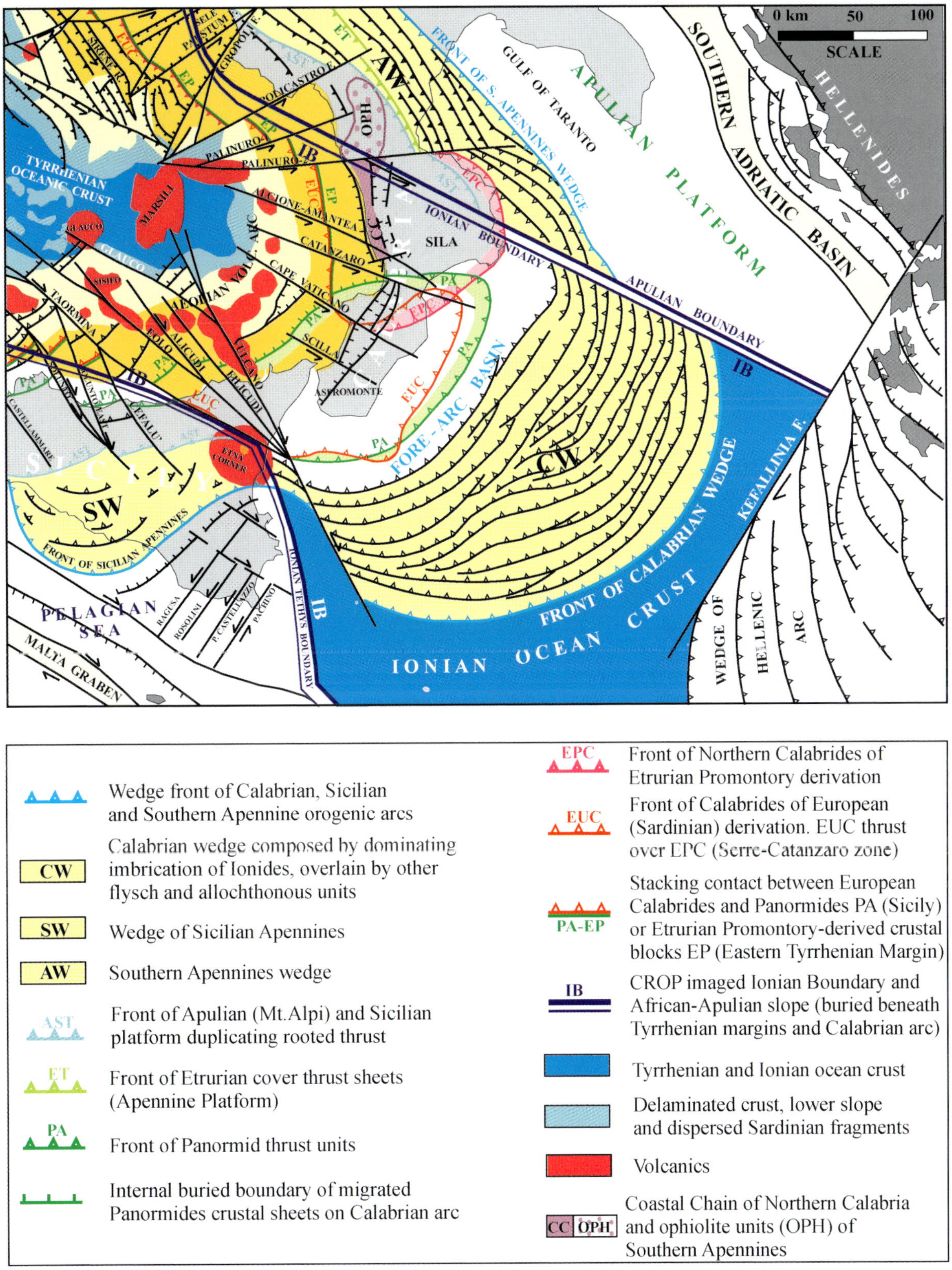

Figure 11: ***Tectonic sketch-map of the Calabrian arc*** **reconstructed from new deep CROP seismic sections and integrating other seismic lines (MS Lines: Finetti and Del Ben, 1986; Finetti et al., this volume, Chapter 1), and lines from the Italian Ministry's public dataset.**

REFERENCES

Alvarez, W., 1976. A former continuation of the Alps. Bull. Geol. Soc. Am. 87, 891–896.

Amodio-Morelli, L., Bonardi, G., Colonna, V., Dietrich, D., Giunta, G., Ippolito, F., Liguori, V., Lorenzoni, S., Paglionico, A., Perrone, V., Piccarreta, G., Russo, M., Scandone, P., Zanettin-Lorenzoni, E., Zuppetta, A., 1976. L'Arco Calabro-Peloritano nell'orogene appenninico-maghrebide. Mem. Soc. Geol. It. 17, 1–60.

Boccaletti, M., Nicolich, R., Tortorici, L., 1984. The Calabrian Arc and the Ionian Sea in the dynamic evolution of the Central Mediterranean. Marine Geology 55, 219–245.

Bonardi, G., Giunta, G., Perrone, V., Russo, M., Zuppetta, A., Ciampo, G., 1980. Osservazioni sull'evoluzione dell'Arco Calabro-Peloritano nel Miocene Inferiore: la formazione di Stilo-Capo d'Orlando. Boll. Soc. Geol. It. 99, 365–393.

Bonardi, G., Cello, G., Perrone, V., Tortorici, L., Turco, E., Zuppetta, A., 1982. The evolution of the northern sector of the Calabria–Peloritani arc in a semiquantitative palynspastic restoration. Boll. Soc. Geol. It. 101, 259–284.

Bonardi, G., Compagnoni, R., Messina, A., Perrone, V., Russo, S., De Francesco, A.M., Del Moro, A., Platt, J., 1992. Sovraimpronta metamorfica alpina nell'unità dell'Aspromonte (settore meridionale dell'Arco Calabro-Peloritano). Boll. Soc. Geol. It. 111, 81–108.

Bonardi, G., De Capoa, P., Fioretti, B., Perrone, V., 1994. Some remarks on the Calabria-Peloritani Arc and its relationships with the Southern Apennines. Boll. Geofis. Teor. Appl. 36, 483–490.

Caputo, M., Panza, G.F., Postpischl, D., 1970. Deep structure of the Mediterranean basin. J. Geophys. Res. 75 (26), 4919–4923.

Cortese, E., 1895. Descrizione geologica della Calabria. In: Mem. descr. Carta Geol. d'Italia, vol. 9, p. 310.

Finetti, I.R., 1976. Mediterranean Ridge: A Young Submerged Chain Associated with the Hellenic Arc. Boll. Geof. Teor. Appl. 28, 31–62.

Finetti, I.R., 1982. Structure, stratigraphy and evolution of Central Mediterranean. Boll. Geof. Teor. Appl. 24, 247–426.

Finetti, I.R., this volume. Understanding the Ionides and their geodynamics. Chapter 10.

Finetti, I.R., Del Ben, A., 1986. Geophysical study of the Tyrrhenian opening. Boll. Geof. Teor. Appl. 28, 75–156.

Finetti, I., Lentini, F., Carbone, S., Catalano, S., Del Ben, A., 1996. Il sistema Appennino Meridionale–Arco Calabro–Sicilia nel Mediterraneo Centrale: studio geologico-geofisico. Boll. Soc. Geol. It. 115, 529–559.

Finetti, I.R., Del Ben, A., Forlin, E., Prizzon, A., this volume. Review of the main results of the pioneering MS seismic exploration programme in the deep water of the Mediterranean Sea (OGS, 1969–1980). Chapter 1.

Finetti, I.R., Lentini, F., Carbone, S., Del Ben, A., Di Stefano, A., Guarnier, P., Pipan, M., Prizzon, A., this volume. Crustal Tectono-stratigraphy and Geodynamics of the Southern Apennines from CROP and other integrated geophysical-geological data. Chapter 12.

Gasparini, C., Iannaccone, G., Scandone, P., Scarpa, R., 1982. Seismotectonics of the Calabrian Arc. Tectonophysics 84, 267–286.

Haccard, D., Lorenz, C., Grandjacquet, C., 1972. Essai sur l'evolution tectonogenetique de la liaison Alpes-Apennins (de la Ligurie a la Calabre). Mem. Soc. Geol. It. 11, 309–341.

Limanowski, M., 1913. Die grosse kalabrische Decke. Bull. Int. Acad. Sc. Cracovie, Cl. Sci. Math. Nat., S. A 6A, 370–385.

Lugeon, M., Argand, E., 1906. La racine de la nappe sicilienne et l'arc de charriage de la Calabre. C. R. Acad. Sci. Paris 142, 1107–1109.

Mele, G., 1998. High-frequency wave propagation from mantle earthquakes in the Tyrrhenian Sea: New constraints for the geometry of the south Tyrrhenian subduction zone. Geophys. Res. Lett. 25 (15), 2877–2880.

Ogniben, L., 1969. Schema introduttivo alla geologia del confine calabro-lucano. Mem. Soc. Geol. It. 8, 435–763.

Ogniben, L., 1973. Schema geologico della Calabria in base ai dati odierni. Geol. Romana 12, 243–585.

Petterschmitt, E., 1956. Quelques donnees nouvelles su les seismes profonds de la Mer Tyrrhenienne. Ann. Geofis. 9, 305–334.

Scandone, P., 1979. Origin of the Tyrrhenian Sea and Calabrian Arc. Boll. Soc. Geol. It. 98, 27–34.

Scandone, P., 1982. Structure and evolution of the Calabrian Arc. Earth Evol. Sci. 3, 172–180.

Van Dijk, J.P., Bello, M., Brancaleoni, G.P., Cantarella, G., Costa, V., Frixa, A., Golfetto, F., Merlini, S., Riva, M., Torricelli, S., Toscano, C., Zerrilli, A., 2000. A regional structural model for the northern sector of the Calabrian Arc (southern Italy). Tectonophysics 324, 267–320.

CROP PROJECT: Deep Seismic Exploration of the Central Mediterranean and Italy
Edited by I.R. Finetti

CHAPTER 18

Crustal Tectono-Stratigraphic Setting and Geodynamics of the Corso-Sardinian Block from New CROP Seismic Data

I.R. Finetti[1,*], A. Del Ben[1], S. Fais[2], E. Forlin[1], E. Klingelé[3], L. Lecca[4], M. Pipan[1], A. Prizzon[1]

ABSTRACT

Based on new CROP lithospheric seismic data integrated with other geophysical and geological data, a well controlled crustal setting for the Corso-Sardinian Block (CSB) has been reconstructed, and an innovative constrained geodynamic evolution is proposed. A continuous marine CROP section passing between Corsica and Sardinia and crossing the CSB from the Balearic Sea to the Tyrrhenian back-arc basin exhibits the tectono-stratigraphy of the block, which is thus fully explored for the first time by transcrustal seismic reflection data. On the western margin, listric faults connected to the Balearic basin opening and CSB rotation are clearly shown. The CSB axial mainland is characterized by thrust faults which thicken a crust previously stretched during the early Balearic stage (Oligocene) and successively recompacted in the late Balearic stage when collisions of Corsica with the NW Adria margin to the north of the block, and of SE Sardinia with the NW Panormid Promontory to the south took place.

The CROP section M10, running from the Tuscan Archipelago to the Ligurian sphenochasm, exhibits an Alpine Tethys slab subducted laterally beneath Adria in the Eo-Alpine stage (Upper Cretaceous–Lower Eocene) and a W-dipping slab subducted beneath the southern European margin (Corsica) in the Balearic stage (Oligocene–Lower Miocene). CROP section M12B illustrates the lithospheric tectono-stratigraphy of the southern Corsica basin where it was closed by eastward-pulled CSB units along the Augustus Line (41°parallel, northern boundary of the Ionian Tethys), just at the NW Adria plate corner (Etrurian Promontory). Clear seismic imaging of the subducted Alpine Tethys slab is visible. On the Adria-verging frontal thrusts of the CSB, at the southern closure of the Corsica basin, Europe-verging thrusts associated with high mag anomalies are present. They are attributed to ophiolite-bearing units of Eo-Alpine obduction. These ophiolites are the southeastern continuation of those outcropping in NE Corsica and, with the exception of some dispersed fragments, are the last to connect those of the Coastal Chain segment in northern Calabria.

The CROP section M28B images the stretched and delaminated eastern Sardinian margin very well. In the Cornaglia terrace, the Sardinian (Europe) crust is four times thinner than on the island's coast. Over the delaminated crustal Sardinian units, thin Europe-verging thrust units associated with high mag anomalies are identified. On the basis of their tectonic framework, associated high mag anomaly, and characteristic seismic facies, we attribute these units to the SE Sardinian segment of the ophiolite-bearing units. With the discovery of this last segment, and by inverting the migration of all known ophiolites to their original position of Eo-Alpine obduction, it is now possible to reassemble an almost complete belt of ophiolites from the Betic zone to the Alps.

From specific new geological-geophysical investigations, a detailed study of Sardinian rift activity has been performed, providing also an explanation of related magmatic activities. The ensemble of all relevant new data allows us to reconstruct a time–space evolution of the CSB. At the end of the Lower Cretaceous, the paleogeography of the central Mediterranean

[1]Department of Geological, Environmental and Marine Sciences, University of Trieste, Via Weiss 1, 34127 Trieste, Italy
[2]Department of Geoengineering and Environment Technologies, University of Cagliari, Piazza d'Armi, 09123 Cagliari, Italy
[3]Geodesy and Geodynamics Laboratory, IGP, ETH Honggerberg, 8093 Zurich, Switzerland
[4]Department of Earth Sciences, University of Cagliari, Via Trentino 51, 09127 Cagliari, Italy
*Corresponding author: Tel.: +39-40-5582276; e-mail: finetti@units.it (I.R. Finetti)

consisted of two intercommunicating ocean basins (Ionian Neotethys opened in Permo-Triassic times, and Alpine Neotethys opened in Middle Jurassic–Lower Cretaceous times), interposed between Europe to the north and Adria–Africa to the south. Two promontories (Etrurian Promontory to the north and Panormid Promontory to the south) delimited the Alpine Tethys crust from the Ionian. In the Eo-Alpine geodynamic stage (Upper Cretaceous–Lower Eocene) the joint convergence of Africa–Adria and interposed Ionian with Europe determined the closure of the Alpine Tethys and continental collision in the sector of the Alps, leaving a wide remnant of the Alpine Tethys west of the Alps unsubducted. In the Balearic stage (Oligocene–Lower Miocene), within the remnant Alpine Tethys, the CSB rotated, closing the Tethyan basin, and opening the Balearic back-arc basin. The Corsica block collided with NW Adria (Etrurian Promontory) and SE Sardinia collided with the NW Panormid Promontory, arresting the CSB rotation. The W-dipping subduction continued without stop in the Ionian Tethys (Tyrrhenian stage), and connected mantle flux dynamics gave origin to the Southern Apennine–Calabrian Arc–Sicilian Apennine thrust belt system, delaminating the Sardinian margin and opening the Tyrrhenian back-arc basin.

Keywords: Corso-Sardinian crustal tectonics, stratigraphy, geodynamics CROP reconstruction

1. INTRODUCTION

In this work, the macro-tectono-stratigraphic setting of the Corso-Sardinian block (CSB) within its complex surrounding context, and its evolving geodynamics are described. The results here presented are based on new CROP seismic data, integrated with other geophysical and geological data obtained from new investigations. Information selected from the literature for the area also contributes greatly to the presented setting.

Geodynamic views of the CSB in the literature

In the Late Oligocene, a regional orogenic volcanism is recognized in Sardinia (Beccaluva et al., 1985, 1987). This event supports the presence of a residual Tethyan oceanic lithosphere east of Sardinia, large enough to subduct towards the northwest. East of the paleo-CSB, Serri et al. (2001) propose the presence of a *Sicilian-Sub-Ligurian oceanic basin* that continued in the Ionian Ocean basin, whose subduction would have activated the orogenetic volcanism in Provence, Sardinia, and the Gulf of Valencia during the Late Oligocene–Early Miocene. A similar paleogeographic and geodynamic scenario has long been proposed by a number of authors (e.g. Alvarez et al., 1974; Tapponnier, 1977 and others). The presence of an Oligo-Miocene ocean basin between the CSB and Adria, as well as the direction of subduction, are highly debated topics that have been proposed in several space–time hypotheses.

The genesis and fragmentation of the paleo-CSB is testimony to a regional geodynamic process, variously described since the 70's in a number of works by both the Italian and the French schools. This process occurred from the Late Oligocene onwards throughout the western Mediterranean and Tyrrhenian. It can be observed in surface geology only partially because most of the structural and lithological products are actually located below sea level. The Tertiary evolution of the area was explained prevalently as being activated by progressive closure of the Africa–Europe gap (e.g. Tapponnier, 1977; Cohen, 1980; Boccaletti et al., 1982, 1990; Dewey et al., 1989). The fragmenting of the CSB was attributed to the shortening that occurred along the Pyrenees, which from the Paleocene to the Oligocene was estimated in the order of 100–150 km in an approximate NE direction (Munoz, 1992; Olivet, 1996). At the same time, to the north, the Alps orogeny (Coward and Dietrich, 1989) induced rotational stresses (Vialon et al., 1989) and an extension in the foreland area, favouring the beginning of a European rifting system (Bois, 1993). Progressively during the Late Oligocene–early Middle Miocene, the formation of the Maghrebian chain and the stacking of the European units (Kabilies, Calabrides) took place (Wildi, 1983; Ben Avraham et al., 1990; Finetti et al., 1996; Torelli et al., 1991).

As regards the problem of oceanic subduction, some authors favour a west-Alpine vergence up to the Eocene-Lower Oligocene, and postulate a later Apenninic vergence from the Eo-Alpine to the Apenninic phases, with Apenninic vergences and retrovergences on Corsica (Bortolotti et al., 2001). Recent CROP seismic data exhibited lateral eastward subduction of the Alpine Tethys slab beneath NW Adria in the Eo-Alpine stage, followed by an eastward migration of the thrust fronts over an Alpine Tethys crust which forms a bulge below Elba island, and shows in the seismic section only the initial uppermost westward subduction below the Corsica basin, where CROP section M12A terminates (Finetti et al., 2001). Other CROP sections now show clearly that a well developed west-dipping Alpine Tethys slab is present below the CSB (Figures 13, 14(b) and 15). The back-arc Balearic basin extension is interpreted as due to roll-back or retreat of the subducted slab (Boccaletti et al., 1974; Ritsema, 1979; Finetti and Del Ben, 1986; Malinverno and Ryan, 1986; Patacca and Scandone, 1989).

The new lithospheric CROP data allow us to reconstruct a new and far better 3D-constrained lithospheric setting and geodynamic evolution of the CSB, which will be described in the following.

2. REGIONAL GEOLOGICAL AND GEOPHYSICAL OUTLINE

The upper crust of the CSB is made up prevalently of metamorphites, derived from terrigenous, volcanic and carbonate successions of Cambrian to Carboniferous age, and of large volumes of granitoids relating to the Hercynian orogeny (Carmignani et al., 1992). Locally, follow thin Permo-Carboniferous terrigenous and volcanic complexes and a Triassic succession in Germanic facies (Costamagna and Barca, 2002) covered by fluvial sediments and by Jurassic and Cretaceous carbonate complexes. In a few zones of southern Sardinia, Eocene and Oligocene terrigenous and shallow-water carbonate sediments are present, covered by Oligo-Miocene to Quaternary volcanic and sedimentary products along the basins of the SR (Cherchi and Montadert, 1982). The north-eastern part of Corsica is made up of deformed, metamorphic complexes derived from the closure of the Ligurian-Piedmontese ocean and generally considered the southward continuation of the western Alps (Durand-Delga, 1984).

This block was considered a residue of a paleo-microplate of European origin, mobile within the Neogene western and central Mediterranean geodynamical system (Alvarez et al., 1974). In its present structure the geology of the block shows mainly tectonic highs and basins that date back to Neogene times, and is characterized by contractional and extensional evolution in Corsica, and in Sardinia by mainly transtensional-extensional structures known as the *Sardinian Rift* (Figure 2).

2.1. Magnetic and Gravity Expressions of the Corso-Sardinian Block

The CSB represents a domain with magnetic anomalies typical of continental crust, interposed between the magnetic domain of the Western Mediterranean and the Tyrrhenian basin. The complex evolution and the diversity of the genetic processes of the lithologies characterized by very different minero-petrographic compositions caused a strongly contrasted magnetic pattern, both in Sardinia and Corsica and surrounding seas. These magnetic patterns have been described in a number of studies (Arisi Rota and Fichera, 1987; Bayer et al., 1973, 1976; Cassano et al., 1979, 1986; Mouge and Galdéano, 1990; Zanolla et al., 1998; Fais et al., 1994, 1996, 2001, 2002; Galdeano and Ciminale, 1987).

In this context, the local ground surveys carried out in Sardinia have been purposely ignored since they are judged inadequate to reconstruct the main deep geological and structural features of the region, but the EGT and AGIP-EMS aeromagnetic data have been used as described below. Similarly for Corsica, mainly aeromagnetic data were employed (Mouge and Galdéano, 1990). The gravity data used to produce the Bouguer anomaly map were taken from the EGT CD-ROM (Klingelé et al., 1989), applying a homogenization procedure on data of different origins as described in the following.

The Airborne Magnetic Data and Main Magnetic Features

The airborne magnetic map reported in Figure 3 was produced with the Airborne Magnetic Anomaly data of the EGT CD-ROM (Wonik et al., 1992), obtained from surveys carried out at different flight altitudes and epochs. Adequate procedures were applied to achieve acceptable coherence in the data. Quantitative interpretations in some sectors of the Sardinian Block where remarkable anomalies are present were carried out using the data collected by the geophysical contractor CGG for AGIP-EMS (Cassano et al., 1979).

In order to obtain a residual anomaly map, a regional field with a N–S gradient of $+3.232$ nT/km and $+0.726$ nT/km from west to east was removed. The parameters of this field were obtained by least-squares fitting of a first order polynomial to the experimental data.

The airborne magnetic map of the CSB shows very interesting anomalies, some of which have already been analyzed and interpreted. In Sardinia and its continental margins, a comparison of this map with the geological

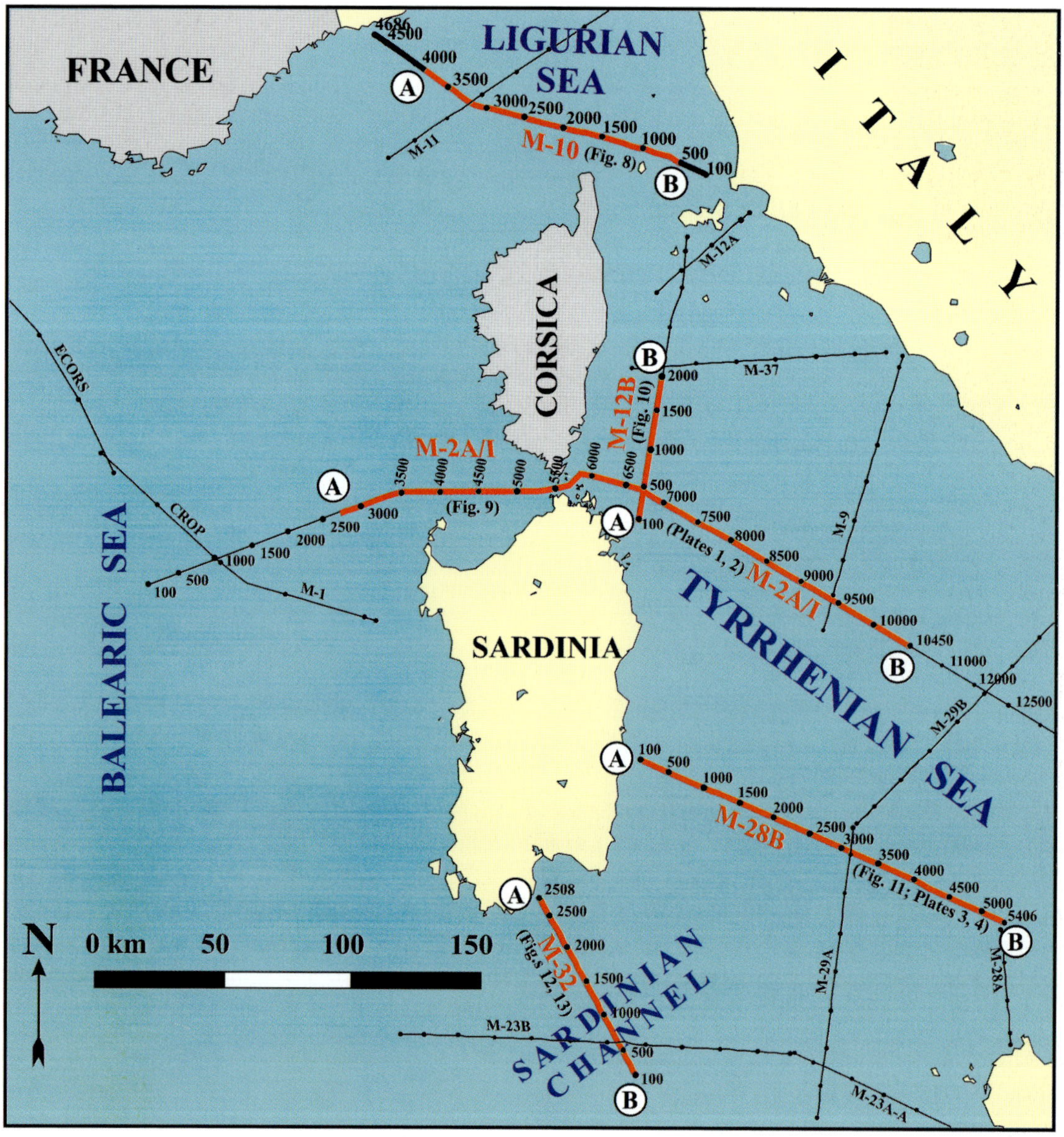

Figure 1: Location map of the CROP seismic sections used in this Chapter 18 (red traces).

map shows a correspondence between the different magnetic patterns and the different lithologies making up the crust. In brief, two main magnetic patterns can be recognized: one characterizing the Palaeozoic crust, the other characterizing the sectors involved in the volcanic processes associated with extensional tectonic phases of the Tertiary.

A magnetic pattern characterized by relative low intensity and low gradient maxima and minima corresponds to the generally integral tectonic blocks made up of Hercynian rocks which are both sedimentary, more or less metamorphic, and intrusive more or less acidic (Carmignani et al., 2001). The anomalies present in these tectonic blocks may be correlated with petrographical or structural Hercynian heterogeneities, such as the one located in north-eastern Sardinia (Gallura), for example (Figure 3). This large and strong magnetic anomaly of more than 400 nT, extending for more than 30 km and with a width of around 15 km, can be referred to a source within the Paleozoic crust. This anomaly has been quantitatively interpreted by Fais et al. (1994) under two hypotheses for explaining its source. In the first hypothesis, the magnetic bodies can be considered as plutonic intruding into a pre-existing granitic complex. In the second hypothesis, the outcropping granites and the magnetic structures are cogenetic but differ in their ferromagnesian mineral content.

Other anomalies are located on the high-angle fault and thrust sheet zones of central and south-eastern Sardinia. Anomalies are also present in correspondence with outcrops of Permo-Carboniferous volcanites, generally rhyo-

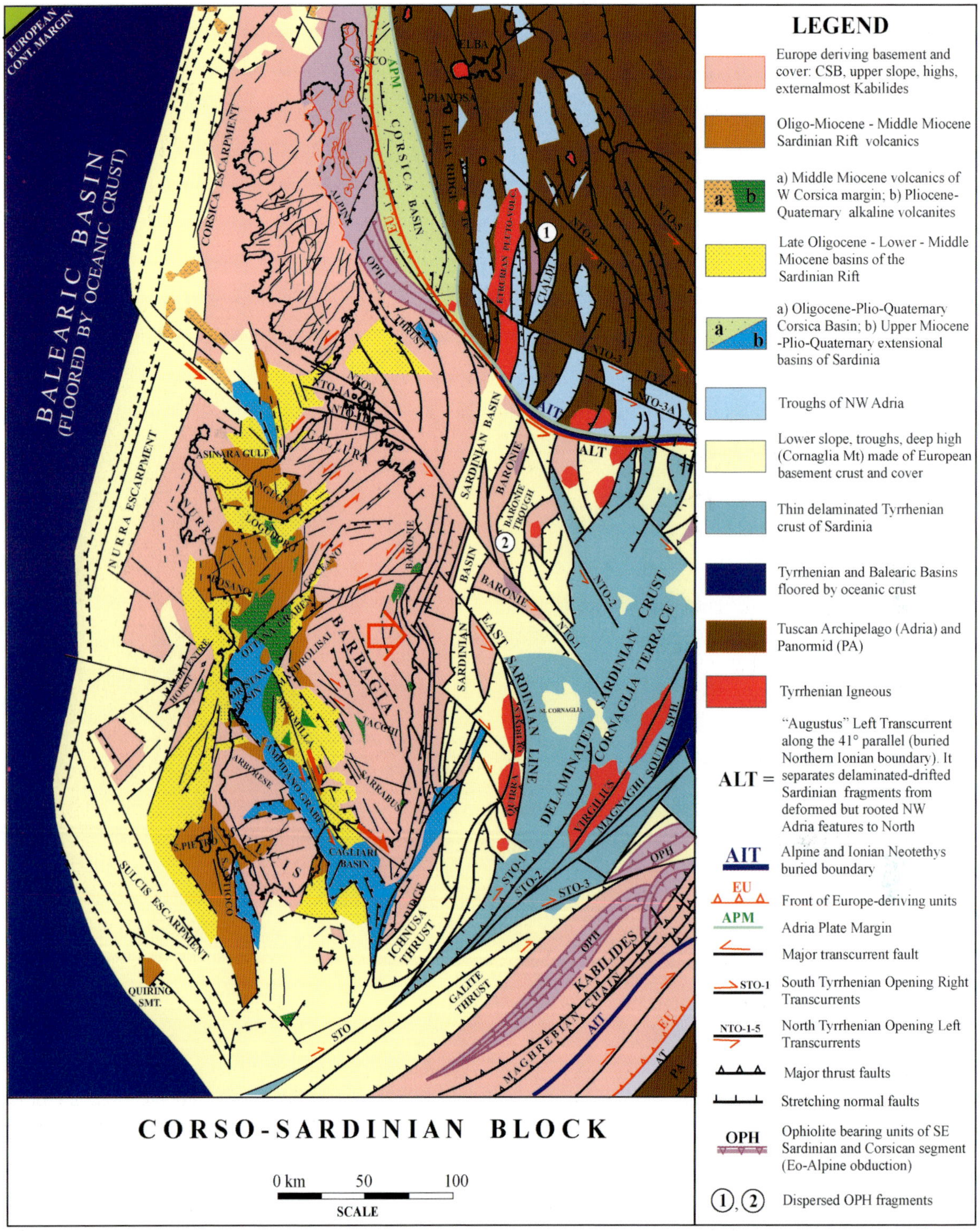

Figure 2: Geological sketch map of the Corso-Sardinian Block (CSB) outlining crustal macro-tectono-dynamics, volcanic and Sardinian rift-basin distribution. The illustrated tectonodynamics are derived from the CROP network interpretation (Finetti, this volume, Chapter 34).

dacitic, and subordinately, andesitic. A magnetic pattern made up of sets of localized high-gradient anomalies corresponds generally to the Oligo-Miocene andesitic calc-alkaline complexes (Brotzu, 1997). These volcanic complexes outcrop in a number of tectonic contexts. The anomalies located along the palaeo-structures of the rift reactivated by the Campidano graben (Monastir, Furtei, Siliqua, Sarroch) are particularly significant from the tectonic point of view.

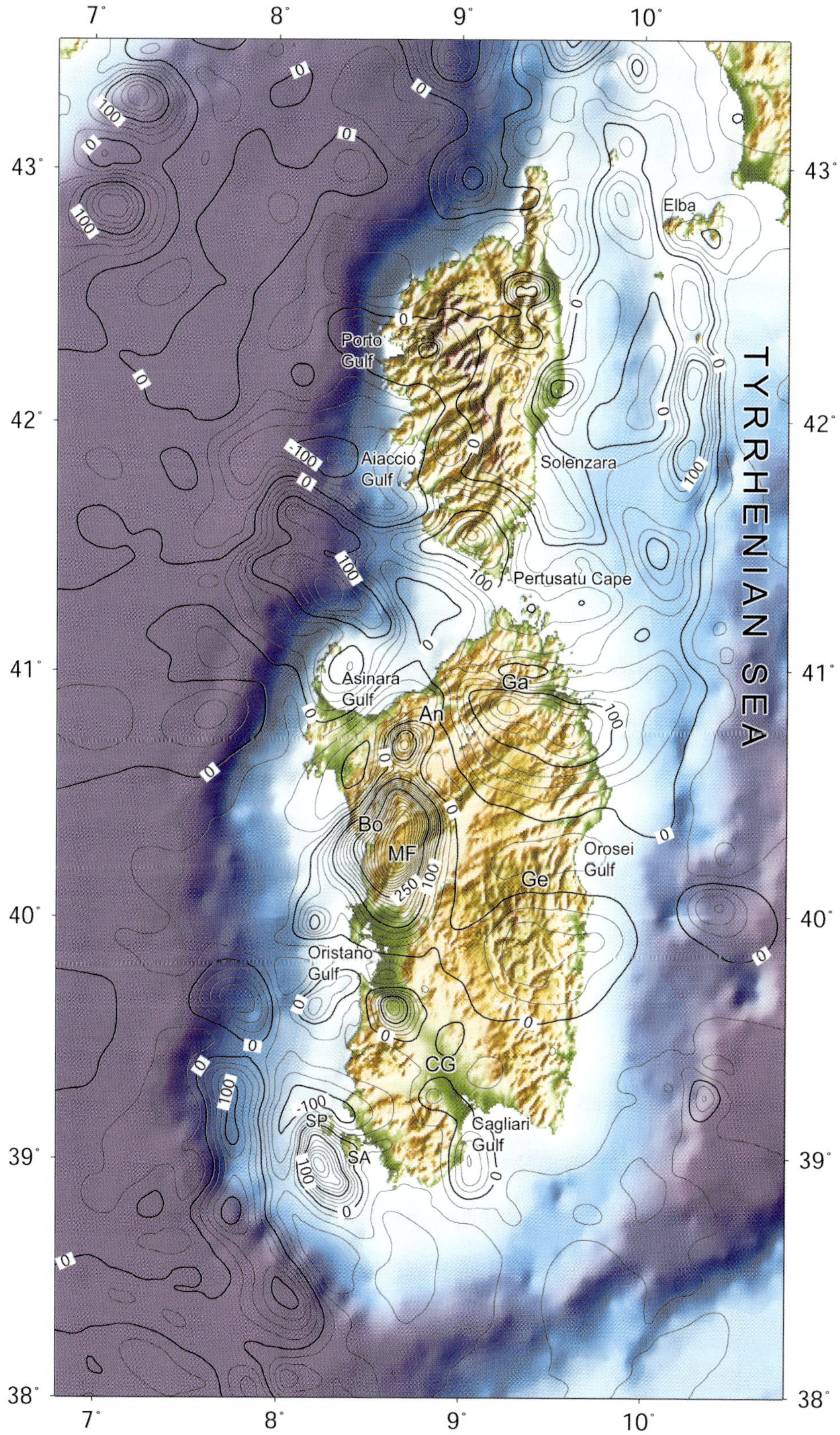

Figure 3: Residual total magnetic field of Corsica and Sardinia and surrounding areas. Units are nT and equidistance between curves is 25 nT. Ga = Gallura; An = Anglona; Bo = Bosano; MF = Monte Ferru; Ge = Gennargentu; CG = Campidano Graben.

Other anomalies have been produced by the Pliocene alkaline volcanism, which is linked with the opening of the Tyrrhenian basin. Among these, the most remarkable feature is the strong positive anomaly northeast of the Gulf of Oristano, which can be attributed to a huge mass of lavas (M. Ferru complex) mainly made up of dacites and basalts (Beccaluva et al., 1977) superimposed on Oligo-Miocene volcanites. Other anomalies are caused by the basaltic plateaux of Campeda and Abbasanta and other small plateaux. Significant anomalies are also present in the submerged sectors prevalently along the western continental margin, and are attributable to Oligo-Miocene volcanites.

In Corsica, two remarkable anomalies can be easily observed (Figure 3). One is located in the southernmost part of the island (Hercynian Corsica) in correspondence to granitoid rocks of different petrographical compositions but mainly represented by biotite leucomonzogranites, biotite monzogranites, granodiorites, tonalo-granodiorites with horneblende-biotite and tonalites (Carmignani et al., 2001). The other is in Alpine Corsica at a latitude of about 42.5 N, close to the eastern coast where ophiolitic formations outcrop. In a general overview of the magnetic pattern of Corsica, some authors (Mauffret et al., 1999) highlight the magnetic positive trend that develops from the Gulf of Solenzara in a SE direction to the south Tyrrhenian basin. This trend is also linked the prosecution of the Eo-Alpine ophiolite-bearing units (Finetti, this volume, Chapter 18). Southwest of the Aiaccio Gulf, a very large two-sided anomaly (around 50 km wide) can be identified, and can be correlated with the Middle Miocene volcanic bodies of the western Corsica margin (Bellaiche et al., 1979; Bellon et al., 1985). For these anomalies in Corsica no quantitative interpretation is available.

In the CSB magnetic panorama, large, strong anomalies coexist with small, weak ones. For this reason, the later are difficult to identify and quantify. In order to overcome this difficulty, a logarithmic transform was applied to the total field map (Figure 4). In this representation, the values are scaled according to their base ten logarithm. The effect of this is to enhance the weak anomalies and downscale the very strong ones.

The Gravity Data and Main Gravity Features

The data used in the map shown in Figure 5, was produced with the Bouguer anomaly data from the EGT CD-ROM (Klingelé et al., 1989). The Corsican data were digitized from the map of Bayer and Lesquer (1976).

The gravity map of the CSB (Figure 5) unlike the magnetic map does not present any very notable features, except in Alpine Corsica. In Sardinia, one of the main features can be observed in the southwestern sector of the Oligo-Miocene rift, where the Plio-Quaternary Campidano graben was involved in a new subsidence cycle. The gravity low is related to the presence of large thicknesses of Late Oligocene–Early Miocene synrift volcanites, continental deposits (Ussana Formation), and marine sediments that infilled the rift which was finally overlain by large-thickness of Plio-Pleistocene continental deposits (Samassi Formation). A prolongation of this structure in the Cagliari basin is clear (Figure 5).

A gravity survey of Sardinia was carried out by Trudu (1962), and later integrated by Balia et al. (1983, 1984) in the southern and central part of the Campidano. On the western Sardinian margin, off the Gulf of Oristano, and in the area of the islands of S. Antioco and S. Pietro, characterized by very interesting magnetic anomalies, the gravity anomalies are weak. The relative gravimetric low in the Gulf of Oristano and in the facing onshore area is linked with the north-western continuation of the Campidano graben. In Corsica, in a general overview (Figure 5), only the central-southern part and the offshore zone east of the Corsica–Sardinia channel show anomalies with notable amplitudes. Alpine Corsica is characterized by a positive anomaly roughly oriented N–S with a maximum amplitude varying between 80 and 110 mGal. The Hercynian Corsica basement contains many anomalies of short wavelength.

Concerning anomalies related to deep structures, it is observed that a strong gradient along the western coast of Corsica marks a sudden decrease in the thickness of the continental crust. A negative anomaly with a minimum in the Solenzara area and facing offshore can be observed. The eastern sector of this anomaly with an approximately N–S trend can be correlated to the large thickness of infilling sediments in the Corsica basin and its coastal underthrusting.

The same logarithmic transformation as applied to the magnetic data was also applied to the gravity data (Figure 6), in order to see more clearly whether a correlation can be found between the magnetic and the gravity anomalies. Except for the Campidano graben, the transformed gravity data do not show any unambiguous correlation with the magnetic data, but they reveal some interesting features. Firstly, a huge NNW–SSE elongated negative anomaly appears in the northeastern part of Sardinia.

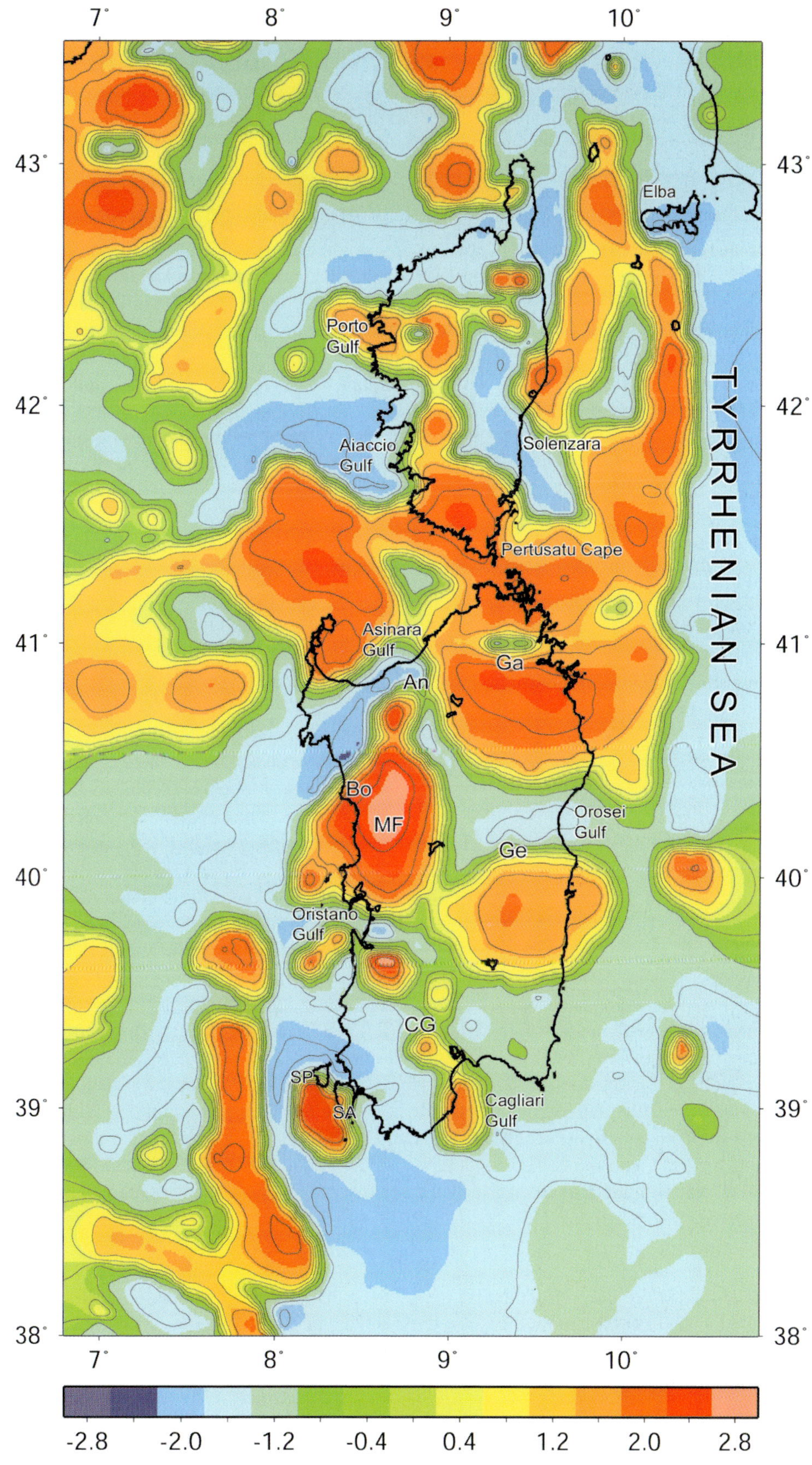

Figure 4: Residual total magnetic field of Corsica and Sardinia and surrounding areas plotted on a base 10 logarithmic scale. Units are $\log_{10}$(nT) and equidistance between curves is 1. For the meaning of the letters see Figure 3.

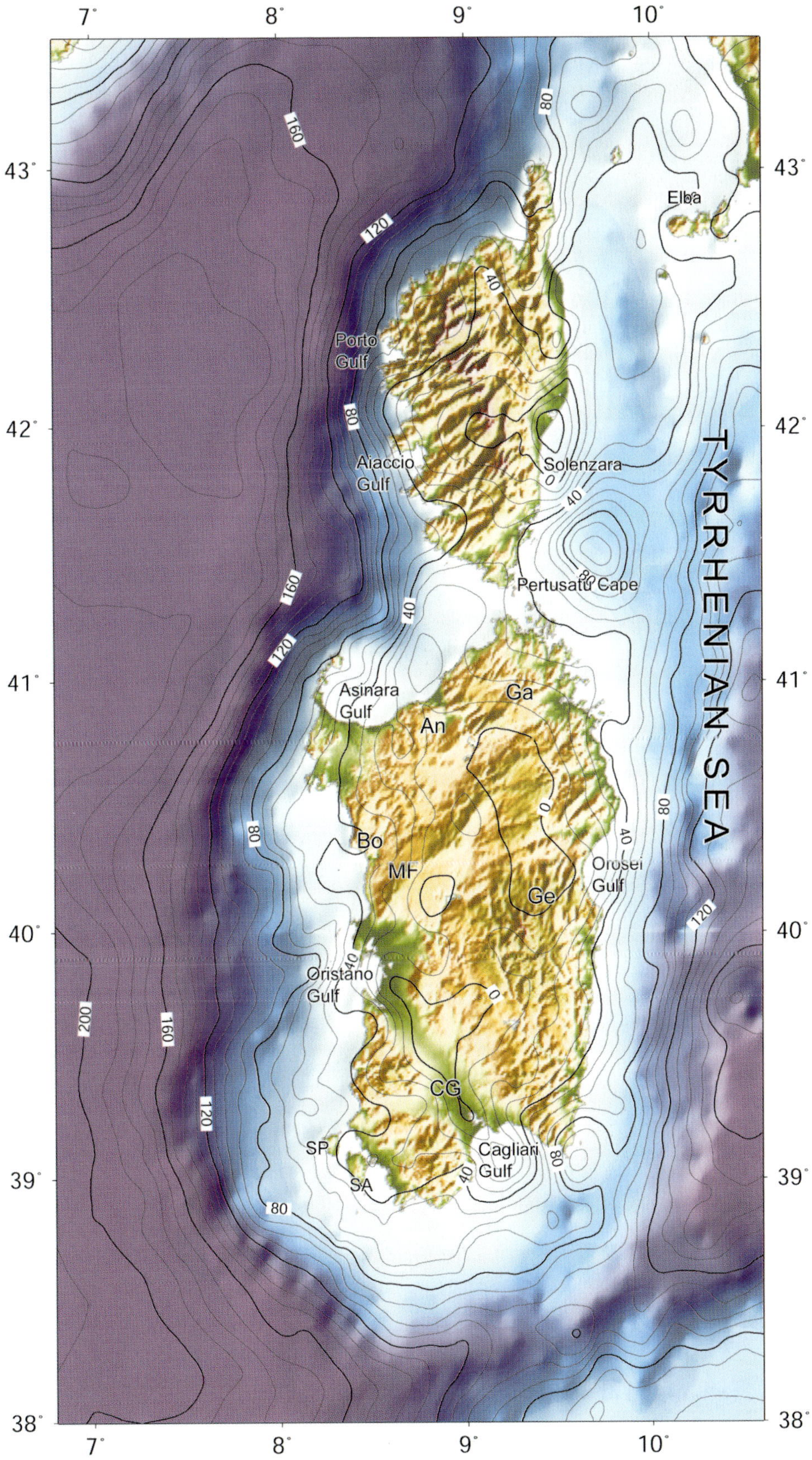

Figure 5: Complete Bouguer anomaly map of Corsica and Sardinia and surrounding areas. Units are mGal (10^{-5} m s^{-2}) and equidistance between curves is 10 mGal. For the meaning of the letters see Figure 3.

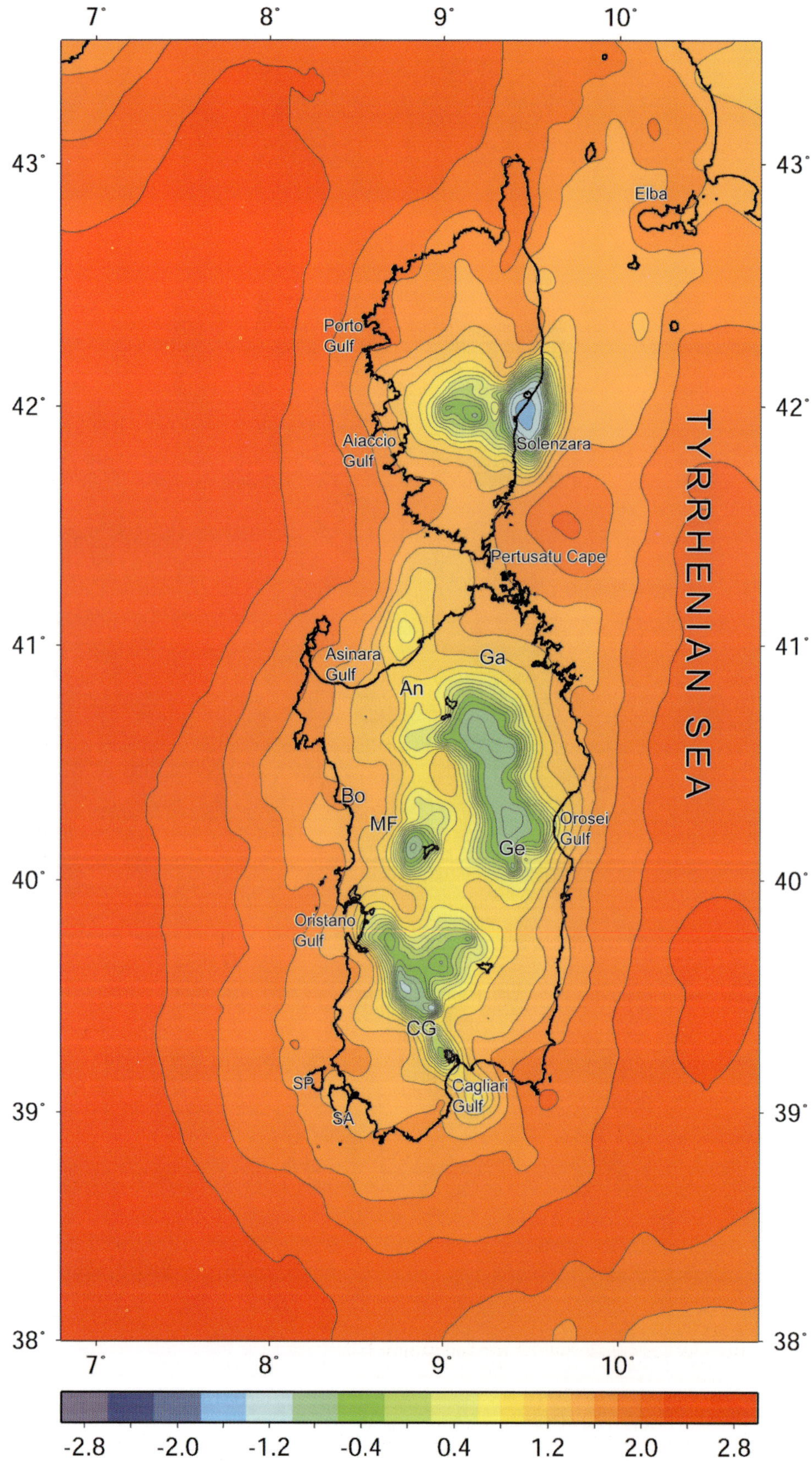

Figure 6: Complete Bouguer anomaly map of Corsica and Sardinia and surrounding areas, plotted on a base 10 logarithmic scale. Units are $\log_{10}$ (mGal) and equidistance between curves is 1. For the meaning of the letters see Figure 3.

The Sardinian Rift System

The SR preserves tectono-stratigraphic imprints of the Balearic and Tyrrhenian geodynamic stages. In the past, the SR was considered monophasic or at most biphasic, and the stratigraphy was believed to be constant from the Gulf of Cagliari to the Gulf of Asinara (Cherchi and Montadert, 1982; Thomas and Gennesseaux, 1986). Over the last decade, a number of works have tended to give polyphasic descriptions.

The basement of Sardinia, as described in many past studies, was involved in the Tertiary reactivation of part of the late Hercynian, Permo-Triassic, and Jurassic tectonic structures. The tectonic events that repeatedly reactivated the extensional movements are various and occur between the Upper Eocene–Oligocene and the Plio-Quaternary. The SR system extends along the entire western part of Sardinia and is segmented by minor transverse depressions, of an approximate east-west direction in the Sulcis-Arburese block, and of a northeast-southwest direction in central and northeastern Sardinia. Other branches of the SR have been located in the western continental margin of Sardinia and in the Gulfs of Asinara and Cagliari. The faults of the rift show evidence of movements that evolve from simple to divergent strike-slip up to extensional. The evolution of the main fault zones produced both a calc-alkaline polyphasic volcanism and a regional system of transtensional-extensional basins, which were filled by continental and epicontinental marine sequences with important volcanoclastic contributions (De Cillia et al., 1992; Lecca et al., 1997; Fais et al., 2000). Late tectonic phases (Middle Miocene to Plio-Quaternary) of the Tyrrhenian stage were superimposed on the Balearic stage rift (Upper Oligocene–Lower Miocene). Moreover, these reactivation phases continued the evolution of the CSB western continental margin and gave rise to new tectonization of the eastern one. Therefore the many rift sub-basins are characterized by the presence of structural elements and stratigraphic sequences belonging to the different phases, not always clearly distinguishable.

The rift faults delimit crustal blocks (widths in the order of a hundred kilometres) whose relative movements have controlled the extensional basins and their associated volcanism. The largest components of the rift are due to extensional movement of the western blocks, which belong to the same tectonic system as the continental margin of western Sardinia, comparable to the main eastern horst block (Figure 2). Inside each second-order block, other minor fault systems (with a spacing on the order of tens of kilometres) subdivide both the horst-blocks and the graben-blocks, giving rise to third-order blocks which are dislocated and/or rotated according to crustal accommodations. These third-order structures involve all the stratigraphic structures, including the Paleozoic basement (Barbagia, Gallura and others), the basement with Mesozoic and Eocenic cover (Tacchi area), the blocks involved in Oligo-Miocene andesitic and ignimbritic volcanism (Anglona, Arburese and others), and finally also the Miocenic arenaceous-calcareous-marly sequences (inside the rift).

A more recent focus of studies on the SR, which is still ongoing, is dedicated to the description of its sub-basins (Thomas and Gennesseaux, 1986; De Cillia et al., 1992; Lecca et al., 1994; Fais et al., 1996, 1997; Funedda et al., 2000; Casula et al., 2001; Spano and Barca, 2002), and to their relationships with the Northern Apennines orogeny (Carmignani et al., 1994; Oggiano et al., 1995) as well as to the interpretation of the rift as an intra-arc (i.e. intra-CSB) rift and to the relationships within its sub-basins of tectonic, volcanic, and sedimentary processes (Assorgia et al., 1995; Brotzu, 1997; Lecca et al., 1997; Sowerbutts and Underhill, 1998; Sowerbutts, 2000; Facenna et al., 2002). The results of a number of biostratigraphic studies (Cherchi, 1985; Iaccarino et al., 1985), field and, particularly, radiochronologic studies converge to support an evolution of the SR ranging between $\sim$ 28 Ma and 15 Ma, with a climax for explosive volcanism at $\sim$ 21–19 Ma, and allow us to distinguish several rifting-related events. A proto-rift phase can be pushed back to the Late Oligocene. During the Upper Oligocene-Aquitanian, a first phase of rifting allowed marine ingression into southern Sardinia and in Asinara Gulf. At the end of the explosive volcanism (upper Burdigalian-Langhian), a clear second phase of rifting was superimposed in northern Sardinia with clear half graben structures, while volcanism in the Sulcis was late and extended up to the Middle Miocene (Morra et al., 1994; Fais et al., 2002). An important reactivation in the Tyrrhenian stage is observed.

2.2. New Investigations–Interpretations of the Sardinian Rift

Integrated geophysical/geological studies have been conducted (Fais et al., 1996, 2001, 2002) to obtain information on the deep features of some of the Sardinian Rift basins. The main results in some key sectors of the Sardinian rift are briefly described here. New geophysical data, especially *reflection seismic* (Ottana basin, Chilivani–Tula basins, offshore Gulf of Oristano), *quantitative interpretation of the available aeromagnetic data* (Campidano graben, Central and Southwestern Sardinian margin), and accurate *seismostratigraphic interpretation* of some sectors (Gulf of Asinara, Gulf of Cagliari) have helped improve our understanding of the SR. Reconstructed sections across the main rift basin are exhibited in Figures 7 to 12.

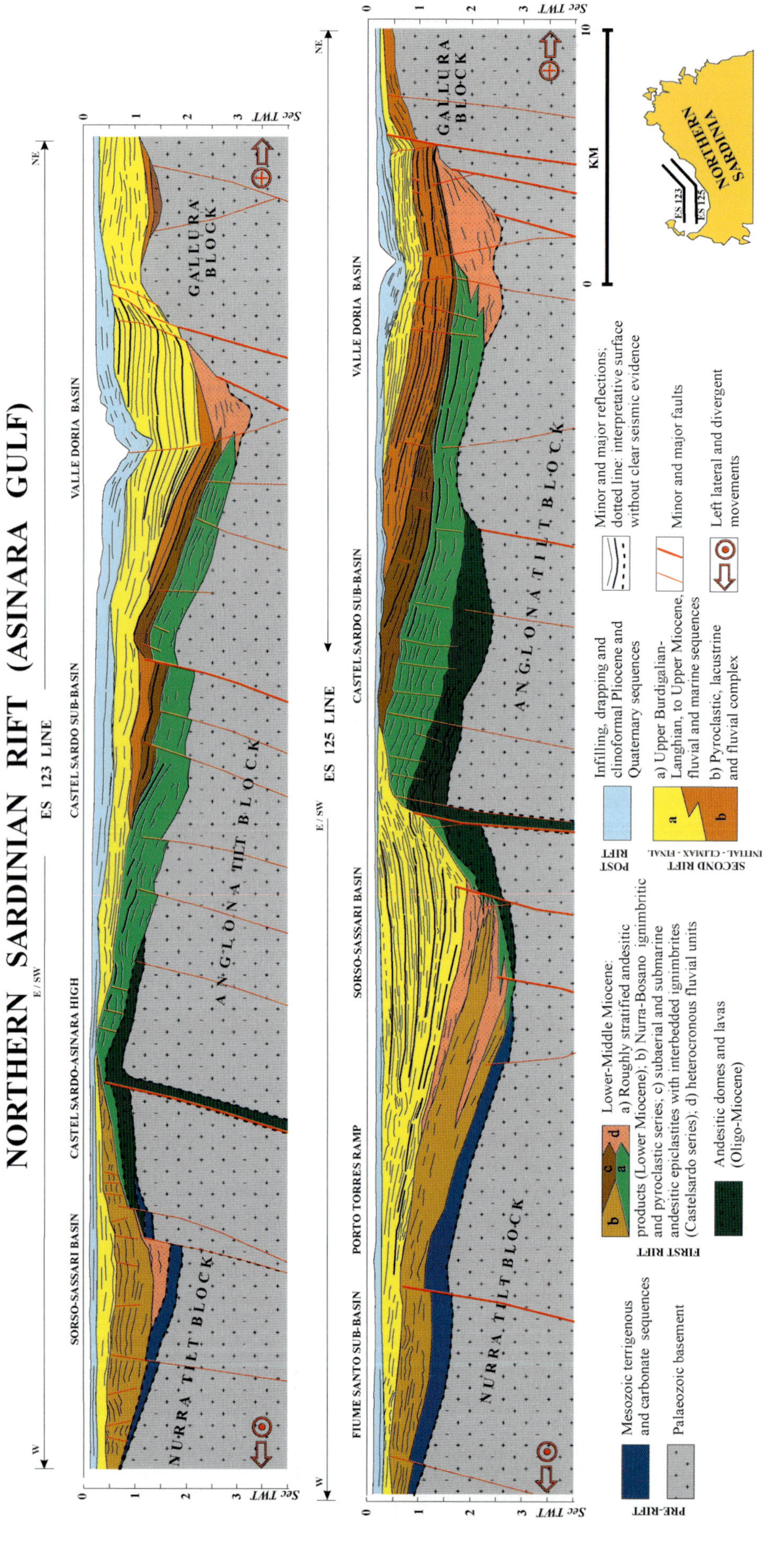

Figure 7: Interpretative seismogeological sections across the Northern Sardinian Rift zone, reconstructed from lines ES123 and ES125 (ENI).

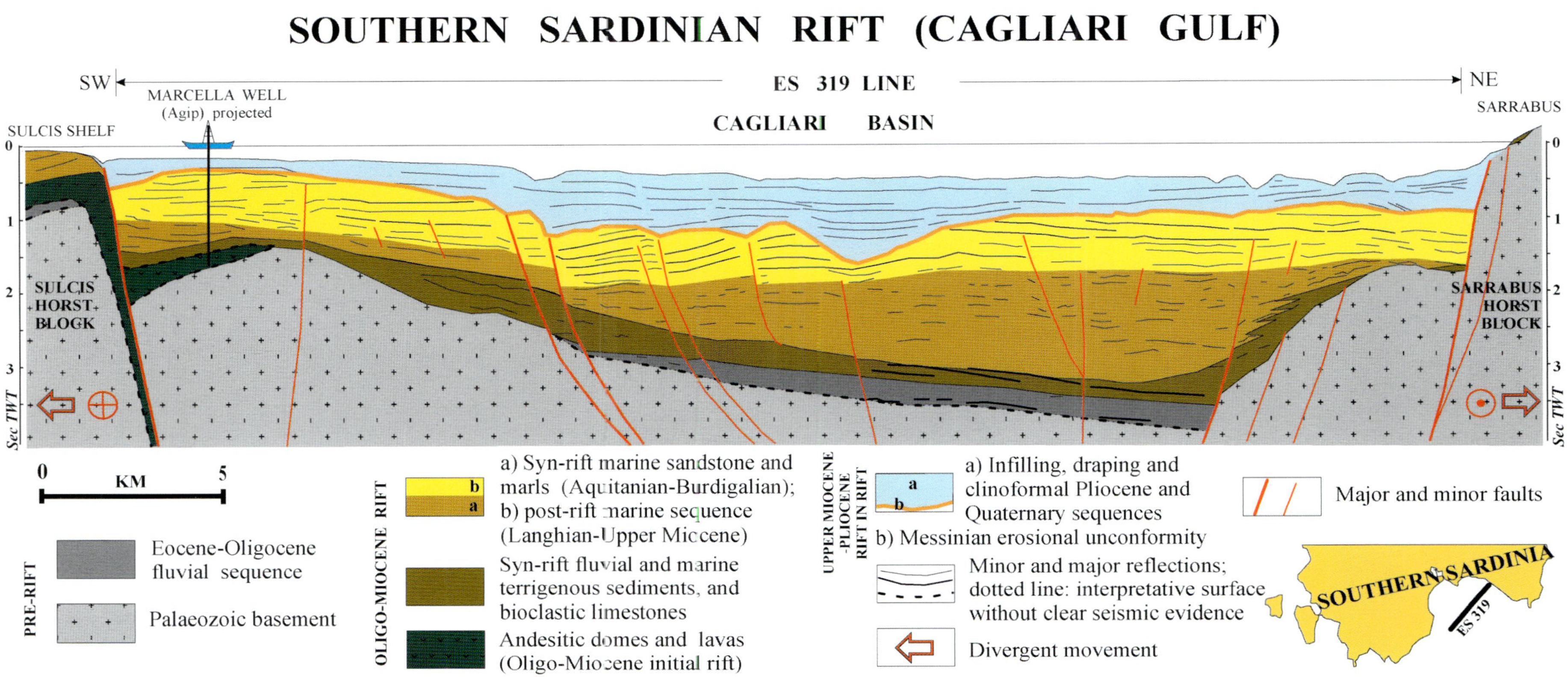

Figure 8: Interpretative seismogeological section across the Southern Sardinian Rift, reconstructed from line ES319 (ENI).

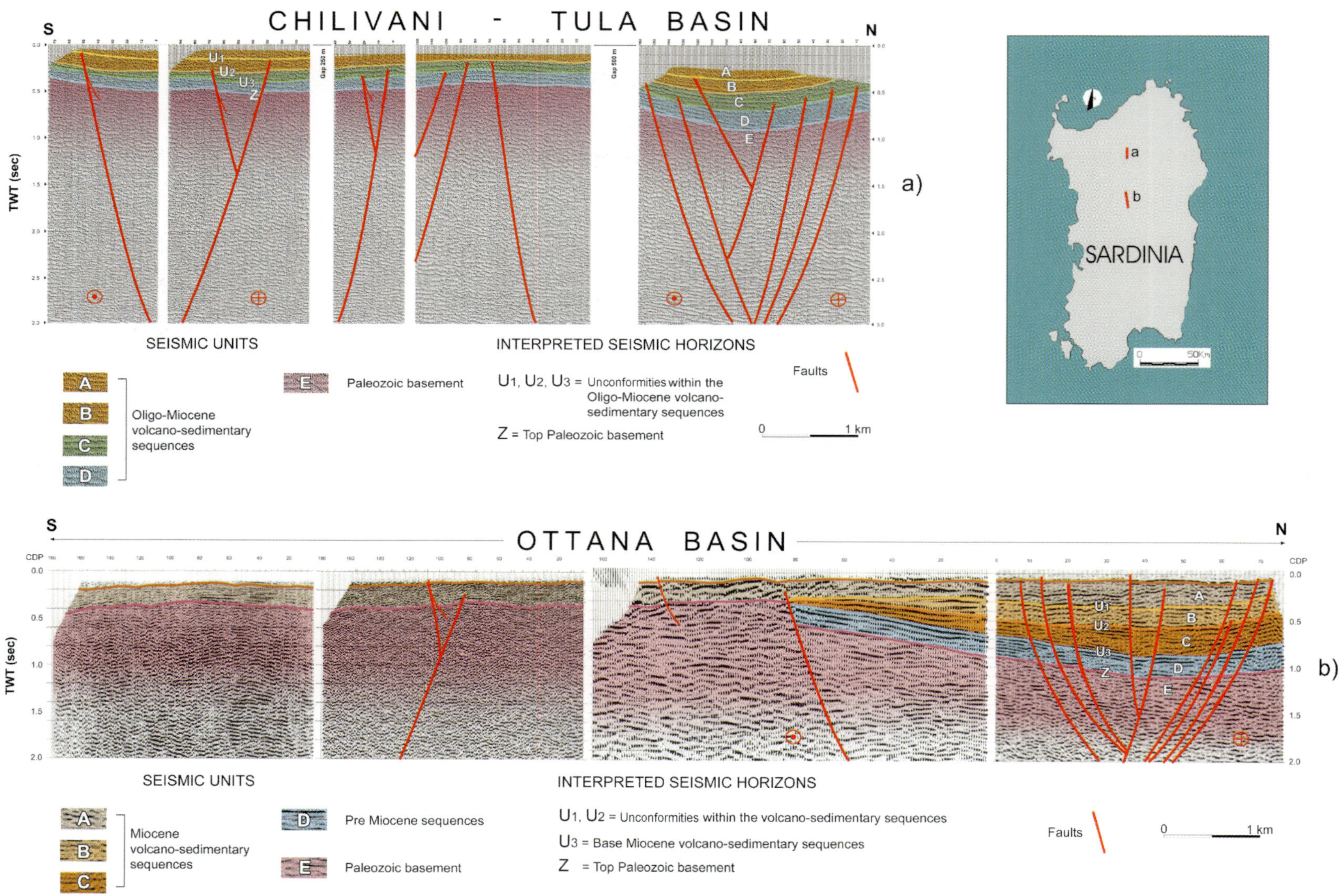

Figure 9: Interpreted seismic sections of the Chilivani–Tula basin (upper part) and the Ottana basin (lower part).

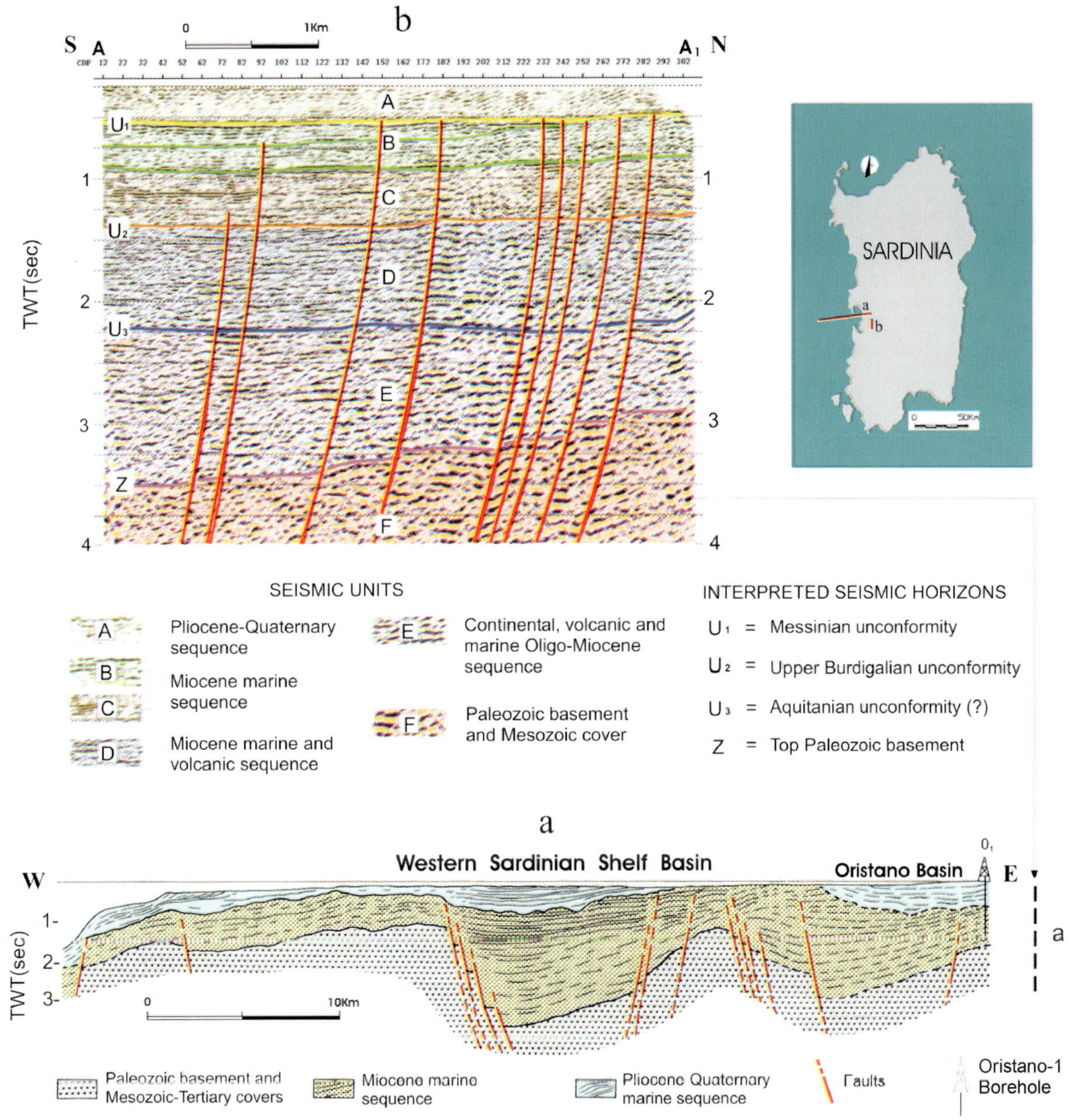

Figure 10: Oristano Basin – western Sardinia shelf: (a) seismogeological section reconstructed from ENI lines; (b) interpreted seismic section of the onshore area of the Oristano basin.

Asinara Gulf. To illustrate the tectono-sedimentary features of this gulf, two seismic lines that cross the basins transversally have been selected and interpreted (Figure 7). Two pre-Oligo-Miocene rifting units are related to the Paleozoic basement and to the Mesozoic sedimentary cover outcropping on the Nurra block (interpretative units without clear distinctive seismic evidence). Andesitic domes and lavas are linked to the initiation of Oligo-Miocene rifting, while roughly stratified andesitic products, subaerial to submarine epiclastites and interbedded ignimbrites, the ignimbrite series of the Nurra and Bosano area and, finally, subaerial and submarine andesitic epiclastites, terrigenous sandstones, and interbedded ignimbrites are linked to the climax (Aquitanian-Burdigalian Castelsardo sequence; Maxia and Pecorini, 1969; Spano and Asunis, 1984). A second rifting initiation shows a pyroclastic, lacustrine and fluvial complex (Lecca et al., 1997), and heterochronous fluvial units. An upper Burdigalian–Upper Miocene fluvial to marine sequence is attributed to a second rifting climax to post-rift. The last unit is an infilling and mud-draping Plio-Quaternary sequence. Along these cross-sections, the tectonic structure of the SR is represented by two half grabens.

Cagliari Gulf. To show the most representative tectono-sedimentary features of this SR zone, a SW–NE line transvers to the basin was interpreted (Figure 8). The Cagliari Gulf represents the southern part of the

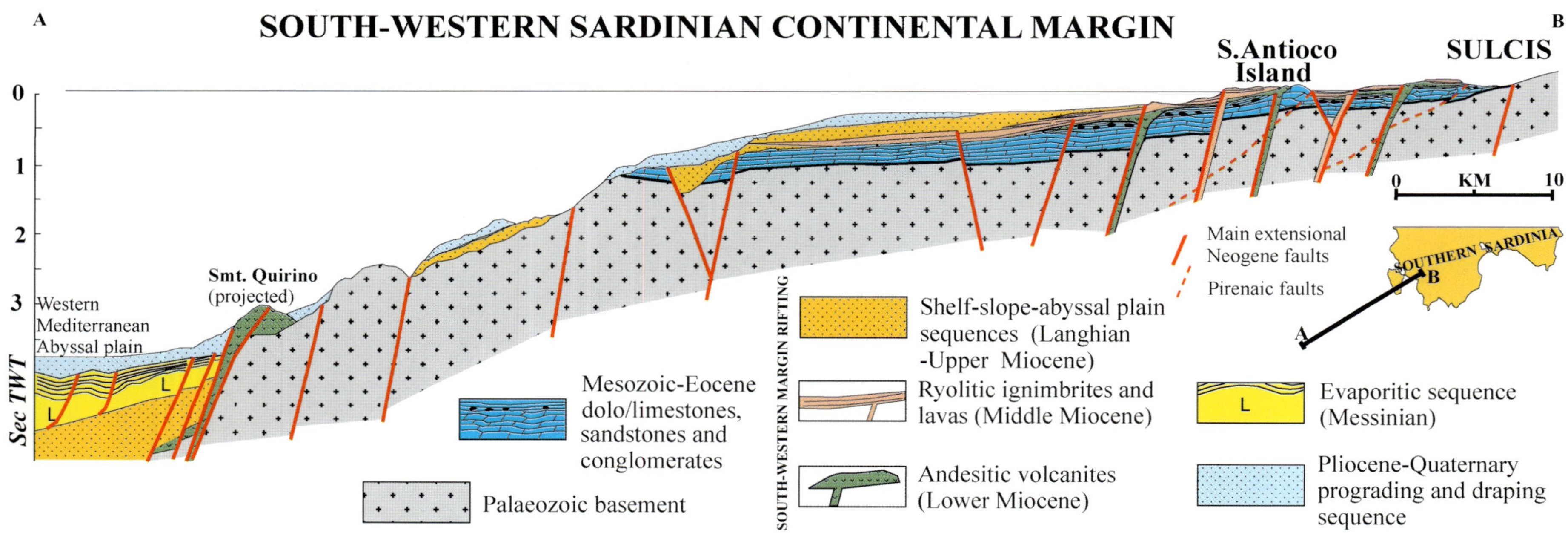

Figure 11: Geological section across the south-western margin of Sardinia, reconstructed from seismic network and magnetic data.

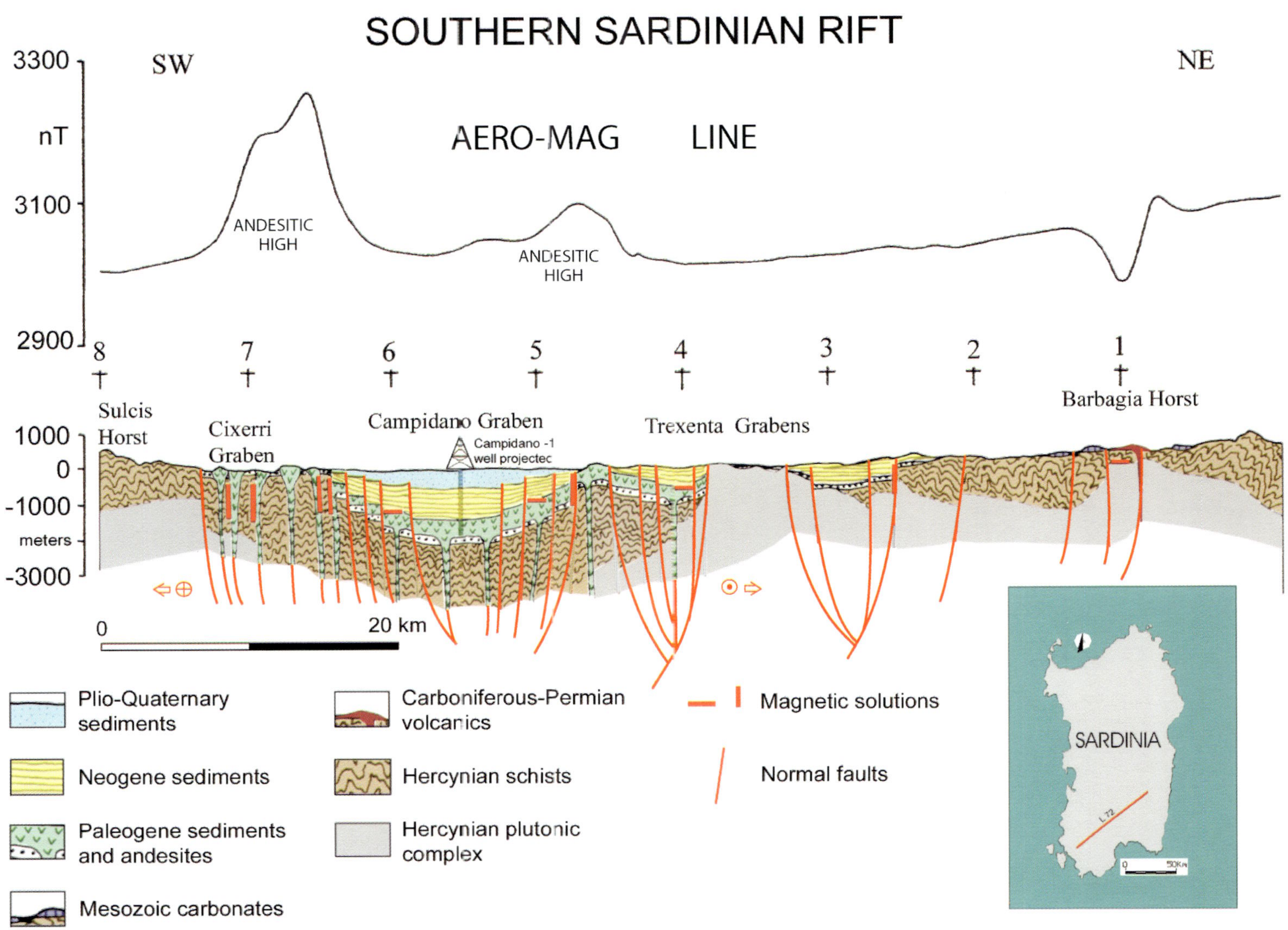

Figure 12: Geological-geophysical section across the Campidano and Trexenta grabens. Transtensional deformation represented by half-grabens, generated by right-shear tectonics, driven by eastward moving Tyrrhenian mantle flux.

SR which here consists of a basin bounded by approximately NW faults. The tectonic structure is made up of a lowered tilted block, a master fault zone towards the NE, an antithetical towards the SW, and other minor parallel antitheticals. A summary of the principal evolutionary stages can be proposed, thanks also to the biostratigraphical and radiochronological information in the literature (Pecorini et al., 1969; Savelli et al., 1979; Cherchi, 1985). The units preceding the Oligo-Miocene rifting phase are represented by the Paleozoic basement and by probable marine Eocene cover, besides fluvial sandstones and claystones of the Cixerri Formation (Upper Eocene–Lower Oligocene).

The initiation of Oligo-Miocene rifting is accompanied by the Upper Oligocene-Aquitanian andesitic volcanism of the southern Sulcis block, penetrated in the lower part of the Marcella-1 well. The beginning of sedimentation inside the rift is characterized by coarse-grained fluvial sediments (Ussana Formation), continental shelf sandstones, and bioclastic-biohermal temperate limestones. During the rotational lowering of the block, the rift climax is represented by a triangular-shaped infilling sequence made up of Aquitanian-Burdigalian marine marls and sandstones. From the upper Burdigalian to Upper Miocene, there was a general lowering without rotation of the block. A transtensional-extensional tectonic phase is mainly attributable to the Upper Miocene–Messinian, and a reactivation of the extensional movements, particularly along the basin margin faults, is underlined by the Plio-Quaternary infilling that clearly testifies to the Tyrrhenian rifting phase, well evident in southern Sardinia as the *Campidano Graben*.

Chilivani–Tula basin. The Chilivani–Tula basin represents a transtensional basin of the Sardinian Rift, developed along sinistral transcurrent faults that prevalently characterize the eastern sector of the northern CSB (Lecca et al., 1994; Oggiano et al., 1995; Fais et al., 1997). The acquisition of new seismic data in the northeastern sector of this basin, and their integrated interpretation with geological data is helping define the basin's structural feature's and tectonic evolution. From field data, the basin appears to be settled on a Paleozoic basement, outcropping on its northern and southern boundaries with middle-high-grade Hercynian metamorphic rocks and subordinately with late-Hercynian granitoids. Continental sediments of Oligo-Aquitanian age, alternating with calc-alkaline volcanic products, outcrop inside the basin and are partly considered syntectonic with its evolution, and partly post-tectonic and of Burdigalian-Langhian age. In Figure 9(a) is exhibited the interpreted final stack of a new seismic section specifically acquired for the present study. The section crosses the Tula basin and the innermost part of the Chilivani–Ardara basin. From integrated interpretation of seismic and geological data, it is possible to correlate the units from D to A with the Oligo-Miocene volcano-sedimentary sequences that represent the infilling of the basin. This basinal setting is the result of the different tectonic activities of the Balearic (Oligocene–Lower Miocene) and Tyrrhenian stages.

Ottana basin. The Ottana rift-basin, as well as other rift-basins of the northern Sardinian Rift, evolved throughout the Oligo-Miocene geodynamic phase mainly as left, transtensive processes (Figure 9(b)). Based on land data, it is acknowledged that the upper part of the filling sequence of the basin is made up of an Early-Middle Miocene succession of cineritic pumiceous pyroclastic banks, volcanic epiclastites, and alluvial conglomeratic sandstones. A silicoclastic-carbonatic continental shelf marine sequence is present only in the uppermost part of the filling succession. Basalt lava flows and Pliocene continental conglomerates cover the oldest units.

Thanks to the new seismic reflection data specifically programmed and acquired (Figure 9(b)), it has been possible to define the structural setting of the NE sector of the Ottana basin. The interpreted cross-section show the presence of a triangular-shaped basin-fill in the northern part of Oligo-Miocene age. It is possible to identify five seismic units (A to E in Figure 9(b)) with boundaries mainly coinciding with unconformities (U_1 to Z in Figure 9(b)). Unit D deposited following extensional tectonics that probably reactivated Hercynian structures. The top of this unit, attributable to pre-Miocene times, is an erosional surface (U_3) due to a tectonic event. This unit lies in angular discordance (Z) on the Paleozoic basement (Unit E). Based on seismic interpretation and geological and petrographical data, the very thick volcano-sedimentary series (units A, B, C), mainly represented by ignimbrites (De Cillia et al., 1992), can be referred to the Lower Miocene (Balearic stage). The entire investigated sector of the Ottana Basin was further reactivated by the Plio-Quaternary extensional tectonics related to the Tyrrhenian stage.

Southwestern Sardinian margin. This Sardinian margin area, structured on a Paleozoic basement, is made up of metamorphic Hercynian rocks and granitoids. The basement is locally covered both by Mesozoic-Tertiary terrigenous and carbonate sequences and by calcalkaline volcanic sequences of the Oligo-Miocene rifting process. This sector of the western Sardinian continental margin–Sardinian rift transition zone is prevalently characterized by listric normal and transfer faults with a prominent NNE and NW trend. In this sector, small half grabens are infilled by syn- and post-rift sequences.

From new reflection seismic data specifically acquired in the onshore area of the Oristano basin, it has been possible to better define the seismo-stratigraphic units than from previous seismic data. Figure 10 show a characteristic cross-section of the western Sardinian shelf basin and the Oristano basin (Figure 10(a)) together with a window five kilometres long of the new seismic data acquired in the onshore area of the Oristano basin (Figure 10(b)). In this seismic example, four main horizons have been recognized, corresponding to as many unconformities, which bound different seismic units. The stratigraphical interpretation of the seismic units is calibrated in the Oristano-1 borehole (Figure 10(a)).

Offshore southwestern Sardinia, there is a complex magnetic pattern consisting of a set of strong magnetic anomalies with amplitude larger than 500 nT, extending for approximately 15 km and with a width of about 5 km. There is a clear relationship between basement highs of the seismic and magnetic results. The section in Figure 11 shows the results of the integrated interpretation of the aeromagnetic and high-resolution seismic data. In the fault zones involved in the Oligo-Miocene rifting, volcanic andesitic extrusions (i.e. Quirino smt) overlain by an ignimbritic sequence, and later by marine Lower-Middle Miocene continental shelf and slope sediments, occur.

Campidano and Trexenta rift-basins. The relationship between the Oligo-Miocene southern Sardinian rift zone and the Plio-Quaternary Campidano graben was analyzed by interpretation of aeromagnetic data (Balia et al., 1991). Using a spectral analysis technique along aeromagnetic SW–NE lines, the depth of the top and, in a few cases, also the bottom of the Oligo-Miocene volcanic formations and, consequently, the thickness of the overlying sedimentary fill were estimated. A significant example of the results obtained with spectral analysis, and a geological interpretation along an aeromagnetic line across the Campidano is shown in Figure 12. The results obtained with spectral analysis are represented by horizontal bars under the centres of corresponding spectral windows; those obtained with the integrated use of data inversion techniques are represented by vertical bars.

A thick sequence of Oligo-Miocene volcanites (andesites) and a thick Miocene marine succession characterize the *Campidano graben*, which was superimposed over a former rift-basin as a consequence of Plio-Quaternary extensional tectonics driven by eastward-moving Tyrrhenian mantle flux (Figures 2 and 19). Minor sub-basins are located on the shoulders of the Campidano basin, as can be observed in the eastern sector (Trexenta region) where small half grabens are present. These minor sub-basins are bounded by normal faults and filled with both volcanites and Miocene marine sediments. On the western shoulder of the southern rift, a significant Oligo-Miocene calc-alkaline volcanism occurred, as is deduced from the outcropping geology and from the strong concentration of magnetic solutions in the Cixerri district (Figure 12).

3. CROP SEISMIC DATA USED

In the present work, five CROP sections or parts of sections are used (Figure 1): CROP M2A/I (SP 2700–10450), M-10 (SP 500–3900), M12B (SP 100–2000), M28B and M32. The CROP section M2A/I (Plates 1, 2; Figure 14) crosses the whole CSB, from the Balearic Basin floored by oceanic crust, to the extended and delaminated Sardinian margin in the Tyrrhenian Sea. This section allows us for the first time to reconstruct a lithospheric seismogeological section across the CSB, fully controlled from the surface to upper mantle. Moreover, since the section is marine, it has the double advantage of lower cost and better seismic response with respect to a land line, or even a combined marine-land section, as was formerly considered (extension of marine ECORS-CROP M1 section across the Sardinian mainland) between M1 and M28B (Figure 1).

The CROP section M10 (part: SP 500–3900; Figure 13) shows the northward prolongation of the Corsican block and of the collisional zone between the rotated block and the NW Adria plate margin. Two subducted Alpine

Table 1
CROP seismic sections used in this work and their respective figure and plate numbers

Crop section	SP	km	Figure N°	Chapter plate N°
M10 (Part)	3900–500	167	13	
M2A/I (Part)	2700–10450	371	14	1–2
M12B (Part)	100–2000	97	15	
M28B	100–5406	267	16	3–4
M32	100–2508	123	17–18	

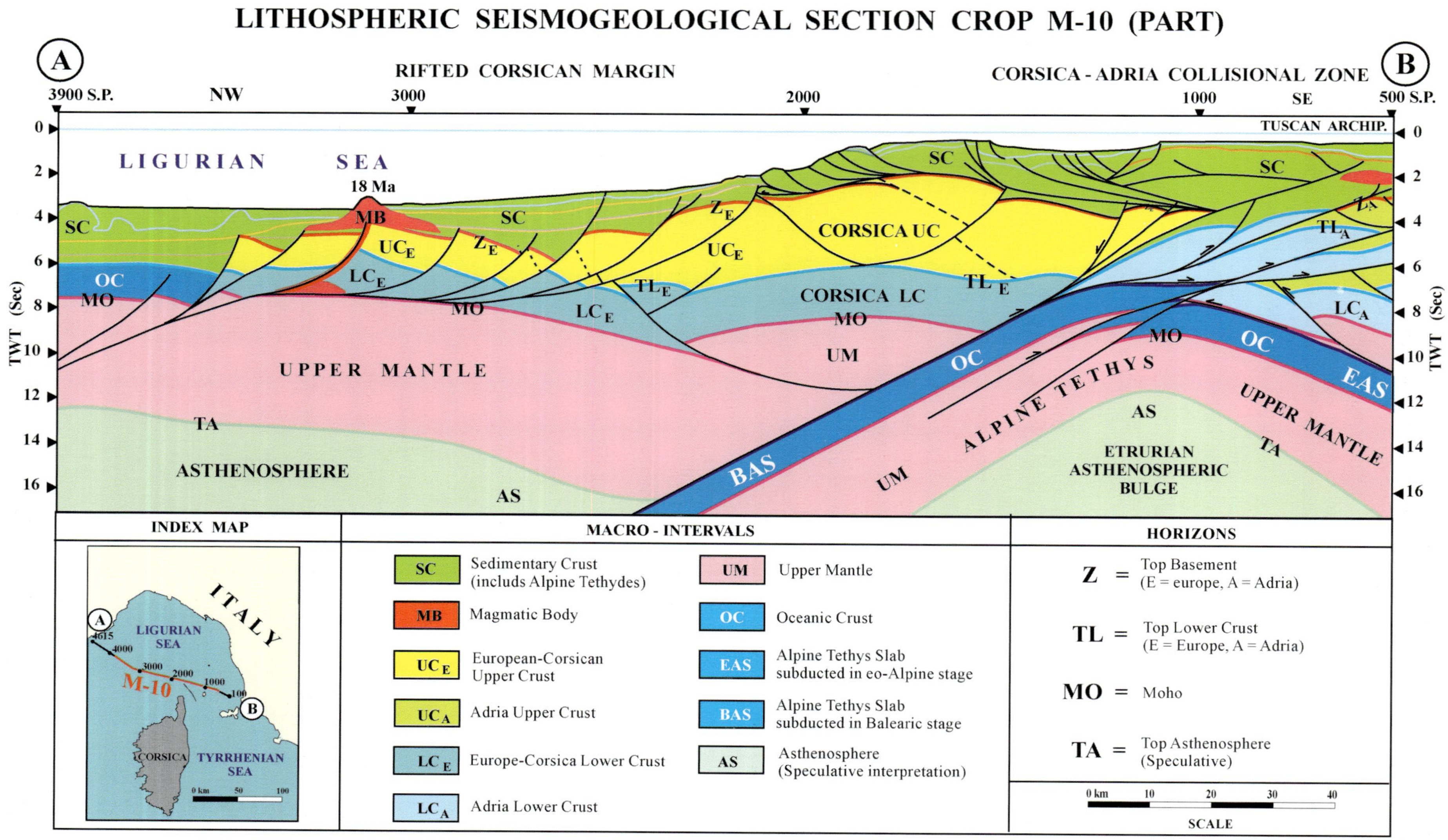

Figure 13: Seismogeological section across the northern extension of the Corsica Block on the Tuscan-Ligurian Sea, reconstructed from CROP section M10 (part: SP 500–3900).

Tethys slabs in two different geodynamic stages are also exhibited. The CROP section M12B (part: SP 100–2000; Figure 15) is located across the frontal thrusts of the CSB and the southern Corsica basin. It controls an important geodynamic sector, strongly debated in the literature, and important for understanding the closure of the Alpine Tethys between Corsica and NW Adria. The section CROP M28B (Plates 3, 4; Figure 16) gives a complete view of the extended and delaminated Sardinian margin and much relevant tectono-dynamic information.

In the Sardinian Channel, the section CROP M32 (Figures 17 and 18) explores the SE Sardinian margin imprinted by a compressive tectonic deformation produced by collision between the rotated block and the NW Panormid Promontory.

Some of the sections (i.e. CROP M2A/I and M12B) whose standard processing furnished poor, almost uninterpretable results, were carefully reprocessed by iterative interpretive reprocessing at the University of Trieste (Exploration Geophysics Group, Faculty of Sciences). The results obtained are markedly better and allow reliable interpretation.

4. SEISMIC INTERPRETATION

In the following, the main results obtained from interpretation of the CROP sections listed in Table 1 are described. These results are framed within a larger tectono-stratigraphic and geodynamic reconstruction based on the whole CROP dataset. In fact, each single studied CROP province does not contain in itself sufficient information for a correct time–space understanding of those events that produced the complex variable local settings we observe now in a given zone. Only a large-scale lithospheric analysis, like that now possible with the CROP network in the central Mediterranean Sea provides a controlled understanding of the examined features within a correct global framework.

For an easier and more immediate tectono dynamic allocation of the seismic imagings described, it is convenient to introduce a general outline of the Corso-Sardinian block and of its regional geodynamics. The new CROP data supply for the first time key lithospheric information on important topics previously discussed on the basis of postulated schemes or hypothetic downward extrapolations of observed shallower facts. The new data image all the most critical features and support controlled innovative reconstructions.

The first important basic new starting point is the pre-orogenic paleogeography of the Central-Western Mediterranean region. There is good CROP control to establish that *at the end of the Lower Cretaceous, before the Eo-Alpine geodynamics, two communicating oceans were interposed between continental plate crusts: the Ionian Neotethys of Permo-Triassic opening and the Alpine Neotethys of Middle Jurassic Lower Cretaceous opening. The first separated the Africa plate from Adria, and the second separated Europe from Adria to the NE (Alps sector), and Europe from Africa to the SW of the Alps region* (Finetti, this volume, Chapter 34; small scheme in Plate 2a). *During the Eo-Alpine geodynamic stage (Upper Cretaceous–Lower Eocene) the NE-Alpine Tethys (Alps sector) subducted the Adria plate, establishing a continental collision. West of the Alps, where the subduction very likely took place as an intra-ocean process, a wide Alpine Tethys remained unsubducted. Ophiolite-bearing units were obducted on the southern European margin from the Betic zone to the Alps. In the Balearic stage (Oligocene–Lower Miocene), within the unsubducted Tethys remnant, the Corso-Sardinian block rotated driven by the W-dipping subduction of the Alpine Tethys slab (Figure* 13). *The Corsica block collided with the NW Adria plate, and the SE Sardinian block collided with the NW Panormid Promontory, stopping the CSB rotational movement* (Finetti, this volume, Chapter 34; Figure 19; Plate 2b). *Between the NW-Adria corner (Etrurian Promontory) and the SW-Panormid Promontory, the W-dipping Alpine Tethys subduction process continued in the Ionian Neotethys (Tyrrhenain stage: Middle Miocene to Present) generating the Southern Apennines–Calabrian Arc–Sicilian Apennines system, while the Northern Apennines continued their compressive deformation. The back-arc Tyrrhenian basin opened within the eastern Sardinian margin, delaminating it and transporting Sardinian crustal fragments and other allochthonous units outward, with oblique thrusting in the Southern Apennines and Sicily and, more cylindrically, on the Calabrian Arc.*

In the following, we describe the results of the seismic interpretation of the CROP sections and the above summarized geodynamic scheme should be a helpful preface for understanding the dynamics related to the crustal tectono-stratigraphic data shown.

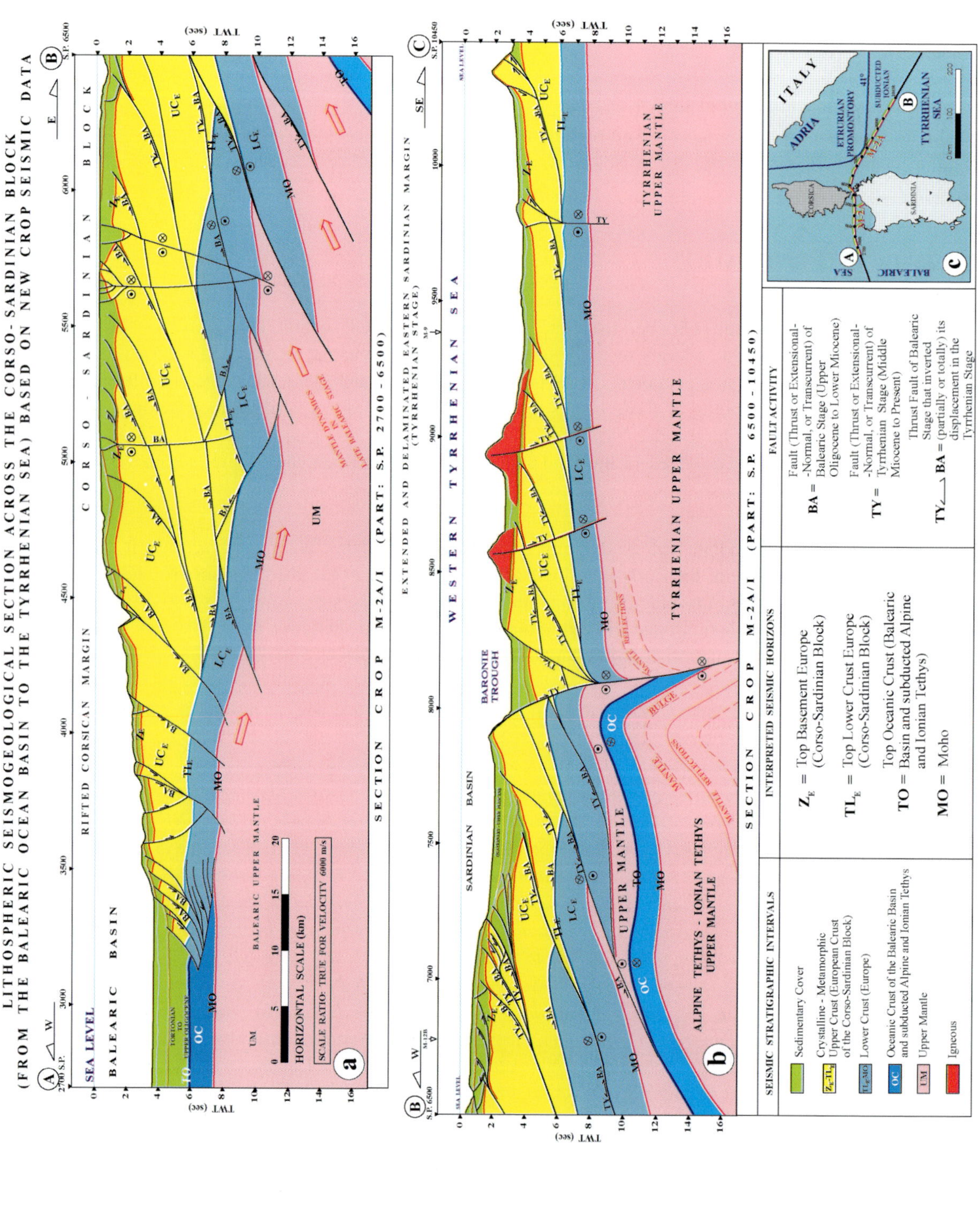
LITHOSPHERIC SEISMOGEOLOGICAL SECTION ACROSS THE CORSO-SARDINIAN BLOCK
(FROM THE BALEARIC OCEAN BASIN TO THE TYRRHENIAN SEA) BASED ON NEW CROP SEISMIC DATA
BALEARIC BASIN
RIFTED CORSICAN MARGIN
CORSO - SARDINIAN BLOCK
SEA LEVEL
BALEARIC UPPER MANTLE
HORIZONTAL SCALE (km)
SCALE RATIO: TRUE FOR VELOCITY 6000 m/s
LATE BALEARIC STAGE MANTLE DYNAMICS
SECTION CROP M-2A/I (PART: S.P. 2700 - 6500)
EXTENDED AND DELAMINATED EASTERN SARDINIAN MARGIN (TYRRHENIAN STAGE)
WESTERN TYRRHENIAN SEA
SARDINIAN BASIN
BARONIE TROUGH
UPPER MANTLE
ALPINE TETHYS - IONIAN TETHYS UPPER MANTLE
TYRRHENIAN UPPER MANTLE
MANTLE BULGE
MANTLE REFLECTIONS
SECTION CROP M-2A/I (PART: S.P. 6500 - 10450)
TWT (sec)
SEISMIC STRATIGRAPHIC INTERVALS
Sedimentary Cover
Crystalline - Metamorphic Upper Crust (European Crust of the Corso-Sardinian Block)
Lower Crust (Europe)
Oceanic Crust of the Balearic Basin and subducted Alpine and Ionian Tethys
Upper Mantle
Igneous
INTERPRETED SEISMIC HORIZONS
Z_E = Top Basement Europe (Corso-Sardinian Block)
TL_E = Top Lower Crust Europe (Corso-Sardinian Block)
TO = Top Oceanic Crust (Balearic Basin and subducted Alpine and Ionian Tethys)
MO = Moho
FAULT ACTIVITY
BA = Fault (Thrust or Extensional-Normal, or Transcurrent) of Balearic Stage (Upper Oligocene to Lower Miocene)
TY = Fault (Thrust or Extensional-Normal, or Transcurrent) of Tyrrhenian Stage (Middle Miocene to Present)
TY BA = Thrust Fault of Balearic Stage that inverted (partially or totally) its displacement in the Tyrrhenian Stage
ITALY
ADRIA
ETRURIAN PROMONTORY
CORSICA
SARDINIA
BALEARIC SEA
TYRRHENIAN SEA
SUBDUCTED IONIAN

Figure 14: Lithospheric seismogeological section across the Corso-Sardinian Block, from the Balearic Basin to the Tyrrhenian Sea, based on new CROP seismic section M2A/I (part: SP 2700–10450).

4.1. Crop Section M10 (part)

In Figure 13, a simplified version of the interpreted lithospheric seismogeological section CROP M10 (part: SP 500–3900) is exhibited. This new section clears up some of the topics widely discussed in the literature and of critical importance for reconstruction of the tectono-dynamics. It crosses: (1) the Ligurian Sea, where a crustal opening occurred in the Balearic stage; (2) the rifted western Corsican margin; (3) the collisional zone between Corsica and the NW Adria plate margin. One point questioned in the literature is whether the Alpine Tethys subducted Europe or the Adria plate. In our interpretation, an E-dipping subduction (EAS), geophysically identified some years ago (Finetti et al., 2001) and attributed to the Eo-Alpine stage, is shown. Here this subduction took place laterally to the almost N–S trending NW-Adria boundary. At the end of the Eo-Alpine stage, west of Adria (which collided with the Europe in the Alps sector) a wide Alpine Tethys remnant remained unsubducted. The

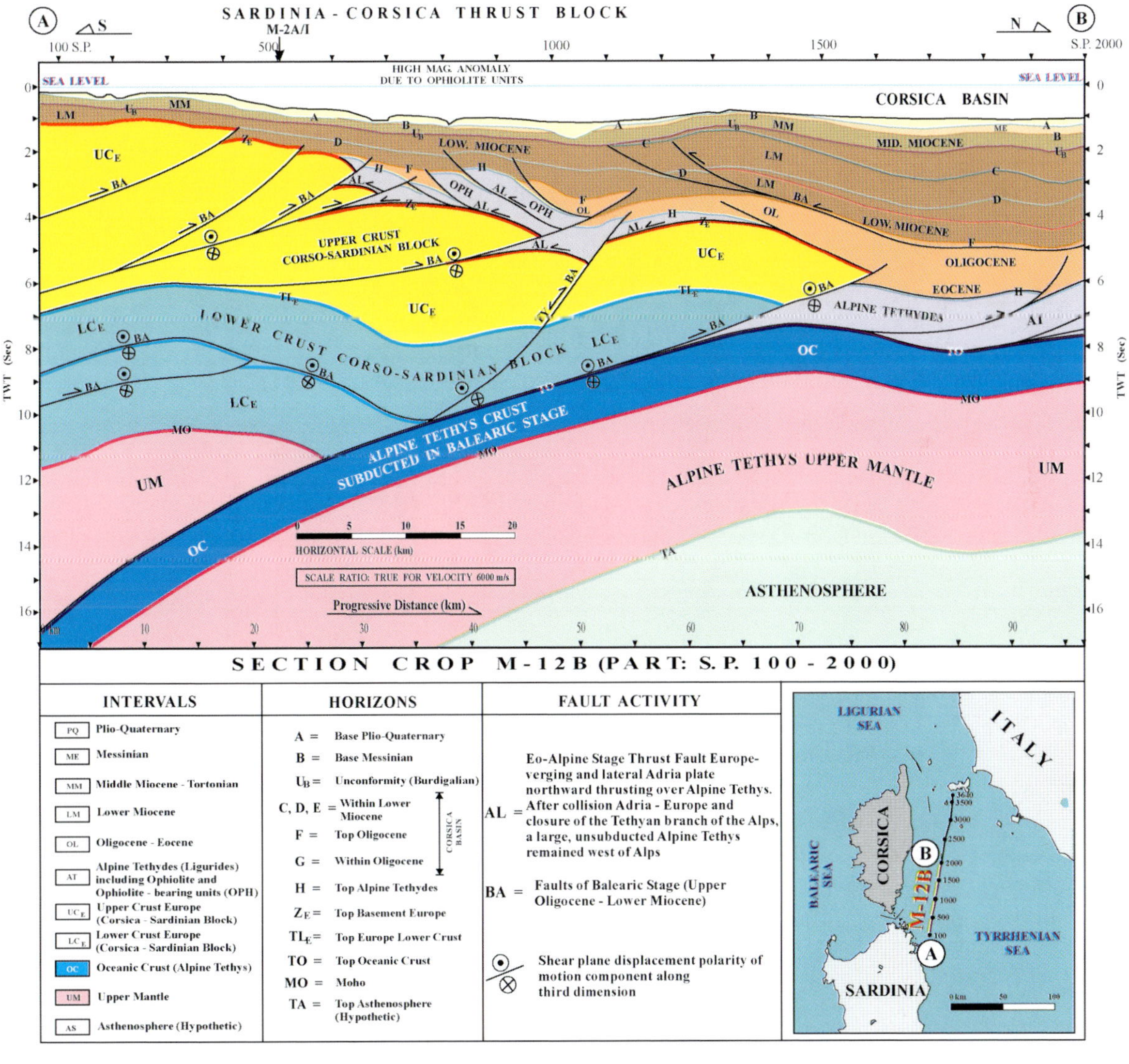

Figure 15: Lithospheric seismogeological section on the eastern Corso-Sardinian thrust front–Corsica Basin. The section is subparallel to the thrust propagation. The buried ophiolite-bearing units of the Alpine Tethydes (OPH) have been obducted over the European (Corso-Sardinian Block) margin in Eo-Alpine stage. The Corsica Basin represents the suture zone between the NW Adria Plate and the Corsican Block whose rotation was driven by NW-dipping subduction of the Alpine Tethys and consequent Balearic mantle dynamics.

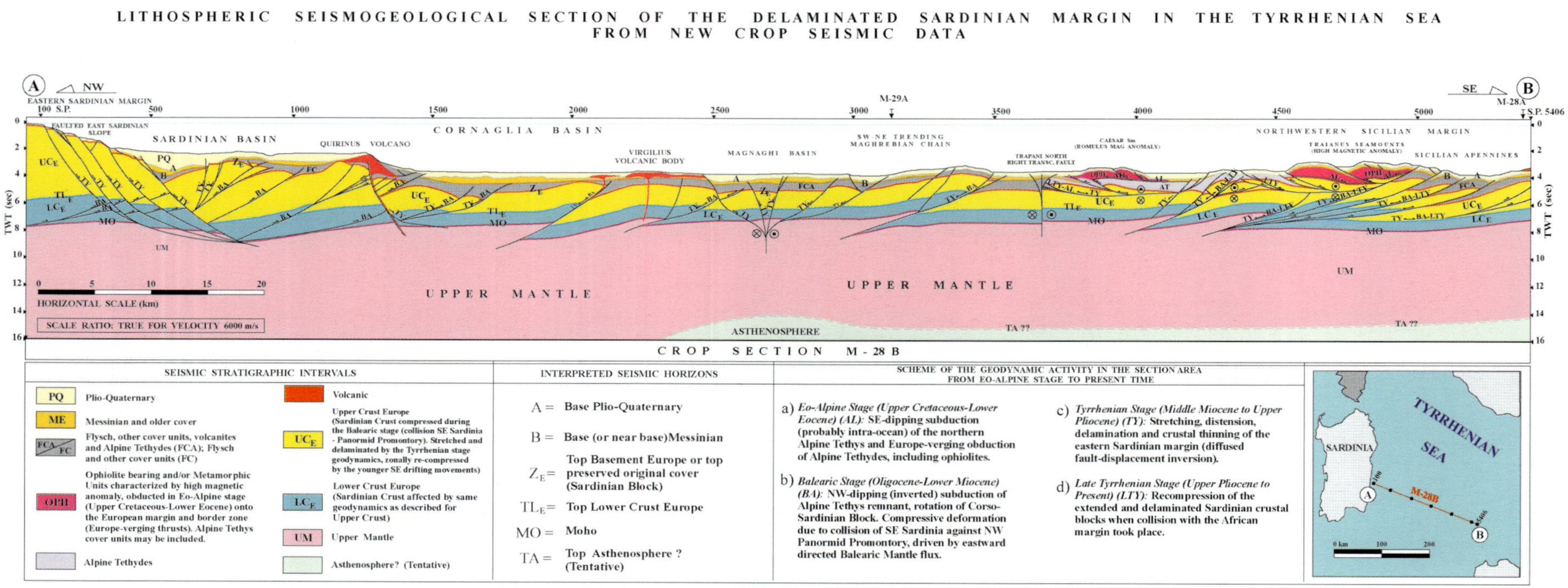

Figure 16: Lithospheric seismogeological section across the extended and delaminated eastern margin of Sardinia, based on new CROP section M28B.

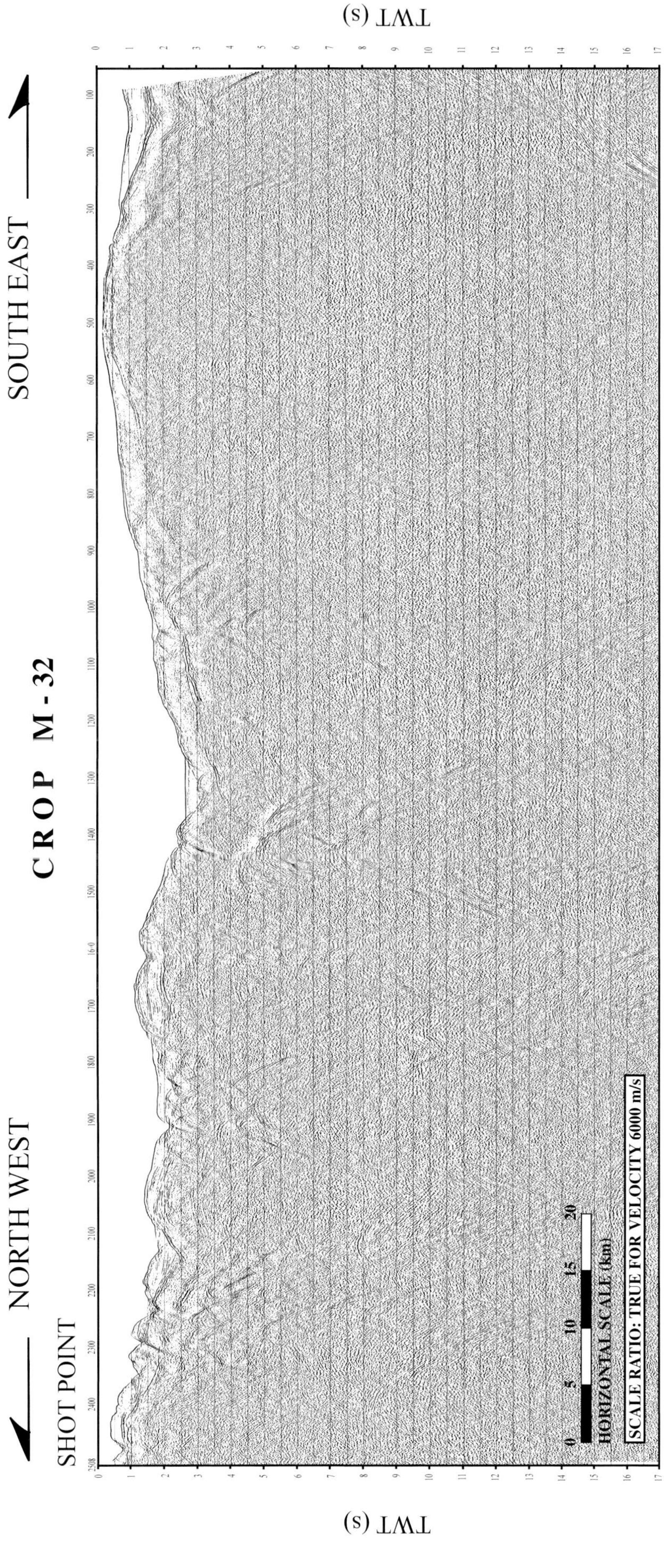

Figure 17: CROP seismic section M32 in the Sardinian Channel, from the south Sardinia coast to the northern Tunisia shelf area.

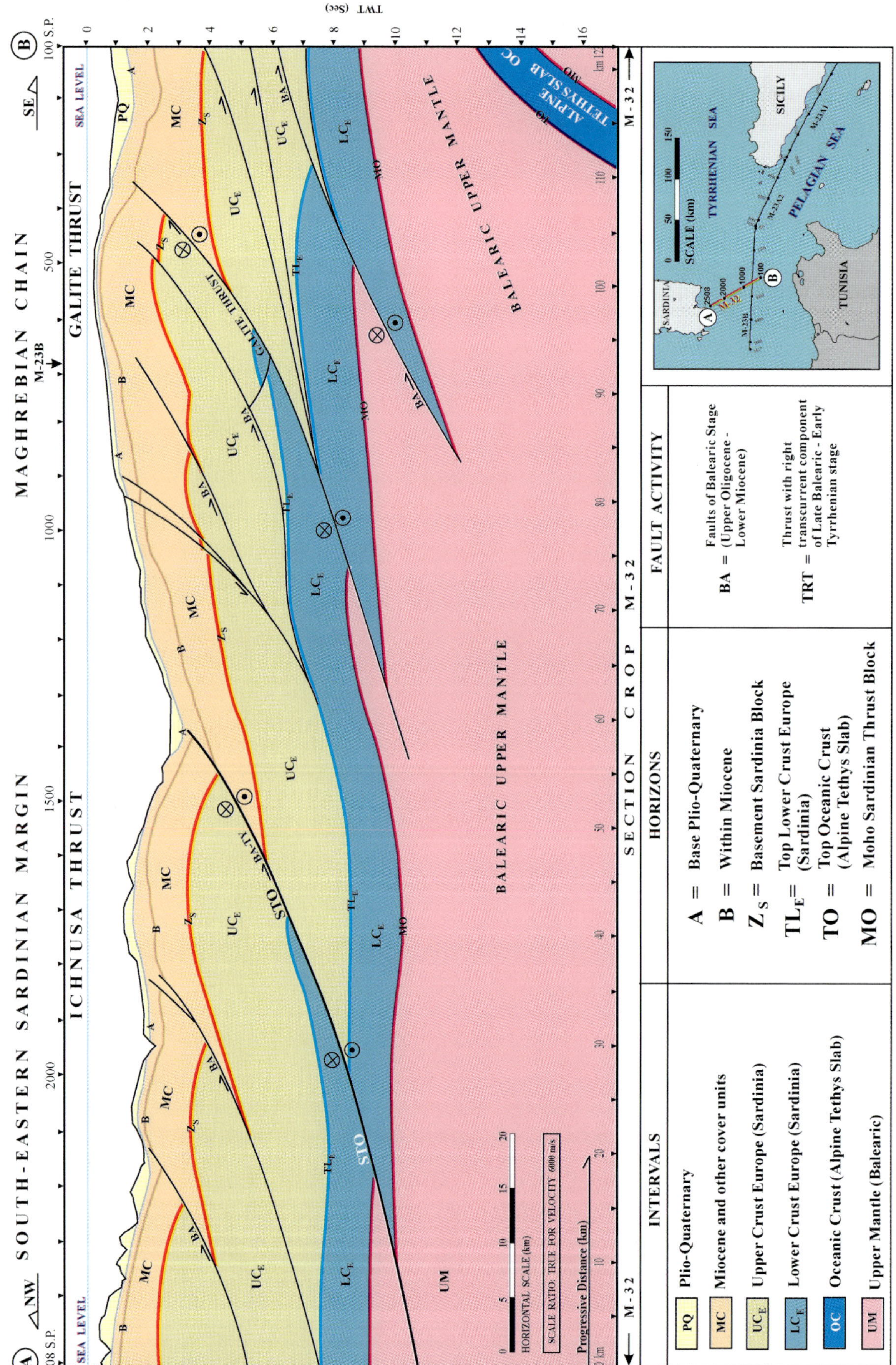
A
2508 S.P.
NW
SOUTH-EASTERN SARDINIAN MARGIN
MAGHREBIAN CHAIN
SE
B
100 S.P.
ICHNUSA THRUST
GALITE THRUST
GALITE THRUST
M-23B
SEA LEVEL
TWT (Sec)
STO
BA-TY
BALEARIC UPPER MANTLE
ALPINE TETHYS SLAB OC
HORIZONTAL SCALE (km)
SCALE RATIO: TRUE FOR VELOCITY 6000 m/s
Progressive Distance (km)
M - 32
SECTION CROP M - 32
INTERVALS
PQ Plio-Quaternary
MC Miocene and other cover units
UCE Upper Crust Europe (Sardinia)
LCE Lower Crust Europe (Sardinia)
OC Oceanic Crust (Alpine Tethys Slab)
UM Upper Mantle (Balearic)
HORIZONS
A = Base Plio-Quaternary
B = Within Miocene
ZS = Basement Sardinia Block
TLE = Top Lower Crust Europe (Sardinia)
TO = Top Oceanic Crust (Alpine Tethys Slab)
MO = Moho Sardinian Thrust Block
FAULT ACTIVITY
BA = Faults of Balearic Stage (Upper Oligocene - Lower Miocene)
TRT = Thrust with right transcurrent component of Late Balearic - Early Tyrrhenian stage
SARDINIA
TYRRHENIAN SEA
SICILY
PELAGIAN SEA
TUNISIA
SCALE (km)

Figure 18: Lithospheric seismogeological section across the Sardinian Channel based on the new CROP section M32. It shows the southeastern Sardinian margin deformed by compression tectonics due to collision with the NW side of the Panormid Promontory in the late Balearic–initial Tyrrhenian stage. The STO transpressive fault, in the Tyrrhenian Sea, splits up into three right-transcurrent faults (Figure 20: STO-1, 2, and 3) which accommodated the back-arc basin opening tectono-dynamics. On the SE extremity is the upper Alpine Tethys slab.

---

interpreted W-dipping slab (BAS) supports the proposal that this Tethyan remnant in the Balearic stage subducted the European margin (BAS), determining the rotation of the CSB and the Corsica–Adria collision, as we see in Figure 13. Between the two slabs, a prominent *asthenospheric bulge* which migrated progressively eastward during the Balearic stage (Upper Oligocene–Lower Miocene) was generated (Finetti, this volume, Chapter 34).

4.2. CROP Section M2A/I across the Corso-Sardinian Block

Up to now CROP M2A/I is the only transcrustal seismic reflection section that crosses uninterrupted the entire CSB, from the Balearic Sea to the Tyrrhenian basin (Plates 1 and 2; Figure 14). The iterated interpretive reprocessing gave a marked signal improvement over the original standard processing. The interpreted seismogeological section here presented can be considered of satisfactory reliability with regard to its fundamental crustal stratigraphic and macro-tectonic setting. A fair to good interpretive guide marker is the well-reflecting lower crust interval, but also the upper crust and major tectonic discontinuities are based on acceptable seismic support.

The CSB presents three distinct tectono-crustal zones: (1) an internal one, rifted, thinned and affected by transcrustal listric faults (RF-1 to 6), connected with the opening of the Balearic back-arc basin (SP 3200–4600); (2) a thicker crustal zone corresponding to the central CSB area (roughly SP 4600–7200); (3) a Tyrrhenian margin zone, prominently stretched and delaminated (SP 7200–10450). This thinned continental zone terminates on the *Selli line* (some 13 km southeast of SP 10450) where the Tyrrhenian Sea floored by oceanic crust opens (Figure 19).

The CROP data provide full 3D evidence what has unanimously been accepted for a long time, that the CSB is formed by a fragment of the European margin, detached during the Balearic geodynamic stage and rotated anticlockwise (Oligocene–Upper Miocene). It is also clear that this European fragment (microplate) was affected by prominent tectonic deformation not simple to understand. In fact, we may clearly observe now that in the section zone (southern Corsica–northern Sardinia) the major tectonic imprint of the central CSB is that of a moderately shortened crust with E-verging thrusts and reverse crustal faults. This is apparently in contradiction with an expected stretching of a subduction-related detachment of the CSB. However, observing that in the Late Balearic stage, to the north of the 41° parallel, the CSB collided with the NW-Adria margin or Etrurian Promontory (Plate 2b), it is evident that, after a stretching and delamination process, the extended Corsican–northern Sardinian margin was recompacted and thickened in Aquitanian–Burdigalian times. The same early Balearic stretching and late Balearic thrusting recompaction occurred also at the SE-Sardinian margin as a consequence of its collision with the NW Panormid Promontory (Figure 19 and Plate 2, c). The total crustal shortening measured along the section of Plate 2 on the top basement (ZE), top lower crust (TL) and Moho (MO) is consistently in the order of 50–55 km, mostly due to the transcrustal thrust faults TF-5, 6 and 7. It is to be noted that on the eastern part of the examined section (Tyrrhenian side) most Balearic thrust-faults partially inverted their displacement during the extensional effects of the Tyrrhenian stage. The CSB thrust fault system also affects the lower crust with the faults TF-2 to TF-8 (Plate 2), while the faults TF-1 and TF-9 to 12 cut only the upper crust. From SP 6300 to 8100, beneath a thin upper mantle attributable to the Balearic mantle flux which drove the CSB to its present-day position, an interval, about 1.2 s TWT thick is interpretable by relatively good and clear deep reflections (Plate 2, Figure 14). After many 3D studies, correlations, and cross-controls of the Alpine Tethys slabs revealed by the CROP network (i.e. CROP M10, M11, M12A, M12B and M37), we reached the well-supported conviction that this reflecting interval belongs to the Alpine Neotethys (Plate 2). Its westward segment (SP 7000–6700) *represents the Alpine Tethys slab subducted in the Balearic stage beneath the CSB* and is quite similar and correlatable with the corresponding slab seen on sections CROP M10 (Figure 13), M11 (Finetti et al., this volume, Chapter 8), and M37 (Finetti et al., this volume, Chapter 9). The eastern part (SP 7950–8200) is truncated by an important lithospheric left-transcurrent fault and *represents an Ionian slab subducted in the early Tyrrhenian stage*. These two slabs are connected by an oceanic crust that is close to the Etrurian Promontory corner (Plate 2, c; Figure 19) and corresponds to the passage from the Alpine Tethys to the Ionian Tethys (point A′ in the Plate 2, a). They form a marked mantle bulge, well imaged by clear mantle reflections (Figure 14).

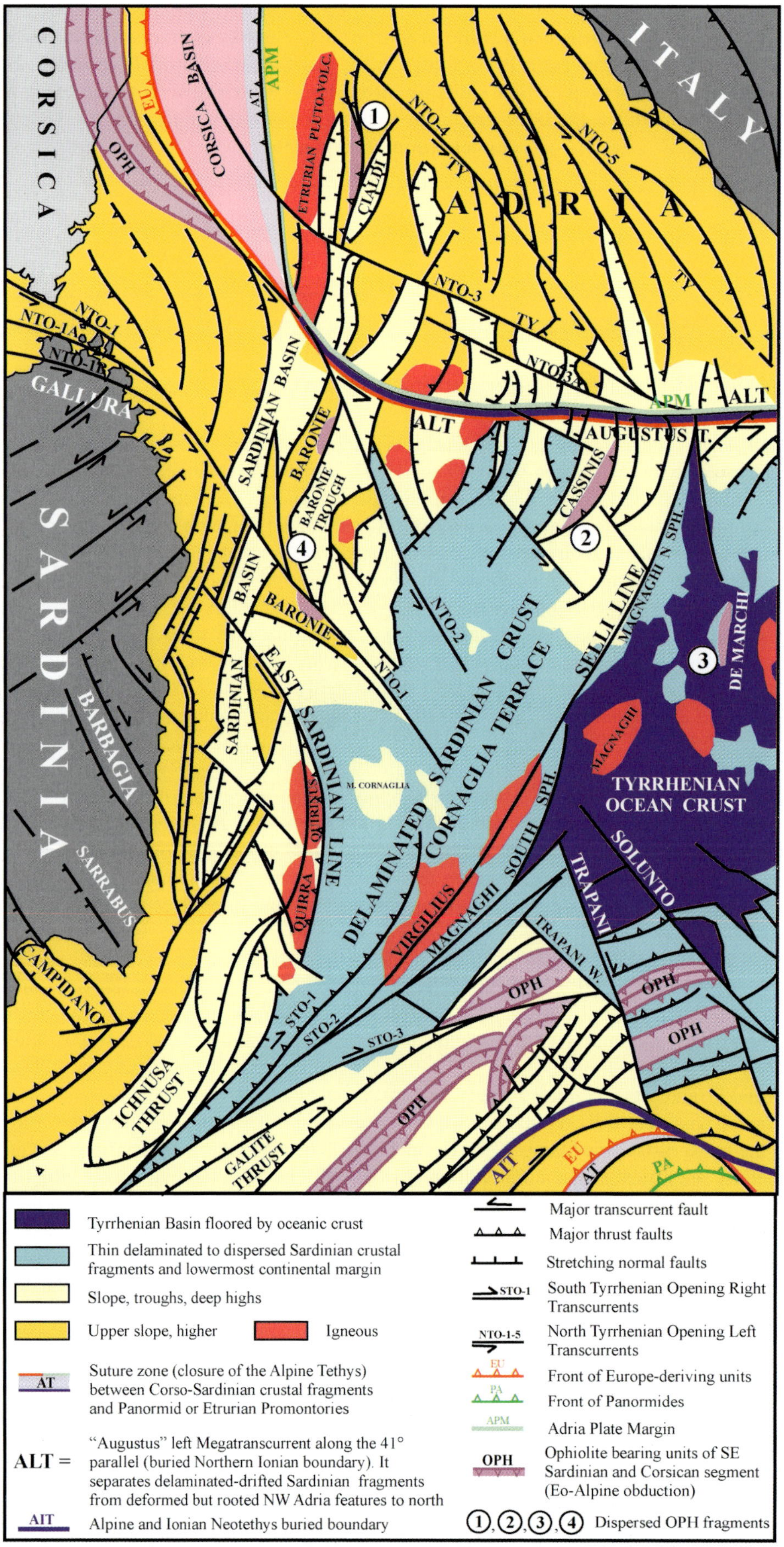
CORSICA
ITALY
ADRIA
SARDINIA
CORSICA BASIN
ETRURIAN PLUTO-VOLC.
CIALDI
NTO-1
NTO-1A
NTO-1B
NTO-2
NTO-3
NTO-3A
NTO-4
NTO-5
TY
EU
AT
APM
OPH
GALLURA
SARDINIAN BASIN
BARONIE
BARONIE TROUGH
ALT
AUGUSTUS T.
CASSINIS
SELLI LINE
MAGNAGHI N SPH.
DE MARCHI
MAGNAGHI
TYRRHENIAN OCEAN CRUST
SOLUNTO
TRAPANI
TRAPANI W.
EAST SARDINIAN LINE
DELAMINATED SARDINIAN CRUST
CORNAGLIA TERRACE
M. CORNAGLIA
MAGNAGHI SOUTH SPH.
VIRGILIUS
QUIRRA
BARBAGIA
SARRABUS
CAMPIDANO
ICHNUSA THRUST
GALITE THRUST
STO-1
STO-2
STO-3
AIT
PA
Tyrrhenian Basin floored by oceanic crust
Thin delaminated to dispersed Sardinian crustal fragments and lowermost continental margin
Slope, troughs, deep highs
Upper slope, higher
Igneous
Suture zone (closure of the Alpine Tethys) between Corso-Sardinian crustal fragments and Panormid or Etrurian Promontories
ALT = "Augustus" left Megatranscurrent along the 41° parallel (buried Northern Ionian boundary). It separates delaminated-drifted Sardinian fragments from deformed but rooted NW Adria features to north
Alpine and Ionian Neotethys buried boundary
Major transcurrent fault
Major thrust faults
Stretching normal faults
South Tyrrhenian Opening Right Transcurrents
North Tyrrhenian Opening Left Transcurrents
Front of Europe-deriving units
Front of Panormides
Adria Plate Margin
Ophiolite bearing units of SE Sardinian and Corsican segment (Eo-Alpine obduction)
①, ②, ③, ④ Dispersed OPH fragments

Figure 19: New CROP-controlled crustal tectono-dynamic sketch illustrating: (1) the collisional zone Sardinia–NW Adria Plate to the north (Etrurian Promontory); (2) the collisional zone SE Sardinia–NW Panormid Promontory to the south; and (3) the Tyrrhenian-opening-related eastern Sardinian margin delamination, driven by the Ionian slab retreat and consequent mantle flux.

As regard to the prominent left transcurrent lithospheric fault (NTO-2 of the *Baronie Trough*), it is an element of an important tectono-dynamic system connected to the *Northern Tyrrhenian opening* (NTO). Other similar faults are NTO-1, NTO-3, NTO-4 and NTO-5 (Figure 19). All these faults are associated with marked magnetic lineaments and frequently determine magmatic effusions–intrusions. The NW–SE directed transcurrent fault NTO-1 (SP 5900), and vicariants NTO-1A (SP 5650) and NTO-1B (SP 5100), activate accommodating NE–SW-directed strike-slip faults in NE Sardinia (Figure 19).

Two mounts, at SP 8100 and 8950, are named here for the first time *M. Marco Polo* and *M. Columbus* respectively, and seem to be affected by volcanic activity which flowed through two transcrustal extensive normal faults.

4.3. CROP Section M12B (Part)

Interpretation of CROP section M12B (Part: SP 100–2000) is exhibited in Figure 15. This is the area of convergence Sardinia–Corsica/NW Adria, here seen on a section that crosses the front of the CSB thrust units and the southern edge of the Corsica basin, along a direction subparallel to the basin axis. The information deduced is important for many questions in this complex area. It is immediately evident that the CSB convergence with NW Adria has its suture zone in the Corsica basin. M12B is the only CROP section that shows the thrust front of Corsica (here, more specifically, southern Corsica–northern Sardinia) into the Corsica basin. This basin represents the foredeep of the Balearic convergence, filled by Alpine Tethydes (Ligurides) and associated flysch units of the Oligocene–Lower Miocene (the time of slab subduction and CSB rotation), followed by thin Middle-Upper Miocene and Plio-Quaternary. The Corsica basin sequence is calibrated by the Martina 1 borehole at the northern extremity of the M12B CROP section (Finetti et al., 2001; Finetti et al., this volume, Chapter 8). When studying the tectono-stratigraphic setting of Figure 15 (M12B) it should be taken into consideration that this is a zone of accommodating deformation between the southern extremity of the Corsica Basin, imprinted by an almost NS-trending tectonics of the Corsica–NW Adria convergence/collision, and the area of extended and delaminated Sardinian margin, along the 41° parallel, characterized by a regional megatranscurrent fault (*"Augustus" left-megatranscurrent*) coincident with the northern boundary of the subducted Ionian Tethys (Figure 19).

On the E-verging frontal thrusts of the CSB, W-verging thin thrust units associated with a high mag anomaly have been identified (Figure 15). It seems logical to connect these units with the nearby outcropping ophiolite-bearing units of Eo-Alpine obduction of NE Corsica. To the south of the Augustus line (41° parallel) these ophiolite-bearing units are found only as a few small dispersed findings on delaminated fragments of the Sardinian margin in the Tyrrhenain Sea (Mt. Cassinis, Mt. De Marchi, Mt. Baronie: sites 2, 3 and 4 of Figure 19) and then, more organized, in the Coastal Chain of northern Calabria. A sea-floor ophiolite unit is also known on Mt. Cialdi (site 1).

4.4. CROP Section M28B across the delaminated Eastern Sardinian Margin

This CROP section M28B (Plate 4) gives an immediate, fairly self-explanatory image of the crustal-lithospheric tectono-stratigraphic setting of the stretched and delaminated eastern Sardinian margin, in the SW Tyrrhenian Sea. Three distinct tectono-dynamic processes can be recognized in the section: (1) a (late) Balearic compressive deformation generated during the collision of SE Sardinia with the NW-Panormid promontory; (2) a Tyrrhenian-stage crustal extension and delamination related to the Ionian slab retreat and Tyrrhenain back-arc basin opening; (3) a late Tyrrhenian stage where the NW-Sicilian margin area, east of the Trapani transcurrent fault, was dissected from the Maghrebian chain by a NW–SE right-transcurrent system, and its previously delaminated units were recompacted on the Sicily margin (Figures 16: SP 4300-5406; and Figure 19). The original compressive tectonics which occurred before the Tyrrhenian opening are still observable as thrust faults: faults TF-1 to TF-3, faults Tf-4 to TF-7. These latter group, during the Tyrrhenian opening stage, partially inverted their displacement (Plate 4). The extensional faults of the Tyrrhenian stage (active since the Middle Miocene) are very impressive on the Sardinian slope and in the Sardinian basin where they truncate previous thrust faults (Plate 4: faults EF-1 to EF-6, from SP 100 to 1400). Connected to the extension fault EF-6 is the effusion pathway of a newly discovered prominent volcano

Table 2
Thinning (elongation) of Pre-Sedimentary crust along the delaminated Tyrrhenian margin of Sardinia (CROP Section M28B). Assumed velocities: UC = 6.0 km/s; LC = 7.0 km/s

Crustal interval	Section zone							
	SP 100		SP 1000		SP 1600–3700		SP 5100	
	TWT (S)	km	TWT (S)	km	TWT (S)	km	TWT (S)	km
UC (Upper Crust)	5.6	16.8	2.3	6.9	1.4	4.2	1.0	3.0
LC (Lower Crust)	2.0	7.0	1.6	5.6	1.0	3.5	0.78	2.8
TC (Total Crust)	7.6	23.8	4.4	12.5	2.4	7.7	1.8	5.8
Thinning (Elongation) Ratio SP 100/ Other site	0		1.90		3.09		4.10	

which we name here the *Quirinus Volcano*. This fault separates a sharply thinned and delaminated eastern crust from a western one, which also stretched, but remained much thicker. Another *newly identified volcanic body* lies from SP 2100 to 2480 and is named *Virgilius*. The extensional fault EF-6 represents a regional tectonic element well evident also in the bathymetry of the area, because it separates two markedly different crustal thicknesses. The position of this prominent tectonic element is reported in Figure 19, with the name of *East Sardinian Line*. Another relevant crustal feature, already recognized in the literature (Finetti and Del Ben, 1986), is the "*Selli Line*" (Figure 19, Plate 4) which also represents a significant boundary between the opened oceanic Tyrrhenian Sea and a stretched Cornaglia terrace zone. In Plate 4 (section M2A/I) the Selli line delimits the *southern sphenochasm of the Magnaghi basin,* where a Tyrrhenian-opening-related right-transcurrent fault passes (STO-2). The *right-transcurrent Trapani North is the westernmost NW–SE directed fault of a system which dissects the Maghrebian chain of SW–NE direction from the Sicilian Apennines Chain.*

A significant point is the identification from good characteristic seismic signals of roof-thrust units of European vergence, associated with high magnetic anomalies (Figure 16, Plate 4: SP 3800–4050 and 4500–5000). After careful analysis and a 3D study of similar seismic facies on other CROP sections (i.e. CROP M28A, Finetti et al., this volume, Chapter 15), *we find it quite convincing to attribute these particular seismic bodies to ophiolite-bearing units which obducted the southern margin of Europe in the Eo-Alpine stage.* Accepting such an interpretation, it is possible to give coherent explanations for the associated high magnetic anomalies and for the uppermost seismic features, otherwise problematic. To the two submarine mounts formed by these interpreted ophiolite-bearing units we give the names of *Caesar Sm.* and *Traianus Sms.* (Plate 4, Figure 16).

In Late (Middle) Tyrrhenian stage, the delaminated crustal units of the Sardinian margin collided with the Sicilian margin and were recompacted with imbricated thrusts (SP 4000–5406).

For a quantitative indication of the crustal stretching and thinning during the Tyrrhenian dynamics in the section area, Table 2 has been compiled, where the thicknesses of the pre-sedimentary upper crust (UC), lower crust (LC) and total crust (TC) of various significant thinned zones are compared (SP 100, 1000, 1600–3700 and 5100). Then, the "TC" ratio (thinning or elongation) with respect to SP 100 was calculated. It is interesting to note that passing from SP 100 to the section segment located between faults EF-5 and EF-6 (SP 1000) the crustal thickness is reduced from nearly 24 km to 12.5 km (elongation from 1 to 1.9). Zone SP 1600–3700 has an average TC thickness of 7.7 km or a thinning (elongation) of 3.09. At SP 5100 the thinning with respect to SP 100 is 4.1 times.

4.5. CROP Section M32 across the Southern Sardinian Margin

The CROP seismic section M32 explores the southern Sardinian margin in the Sardinian Channel area. Figure 17 shows a reduced version of the seismic section and Figure 18 its interpretation. The whole section is imprinted by compressive crustal tectonics represented by thrust faults, some involving the lower crust (transcrustal thrusts) and some limited to the upper crust. This tectonic style is coherent throughout the CROP explored Maghrebian chain area and, in this specific zone, is the effect of the continental collision between rotating Sardinia and the NW Panormid Promontory. One of the major transcrustal faults is the *Ichnusa thrust* (Figure 18). This thrust fault (STO), during the late Balearic–early Tyrrhenian stage, was accompanied by a marked right-transcurrent motion component and, going into the Tyrrhenian Sea, it splits up into three important faults dominated by right-transcurrent motion that accommodated the back-arc basin opening and crustal delamination (Figure 19: faults STO-1, 2, and 3).

Another transcrustal compressive fault is the *Galite thrust* (Figure 18), it too evidently affected by minor right-transcurrent motion. The third lithospheric fault lying at the SE extremity of the section is a major one of the Maghrebian chain sector collision with the Panormid Promontory. It is to observe that the involved crustal thicknesses, removed the thrusting, is in the order of about 2.8 s TWT in the pre-sedimentary upper crust (UC) or about 8.4 km, and 1.7 s TWT in the lower crust (LC), or about 5.5 km. We are therefore dealing with a thinned crust. It seems very likely that the thinning process occurred in the Balearic stage before the collision between the SE Sardinian margin and the NW Panormid Promontory.

Around SP 100 of the section, in the reflection time interval of 12.6 to 14.5, a characteristic band of good reflections dipping northwestward is present (Figure 18). The 3D CROP network data supports this being the imaging of a subducted slab and, more precisely, the upper part of the Alpine Tethys slab, close to the NW side of the Panormid Promontory boundary.

CONCLUDING REMARKS

From interpretation of the CROP seismic data network are derived important control data and a new understanding of the tectono-dynamics of the Corso-Sardinian Block:

(a) *The CSB detached in the Balearic stage from Europe and rotated counterclockwise, driven by W-dipping subduction of a wide western Alpine Tethys lithospheric remnant preserved after the Eo-Alpine stage that determined the closure of the eastern Alpine Tethys in the sector of the Alps (Adria-Europe collision and formation of the Eo-Alps).*

(b) *The CSB rotation terminated at the end of the Balearic stage when the Corsica block collided with NW Adria (Etrurian Promontory), and SE Sardinian block collided with the Panormid Promontory.*

(c) *The eastern-moving roll-back sinking process of the Alpine Tethys lithosphere continued into the openly connected Ionian Tethys during the Tyrrhenian stage.*

(d) *The Tyrrhenian opening took place within eastern Sardinia, south of the 41° parallel, with severe delamination of the eastern Sardinian margin and tectonic transport of CSB fragments over complex allochthonous crustal stacks of the Southern Apennines and Sicilian Apennines Tyrrhenian margins and on the Calabrian Arc.*

(e) *Both the Balearic (Upper Oligocene–Lower Miocene) and Tyrrhenian (Middle Miocene to Present) geodynamic stages have imprinted the Sardinian block with transtensive tectonics of dominant left polarity to the north and right to the south (Sardinian rift), accompanied by significant magmatic activity.*

ACKNOWLEDGEMENTS

This work was supported by MIUR, COFIN 1998, 2000 and 2002 with fundings to I.R. Finetti, University of Trieste, as project coordinator and to S. Fais, as Cagliari University unit coordinator. Many thanks are due to OGS, Seismic Division and to OGS-Explora's seismic crews for the CROP seismic data acquisition, to ENEL-Ismes for processing the sections CROP M28A and M32. Our gratitude is expressed to the EGG (Exploration Geophysics Group) of the University of Trieste, Faculty of Sciences, for the invaluable contribution of the iterated interpretative reprocessing of the CROP Sections M2A/I, M10 and M12B. The signal quality improvement obtained was determinant for understanding this complex area.

REFERENCES

Alvarez, W., Cocozza, T., Forese Wezel, C., 1974. Fragmentation of the Alpine orogenic belt by microplate dispersal. Nature 248, 209–314.

Arisi Rota, F., Fichera, R., 1987. Magnetic interpretation related to geo-magnetic provinces: the Italian case history. Tectonophysics 138, 179–196.

Assorgia, A., Balogh, K., Lecca, L., Ibba, A., Porcu, A., Secchi, F., Tilocca, G., 1995. Volcanological characters and structural context of Oligo-Miocene volcanic succession from Central Sardinia (Italy). Acc. Naz. Sc. XL XIV (Rapporti Alpi-Appennino), 397–424.
Balia, R., Fais, S., Trudu, R., 1983. Studio gravimetrico del Campidano meridionale. In: Atti II Conv. Naz. Geof. Terra Solida, pp. 449–457.
Balia, R., Ciminale, M., Loddo, M., Pecorini, G., Ruina, G., Trudu, R., 1984. Gravity survey and interpretation of Bouguer anomalies in the Campidano geothermal area (Sardinia, Italy). Geothermics 13 (4), 333–347.
Balia, R., Fais, S., Klingelé, E., Marson, I., Porcu, A., 1991. Aeromagnetic constraints on the geostructural interpretation of the southern part of the Sardinian Rift, Italy. Tectonophysics 195, 347–358.
Bayer, R., Le Mouel, J.L., Le Pichon, X., 1973. Magnetic anomaly pattern in the Western Mediterranean. Earth Planet. Sci. Lett. 19, 168–176.
Bayer, M., Bayer, R., Lesquer, A., 1976. Quelques remarques sur la structure gèologique de la Corse d'après la gravimetrie et le magnétisme. Bull. Soc. Geol. France XVIII (5), 1189–1194.
Bayer, R., Lesquer, A., 1976. Carte gravimétrique de la Corse. Cent. Géol. Géophys., CNRS, Montpellier.
Beccaluva, L., Deriu, M., Macciotta, G., Savelli, C., Venturelli, G., 1977. Geocronology and magmatic character of the Pliocene-Pleistocene volcanism in Sardinia (Italy). Bull. Volcanol. 40 (3), 1–16.
Beccaluva, L., Civetta, L., Macciotta, G., Ricci, C.A., 1985. Geochronology in Sardinia: results and problems. Rend. Soc. It. Min. Petr. 40, 57–72.
Beccaluva, L., Brotzu, P., Macciotta, G., Morbidelli, L., Serri, G., Traversa, G., 1987. Cainozoic tectono-magmatic evolution and inferred mantle sources in the Sardo-Tyrrhenian area. In: The Lithosphere in Italy: Advances in Earth Science Research. Accad. Naz. Lincei 80, 229–248.
Bellaiche, G., Rehault, J.P., Robin, C., Gennesseaux, M., Auzende, J.M., Boivin, M.C., 1979. Petrological evidence of typical oceanic crust in the western Mediterranean Sea. Marine Geology 32, M11–M18.
Bellon, H., Maury, R., Bellaiche, G., Rehault, J.P., Mermet, J.F., Auzende, J.M., 1985. Âge et nature des formations volcaniques prismée observées et prelevées dans le Canyon des Moines (sud-ouest Corse) pendant la campagne Cyaligure. Marine Geology 67, 163–176.
Ben Avraham, Z., Boccaletti, M., Cello, G., Grasso, M., Lentini, F., Torelli, L., Tortorici, L., 1990. Principali domini strutturali originatisi dalla collisione neogenico-quaternaria nel Mediterraneo centrale. Mem. Soc. Geol. It. 45, 453–462.
Boccaletti, M., Guazzone, G., Manetti, P., 1974. Evoluzione paleogeografia e geodinamica del Mediterraneo: i bacini marginali. Mem. Soc. Geol. It., Suppl. 2 13, 162–199.
Boccaletti, M., Guazzone, G., Dainelli, P., 1982. Il sistema regmatico neogenico-quaternario nell'area mediterranea: esempio di deformazione plastico-rigida post-collisionale. Mem. Soc. Geol. It. 24, 465–482.
Boccaletti, M., Calamita, F., Deiana, G., Gelati, R., Massari, F., Moratti, G., Ricci Lucchi, F., 1990. Migrating foredeep-thrust belt system in the northern Apennines and southern Alps. Paleogeogr. Paleoclimatol. Paleoecol. 77, 3–14.
Bois, C., 1993. Initiation and evolution of the Oligo-Miocene rift basins of south-western Europe: a contribution of deep seismic reflection profiling. Tectonophysics 226, 227–252.
Bortolotti, V., Principi, G., Treves, B., 2001. Ophiolites, Ligurides and the tectonic evolution from spreading to convergence of a Mesozoic western tethys segment. In: Vai, G.B., Martini, J.P. (Eds.), Anatomy of an Orogen: the Apennines and Adjacent Mediterranean Basins. Kluwer Academic, pp. 151–164.
Brotzu, P. (Ed.), 1997. The Tertiary calcalkaline volcanism of Sardinia. Periodico di Mineralogia, vol. 66, pp. 1–232.
Carmignani, L., Pertusati, P.C., Barca, S., Carosi, R., Di Pisa, A., Gattiglio, M., Musumeci, G., Oggiano, G., 1992. Struttura della catena ercinica della Sardegna. Guida all'escursione del Gruppo Inf. Geol. Strutt., Siena, pp. 1–177.
Carmignani, L., Barca, S., Disperati, L., Fantozzi, P., Funedda, A., Oggiano, G., Pasci, S., 1994. Tertiary compression and extension in Sardinian Basement. Boll. Geof. Teor. Appl. 36 (141–144), 45–62.
Carmignani, L., Oggiano, G., Barca, S., Conti, P., Salvadori, I., Eltrudis, A., Funedda, A., Pasci, S., 2001. Geologia della Sardegna. Note illustrative della Carta Geologica della Sardegna a scala 1:200.000, con carta della Sardegna e della Sardegna-Corsica. Ist. Pol. Zecca dello Stato, Roma, pp. 1–283.
Cassano, E., Marcello, A., Nannini, R., Pretti, S., Ranieri, G., Salvaderi, R., Salvadori, I., 1979. Rilievo aeromagnetico della Sardegna e del mare circostante. Boll. Serv. Geol. Ital. C, pp. 199–234.
Cassano, E., Fichera, R., Arisi Rota, F., 1986. Rilievo aeromagnetico d'Italia alcuni risultati interpretativi. In: Atti V Conv. Gruppo Naz. Geof. Terra Solida, pp. 939–962.
Casula, G., Cherchi, A., Montadert, I., Murru, M., Sarria, E., 2001. The Cenozoic graben system of Sardinia (Italy): geodynamic evolution from new seismic and field data. Marine and Petroleum Geology 18, 863–888.
Cherchi, A., Montadert, L., 1982. The Oligo-Miocene Rift of Sardinia and early history of the western Mediterranean basin. Nature 298, 736–739.

Cherchi, A. (Ed.), 1985. 19th European micropaleontological colloquium. Sardinia, 1985. Micropaleontological researches in Sardinia – Guidebook. AGIP, pp. 1–338.
Cohen, C.R., 1980. Plate tectonic model for the Oligo-Miocene evolution of the western Mediterranean. Tectonophysics 68, 283–311.
Costamagna, L., Barca, S., 2002. The "Germanic" Triassic of Sardinia (Italy): a stratigraphic, depositional and palaeogeographic review. Riv. It. Pal. Strat. 108, 67–100.
Coward, M.P., Dietrich, D., 1989. Alpine tectonics – an overview. In: Coward, M.P., Dietrich, D., Park, R.G. (Eds.), Alpine Tectonics. Geol. Soc. Spec. Publ. 45, 1–29.
De Cillia, C., Fais, S., Tocco, R., 1992. Geostructural features of de middle Tirso valley from seismic reflection data. Boll. Geof. Teor. Appl. 34 (136), 273–286.
Dewey, F.J., Helman, F.L., Turco, E., Hutton, D.H.W., Knot, D.S., 1989. Kinematics of the western Mediterranean. In: Coward, M.P., Dietrich, D., Park, R.G. (Eds.), Alpine Tectonics. Geol. Soc. Spec. Publ. 45, 265–283.
Durand-Delga, M., 1984. Principaux traits de la Corse alpine et correlations avec les Alpes ligures. Mem. Soc. Geol. It. 28, 285–329.
Facenna, C., Speranza, F., Caracciolo, F.D., Mattei, M., Oggiano, G., 2002. Extensional tectonics on Sardinia (Italy): insights into the arc-back-arc transitional regime. Tectonophysics 356, 213–232.
Fais, S., Klingelé, E.E., Tocco, R., 1994. Geophysical interpretation of the Gallura magnetic anomaly (northeast Sardinia, Italy). Tectonophysics 233, 125–144.
Fais, S., Klingelé, E.E., Lecca, L., 1996. Oligo-Miocene half graben structure in Western Sardinia Shelf (western Mediterranean): reflection seismic and aeromagnetic data comparison. Marine Geology 133, 203–222.
Fais, S., Barca, S., Porcu, A., Tocco, R., 1997. Stratigraphic-structural and seiosmic reflection investigations on cenozoic basins of Sardinia, Italy. In: 59rd EAGE Ext. Abstr., vol. 2, pp. 1–4.
Fais, S., Lecca, L., Tocco, R., 2000. Reflection seismic and geological data in a central sector of the Sardinian Rift (western Mediterranean). In: IX Cong. Geol. Chileno, Puerto Varas, Chile, pp. 1–4.
Fais, S., Klingelé, E.E., Lecca, L., 2001. Interpretation of geophysical data in the SW Sardinian shelf (western Mediterranean). In: 63rd EAGE Conf. Ext. Abstr., vol. 2, pp. 1–4.
Fais, S., Klingelé, E.E., Lecca, L., 2002. Structural features of the south-western Sardinian shelf (western Mediterranean) deduced from aeromagnetic and high-resolution reflection seismic data. Hecl. Geol. Helv. 95 (2), 169–182.
Finetti, I.R., this volume. Geodynamic evolution of the Mediterranean Region, from the Permo-Triassic Ionian opening to the Present, constrained by new lithospheric CROP seismic data. Chapter 34.
Finetti, I.R., Del Ben, A., 1986. Geophysical study of the Tyrrhenian opening. Boll. Geof. Teor. Appl. 28, 75–156.
Finetti, I.R., Lentini, F., Carbone, S., Catalano, S., Del Ben, A., 1996. Il sistema Appennino meridionale-Sicilia nel Mediterraneo centrale: studio geologico-geofisico. Boll. Soc. Geol. It. 115, 529–559.
Finetti, I.R., Boccaletti, M., Bonini, M., Del Ben, A., Pipan, M., Prizzon, A., Sani, F., this volume. Lithospheric tectono-stratigraphic setting of the Ligurian Sea Northern Apennines–Adriatic foreland from integrated CROP seismic data. Chapter 8.
Finetti, I.R., Del Ben, A., Forlin, E., Pipan, M., Prizzon, A., Calamita, F., Crescenti, U., Rusciadelli, G., Scisciani, V., this volume. Crustal geologic section across Central Italy from the Corsica basin to the Adriatic Sea, based on geological and CROP seismic data. Chapter 9.
Finetti, I.R., Lentini, F., Carbone, S., Del Ben, A., Di Stefano, A., Forlin, E., Guarnieri, P., Pipan, M., Prizzon, A., this volume. Geological outline of Sicily and lithospheric tectono-dynamics of its Tyrrhenian margin from new CROP seismic data. Chapter 15.
Finetti, I.R., Boccaletti, M., Bonini, M., Del Ben, A., Geletti, R., Pipan, M., Sani, F., 2001. Crustal section based on CROP seismic data across the North Tyrrhenian–Northern Apennines–Adriatic Sea. Tectonophysics 343, 135–163.
Funedda, A., Oggiano, G., Pasci, S., 2000. The Logudoro basin: a key area for the Tertiary tectono-sedimentary evolution of North Sardinia. Boll. Soc. Geol. It. 119, 31–38.
Galdeano, A., Ciminale, M., 1987. Aeromagnetic evidence for the rotation of Sardinia (Mediterranean Sea): comparison with the paleomagnetic measurements. Earth Planet. Sci. Lett. 82, 193–205.
Iaccarino, S., D'Onofrio, S., Murru, M., 1985. Miocene foraminifera of several sections of the Marmilla area (central western Sardinia). Boll. Soc. Paleontol. It. 23, 395–412.
Klingelé, E., Lahmeyer, B., Marson, I., Schwarz, G., 1989. A 2-D gravity model of the seismic refraction profile of the EGT southern segment. In: Freeman, R., Mueller, St. (Eds.), Proceedings of the Sixth EGT: Data Compilation and Synoptic Interpretation. European Science Foundation, pp. 271–278.
Lecca, L., Pala, A., Tilocca, G., 1994. L'half graben di Chilivani-Ardara come modello tettonico-sedimentario e idrogeologico di una struttura secondaria miocenica del rifting della Sardegna. Rend. Sem. Fac. Sc. Univ. Cagliari 64 (2), 267–285.

Lecca, L., Lonis, R., Luxoro, S., Melis, E., Sechi, F., Brotzu, P., 1997. Oligo-Miocene volcanic sequences and rifting stages in Sardinia: a review. Periodico di Mineralogia 66, 7–61.

Malinverno, A., Ryan, W.B.F., 1986. Extension in the Tyrrhenian sea and shortening in the Apennines as a result of arc migration driven by sinking of the lithosphere. Tectonics 5, 227–245.

Mauffret, A., Contrucci, I., Brunet, C., 1999. Structural evolution of the Northern Tyrrhenian Sea from new seismic data. Mar. Petr. Geol. 16, 381–407.

Maxia, C., Pecorini, G., 1969. La zona di Castelsardo: la meno incompleta serie miocenica della Sardegna. In: Comm. Medit. Neog. Strat., Proc. IV Sess. Giorn. Geol. 35, 345–357.

Morra, V., Secchi, F., Assorgia, A., 1994. Petrogenetic significance of peralkaline rocks from cenozoic calcalkaline volcanism from SW Sardinia (Italy). Chem. Geol. 118, 109–142.

Mouge, P., Galdéano, A., 1990. Aeromagnetic compilation over the southern traverse of EGT. In: Freeman, R., Mueller, St. (Eds.), Proceedings of the Sixth EGT Workshop: Data Compilation and Synoptic Interpretation. European Science Foundation, pp. 237–246.

Munoz, J.A., 1992. Evolution of a continental collision belt: ECORS-Pyrenees crustal balanced cross-section. In: Mc Clay, K.R. (Ed.), Thrust Tectonics. Chapman & Hall, pp. 235–246.

Oggiano, G., Pasci, S., Funedda, A., 1995. Il bacino di Chilivani–Berchidda: un esempio di struttura transtensiva. Possibili relazioni con la geodinamica cenozoica del Mediteraneo occidentale. Boll. Soc. Geol. It. 114, 465–475.

Olivet, J.L., 1996. La cinématique de la plaque Ibérique. Bull. Centres Rech. Explor. Prod. ELF Aquit. 20 (1), 131–195.

Patacca, E., Scandone, P., 1989. Post-Tortonian mountain building in the Apennines. The role of the passive sinking of a relic lithospheric slab. In: The Lithosphere in Italy: Advances in Earth Science Research. Acc. Naz. Lincei, 157–176.

Pecorini, G., Pomesano Cherchi, A., 1969. Ricerche geologiche e biostratigrafiche sul Campidano meridionale (Sardegna). Mem. Soc. Geol. It. 8, 421–451.

Ritsema, A.R., 1979. Active or passive subduction at the Calabrian arc. Geologie en Mijnbouw 58 (2), 127–134.

Savelli, C., Beccaluva, L., Deriu, M., Macciotta, G., Maccioni, L., 1979. K–Ar geochronology and evolution of the tertiary "calc-alkalic" volcanism of Sardinia (Italy). J. Volcanol. Geotherm. Res. 5, 257–269.

Serri, G., Innocenti, F., Manetti, P., 2001. Magmatism from Mesozoic to present: petrogenesis, time–space distribution and geodynamic implications. In: Vai, G.B., Martini, J.P. (Eds.), Anatomy of an Orogen: The Apennines and Adjacent Mediterranean Basins. Kluwer Academic, pp. 77–104.

Sowerbutts, A., Underhill, J.R., 1998. Sedimentary response to intra-arc extension: controls on Oligo-Miocene deposition, Sarcidano sub-basin, Sardinia. J. Geol. Soc. 155, 491–508.

Sowerbutts, A., 2000. Sedimentation and volcanism linked to multiphase rifting in an Oligo-Miocene intra-arc basin, Anglona. Sardinia. Geol. Mag. 137 (4), 395–418.

Spano, C., Asunis, M.I., 1984. Ricerche biostratigrafiche nel settore di Castelsardo (Sardegna settentrionale). Boll. Soc. Sarda Sc. Nat. 23, 45–75.

Spano, C., Barca, S., 2002. Ecobiostratigraphic, lithostratigraphic, depositional and synthemic setting of Cenozoic units in Southern Sardinia (Italy). Boll. Soc. Geol. It. 121, 19–34.

Tapponnier, P., 1977. Evolution tectonique du système alpin en Méditerranée: poinçonnement et écracement rigide-plastique. Bull. Soc. Geol. France 19, 437–460.

Thomas, B., Gennesseaux, M., 1986. A two-stage rifting in the basins of the Corsica–Sardinia straits. Mar. Geol. 72, 225–239.

Torelli, L., Zitellini, N., Argnani, A., Brancolini, G., De Cillia, C., Peis, D., Tricart, P., 1991. Sezione gelogica crostale dell'avanpaese pelagiano al bacino di retroarco tirrenico (Mediterraneo centrale). Mem. Soc. Geol. It. 47, 385–399.

Trudu, R., 1962. Rilevamento gravimetrico della Sardegna. Boll. Geof. Teor. Appl. 4 (15), 299–339.

Vialon, P., Rochette, P., Ménard, G., 1989. Indentation and rotation in the western Alpine arc. In: Coward, M.P., Dietrich, D., Park, R.G. (Eds.), Alpine Tectonics. Geol. Soc. Spec. Publ. 45, 329–338.

Wildi, W., 1983. La chaîne tello-rifaine (Algérie, Maroc, Tunisie) : structure, stratigraphie et évolution deu Trias au Miocéne. Rev. Géol. Dyn. Géogr. Phys. 24 (3), 201–297.

Wonik, T., Galdeano, A., Hahn, A., Mouge, P., 1992. Magnetic anomalies. In: Freeman, R., Mueller, S. (Eds.), A Continental Revealed – The European Geotraverse. Atlas of Compiled Data. Cambridge University Press, Cambridge, pp. 31–34.

Zanolla, C., Morelli, C., Marson, I., 1998. The magnetic anomalies of the Mediterranean Sea (IBCM-M). Boll. Geof. Teor. Appl. 39 (1), 1–36.

CROP PROJECT: Deep Seismic Exploration of the Central Mediterranean and Italy
Edited by I.R. Finetti

CHAPTER 19

Crustal Tectono-Stratigraphy of the Ionian Sea from New Integrated CROP Seismic Data

I.R. Finetti[1,*], A. Del Ben[1]

ABSTRACT

The CROP seismic data indicate very clearly that the Ionian Sea is floored by the old oceanic crust of a Tethyan basin which opened in the Permo-Triassic, of which only a remnant is now seen because the major part subducted during the Tyrrhenian geodynamic stage (Middle Miocene to Present). Buried beneath complex tectonic stacks of allochthonous units of the peri-Tyrrhenian margins of the Southern Apennines–Calabrian Arc–Sicilian Apennines, the CROP sections disclosed for the first time clear seismic imagings of the upper Ionian slab and its continental boundary. Such imagings have allowed us to reconstruct a well-constrained paleogeography of the pre-orogenic Ionian Tethys. It emerges that this oceanic basin formed by detachment of the Adria plate from Africa, whose CROP defined Ionian boundaries are still reconnectable by clockwise back-rotation of Adria by about 40°. From CROP data integrated with geophysical-geological information, it is evident that the Ionian Tethys and the Alpine Tethys of Middle Jurassic opening, delimited by the Etrurian and Panormid promontories, were communicating and played a major role in the successive geodynamic process.
The Ionian seismic stratigraphy is characterized by a condensed sequence deposited from the Ionian opening (Upper Permian in the initial rift trough, to Ladinian in the drift area) to the Oligocene, followed by a Neogene of higher deposition rate. In the Eo-Alpine geodynamic stage (Upper Cretaceous–Eocene), the Alpine Tethys sector of the Alps closed (collision Adria–Europe), preserving a reduced western Alpine Tethys which closed in the Balearic stage (Oligocene–Lower Miocene) with W-dipping subduction of the Alpine Tethys remnant and rotation of the Corso-Sardinian block. At the end of the Balearic stage, a large eastward migrating orogenic arc included the Northern Apennines, the Southern Apennines, the Calabrian Arc and the Sicilian–Maghrebian chain. The Corso-Sardinian rotation stopped due to collision of Corsica with the Etrurian promontory, and of SE Sardinia with the Panormid promontory, but the *eastward roll-back sinking continued within the Ionian Sea* as the basic driving force of the Tyrrhenian geodynamic stage (Burdigalian to Present) which generated the Southern Apennine–Calabrian Arc–Sicilian Apennine thrust belt.
The Ionian cover was stripped off from its subducting oceanic basement, and in part thrust obliquely to form the *Ionides allochthonous units* in the Southern Apennines (Lagonegrese) and in the Sicilian Apennines (Imerese, Sicanian). The major part stacked in the imbricated Calabrian wedge. Other Ionides were imbricated in the western Hellenic wedge. Tectono-stratigraphic units coming from different realms (Sardinia, Alpine Tethys, Etrurian and Panormid promontories and internal-external flysch) took part in the construction of the tectonic stacks which are well defined by the CROP data.

Keywords: CROP exploration Ionian Sea, tectonics, seismic stratigraphy

1. INTRODUCTION

The *Ionian Sea–Eastern Mediterranean basin*, before the recent intensive geophysical investigation, was interpreted under various hypotheses, involving quite different models. The first fundamental question was the defin-

[1]Department of Geological, Environmental and Marine Sciences, Exploration Geophysics Group, University of Trieste, Via Weiss 1, 34127 Trieste, Italy
*Corresponding author. Tel.: +39 40 5582276; e-mail: finetti@units.it (I.R. Finetti)

ition of the type of crust: continental or oceanic? Around 1970, a decade of prolific geophysical exploration in various parts of the Mediterranean Sea, including the Ionian–Eastern Mediterranean started. From extensive regional seismic exploration (Finetti, 1982), it was observed that *the Ionian Sea* (in its undeformed area, beyond the Calabrian and Hellenic wedges) *is floored by a rather flat basement crust of relatively good reflectivity. This seismic reflection data also imaged for the first time an old Ionian oceanic crust* (Finetti, 1982). It was also noted that the Ionian cover sequence is thin in its Mesozoic-Paleogenic succession and that this crust is underthrusting the Calabrian and Hellenic arcs.

Further investigations advanced our knowledge of the Ionian. Correlating the Ionian sedimentary cover and the Southern Apennines thrust-stack across the gulf of Taranto, it was concluded that the Lagonegrese (condensed deep deposition sequence of Triassic to Oligocene age) represented allochthonous units detached from the subducted Ionian slab and tectonically transported into the Southern Apennines, over the Apulian platform in the Tyrrhenian stage (Middle Miocene to Present). More specifically, it was observed that the allochthonous cover units of Ionian origin thrust over the Upper Miocene in the internal Southern Apennines and over the Lower Pliocene in the external sector, indicating that the thrusting of the Ionian (Lagonegrese) units took place from the Lower to Middle Pliocene (Finetti et al., 1996). This conclusion led the authors also to extend the Ionian paleogeography northward (area of the present Tyrrhenian basin) with no precise boundaries. The checked age of the older Lagonegrese units established the timing of the Ionian opening. *The deep seismic exploration CROP project now provides new data that for the first time constrain a well-defined paleogeography of the Ionian Neotethys.* In this work, we exhibit and comment on the interpretation of some significant CROP sections in the Ionian Sea, and derive conclusions, referring the reader to other chapters of this volume for information on sections or parts of sections included in other works.

2. CROP SEISMIC DATA EXHIBITED AND DISCUSSED

A large part of the CROP seismic network was acquired in the Ionian Sea (Figure 1). The northeastern part of the CROP sections reported in the location map of Figure 1 (CROP M8, M38 and parts of M5 and M34) explore the SE-prolongation of the Apulian platform, described in Finetti and Del Ben (this volume, Chapter 23). Parts of CROP M5, M2B, M4 and M27 explore the wedge of the Calabrian arc Finetti (this volume, Chapter 17). In this chapter we exhibit and describe the interpretation of the CROP sections listed in Table 1.

Two other good seismic examples selected from the MS lines (OGS regional seismic network covering the whole Mediterranean; Finetti et al., this volume, Chapter 1) show our interpreted tectono-stratigraphy and crustal setting of the Ionian Sea to the front of the Hellenic arc (Figures 10 and 11) which are quite similar to those observed to the front of the Calabrian arc.

Table 1
CROP seismic sections exhibited and described in this work (Chapter 19)

CROP section	SP	Length (km)	Chapter 19			
			Figure N°		Plate N°	
			Seismic	Int.	Seismic	Int.
M3 (close up)	1600–2000	26	2a	2b	–	–
M2B	100–1300	72		3	–	–
M2B	1300–2800	92		4	–	–
M3	100–5760	356		5	1	2
M4 (Part)	100–1820	110		6	–	–
M22	100–3680	182		7	3	4
M23A (Part)	100–4600	226		8	5	6
M34 (Part)	2200–7178	248		9	7	8

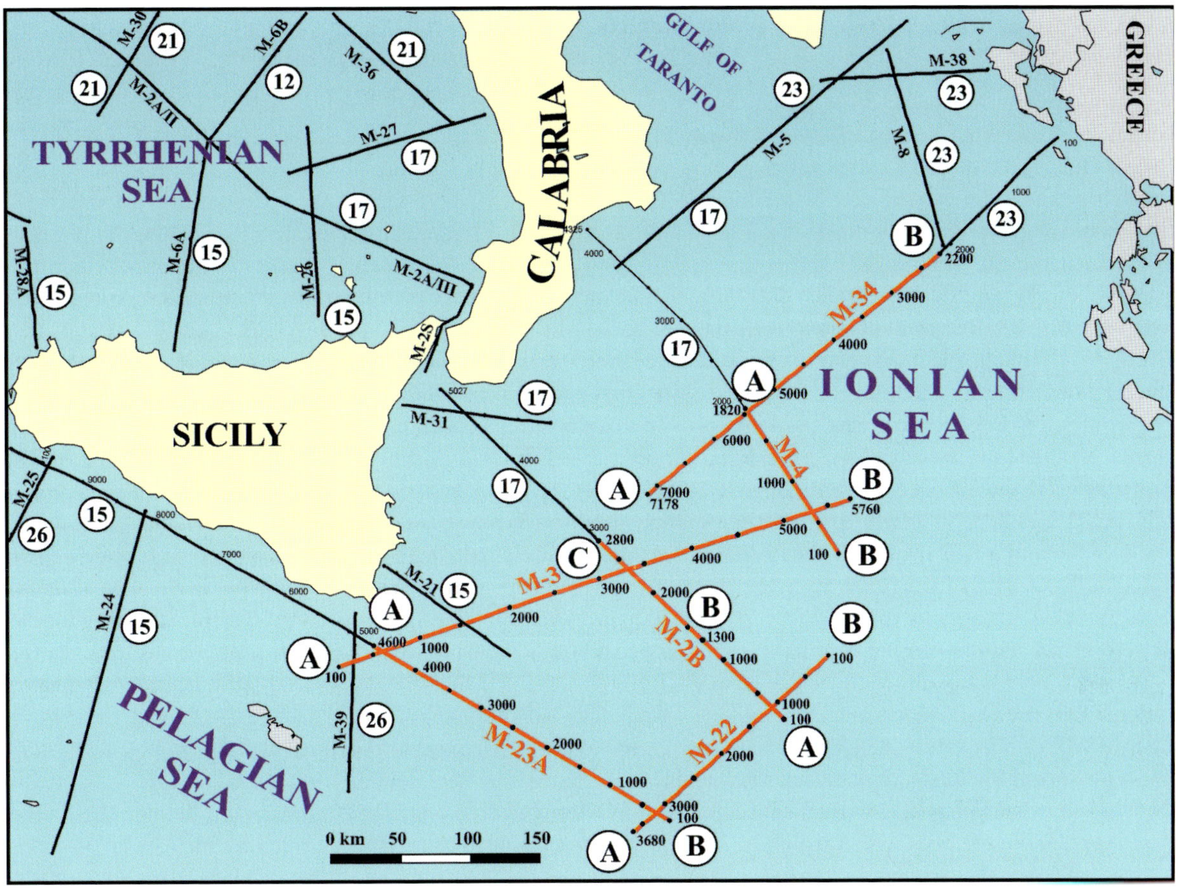

Figure 1: Location map of the interpreted CROP seismic sections exhibited and commented on in this work (red traces) (Chapter 19). Black traced CROP sections are exhibited and commented on in chapters indicated within white circles. Chapter 15 relates to Sicily, Chapter 17 to the Calabrian arc, Chapter 23 to the Adriatic–Apulian platform, and 26 to the Pelagian Sea.

3. SEISMIC INTERPRETATION

The Ionian Sea is a key area for the understanding of the Mediterranean geodynamics. The most important questions posed since the period of the pioneering scientific investigations were that of the nature of its crust and age of the cover succession. A hypothetical reconstruction considered the Ionian Sea as a continental basin which foundered in Neogene times (Scandone, 1979). *On seismic reflection data, over two decades ago, beneath a thick sedimentary succession, a thin, flat basement interval (overlain by a Mesozoic succession) and attributed to an old ocean crust, was imaged and identified in the Ionian Sea* (Finetti, 1982). It was also seismically evident that the Ionian oceanic crust continued northwestward beneath the Calabrian arc. Later on, in a geophysical-geological reconstruction of the paleogeographic setting of the central Mediterranean, a full connection between the Ionian Tethys and the Alpine Tethys which opened in different times, was proposed (Finetti et al., 1996). *New CROP data confirm that the Ionian Sea is floored by a typical thin oceanic crust and show that it opened in Permo-Triassic times, as indicated by its Mesozoic-Cenozoic sequence* (Finetti, this volume, Chapter 10). *On both the Apulian and Pelagian sides, the deep Ionian basin cover is delimited by steep slopes whose continental margins show listric faults generated during the Permo-Triassic opening, some of which reactivated later on* (Figures 5(a), 8(a) and 9(b)). *These Africa and Adria continental-slope-Ionian boundaries have been clearly identified from CROP data for the first time, even where they are buried beneath the allochthonous crustal blocks of the Tyrrhenian margins* (Figure 14). *Under a complex tectonic stack of migrated units, coming from different domains, the Tyrrhenian CROP sections allow us to discriminate in a clear, reliable manner the upper Ionian slab and connected Apulian or African continental paleo-margin* (Finetti and Del Ben, this volume, Chapter 21; Finetti et al., this volume, Chapter 15; Finetti, this volume, Chapter 17).

In the following, a description of the more significant results obtained from interpretation of the CROP seismic sections listed in Table 1 is given.

3.1. Interpreted seismic stratigraphy of the Ionian Sea

The stratigraphic sequence of the Ionian Sea is explored by several CROP sections and MS seismic lines. The best place to conduct a seismic stratigraphic analysis and identification of reflecting intervals is that where the Ionian crust is tectonically undeformed. Such a condition is present beyond the Calabrian and Hellenic wedges. In Figure 2(a) a close-up of CROP section M3 (SP 1600–2000) is exhibited, and in Figure 2(b) its interpretation. Beneath a thin Plio-Quaternary (interval A_0–A) a thick, rather seismically transparent interval (AB) is clearly evident, characterized by diffuse diffraction noise. This is the typical Messinian evaporite sequence which progressively thickened during the outward migration of the Calabrian arc, due to a sort of snowplough-like action over horizon "B". Between the interpreted seismic horizons "B" and "I", and "I" and "TO", a well reflecting, relatively thin sequence is imaged. From wide-scale seismic sequence analysis, specific seismic stratigraphic study, and careful seismic correlation between deep basin units (outcropping and/or penetrated by boreholes) of the southern Apennines (Lagonegrese) and the *Ionian cover sequence*, it has been possible to show that *the interval I–TO corresponds to the condensed sequence of the Lagonegrese thrust units, now called more appropriately "Ionian Tethydes" (IT) or "Ionides"* (Finetti, this volume, Chapter 10). It was also concluded that *this sequence deposited from the Permo-Triassic Ionian opening* (Finetti, this volume, Chapter 6) *to the Oligocene in a quiet basin at a very low sedimentation rate.* In the Upper Permian–Anisian, the initial rift stage of the Ionian opening, clastic deposition (M. Facito) very likely took place on the slopes of the created trough. After the Anisian, in the spreading and post-spreading phases, the condensed deep water sequence deposited on a flat oceanic basement, confining the earlier Upper Permian to Anisian deposition to the slope and immediate proximal basin position. Over the condensed I–TO interval of low sedimentation rate (1 to 1.6 km in about 200 Ma at the distal zone; about 1 mm/200–125 a) the post-Oligocene–Pre-Messinian B–I interval (Aquitanian to Tortonian) of higher sedimentation rate (from 0.4 to 1.25 s TWT, or about from 0.8 to 2.50 km in 13 Ma; about 1 mm/16.2 to 5.2 a) deposited. The Miocene deposition thickens toward the Hellenic and Calabrian wedges. These average sedimentation rates and thicknesses obtained from seismic interpretation of the Ionian interval I–TO are roughly comparable to observed data of the Ionides (Lagonegrese) in the Southern Apennines. A flat undeformed Ionian crust characterized by its oceanic basement interval and overlying condensed sequence (Trias to Oligocene) can be observed in several CROP sections or parts of sections: CROP M3 (Figures 2 and 5 and Plates 1–2); M2B (Figures 3 and 4); M4 (Figure 6); M22 (Figure 7 and Plates 3–4) and M23A1 (Figure 8 and Plates 5–6). Within the deep basinal Ionian cover various igneous intervals (Figures 5(a), 7, 8 and 10) associated with magnetic anomalies (Figure 13) have been identified. Prominent repeated basaltic intercalations were penetrated by boreholes on the Pelagian border zone, south of the Hyblean plateau. On the Hyblean plateau (HP), Hyblean–Maltese escarpment (HME), Medina plateau (MP), Medina Mts. (MM), Archimedes Sm. (A), and other zones of the eastern Pelagian Sea–Hyblean/Maltese escarpment, major magnetic anomalies (Figure 13), associated with magmatic intrusions–extrusions are located.

The *Ionian Oceanic crust* is 1.2–1.6 s (TWT) thick or about 4.2–5.6 km for an assumed conversion velocity of 7.0 km/s. It is interesting to note that there is good correlation between the Bouguer gravity field (Figure 12) and the Moho time-depth. The Ionian oceanic crust and its cover has been identified and interpreted also outside the CROP seismic network using the MS seismic lines (Figures 10 and 11). These lines (Finetti et al., this volume, Chapter 1), having a record length of 10 s, in the Ionian and Tyrrhenian Seas give as complete transcrustal information as the CROP sections.

3.2. The wedge of the Calabrian Arc

Most of the Ionian Sea is covered by wedges of the Calabrian and Hellenic arcs, while flat, undeformed ocean basin is represented only by a minor remnant (Figure 14). The *Calabrian wedge* is confined by a steep Apulian escarpment to the NE (SE-extension of the Apulian platform) (Figure 9(b) and Plates 7–8), and by the Hyblean–Maltese escarpment to the SW, and is also very steep (Figure 5(a) and Plates 1–2). More precisely, the Hyblean–Maltese escarpment delimits a plastic thickening of the outer zone of the wedge, while the southeastward-propagating thrust-deformation of the wedge is substantially delimited by the Vulcano Fault (Figure 5(a)). *This very active fault dissects the Sicilian Apennines from the Calabrian arc, transferring the latter southeastward along a direction subparallel to the Hyblean–Maltese escarpment.* In the Ionian sedimentary thrust-pile, the Vulcano fault is acting as a right-lateral strike-slip, and on the subducting oceanic Ionian basement crust, as a dip-slip of inward-progressive vertical displacement increase. *The Vulcano fault, from the seismogenic viewpoint, is certainly that of highest friction force in Italy and, very likely, must be associated with the seismic risk number one in the country and responsible for the catastrophic earthquake of Messina in 1908.*

CLOSE-UP OF CROP SEISMIC SECTION M3 IN WESTERN IONIAN SEA (SP 1600-2000)

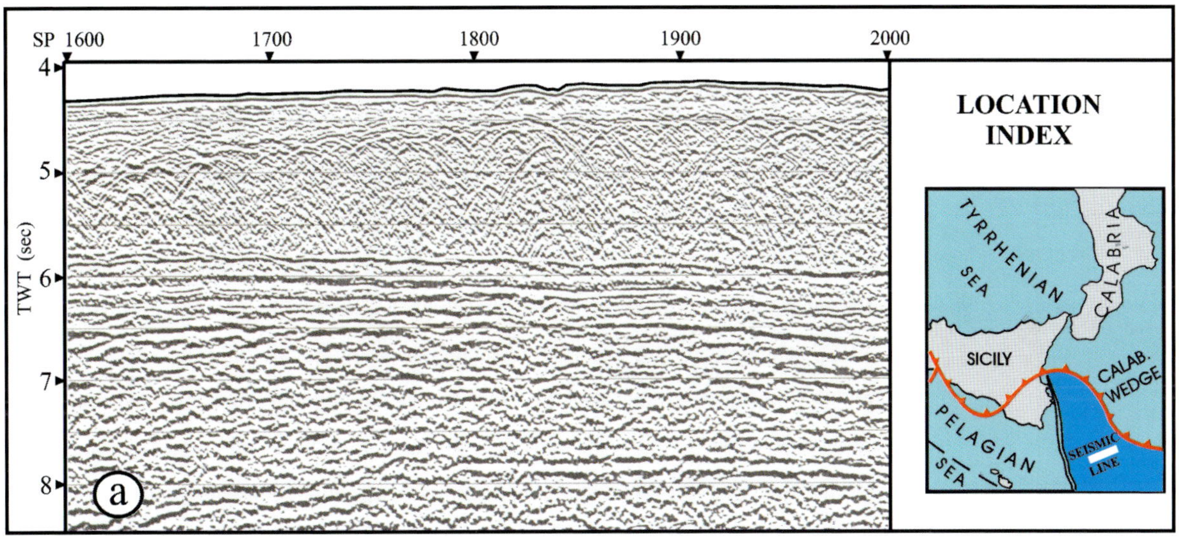

INTERPRETED CRUSTAL SEISMIC STRATIGRAPHY OF THE IONIAN SEA

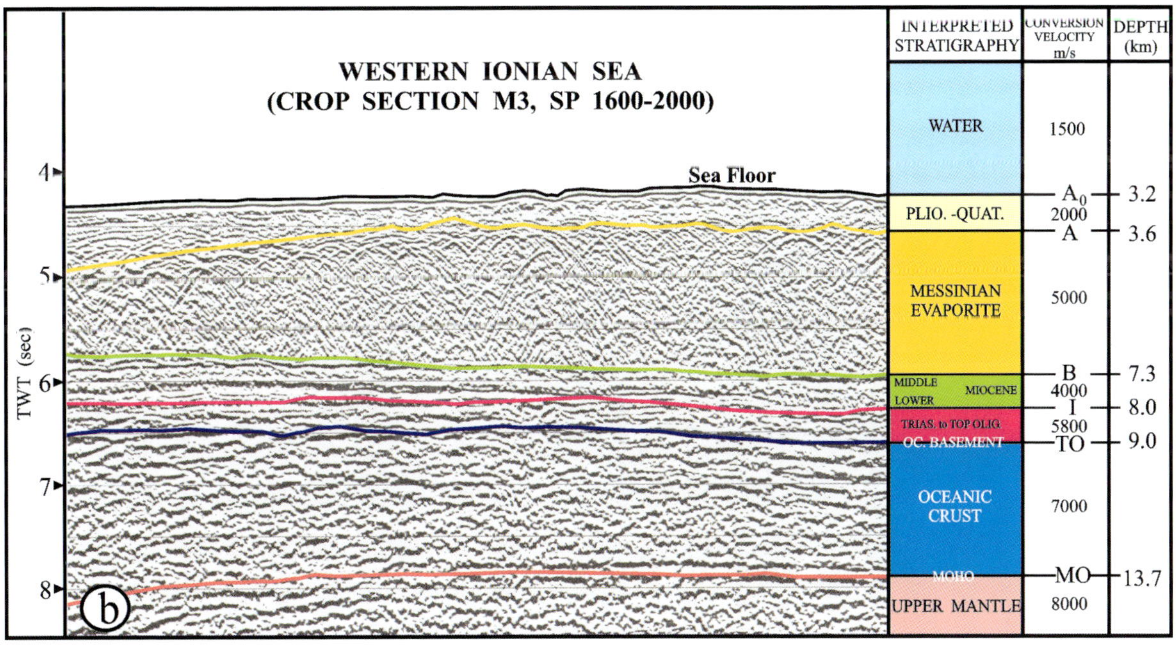

Figure 2: Seismic example (a) and interpreted seismostratigraphy (b) of the oceanic crust of the Ionian Sea and its cover. This seismic close-up is taken in the Western Ionian Sea, where the crust is very thin (see Plate 2). The oceanic basement (TO) is covered by a continuous deep-water condensed sedimentation from the Trias to Oligocene (interval I–TO), overlain by Lower-Middle Miocene (B–I), thick Messinian salt (A–B) and thin Plio-Quaternary (A_0–A). The condensed sequence I–TO (*Ionian Tethydes* or *Ionides*) was stripped off its subducting Ionian basement crust and tectonically transported, by oblique thrusting, onto the Southern Apennines (*Lagonegro units*) and Sicilian Apennines (*Imerese–Sicanian units*). A major part of the Ionides is stacked in the wedges of the Calabrian and Hellenic arcs.

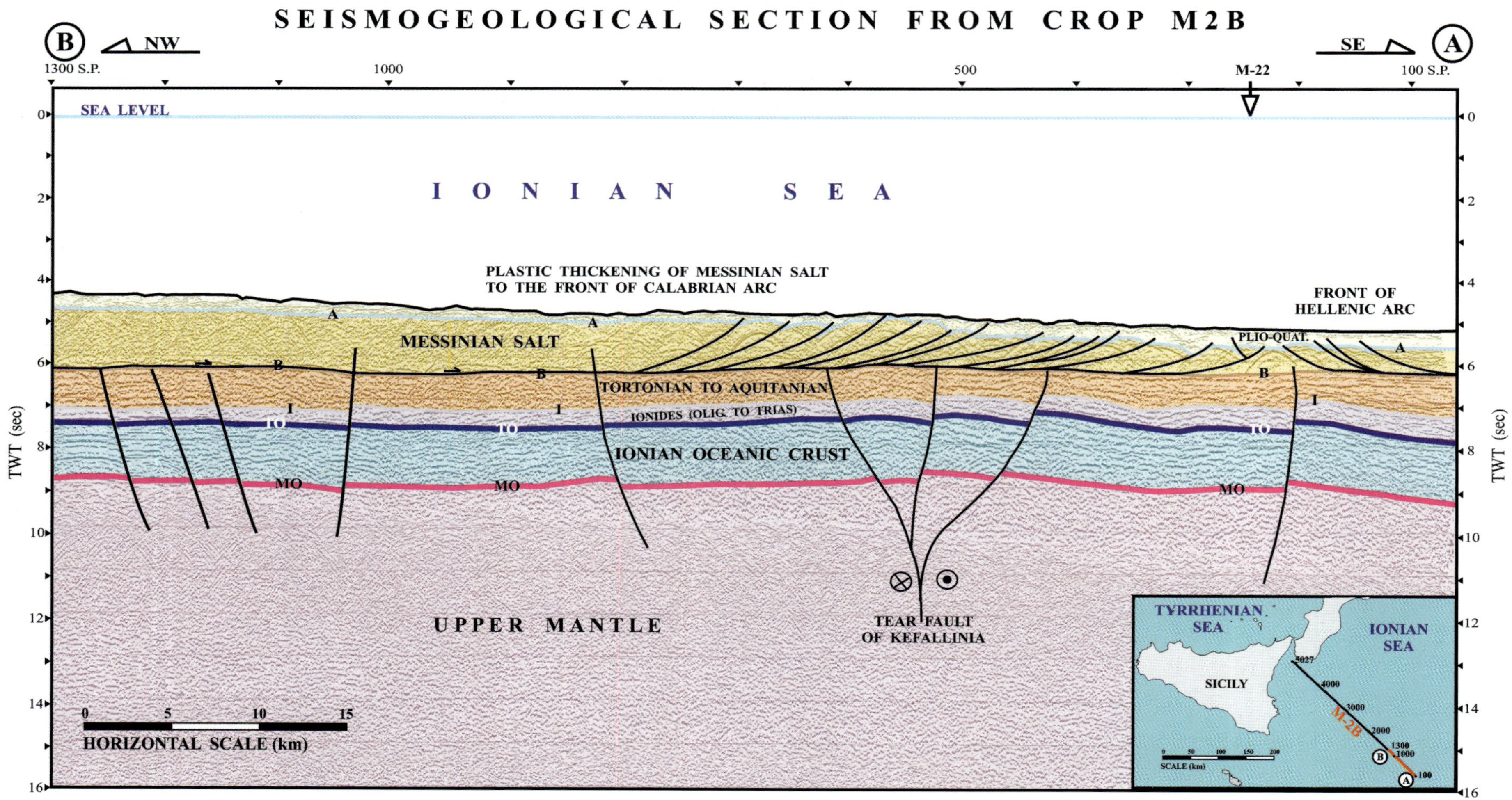

Figure 3: Lithospheric seismogeological section in the Ionian Sea, in front of the Calabrian arc. Plastic thickening of the Messinian salt is due to the thrust propagation of the Calabrian arc. A right-transcurrent fault, a tear element of the lateral Kefallinia transcurrent, is seismically imaged. A = base Plio-Quaternary; B = base Messinian evaporite; I = top Ionian condensed sequence (Ionides); TO = top oceanic basement crust; MO = Moho.

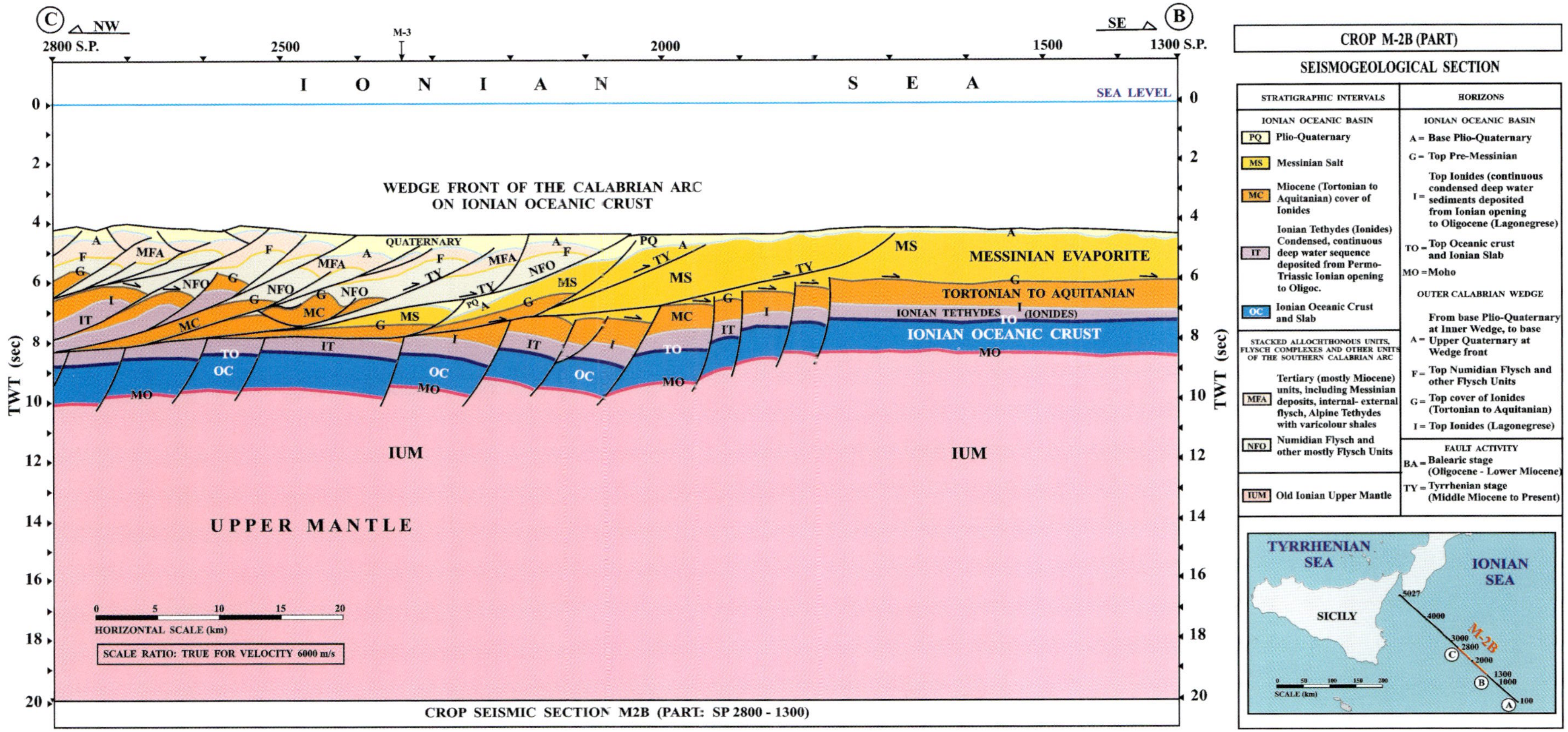

Figure 4: Seismogeological section across the outermost wedge of the Calabrian arc over the oceanic Ionian crust (CROP M2B, part). A strong crustal reaction to the wedge leading is clearly evident.

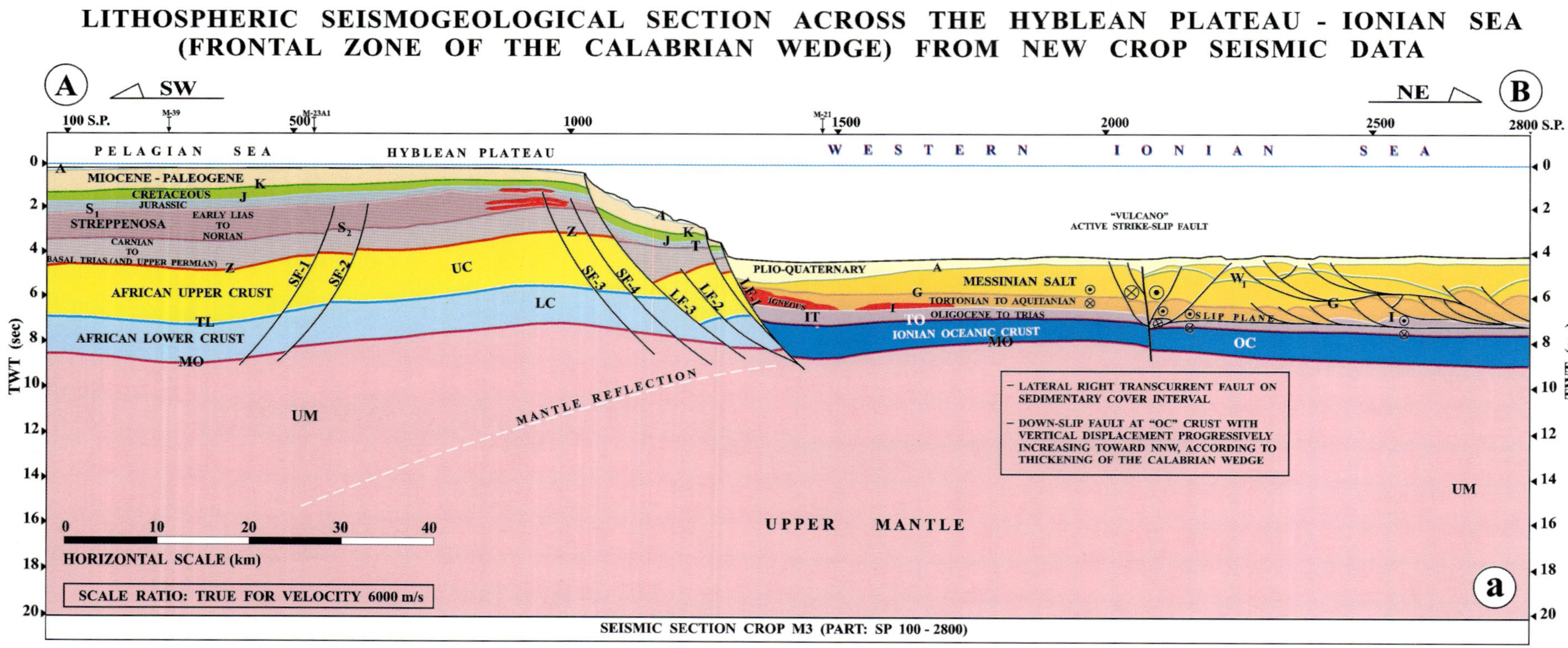

Figure 5: Lithospheric seismogeological setting across the eastern Hyblean plateau–Ionian oceanic crust (CROP M3, part). Permo-Triassic rift faults of Ionian opening, some of which reactivated successively, are evident. Typical Ionian oceanic crust and its cover is also evident and self-explanatory. The *Vulcano strike-slip* fault is shown around SP 2060.

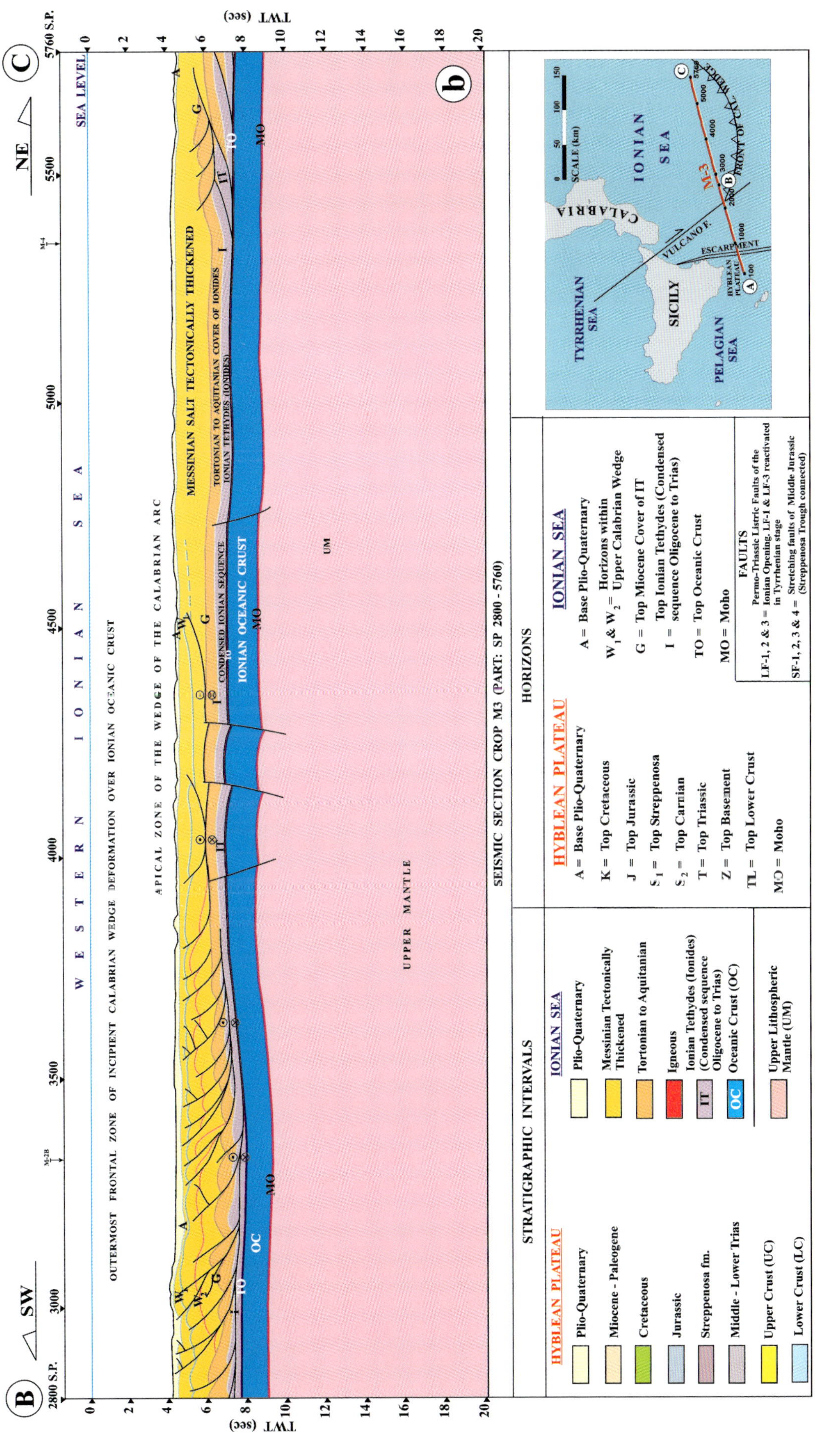

Figure 5: *Continued.*

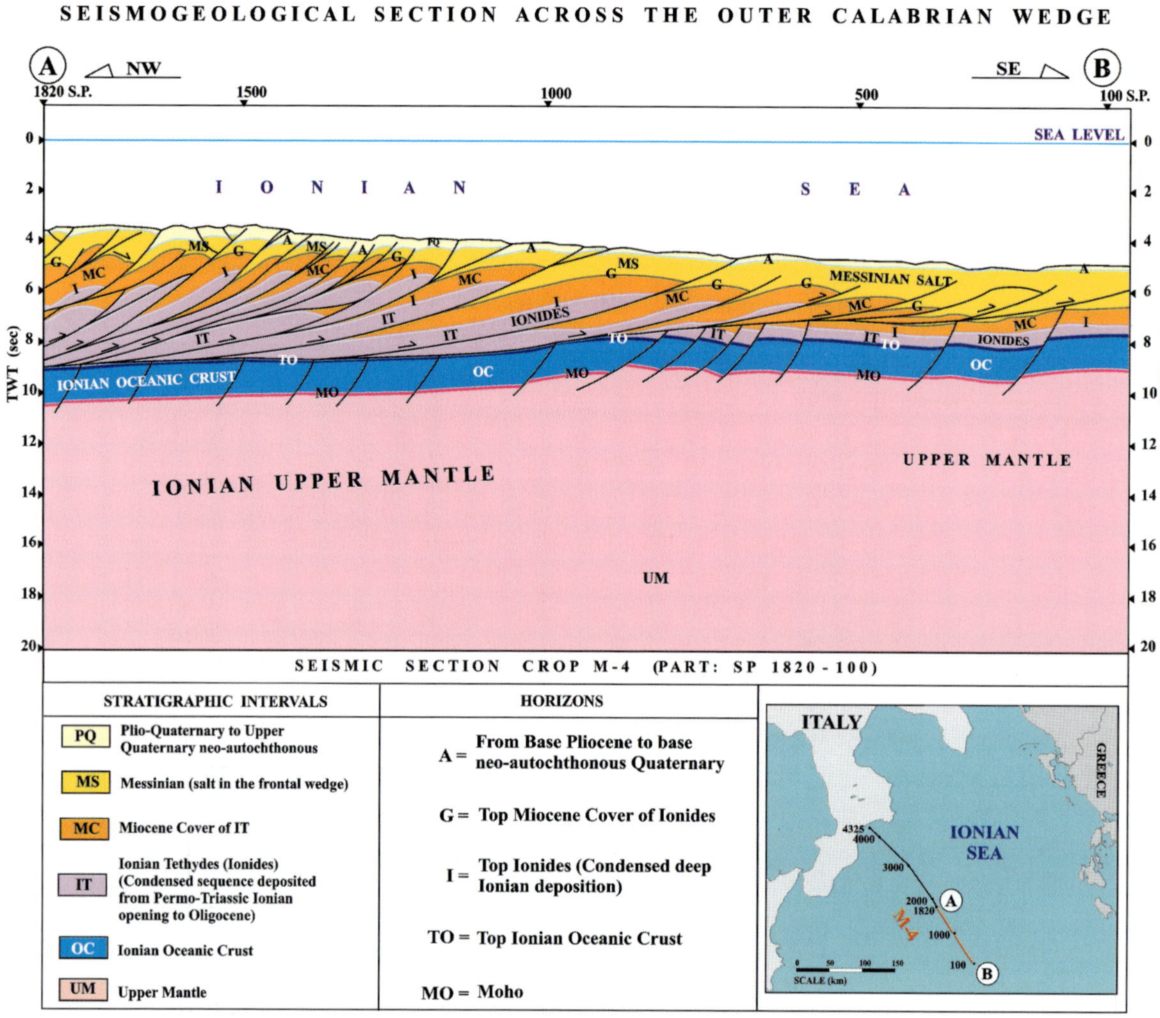

Figure 6: Lithospheric seismogeological section showing the outer wedge of the Calabrian arc over the oceanic Ionian crust. Thickening of the Messinian evaporite is produced by a snowplough-like action of the outward arc propagation.

In Figure 3, a flat Ionian sequence overlain by a thickened Messinian evaporite is shown, which represents the outermost shallow effect of the wedge propagation of the Calabrian arc, similar to that of the Hellenic arc (Figures 10 and 11). Continuing inward, CROP section M2B (Figure 4) exhibits the wedge front with its thrust-pile over the condensed deep depositional Ionian sequence (Ionides). *It is interesting to observe that a relatively modest tectonic thickening produces a remarkable load-faulting effect on the Ionian crust. This means that the old Ionian basement crust (Permo-Triassic) has a density close to the buoyancy limit over its mantle.* On CROP section M4 (SP 100–1820), (Figure 6) a repeated thrust imbrication of the Ionides and their pre-Messinian Miocene cover is clearly visible. These thrust units are stripped off from the subducting oceanic basement crust. *It is evident from CROP data that no significant amount of the Ionides subducted* (Finetti and Del Ben, this volume, Chapter 21). The Calabrian wedge is seen also on CROP sections subperpendicular to the thrust propagation (Figures 5 and 9). CROP section M3 explores the Calabrian wedge on its outermost zone (Figure 5 and Plates 1–2) and shows the apical deformation of the arc (Figure 5(b)) where the thrusting tectonics is close to terminating. CROP section M34 crosses the apical zone of the arc more internally, where the arcuate thrust stack is prominently thicker (Figure 9(a) and Plates 7–8). On the innermost wedge, the pile of the Ionides and cover arrives at 5–6 imbricated thrust units, and each single thrust component is generally well discriminated seismically. Repeated thrust units of the Ionides have also been found in boreholes of the Southern Apennines.

The interpreted Trias to Oligocene condensed sequence of the Ionian Tethydes or Ionides (IT) indicates a rather uniform thickness of about 0.8–1.2 km in the distal zones (Figures 2 to 11). On some zones proximal to the steep escarpment, it is possible to identify thicker intervals of Ionides. The relatively thin interval can be in part explained by the observed crustal setting of the two continental border zones of the Ionian basin (Hyblean–Maltese

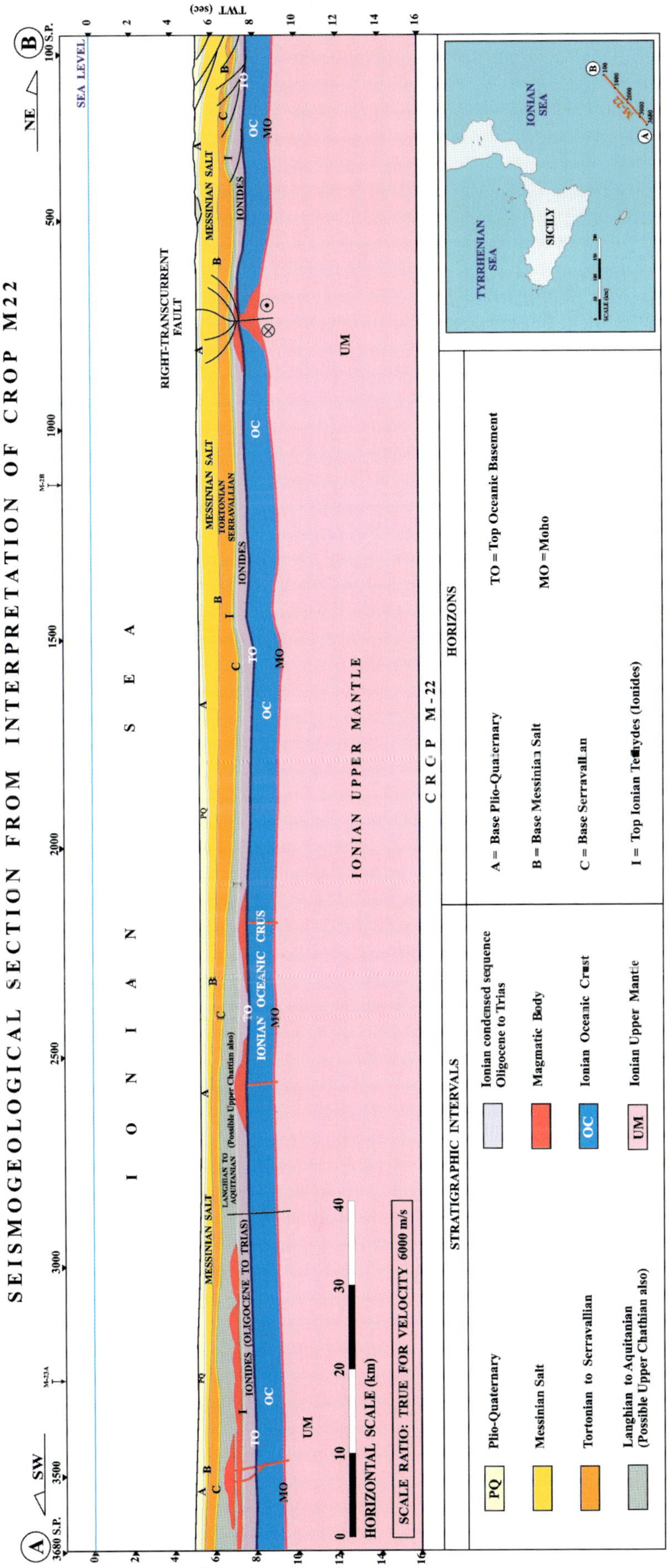

Figure 7: Lithospheric seismogeological section on the undeformed southwestern Ionian Sea obtained from interpretation of CROP section M22 (Plates 3 and 4). It exhibits the flat Ionian oceanic crust and overlying cover.

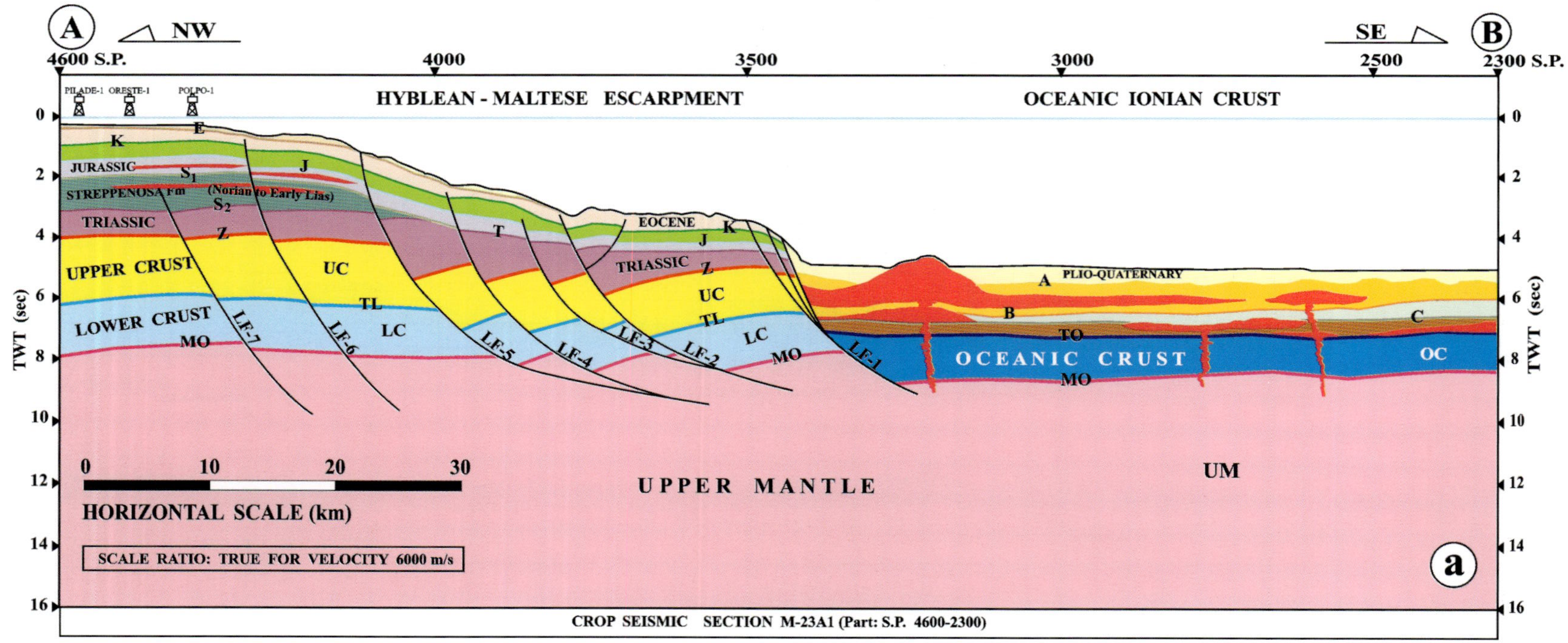

Figure 8: Interpreted lithospheric seismogeological CROP section (M23A1, part) crossing the eastern Hyblean plateau (a) and the western Ionian Sea floored by a very thin oceanic crust (a), (b). There is prominent igneous activity.

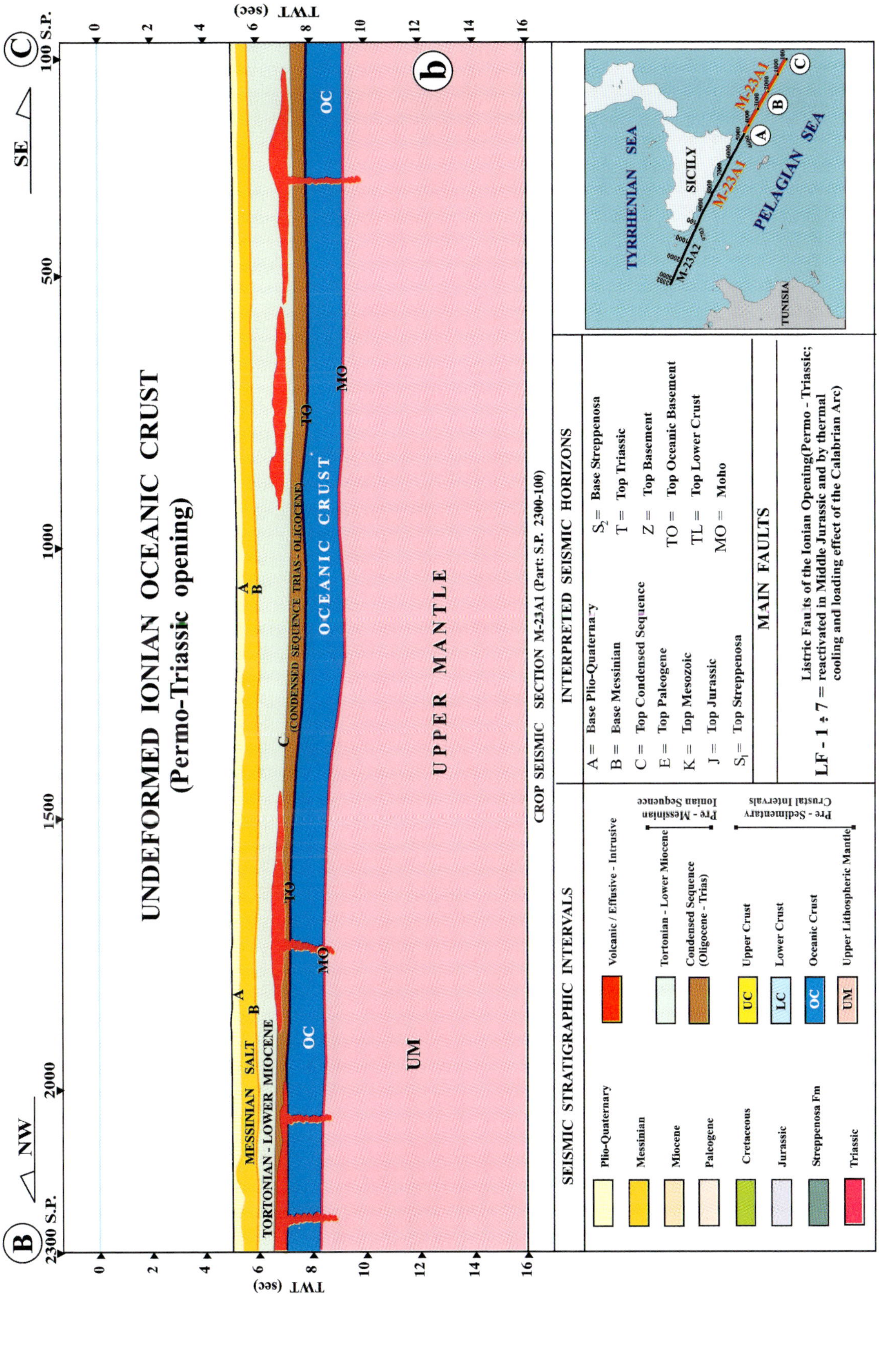

Figure 8: *Continued.*

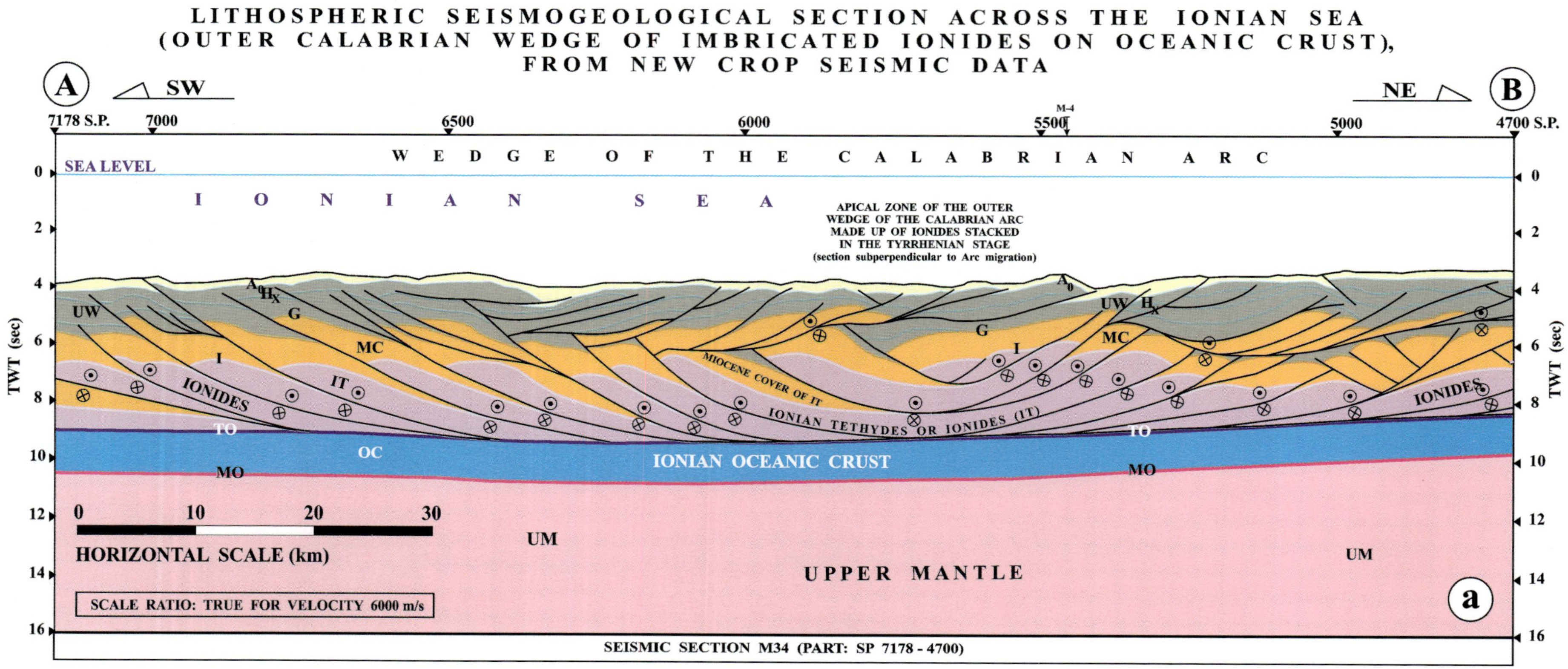

Figure 9: Interpreted seismogeological section (M34, part) of the Calabrian wedge over the oceanic Ionian crust (a) and of the boundary area between the Apulian platform and Calabrian wedge (b). The section visualizes the crustal setting perpendicular to the wedge propagation formed by imbricated Ionides and cover. Permo-Triassic rift-faults of Ionian opening are clearly exhibited (b).

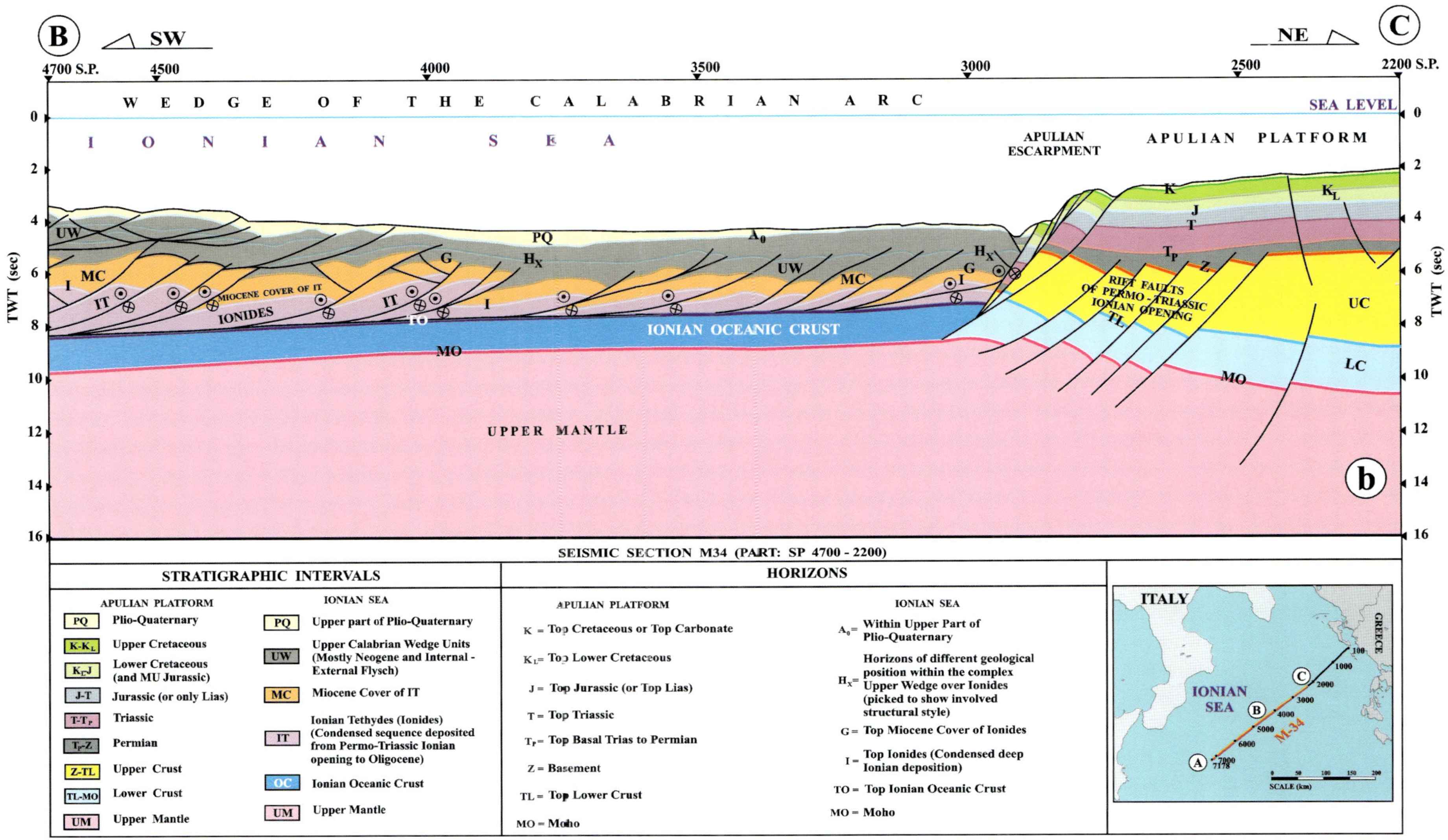

Figure 9: *Continued.*

and Apulian shelf areas). It is here evident that a marked back-tilting occurred during the Ionian opening, which has persisted up to now (Figures 5(a), 8(a) and 9(b) and Plates 2, 6 and 8): such continental-border highs may have prevented the basinward sedimentary flow.

Only in the Southern Apennines and Sicily do we find thicker intervals of basal deposition (Late Permian to Anisian) evidently connected to the earlier rift activity with deposition confined to the slope and immediate proximal deep trough zone.

3.3. The Hellenic arc wedge in the Ionian Sea

Based on MS seismic lines, a tectonic outline of the Hellenic wedge into the Ionian Sea has been reconstructed (Figures 10, 11 and 14). It is evident that also in the outer Hellenic wedge is the tectono-stratigraphic setting dominated by a Messinian evaporite interval thickened by the outward propagation of the arc over the horizon "B" (top Tortonian) which remains flat and undeformed (Figures 10 and 11). Moving internally, the thrust pile of the wedge includes progressively older cover units up to Ionides (interval I–TO) and Miocene cover (interval B–I).

The Mediterranean ridge, the old name of the Hellenic wedge, in the past, was interpreted by various authors in different ways (Rabinowitz and Ryan, 1970; Mulder, 1973). The MS seismic line network helped show that this prominent feature represents the outer sector of the Hellenic-arc-related tectonodynamics of Neogene-Quaternary age (Finetti, 1976). From CROP and MS section information, we now see that the Hellenic wedge is formed by a dominating imbrication of Ionian–East Mediterranean cover detached from the subducted oceanic slab.

To the north, the Hellenic wedge (Figure 14: HW) is delimited by the Kefallinia right-lateral transcurrent fault. This important regional fault, seismically identified and defined many years ago (Finetti, 1982) dissects the Hellenides and accommodates slip tectonics related to the Anatolian fault dynamics and Aegean back-arc basin opening.

4. GEODYNAMIC EVOLUTION

Using the CROP seismic dataset integrated with other geophysical-geological information, it is possible to reconstruct a much better constrained geodynamic evolution of the Ionian Sea than in the past. This is mainly due to the fact that the CROP sections exhibit for the first time very clear imaging of the upper Ionian slab beneath buried allochthonous stacked crustal units around the Tyrrhenian margins of the Southern Apennines–Calabrian arc–Sicilian Apennines. *This Ionian slab seismic imaging allows us to reconstruct the paleogeography of the Ionian Sea, already known on the Ionian ocean remnant* (Finetti, 1982), *also where the Tethyan basin has disappeared due to subduction.*

The CROP seismic sections in the Tyrrhenian Sea show that the steep NW–SE-directed Apulian slope and adjacent Ionian oceanic crust, clearly visible in the present Ionian area (Figures 9(b) and 14), continue with the same trend northwestward, across the Calabrian arc, on the Campanian-Latial Tyrrhenian margin, up to the 41° parallel (Figure 15(a) segment B′C′). Along the present day 41° parallel, the northern boundary of the subducted Ionian Sea passed (segment A′B′ of Figure 15(a)). North of the Sicilian platform, buried beneath the Tyrrhenian margin have been imaged the Ionian boundary segments AB and BC (Figure 15(a)). From this very important new data it is evident that the Ionian Sea opened as a consequence of detachment and anticlockwise rotation of the Adria plate. Rotating Adria back (clockwise) by about 40° it is possible to reconnect it to Africa. Regarding the time of opening of the Ionian Sea, the age of the initial deposition into the earlier Ionian trough is indicative. In the Ionides (Lagonegrese) of the Southern Apennines, the first recognized deposition (M. Facito Fm.) is dated Upper Permian

---

Figure 10: Example of interpreted MS line (MS 33, SP 1722–1425) across the Ionian basin-front of the Hellenic wedge. A = base Plio-Quaternary; B = base Messinian; I = top Ionian Tethydes (Ionides): a condensed deep basin sequence sedimented over the oceanic crust from the Trias to Oligocene; TO = top oceanic basement; MO = Moho. The pre-Messinian Miocene sequence is characterized by a marked increase in sedimentation rate due to the orogenic activity of the Hellenic arc. BT = backthrust due to lateral extrusion of the Messinian induced by collision of the Hellenic wedge with the Cyrenaican slope.

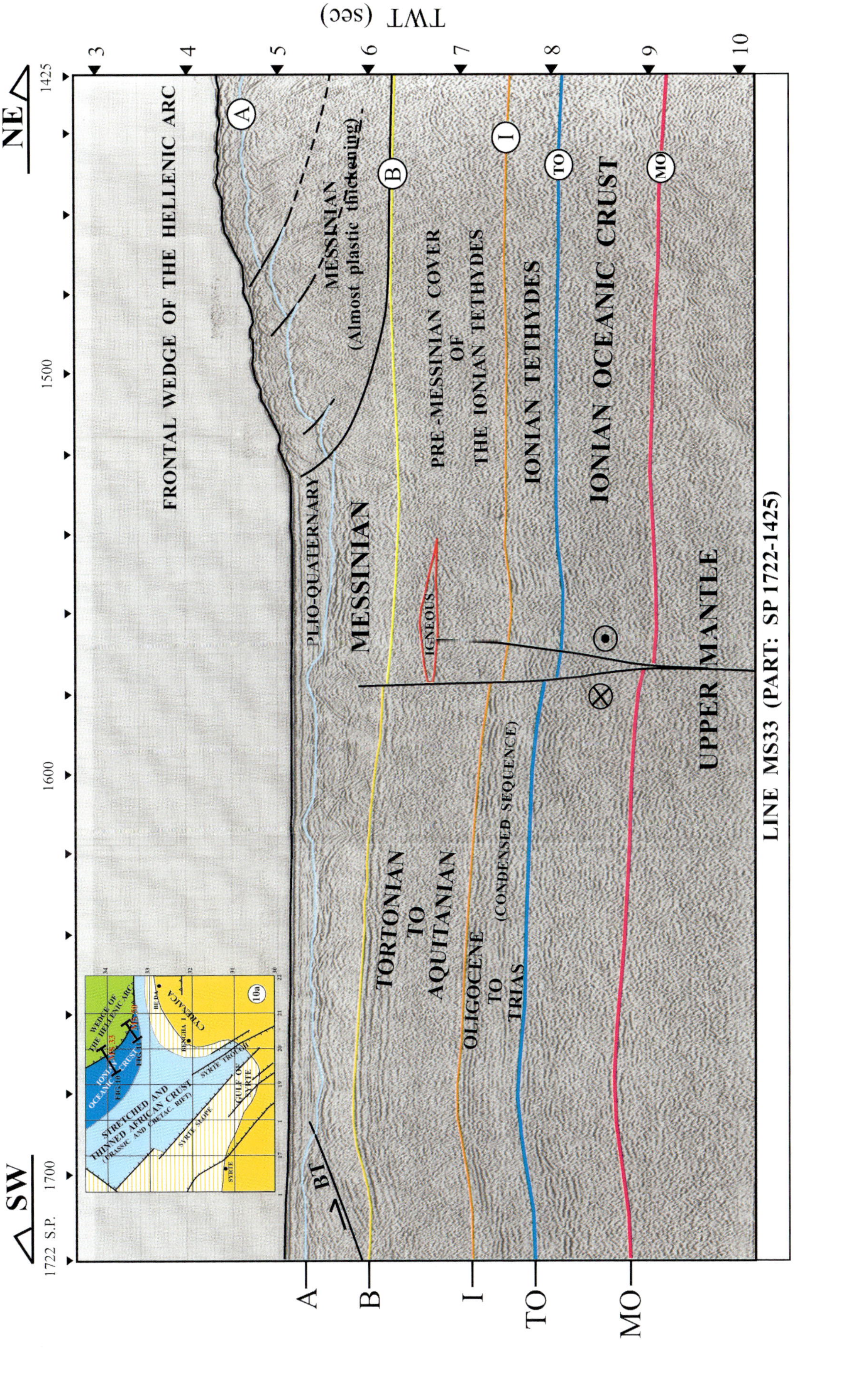

NE
SW
1425
1500
1600
1700
1722 S.P.
TWT (sec)
3
4
5
6
7
8
9
10
FRONTAL WEDGE OF THE HELLENIC ARC
MESSINIAN
(Almost plastic thickening)
PLIO-QUATERNARY
MESSINIAN
PRE-MESSINIAN COVER
OF
THE IONIAN TETHYDES
IONIAN TETHYDES
IONIAN OCEANIC CRUST
IGNEOUS
UPPER MANTLE
TORTONIAN
TO
AQUITANIAN
OLIGOCENE
TO
TRIAS
(CONDENSED SEQUENCE)
BT
A
B
I
TO
MO
LINE MS33 (PART: SP 1722-1425)
10a

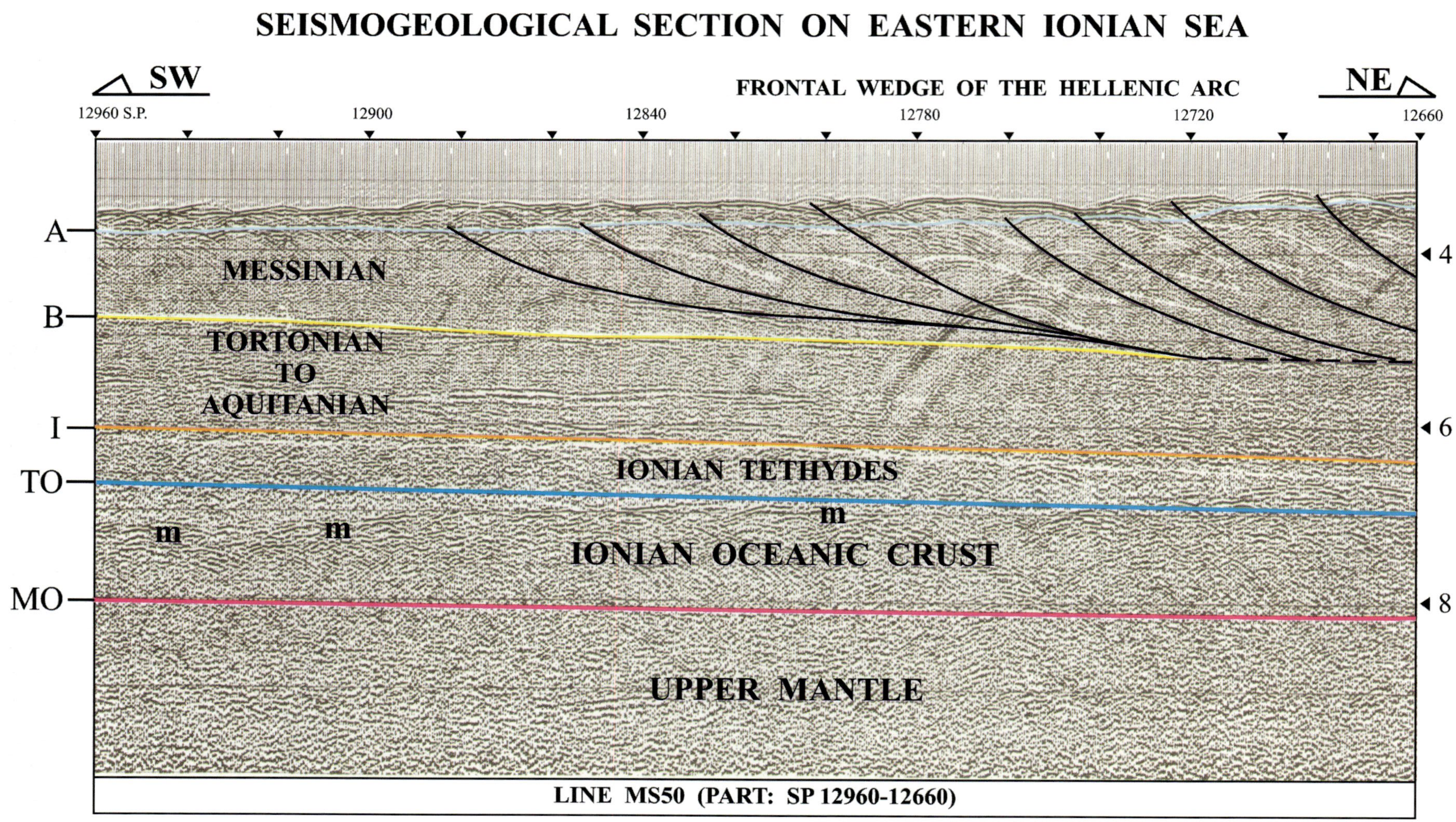

Figure 11: Interpreted seismic reflection line (MS 50 part: SP 12 960–12 660) in the Ionian Sea, floored by oceanic crust, showing that the outermost deformation of the Hellenic arc is affecting the Messinian interval and covering Plio-Quaternary. A = base Plio-Quaternary; B = base Messinian; I = top Ionian Tethydes: a condensed Oligocene to Trias sequence, covered by the BI interval of higher deposition rate (Tortonian to Aquitanian); TO = top oceanic crust, (m = multiple); M = Moho. For location, see previous Figure 10(a). The Ionian Tethydes sequence constitutes a large part of the imbricated tectonic units of the Hellenic wedge.

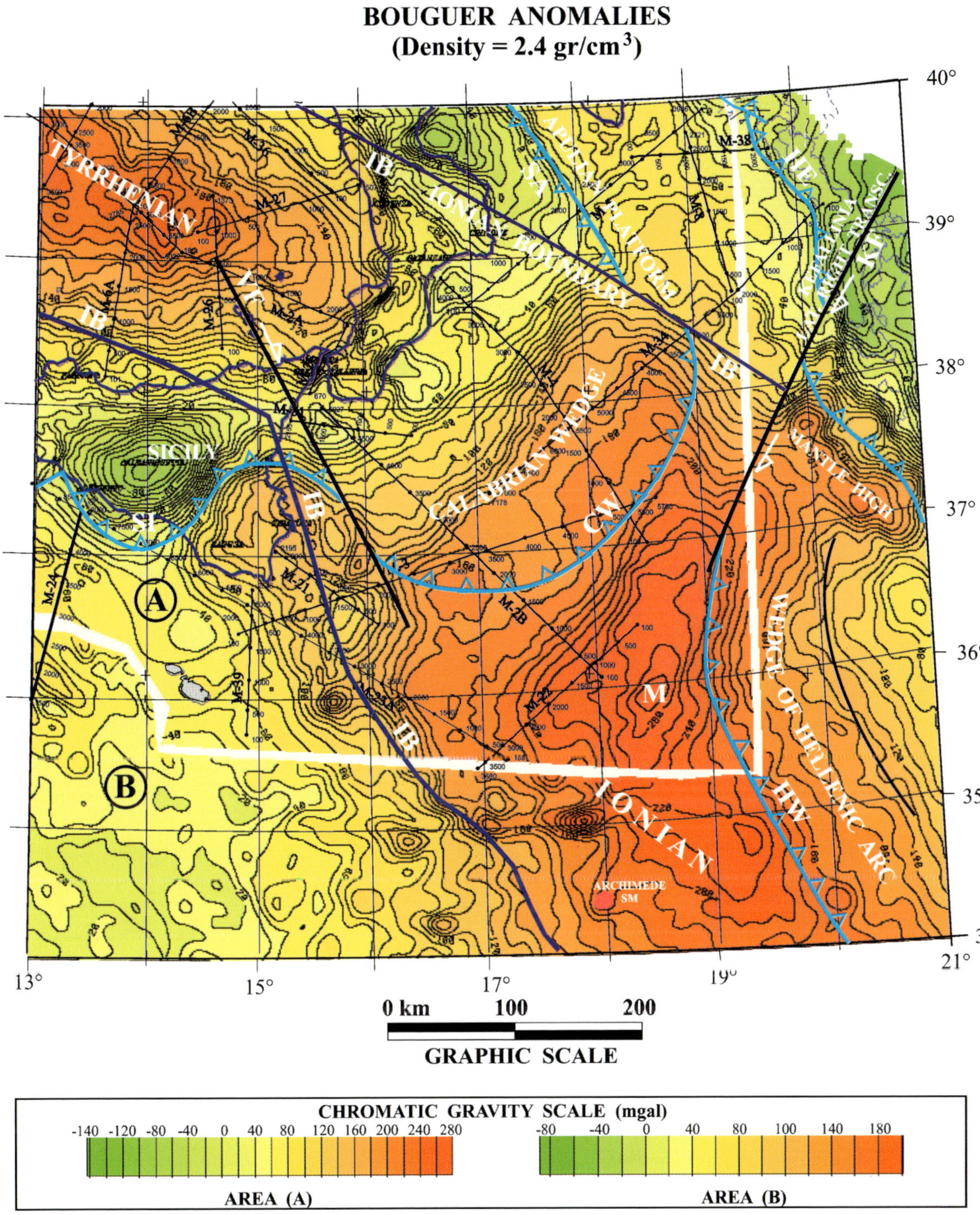

Figure 12: Crustal expression of the regional Bouguer gravity macro-anomalies in the Ionian Sea and surrounding area, defined by the CROP seismic dataset. (Gravity data from Bernardelli et al., this volume, Chapter 3). IB = Ionian boundary (buried beneath the Southern Apennines and Sicilian Apennines Tyrrhenian margins); CW = front of the Calabrian wedge; HW = front of the Hellenic wedge; SA = front of the Southern Apennines; SI = front of the Sicilian Apennines; HE = front of the Hellenides; KF = Kefallinia right-lateral transcurrent fault; VF = Vulcano right-lateral transcurrent fault. M = Marconi gravity anomaly in the Ionian Sea (over 250 mgal).

(Ciarapica and Passeri, this volume, Chapter 11). Permian depositions of the Ionian basin have been recognized also in Sicily (Finetti et al., this volume, Chapter 15). More to the southeast, Permian (and older) deposits of the Eastern Mediterranean–Neotethyan ocean basin are assumed to be early deposition in the newly opened basin (Stampfli, 2000). All these data lead to the conclusion that the *Ionian–Eastern Mediterranean opening initiated*

RESIDUAL MAGNETIC FIELD (REDUCED TO THE POLE)

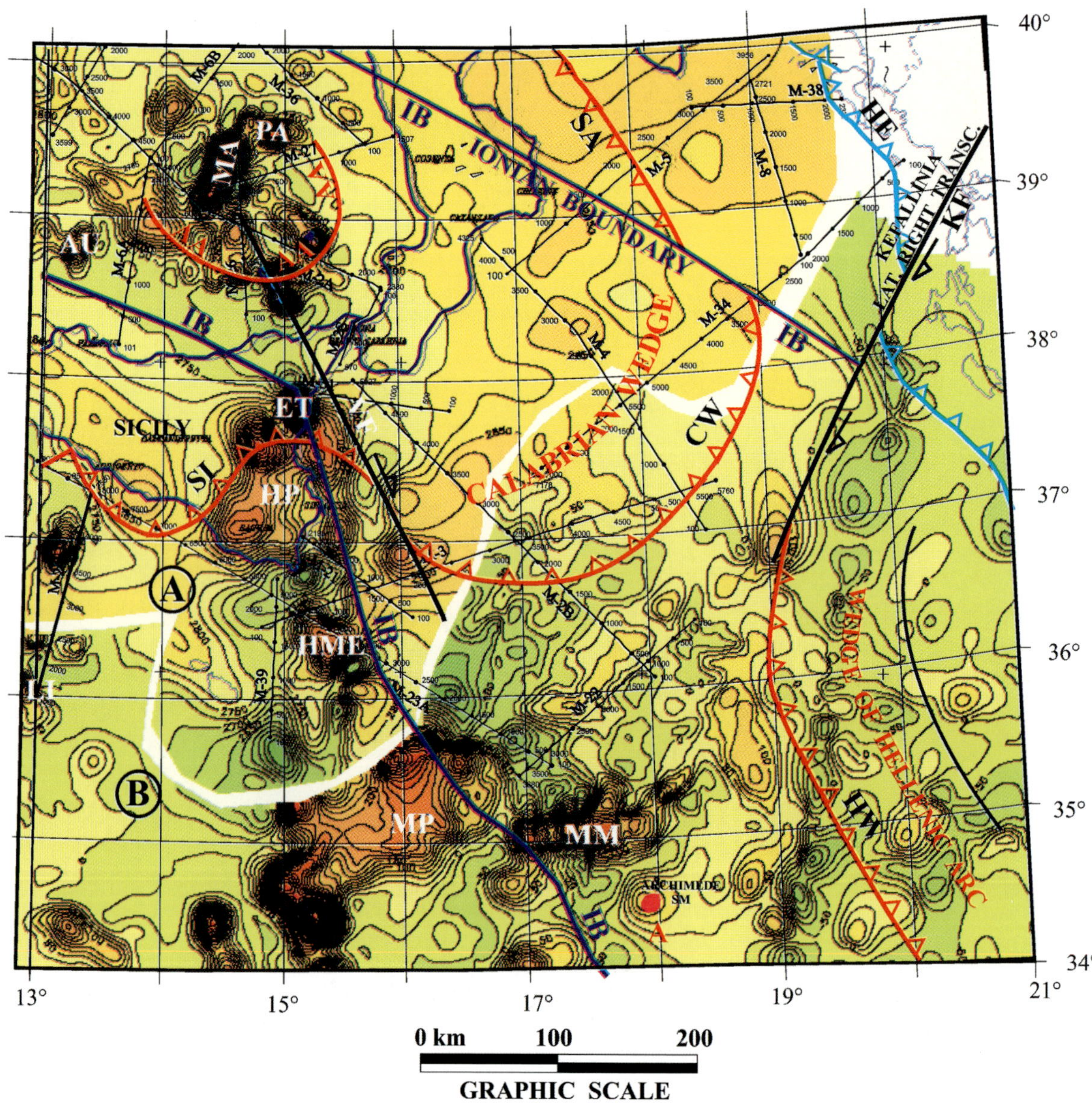

Figure 13: Residual magnetic anomalies (reduced to the pole) in the Ionian Sea and surrounding areas (after Bernardelli et al., this volume, Chapter 3). Reported crustal boundaries and tectonic features are the same as in Figure 12. Some major anomalies: (a) *Tyrrhenian* (AA = Aeolian volcanic arc; MA = Marsili; AU = Anchise–Ustica; PA = Palinuro); (b) *Etna–Hyblean plateau–Pelagian Sea/western Ionian* (ET = Etna; HP = Hyblean plateau; HME = Hyblean–Maltese escarpment; MM = Medina Mounts; MP = Medina plateau; A = Archimede Mt.; LI = Linosa). Crustal Boundaries: IB = Ionian ocean boundary (buried beneath the Calabrian arc and Tyrrhenian margins). Tectonic Features: CW = front of Calabrian wedge; SA = front of Southern Apennines; SI = front of Sicilian Apennines; HW = front of Hellenic wedge; HE = front of Hellenides; VF = Vulcano fault; KF = Kefallina fault.

in the (*Upper*) *Permian*. The fact that most of the Ionides (Lagonegrese) of the Southern Apennines initiate with Anisian deposits and continue later with deep sediments (cherty limestone) can be explained by an earlier Upper Permian limited to the initial Ionian rift basin (located on the slopes after spreading). The deep persistent deposition followed the initial stage with broad sedimentation starting from the cherty limestone. Moreover, it is possible that the basal clastic layer has been dispersed by tectonic processes. CROP sections across the Tuscan Archipelago–inner Northern Apennines belt imaged an E-dipping Alpine Tethys basement crust recognized as a slab of Eo-Alpine subduction (Finetti et al., 2001). Other CROP sections crossing the Tuscan Archipelago–Ligurian Sea and

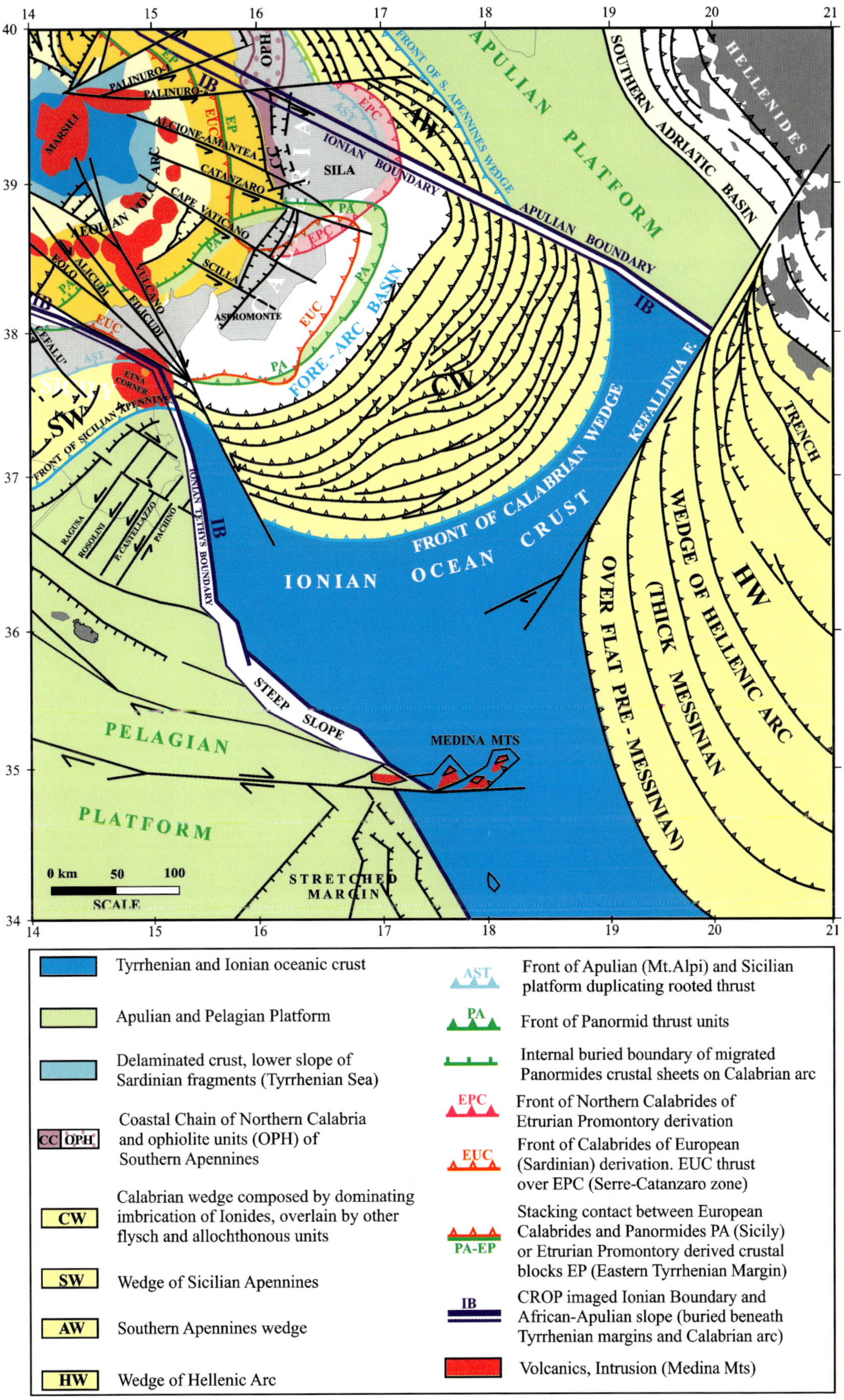

Figure 14: Tectonic sketchmap of the Ionian Sea and surrounding provinces, and crustal boundaries of the subducted and remnant Ionian Tethys.

EVOLUTIVE PALEOGEOGRAPHY OF THE CENTRAL MEDITERRANEAN

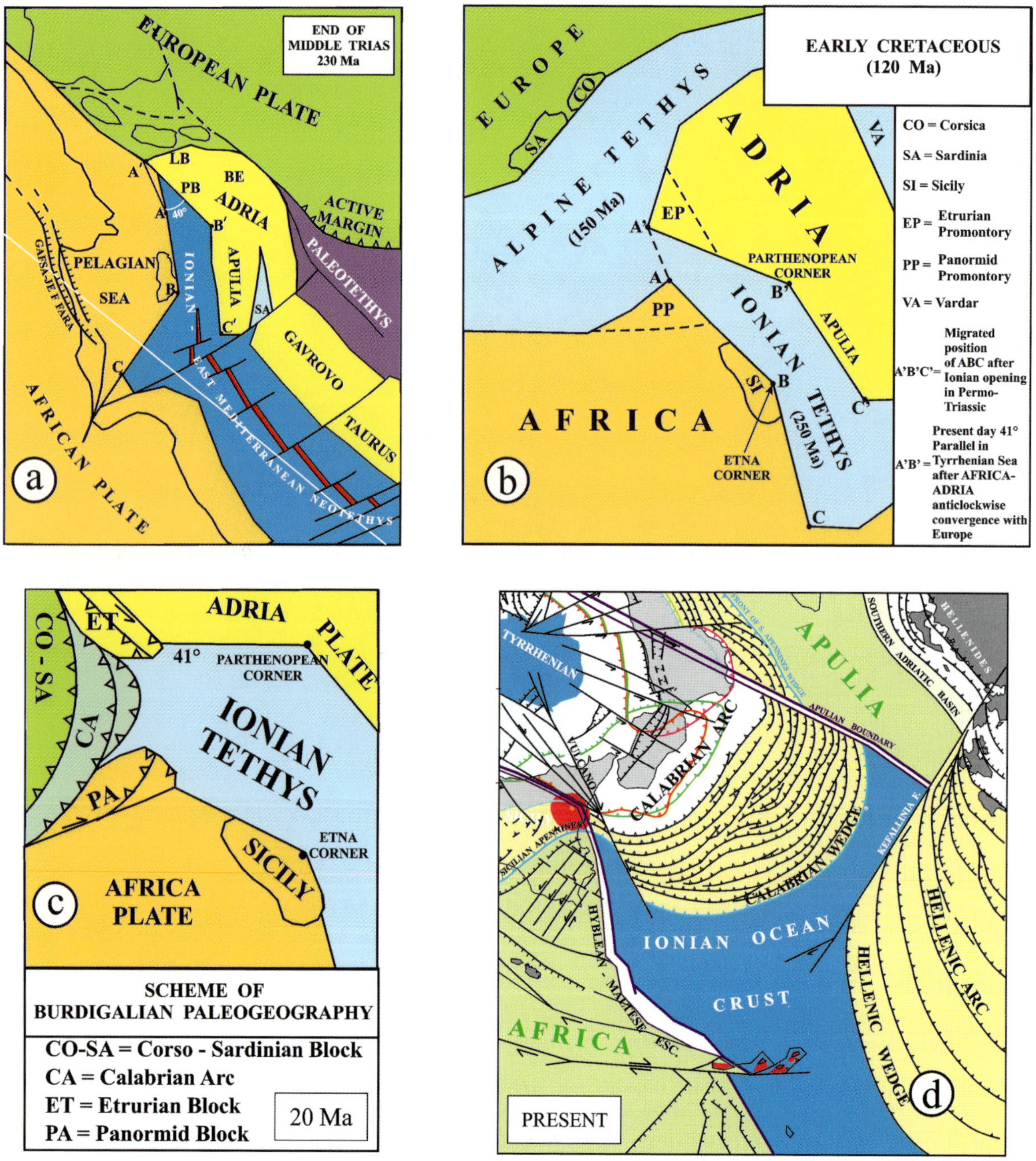

Figure 15: CROP-constrained paleogeographic schemes for the central Mediterranean region at different the stages, from the Permo-Triassic Ionian opening (a) to the Present (d).

Corsica basin showed a prominent bulge of the Alpine Tethys mantle and asthenosphere and a W-dipping slab subducted beneath the Corso-Sardinian block in the Balearic stage (Finetti et al, this volume, Chapter 8; Finetti et al., this volume, Chapter 18). These CROP data on the subducted Alpine Tethys crust together with those shown on the same Tethyan ocean by other transcrustal seismic sections of the Alps (CROP-ECORS, CROP-NPF-20 and TRANSALP), constrain the paleogeographic reconstruction of the Alpine Tethys as outlined in Figure 15(b). The CROP network data clearly show that the Ionian Tethys of Permo-Triassic opening, and the Alpine Tethys of Middle Jurassic–Early Cretaceous opening, were fully communicating. Two promontories delimited the passage from one basin to the other: the *Etrurian promontory* (Adria plate corner), and the *Panormid promontory* (Africa plate corner). The continental plate–ocean basin context, as well as these two promontories, played a prominent role in the geodynamic evolution from the Eo-Alpine stage to the Present.

In the *Eo-Alpine stage* (Upper Cretaceous to Early Eocene), Africa–Adria and the interposed Ionian Tethys converged with Europe causing an Adria–Europe collision and forming the Eo-Alps, leaving unsubducted a restricted, but still wide Alpine Tethys remnant west of the Alps (Finetti et al, this volume, Chapter 8; Finetti, this volume, Chapter 34). The Eo-Alpine subduction of the Alpine Tethys crust was S to SE-dipping (beneath Adria) and, west of Adria, took place as a SE-dipping intra-ocean subduction. In the Balearic stage (Oligocene–Lower Miocene), and probably starting from an intra-ocean back-thrust of the Eo-Alpine stage close to the European margin, a W-dipping subduction of the western Alpine Tethys remnant and its closure occurred, accompanied by the detachment and rotation of the Corso-Sardinian block. This rotational movement stopped almost completely after the end of the Aquitanian when Corsica collided with the Etrurian promontory and southeastern Sardinia collided with the Panormid promontory (Figure 15(c)).

The W-dipping subduction, after the total underthrusting of the *Alpine Tethys*, continued without stop into the *Ionian Tethys*, giving rise to the *Tyrrhenian stage*. Over the Ionian Tethys, the back-arc Tyrrhenian opening was generated, driven by a fast southeastward slab retreat of the Ionian oceanic lithosphere. The Ionian cover was stripped off the subducting basement and went to form a prominent part of the Calabrian wedge, the *Ionides* (Lagonegrese) allochthonous units on the Southern Apennines, and the Ionides (Imerese and Sicanian) allochthonous units of the Sicilian Apennines. Crustal and cover units of the Etrurian and Panormid promontories were transported onto the Tyrrhenian margins and belts of the Southern Apennines, Calabrian Arc and Sicilian Apennines, followed in turn by Calabrides, forming, jointly with flysch units, a complex orogenic system (Figure 15(d)). *The Balearic and Tyrrhenian geodynamic stages were almost only driven by a fast roll-back sinking (up to 10 cm/a) of the Alpine Tethys and Ionian lithosphere, accompanied by a slower Africa–Adria/Europe convergence (about 0.4 cm/a) of incomparable tectonic significance in the Apennines–Calabrian Arc–Sicilian Apennines–Maghrebian orogeny.*

ACKNOWLEDGEMENTS

This work was supported by MIUR, COFIN 1998, 2000 and 2002 with fundings to I.R. Finetti, University of Trieste. The University of Trieste contributed to the interpretative seismic reprocessing and interpretation. Many thanks are due to OGS, Seismic Division, and to the OGS-Explora's seismic crews for the CROP seismic data acquisition, to ENI-AGIP for processing of CROP sections M22, 23A1 and 34. Our gratitude is expressed to the EGG (Exploration Geophysics Group) of the University of Trieste, Faculty of Sciences, for the invaluable contribution made in the iterated interpretative reprocessing of the CROP sections M2B, 3 and 4. The improvement in signal quality obtained was determinant for understanding this complex area of the Calabrian wedge.

REFERENCES

Bernardelli, P., Cavalli, C., Longoni, R., Giori, I., this volume. Gravity and magnetic fields of the Central Mediterranean Region. Chapter 3.

Ciarapica, G., Passeri, L., this volume. Ionian Tethydes in the Southern Apennines. Chapter 11.

Finetti, I., 1976. Mediterranean Ridge: a young submerged chain associated with the Hellenic Arc. Boll. Geof. Teor. Appl. 28, 31–62.

Finetti, I.R., 1982. Structure, stratigraphy and evolution of Central Mediterranean. Boll. Geof. Teor. Appl. 24, 247–426.

Finetti, I., Lentini, F., Carbone, S., Catalano, S., Del Ben, A., 1996. Il sistema Appennino Meridionale–Arco Calabro–Sicilia nel Mediterraneo Centrale: studio geologico-geofisico. Boll. Soc. Geol. It. 115, 529–559.

Finetti, I.R., this volume. Ionian and Alpine Neotethyan oceans opening. Chapter 6.

Finetti, I.R., this volume. Understanding the Ionides and their geodynamics. Chapter 10.

Finetti, I.R., this volume. The Calabrian arc and subducting Ionian slab from new CROP seismic data. Chapter 17.

Finetti, I.R., this volume. Geodynamic evolution of the Mediterranean Region, from the Permo-Triassic Ionian opening to the Present, constrained by new lithospheric CROP seismic data. Chapter 34.

Finetti, I.R., Del Ben, A., this volume. Ionian Tethys lithosphere roll-back sinking and back-arc Tyrrhenian opening from new CROP seismic data. Chapter 21.

Finetti, I.R., Del Ben, A., this volume. Crustal tectono-stratigraphic setting of the Adriatic Sea from new CROP seismic data. Chapter 23.

Finetti, I.R., Del Ben, A., Forlin, E., Prizzon, A., this volume. Review of the main results of the pioneering MS seismic exploration programme in the deep water of the Mediterranean Sea (OGS, 1969–1980). Chapter 1.

Finetti, I.R., Boccaletti, M., Bonini, M., Del Ben, A., Geletti, R., Pipan, M., Sani, F., 2001. Crustal section based on CROP seismic data across the North Tyrrhenian–Northern Apennines–Adriatic Sea. Tectonophysics 343, 135–163.

Finetti, I.R., Boccaletti, M., Bonini, M., Del Ben, A., Pipan, M., Prizzon, A., Sani, F., this volume. Lithospheric tectono-stratigraphic setting of the Ligurian Sea–Northern Apennines–Adriatic foreland from integrated CROP seismic data. Chapter 8.

Finetti, I.R., Lentini, F., Carbone, S., Del Ben, A., Di Stefano, A., Forlin, E., Guarnieri, P., Pipan, M., Prizzon, A., this volume. Geological outline of Sicily and lithospheric tectono-dynamics of its Tyrrhenian margin from new CROP seismic data. Chapter 15.

Finetti, I.R., Del Ben, A., Fais, S., Forlin, E., Klingelé, E.E., Lecca, L., Pipan, M., Prizzon, A., this volume. Crustal tectono-stratigraphic setting and geodynamics of the Corso-Sardinian Block from new CROP seismic data. Chapter 18.

Mulder, C.J., 1973. Tectonic framework and distribution of Miocene evaporates in the Mediterranean. In: Drooger, C.W. (Ed.), Messinian Events in the Mediterranean. North-Holland, Amsterdam, pp. 44–59.

Rabinowitz, P.R., Ryan, W.B.F., 1970. Gravity anomalies and crustal shortening in the eastern Mediterranean. Tectonophysics 10, 585–608.

Scandone, P., 1979. Origin of the Tyrrhenian Sea and Calabrian Arc. Boll. Soc. Geol. It. 98, 27–34.

Stampfli, G., 2000. Tethyan oceans. In: Bozkurt, E., Winchester, J.A., Piper, J.D.A. (Eds.), Tectonic and Magmatism in Turkey and the Surrounding Area. Geol. Soc. London Spec. Publ. 173, 1–23.

CROP PROJECT: Deep Seismic Exploration of the Central Mediterranean and Italy
Edited by I.R. Finetti

CHAPTER 20

Lithospheric Tectono-Dynamics of the Balearic Basin Opening from CROP-ECORS Seismic Data

I.R. Finetti[1,*], E. Forlin[1], M. Pipan[1]

ABSTRACT

We reconstruct the structure and geodynamics of the Balearic basin based on deep seismic reflection data from the CROP and CROP-ECORS projects. An interpretative processing (IP) sequence and innovative velocity filtering techniques were exploited to reprocess most of the CROP seismic lines from the area, and to enhance deeper crustal events. The resulting good quality data now image the slab of the Alpine-Tethys oceanic crust, subducted in the Balearic stage (Oligocene–Lower Miocene). Such an interpretation supports the idea that rotation of the Corso-Sardinia block in the Balearic geodynamic stage was driven by W-dipping subduction of the Alpine-Tethys lithosphere. In all the seismic sections examined in this work, the rift tectonics of both the margins of the Balearic Basin and the ocean–continent boundary are well evident.

1. INTRODUCTION

The West Mediterranean Sea has been interpreted as a back-arc basin which opened as consequence of NW-dipping subduction of the African plate (Réhault et al., 1984; Burrus, 1984). The opening of the Western Mediterranean basins initiated at about 30 Ma (Jolivet and Facenna, 2000) with the first rifting phase. The drifting phase was coeval with the anti-clockwise rotation of the Corsica–Sardinia block between 21 and 19 Ma (Réhault, 1981). The duration of the drifting phase is still controversial and the end of the rotation could be younger (15 Ma, Vigliotti and Langenheim, 1995).

Rift-related volcanism is widespread in the Western Mediterranean basins. The oldest samples recognized (Oligocene in age) are located in the Valencia trough, in the Sardinian rift zone and along the Liguro-Provençal margins (Coulon, 1977; Bellon, 1981). According to kinematic reconstructions (Réhault, 1981; Burrus, 1984), this volcanism roughly trended NE–SW, parallel to the paleo-subduction zone, before the rotation of the Corso-Sardinian block; therefore this volcanic activity is very likely linked to the subduction process.

The oceanic domain is characterized by narrow, isolated magnetic anomalies that are interpreted as being due to tholeitic volcanics. Calcalkaline volcanism, very likely related to subduction, appears to be discontinuously distributed in space and time, from the beginning of rifting on the Ligurian margin (~30 Ma), to the end of the Corsica–Sardinia block rotation (~16 Ma) toward the Corsica margin. This space and time shift reveals the importance of both the rollback of the Alpine Tethys slab and the migration of the Alpine-Apenninic thrust-front toward the ESE during the basin formation (Rollet et al., 2002). According to these authors, the initial rheological conditions and inherited crustal fabric would induce important changes in the styles of deformation observed along margins and between conjugate margins. Therefore the mode of opening and the margin structures mainly depend on the balance between intrinsic, inherited crustal heterogeneity (fabric and rheological changes) and external conditions imposed by rollback of the subducting lithosphere.

It is widely accepted that the traction force induced by the subducting (Adria) slab is the most important cause for the Tyrrhenian back arc extension, which occurred in continuity with the Ligurian basin opening (Malinverno and Ryan, 1986; Facenna et al., 1996, 1997). The rollback of the subducting zone hinge produces an extensional stress

[1]Department of Geological, Environmental and Marine Sciences, Exploration Geophysics Group, University of Trieste, Via Weiss 1, 34127 Trieste, Italy
*Corresponding author. Tel.: +39 40 5582276; e-mail: finetti@units.it (I.R. Finetti)

Table 1
CROP seismic sections interpreted and described in the present chapter (Balearic Ligurian Sea, Chapter 20)

CROP seismic section	SP	Length (km)
M2A/I	100–3600	175
M10	2500–4686	109
M11	100–3000	145
M23B	3000–5417	121

field in the upper plate, which is believed to generate an asymmetrical mechanism of rifting (Martinez et al., 1995). In this model, back arc extension forms a narrow margin on the side of the volcanic arc, close to the subduction zone, whereas the conjugate margin is widening by tectonic extension and is accompanied by volcanism in the first stage of opening. This geometry is observed in the Provençal basin, where the Gulf of Lion margin is wider than its west Sardinian conjugate (Burrus, 1984; de Voogd et al., 1991).

In the Western Mediterranean basin, various geophysical methods have been employed to study the crustal setting, including advanced seismic profiling, wide-angle seismics, heat flow measurements, magnetic and gravimetric surveying. However, some parts of this area are still poorly controlled. The interpreted new CROP sections described in this work, together with the CROP-ECORS section, place the crustal tectono-stratigraphic setting of the Balearic basin within the wider context of the Mediterranean region investigated by the CROP project. Reconstruction of the Balearic stage geodynamics is essential for understanding the multi-staged Apenninic-Maghrebian orogeny.

2. GEOPHYSICAL DATASET

This work is based on: (a) new CROP seismic data (Table 1); (b) the CROP-ECORS section "Gulf of Lion–Sardinia", already presented and discussed in the literature; (c) MS seismic lines (Finetti et al., this volume, Chapter 1); and (d) gravity, magnetic and other geophysical-geological data available in the literature. Figure 1 shows the location map of the CROP and CROP-ECORS sections.

The acquisition parameters of the new CROP sections M2A/I, M10, M11 and M23B are reported by Finetti et al. (this volume, Chapter 5).

Most of the listed sections (M2A/I-M10-M11) have been reprocessed (interpretative reprocessing) by the University of Trieste to enhance signals from deeper crustal events. The choice of processing flow was constrained by the need to cope with the characteristics of a quickly varying structural environment. In such seismic sections, continental and oceanic crust are present, implying completely different problems, which need to be solved using specific processing techniques and parameters. The concurrent problems of persistent multiples and deep events characterized by low reflectivity were addressed using interpretative processing (IP) and innovative velocity filtering techniques. The signal quality thus obtained is markedly better than that of the standard processing (Finetti et al., this volume, Chapter 5). The CROP section M23-B was processed by ENI-AGIP.

The interpretation of the CROP data was done to a working scale 1:50,000 and presented as reduced figures, where the vertical scale is true for a velocity of 6,000 m/s.

On the basis of the acquired knowledge of the tectono-stratigraphic setting of the area, we have also re-examined and included in our database the CROP-ECORS profile taken from the literature (de Voogd et al., 1991; Fanucci and Morelli, 2001).

3. SEISMIC INTERPRETATION

In the Ligurian Gulf two CROP sections illustrate very clearly the transcrustal tectono-stratigraphic setting of the area: sections M10 and M11. Section M10 exhibits the European and Corsican rifted margins and the interpreted

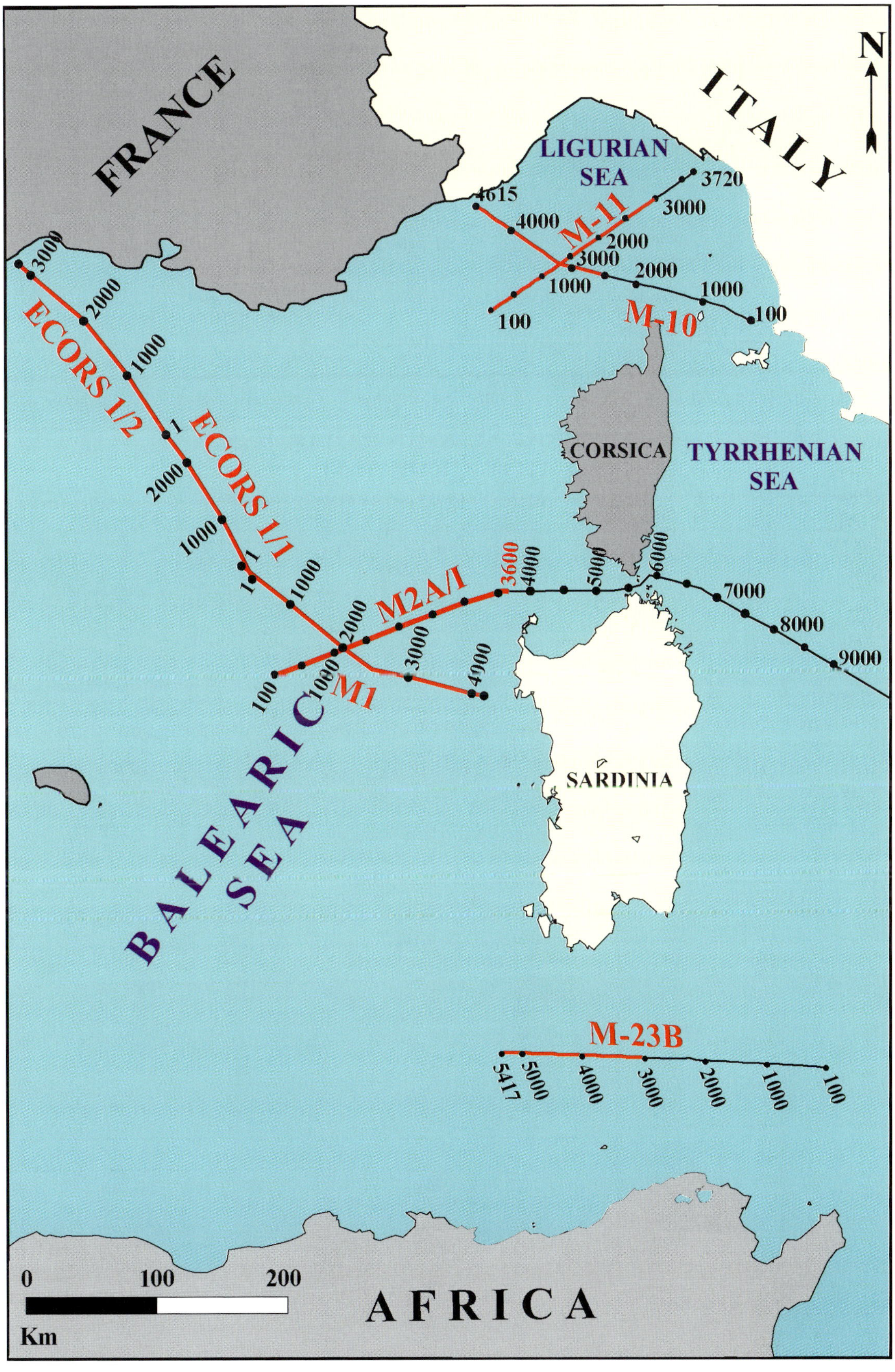

Figure 1: Location map for the CROP seismic sections used in this chapter.

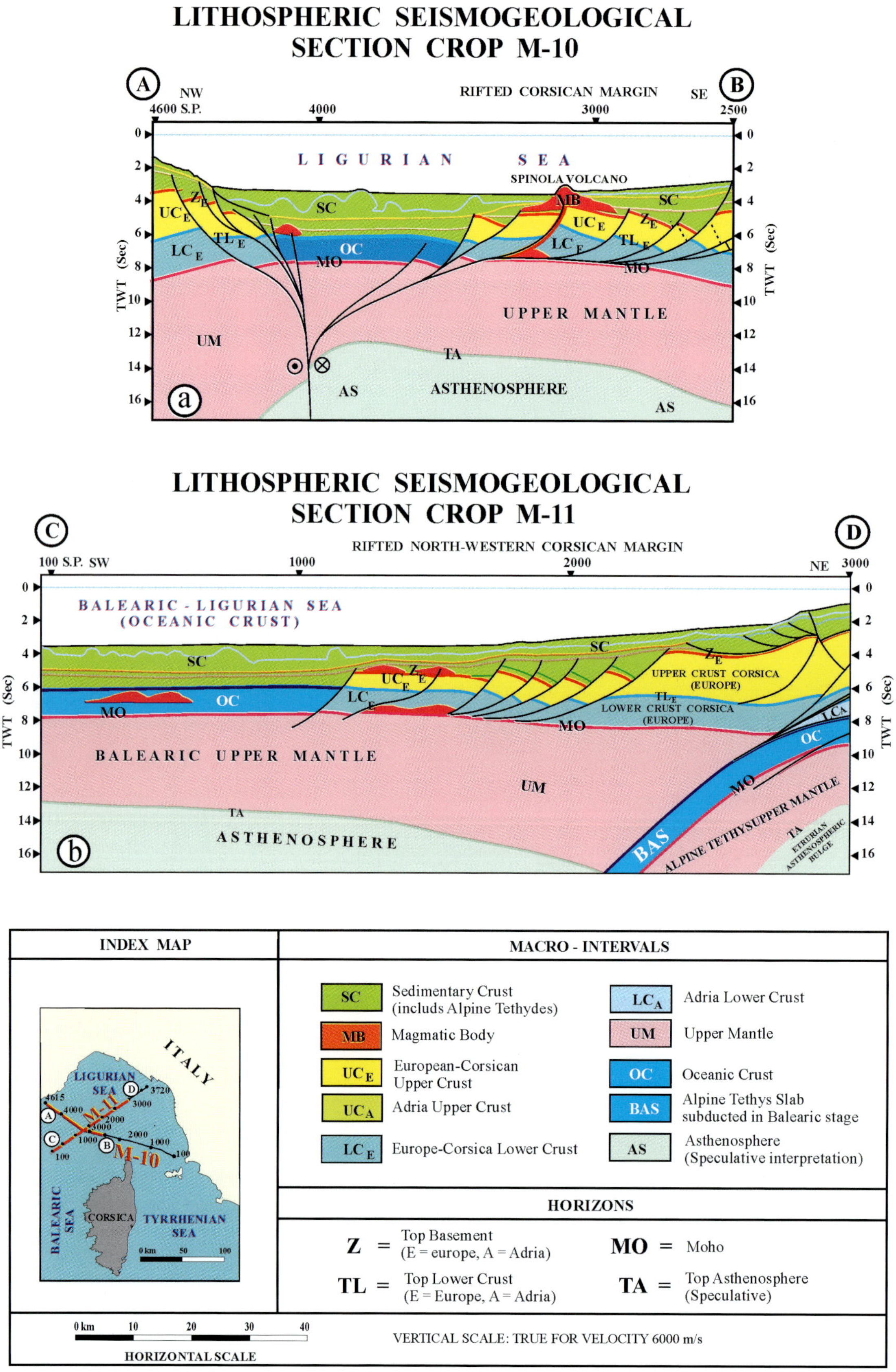

Figure 2: Simplified version of the interpreted seismic sections M10 and M11 (part). Seismic section M10 shows the Ligurian basin floored by oceanic crust, the rifted Corsican and Ligurian margins. Seismic section M11 shows also the rift tectonics of the Corsica margin and the Alpine Tethys slab subducted in the Balearic stage.

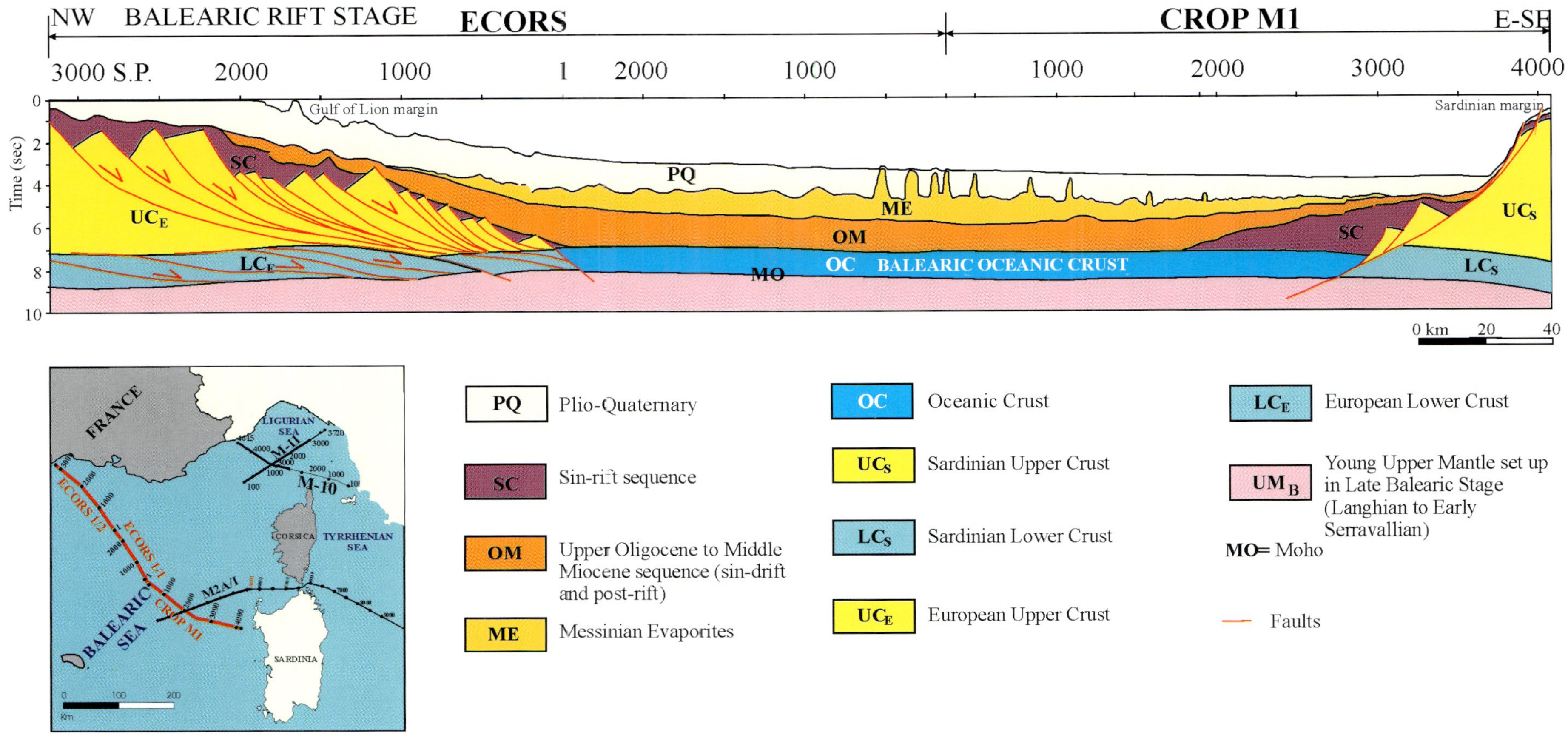

Figure 3: Interpreted seismic profile CROP-ECORS (modified from (Fanucci and Morelli, 2001)). This section clearly shows the different rift-tectonics of the two margins of the Balearic basin (Sardinian and Gulf of Lion). This differentiation is connected to the fact that the Alpine Tethys slab retreat-related eastward mantle flux delaminates and thins the internal margin, and transports, the migrating block (Sardinia) outward without much stretching. The same unusual tectonics can be observed in the Tyrrhenian back arc basin.

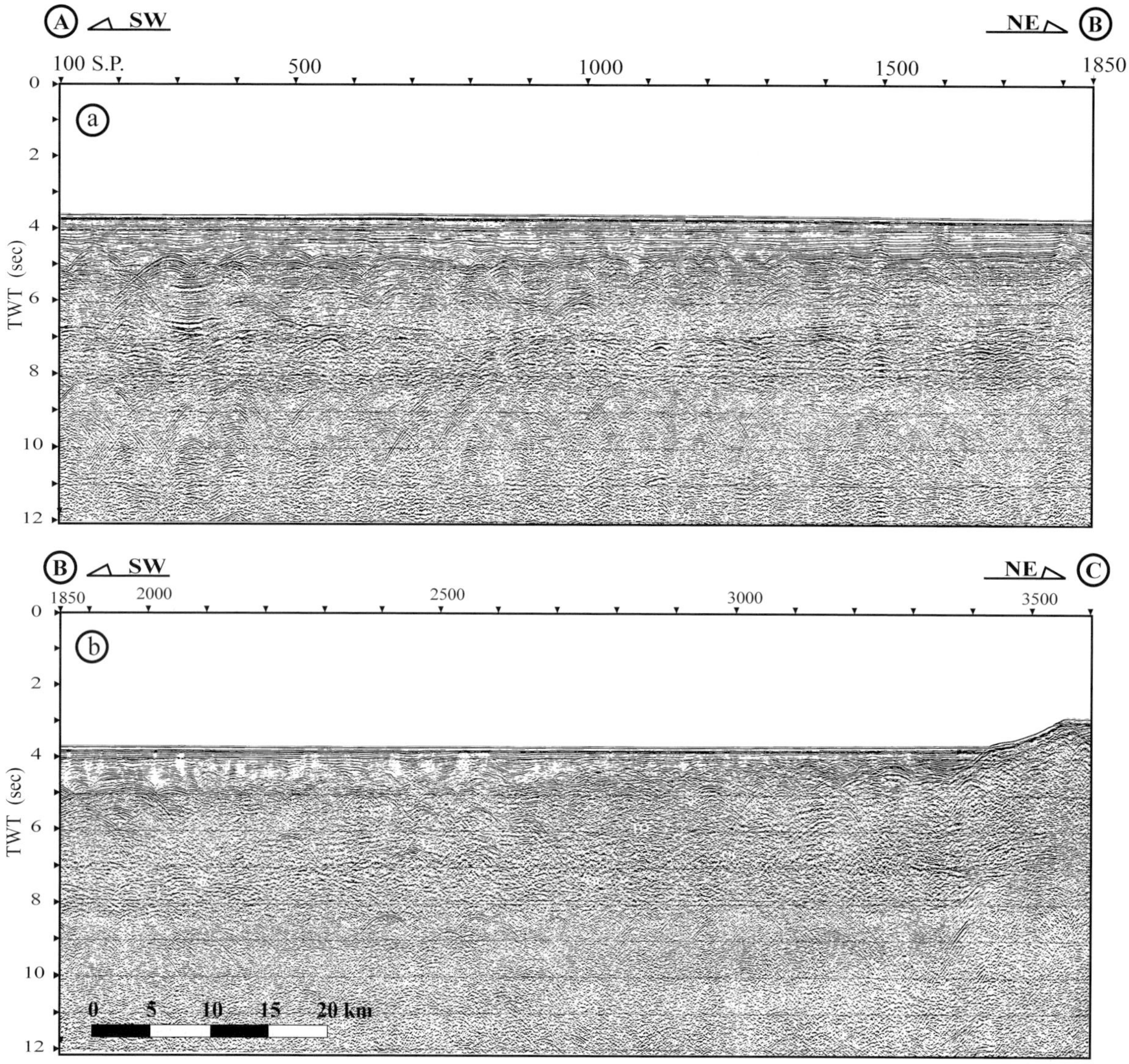

Figure 4: Reduced version of CROP seismic section M2A/I (part: SP 100–3600) in the Balearic basin. Scale ratio true for $V_p = 6000$ m/s. For location see Figure 5.

segment of oceanic crust (Figure 2(a)). The rift tectonics of the Corsican margin are well evident also in section M11 (Figure 2(b)). At the western edge of the rifted Corsican margin the Volcano Spinola (20 Ma) is imaged, which testifies to the age of the related extensional tectonics. As usual, the lower crust can be discriminated as an interval of higher reflectivity than the overlying upper basement crust. At the NE extremity of Figure 2(b) (section M11) the interpreted slab of the Alpine Tethys oceanic crust is visible, subducted in the Balearic stage (Oligocene–Lower Miocene). Such an interpretation supports the idea that the rotation of the Corso-Sardinian block in the Balearic geodynamic stage was driven by W-dipping subduction of the Alpine Tethys lithosphere. This explains how a post-Eo-Alpine Tethys existed. In other chapters of this volume the problem is more extensively treated (Finetti et al., this volume, Chapter 8; Finetti, this volume, Chapter 34). Here we give a summary only. The Eo-Alpine stage gave rise to Adria-Europe collision in the Alps sector, leaving unsubducted a restricted but still wide Alpine Tethys remnant west of the Alps. In the western Alpine Tethys, the SE-dipping subduction of the Eo-Alpine stage occurred as an intra-ocean process. In the Balearic stage (Oligocene–Lower Miocene) the Alpine Tethys remnant subducted (NW to W-dipping) under the European plate, giving rise to the back arc Balearic basin opening. Figure 3 shows our interpretation of the seismic profile ECORS-CROP running from the Gulf of

---

Figure 5: Interpreted seismological CROP section M2A/I (part: 100–3600) crossing the western margin of the Corso-Sardinian block and the Balearic basin.

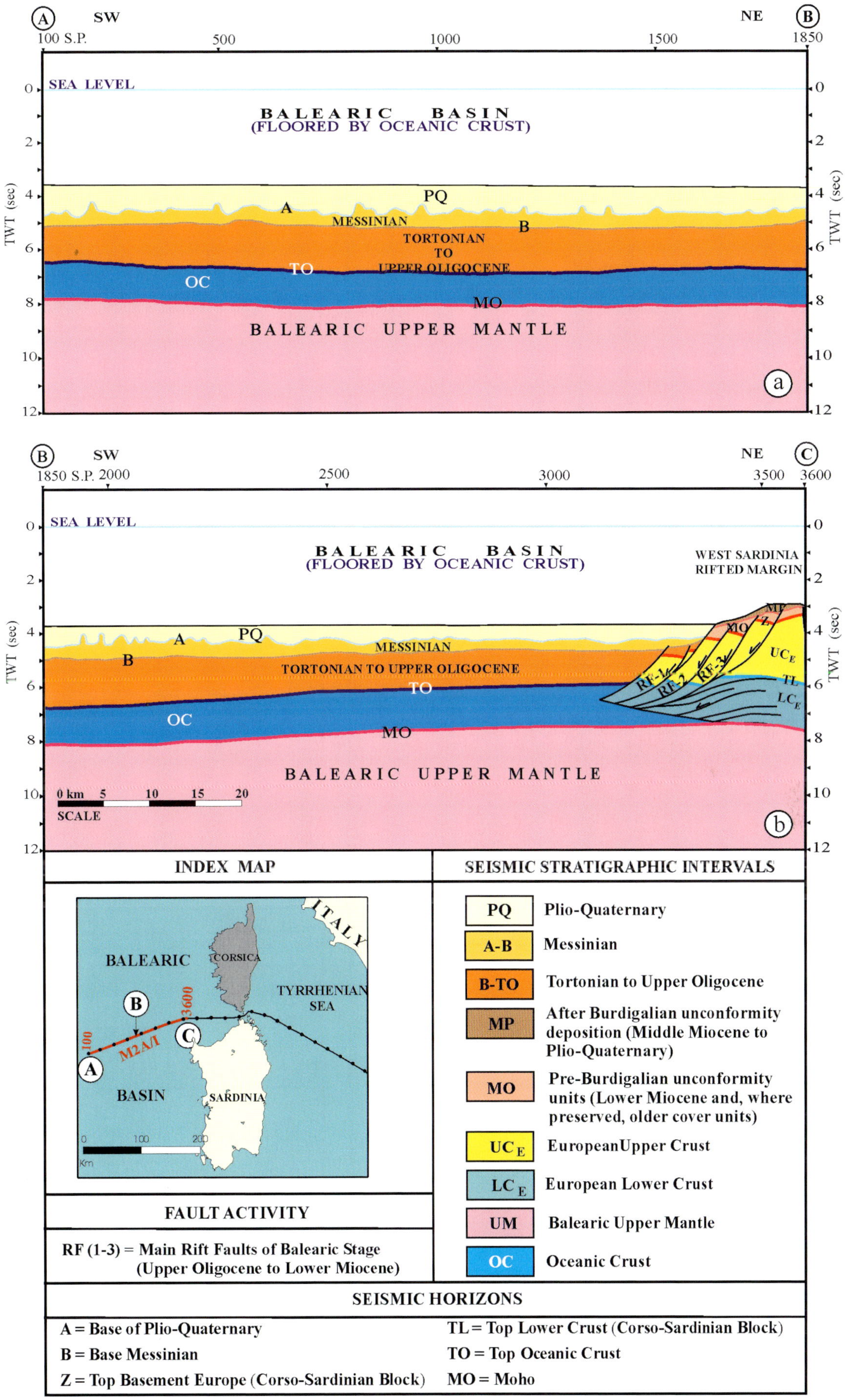

A SW NE B
100 S.P. 500 1000 1500 1850
SEA LEVEL
BALEARIC BASIN
(FLOORED BY OCEANIC CRUST)
PQ
A
MESSINIAN
B
TORTONIAN TO UPPER OLIGOCENE
TO
OC
MO
BALEARIC UPPER MANTLE
TWT (sec)
a
B SW NE C
1850 S.P. 2000 2500 3000 3500 3600
SEA LEVEL
BALEARIC BASIN
(FLOORED BY OCEANIC CRUST)
WEST SARDINIA RIFTED MARGIN
MP
MO
Z
UCE
TL
LCE
RF-1
RF-2
RF-3
A
PQ
B
MESSINIAN
TORTONIAN TO UPPER OLIGOCENE
TO
OC
MO
BALEARIC UPPER MANTLE
0 km 5 10 15 20
SCALE
b
INDEX MAP
ITALY
BALEARIC
CORSICA
TYRRHENIAN SEA
BASIN
SARDINIA
M2A/1
100
3600
0 100 200 Km
FAULT ACTIVITY
RF (1-3) = Main Rift Faults of Balearic Stage (Upper Oligocene to Lower Miocene)
SEISMIC STRATIGRAPHIC INTERVALS
PQ Plio-Quaternary
A-B Messinian
B-TO Tortonian to Upper Oligocene
MP After Burdigalian unconformity deposition (Middle Miocene to Plio-Quaternary)
MO Pre-Burdigalian unconformity units (Lower Miocene and, where preserved, older cover units)
UCE EuropeanUpper Crust
LCE European Lower Crust
UM Balearic Upper Mantle
OC Oceanic Crust
SEISMIC HORIZONS
A = Base of Plio-Quaternary
B = Base Messinian
Z = Top Basement Europe (Corso-Sardinian Block)
TL = Top Lower Crust (Corso-Sardinian Block)
TO = Top Oceanic Crust
MO = Moho

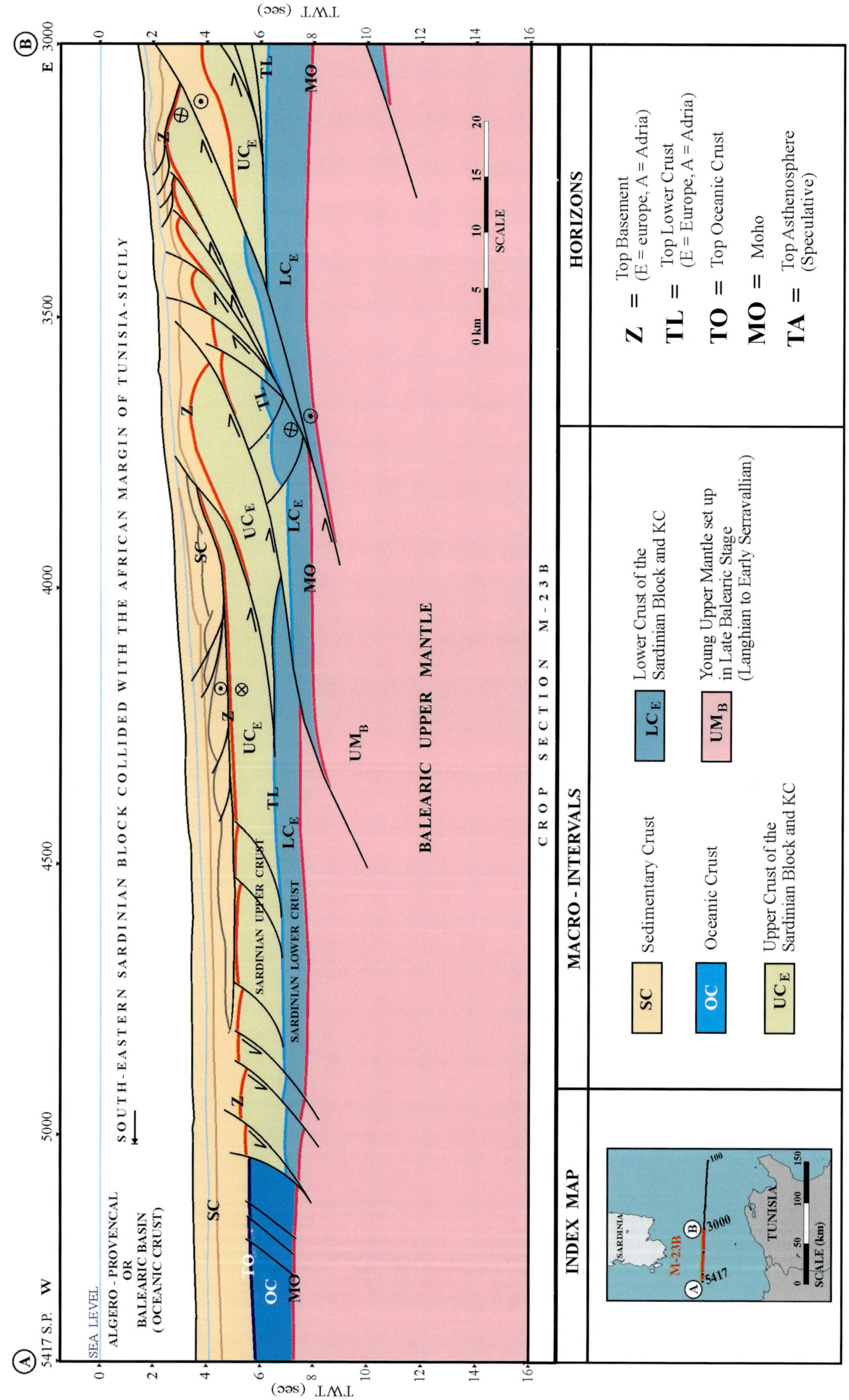

Figure 6: Simplified seismological CROP section M23B (part) showing the SE edge of the Balearic basin and the southern extremity of the rotated Sardinian block.

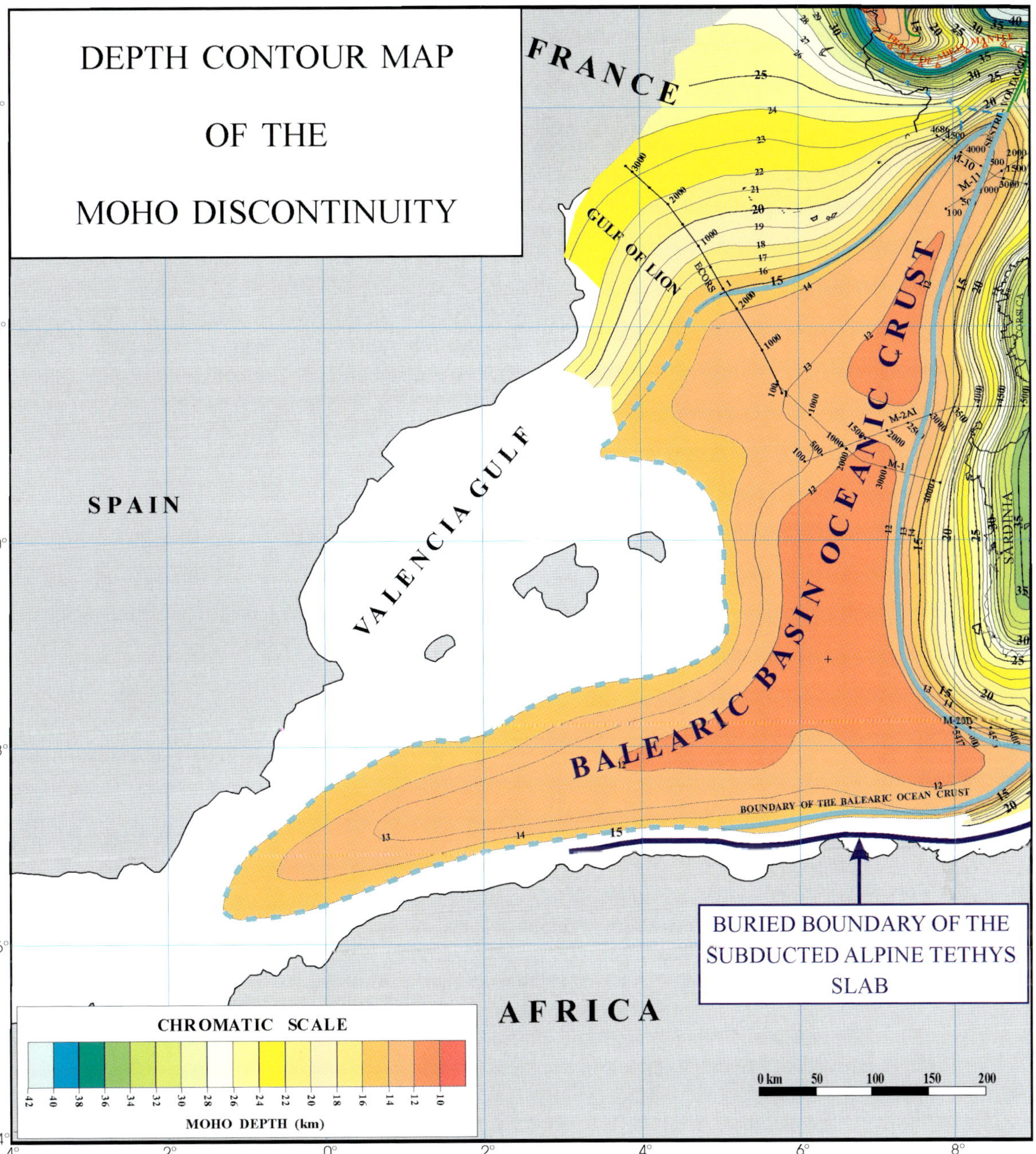

DEPTH CONVERSION VELOCITY (m/s)

- Water	1500
- Plio-Quaternary	2000
- Terrigenous Tertiary and various other units	2400 - 3600
- Alpine Tethydes and Ionides	4000
- Messinian Evaporite	5000
- Carbonate sequences	5500 - 6000
- Basement Upper Crust	6000
- Lower Crust and Oceanic Basement Crust	7000

Moho depth contour line. Contour interval: 1 km

Oceanic crust limit of the Balearic basin.

CROP AND ECORS SEISMIC SECTION POSITION

Figure 7: Depth contour map of the Moho discontinuity in the Balearic basin based on new CROP seismic data, CROP-ECORS section "Gulf of Lion–Sardinia", MS seismic sections, gravity data and data from the literature. Interpolation among the CROP sections was aided by gravity data and other geophysical information.

Lion to the NW-Sardinian margin. This interpretation is basically in line with other interpretations presented in the literature (Fanucci and Morelli, 2001). We consider identifiable the lower crust also in the European margin. Rift tectonics are well evident in the upper crust of the European margin while in the lower crust the tectonics are mostly represented by delamination. A more drastic passage from continental to oceanic crust occurs at the Sardinian margin.

Figures 4 and 5 show a reduced version of CROP section M2A/I (part) and its interpretation, respectively. This section basically shows the oceanic crust of the Balearic basin and its cover, which are fully self-explanatory. At the eastern extremity, the lower Sardinian margin is shown.

Most of Figure 6 images the rifted and delaminated margin of SW Sardinia, where the rifting and delaminating tectonics of the Balearic stage occurred and were followed at the end compressive tectonics, when the Sardinia block collided with the Panormid Promontory.

Finally Figure 7 shows the depth-contour map of the Moho discontinuity of the Balearic basin and the Ligurian Sea reconstructed with all the data referred to in Section 2. A complete Moho depth map of the central-Western Mediterranean is presented by Finetti (this volume, Chapter 27).

ACKNOWLEDGEMENTS

This work was funded by MIUR, COFIN 1998, 2000 and 2002 with grants to I.R. Finetti, University of Trieste. The Authors are grateful to OGS for the CROP data acquisition and to Eni-AGIP for the processing of CROP section M23-B. We sincerely thank the Exploration Geophysics Group (EGG) of the Faculty of Science of the University of Trieste for the extensive work of interpretative reprocessing of most of the data used. The signal enhancement obtained was very helpful for the interpretation illustrated in this chapter.

REFERENCES

Bellon, H., 1981. Chronologie radiométrique (K–Ar) des manifestations magmatiques autor de la Méditerranée occidentale entre 33 et 1 Ma. In: Wezel, F.C. (Ed.), Sedimentary Basins of Mediterranean Margins. Technoprint, Bologna, Italy, pp. 341–360.

Burrus, J., 1984. Contribution to a geodynamic synthesis of the Provencal Basin (north-western Mediterranean). Marine Geology 55, 247–269.

Coulon, C., 1977. Le volcanisme calco-alcaline cénozoique de Sardaigne (Italie) : Pétrologie, géochimie et genèse des laves andésitique et des ignimbeites : Signification géodinamique. Thèse de doctorat, Univ. D'Aix-Marseille III, Aix en Provence, France.

de Voogd, B., Nicolich, R., Olivet, J.L., Fanucci, F., Burrus, J., Mauffret, A., Pascal, G., ECORS-CROP Profile Group, 1991. First deep seismic reflection transect from the Gulf of Lions to Sardinia (ECORS-CROP profiles in western Mediterranean). In: Meissner, R., et al. (Eds.), Continental Lithosphere: Deep Seismic Reflection. In: Geodyn. Ser., vol. 22. AGU, Washington, DC, pp. 265–274.

Facenna, C., Davy, P., Brun, J.P., Funicello, R., Giardini, D., Mattei, M., Nalpas, T., 1996. The dynamics of back-arc extension: An experimental approach to the opening of the Tyrrhenian Sea. Geophys. J. Internat. 126, 781–795.

Facenna, C., Mattei, M., Funicello, R., Jolivet, L., 1997. Styles of back-arc extension in the central Mediterranean. Terra Nova 9, 126–130.

Fanucci, F., Morelli, D., 2001. Modalità e cause della deriva del Blocco Sardo-Corso. Studi Trentini di Scienze Naturali – Acta Geologica, Trento 77 (2000), 5–14.

Finetti, I.R., this volume. Depth contour map of the Moho discontinuity in the Central Mediterranean Region from new CROP seismic data. Chapter 27.

Finetti, I.R., this volume. Geodynamic evolution of the Mediterranean Region from Permo-Triassic Ionian opening to Present, constrained by new lithospheric CROP seismic data. Chapter 34.

Finetti, I.R., Del Ben, A., Forlin, E., Prizzon, A., this volume. Review of the main results of pioneering MS seismic exploration programme in the deep water of the Mediterranean Sea (OGS, 1969–1980). Chapter 1.

Finetti, I.R., Forlin, E., Pipan, M., this volume. CROP seismic data acquisition, processing and interpretative re-processing. Chapter 5.

Finetti, I.R., Boccaletti, M., Bovini, M., Del Ben, A., Pipan, M., Prizzon, A., Sani, F., this volume. Lithospheric tectono-stratigraphic setting of the Ligurian Sea–Northern Apennines–Adriatic Foreland from integrated CROP seismic data. Chapter 8.

Jolivet, L., Facenna, C.D., 2000. Mediterranean extension and the Africa–Eurasia collision. Tectonics 19, 1095–1107.

Malinverno, A., Ryan, W.B.F., 1986. Extension in the Tyrrhenian Sea and shortening in the Apennines as result of arc migration driven by sinking of the lithosphere. Tectonics 5, 227–245.

Martinez, F.P., Fryer, P., Baker, N.A., Yamazaki, T., 1995. Evolution of back-arc rifting: Mariana Trough, 20°–24° N. J. Geophys. Research 100, 3807–3827.

Réhault, J.-P., 1981. Evolution tectonique et sédimentaire du bassin Ligure (Méditerranée Occidentale). Thèse d'ètat, Univ. Pierre et Marie Curie, Paris VI.

Réhault, J.-P., Boillot, G., Mauffret, A., 1984. The western mediterranean basin geological evolution. Marine Geology 55, 447–477.

Rollet, N., Déverchère, J., Beslier, M.-O., Guennoc, P., Réhault, J.-P., Sosson, M., Truffert, C., 2002. Back arc extension, tectonic inheritance, and volcanism in the Ligurian Sea, Western Mediterranean. Tectonics 21 (3), 1015.

Vigliotti, L., Langenheim, V.E., 1995. When did Sardinia stop rotating?: New paleomagnetic results. Terra Nova 7, 424–435.

CROP PROJECT: Deep Seismic Exploration of the Central Mediterranean and Italy
Edited by I.R. Finetti

CHAPTER 21

Ionian Tethys Lithosphere Roll-Back Sinking and Back-Arc Tyrrhenian Opening from New CROP Seismic Data

I.R. Finetti[1,*], A. Del Ben[1]

ABSTRACT

New lithospheric information of major importance for understanding Tyrrhenian geodynamics has been obtained from the CROP deep seismic exploration project in the central Mediterranean region. Notable seismic exploration results are the clear imaging of upper slab segments of subducted Ionian Tethys, buried beneath circum-Tyrrhenian margins, and of the Alpine Tethys slab, laterally subducted beneath NW Adria in the Eo-Alpine stage (Upper Cretaceous–Lower Eocene) and beneath the Corso-Sardinian block in the Balearic stage (Oligocene–Lower Miocene). These new seismic images now allow us to reconstruct a well-constrained pre-orogenic paleogeographic framework, consisting of two communicating ocean basins: the Ionian Tethys of Permo-Triassic opening, interposed between Africa and Adria, and the Alpine Tethys, of Middle Jurassic–Lower Cretaceous opening, interposed, in the Alps sector, between Adria and Europe, and, west of the Alps, between Africa and Europe. Two well-defined promontories of great importance for the evolving tectono-dynamics delimited the Alpine Tethys from the Ionian: the Etrurian promontory to the north (NW Adria) and the Panormid promontory to the south (Africa).

At the end of the Eo-Alpine geodynamic stage, the NE-Alpine Tethys closed with the Adria–Europe collision in the Alps sector, but west of the Alps, a wide Alpine Tethys remnant persisted. This Tethys remnant in the Balearic stage subducted beneath the European margin, giving rise to rotation of the Corso-Sardinian block, closure of the Alpine Tethys, and formation of a large migrating orogenic arc, which included earlier settings of the Northern Apennines, of the Southern Apennines, of the Calabrian arc, of the Sicilian Apennines and of the Maghrebian chain. The Etrurian promontory, at the start of the Tyrrhenian stage (Middle Miocene), dissected the Northern Apennines from the Southern, and the Etrurian promontory dissected the Maghrebian chain from the Sicilian Apennines. Driven by fast roll-back sinking of the Ionian Tethys lithosphere, in the Tyrrhenian stage (Middle Miocene to Present), the geodynamics that generated the Southern Apennines–Calabrian arc–Sicilian Apennines thrust belt system, whose CROP reconstructed macro-tectono-stratigraphic setting we exhibit and discuss in this work, took place.

The eastern Tyrrhenian margin is composed of migrated crustal fragments coming from the Etrurian promontory which collided with the Adria slope, south of the 41° parallel (northern boundary of the Ionian Tethys). Tectonically stacked with the Etrurian fragments, there are European crustal fragments detached from Sardinia–southern Corsica and carried eastward after the Tyrrhenian opening, occurring within the Sardinian margin. Etrurian promontory cover sheets were thrust onto the roof of the Southern Apennines nappe, where they are known as Apennine platform units. The condensed deep basin sequence that covers the Ionian Tethys was stripped off from the subducting basement and went to form allochthonous units in the Southern Apennines and Sicily. In our CROP reconstruction, it is very clear that the basinal units of the Southern Apennines, widely discussed in the literature and known as the Lagonegrese, are formed from Ionian cover units, as well as for Imerese units of Sicily. It is also evident that most of these Ionian cover units are imbricated in the frontal wedge of the Calabrian arc in the Ionian Sea. For all these Ionian-derived units we use the unifying and self-explanatory name Ionides. Also the southern Tyrrhenian margin is formed from a stack of two different allochthonous crustal fragments, like the eastern

[1]Department of Geological, Environmental and Marine Sciences, University of Trieste, Via Weiss 1, 34127 Trieste, Italy
*Corresponding author. Tel.: +39 40 5582276; e-mail: finetti@units.it (I.R. Finetti)

margin, but here the block colliding with the buried Sicilian platform slope came from the Panormid promontory. In the Calabrian arc, tectonic units coming from both promontories, followed by the Calabrides, are stacked together with Alpine Tethydes, Flysch units and a huge wedge of Ionides. Prominent transcurrent faults (leftt in the Southern Apennines and Calabria; right in Sicily) intersect the orogenic arcs.

Keywords: Ionian subduction, Tyrrhenian opening, tectono-dynamics

1. INTRODUCTION

In 1965, when the Mediterranean Sea basin was poorly known, some initial geophysical exploration campaigns were started. Soon after, many geoscientists and institutions of various countries (Italy, France, UK, Germany, Russia, USA and others) took part in pioneering geophysical investigations of various zones of this complex basin. OGS of Trieste, supported by the CNR (National Research Council of Italy) in 1968 programmed a pluriannual (1969–1980) regional exploration of the whole Mediterranean, from Gibraltar to the Levantian Sea, employing the new multichannel-multifold digital seismic techniques just appearing at that time in petroleum exploration. The results obtained, together with those of other investigations, contributed markedly to providing the international geoscience community with new information of relevant importance for a progressively better understanding of the area. It was immediately clear that the Tyrrhenian Sea constituted a key geological part of the Mediterranean for reconstructing the geodynamic evolution of this complex region.

A fast growing literature proposed continually different geodynamic models based on emerging new geophysical-geological data and its interpretations. The impressive young extensional tectonics of the Tyrrhenian Sea were interpreted in different ways: (a) *Back-arc-related extension in subduction-slab retreat models* (Boccaletti et al., 1971; Malinverno and Ryan, 1986; Finetti and Del Ben, 1986; Royden et al., 1987; Wortel and Spakman, 1992; Jolivet and Faccenna, 2000); (b) *Lithospheric delamination* (Roeder, 1980); (c) *Mantle uprise determining gravity-sliding effects* (Van Bemmelen, 1972; Locardi, 1982; Wezel, 1982; and, *with variations*: Lavecchia and Stoppa, 1996; D'Agostino et al., 2001); (d) *Lateral extrusion connected to Africa–Europe convergence* (Tapponnier, 1977; Boccaletti et al., 1982; Faccenna et al., 1996; Mantovani et al., 1997); and (e) *Late orogenic collapse of an overthickened accretionary wedge* (Carmignani and Kligfield, 1990; Carmignani et al., 1994). A regional study of the whole Tyrrhenian basin and surrounding geological provinces, totally controlled by a significant amount of seismic reflection data, was presented in the literature several years ago (Finetti and Del Ben, 1986). This provided a first framework for the Tyrrhenian opening-related tectono-dynamics completely based on new geophysical data, MS seismic network, specifically acquired for the task. These pioneering exploration data have been successively augmented by various other local seismic explorations but, until now, this is the only seismic project which covers the Mediterranean Sea entirely with a regional grid. The MS (Mediterranean Sea) seismic network (Finetti et al., this volume, Chapter 1) was acquired with a record length of 8–10 s TWT, or with deep but not transcrustal penetration in continental areas. In the deep basins preserved from subduction, like the Balearic, Tyrrhenian and Ionian Sea, such penetration was sufficient to explore the entire, mostly oceanic crust, where the Moho ranges from about 7.0–7.5 s (Tyrrhenian) to 8.0–9.0 s (undeformed oceanic crust of the Ionian Sea). Therefore, the MS seismic lines are of major geophysical importance, and reprocessing still continues to refine the information.

But in the time it became clear that for further advances in knowledge of the Mediterranean, it would be necessary to explore its lithosphere with deeper seismic penetration. For this reason, in line with similar national projects like COCORP (USA), ECORS (France), BIRPS (UK), DEKORP (Germany) and others, the Italian CROP programme was projected and performed, of which nearly 90% is located offshore.

In this work, we present and describe the interpretation of significant CROP-based reconstructions of the pre-Tyrrhenian paleogeographic framework, and of Tyrrhenian opening-related geodynamics. Our description is limited to a specific Tyrrhenian part of the CROP network, but the most relevant conclusions are based on the whole network.

2. CROP SEISMIC DATA DESCRIBED

Since previous investigations showed us that several key elements for understanding the Mediterranean dynamics are imprinted in the Tyrrhenian Sea, a large part of the CROP seismic network was acquired in this basin. In this

Table 1
CROP seismic sections exhibited and described in this work

CROP section	SP	km	Figure N°	Chapter plate N°
M2A/I (Part)	6500–10450	188	2	
M2A/II	10450–16022	278	3	1–2
M2A/III (Part)	100–2000	120	4	
M9	100–3635	178	5	3–4
M29B	100–5545	300 (Extended 25 km)	6	5–6
M30	100–3599	180	7	7–8
M36	100–6721	338	8	9–10
M28B (Part)	2000–4000	101	9	
M6B	100–2730	140	10	

work we describe the major features, and provide lithospheric seismic interpretations of the CROP sections listed in Table 1, framed within a larger setting based on interpretation of the whole Italian deep seismic exploration project (Figure 1).

The Tyrrhenian CROP sections (Figure 1) not listed in Table 1 are specifically described and exhibited in plates and figures of other chapters in this volume (M12A and M12B in Chapter 8; M37 in Chapter 9; M6B in Chapter 12; M6A, M23, M26, M28A and M29A in Chapter 15; M2A/III and M27 in Chapter 17; M2A/I, M28B and M32 in Chapter 18).

Several other important data were utilized in the interpretation of the complex tectonics of the Tyrrhenian basin and its margins. The MS seismic lines (Finetti and Del Ben, 1986; Finetti et al., this volume, Chapter 1) that cover the Tyrrhenian Sea with a more dense network were very useful in many studied zones. Various other spot seismic data, and the seismic dataset available from the Italian Ministry of Industry in the shelf areas also contributed in reconstructing some important features. The ODP (Ocean Drilling Programme) deep boreholes and some hydrocarbon boreholes were used for calibration of interpreted seismic horizons in specific zones. But the most important macrostratigraphic identification was done on the basis of the regional knowledge of characteristic reflecting horizons and crustal intervals that has matured over the last few decades. Such characteristic horizons and seismic reflecting intervals were submitted to a careful, reasoned 3D mutual cross-control among the CROP section network. The CROP sections M2A/I, M2A/II, M2A/III, M9 and M30 were carefully reprocessed (interpretative reprocessing) by the processing centre of the University of Trieste, Faculty of Sciences, obtaining a satisfactory improvement in signal over the original poor sections of the standard processing.

3. CROP SEISMIC SECTIONS INTERPRETATION

The deep seismic exploration data of the Italian national CROP project extend downward the lithospheric information previously provided by the Italian MS lines seismic dataset for the whole Mediterranean acquired by OGS (Finetti et al., 1970; Finetti and Morelli, 1973; Finetti, 1982; Finetti and Del Ben, 1986; Finetti et al., this volume, Chapter 1). Downward extension of the medium penetration MS lines was of fundamental importance because it allowed us to study key deep lithospheric data related to the evolving geodynamics of the region. *One of the most important new geophysical imagings of the CROP data in the Tyrrhenian Sea is surely that of the subducted upper Ionian slab, buried beneath allochthonous crustal blocks of the Tyrrhenian margin of the Southern Apennines–Calabrian arc–Sicilian Apennines system. The same importance can be attributed to identification of the upper slab of the Alpine Tethys. These new data allow us to reconstruct a very well constrained pre-orogenic Mediterranean paleogeography and to observe that this setting played a determinant role in the evolving tectono-dynamics of the area.*

It is to be remarked that only with a careful, patient, interpretation of a 3D network, difficult even for seismic interpreters of world-wide experience, is it possible to extract from a single transcrustal section of a tectonically complex area so much information as here exhibited. This can be done using even weak reflections of low to very low amplitude, by verifying their coherence within each delimited tectono-stratigraphic body. Moreover, the tectono-stratigraphic reconstruction is obtained progressively, brick by brick, starting from those elements most

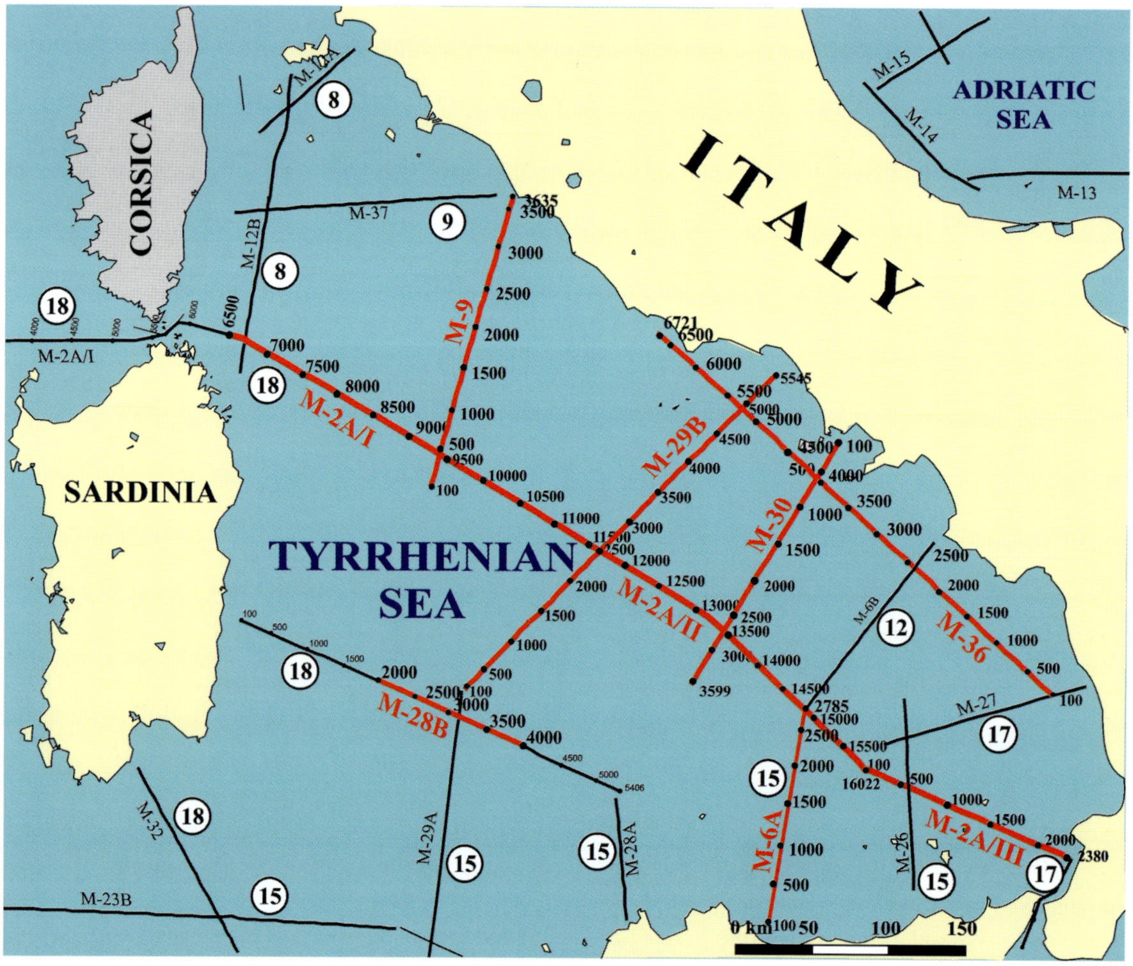

Figure 1: Location map for the basic CROP sections described in this chapter (red-traced lines). Reconstructed tectono-dynamics benefit also from results of the other (black-traced) CROP sections, described in other chapters of the present volume. Within white circles are the chapter numbers where the CROP sections are fully described and presented on plates as the basic specific document for that chapter.

reliably identifiable and extending the interpretation with continuously cross-correlated control. In such a manner, an experienced seismic interpreter may obtain results of whose validity he is aware, while a rash external judgment may think them to be an over-interpretation.

3.1. CROP Section M2A across the Corso-Sardinian Block and Tyrrhenian Sea

The section M2 is the longest one of the CROP seismic dataset. It is composed of two main parts: one from the Balearic basin to the SE Tyrrhenian (M2A), and the other in the Ionian Sea (M2B). CROP M2A starts to the NW in the Balearic basin and crosses the Corso-Sardinian block, passing between the two islands; the delaminated Sardinian margin in the Tyrrhenian Sea; the axial zone of the deep Tyrrhenian floored by oceanic crust, and terminates on the Tyrrhenian margin of the Calabrian arc, composed of stretched drifted crustal fragments of the eastern Sardinian margin. Section M2A is divided into three parts: M2A/I, M2A/II and M2A/III (Figure 1). CROP M2A/I (SP 100–10450), crossing the Corso-Sardinian block, is included in the chapter relative to that block (Finetti et al., this volume, Chapter 18). But its eastern part (SP 6500–10450), important for understanding the Tyrrhenian opening, is exhibited also in the present chapter (Figure 2). M2A/II, from SP 10450 to 16022 (Figure 3; Plates 1 and 2), explores most of the axial zone of the Tyrrhenian ocean. M2A/III (SP 100–2380) is included in the chapter on the Calabrian arc (Finetti, this volume, Chapter 17) but exhibited (Figure 4) and commented on in this chapter also as example of the crustal setting of the Tyrrhenian margin of the Calabrian arc segment that migrated over the subducting Ionian area without colliding with continental margins.

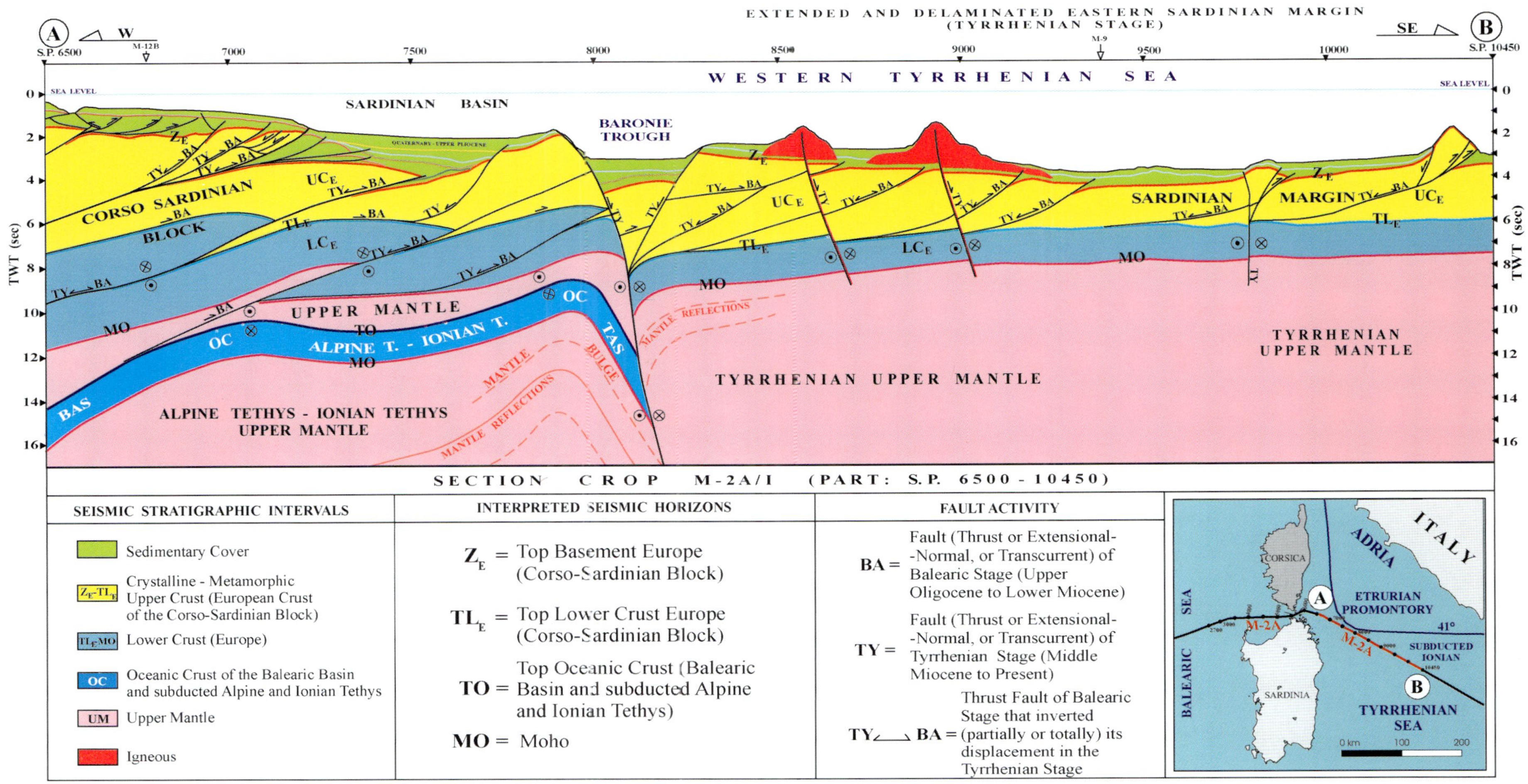

Figure 2: Reduced version of the lithospheric seismogeological section CROP M2A/I (SP 6500–10450) across the stretched eastern Sardinian margin, in the Tyrrhenian Sea. It images a segment of oceanic crust at the passage between Alpine to Ionian Tethys, the upper slab of the Alpine Tethys subducted during the Balearic stage (BAS), and the Ionian slab subducted and truncated in the Tyrrhenian stage (TAS). A prominent left-transcurrent fault across the Baronie trough and a mantle bulge are shown. To the east of the Baronie trough is stretched-delaminated crust of the Sardinian margin.

After much coordinated effort of iterative reprocessing and progressive, 3D, cross-controlled identification of the reflecting layers and crustal intervals, the complex lithospheric tectono-stratigraphic setting of Figure 2 has been reconstructed from CROP section M2A/I (SP 6500–10450).

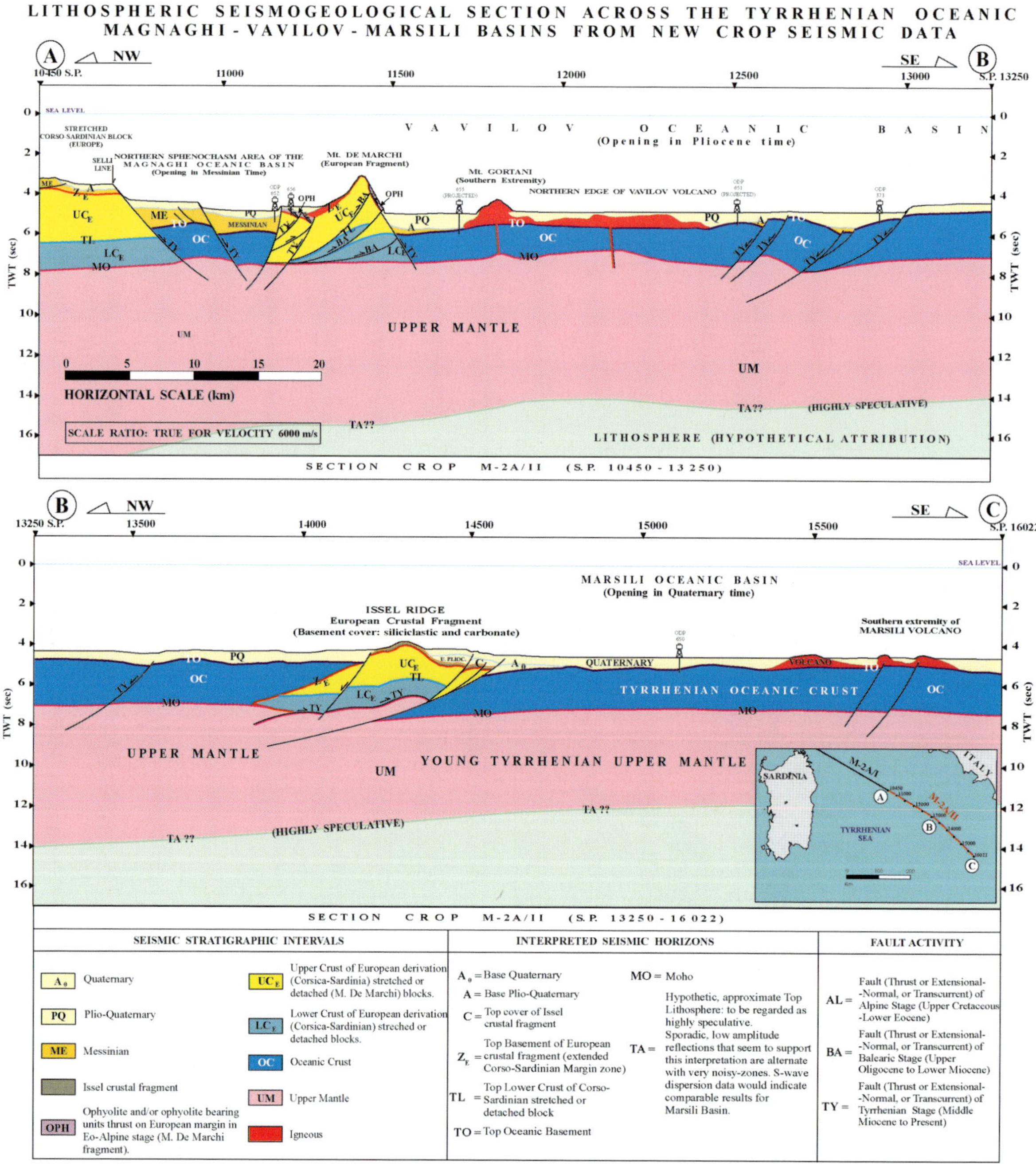

Figure 3: Reduced version of the lithospheric seismogeological section CROP M2A/II, across the Tyrrhenian Vavilov and Marsili oceanic basins.

Figure 4: Lithospheric seismogeological section of the Tyrrhenian margin of the Calabrian arc (part of CROP section M2A/III: SP 100 to 2000). This part of the internal Calabrian margin is made up of a stretched and delaminated crustal fragment from the eastern Sardinian margin. Only at the SE extremity (SP 1800–2000), beneath the Calabrides, does a layer of Panormid lower crust units appear.

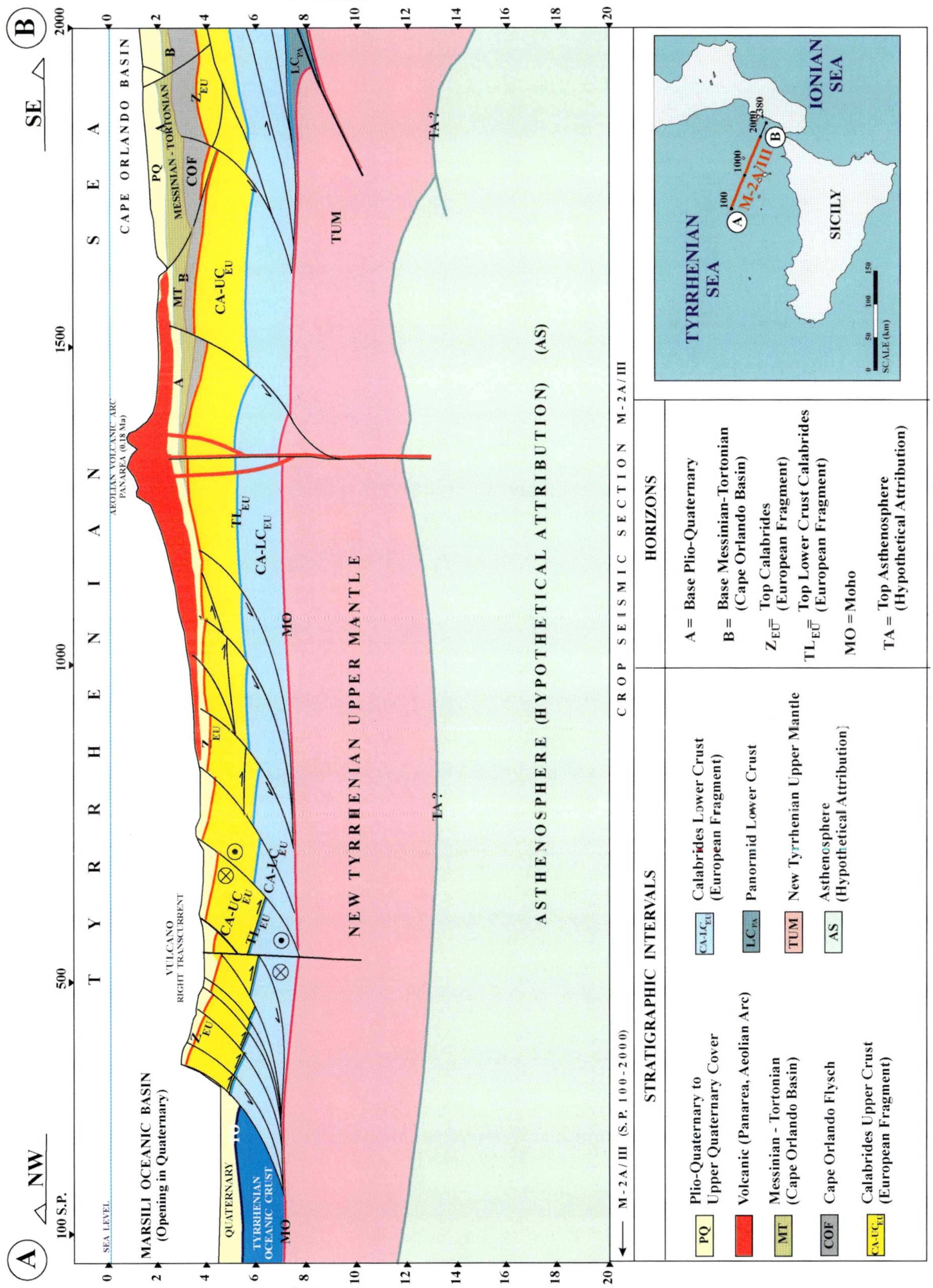
A NW
B SE
100 S.P.
500
1000
1500
2000
TWT (sec)
SEA LEVEL
MARSILI OCEANIC BASIN
(Opening in Quaternary)
QUATERNARY
TYRRHENIAN OCEANIC CRUST
T Y R R H E N I A N S E A
CAPE ORLANDO BASIN
VULCANO RIGHT TRANSCURRENT
AEOLIAN VOLCANIC ARC PANAREA (0.18 Ma)
PQ
MESSINIAN - TORTONIAN
COF
MT
CA-UC
CA-LC
TUM
NEW TYRRHENIAN UPPER MANTLE
ASTHENOSPHERE (HYPOTHETICAL ATTRIBUTION) (AS)
TA?
MO
M-2A/III (S.P. 100-2000)
CROP SEISMIC SECTION M-2A/III
STRATIGRAPHIC INTERVALS
PQ Plio-Quaternary to Upper Quaternary Cover
Volcanic (Panarea, Aeolian Arc)
MT Messinian - Tortonian (Cape Orlando Basin)
COF Cape Orlando Flysch
CA-UC EU Calabrides Upper Crust (European Fragment)
CA-LC EU Calabrides Lower Crust (European Fragment)
LC PA Panormid Lower Crust
TUM New Tyrrhenian Upper Mantle
AS Asthenosphere (Hypothetical Attribution)
HORIZONS
A = Base Plio-Quaternary
B = Base Messinian-Tortonian (Cape Orlando Basin)
Z EU = Top Calabrides (European Fragment)
TL EU = Top Lower Crust Calabrides (European Fragment)
MO = Moho
TA = Top Asthenosphere (Hypothetical Attribution)
TYRRHENIAN SEA
IONIAN SEA
SICILY
M-2A/III
SCALE (km)

To understand the features in this section, it is necessary to note in which particular paleogeographic position each part of the section is located. Looking at the small figures "a" and "b" in Plate 2 and the section position (Figure 1), it is seen that M2A/I of Figure 2 is close to the paleogeographic corner of the Etrurian promontory (NW Adria plate). There is CROP evidence that in the Balearic stage (Oligocene–Lower Miocene) the Corso-Sardinian block, driven by W-dipping subduction of the Tethys remnant (after the Eo-Alpine stage), rotated and Sardinia collided with the Etrurian promontory. In Figure 2, a deep reflecting interval is imaged (0.9–1.2 s TWT thick) attributed to the oceanic crust at the passage between the Alpine and the Ionian Tethys, around the Etrurian promontory corner. This oceanic crustal interval and its Alpine–Ionian Tethys mantle form a prominent bulge beneath Baronie Mt. and shows an E-dipping Ionian slab (TYS), truncated by a prominent left transcurrent thrust of the Tyrrhenian stage. From SP 7000 to 6500, a W-dipping subduction of the Alpine Tethys of evident Balearic age (BAS) is shown.

Over the Alpine–Ionian Tethys crust and to the east of the Baronie trough, a crustal seismic imaging of the Corso-Sardinian block and of its stretched Tyrrhenian margin is exhibited (Figure 2). To the west of the Baronie trough the section images compressive crustal faults connected to the collision of Sardinia with the Etrurian promontory in the Late Balearic stage (Lower Miocene). These faults partially inverted their displacement in the Tyrrhenian stage (Middle Miocene to Present) when the Tyrrhenian opening within the Sardinian margin took place. The stretched and delaminated crust of the Sardinian units lying to the east of the Baronie trough is related to easternward flux of the Tyrrhenian mantle along the 41° parallel, which is the northern margin of the Ionian Tethys (Figures "a" and "b" in Plate 2).

Plate 2 and Figure 3 show a lithospheric section of the young oceanic Tyrrhenian basin seen along its axial zone, from the Selli line to the southern foot of the Marsili volcano (section CROP M2A/II, SP 10450–16022). Two crustal fragments are dispersed within the ocean crust: Mt. De Marchi and the Issel ridge. The Selli line and Mt. De Marchi border the northern sphenochasm of the Magnaghi basin floored by oceanic crust. The Issel ridge separates the Vavilov basin from that of Marsili. Basaltic bodies of Mt. Gortani, of the northern foot of Vavilov volcano, and of the southern edge of Marsili volcano are crossed by the section. The deep sea drilling project and the ODP (Ocean Drilling Programme) boreholes calibrate 9 sites in the Tyrrhenian basin (from NW to SE: sites N. 654, 132–653, 652, 656, 655, 651, 373 and 650. See Figure 2 and Plate 2 for their projected locations on CROP sections M2A/I and M2A/II). Boreholes 654, 653 and 132, located west of the Selli line, and 652 on the Magnaghi sphenochasm border, encountered the Messinian, and Messinian evaporite the first three: this means that the crustal stretching of the Sardinian margin was active before the Upper Miocene. ODP borehole 656 on Mt. De Marchi found ophiolite units and their Lower Cretaceous–Upper Jurassic cover beneath Plio-Quaternary and a thin Messinian layer. This is one of the dispersed ophiolite-bearing units reported in Figure 11. The other boreholes 655, 651, 373 and 650 found the oceanic basaltic basement after crossing a progressively younger Plio-Quaternary cover, in accord with the time–space opening dynamics.

CROP section M2A/III in the southeastern Tyrrhenian (Figure 4) shows the edge of the Marsili basin which opened in Quaternary, and the inner margin of the Calabrian arc, formed from stretched crustal units of Calabrides (Sardinian margin coming). There is major thinning of the upper crust which is here only 1.7–1.8 s thick. The young volcano *Panarea* (0.18 Ma), as well as the very active *Vulcano right transcurrent fault*, are crossed and imaged by the section. This fault is a lateral megashear that dissects the Calabrian arc from the Sicilian Apennines, transferring the thick tectonic stack and the wedge of the Ionides sharply southward, along the Hyblean-Maltese escarpment (Figure 11).

3.2. CROP Section M9 across the Northern Ionian Tethys Boundary

This CROP section crosses three different crustal sectors each of a specific geological significance. In the northern part (SP 1600/1700 to 3635) the Latial continental platform area is shown, to the north of the 41° parallel (Figure 5; Plate 3 and 4). The interpreted seismogeological section in this sector images a crust affected by E-verging high-angle thrust-faults of relatively modest horizontal displacement, which, for the most part, are confined to the upper crust. This contractional deformation is associated with the Late Balearic–Early Tyrrhenian geodynamic stage (Miocene–Early Pliocene), when the W-dipping subduction of the western Alpine Tethys (Tethys remained unsubducted to the west of the Alps, after the Eo-Alpine stage) drove the rotation of the Corso-Sardinian block, determining the Corsica–Adria collision to the north of the 41° parallel (Plate 4; scheme "c"). All the structures of this Adria plate zone remain substantially rooted in their basement. Over the platform cover (AP) lies a roof-nappe of Alpine Tethydes (Ligurides) and flysch units of Miocene age. Within depressions, Messinian units are identified. Plio-Quaternary sediments cover the underlying tectono-stratigraphic sequence.

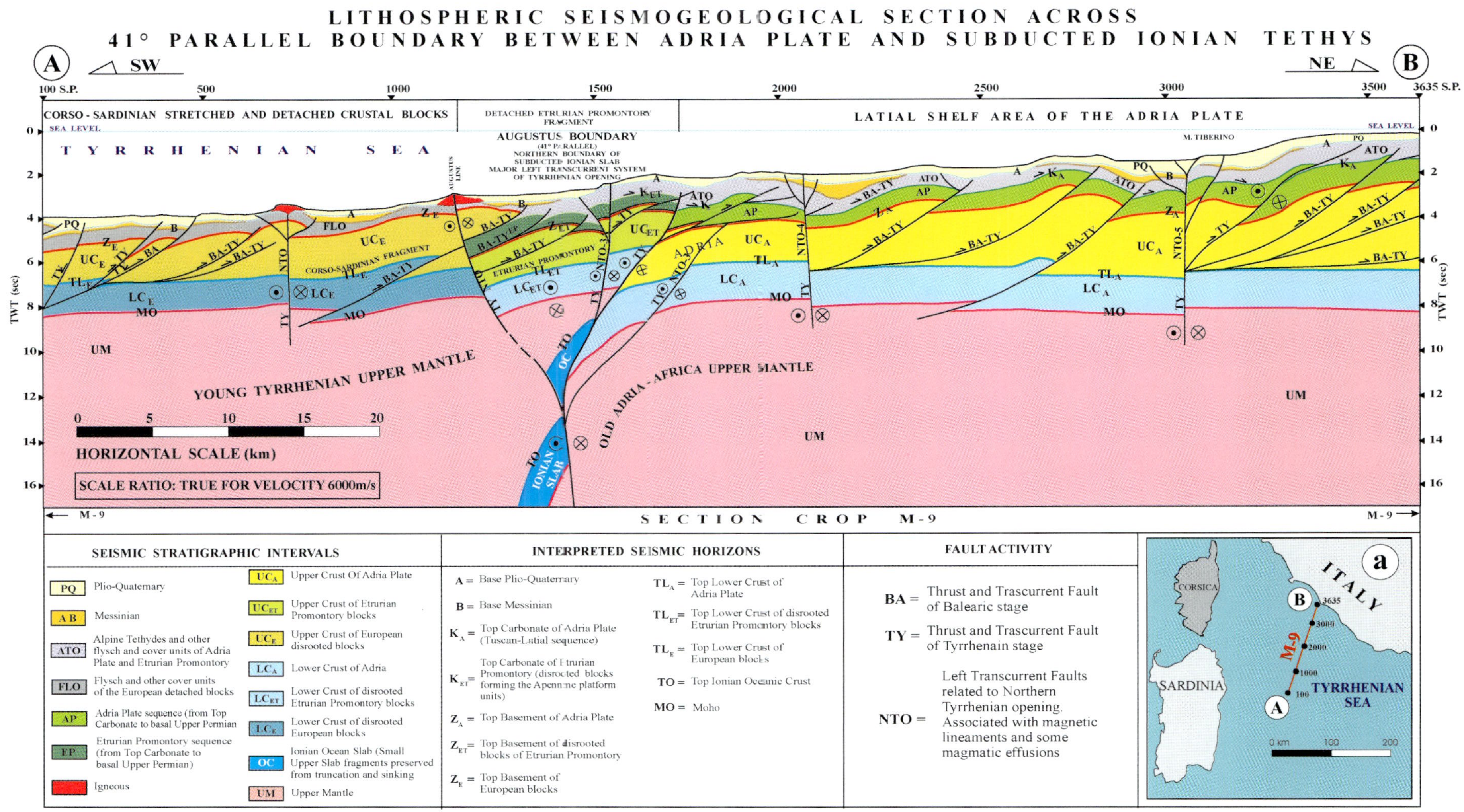

Figure 5: Reduced version of the lithospheric seismogeological section CROP M-9, across the Adria plate margin-truncated Ionian slab (41° parallel Augustus line) – migrated Sardinian crustal fragments.

After the Late-Balearic collisional contraction, when in the Early Tyrrhenian stage (Middle Miocene), driven by the east-retreating dynamics of the Ionian slab, the Tyrrhenian Sea started to open, the NW Adria plate area was affected by NW–SE-directed left-transcurrent faults. These Northern Tyrrhenian opening-related faults (NTO) also helped to reactivate previous thrust faults and to create new flower features, in some cases accompanied by magmatic intrusion–extrusion. The major NTO faults are exhibited in Plate 2 (NTO-4, NTO-5) and in Figure 11 (NTO-1, NTO-2 and NTO-3). These left-transcurrent faults are associated with clear magnetic lineaments indicative of the mantle kinematics involved. By far the most impressive NTO fault is that connected to the 41° parallel boundary of the Ionian Sea, and named the *Augustus boundary*, which is also a prominent tectono-kinematic line (*Augustus line*) (Finetti, this volume, Chapter 34). This paleogeographic boundary, as well as the whole pre-orogenic paleogeographic framework, played a major role in the geodynamic evolution of the area. A sharp compositional distinction between the igneous bodies lying to the north of the 41° parallel and those located to the south has already been noted (Serri, 1990). Now we have clear explanation why.

The Tyrrhenian opening-related mantle dynamics detached crustal fragments of the Etrurian promontory corner area, transporting them eastward along the Augustus line, and truncating the Ionian slab (Plate 2; Figure 5). The Etrurian promontory crustal units are followed by eastward migrated crustal fragments of the Corso-Sardinian block.

3.3. CROP Section M29B across the Campanian Margin-Tyrrhenian Ocean

The NE part of CROP section M29B crosses the *Northern Campanian margin* and the *Augustus megashear fault* (SP 3500–5545). On the basis of available seismic data of middle penetration and of borehole data, the section has been extrapolated landward some 25 km (Plates 5 and 6; Figure 6). The major part of the section is located in the Tyrrhenian oceanic area, across the *Magnaghi* and *Vavilov* basins (SP 600–3500). The SW extremity (SP 100–600) images the delaminated and thinned eastern margin of Sardinia.

An important contribution of this section to reconstruction of the pre-orogenic paleogeographic framework is the clear imaging of the bent *upper Ionian slab* in a zone which is close to the *Parthenopean corner* of the Ionian Tethys, formed by the 41° parallel northern limit of the subducted basin (*Augustus boundary*) and the NW–SE-trending Ionian Tethys boundary (Figure 11). The section M28A crosses the 41° parallel side of the corner. The Adria plate slope has collided with a crustal block detached from the Etrurian promontory and migrated along the Augustus boundary. This block is deformed by contractional tectonics. *The platform cover of the Etrurian promontory derived crustal blocks were frequently detached and thrust obliquely over the Apulian margin, forming roof-sited allochthonous sheets units in the Southern Apennines (Apennine platform units). So, the CROP data clarifies the debated provenance of these allochthonous platform units of the Southern Apennines.*

The Ionian subduction process is accompanied by detachment of its condensed deep water sequence, ranging in time from the Ionian opening (Permo-Triassic) to the Oligocene. These cover units were tectonically transported outward and, in part, thrust obliquely onto the Southern Apennine platform sequence (over the Lower Pliocene), in part onto the Sicilian Apennines, but most form a prominent wedge of the Calabrian arc in the Ionian Sea. *The allochthonous units of the Southern Apennines known in the literature as "Lagonegrese units" and those of the Sicilian Apennines known as "Imerese units", as clearly shown by CROP data* (Finetti, this volume, Chapter 10), *are interpreted as Ionian cover units* (*Ionian Tethydes "IT" or Ionides*). Most of the Ionides forms a prominent wedge of the Calabrian arc in the Ionian Sea. In section M29B (Plate 6), which is the first northern section affected by Ionides detachment thrust over the Southern Apennines, the position of the IT is purely indicative.

The allochthonous crustal units coming from the (Corso)-Sardinian margin (Europe) closed the Alpine Tethys in the Balearic stage (Oligocene–Lower Miocene), and in the Tyrrhenian stage (Middle Miocene–Pliocene) *were detached by the Tyrrhenian opening and migrated along the Augustus Line to reach their present position together with the Etrurian crustal fragment.* The innermost European fragments are affected by impressive listric faults and block rotation. The flat oceanic crust of the Magnaghi and Vavilov basins is covered by thin Upper Miocene and Plio-Quaterary sediments, which date the first crustal opening of this northwestern part of the basin. Some evident volcanic bodies are exhibited (Plate 6: *D'Ancona*, *Gortani*). The edge of the stretched and delaminated Sardinian margin, and two major right-transcurrent faults connected to the Tyrrhenian opening (i.e. the *Trapani* and *Solunto* faults) are self-explanatory in the section (Plate 6) and in the tectonic map (Figure 11).

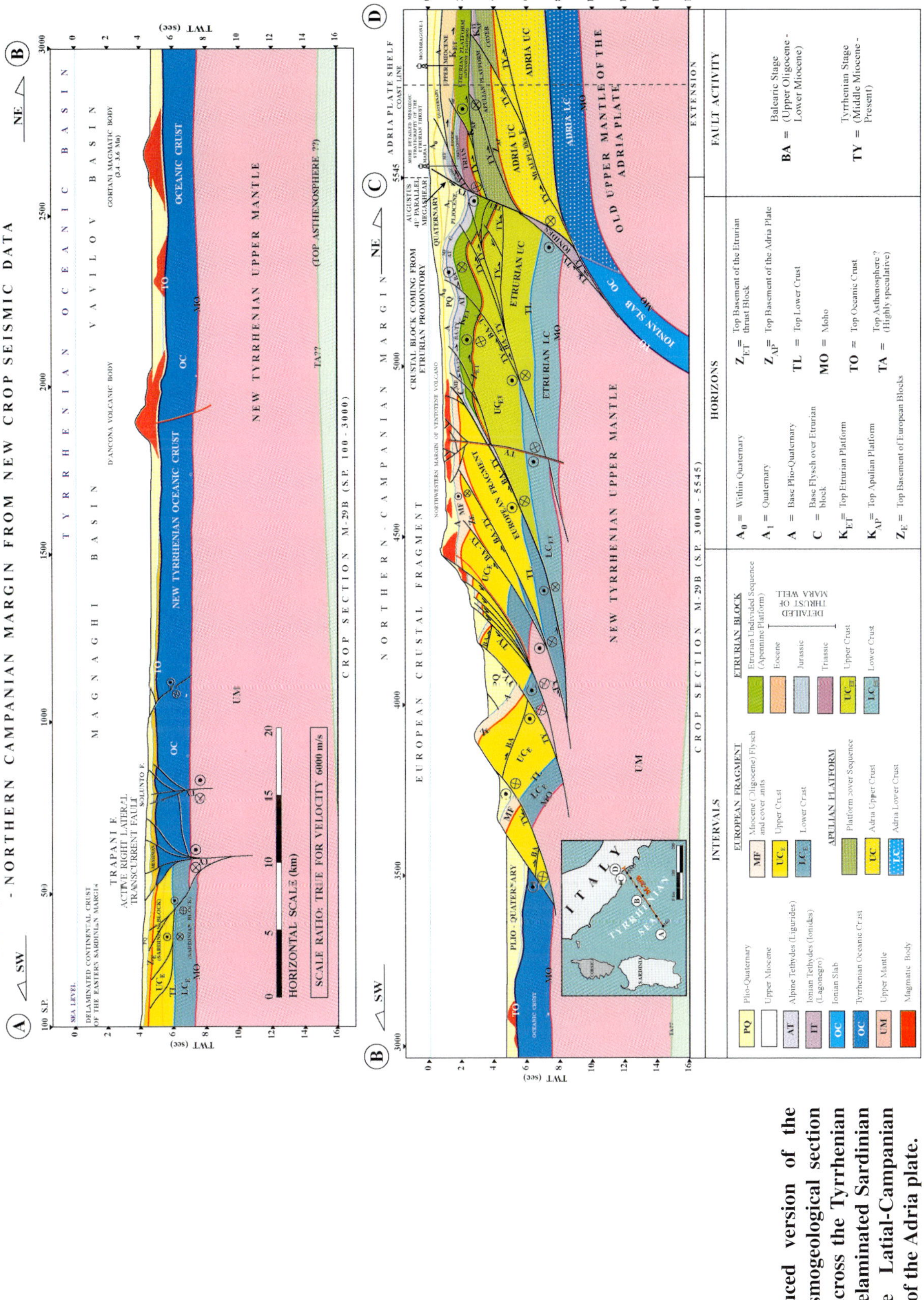

Figure 6: Reduced version of the lithospheric seismogeological section CROP M29B, across the Tyrrhenian Sea, from the delaminated Sardinian margin to the Latial-Campanian coastal area of the Adria plate.

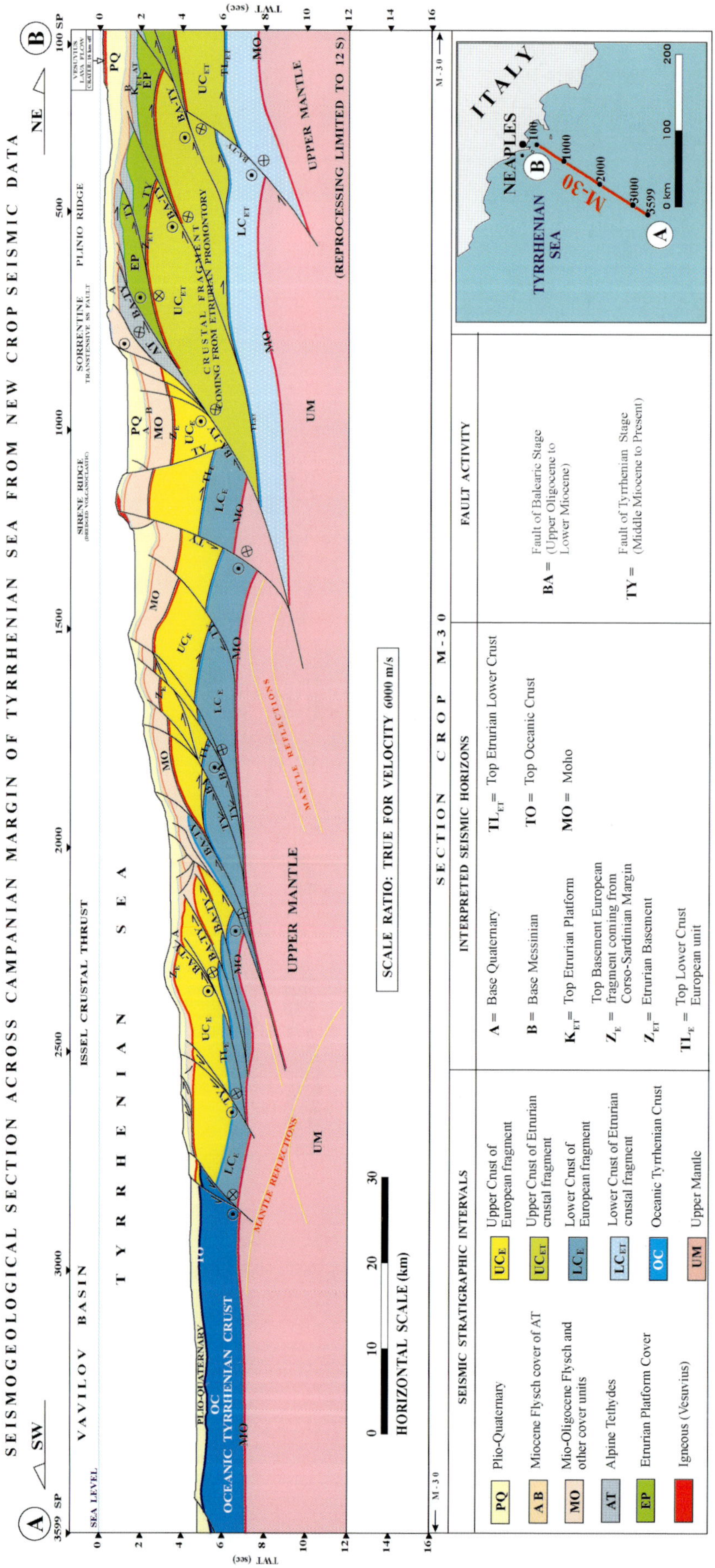

Figure 7: Reduced version of the seismogeological section CROP M30 from the Naples coast to the Tyrrhenian ocean (Vavilov basin).

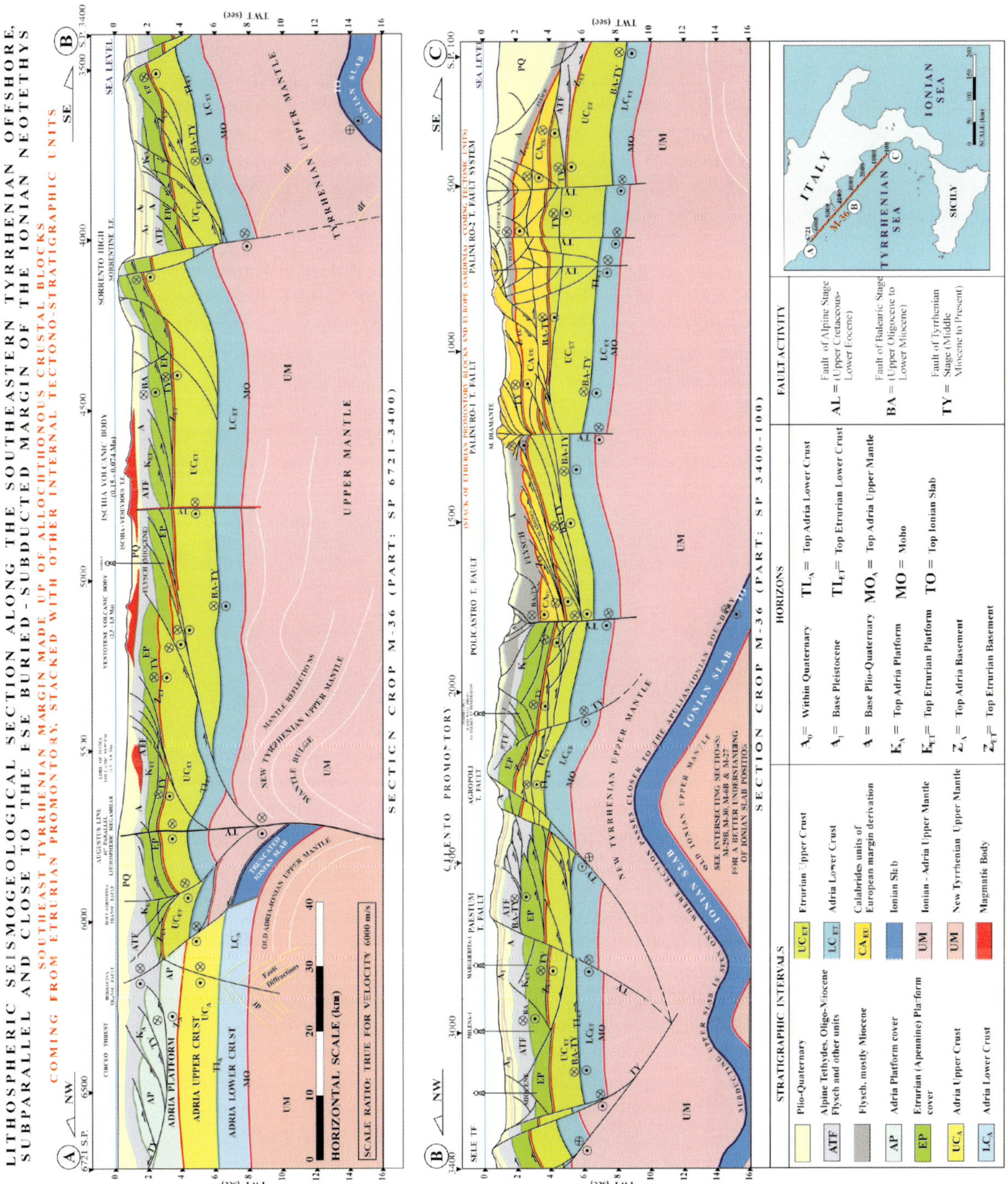

Figure 8: Reduced version of the lithospheric seismogeological section CROP M36 along the Latial-Campanian coastal area of the Tyrrhenian Sea.

3.4. CROP Section M30 across the Campanian Margin of the Tyrrhenian Sea

CROP section M30 (Plate 7) crosses the Campanian margin of the Tyrrhenian Sea, terminating in the Gulf of Naples, in front of the Vesuvian-Phlegrean volcanic area. The interpreted seismogeological section (Plate 8) shows that two tectonically stacked crustal blocks, coming from different paleogeographic positions (Sardinian margin and Etrurian promontory) form the crust of this margin area over a Tyrrhenian mantle of new settlement. In this coastal zone, evidently, the bending truncated upper Ionian slab is not imaged because it lies a little more to the east, beneath the Vesuvious-Phlegrean volcanic zone.

The more internal crustal fragment facing the Tyrrhenian ocean comes from the Sardinian margin, from which was detached during the Tyrrhenian opening. Further back in time, before the Balearic geodynamic stage, this block, jointly with the Corso-Sardinian block, belonged to the southern European margin. In the Balearic stage, the Corso-Sardinian block rotated about 45°, closing the subducted Tethyan ocean basin with tectonic transport of Alpine Tethydes over the Adria (Etrurian platform) cover. In the Tyrrhenian stage, this fragment and the fragment coming from the Etrurian promontory (Plate 8, SP 100–1000; Figure 7) were driven to their present position by collision with the Campanian margin of the Adria plate. The section (Plate 8) is fairly self-explanatory with regard to the macro-tectonic and stratigraphic setting and the fault time-analysis. We want only to draw attention to the Sirene ridge which constitutes a long narrow characteristic feature of the margin due to a complex combination of thrusting and perpendicular transtensive effects of the Sorrentine left transcurrent fault (Figure 11). On the NE extremity, a flat igneous layer representing a lava flow from the Vesuvious volcanic apparatus is visible.

3.5. CROP Section M36, along the Eastern Tyrrhenian Coast

CROP seismic section M36 (Plate 9) was acquired to control the tectono-stratigraphic features of the eastern Tyrrhenian margin, which also on the MS lines (Finetti and Del Ben, 1986) revealed clear tectonic trends perpendicular to the Southern Apennine system. The interpretation given in seismogeological section M36 (Plate 10; Figure 8) shows that the new lithospheric information supplied is of great importance, even more than expected. At the NW extremity, the section crosses the *Augustus Line* (northern boundary of the Ionian Tethys) where lithospheric megashear-related tectonics started with the early Tyrrhenian opening process. The truncation of the Ionian slab by the Augustus megashear fault, and the connected mantle tectono-dynamics imaged by good reflection data even in the mantle, is very impressive. A bulge of the new Tyrrhenian mantle and the bending upper Ionian slab truncated by the Augustus left transcurrent fault are also evident. From the Augustus shear zone to SP 3800, the subducted Ionian slab is deeper than section bottom; from SP 3800 to 1650, where the slab is closer to its boundary, it is visible as an undulating reflecting oceanic crustal interval; from SP 1650 to the SE extremity of the section, again it is deeper than section bottom. To better understand the position of the bent upper Ionian slab, it help to look at the intersecting CROP sections M9, M29B and M30 in this work, together with M6B (Finetti et al., this volume, Chapter 12), and M27 (Finetti, this volume, Chapter 17). From these CROP sections perpendicular to the coast, it is also easier to understand the tectono-dynamics of the allochthonous crustal fragments which detached from their original crustal setting and migrated eastward and southeastward to collide with the Apulian margin in the Ionian Tethys. From careful cross-controlled 3D seismic interpretation, it is evident that the crustal blocks we see on section M36 from the Augustus Line to the *Policastro* transcurrent fault (Plate 10, 10b, 10c and Figure 8) are crustal fragments coming from the Etrurian promontory area, detached and migrated in the Tyrrhenian geodynamic stage (Middle Miocene to Pleistocene). Such Etrurian crustal blocks include the lower and upper crust and the cover which are detached from basement and driven eastward. As reported above in Section 3.3, the *cover of Etrurian promontory-derived fragments constitute the roof-thrust carbonate units of the Southern Apennines, known in literature the as Apennine platform units*, and of questioned provenance.

From the Policastro transcurrent fault to the SE extremity of the section, the cover units of the Etrurian crustal block thrust eastward, are replaced by Calabrides units (Plate 10; Figure 8) coming from the eastern margin of Sardinia. *The routes followed by the main crustal fragments of the tectonic stack which collided with the Apulian slope are imprinted as magnetic lineaments across the Tyrrhenian Sea.* Over the platform cover units, the Alpine Tethydes

---

Figure 9: Lithospheric seismogeological section across the delaminated Sardinian margin in the southwestern Tyrrhenian Sea (part of CROP section M28B: SP 2000 to 4000).

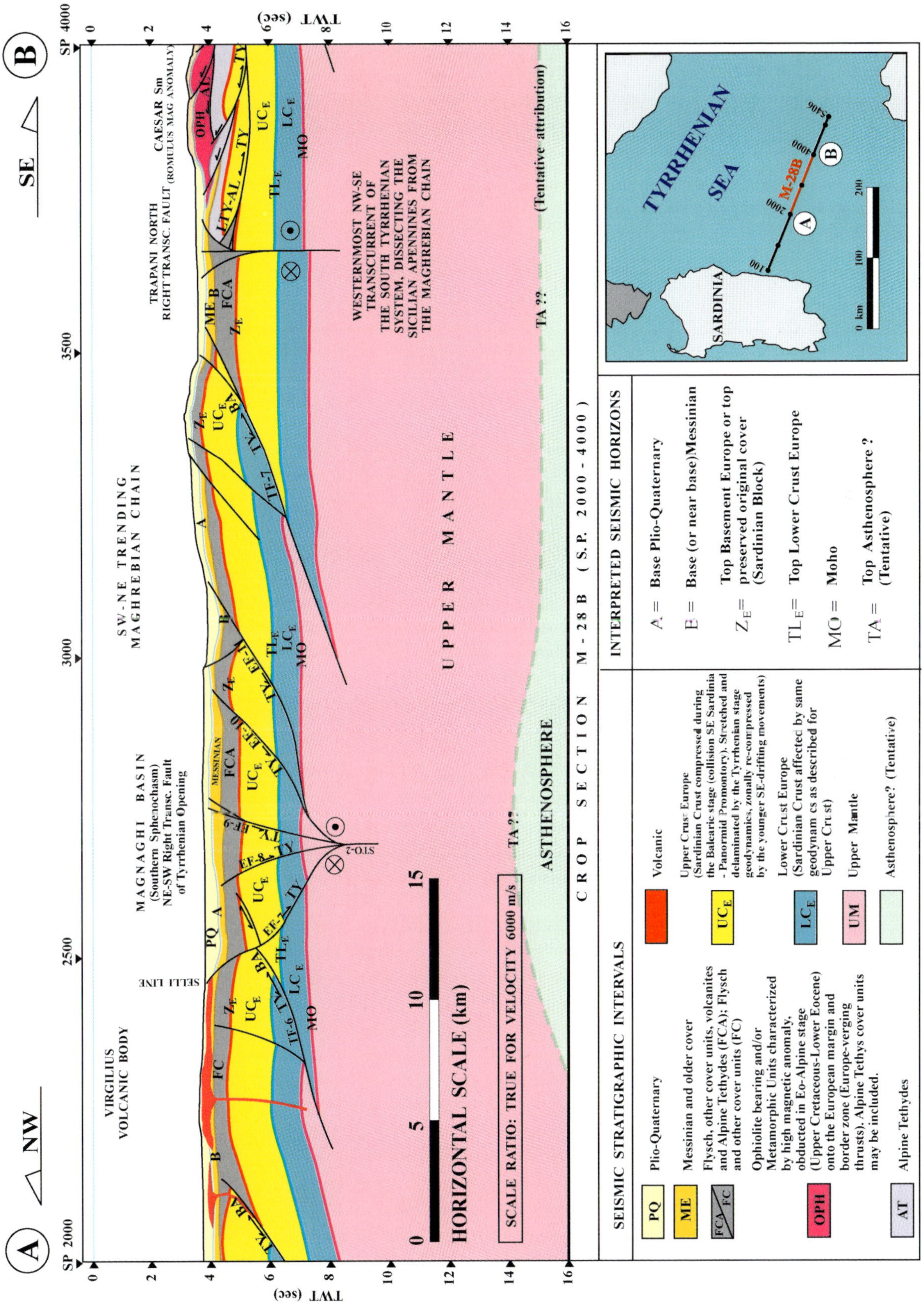
A
B
NW
SE
SP 2000
2500
3000
3500
SP 4000
TWT (sec)
VIRGILIUS VOLCANIC BODY
SELLI LINE
MAGNAGHI BASIN (Southern Sphenochasm) NE-SW Right Transc. Fault of Tyrrhenian Opening
SW-NE TRENDING MAGHREBIAN CHAIN
TRAPANI NORTH RIGHT TRANSC. FAULT
CAESAR Sm (ROMULUS MAG ANOMALY)
WESTERNMOST NW-SE TRANSCURRENT OF THE SOUTH TYRRHENIAN SYSTEM, DISSECTING THE SICILIAN APENNINES FROM THE MAGHREBIAN CHAIN
MESSINIAN
STO-2
UPPER MANTLE
ASTHENOSPHERE
TA ??
(Tentative attribution)
HORIZONTAL SCALE (km)
SCALE RATIO: TRUE FOR VELOCITY 6000 m/s
CROP SECTION M-28B (S.P. 2000-4000)
SEISMIC STRATIGRAPHIC INTERVALS
PQ Plio-Quaternary
ME Messinian and older cover
FCA/FC Flysch, other cover units, volcanites and Alpine Tethydes (FCA); Flysch and other cover units (FC)
OPH Ophiolite bearing and/or Metamorphic Units characterized by high magnetic anomaly, obducted in Eo-Alpine stage (Upper Cretaceous-Lower Eocene) onto the European margin and border zone (Europe-verging thrusts). Alpine Tethys cover units may be included.
AT Alpine Tethydes
Volcanic
UC_E Upper Crust Europe (Sardinian Crust compressed during the Balearic stage (collision SE Sardinia - Panormid Promontory). Stretched and delaminated by the Tyrrhenian stage geodynamics, zonally re-compressed by the younger SE-drifting movements)
LC_E Lower Crust Europe (Sardinian Crust affected by same geodynamics as described for Upper Crust)
UM Upper Mantle
Asthenosphere? (Tentative)
INTERPRETED SEISMIC HORIZONS
A = Base Plio-Quaternary
B = Base (or near base) Messinian
Z_E = Top Basement Europe or top preserved original cover (Sardinian Block)
TL_E = Top Lower Crust Europe
MO = Moho
TA = Top Asthenosphere ? (Tentative)
TYRRHENIAN SEA
SARDINIA
M-28B
0 km 100 200

(Ligurides) units and flysch units (ATF) were thrust during the Balearic geodynamic stage. Over the Calabrides, we find Oligo-Miocene flysch.

From the tectonic viewpoint, it is interesting to observe how the prominent lithospheric slices were produced on section M36 by *left-transcurrent faults, mostly displacing the stacked thrust-pile*. These are the *Sorrentine, Sele, Paestum, Agropoli, Policastro, Palinuro 1* and *Palinuro 2 systems*. The Sorrentine fault seems to also affect the Apulian crust and to extend as far as the Vulture volcano (Figure 11). *The Policastro left transcurrent fault transfers ophiolite-bearing units into the Southern Apennines*. Further, more detailed information can be obtained directly from the self-explanatory annotation of the section.

3.6. CROP Section M28B on Stretched-Delaminated Sardinian Margin

Part of interpreted section CROP M29B (SP 2000–4000) is plotted in Figure 9 and *images the crustal characteristics of a stretched and delaminated Sardinian margin*. This marginal area is more completely described in another part of this volume (Finetti et al., this volume, Chapter 18) dedicated to the Corso-Sardinian block. In this paragraph we examine this very thin continental crustal zone of the Tyrrhenian Sea for a better understanding of the delaminating effects produced by the Ionian slab subduction and dynamics of the connected Tyrrhenian mantle flux. *Opening of the Tyrrhenian back-arc basin took place within the eastern Sardinian crust, starting from Messinian times* (although stretching initiated earlier in the Burdigalian) as indicated by the sedimentary cover distribution on the newly created oceanic crust. From the Selli line, the opening process progressively expanded the back-arc basin to its present size (Figure 11). The *Trapani right lateral transcurrent fault (Figure 9), at the end of the Balearic geodynamic stage (Lower Miocene), dissected a continuous wide eastward migrating Apennino-Maghrebian arc, separating off the Sicilian Apennines which developed in the Tyrrhenian stage (Middle Miocene to Present) as an orogenic system characterized by oblique thrusting and prominent right-transcurrent tectonics. Seismically characteristic, thin uppermost, Europe-verging thrust units, associated with prominent magnetic anomalies has been found on the SE margin of Sardinia*. In our analysis *there is good evidence that such units represent ophiolite-bearing rocks thrust over the southern European (Sardinian) margin in the Eo-Alpine stage (Upper Cretaceous–Lower Eocene)*. One of these characteristic thrust elements is the *Caesar Sm.* site (Figure 9). *The discovery of these ophiolite-bearing units of the SE Sardinian margin provides that previously missing segment of ophiolite needed for the reconstruction of a continuous ophiolite belt from the Betic zone to the Alps.*

3.7. CROP Section M6A across the Sicilian Margin

The northern Sicilian margin in the Tyrrhenian Sea is dominated by a crustal stack of allochthonous tectonic units coming from different paleogeographic sectors. From the inner to outer zone we find thrust crustal blocks of Calabrides coming from the Sardinian margin, stacked with crustal blocks coming from the Panormid promontory. The Calabrides in the Early Balearic stage (Oligocene) were stretched during the W-dipping subduction of the Alpine Tethys and successively, in the Late Balearic stage (Lower Miocene), due to collision of the rotating SE-Sardinian margin with the NE Panormid promontory, were recompacted. Stretching also affected the Calabrides around the Panormid corner area, in the Early Tyrrhenian stage (Middle Miocene–Lower Pliocene), followed by recompaction during their oblique thrusting over the Sicilian slope (Middle Upper Pliocene).

Two different deep basin units are associated with the complex tectonic stack which collided with the northern Sicily slope: the Alpine Tethydes (Ligurides, Sicilides) and the Ionian Tethydes or Ionides (Imerese) units. The Alpine Tethydes (AT) were stripped off from their oceanic basement (Alpine Tethys remnant conserved after the Eo-Alpine stage) in the Balearic stage, and the Ionides (IT) in the Tyrrhenian stage.

CROP section M6A (Figure 10) images very clearly the Ionian slab and transcrustal right transcurrent faults (Solunto, Solunto West and Taormina). Between the bent Sicilian-Ionian crust and overriding thrust-stack, well reflecting disjointed units are imaged, which we attribute to Ionides trapped in the orogenic system. Among the several CROP section examples that exhibit converging zones on subducted Apulia and Sicily margin, this in Figure 10 is the only one that shows such characteristic reflecting units, which evidently detached from the subducted Ionian slab and could not thrust outward because they were trapped by the stack process.

CROP section M6A has been more extensively described in Chapter 15 on Sicily.

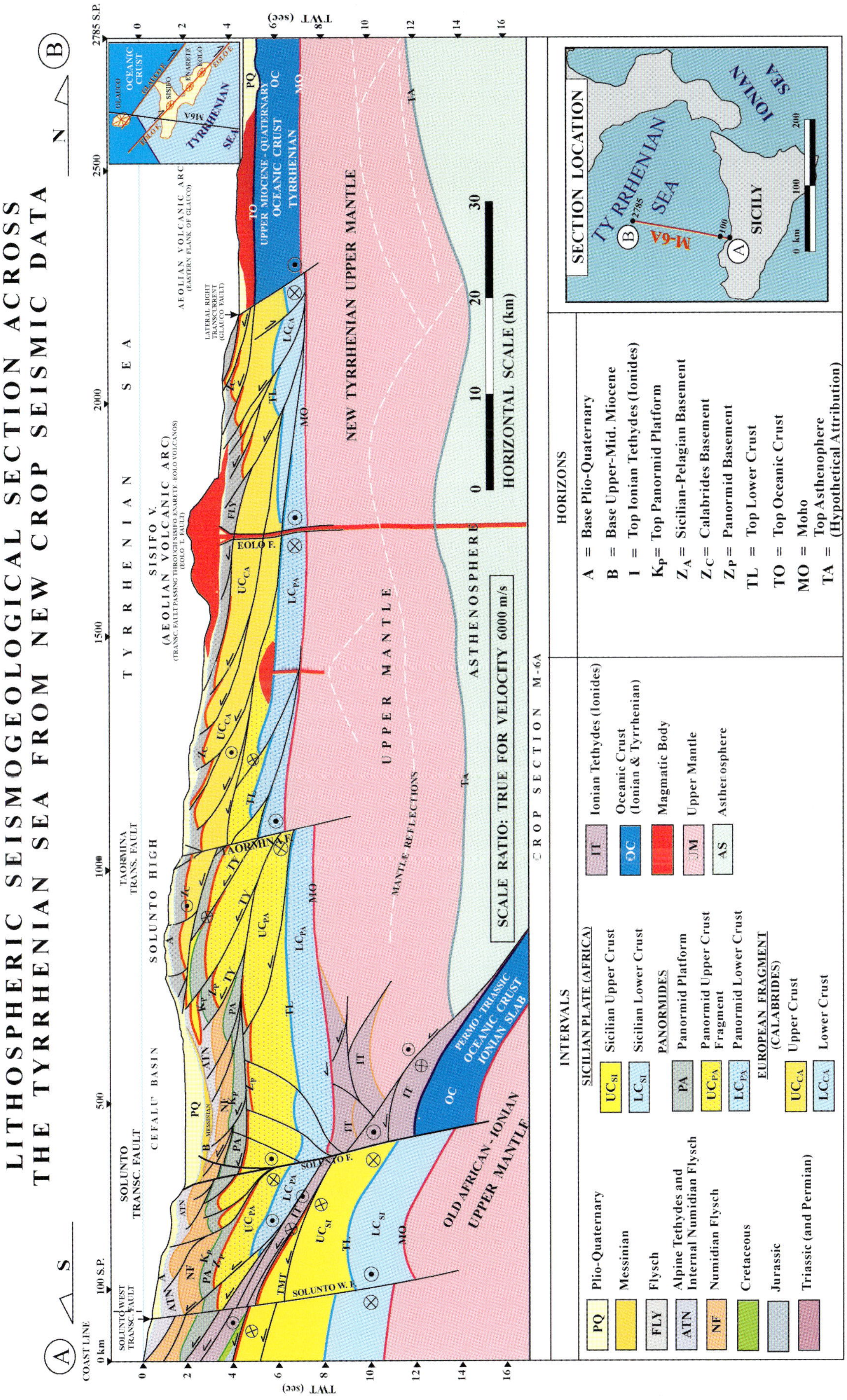

Figure 10: Example of seismogeological section on the Tyrrhenian margin of Sicily (M6A, SP 100–1200).

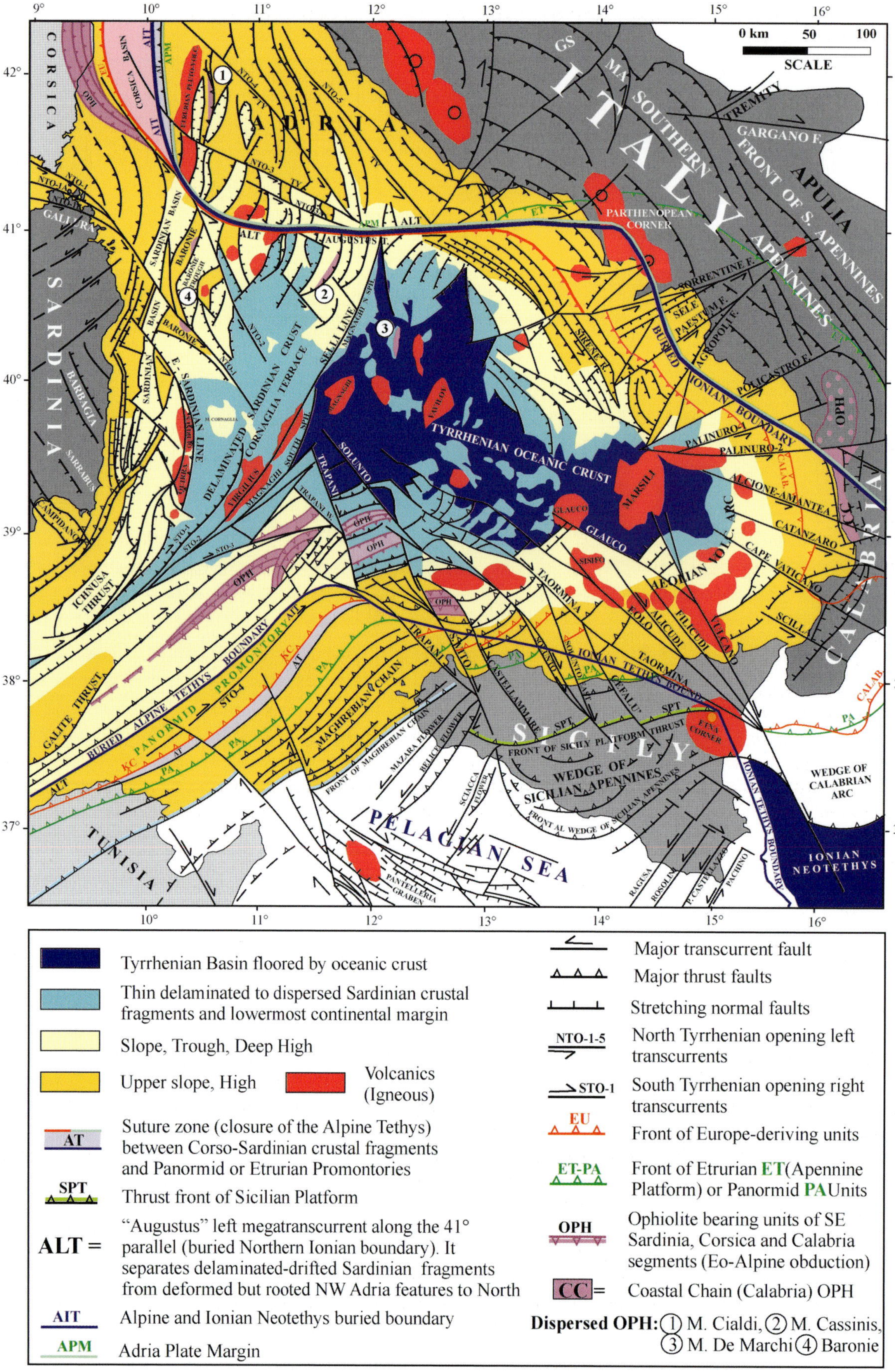

Figure 11: Sketch map of the Tyrrhenian Sea and surrounding area, showing major tectono-dynamic features related to Alpine–Ionian Tethys lithosphere roll-back sinking and paleogeographic setting.

4. TYRRHENIAN GEODYNAMICS DERIVED FROM NEW LITHOSPHERIC CROP SEISMIC CONSTRAINTS

As mentioned in the introduction, various models have been proposed in the literature for the Tyrrhenian opening. A detailed tectono-stratigraphic setting and a paleogeographic evolution of the Tyrrhenian opening interpreted as a subduction-related back-arc basin, supported by a significant multichannel-multifold seismic dataset, was presented years ago (Finetti and Del Ben, 1986). In the same year, other authors also (Malinverno and Ryan, 1986) proposed a subduction-related model with the introduction of a schematic context that now is seen to have been very intuitive. Other geoscientists also favoured this model (Boccaletti et al., 1971, successively abandoned; Royden et al., 1987; Wortel and Spakman, 1992; Jolivet and Faccenna, 2000).

The new CROP seismic data described here and in other chapters of this volume, supply new seismic information extending through the whole crust-upper mantle/lithosphere on those key features that provide constraints for a realistic reconstruction.

Firstly *the CROP data imaged the upper Ionian slab buried beneath the Tyrrhenian margins and the Alpine Tethys slab, allowing a new, well-controlled reconstruction of the pre-orogenic paleogeography which was determinant for understanding the successive evolution.* Starting from the Lower Cretaceous paleogeographic setting, it was possible to establish with satisfactory reliability that *in the Eo-Alpine geodynamic stage (Upper Cretaceous–Lower Eocene) the Africa–Adria and interposed Ionian Tethys (Permo-Triassic opening) converged with Europe as a joint block. This convergence determined a S-dipping subduction of the Alpine Tethys beneath Adria and the Adria–Europe collision in the Alps sector. To the west of the Alps a wide Alpine Tethys remnant persisted, even if restricted with respect to its original Lower Cretaceous extent. This is proven both by the new reconstructed Lower Cretaceous paleogeography which leaves Adria to collide with Europe without closing the western Alpine Tethys, and by the CROP imaging of a W-dipping Alpine Tethys slab that subducted under the Corso-Sardinian block in the Balearic stage (Oligocene–Lower Miocene). At the end of the Balearic stage, a single ample continuous orogenic arc was formed which included earlier elements of the Northern and Southern Apennines, and of the Calabrian Arc, Sicilian Apennines and Maghrebian chain. The rotated Corso-Sardinian block (eastern margin) was facing three different situations*: (1) *Corsica, after closure of the interposed Alpine Tethys, was colliding with NW Adria (Etrurian promontory)*; (2) *SE Sardinia, after rotation and closure of the wide interposed Alpine Tethys was colliding with the NW Panormid promontory*; (3) *the southern extremity of Corsica and that major part of Sardinia lying to the north of the Panormid promontory, jointly with their frontal wedge formed of Alpine Tethydes and flysch, were entering into the Ionian Tethys ocean of Permo-Triassic opening with no continental obstacle in front.*

From this evolutionary stage, driven by the lithospheric roll-back and sinking of the Ionian Tethys with consequent SE-directed flux of Tyrrhenian mantle asthenosphere, commenced the impressive geodynamic evolution of the Tyrrhenian stage (Middle Miocene to Present).

To the north, the Etrurian promontory bisected the wide migrating continuous arc of the Balearic stage, separating the Northern Apennines from the Southern Apennines. To the south, the Panormid promontory bisected the Maghrebian chain from the Sicilian Apennines.

The Northern Apennine contraction continued driven by a huge eastward-migrating asthenospheric bulge formed under the Tuscan Archipelago, and the eastward-moving Tyrrhenian mantle flux along the Augustus Line (41° *parallel northern Ionian boundary) contributed to its connected counterclockwise rotation. The Southern Apennine–Calabrian arc–Sicilian Apennines great arc developed progressively from the Middle Miocene to the Present with a fast Ionian slab retreat (over 10 cm/year during the Messinian–Lower Pliocene). The condensed deep basinal cover sequence of the Ionian Tethys, deposited from the Permo-Triassic basin opening to the Oligocene, detached from its subducting basement (Ionian slab) and thrust obliquely onto the Apulian platform in the Southern Apennines, and onto the Sicilian Platform in Sicily, while on the southeastward migrating Calabrian Arc, it stacked to form a prominent wedge of imbricated sheets. These Ionian cover units go to form the Lagonegro thrust units in the Southern Apennines and the Imerese-Sicanian in Sicily. To simplify the origin of these deep basinal units of the Southern Apennines and Sicily, and their nomenclature, we adopt the unifying, self-explanatory name of Ionian Tethydes or Ionides* (Finetti, this volume, Chapter 10).

The Tyrrhenian Sea opened as a back-arc basin within the eastern margin of Sardinia, with a prominent delamination of the area which remained to become the present Sardinian margin (Cornaglia Terrace), while the detached crustal Sardinian units migrated outward on the whole great circum-Tyrrhenian arc. Crustal fragments detached also from the Etrurian and Panormid promontories and stacked in the circum-Tyrrhenian margin of the arc. This great arc was affected by prominent right-transcurrent faults on the Apulian margin and Calabria, and left-transcurrents on the Sicilian margin.

During the fast oceanic lithosphere roll-back/sinking dynamics of the Balearic and Tyrrhenian stages, the Africa–Europe convergence, from the Oligocene, acted in the area as a much slower process with the most evident effects along the right-transcurrent Insubric flower and along right-transcurrent faults of the Adriatic and Pelagian seas (Finetti, this volume, Chapter 34).

ACKNOWLEDGEMENTS

This work was partially funded by CROP Partners (CNR, ENEL, ENI) with CNR grants to I.R. Finetti: CTR 96.00797; CTB 97.04525.CT05. Major financial support was given by MIUR: Italian programmes; grants COFIN 98, COFIN 2000 and COFIN 2002 to I.R. Finetti (University of Trieste). Relevant financial support was granted also by the University of Trieste for reprocessing and interpretation. The Authors express their gratitude to all supporters. Many thanks are due to OGS, Seismic Division, to all Teams of the OGS-Explora Seismic Vessel for the CROP seismic data acquisition; to ENEL for the processing of part of the CROP seismic sections used; and to the EGG (Exploration Geophysics Group) of the Faculty of Sciences of the University of Trieste for iterated interpretative reprocessings of the major part of the CROP data used, where signal improvement was crucial for a better understanding of the complex features involved. We wish to thank William B.F. Ryan of the Lamont-Doherty Earth Observatory, Columbia University, USA, for reviewing the manuscript and for much appreciated comments.

REFERENCES

Boccaletti, M., Elter, P., Guazzone, G., 1971. Plate tectonic models for the development of the Western Alps and Northern Apennines. Nature, Phys. Sci. 234, 108–111.

Boccaletti, M., Conedera, C., Dainelli, P., Gocev, P., 1982. The recent (Neogene-Quaternary) regmatic system of the Western Mediterranean region. J. Petr. Geol. 5, 31–49.

Carmignani, L., Kligfield, R., 1990. Crustal extension in the Northern Apennines: the transition from compression to extension in the Alpi Apuane core complex. Tectonics 9, 1275–1303.

Carmignani, L., Decandia, F.A., Fantozzi, P.L., Lazzarotto, A., Liotta, D., Meccheri, M., 1994. Tertiary extensional tectonics in Tuscany (Northern Apennines). Tectonophysics 238, 295–315.

D'Agostino, N., Jackson, J.A., Dramis, F., Funiciello, R., 2001. Interactions between mantle upwelling, drainage evolution and active normal faulting: an example from the central Apennines (Italy). Geophys. J. Int. 147, 475–497.

Faccenna, C., Davy, P., Brun, J.P., Funiciello, R., Giardini, D., Mattei, M., Nalpas, T., 1996. The dynamics of back-arc extension: an experimental approach to the opening of the Tyrrhenian Sea. Geophys. J. Int. 126, 781–795.

Finetti, I.R., 1982. Structure, stratigraphy and evolution of Central Mediterranean. Boll. Geof. Teor. Appl. 24, 247–426.

Finetti, I.R., this volume. Understanding the Ionides and their geodynamics. Chapter 10.

Finetti, I.R., this volume. The Calabrian arc and subducting Ionian slab from new CROP seismic data. Chapter 17.

Finetti, I.R., this volume. Geodynamic evolution of the Mediterranean Region, from the Permo-Triassic Ionian opening to the Present, constrained by new lithospheric CROP seismic data. Chapter 34.

Finetti, I.R., Del Ben, A., 1986. Geophysical study of the Tyrrhenian opening. Boll. Geof. Teor. Appl. 28, 75–156.

Finetti, I.R., Morelli, C., 1973. Geophysical exploration of the Mediterranean Sea. Boll. Geof. Teor. Appl. 15, 263–341.

Finetti, I.R., Morelli, C., Zarudzki, 1970. Reflection seismic study of the Tyrrhenian Sea. Boll. Geof. Teor. Appl. 12, 311–345.

Finetti, I.R., Del Ben, A., Forlin, E., Prizzon, A., this volume. Review of the main results of the pioneering MS seismic exploration programme in the deep water of the Mediterranean Sea (OGS, 1969–1980). Chapter 1.

Finetti, I.R., Lentini, F., Carbone, S., Del Ben, A., Di Stefano, A., Guarnier, P., Pipan, M., Prizzon, A., this volume. Crustal tectono-stratigraphy and geodynamics of the Southern Apennines from CROP and other integrated geophysical-geological data. Chapter 12.

Finetti, I.R., Del Ben, A., Fais, S., Forlin, E., Klingelé, E., Lecca, L., Pipan, M., Prizzon, A., this volume. Crustal tectono-stratigraphic setting and geodynamics of the Corso-Sardinian block from new CROP seismic data. Chapter 18.

Jolivet, L., Faccenna, C., 2000. Mediterranean extension and the Africa–Eurasia collision. Tectonics 19, 1095–1106.
Lavecchia, G., Stoppa, F., 1996. The tectonic significance of Italian magmatism: an alternative view to the popular interpretation. Terra Nova 8, 435–446.
Locardi, E., 1982. Individuazione di strutture sismogenetiche dall'esame dell'evoluzione vulcano-tettonica dell'Appennino e del Tirreno. Mem. Soc. Geol. It. 24, 569–596.
Malinverno, A., Ryan, W.B.F., 1986. Extension in the Tyrrhenian Sea and shortening in the Apennines as a result of arc migration driven by sinking of the lithosphere. Tectonics 5, 227–245.
Mantovani, E., Albarello, D., Tamburelli, C., Babbucci, D., Viti, M., 1997. Plate convergence, crustal delamination, extrusion tectonics and minimization of shortening work as main controlling factors of the recent Mediterranean deformation pattern. Ann. Geofis. 40, 611–643.
Roeder, D., 1980. Geodynamics of the Alpine-Mediterranean system – a synthesis. Eclog. Geol. Helv. 73, 353–377.
Royden, L.H., Patacca, E., Scandone, P., 1987. Segmentation and configuration of subducted lithosphere in Italy: an important control on thrust-belt and foredeep-basin evolution. Geology 15, 714–717.
Serri, G., 1990. Neogene-Quaternary magmatism of the Tyrrhenian region: characterization of the magma sources and geodynamic implication. Mem. Soc. Geol. It. 41, 219–242.
Tapponnier, P., 1977. Evolution tectonique du système alpin en Méditerranée: poinçonnement et écrasement rigide-plastique. Bull. Soc. Géol. Fr. 7, 437–460.
Van Bemmelen, R.W., 1972. Driving forces of Mediterranean orogeny (Tyrrhenian test-case). Geol. Mjinbouw 51, 548–573.
Wezel, F.C., 1982. The Tyrrhenian Sea: a rifted krikogenic-swell basin. Mem. Soc. Geol. It. 24, 531–568.
Wortel, M.J.R., Spakman, W., 1992. Structure and dynamics of subducted lithosphere in the Mediterranean region. Proc. K. Ned. Akad. Wet., Biol. Chem. Geol. Phys. Med. Sci. 95, 325–347.

CHAPTER 22

Geodynamic Significance of Tyrrhenian Seafloor Morphology

M. Marani[1]

ABSTRACT

Based on the morphology of the Tyrrhenian seafloor, four discrete sectors with distinctive topography can be identified. Given the recent development of the Tyrrhenian back-arc basin, its sea-bottom terrain chronicles the geologic events that led to its formation. The morphological sectors that distinguish the Tyrrhenian Sea are shown to be related to four geodynamic provinces closely linked to the evolving back-arc basin. Furthermore, in some cases, seafloor topography results from the effects of deeper geological processes that act at the lithospheric level, giving key information concerning mantle properties and dynamics. Detailed swath bathymetry therefore develops into a significant additional research tool in geologically active marine regions, to be fully exploited along with deep-sounding geophysical surveys.

Keywords: Tyrrhenian seafloor multibeam, geodynamic significance

1. INTRODUCTION

The Tyrrhenian Sea, the youngest back-arc basin of the Mediterranean, has been entirely surveyed by means of multibeam swath bathymetry, from its deeper portions up to 400 metres water depth. The resulting data set provided baseline morphological information to correlate research in portions of the Tyrrhenian involving rock sample collection and deep-towed side-scan sonar surveys.

Detailed seafloor mapping in evolving arc/back-arc settings can reveal surface features that, for the most part, result from the deeper geological processes acting in the region. It is evident that deep-seated events influence the pattern of structural styles, the development of volcanism and even the occurrence of large sedimentary transport and depositional systems. However, the comprehensive characterization of the surface expression and effects of these processes provides a means to define the properties, structure and dynamics of their source region, the mantle.

In some cases, as this chapter attempts to show, the effects of events occurring as deep as within the asthenosphere or at the asthenosphere–lithosphere boundary are well revealed by surface morphology.

2. REGIONAL SETTING

The formation of the Tyrrhenian Sea occurs within the overall context of slow convergence between the African and Eurasian plates, which currently characterizes the Mediterranean (Dewey et al., 1989; Argus et al., 1989; De Mets et al., 1990; Ward, 1994). Bordered to the east and south by the Apennine–Maghrebid mountain belt, and to the west by the passive Sardinian margin, the Tyrrhenian basin formed as a consequence of rifting and back-arc extension of the Alpine/Apennine suture above the currently northwest-subducting Ionian oceanic slab (Kastens et al., 1988, 1990; Sartori, 1990; Jolivet, 1991).

E–W directed rifting in the northern Tyrrhenian in the Lower-Mid Miocene (~ 15 Myr) and along the western margin of Sardinia in the Tortonian (~ 11 Myr) (Zitellini et al., 1986; Kastens et al., 1990) marks the initial opening of the Tyrrhenian basin leading to the formation of oceanic domains in the Southern Tyrrhenian. First, production of ocean crust occurred westward, during the Pliocene spreading of the Vavilov basin (4.3–2.6 Myr), (Kastens et al., 1990) accompanied by the thermal subsidence of the thinned western margin crust. A subsequent

[1]Institute of Marine Geology, CNR, Via P. Gobetti 101, 40129 Bologna, Italy. E-mail: michael.marani@bo.ismar.cnr.it

change to ESE-directed extensional stress in Late Pliocene–Quaternary resulted in the emplacement of basaltic crust southeastwards, generating the Marsili back-arc basin (2 Myr) (Kastens et al., 1990).

The most widely accepted mechanism to account for the migration in space and time of rifting, volcanism and ocean crust emplacement links the eastward migration of crustal thinning and oceanic accretion to the passive rollback of the Ionian plate (Malinverno and Ryan, 1986; Savelli, 1988; Argnani and Savelli, 1999). Since the Early Miocene, the time of the Alpine/Apennine collisional suture in the future Tyrrhenian area (Beccaluva et al., 1994), most of the pre-existing Mesozoic oceanic lithosphere in the western and central Mediterranean had been consumed, with the exception of the remnant Ionian ocean, delimited to the southwest by its ancient margin, the Malta escarpment.

The Tyrrhenian opening, induced by the rollback of the remnant slab, was matched by the eastward and southward radial growth of the Apennine–Maghrebid fold and thrust belt on the Italian peninsula and Sicily (Sartori, 1989). In step with back-arc basin development, the subduction-related island arc volcanism of the southern Tyrrhenian basin migrated from west to southeast, from Sardinia (32–13 Ma) to the currently active Aeolian island arc (Serri, 1997, and references therein), developing the present-day arc/back-arc configuration of the southern Tyrrhenian region.

The deep-water, ocean crust floored, central and southern Tyrrhenian Sea is surrounded by different geodynamic margin settings. The western Tyrrhenian margin represents a typical passive continental margin, while the eastern and southern margins are associated with high seismicity, active volcanism and elevated rates of uplift of land areas represented by the Apennine–Maghrebid mountain chain. In these latter areas, the deposits deriving from both onshore and marine volcanic activity make an important contribution to material entering the basin. Several submarine canyons dissect the continental slopes, representing major sediment transport pathways from the margins to the deeper parts of the basin, the sites of mixed terrigenous and volcanoclastic turbititic accumulation.

3. GEODYNAMIC PROVINCES OF THE TYRRHENIAN BASIN

Due to the young age of the Tyrrhenian Sea, tectonic and associated volcanic processes exert a strong control on the seafloor make-up. Consequently, simple examination of the bathymetric characteristics of the basin allows a primary subdivision (Figure 1) of the region on the basis of its morphology: the northern Tyrrhenian zone; the western Sardinia margin; the eastern Tyrrhenian margin of peninsular Italy and Sicily, and the deep, central and southeastern abyssal plains.

However, this classification based on purely morphological features, which undeniably derives from the formational history of the basin, is substantiated by distinctive geological and geophysical information that characterizes each zone. As a result, the morphology of the Tyrrhenian seafloor essentially distinguishes four principal geodynamic provinces related to its formation and development.

3.1. Basin and Range Province: Northern Tyrrhenian

Bordered to the north by the Tuscan Archipelago and to the south by the conjunction of the Baronie seamount and Pontine Islands, this sector is roughly triangle-shaped and extends between 42°N, 11°W (northern edge of survey) and 40°30′N (Figure 2).

Bathymetrically, the northern Tyrrhenian principally develops as a series of N–S and NNW–SSE-trending structural highs and adjacent, relatively flat-lying basins. On average, the structures have a length of $\sim$40 km and are set 20/25 km apart. The bathymetric base level of the region, corresponding to the basin depths, is $\sim$1500 meters in the northern portion of the region and reaches 2000 meters only in the southernmost part of it. The major structures (Figure 2), such as the Etruschi seamount in the west, and the Civitavecchia valley in the east, have a continuity of more than 70 km, and the average elevation of the basin highs, with respect to the lows, is more than 1000 meters, with the summits reaching 300/400 meters below sea level.

Previous work in the region, based on single channel seismic reflection profiles, has shown the extensional nature of the northern Tyrrhenian terrain (Zitellini et al., 1986; Marani and Zitellini, 1986; Bartole, 1995), which is composed of a series of rotational normal faults resulting in a tectonic framework of tilted blocks, half-graben and horst structures.

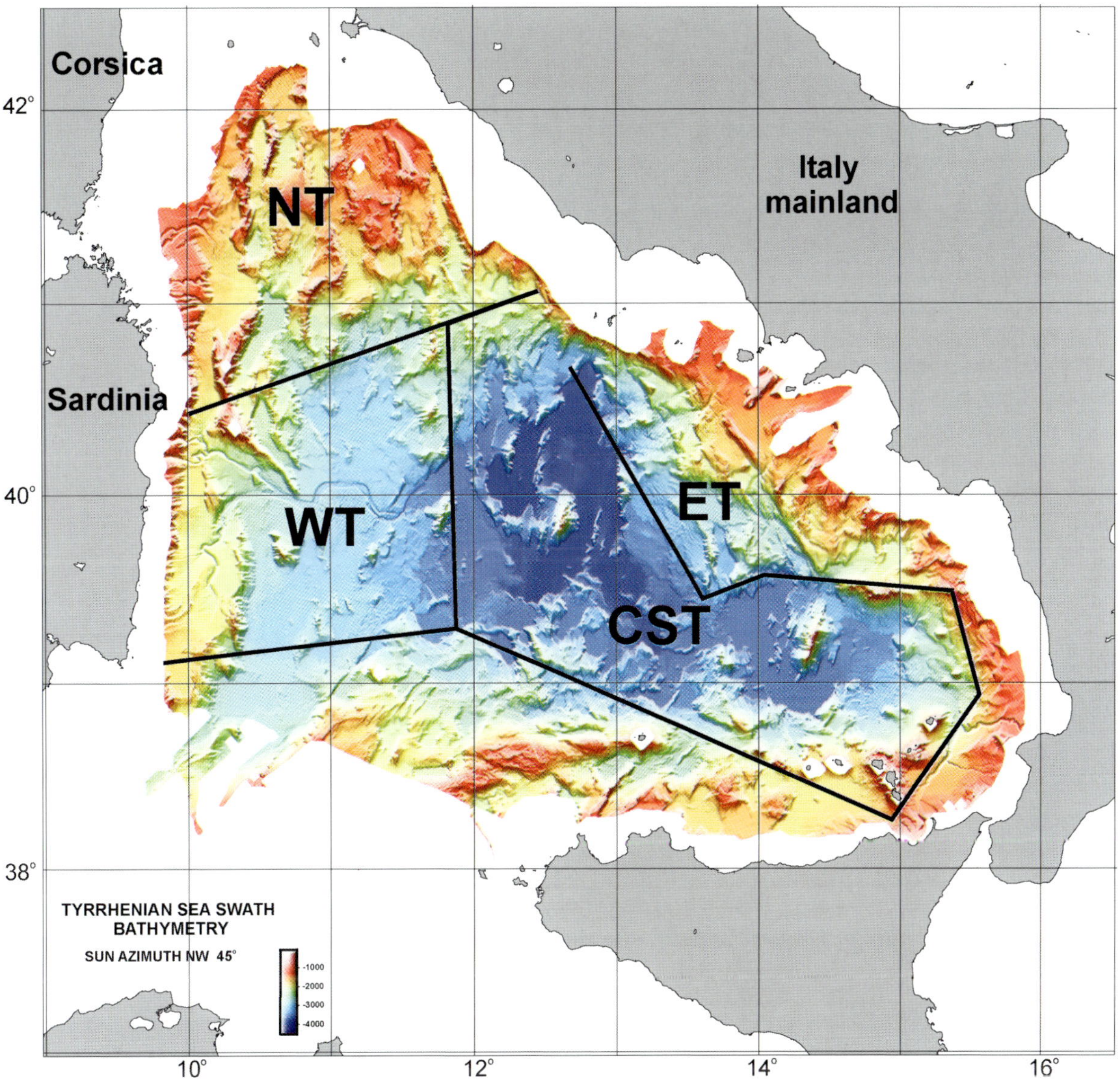

Figure 1: Shaded relief bathymetric map of the Tyrrhenian Sea showing the surveyed region. Depths are colour-coded, illumination from the NW. Black lines divide the four geodynamic provinces discussed in the text: NT, Northern Tyrrhenian Basin and Range province; WT, Western Tyrrhenian passive margin province; ET, Eastern Tyrrhenian active margin province and CST, Central and Southern Tyrrhenian ocean-crust-floored province.

Extensional thinning of the significantly thickened crust resulting from the Alpine/Apennine collisional suture that previously occupied this region began in the Early-Mid Miocene (Carmignani and Kligfield, 1990; Jolivet, 1991; Pascucci et al., 1999). Extension with accompanied magmatic activity gets younger from west to east (Beccaluva et al., 1989; Serri et al., 1993; Carmignani et al., 1995), from ~15/20 Myr in the Corsica basin to the presently active normal faults in the northern Apennines. The northern Tyrrhenian is currently characterized by relatively thin crust (Scarascia et al., 1994) of the order of 22 km, and a shallow asthenosphere, 50 km deep (Suhaldoc and Panza, 1989; Serri et al., 1993; Finetti et al., 2001; Finetti et al., this volume, Chapter 8).

Detailed seafloor bathymetry thus furnishes a comprehensive picture of the array of extensional features that affected a thickened crustal wedge in the 200-km-wide northern Tyrrhenian marine region. The overall surface topography of this broad extensional system reveals a marked similarity to areas of diffuse continental extension, such as the Basin and Range province of the western United States. This comparison between structural styles, based only on morphological grounds, seems supported however by the finding of low-angle detachment faults and core complexes in Corsica and Tuscany (Carmignani and Kligfield, 1990; Jolivet, 1991), by the extensive block faulting, and by the diachronous development of extension and magmatism. In effect, all these traits are common to the

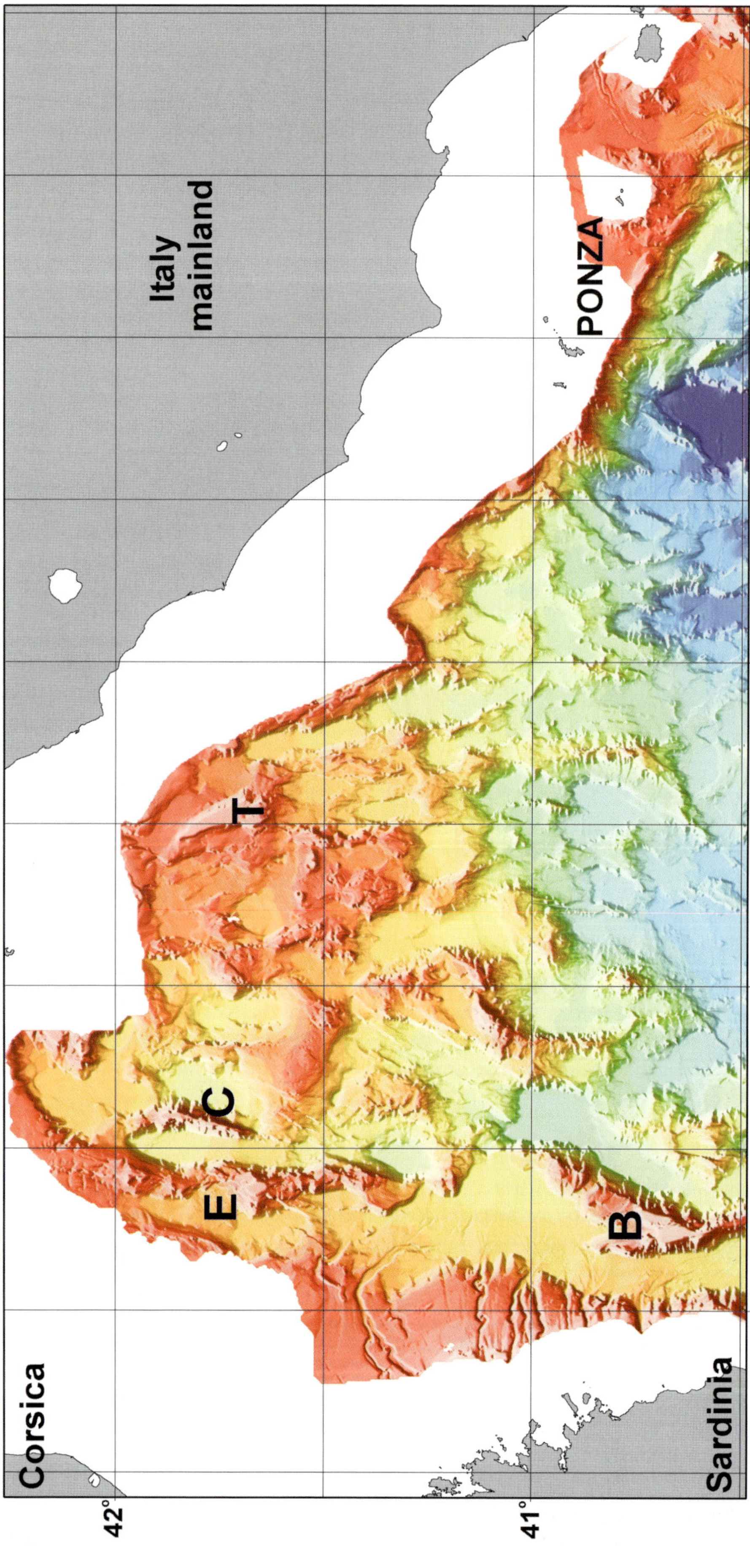

Figure 2: North Tyrrhenian Province. Colour code and illumination as in Figure 1. The relatively shallow base-level of the seafloor is interrupted by mostly N–S trending structural highs and intervening basins. Major seamounts are Etruschi (E), Cialdi (C) and Tiberino (T). Baronie seamount (B) to the west and Ponza offshore structures to the east delimit the region to the south. Refer to the text for further discussion.

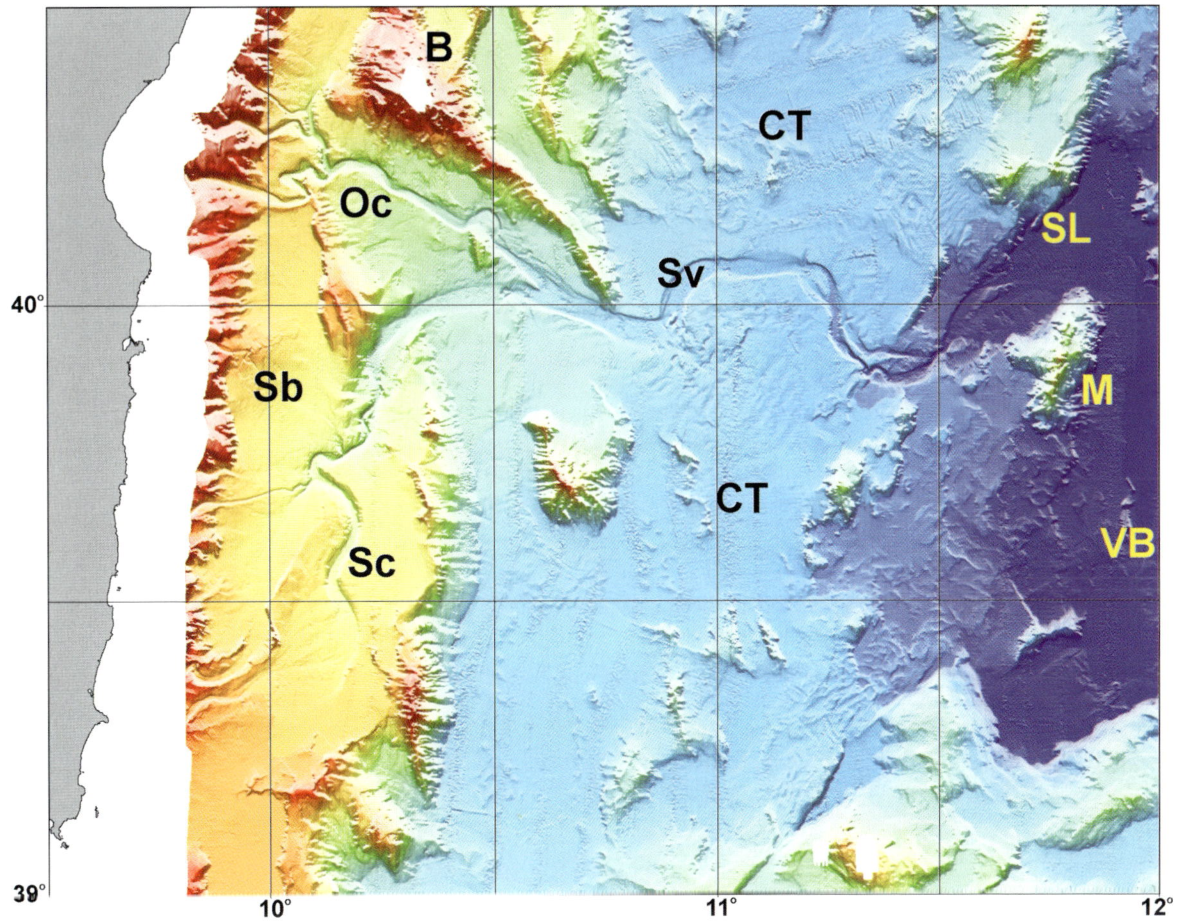

Figure 3: Western Tyrrhenian Province. Colour code and illumination as in Figure 1. The region is subdivided into two separate physiographic zones by the outcropping morphological highs: the Sardinia basin (Sb) and the Cornaglia terrace (CT). The trend of the southern portion of Baronie seamount (B) allows breaching of the Sarrabus (Sc) and Orosei (Oc) canyons seaward where they merge to form the Sardinia valley (Sv). The western province is delimited eastwards by the Selli Line (SL) which drops down into the deep Vavilov basin (VB). Magnaghi seamount is one of the two volcanoes that characterize the VB. Refer to the text for further discussion.

Basin and Range province; namely extension involving low-angle extensional faults coupled to high-angle, rotational block faults, and the documented migration in time and space of extension and magmatism (Wernicke, 1981; Lister and Davis, 1989; Surpless et al., 2002).

3.2. Passive Margin Province: The Western Sardinia Margin

The morphological link between the northern Tyrrhenian Basin and Range province and the passive margin province of western Sardinia is the Baronie seamount, the largest structural high in the Tyrrhenian Sea, with a length of over 120 km. Moreover, from a geodynamic point of view, the Baronie seamount represents the tectonic connection between the two provinces, being made up of Alpine units, analogous to those outcropping in eastern Corsica, and Alpine foreland units that make up most of the islands of Corsica–Sardinia (Structural model of Italy, 1991). South of the Baronie seamount, in fact, only the latter units seem to persist, according to rather extensive seafloor sampling.

The western Sardinia province is thus is bound to the north by the Baronie seamount at 40°30′N and to the south at 39°N, covering the marine area eastwards to 11°30′E (Figure 3).

The margin consists of two distinct physiographic belts parallel to the Sardinia coastline and with increasing water depth. At a regional depth of ~ 1000–1500 meters, an upper slope belt, about 50 km wide, of sediment-filled, flat-lying hanging basins develops, bounded seawards by a series of structural highs. The basins, together generally termed the Sardinia basin (SB) are delimited landwards by the outer continental shelf and slope, dissected by numerous, small rectilinear canyons. These promote basin filling but also contribute to the development of the larger-scale Sarrabus and Orosei canyon systems that merge at a breach in the bounding structural highs, in proximity to the southern Baronie seamount, and continue to the deeper ocean as the Valley of Sardinia.

The Valley of Sardinia extends within the second belt, which is also bathymetrically well-defined. This consists of a relatively flat lying deep-water plain ($\sim$ 2500–2800 meters water depth), extending about 70 km seawards, called the Cornaglia Terrace (CT). The plain is bounded eastwards by a NE–SW-directed fault scarp, the Selli Line that, in effect, delimits the Sardinia margin from the Tyrrhenian abyssal plain.

Apart from the Baronie seamount, the structural highs that bound the SB display a subdued topography. Numerous reflection seismic studies in the region show that the structural highs are actually the footwall leading edges of large tilted blocks, formed by the development of rotational faults dipping eastwards, with the half-grabens formed by block tilting now practically filled in by the SB sediments. Similar structures underlie the CT, although the thinner sediment cover (most of it being prevented from reaching the CT by the hanging basins of the SB) does not register block tilting as adequately as in the SB.

The initiation of activity of the structures has been established by the drilling results of ODP Leg 107 (Kastens et al., 1990). A date of Mid-Upper Miocene (Tortonian $\sim$ 10 Myr) is established from hole 654. However, given the fact that un-datable conglomerates make up the base of the sequence, one cannot rule out that the age of inception of the activity is earlier.

Thus, the Sardinia slope area morphology records the rifting episode that resulted in the subsequent formation of the ocean-crust-floored deep Tyrrhenian basin domain. During this latter period, the Sardinia slope area began to represent the passive margin geodynamic province of the basin. Crust thickness is a clear record of this process, thinning from $\sim$30 km beneath Sardinia to less than 10 km in the Vavilov basin (Suhaldoc and Panza, 1989; Scarascia et al., 1994).

3.3. Active Margin Province: The Eastern Tyrrhenian

For the purpose of this chapter, the eastern Tyrrhenian margin is defined as the continental slope region between the Pontine Archipelago and the Palinuro seamount. It has variable width, between a minimum of a few kilometers to about 60 km, and runs parallel to peninsular Italy in a NW–SE direction (Figure 4). The southern Apennine chain represents the emergent area to the east of the margin.

Morphologically, the limits of the area are well revealed. To the north, a series of $\sim$ 25-km-long fault scarps that develop in a NE–SW direction west and south of Ponza Island are seen to clearly separate the structural trends of the eastern Tyrrhenian margin from the northern Tyrrhenian Basin and Range province. Similarly, the E–W trending, 50-km-long Palinuro volcanic complex defines the boundary between the Aeolian arc region and the eastern Tyrrhenian. The western limits of the area are the flat lying, deep-water, Vavilov and Marsili abyssal plains.

In contrast to the passive margin geodynamic province, the structures in this region are morphologically distinct, despite the undoubtedly large sediment supply due to proximity to the Apennine mountain chain, implying a very recent or on-going tectonic activity. On the basis of its morphology, the eastern Tyrrhenian margin can be morphologically and structurally sub-divided into two sectors roughly corresponding to the upper ($<$ 2000 m depth) and lower ($>$ 2000 m depth) continental slope regions.

Structural partitioning of the margin is brought about by a 330°/340°-trending, 80-km-long normal fault system. The system basically consists of a 330°-directed, west-dipping master fault which displays increasing throw southwards, from $\sim$400 meters to over 1 km. An east-dipping 60 km-long conjugate fault develops with a 340° trend and a throw of the order of 300 meters. The resulting graben consequently has southward increasing widths, from $\sim$5 km in the north to $\sim$ 12 km in the south. Between the two faults, three NW–SE faults develop with 200-meter throws, producing three steps, distanced roughly 15 km, within the graben. Based on its overall morphological characteristics, the fault system, named the Eastern Tyrrhenian Margin Fault (ETMF) has been interpreted as a left-lateral transtensive system (Musacchio et al., 1999). The southern termination of the ETMF is in proximity to the conjunction between the Palinuro and Marsili volcanic seamounts.

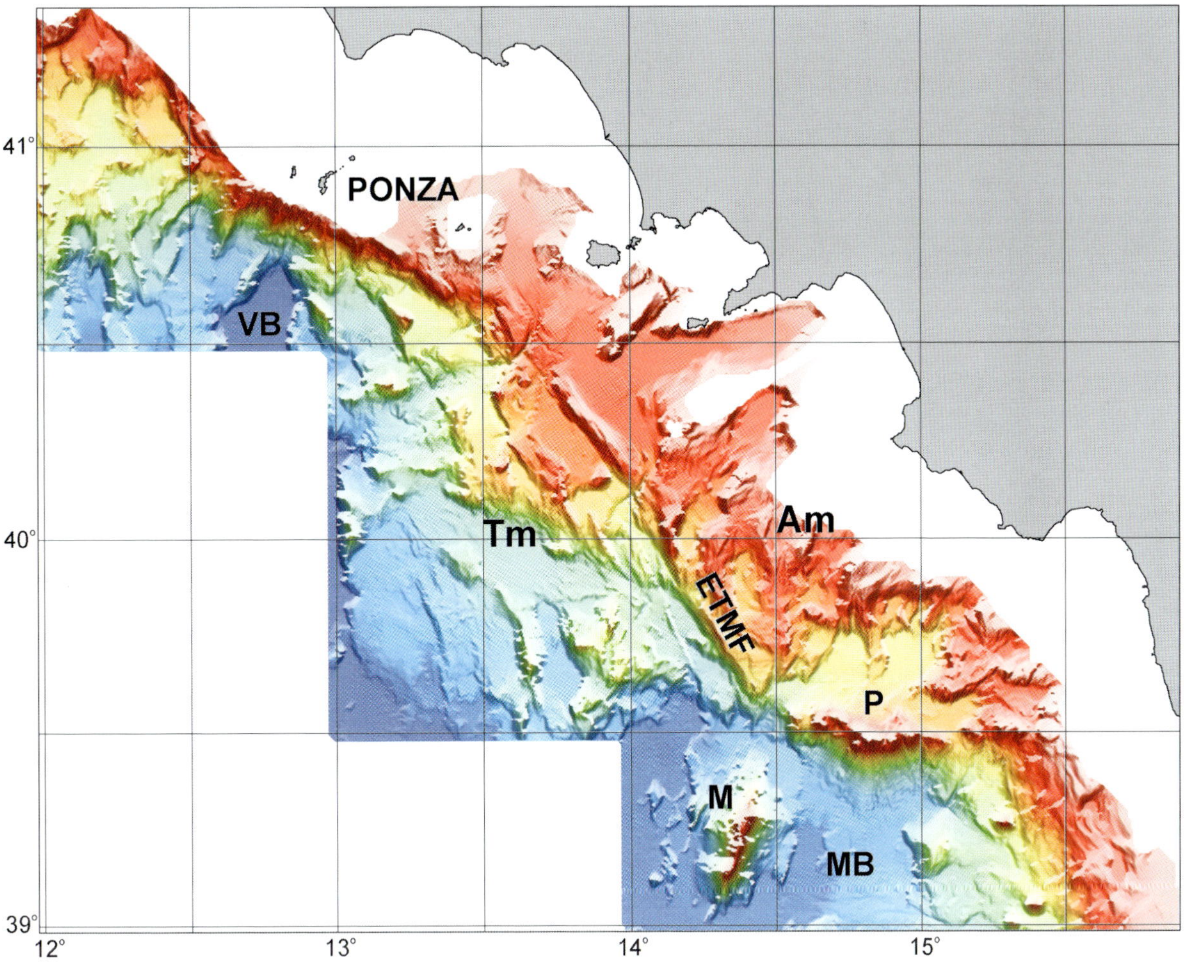

Figure 4: Eastern Tyrrhenian Province. Colour code and illumination as in Figure 1. The outstanding feature of the region is the Eastern Tyrrhenian margin fault (ETMF) that separates the structurally diverse upper ("Apenninic"-type margin, Am) and lower ("Tyrrhenian"-type margin, Tm) slopes of the area. Also visible are the Palinuro volcanic complex (P) forming the southern boundary to the province, the Marsili volcano (M) within the Marsili basin (MB) and the northern extension of the Vavilov basin (VB) offshore Ponza Island. Refer to the text for further discussion.

The upper slope region is characterized by the development of numerous relatively short, 100 to 250 meter fault scarps. Fault orientation falls within the N–S quadrant mostly in the seaward portion of the region, while mostly NE–SW trends are present to the east. The scarps extend on average 20 km and dissect the region into several small, N–S/NE–SW-trending horsts and grabens, each on average about 8 km wide.

The development of these upper slope structures occurs exclusively to the north of the ETM; they are not present in the Sirene seamount region or to the east of the ETMF's southern termination. Moreover, the southern terminations of the structures are directly juxtaposed with the major ETMF hanging wall. This seems to indicate that there may well be a structural connection between these two tectonic regimes.

A distinctive morphology prevails within the lower slope of the eastern Tyrrhenian margin, west of the ETMF system, from offshore its southern termination to south of the NE–SW faults adjacent to Ponza. The striking morphological differentiation from the upper slope region is brought about by the development of a number of linear fault scarps that closely follow the trend of the ETMF. The faults affect the entire extension of the lower slope area, up to the limits of the deep-water abyssal plain. Average fault length is in the order of 35 km with down-to-the-west variable throw between 100 and 400 meters. The ETMF trend is dominant, with virtually the only exception being the N–S fault bounding Flavio Gioia seamount.

The ETMF in point of fact defines a lower slope affected by extensional structures directly linked to the generation of the deeper portion of the Tyrrhenian Sea, and an upper slope area that can only be loosely associated with this event; the ETMF seems to divide a "Tyrrhenian-type" of lower slope from an "Apenninic-type" upper slope. This latter structural style is confined to the upper slope area north of the ETMF-Palinuro seamount alignment. In fact, south of Palinuro seamount, a remarkable change occurs in morphology and structural style, since it is both the northern limit of the ocean-floored Marsili basin and the associated active Aeolian volcanic arc

A deep-seated cause can be advanced to better comprehend these structural differences. Considering that the seismicity related to subduction beneath the southeastern Tyrrhenian Sea is rather well delimited by the entire length of the ETMF, occurring practically only to the west of it, mantle stress gradients generated by slab rollback and sinking may have propagated to superficial crustal levels to form structural divides, such as the ETMF, approximately aligned along the deep slab edge (Marani and Trua, 2002). In this sense, this region is defined as the active margin geodynamic province, notwithstanding that, since at least the Mid-Pleistocene, the inner portions of the Apennine fold and thrust belt bordering this sector have been affected by normal faulting, the compressional front having migrated to the Adriatic Sea.

3.4. Oceanic Terrain Province: Central/Southern Tyrrhenian Back-Arc and Arc

The central and southeastern, deep-water (> 3000 m) portions of the Tyrrhenian Sea are floored by oceanic crust produced during two distinct episodes of accretion: first in the central Tyrrhenian Vavilov back-arc basin (VB), followed by a shift to the south-east to the Marsili back-arc basin (MB). Generation of the VB and MB are dated 4.3–2.6 Myr and < 2 Myr respectively. Regional lithosphere models for both basins concur with their oceanic nature, giving crustal thicknesses of < 10 km and a 30 km LID.

Outstanding morphological features of the otherwise flat lying, turbidite-filled abyssal plains are the large seamounts that occupy the central parts of the VB and MB (Figures 5, 6). The submerged portion the Aeolian Island volcanic arc associated with the development of the MB, moreover, is well developed both to the west and east of the islands.

The deep plains of the Vavilov basin have roughly a triangular shape, delimited to the west by the Selli Line and to the east by the lower slope of the southeastern Tyrrhenian active margin and the MB (Figure 5). The vertex of the triangle, the northernmost limit of the VB, occurs only a few km south of Ponza Island, forming one of the highest gradient slopes (from 0 to -3500 m) in the Tyrrhenian Sea. The southern border of the VB occurs in a region of complex, sediment-capped topography north of western Sicily.

Several large seamounts, along with seamount chains and linear-trending fault scarps complicate the morphology of the VB. Extensive older (Structural model of Italy, 1991) and recent seafloor sampling shows that the structural features are related either to basement terrains or to the development of large submarine volcanoes.

Outcropping "Corsica-type" Alpine units and Upper Miocene rocks characterize the scarp of the NE–SW Selli Line (SL). It is morphologically distinct for about 80 km and has on average 250/300 m throw, dropping down to the VB abyssal plain. Fault throw is variably distributed along closely spaced splays, in some cases connected by relay ramps. The SL is a fault system of significant structural importance, dividing the passive margin province from the Tyrrhenian oceanic domain.

Within the northern part of the VB, two seamounts represent a second and probably definite boundary between the oceanic crust of the basin and the surrounding thinned crustal blocks. These are the western De Marchi, and the eastern Flavio Gioa seamounts. They are 80 km apart, both are N–S trending and characterized by a distinct asymmetry, the former presenting a steep eastwards-dipping fault scarp and gentler western slope, the latter is a mirror image, with a steep westwards-dipping fault scarp and gentler eastern slope. Both have an elevation of 1200 m, rising from the ~ 3600 m deep VB plain. The seamounts, composed of metamorphic units related to Alpine Corsica in the case of De Marchi and to Calabride units in that of Flavio Gioia, in effect represent the final trace of rotational crustal blocks within the VB. The Gortani Ridge, a ~ 40 km linear morphological high, with a maximum elevation in the order of 300/400 meters, is positioned between the two tilted blocks, about 20 km from De Marchi seamount and 60 km from Flavio Gioia seamount. ODP drill-site 655 showed this feature to be made up of T-MORB basalt flows, considered by Kastens et al. (1990), to be the earliest evidence (~ 4 Myr) for emplacement of oceanic crust in the VB.

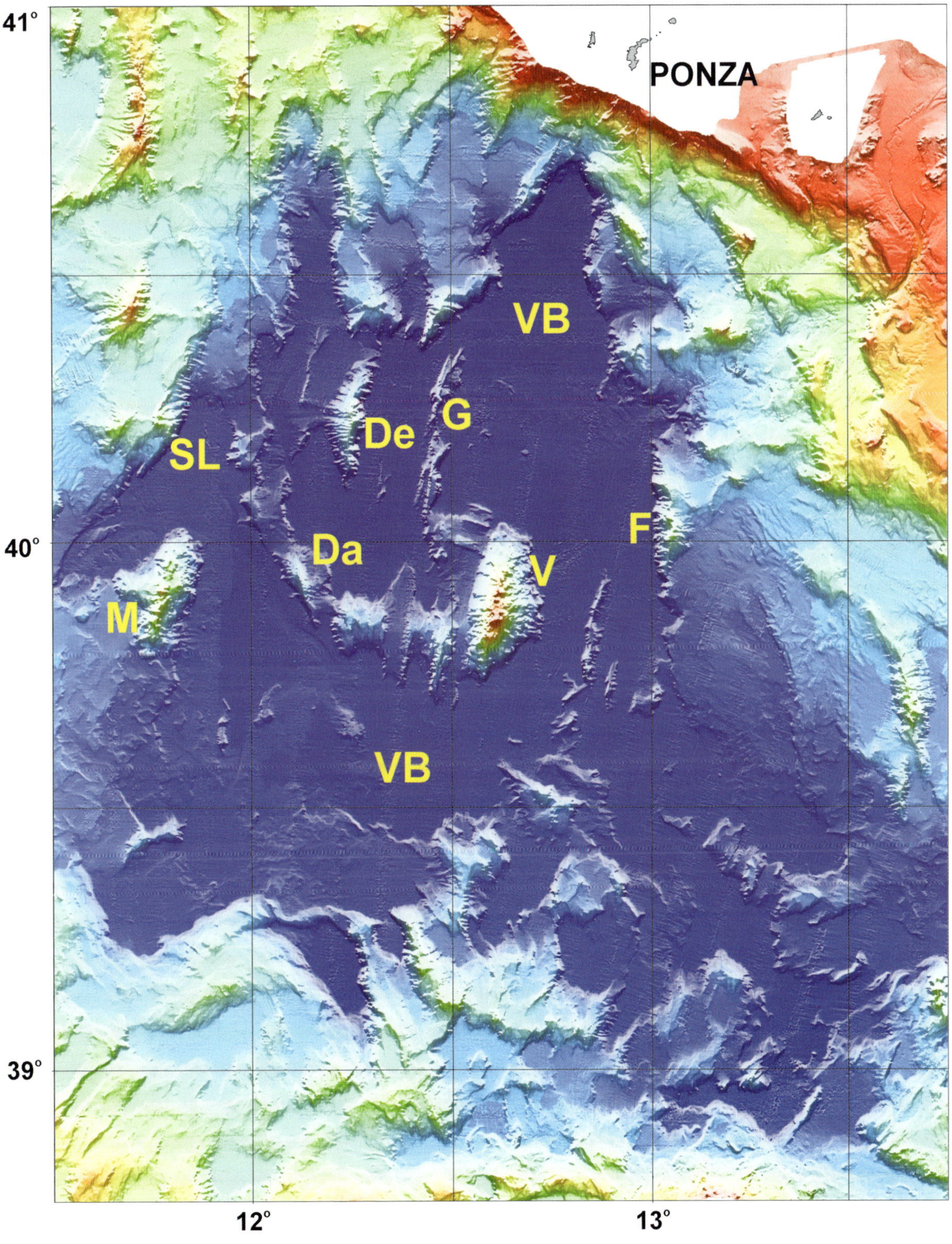

Figure 5: Vavilov Basin Province. Colour code and illumination as in Figure 1. The deep Vavilov basin (VB) plain is interspersed with diverse morphological elements. These are the Selli Line (SL) fault, the De Marchi (De) and Flavio Gioia (F) tilted blocks, the Gortani (G) and D'Ancona (Da) ridges and the large volcanoes Vavilov (V) and Magnaghi (M). Refer to the text for detailed discussion of each feature.

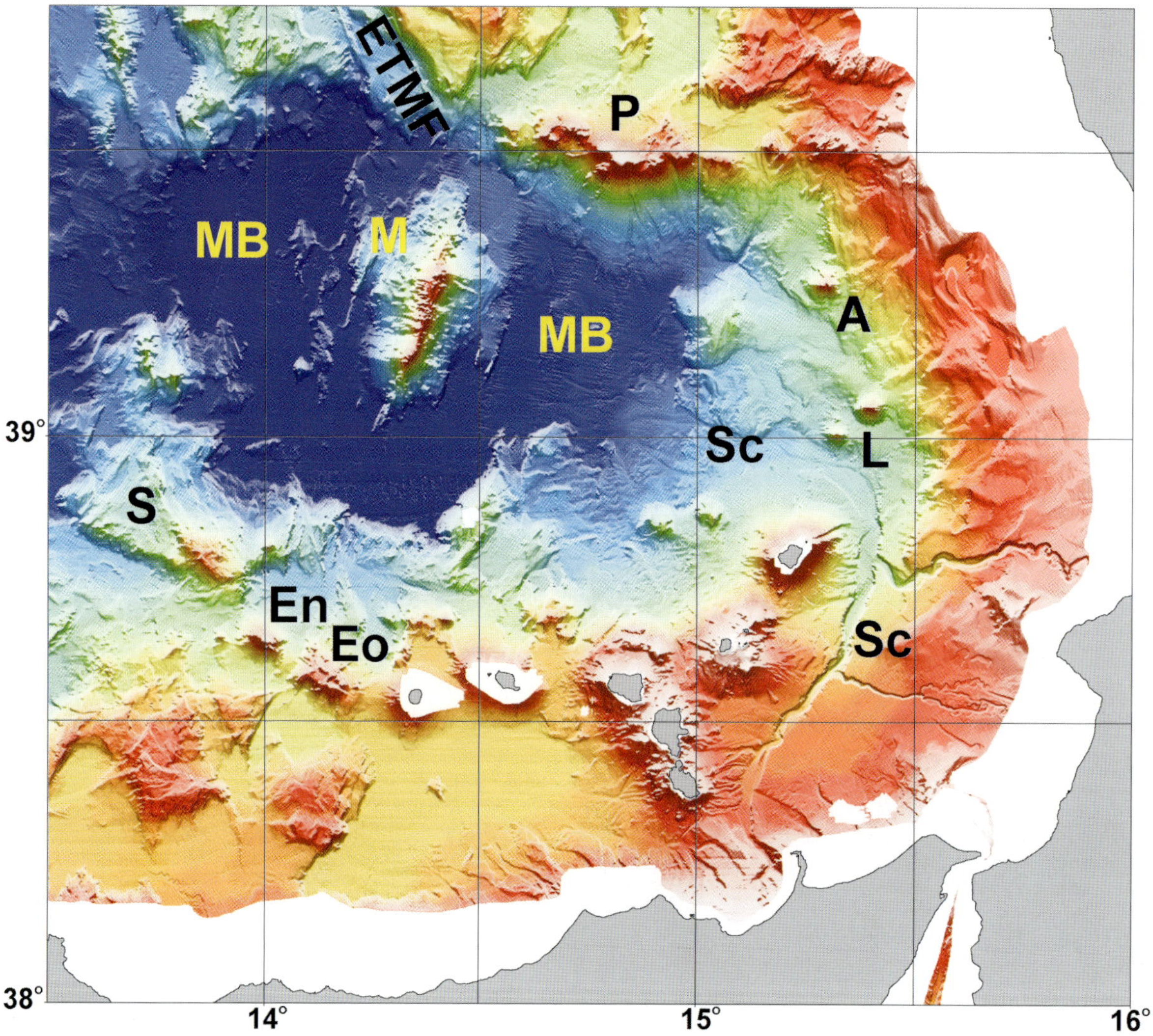

Figure 6: Marsili Basin Province. Colour code and illumination as in Figure 1. Marsili basin (MB) is the recent-most ocean-crust-floored basin, characterized by the large, axially located Marsili volcano (M). The deep basin is delimited by the Palinuro volcanic complex (P) and by the Aeolian volcanic arc located on the upper slope, which is cut by the Stromboli canyon (Sc). Submarine extensions to the arc are represented by the Lametini seamounts (L) and Alcione volcano (A) to the east and Eolo (Eo), Enarete (En) volcanoes and the Sisifo volcanic (S) ridge to the west. Refer to the text for detailed discussion of each feature.

The central VB is more intricate from a morphological point of view. The major feature is the axially located Vavilov volcano. It is positioned just south and midway between the De Marchi and Flavio Gioia seamounts. Vavilov volcano is elongated NNE/SSW for 40 km, is on average about 10 km wide, and stands 2800 meters above the abyssal plain, with summit at 730 meters depth. Its distinctive asymmetry, displayed by a steep, smooth western flank and a gentler, satellite-cone-affected eastern flank, presupposes a probable collapse of the volcano's western flank. Deep-tow sidescan sonar data along the base of the western flank do not definitely support this hypothesis, since the areal extension of the isolated patches of large blocks that have been found to outcrop cannot be verified due to the area being mostly sediment covered. However, reflection seismic data (Marani et al., in press) along the western base region show a very shallow acoustic basement compared to the eastern base, which could be related to infilling by rock avalanche deposits.

One other large volcano, Magnaghi seamount, dominates the western part of the central VB, in proximity to the SL. It has approximately the same trend as Vavilov, with a length of ~ 25 km and a summit at 1470 meters.

Both Vavilov and Magnaghi volcanoes are made up of tholeitic to alkalic basalts (Robin et al., 1987; Savelli, 1988) which erupted in the Late Pliocene (Kastens et al., 1990; Savelli, 1988). These authors therefore conclude that the large volcanoes of the VB were formed after the bulk of the low-standing basaltic crust of the basin had been generated.

The final morphological feature of the central VB casts some doubt on the actual position of the boundary between continental and oceanic crust in this portion of the basin. The D'Ancona ridge is an arcuate high-standing feature that initiates in the region between the SL and the De Marchi seamount, and terminates against the southern tip of the Vavilov volcano. The structure is made up of a series of highs, with elevations reaching even 800 meters but an average between 200 and 400 meters. In proximity to the Vavilov volcano, the D'Ancona ridge is formed by sharply defined N–S striking linear features, similar to the trend of the volcano and to the trend of the linear, subdued ridge to the east of the Vavilov seamount. Recently acquired shallow air-gun seismic reflection data (Marani et al., in press) show that the central portion of the D'Ancona ridge, positioned between the Magnaghi and Vavilov seamounts, possesses a sedimented ($\sim$ 250 meters cover), faulted basement with a seismic facies very different from the strongly reflective, basaltic crust, and more akin to continental basement. Although Vavilov seamount has recently been extensively sampled, no samples were collected from the nearby linear part of the D'Ancona ridge which, due its morphological character and proximity to the volcano, could be of a volcanic nature. However, the continental-basement-type character of that portion of the D'Ancona ridge lying between the Magnaghi and Vavilov volcanoes demonstrates that the alleged geodynamic setting of the northern VB does not apply to the central VB. The differentiation could simply be brought about by the presence of isolated rafted blocks of continental crust within the basaltic basement of the central VB, which did not develop in its northern portion. Alternatively, the D'Ancona ridge could play a more important role as a strong demarcation in crustal nature in some way involving the development of the two large volcanoes that characterize this region.

To the southeast, across a region of subdued seafloor topography corresponding to a saddle of $\sim$ 15 km thick crust (Scarascia et al., 1994; Suhaldoc and Panza, 1989), the Vavilov basin passes to the more recent Marsili basin (Figure 6) where crustal thickness returns to less than 10 km. This small, $\sim$ 2 Myr-old near-circular back-arc basin stands at 3500 meters water depth. Kilometre-scale fault scarps, which develop at the southern part of the basin, delimit the upper slope area in which the Aeolian volcanic arc develops. To the north, the Palinuro volcanic complex bounds the deep basin plain. More subdued topography characterizes the eastern boundary of the MB along the course of the Stromboli canyon as it reaches the abyssal plain.

Most of the MB is occupied by the Marsili volcano, which is practically the only morphological element of relevance. The volcano rises 3000 metres from the deep plain of the Marsili basin to a minimum depth of 489 metres, and is elongated NNE–SSW, with a length in the order of 50 km and a mean width of 16 km. It is flanked by a series of linear basin-floor fault scarps that form a symmetrical horst and graben structure at its sides. In its summit region, the volcano is characterized by a marked linearity formed by coalesced or single elongated cones. The development of fields of small volcanoes with mostly flat tops, and several with breached craters, characterizes its lower flanks.

Sampling data (Marani et al., 1999; Trua et al., 2002) show that the volcanoes of the cone fields and lower flanks of the volcano are composed of calc-alkaline basalts, while evolved high-K andesites were recovered only on the summit cones. Moreover, modelling results (Trua et al., 2002) indicate that the formation of the andesites is compatible with the differentiation of the type of basalts found in the lower reaches of the volcano, indicating a possible upper magma chamber within the volcano. The age of the volcano is estimated of $\sim$ 0.7 Ma, based on the magnetic anomaly pattern of the edifice (Faggioni et al., 1995) which indicates that the bulk of the volcano erupted within the normal polarity geomagnetic chron C1 (Bruhnes). Radiometric dating of rocks collected from near the summit area (Selli et al., 1977) furnish ages of $\sim$ 0.1 Ma. The volcano has been interpreted to result from a strong thermal pulse of asthenospheric material produced by a rapid phase of slab rollback during the Mid Pleistocene (Marani and Trua, 2002).

The islands and submarine volcanoes comprising the active volcanic arc associated with the Marsili back-arc basin occupy the upper slope of the southeastern Tyrrhenian Sea. Submarine volcanoes, developed since 0.8 Ma (Beccaluva et al., 1985), form an extension to the emerged islands both eastwards and westwards. The arc is well delimited to the north by the 50 km E–W development of the $\sim$0.35 Ma old (Colantoni et al., 1981) Palinuro composite basaltic andesite volcano. The volcano's morphology is made up of at least 6 distinct volcanic cones developed at a base level around 1400 metres in the slope margin between the active margin geodynamic province to the north and the MB oceanic terrain to the south. In fact, the southern flank of the Palinuro volcanic complex drops more than 3 km to the MB abyssal plain. Elevations of the cones range from 800 to more than 1200 meters, the central cone reaching $\sim$ 70 meters water depth. This cone has a wide, flat top presumably due to erosion during the last glacial 125-meters lowstand. Whereas one of the cones, east of the central cone, displays a clearly breached crater, the westernmost portion of Palinuro is characterized by a caldera structure on whose rim develop a series of small domes.

The complex morphology, and the key location of Palinuro volcano at the boundary between two wholly distinct geodynamic provinces, point to the important role played by this structure during the recent evolution of the south-eastern Tyrrhenian Sea. Its structural importance may be followed also through its extension onto the emerged areas of the Italian peninsula, where the boundary between the Apennine terrains and the Calabride complex occurs.

South of Palinuro volcano, the three submarine volcanoes that form the eastern extension of the Aeolian arc are morphologically distinct. The southernmost features are the two twin volcanoes of the Lametini seamounts which display regular cone shapes with circular bases, with heights of 650 and 1000 meters. They are about 5 km apart and trend NE–SW along the slope area at a base level of 2000 meters. Alcione volcano is positioned midway between the Lametini volcanoes and Palinuro. Although generally cone-shaped with 800 meter height, it displays a striking morphology, the cone being entirely dissected by a central NNW–SSE trending scarp that divides two summit craters. The scarp is slightly curved and drops the western crater 50 meters seawards to the west. The morphology of the surrounding seabed does not furnish evidence of similarly trending fault scarps, indicating that the structure is limited exclusively to the volcano. The formation of the scarp is due to limited seaward failure of part of the edifice. In general, the properties and dip of the basement on which a submarine volcano is positioned together with its intrinsic growth dynamics are joint causes of instability at its base. In the same way, some degree of basal gravity sliding could be the mechanism for formation of the unique morphology of Alcione volcano.

The submerged western extension of the arc is formed by the NW–SE alignment of the Eolo and Enarete volcanoes and the Sisifo volcanic ridge. Eolo and Enarete are isolated edifices located about 30 km northwest of Alicudi Island. The morphology of Eolo volcano is singular. It has a base of ~ 12 km and height of 900 metres. The summit (950 m water depth) is made up of a 6-km-wide flat surface surrounded by a series of small, 100-m-high cones. The cones become more developed, reaching 250 m to the southeast where they encircle a more depressed area of the volcano. Enarete volcano, on the other hand, is a near perfect cone, slightly elongated NW–SE, with a base diameter of ~ 15 km and a height of 1600 metres. Sisifo volcanic ridge develops in a NW–SE direction also, for ~ 30 km. It is formed by a southern, southwest-dipping 500-meter scarp and a more gently dipping northern part. Volcanic rocks have been collected from both its southeastern and northwestern extremities and from a small cone in its summit area.

4. QUATERNARY MANTLE DYNAMICS IN THE SOUTHEASTERN TYRRHENIAN SEA

The preceding account has attempted to show the geological development of the Tyrrhenian Sea by way of the present-day seafloor morphology. It is evident that seafloor makeup furnishes more comprehensible information the more recent are the geological events that contribute to its creation. The Marsili basin is thus the region of the Tyrrhenian Sea where a possible link between seafloor morphology and deep structure can be established. In fact this recent back-arc basin is underlain by a steeply dipping subducting slab of Mesozoic oceanic crust and, superficially contains or is surrounded by some of the major structures described in this chapter. If one extends a view to the surrounding southern Apennine chain and Calabrian arc, then the geological overall picture becomes of geodynamic importance.

Recently, various authors have put forward models relating deep mantle dynamics induced by rollback of the Ionian slab to the coupled observable surficial effects. Gvirtzman and Nur (1999), based on crustal modelling studies, link the formation of Etna volcano to sideways asthenospheric flow induced by slab decoupling beneath Calabria, giving rise also to the uplift of Calabria itself. Based on the morphology of the southern Tyrrhenian seafloor, its structural makeup, and with the support of petrographical studies of recently sampled rocks (Trua et al., in press; Marani and Trua, 2002) suggest that dip-directed tears at the sides of the slab induce the rise of deep asthenospheric material to the Marsili basin area and surroundings, and recognize the surface traces of the tears.

This chapter has endeavoured to show that seafloor morphology can serve a variety of functions in regions of recent formation. Particularly in cases where detailed topographic data may indirectly reveal the surficial effects of deep-seated processes, the same information becomes the baseline data for subsequent research, for instance seafloor sampling and petrological studies or geophysical surveys, such as the CROP project, that can be undertaken to substantiate the surface observations.

REFERENCES

Argnani, A., Savelli, C., 1999. Cenozoic volcanism and tectonics in the southern Tyrrhenian Sea: space–time distribution and geodynamic significance. Geodynamics 27, 409–432.

Argus, D.F., Gordon, R.G., De Mets, C., Stein, S., 1989. Closure of the Africa–Eurasia–North America plate motion circuit and tectonics of the Gloria Fault. J. Geophys. Res. 94, 5,585–5,602.

Bartole, R., 1995. The north Tyrrhenian–Northern Apennines post-collisional system: constrain for a geodynamic model. Terra Nova, 7–30.

Beccaluva, L., Brotzu, P., Macciota, G., Morbidelli, L., Serri, G., Traversa, G., 1989. Cainozoic tectono-magmatic evolution and inferred mantle sources in the Sardo-Tyrrhenian area. In: Boriani, A., Bonafede, M., Piccardo, G.B., Vai, G.B. (Eds.), The Lithosphere in Italy. Accad. Naz. Lincei, Rome, pp. 229–248.

Beccaluva, L., Coltorti, M., Galassi, B., Maciotta, G., Siena, F., 1994. The Cainozoic calcalkaline magmatism of the western Mediterranean and its geodynamic significance. Boll. Geof. Teor. Appl. XXXVI (141–144), 293–308.

Beccaluva, L., Gabbianelli, G., Lucchini, F., Rossi, P.L., Savelli, C., 1985. Petrology and K/Ar ages of volcanics dredged from the Eolian seamounts: implications for geodynamic evolution of the southern Tyrrhenian basin. Earth Planet. Sci. Lett. 74, 187–208.

Carmignani, L., Decandia, F.A., Disperati, L., Fantozzi, P.L., Lazzarotto, A., Liotta, D., Oggiano, G., 1995. Relationships between the Tertiary structural evolution of the Sardinia–Corsica–Provençal domain and the northern Apennines. Terra Nova 7, 128–137.

Carmignani, L., Kligfield, R., 1990. Crustal extension in the northern Apennines: the transition from compression to extension in the Alpi Apuane complex. Tectonics 9, 1275–1305.

Colantoni, P., Lucchini, F., Rossi, P.L., Sartori, R., Savelli, C., 1981. The Palinuro Volcano and magmatism of the south-eastern Tyrrhenian Sea (Mediterranean). Mar. Geol. 39, M1–M12.

De Mets, C., Gordon, R.G., Argus, D.F., Stein, S., 1990. Current plate motions. Geophys. J. Int. 101, 425–478.

Dewey, J.F., Helman, M.L., Turco, E., Hutton, D.H.W., Knott, S.D., 1989. Kinematics of the western Mediterranean. In: Coward, M.P., Dietrich, D. (Eds.), Alpine Tectonics. Geol. Soc. Spec. Publ. 45, 265–283.

Faggioni, O., Pinna, E., Savelli, C., Schreider, A.A., 1995. Geomagnetism and age study of Tyrrhenian seamounts. Geophys. J. Int. 123, 915–930.

Finetti, I.R., Boccaletti, M., Bonini, M., Del Ben, A., Geletti, R., Pipan, M., Sani, F., 2001. Crustal section based on CROP seismic data across the North Tyrrhenian–Northern Apennines–Adriatic Sea. Tectonophysics 343, 135–163.

Finetti, I., Boccaletti, M., Bonini, M., Del Ben, A., Pipan, M., Prizzon, A., Sani, F., this volume. Lithospheric tectono-stratigraphic setting of the Ligurian Sea–Northern Apennines–Adriatic foreland from integrated CROP seismic data. Chapter 8.

Gvirtzman, Z., Nur, A., 1999. The formation of Mount Etna as the consequence of slab rollback. Nature 401, 782–785.

Jolivet, L., 1991. Extension of thickened continental crust, from brittle to ductile deformation: examples from Alpine Corsica and Aegean Sea. Ann. Geofis. 36, 139–153.

Kastens, K.A., et al., 1988. ODP Leg 107 in the Tyrrhenian Sea: insight into passive margin and back-arc basin evolution. Geol. Soc. Amer. Bull. 100, 1,140–1,156.

Kastens, K.A., et al., 1990. The geological evolution of the Tyrrhenian Sea: an introduction to the scientific results of ODP Leg 107. In: Kastens, K.A., et al. (Eds.), Proceedings of the ODP. Scientific Results, vol. 107, pp. 3–26.

Lister, G.S., Davis, G.A., 1989. The origin of metamorphic core complexes and detachment faults formed during Tertiary continental extension in the northern Colorado River region, U.S.A. J. Struct. Geol. 1/2, 65–94.

Malinverno, A., Ryan, W.B.F., 1986. Extension in the Tyrrhenian Sea and shortening in the Apennines as result of arc migration driven by slab sinking in the lithosphere. Tectonics 5, 227–245.

Marani, M.P., Gamberi, F., Ivanov, M., ship-board party, in press. Introduction and main objectives of TTR-12 Leg IV–Tyrrhenian Sea.

Marani, M.P., Gamberi, F., Casoni, L., Carrara, G., Landuzzi, V., Musacchio, M., Penitenti, D., Rossi, L., Trua, T., 1999. New rock and hydrothermal samples from the southern Tyrrhenian sea: the MAR-98 research cruise. Giorn. Geol. 61, 3–24.

Marani, M.P., Trua, T., 2002. Thermal constriction and slab tearing at the origin of a superinflated spreading ridge: Marsili volcano (Tyrrhenian Sea). J. Geophys. Res. 107 (B9), 2188, 10.1029/2001JB000285.

Marani, M., Zitellini, N., 1986. Rift structures and wrench tectonics along the continental slope between Civitavecchia and C. Circeo. Mem. Soc. Geol. Ital. 35, 453–457.

Musacchio, M., Carrarra, G., Gamberi, F., Marani, M., 1999. Tectonic setting of the eastern Tyrrhenian margin. In: Geoitalia, 2° Forum FIST, Riassunti, vol. 1, pp. 184–185.

Pascucci, V., Merlini, S., Martini, I.P., 1999. Seismic stratigraphy of the Miocene-Pleistocene sedimentary basins of the Northern Tyrrhenian Sea and western Tuscany (Italy). Basin Res. 11, 337–356.

Robin, C., Colantoni, P., Gennesseaux, M., Rehault, J.P., 1987. Vavilov seamount: a mildly alkaline Quaternary volcano in the Tyrrhenian basin. Mar. Geol. 78, 122–136.

Sartori, R., 1990. The main results of ODP Leg 107 in the frame of Neogene to Recent geology of peri-Tyrrhenian areas. In: Kastens, K.A., et al. (Eds.), Proceedings of the ODP. Scientific Results, vol. 107, pp. 715–730.

Sartori, R., ODP Leg 107 Scientific Staff, 1989. Drillings of ODP Leg 107 in the Tyrrhenian Sea: tentative basin evolution compared to deformations in the surrounding chains. In: Boriani, A., Bonafede, M., Piccardo, G.B., Vai, G.B. (Eds.), The Lithosphere in Italy. Accad. Naz. Lincei, Rome, pp. 139–156.

Savelli, C., 1988. Late Oligocene to recent episodes of magmatism in and around the Tyrrhenian Sea: implications for the processes of opening in a young inter-arc basin of intra-orogenic (Mediterranean) type. Tectonophysics 146, 163–181.

Scarascia, S., Lozej, A., Cassinis, R., 1994. Crustal structures of the Ligurian, Tyrrhenian and Ionian Seas and adjacent onshore areas interpreted from wide-angle seismic profiles. Boll. Geof. Teor. Appl. 36 (141–144), 4–19.

Selli, R., Lucchini, F., Rossi, P.L., Savelli, C., Del Monte, M., 1977. Dati geologici, petrochimici e radiometrici sui vulcani centro-tirrenici. Giorn. Geol. XLII, 221–246.

Serri, G., 1997. Neogene-Quaternary magmatic activity and its geodynamic implications in the central Mediterranean region. Ann. Geophys. 40 (3), 681–703.

Serri, G., Innocenti, F., Manetti, P., 1993. Geochemical and petrological evidence of the subduction of delaminated Adriatic continental lithosphere in the genesis of the Neogene-Quaternary magmatism of central Italy. Tectonophysics 223, 117–147.

Structural model of Italy, 1991. C.N.R. 114, vol. 3.

Suhaldoc, P., Panza, G.F., 1989. Physical properties of the lithosphere–asthenosphere system in Europe from geophysical data. In: Boriani, A., Bonafede, M., Piccardo, G.B., Vai, G.B. (Eds.), The Lithosphere in Italy. Accad. Naz. Lincei, Rome, pp. 15–40.

Surpless, B.E., Stockli, D.K., Dumitru, T.A., Miller, E.L., 2002. Two-phase westward encroachment of Basin and Range extension into the northern Sierra Nevada. Tectonics 21 (1), 10.1029/2000TC001257.

Trua, T., Serri, G., Marani, M., in press. Lateral flow of African mantle below the nearby Tyrrhenian plate: geochemical and isotopic evidence. Terra Nova.

Wernicke, B., 1981. Low-angle normal faults in the Basin and Range province: nappe tectonics in an extending orogen. Nature 291, 645–648.

Trua, T., Serri, G., Renzulli, A., Marani, M., Gamberi, F., 2002. Volcanological and petrological evolution of Marsili seamount (southern Tyrrhenian Sea). J. Volcanol. Geotherm. Res. 114, 441–464.

Ward, S.N., 1994. Constraints on the seismotectonics of the central Mediterranean from Very Long Baseline Interferometry. Geophys. J. Int. 117, 441–452.

Zitellini, N., Trincardi, F., Marani, M., Fabbri, A., 1986. Neogene tectonics of the northern Tyrrhenian Sea. Giorn. Geol. 48 (1/2), 25–40.

CHAPTER 23

Crustal Tectono-Stratigraphic Setting of the Adriatic Sea from New CROP Seismic Data

I.R. Finetti[1,*], A. Del Ben[1]

ABSTRACT

Eight CROP seismic sections in the Adriatic Sea and four on the southeastern Apulian extension area provide for the first time a well-controlled regional transcrustal tectono-stratigraphic setting for the offshore Adria plate. In Permo-Triassic times the Adria continent detached from Africa generating the interposed Ionian Neotethys basin, floored by oceanic crust. Rift faults, related to Adria detachment and anticlockwise rotational-drifting, can be observed on CROP sections of the Adriatic Sea and Apulian platform. Stratigraphic (borehole) data, supported by geophysics, indicate that sedimentation commenced in paleo-depressions of the Adria plate's Hercynian basement in the Permian, when the above mentioned rift-tectonics began to be active. Successive rift stages (Trias, Lias) gave rise to a composite space–time distribution of basins and platforms, well defined in detail by the hydrocarbon industry's exploration.

The Adriatic cover sequence is calibrated by numerous wells and is seismically outlined by various interpreted reflecting horizons, generally of good quality and regionally of high reliability. Three different evaporitic cycles in the Permo-Triassic have been recognized in the Southern Adriatic. Prominent Triassic batholithic bodies have been recognized in the Mid-Adriatic. Compressive, extensive and diffused shear tectonics affect different provinces of this wide area. The Moho discontinuity and the top of the lower crust are generally identified as lower and upper boundaries of a fairly well discriminated reflective lower crustal interval.

Keywords: Adriatic crustal tectono-stratigraphy CROP controlled

1. INTRODUCTION

The Padano-Adriatic foreland is formed from a thick depositional sequence which started in Late Paleozoic times (Permian) and has continued with time–space hiatuses until the present. This sedimentary cover is lying over a Hercynian crystalline-metamorphic basement belonging to the Adria plate, detached from Africa in Permo-Triassic times, when opening of the interposed Ionian Neotethys took place. The Po Valley and the Adriatic Sea floor are rather flat and constituted by undeformed young sediments that cover tectonically deformed features. Only geophysical exploration can image such deformed subsurface layers. The geoscience community should acknowledge that hydrocarbon exploration was determinant in revealing the Padano-Adriatic tectono-stratigraphy and its geological evolution. From the early fifties extensive exploration campaigns began, and then continued with detailed investigations and successful exploration results. The international oil industry, mainly ENI-AGIP, collected significant subsurface geophysical and geological data for a correct and careful reconstruction of the tectono-stratigraphy, and from the early eighties began to disclose basic information to the scientific community (Pieri and Groppi, 1981; Dragasevic, 1983; Cati et al., 1987).

In the deep water of the Southern Adriatic, the MS (Mediterranean Sea) seismic lines began in the early seventies, discovering a thick Tertiary sedimentary basin in front of the Helleno-Dinaric thrust belt (Finetti and Morelli, 1973). Thus on the basis of seismic data acquired for industrial (hydrocarbon exploration) or scientific (MS lines) purposes, it was possible to reconstruct a structural sketch-map of the whole Padano-Adriatic

[1]Department of Geological, Environmental and Marine Sciences, Exploration Geophysics Group, University of Trieste, Via Weiss 1, 34127 Trieste, Italy
*Corresponding author. Tel.: +39 40 5582276; e-mail: finetti@units.it (I.R. Finetti)

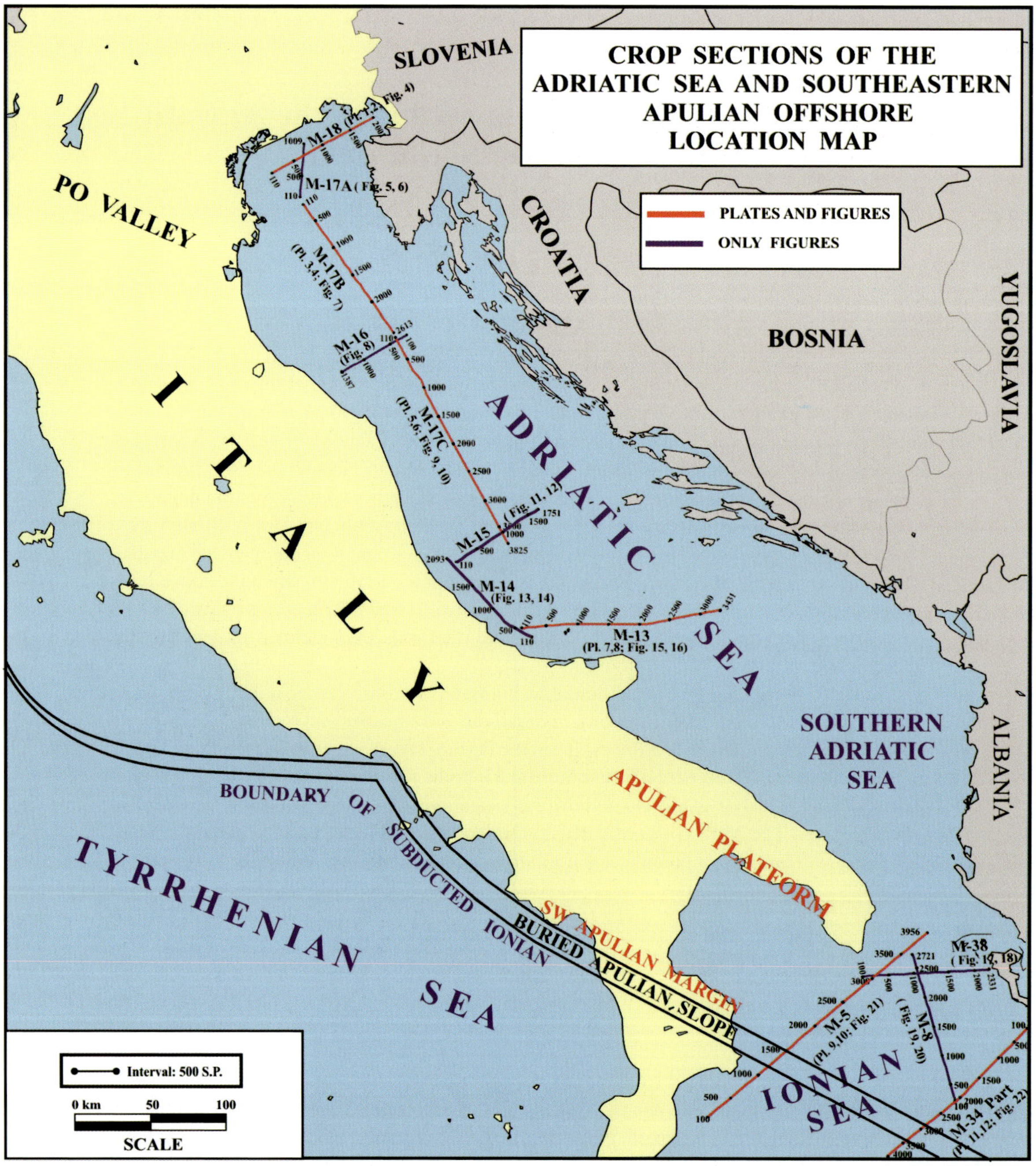

Figure 1: Location Map of the CROP seismic sections of the Adriatic Sea (M-13, M-14, M-15, M-16, M-17A, M-17B, M-17C and M-18), and of the Southern Apulian platform (M-5, M-8, M-34 Part and M-38). Red traced sections are exhibited in the plates and in the figures at reduced scale. For each red section, the plate number (PL) for the blank seismic sections and respective interpreted seismogeological section, plus the figure number (Fig.) at reduced scale exhibited in the chapter text is indicated. The shorter blue sections are exhibited only in the figures at reduced scale.

foreland area (Finetti et al., 1987). Seismic-reflection-based structural reconstructions of the Southern Adriatic basin and southeastward extension of the Apulian Platform have been presented in the literature (Finetti, 1982; Finetti and Del Ben, 1986; Del Ben et al., 1994). This chapter exhibits a crustal interpretation of the new CROP seismic sections of the Adriatic and southeast continuation of the Apulian platform (Figure 1). The red-traced CROP sections in Figure 1 are presented both as Plates (1:300,000 horizontal scale) and as reduced figures in the text. The blue-traced shorter CROP sections of Figure 1 are presented only as reduced figures in the text. When a Plate is presented, the related black and white seismic section is not included as a reduced figure, but only its interpretation. Plate and figure numbers are reported also on the location map of Figure 1. Two CROP sections (M5 and M34) exhibit the crustal passage from the Apulian platform to the Ionian basin floored by oceanic crust.

The CROP network of the Adriatic basin is composed of a long axial section (M17A–M17B–M17C) and four transverse sections (M-18, M16, M15 and M13). CROP section M14 explores the coastal zone of the deep Pescara basin. The CROP programme in the Southern Adriatic was not done due to temporary navigation restriction during the acquisition period (Bosnian crisis). Three of four Apulian CROP sections are located transversally to the thick platform crust (M5, M34 and M38) and one along the axial zone (MS).

The Padano-Adriatic foreland area of the Adriatic plate is bounded by active Adria-verging thrust-fronts that prominently deformed the Adria cover. To the northeast, the foreland is delimited by the Dinaric orogenic system, to the north by the Southalpine backthrusting tectonics of the Alpine system, and to the southwest by the Apennines thrust belt. The crustal tectono-stratigraphy of the Adriatic foreland area is well imaged seismically by the interpreted CROP sections. Compared to the bounding thrust-belts, the Adriatic and Southern Apulian CROP sections exhibit relatively modest and simpler tectonization. All Adriatic-Apulian CROP sections are clearly self-explanatory. Therefore, we limit their description to a minimum, leaving the reader to gather directly for himself the immediately visible transcrustal information provided by each seismogeological CROP section.

2. SEISMIC STRATIGRAPHY

For the CROP seismic profiles of this chapter, the sedimentary cover is calibrated by numerous boreholes, only in part indicated on the interpreted seismogeological sections. From a wide, pluriannual seismic exploration of the Adriatic, various characteristic reflectors have been identified and can be reliably recognized at large scale even without direct calibration. Note that, for the scientific purposes of the CROP programme, is not so important to reconstruct a careful detailed stratigraphy of the sedimentary cover, but rather to identify and define correctly the basic transcrustal macro-tectono-stratigraphy and most significant events.

In Figure 2 a generalized stratigraphy of the Padano-Adriatic cover deposits overlying the Hercynian basement and main provincial or regional variations is reported. In Figure 3 the stratigraphy encountered by the Amanda 1 bis deep borehole (TD = 7305 m) of the Northern Adriatic Sea is exhibited. This well stopped not far from the interpreted basement reflector "Z" (Figure 4). The well Assunta-1, near Venice, found the basement on a structural high covered by Norian sediments. Another deep borehole drilled on the Northern Apulian swell (well Puglia 1) stopped at 7070 m in the Permian (having crossed about 660 m of Permian sandstone with siliceous-calcareous cement beneath Scythian deposits). Also in this case the borehole's total depth is not far from the interpreted seismic basement. These data for the northern and southern Adriatic foreland basal cover coherently indicate that the Hercynian basement (or pre-Permian preserved zone) began to be covered (on existing paleo-depressions) in Permian times when rift activity of the Adria plate detachment initiated.

The interpreted seismic horizons are plotted and defined on every Plate and Figure of the CROP sections of this chapter. Various horizons are more or less seismically identifiable after interpreting experience of the reflectivity in the area has matured. Horizons like "A" (*base Plio-Quaternary*), "K" (*top Carbonate*), "M" (*top Maiolica Fm.*), "J" (*top Jurassic*), "T_1" (*top Dolomia Principale Fm.*), "T_X" (*Middle Trias unconformity*, or *base of the Dolomia Principale Fm.*) can be identified and picked with good to fair reliability and acceptable correctness for CROP regional purposes. Other horizons of less regional or of local significance are also evident, or are calibrated by boreholes, or are of lower reliability. Basement horizon "Z" is evident only in given parts of the studied area, and in other major parts must be considered as only indicative, but acceptable for the purpose. The horizons "TL" (top lower crust) and "MO" (Moho), in general, can be identified as top and base of the lower crust interval, and more reflective than the upper crust and Mantle. In various poorly reflective zones, "TL" and "MO" mostly represent an acceptable interpolation among other well characterized zones where this interval is more evident.

3. SEISMIC INTERPRETATION OF THE ADRIATIC–SOUTHEASTERN APULIAN CROP SECTIONS

The CROP seismic sections of the Adriatic–Southeastern Apulian foreland exhibit for the first time the deep crustal setting of the area, while accurate knowledge of its sedimentary cover has been acquired over several years by the petroleum exploration industry. In this chapter, the most significant regional crustal settings deduced from the many CROP sections interpreted are described in summary (Table 1; Figures 4 to 22; Plates 1 to 12). The upper basement crust, the lower crust and deep tectonics are the more innovative aspects of the interpretation.

Table 1
CROP seismic sections of the Adriatic foreland and Southeastern Apulia exhibited and commented on in this work (Chapter 23)

CROP section	SP	Length (km)	Chapter 23			
			Figure N°		Plate N°	
			Seismic	Int.	Seismic	Int.
Adriatic CROP sections						
M18	100–2001	97	–	4	1	2
M17A	100–1009	48	5	6	–	–
M17B	100–2613	128	–	7	3	4
M16	100–1387	65	–	8	–	–
M17C (AB)	100–2000	189	–	9	5	6
M17C (BC)	2000–3825			10		
M15	100–1751	85	11	12	–	–
M14	100–2093	102	13	14	–	–
M13 (AB)	100–1800	169	–	15	7	8
M13 (BC)	1800–3431			16		
Southeastern Apulian CROP sections						
M38	100–2331	114	17	18	–	–
M8	100–2721	133	19	20	–	–
MS	100–3956	243	–	21	9	10
M34 (Part)	100–4000	198	–	22	11	12

3.1. CROP Seismic Sections of the Adriatic Sea

Interpretation of the eight CROP sections of the Adriatic Sea (M18, M17A, M17B, M16, M17C, M15, M14 and M13) provides a clear regional transcrustal setting for the explored area. The pre-orogenic sedimentary sequence (Permo-Mesozoic to Lower Paleogene) is regulated by a basin-platform distribution that evolved in time and space as a consequence of the rift-stages. During the Permo-Triassic detachment of the Adria plate from Africa and opening of the interposed Ionian Neotethys, prominent rift tectonic processes affected Adria. On its northern margin, prominent rift-related tectono-magmatic and sedimentary events testify to extensional tectonics (i.e. Permian sandstone of Val Gardena and Lombardy; igneous intrusion–effusion in Permian times on the Western Alps and Ivrea body).

The sedimentary crust of the Adriatic Sea, which includes thrust-thickening, is everywhere rather thick and varies from a minimum of 3.0 sec (TWT) on CROP section M18 to a maximum of 6.3 sec on section M13 (Table 2). Upper basement crust thickness mostly ranges between 5.0 sec (about 12 km) and 6.0 sec (about 15 km), with a maximum of 6.8 sec (about 20 km) on section CROP M18 (Istrian area, Trieste offshore). The Adriatic lower crust interval (Table 2) is much more regular than the upper crust and ranges from 2.0 sec (about 7 km) to 3.0 sec reflection time (about 10.5 km).

CROP Section M18. On interpreted seismogeological section M18, the transcrustal tectono-stratigraphic setting of the Northern Adriatic region is very clearly exhibited (Figure 4; Plates 1 and 2). The sedimentary cover in the middle of the section is calibrated by the deep borehole Amanda-1 bis (Figure 3). This well, in fact, stopped in Permian strata (7,305 m TD), near the interpreted basement (Figure 4). The sedimentary sequence is well defined seismically by the many calibrated and/or identified and interpreted horizons (A_0, A, C, E, K, M, J, T_1, T_X, and Z) whose stratigraphic position is reported in the section legend. While the sedimentary sequence thickens progressively southwestward, the basement upper crust (UC) and the lower crust (LC) become thicker northwestward on the Istrian–Friulian platform area.

Some Permo-Triassic and Liassic rift faults have been interpreted. These faults, on the basis of extensive analysis of CROP data, are interpreted as being connected to left-shear motion. Around SP 1600, at the basin-platform

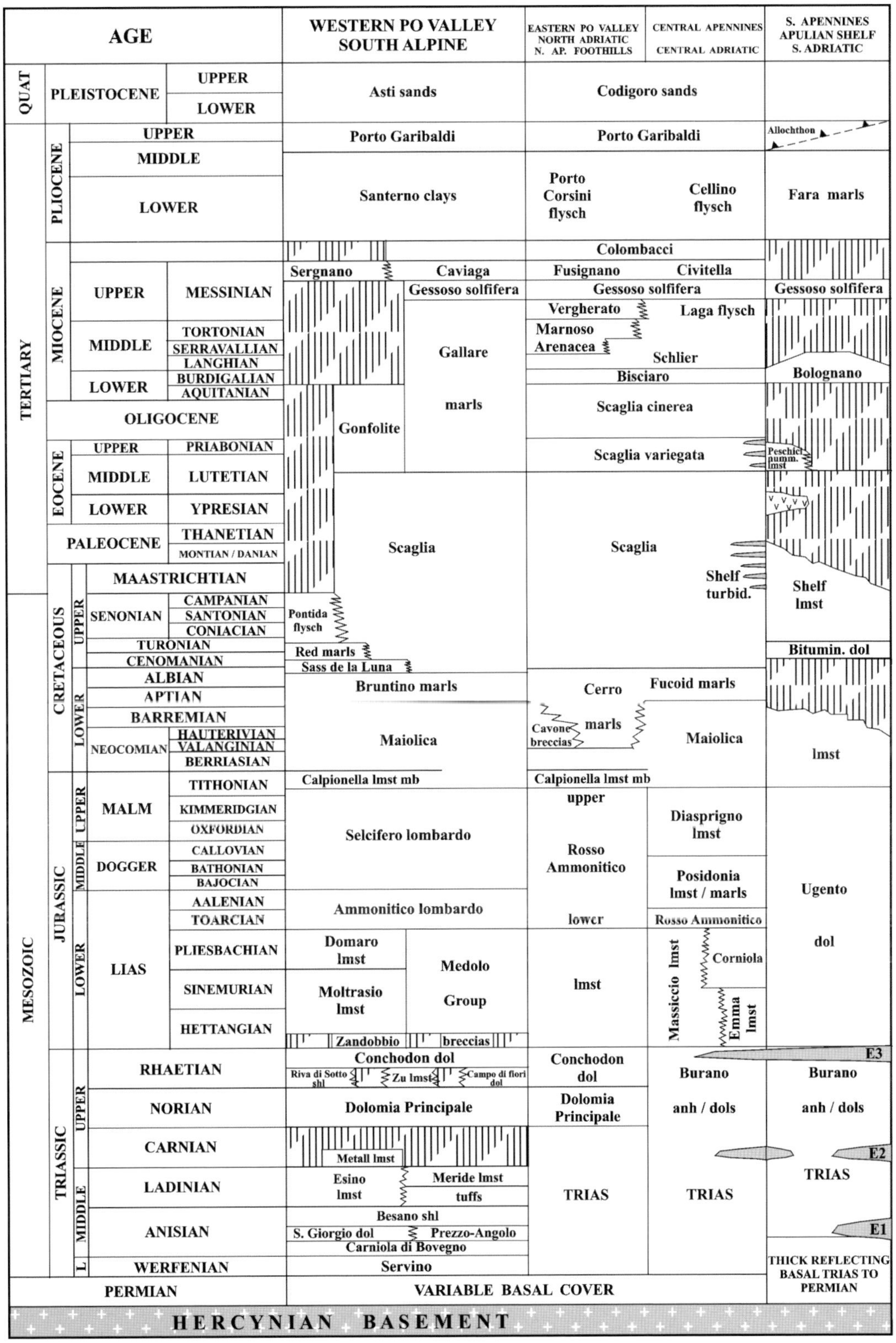

Figure 2: Indicative stratigraphic chart of the Adriatic–Po Valley–Apulian foreland and adjacent Apenninic front areas (after Casero, 2004, modified), controlled by numerous hydrocarbon boreholes. E1, E2, E3 = Permo-Triassic evaporitic cycles of the Southern Adriatic (see Figure 16). Pre-Permian Paleozoic depositions on local old depressions are possible.

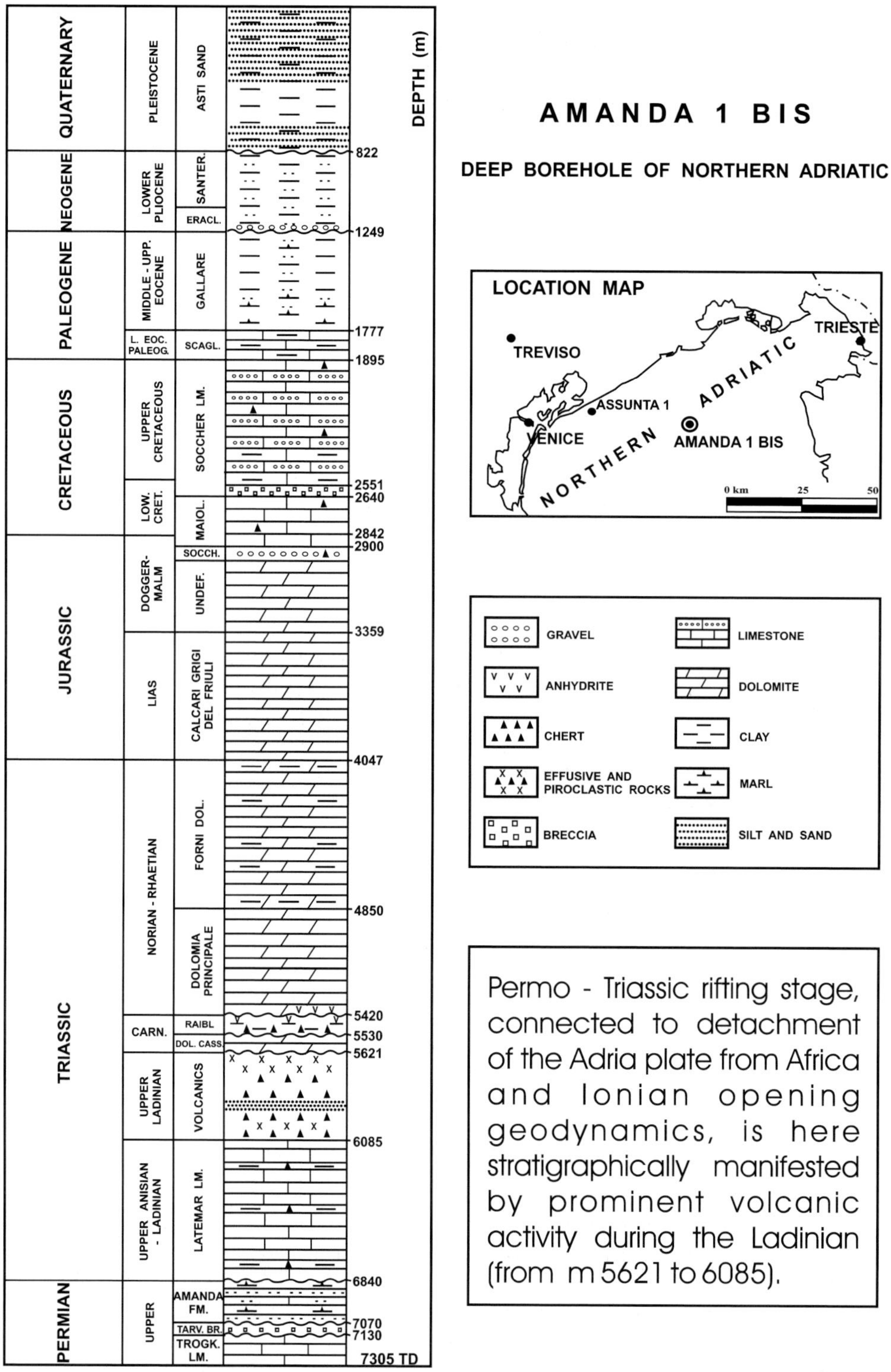

Figure 3: Stratigraphy of the Amanda 1 Bis, deep borehole (TD = 7305 m) of the Northern Adriatic Sea.

passage, the Dinaric geodynamics produced a small compressive deformation within a flat regional undeformed context (Figure 4, Plate 2).

CROP Section M17A. The reduced version of the interpreted CROP section M17A, in the Northern Adriatic, is shown in Figure 5. Its seismogeological interpretation, done to the large working scale of 1:50,000, as were all other CROP sections, is exhibited at a reduced scale in Figure 6. This section repeats the tectono-stratigraphic characteristics seen on previous CROP sections in the area of the Amanda-1 bis deep borehole. Also here, rift faults of Permo-Lower Triassic age have been interpreted, and attributed to left-shear tectonics. Excluding modest

Table 2
Minimum and maximum thickness of the sedimentary cover interval (SC), upper crust (UC) and lower crust (LC) interpreted along with the Adriatic CROP sections

CROP section	Interval thickness (TWT sec)					
	SC		UC		LC	
	Min	Max	Min	Max	Min	Max
M18	3.0	5.0	3.7	6.8	2.4	3.0
M17A	4.2	4.8	3.4	4.3	2.0	2.4
M17B	4.2	4.7	4.2	6.0	2.2	3.0
M16	4.2	5.6	4.2	5.8	2.7	3.1
M17C	4.6	6.6	4.4	5.9	2.5	2.8
M15	5.0	6.4	4.1	6.0	2.3	2.4
M14	4.3	6.2	4.3	5.2	2.3	2.5
M13	3.6	6.3	3.7	5.2	2.2	2.6

Table 3
Minimum and maximum thickness of the sedimentary cover interval (SC), upper crust (UC) and lower crust (LC) interpreted along with the southeastern Apulian platform CROP sections

CROP section	Interval thickness (TWT sec)					
	SC		UC		LC	
	Min	Max	Min	Max	Min	Max
M38	4.7	7.1	3.6	4.2	2.1	3.0
M8	3.1	4.1	3.8	4.2	1.9	3.1
M5*	2.8	4.6	3.7	4.1	2.5	3.1
M34*	3.0	5.5	3.2	4.2	1.5	1.8

*Only the Apulian platform crust was considered.

deformation effects of the old rift tectonics, the structural setting is almost flat. The interpreted horizon "T_X" is connected to a *Middle Trias unconformity at the base of the Dolomia Principale Fm.*

CROP Section M17B. This section, located on the axial zone of the Northern Adriatic Sea, is composed of a thick platform zone to the north (Friulian-Istrian platform) and by a basinal zone to the south (Bellunese basin) generated by Mesozoic (Trias, Lias) rift tectonics, which are well evident on the seismic data (Figure 7; Plates 3 and 4). The basement upper crust, in the platform area, is markedly thicker (up to 6 sec TWT, about 18 km) than in the basinal area (about 4.5 sec, 13 km). Lower crust thickness is between 2.2 sec (about 7.5 km) and 3.0 sec (about 10 km), and the Moho discontinuity, from a depth of 29–30 km to the north, deepens southward to 38 km at the section extremity (SP 2613).

CROP Section M16. This CROP section has been carefully reprocessed with the specific purpose of improving the pre-TX-horizon crustal reflection signals in order to better interpret the Mesozoic rift faults. After a long iterative reprocessing, a markedly improved section was obtained that allowed much better the rift fault identification (Figure 8). The reconstructed tectono-stratigraphy of the section gives a realistic image of the rift tectonics (Triassic, Lias) connected to the generation of the wide basin known in the literature as the *Umbria–Marche basin*. The horizon T_X (Middle Trias unconformity) is well identified below a poorly reflective interval T_I–T_X. It was possible to identify and continuously follow also horizons within the upper basement crust (UC_X, UC_Y) and lower crust (LC_X). Top lower crust (TL) and Moho (MO) are identifiable with satisfactory reliability. At SP 350, the passage between the Istrian-Dalmatian platform and basinal successions begins. The platform forms an island (Barbara Is-

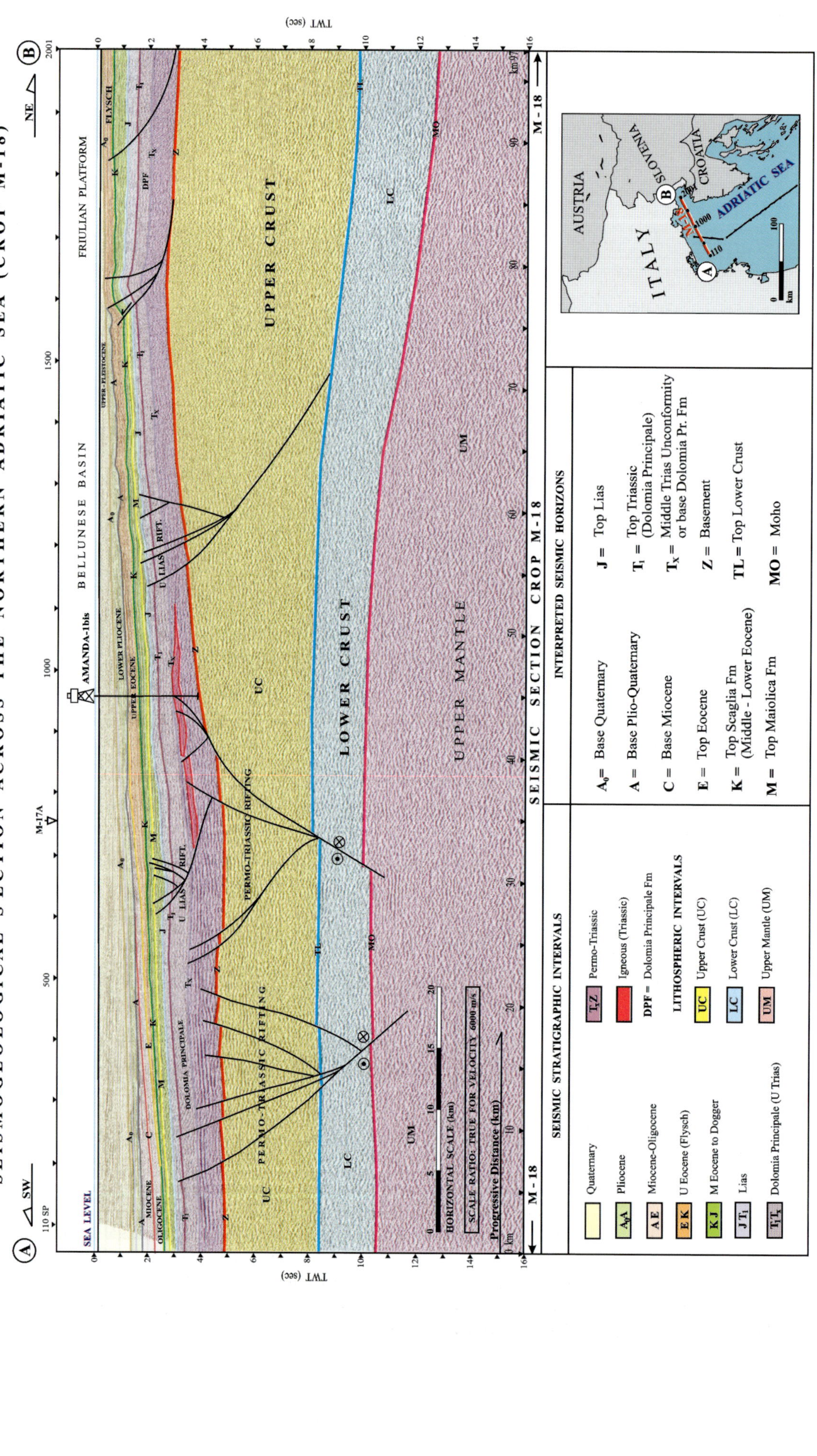
SEISMOGEOLOGICAL SECTION ACROSS THE NORTHERN ADRIATIC SEA (CROP M-18)
A SW
NE B
110 SP
500
1000
1500
2001
M-17A
AMANDA-1bis
SEA LEVEL
BELLUNESE BASIN
FRIULIAN PLATFORM
UPPER CRUST
LOWER CRUST
UPPER MANTLE
PERMO-TRIASSIC RIFTING
DOLOMIA PRINCIPALE
MIOCENE
OLIGOCENE
LOWER PLIOCENE
UPPER EOCENE
UPPER - PLEISTOCENE
FLYSCH
U LIAS
RIFT.
DPF
UC
LC
UM
TL
MO
TWT (sec)
HORIZONTAL SCALE (km)
SCALE RATIO: TRUE FOR VELOCITY 6000 m/s
Progressive Distance (km)
SEISMIC SECTION CROP M-18
M - 18
AUSTRIA
SLOVENIA
CROATIA
ITALY
ADRIATIC SEA
SEISMIC STRATIGRAPHIC INTERVALS
Quaternary
Pliocene
Miocene-Oligocene
U Eocene (Flysch)
M Eocene to Dogger
Lias
Dolomia Principale (U Trias)
Permo-Triassic
Igneous (Triassic)
DPF = Dolomia Principale Fm
LITHOSPHERIC INTERVALS
Upper Crust (UC)
Lower Crust (LC)
Upper Mantle (UM)
INTERPRETED SEISMIC HORIZONS
A0 = Base Quaternary
A = Base Plio-Quaternary
C = Base Miocene
E = Top Eocene
K = Top Scaglia Fm (Middle - Lower Eocene)
M = Top Maiolica Fm
J = Top Lias
T1 = Top Triassic (Dolomia Principale)
TX = Middle Trias Unconformity or base Dolomia Pr. Fm
Z = Basement
TL = Top Lower Crust
MO = Moho

Figure 4: Seismogeological section crossing the Northern Adriatic Sea, reconstructed from interpretation of CROP section M-18. Major evidence of Late-Permian–Early Triassic rifting as well as the main Triassic igneous layer is also shown. Upper Lias rift faults are locally also evident. These transtensive rift tectonics, only partially indicated, are associated with left-strike-slip motion accompanied by counterclockwise rotation of the Adria plate with respect to the African mother plate.

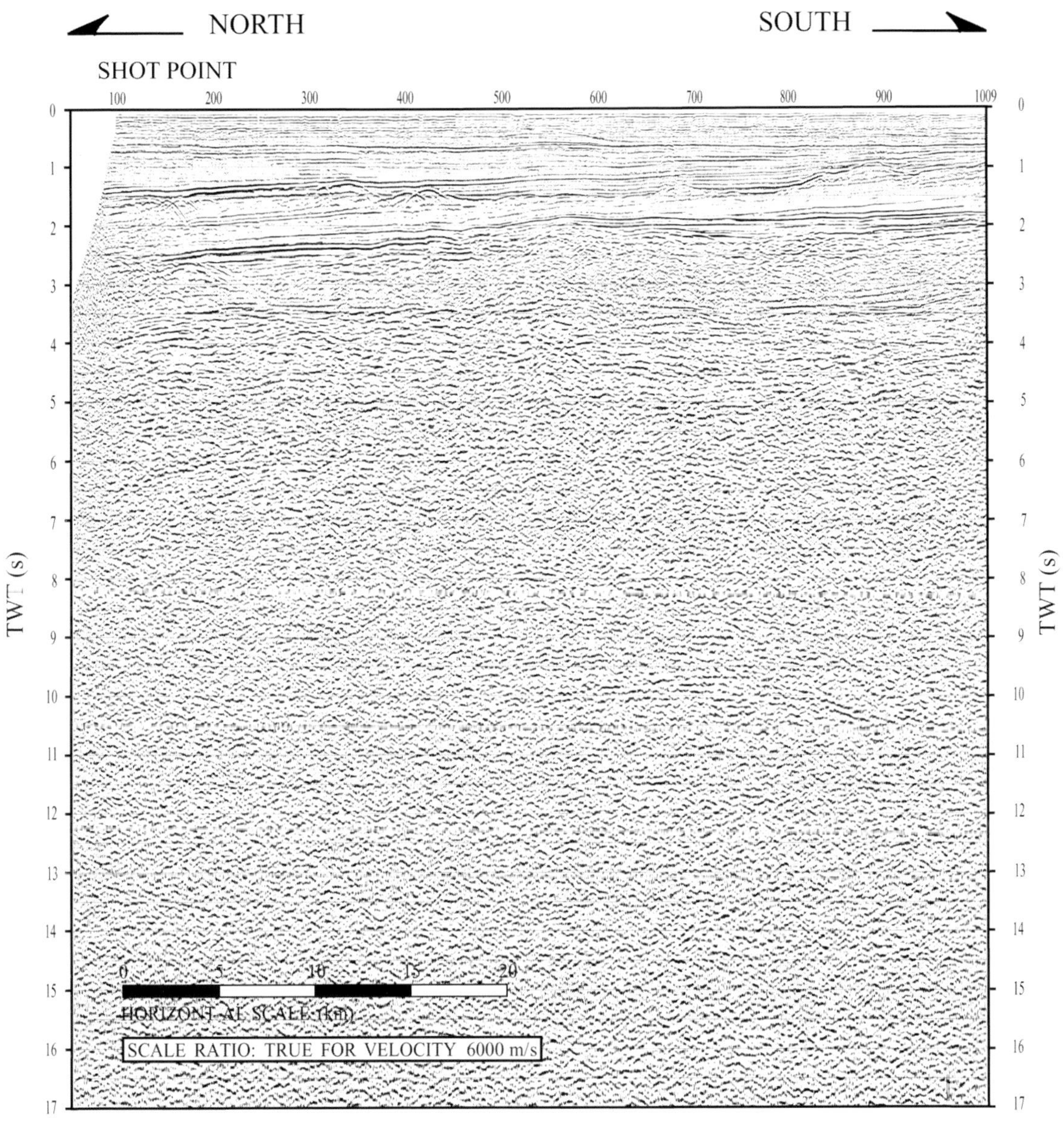

Figure 5: CROP seismic section M-17A along the axial zone of the Northern Adriatic Sea.

land) between SP 500 and 750. The Northern Apennine thrust front is deforming the upper sedimentary cover at the line's SW extremity. As in section M18, the Istrian platform is slightly thrust over the adjacent basinal sequence.

CROP Section 17C. The CROP section M17C (Plates 5 and 6; Figure 9 northern part; Figure 10 southern part) reveals prominent igneous data previously completely unknown. On the northern part, between SP 630 and 2000 (Figure 9; Plate 6), two huge batholithic bodies are intruded at different depths into the upper basement crust. Time analysis of the overlying cover deformation indicates that the intrusion took place in Triassic times as consequence of rift activity and was, successively, reactivated during the Dinaric geodynamic stage (Paleogene). On CROP section M17C (Figure 9) the *Batholith "A"*, very probably elongated along the Adriatic axis, has a length of about

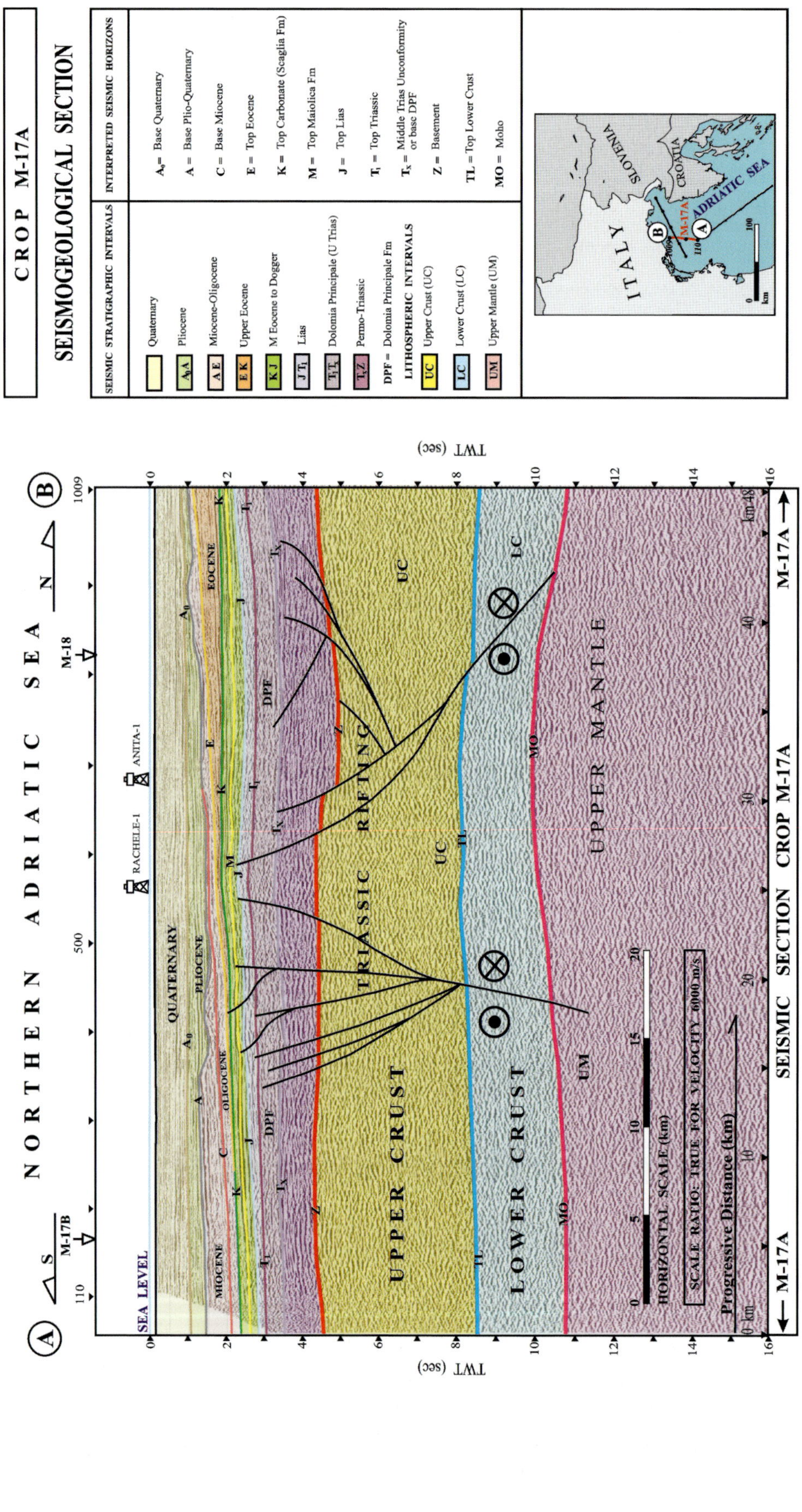

Figure 6: Crustal seismogeological section along the axial zone of the Northern Adriatic Sea, from CROP M-17A.

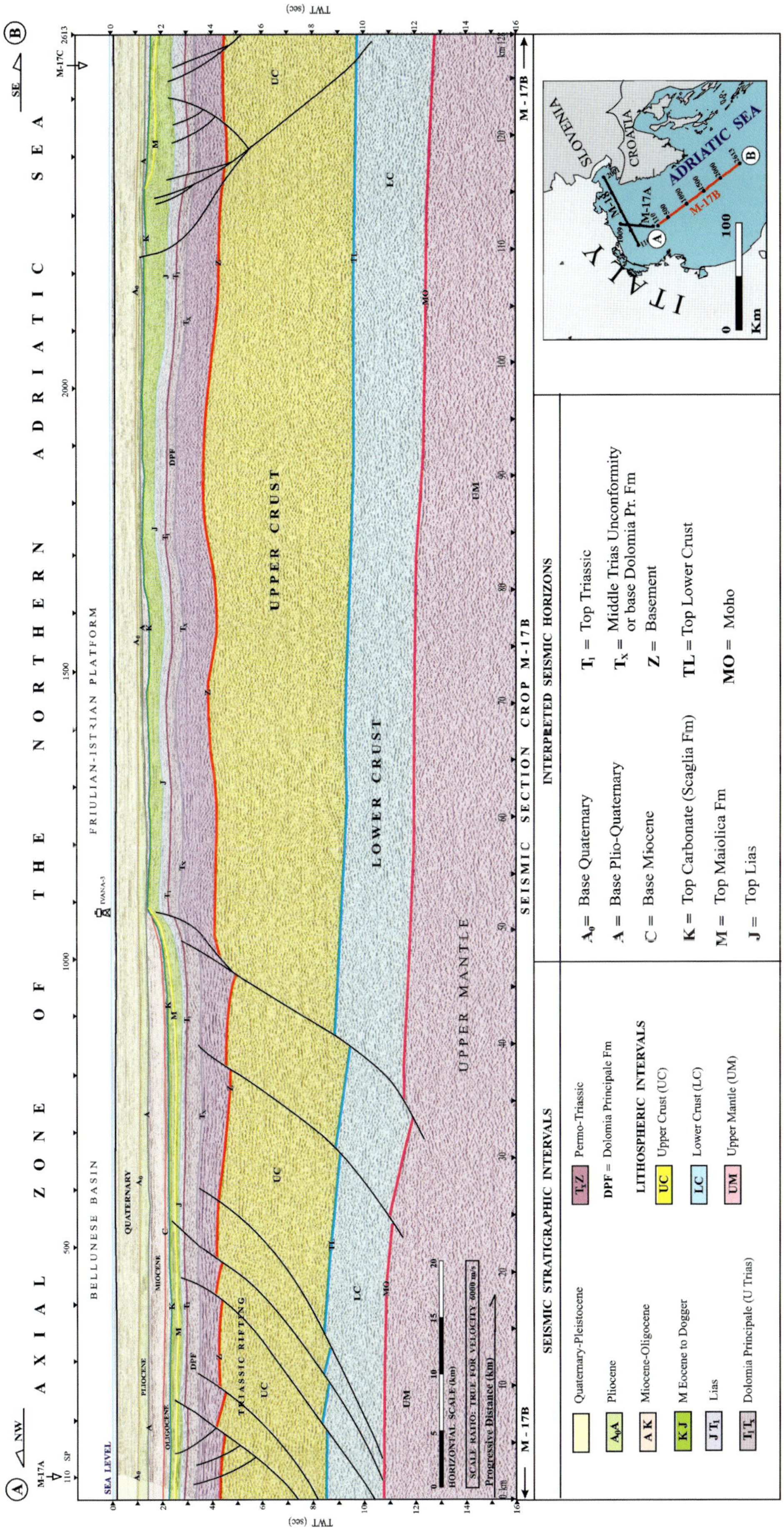

Figure 7: Crustal seismogeological section along the axial zone of the Northern Adriatic Sea, from CROP M-17B.

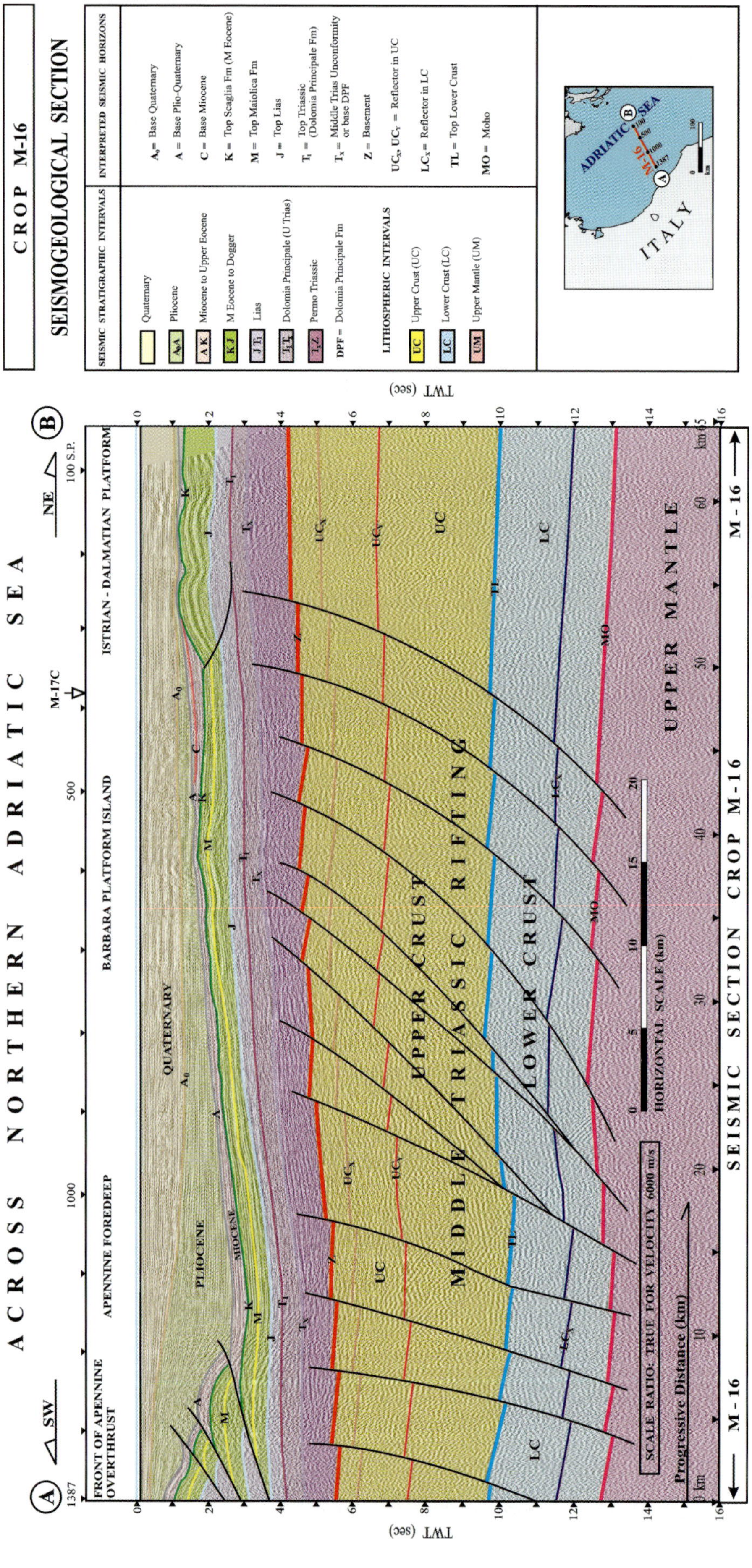

Figure 8: Crustal seismogeological section across the Northern Adriatic Sea, from CROP M-16.

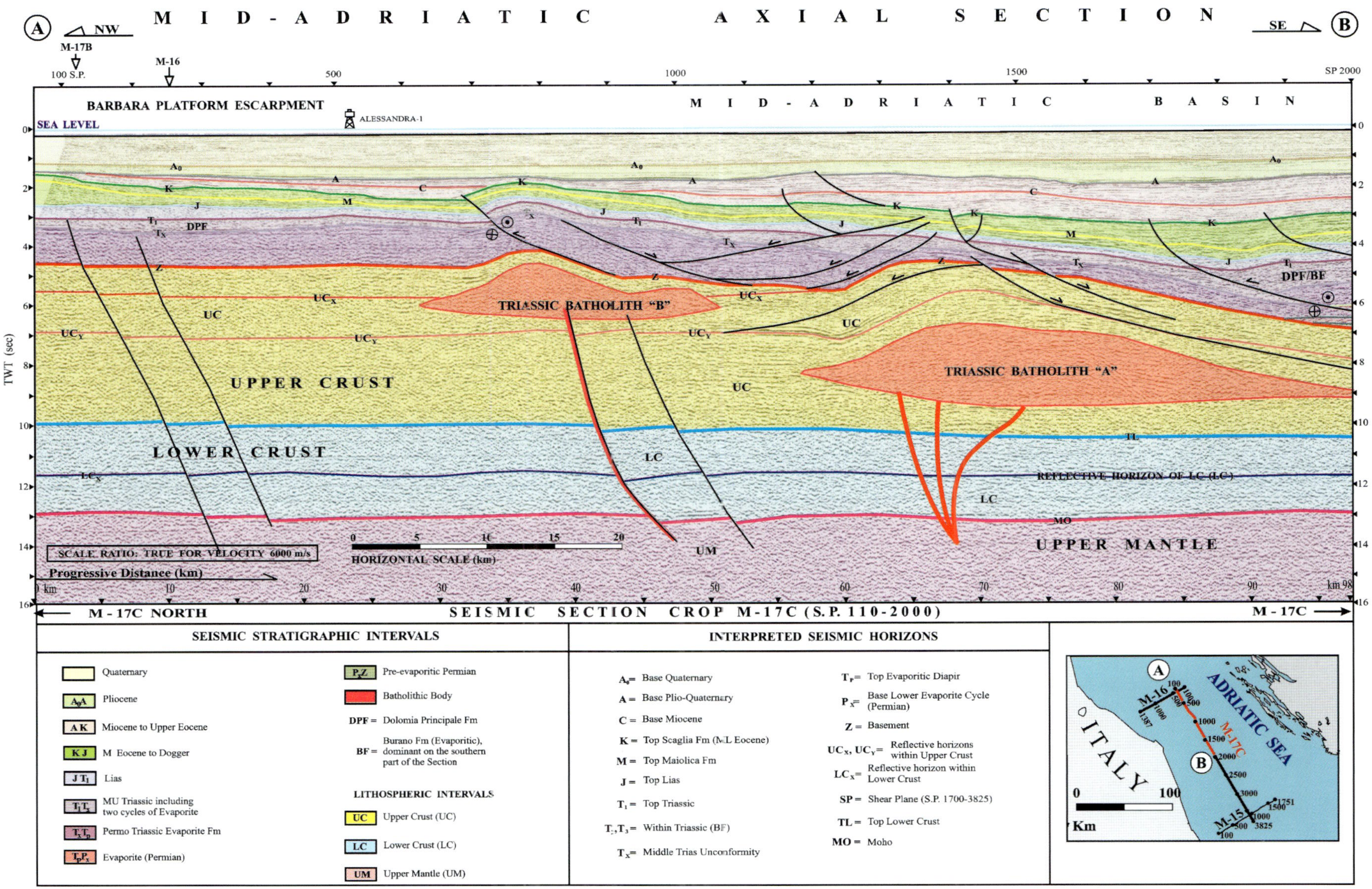

Figure 9: Crustal seismogeological section along the axial zone of the central Adriatic Sea, from CROP M-17C (part: SP 100–2000).

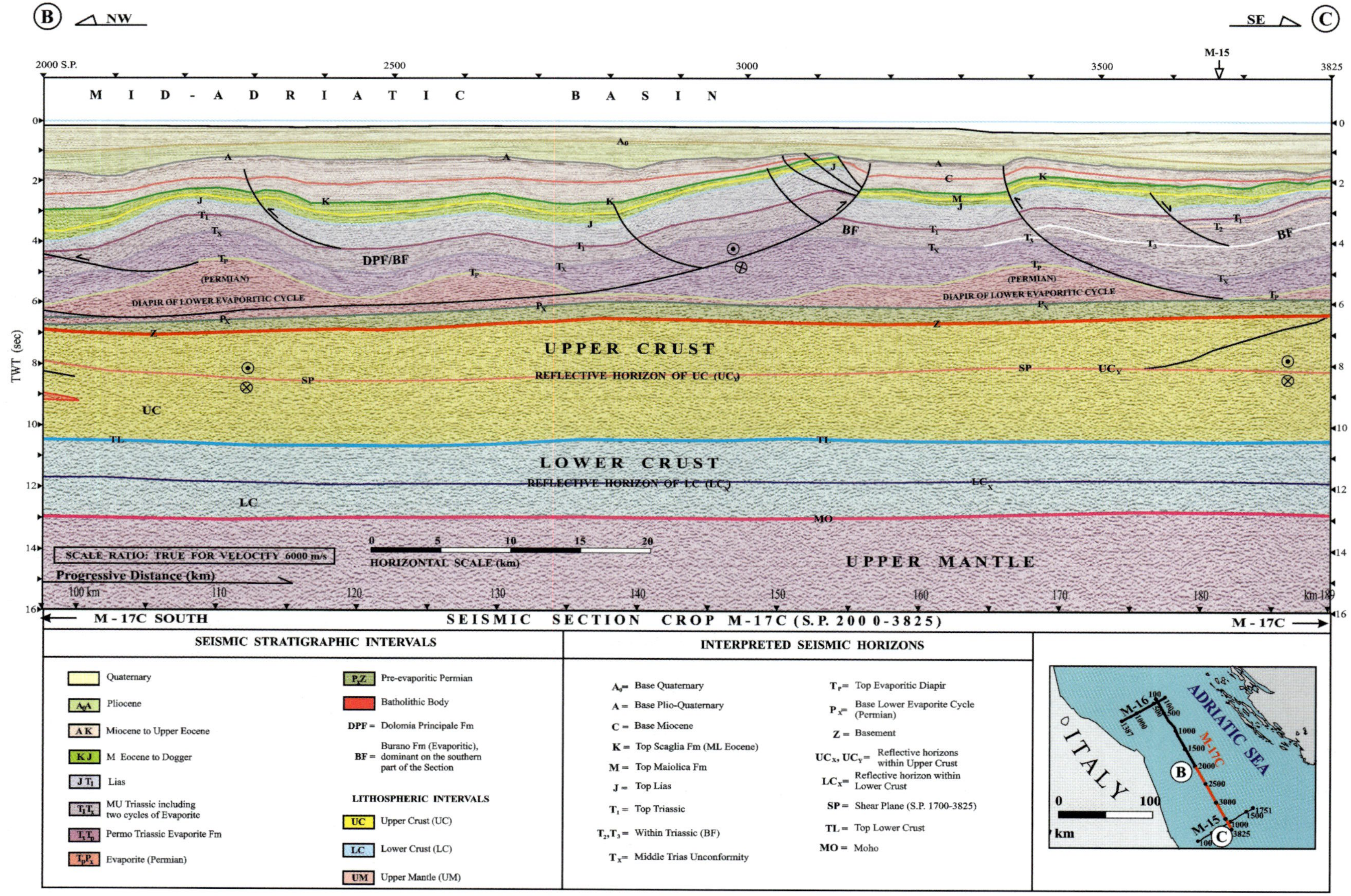

Figure 10: Crustal seismogeological section along the axial zone of the central Adriatic Sea, from CROP M-17C (part: SP 2000–3825).

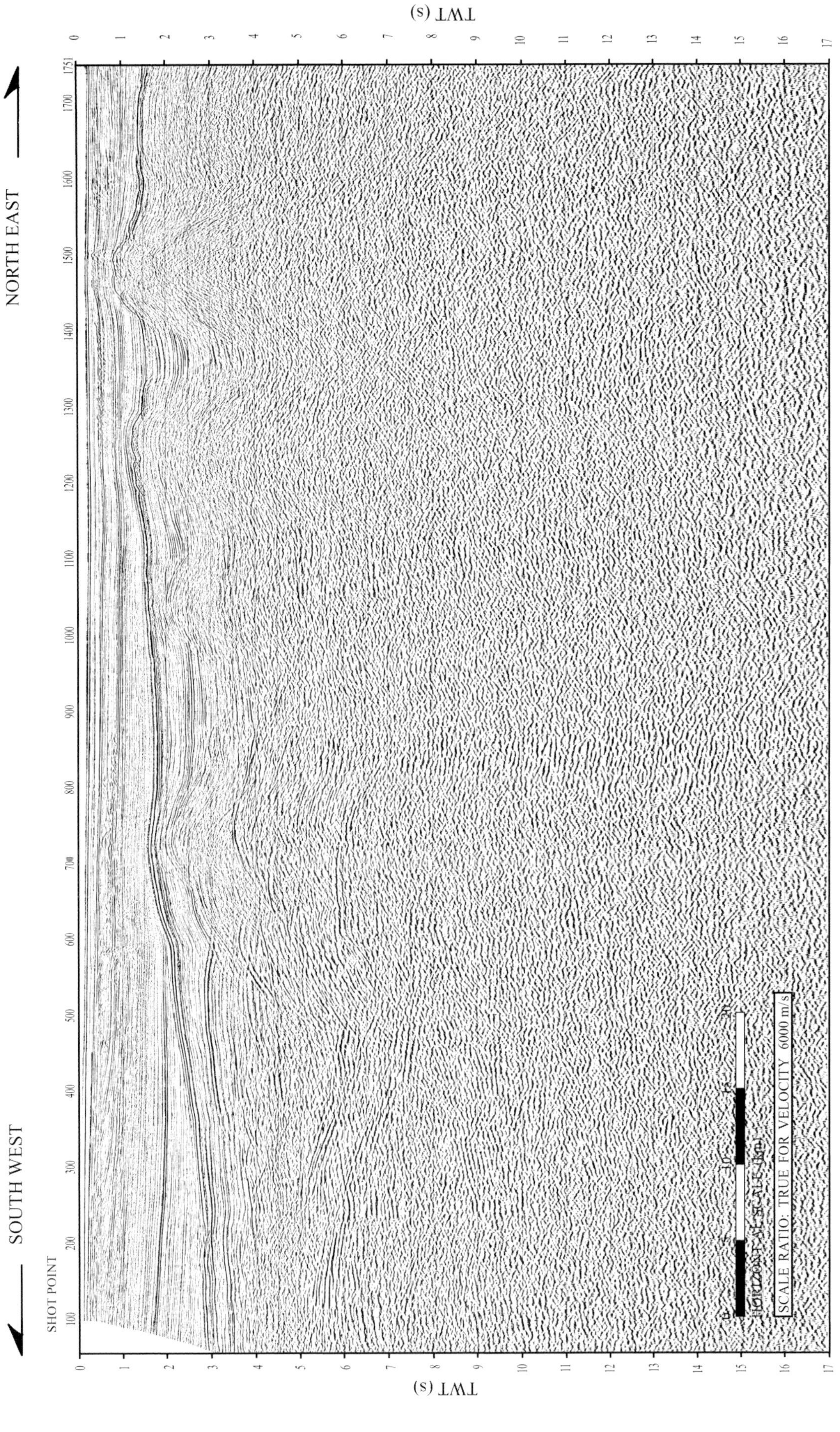

Figure 11: CROP seismic section M-15, across the central Adriatic Sea. This section at larger scale (1:300,000) can be seen in Plate 1 of Chapter 9 of the present volume.

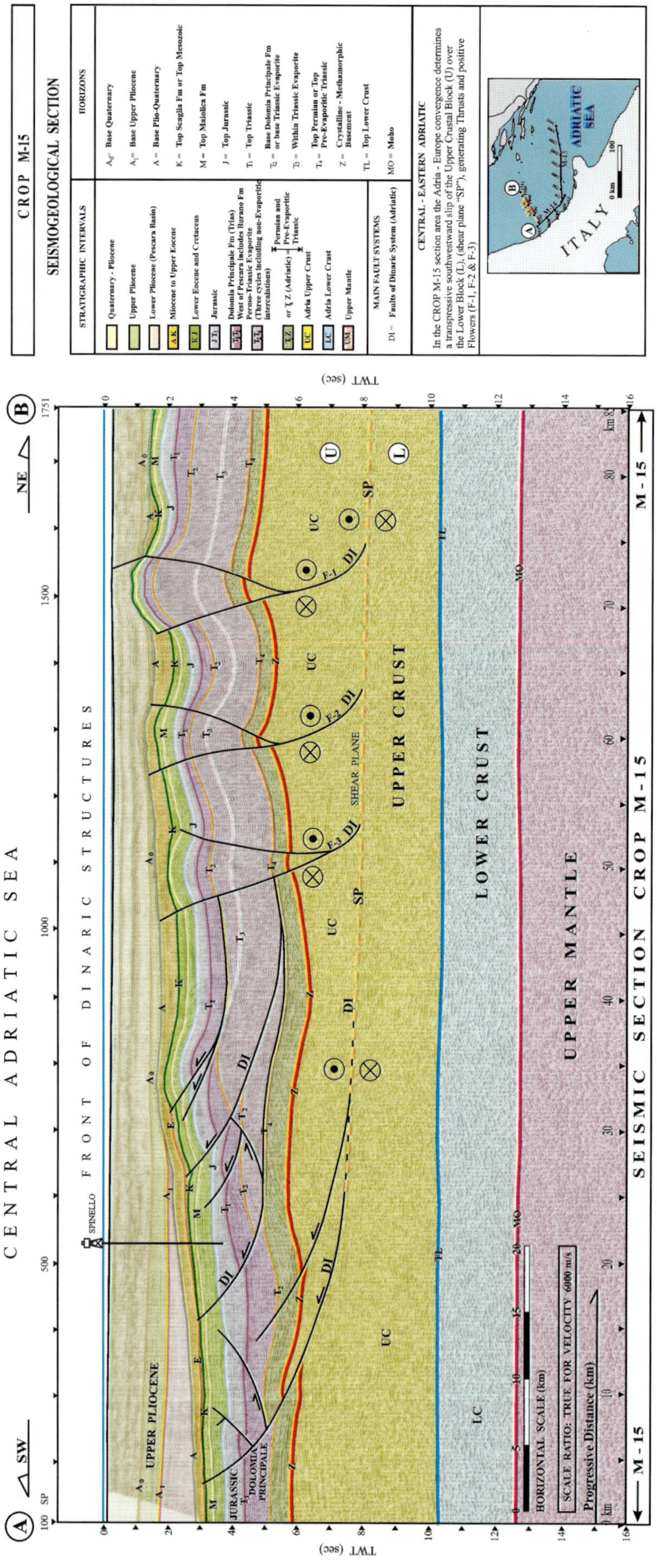

Figure 12: Crustal seismogeological section across the Central Adriatic Sea, from CROP M-15.

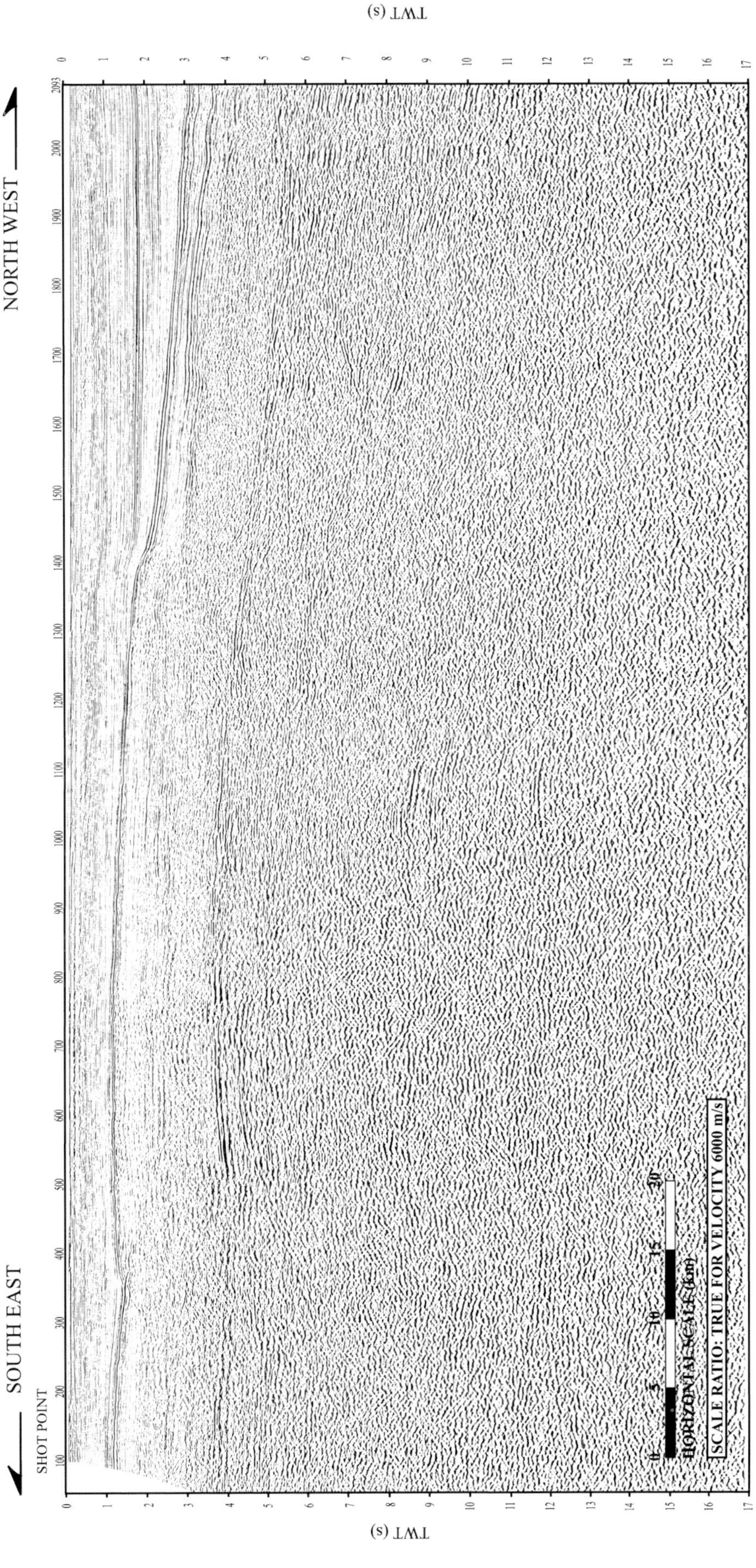

Figure 13: CROP seismic section M-14 along the Abruzzese–Molisian coastal zone of the Adriatic Sea.

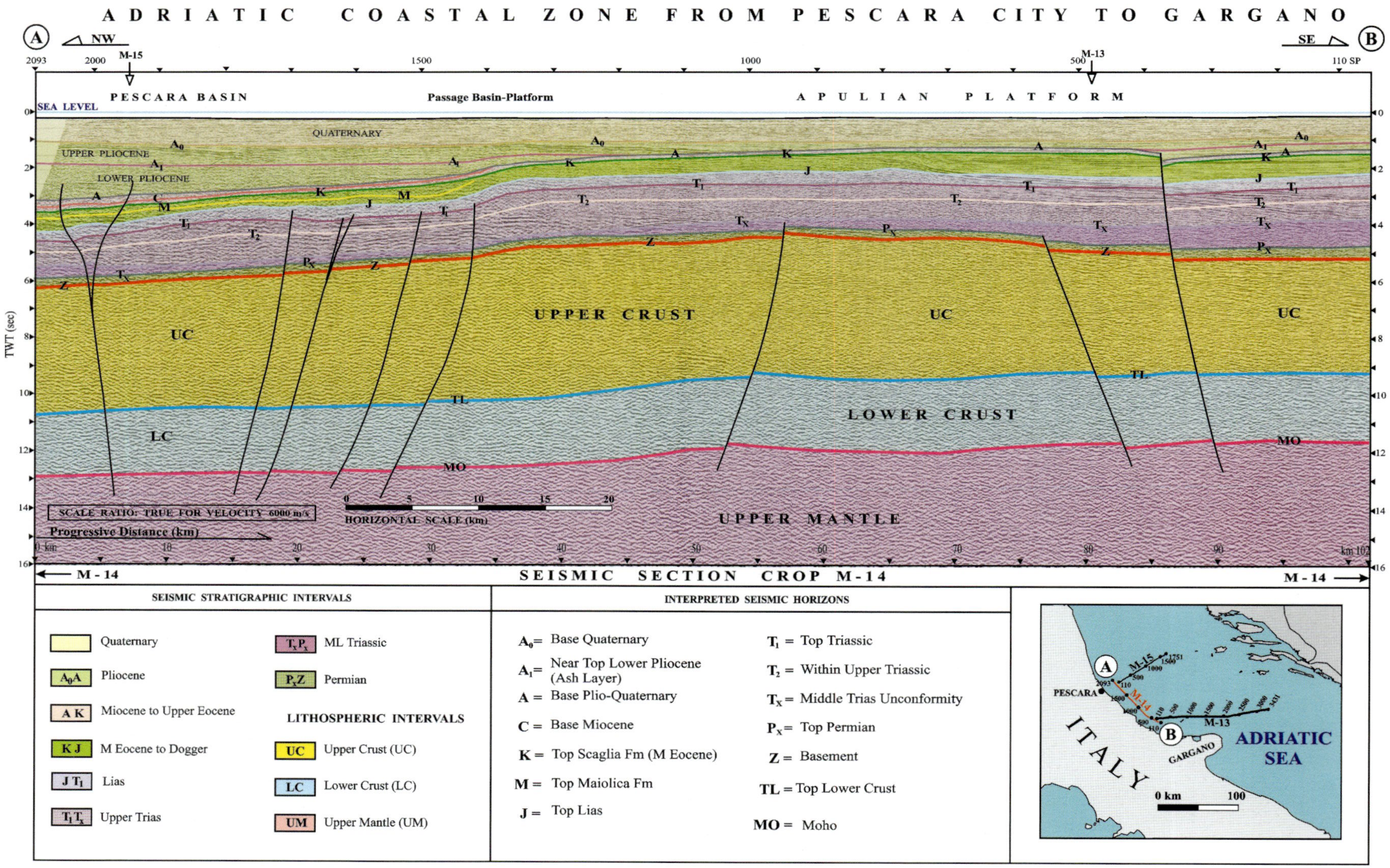

Figure 14: Crustal seismogeological section along the Abruzzese–Molisian coastal zone of the Adriatic, from CROP M-14. This section crosses the Pescara basin–Apulian platform boundary.

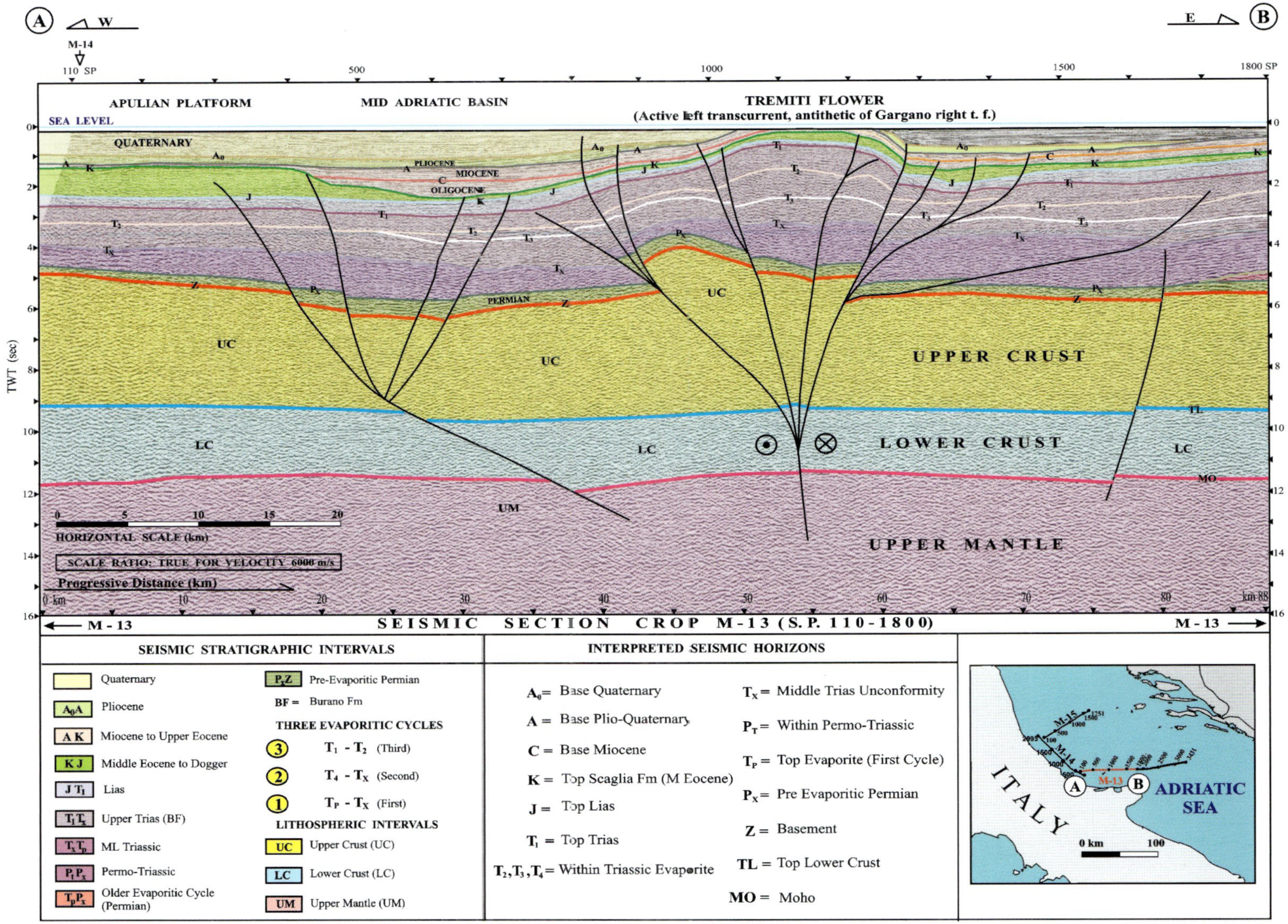

Figure 15: Crustal seismogeological section across the Southern Adriatic Sea, from CROP M-13 (part: S.P. 110–1800). There is impressive seismic imaging of the Tremiti flower generated by an active left-strike slip, antithetic to the "Gargano" right transcurrent driven by the Africa–Europe convergence.

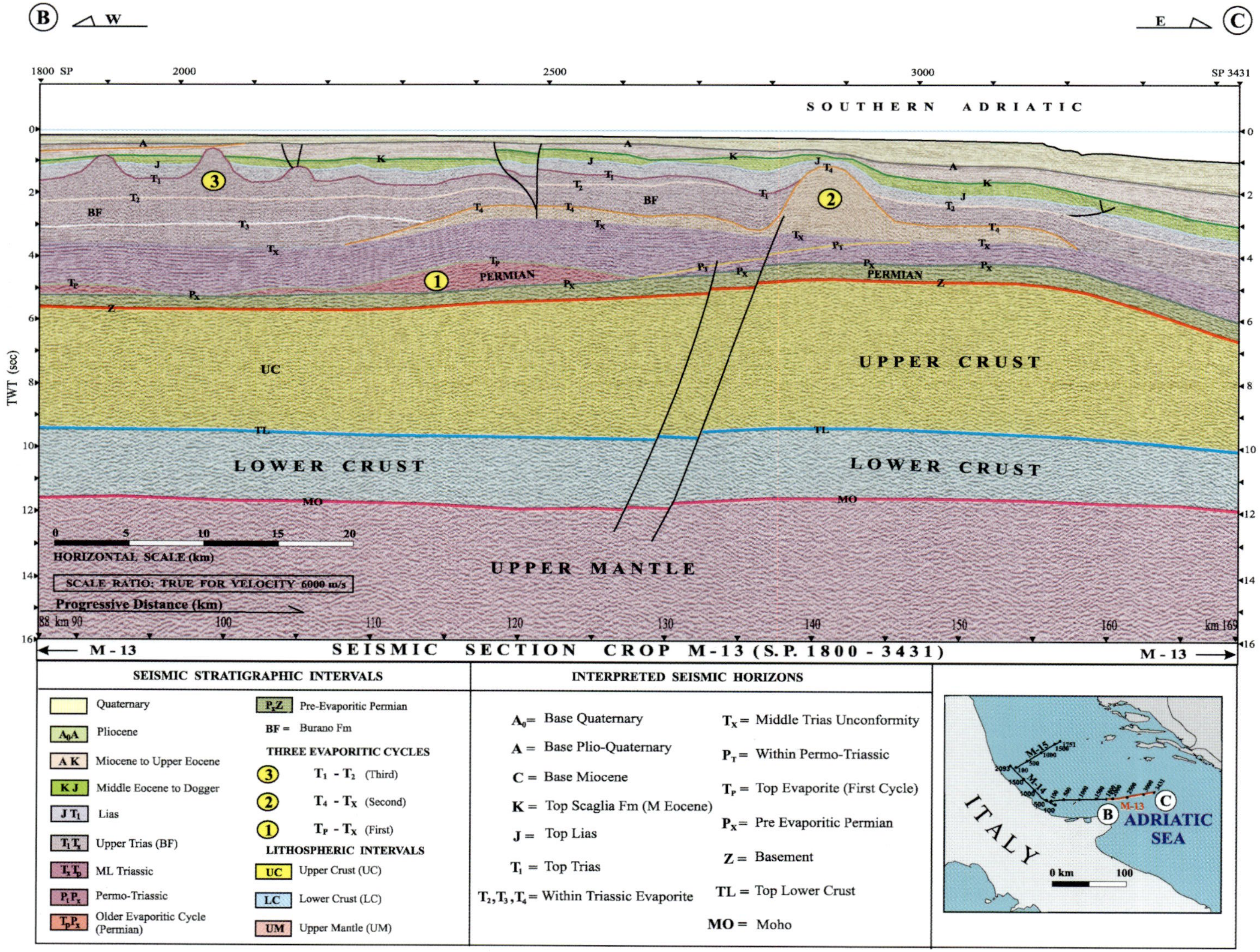

Figure 16: Crustal seismogeological section across the Southern Adriatic Sea, from CROP M-13 (part: S.P. 1800–3431).

Figure 17: CROP seismic section M-38.

45 km and a maximum height of 2.7 sec (about 8 km). The *Batholith "B"* is about 23 km long with a maximum height of 5.8 km.

The thick sedimentary cover is affected by Dinaric tectonics that over the batholiths amplified their effects within the Triassic evaporite (Burano Fm.). In the southern part of the section (Figure 10) evaporitic domes and compressive structure of the frontal dinaric deformation are clearly visible, subparallel to the regional trends. More extensive reconstruction of the evaporitic interval's distribution (see Figure 16) indicates that the domes of Figure 10 very probably belong to the deeper cycle, attributable to the Permian. Within the upper crust a continuous well-evident seismic horizon (UC_X) is picked. This horizon is connected to a shear plane of the westward-displacing dinaric thrust tectonics.

CROP Section M15. Buried beneath flat Quaternary sediments, a significant tectonic deformation of difficult interpretation is seismically evident. From SP 200 to 1000 west-verging thrust-faults attributable to the front of the Dinaric structure are exhibited. These tectonics deform an area which extends westward to the NE flank of the Pescara basin (Figure 12). The age of deformation is in line with the Dinaric stage. From SP 1000 to the NE extremity of the section, three transpressive flower features (F1, F2 and F3) characterize the tectonics of this part of the section. Such faults, together with the frontal thrust faults at SP 200, are rooted on a shear plane (SP) located in the basement upper crust, at a depth of about 8 sec TWT, where a reflector can be identified. In this interpretation the compressive tectonic deformation is driven by northward motion of the Adria plate (Adria–Europe convergence) accompanied by transpressive right-transcurrent and thrust faults over the shearing plane SP (Figure 12). Going more internally to the Dinaric orogen, the compressive and shear tectonics involve progressively deeper crustal layers until transcrustal features are produced. Shear tectonics, especially in the central Adriatic, assume an important role with exhumation of deep-seated units and igneous rocks (i.e. Palagruza gabbros).

CROP Section M14. As for all other Adriatic CROP sections, the interpreted seismic horizons are stratigraphically defined in the section legend. CROP section M14 (Figures 13 and 14) is located across the *Pescara basin–Apulian platform.* The interpreted seismogeological section (Figure 14) is self-explanatory. Thickening of the Pliocene sequence toward the north clearly dates the prominent Apennine overthrusting process.

CROP Section M13. This section in the Southern Adriatic is tectonically and stratigraphically very interesting (Plates 7 and 8; Figures 15 and 16). The western part (Figure 15) shows the huge positive *Tremiti Flower which is the antithetic left-shear fault of the Gargano Right-Transcurrent Fault* (Figure 23). This type of shear fault system is rather recurrent in the Adriatic foreland (as well as in the Pelagian foreland (Finetti and Del Ben, this volume, Chapter 26). The Gargano-Tremiti transcurrent system, identified and defined seismically several years ago (Finetti, 1982), is still very active. In fact, it recently (2002) produced a large earthquake in the city of S. Giuliano di Puglia, where children of a local school which collapsed (very likely sited close to the fault displacement zone) lost their life.

At SP 450, the passage from the Apulian platform to the Adriatic basin starts (Figure 15). On the eastern part of the section (Figure 16) it is possible to observe the halokinetic structures of various evaporitic cycles. In our interpretation, three main cycles can be recognized: the first (1 within yellow circle) delimited by seismic horizons T_Y–T_X, and attributed to the Permian; a second (2) lying between horizons T_4–T_X which form a large dome at SP 2880, of imprecise Triassic age; and a third one (3) of Upper Triassic age (Figure 16).

3.2. CROP Seismic Sections of Southeastern Apulia

All four sections of the Southeastern Apulia extension explore the thicker platform zone. Two sections extend eastward to the Greek islands of Corfu (M38) and Paxos (M34). One section (M8) runs subparallel to the Apulian axis, near the top of the platform swell. Two CROP sections (M5 and M34) exhibit the crustal setting of the Apulian platform and its passage to the Ionian Neotethys, floored by oceanic crust.

Figure 18: Interpreted CROP section M-38, crossing the Apulian platform, the Otranto Channel and the Western Corfu shelf. It shows the complex structural setting of the Corfu shelf area where a tectonic thickening of the Triassic Evaporite and an overlying thrusting Tertiary wedge are evident. The Apulian platform–Otranto Channel basin transition is seismically evident.

SEISMOGEOLOGICAL LITHOSPHERIC SECTION FROM APULIAN PLATFORM TO CORFU FROM NEW CROP SEISMIC DATA (SECTION M-38)

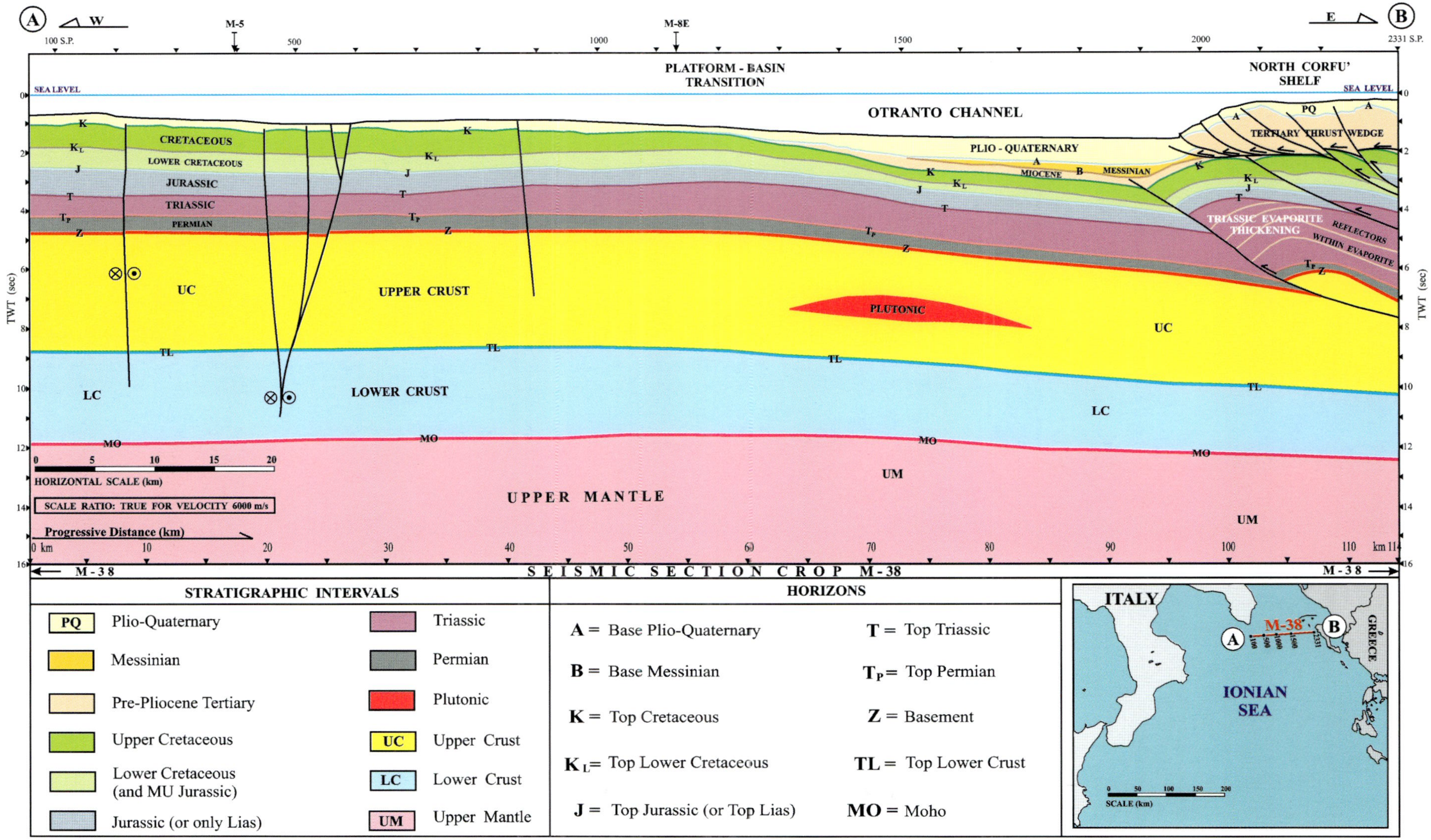

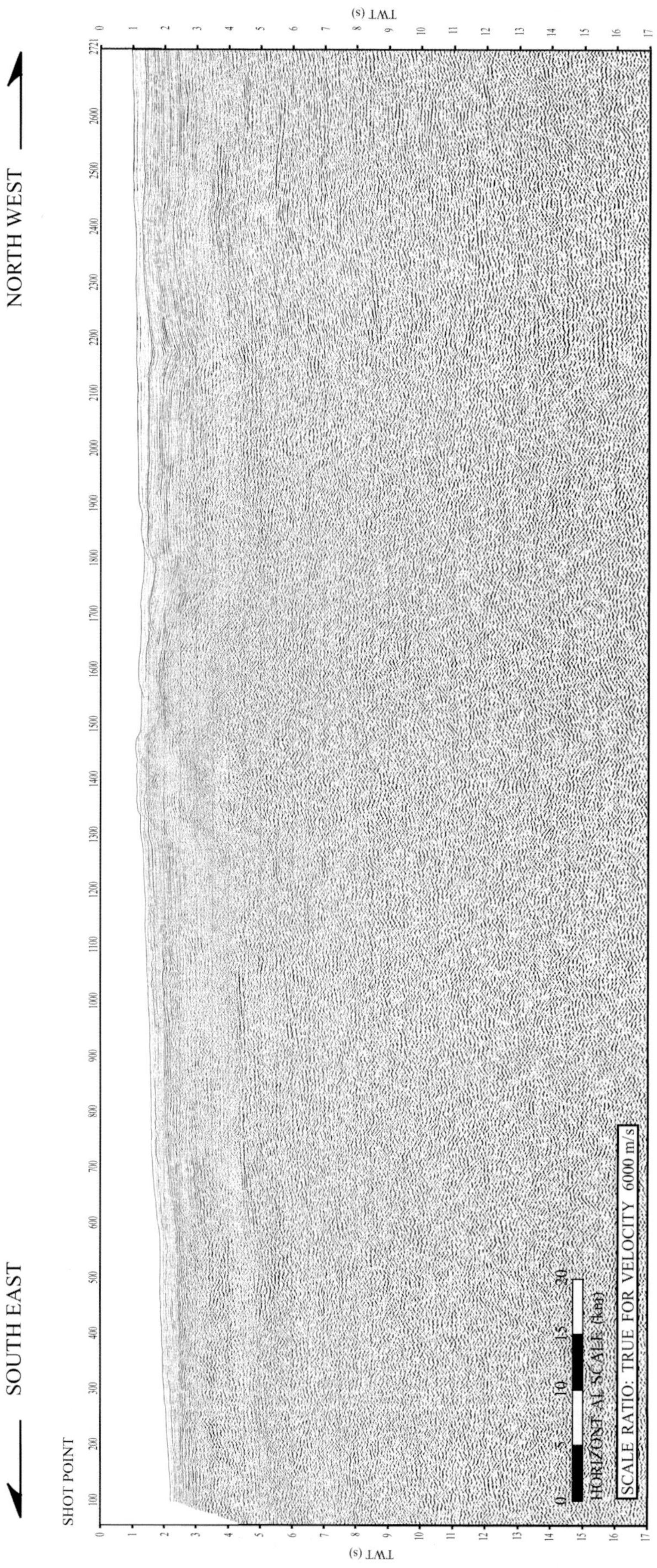

Figure 19: CROP seismic section M-8.

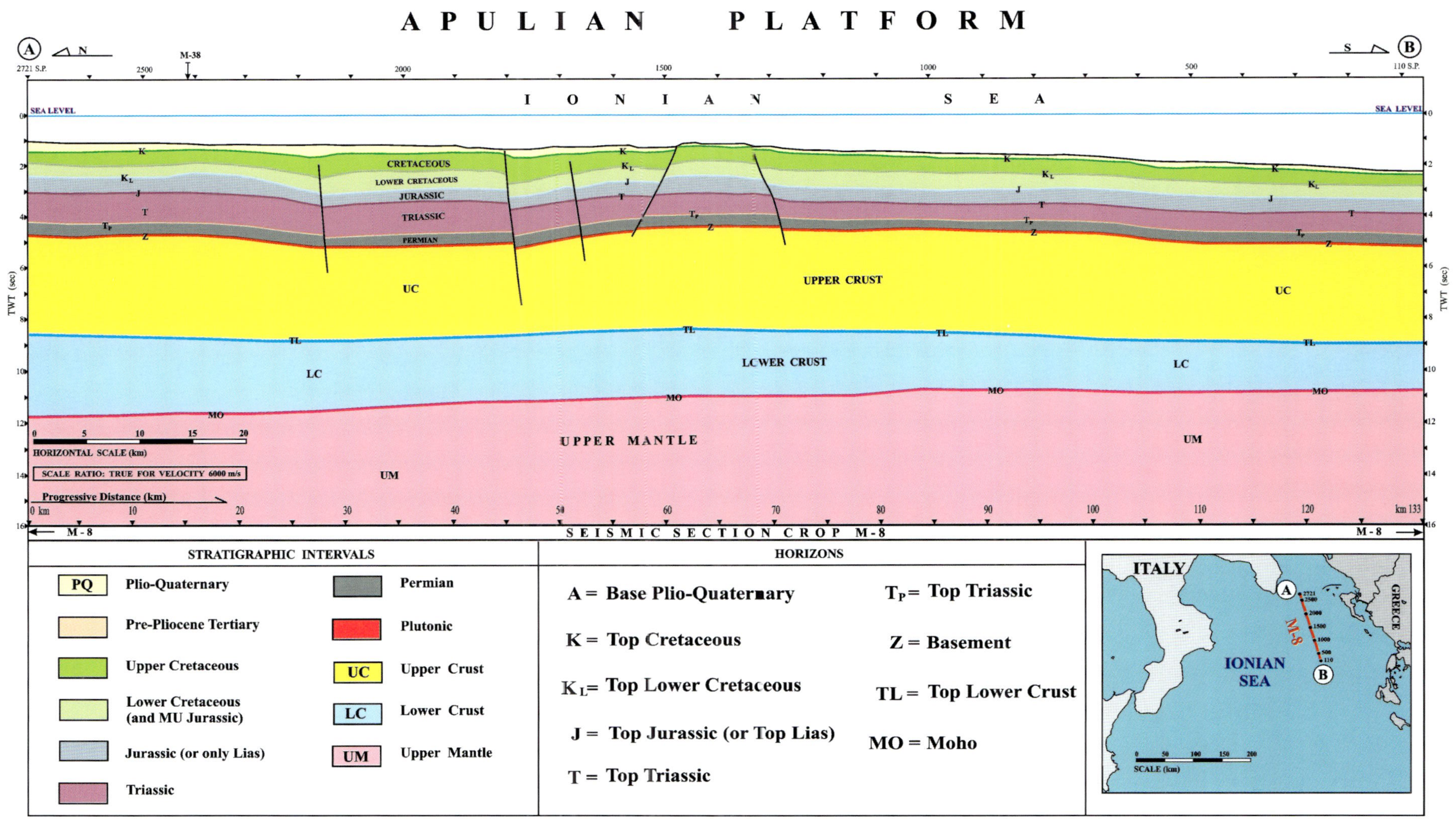

Figure 20: Interpreted CROP section M-8 on the southeastern extension of the Apulian platform, characterized by a flat setting, and interrupted by a few minor faults, probably connected to small shear movements.

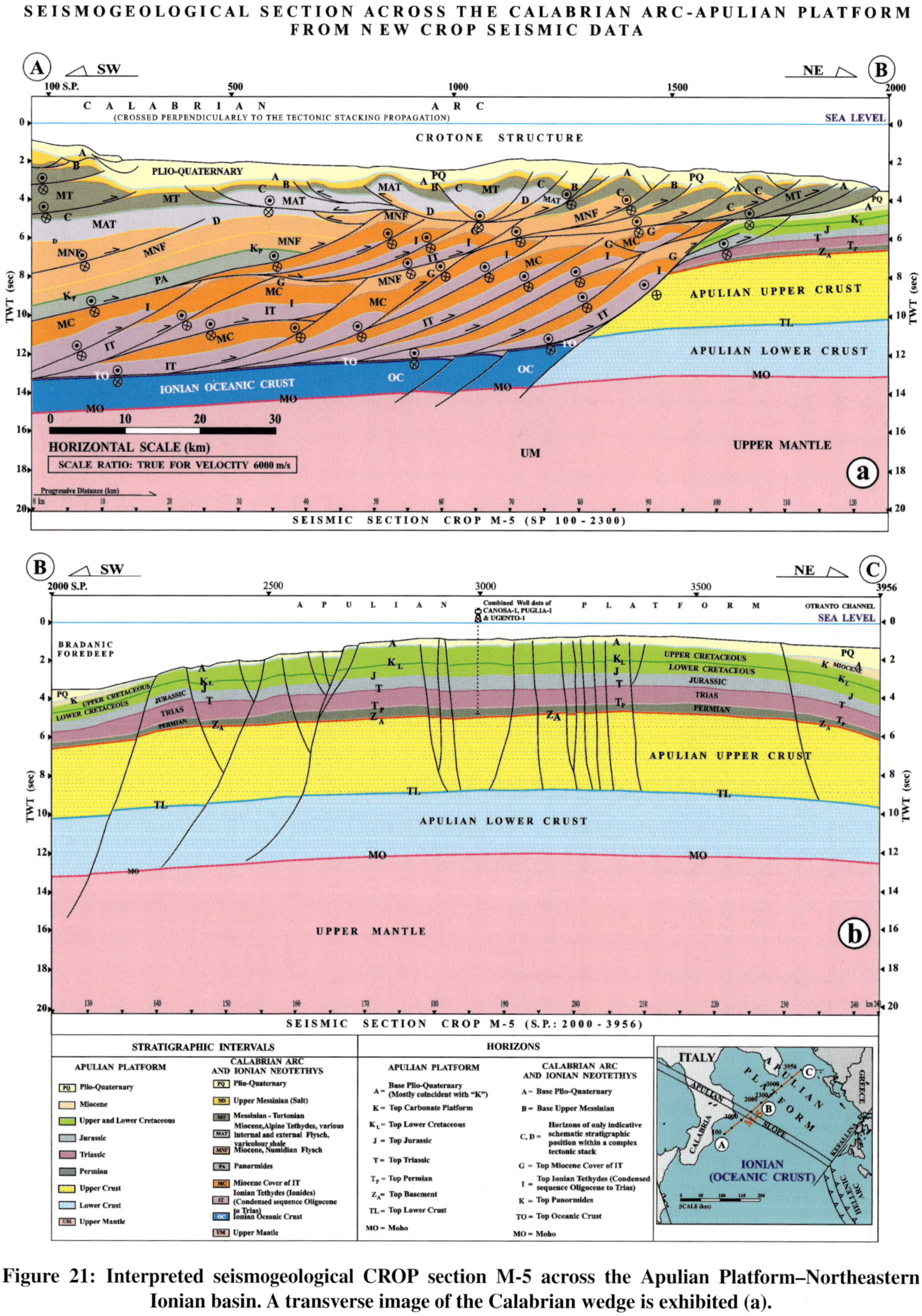

Figure 21: Interpreted seismogeological CROP section M-5 across the Apulian Platform–Northeastern Ionian basin. A transverse image of the Calabrian wedge is exhibited (a).

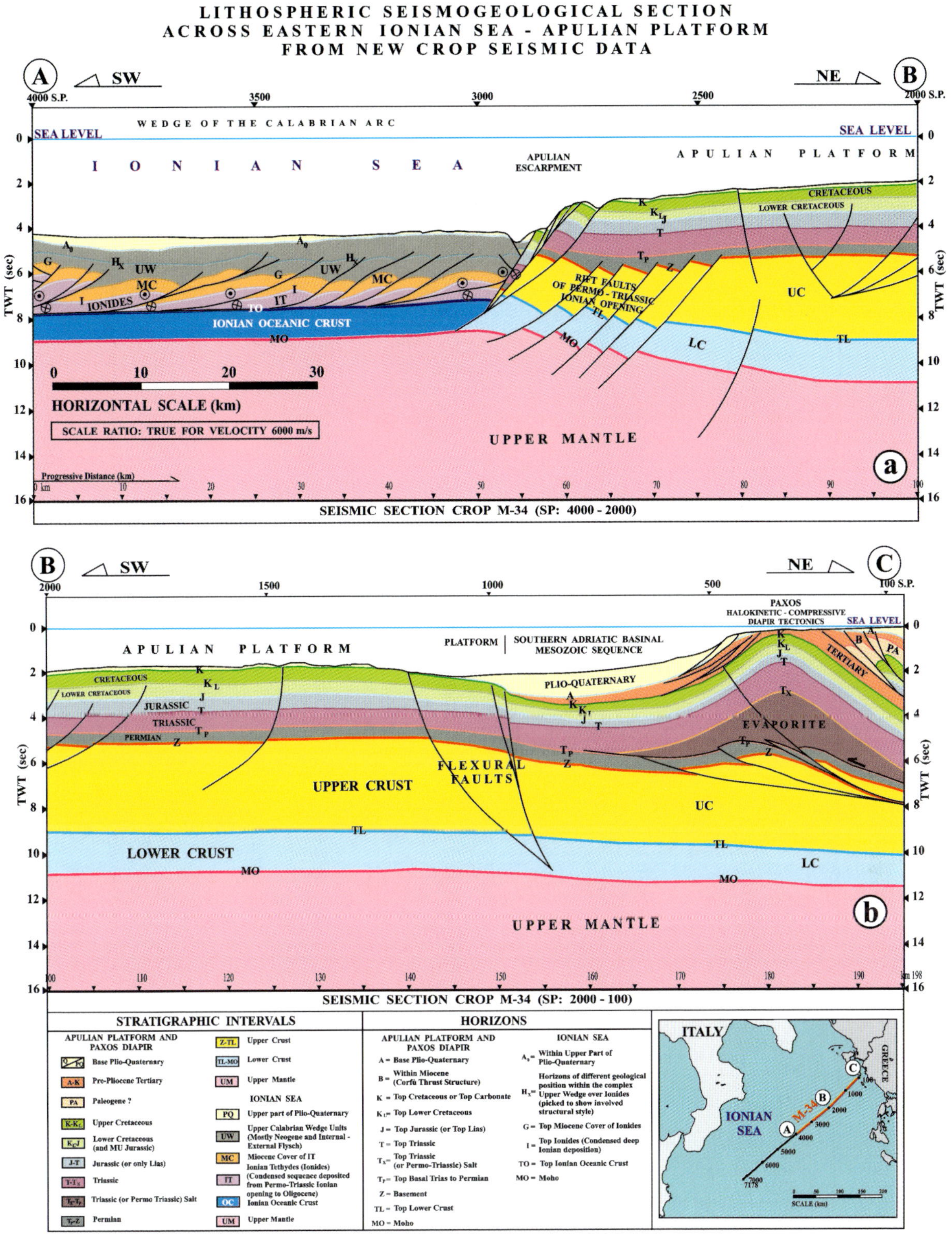

Figure 22: Interpreted seismogeological CROP section M-34 (part) across the northeastern Ionian basin–Apulian margin–Paxos Island.

CROP Section M38. This section crosses the Apulian Platform, the Otranto channel basin, and the western margin of Corfu Island (Figure 17: seismics; Figure 18: interpreted seismogeological section). The tabular platform setting of Apulia is interrupted by a few small normal faults, probably generated by modest right-shear motion (Adria–Europe convergence). From SP 1250 to 1400, the platform passes to basinal deposition, where the Cretaceous and Jurassic interval marks an evident layer thinning (Figure 18). From SP 1900 to the eastern extremity of the section, the Corfu thrust tectonics is seismically well evident.

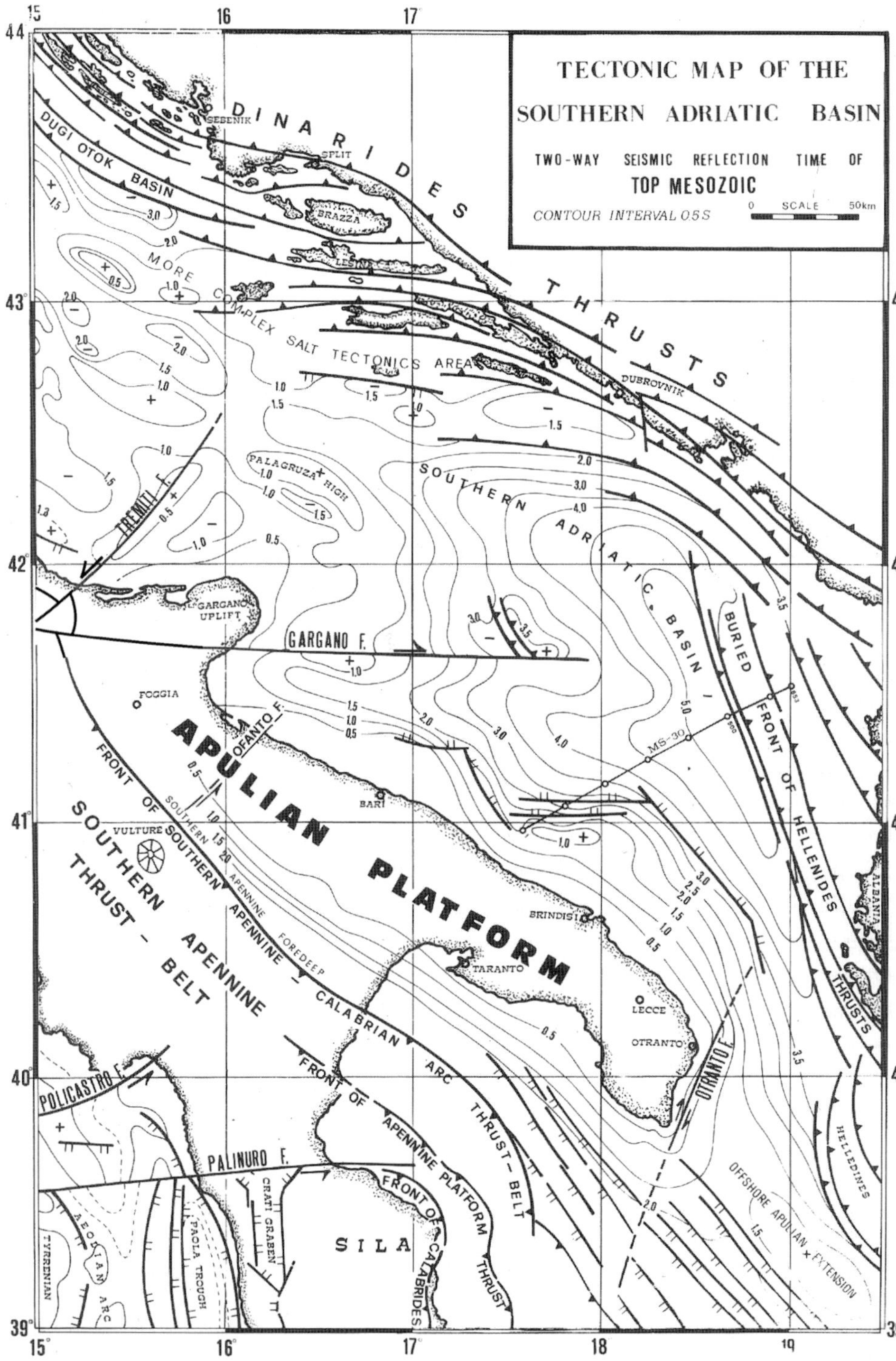

Figure 23: Structural map of the Southern Adriatic basin. Contour lines refer to seismic reflection time (sec. TWT) of the top Mesozoic.

CROP Section M8. This section exhibits the typical flat crustal stratigraphy of the axial zone of the Apulian platform (Figures 19 and 20). The crust-mantle interface (Moho) is at a depth of 11.8 sec TWT (about 32.5 km) to the northern extremity, and 10.8 sec (about 27 km) to the southern end (Figure 20). From north to south, the crustal thickness (water excluded) decreases progressively from about 31 km to 25.4 km.

CROP Section M5. The CROP section M5 shows the very clear and interesting transcrustal setting of the Apulian platform, of its buried slope on the Ionian Tethys and of the Calabrian wedge, stacked on the oceanic crust (Plates 9 and 10; Figures 21 and 22). On the *Bradanic foredeep* side, the platform is affected by transcrustal flexural faults

(Figure 21(b)). From SP 1200 to 1700, there is the Apulian slope over which the wedge of imbricated Ionian Tethys (IT) cover units (Ionides) is thrust, and seen transversally to the Calabrian Arc tectonic propagation. Over the Ionides, there is their Miocene cover (MC) and other thrust units (P = Panormides; Numidian flysch and Miocene MNF; Alpine Tethydes and Miocene MAT; Messinian-Tortonian).

CROP Section M34 (Part). The section M34 crosses a large part of the Ionian Sea with its oceanic crust, the entire Apulian platform, the southern Adriatic basin, and the Paxos Island compressive feature. Here M34 is exhibited from SP 100 to 4000 (Plates 11 and 12; Figure 22), leaving part of its Ionian segment to Chapter 19 by Finetti and Del Ben (this volume). The M34 section crosses the Ionian Sea where the Calabrian wedge is thin and does not cover the whole Apulian slope. Note that the Apulian margin is affected by Permo-Triassic rift-faults connected with the Ionian opening (Figure 22(a)). It is not clear why similar faults are not evident on the Apulian margin in section M5.

4. STRUCTURAL MAP OF THE SOUTHERN ADRIATIC

To connect the geophysical information for the Adriatic CROP sections with that of the southeastern Apulian extension on the southern Adriatic where no CROP sections could be acquired, a structural sketch-map based on MS seismic lines (Finetti et al., this volume, Chapter 1) and other geophysical and borehole data is presented (Figure 23). This map exhibits the areal distribution of major tectonic features and the contour lines of the top Mesozoic. It should be noted that the available seismic data, borehole data, gravity data and crustal interpolations among CROP data suggest that the southern Adriatic crust underthrusting the southern Dinarides and Hellenides is markedly thinner than that in the northern sector. A possible interpretation is that the southern Adriatic crust was stretched and thinned during the Permo-Triassic Ionian opening.

ACKNOWLEDGEMENTS

The Authors are very grateful to OGS for the CROP data acquisition and to ENI-AGIP for processing. Many thanks are due to the Processing Center of the University of Trieste for the reprocessing of section M16. This work was funded by MIUR with grants to I.R. Finetti (COFIN 1998, 2000 and 2002).

REFERENCES

Casero, P., Structural setting of petroleum exploration plays in Italy. In: Crescenti, V., D'Offizi, S., Merlino, S., Sacchi, L. (Eds.), Special Volume of the Italian Geological Society for the IGC 32, Florence – 2004.

Cati, A., Sartorio, D., Venturini, S., 1987. Carbonate platforms in the subsurface of the Northern Adriatic area. Mem. Soc. Geol. It. 40, 295–308.

Del Ben, A., Finetti, I.R., Mongelli, F., Zito, G., 1994. Seismic and heat flow study of the Southern Adriatic basin. Boll. Geof. Teor. Appl. XXXVI, 29–43.

Dragasevic, T., 1983. Oil geological exploration in the Montenegro offshore in Yugoslavia. Journal of Jugoslav Committee of the World Petroleum Congresses XXXIV (7–8), 397–403.

Finetti, I.R., 1982. Structure, stratigraphy and evolution of Central Mediterranean. Boll. Geof. Teor. Appl. 24, 247–426.

Finetti, I.R., Del Ben, A., 1986. Geophysical study of the Tyrrhenian opening. Boll. Geof. Teor. Appl. 28, 75–156.

Finetti, I.R., Del Ben, A., this volume. Crustal tectono-stratigraphic setting of the Pelagian foreland from new CROP seismic data. Chapter 26.

Finetti, I.R., Del Ben, A., this volume. Crustal tectono-stratigraphic setting of the Ionian Sea from new integrated CROP seismic data. Chapter 19.

Finetti, I.R., Morelli, C., 1973. Geophysical exploration of the Mediterranean Sea. Boll. Geof. Teor. Appl. 15, 263–341.

Finetti, I.R., Del Ben, A., Forlin, E., Prizzon, A., this volume. Review of the main results of pioneering MS seismic exploration programme in the deep water of the Mediterranean Sea (OGS, 1969–1980). Chapter 1.

Finetti, I.R., Bricchi, G., Del Ben, A., Pipan, M., Xuan, A., 1987. Geophysical study of the Adria plate. Mem. Soc. Geol. It. 40, 335–344.

Pieri, M., Groppi, G., 1981. Subsurface geological structure of the Po Plain, Italy. Pubb. N° 414 Progetto Finalizzato Geodinamica.

CHAPTER 24

Seismic Modelling Applied to CROP Crustal Section in the Adriatic Sea

J.M. Carcione[1,*], I.R. Finetti[2], D. Gei[1]

ABSTRACT

We use a seismic-modelling methodology to validate the interpretation of a deep-crust seismic survey. We compute a zero-offset stacked section, with a single simulation, by using the exploding-reflector concept and the non-reflecting wave equation. The geological model is verified by computing exploding-reflector experiments and comparing them with the real stacked section. In the final phase of the modelling study, we use the variable-density acoustic wave equation to compute common-shot and common-offset synthetic surveys, which are used to obtain the synthetic stacked time section using the standard processing sequence. These simulations and verification of the geological model by post-stacked migration constitute an additional test. The methodology is applied to a seismic line from the Italian deep-crust exploration project CROP, acquired in the Adriatic Sea.

Keywords: seismic modelling, CROP, crustal section, Adriatic

1. INTRODUCTION

Seismic numerical modelling is a valuable tool for validation of seismic interpretations. The method has been used in hydrocarbon exploration (Fagin, 1992), earthquake-seismology (Priolo, 1999) and crustal studies (Yarnold et al., 1993; Morgante et al., 1998). Here, we develop a methodology for investigating the seismic response of the Earth's crust at the large-scale, with the purpose of interpreting the main geological features from the surface to the upper mantle.

The main deep crust exploration project in Italy is CROP (Finetti, 1994; Finetti et al., 2001). It is the equivalent of DEKORP (Germany) (Rabbel and Gajewski, 1999), COCORP (USA), BIRPS (UK) and ECORS (France). The data was acquired on land and offshore in the Adriatic and throughout the central Mediterranean. In particular, the CROP M-15 seismic section was acquired offshore in the Adriatic Sea. The source and acquisition parameters for this deep seismic section are given in Table 1. Land CROP data obtained nearby in the Southern Apennines are exhibited in the literature (Mazzotti et al., 2000). Amplitude information is relatively important, but a precise determination of the interval velocities is difficult because the residual NMO of reflection events beyond 4–5 s becomes critically small. This fact precludes the use of accurate velocity analyses, reflection tomography, and techniques such as prestack depth migration. Therefore, we deal with almost zero offset P-wave data, and the model design is based on the interpretation of the stacked time section, since well data are scarce, and only oil exploration wells down to 4 km depth are available. We do not consider mode conversion (i.e., S waves) and intrinsic attenuation, which, in this situation, constitute second order effects. Anisotropy of the P wave is modelled with an elliptical anisotropic rheology, since it can be important in the upper mantle (Guest and Thomson, 1992).

The modelling method is based on the 2-D acoustic wave equation with variable density. This choice allows us to define the acoustic impedance of each macro-layer. The velocity field and density of each stratum can be defined on the basis of the reflector strength and global geological information for the study area. The data in the Adriatic Sea has a good signal-to-noise ratio, unlike onshore data, where strong tectonism in the Apennines region has created highly discontinuous reflectors. However, the signal is degraded by the heterogeneous nature of the crust

[1]Istituto Nazionale di Oceanografia e di Geofisica Sperimentale (OGS), Borgo Grotta Gigante 42c, 34010 Sgonico, Trieste, Italy
[2]EGG (Exploration Geophysics Group), Department of Geological, Environmental and Marine Sciences, Faculty of Sciences, University of Trieste, via Weiss 1, 34127 Trieste, Italy
*Corresponding author. Fax: +39 040 327521; e-mail: jcarcione@ogs.trieste.it (J.M. Carcione)

Table 1
Seismic vessel and acquisition parameters of CROP M-15 seismic survey

Seismic vessel	OGS-Explora
Source	Air-guns (80.4 l)
Recording length	17 s
Sampling interval	4 ms
Streamer length	4500 m
Group spacing	25 m
Shot interval	50 m
Channels	180
Coverage	4500%

Figure 1: Location of the CROP M-15 seismic line in the Adriatic Sea.

at small scales. Constraints on the velocity variations of these small-scale heterogeneities can be estimated from P-wave sonic logs. We introduce these inhomogeneities by using a spatially isotropic von Kármán autocovariance probability function of high fractal dimension, which simulates scattering-Q effects (Holliger, 1997). A typical correlation length is 100 m and the standard deviation of the velocity fluctuations ranges from 200 to 400 m/s.

As the first step, a ray-tracing algorithm is used to obtain a preliminary geological model by comparing the synthetic (zero-offset) two-way travel times to those of the real stacked section. Simulation of a stacked seismic section requires the calculation of a set of common-shot experiments and application of the standard processing sequence. An approximation of the stacked section is obtained with the non-reflecting wave equation and the exploding-reflector approach. These methods allow us to reduce computing time (Baysal et al., 1984; Carcione et al., 1994). The non-reflecting wave equation implies a constant impedance model to avoid multiple reflections, which are, in principle, absent from stacked sections and constitute unwanted artifacts in migration processes. In the exploding-reflector method, each reflection point in the subsurface explodes at $t = 0$ with a magnitude proportional to the normal-incidence reflection coefficient. The non-reflecting equation is a modification of the wave equation, where the impedance is constant over the whole model space. In this way, non-physical multiple reflections are avoided and the recorded events are primary reflections. The density is used as a free parameter to obtain a constant impedance model and avoid multiple reflections. The reflection strength is then implicit in the source strength. Moreover, the method generates normal-incidence reflections, i.e., those having identical downgoing and upgoing wave paths. In order to obtain the two-way travel time, the phase velocities are halved. Due to sampling constraints, halving the velocities implies doubling the number of grid points. After a set of exploding-reflector experiments and comparison with the real stacked section, we obtain the final geological model. We do not consider anisotropy and scattering in this process. The "exact" synthetic stacked section is then obtained by applying the standard

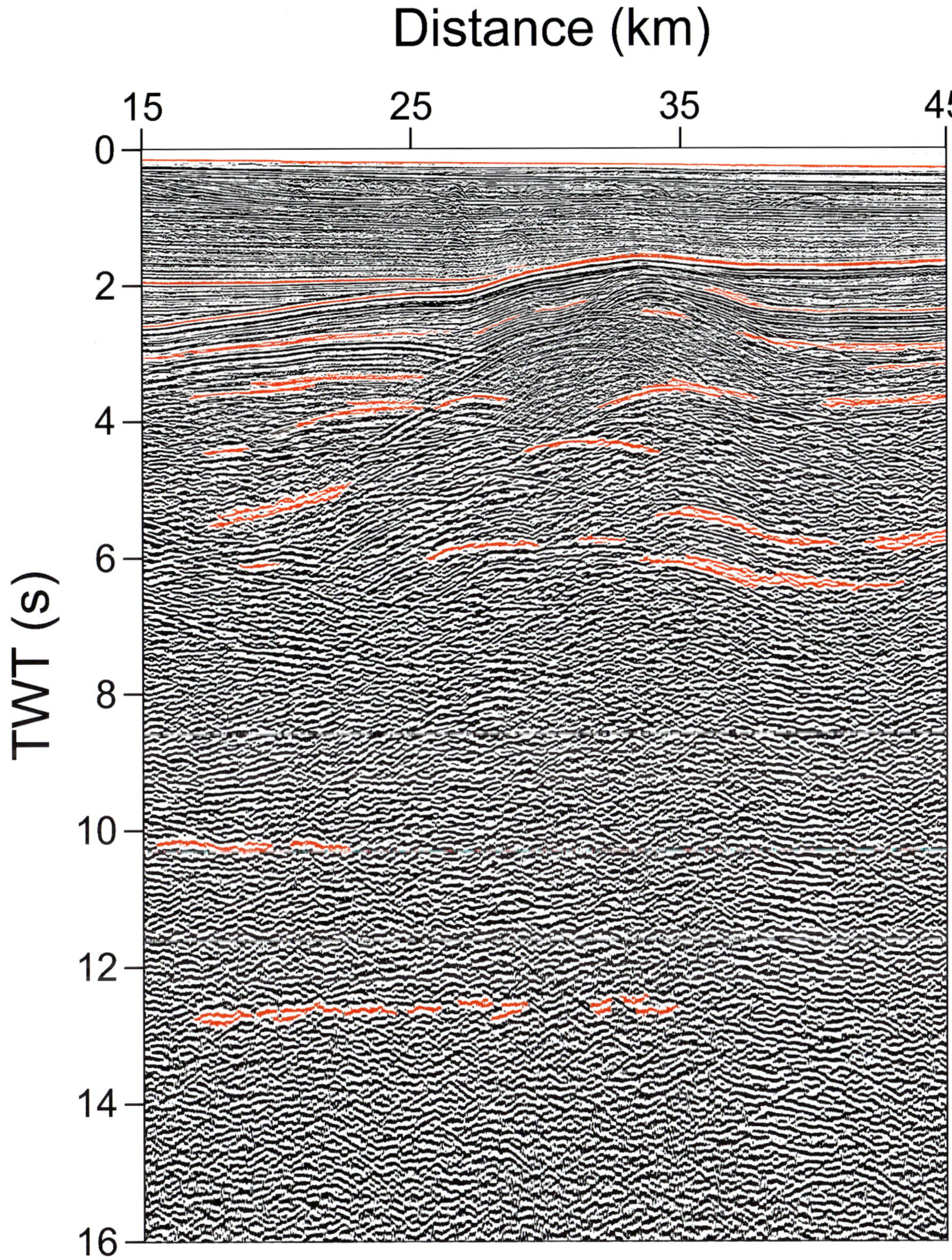

Figure 2: CROP M-15 seismic line (part).

processing sequence to a set of synthetic common-shot profiles computed with the variable-density acoustic wave equation.

The numerical solver consists of the pseudospectral Fourier method for computing the spatial derivatives, and a 2nd-order leap-frog method for time integration (Carcione et al., 1994). An averaging method, developed by Zeng and West (1996), is used to reduce spurious diffractions arising from an inappropriate modelling of curved and dipping interfaces (the so-called staircase effect). It is based on a spatially weighted averaging of the model properties.

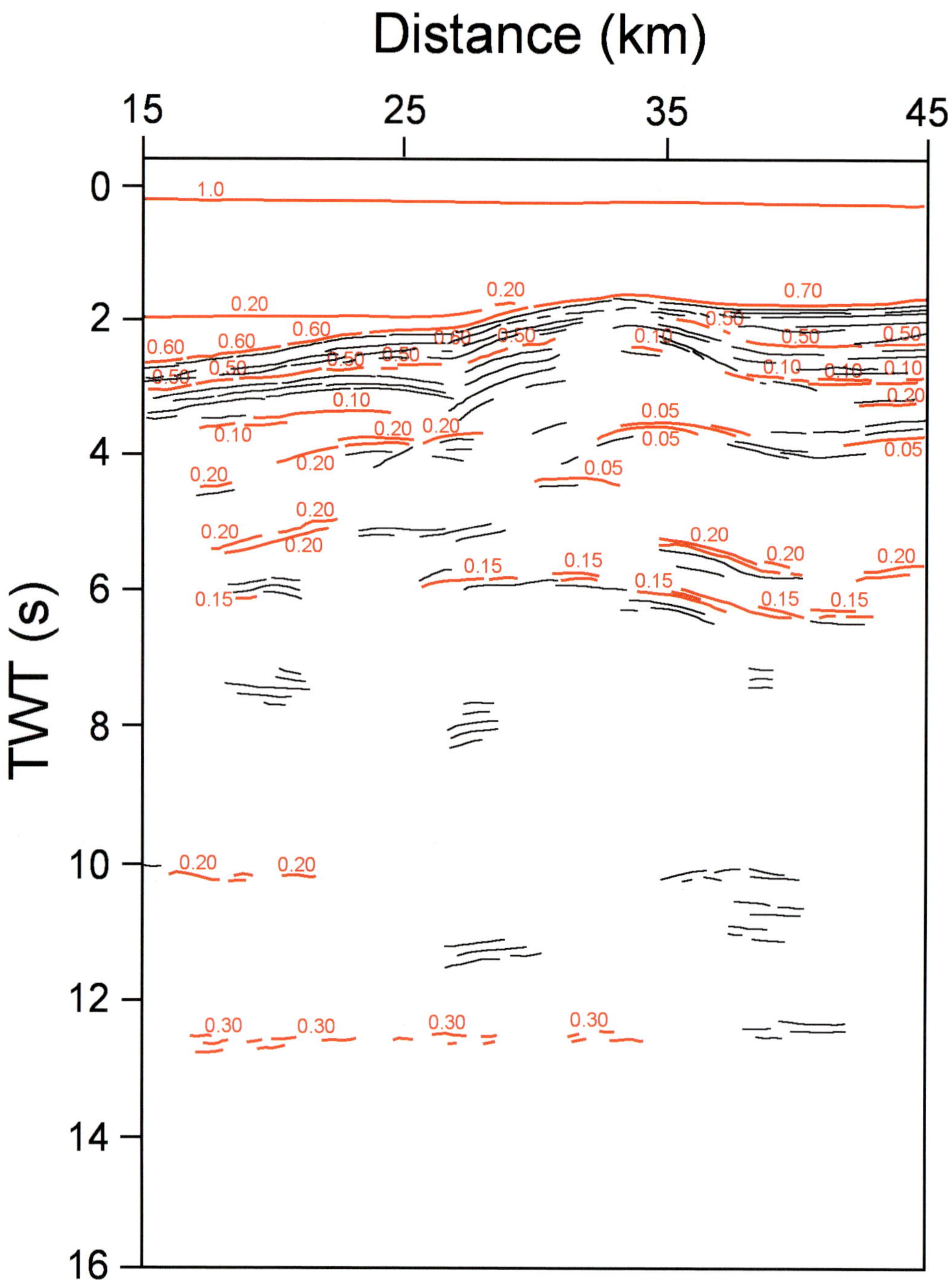

Figure 3: Schematic line drawing of CROP M-15 seismic line. The numbers indicate the relative amplitude of the events highlighted in Figure 2.

2. THE MODELLING TECHNIQUE

The variable-density wave equation for elliptically anisotropic media is

$$\rho c^2 \frac{\partial}{\partial x}\left(\frac{1}{\rho}\frac{\partial p}{\partial x}\right) + \rho c_3^2 \frac{\partial}{\partial z}\left(\frac{1}{\rho}\frac{\partial p}{\partial z}\right) = \frac{\partial^2 p}{\partial t^2} + s, \tag{1}$$

where $p(x, z, t)$ is the pressure field, $\rho(x, z)$ is the material density, $c(x, z)$ is the horizontal P-wave velocity, $c_3(x, z)$ is the vertical P-wave velocity, and $s(x, z, t)$ is the source. Assuming a plane wave kernel

$$\exp\left[i\omega(t - s_x x - s_z z)\right],$$

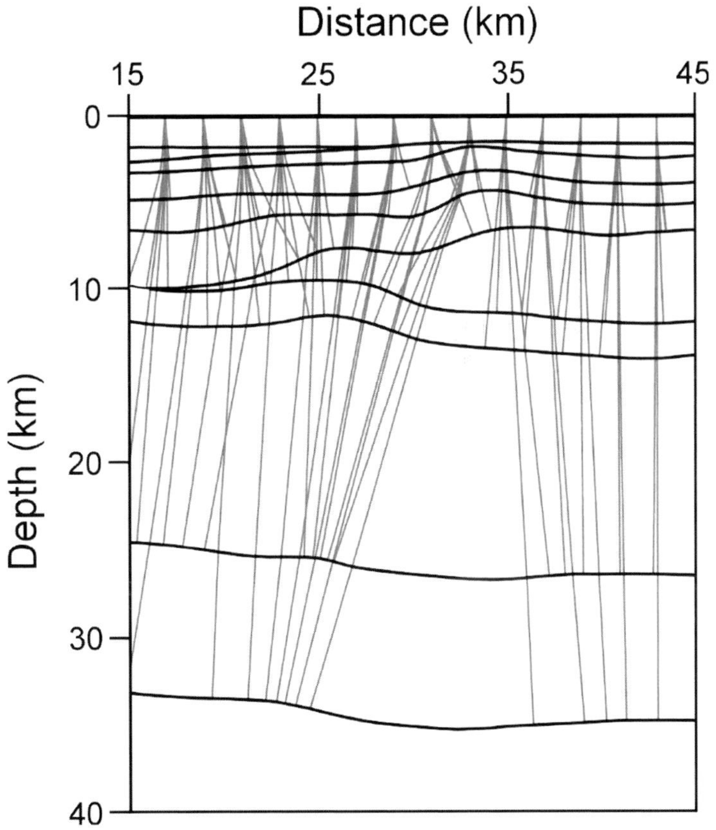

Figure 4: First version of the geological model obtained from zero-offset ray tracing. The zero-offset ray paths are indicated.

where ω is the angular frequency and s_i are the slowness components, we obtain the dispersion relation

$$\frac{s_x^2}{1/c^2} + \frac{s_z^2}{1/c_3^2} = 1. \tag{2}$$

This is an ellipse with semiaxes $1/c$ and $1/c_3$. We assume that the major semiaxis of the ellipse is vertical, that is $c_3 \leqslant c$. Seismic anisotropy is usually reported as a percent,

$$A = 100\left(1 - \frac{c_3}{c}\right) \tag{3}$$

(e.g., Rudnick and Fountain, 1995). The isotropic case is obtained for $c_3 \equiv c$.

We consider the source term

$$s(x, z, t) = \delta(x - x_0)\delta(z - z_0)h(t), \tag{4}$$

where δ is Dirac's delta, (x_0, z_0) is the source location and $h(t)$ is the source time-history (a Ricker-type function in our simulations).

To derive the exploding-reflector isotropic wave equation, with the non-reflecting – constant impedance – condition, we assume that the acoustic impedance

$$I = \rho c \tag{5}$$

is constant throughout the model space. Using this condition, Equation (1) becomes

$$c\frac{\partial}{\partial x}\left(c\frac{\partial p}{\partial x}\right) + c\frac{\partial}{\partial z}\left(c\frac{\partial p}{\partial z}\right) = \frac{\partial^2 p}{\partial t^2} + s \tag{6}$$

(Baysal et al., 1984; Carcione et al., 1994). The normal-incidence reflection coefficient is zero for this equation, and it becomes the constant density wave equation when the velocity is constant. We place a source on each grid

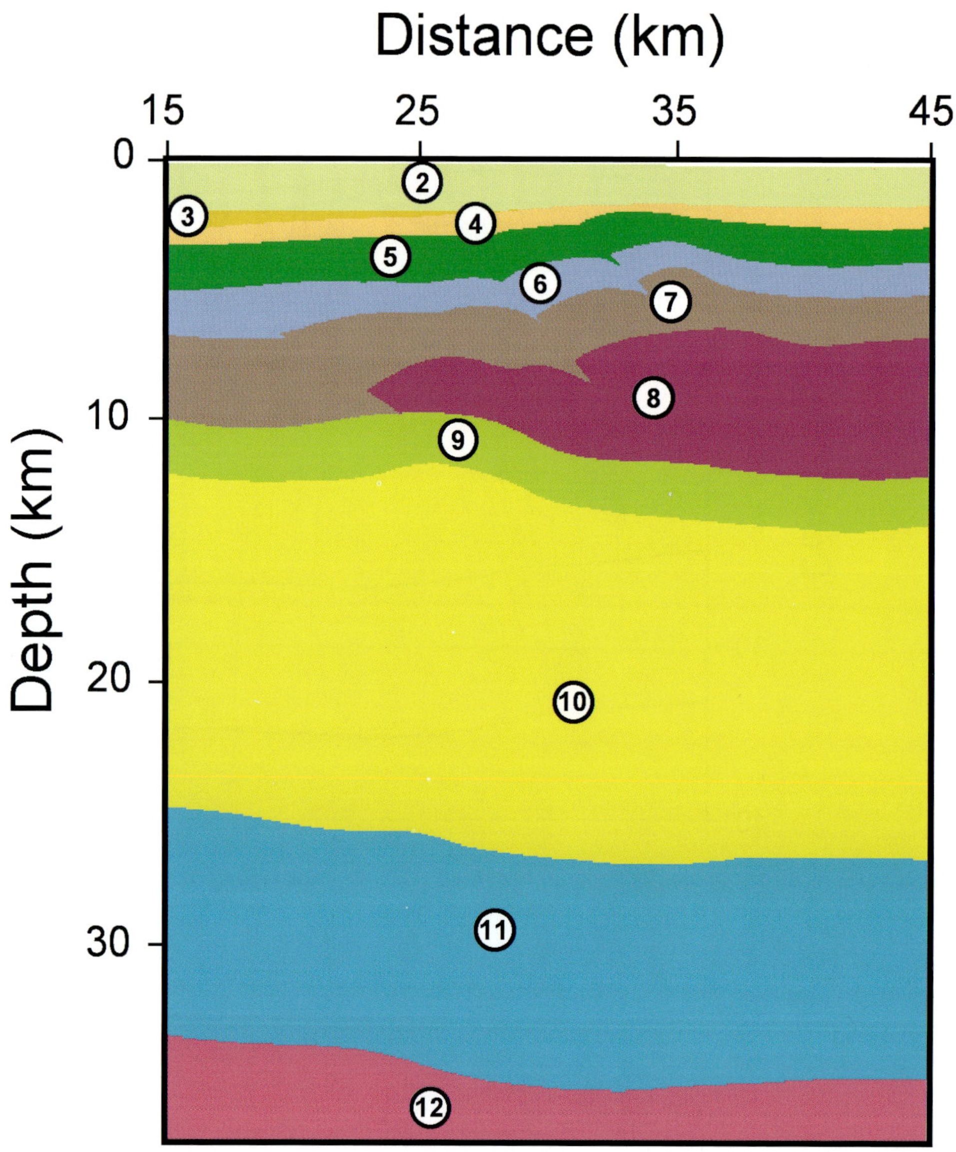

Figure 5: Geological model in terms of seismic velocity and mass density, used for the exploding-reflector simulations. The numbers denote the layers indicated in Table 2.

point (i, j) defining the interfaces,

$$s(x, z, t) = R\delta(x - x_i)\delta(z - z_j)h(t), \tag{7}$$

where R is the normal-incidence reflection coefficient, and (x_i, z_j) is the source location. The normal-incidence reflection coefficient is

$$R = \frac{I_2 - I_1}{I_1 + I_2} \tag{8}$$

(DeSanto, 1992, p. 5), where 1 and 2 denote the upper and lower media.

Because the wave equation is linear, seismograms with different dominant frequencies – and time histories – can be implemented by convolving $h(t)$ with only one simulation obtained with $\delta(t)$ as a source (a discrete delta with strength $1/dt$).

Scattering due to heterogeneities in the crust and mantle is considered.

Table 2
Acoustic properties. P-wave velocities, anisotropy parameter, density, autocorrelation distance, maximum velocity perturbation and fractal number

Layer	Medium	c (km/s)	A (%)	ρ (g/cm^3)	a (m)	Δc_0 (m/s)	ν
1	Adriatic Sea	1.5	0	1.03	–	–	–
2	Quaternary and Upper Pliocene	2	0	2	–	–	–
3	Lower Pliocene	2.3	0	2.1	–	–	–
4	Miocene to Upper Eocene	3.3	0	2.4	–	–	–
5	Lower Eocene and Cretaceous	5	0	2.5	–	–	–
6	Jurassic	5.4	0	2.55	–	–	–
7	Dolomia Principale	6.3	0	2.55	–	–	–
8	Permo-Triassic Evaporite	6	0	2.7	–	–	–
9	Permian and Pre-Evaporite Triassic	5.5	0	2.5	–	–	–
10	Upper crust	6	3	2.67	200	480	0.15
11	Lower crust	7	5	2.69	150	420	0.2
12	Upper mantle	8	6	3.28	300	640	0.18

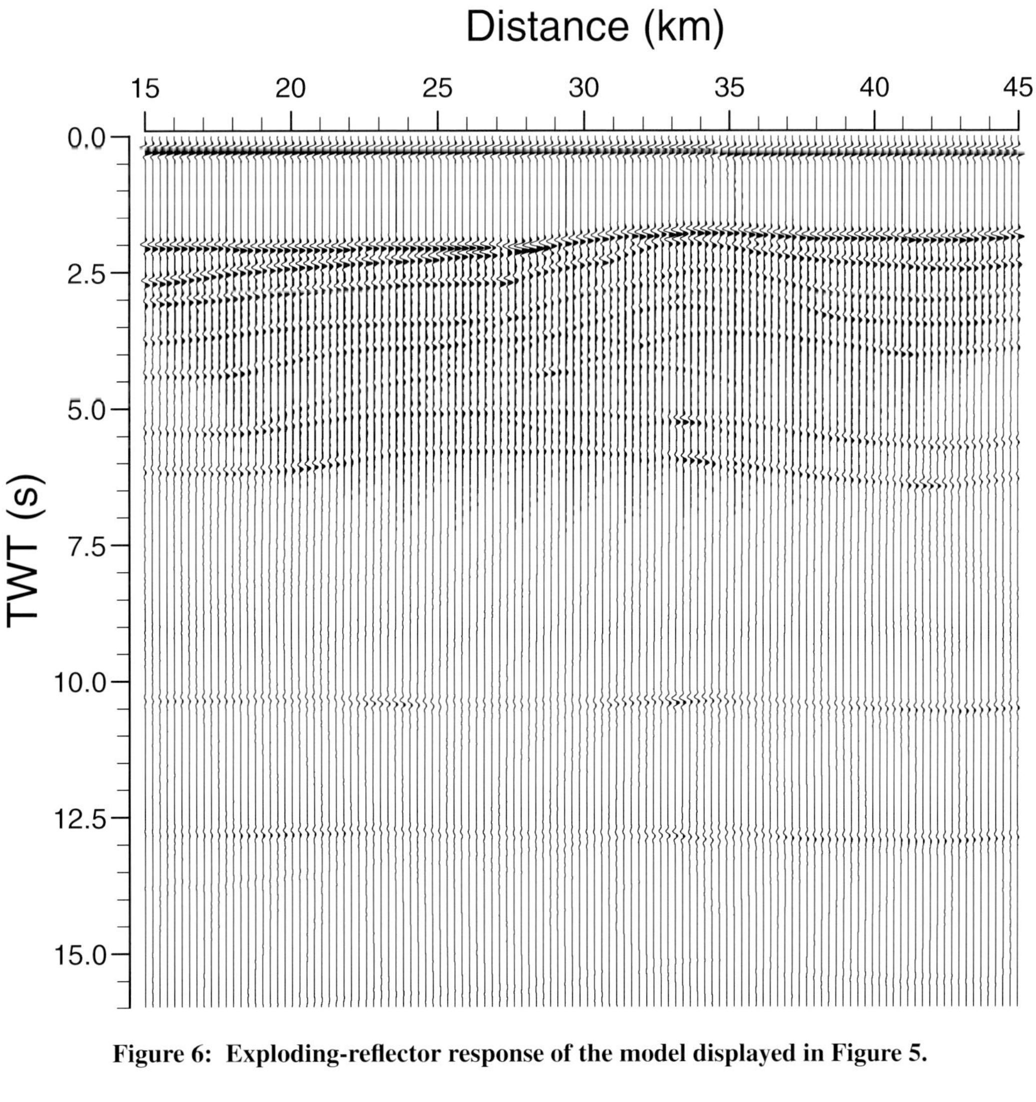

Figure 6: Exploding-reflector response of the model displayed in Figure 5.

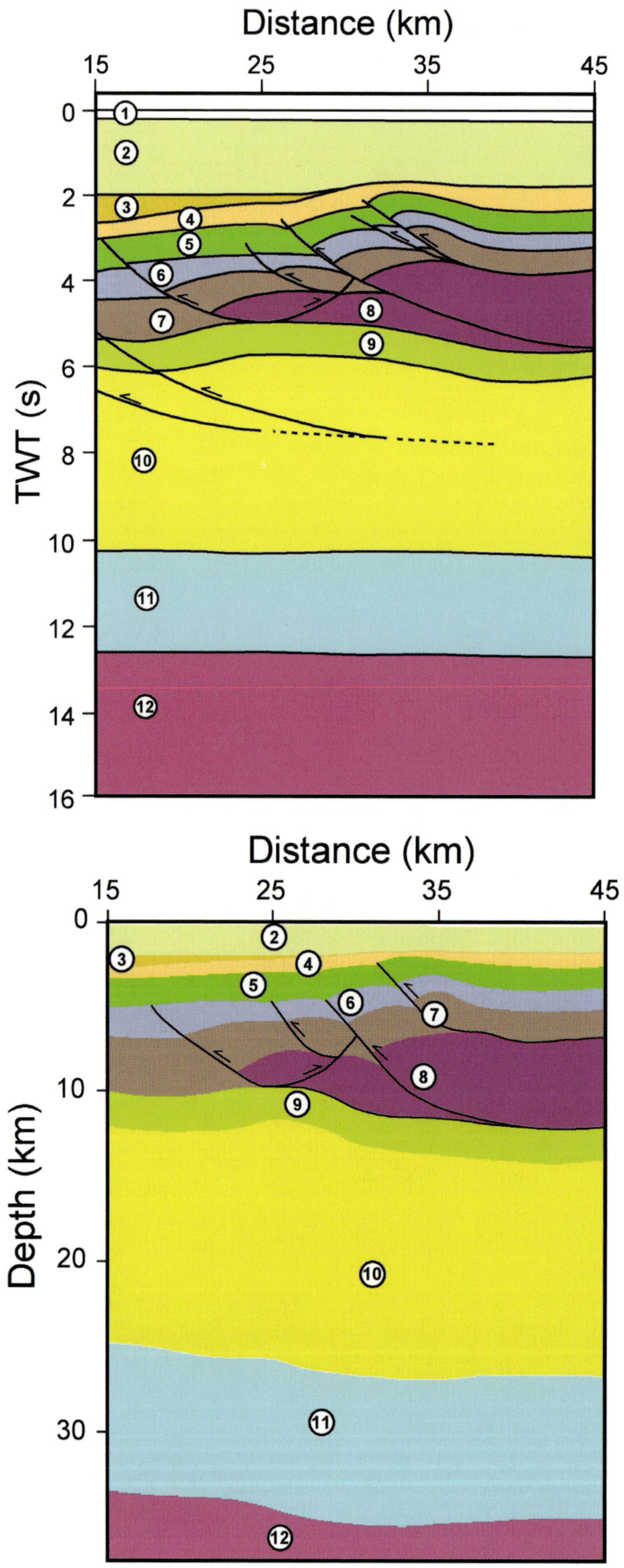

Figure 7: Seismogeological section – in two-way travel time (upper picture) and depth (lower picture) – after a re-interpretation on the basis of the ray-tracing results, the exploding-reflector simulations and a priori geological information.

Let Δc_0 be the maximum deviation of the velocity field from the background value c_0. The velocity field at (x, z) is first subjected to the variations $(\Delta c)^r$, such that

$$-\Delta c_0 \leqslant (\Delta c)^r \leqslant \Delta c_0, \tag{9}$$

where $(\Delta c)^r$ is obtained from a 2-D random generator, and the superindex "r" denotes random. (Random numbers between 0 and 1 are generated and then scaled to the interval $[-1, 1]\Delta c_0$.)

Small-scale P-wave velocity variations in the lithosphere are well described by the von Kármán autocovariance function (Frankel and Clayton, 1986; Holliger, 1997). The corresponding wavenumber-domain power spectrum is

$$C(k_x, k_z) = K(1 + k^2a^2)^{-(\nu+N/2)}, \tag{10}$$

where $k = \sqrt{k_x^2 + k_z^2}$ is the wavenumber, a is the correlation length, ν $(0 < \nu < 1)$ is a self-similarity coefficient, K is a normalization constant, and N is the Euclidean dimension. The von Kármán correlation function describes self-affine, fractal processes of fractal dimension $N + 1 - \nu$ at a scale smaller than a.

The velocity is then calculated as

$$c(x, z) = c_0 \pm \Delta c(x, z), \tag{11}$$

where

$$\widetilde{\Delta c}(k_x, k_z) = \widetilde{(\Delta c)}^r(k_x, k_z)C(k_x, k_z), \tag{12}$$

with $\widetilde{(\Delta c)}^r(k_x, k_z)$ being the Fourier transform of $(\Delta c)^r(x, z)$. (The tilde denotes the space Fourier transform.)

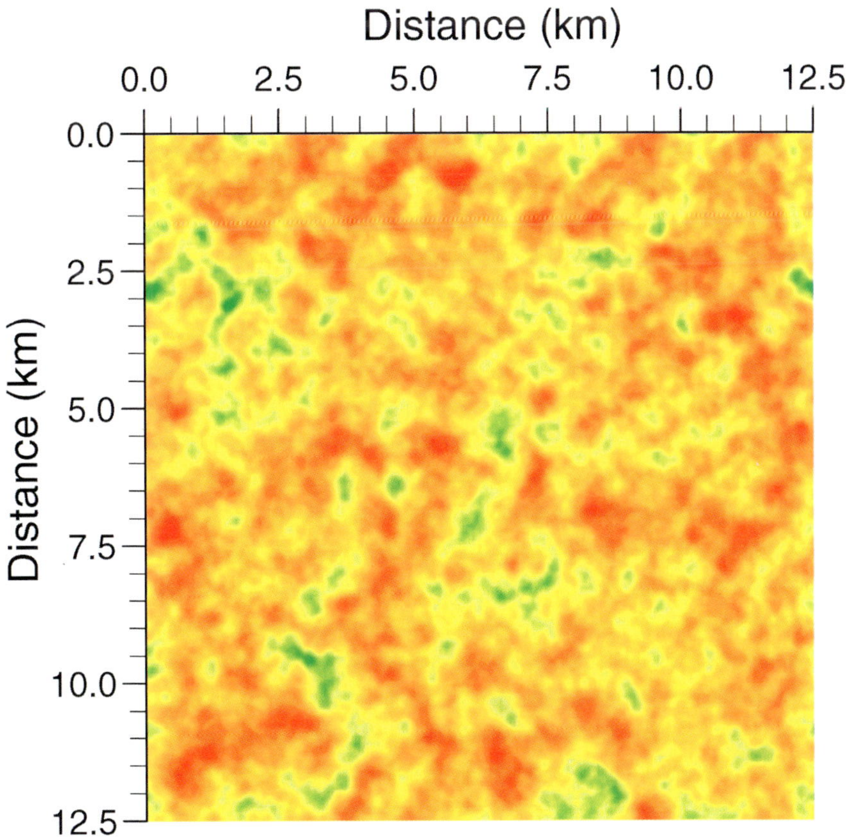

Figure 8: Velocity field perturbations for part of the upper mantle obtained from the von Kármán autocovariance function (see Table 2).

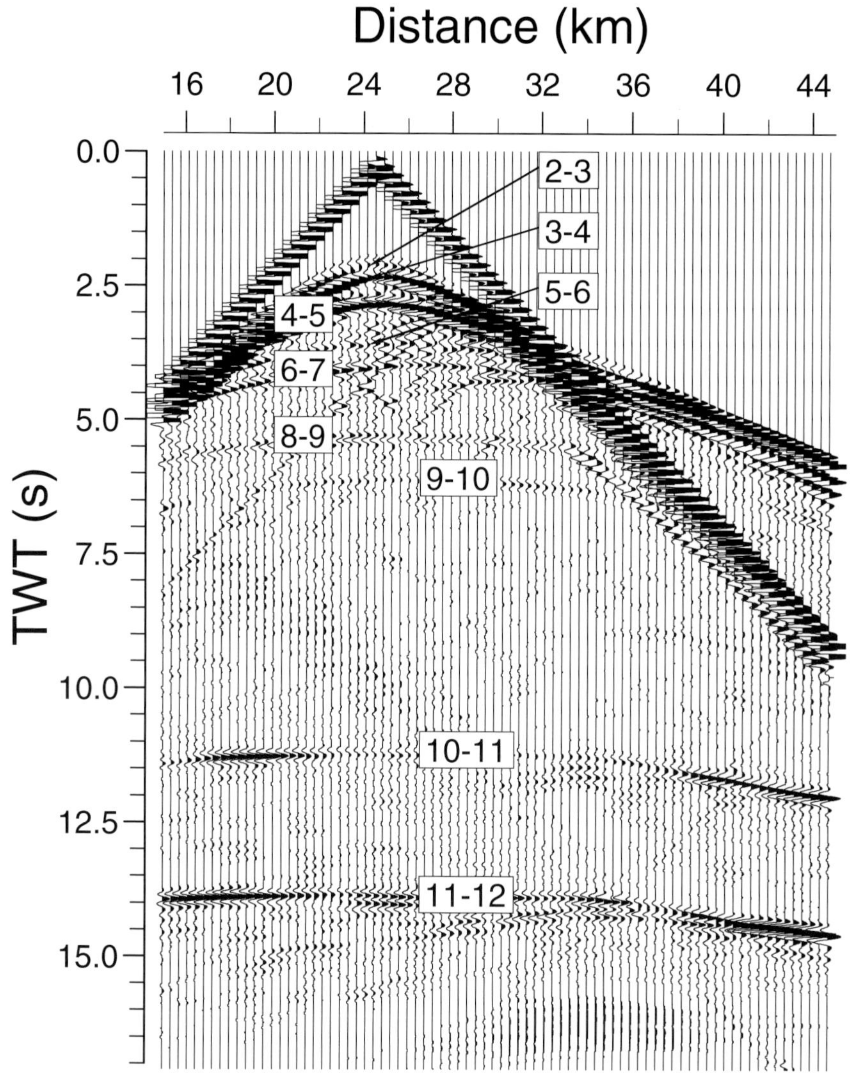

Figure 9: Common-shot gather corresponding to the model shown in Figure 7. The labels denote the reflection events generated at the interfaces indicated in Figure 7.

The spatial derivatives are computed by using the Fourier method. The spectral coefficients are calculated with the fast Fourier transform (FFT) (Fornberg, 1996; Carcione, 2001, p. 303). The time integration of Equations (1) and (6) is performed with a standard second-order differencing scheme (e.g., Celia and Gray, 1992).

3. THE DEEP SEISMIC EXPLORATION SECTION CROP M-15

The location of the CROP M-15 seismic section in the Adriatic Sea is shown in Figure 1. A time section is displayed in Figure 2 (the processing technique is standard). The stronger seismic events are plotted in Figure 3 (line drawing). In general, the numbers indicate the relative amplitudes of the events. In this area, we observe a thick undeformed continental crust, where the sediment column, and upper and lower crust macrointervals can be reliably identified. The section shows the gentle compressive deformation of the frontal Dinaric belt, where the platform slightly thrusts along a high-angle fault over the basinal succession in Oligocene-Miocene times (Finetti and Del Ben, this volume, Chapter 22).

The first phase of the modelling approach is to use a zero-offset ray-tracing algorithm (iteratively) to obtain the location of the geological interfaces. A first estimation of the seismic velocities is obtained from a priori geological and geophysical information for the study area. P-wave seismic-velocity values are based on the data published by Christensen (1989), Rudnick and Fountain (1995) and Brittan and Warner (1996). A first version of the geological model is shown in Figure 4, where the main zero-offset ray trajectories are indicated. Due to the limitations of the ray-tracing algorithm, structures such as faults planes cannot be simulated.

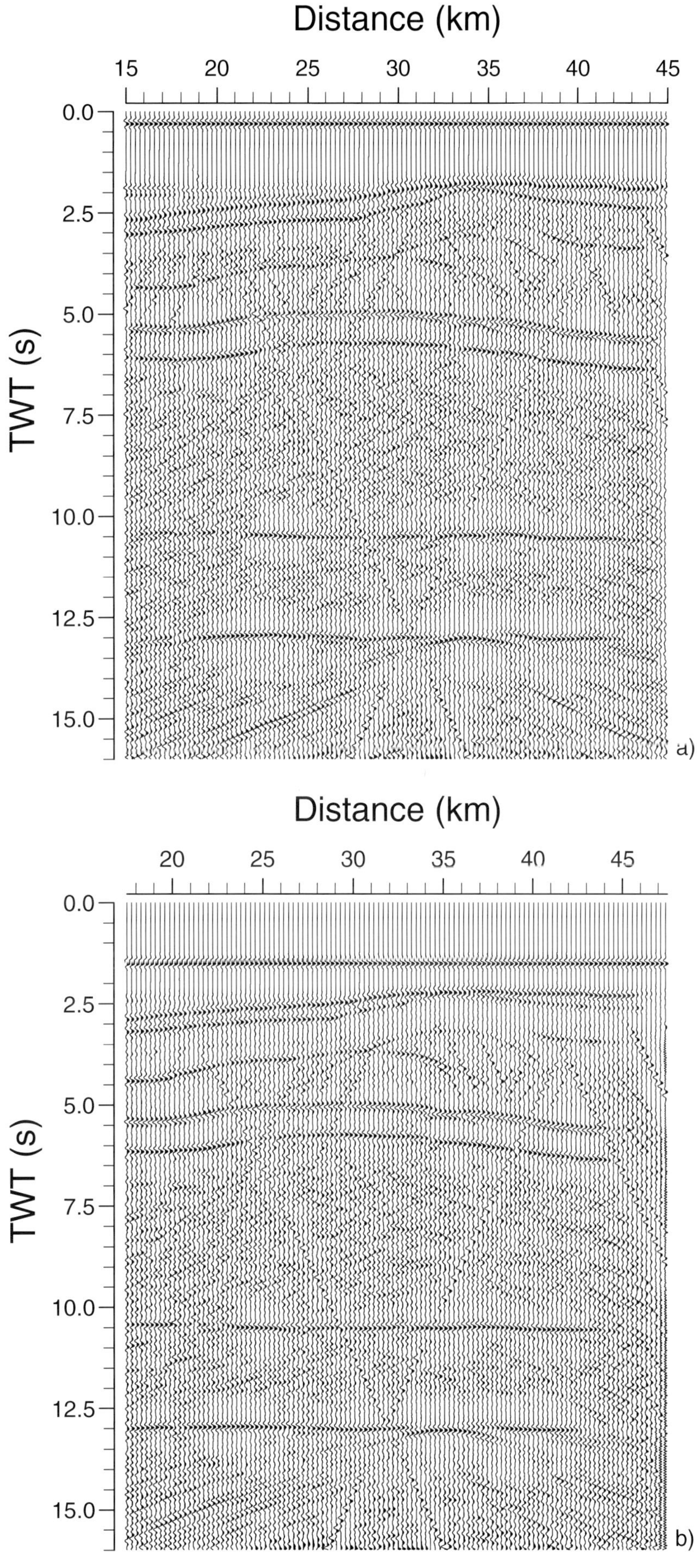

Figure 10: Near-offset section (a) and far-offset section (b) corresponding to the model shown in Figure 7.

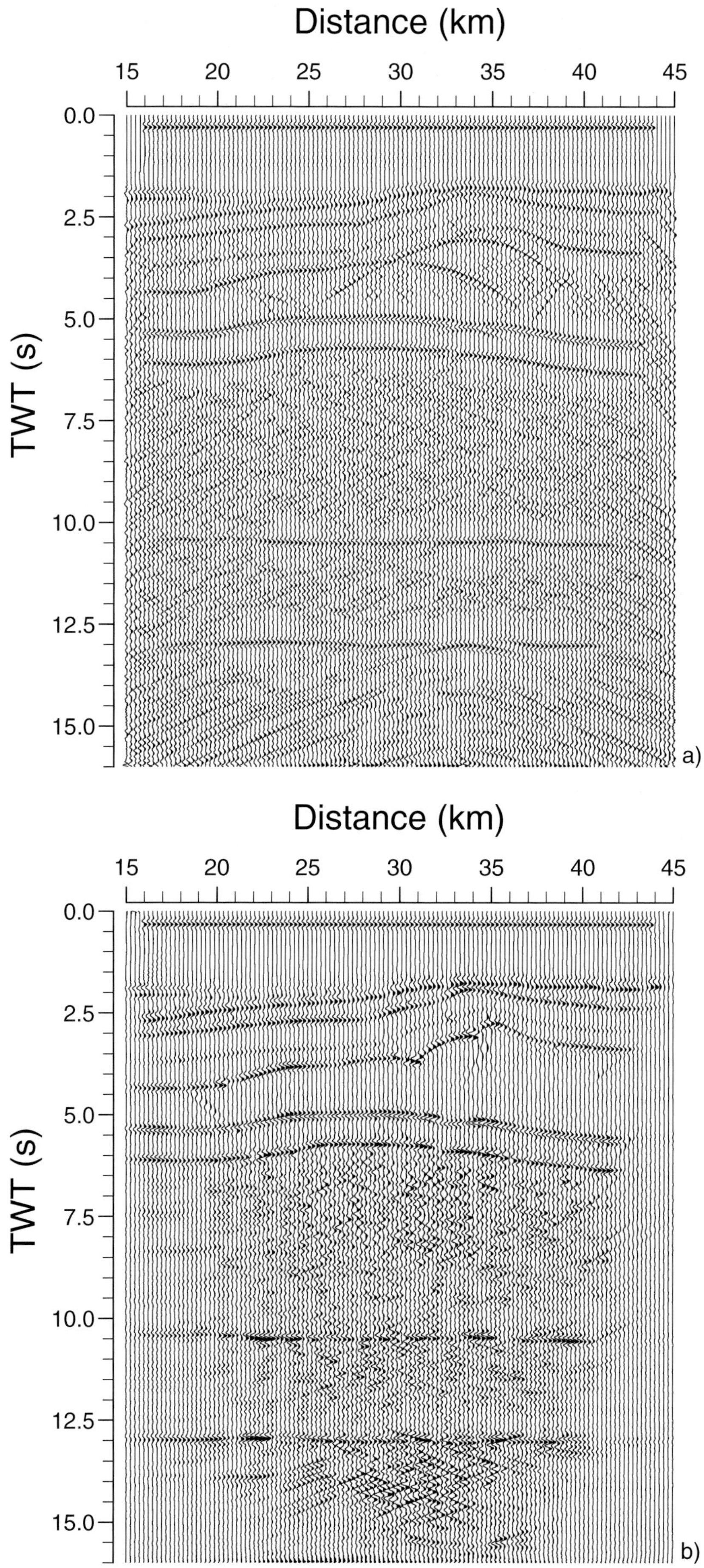

Figure 11: Stacked section (a) and time-migrated section (b) corresponding to the model shown in Figure 7.

The model is refined by using exploding-reflector simulations. In the exploding-reflector algorithm the source strength is proportional to the reflection coefficient, which is proportional to the acoustic-impedance contrast. Thus, the density of each layer is obtained from data provided by Rudnick and Fountain (1995), and using the relative amplitudes indicated in Figure 3. The mesh has 1080×1200 points, with a grid spacing of 31 m. In order to avoid wraparound, absorbing strips of length 60 grid points are implemented at the boundaries of the numerical mesh. The dominant frequency of the source is 6 Hz, and the wavefield is computed by using a time step of 1 ms. The final geological model (see Figure 5 and Table 2) is obtained after computing many exploding-reflector sections and comparing the simulations with the real stacked section (anisotropy and scattering effects are not taken into account in this phase). Figure 6 shows the exploding-reflector seismic section. The improvement over the model shown in Figure 4 consists in the inclusion of fault planes based on the presence of diffraction events. The interpreter plays an important role in this phase, integrating a priori knowledge of the regional tectonic features. Figure 7 shows the geological section in two-way travel time (a) and depth (b), after a re-interpretation on the basis of the preceding information and a priori regional geological information.

In the following, we consider anisotropy and scattering effects according to the values given in Table 2. Since the wavefront is elliptical, the moveout velocity is the velocity for a wave traveling in the horizontal direction. This velocity, c, is obtained from surface measurements (Levin, 1978). Figure 8 shows the perturbation of the velocity field ($\Delta c(x, z)$) of a representative part of the upper mantle. In order to simulate ten-fold CMP acquisition, 120 common shots with 80 split-symmetric channels are computed. The shot interval is 248 m, and the group spacing is 62 m. The mesh has 600×720 points, a grid spacing of 62 m, and 60 grid points for each absorbing strip. The positions of the first and last shots are $x = 15$, and $x = 44.76$ km, respectively. Figure 9 shows a common-shot gather, where the indices denote the reflection events generated at the different interfaces indicated in Figures 5 and 7. Figure 10 shows two common-offset sections, corresponding to the near offsets (a) and far offsets (b). They resemble the stacked section, in particular the near-offset section. The stacked section and its post-stack time migration are displayed in Figures 11(a) and 11(b), respectively. The strength of the events in the unmigrated section agree fairly well with the values indicated in Figure 2, and the migrated section is in good agreement with the model shown in Figure 7.

4. CONCLUSIONS

The procedure used to simulate the CROP M-15 seismic line involves: (i) line drawing, to identify the location and strength of the main events, (ii) ray tracing, to generate a first version of the geological model in terms of seismic velocities; (iii) exploding-reflector experiments, to generate a geological model in terms of seismic velocity and mass density; (iv) refining of the model by the interpreter on the basis of a priori geological and geophysical information (inclusions of fault planes, etc.).

The model is further improved by considering random heterogeneities, which characterize the seismic response of the crust and mantle at different scales. The result is a complete characterization of the geological setting, with the possibility of calculating realistic common-shot seismograms to further investigate and improve the interpretation of the different geological structures.

It is important to point out that the modelling phase is a validating tool of the interpretation process. A substantial role is played by the ability and knowledge of the interpreter, who integrates a priori knowledge of the main tectonic characteristics of the area under study.

ACKNOWLEDGEMENTS

This work was funded in part by MURST (Ministry of University, Scientific Research and Technology) under the framework of COFIN-2000 granted to I.R. Finetti.

REFERENCES

Baysal, E., Kosloff, D.D., Sherwood, J.W.C., 1984. A two-way nonreflecting wave equation. Geophysics 49, 132–141.

Brittan, J., Warner, M., 1996. Seismic velocity, heterogeneity, and the composition of the lower crust. Tectonophysics 264, 249–259.

Carcione, J.M., 2001. Wave Fields in Real Media. Wave Propagation in Anisotropic, Anelastic and Porous Media. In: Handbook of Geophysical Exploration, vol. 31. Pergamon Press.

Carcione, J.M., Böhm, G., Marchetti, A., 1994. Simulation of a CMP seismic section. J. Seis. Expl. 3, 381–396.

Celia, M.A., Gray, W.G., 1992. Numerical Methods for Differential Equations. Fundamental Concepts for Scientific and Engineering Applications. Prentice-Hall.

Christensen, N.I., 1989. Reflectivity and seismic properties of the deep continental crust. J. Geophys. Res. 94 (B12), 17793–17804.

DeSanto, J.A., 1992. Scalar Wave Theory. Springer Series on Wave Phenomena. Springer-Verlag, Berlin.

Fagin, S.W., 1992. Seismic Modeling of Geological Structures: Applications to Exploration Problems. Geophysical Development Series, vol. 2. Society of Exploration Geophysicists.

Finetti, I. (Ed.), 1994. CROP project, offshore crustal seismic profiling in the central Mediterranean. Boll. Geof. Teor. Appl. 36 (1994) 1–536.

Finetti, I.R., Boccaletti, M., Bonini, M., Del Ben, A., Geletti, R., Pipan, M., Sani, F., 2001. Crustal section based on CROP seismic data across the North Tyrrhenian–Northern Apennines–Adriatic Sea. Tectonophysics 343, 135–163.

Finetti, I.R., Del Ben, A., this volume. Crustal tectono-stratigraphic setting of the Adriatic sea from new CROP seismic data. Chapter 22.

Fornberg, B., 1996. A Practical Guide to Pseudospectral Methods. Cambridge University Press.

Frankel, A., Clayton, R.W., 1986. Finite difference simulations of seismic scattering: implications for the propagation of short-period seismic waves in the crust and models of crustal heterogeneity. J. Geophys. Res. 91 (B6), 6465–6489.

Guest, W.S., Thomson, C.J., 1992. A source of significant transverse isotropy arrivals from an isotropic–anisotropic interface, e.g., the Moho. Geophys. J. Internat. 111, 309–318.

Holliger, K., 1997. Seismic scattering in the upper crystalline crust based on evidence from sonic logs. Geophys. J. Internat. 128, 65–72.

Levin, F., 1978. The reflection, refraction, and diffraction of waves in media with an elliptical velocity dependence. Geophysics 43, 528–537.

Mazzotti, A.P., Stucchi, E., Fradelizio, G.L., Zanzi, L., Scandone, P., 2000. Seismic exploration in complex terrains: a processing experience in the Southern Apennines. Geophysics 65, 1402–1417.

Morgante, A., Barchi, M.R., D'Offizi, S., Minelli, G., Pialli, G., 1998. The contribution of seismic modeling to the interpretation of the CROP-03 line. Mem. Soc. Geol. It. 52, 91–100.

Priolo, E., 1999. 2-D spectral element simulation of destructive ground shaking in Catania (Italy). J. Seismology 3, 289–309.

Rabbel, W., Gajewski, D. (Eds.), 1999. Seismic exploration of the deep continental crust. Pure Appl. Geophys. 156 (1999) 1–370.

Rudnick, R.L., Fountain, D.M., 1995. Nature and composition of the continental crust: a lower crustal perspective. Rev. Geophys. 33 (3), 267–309.

Yarnold, J.C., Johnson, R.A., Sorensen, L.S., 1993. Identification of multiple generations of crosscutting "domino"-style faults: insights from seismic modeling. Tectonics 12, 159–168.

Zeng, X., West, G.F., 1996. Reducing spurious diffractions in elastic wavefield calculations. Geophysics 61, 1436–1439.

CHAPTER 25

Mechanism and Timing of the Pliocene-Pleistocene Foredeep Migration in the Abruzzo Offshore Area (Central Adriatic)

S. Carruba[1], R. Casnedi[1,*], C.R. Perotti[1]

ABSTRACT

An elongated foredeep basin (Periadriatic foredeep) is present at the transition zone between the Apennine chain and the Adriatic foreland in Italy.
The flexure of the Adriatic lithosphere underneath the Apennines has caused the accumulation of a thick clastic succession, composed mostly of turbiditic sediments, from the Oligocene to the present. Both the orogenic front and the basin depocentres migrated towards the foreland during the Apennine orogenesis, leading the Apennine thrusts to incorporate progressively younger foredeep deposits.
In the Abruzzo offshore area (central Adriatic Sea), seismic profiles have imaged the outermost part of the basin, here filled with Plio-Pleistocene sediments, revealing its present architecture. The most noteworthy features are the progressive pinch-out of the Pliocene foredeep deposits on to the westward-dipping foreland ramp, the Upper Pliocene compressional deformations mainly affecting the foredeep substratum, and the occurrence of a deltaic progradational system replacing turbidites from the Pleistocene.
Interpretation and restoration of the seismic sections have allowed a reconstruction of this sector of the foredeep. During the Lower and Middle Pliocene, turbidites accumulated in a wedge-shaped basin, onlapping onto a westward-dipping undeformed foreland ramp. The onlap migrated gradually eastwards. In the Upper Pliocene, folding and faulting involved the foreland ramp and, locally, the foredeep deposits, and syntectonic sedimentation was superimposed on the overall foredeep migration towards the foreland. At the beginning of the Pleistocene, the westward tilting of the foreland plate ceased, and the basin was filled by shallow-water and deltaic sediments prograding eastward.
The migration rate of the foredeep, calculated from the base of the pinch-out shifting, was about 11–17 mm/yr from 4.15 to 1.50 Ma, with a variation from 4 to 34 mm/yr depending on the considered time span and on the area.
The geometry of the foredeep fill and its migration velocity fit into a geodynamic model involving the eastwards roll-back of the subducting Adria plate under the Apennine chain as the major mechanism of the development of the Periadriatic foredeep.

Keywords: Plio-Pleistocene, Abruzzo, foredeep migration

1. INTRODUCTION

The Periadriatic foredeep developed along the boundary between the uplifting Apennines and the Adriatic foreland during the Apennine orogenesis (Ricci Lucchi, 1986).

It is already known, on the basis of onshore subsurface data (Casnedi et al., 1982), that the foredeep migrated eastward with a southern trend from the Northern Apennines to the Taranto Gulf. This migration was due to the collision of the Adria plate with the European plate (Finetti et al., 1987), and to the flexure of the Adria plate underneath the thrust belt (Royden and Karner, 1984; Moretti and Royden, 1988).

The coeval migration of the orogenic thrust front caused the progressive deformation of the foredeep deposits; the older ones are now incorporated in the chain, while the younger are buried in the external part of the foredeep and are affected by weaker thrusting and folding.

[1]Dipartimento di Scienze della Terra, Università di Pavia, Via Ferrata, 1, 27100 Pavia, Italy
*Corresponding author. Fax: +39 382 985890; e-mail: rcasnedi@unipv.it (R. Casnedi)

The chain–foredeep–foreland system in the Abruzzo Apennines started during Messinian times and continued eastward during the Pliocene. Seismic profiles acquired in the Abruzzo offshore area for hydrocarbon exploration give evidence of the last phases of the central Adriatic foredeep migration. These data have been studied with the aim of reconstructing the tectono-sedimentary evolution and migration of this foredeep from the Lower Pliocene to the present.

Identification of the easternmost margin of the basin at different stages allows the calculation of the migration velocity of the foredeep, which can be compared with the velocity of the Adria plate subduction and the Apennine deformation estimated using different methods.

2. GEOLOGICAL SETTING

The Periadriatic foredeep is a deep (up to 8 km) sedimentary basin lying between the Apennine thrust belt and the Adriatic foreland, from the Po Plain in the north to the Taranto Gulf in the south. The development of this foredeep is related to the collision of the European plate against the Adria plate (African continental margin), begun in the early Tertiary, and connected with the flexure and subduction of the latter under the Apennine thrust belt (Kligfield, 1979; Boccaletti et al., 1980; Casnedi et al., 1982; Paltrinieri et al., 1982; Royden and Karner, 1984; Finetti et al., 1987; Patacca and Scandone, 1989; Doglioni, 1991; Finetti et al., 2001).

The stratigraphic record of the foredeep deposits is not represented by a single event but, due to the basin migration, by individual sedimentary sequences, deposited at different times in different sectors of the foredeep (Ricci Lucchi, 1986).

The age of the foredeep sediments indicates that the basin migrated towards the east, but with a southward deepening. So the older deposits of the foredeep are found in the Northern Apennines (Macigno, Oligocene) and they gradually became younger towards the SE, up to the present-day clastic sedimentation in the Taranto Gulf. This basin migration is consistent with an oblique convergence between the two colliding continental margins, with an angle open to the south (Casnedi, 1991).

The deposits of the foredeep were progressively affected by compressional events associated with the migration of the Apennine overthrusting towards the foreland: the inner and older sediments were uplifted and at present crop out in the Apennine chain, while the more external and younger ones are buried and located in a more external position, between the more recent Apennine thrust front and the Adriatic foreland.

The basin fill sediments were generally carried in by turbidity currents of northern origin.

The investigated area comprises the external part of the Abruzzo foredeep, in which the deposition of a thick foredeep sequence took place in the Plio-Pleistocene at the front of the emerging chain (Crescenti et al., 1980). In this area, Pliocene siliciclastic deposits, composed mainly of turbiditic sediments, were deposited on to the sedimentary cover of the Adria plate, represented by the Umbria–Marche carbonatic succession, which ranges from Upper Triassic dolostones and evaporites to the Messinian evaporites (Gessoso-Solfifera Formation) (Finetti et al., 1987). Shallow-water and deltaic sediments replaced turbidites during the Pleistocene (Ori et al., 1986). As occurred throughout the Adriatic foredeep, here the synorogenic deposits were in turn affected by the compressional events, mainly Pliocene in age, related to the migration of the Apennine thrust belt front.

Crustal shortening caused the development of structural highs, already explored for hydrocarbon exploitation (Figure 1). In the present Abruzzo onshore area, an Internal Structure and a Coastal Structure are recognized: they are buried thrust systems, trending in a roughly N–S direction, and of Lower Pliocene and Middle Pliocene age, respectively. Between these, the Morro d'Oro Structure, formed in the Lower Pliocene, is present (Bolis et al., 2003).

In the study area, starting from the Coastal Structure to the offshore area, the Emilio, the Giovanna–Clorinda and the Edmondo–Patrizia structures follow outward. Each corresponds to an individual basin, active at different times, from the Late Messinian to the Pleistocene, up to the Central Adriatic divide, where a regional structural high (Mid-Adriatic ridge of Finetti et al., 1987) seems to represent the peripheral bulge of the doubly subducted Adriatic lithosphere (Moretti and Royden, 1988; De Alteriis, 1995).

Generally speaking, from the Internal Structure seawards, the deformation and consequent overall configuration of the basins progressively become younger.

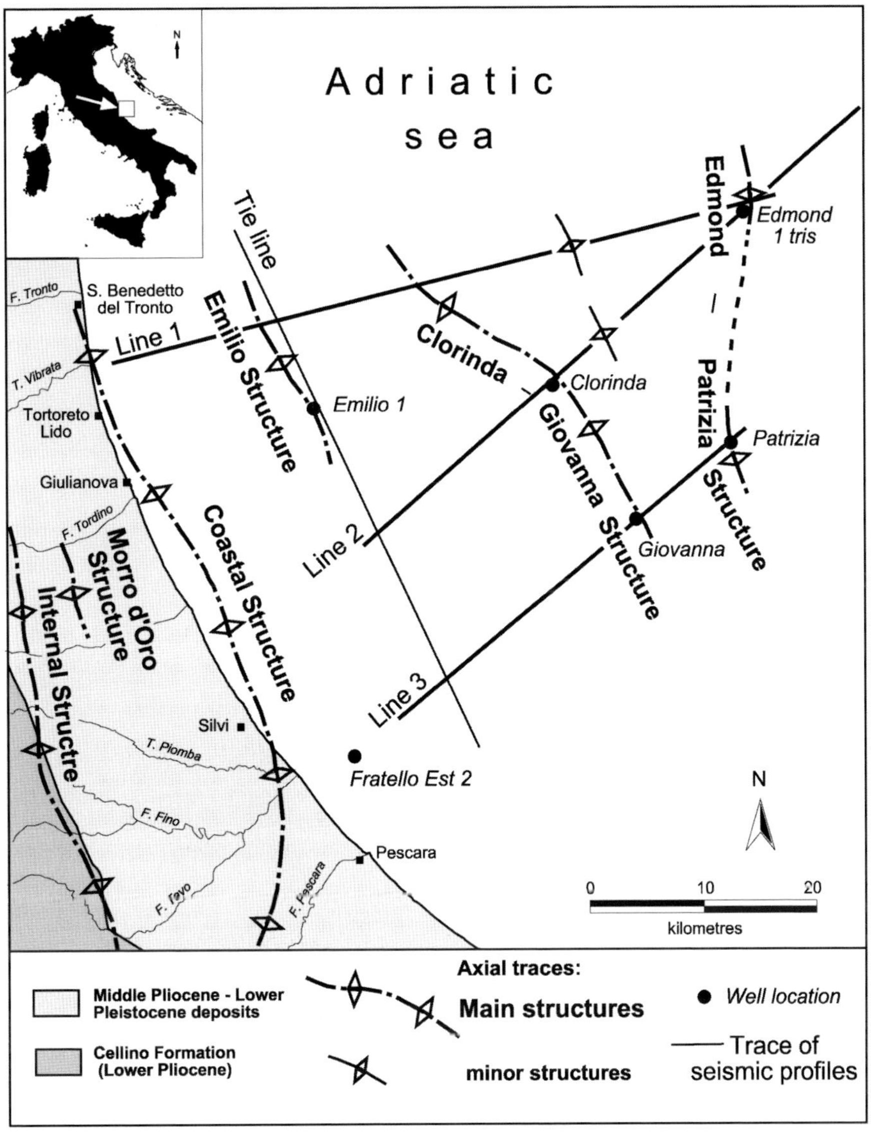

Figure 1: Sketch map of the study area, with the location of the three seismic profiles illustrated in the chapter, the tie line used for horizon correlation, the available wells and the axes of major tectonic structures.

3. SEDIMENTARY AND STRUCTURAL OUTLINE OF THE PLIO-PLEISTOCENE SUCCESSION

Seismic profiles crossing the foredeep in the Abruzzo offshore area up to the outer rise in the central Adriatic Sea are here interpreted (Figures 3, 4 and 5). Their orientation (SW–NE) is approximately perpendicular both to the foredeep axis and to the tectonic structures.

In the eastern parts of sections 2 and 3 (about 60 km offshore), the Edmond 1-tris well (Figure 2) penetrates the foredeep deposits where they show minimum thickness on a structural high, reaching the foreland Meso-Cenozoic carbonates. This well allows the recognition, using stratigraphy and interval velocities, of the Messinian (top evaporites), Lower Pliocene, Middle Pliocene and basal Pleistocene deposits (*Hyalinea baltica* F.O.).

The clastic sediments thin outward, with basal onlap, from the Pliocene to the Pleistocene, over the Messinian basement, which forms the foreland ramp. This dips westwards under the Apennine chain, being considered the top of the subducting Adriatic platform (Paltrinieri et al., 1982; Bally et al., 1986; Royden et al., 1987; Casnedi, 1991; Doglioni, 1991; Ori et al., 1993).

The wedge of Pliocene sediments onlapping the foreland ramp corresponds to the seismic depositional unit 1 of Ori et al. (1986), and is composed of turbidites of northern provenance.

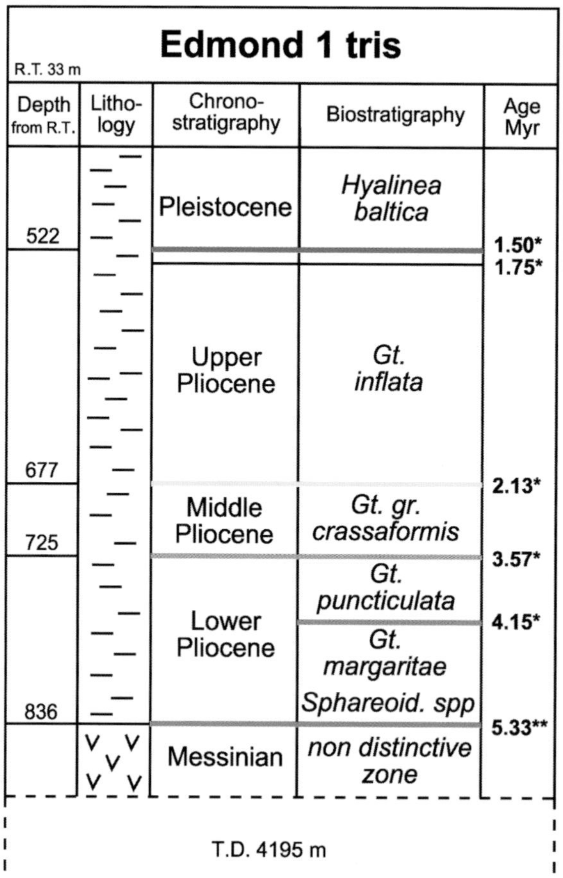

Figure 2: Stratigraphy of the upper part of the Edmond 1-tris well. Ages from Sprovieri (1993) (*) and Van Couvering et al. (2000) (**). Due to its external position, the well penetrates a thin pelitic Plio-Pleistocene sequence.

The occurrence of structural highs allows us to distinguish three main basins and sites of turbidite sedimentation:

(1) The Coastal Basin (between the Coastal Structure to the W and the Emilio Structure, where present, to the E).

The Coastal Basin started in the latest Middle Pliocene, with flat turbiditic deposition onlapping eastward the Emilio Structure, active at this time in the northern part of the study area (line 1). In the Late Pliocene the structure continued its growth, and the turbiditic basinal beds appear tilted. By the end of the Pliocene the structure had already grown, and flat onlapping of turbiditic beds occurred. On lines 2 and 3, the Emilio Structure does not appear: therefore in the southernmost sector, the basin was open to the Giovanna Structure. The layers appear flat even in the western area, near the Fratello Est 2 well.

(2) The Giovanna Basin (between the Emilio Structure, where present, and the Giovanna–Clorinda Structure).

The Giovanna Structure began its growth in the Late Pliocene. Turbiditic beds of this age thin eastwards and appear tilted in eastern onlap against the structure. The basinal sedimentation of distal turbidites is fairly thin, the basin being open eastward. The closure of the basin caused thick turbidite sedimentation with onlap on both sides of the Giovanna Structure (line 1). This line shows flat onlap westwards, against the Emilio Structure, and tilted onlap eastwards, against the Giovanna Structure.

In the southern part (lines 2 and 3) during the late Pliocene, the Clorinda–Giovanna Structure was less active: the turbiditic bodies developed above it in sheet-like continuous reflectors.

(3) The Edmond Basin (between the Giovanna–Clorinda Structure and the Edmond–Patrizia Structure).

The Early-Middle Pliocene sedimentation is represented by thin clays, the basin being open to the east and filled with distal turbidites. During the Late Pliocene the Edmond Structure underwent its main deformation, defining a

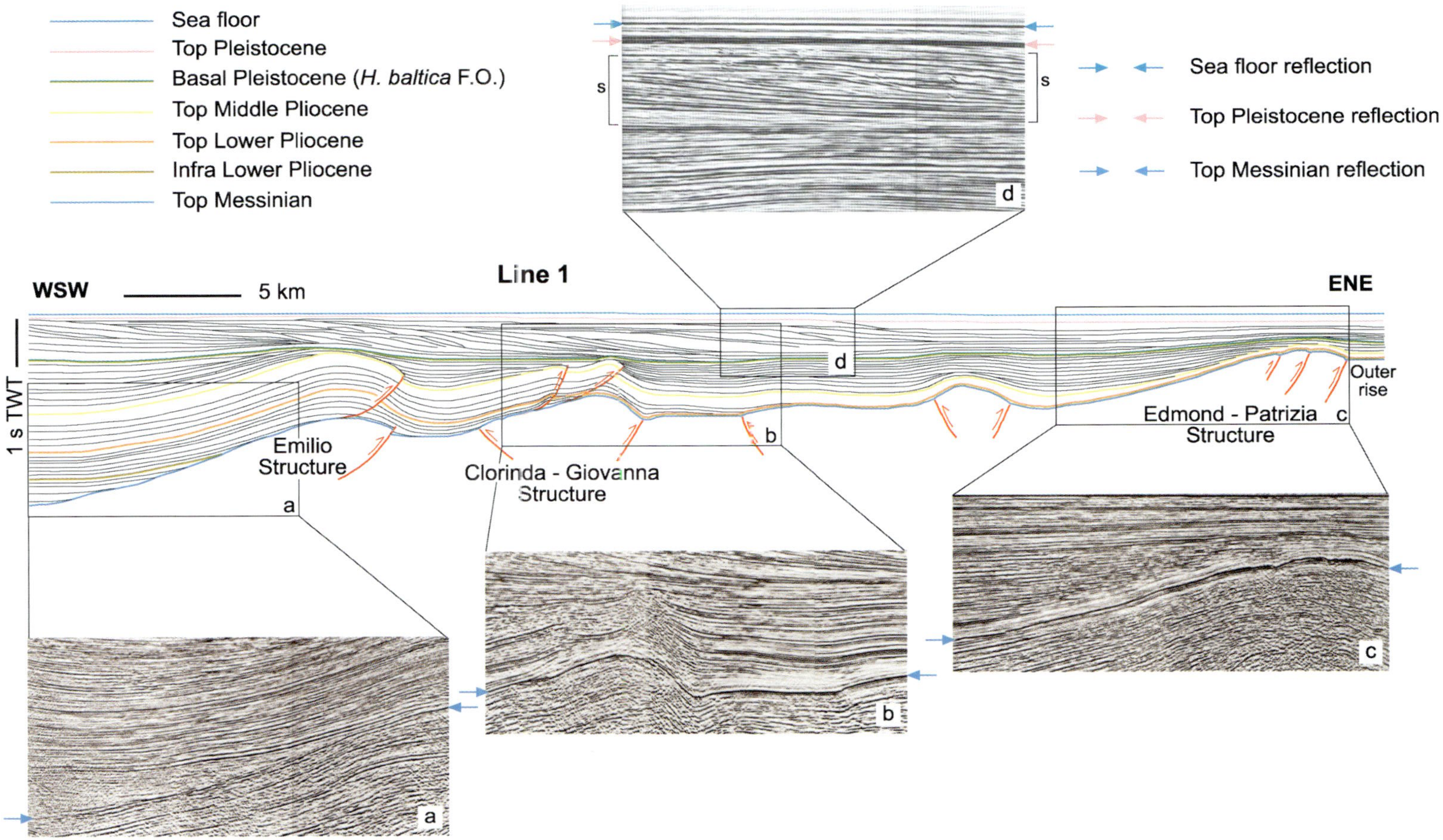

Figure 3: Geological interpretation of seismic line 1 (time version). Close-ups a, b and c show the seismic features of the onlap of siliciclastic foredeep deposits onto the Messinian evaporites (foredeep substratum). a = onlap of the basal Pliocene; b = onlap of Lower Pliocene (affected by thrusting) and Middle Pliocene; c = onlap of Upper Pliocene. Close-up d displays the complex sigmoidal pattern (s) of the Pleistocene sequence, overlain by flat-lying Holocene deposits, gradually thinning westward (see text).

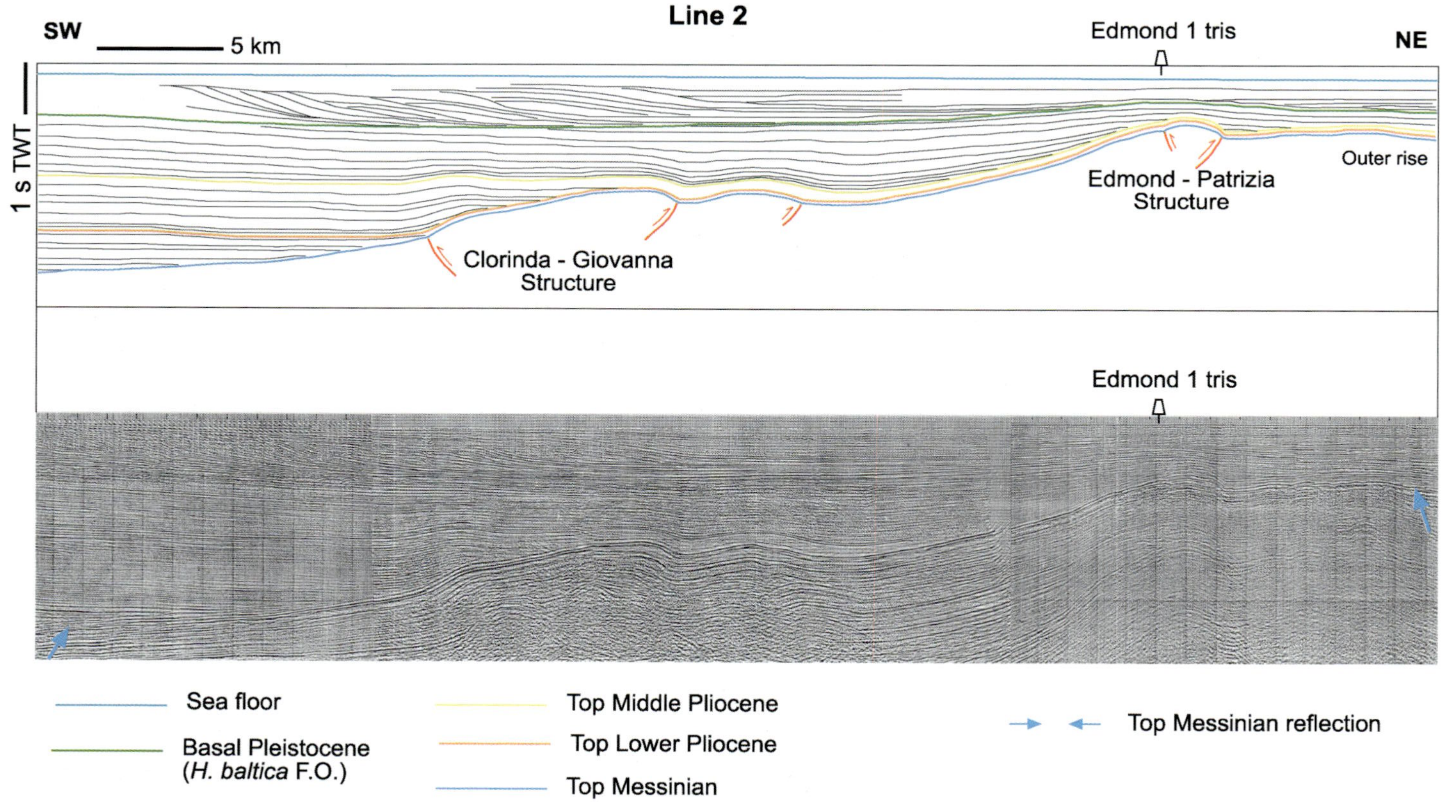

Figure 4: Geological interpretation of seismic line 2 (time version). Along this transect, located south of line 1, faulting did not affect the Plio-Pleistocene sequence.

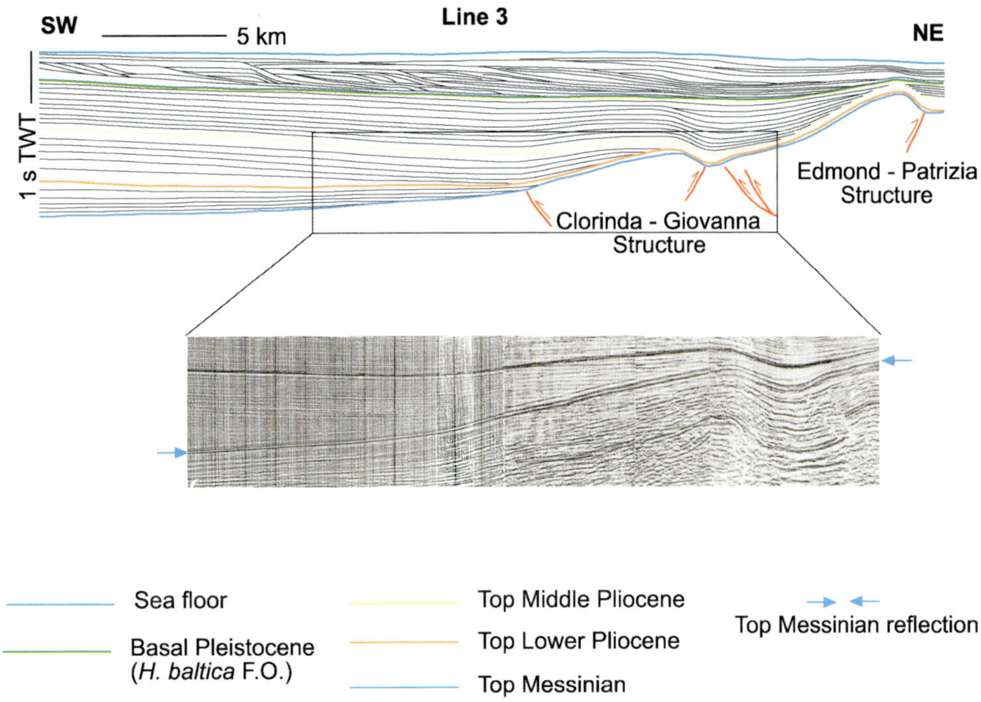

Figure 5: Geological interpretation of seismic line 3 (time version). The close-up clearly shows the onlap of the Lower-Middle Pliocene turbidites against the foredeep substratum (Messinian evaporites).

basin included between the features of the Abruzzo offshore area. This basin, with an almost flat base, but slightly undulating at the bottom indicating minor pop-up of the subsurface, terminates sharply against the Edmond high. In the northern sector of the study area the basin developed transversally for 30 km, with a secondary high in the middle (line 1), but strongly reduced in width to the south (the Giovanna–Patrizia distance in line 3 is less than 10 km). The Edmond ramp, extending up to 10 km, forms the main feature of the Late Pliocene sedimentation. Onlap terminations of the beds of clay are tilted by the growth of the synsedimentary structure.

On top of this structure the Pliocene sedimentation is condensed into only 300 m (Figure 2): in the Edmondo 1-tris well, very near to the top of the high, 150 m of Upper Pliocene clays have been drilled. Erosion surfaces are present both at the top and bottom of the Late Pliocene, and in the uppermost Middle Pliocene. Erosion was therefore active throughout the growth of the structure. The base of the Pleistocene appears deformed into a gentle anticline, showing that the Edmond Structure was still active at this time.

An overlying Pleistocene sequence with clinoform-sigmoidal geometry thins slightly to the east. This feature is peculiar to deltaic progradational systems (units 2 and 3 of Ori et al., 1986) with a transverse source. An overall regression, due to the growth of the Apennine chain, caused a progressive shifting of the coastline, and the progradation of the related deltaic deposition, as a product of coalescence of many fan deltas (Ori et al., 1986, 1993). Eustatic movements linked to the Pleistocene glaciations certainly played an important role in the development of the complex internal architecture of this system. On seismic profiles, packages of parallel and clinoform sigmoidal reflectors, wedging seawards, indicate the eastwards progradation of deltaic deposits. Sets are laterally continuous at the bottom where they display tangential cover on the sub-bottom layers. There is evidence for this facies for about 40 km seaward of the present coastline. Pelagic clays follow, onlapping onto the more external dipping beds. A Pleistocene basin filled with clays, 20 km wide, is therefore located between the termination of the deltaic progradational system and the Edmond–Patrizia Structure.

The prograding sequence is overlain by a thin (80–100 m) packet of parallel reflections gradually thinning toward the coast (onlap). This coastal onlap can be attributed to sea-level rise after the last glacial episode.

Compressional movements, related to the Apennine orogenesis, occurred mainly during the Upper Pliocene and gently deformed the foredeep. This produced faulted anticlines with an Apennine transport direction and associated back thrust (Edmond–Patrizia Structure), as well as pop-up structures (Giovanna–Clorinda and other minor intermediate structures).

Faults die out in the Messinian evaporites, the overlying siliciclastic sediments being affected only by folds. In the northern profile (Figure 3) the Pliocene sequence is deformed by east-verging fault-related folds detached over the Emilio and Clorinda–Giovanna structures, onto the Messinian evaporites. These deformations are clearly disharmonic with respect to the underlying structures in the foredeep substratum and are longitudinally discontinuous, and missing in the southern part of the study area.

Moreover, the base of the Pleistocene is clearly deformed in a gentle anticline related to the Edmond and Emilio structural highs, which were active up to this time. The present sea floor is flat lying, and undeformed by these structural highs.

4. PLIOCENE EVOLUTION OF THE FOREDEEP

In order to determine the evolution of the central Adriatic foredeep the interpreted seismic sections were at first depth-converted using the data from the available wells. Subsequently depth profiles were restored to their unstrained state at different time stages: the base of the Pleistocene (*Hyalinea baltica* F.O.), the top of the Middle Pliocene and the top of the Lower Pliocene (Figures 6, 7 and 8). Three transverse geological sections of the foredeep basin at the above-mentioned times were obtained by unfolding the succession, and returning the reflectors corresponding to these ages to the horizontal. The relative map view of the foredeep evolution is depicted in Figure 9. Compaction of deposits was not considered. Throughout the whole Pliocene the sedimentary succession comprised turbiditic bodies of northern origin.

Starting from the base of Pliocene, the depositional and structural evolution of the foredeep can be synthesized as follows:

– Lower Pliocene (5.33–3.57 Ma): the layered turbiditic sediments of the foredeep onlap onto a 1°–2° inclined foreland ramp. In lines 2 and 3 a progressive hinterland thickening of the strata is visible. The pinch-out of the turbidites (black arrows in Figures 6, 7 and 8) marks the flexure point of the foreland ramp. Eastward of this point the ramp is flat (outer ramp) and covered only by a thin pelitic drape of sediments (Edmond 1-tris). This flexure point thus represents the external margin of the foredeep basin, which has a well-defined wedge-shaped geometry.

Seismic data available between the restored profiles (lines 1, 2 and 3) and the coastline allow a reconstruction of the extent of the foredeep at basal Lower Pliocene times, mapping a seismic marker horizon next to the *Globorotalia puncticulata* F.O. (Carruba, 2001), a biostratigraphical event dated to about 3.71 Ma (Sprovieri, 1993). The extent of this seismic marker shows that the foredeep, during the basal Lower Pliocene, is abruptly restricted north of Silvi owing to a structural high of the substratum (Figure 9(a)).

– Middle Pliocene (3.6–2.13 Ma): deepening of the foredeep induces the sedimentation of turbidites which onlap onto a 2°–3° inclined foreland ramp. The flexure point of the ramp and consequently the eastern border of the foredeep basin migrate 5–6 km eastwards. The foredeep geometry is wedge-shaped with horizontal strata and without appreciable thickness variations (conformable onlaps). Shaley drapes cover the outer ramp.

At the end of the Middle Pliocene, reverse faults initiate in the area near the flexure point. These deformations will lead to the growth of the Giovanna–Clorinda Structure in the Upper Pliocene.

– Upper Pliocene to basal Pleistocene (2.13–1.50 Ma): the Edmond ramp forms the main feature of the Late Pliocene sedimentation.

The ramp was deformed by compressional structures, but the overall wedge geometry of the basin and its migration towards the foreland is still evident. Onlap terminations of the beds of clay were tilted by the growth of the synsedimentary structure. The western thick turbiditic bodies thin towards the Edmond 1-tris well, very near to the top of the Edmond–Patrizia Structure, where the Upper Pliocene is made up of a thin shaley layer (about 150 m).

Thickness is reduced by combination of growth and erosion. The Edmond–Patrizia anticline was active during the Late Pliocene sedimentation; therefore the beds of clay onlap on both sides of this structure. The thickness of the Upper Pliocene sediments increases very little on the eastern limb of the anticline and then it thins at once: the flexure point of the foreland ramp should be located to the west of the Edmond 1-tris well, and close to it. The ramp was therefore a depositional ramp up to the Middle Pliocene, and subsequently gently deformed during the Upper Pliocene.

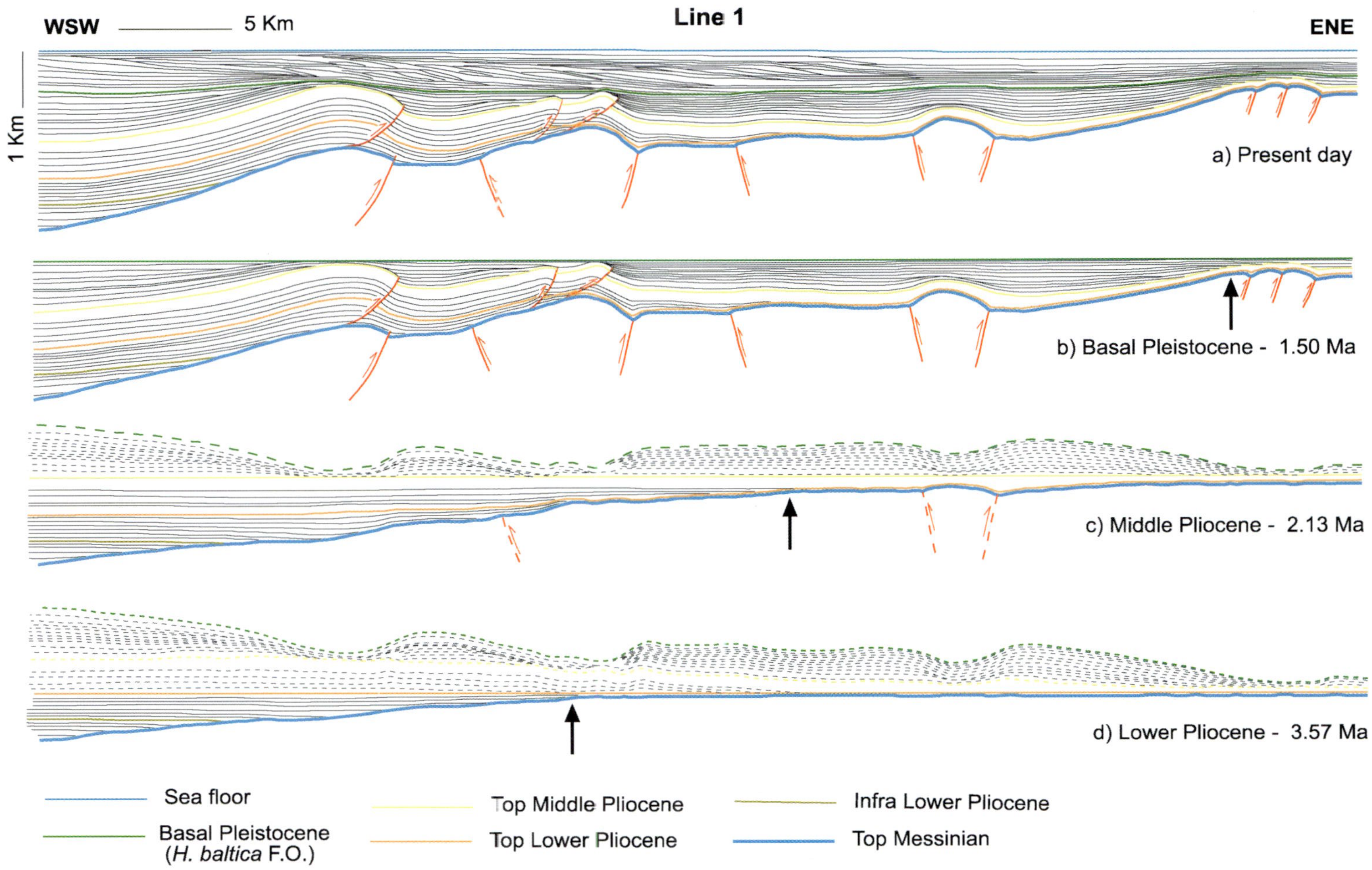

Figure 6: Depth conversion of Line 1 (a), and retro-deformation steps, showing the foredeep geometry at different times: (b) basal Pleistocene; (c) top of the Middle Pliocene; and (d) top of the Lower Pliocene. Up to the end of the Middle Pliocene the foredeep was a simple wedge-shaped basin thinning towards the foreland; in the Upper Pliocene this overall geometry was complicated by thrusting.

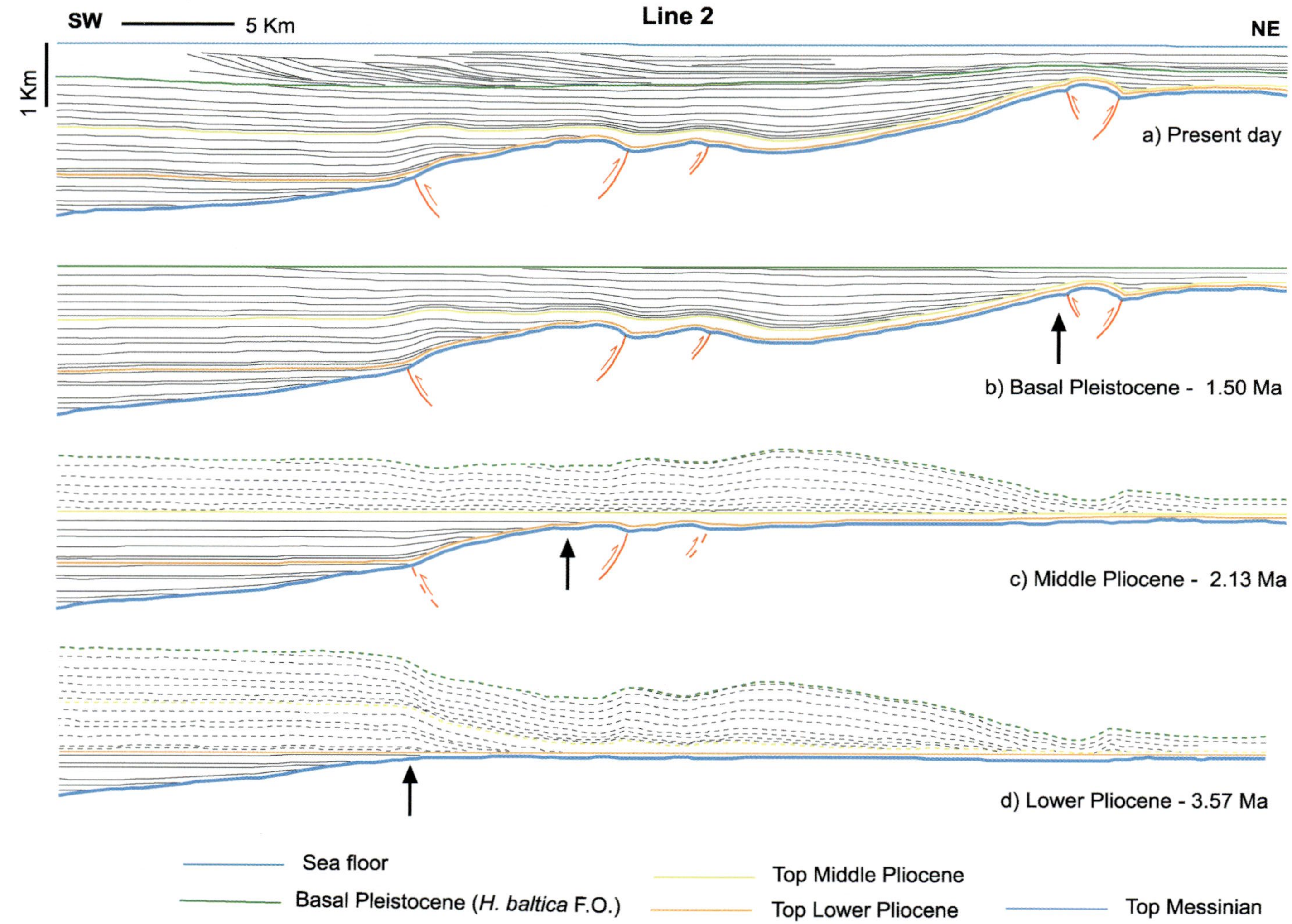

Figure 7: Depth conversion of Line 2 (a), and foredeep geometry at (b) basal Pleistocene, (c) top Middle Pliocene, and (d) top Lower Pliocene, achieved by retro-deformation.

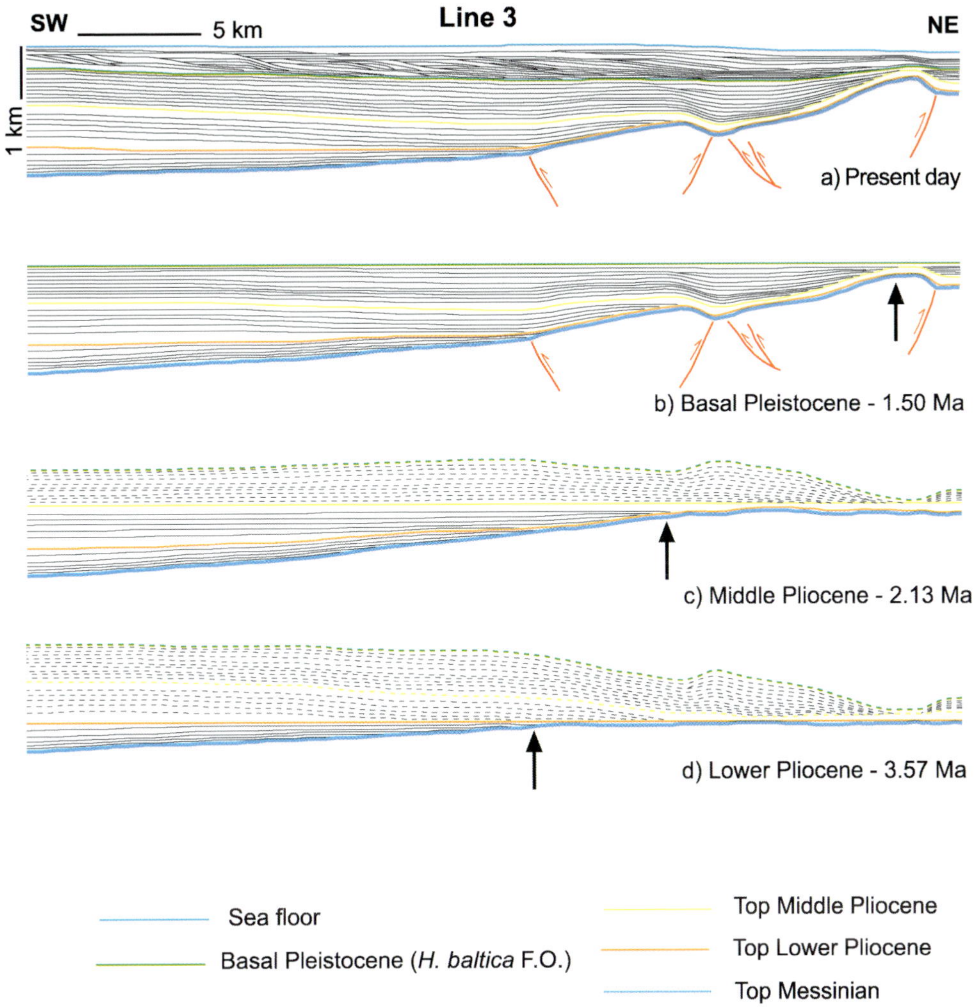

Figure 8: Depth conversion of Line 3 (a), and foredeep geometry at (b) basal Pleistocene, (c) top Middle Pliocene, and (d) top Lower Pliocene.

Flexure points corresponding to the top of the Middle Pliocene and basal Pleistocene are shifted by 20 km in the north and 10 km in the south, marking a southward-narrowing basin. During this time, onlapping phenomena are evident on both limbs of the growing structures, which in turn tilted the older turbiditic beds. Minor "open" piggy-back basins (*sensu* Ori et al., 1993) were thus formed.

Compressional deformations affected the foredeep deposits and their substratum in the Upper Pliocene, causing the growth of the Emilio, Giovanna–Clorinda and Edmond–Patrizia structures.

– Pleistocene (from 1.50 Ma): the whole area was subjected to progradational delta deposition from a western source; the westward tilting of the foredeep substratum and the consequent outward migration of the wedge-shaped basin ceased. This leads us to think that in this area no flexure or subduction of the Adriatic lithosphere took place from the basal Pleistocene. The end of subduction was preceded by Upper Pliocene deformation in the foreland plate.

Restorations of the interpreted seismic section performed at different time intervals (Figures 6, 7 and 8), show that thrusting and folding of the foreland ramp developed mainly near the flexure point of the foreland ramp, which seems to be a weak point where compressional strains are centred, and therefore a likely source of foreland thrusts (see De Alteriis, 1995; Argnani, 1998; Bertotti et al., 2001; for a discussion about the significance of the Edmond–Patrizia Structure).

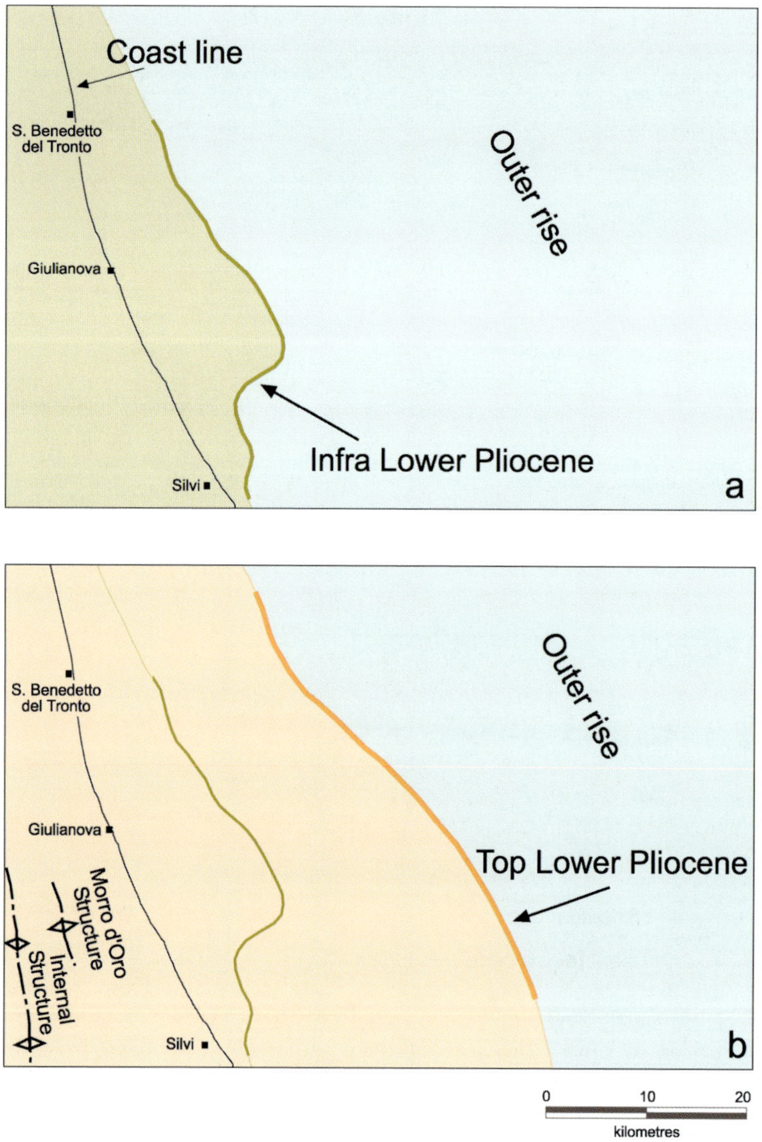

Figure 9: Map view of the terminal onlap of the foredeep deposits at different times: (a) basal Lower Pliocene; (b) Lower Pliocene; (c) top Middle Pliocene; and (d) basal Pleistocene (*H. baltica* F.O.). Present coast-line is shown for reference. The line marking the terminal onlap is thinner where its position is extrapolated.

5. THE FOREDEEP MIGRATION RATE

Analysis of seismic and well data in the offshore Abruzzo area led to the recognition of the progressive eastern migration of the external margin of the Pliocene foredeep, where the turbiditic deposits pinch out onto an inclined pre-Pliocene substratum. Restoration of the interpreted seismic profiles shows that the pinch-out ended at the flexure point of the foreland ramp, and that this point progressively shifted towards the east during Pliocene times (pull-back); it has therefore been possible to measure the relative distance of the flexure point at different stages during the Pliocene.

Ages provided by Sprovieri (1992, 1993) have been applied to date these migration steps:

- Basal Pliocene (*Globorotalia puncticulata* F.O.): 4.15 Ma
- Top Lower Pliocene: 3.57 Ma
- Top Middle Pliocene: 2.13 Ma
- Basal Pleistocene level: 1.50 Ma (*Hyalinea baltica* F.O.)

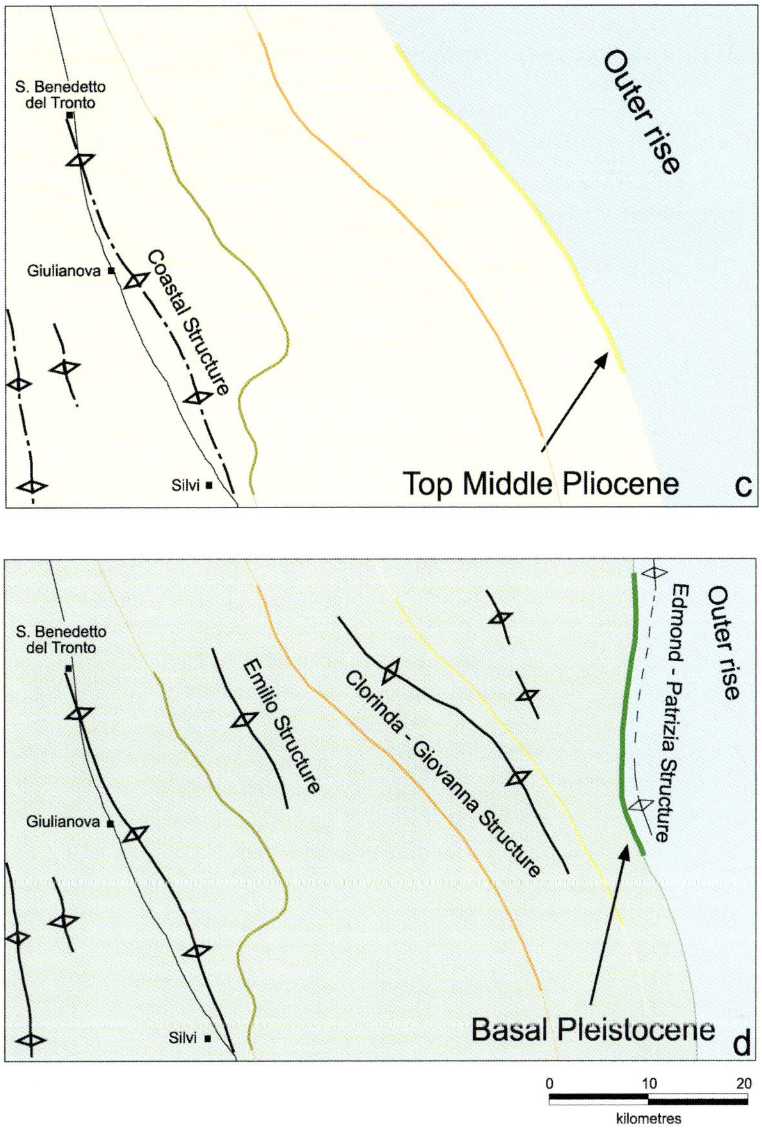

Figure 9: ***Continued.***

These data thus yielded the timing of migration of the flexure point of the foreland ramp.

The flexure point shifted about 15 km towards the foreland during the upper part of the Lower Pliocene, 6–10 km in the Middle Pliocene and 10–21 km in the Upper Pliocene. The lowest values are found in the southern part of the study area, the highest in the northern part.

The velocities of the pull-back of the flexure ramp can thus be evaluated as 25 mm/yr during the upper part of the Lower Pliocene, 4–7 mm/yr in the Middle Pliocene, and 16–34 mm/yr during the Upper Pliocene to basal Pleistocene time-span (without calculating the negligible shortening due to folding and faulting affecting the foredeep). As a whole, (from basal Pliocene to basal Pleistocene) the migration rate is about 11–17 mm/yr.

6. GEODYNAMIC IMPLICATIONS AND DISCUSSION

As shown before, the eastwards migration velocity of the Abruzzo foredeep during Pliocene times ranges from 5 to 34 mm/yr, with an average rate of about 11–17 mm/yr. The variation seems to depend on the area of evaluation (the maximum rates occurred towards the N, in the S. Benedetto del Tronto offshore area) and the considered time-span (the minimum values occurred during the Middle Pliocene).

It is useful to compare the rate of foreland migration of the foredeep in the Abruzzo-Adriatic region with the roll-back velocities (subduction velocities) of the Adria lithosphere dipping under the Apennines, calculated using different methods.

Doglioni et al. (1994) estimated a very large variation (10–70 mm/yr) in the rate of subduction of the Adria plate under the Apennine chain, from Miocene to present times; the lower values relate to the Northern Apennines and Sicily and the higher to the Ionian area.

Along the Provençal Basin–Sardinia Block–Ionian Sea transect, Gueguen et al. (1997, 1998) computed a roll-back velocity of 34 mm/yr during the last 23 Ma, and 32 mm/yr from 5 Ma to the present, on the basis of lithospheric balance. For the Abruzzo transect during the last 5 Ma, the same authors evaluated an eastwards migration rate of the Apennine arc of about 7.5 mm/yr.

Other authors calculated the shortening rate in the central Apennines from balanced and retro-deformable geological sections. Calamita et al. (1991) found a rate of deformation of about 15 mm/yr from the Upper Messinian to Upper Pliocene, in a section through M. Utero–Tortoreto Lido. Lavecchia et al. (1994) estimated the shift in the front of deformation in the central Apennines (balanced cross-section along the Tiberina valley–Ascoli Piceno–Tortoreto Lido transect) of 20–25 mm/yr from the late Miocene to the present. According to Bigi et al. (1995), the central Apennines compressional front migrated eastwards at a rate of 11 mm/yr in the Messinian and 30 mm/yr in the Lower Pliocene, while during the Messinian the depocentre of the Laga basin migrated eastwards at a velocity of 40 mm/yr. Finally, Barchi et al. (1997) estimated a total shortening along the CROP 03 seismic line, across the central Apennines, as 200 km, with a deformation rate of 8.5–9 mm/yr during the Miocene–present time interval.

A comparison between the central Apennines shortening rate and the rate of eastwards migration of the Abruzzo foredeep indicates that the two velocities are fully compatible. In particular they are almost the same for the most recent time period, from the Lower Pliocene to the present. This coincidence is much more significant considering the different methods used in the evaluations. The similar rate of Apennine shortening and foredeep migration shows that the two phenomena are induced by the same geodynamic mechanism, i.e. the eastwards roll-back of the Adria plate under the Apennine chain (Malinverno and Ryan, 1986; Royden et al., 1987; Doglioni, 1991; Gueguen et al., 1997, 1998, and references therein). If the eastwards foredeep migration were caused by the flexural response to loading by sediment deposition or thrust nappes over the Adriatic lithosphere, its rate would not necessarily be the same as the Apennine total shortening rate (Flemings and Jordan, 1989; Sinclair et al., 1991); rather, it would probably be smaller. In fact, the rate of crustal shortening causing the thickening of the orogenic wedge would exceed (without considering other factors like climate, erosion rate, sediment transport coefficient or the slope angle of the wedge) the flexural response inducing the slab retreat of the lower elastic plate. Only the rate of propagation of the orogenic compressional front into the foredeep can be successfully compared in all cases to the progressive migration of the basin, at least for underfilled foredeeps like in the French and Swiss Alps (Sinclair, 1997).

On the other hand Royden and Karner (1984), Royden (1988) and Royden and Horvath (1988) show that the topographic load is insufficient to explain the deep foreland basins at the front of the Apennine and Carpathian belts.

The roll-back of the subduction hinge to the front of the central Apennines appears to have caused the progressive eastwards migration of the foredeep. In such a geodynamic context, the different sedimentary units that infill the foredeep do not necessarily show an internal architecture characterized by laterally shifted thicknesses (hinterland thickening). Even if a westwards thickening of the stratigraphic units of the Lower Pliocene is visible in line 3 of Figure 8, generally the Pliocene Abruzzo foredeep shows basin-fill geometries characterized by conformable onlap phenomena, without appreciable lateral variations in the thickness of the stratigraphic succession. The progressive retreat of the Adriatic slab and the consequent creation of extra space seem to precede the emplacement of new allochthonous bodies. This can be explained by considering that the motor for the foredeep subsidence is not the increasing load of the thickened wedge, but the roll-back of the subduction hinge, with a probable parallel geometry (at least in the foredeep area) of the descending slab (Figure 10).

The hypothesis that roll-back of the subduction hinge, associated with the progressive retreat of the Adriatic slab, is the active geodynamic mechanism in the Adriatic area finds other particular indications (Doglioni, 1993): (a) the Apennine chain, adjacent to the Adriatic foredeep, has anomalously low relief; (b) the belt is characterized by extensional tectonics in its western (inner) part: the eastern boundary of the area subject to extension is very near

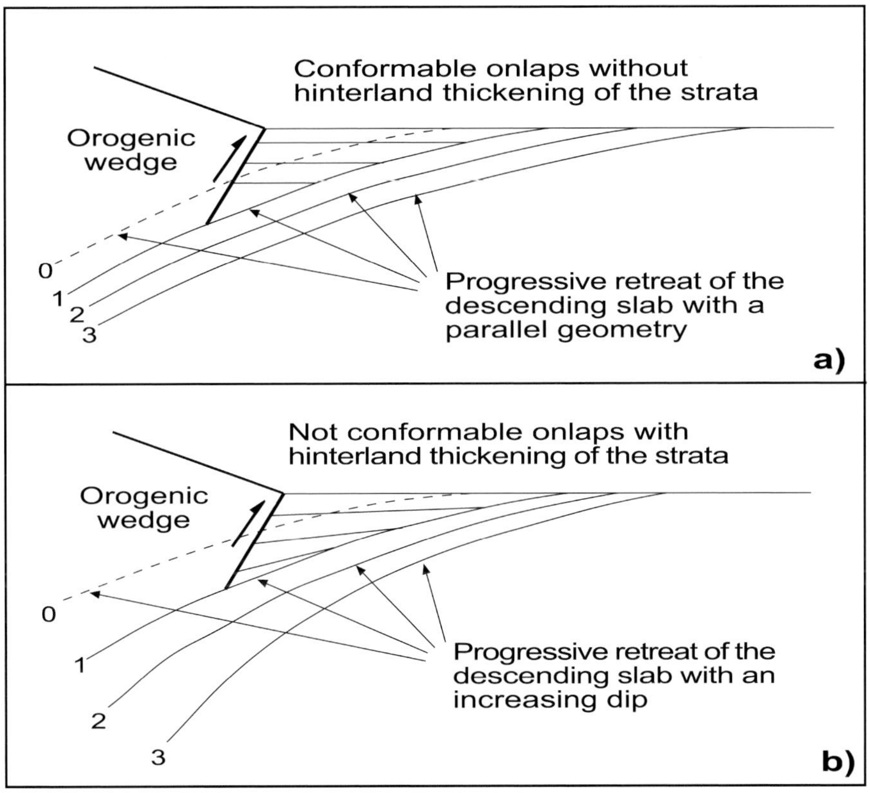

Figure 10: In a foredeep caused by the progressive retreat of the descending slab with a parallel geometry (roll-back of the subduction hinge) the strata show conformable onlaps (a); if the dip of the slab were increased by the orogenic load, the strata would thicken towards the hinterland, with an upwardly decreasing angle (b).

to the active compressional front; (c) the foredeep adjacent to the chain is very deep (several kilometres) and with a high rate of subsidence (up to 1000 m/Ma).

Recently, Finetti et al. (2001) proposed an alternative geodynamic interpretation of the Apennine chain. According to these authors, during the Late Miocene to present continental collision, east-vergent thrust faults affected the sedimentary cover of the Adriatic margin down to the lower crust and mantle, leading the Adriatic lithosphere to underthrust the Apennines. In this model, the inner crustal extension represents a subordinate geodynamic event related to the gravity sliding of sedimentary blocks over the basement high due to the Neogene lithospheric thrusting.

The comparable rate of Apennine shortening and foredeep migration does not fully exclude this last interpretation, as long as the underthrusting Adriatic lithosphere is affected by a retreat mechanism with a fairly parallel geometry.

The compressional deformation causing lithospheric or crustal folding (buckling) of the cratonic area some distance away from the collisional belt is a geodynamic mechanism recognized in various areas (Cloething et al., 1999, and references therein), and is the result of horizontal stresses transmitted through the lithosphere (Letouzey, 1986; Karner et al., 1993; Heller et al., 1993; Cloething et al., 1999). Recently it has been proposed for the Puglia area (Doglioni et al., 1994) and for the whole Adria Plate (Bertotti et al., 2001). However, in the Pliocene Abruzzo foredeep, this mechanism does not seem applicable due to the presence of progressively migrating onlap towards the foreland, and because of the occurrence of consistent extensional tectonics in hinterland areas, contrasting with the generally compressional state necessary for lithospheric buckling.

7. CONCLUSIONS

The topographic load of the Apennine thrust belt cannot account for the development and subsidence of the Periadriatic foredeep (Royden and Karner, 1984; Royden, 1988). Several geological and geophysical elements, including the probable presence of a steeply dipping lithospheric slab under the Apennines, the low relief of the Apennine chain associated with extensional tectonics close to the active compressional front, the great depth of the fore deep, and the high rate of subsidence, support the hypothesis that the geodynamic mechanism which led to the formation of the chain-foredeep system in the Apennines was the roll-back of the subduction hinge combined with the retreat of the Adriatic slab (Malinverno and Ryan, 1986; Doglioni, 1991; Lucente et al., 1999).

Seismic profiles across the outermost sector of the central Adriatic foredeep show the evolution of this basin during Plio-Pleistocene times. The migration of the foredeep towards the foreland was caused by the eastwards shifting of the flexure point of the foreland ramp. The migration rate has been calculated on the basis of subsequent time intervals, reported on horizontal datum, of the terminal onlaps of the foredeep sediments. In the study area the estimate is 11–17 mm/yr, for the interval from basal Pliocene (4.15 Ma) to basal Pleistocene (1.50 Ma).

These values fall in the same range as those determined for the roll-back velocity of the Adriatic lithosphere under the Apennine chain. This fact, and the basin-fill geometry being generally marked by conformable onlap, indicate that the eastwards migration of the foredeep is consistent with a progressive and parallel retreat of a descending slab; that is, with the roll-back of the subduction hinge of the Adria plate. The similar rates of Apennine chain shortening and foredeep migration corroborate the roll-back hypothesis, in that the rate of flexural response of the Adria lithosphere does not necessarily have to be the same as the chain shortening in the case of loading of sediments and thrust sheets.

A major change in the depositional pattern associated with the end of westward tilting of the foreland plate, and of the associated development of the foredeep basin, seems to indicate that the roll-back of the subduction hinge stopped at the beginning of the Pleistocene in the Abruzzo area, while it has continued up to the present in the Bradano Trough and the Taranto Gulf.

ACKNOWLEDGEMENTS

We are grateful to Edison Gas S.p.A. and ENI-Agip S.p.A., who kindly supplied the subsurface data. Their support and helpful collaboration has made this work possible.

REFERENCES

Argnani, A., 1998. Structural elements of the Adriatic foreland and their relationships with the front of the Apennine fold-and-thrust belt. Mem. Soc. Geol. It. 52, 647–654.

Bally, A.W., Burbi, L., Cooper, C., Ghelardoni, R., 1986. Balanced sections and seismic reflection profiles across the Central Apennines. Mem. Soc. Geol. It. 35, 257–310.

Barchi, M., Minelli, G., Pialli, G., 1997. The Crop 03 profile: a synthesis of results on deep structures of the Northern Apennines. Mem. Soc. Geol. It. 52, 383–400.

Bertotti, G., Ricotti, V., Chilovi, C., Fantoni, R., Merlini, S., Mosconi, A., 2001. Neogene to Quaternary sedimentary basins in the south Adriatic (Central Mediterranean): Foredeeps and lithosperic buckling. Tectonics 20 (5), 771–787.

Bigi, S., Calamita, F., Cello, G., Centamore, E., Deiana, G., Paltrinieri, W., Ridolfi, M., 1995. Evoluzione messiniano-pliocenica del sistema catena-avanfossa nell'area marchigiano-abruzzese esterna. Studi Geologici Camerti spec. vol. 1995/1, 29–35.

Boccaletti, M., Coli, M., Decandia, F.A., Giannini, E., Lazzaretto, A., 1980. Evoluzione dell'Appennino Settentrionale secondo un nuovo modello strutturale. Mem. Soc. Geol. It. 21, 359–373.

Bolis, G., Carruba, S., Casnedi, R., Perotti, C.R., Ravaglia, A., Tornaghi, M., 2003. Compressional tectonics overprinting extensional structures in the Abruzzo Periadriatic Foredeep (Central Italy) during Pliocene times. Boll. Soc. Geol. It. 122, 251–266.

Calamita, F., Cello, G., Centamore, E., Deiana, G., Micarelli, A., Paltrinieri, W., Ridolfi, M., 1991. Stile deformativo e cronologia della deformazione lungo tre sezioni bilanciate dall'Appennino umbro-marchigiano alla costa adriatica. Studi Geologici Camerti spec. vol. 1991/1, 295–314.

Carruba, S., 2001. Evoluzione tettonica e deposizionale dell'avanfossa pliocenica abruzzese. PhD thesis, Università di Pavia, Italy, 128 p.

Casnedi, R., 1991. Hydrocarbon accumulation in turbidites in migrating basins of the southern adriatic foredeep (Italy). In: Bouma, A.H., Carter, R.M. (Eds.), Facies Models in Exploration and Development of Hydrocarbon and Ore Deposits, Proceedings of the 28th International Geology Congress. VSP, Utrecht, pp. 219–233.

Casnedi, R., Crescenti, U., Tonna, M., 1982. Evoluzione dell'Avanfossa adriatica meridionale nel Plio-Pleistocene, sulla base di dati di sottosuolo. Mem. Soc. Geol. It. 24, 243–260.

Cloething, S., Burov, E., Poliakov, A., 1999. Lithospheric folding: primary response to compression? (from central Asia to Paris Basin). Tectonics 18, 1064–1083.

Crescenti, U., D'Amato, C., Balduzzi, A., Tonna, M., 1980. Il Plio-Pleistocene del sottosuolo abruzzese-marchigiano tra Ascoli Piceno e Pescara. Geologica Romana 19, 63–84.

De Alteriis, G., 1995. Different foreland basins in Italy: examples from the central and southern Adriatic Sea. Tectonophysics 252, 349–373.

Doglioni, C., 1991. A proposal for the kinematic modelling of W-dipping subductions – possible applications to the Tyrrhenian-Apennines system. Terra Nova 3, 423–434.

Doglioni, C., 1993. Some remarks on the origin of foredeeps. Tectonophysics 228, 1–20.

Doglioni, C., Monelli, F., Pieri, P., 1994. The Puglia uplift (SE Italy): an anomaly in the foreland of the Apenninic subduction due to buckling of a thick continental lithosphere. Tectonics 13, 1309–1321.

Finetti, I.R., Bricchi, G., Del Ben, A., Pipan, M., Xuan, Z., 1987. Geophysical study of the Adria Plate. Mem. Soc. Geol. It. 40, 335–344.

Finetti, I.R., Boccaletti, M., Bovini, M., Del Ben, A., Geletti, R., Pipan, M., Sani, F., 2001. Crustal section based on CROP seismic data across the North Tyrrhenian–Northern Apennines–Adriatic Sea. Tectonophysics 343, 135–163.

Flemings, P.B., Jordan, T.E., 1989. A synthetic stratigraphic model of foreland basin development. J. Geophys. Res. 94 (B4), 3851–3866.

Gueguen, E., Doglioni, C., Fernandez, M., 1997. Lithospheric boudinage in the western Mediterranean back-arc basin. Terra Nova 9, 184–187.

Gueguen, E., Doglioni, C., Fernandez, M., 1998. On the post-25 Ma geodynamic evolution of the western Mediterranean. Tectonophysics 298, 259–269.

Heller, P.L., Beekman, F., Angevine, C.L., Cloething, S., 1993. Cause of tectonic reactivation and subtle uplifts in the Rocky Mountain region and its effect on the stratigraphic record. Geology 21, 1003–1006.

Karner, G.D., Driscoll, N.W., Weissel, J.K., 1993. Response of the lithosphere to in-plane force variations. Earth Planet. Sci. Lett. 114, 397–588

Kligfield, R., 1979. The northern Apennines as a collisional orogen. Am. J. Sci. 279, 676–691.

Lavecchia, G., Brozzetti, F., Barchi, M., Menichetti, M., Keller, J.V.A., 1994. Seismotectonic zoning in east-central Italy deduced from an analysis of the Neogene to present deformations and related stress fields. Geol. Soc. Am. Bull. 106, 1107–1120.

Letouzey, J., 1986. Cenozoic paleo-stress pattern in the Alpine foreland and structural interpretation in a platform basin. Tectonophysics 132, 215–231.

Lucente, F.P., Chiarabba, C., Cimini, G.B., 1999. Tomographic constraints on the geodynamic evolution of the Italian region. J. Geophys. Res. 104 (B9), 20,307–20,327.

Malinverno, A., Ryan, W.B.F., 1986. Extension in the Tyrrhenian Sea and shortening in the Apennines as a result of arc migration driven by sinking of the lithosphere. Tectonics 5, 227–245.

Moretti, I., Royden, L., 1988. Deflection, gravity anomalies and tectonics of doubly subducted continental lithosphere: adriatic and ionian seas. Tectonics 7 (4), 875–893.

Ori, G.G., Roveri, M., Vannoni, F., 1986. Plio-Pleistocene sedimentation in the Apenninic-Adriatic Foredeep (Central Adriatic Sea, Italy). In: Allen, P.A., Homewood, P. (Eds.), Foreland Basins. IAS Spec. Publ. 8, 183–198.

Ori, G.G., Serafini, G., Visentin, C., Ricci Lucchi, F., Casnedi, R., Colalongo, M.L., Mosna, S., 1993. Depositional history of the Pliocene-Pleistocene Adriatic Foredeep (Central Italy) from surface and subsurface data. In: Spencer, A.M. (Ed.), Generation, Accumulation and Production of Europe's Hydrocarbons. Spec. Publ. Eur. Assoc. Pet. Geosci. 3, 233–258.

Paltrinieri, W., Zanchini, G., Martini, N., Roccia, L., 1982. Evoluzione del bacino torbiditico marchigiano-abruzzese a partire dal Messiniano in base a lineazioni profonde. Mem. Soc. Geol. It. 24, 233–241.

Patacca, E., Scandone, P., 1989. Post-Tortonian mountain building in the Apennines. The role of the passive sinking of a relic lithospheric slab. In: Boriani, A., Bonafede, M., Piccardo, G.B., Vai, G.B. (Eds.), The Lithosphere in Italy. Atti Acc. Naz. Lincei 80, 157–176.

Ricci Lucchi, F., 1986. The Oligocene to Recent foreland basins of the northern Apennines. In: Allen, P.A., Homewood, P. (Eds.), Foreland Basins. IAS Spec. Publ. 8, 105–139.

Royden, L.H., 1988. Flexural behaviour of the continental in Italy: constraints imposed by gravity and deflection data. J. Geophys. Res. 93 (B7), 7747–7766.

Royden, L.H., Horvath, F., 1988. The Pannonian basin. Mem. Am. Ass. Petrol. Geol. 45, 1–394.

Royden, L.H., Karner, G.D., 1984. Flexure of lithosphere beneath Apennine and Carpathian Foredeep basins: evidence for an insufficient topographic load. Am. Ass. Petrol. Geol. Bull. 68 (6), 704–712.

Royden, L.H., Patacca, E., Scandone, P., 1987. Segmentation and configuration of subducted lithosphere in Italy: an important control on thrust-belt and foredeep evolution. Geology 15, 714–717.
Sinclair, H.D., 1997. Tectonostratigraphic model for underfilled peripheral foreland basins: an Alpine perspective. Geol. Soc. Am. Bull. 109, 324–346.
Sinclair, H.D., Coakley, B.J., Allen, P.A., Watts, A.B., 1991. Simulation of foreland basin stratigraphy using a diffusion model of mountain belt uplift and erosion: an example from the central Alps, Switzerland. Tectonics 10 (3), 599–620.
Sprovieri, R., 1992. Mediterranean Pliocene biochronology: an high resolution record based on quantitative planktonic foraminifera distribution. Riv. It. Paleont. Strat. 98 (1), 61–100.
Sprovieri, R., 1993. Pliocene–Early Pleistocene astronomically forced planktonic foraminifera abundance fluctuations and chronology of Mediterranean calcareous plankton bio-events. Riv. It. Paleont. Strat. 99, 371–414.
Van Couvering, A., Castratori, D., Cita, M.B., Hilgen, F.J., Rio, D., 2000. The base of the Zanclean Stage and of the Pliocene series. Episodes 23 (3), 179–187.

CROP PROJECT: Deep Seismic Exploration of the Central Mediterranean and Italy
Edited by I.R. Finetti

CHAPTER 26

Crustal Tectono-Stratigraphic Setting of the Pelagian Foreland from New CROP Seismic Data

I.R. Finetti[1,*], A. Del Ben[1]

ABSTRACT

New deep CROP seismic data have allowed us to reconstruct a regional transcrustal tectono-stratigraphic setting for the Pelagian Sea. This area is delimited to the NW by the Maghrebian chain, to the N by the Sicilian Apennines, to the E by the Ionian Sea floored by Permo-Triassic oceanic crust, and to the W and S by the northern Africa mainland. Cover strata very likely became widespread in the Permian, with probable older local depositions into rift-related depressions. Thick Mesozoic-Tertiary successions characterize wide platform areas of the Pelagian Sea and interposed rift-basin sequences. Tectonics are dominantly represented by a tabular setting interrupted by impressive, prominent shear-related transtensional deformations. Two transcurrent systems imprint the structural setting of the Pelagian Sea: a NW–SE-directed dextral main one and a left antithetic NE–SW-trending one. Along the axial zone of the Pelagian Sea, the NW–SE main transcurrent system gave rise to significant rifting processes accompanied by crustal thinning and huge mantle uplift. Major mantle culminations in the Pantelleria and Linosa graben areas are at a depth of only 17 km, or about 10 km higher than on lateral tabular zones. Diffuse igneous shows are directly and/or geophysically evident in the Pelagian province, all connected to the shear-tectonics.

Keywords: Pelagian Sea, CROP transcrustal seismic exploration

1. INTRODUCTION

The Pelagian foreland is a mostly flat area of the African plate, covered by shallow water, and located between the African coast and Sicily, where the tectonic deformation is dominated by a NW–SE-trending system of transcurrent to transtensive strike-slip motions of right polarity, combined with antithetic NNE–SSW-directed left-transcurrent faults. Geophysical exploration campaigns were conducted on the Pelagian Sea many years ago (Finetti and Morelli, 1973; Winnock, 1981; Finetti, 1982, 1984; Finetti and Del Ben, 1986). From these pioneering investigations, it became clear that the Pelagian Sea is dominated by a mostly tabular setting, interrupted by strike-slip faults, and transtensive (negative flower and pullapart troughs) or transpressive (positive flower) tectonics. The new geophysical-geological data collected in the initial exploration activity of the Mediterranean stimulated the geoscience community to place it in a large-scale plate tectonic framework (Biju-Duval, 1980; and many other).

The stratigraphy of the Pelagian Sea foreland is represented by a relatively thick sedimentary deposition, mostly of Mesozoic age. In accordance with the widely observed basal cover of the explored Northern Africa region, on the crystalline-metamorphic basement, a more or less thick interval of quartzitic sandstone of Cambro-Ordovician age and successive Paleozoic depositions are expected. In the areas of the Panormid and Etrurian promontories the quartzitic interval was tectonically removed and went to feed the future Numidian flysch. According to the location, at varying epochs of the Paleozoic era, local deposition started and later widespread cover continued in the Meso-Cenozoic. The oldest Paleozoic sedimentary units were recognized to the north of the Jeffara extensional fault.

The interpreted CROP sections described in this work illustrate very well the crustal tectono-stratigraphic setting of the Pelagian Sea (Figures 2, 3, 5, 6 and 7; Plates 2 and 4).

[1]Department of Geological, Environmental and Marine Sciences, Exploration Geophysics Group, University of Trieste, Via Weiss 1, 34127 Trieste, Italy
[*]Corresponding author. Tel.: +39 40 5582276; e-mail: finetti@units.it (I.R. Finetti)

2. CROP SEISMIC DATA INTERPRETED AND DESCRIBED

Four long CROP sections explore the Pelagian Sea (Figure 1; Table 1). Three of them (M25, M24 and M39) cross transversally to the main rift tectonic features, and one (M23A1) runs close and subparallel to the southern Sicilian coast line. The section M23A1 (part) exhibits various flower features and transtensive faults, which slice the whole crust. Moreover, its southeast extremity images very clearly the listric faults associated with the Permo-Triassic Ionian opening (Figure 2; Plates 1 and 2). The interpretation of the CROP data was done on sections to the scale

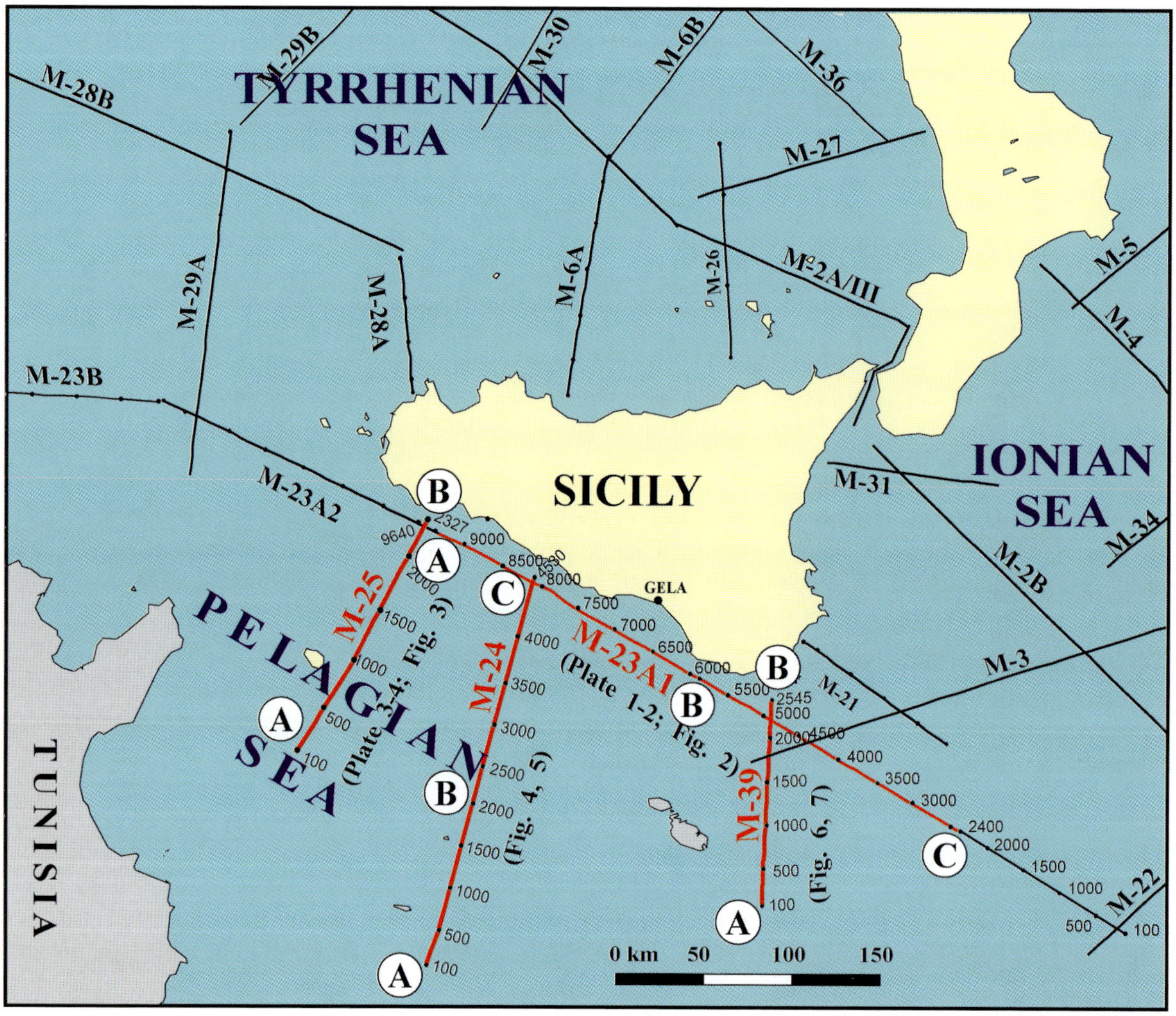

Figure 1: Location map (red traces) of the interpreted and described CROP seismic sections of this work (Chapter 26).

Table 1
CROP seismic sections exhibited and described in this work on the Pelagian Sea (Chapter 26)

Crop section	SP	Length (km)	Chapter 26			
			Figure N°		Plate N°	
			Seismic	Int.	Seismic	Int.
M23A1(P)	2400–9400	349	–	2	1	2
M25	100–2327	157	–	3	3	4
M24	100–4530	224	4	5	–	–
M39	100–2545	125	6	7	–	–

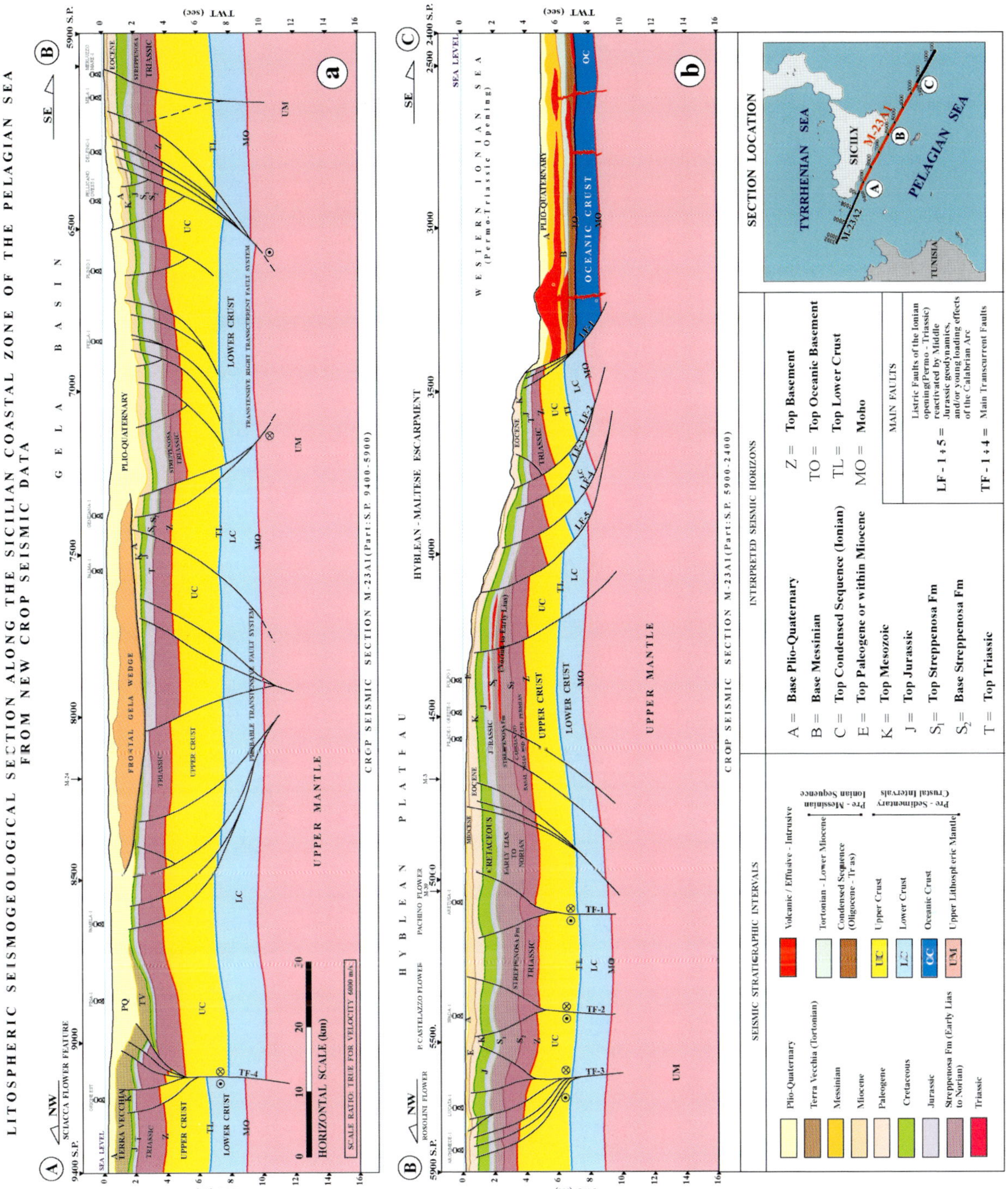

Figure 2: Reduced version of the interpreted crustal seismogeological section of Plate 2: CROP M23A1 (part).

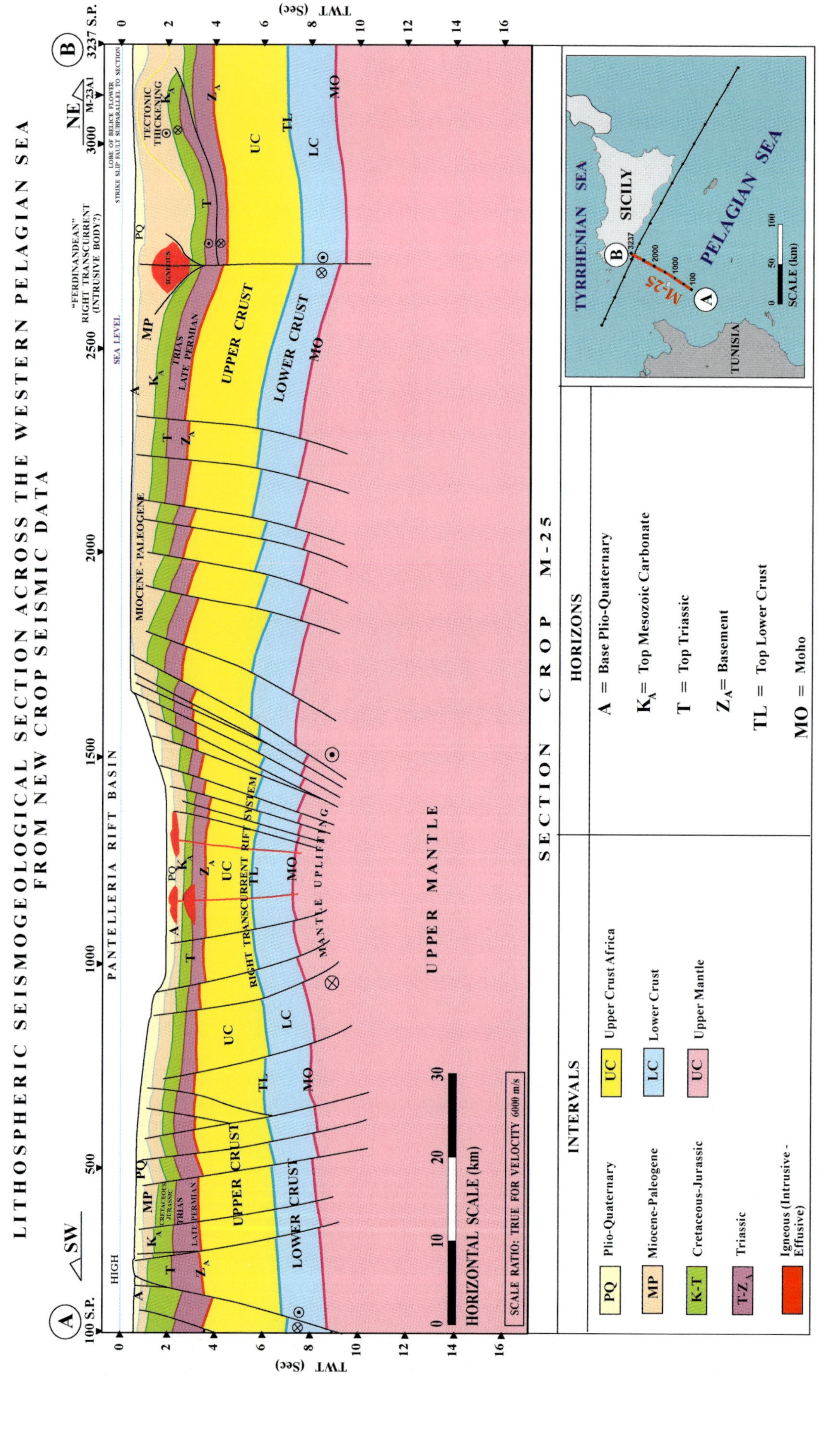

Figure 3: Reduced version of the interpreted crustal seismogeological section CROP M25 of Plate 4.

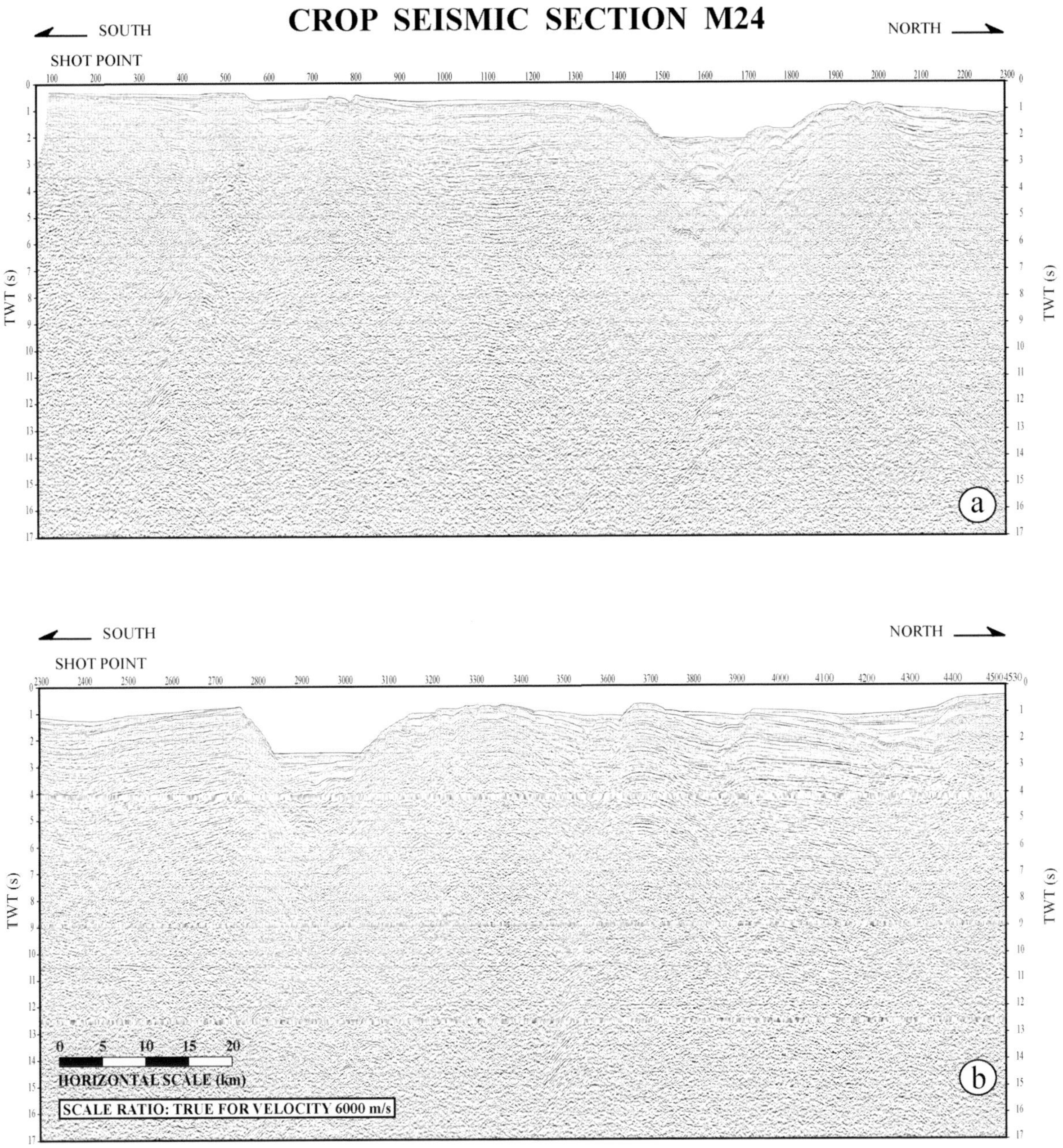

Figure 4: CROP seismic section M24 across the Pelagian Sea. Its location is shown in Figure 1, and its interpretation in Figure 5.

1:50 000 and the results are presented on annexed Plates using the horizontal scale 1:300 000, and vertical scale true for a velocity of 6.000 m/s. At the same scale, the uninterpreted seismic sections are also provided. This scale is reserved for the longer CROP lines (M23A1 part and M25), while the shorter sections (M24 and M39) are exhibited only as reduced figures (Figures 4 and 6 for the seismic section, and 5 and 7 for the respective interpreted seismogeological sections).

Other geophysical information is also integrated with the CROP data: the *MS seismic lines* (Finetti et al., this volume, Chapter 1); seismic data from the *public dataset* of the Italian Ministry of Industry; *gravity* and *magnetic data*; and *borehole data*.

3. SEISMIC INTERPRETATION

The sedimentary cover stratigraphy of the interpreted CROP seismic sections in the Pelagian Sea is calibrated by much borehole data from the pluridecennial hydrocarbon exploration activity. The interpreted horizons are all listed

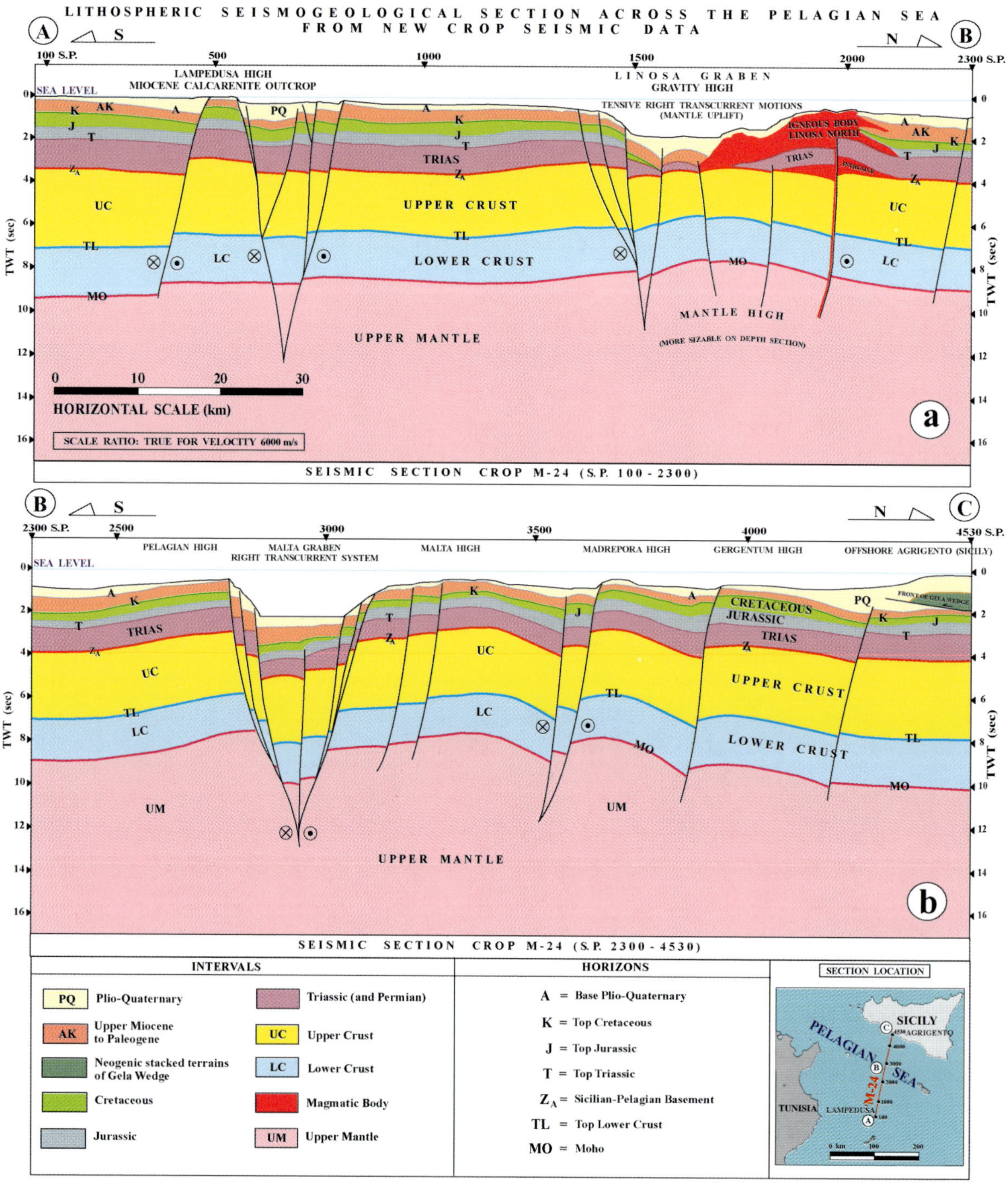

Figure 5: Interpreted crustal seismogeological section CROP M24 across the Pelagian Sea.

and specified in the legend of the interpreted seismogeological sections (Figures 2, 3, 5 and 7; Plates 2 and 4). The horizons "A" (base Plio-Quaternary), "B" (base Messinian), and "C" (top Carbonate) are normally well seismically characterized and of clear identification. The other picked horizons of the sedimentary cover are calibrated at various points. The basement (horizon "Z") is everywhere interpreted and must be regarded as an approximate, but realistic indication, locally well evident seismically on the sections. The horizons "TL" (top lower crust) and "MO" (moho) are of regionally acceptable reliability.

All picked horizons of the Ionian Sea are well characterized seismically and of good reliability. On the Hyblean plateau and its escarpment, as well as on some other local areas of the Pelagian rift zone (*Sicily Channel rift zone*) several igneous layers, impossible to exhibit in the reduced sections of the present work, are evident and in part also calibrated by boreholes. The stratigraphic cover began to deposit on existing late Paleozoic depressions

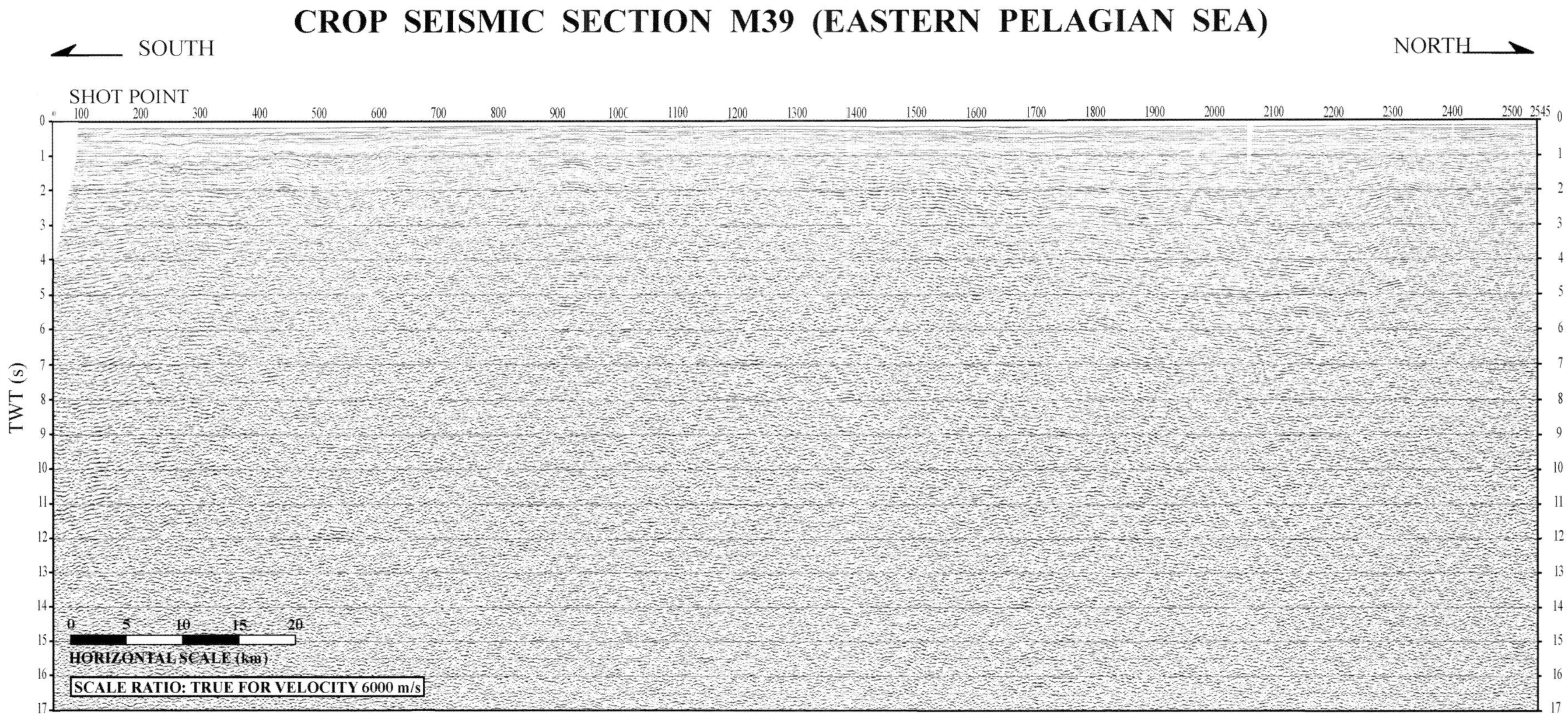

Figure 6: CROP seismic section M39 across the eastern edge of the Pelagian platform, to the south of the Hyblean plateau. See Figure 1 for location. Its interpretation is exhibited in Figure 7.

CRUSTAL SEISMOGEOLOGICAL SECTION ACROSS THE EASTERN MARGIN OF THE PELAGIAN SEA
(FROM NEW CROP SEISMIC DATA: SECTION M-39)

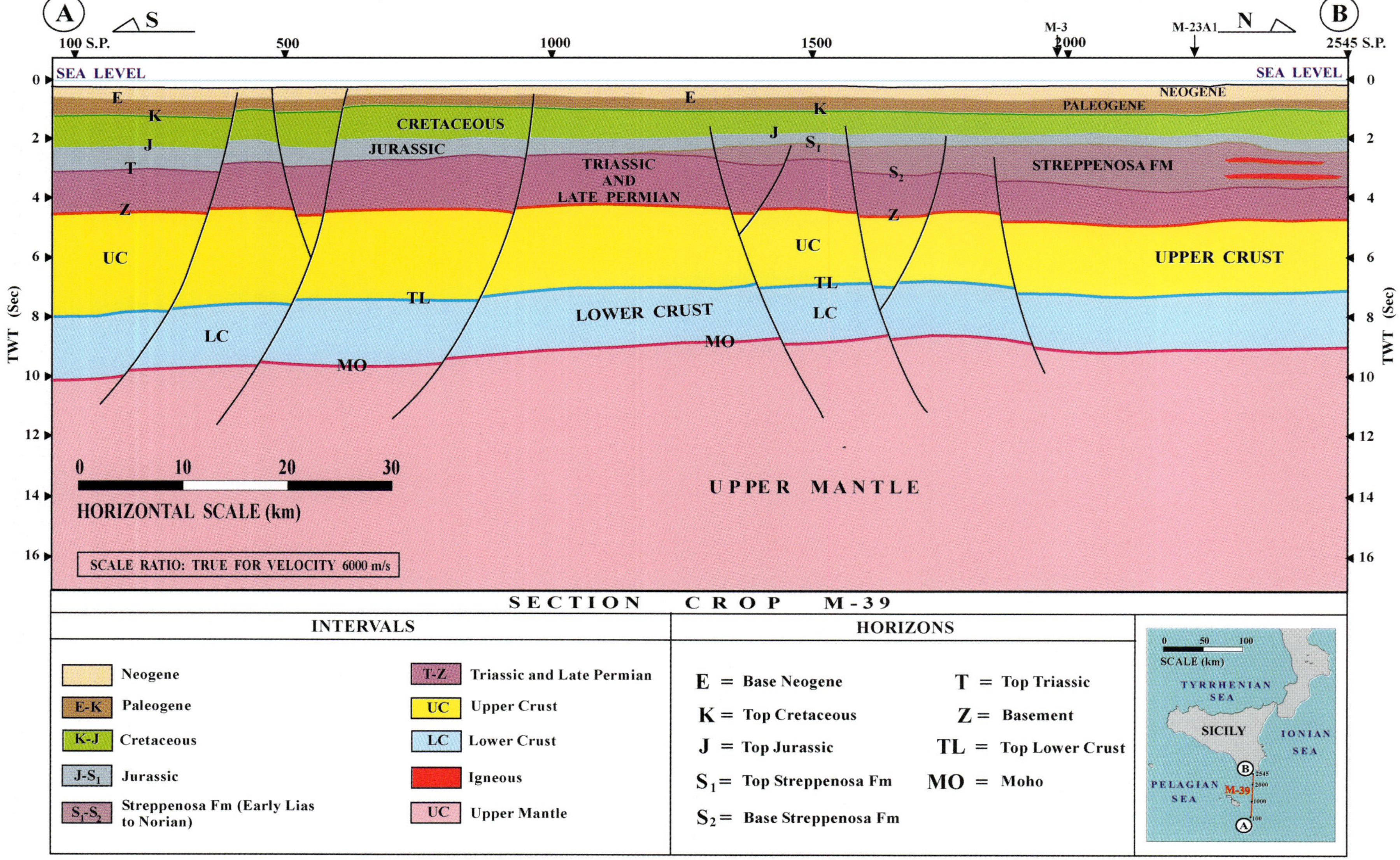

Figure 7: Interpreted crustal seismogeological section CROP M39 across the eastern edge of the Pelagian platform, to the south of the Hyblean plateau.

at different epochs. The oldest known depositional cover is that of the Jeffara-Tripolitania basin, opened by the Jeffara rift fault which determined a marked subsidence from the Permian. The presence of other old rift-related depressions in the Pelagian Sea can be assumed. It is possible that in Late Permian times most of the Pelagian Sea was under sedimentary deposition. A thick Triassic, Jurassic and Cretaceous succession characterizes the rather tabular stratigraphy of this sea, followed by variable intervals of Tertiary and Quaternary.

The regional CROP seismic data explore deep crustal upper mantle lithospheric intervals where the information essential for a correct understanding and reconstruction of the tectonic deformation and its driving dynamics is imprinted. The interpretation should be self-explanatory; therefore we concentrate our description mostly on the tectono-dynamics than on the stratigraphy. Thrust-related deformation of the Pelagian foreland at the outer zone of the Maghrebian-Sicilian Apennines chain was recently interpreted from new seismic data (Bello et al., 2000; Catalano et al., 2000). These new seismic data for Sicily have been reinterpreted and assembled with the CROP data (Finetti et al., this volume, Chapter 15).

3.1. Basic Tectono-Dynamics of the Pelagian Sea

Before describing the major crustal features reconstructed from interpretation of the CROP sections in the Pelagian Sea, it is convenient to first summarize which basic transcrustal dynamics imprinted the tectonic deformation of the area. From CROP and other geophysical data (MS seismic lines; public seismic dataset of the Italian Ministry of Industry; gravity and magnetic data) it is seen that shear tectonics dominate the deformation pattern of the Pelagian Sea in the area lying to the south of the Sicilian Apennines and to the east of the Maghrebian chain.

Large-scale seismic reconstructions of tensive right-transcurrent tectonics in the Pelagian area have been reported in the literature for over twenty years (Finetti, 1984). The new transcrustal CROP seismic data show that two different interconnected systems of strike-slip faults were and are still active throughout the region (Figure 8). The axial zone

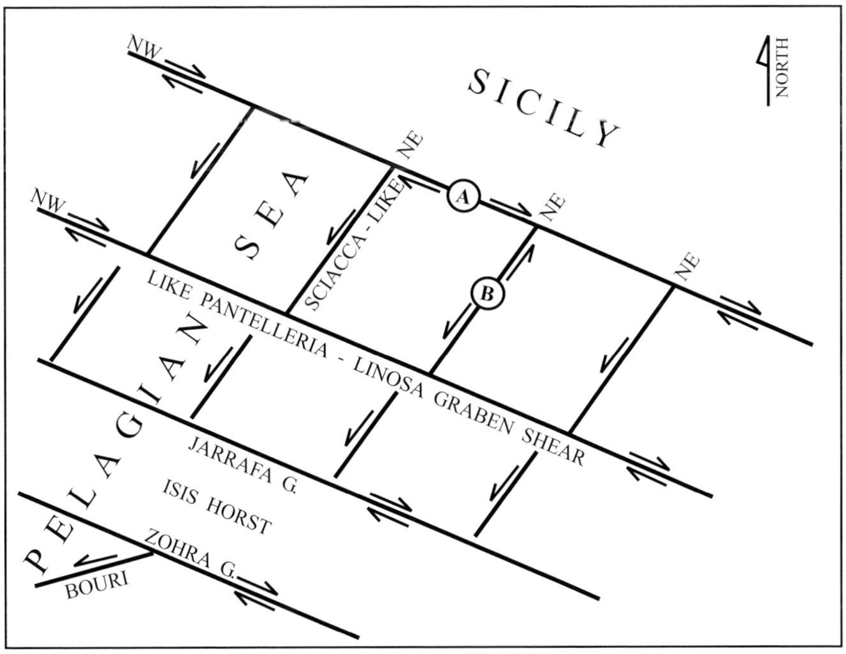

Figure 8: Scheme of the dominating shear tectonics of the Pelagian Sea: *NW–SE right-transcurrents* (type "A") and antithetic *NE–SW left-transcurrents* (type "B"). South of the *Isis horst* and *Zohra graben* (Figures 10 and 12), in the *Tripolitania basin*, the left-transcurrency-related deformation generates a series of ENE–WSW to NE–SW-directed structures where the Mesozoic evaporite sequence underwent its halokinetic reaction (i.e. *Bouri oilfield structure* and other subparallel hydrocarbon structures in the area).

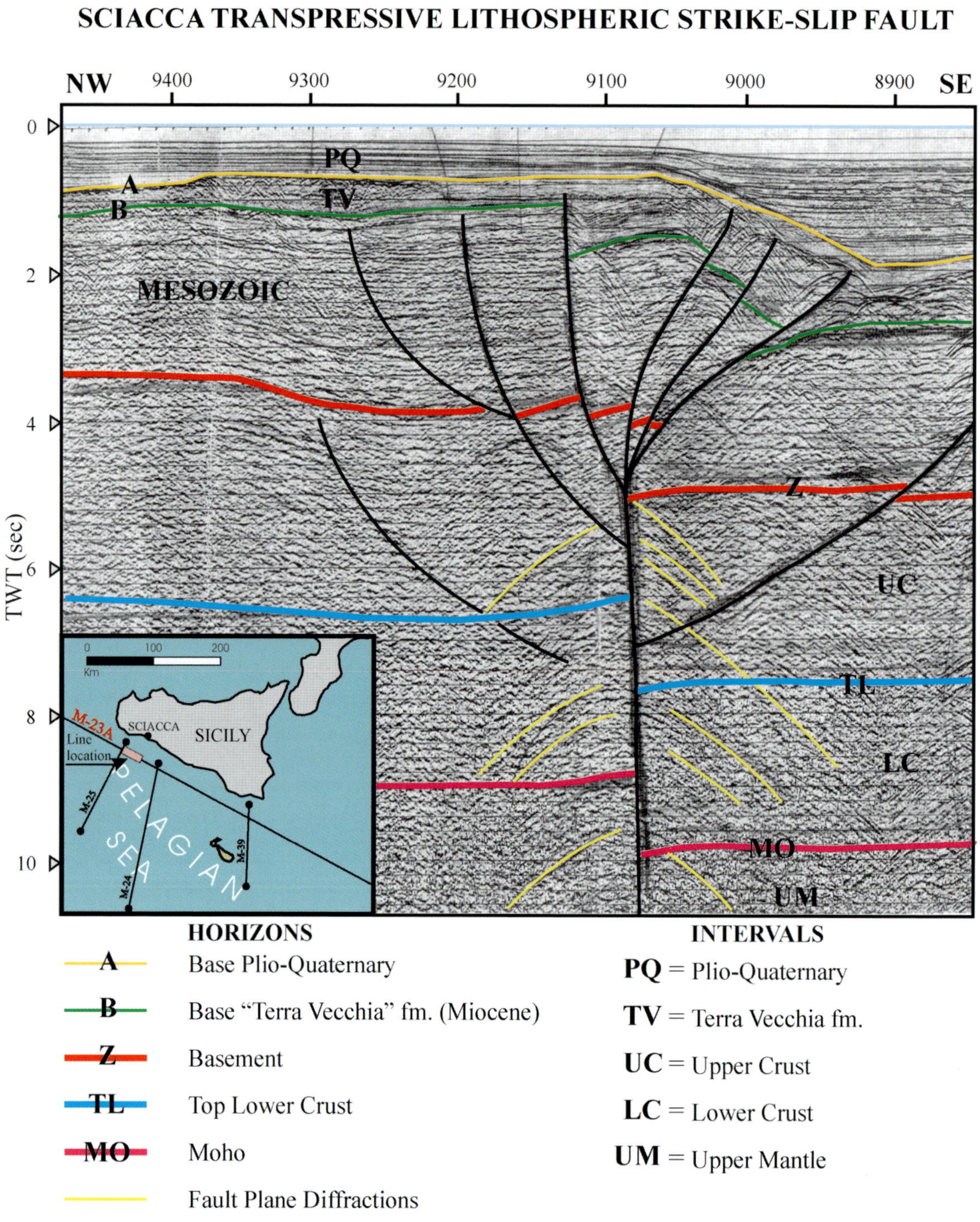

Figure 9: Seismic example of a NE–SW-directed left-strike-slip fault (*Sciacca Fault*), antithetic to the main NW–SE right-transcurrent system of the Sicily Channel rift zone in the Pelagian Sea. This figure, part of interpreted CROP section M23A, exhibits a flower feature modelled as type "B" in the previous Figure 8. Other recognized transpressive left-strike-slip faults of this type are located in Figure 12: *Mazara*; *Belice*; *Ragusa*; *Rosolini*; *P. Castellazzo* and *Pachino*. In Figure 2 and Plate 2, the crustal section of the last four mentioned faults is shown.

of the Pelagian Sea is affected by a prominent NW–SE-directed right-transcurrent fault system (fault type "A" in the model of Figure 8). A NE–SW trending system is connected as antithetic left-transcurrent faults (fault type "B" in the model of Figure 8, and Figure 9 for a real example). *Prominent graben depressions and mantle uplift are associated with the right-transcurrent system* (i.e. the *Pantelleria* and *Linosa grabens*). The *Medina graben* is the continuation of the axial shear zone of the Pantelleria and Linosa grabens (grey dashed area of Figure 10), but does not gives rise to significant mantle uplift because we are dealing with a substantial crustal shear with no

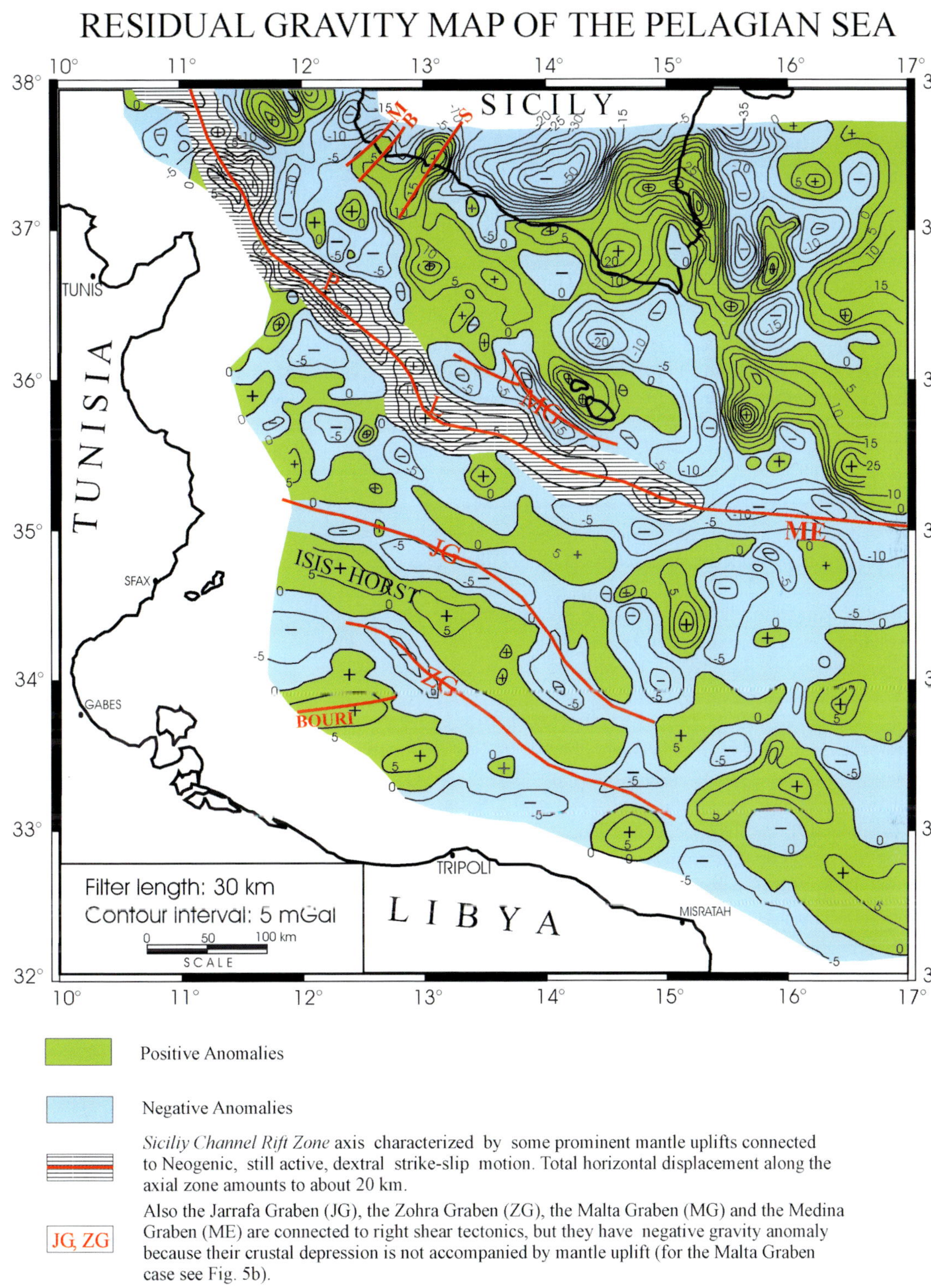

Figure 10: Residual gravity map of the Pelagian Sea showing anomalies connected to the shear tectonics. The axial grey-dashed zone indicates local prominent mantle uplift (gravity high) which in the Pantelleria (P) and Linosa (L) grabens arrives at about 17–18 km: about 10 km higher than in the Sicilian coast area. Also the left-transcurrent faults may be associated with a positive gravity anomaly and mantle uplift as are the Sciacca fault (S) and the Belice (B) and Mazara (M) faults.

vertical mantle displacement. The left-transcurrent system mostly forms transpressive features (positive flowers), not only in the Pelagian Sea (example in Figure 9), but also in Sicily. In Figure 2, the CROP section M23A1 (part) which crosses various positive flowers located across the southeastern Sicily onshore-offshore zone is exhibited (i.e. *Pachino, P. Castellazzo, Rosolini* and others).

Frequent igneous intrusion–effusion shows accompany the above sketched shear tectonics. Major impressive examples are the *Pantelleria and Linosa volcanoes*. But many other igneous bodies of various ages are found in boreholes and/or inferred from seismic and/or magnetic data (Figure 12). It seems possible that shear tectonic deformations and related igneous activities have occurred since the early Mesozoic at various epochs. It seems also possible to propose inversion of displacement polarity between a postulated pre-Eo-Alpine NW–SE left-transcurrency and the observed right-transcurrency of the same system after the Adria–Europe collision in the sector of the Alps.

3.2. Interpreted seismogeological CROP sections of the Pelagian Sea

The transcrustal seismogeological section of Figure 2 (and Plate 2), reconstructed from interpretation of **CROP section M23A1** (part) shown in Plate 1, being NW–SE directed, is obviously the best one for control of the NE–SW-trending left-transcurrent faults. On its northwest extremity (Figure 2(a)), the *Sciacca fault* is shown, of which a seismic close-up is given in Figure 9. This section in Figure 2(a), from the Sciacca fault to the Ragusa left-transcurrent fault, shows various rift faults that can be connected to the same basic shear tectonics schematized in Figure 8. In Figure 2(b), the positive flower generated by the *Rosolini, P. Castellazzo* and *Pachino left-transcurrent faults* of the Hyblean plateau and its offshore is fully evident. At the eastern edge of the Hyblean plateau, the crust thins, and on the escarpment, the Permo-Triassic rift faults of the Ionian opening are visible, more or less reactivated in the Neogene-Quaternary by the loading effects of the Calabrian wedge.

From the Gela basin to the Hyblean–Maltese escarpment, the platform cover is interbedded by an early Lias to Norian basinal deposition (euxinic succession of the *Streppenosa Fm.*, well described in Frixa et al., 2000). Several Mesozoic igneous layers, impossible to draw on the reduced scale of the section, testify to the prominent rift tectonics in the Triassic-Jurassic.

The **CROP section M25** (Figure 3; Plates 3 and 4) crosses one of the most prominent and impressive rift zones of the Pelagian Sea: the *Pantelleria rift-basin* (or *Pantelleria graben*: segment of the *Sicily Channel rift zone*; Figure 10; Finetti, 1984). The section passes several kilometers to the southeast of the Pantelleria volcano, and therefore the igneous bodies we see, even if connected to the same crustal extension process, are of smaller dimension. It is evident that crustal thinning and mantle uplift are here remarkable. In a depth converted section (water effect removed) the crustal-mantle boundary rises to 17 km, about 10 km higher than at the two extremities of the section (Figures 3 and 11). Strong positive gravity and magnetic anomalies characterize the Pantelleria graben zone (Figures 10 and 12). Rift-faults, buried beneath the flat Tertiary cover are present also to the north of the Pantelleria basin, from SP 1800 to 2800. *Time analysis of the Pantelleria rift and igneous activity indicates that the main shear-transtensional phase initiated in the early Tyrrhenian stage* (12–14 *Ma*) *and continued into historical times.* A young plutonic intrusion can be seen across a right-transcurrent fault named the *Ferdinandean fault* (Figure 3, SP 2700). This fault also crosses the igneous zone of the Graham bank.

The northern extremity of the CROP M25 section exhibits a subparallel transpressive thrust, connected to the left-transcurrent motion of the *Belice fault* (Figure 3).

The **CROP section M24** (Figures 4 and 5) crosses the Pelagian Sea from Lampedusa island to the frontal wedge of the *Sicilian Apennines*. Two main rift-graben zones are well shown: the *Linosa graben* (Figure 5(a)) and the *Malta graben* (Figure 5(b)). The Lionosa graben formed as a consequence of a prominent right-transtensive shear motion accompanied by a significant crustal thinning and mantle uplift. In the mantle culmination, the moho is at a depth of only about 18 km (Figure 11). The Linosa graben zone explored by this section lies some 4 km southeast of the main *Linosa volcano*, but other huge igneous bodies of the Linosa magmatic apparatus are shown (Figure 5(a)). On the northern flank of the graben, a nearly two seconds thick, complex intrusive-effusive magmatic body is evident, which seems to have partly digested intervals of the Mesozoic cover. We name this volcano-plutonic show *Linosa north*. Prominent gravity highs and magnetic anomalies are associated with the Linosa graben feature (Figures 10 and 12), as is also the case for the Pantelleria graben.

The *Malta graben* is formed by a negative flower generated by a right-transcurrent transtensive fault (Figure 5(b)). But, contrary to the Pantelleria and Linosa grabens, in the Malta graben zone, there is collapse of lower crust and

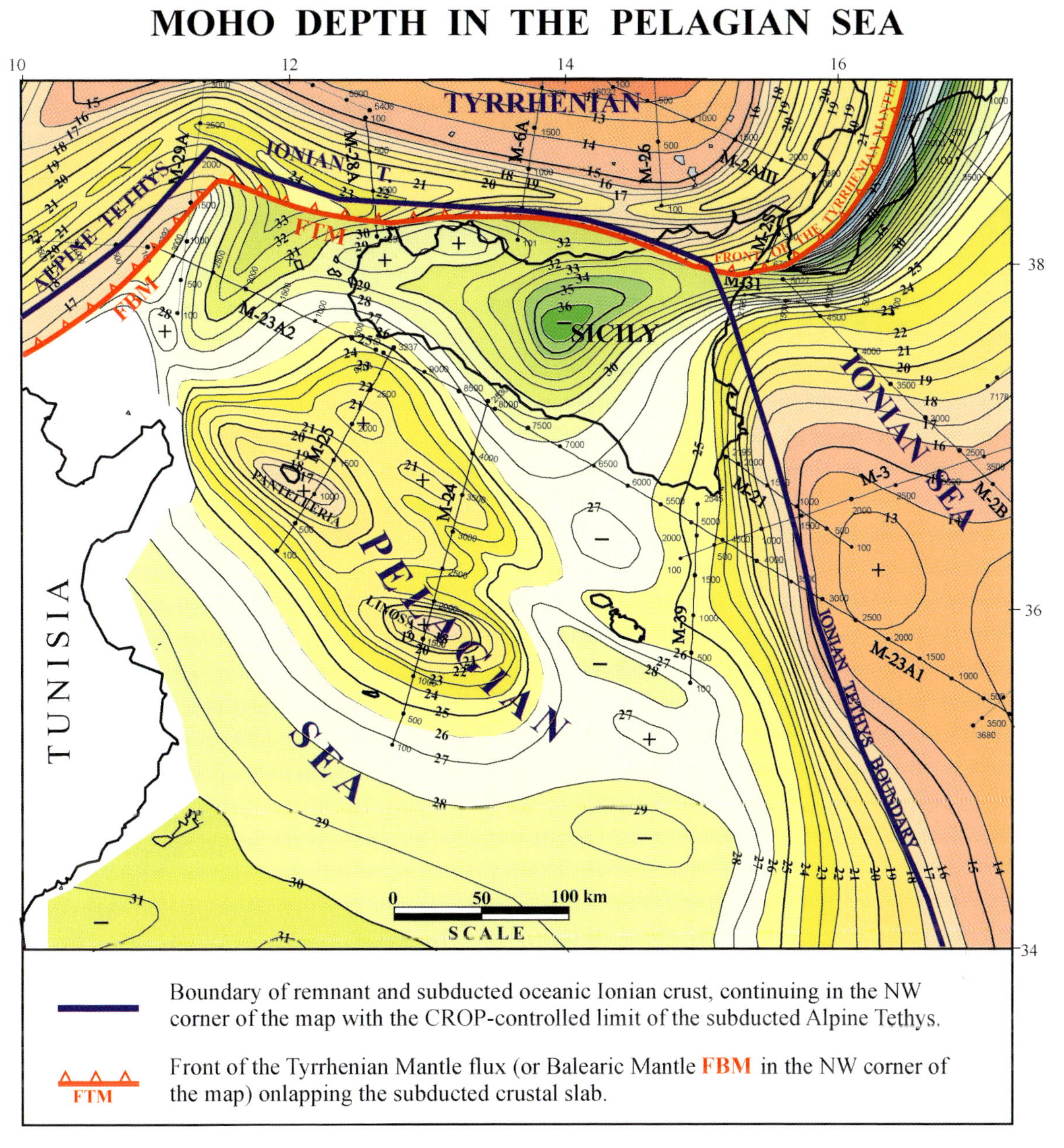

Figure 11: Depth-contour map of the moho discontinuity in the Pelagian Sea, reconstructed from new CROP seismic data. Contour interval 1 km. The mantle uplift of the Pantelleria and Linosa areas is generated by a NW–SE-trending transcrustal right-shear fault system of regional magnitude.

basement upper crust, no mantle uplift, negative gravity anomaly, and strike-slip motion predominates over the tensive deformation component.

The NW prolongation of the *Malta horst* features, on the M24 CROP section, gives rise to a rift-high delimited by two composed normal faults of which the southern one uplifts the Malta-correlated block while the northern one down-slips this block (Figure 5(b)). Analogous rifted crustal blocks to that of the NW prolongation of the *Malta high* are those of the *Madrepora high* and *Gergentum high.* To the northern extremity, the M25 CROP section shows the frontal thrust of the Sicilian Apennines wedge.

On the eastern edge of the Pelagian platform is located the **CROP section M39** (Figure 6 seismics; Figure 7 interpretation). This section exhibits a rather tabular crustal structure cut by transcrustal direct faults of relatively insignificant displacement. From SP 1300 to 1800 are present rift-faults connected to the Mesozoic stretching tectonics that generated the *Streppenosa basin* of which these faults constitute the southern flank. The gentle high from SP 600 to 1300 is the southeast prolongation of the *Malta high.* High magnetic anomalies and frequent effusive-

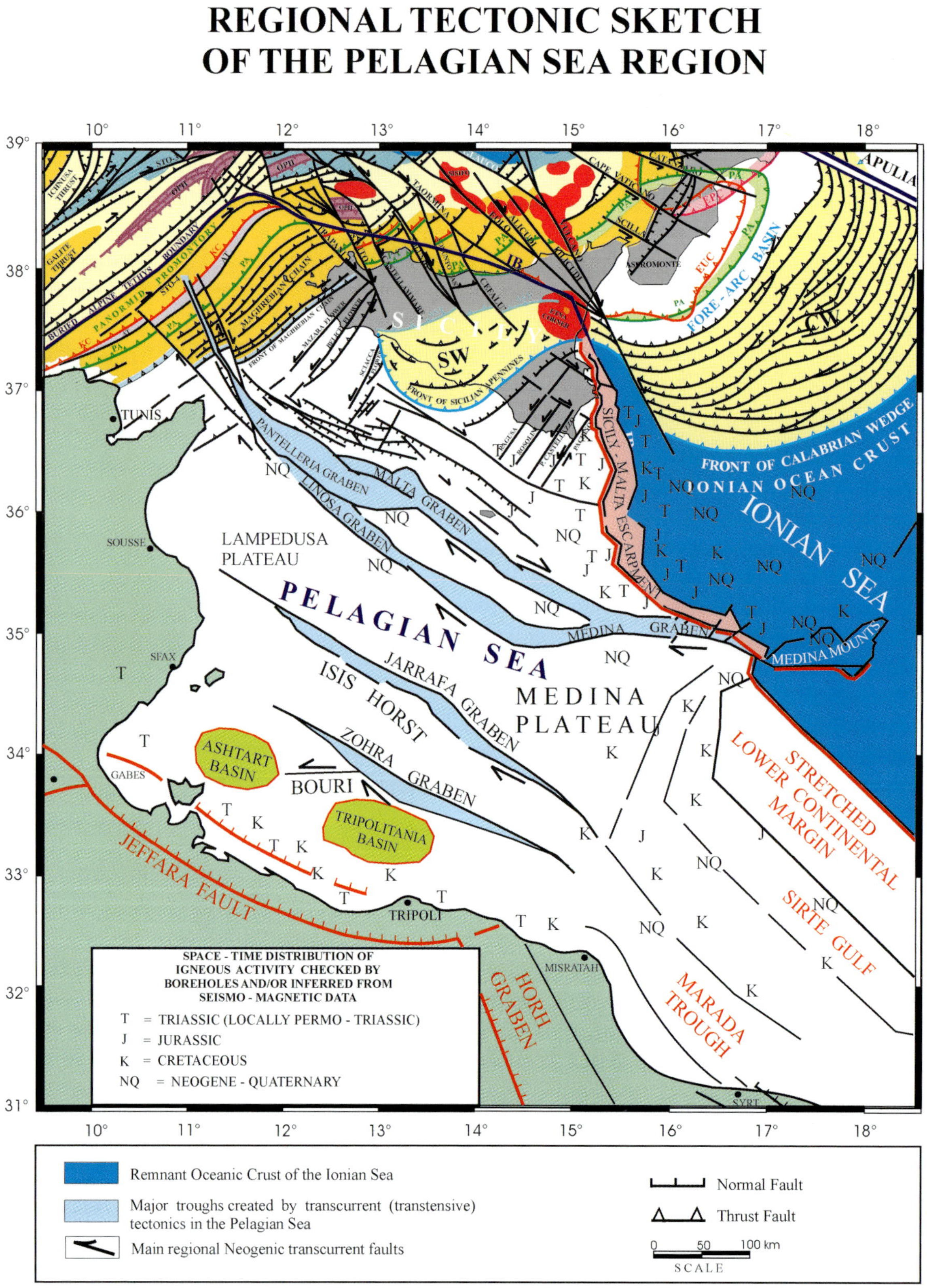

Figure 12: Regional tectonic sketch map of the Pelagian Sea. The complex tectonic reconstruction of Sicily and the Calabrian Arc is reported only to show the northern boundary of the tabular tectonics of the Pelagian Sea. For the Sicilian and Calabrian Arc legends the reader is addressed to Chapters 15 (Sicily) and 17 (Calabrian Arc) of this CROP volume.

intrusive magmatic layers (checked by boreholes or inferred from seismic and/or magnetic data) characterize the eastern border zone of the Pelagian Sea and of the Hyblean–Maltese escarpment. In Figure 12, jointly with the tectonic sketch of the major features of the Pelagian Sea and surrounding provinces, the distribution of major igneous shows, checked or inferred, is reported.

ACKNOWLEDGEMENTS

This work benefited from funding by CROP supporters (CNR, ENEL, ENI) and MIUR (COFIN 1998, 2000 and 2002 to I.R. Finetti): the authors express their gratitude. We wish to thank OGS for recording and ENI-AGIP for processing the CROP data used.

REFERENCES

Bello, M., Franchino, A., Merlini, S., 2000. Structural model of eastern Sicily. Mem. Soc. Geol. It. 55, 61–70.

Biju-Duval, B., 1980. De la Tethys aux Mers Intra-alpines Actuelles: Introduction. In: Mem. B.R.G.M., vol. 115, Coll. C5: "Géologie des chains alpines issues de la Tethys", 26° CGI, pp. 231–235.

Catalano, R., Franchino, A., Merlini, S., Sulli, A., 2000. Central western Sicily structural setting interpreted from seismic reflection profiles. Mem. Soc. Geol. It. 55, 5–16.

Finetti, I.R., 1982. Structure, stratigraphy and evolution of Central Mediterranean. Boll. Geof. Teor. Appl. 24, 247–426.

Finetti, I.R., 1984. Geophysical study of the Sicily Channel rift zone. Boll. Geof. Teor Appl. 12, 263–341.

Finetti, I.R., Del Ben, A., 1986. Geophysical study of the Tyrrhenian opening. Boll. Geof. Teor. Appl. 28, 75–156.

Finetti, I.R., Morelli, C., 1973. Geophysical exploration of the Mediterranean Sea. Boll. Geof. Teor. Appl. 15, 263–341.

Finetti, I.R., Del Ben, A., Forlin, E., Prizzon, A., this volume. Review of the main results of the pioneering MS seismic exploration programme in the deep water of the Mediterranean Sea (OGS, 1969–1980). Chapter 1.

Finetti, I.R., Lentini, F., Carbone, S., Del Ben, A., Di Stefano, A., Forlin, E., Guarnieri, P., Pipan, M., Prizzon, A., this volume. Geological outline of Sicily and lithospheric tectono-dynamics of its Tyrrhenian margin from new CROP Seismic data. Chapter 15.

Frixa, A., Bertamoni, M., Catrullo, D., Trincianti, E., Miuccio, G., 2000. Late Norian-Hettangian paleogeography in the area between wells Noto 1 and Polpo 1 (SE Sicily). Mem. Soc. Geol. It. 55, 279–284.

Winnock, E., 1981. Structure du Bloc Pelagien. In: Wezel, F.C. (Ed.), Sedimentary Basins of Mediterranean Margins. Technoprint, Bologna, pp. 445–464.

CHAPTER 27

Depth Contour Map of the Moho Discontinuity in the Central Mediterranean Region from New CROP Seismic Data

I.R. Finetti[1,*]

ABSTRACT

From interpretation of new CROP transcrustal seismic data, integrated with other seismic sections, a Moho depth contour map of the central Mediterranean region, from southern Europe to Africa, across Italy and the surrounding seas, has been reconstructed. This map, the first for the Mediterranean area based on advanced near-vertical deep seismic profiling, supplies the geoscience community with wide-scale lithospheric information, which is much better regionally controlled than previously. It also allows a reliable identification, delimiting and definition of the crustal provinces that compose the present complex geological setting of the explored region, so helping to resolve controversies about the evolving tectonics.

The obtained Moho-related data are very useful for understanding the major macro-tectono-dynamic events in which the crust-mantle interface took part (thrusting, rifting, shear-slip, extensional and compressive delamination, subduction, exhumation, etc.). In the Alps, the subduction of the southern European margin beneath the overthrusting Adria plate (and closure of the NE-Alpine Tethys) is well imaged. In the Northern Apennines, Southern Apennines, Calabrian arc, Sicilian Apennines and Maghrebian chain, the smoothed Moho tectonic setting is regionally reconstructed, together with that of the Corso-Sardinian block. Moho isobaths for the major basins (Balearic, Tyrrhenian and Ionian) and foreland areas (Po Valley–Adriatic Sea and Pelagian Sea) are reconstructed with very good reliability.

Keywords: Moho depth contour map Central Mediterranean CROP controlled

1. INTRODUCTION

Analog wide-angle seismic profiling in the Italian mainland was started in 1966 to study the crustal setting, at that time almost completely unknown. DSS geotraverses performed under European cooperation continued and proliferated subsequently in various parts of Europe, in Italy and the Mediterranean region. The adoption of digital techniques in the late eighties helped improve the quality of the WARR (Wide-Angle Reflection–Refraction) seismic results obtained. A rich literature has been produced on the WARR profiling over the last 30 years. A review of the results obtained up to the late seventies in the Alps was performed by Giese and Prodelh (1976). A complete up-to-date review of the WARR profiles in the central Mediterranean region from the Alps to Africa is presented by Cassinis et al. (this volume, Chapter 2).

In this chapter, a *new depth-contour map of the Moho discontinuity, wholly controlled for the first time by near vertical multifold profiling is presented and discussed.* This map is based on over 10,000 km of CROP seismic sections integrated with other transcrustal seismic data from the literature (ECORS, TRANSALP) (Figure 1) and seismic lines of middle-to-deep crustal penetration. In the Ionian and Tyrrhenian Seas, where a thinner crust floored by oceanic basement type occurs, the MS lines, programmed in an initial pioneering exploration of the whole Mediterranean area, exhibited the Moho discontinuity many years ago (Finetti, 1982; Finetti and Del Ben, 1986).

[1]Department of Geological, Environmental and Marine Sciences, Exploration Geophysics Group, University of Trieste, Via Weiss 1, 34127 Trieste, Italy
[*]Corresponding author. Tel.: +39 40 5582276; e-mail: finetti@units.it (I.R. Finetti)

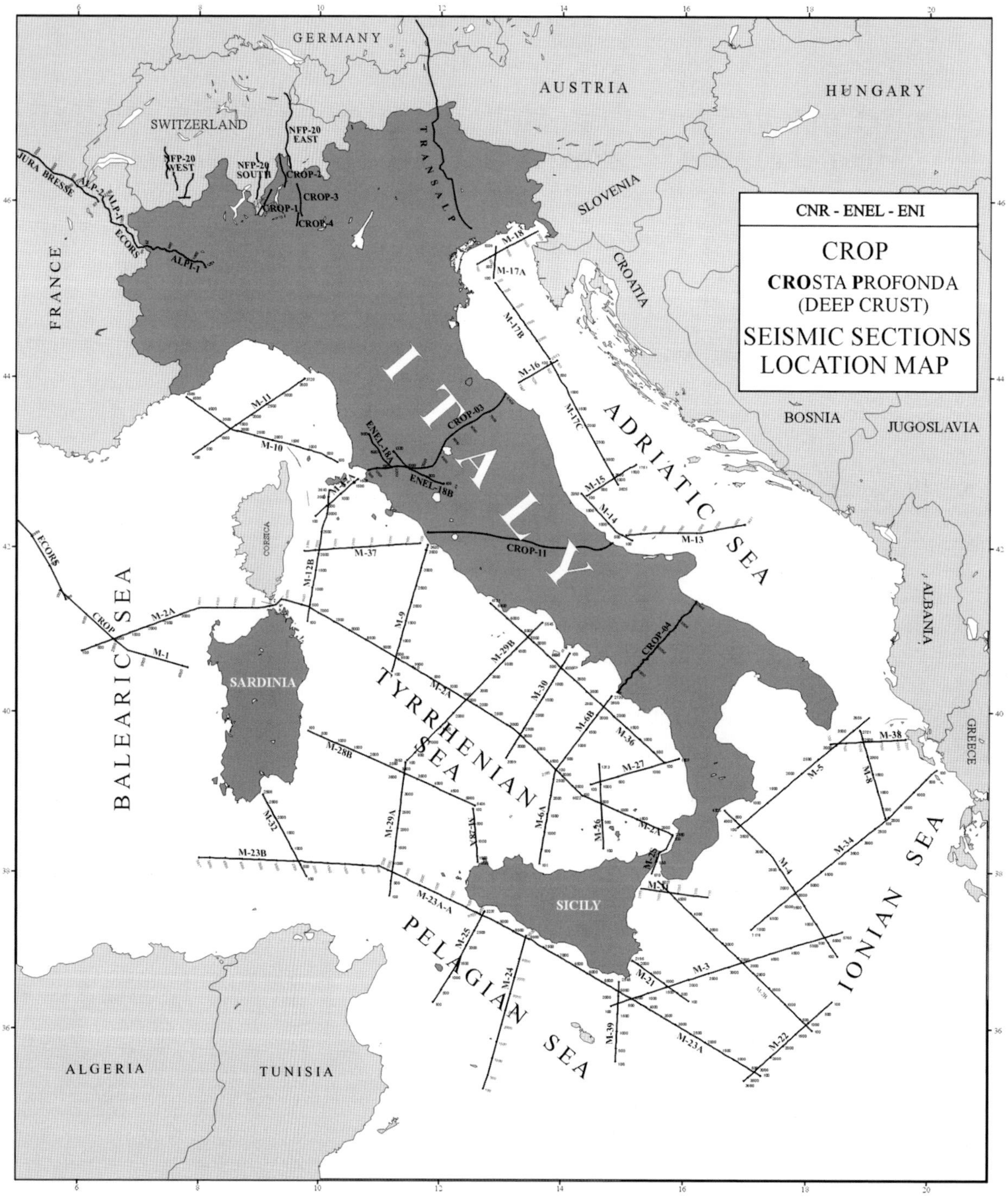

Figure 1: Location map for the CROP seismic network and other transcrustal seismic profiles whose data have been utilized for the reconstruction of the depth-contour lines in Figure 5 and the annexed plate of this chapter. In the Tyrrhenian Sea, Ionian Sea, Balearic Sea (and Southern Adriatic), various MS seismic lines (see Finetti et al. (this volume, Chapter 1)) integrated the CROP geophysical control of the crust-mantle interface.

Interpretation of the CROP sections is exhibited and extensively described in various chapters of this CROP volume (Chapters 7–10, 12, 15–21, 23 and 26). *These interpreted CROP sections constitute the basic data on which the depth-contour map of the Moho of Plate 1 and Figure 5 has been reconstructed.*

Among the CROP seismic sections, interpolating contour lines do not represent a pure geometric correlation, but are conditioned by a reasoned integration of seismic and Bouguer gravity data. Only the most significant Moho faults have been mapped, while minor faults were smoothed by local contour line gradient increase.

Table 1
Depth conversion velocities of the crustal stratigraphic intervals overlying the Moho discontinuity

Crustal intervals	P-wave velocity (m/s)
Water	1500
Plio-Quaternary	2000
Terrigenous Tertiary and various other units	2400–3600
Alpine Tehydes and Ionides	4000
Messinian Evaporite	5000
Carbonate Sequences	5500–6000
Basement Upper Crust	6000
Lower Crust and Oceanic Basement Crust	7000
Thrust Upper Mantle	8000

2. MOHO DEPTH MAP RECONSTRUCTION PROCEDURE

Interpretation of the CROP seismic sections was conducted at the working scale 1:50,000, considered adequate to pick up all significant details. This scale allows one to identify and follow reflectors and chosen horizons much better than does the reduced 1:300,000 scale of the plates in this volume relative to the attached seismic sections. In fact, at the larger scale, it is possible to discriminate low to very low signal amplitudes of primary reflections not visible on the reduced scale.

For specific comments the reader is addressed to the chapters of this CROP volume listed in the introduction. As a general comment it can be said that the fundamental 2D setting of the crust-mantle boundary reconstructed along the CROP sections integrated with other seismic data is in general of good-to-fair reliability, and its interpolated 3D surface setting of good regional validity. Moreover the mantle-type definition obtained is of major importance for a correct understanding of the complex geodynamics and related driving forces.

Only the major, most significant faults of the Moho surface have been reported in the map of Plate 1, while minor vertical displacement have been smoothed during contouring. Line contouring was done to the working scale 1:1,000,000, helped by the Bouguer gravity field for interpolation between the 2D CROP sections. In this work the Moho depth-map is presented to the reduced scale 1:2,000,000 (Figure 5, Plate 1).

Time-Depth Conversion Velocities

Based on numerous velocity functions obtained by processing and interpretative reprocessing of the CROP seismic data, the Moho reflection times of the interpreted sections were converted into depth assuming the stratigraphically simplified interval velocities reported in Table 1.

Moho depth was carefully calculated according to the interpretation of the CROP and other seismic sections on 3196 points, averaging one every about 3.2 km of interpreted section. Contouring of the obtained Moho depth dataset shows clear coherent crustal provinces, each one having particular characteristics.

3. CRUSTAL PROVINCES OF THE CENTRAL MEDITERRANEAN REGION AND THEIR GEOLOGICAL HISTORY

The new contoured depth map of the Moho discontinuity of the central Mediterranean region allows us to better understand the crustal characteristics and major tectonic features of a vast complex composit area running from Southern Europe–Alps to the Northern African plate. Where one Moho interface is superimposed on another (lithospheric thrusting), the map (Figure 5, Plate 1) reports the upper one, ignoring, for simplicity, the underthrusting (i.e. Alps, Northern and Southern Apennines, Calabrian arc, Sicilian Apennines, Maghrebian chain and Corso-Sardinian block).

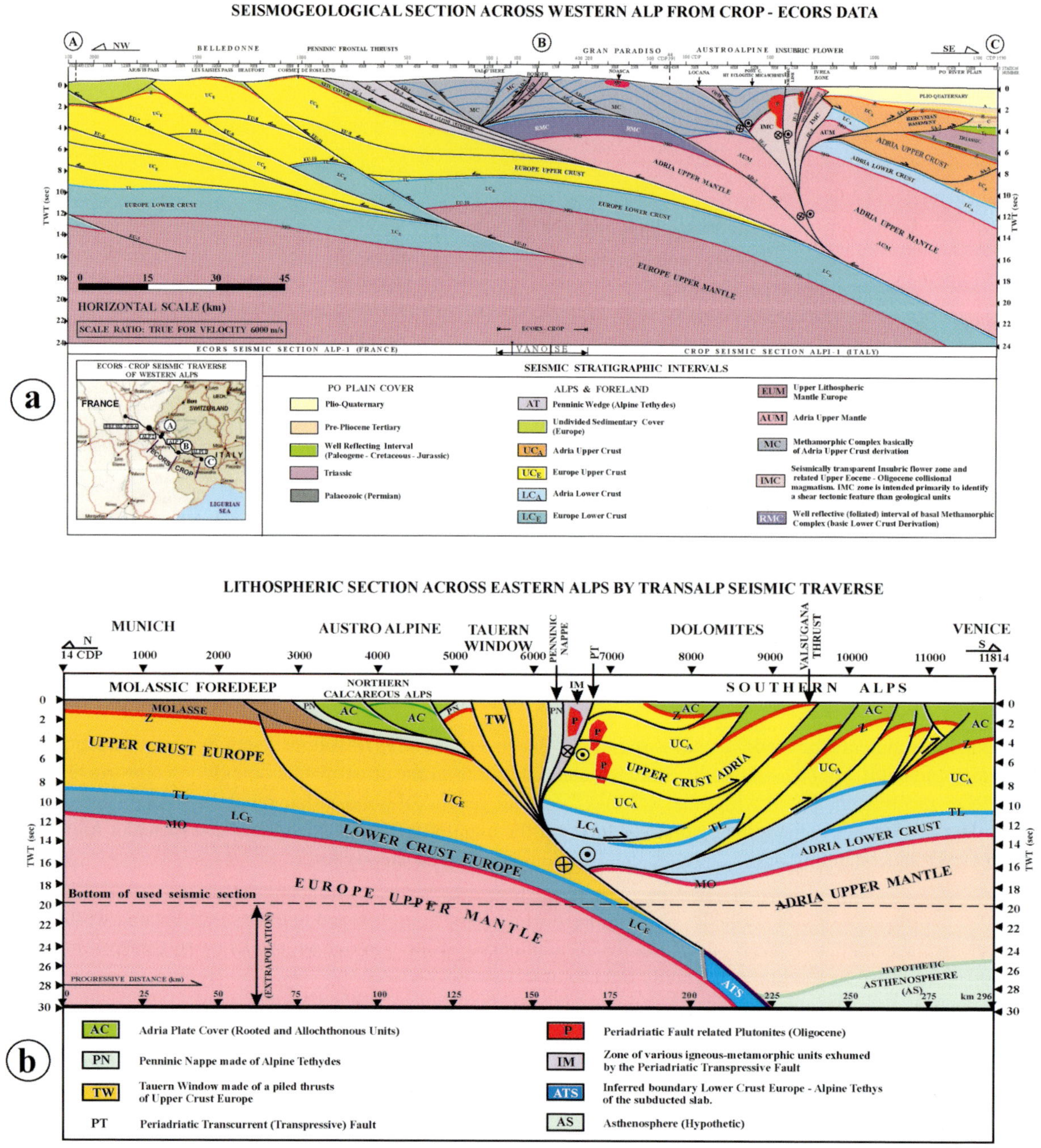

Figure 2: Reduced version of reinterpreted seismic sections crossing the Western Alps (a) and Eastern Alps (b). These sections are extensively described in Finetti (this volume, Chapter 7) but are exhibited here for a better understanding of the Moho map in the Alpine sector of the annexed Plate.

3.1. Southern Europe–Alps and Related Lithospheric Tectono-Dynamics

The subduction of the Southern European plate beneath the Adria plate (Alpine Tethys closure in the Alps sector) and formation of the Alpine chain has been investigated by transcrustal seismic profiles acquired under international cooperation: Italy–France (Western Alps), Italy–Switzerland (Central Alps) and Italy–Germany–Austria (Eastern Alps: Transalp). In the Western Alps, the Alpi-1 (CROP)–Alp-1 (ECORS) profile is continued to the northwest by the Bresse–Jura ECORS profiles as far as the Massif Central. Bresse–Jura–Alp-1–Alpi-1 and Transalp profiles cross the whole Alpine chain and large parts of the European plate (Figures 1 and 5).

The Plate 1 shows clearly the front of the Adria plate, the related mega-thrust and the subducting European plate. On the Transalp profile, the Moho of the subducting European plate, below the frontal limit of the Adria mantle thrust, has foundered to a depth of about 60 km, while the onlapping crust-mantle boundary of the overthrusting Adria plate, at the same point, is at a depth of 44 km. In the Western Alps, the Alpi-1 (CROP)–Alp-1 (ECORS) profiles show a prominent, late orogenic westward thrust-displacement of the Adria plate over the European crust. These regional dynamics took place in the *Meso-Alpine stage* (Oligocene–Lower Miocene) when the N–S component of the Adria–Europe convergency, after the *Eo-Alpine stage* (Upper Cretaceous–Eocene), encountered progressively increasing friction forces, and the contractional deformation found its easiest escape westward. The *Insubric right-transcurrent regional fault* slices this Adria plate motion to the north with huge impressive transpressional flowers features which form a large part of the Southalpine and Austroalpine tectonics. In the Ivrea zone, the Insubric flower produced the highest Adria mantle uplift (and exhumation) of the Alpine chain (Figure 2(a); Plate 1). In front of the Adria mantle thrust, along the Alpi-1–Alp-1 profile, the Moho of the subducting European plate is at a depth of about 40 km (60 km on the Transalp) and the Adria crust-mantle boundary, at the same point, is only 10–11 km deep (44 km on the Transalp). The European Moho, in the area shown in Plate 1, deepens progressively toward the Alpine chain, starting from a depth of nearly 20 km in the Central Massif zone.

On the sector of the map (Plate 1) relative to the Alpine tectono-dynamics, besides the Moho bathymetry, are plotted:

(a) The front of the Adria plate thrusting (including uprooted tectonic units of Europe as for the Tauern Window);
(b) The front of the Adria plate mantle which onlaps the subducting European plate;
(c) The Insubric shear line as crustal slicing element within the Adria plate thrust;
(d) The front of the Southern Alps;
(e) The estimated buried limit of the subducted European continental crust bordering the subducted Alpine Tethys (yellow line). In the Transalp profile, this boundary is inferred to be at a depth of about 80 km (see Chapter 7 of this volume).

The significant Moho highs of the Lessini and Euganei mounts are very likely connected to crustal shear tectonics (i.e. Schio-Vicenza transcurrent fault), combined with Southalpine deformation. This mantle high continues in the Northern Adriatic-Istrian area (CROP M17A, M17B and M18).

3.2. Balearic Back-Arc Basin, Corso-Sardinian Block, and Balearic Mantle Dynamics-Related Apenninic-Maghrebian Arc

After the Eo-Alpine stage that closed the NE-Alpine Tethys in the Alps sector, leaving unsubducted a wide SW-Alpine Tethys remnant, the Balearic geodynamic evolution (Oligocene–Lower Miocene) took place. This stage, in the author's simplified scheme for the central Mediterranean context, is considered the Meso-Alpine. Eastward-migrating lithospheric roll-back of the SW-Alpine Tethys remnant drove the Balearic stage geodynamics with: (a) Corso-Sardinian block rotation; (b) back-arc opening of the Balearic Sea; (c) closure of the Alpine Tethys remnant to the north of the 41° parallel, in the Corsica basin zone, with the *Corsica–NW Adria* (*Etrurian promontory*) *collision*; (d) closure to the south of the Alpine Tethys with the *SE-Sardinia–Panormid promontory collision*; *formation of a single great orogenic arc including the early tectonic stack units of the Northern Apennines–Southern Apennines–Calabrian arc–Sicilian Apennines–Maghrebian chain.* In the subsequent stage (Middle Miocene to Present), as briefly commented on in the paragraph 3.3, Ionian lithosphere roll-back, continuing from the Balearic stage (the Alpine and Ionian Tethys were openly connected), drove the geodynamics that formed the Southern Apennine–Calabrian arc–Sicilian Apennines (*Tyrrhenian stage*, or Neo-Alpine stage) and completed the Apennines tectonic deformation (Finetti, this volume, Chapter 34).

The above summarized basic events are helpful for understanding the various crustal-mantle elements that compose the ensemble of the complex investigated region.

Balearic Back-Arc Basin

The Moho depth tops at about 12 km and is 14–15 km around the oceanic-continental crust boundary. The Messinian halokinetic structuring creates severe noise in the pre-Messinian sequence reflections, but the oceanic basement and the Moho reflector have been identified with good reliability on all CROP sections (M10, M11, M1, M2A1 and M23B). Other MS seismic lines (see Chapter 1 of this volume) contributed to control the Moho in this basin. Rifted European and Corso-Sardinian margins of the Balearic basin have been clearly defined together with the pre-rift, syn-rift and post-rift sequences that define the basin opening time: Upper Oligocene–Lower Miocene.

Corso-Sardinian Block

The depth-contour lines for the Moho of the Corso-Sardinian block (see Plate) have been smoothed across transcrustal faults. In fact, this continental crustal block is affected by rift-faults and thrust-faults generated during the late Balearic stage collision with the Etrurian promontory to the north and with the Panormid promontory to the south. From north to south the Corso-Sardinian Moho is controlled by CROP sections M10, M11, M2A1, M1, M28B, M32 and M23B. The CROP section M2A1 was specifically programmed to cross the block transversally as a marine section with the expected double advantage of lower cost and better results compared to a land section, as initially conceived as a continuation to the French–Italian cooperative section ECORS-CROP M1. The CROP section M2A1 contributed successfully to reconstruction for the first time of a transcrustal section in the Corso-Sardinian block, fully controlled by seismic data of satisfactory quality (Finetti et al., this volume, Chapter 18). In its axial zone, the Corso-Sardinian block has a crustal thickness of about 34–35 km, clearly thicker than the Adria plate in the Tuscan Archipelago, and the Southern European plate. Going from the thicker axial zone to the Balearic Sea on one side, and to the Tyrrhenian Sea on the other, the Corso-Sardinian block, stretched and delaminated, thins progressively to less than 15 km.

Balearic Mantle-Related Apenninic-Maghrebian Arc

Driven by eastward-moving roll-back (W-dipping Alpine Tethys slab) of that Alpine Tethys which remained unsubducted after the Eo-Alpine, Adria–Europe collision in the Alps sector, a new Balearic mantle flux (Balearic stage: Oligocene–Lower Miocene) transported the Corso-Sardinian block to collide with the Etrurian and Panormid promontories, stopping rotation at the beginning of the Middle Miocene. The frontal limit of the Balearic mantle to the north faces the Corsica basin, and to the south thrusts onto the Panormid promontory-Maghrebian sector of the Alpine Tethys slab (see annexed Plate 1 and Figure 5). The Corsica basin area is a key zone for understanding the geodynamic evolution and identifying the crustal provinces. It constitutes a window on the Alpine Tethys oceanic basement, located near a culmination of a prominent asthenospheric high (Figures 3(a), 3(b) and 3(c)) called the *Etrurian Asthenospheric Bulge* (Plate 1 and Figure 5). The Alpine Tethys basement of the Corsica basin is covered by a wedge of piled flysch and Alpine Tehydes units (Figures 3(b) and 3(c)) sandwiched between the underthrusting Tethyan crust–Adria plate boundary (Plate 1, green line) and the overthrusting Corsica-Balearic mantle. At the end of the Balearic stage, a single wide, continuous orogenic arc, including the Northern Apennines, Southern Apennines, Calabrian arc, Sicilian Apennines and Maghrebian chain was active. Slab retreat continued without stop to the east of the Etrurian and Panormid promontories with subduction of the Ionian Tethys crust (Middle Miocene to Present): this was the *Tyrrhenian stage*.

3.3. Tyrrhenian Mantle-Related Tectonodynamics

The annexed Plate 1 (and Figure 5), in the Tyrrhenian area, shows the boundary of subducted Ionian slab buried beneath drifted-thrust units. The Ionian Tethys boundary (Permo-Triassic opening) is shown also on the unsubducted area between the Apulian platform and the Hyblean–Maltese escarpment (Plate 1: blue line). To the north, the Ionian ocean limit coincides with an east-west line running along the 41° parallel, named the *Augustus Line* (Figure 4(a)). The CROP seismic network data allowed a clear identification and definition of the Ionian slab for the first time, giving an invaluable contribution to progress in knowledge of this complex Mediterranean geo-province. Two continent-ocean crust boundary corner zones delimit the Alpine Tethys from the Ionian ocean paleo-basins: the *Etrurian promontory* to the north (NW Adria corner) and the *Panormid promontory* to the south (African corner). The buried boundary of the SE-Alpine Tethys, west of the Panormid promontory corner has also been identified (Plate 1). The reconstructed CROP-controlled paleogeography indicates very clearly that the Alpine Tethys of Middle Jurassic–Lower Cretaceous opening and the Ionian Tethys of Permo-Triassic opening formed a single continuous oceanic crust, roughly delimited, with respect to their different opening times and cover sequences, by a line connecting the two above-mentioned promontories. At the end of the Balearic stage (end of the Lower Miocene), driven by Ionian ocean lithospheric roll-back (an uninterrupted continuation of the Alpine Tethys lithospheric roll-back), the Tyrrhenian back-arc basin opened within eastern Sardinia, delaminating its present margin and transporting crustal fragments and other stacked allochthonous units over the subducting Ionian slab. The Etrurian promontory dissected the Northern Apennines from the Southern Apennines and the Panormid promontory dissected the Maghrebian chain from the Sicilian Apennines. The Tyrrhenian mantle front is well defined in the Southern Apennines–Calabrian Arc–Sicilian Apennines thrust belt by the CROP data (Figures 4(a), 4(b), 4(d); Plate 1).

The Moho of the Tyrrhenian Sea tops at around 10 km and deepens to some 20–25 km at thrust-front zones.

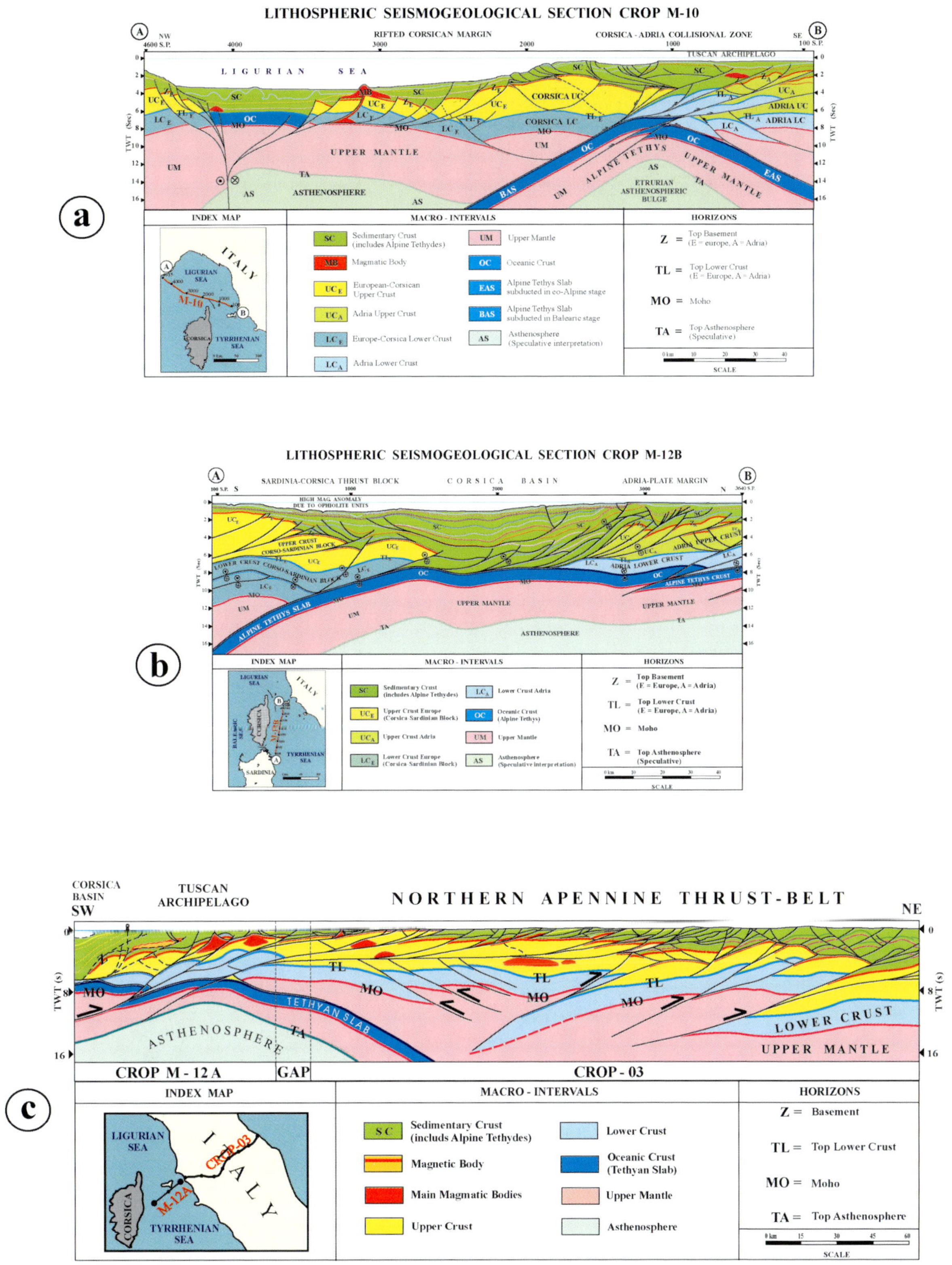

Figure 3: Reduced version of interpreted CROP seismic sections exhibited to better understand the Moho map of the annexed Plate. These sections are extensively described in Chapter 8 of this volume.

3.4. Ionian Sea

The Ionian Sea has been known to have an old oceanic crust since over twenty years (Finetti, 1982). Now the CROP seismic network allows the reconstruction of a well-controlled bathymetric setting for the Ionian Neotethys remnant (Plate 1). In Chapters 17, 19 and 21 of this volume, the Ionian ocean crust is seismically well-imaged

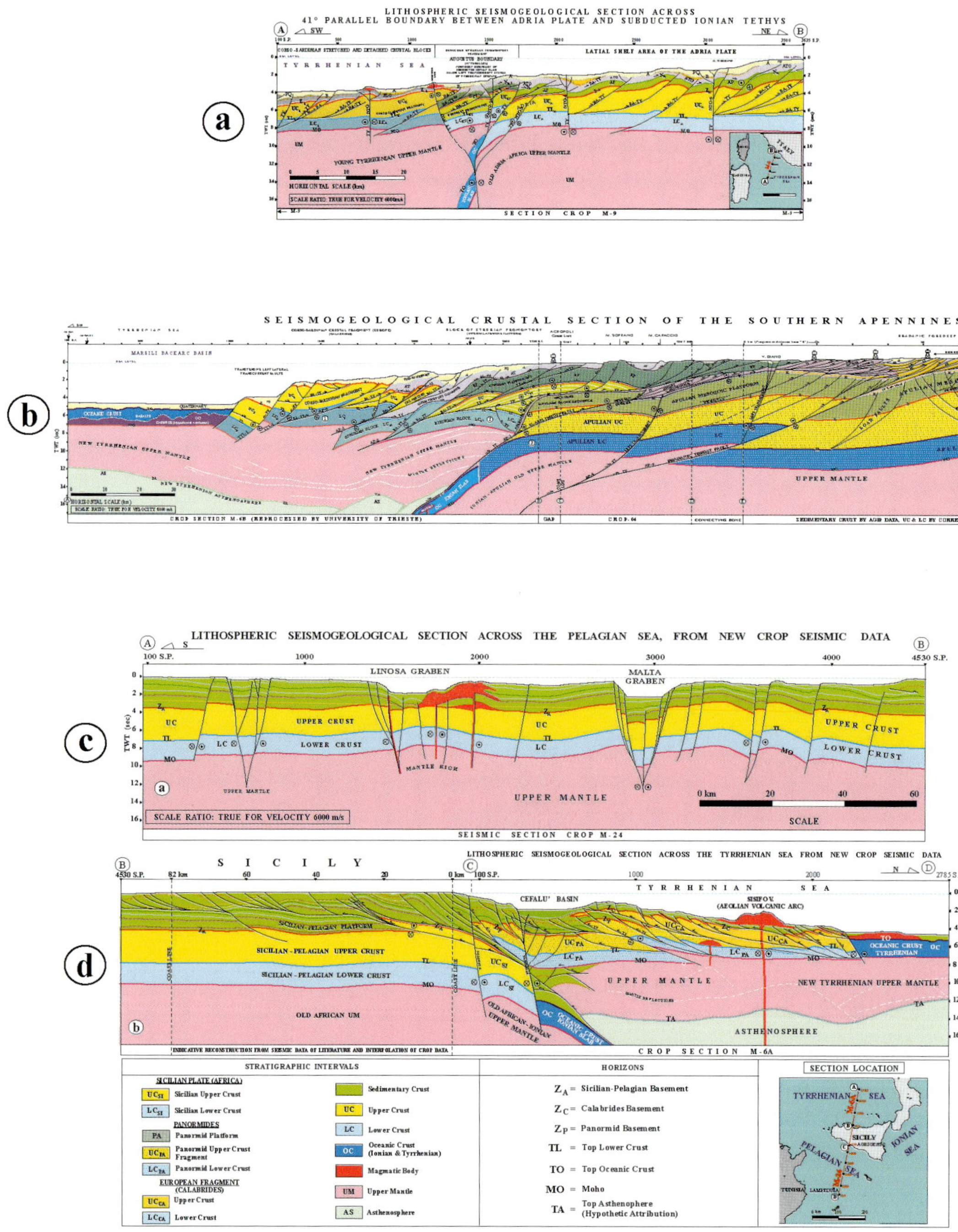

Figure 4: Reduced versions of interpreted CROP seismic sections extensively described in other chapters of this volume ("a" in Chapter 21; "b" in Chapter 12; "c" and "d" in Chapter 15). These sections help understand the annexed Moho map, fully controlled by new transcrustal near-vertical seismic reflection data.

and described also in its subducted zones (upper slab). Beneath the Calabrian arc, the Ionian slab has been clearly imaged by careful reprocessing to a depth of over 150 km (Finetti, this volume, Chapter 17).

In the annexed map (Plate 1) the Ionian Sea Moho tops at about 11 km, over the strong Bouguer anomaly zone (Marconi Seamount area). Where no tectonic-wedge load effects occur, the crustal setting is substantially flat. At the two undeformed ocean basin boundaries (Southern Apulian and Hyblean-Maltese continental crust boundaries) the Ionian Moho is at a depth of 14–16 km. Beneath the Calabrian and Hellenic arcs, it deepens progressively with the wedge thickness (subduction and load effect). In the frontal zone of the Calabrian arc, where the Tyrrhenian mantle onlaps the subducting Ionian slab, at a calculated depth of 45–50 km, the Moho is clearly imaged.

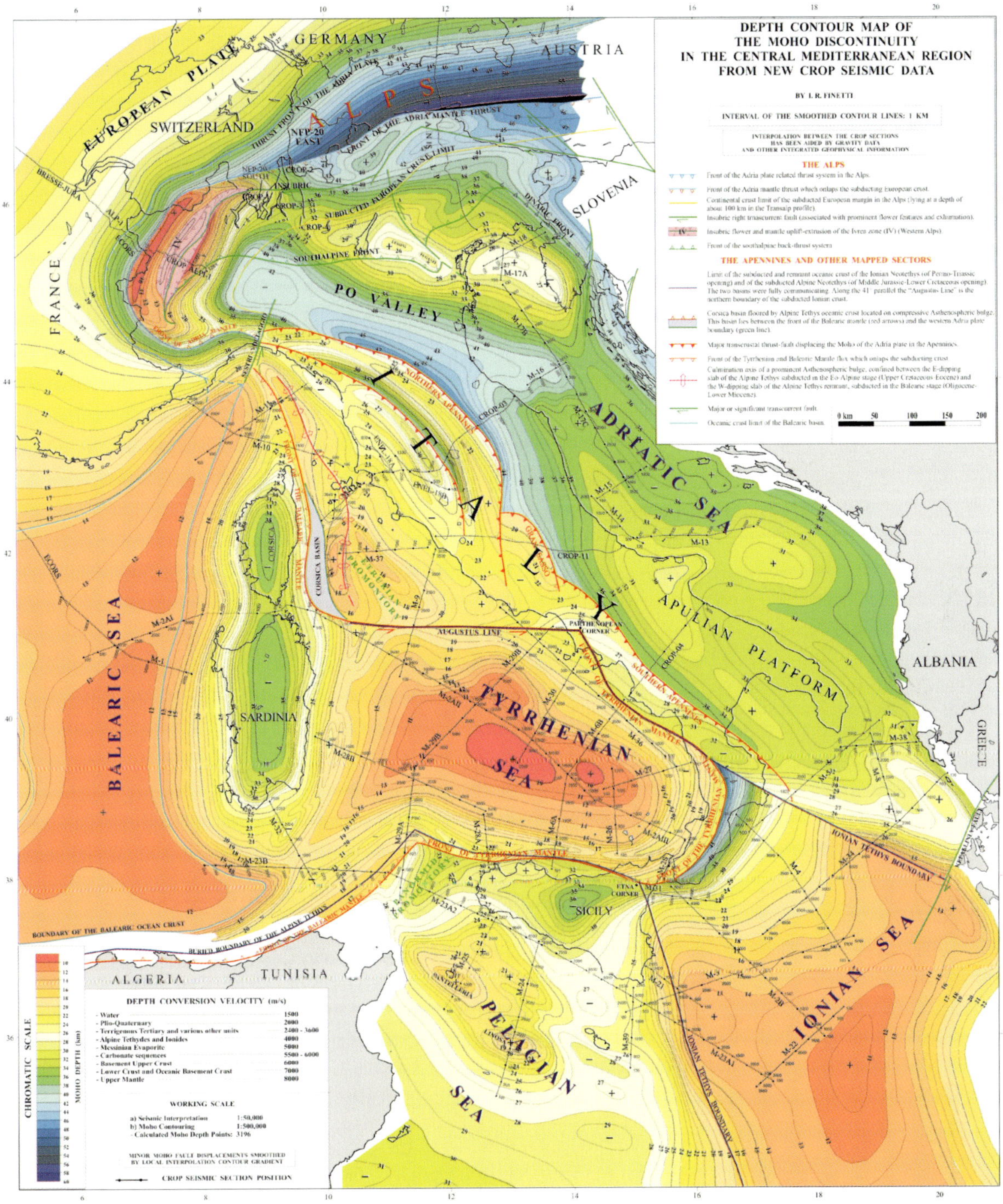

Figure 5: Reduced version of the 1:2,000,000 scale Moho depth-contour map of the annexed plate.

3.5. Padan-Adriatic Foreland

The Moho in the Po valley and Adriatic foreland area is sufficiently self-explanatory (Plate 1, Figure 5) to not need much comment. In the Po valley, the mantle uprises on the Lessini-Euganei mounts (26 km) and on the area lying between the Venice lagune and Istrian peninsula (27 km). The latter of these two mantle high areas is controlled by CROP sections M17A, M17B and M18. The Bouguer gravity field (see Bernardelli et al. (this volume, Chapter 3)) correlates very well with the depth contour lines of these highs. The Lessini-Euganei mantle uplift seems to be connected with transcrustal shear tectonics (i.e. Schio-Vicenza transcurrent fault). To the south of these mantle uplifts, the Moho discontinuity of the Po valley deepens progressively beneath the Northern Apennines thrust belt, to a depth of 47 km beneath the front of the externalmost Moho thrust of the Northern Apennines (Plate 1). Two main Moho thrusts affect the Adria plate in the Northern Apennines thrust belt (Figure 3c, Plate 1). The Adriatic–Apulian area shows a Moho that deepens beneath the Gran Sasso–Southern Apennines thrust belt and beneath the Dinarides. In the Gargano and East-Gargano offshore zone, there is an east-west trending Moho high (30–31 km) which is very likely generated by transcrustal shear tectonics.

3.6. Pelagian Sea

In central Sicily, the Moho forms a marked low (36 km) filled by a thick thrust pile of the Sicilian Apennines wedge. To the south of the Sicilian coast, in the Pelagian Sea, the Moho is characterized by prominent mantle uplifts (*Pantelleria* and *Linosa* highs) which are clearly connected to the NW–SE-directed right-transcurrent tectonics of the Sicily channel rift zone (Finetti, 1984). These mantle highs top at a depth of only 17–18 km, about 10 km higher than in the surrounding areas. The eastern edge of the Pelagian Sea delimits a sharp passage from the African plate continental crust (20–25 km) to the Ionian oceanic crust (14–16 km).

ACKNOWLEDGEMENTS

This work was supported by MIUR, COFIN 1998, 2000 and 2002 to I.R. Finetti, University of Trieste. Particular thanks are due to Dr. E. Forlin for his much appreciated contribution to the time-depth conversion work on Moho reflection data. Dr. A. Prizzon is also acknowledged for the graphical collaboration granted.

REFERENCES

Bernardelli, P., Cavalli, C., Longoni, R., Giori, I., this volume. Gravity and magnetic fields of the Central Mediterranean region. Chapter 3.

Cassinis, R., Scarascia, S., Lozej, A., this volume. Review of seismic wide-angle reflection–refraction (WARR) results in the Italian region (1956–1987). Chapter 2.

Finetti, I.R., 1982. Structure, stratigraphy and evolution of Central Mediterranean. Boll. Geof. Teor. Appl. 24, 247–426.

Finetti, I.R., 1984. Geophysical study of the Sicily Channel rift zone. Boll. Geof. Teor Appl. 12, 263–341.

Finetti, I.R., this volume. Crustal tectono-stratigraphic sections across the Western and Eastern Alps from ECORS-CROP and Transalp seismic data. Chapter 7.

Finetti, I.R., this volume. The Calabrian arc and subducting Ionian slab from new CROP seismic data. Chapter 17.

Finetti, I.R., this volume. Geodynamic evolution of the Mediterranean Region, from the Permo-Triassic Ionian opening to the Present, constrained by new lithospheric CROP seismic data. Chapter 34.

Finetti, I.R., Del Ben, A., 1986. Geophysical study of the Tyrrhenian opening. Boll. Geof. Teor. Appl. 28, 75–156.

Finetti, I.R., Del Ben, A., Forlin, E., Prizzon, A., this volume. Review of the main results of pioneering MS seismic exploration programme in the deep water of the Mediterranean Sea (OGS, 1969–1980). Chapter 1.

Finetti, I.R., Del Ben, A., Fais, S., Forlin, E., Klingelé, E.E., Lecca, L., Pipan, M., Prizzon, A., this volume. Crustal tectono-stratigraphic setting and geodynamics of the Corso-Sardinian block from new CROP seismic data. Chapter 18.

Giese, P., Prodelh, C., 1976. Main features of crustal structures of the Alps. In: Giese, P., Prodehl, C., Stein, A. (Eds.), Explosion Seismology in Central Europe. Deutsch. Geophys. Gesell. Springer, pp. 347–375.

CROP PROJECT: Deep Seismic Exploration of the Central Mediterranean and Italy
Edited by I.R. Finetti

CHAPTER 28

Triassic Magmatism and Jurassic Ophiolites at the Margins of the Adria Plate

L. Beccaluva[1], M. Coltorti[1], E. Saccani[1], F. Siena[1], O. Zeda[2]

ABSTRACT

Triassic magmatic rocks, spatially associated with Jurassic ophiolites, are distributed along the Apennine–Alpine–Dinaride–Hellenide orogenic belts around the Adria continental plate. They can be related to a rifting phase which was a precursor to the Jurassic oceanization, and show various affinities. Alkaline to transitional anorogenic magmatism characterizes the western margin of the Adria plate, from Calabria to Tuscany, while rocks of the calc-alkaline/shoshonite orogenic series are widespread at the northern and eastern margins, from the Alps to Dinarides and Hellenides. The apparent discrepancy between the "orogenic" character of the latter and the general "anorogenic" setting of the former can be explained if mantle sources at the northern and eastern Adria margins inherited subduction-related geochemical components from the Hercynian orogenic cycle.

The subsequent Jurassic oceanization produced multiple oceanic basins: the Western Tethyan (Ligure–Piemontese) basin at the western margin, and the Serbian, Mirdita–Pindos basins of the Subpelagonian Zone at the eastern Adria margin. The western ophiolites of the Alpine–Apennine orogenic belt display exclusively MORB affinities, and structural features which indicate discontinuous oceanic crust generation along a "slow-spreading" system characterized by passive lithospheric extension and tectonic denudation of large sectors of subcontinental mantle peridotites. In contrast, the Subpelagonian Zone ophiolitic complexes are characterized by the juxtaposition of two subparallel belts, of MORB-type to the west, and of suprasubduction-type to the east, implying the development of intra-oceanic subduction processes within a pristine oceanic basin.

The hypothesized original location of this basin between the Adria and Pelagonian continental blocks is to be preferred over a possible location in the Vardar Almopias oceanic basin.

Keywords: Triassic magmatism, Jurassic Ophiolites, Adria Plate

1. INTRODUCTION

Triassic magmatic episodes were widespread during the rifting processes and continental stretching related to first-order sinistral transcurrent movements between the European and African Plates (Abbate et al., 1986). The main magmatic occurrences, mostly intercalated in Triassic sedimentary sequences, are documented at the borders of the Adria plate (Figure 1): in Sicily, Calabria, Sardinia, Tuscany and Liguria (to the west); in the Dolomitic area, southern Alps (to the north); and in the Dinarides and Hellenides orogenic belts (to the east). The subsequent Jurassic oceanization led to the generation of multiple oceanic basins on the eastern, western and northern sides of the Adria Plate, ultimately resulting in ophiolitic relics cropping out in the orogenic Mediterranean chains.

This chapter presents new petrological results on the Triassic and Jurassic magmatic rocks, and a review of the available data in the literature, with the aim of defining the affinities and original tectonic settings of the various magmatic associations, and providing a contribution to the paleogeographic reconstruction and geodynamics of the Adria plate during the Mesozoic.

[1]Dipartimento di Scienze della Terra, Università di Ferrara, Via Saragat, 1, 44100 Ferrara, Italy
[2]Dipartimento di Scienze della Terra, Università di Parma, Parco Area delle Scienze, 43100 Parma, Italy

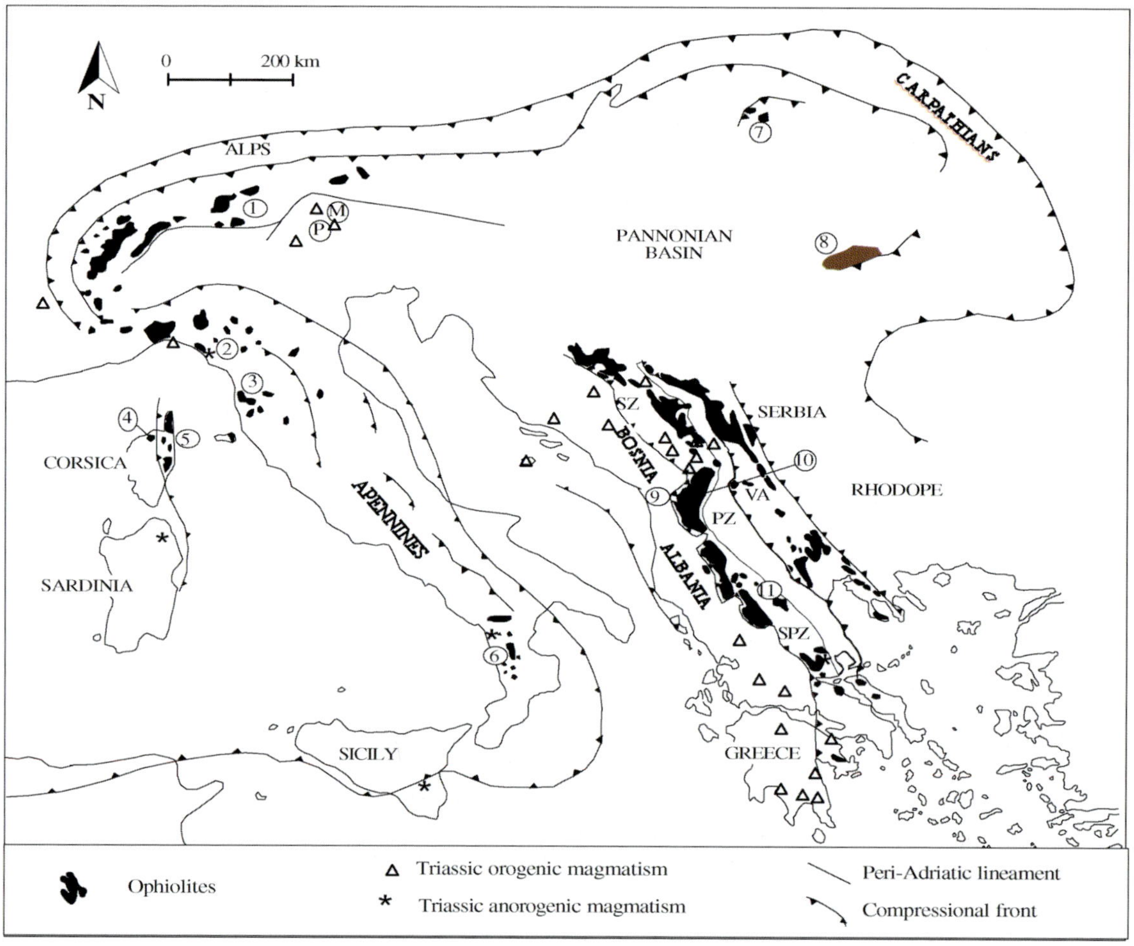

Figure 1: Tectonic sketch map of the Alpine–Apennine and Dinarides–Hellenides orogenic system, showing the main Triassic magmatic suites and the Jurassic ophiolitic complexes. Abbreviations: SZ = Serbian Zone; SPZ = Sub-Pelagonian Zone; PZ = Pelagonian Zone; M = Monzoni Complex; P = Predazzo Complex; 1 = Engadine; 2 = Eastern Liguria; 3 = Tuscany; 4 = Balagne; 5 = Alpine Corsica; 6 = Calabria; 7 = Bukk Mts.; 8 = South Apuseni Mts.; 9 = Western Mirdita; 10 = Eastern Mirdita; 11 = Pindos–Vourinos.

2. THE RIFTING STAGE

Permo-Triassic Precursors

Tectono-magmatic precursors to the rifting processes have been reported by several authors, who propose a gradual, asymmetric, passive extension of the continental lithosphere in the Europe–Africa domain (Lemoine et al., 1987; Piccardo et al., 1994). Permian intrusions such as the Mt. Cervino and Mt. Collon gabbros are the products of extension-related mantle partial melting and thinning of the continental lithosphere displaced along a master detachment fault at the northwestern margin of the future Adria plate (Dal Piaz, 1993). This was probably related to the activation of a brittle/ductile intra-lithospheric shear zone and thermal perturbation connected with the intrusion of mantle-derived basic magma into the lower crust, and acidic anatectic magmatism in the upper crust. A likely example was the Early Permian Ivrea mafic-ultramafic intrusion which can be connected to the synchronous extension of the lower crust, followed by a progressive migration of the magmatism toward higher crustal levels (Sinigoi et al., 1994; Mayer and Sinigoi, 1997).

Passive mechanisms for the extension of the subcontinental lithospheric mantle have also been proposed for the onset of the Ligure–Piemontese oceanic basin (Beccaluva et al., 1984a), associated with precursor Permian magmatic episodes (Rampone and Piccardo, 2000).

In conclusion, a general tectono-magmatic scheme can be outlined for the intervening Europe–Africa (Adria) boundary, characterized by the asymmetric passive rifting of continental lithosphere from the Early Permian on-

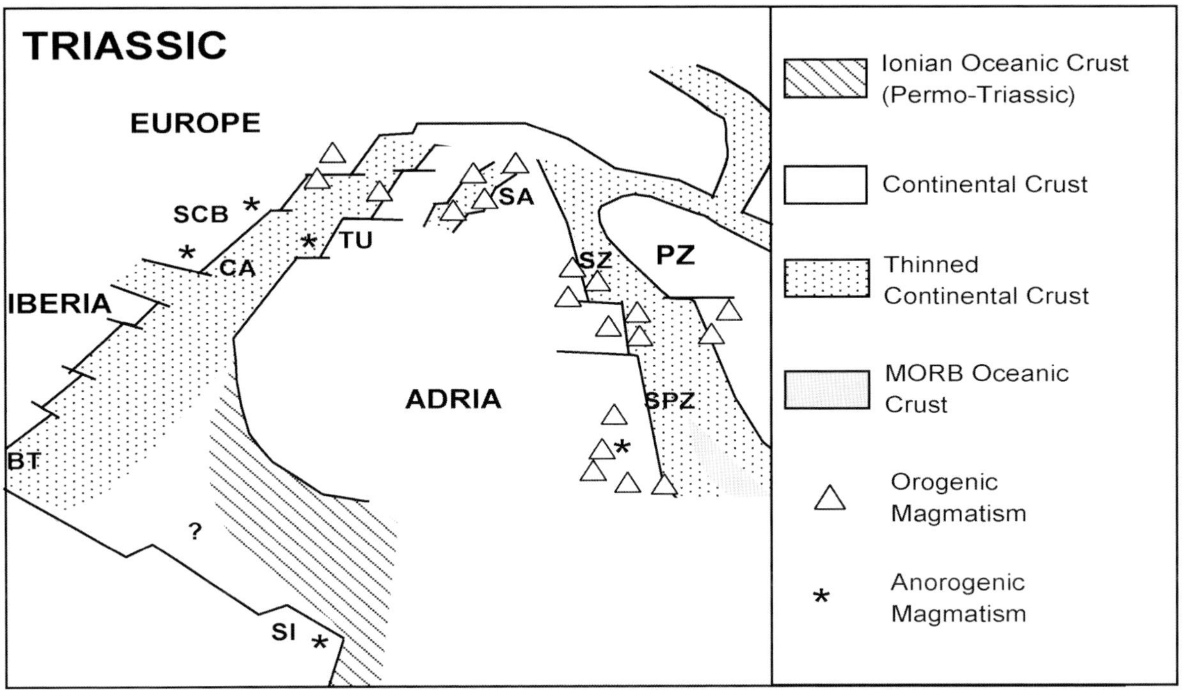

Figure 2: Tentative sketch of the Triassic rift systems in the Adria domain, showing the main orogenic and anorogenic magmatic events (not to scale). Abbreviations: SI = Sicily; Ca = Calabria; SCB = Sardinia–Corsica Block; TU = Tuscany; SA = Southern Alps; SZ = Serbian Zone; SPZ = Sub-Pelagonian Zone; PZ = Pelagonian Zone. The Permo-Triassic Ionian oceanic crust is inferred from Finetti et al. (1996) and Finetti and Del Ben (2000).

ward, with shear zones, passive asthenospheric upwelling and thermal perturbation, and generation of basic magmas in the subcontinental mantle and their intrusion into the overlying crust.

Middle-Late Triassic Magmatism

The mature stage of continental rifting is marked by widespread deposition of shallow-water carbonates and the onset of narrow seaways in the Middle-Late Triassic (Scandone, 1975). The main tectonic phase took place during the Middle Triassic, producing a complex system of extensional lithospheric faults, sinking blocks, and marked vertical movements which ultimately lead to the formation of the Adria plate (Figure 2).

The main magmatic occurrences, mostly intercalated in Triassic sedimentary sequences, are documented at the margins of the Adria plate as follows (Figure 2): in Sicily, Calabria, Sardinia, Tuscany and Liguria, to the west; in the Dolomitic area (southern Alps), to the north; and in the Dinarides and Hellenides orogenic belts, to the east. Different magmatic affinities characterize the western and north-eastern sectors: basaltic episodes of transitional to alkaline anorogenic affinity in the west (from Sicily to Tuscany); calc-alkaline to shoshonitic basic to acid rocks.

Anorogenic magmatism. Anorogenic magmatism has been studied in several localities.

- The south-westernmost anorogenic magmatic occurrence lies in the Iblean foreland of southeastern Sicily, where alkali basalts and basanites, intercalated in Late Triassic dolomites, have been recovered during drilling (Cristofolini, 1966; Longaretti and Rocchi, 1992).
- Metabasalts of transitional affinity have been reported in Calabria, interlayered with the Middle Triassic calc-schists of the San Donato Unit (Macciotta et al., 1986).
- On the paleo-European margin, in Sardinia, two sets of coeval basaltic dikes cut the Paleozoic crystalline basement: one with transitional, the other with alkaline affinity (Traversa and Vaccaro, 1992). These are dated at 230 ± 10 Ma by Ar/Ar, K/Ar and Rb/Sr methods on mineral separates (Vaccaro et al., 1991).
- Metabasaltic lenses of transitional affinity have been reported within the Middle Triassic sedimentary sequences of Monte Brugiana and Punta Bianca in Tuscany (Ricci and Serri, 1975).

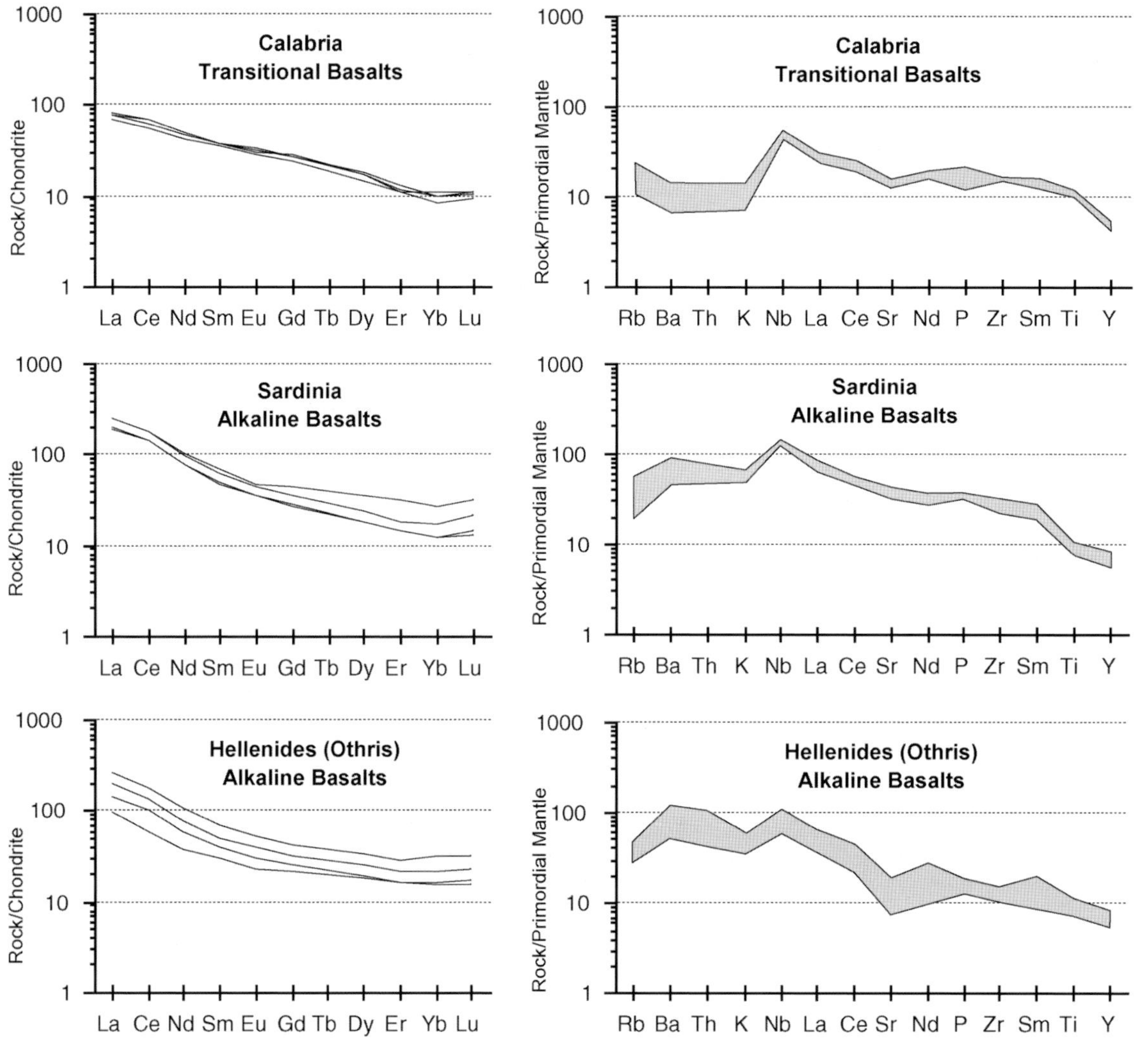

Figure 3: Chondrite- and primordial mantle-normalized REE and incompatible element patterns of Triassic anorogenic rocks from Calabria (Macciotta et al., 1986), Sardinia (Traversa and Vaccaro, 1992), and Hellenides (this work). Normalizing factors are from Sun and McDonough (1989).

- The only significant episode of anorogenic magmatism not located in the western Adria margin is represented by the Othris alkali basalts of the Hellenides (Middle Triassic age: Pe-Piper, 1982).

As shown by the primordial mantle-normalized incompatible element distributions in Figure 3, all the above basaltic episodes share geochemical affinities with anorogenic intra-plate magmatism, i.e. constantly high contents of high field strength elements (HFSE) such as Zr, Ti, Y, Nb, and P, associated with a marked positive Nb anomaly. These basalts show variable enrichments in the low field strength elements (LFSE) such as Rb, Ba, Th and K, clearly related to the degree of alkalinity. This is also shown by the parallel light rare earth element (LREE) enrichment in the chondrite-normalized REE patterns (Figure 3), with La/Yb_N ranging from ca. 7 to 17, characteristic of transitional to alkaline basalts, respectively.

The available isotopic data for $t = 230$ Ma indicate depleted mantle sources with respect to CHUR for the Sardinia alkali and transitional basalts (Vaccaro et al., 1991). This apparent geochemical decoupling between the overall depleted isotopic compositions of magmas and their incompatible element enrichment, which is particularly marked in alkali basalts, suggests a relatively recent elemental enrichment of the mantle sources, which possibly occurred during the Permo-Triassic. This feature is commonly observed in recent anorogenic volcanic associations and related mantle xenoliths (Siena et al., 1991; Beccaluva et al., 1998a).

Orogenic magmatism. Middle-Late Triassic magmatism with orogenic affinities is found along an arcuate belt on the northern and eastern Adria margins extending from Liguria, through the Southern Alps, to the Dinarides and Hellenides (Figure 1).

In Liguria, volcanoclastites of calc-alkaline andesitic compositions are intercalated in the Middle-Triassic carbonate sedimentary sequence of the Voltri Group (Cortesogno et al., 1982).

Intensive magmatic activity developed during the Middle Triassic in the Dolomites area with effusive and volcanoclastic products, as well as the intrusive complexes of Predazzo and Monzoni, which are dated at 238–232 and 230 ± 8 Ma, respectively (Lucchini et al., 1982; Petersen et al., 1980; Laurenzi et al., 1994; Visonà, 1997; Bonadiman et al., 1994, and references therein). These magmatic products mainly belong to the shoshonitic series, ranging in composition from basalts to latites with rare trachytes in the lavas, and from gabbros to monzonites and syenites in their plutonic equivalents. Rocks of the high-K calc-alkaline series are less abundant, and are represented by basaltic and basaltic andesitic lavas, as well as the granitic Predazzo intrusion (Beccaluva et al., 1996; Coltorti, 1996).

In the Dinarides, widespread Middle Triassic calc-alkaline magmatism comprising basalts, andesites, and rhyolites has been reported from the Dalmatian to the Serbian Zone (Bébien et al., 1978).

In the Hellenides, widespread Middle Triassic calc-alkaline volcanics and less abundant shoshonitic basalts and shoshonites have been reported mainly from the Peloponnese region (Pe-Piper, 1983, 1998).

The orogenic affinities of all these magmatic associations is clearly demonstrated by the primordial mantle-normalized incompatible element patterns given in Figure 4. These are characterized by marked Nb and Ti negative anomalies, coupled with overall enrichment (from calc-alkaline to shoshonitic rocks) of LFSE, commonly considered a characteristic of subduction-related components. The chondrite-normalized REE distributions in Figure 4 show nearly-flat or slightly positive fractionated heavy-REE (HREE) patterns, as well as the expected gradual enrichment in LREE from the calc-alkaline to the shoshonitic series.

Overall, the apparent regional discrepancy between the orogenic character of this magmatism and the Triassic anorogenic setting may be explained if the mantle sources inherited subduction-related geochemical components from the Hercynian orogenic cycle, as already observed by Bonadiman et al. (1994) for the Monzoni complex, and Pe-Piper (1998) for the Hellenides. It should also be noted that the available Sr–Nd isotopic data on the Triassic magmatism from the Southern Alps compare favourably with Permian orogenic magmatic events (Macera et al., 1994, and references therein; Rottura et al., 1998).

3. THE JURASSIC OCEANIZATION

Continuous rifting ultimately led to the opening of several oceanic basins bordering the Adria Plate during the Jurassic (Figure 5). Oceanization was apparently related to the initial opening of the central Atlantic at about 190–180 Ma, and resulted from of interaction of the European and African plates with the various micro-plates (Iberia, Adria, and Pelagonia) located between them. Tethyan ophiolites from the Alpine–Apennine and Dinarides–Albanides–Hellenides systems thus represent remnants of the oceanic basins which developed during the Jurassic between Europe/Iberia and Adria (western Tethys), and Adria and Pelagonia (eastern Tethys), respectively.

Western Adria ophiolites. The oldest absolute ages for the ophiolites located west of Adria are 180–170 Ma (Beccaluva et al., 1981, and references therein), whereas biostratigraphic ages for the Apennine ophiolites cluster in the interval 162–158 Ma (Chiari et al., 1997).

Ophiolitic complexes along the western Adria margin, in Calabria, the northern Apennines, Alpine Corsica, and the western and central Alps, invariably show Mid-Ocean Ridge (MOR) affinity, and consist of lherzolitic mantle tectonites and cumulate sequences of plagioclase-dunites, troctolites, gabbros, Fe-gabbros, and very scarce plagiogranites (Figure 6) (Beccaluva and Piccardo, 1978; Beccaluva et al., 1977, 1980, 1982, 1984a; Serri, 1980; Puga, this volume). The associated volcanics consist of high-Ti basalts displaying Ti/V ratios > 20 (Figure 7). They have LREE-depleted (Engadine, Inzecca, Pineto, Cap Corse, and Internal Ligurides) or flat to slightly enriched (Calabria, Betic Cordillera, Balagne and Nebbio Units, and External Ligurides) chondrite-normalized patterns (Figure 8), similar to those of normal and transitional MOR basalts, respectively (Beccaluva et al., 1979, 1989; Rampone et al., 1998; Padoa et al., 2001; Venturelli et al., 1981; Puga, this volume). The original tectonic setting of these basalts is therefore represented by an oceanic spreading system without any influence of subduction-related processes in magma generation. The transitional MORB affinity of most of the external ophiolitic units from the Alpine–Apennine belt can be related to the early stages of opening of this

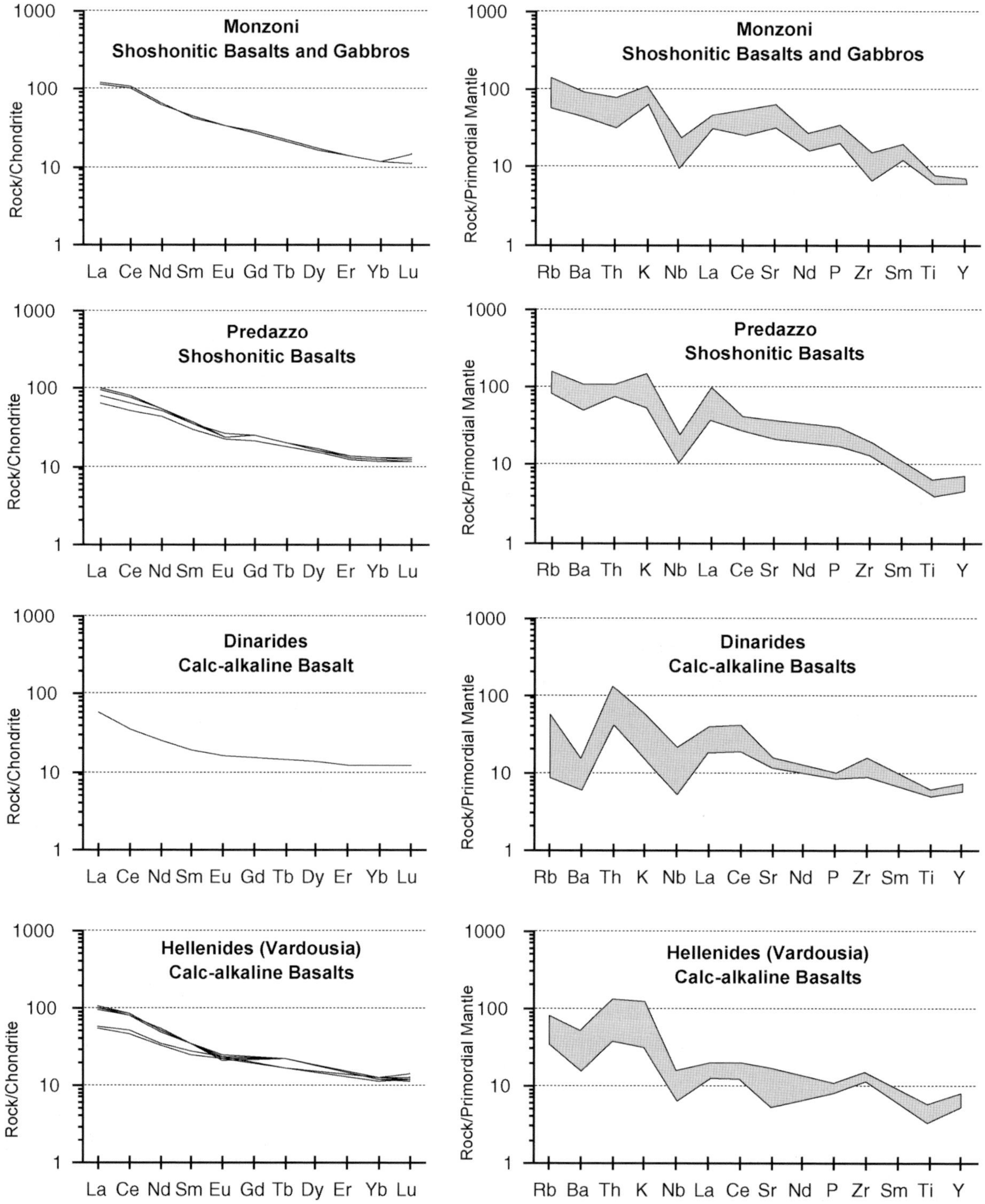

Figure 4: Chondrite- and primordial mantle-normalized REE and incompatible element patterns of Triassic orogenic rocks from the Monzoni Complex (Bonadiman et al., 1994), Predazzo Complex (Marrocchino, 2001), Dinarides (this work), and Hellenides (incompatible elements, this work; REE from Pe-Piper and Mavronichi, 1990). Normalizing factors are from Sun and McDonough (1989).

basin; the proximity of the transitional MORB magmatism to the original continental margins is also indicated by the close association of these units with continent-derived debris (Beccaluva et al., 1980, 1984b; Marroni et al., 1998). On the whole, the peculiarities of the western Adria ophiolites – such as the lack of sheeted dike complexes, reduced and variable crustal thickness (up to 2–3 km), and evidence of early oceanic tectonism (e.g., oceanic breccias, deposition of cherts on the top of gabbros and serpentinized mantle peridotites), as well as the characteristics of primary magmas – suggest that the original tectonic setting was a narrow oceanic

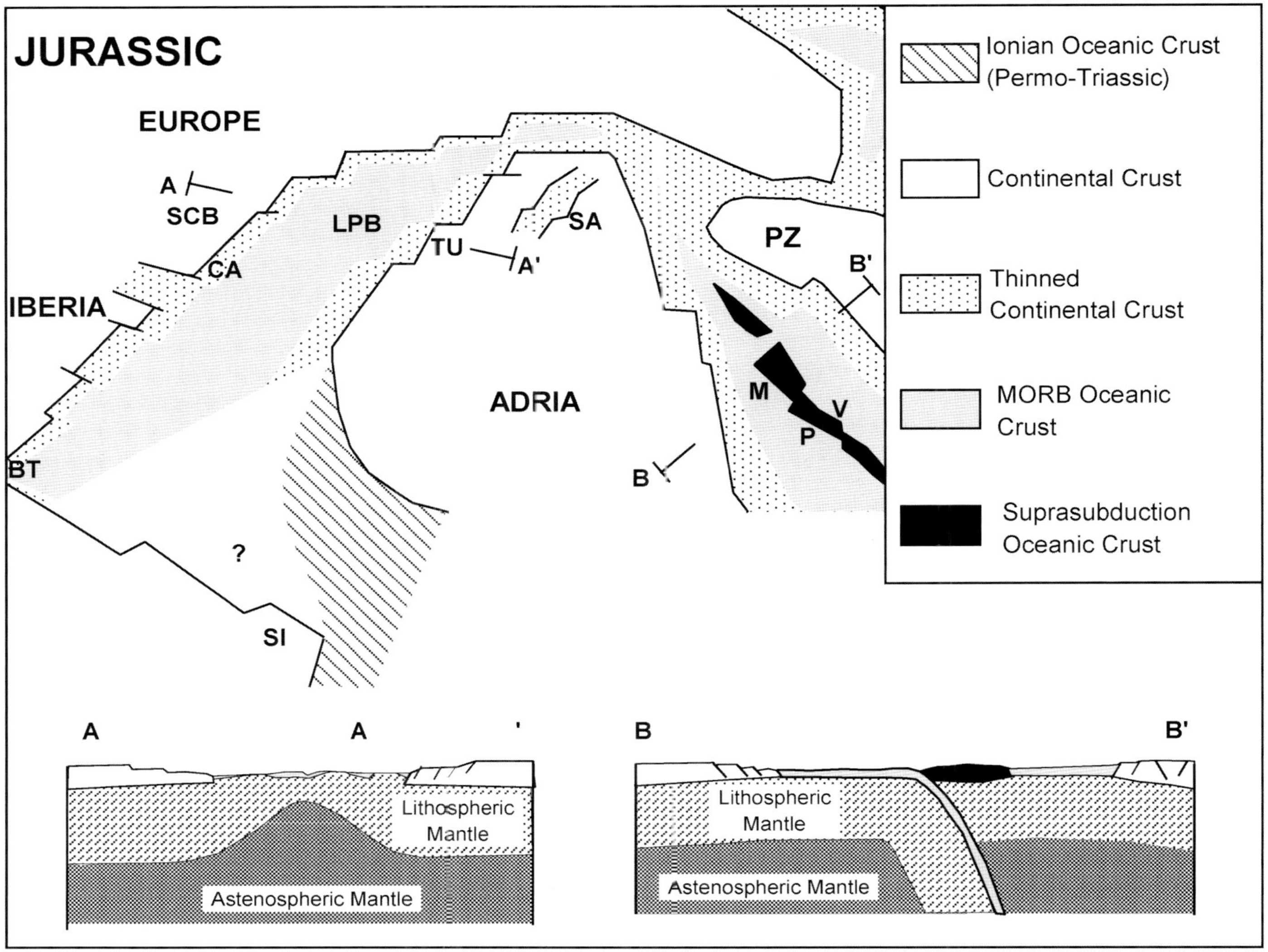

Figure 5: Tentative sketch and cross-sections (A–A′, B–B′) of the Jurassic oceanic basins around the Adria plate (not to scale). Abbreviations: SI = Sicily; Ca = Calabria; SCB = Sardinia–Corsica Block; LPB = Ligure–Piemontese Basin; TU = Tuscany; SA = Southern Alps; M = Mirdita; P = Pindos; V = Vourinos; PZ = Pelagonian Zone. The question mark denotes the uncertain relationships between Jurassic and Permo-Triassic oceanic crusts, as well as the possible occurrence of a continental platform (Panormide) in between (Finetti et al., 1996; Muttoni et al., 2001, and references therein).

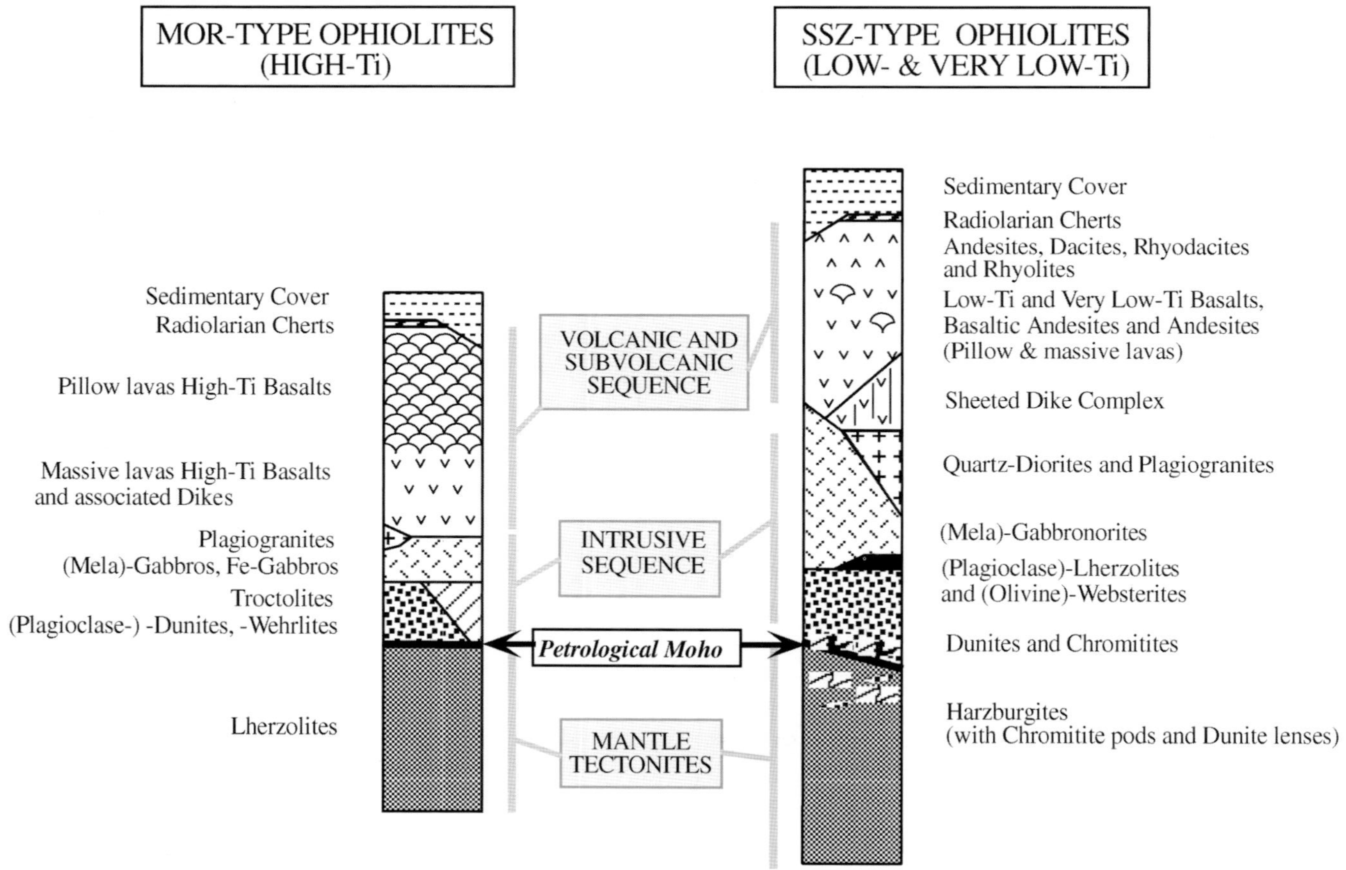

Figure 6: Generalized stratigraphical schemes (not to scale) for MOR- and SSZ-type ophiolites from the peri-Adriatic region.

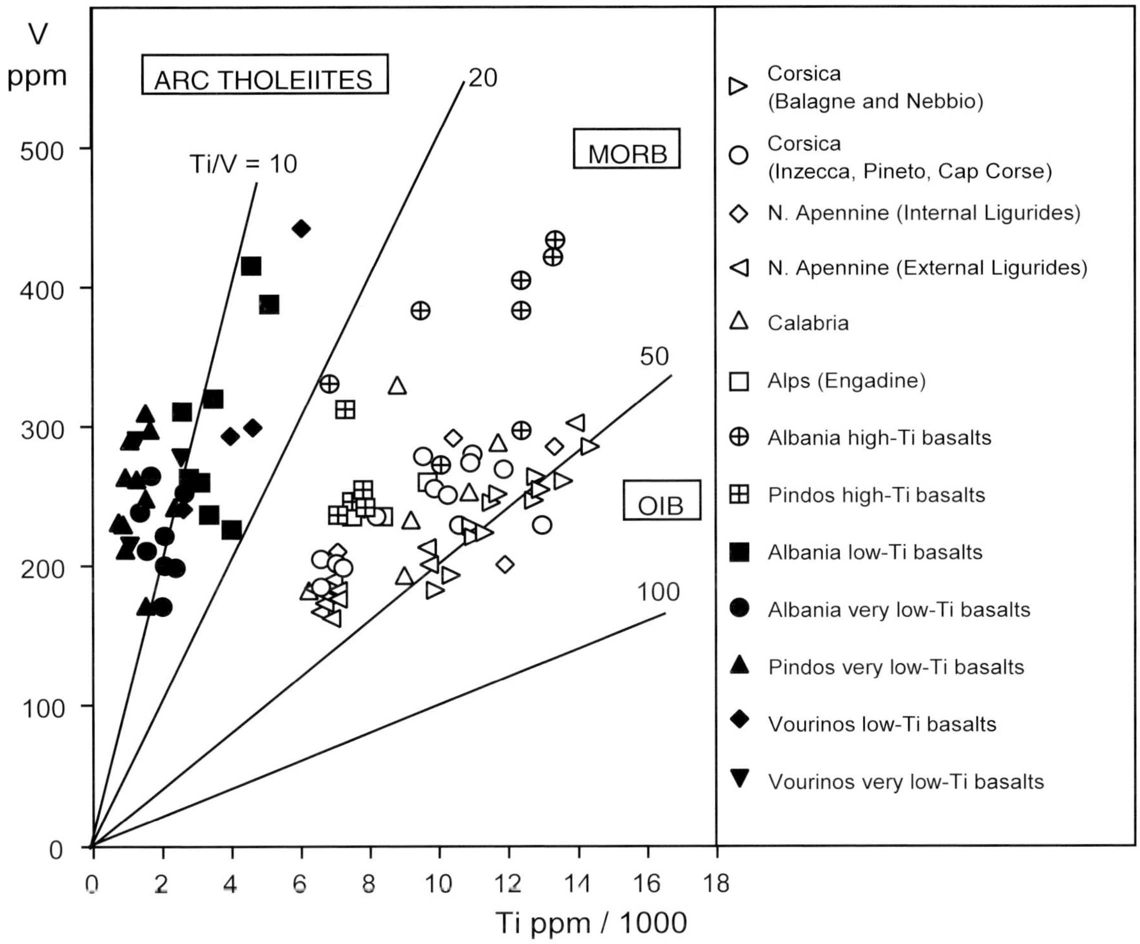

Figure 7: Ti/1000 versus V diagram (after Shervais, 1982) for basaltic rocks from the peri-Adriatic ophiolitic complexes.

basin characterized by a "slow-spreading" system and passive lithospheric extension (Decandia and Elter, 1972; Beccaluva and Piccardo, 1978; Beccaluva et al., 1984a, 1984b; Hébert et al., 1989; Bortolotti et al., 1990; Abbate et al., 1994; Martin et al., 1994; Lagabrielle and Lemoine, 1997). The tectonic denudation of large sectors of subcontinental mantle peridotites, genetically unrelated to the Jurassic oceanic crust generation, has recently been highlighted by Rampone et al. (1998), Rampone and Piccardo (2000) and Schaltegger et al. (2002).

Eastern Adria ophiolites. The ophiolitic complexes between the Pelagonian and Adria continental blocks belong to the Serbian (Slatibor, Banja Luka, etc.), Mirdita, and Subpelagonian (e.g. Vourinos and Pindos) zones of the Dinarides, Albanides and Hellenides, respectively (Figure 1). Absolute ages of the amphibolitic soles beneath the Hellenides and Dinarides ophiolites date back to between 162 and 180 Ma (Spray et al., 1984; Vergély et al., 1998), while accurate biostratigraphic ages on the radiolarites of the Albanian ophiolites span from the late Bajocian (170 Ma) to mid Callovian (160 Ma) (Marcucci and Prela, 1996). It should, however, be noted that Late Triassic oceanization processes are indicated by MOR-type basalts in northern Greece and the Argolis Peninsula (Jones et al., 1991; Saccani et al., 2002) (see also Figure 2).

Eastern Adria ophiolites are characterized by the peculiar juxtaposition of MOR-type complexes on their western side, and suprasubduction-type (SSZ) on the eastern side. The latter ophiolitic type is generally considered as having formed above an intra-oceanic subduction zone (Pearce et al., 1984). The SSZ ophiolites, such as those of eastern Mirdita, Vourinos, and Pindos (Figure 6), are characterized by predominantly harzburgitic mantle tectonites, a cumulitic plutonic sequence made up of dunites (sometimes with important chromititic bodies), (plagioclase)-lherzolites, (olivine)-websterites, and gabbro-norites, as well as isotropic gabbros, quartz-diorites, and plagiogranites (Beccaluva et al., 1984c, 1994, 1997, 1998b; Bébien et al., 1998, 2000). The associated sheeted dike complexes and volcanics consist of low-Ti basalts, basaltic andesites, andesites, dacites, and rhyolites with island-arc tholeiitic (IAT) affinity. Occasional, very low-Ti lavas with boninitic affinity occur, usually later in the sequence (Beccaluva

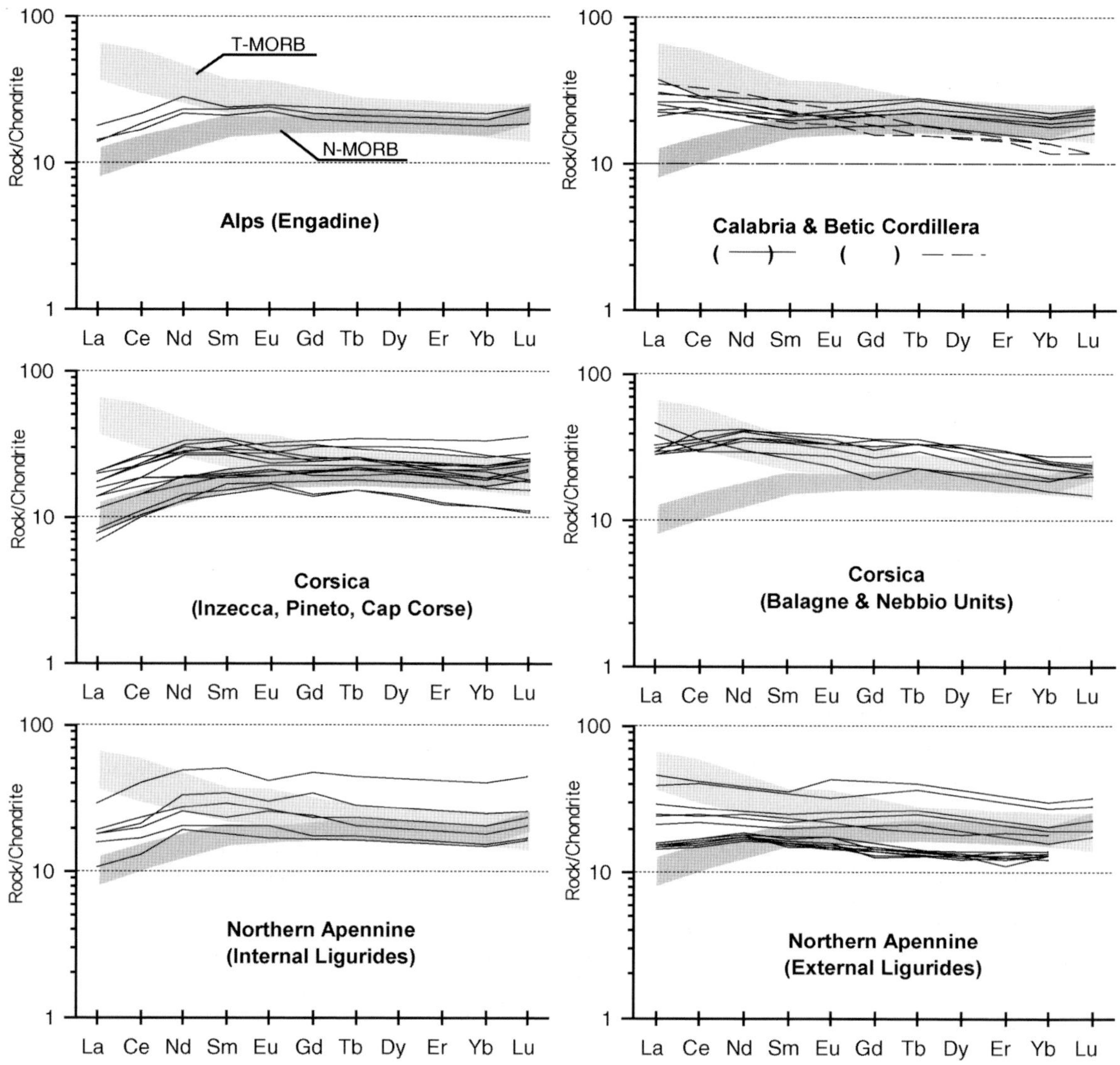

Figure 8: Chondrite-normalized REE patterns of Jurassic ophiolites from the Western Tethys. Data sources: Engadine, Venturelli et al. (1981); Calabria, Beccaluva et al. (1982, 1983); Betic Cordillera, Puga (this volume); Inzecca, Venturelli et al. (1981); Pineto, Saccani et al. (2000); Cap Corse, this work; Balagne, Venturelli et al. (1979); Nebbio, Saccani et al. (2000); Internal Ligurides, Venturelli et al. (1981), and this work; External Ligurides, Vannucci et al. (1993) and this work. Normalizing factors are from Sun and McDonough (1989). Normal (N-) and Transitional (T-) MORB fields after Le Roex (1987).

and Serri, 1988; Beccaluva et al., 1984c; Bortolotti et al., 2002). The Ti/V ratios of SSZ ophiolitic basalts are constantly less than 20 (Figure 7), while their chondrite-normalized REE distributions (Figure 9) show an overall depletion with respect to N-MORB ophiolites. In particular, the very low-Ti basaltic rocks show the strongest overall REE depletion, which is typical of boninitic magmatism. Consequently, the petrological characteristics of low- and very low-Ti ophiolites indicate a suprasubduction origin of the parental magmas, and imply the development of an intraoceanic subduction within a pristine oceanic domain.

A paleogeographic reconstruction (Beccaluva et al., 1994; Bortolotti et al., 2002) of the multiple oceanic basins that developed during the Jurassic between the Adria and Pelagonian blocks (Figure 5) should therefore indicate: (1) the early creation of MORB-type oceanic lithosphere between the newly-formed continental margins after the Triassic rifting stage; (2) a subsequent intra-oceanic, converging plate system which developed on the eastern side of this basin, with the generation of a suprasubduction-type oceanic lithosphere.

In the model of Figure 5, a north-eastward dipping subduction is assumed, based on the general structure of the Albanides characterized by westward thrusting. However, on the basis of early deformation patterns in the Mirdita and Vourinos ophiolites, Robertson and Shallo (2000) proposed a generalized, Jurassic, south-westward dipping

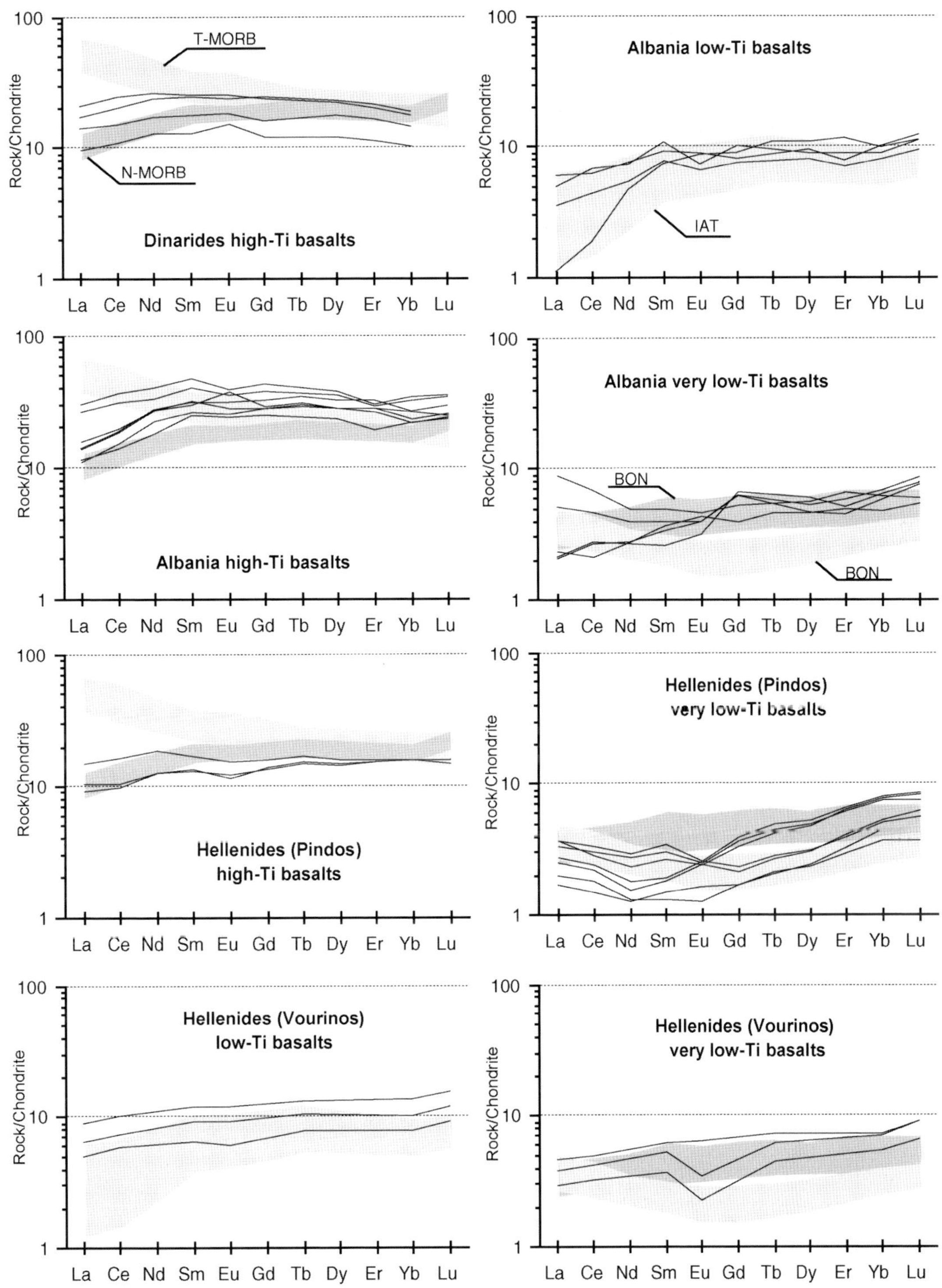

Figure 9: Chondrite-normalized REE patterns of Jurassic ophiolites from the east of the Adria plate. Data sources: Dinarides, Lugovic et al. (1991); Albania high-Ti, low-Ti, and very low-Ti basalts, Bortolotti et al. (2002), and this work; Pindos high-Ti, and very low-Ti basalts, this work; Vourinos low-Ti, and very low-Ti basalts, Beccaluva et al. (1984c). Normalizing factors are from Sun and McDonough (1989). Normal (N-) and Transitional (T-) MORB fields after Le Roex (1987); Island Arc low-Ti basalt (IAT) and Boninite (Bon) fields after Beccaluva and Serri (1988).

subduction of the oceanic lithosphere, followed by a north-eastward flipping during the Tertiary collision. The Jurassic age, the south-westward dips, the considerable thickness of the Eastern Belt ophiolites (up to 14 km in Albania), the age of transgressive sediments (Robertson and Shallo, 2000), and the characteristic presence of boninites and absence of post-ophiolite granites, suggest that the Mirdita–Pindos complexes originated between the Adria and Pelagonia continental blocks, and argue against a derivation from the Vardar–Almopias oceanic basin.

CONCLUSIONS

The review of available data on the Triassic magmatic rocks and Jurassic ophiolites at the margins of the Adria Plate presented in this chapter leads to the following conclusions:

1. The Triassic rifting phase (Figure 2), precursor to Jurassic oceanization, produced alkaline to transitional anorogenic magmatism to the west of the Adria plate (from Calabria to Tuscany), and calc-alkaline/shoshonitic magmas at its northern and eastern margins (from the Alps to Dinarides and Hellenides). The orogenic character of the latter – unusual for a rift setting – can be explained if mantle sources at the northern and eastern Adria margins inherited subduction-related geochemical components from the Hercynian orogenic cycle.
2. Oceanization began in the Middle-Late Triassic at the southeastern Adria margin, giving rise during the Middle-Late Jurassic to multiple oceanic basins at both the western (Western Tethys) and eastern (Serbian, Mirdita, and Pindos basins) Adria Plate margins (Figure 5).
3. The ophiolites of the Alpine–Apennine belt (Figure 1), representing remnants of the western Tethys, display exclusively MORB affinities, and are related to discontinuous oceanic crust generation along a "slow-spreading" system characterized by the passive lithospheric extension and widespread tectonic denudation of subcontinental mantle peridotites.
4. The ophiolitic complexes of the Subpelagonian Zone (Figure 1), representing remnants of the Serbian, Mirdita, and Pindos basins, are characterized by the juxtaposition of two subparallel belts: MORB-type to the west, and suprasubduction-type to the east. This implies the development of intra-oceanic subduction processes during the Middle-Late Jurassic within a pristine MORB-type oceanic basin located between the Adria and Pelagonian continental blocks (Figure 5).

ACKNOWLEDGEMENTS

We gratefully acknowledge J.A. Pearce and an anonymous referee for very constructive reviews. This work was financially supported by the MIUR-COFIN project 2000.

REFERENCES

Abbate, E., Bortolotti, V., Conti, M., Marcucci, M., Principi, G., Passerini, P., Treves, B., 1986. Apennines and Alps ophiolites and the evolution of the western Tethys. Mem. Soc. Geol. It. 31, 23–44.

Abbate, E., Bortolotti, V., Marcucci, M., Passerini, P., Principi, G., 1994. Genetic models for the northern Apennines ophiolites: a discussion in the light of radiolarian biostratigraphy. Ofioliti 19, 333–347.

Bébien, J., Blanchet, R., Cadet, J.P., Charvet, J., Chorowicz, J., Lapierre, H., Rampnoux, J.P., 1978. Le volcanisme Triassique des Dinarides en Yugoslevie : sa place dans l'évolution geotectonique peri-mediterraneenne. Tectonophysics 47, 159–176.

Bébien, J., Shallo, M., Manika, K., Gega, D., 1998. The Shebenik Massif (Albania): a link between MOR- and SSZ-type ophiolites? Ofioliti 23, 7–15.

Bébien, J., Dimo-Lahitte, A., Vergély, P., Insergueix-Filippi, D., Dupeyrat, L., 2000. Albanian ophiolites. I – Magmatic and metamorphic processes associated with the initiation of a subduction. Ofioliti 25, 39–45.

Beccaluva, L., Ohnenstetter, D., Ohnenstetter, M., Venturelli, G., 1977. The trace element geochemistry of Corsican ophiolites. Contrib. Mineral. Petrol. 64, 11–31.

Beccaluva, L., Piccardo, G.B., 1978. Petrology of the Northern Apennine ophiolites and their significance in the Western Mediterranean area. In: Closs, H., Roeder, D., Schmidt, K. (Eds.), Alps, Apennines, Hellenides. In: Inter-Union Commission on Geodynamics, Scientific Report, vol. 38. E. Schweizerbart'sche Verlagsbuchhandlung, Stuttgart, pp. 243–254.

Beccaluva, L., Ohnenstetter, D., Ohnenstetter, M., 1979. Geochemical discrimination between ocean floor and island arc tholeiites – Application to some ophiolites. Canad. J. Earth Sci. 16, 1874–1882.

Beccaluva, L., Piccardo, G.B., Serri, G., 1980. Petrology of northern Apennine ophiolites and comparison with other Tethyan ophiolites. Ophiolites. Int. Oph. Symp., Cyprus 1979, 314–331.

Beccaluva, L., Chiesa, S., Delaloye, M., 1981. K/Ar age determinations on some tethyan ophiolites. Rend. Soc. It. Min. Petr. 37, 869–880.

Beccaluva, L., Macciotta, M., Spadea, P., 1982. Petrology and geodynamic significance of the Calabria–Lucania Ophiolites. Rend. Soc. It. Min. Petr. 38, 973–987.

Beccaluva, L., Dostal, J., Macciotta, G., Zerbi, M., 1983. Trace element geochemistry of some ophiolites from Calabria (Southern Italy). Ofioliti 8, 325–332.

Beccaluva, L., Macciotta, G., Piccardo, G.B., Zeda, O., 1984a. Petrology of lherzolitic rocks from the Northern Apennine ophiolites. Lithos 17, 299–316.

Beccaluva, L., Dal Piaz, G.V., Macciotta, G., 1984b. Transitional-to normal-MORB affinity in ophiolitic metabasites from the Zermatt-Saas, Combin and Antrona units, Western Alps: implications for the paleogeographic evolution of the Western Tethyan basin. Geol. en Mijnb. 16, 165–177.

Beccaluva, L., Ohnenstetter, D., Ohnenstetter, M., Paupy, A., 1984c. Two magmatic series with island arc affinities within the Vourinos ophiolite. Contrib. Mineral. Petrol. 85, 253–271.

Beccaluva, L., Serri, G., 1988. Boninitic and low-Ti subduction-related lavas from intraoceanic arc-backarc systems and low-Ti ophiolites: a reappraisal of their petrogenesis and original tectonic setting. Tectonophysics 146, 291–315.

Beccaluva, L., Macciotta, G., Piccardo, G.B., Zeda, O., 1989. Clinopyroxene composition of ophiolite basalts as petrogenetic indicator. Chem. Geol. 77, 165–182.

Beccaluva, L., Coltorti, M., Premti, I., Saccani, E., Siena, F., Zeda, O., 1994. Mid-Ocean ridge and suprasubduction affinities in the ophiolitic belts of Albania. Ofioliti 19, 77–96.

Beccaluva, L., Coltorti, M., Siena, F., 1996. Il magmatismo Medio-Triassico delle Dolomiti. In: Cassiano, S. (Bz), La Geologia delle Dolomiti, 78° Riunione Estiva della Società Geologica, 16–18 Settembre 1996. Introduzione Geologica e Guida alle escursioni, pp. 34–36.

Beccaluva, L., Shallo, M., Coltorti, M., Premti, I., Siena, F., 1997. Albania. In: Moores, E.M., Fairbridge, R.W. (Eds.), Encyclopedia of European and Asian Regional Geology. Chapman and Hall, London, pp. 7–16.

Beccaluva, L., Siena, F., Coltorti, M., Di Grande, A., Lo Giudice, A., Macciotta, G., Tassinari, R., Vaccaro, C., 1998a. Nephelinitic to tholeiitic magma generation in a transtensional tectonic setting: an integrated model for the Iblean volcanism. Sicily. J. Petrol. 39, 1547–1576.

Beccaluva, L., Coltorti, M., Ferrini, V., Saccani, E., Siena, F., Zeda, O., 1998b. Petrological modelling of Albanian ophiolites with particular regard to the Bulqiza chromite ore deposits. Periodico di Mineralogia 67, 7–23.

Bonadiman, C., Coltorti, M., Siena, F., 1994. Petrogenesis and T-fO_2 estimates of Mt. Monzoni complex (Central Dolomites, Southern Alps): a Triassic shoshonitic intrusion in a transcurrent geodynamic setting. Eur. J. Mineral. 6, 943–966.

Bortolotti, V., Principi, G., Treves, B., 1990. Mesozoic evolution of the Western Tethys and the Europe/Iberia/Adria plate junction. Mem. Soc. Geol. It. 45, 393–407.

Bortolotti, V., Marroni, M., Pandolfi, L., Principi, G., Saccani, E., 2002. Interaction between mid-ocean and subduction magmatism in Albanian ophiolites. J. Geol. 110, 561–576.

Chiari, M., Cortese, G., Marcucci, M., Nozzoli, N., 1997. Radiolarian biostratigraphy in the sedimentary cover of the ophiolites of south-western Tuscany, central Italy. Eclog. Geol. Helv. 90, 55–77.

Coltorti, M., 1996. Il magmatismo ladinico di Predazzo. In: Geologia delle Dolomiti, 78° Riunione SGI, San Cassiano, 16–18 Settembre 1996. Introduzione Geologica Guida alla escursione generale.

Cortesogno, L., Rossi, P.M., Vannucci, R., 1982. Episodi vulcanici in successioni carbonatiche medio-Triassiche pre-piemontesi (Cogoleto, Liguria). Riv. Ital. Paleont. 88, 21–44.

Cristofolini, R., 1966. Le manifestazioni eruttive basiche del Trias superiore nel sottosuolo di Ragusa (Sicilia sud-orientale). Periodico di Mineralogia 1, 1–28.

Dal Piaz, G.V., 1993. Evolution of Austro-Alpine and Upper Pennine basement in the northwestern Alps from Variscan convergence to post-Variscan extension. In: Von Raumer, J.F., Neubauer, F. (Eds.), Pre-Mesozoic Geology in the Alps. Springer-Verlag, pp. 327–344.

Decandia, F.A., Elter, P., 1972. La zona ofiolitifera del Bracco nel settore compreso fra Levanto e la Val Graveglia (Appennino Ligure). Mem. Soc. Geol. It. 11, 503–530.

Finetti, I., Lentini, F., Carbone, S., Catalano, S., Del Ben, A., 1996. Il sistema Appenninico Meridionale–Arco Calabro–Sicilia nel Mediterraneo Centrale: studio geofisico-geologico. Boll. Soc. Geol. It. 115, 529–559.

Finetti, I., Del Ben, A., 2000. Crustal stratigraphy and tectono-dynamics of the Pelagian Sea Region from new "CROP" seismic data. In: Second Symposium of Geology of Northwest Libya, Tripoli, 6–8 November 2000, Abstract volume, p. 42.

Hébert, R., Serri, G., Hekinian, R., 1989. Mineral chemistry of ultramafic tectonites and ultramafic to gabbroic cumulates from the major oceanic basins and northern Apennines ophiolites (Italy): a comparison. Chem. Geol. 77, 183–207.

Jones, G., Robertson, A.H.F., Cann, J.R., 1991. Genesis and emplacement of the supra-subduction zone Pindos ophiolite, northwestern Greece. In: Peters, Tj., et al. (Eds.), Ophiolite Genesis and Evolution of the Oceanic Lithosphere. Ministry of Petroleum and Minerals, Sultanate of Oman, pp. 771–799.

Lagabrielle, Y., Lemoine, M., 1997. Alpine, Corsican and Apennine ophiolites – The slow-spreading ridge model. C. R. Acad. Sci. Paris 325, 909–920.

Laurenzi, M.A., Visonà, D., Zantedeschi, C., 1994. High resolution Ar/Ar chronology of Predazzo Magmatic Complex (Southern Alps, Italy). In: ICOG 8, 5–11 July 1994, Abstract volume, p. 186.

Lemoine, M., Tricart, P., Boillot, G., 1987. Ultramafic and gabbroic ocean floor of the Ligurian Tethys (Alps, Corsica, Apennines): in search of a genetic model. Geology 15, 622–625.

Le Roex, A.P., 1987. Source region of mid-ocean ridge basalts: evidence for enrichment processes. In: Menzies, M.A., Hawkesworth, C.J. (Eds.), Mantle Metasomatism. Academic Press, London, pp. 389–419.

Longaretti, G., Rocchi, S., 1992. Il magmatismo dell'Avampaese Ibleo (Sicilia orientale) tra il Trias e il Quaternario: dati stratigrafici e petrologici di sottosuolo. Mem. Soc. Geol. It. 45, 911–925.

Lucchini, F., Rossi, P.L., Simboli, G., 1982. Il magmatismo triassico dell'area di Predazzo (Alpi Meridionali, Italia). In: Castellarin, A., Vai, G.B. (Eds.), Guida alla Geologia del Sudalpino Centro-Orientale. Soc. Geol. It., pp. 211–220.

Lugovic, B., Altherr, R., Raczek, I., Hofmann, A.W., Majer, V., 1991. Geochemistry of peridotites and mafic igneous rocks from the Central Dinaric Ophiolite Belt, Yugoslavia. Contrib. Mineral. Petrol. 106, 201–216.

Macciotta, G., Zuppetta, A., Zeda, O., 1986. Caratteri petrografici e significato geotettonico delle metavulcaniti basiche triassiche dell'Unità di S. Donato (Calabria). Boll. Soc. Geol. It. 105, 195–202.

Macera, P., Del Moro, A., Bargossi, G.B., Campana, R., Rottura, A., 1994. Poligenetic nature of the Cima d'Asta intrusive complex, Southern Alps, Italy. Inferences from petrological, geochemical and isotopic (Sr and Nd) data. Lithos 32, 47–62.

Marcucci, M., Prela, M., 1996. The Lumi i Zi (Puke) section of the Kalur cherts: radiolarian assemblages and comparison with other sections in northern Albania. Ofioliti 21, 71–76.

Marrocchino, E., 2001. Caratteri petrologici del Complesso vulcano-plutonico di Predazzo (Trento). Tesi di Dottorato, XIII Ciclo. Università di Modena, 51 p.

Marroni, M., Molli, G., Montanini, A., Tribuzio, R., 1998. The association of continental crust rocks with ophiolites in the Northern Apennines (Italy): implications for the continent–ocean transition in the Western Tethys. Tectonophysics 292, 43–66.

Martin, S., Tartarotti, P., Dal Piaz, G.V., 1994. The Mesozoic ophiolites of the Alps: a review. Boll. Geof. Teor. Appl. XXXVI (141–144), 175–219.

Mayer, A., Sinigoi, S., 1997. Is the Ivrea mafic intrusion a lower-crust precursory event of the Alpine Tethys? Plinius 18, 142–143.

Muttoni, G., Garzanti, E., Alfonsi, L., Cirilli, S., Germani, D., Lowrie, W., 2001. Motion of Africa and Adria since the Permian: paleomagnetic and paleoclimatic constraints from northern Libya. Earth Planet. Sci. Lett. 192, 159–174.

Padoa, E., Saccani, E., Durand-Delga, M., 2001. Structural and geochemical data on the Rio Magno Unit: evidences for a new "Apenninic" ophiolitic Unit in Alpine Corsica and its geodynamic implications. Terra Nova 13, 135–142.

Pearce, J.A., Lippard, S.J., Roberts, S., 1984. Characteristics and tectonic significance of supra-subduction zone ophiolites. In: Kokelaar, B.P., Howells, M.F. (Eds.), Marginal Basin Geology. Geol. Soc. London Spec. Publ. 16, 77–94.

Pe-Piper, G., 1982. Geochemistry, tectonic setting and metamorphism of mid-Triassic volcanic rocks of Greece. Tectonophysics 85, 253–272.

Pe-Piper, G., 1983. The Triassic volcanic rocks of Tyros, Zarouhla, Kalamae and Epidavros, Peloponnese, Greece. Schweiz. Mineral. Petrogr. Mitt. 63, 249–266.

Pe-Piper, G., 1998. The nature of Triassic extension-related magmatism in Greece: evidence from Nd and Pb isotope geochemistry. Geological Magazine 135, 331–348.

Pe-Piper, G., Mavronichi, M., 1990. Petrology, geochemistry and regional significance of the Triassic volcanic rocks of the Western Parnassos isopic zone of Greece. Ofioliti 15, 269–285.

Petersen, J.S., Morten, L., Simboli, G., Lucchini, F., 1980. REE abundances in the Predazzo-Monzoni intrusive complex, Dolomites, North Italy. Riv. It. Paleont. Strat. 85, 3–4.

Piccardo, G.B., Rampone, E., Vannucci, R., Cimmino, F., 1994. Upper mantle evolution of ophiolitic peridotites from the Northern Apennine: petrological constraints to the geodynamic processes. Mem. Soc. Geol. It. 48, 137–148.

Puga, E., this volume. A reappraisal of the Betic Ophiolitic Association: the westernmost relic of the Alpine Tethys. Chapter 31.

Rampone, E., Hofmann, A.W., Raczek, I., 1998. Isotopic contrast within the internal Liguride ophiolite (N. Italy): the lack of a genetic mantle-crust link. Earth Planet. Sci. Lett. 163, 175–189.
Rampone, E., Piccardo, G., 2000. The ophiolite-oceanic lithosphere analogue: new insights from the Northern Apennines (Italy). In: Dilek, Y., Moores, E.M., Elthon, D., Nicolas, A. (Eds.), Ophiolites and Oceanic Crust: New Insights from Field Studies and the Ocean Drilling Program. Geological Society of America, Special paper, vol. 349, pp. 21–34.
Ricci, C.A., Serri, G., 1975. Evidenze geochimiche sulla diversa affinità petrogentica delle rocce basiche comprese nelle serie a facies toscana. Boll. Soc. Geol. It. 94, 1187–1198.
Robertson, A.H.F., Shallo, M., 2000. Mesozoic-Tertiary tectonic evolution of Albania in its regional Eastern Mediterranean context. Tectonophysics 316, 197–254.
Rottura, A., Bargossi, G.M., Caggianelli, A., Del Moro, A., Visonà, D., Tranne, C.A., 1998. Origin and significance of the Permian high-K calc-alkaline magmatism in the central-eastern Southern Alps, Italy. Lithos 45, 329–348.
Saccani, E., Padoa, E., Tassinari, R., 2000. Preliminary data on the Pineto gabbroic massif and Nebbio Basalts: progress toward the geochemical characterization of Alpine Corsica Ophiolites. Ofioliti 25, 75–85.
Saccani, E., Padoa, E., Photiades, A., 2002. Tectono-magmatic significance of Triassic MORBs from the Argolis Peninsula (Greece): implication for the origin of the Pindos ocean. Ofioliti 27, 73.
Scandone, P., 1975. Triassic seaways and the Jurassic Tethys Ocean in the central Mediterranean area. Nature 256, 117–118.
Schaltegger, U., Desmurs, L., Manatschal, G., Muentener, O., Meier, M., Frank, M., Bernoulli, D., 2002. The transition from rifting to sea-floor spreading within a magma-poor rifted margin: field and isotopic constraints. Terra Nova 14, 156–162.
Serri, G., 1980. Chemistry and petrology of gabbroic complexes from the northern Apennine ophiolites. Ophiolites, Int. Oph. Symp., Cyprus 1979, 296–313.
Shervais, J.W., 1982. Ti–V plots and the petrogenesis of modern ophiolitic lavas. Earth Planet. Sci. Lett. 59, 101–118.
Siena, F., Beccaluva, L., Coltorti, M., Marchesi, S., Morra, E., 1991. Ridge to hot-spot evolution of the Atlantic lithospheric mantle: evidence from Lanzarote peridotite xenoliths (Canary Islands). J. Petrol. Spec. Issue, 271–290.
Sinigoi, S., Quick, J.E., Clemens-Knott, D., Mayer, A., Demarchi, G., Mazzucchelli, M., Negrini, L., Rivalenti, G., 1994. Chemical evolution of a large mafic intrusion in the lower crust, Ivre-Verbano Zone, Northern Italy. J. Geophys. Res. 99, 21575–21590.
Spray, J.G., Bébien, J., Rex, D.C., Roddick, J.C., 1984. Age constraints on the igneous and metamorphic evolution of Hellenic-Dinaric ophiolites. In: Dixon, J.E., Robertson, A.H.F. (Eds.), The Geological Evolution of the Eastern Mediterranean. In: Geol. Soc. London Spec. Publ., vol. 17. Blackwell, Oxford, pp. 619–628.
Sun, S.S., McDonough, W.F., 1989. Chemical and isotopic systematics of ocean basalts: implications for mantle composition and processes. In: Saunders, A.D., Norry, M.J. (Eds.), Magmatism in the Ocean Basins. In: Geol. Soc. Spec. Publ., vol. 42. Blackwell, Oxford, pp. 313–346.
Traversa, G., Vaccaro, C., REE Distribution in the Late Hercynian Dykes from Sardinia. In: IGCP Project No. 276. Newsletter, vol. 5, pp. 215–226.
Vaccaro, C., Atzori, P., Del Moro, A., Oddone, M., Traversa, G., Villa, I.M., 1991. Geochronology and Sr isotopes geochemistry of late-Hercyian dykes from Sardinia. Schweiz. Miner. Petrogr. Mitt. 71, 221–230.
Vannucci, R., Rampone, E., Piccardo, G.B., Ottolini, L., Bottazzi, P., 1993. Ophiolitic magmatism in the Ligurian Tethys: an ion microprobe study of basaltic clinopyroxenes. Contrib. Mineral. Petrol. 115, 123–137.
Venturelli, G., Capedri, S., Thorpe, R.S., Potts, P.J., 1979. Rare-earth and other element distribution in some ophiolitic metabasalts of Corsica, Western Mediterranean. Chem. Geol. 24, 339–353.
Venturelli, G., Thorpe, R.S., Potts, P.J., 1981. Rare earth and trace element characteristics of ophiolitic metabasalts from the Alpine–Apennine belt. Earth Planet. Sci. Lett. 53, 109–123.
Vergély, P., Dimo, A., Monie, P., 1998. Datation des semelles métamorphiques ophiolitiques d'Albanie par la methode $^{40}Ar/^{39}Ar$: conséquences sur le mécanisme de leur mise en place. C. R. Acad. Sci. Paris 326, 717–722.
Visonà, D., 1997. The Predazzo multipulse intrusive body (Western Dolomites, Italy), field and mineralogical studies. Mem. Sci. Geol. Padova 49, 117–125.

CHAPTER 29

Cenozoic Tectono-Magmatic Evolution of the Central-Western Mediterranean: Migration of an Arc-Interarc Basin System and Variations in the Mode of Subduction

L. Beccaluva[1], G. Bianchini[1], M. Coltorti[1], F. Siena[1], M. Verde[1]

ABSTRACT

The space–time distribution, petrogenetic affinity and mantle source characteristics of Cenozoic orogenic magmatism in the central-western Mediterranean area, from Provence to the Eolian Arc, are reviewed here. The best-fit model to account for the tectono-magmatic evolution of the area is based on a single evolutionary process of subduction starting from the Middle-Late Eocene beneath the paleo-European continental margin, migrating southeastward with time, intercalated by the opening of two interarc oceanic basins: the Ligurian-Balearic and the Tyrrhenian. The northwest-dipping subduction of the Ionian oceanic lithosphere produced tholeiitic/calcalkaline magmas in Provence and Sardinia (34–26 Ma), with petrological and isotopic characteristics typical of the initial stages of arc magmatism related to the subduction of pure oceanic lithosphere. Interarc rifting of the paleo-European continental margin developed at least from the Early Miocene and resulted in the opening of the Ligurian-Balearic oceanic basin, concomitantly with the southeastward drifting and rotation of the Sardinia–Corsica block. Since the Burdigalian, the Sardinia orogenic magmatism has displayed a more mature character, with both calcalkaline and shoshonite products, likely in connection with steepening of the subducted slab, continental collision in the Corsica–Apennine sector, and local involvement of continental-derived terrigenous sediments in the subduction. After a period of quiescence, a new tectono-magmatic phase began in the Late Miocene, with rifting along the Apennines and the Calabrian Alps, which ultimately led to the diachronous opening of the Tyrrhenian basin (Vavilov 4–3.4 Ma; Marsili 1.9–1.6 Ma) and the southeastward migration of the Calabrian Arc to its present position. The petrological and isotopic features of the Quaternary orogenic magmas along the eastern peri-Tyrrhenian border – namely their general evolution from calcakaline to shoshonitic and ultra-potassic products – are related to an accentuated steepening of the subducted lithosphere, as well as the progressive involvement of terrigenous sediments, increasing from the Eolian area to the Campanian and Roman Provinces. In the proposed evolutionary model the modes of subduction developed distinct features over time: (1) the Eocene-Oligocene subduction is considered an "active" process with shallow dip beneath the paleo-European continental margin and moderate decoupling between the converging plates; (2) the Neogene-Quaternary subduction, instead, became increasingly "passive", characterized by accelerated slab sinking and retreat, leading to progressive decoupling of the upper and lower plates, intense asthenospheric diapirism and interarc opening in the upper plate. The ca. 800-km long lithospheric slab consumed during the entire subduction process predicted by the model is in excellent agreement with recent tomographic data which depict the Ionian lithosphere currently dipping beneath the Aeolian-Calabrian arc as a single body flattening in the upper mantle at around 670 km beneath the Tyrrhenian basin and adjacent Sardinia block.

Keywords: Cenozoic tectono-magmatic evolution, central-western Mediterranean

[1]Dipartimento di Scienze della Terra, Università di Ferrara, Via Saragat, 1, 44100 Ferrara, Italy

1. INTRODUCTION

In this chapter we present a review and new data, including rare earth elements (REE) and Sr–Nd isotopes, on the orogenic magmatic events which developed in the central-western Mediterranean area during the Cenozoic. The petrogenetic affinity and mutual relationships of these magmatic associations, which appear progressively younger southeastward, represent a key-point for interpreting the Mediterranean convergent system between Europe and Africa (Dewey et al., 1989; Mantovani et al., 1997) and its tectono-magmatic evolution.

The aim of this chapter is to define the petrological significance of the magmatic associations and their mantle sources, as well as their Cenozoic spatio-temporal evolution. We shall focus attention on the central-western Mediterranean sector, considering particularly an ideal NW–SE profile from Provence to the Aeolian-Calabrian Arc. This allows us to place further constraints on the geodynamic history of the area (Beccaluva et al., 1987 and references therein), and to propose a model which may accommodate the tectono-magmatic events and the mode of subduction processes in a single, coherent scheme.

2. SPATIO-TEMPORAL EVOLUTION OF THE OROGENIC MAGMATISM

In the central-western Mediterranean area, the Cenozoic magmatic activity with orogenic affinity developed in two main cycles separated in space and time: (a) in Provence, the Ligurian-Provençal sea, the Valencia Trough, Sardinia and, to a lesser extent in Corsica, within the time span 34–13 Ma; (b) in the eastern peri-Tyrrhenian domain, from the Aeolian Arc to the Roman Province, during the Pliocene and Quaternary (Figure 1).

In ***Provence***, the Cenozoic orogenic subvolcanic and volcanic rocks are mainly represented by microdiorites, basalts, andesites and dacites from Estèrel (34–20 Ma: Bellon, 1981), and andesitic clasts in conglomerates near Nice. Their tholeiitic/calcalkaline affinity is typical of the initial orogenic magmatism in active continental margins (Beccaluva et al., 1994). Therefore, the age and serial affinity of this magmatism coherently indicate that the paleo-European continental margin was affected by subduction of oceanic lithosphere at least from the Late Eocene to Early Miocene. A southwestward extension of this volcanic district is attested to by calcalkaline volcanics (andesites, dacites and rhyolites) outcropping in Majorca and offshore in the ***Valencia Trough*** (24–18.6 Ma; Martì et al., 1992). Calcalkaline (to shoshonitic) basalts, andesites and dacites have also been recovered in the ***Ligurian-Provençal sea***, with K–Ar ages spanning between 30 and 12 Ma (Sosson et al., 1998).

Orogenic volcanic activity in ***Sardinia*** started at about 32 Ma, with basaltic-andesitic lavas locally occurring along the western graben of the island. Since about 23 Ma, large eruptions of rhyodacitic ignimbrites have been recorded in the same area, alternating, and partly contemporary with basaltic-andesitic lavas, until about 13 Ma, which marks the end of volcanic activity in Sardinia (Beccaluva et al., 1985a and references therein). The calcalkaline *s.l.* character of this magmatism further supports the existence of oceanic lithospere subduction below the paleo-European continental margin. The climax of the ignimbrite emplacement at 21–19 Ma in both Provence and Sardinia (Bellon, 1981; Beccaluva et al., 1987), and the eruption of unusually high-Mg calcalkaline basalts in Sardinia (18 Ma; Lecca et al., 1997; Morra et al., 1997) suggest paroxystic extensional tectonics accompanying the opening of the Balearic backarc basin and the anticlockwise rotation of the Sardinia–Corsica microplate (Montigny et al., 1981; Cherchi and Montadert, 1982; Burrus, 1984; Rehault et al., 1984; Beccaluva et al., 1987). Since then, the occurrence of high-K calcalkaline and shoshonitic lavas (Mameli, 1996) has indicated a more mature stage of arc magmatism. This phase ended in the Langhian (Vigliotti and Langenheim, 1995; Chamot-Rooke et al., 1999). Miocene volcanic activity also developed offshore from Sardinia, with calcalkaline (s.l.) andesites in the north (17–16 Ma; Rossi et al., 1998) and shoshonitic products in the Sardinia Channel (13–12 Ma; Mascle et al., 2001). Rhyolitic ignimbrites (18.9–19.3 Ma) are reported by Ottaviani-Spella et al. (1996) in southern ***Corsica***, while lamproites (14 Ma) occur at Sisco in the northern part of the island (Civetta et al., 1978; Serri et al., 1993).

Figure 1: Simplified sketch map of the studied area with the main occurrences of Cenozoic orogenic volcanics and a generalized cross-section from Provence and Sardinia to the Aelioan–Calabrian arc (AA′), modified after Beccaluva et al. (1994). Legend: 1 = Compressional front of the Alpine and Apenninic Chains; 2 = Areas with oceanic crust; 3 = Cenozoic orogenic magmatism; dots indicate hypocenters. Abbreviations: Th = tholeiitic, CA = calcalkaline and high potassium calcalkaline, SHO = shoshonitic, UK = ultrapotassic.

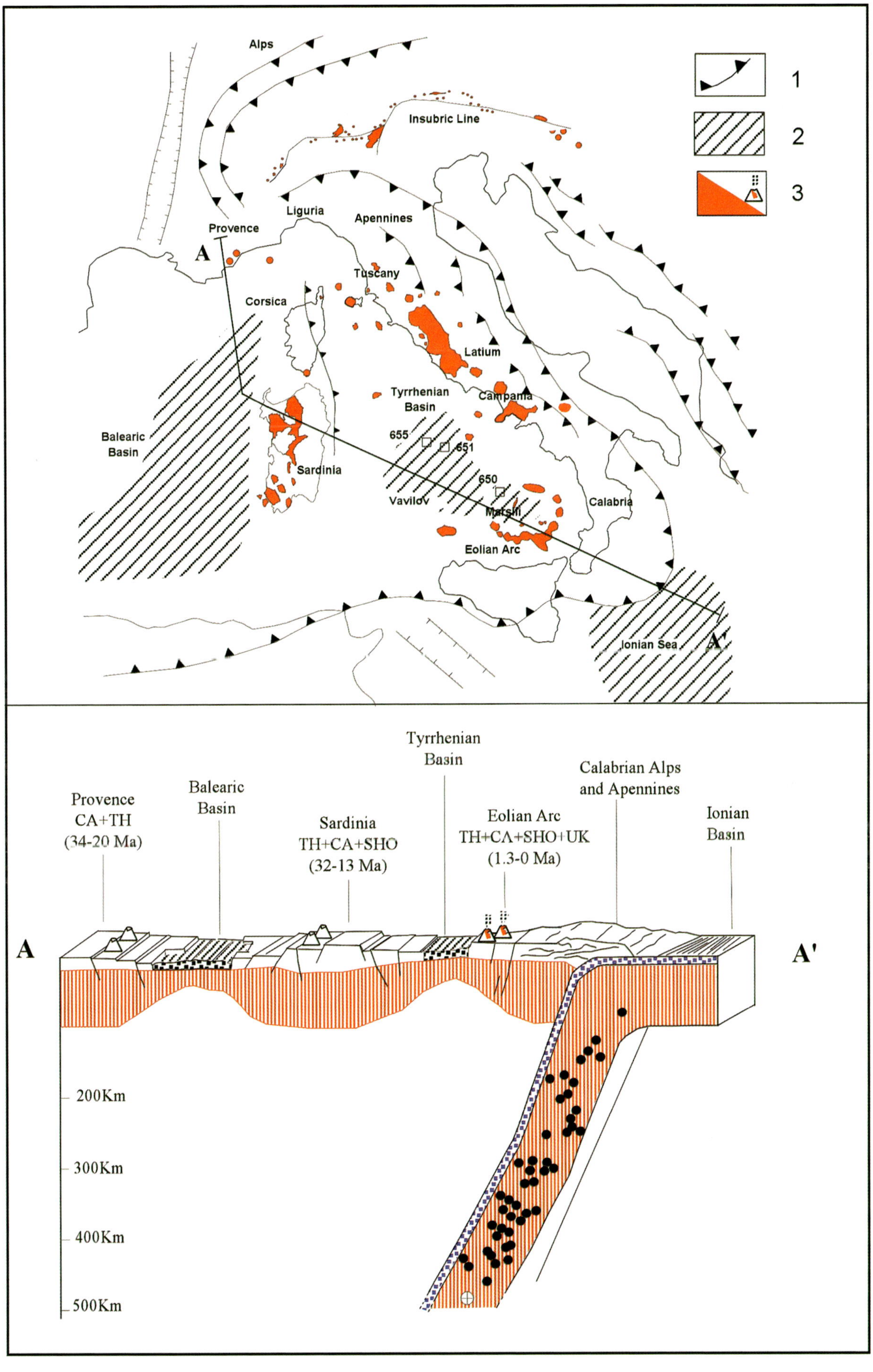
Alps
Insubric Line
Liguria
Apennines
Provence
A
Tuscany
Corsica
Latium
Tyrrhenian Basin
Campania
Balearic Basin
655
651
Sardinia
650
Vavilov
Marsili
Calabria
Eolian Arc
Ionian Sea
A'
1
2
3
Tyrrhenian Basin
Calabrian Alps and Apennines
Provence CA+TH (34-20 Ma)
Balearic Basin
Sardinia TH+CA+SHO (32-13 Ma)
Eolian Arc TH+CA+SHO+UK (1.3-0 Ma)
Ionian Basin
A
A'
200Km
300Km
400Km
500Km

Further evidence of Oligo-Miocene orogenic volcanism is provided by the calcalkaline (s.l.) volcanoclastic levels in sedimentary formations of the s thrust-belt: in the upper part of the Ranzano Formation (32–30 Ma; Cibin et al., 1998), and in the Aveto-Petrignaccola Formation (29 Ma; Mattioli, 1997) from the Northern Apennines; in the Bisciaro Formation (Umbria-Marche) (26.8–17.1 Ma; Balogh et al., 1993), and in several other localities from the Southern Apennines (Crisci et al., 1988; Critelli, 1993).

To the southeast of the Sardinia district, the ***Tyrrhenian Sea*** plays the role of an interarc basin between the Oligo-Miocene Sardinian arc and the still active Aeolian arc. This basin can be subdivided in two sectors, north and south of the 41 °N parallel respectively. The tectono-magmatic evolution of the northern sector reflects the geological history of the adjacent Tuscan area, with E–W extension associated with widespread acidic magmatism, starting at $\sim$ 7–9 Ma (Serri et al., 1993). The southern sector was affected by more recent E–W/NW–SE extensional tectonics associated with widespread basic magmatism and the formation of new oceanic crust in two different sub-basins, the Vavilov and the Marsili basins (ODP Leg 107: sites 650, 651, 655;). Sea-floor spreading occurred in the Vavilov Basin around 4–3.5 Ma and in the Marsili Basin around 1.9–1.7 Ma (Kastens et al., 1986). Geochemical characteristics of igneous rocks from Leg 107 evidence a clear subduction-related affinity (calcalkaline to high-K calcalkaline) for sites 650 (Marsili Basin) and 651 (Vavilov Basin), in contrast with transitional MORB affinity recorded at site 655 in the Vavilov basin (Beccaluva et al., 1990).

Further evidence of subduction-related components within the mantle underlying the Tyrrhenian crust is provided by the recent products of the Marsili seamounts, characterized by a marked calcalkaline signature (Beccaluva et al., 1985b; Savelli and Gasparotto, 1994).

The volcanic activity of the ***Aeolian Archipelago*** is entirely Quaternary in age, with the oldest products, dated at 1.3 Ma, dredged from the westernmost seamounts (Beccaluva et al., 1985b). The Aeolian volcanic arc is underlain by an almost continuous, 250–300 km wide and 550–650 km long, northwest-dipping Benioff zone, as indicated by seismic data (Gasparini et al., 1982). The activity mainly consists of basaltic-andesitic to rhyolitic volcanics, belonging to the calcalkaline and high-K calcalkaline series, and subordinate shoshonite and leucite-tephrite products; the latter are chemically comparable with some leucite-bearing lavas of the Roman province. A few arc tholeiites have also been dredged from the submerged Aeolian slope and north Lametini seamount (Beccaluva et al., 1985b). The evolutionary tendency of the Aeolian magmatism is to be noted: after an initial phase in which tholeiitic volcanics have been recorded, calcalkaline (s.l.) products appear to predominate in most eruptive centers, together with potassic rocks mainly recovered in the eastern part of the Archipelago, particularly in the still active volcanos of Stromboli and Vulcano (Beccaluva et al., 1982 and references therein).

Aeolian potassic products strongly resemble those of the ***Roman Province*** of Campania, which can therefore be considered the northward expression of the same orogenic magmatism. Significantly, about 1550 m of high-K calcalkaline andesites (ca. 2 Ma: Barbieri et al., 1979) have been drilled in the northwestern border of the Phlegrean Fields (Campania), where they are buried under the potassic volcanics (Di Girolamo, 1978; Beccaluva et al., 1991). The existence of these calcalkaline lavas, volumetrically as important as the younger shoshonitic and leucite-bearing volcanics of the same area, testifies to a complete spectrum of orogenic series, as recorded in the Aeolian volcanism.

3. GEOCHEMICAL AND PETROLOGICAL CONSTRAINTS FOR MAGMA GENESIS

Primordial mantle-normalized incompatible element and chondrite-normalized REE distributions of orogenic basic magmas from the central-western Mediterranean area are reported in Figures 2 and 3: all patterns display the typical features of subduction-related magmas, such as negative anomalies of high field strength elements (HFSE; e.g. Nb, Ti), coupled with enrichments of low field strength elements (LFSE) and light rare earth elements (LREE). A general similarity in geochemical patterns is to be remarked between calcalkaline and high-K calcalkaline series from Sardinia, the Aeolian Islands, and Campania. A similar subduction-related geochemical imprint is also shown by basic rocks from sites 650 and 651 (LEG107) in the Tyrrhenian basin (Figure 3).

$^{87}Sr/^{86}Sr$ and $^{143}Nd/^{144}Nd$ analyses of these orogenic products are reported in Table 1, and compared with available data on the Mediterranean magmatic associations in Figure 4. The orogenic magmas show only a limited overlap with the anorogenic mantle array of the area, and a general displacement toward higher $^{87}Sr/^{86}Sr$ and lower $^{143}Nd/^{144}Nd$ compositions. This diagram is useful to highlight the nature of fluids/melts released by the subducted slab into the overlying mantle wedge, and to unravel the possible involvement of continental crust materials

Table 1
Sr–Nd isotope data of selected orogenic (subduction-related) volcanics from the central-western Mediterranean area (after Galassi, 1995). Abbreviations: Th = tholeiitic, CA = calcalkaline, HK–CA = high potassium calcalkaline, SHO = shoshonitic

Locality	Sample label	Lithotype	Serial affinity	$^{87}Sr/^{86}Sr$	$^{143}Nd/^{144}Nd$
Provence					
Villenueve-Loubet	PR1	Andesite	Th/CA	0.70579	0.51266
Villenueve-Loubet	PR2	Andesite	Th/CA	0.70543	0.51266
Esterel-Agay	PR4	Dacite	Th/CA	0.70478	0.51292
Esterel-Agay	PR6	Hb-Gabbro	Th/CA	0.70453	0.51268
Esterel-Agay	PR8	Dacite	Th/CA	0.70470	0.51291
Sardinia					
Logudoro-Bosano	A501	Basalt	CA	0.70605	0.51259
Logudoro-Bosano	A519	Basalt	CA	0.70455	0.51264
Logudoro-Bosano	A577	Basalt	CA	0.70680	0.51246
Logudoro-Bosano	A503	Basalt-Andesite	CA	0.70602	0.51251
Logudoro-Bosano	A530	Basalt-Andesite	CA	0.70666	0.51254
Logudoro-Bosano	A540	Andesite	CA	0.70653	0.51249
Logudoro-Bosano	A567	Andesite	HK-CA	0.70676	0.51255
Anglona	AL1	Latite	SHO	0.70605	0.51244
Anglona	AL3	Trachyte	SHO	0.70618	
Arcuentu	AR2B	Basalt-Andesite	CA	0.70993	0.51229
Arcuentu	AR4	Basalt-Andesite	CA	0.70959	0.51224
Marmilla	A714	Basalt	Th/CA	0.70654	0.51237
Tyrrhenian oceanic crust (LEG 107; sites 650 and 651)					
Marsili basin	650 66-2	Basalt-Andesite	CA	0.70537	0.51277
Marsili basin	650 67-1	Basalt-Andesite	CA	0.70505	
Vavilov basin	651 43-1	Basalt-Andesite	CA	0.70559	0.51269
Vavilov basin	651 43-2	Basalt-Andesite	CA	0.70558	0.51269
Vavilov basin	651 49-1	Basalt-Andesite	CA	0.70438	0.51294
Aeolian Islands (Is) and Seamounts (Smt)					
Sisifo Smt	T76/27 5a1	Basalt	CA	0.70369	
Sisifo Smt	T76/27 3b5	Basalt	CA	0.70358	0.51280
Sisifo Smt	T78/62 3	Trachyte	SHO	0.70431	
Enarete Smt	T76/25 18a	Latite	SHO	0.70499	0.51259
Enarete Smt	T76/25 20a	Latite	SHO	0.70503	0.51260
Alcione Smt	T72/28 c2	Basalt	HK-CA	0.70558	0.51260
Lamentini Smt	T79/41	Basalt	Th/CA-B	0.70484	0.51266
Marsili Smt	M1/1	Basalt	CA	0.70454	0.51275
Marsili Smt	M2/1	Andesite	CA	0.70465	0.51275
Marsili Smt	M1/4	Basalt	CA	0.70530	0.51265
Lipari Is	LP1	Basalt-Andesite	CA	0.70449	
Lipari Is	LP5	Basalt-Andesite	CA	0.70444	
Lipari Is	LP9	Basalt-Andesite	CA	0.70472	0.51262
Lipari Is	LP13	Basalt-Andesite	CA	0.70495	
Lipari Is	LP21	Basalt-Andesite	CA	0.70525	
Lipari Is	LP61	Basalt-Andesite	CA	0.70637	0.51261
Lipari Is	LP10	Andesite	CA	0.70930	0.51236
Lipari Is	LP20	Basalt-Andesite	HK-CA	0.70540	0.51241
Lipari Is	LP23	Basalt-Andesite	HK-CA	0.70445	
Lipari Is	LP2	Andesite	HK-CA	0.70590	0.51255
Lipari Is	LP37	Andesite	HK-CA	0.70564	

Table 1
Continued

Locality	Sample label	Lithotype	Serial affinity	$^{87}Sr/^{86}Sr$	$^{143}Nd/^{144}Nd$
Campania (Volturno plain boreholes)					
Lipari Is	LP33	Latite	SHO	0.70479	0.51256
Volturno	P2-1800	Basalt-Andesite	HK-CA	0.70855	0.51218
Volturno	CV2-1440	Basalt-Andesite	HK-CA	0.70811	0.51222
Volturno	CV3-1889	Basalt-Andesite	HK-CA	0.70581	0.51245

in the subduction. In fact, fluids released by subduction of oceanic crust are characterized by $^{87}Sr/^{86}Sr$ enrichment trends (without significant $^{143}Nd/^{144}Nd$ variation) caused by ocean-floor metamorphism, as shown by the Ligurian ophiolite field in Figure 4. This implies that magmas generated above subducted oceanic lithosphere necessarily reflect this isotopic signature, in addition to that of the pristine mantle wedge.

Orogenic rocks from Provence – characterized by a $^{87}Sr/^{86}Sr$ between 0.70453 and 0.70579, and a $^{143}Nd/^{144}Nd$ between 0.51292 and 0.51265 – are thus consistent with generation from mantle sources enriched solely by fluids driven out of oceanic crust (Wilson, 1989 and reference therein).

The Sr–Nd isotopic systematics of orogenic rocks from Sardinia indicate a more complex framework: the compositions of the older products are generally characterized by a relatively low $^{87}Sr/^{86}Sr$ ($\leqslant 0.706$) and a high $^{143}Nd/^{144}Nd$ (> 0.5125), which are compatible with the subduction of pure oceanic lithosphere (as for Provence); surprisingly some compositions of more recent products show a very high $^{87}Sr/^{86}Sr$ (up to 0.7113) and low $^{143}Nd/^{144}Nd$ (down to 0.51219), requiring additional components. In fact, these extreme isotopic values cannot be merely ascribed to fluids released by oceanic crust, nor to shallow-level crustal contamination, but imply source contamination by continental crust components, possibly via partial fusion of subducted terrigenous sediments (Ellam et al., 1989; Beccaluva et al., 1991). Analogously, Downes et al. (2000), in their detailed study on the Sr–Nd-oxygen isotopic features of Mt. Arcuentu volcanics (south Sardinia), envisaged a remarkable enrichment of mantle sources by subducted siliceous sediments. The occurrence of terrigenous sedimentation along the Corsica–Apennines sector provides a suitable source for crustal components to be carried to depth by the ongoing subduction processes.

Isotopic compositions of mafic rocks in the Tyrrhenian oceanic crust (Vavilov and Marsili basins) and most of the Aeolian volcanics (only slightly displaced from the anorogenic mantle array; Figure 4) are compatible with mantle sources enriched solely by fluid components from subducted oceanic crust, as further confirmed by Boron isotopic data (Tonarini et al., 2001). However, it is interesting to note that shoshonites from the younger series of Stromboli display a distinctly higher $^{87}Sr/^{86}Sr$ (up to 0.7075) and lower $^{143}Nd/^{144}Nd$ composition (down to 0.51242), as well as relatively low $\delta^{11}B$, compared with the other Aeolian magmas, thus again requiring continental crust components in their mantle source (Beccaluva et al., 1985b; Ellam et al., 1989; Francalanci et al., 1993; Francalanci and Manetti, 1994; Tonarini et al., 2001).

The influence of these continental crust-derived components appears even more important in the mantle sources of the Campanian volcanics, where extreme Sr–Nd isotopic compositions are recorded ($^{87}Sr/^{86}Sr$ up to 0.7097; $^{143}Nd/^{144}Nd$ down to 0.5122). Moreover, Sr–Nd isotope compositions of Campanian high-K calcalkaline andesites, which approach the Roccamonfina shoshonitic and leucititic fields (Figure 4), suggest that their magma sources were significantly contaminated by subducted terrigenous sediments, irrespective of K-enrichment (Serri, 1990; Beccaluva et al., 1991; Conticelli et al., 2002). Boron isotopic data also support continental component contamination in mantle sources from which Campanian magmas were generated (Tonarini et al., 2004).

From the above it may be inferred that the involvement – via subduction – of continental crust materials (most probably terrigenous sediments) in the mantle sources of Miocene (Sardinia) and Plio-Quaternary orogenic magmas is likely related to the post-collisional stages of convergence. In particular, the transition from calcalkaline to potassic products at Stromboli may be connected with a progressive steepening of the subducted oceanic lithosphere and the involvement of continental crust materials, plausibly in the form of terrigenous sediments (Beccaluva et al., 1985b; Ellam et al., 1988, 1989; Ellam and Harmon, 1990).

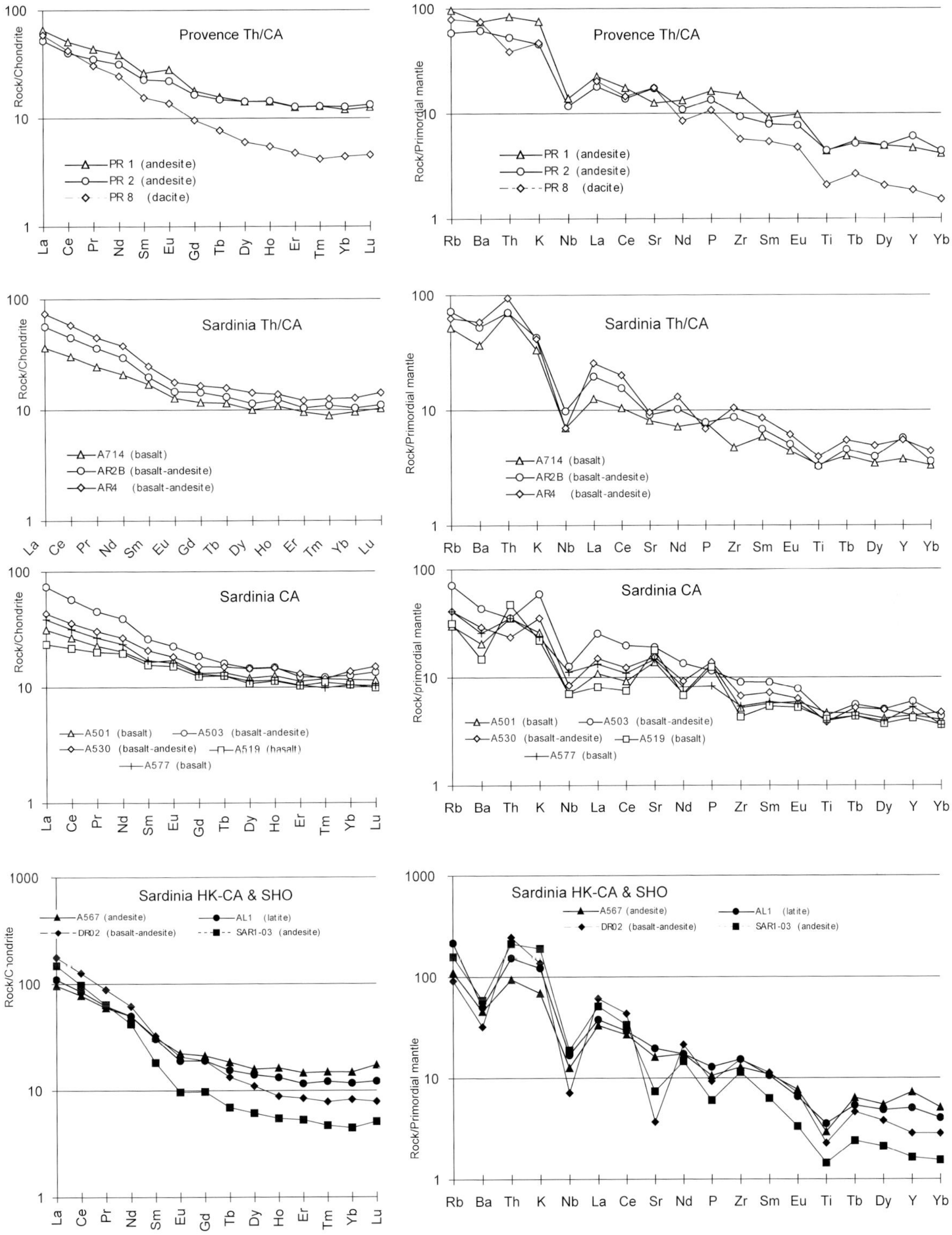

Figure 2: Chondrite-normalized REE and primordial mantle-normalized incompatible element patterns of orogenic volcanics from Provence and Sardinia. Data after Beccaluva et al. (1994) and Galassi (1995). Submarine volcanics DR02 from the western Corsica margin (Rossi et al., 1998) and SAR 1-03 from the Sardinia Channel (Mascle et al., 2001) are also reported. Normalizing factors after Sun and McDonough (1989). Abbreviations as in Figure 1.

As concerns the lead isotopic composition, the available literature data reported in the $^{207}Pb/^{204}Pb$ vs $^{206}Pb/^{204}Pb$ diagram (Figure 5) show that the orogenic magmas, although partially overlapping with the Mediterranean anorogenic mantle array, are displaced toward higher $^{207}Pb/^{204}Pb$ values, as commonly observed in magmatic suites from subduction related settings (Ellam et al., 1989; Sun et al., 1998; Huang et al., 2000). Nevertheless, it is diffi-

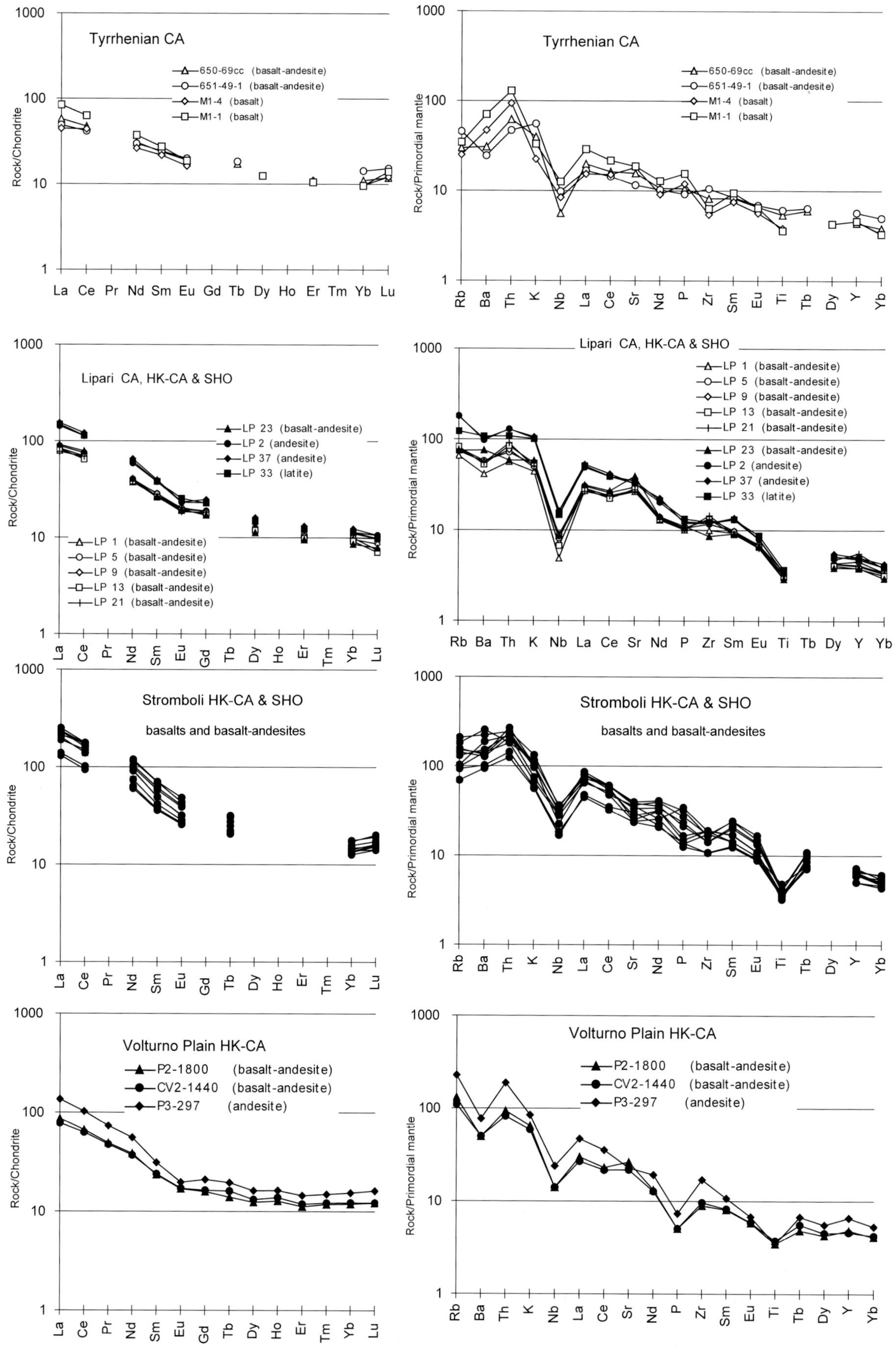
Tyrrhenian CA
650-69cc (basalt-andesite)
651-49-1 (basalt-andesite)
M1-4 (basalt)
M1-1 (basalt)
Rock/Chondrite
Rock/Primordial mantle
Lipari CA, HK-CA & SHO
LP 1 (basalt-andesite)
LP 5 (basalt-andesite)
LP 9 (basalt-andesite)
LP 13 (basalt-andesite)
LP 21 (basalt-andesite)
LP 23 (basalt-andesite)
LP 2 (andesite)
LP 37 (andesite)
LP 33 (latite)
Stromboli HK-CA & SHO
basalts and basalt-andesites
Volturno Plain HK-CA
P2-1800 (basalt-andesite)
CV2-1440 (basalt-andesite)
P3-297 (andesite)
La Ce Pr Nd Sm Eu Gd Tb Dy Ho Er Tm Yb Lu
Rb Ba Th K Nb La Ce Sr Nd P Zr Sm Eu Ti Tb Dy Y Yb

Figure 3: Chondrite-normalized REE and primordial mantle-normalized incompatible element patterns of orogenic volcanics from the Tyrrhenian basin (LEG 107: sites 650 and 651) from the Aeolian Islands and Seamounts, and from Campania (Volturno plain boreholes). Data after Beccaluva et al. (1994) and Galassi (1995). Data for Stromboli volcanics are from Ellam et al. (1989). Normalizing factors after Sun and McDonough (1989). Abbreviations as in Figure 1.

---

cult to explain the wide $^{206}Pb/^{204}Pb$ compositional spectrum only in terms of orogenic components, thus implying significant compositional differences in the mantle regions prior to subduction-related metasomatism. In particular, the pre-subduction geochemical imprint of Aeolian and Campanian lavas (with $^{206}Pb/^{204}Pb$ in the ranges 19.0–19.8 and 18.9–19.2, respectively) seems to correspond to OIB-type alkaline mantle sources, whereas OIB-tholeiitic mantle sources appear to be more appropriate for Latium and Sardinia volcanics (with $^{206}Pb/^{204}Pb$ in the ranges 18.7–18.8 and 18.5–18.7, respectively), as already suggested by Beccaluva et al. (1991).

On the whole, the complex petrological and geochemical variations of orogenic magmatism in the central-western Mediterranean can be satisfactorily accounted for by the interplay of a number of factors: (1) original (pre-subduction) regional differences in mantle sources, i.e. OIB-tholeiitic for Provence, Sardinia and Latium, and OIB-alkaline for Aeolian and Campanian districts; (2) heterogeneous subduction-related metasomatic processes induced in the mantle wedge by fluids driven from subducted oceanic lithosphere in the early stages (tholeiitic and calcalkaline magmatism), accompanied by continental crust-derived components in the late stages of convergence and continental collision (more potassic magmas); (3) variable degrees of partial melting, which preferentially

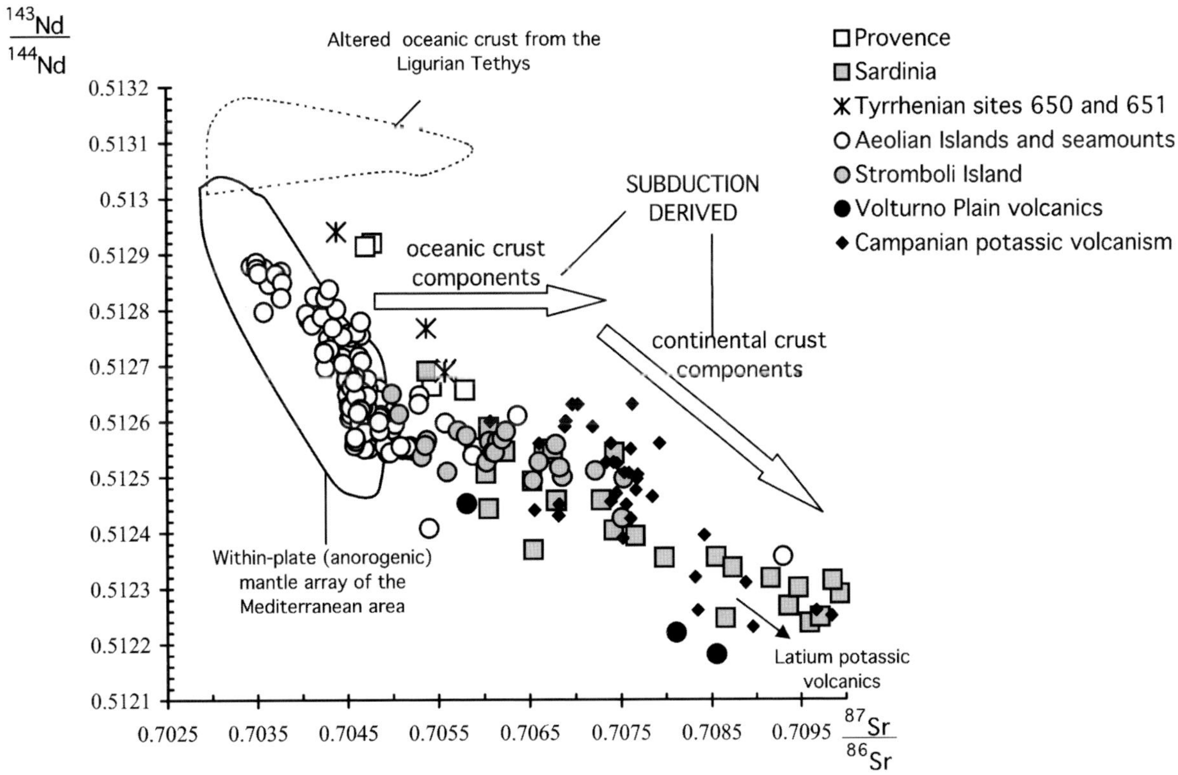

Figure 4: Variation of $^{143}Nd/^{144}Nd$ versus $^{87}Sr/^{86}Sr$ for orogenic volcanics from the central-western Mediterranean area, based on available data from the literature: Hawkesworth and Vollmer (1979, and references therein), Ellam et al. (1989), Francalanci et al. (1993), Galassi (1995), Del Moro et al. (1998), De Astis et al. (2000), Downes et al. (2000). The anorogenic non-subduction-related mantle array is based on an extensive data set of alkaline lavas and associated mantle xenoliths from the Mediterranean area (Beccaluva et al., 2001). A compositional field of basalts and gabbros from Ligurian ophiolites of the northern Apennines (Rampone et al., 1988) is reported for comparison (see text for explanation). Arrows indicate the expected compositional variations of mantle sources (and related magmas) induced by oceanic crust vs. continental crust metasomatic components, driven off a subducted slab.

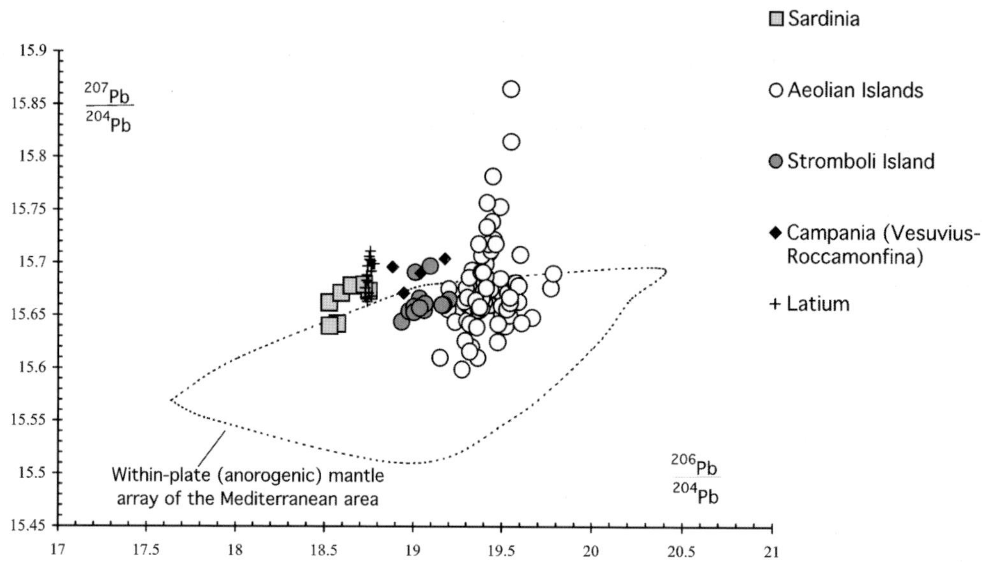

Figure 5: Variation of $^{207}Pb/^{204}Pb$ versus $^{206}Pb/^{204}Pb$ for orogenic volcanics from the central-western Mediterranean area based on available data from the literature: Vollmer (1976), Ellam et al. (1989), Francalanci et al. (1993), Caron and Orgeval (1995) Del Moro et al. (1998), De Astis et al. (2000). The anorogenic non-subduction-related mantle array is based on an extensive data set of alkaline lavas and associated mantle xenoliths from the Mediterranean area (Beccaluva et al., 2001).

tapped low-solidus highly metasomatized (and phlogopite bearing) regions of the mantle, producing K-rich magmas at relatively low melting degrees; and (4) the variable depths at which partial melting occurred, with a tendency for more potassic (and silica undersaturated) products to be more abundant in the late stage of each magmatic cycle, in relation to steepening of the subducted slab and deepening of the related mantle sources.

4. TECTONO-MAGMATIC MODEL

In any discussion of the general tectono-magmatic model of the central-western Mediterranean area, the south-eastward migration of the orogenic arc magmatism, and subsequent opening of the Balearic and Tyrrhenian oceanic basins, as well as the lithosphere structure, must first be taken into consideration (Beccaluva et al., 1985b, 1987, 1994; Malinverno and Ryan, 1986; Doglioni et al., 1999; Jolivet et al., 1999; Wilson and Bianchini, 1999; Serri et al., 2001).

As seen in the previous sections, the oldest magmatic events developed along the paleo-European continental margin of Provence and Sardinia, starting from 34–32 Ma; they were characterized by tholeiitic/calcalkaline magmas, typical of initial stages of orogenic magmatism related to subduction of oceanic lithosphere. These events induced opening and oceanization of the Ligurian–Balearic basin (21–18 Ma), as well as drifting and anticlockwise rotation of the Sardinia–Corsica microplate relative to stable Europe. This suggests prolonged subduction (at least from the Late Eocene) of NW-dipping oceanic lithosphere below the paleo-European continental margin. SE drifting of the Sardinia–Corsica microplate is probably marked by the earliest ignimbrites of Aquitanian age in both Provence and Sardinia (Bellon, 1981; Beccaluva et al., 1985a). The paleogeographic position of the oceanic lithosphere subducting below the paleo-European margin is necessarily to be assigned to the Sicilide-Canetolo or Sicilian-Subligurian oceanic basin (Beccaluva et al., 1987; Serri et al., 2001), corresponding to the northwestward extension of the Ionian domain (Finetti et al., 1996; Finetti et al., 2001).

In the proposed model (Figure 6), this oceanic sector is bounded, to the north-west, by the Alpine chain of Corsica and Calabria and, to the north-east, by the Adriatic continental domain. The distribution of the orogenic volcanism, the opening of the Ligurian-Balearic basin, and the related anticlockwise rotation of the Sardinia–Corsica block imply a northward decrease in the width of the Ionian oceanic basin and its subsequent diachronous closure (Figure 6A–B). From the Burdigalian, the Sardinia orogenic magmatism displays a more mature character, with both calcalkaline (s.l.) and shoshonite products, probably in connection with a steepening of the subducted slab and the local involvement of continental crust-derived materials in the subduction (Figure 6B). Concomitantly, the volcanic activity gradually decreased in Sardinia, ending at around 13 Ma, in relation to a marked decrease in the subduction rate, in turn induced by the intervening continental collision. Diachronous continental collision between the Alpine chain of Corsica (and Calabria) and the Apennine domains of the Adriatic margin is in accordance with the Early Oligocene and Aquitanian ages of tectonic transport onto the Apennine domain of the Canetolo and Sicilide units, respectively (Beccaluva et al., 1987 and references therein).

During the Serravalian and Tortonian, orogenic magmatism was substantially absent in the central Mediterranean area, coinciding with the last main compressional phase in the Northern Apennines (Carmignani et al., 1978). In this sector, therefore, active subduction diminished and ultimately ceased, leaving a subducted slab plunging almost vertically beneath the central Apennines (Panza et al., 1980), as also recently confirmed by tomographic data (Spakman, 1990; Wortel and Spakman, 2000).

A new tectono-magmatic phase started around 7 Ma, with rifting along the Apennines and Calabrian Alps, and production of the Tuscan Province magmatism, characterized by a complex interaction between mantle-derived potassic magmas and crustal anatectic melts (Serri et al., 1993 and references therein). Southeastward migration of the rifting subsequently resulted in the opening of the Tyrrhenian basin, with the diachronous development of the Vavilov (4–3.5 Ma) and Marsili (1.9–1.6 Ma) oceanic crusts, as demonstrated by Leg 107 results (Kastens et al., 1986). A relict Pliocene volcanic arc (mostly submerged) between these two oceanic domains may be envisaged as an arcuate structure extending from the Pontine Islands to the Anchise seamounts (Savelli and Gasparotto, 1994; Francalanci and Manetti, 1994).

A new arc magmatism developed during the Quaternary in the Aeolian Archipelago and Campania, eastward of the Tyrrhenian basin, which therefore plays the role of an interarc basin between Sardinia and the eastern peri-Tyrrhenian Adriatic margin (Figure 6C). These tectono-magmatic features are totally coherent with NW-dipping of the Ionian oceanic lithosphere, plunging below the Calabrian-Aeolian area with a still seismically active Benioff zone. In our opinion, active mantle diapirism above the subduction zone, together with subduction retreat, not only supplied the basaltic magmas of the Tyrrhenian oceanic crust, but also favoured the counterclockwise rotation of Southern Italy (Scheepers et al., 1993), and pushed the Calabrian arc into its present position.

It should be noted that a significantly different interpretation of the Aeolian-Tyrrhenian area is favoured by Patacca et al. (1990) and Francalanci and Manetti (1994), who consider this system to have been generated by a recent subduction process started at the end of the Miocene, on the basis of the depth reached by the lithospheric slab and the estimated opening rate in the Tyrrhenian Sea of 6 cm/year for the last 8 Ma (Patacca et al., 1990). This view should be rejected at least for the following reasons: (a) the age of any subduction process cannot be estimated on the basis of the maximum depth reached by the lithospheric slab at a given moment, since we cannot know the extent of the lithosphere which disappeared once it recycled in the transition zone of the mantle or flattened toward it; (b) in the hypothesis of a short-lived Late Miocene subduction, deep lithospheric relics of an old oceanic slab – subducted during the Oligocene beneath the paleo-European margin – should be geophysically detectable below Sardinia, as in the general case of the Alpine–Apennine–Maghrebian chain (Spakman, 1990; Wortel and Spakman, 2000); on the contrary, tomographic data reported by Lucente et al. (1999) depict the slab currently dipping beneath the Aeolian-Calabrian arc as a unique body flattening in the upper mantle at around 670 km beneath the Tyrrhenian basin and adjacent Sardinia block (Figure 6C).

In conclusion we maintain that the Cenozoic tectono-magmatic evolution of the central-western Mediterranean (Beccaluva et al., 1985b, 1987) bears some analogies with that of the western Pacific arc/backarc systems, where arc magmatism linked to a unique subduction process migrates eastward with time, and is diachronously intercalated by the opening of several interarc oceanic basins (Beccaluva et al., 1980; Crawford et al., 1981; Gribble et al., 1998). Consequently, the best-fit model accounting for the space–time tectono-magmatic evolution and plate dynamics of the studied area relies on the following points:

1 – A NW-dipping subduction of oceanic lithosphere developed, in the Late Eocene, below the paleo-European continental margin, producing tholeiitic/calcalkaline magmas in Provence and Sardinia (34–26 Ma), typical of the initial stages of orogenic-arc magmatism related to subduction of pure oceanic lithosphere (Figure 6A).

2 – Interarc rifting of the Paleo-European continental margin developed later, resulting in the opening of the Ligurian–Balearic oceanic basin and the southeastward drifting and rotation of the Sardinia–Corsica block (Figure 6B). This evolutionary stage bears analogies with that of Japan, and can be satisfactorily accounted for – according to geodynamic modelling (Uyeda and Kanamori, 1979; Dewey, 1980) – by a progressive sinking of the subducted lithosphere and decoupling of the converging plates, with concomitant asthenospheric diapirism and extension in the overriding plate.

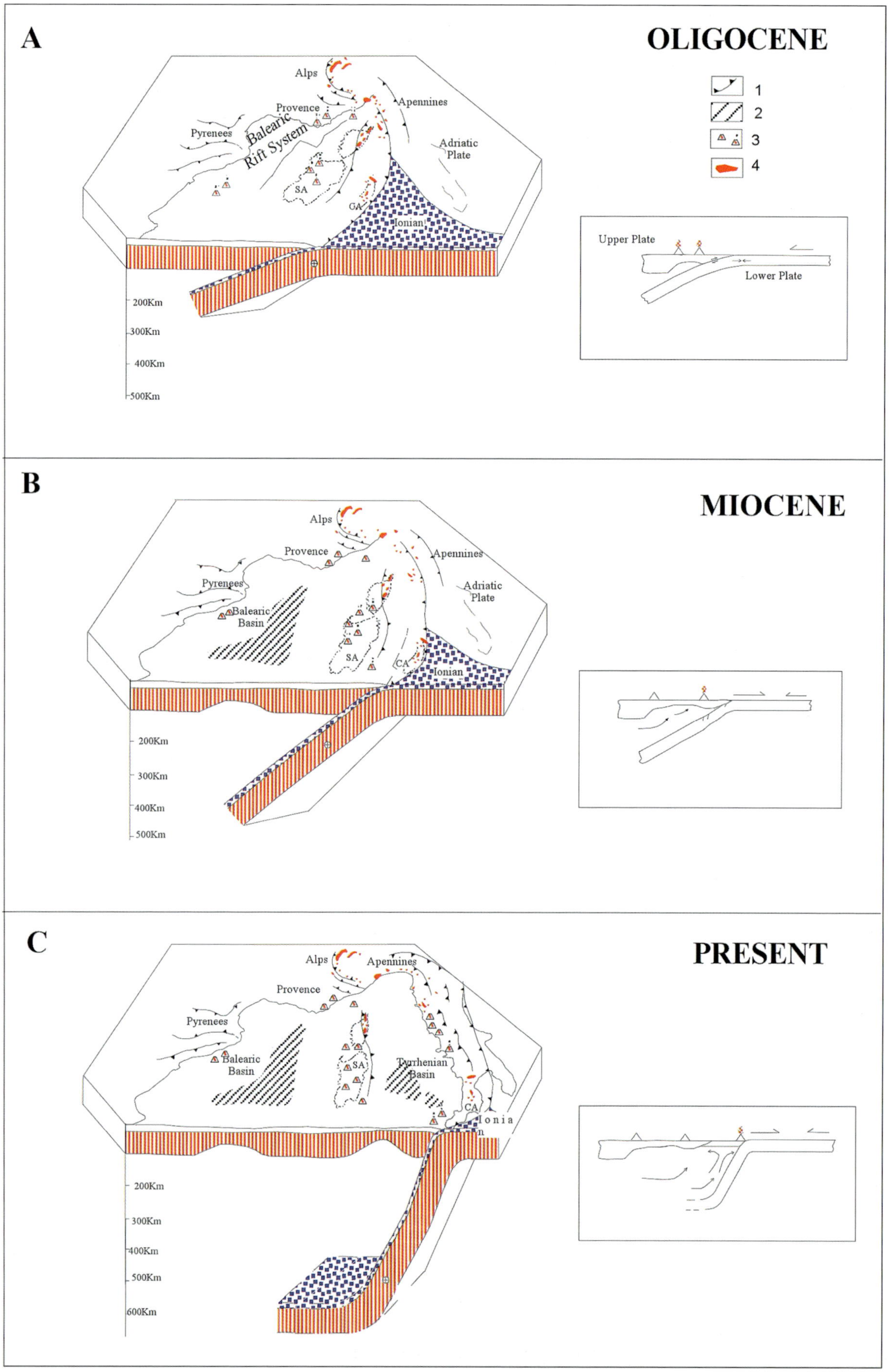

A
OLIGOCENE
1
2
3
4
Alps
Provence
Apennines
Pyrenees
Balearic
Rift System
Adriatic
Plate
SA
GA
Ionian
200Km
300Km
400Km
500Km
Upper Plate
Lower Plate
B
MIOCENE
Alps
Provence
Apennines
Pyrenees
Adriatic
Plate
Balearic
Basin
SA
CA
Ionian
200Km
300Km
400Km
500Km
C
PRESENT
Alps
Apennines
Provence
Pyrenees
Balearic
Basin
SA
Tyrrhenian
Basin
CA
200Km
300Km
400Km
500Km
600Km

Figure 6: Schematic model depicting the geodynamic evolution of the central-western Mediterranean area from the Oligocene (A), through the Miocene (B) up to the Present configuration (C). For comparison, the insets show progressive decoupling of the upper and lower plates, according to the evolutionary model of Uyeda and Kanamori (1979): in (A), convergence between the upper plate and subduction boundary results in a limited decoupling (Japan-type); in (C), sinking of the lower plate and retreat of the subduction boundary from the upper plate result in further decoupling, allowing accentuated asthenospheric diapirism and backarc opening (Mariana-type). The overall consumption of ca. 800 km of oceanic lithosphere, as reported in (C), is in excellent agreement with tomographic data (Lucente et al., 1999) depicting the Ionian slab subducted and flattened at around 670 km beneath the Tyrrhenian-Sardinia region. Legend: 1 = Compressional front of the Alpine and Apennine Chains; 2 = Oceanic crust of inter-arc basins; 3 = Cenozoic orogenic magmatism; 4 = Tethyan Jurassic ophiolites.

---

3 – The arc volcanism in Sardinia strongly decreased from 18 Ma and finished around 13 Ma, in relation to the steepening of the oceanic lithospheric slab, whereas further north, in Tuscany, continent-continent collision had already occurred (Jolivet et al., 1999 and reference therein). Rifting and splitting of the Calabrian nappes in a forearc position east of Sardinia started in the Late Miocene as a precursor to the Tyrrhenian oceanization (Figure 6B).

4 – Opening of the Tyrrhenian basin took place concomitantly with the anticlockwise rotation of Southern Italy, carrying the Calabrian arc southeastward to its present position (Figure 6C). During this phase, subduction-related magmatism migrated to the oceanic extension axis, as indicated by the calcalkaline signature of igneous rocks from Tyrrhenian sites 650 and 651. A similar tectono-magmatic evolution has been observed in other backarc basins, such as the Mariana Trough (Gribble et al., 1998) and is predicted by the Uyeda and Kanamori (1979) evolutionary model, where accentuated slab sinking ultimately results in retreating subduction, intense mantle diapirism, and opening of a new backarc basin.

5 – Geophysical modelling based on density and viscosity contrasts (Verde, 1996; Giunchi et al., 1996) and laboratory simulations (scaled analogue models; Faccenna et al., 2001) show excellent correspondence with seismic tomography (Spakman, 1990; Lucente et al., 1999; Wortel and Spakman, 2000), depicting nearly vertical, still seismically active subducted slab, down to 500 km, and its bending and flattening at greater depths. All evidence coherently indicates that the present tectono-magmatic phase is controlled by accentuated sinking of the subducted oceanic lithosphere, resulting in its further retreat, verticalization and incipient segmentation, as well as in subsidence and uplift (1–2 mm/y) in the Marsili basin and Calabrian arc, respectively.

6 – The Quaternary orogenic volcanism, from the Aeolian arc to the Roman Province, appears to be mainly related to extensional tectonics and deep lithospheric rifting of the eastern peri-Tyrrhenian margin, which allowed the generation and eruption of magmas from mantle sources previously modified by subduction-derived geochemical components.

On the whole, the central-western Mediterranean convergent system, although related to a single subduction process (from the Eocene to the Present), developed distinct features over time. The Eocene-Oligocene subduction phase was an "active" process characterized by moderate, though progressively increasing, decoupling between the converging plates, with relatively shallow dip of the Ionian oceanic lithosphere beneath the paleo-European continental margin. The Neogene-Quaternary subduction phase, instead, became increasingly "passive", characterized by accelerated slab sinking and deepening, and gradual retreat which produced intense asthenospheric diapirism and interarc opening in the over-riding plate.

REFERENCES

Balogh, K., Delle Rose, M., Guerrera, F., Ravasz-Baranyay, Veneri, F., 1993. New data concerning the inframiocenic "Bisciaro volcanoclastic event" (Umbro-Marche Apennine) and comparison with similar occurrences. Giornale di Geologia, ser. 3a, 55 2, 83–104.

Barbieri, M., Di Girolamo, P., Locardi, P., Lombardi, G., Stanzione, D., 1979. Petrology of the calc-alkaline volcanics of the Prete 2 well (Campania, Italy). Per. Mineral. 48, 53–74.

Beccaluva, L., Macciotta, G., Savelli, C., Serri, G., Zeda, O., 1980. Geochemistry and K/Ar ages of volcanics dredged in the Philippine Sea (Mariana, Yap and Palau trenches and Parece Vela basin). In: Hayes (Ed.), The Tectonic and Geological Evolution of Southern Asian Seas and Islands. AGU – Geophysical Monograph, vol. 23, pp. 247–268.

Beccaluva, L., Rossi, P.L., Serri, G., 1982. Neogene to recent volcanism of the Southern Tyrrhenian–Sicilian area: implication for the geodynamic evolution of the Calabrian arc. In: Monograph Structure, Evolution and Present Dynamics of the Calabrian Arc. Earth Evol. Sci. 3, 222–238.

Beccaluva, L., Civetta, L., Macciotta, G., Ricci, C.A., 1985a. Geochronology in Sardinia: results and problems. Rend. Soc. It. Min. Petr. 40, 57–72.

Beccaluva, L., Gabbianelli, G., Lucchini, F., Rossi, P.L., Savelli, C., 1985b. Petrology and K/Ar ages of volcanics dredged from the Eolian Seamounts: implications for geodynamic evolution of the southern Tyrrhenian basin. Earth Planet. Sci. Lett. 74, 187–208.

Beccaluva, L., Brotzu, P., Macciotta, G., Morbidelli, L., Serri, G., Traversa, G., 1987. Cainozoic tectono-magmatic evolution and inferred mantle in the Sardo-Tyrrhenian area. The lithosphere in Italy – advances in Earth science research. Accad. Naz. Lincei 80, 229–248.

Beccaluva, L., Bonatti, E., Dupuy, C., Ferrara, G., Innocenti, F., Lucchini, F., Macera, P., Petrini, R., Rossi, P.L., Serri, G., Seyler, M., Siena, F., 1990. Geochemistry and mineralogy of volcanic rocks from ODP Sites 650, 651, 655, and 654 in the Tyrrhenian Sea. In: Kastens, K.A., Mascle, J., et al. (Eds.), 1990 Proc. ODP, Scientific Results, vol. 107, pp. 49–74.

Beccaluva, L., Di Girolamo, P., Serri, G., 1991. Petrogenesis and tectonic setting of the Roman Volcanic Province, Italy. Lithos 26, 191–221.

Beccaluva, L., Coltorti, M., Galassi, B., Macciotta, G., Siena, F., 1994. The Cainozoic calcalkaline magmatism of the western Mediterranean and its geodynamic significance. Boll. Geof. Teor. Appl. 36 (141–144), 293–308.

Beccaluva, L., Bianchini, G., Coltorti, M., Siena, F., Vaccaro, C., 2001. Distinct metasomatic processes in the European and African lithospheric mantle: evidence from alkaline lavas and entrained peridotite xenoliths. In: Geophysical Research Abstract, vol. 3. 26th General Assembly of the European Geophysical Society, Nice, March 2001.

Bellon, H., 1981. Chronologie radiometrique (K–Ar) des manifestations magmatiques autour de la Mediterranee occidentale entre 33 et 1 Ma. In: Wezel, F.C. (Ed.), Sedimentary Basins of Mediterranean Margins, C.N.R. Italian Project of Oceanography. Technoprint, Bologna, pp. 341–360.

Burrus, J., 1984. Contribution to a geodynamic synthesis of the Provençal basin (N–W Mediterranean). Marine Geology 55, 247–269.

Carmignani, L., Giglia, G., Kligfield, R., 1978. Structural evolution of the Apuane Alps: an example of continental margin deformation. J. Geology 86, 487–504.

Caron, C., Orgeval, J.J., 1995. Origine du plomb des minéralisations à métaux de base et métaux précieux de Sardaigne occidentale et leur encaissant volcanique tertiaire. C. R. Acad. Sci. Paris, Sér. IIa 323, 41–47.

Chamot-Rooke, N., Gaulier, J.-M., Jestin, F., 1999. Constraints on Moho depth and crustal thickness in the Liguro-Provençal basin from a 3D gravity inversion: geodynamic implications. In: Durand, et al. (Eds.), The Mediterranean Basin: Tertiary Extension within the Alpine Orogen. Geol. Soc. Spec. Publ. 156, 37–62.

Cherchi, A., Montadert, L., 1982. Oligo-Miocene rift of Sardinia and the early history of the western Mediterranean basin. Nature 298, 736–739.

Cibin, U., Tateo, F., Catanzariti, R., Martelli, L., Rio, D., 1998. Composizione, origine ed età del vulcanismo andesitico oligocenico inferiore dell'Appennino Settentrionale: le intercalazioni vulcano-derivate nella formazione di Ranzano. Boll. Soc. Geol. It. 117, 569–581.

Civetta, L., Orsi, G., Scandone, P., Pece, R., 1978. East-wards migration of the Tuscan anatectic magmatism due to anti-clockwise rotation of the Apennines. Nature 276, 604–606.

Conticelli, S., D'Antonio, M., Pinarelli, L., Civetta, L., 2002. Source contamination and mantle heterogeneity in the genesis of Italian potassic and ultrapotassic volcanic rocks: Sr–Nd–Pb isotope data. Mineral. and Petrol. 74, 189–222.

Crawford, A.J., Beccaluva, L., Serri, G., 1981. Tectono-magmatic evolution of the west Philippine-Mariana region and the origin of boninites. Earth Planet. Sci. Lett. 54, 346–356.

Crisci, G.M., Critelli, S., De Rosa, R., 1988. Vulcanismo sinsedimentario nella successione terrigena della Formazione di San Mauro (Miocene Inferiore, Unità del Cilento), Appennino Meridionale. Miner. Petr. Acta XXXI, 159–178.

Critelli, S., 1993. Sandstone detrital modes in the Paleogene Liguride Complex, accretionary wedge of the southern Apennines (Italy). J. Sedimentary Petrol. 63 (3), 464–476.

De Astis, G., Peccerillo, A., Kempton, P., La Volpe, L., Wu, T.W., 2000. Transition from calc-alkaline to potassium-rich magmatism in subduction environments: geochemical and Sr, Nd, Pb isotopic constraints from the island of Vulcano (Aeolian arc). Contrib. Mineral. Petrol. 139, 684–703.

Del Moro, A., Gioncada, A., Pinarelli, L., Sbrana, A., Joron, J.L., 1998. Sr, Nd, and Pb isotope evidence for open system evolution at Vulcano, Aeolian Arc, Italy. Lithos 43, 81–106.

Dewey, J.F., 1980. Episodicity, sequence and style at convergent plate boundaries. In: Stranway, D.W. (Ed.), The Continental Crust and its Mineral Deposits. Spec. Pap. Geol. Assoc. Can. 20, 553–573.

Dewey, J.F., Helman, M.L., Torco, E., Hutton, D.H.W., Knott, S.D., 1989. Kinematics of the Western Mediterranean. In: Coward, et al. (Eds.), Alpine Tectonics. Geol. Soc. Spec. Publ. 45, 265–283.

Di Girolamo, P., 1978. Geotectonic setting of Miocene-Quaternary volcanism in and around the eastern Tyrrhenian Sea border (Italy) as deduced from major elements geochemistry. Bull. Volcanol. 41 (3), 1–22.

Doglioni, C., Gueguen, E., Harabaglia, P., Mongelli, F., 1999. On the origin of west-directed subduction zones and applications to the western Mediterranean. In: Durand, et al. (Eds.), The Mediterranean Basin: Tertiary Extension within the Alpine Orogen. Geol. Soc. Spec. Publ. 156, 541–561.

Downes, H., Thirlwall, M.F., Trayhorn, S.C., 2000. Miocene subduction-related magmatism in southern Sardinia: Sr–Nd-and oxygen isotopic evidence for mantle source enrichment. J. Volcanol. Geoth. Res. 106, 1–21.

Ellam, R.M., Menzies, M.A., Hawkesworth, C.J., Leeman, W.P., Rosi, M., Serri, G., 1988. The transition from calc-alkaline to potassic orogenic magmatism in the Aeolian Islands, Southern Italy. Bull. Vulcanol. 50, 386–398.

Ellam, R.M., Hawkesworth, C.J., Menzies, M.A., Rogers, N.W., 1989. The volcanism of Southern Italy: Role of subduction and the relationship between potassic and sodic alkaline magmatism. J. Geophys. Res. 94, 4589–4601.

Ellam, R.M., Harmon, R.S., 1990. Oxygen isotope constraints on the crustal contribution to the subduction-related magmatism of the Aeolian Islands, Southern Italy. J. Volcanol. Geoth. Res. 44, 105–122.

Faccenna, C., Funiciello, F., Giardini, D., Lucente, P., 2001. Episodic back-arc extension during restricted mantle convection in the Central Mediterranean. Earth Planet. Sci. Lett. 187 (1–2), 105–116.

Finetti, I., Lentini, F., Carbone, S., Catalano, S., Del Ben, A., 1996. Il sistema Appennino Meridionale–Arco Calabro–Sicilia nel Mediterraneo Centrale: studio geologico e geofisico. Boll. Soc. Geol. It. 115, 529–559.

Finetti, I., Boccaletti, M., Bonini, M., Del Ben, A., Geletti, R., Pipan, M., Sani, F., 2001. Crustal section based on CROP seismic data across the North Tyrrhenian–Northern Apennines–Adriatic Sea. Tectonophysics 343, 135–163.

Francalanci, L., Taylor, S.R., McCulloch, M.T., Woodhead, J.D., 1993. Geochemical and isotopic variations in the calc-alkaline rocks of Aeolian arc, southern Tyrrhenian Sea, Italy: constraints on magma genesis. Contrib. Mineral. Petrol. 113, 300–313.

Francalanci, L., Manetti, P., 1994. Geodynamic models of the southern Tyrrhenian region: constraints from the petrology and geochemistry of the Aeolian volcanic rocks. Boll. Geof. Teor. Appl. 36, 283–292.

Galassi, B., 1995. Evoluzione magmatologica delle associazioni orogeniche cenozoiche del Mediterraneo Centro-Occidentale. PhD thesis, University of Ferrara.

Gasparini, C., Iannaccone, G., Scandone, P., Scarpa, R., 1982. Seismotectonics of the Calabrian Arc. Tectonophysics 84, 267–286.

Giunchi, C., Sabadini, R., Boschi, E., Gasperini, 1996. Dynamic models of Subduction – Geophysical and geological evidence in the Tyrrhenian sea. Geophys. J. Internat. 126, 555–578.

Gribble, R.F., Stern, R.J., Newman, S., Bloomer, S.H., O'Hearn, T., 1998. Chemical and isotopic composition of lavas from the Northern Marianna Trough: implication for magmagenesis in back-arc basin. J. Petrol. 39 (1), 125–154.

Hawkesworth, C.J., Vollmer, R., 1979. Crustal contamination versus enriched mantle: $^{143}Nd/^{144}Nd$ and $^{87}Sr/^{86}Sr$ evidence from the Italian volcanics. Contrib. Mineral. Petrol. 69, 151–165.

Huang, Y.M., Hawkesworth, C., Smith, I., Van Calsteren, P., Black, P., 2000. Geochemistry of late Cenozoic basaltic volcanism in Northland and Coromandel, New Zealand: implication for mantle processes. Chem. Geol. 164, 219–238.

Jolivet, L., Frizon, De La Motte, D., Mascle, A., Seranne, M., 1999. The Mediterranean Basins: Tertiary extension within the Alpine Orogen – an introduction. In: Durand, et al. (Eds.), The Mediterranean Basin: Tertiary Extension within the Alpine Orogen. Geol. Soc. Spec. Publ. 156, 1–14.

Kastens, K., Mascle, J., Aurox, C., Bonatti, E., Broglia, C., Channel, J., Curzi, P., Emeis, K.C., Glaçcon, G., Hasegawa, S., Hieke, W., Mascle, G., McCoy, F., McKenzie, J., Mendelson, J., Mueller, C., Rehault, J.P., Robertson, A., Sartori, R., Sprovieri, R., Torii, M., 1986. Young Tyrrhenian sea evolve very quickly. Geotimes 31.

Lecca, L., Lonis, R., Luxoro, S., Melis, E., Secchi, F., Brotzu, P., 1997. Oligo-Miocene volcanic sequences and rifting stages in Sardinia: a review. Per. Mineral. LXVI (1–2), 7–62.

Lucente, P.F., Chiarabba, C., Cimini, G.B., Giardini, D., 1999. Tomographic constraints on the geodynamic evolution of the Italian region. J. Geophys. Res. 104, 20307–20327.

Malinverno, A., Ryan, W.B.F., 1986. Extension in the Tyrrhenian Sea and shortening in the Apennines as results of arc migration driven by sinking of the lithosphere. Tectonics 5, 227–245.

Mameli, P., 1996 Studio petrografico e geochimico delle vulcaniti oligo-mioceniche dell'Anglona (Sardegna Settentrionale). Degree Thesis, University of Ferrara.

Mantovani, E., Albarello, D., Tamburelli, C., Babbucci, D., Viti, M., 1997. Plate convergence, crustal delamination, extrusion tectonics and minimization of shortening work as main controlling factors of the recent Mediterranean deformation pattern. Ann. Geofis. 40, 611–643.
Martì, J., Mitjavila, J., Roca, E., Aparicio, A., 1992. Cenozoic magmatism in the Valencia trough (western Mediterranean): relationships between structural evolution and volcanism. Tectonophysics 203, 145–165.
Mascle, G.H., Tricart, P., Torelli, L., Boullin, J.-P., Rolfo, F., Lapierre, H., Monie, P., Depardon, S., Mascle, J., Peis, D., 2001. New constraints for the opening of the Tyrrhenian sea: structure of the Sardinia Channel, age, petrology and geochemistry of the Cornacya seamount. Marine Geology 179, 179–201.
Mattioli, M., 1997. Tertiary volcanism in the Northern Apennines (Italy): evidences from the Canetolo unit and buried volcanic bodies. Plinius – Eur. J. Miner. Italian Supplement 17, 150–154.
Montigny, R., Edel, J.B., Thuizat, R., 1981. Oligo-Miocene rotation of Sardinia: K/Ar ages and paleomagnetic data of Tertiary volcanics. Earth Planet. Sci. Lett. 54, 261–271.
Morra, V., Secchi, F.A.G., Melluso, L., Franciosi, L., 1997. High-Mg subduction related Tertiary basalts in Sardinia, Italy. Lithos 40, 69–91.
Ottaviani-Spella, M., Girad, M., Cheilletz, A., 1996. Les ignimbrites burdigaliennes du Sud de la Corse. C. R. Acad. Sci. Paris 323, 1–3.
Panza, G.F., Calcagnile, G., Scandone, P., Mueller, S., 1980. La struttura profonda dell'area mediterranea. Le Scienze 141, 60–69.
Patacca, E., Sartori, R., Scandone, P., 1990. Tyrrhenian basin and Apenninic arcs. Kinematic relations since Late Tortonian times. Mem. Soc. Geol. It. 45, 425–451.
Rampone, E., Hofmann, A.W., Raczeck, I., 1988. Isotopic contrasts within the Internal Liguride ophiolite (N. Italy): the lack of a genetic mantle-crust link. Earth Planet. Sci. Lett. 163, 175–189.
Rehault, J.P., Mascle, J., Boillot, G., 1984. Evolution geodynamique de la Mediterranee depuis l'Oligocene. Mem. Soc. Geol. It. 27, 85–96.
Rossi, P., Guennoc, P., Rehault, J.P., Arnaud, N., Jakni, B., Poupeau, G., Tegyey, M., Ferrandini, J., Sosson, M., Beslier, M.-O., Rollet, N., Gloaguen, R., 1998. Importance du volcanisme calco-alcalin miocène sur la marge sud-ouest de la Corse (campagne MARCO). C. R. Acad. Sci. Paris, Sciences de la terre et des planètes 327, 369–376.
Savelli, C., Gasparotto, G., 1994. Calc-alkaline magmatism and rifting of the deep-water volcano of Marsili (Aeolian back-arc, Tyrrhenian sea). Marine Geology 119, 137–157.
Scheepers, P.J.J., Langereis, C.G., Hilgen, F.J., 1993. Counter-clockwise rotations in the southern Apennines during the Pleistocene: paleomagnetic evidences from the Matera area. Tectonophysics 225, 379–410.
Serri, G., 1990. Neogene-Quaternary magmatism of the Tyrrhenian region: characterization of the magma sources and geodynamic implications. Mem. Soc. Geol. It. 41, 219–242.
Serri, G., Innocenti, F., Manetti, P., 1993. Geochemical and petrological evidence of the subduction of delaminated Adriatic continental lithosphere in the genesis of the Neogene-Quaternary magmatism of Central Italy. Tectonophysics 223, 117–147.
Serri, G., Innocenti, F., Manetti, P., 2001. Magmatism from Mesozoic to Present: Petrogenesis, time–space distribution and geodynamic implications. In: Vai, G.B., Martini, I.P. (Eds.), An Anatomy of an Orogen: The Apennines and Adjacent Mediterranean Basins. Kluwer Academic, pp. 77–104.
Sosson, M., Rollet, N., Deverchere, J., 1998. New constraints on the Ligurian sea formation and evolution. In: Insight on the Formation and Evolution of Mediterranean Basins, MAST Advanced Study Course (Villefranche sur Mer, France, July 1998) Abstracts-Book of the Lecturers, pp. 33–34.
Spakman, W., 1990. Tomographic images of the upper mantle below central Europe and the Mediterranean. Terra Nova 2, 542–553.
Sun, S.-S., McDonough, W.F., 1989. Chemical and isotopic systematics of oceanic basalts: implications for mantle composition and process. In: Saunders, Norry (Eds.), Magmatism in the Oceanic Basins. Geol. Soc. Spec. Publ. 42, 313–346.
Sun, C.-H., Stern, R., Yoshida, T., Kimura, J.-I., 1998. Fukutoku-oka-no-ba Volcano: A new perspective on the alkalic volcano province in the Izu-Bonin-Mariana Arc. The Island Arc 7, 432–442.
Tonarini, S., Leeman, W.P., Ferrara, G., 2001. Boron isotopic variations in lavas of the Aeolian volcanic arc, South Italy. J. Volcanol. Geoth. Res. 110, 155–170.
Tonarini, S., Leeman, W.P., Civetta, L., D'Antonio, M., Ferrara, G., Necco, A., 2004. B/Nb and δ^{11}B systematics in the Phlegrean volcanic district (PVD). J. Volcanol. Geoth. Res. 133, 123–139.
Uyeda, S., Kanamori, H., 1979. Back-arc opening and the mode of subduction. J. Geophys. Res. 84, 1049–1106.
Verde, M., 1996. Aspetti petrologici e reologici del mantello superiore nell'area mediterranea mediante lo studio di xenoliti di mantello inclusi in vulcaniti basiche alcaline. Ph.D. Thesis, University of Ferrara.
Vigliotti, L., Langenheim, V.F., 1995. When Sardinia stop rotating? New Paleomagnetic results. Terra Nova 7, 424–435.

Vollmer, R., 1976. Rb–Sr and U–Th–Pb systematics of alkaline rocks: the alkaline rocks from Italy. Geochim. Cosmochim. Acta 40, 283–295.

Wilson, M., 1989. Igneous Petrogenesis. International Thompson, London.

Wilson, M., Bianchini, G., 1999. Tertiary-Quaternary magmatism within the Mediterranean and surrounding regions. In: Durand, et al. (Eds.), The Mediterranean Basin: Tertiary Extension within the Alpine Orogen. Geol. Soc. Spec. Publ. 156, 141–168.

Wortel, M.J.R., Spakman, W., 2000. Subduction and slab detachment in the Mediterranean-Carpatian region. Science 290, 1910–1917.

CROP PROJECT: Deep Seismic Exploration of the Central Mediterranean and Italy
Edited by I.R. Finetti

CHAPTER 30

Within-Plate Cenozoic Volcanism and Lithospheric Mantle Evolution in the Western-Central Mediterranean Area

L. Beccaluva[1], G. Bianchini[1], C. Bonadiman[1], M. Coltorti[1], G. Macciotta[2], F. Siena[1], C. Vaccaro[1]

ABSTRACT

Extension-related within-plate Cenozoic volcanism in the central-western Mediterranean mostly occurs in three provinces belonging to distinct lithospheric domains: the Veneto, Adria plate, i.e. African northern promontory; the Iblei (Sicily), African Plate; and Sardinia, a drifted fragment of the European Plate. In Veneto (Paleogene) and Iblei (Neogene-Quaternary), transtensional rift volcanism developed as foreland reaction to collisional processes along the Alpine and Maghrebian chains respectively, generating prevailingly basic magmas – from tholeiites to Na-alkali basalts, basanites and nephelinites. In Sardinia, the Pliocene-Quaternary volcanism – related to general tensional tectonics in the central Mediterranean – produced magmas comparatively richer in potassium: from subalkaline basalts, alkali basalts/trachybasalts to basanites, locally associated with rhyolitic and phonolitic differentiates.

An integrated petrogenetic study, based on incompatible element and Sr–Nd–Pb isotope systematics for both lavas and associated mantle xenoliths for the three volcanic provinces, leads to the following constraints: (1) the primary magmas, from tholeiites, alkali basalts to basanites and nephelinites, were generated by decreasing melting degrees of progressively deeper lithospheric mantle sources (ca. 30 to 100 km depth); (2) extensive and multiple enrichment processes by alkali-silicate metasomatizing components widely affected all mantle sources, which invariably require 5–10% metasomatic volatile-bearing phases, mainly amphibole for Veneto and Iblei, and phlogopite + amphibole for Sardinia; (3) a previously depleted lithospheric mantle (DM) is enriched by prevailing HIMU (with subordinate EMII) geochemical components in both the Veneto and Iblean Provinces, whereas in Sardinia a distinctly more potassic EMI signature predominates, in addition to HIMU; (4) in all the investigated provinces, the more alkaline deeper basic magmas (compared to subalkaline basalts) show relatively more marked HIMU signatures, also with involvement of carbonatitic components in the lowest lithospheric sources where nephelinites were generated.

A comparison at a circum-Mediterranean scale reveals that the geochemical features of Sardinia are also observed in the Cenozoic anorogenic magmas and associated mantle xenoliths located within the Variscan basement of central-western Europe. Peridotite massifs, tectonically emplaced during the Alpine orogenesis, are similarly characterized by a prevalent EMI signature, which could thus have been effective at least since the pre-middle Mesozoic. By contrast, EMI signature is lacking in Veneto and Iblei, as well as in the mantle xenoliths and host alkaline lavas from other occurrences of the stable North African domain, where the HIMU component predominates. The ubiquitous presence of HIMU in the European/North African domains lends support to the existence of a Cenozoic asthenospheric mantle plume (thought to extend from the eastern Atlantic to Central Europe and the western Mediterranean), which could therefore represent the provenance of this component. Consequently, the more recent HIMU metasomatizing agents rising from the convecting sub-lithospheric mantle appear to have been effectively accumulated in the lower lithospheric portions (i.e. Thermal Boundary Layer), whereas older metasomatic components (e.g. EMI for Europe) may have been better preserved in the upper, more rigid lithospheric mantle (i.e. Mechanical Boundary Layer).

[1]Department of Earth Sciences, University of Ferrara, Ferrara, Italy
[2]Department of Earth Sciences, University of Cagliari, Cagliari, Italy

Keywords: Mediterranean area, within-plate Cenozoic volcanism, lithospheric mantle evolution

1. INTRODUCTION

The composition of basic magmas generated in the different intra-plate geodynamic settings – such as active and passive continental rifts, transtensive systems and "plume regions" – is strongly constrained by the geochemical evolution, particularly the enrichment processes, which affected their mantle sources. Therefore, the geochemical components of magmas basically reflect those of the source from which they were generated. The associated mantle xenoliths, in turn, represent a further opportunity to study the compositional evolution of the lithospheric mantle related to: (i) depletion events by partial melting and extraction of basic magmas; (ii) enrichment processes due to the interaction between the pristine mantle parageneses and percolating metasomatizing melts of sublithospheric provenance. Previous works have shown the complex evolution of the lithospheric mantle in the central-western Mediterranean, whose geochemical components seem to be attributable to multiple enrichment events (Beccaluva et al., 2001a, 2001b; Downes, 2001 and references therein), including those related to the influence of a Cenozoic mantle plume supposed to extend from the eastern Atlantic to central Europe (Hoernle et al., 1995; Granet et al., 1995).

In this chapter, we review the petrological characteristics of the three main Cenozoic anorogenic volcanic provinces (and associated mantle xenoliths) of the central-western Mediterranean area: the Pliocene-Quaternary volcanism of Sardinia (European lithosphere); the Neogene-Quaternary volcanism of the Iblean area, eastern Sicily (African lithosphere); and the Paleogene Veneto Volcanic Province (Adriatic lithosphere, African northern promontory). This integrated study is aimed at defining the petrological and geochemical characteristics of basic parental magmas and their mantle sources, as well as the evolution of the lithospheric mantle beneath the investigated area. This may also contribute to highlighting regional analogies and differences, in terms of geochemical components, of important lithospheric sections belonging to the European and North African domains and facing one another in the central Mediterranean.

2. PETROLOGY OF BASIC MAGMAS

In the central-western Mediterranean area, the volcanic provinces of Veneto, Iblei and Sardinia record the major within-plate magmatic events during the Cenozoic over an area of more than 4000 km^2 (Figure 1). The first two provinces lay on the African lithosphere (Iblei) and its northern promontory represented by the Adria micro plate (Veneto), whereas the latter rests on the Sardinian continental block, which drifted and rotated during the Oligocene-Miocene away form the palaeo-European continental margin (Beccaluva et al., 1987, and this volume). For all the investigated areas, geophysical data indicate a Moho depth of 28–35 km, with maximum lithospheric thickness of 90–100 km. The overall tectono-magmatic features of the three provinces are typical of within-plate settings, with Veneto and Iblei most favourably comparing with impactogenic low-volcanicity rift, where magmatism is triggered as foreland reaction of orogenic collisional processes. In the following sections, the geochemical characteristics of basic magmas and associated xenoliths are discussed in terms of the major mantle components defined for Ocean Island Basalts (OIB) (Zindler and Hart, 1986; Weaver, 1991; Hofmann, 1997).

Veneto Volcanic Province

The Veneto Volcanic Province (VVP), of late Palaeocene to late Oligocene age (De Vecchi and Sedea, 1995), extends over an area of about 1500 km^2, with numerous eruptive centres mostly aligned along NNW–SSE tectonic trends. Magma generation appears to have been triggered by decompression events related to tensional tectonics which affected the South Alpine foreland in response to general NS compression during the Alpine orogenesis (Beccaluva et al., 1999a; Bonadiman et al., 2001; Milani et al., 1999). Geophysical data indicate a Moho peak of 28 km below the area, and a lithosphere thickness of about 100 km (Panza and Suhadolc, 1990).

VVP lavas are mostly basic in composition, and encompass a wide range of serial affinities comprising mela(M)-nephelinites, basanites, alkaline and transitional basalts, olivine(ol)- and quartz(qz)-tholeiites (Figure 2), as observed in low-volcanicity rifts (Barberi et al., 1982). Nephelinites and basanites often carry spinel-peridotite mantle xenoliths. The relative abundance of silica-undersaturated to silica-oversaturated products shows a regional variation with abundant nephelinites and basanites to the west (Val d'Adige and western Lessini Mts.) and predominant

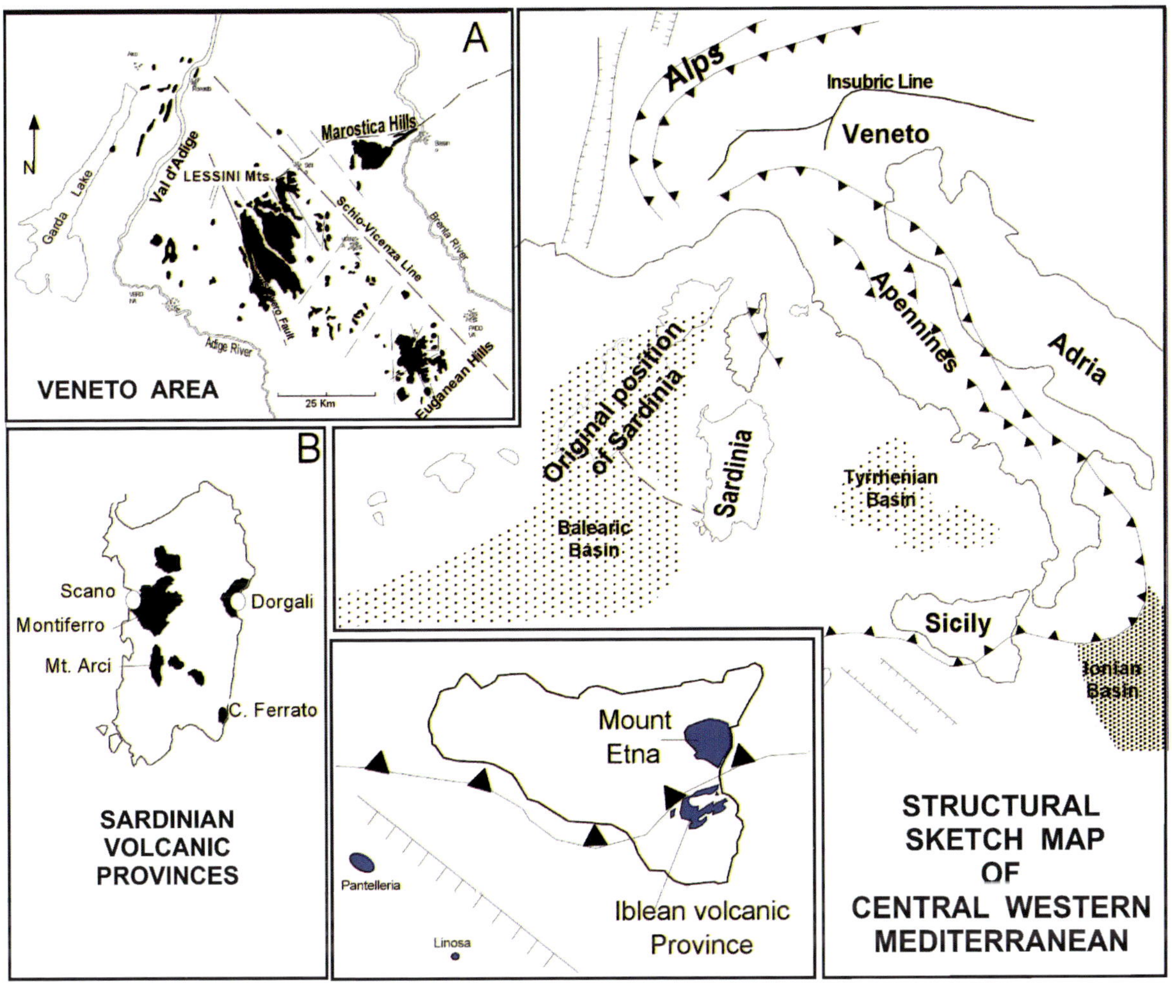

Figure 1: Structural sketch map of the central-western Mediterranean area. In the insets, the Veneto (A), Sardinian (B), and Iblean (C) intra-plate volcanic provinces are reported.

alkali and transitional basalts, and tholeiites to the east (eastern Lessini and Marostica Hills). Differentiated products only occur in the Euganean complex, where transitional basalts to quartz-trachytes and rhyolites predominate (Milani et al., 1999).

Primitive mantle-normalized incompatible element distribution is reported for the different basic magmas in Figure 3, where sub-parallel patterns are observed, becoming gradually more enriched from tholeiites to basanites. These patterns show intermediate characteristics between HIMU (High U/Pb) and EMII (Enriched Mantle II) of OIB end-members. It should be noted that M-nephelinites show marked differences in incompatible element distribution compared to the other basic magmas, as they are characterized by much higher values of Ba, and by the appearance of significant Sr, Zr, and Ti negative anomalies. This suggests that their mantle sources were markedly different from those of tholeiites to basanites. The petrological modelling of primary magmas indicates that tholeiites to M-nephelinites may have been generated by decreasing degrees of partial melting (~ 25 to $\sim 3\%$) of spinel-peridotite mantle sources at increasing depths (~ 30 to ~ 100 km) (Beccaluva et al., 1999a, 2000; Bonadiman et al., 2001). The magma sources have to be lherzolites bearing metasomatic amphibole ($\pm$ phlogopite) for tholeiites to basanites, whereas cpx-rich lherzolites (or even wehrlites), metasomatized also by carbonatitic components, are required to generate the M-nephelinites.

Sr, Nd, and Pb isotopic data (Figure 4) indicate that a previously depleted mantle (DM) was subsequently affected by metasomatic enrichment(s) with a prevalent HIMU signature and a subordinate contribution of the EMII component, particularly for tholeiitic magmas. This is evidenced by significantly different isotopic ranges between alkaline ($^{87}Sr/^{86}Sr$ 0.70315–0.70344, $^{143}Nd/^{144}Nd$ 0.512902–0.512976, and $^{206}Pb/^{204}Pb$ 19.348–19.789) and tholeiitic lavas ($^{87}Sr/^{86}Sr$ 0.70325–0.70387, $^{143}Nd/^{144}Nd$ 0.512858–0.512894, and $^{206}Pb/^{204}Pb$ 19.202–19.219).

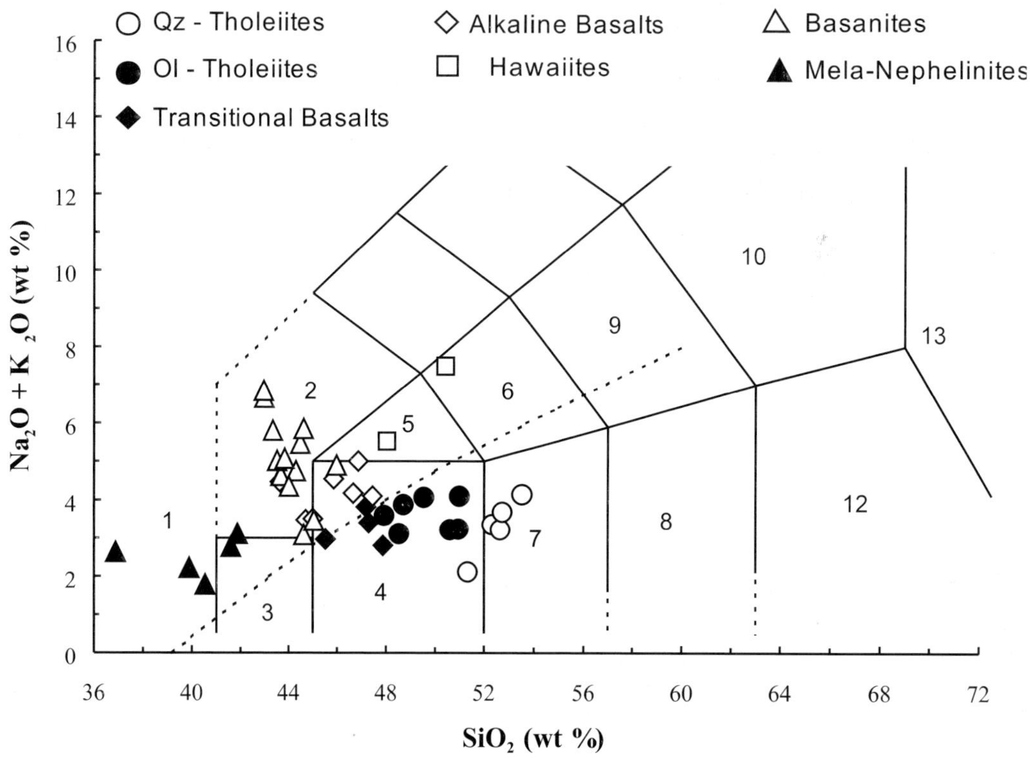

Figure 2: Total alkali-silica diagram for the Veneto basic lavas. Classification fields after Le Bas et al. (1992): 1 – nephelinites; 2 – basanites and tephrites; 3 – picritic basalts; 4 – basalts; 5 – trachybasalts; 6 – basaltic trachyandesites; 7 – basaltic andesites; 8 – andesites; 9 – trachyandesites; 10 – trachytes; 11 – phonolites; 12 – dacites; 13 – rhyolites. The dividing line between alkalic and subalkalic series is after Irvine and Baragar (1971).

Iblean Volcanic Province

In south-eastern Sicily, south of Etna, Miocene and Pliocene-Pleistocene volcanics cover an area of about 500 km^2. Fissural activity, with subaerial and submarine eruptions, produced tholeiitic to nephelinitic lavas along a regional NE–SW lithospheric wrench fault system (Beccaluva et al., 1993, 1998; Trua et al., 1998; Di Grande et al., 2002). The Miocene (mainly Tortonian) magmatic phase, predominantly alkaline in composition, was characterized by a low melt production rate. After a period of quiescence from about 6.5 to 4 Ma (according to biostratigraphical data), volcanism resumed in the Early-Middle Pliocene and continued until the Early Pleistocene. No definite spatial-temporal zonation has been observed; however, the oldest and volumetrically predominant products of Pliocene activity are tholeiites, followed in order of decreasing abundance by basanites, alkali-basalts + hawaiites, transitional basalts and nephelinites (Figure 5).

Primitive mantle-normalized incompatible element patterns for the Iblean basic lavas (Figure 6) display intermediate characteristics between HIMU and EMII OIB end-members, and are closely comparable to those observed in lavas from the VVP area. The available Sr–Nd–Pb isotope data (Figure 4) plot from the depleted mantle (DM) to the HIMU component (Beccaluva et al., 1998; Bianchini et al., 1998, 1999). In particular, subalkaline lavas display a more depleted character ($^{87}Sr/^{86}Sr$ 0.70271–0.70302 and $^{143}Nd/^{144}Nd$ 0.51325–0.51299), whereas alkaline lavas are more enriched ($^{87}Sr/^{86}Sr$ 0.70287–0.70327 and $^{143}Nd/^{144}Nd$ 0.51302–0.51291). Beccaluva et al. (1998) presented a petrogenetic model where most of the Iblean magmas were generated within the spinel peridotite lithospheric mantle (30 to ca. 90 km depth) from progressively deeper sources, from tholeiites to nephelinites, with a parallel decrease in the degree of melting (≈ 30 to $\approx 3\%$). The modelled mantle sources were: (i) lherzolites bearing amphibole $\pm$ phlogopite for the generation of tholeiites, alkali basalts and basanites; (ii) clinopyroxene-rich lherzolites (or even wehrlites) bearing amphibole + phlogopite + carbonatitic components for nephelinitic magmas.

There are many similarities between the Iblean volcanism and that of the Sicily Channel, particularly on the islands of Pantelleria and Linosa. Stratigraphic evidence and K–Ar dating indicate that volcanic activity com-

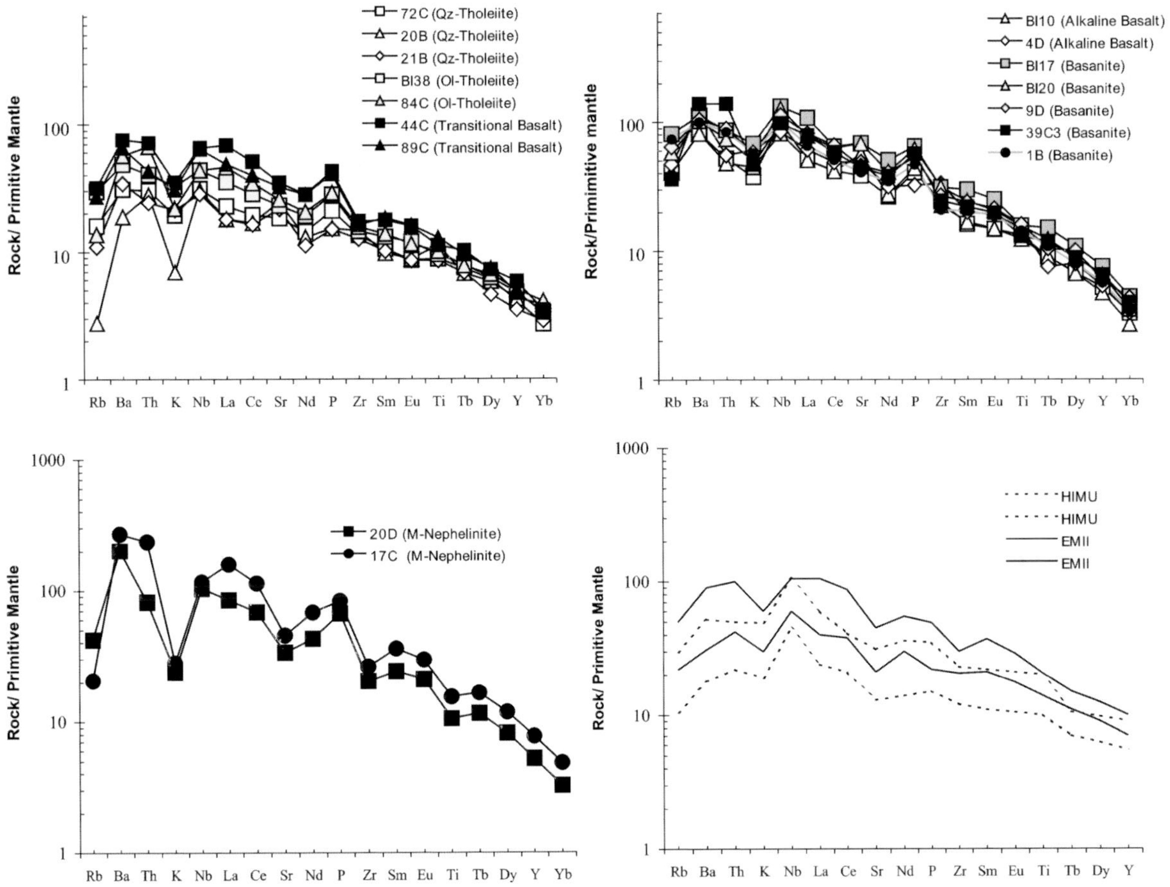

Figure 3: Primitive mantle-normalized incompatible element patterns for the Veneto basic lavas. Normalizing factors from Sun and McDonough (1989). The typical patterns of HIMU and EMII ocean island basalts, after Weaver (1991), are also reported.

menced here in the Upper Miocene (ca. 10 Ma; Beccaluva et al., 1981) and continued until the Present, with the last observed manifestations in 1891 (ephemeral Ferdinandea Island). The most primitive volcanic rocks on Pantelleria are transitional-mildly alkaline basalts and hawaiites. Acidic magmas (trachytes and peralkaline rhyolites) are widespread, and are usually interpreted as differentiation products of parental transitional basalts (Civetta et al., 1998). Linosa is predominantly composed of alkali-basalts with subordinate hawaiites. Beccaluva et al. (1981) and Calanchi et al. (1989) studied submarine volcanic rocks dredged from different locations in the Sicily Channel, reporting a wide range of basic magmas from tholeiites to nepheline basanites. It should be emphasized that the trace element and Sr–Nd–Pb isotopic compositions of volcanic rocks from the Sicily Channel show striking analogies with comparable Iblean lavas, consistently indicating the prevalence of the HIMU metasomatic components in magma sources of this sector of the African Plate (Esperança and Crisci, 1995; Civetta et al., 1998).

To the north, the eruptive activity of Mount Etna displays analogies with Iblean volcanism. The oldest Etnean products (ca. 600,000 years BP; Romano, 1983) are tholeiitic, whereas alkaline mafic magmatism commenced around 220,000 years BP (Condomines et al., 1982; Gillot et al., 1994), alternating for a short period with the tholeiitic lavas. More recently, magmatism has shown exclusively alkaline affinity and the erupted lavas have become hawaiitic in composition, consistently with fractionation processes taking place in a high-level magma chamber beneath the volcano (Clocchiatti et al., 1988). Melt inclusion data in olivine phenocrysts from the Etnean rocks (Schiano et al., 2001) record a wide range of near-primary OIB types, although some of these also reveal a significant imprint of arc magmas in their trace element distribution. As already observed by Beccaluva et al. (1982), the Etnean lavas generally display some enrichment of low field strength elements (LFSE, such as: Cs, Rb, Ba, K, Th, U, Sr) as well as light rare earth elements (LREE) with respect to Iblean products. Detailed Sr–Nd–B isotopic analyses, carried out on both whole rocks and clinopyroxene separates by Tonarini et al. (2001), have shown that such enrichment cannot be induced merely by shallow-level crustal contamination. In order to explain the temporal evolution to higher concentrations of LFSE, higher δ 11B and $^{87}Sr/^{86}Sr$ values and lower $^{143}Nd/^{144}Nd$,

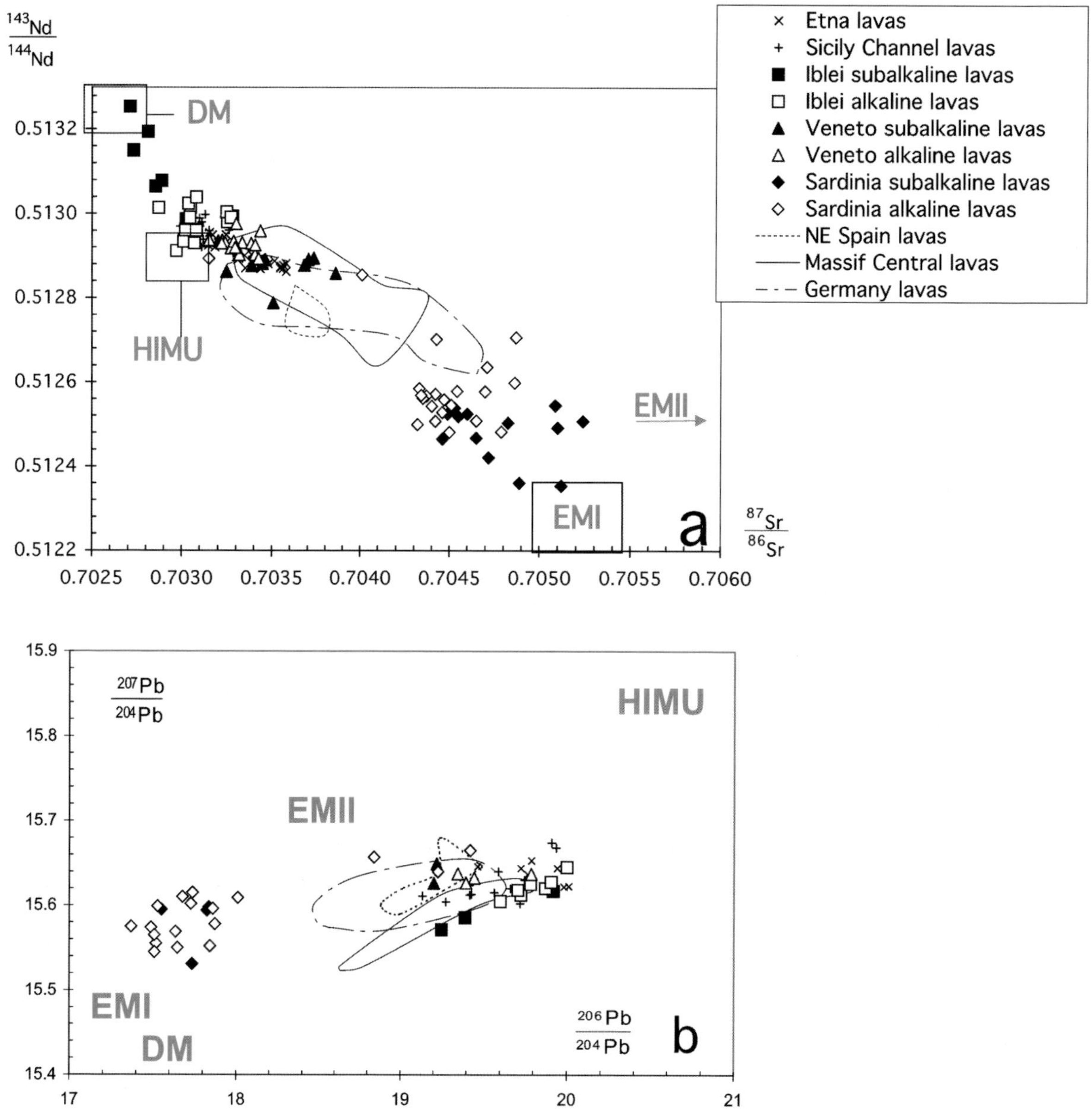

Figure 4: $^{87}Sr/^{86}Sr$ vs $^{143}Nd/^{144}Nd$ (a) and $^{206}Pb/^{204}Pb$ vs $^{207}Pb/^{204}Pb$ (b) diagrams for the Veneto, Iblean and Sardinian basic lavas. Data sources: Beccaluva et al. (2000) and unpublished data for Veneto: Beccaluva et al. (1998) and Bianchini et al. (1999) for Iblei; Lustrino et al. (2000, 2002), Gasperini et al. (2000), and unpublished data for Sardinia. Data for Etna and the Sicily Channel (Carter and Civetta, 1977; Civetta et al., 1998; Tonarini et al., 2001) are also reported for comparison, as well as compositional fields for Massif Central (Chauvel and Jahn, 1984; Briot et al., 1991; Wilson et al., 1995a), Garrotxa, NE Spain (Cebrià et al., 2000) and mafic lavas from Germany (Worner et al., 1986; Kramm and Wedepohl, 1990; Wedepohl et al., 1994). Mantle end member components DM, HIMU, EMI, and EMII after Zindler and Hart (1986).

the same authors proposed a geochemical involvement of fluids driven off the subducted Ionian lithosphere through a slab window gradually opening along the Sicilian margin. In conclusion, though Etna magmas share with the Iblean lavas most of the petrological and geochemical characteristics typical of within-plate settings, it can be confirmed that their mantle sources have been slightly contaminated by fluids related to the subduction of the neighbouring Ionian oceanic slab, as previously suggested by Beccaluva et al. (1982).

Sardinia Volcanic Province

After extensive subduction-related magmatism (32–13 Ma), within-plate Pliocene-Quaternary fissural volcanism in Sardinia took place concomitantly with tensional tectonics which also involved the adjoining Tyrrhenian area

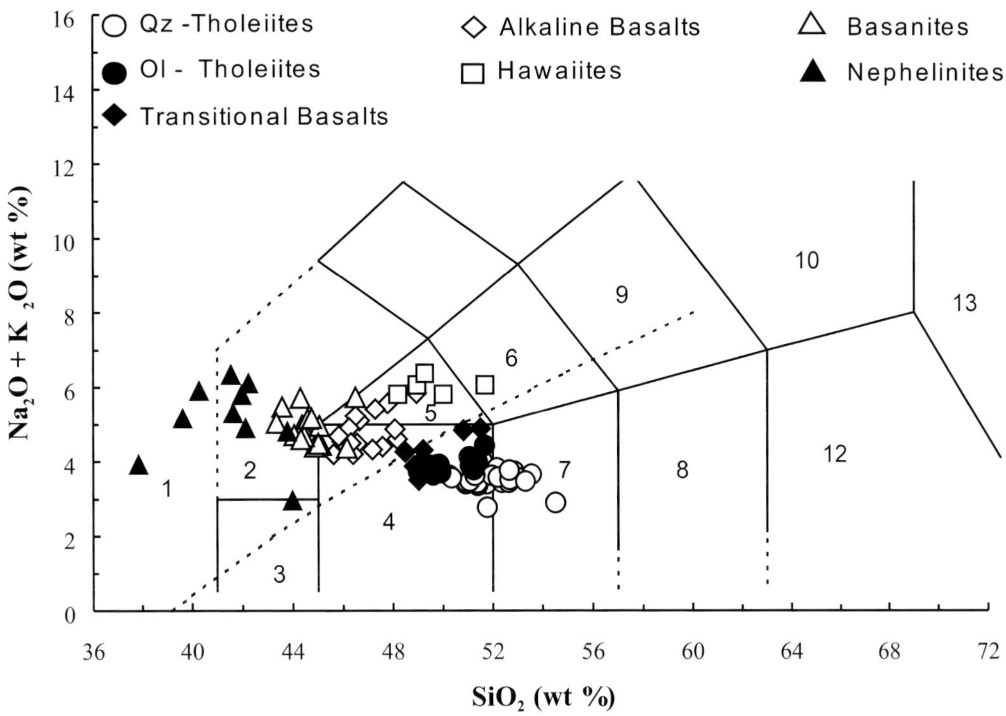

Figure 5: Total alkali-silica classification diagram for the Iblean lavas. Classification fields and dividing line between alkalic and subalkalic series as in Figure 2. Data sources: Beccaluva et al. (1998), Bianchini et al. (1998).

(Beccaluva et al., 1977, 1983, 1987; Macciotta et al., 1978). A wide range of alkaline to subalkaline basic lavas was extruded on the Island during the same time span (4 to < 0.2 Ma), though the climax of subalkaline activity occurred slightly earlier (3.5–3.0 Ma) than that of alkaline volcanism (Beccaluva et al., 1985). Alkaline, transitional and subalkaline differentiation series can be observed in three representative volcanic complexes: alkali basalts/trachybasalts and basanites, to trachyphonolites and phonolites at Montiferro; transitional basalts to qztrachytes at Capo Ferrato and Mt. Arci; and subalkaline basalts to rhyolites at Mt. Arci (Beccaluva et al., 1984) (Figure 7). Major and trace element variations and mineral chemistry suggest that differentiation in the various complexes may be due to distinct fractional crystallization processes of the main silicate phases and oxides, although magma mixing could better account for some of the intermediate lavas at Mt. Arci (Beccaluva et al., 1984; Montanini et al., 1994).

The incompatible element abundances in the most basic lavas of the various volcanic series have been plotted in Figure 8. All magmas exhibit a distinct, positive Ba anomaly which – together with higher Ba/Nb and Ba/La ratios than those of either Veneto or Iblei – corresponds to typical EMI OIB patterns (Weaver, 1991). Petrological modelling (Beccaluva et al., 1984) indicates that basanitic to subalkaline basaltic magmas were probably derived from lherzolitic lithospheric mantle sources, which underwent increasing degrees of partial melting (6–25%) at decreasing depths. Calculations invariably require metasomatic phlogopite $\pm$ amphibole in all modelled sources, as well as possible residual garnet in the source of most alkaline magmas, to account for their pronounced positive fractionation of HREE.

Sr–Nd–Pb isotopic data show a wide compositional range (Lustrino et al., 2000, 2002; Gasperini et al., 2000, and this work) ($^{87}Sr/^{86}Sr$: 0.70315–0.70534; $^{143}Nd/^{144}Nd$: 0.51289–0.51235; $^{206}Pb/^{204}Pb$: 17.5–18.0) which is intermediate between HIMU and EMI components (Figure 4). Most Sardinian samples plot toward EMI, while a predominant HIMU component has been recorded in only three alkaline lavas (Lustrino et al., 2000). More in detail, alkaline basalts and basanites display $^{87}Sr/^{86}Sr$ between 0.70315 and 0.70514, and $^{143}Nd/^{144}Nd$ between 0.51289 and 0.51251, while subalkaline basalts are more displaced toward the EMI component, with $^{87}Sr/^{86}Sr$ between 0.70453 and 0.70534, and $^{143}Nd/^{144}Nd$ between 0.51254 and 0.51235. This suggests that the EMI metasomatic component was preferentially preserved in the shallow lithospheric mantle ($<$ 50–60 km depth) where subalkaline basalts were generated. On the other hand, the deeper mantle sources of alkaline basic magmas were

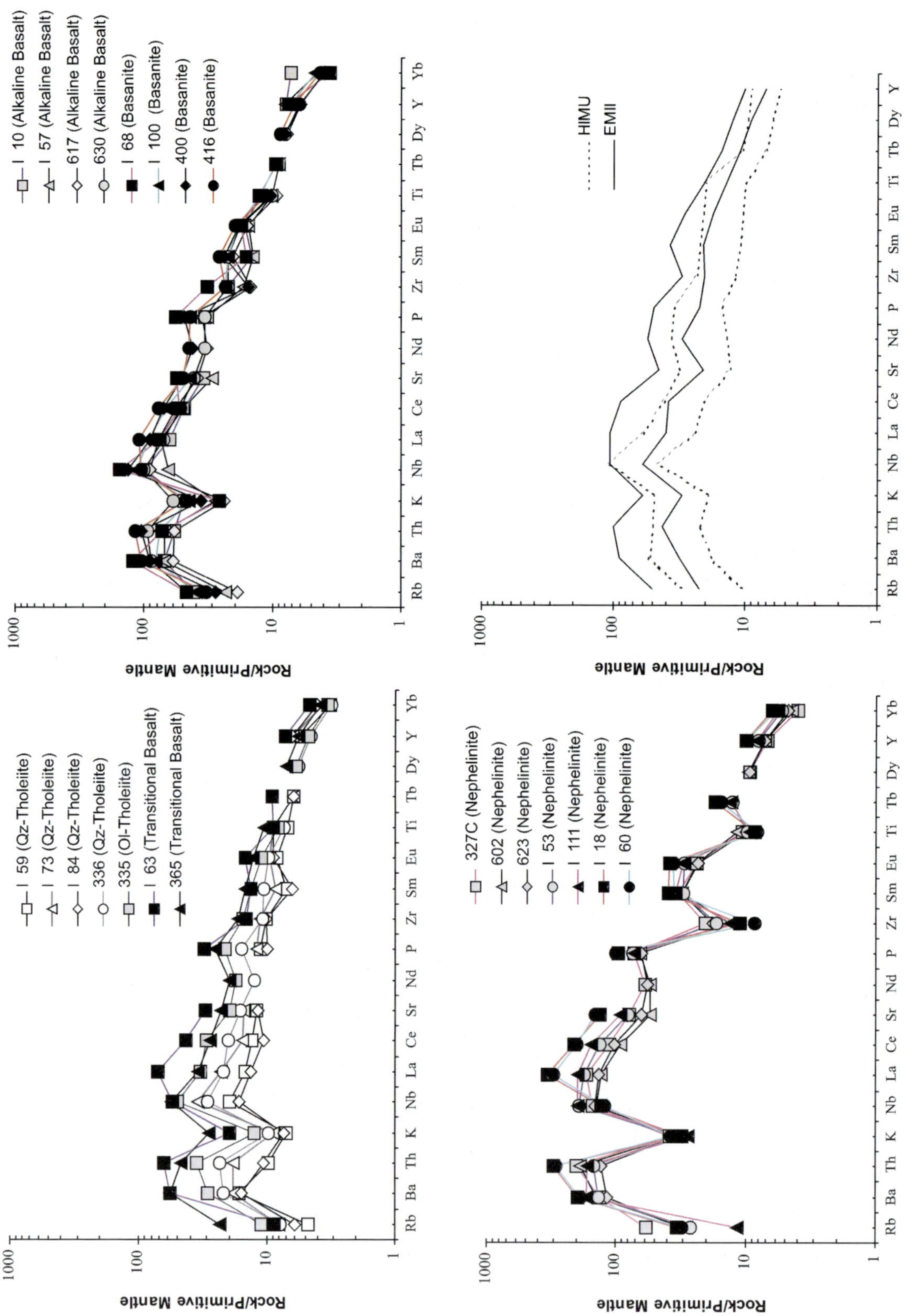
Rock/Primitive Mantle
1000
100
10
1
Rb Ba Th K Nb La Ce Sr Nd P Zr Sm Eu Ti Tb Dy Y Yb
I 59 (Qz-Tholeiite)
I 73 (Qz-Tholeiite)
I 84 (Qz-Tholeiite)
336 (Qz-Tholeiite)
335 (Ol-Tholeiite)
I 63 (Transitional Basalt)
365 (Transitional Basalt)
I 10 (Alkaline Basalt)
I 57 (Alkaline Basalt)
617 (Alkaline Basalt)
630 (Alkaline Basalt)
I 68 (Basanite)
I 100 (Basanite)
400 (Basanite)
416 (Basanite)
327C (Nephelinite)
602 (Nephelinite)
623 (Nephelinite)
I 53 (Nephelinite)
I 111 (Nephelinite)
I 18 (Nephelinite)
I 60 (Nephelinite)
HIMU
EMII

Figure 6: Primitive mantle-normalized incompatible element patterns for the Iblean lavas. Normalizing factors from Sun and McDonough (1989). The typical patterns of HIMU and EMII ocean island basalts, after Weaver (1991), are also reported. Data sources: Beccaluva et al. (1998), Bianchini et al. (1998).

more subjected to rejuvenation by additional sub-lithospheric metasomatizing agents (i.e. the HIMU component) leading to the observed isotopic signature in the generated magmas.

An intra-plate volcanism similar to that of Sardinia occurred during the Neogene-Quaternary in Spain: an association of leucitite, nephelinite, melilitite, and olivine basaltic lavas and pyroclastites is widespread over an area of about 4000 km^2 in the Calatrava Province (central Spain: Cebrià and Lòpez-Ruiz, 1995), whereas in the area of Garrotxa (north-east Spain) a number of leucite-, nepheline-basanite, and alkali basaltic cones and lava flows are distributed over an area of ca. 2500 km^2 (Cebrià et al., 2000). The geochemical features of all these lavas can be closely compared with those of Sardinia and the Cenozoic intra-plate volcanics of Europe in general, as they are variably characterized by HIMU and EMI geochemical components (Cebrià et al., 2000).

3. PETROLOGY OF MANTLE XENOLITHS

Veneto Volcanic Province

In the VVP, mantle xenoliths consist of predominant spinel lherzolites and minor harzburgites (Figure 9), mostly protogranular textured. The mineral paragenesis is formed by large crystals of olivine (ol) and orthopyroxene (opx) (up to 2 mm across), smaller clinopyroxene (cpx) (0.5–1.0 mm), and spinel (sp) with typical lobate shape. Metasomatic (pyrometamorphic) textures are superimposed on the primary parageneses in both lherzolites and harzburgites, and consist of secondary minerals (ol, cpx ± feldspar), spongy clinopyroxene, and variably recrystallized glassy patches (Siena and Coltorti, 1989, 1993; Coltorti et al., 2000).

A continuous depletion trend is recorded from clinopyroxene-rich lherzolites to harzburgites, which is chemically reflected in the gradual decrease in the most fusible elements such as Al_2O_3, CaO and TiO_2, and the parallel

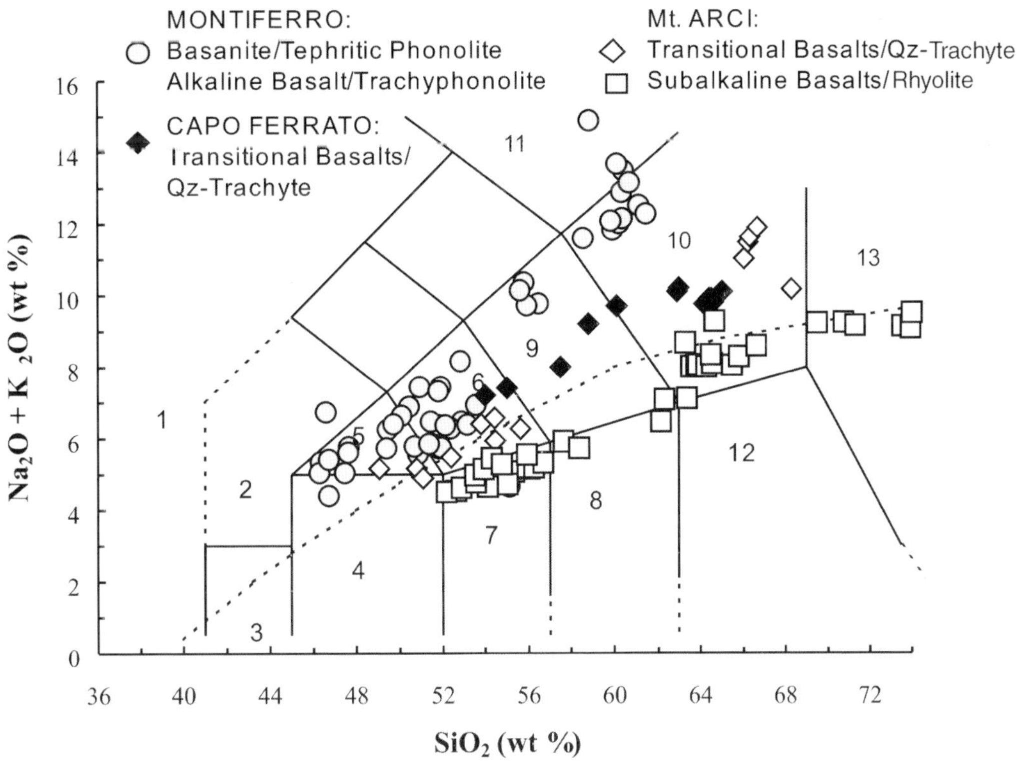

Figure 7: Total alkali-silica diagram for the Sardinian lavas. Classification fields and dividing line between alkalic and subalkalic series as in Figure 2. Data sources: Beccaluva et al. (1984), and unpublished data.

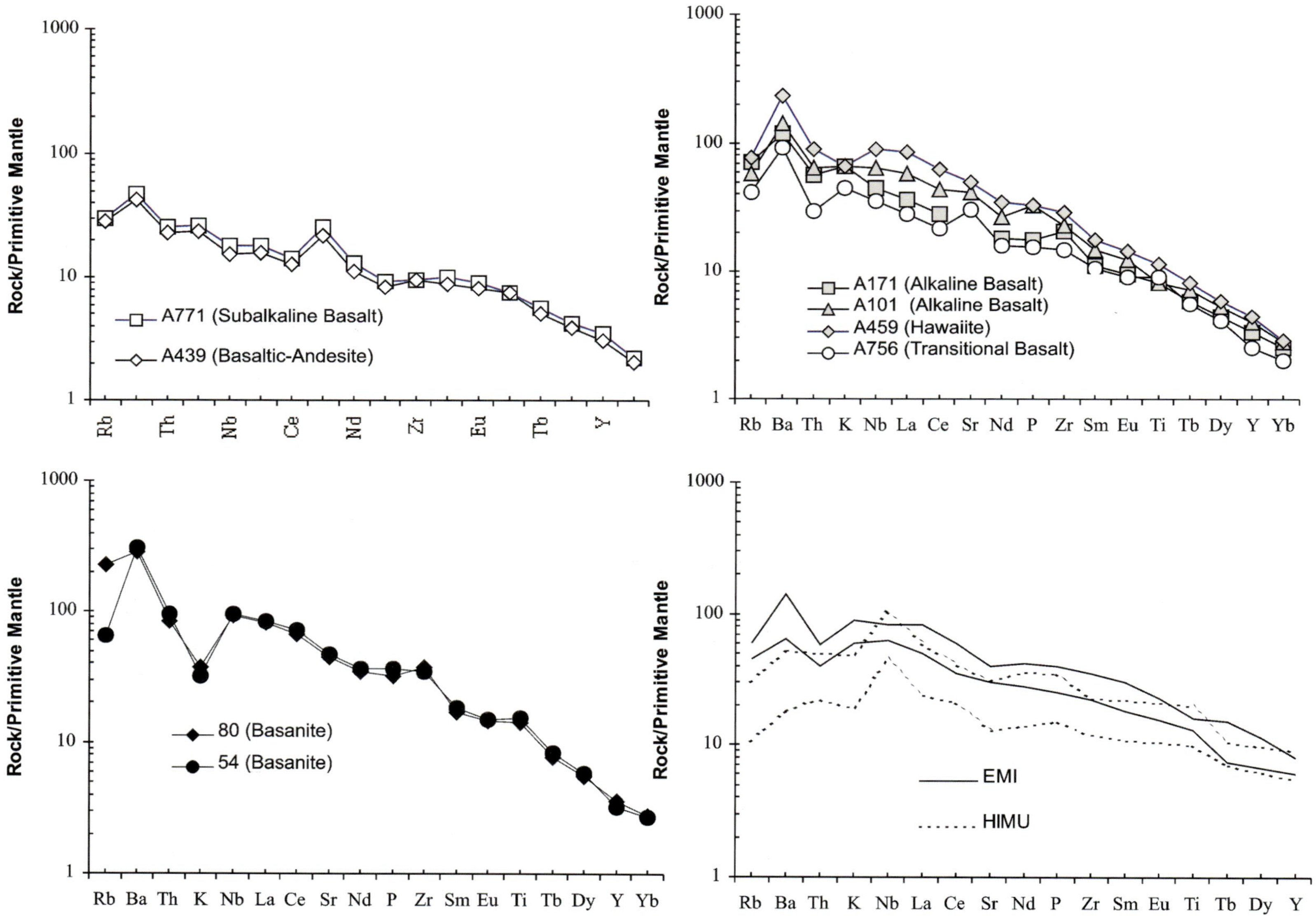

Figure 8: Primitive mantle-normalized incompatible element patterns for the Sardinian basic lavas. Normalizing factors from Sun and McDonough (1989). The typical pattern of HIMU and EMI Ocean Island Basalts, after Weaver (1991), are also reported. Data sources: Beccaluva et al. (1984), and unpublished data.

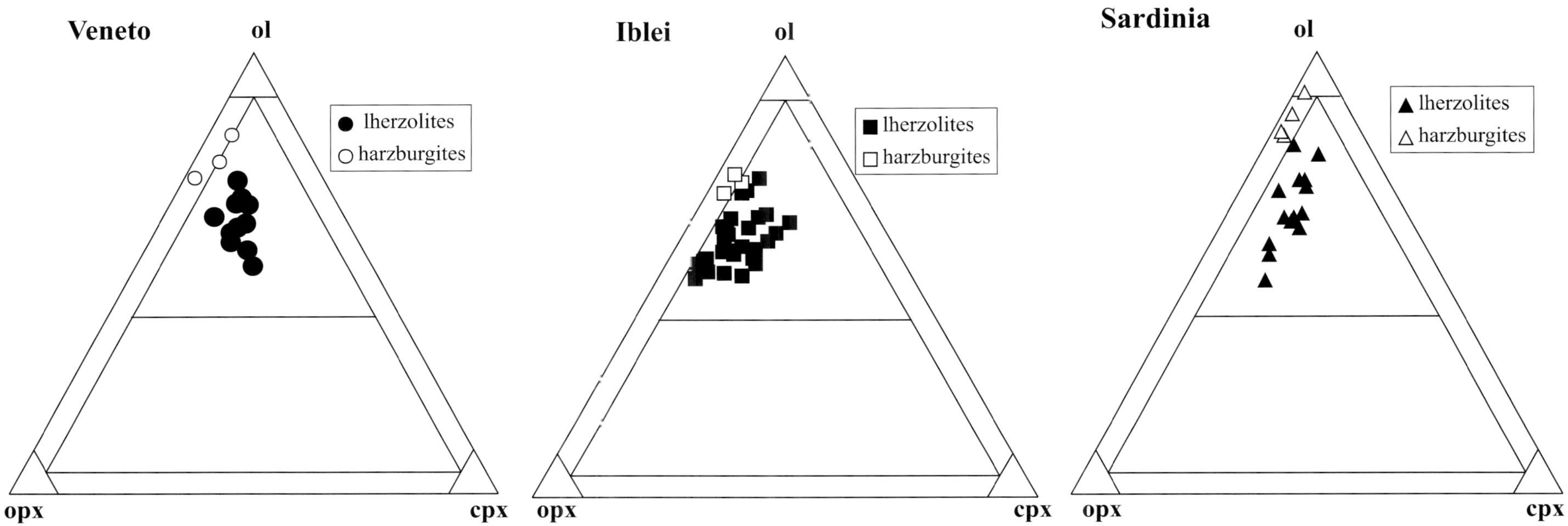

Figure 9: Calculated modal compositions in terms of olivine (ol), orthopyroxene (opx), and clinopyroxene (cpx) of peridotite mantle xenoliths entrained by alkaline lavas from Veneto, Iblei and Sardinia.

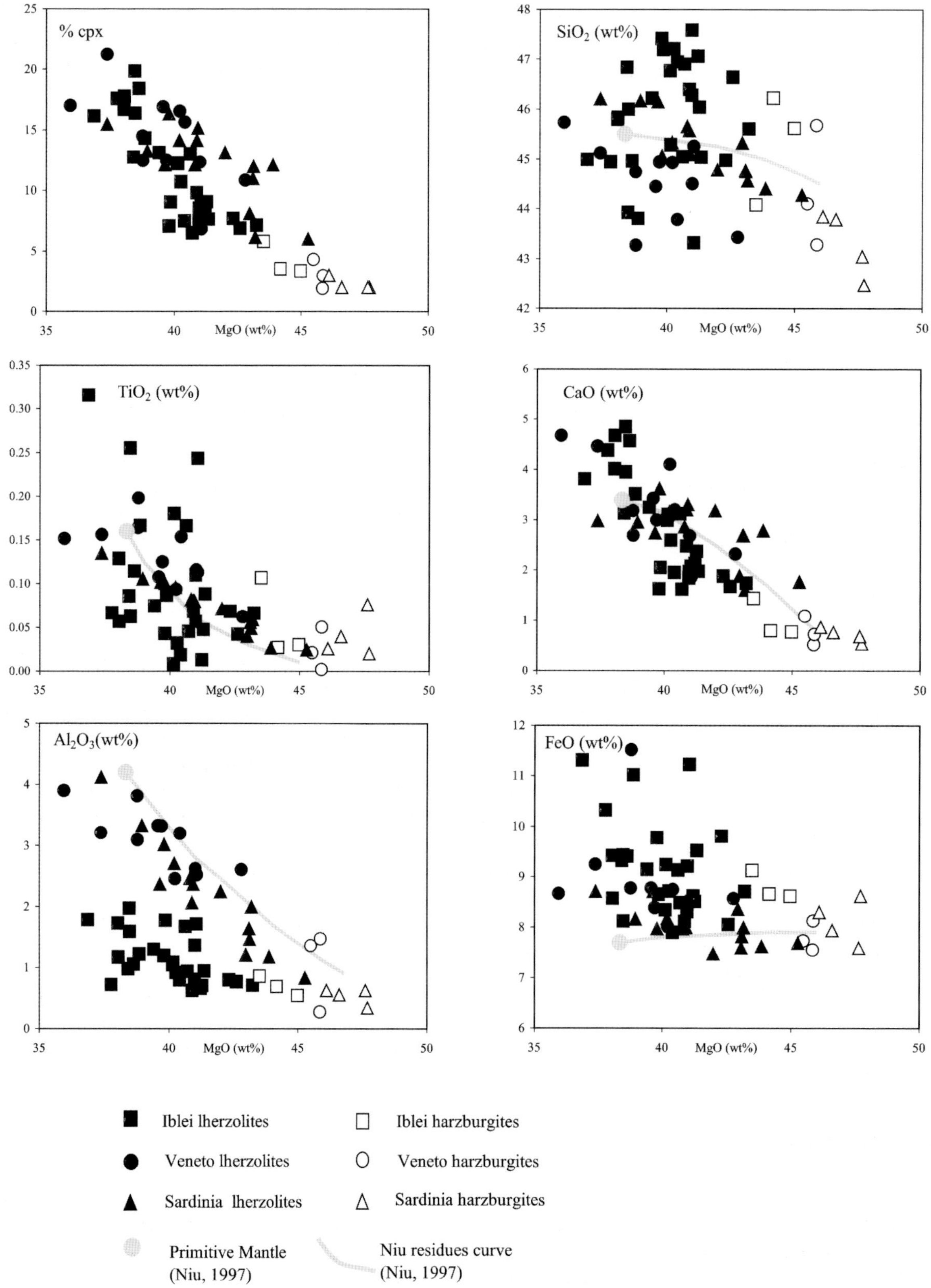

Figure 10: Variation diagrams of SiO_2, CaO, FeO_{tot}, Al_2O_3, TiO_2, and modal cpx vs MgO for peridotite mantle xenoliths entrained by alkaline lavas from Veneto, Iblei and Sardinia. The model fertile lherzolite and the residual composition trend by partial melting, after Niu (1997), are also reported.

increase in Ni and mg# (Mg/(Mg + Fe)*100), as commonly observed in mantle rocks progressively depleted by extraction of basaltic melts (Figure 10). The heavy (H)-REE distribution varies accordingly from 3 times chondritic in cpx-rich lherzolites (cpx up to 20.9%) to 0.2 times chondritic in harzburgites (cpx = 4.3–1.9%) (Figure 11). Moreover, both lherzolites and harzburgites are variably enriched in light (L)-REE (La_N /Yb_N up to 12.6 and 19.2 respectively), suggesting post-depletion enrichments related to metasomatic processes (Beccaluva et al., 2001a) (Figure 11).

Protogranular-textured clinopyroxenes from lherzolites show HREE contents about 10 times chondritic and strong LREE-depletion (La_N/Yb_N = 0.002–0.025), while those from harzburgites vary from slightly LREE-depleted to LREE-enriched (La_N/Yb_N = 0.64–4.00), with much lower HREE contents (ca. 3 times chondritic). Most HREE patterns can be reproduced by a simple fractional melting model and extraction of ca. 5% to 22% basic melts, starting from the most fertile VVP spinel lherzolite. The slightly LREE-depleted to LREE-enriched patterns of protogranular-textured clinopyroxenes in harzburgites, and spongy clinopyroxenes in lherzolites (La_N/Yb_N = 0.27–5.67) can be accounted for by metasomatic enrichment events (Figure 12).

Major and trace element mass balance calculations indicate that the metasomatic parageneses may be accounted for by the addition of 1–6% of Na-alkaline basic melt/s to the primary mineral assemblages (Beccaluva et al., 2001a). The modelled metasomatic agents strongly resemble the late Cretaceous lamprophyric dikes of the south-Alpine domain, as well as some VVP alkaline basic lavas. The presence of glassy patches, as well as the persistent REE zoning of clinopyroxenes, indicate that the most recent metasomatic processes occurred shortly before the entrainment of mantle xenoliths by the host lavas. The available Sr–Nd isotope compositions (whole rocks and cpx separates: Figure 13) plot between the depleted mantle (DM) and the HIMU components, suggesting that the latter represents the isotopic signature of the metasomatizing agent.

Iblean Volcanic Province

In the Iblean volcanic province, spinel-peridotite mantle xenoliths with a typical four phase mineral assemblage are commonly included in the alkaline lavas, the largest-sized (up to 10 cm) occurring in nephelinitic Miocene diatremes of the north-eastern sector (Di Grande et al., 2002, and references therein). CO_2 fluid inclusion data (Bergamini, 1992) and crystal structure modelling of clinopyroxenes (Nimis, 1995) indicate a pressure range of 10–15 kb, corresponding to a maximum provenance depth for these xenoliths of ca. 50 km. Their composition varies from lherzolites to rare harzburgites (Figure 9), and is reflected in a parallel gradual depletion in the most fusible elements (Figure 10). Textures vary from protogranular to porphyroclastic, with superimposed pyrometamorphic features – consisting of secondary phases (including rare phlogopite and feldspars), spongy borders in pyroxenes and variably recrystallized glassy patches which testify to metasomatic processes.

Chondrite-normalized REE distributions of bulk rock and clinopyroxenes suggest widespread reactions between metasomatic agent(s) and the mantle peridotite matrix (Figures 11 and 12). Accordingly, LREE enrichment varies from $(La/Yb)_N$ 3.8 to 10.5 in whole rock, and from 1.5 to 4.0 in clinopyroxene separates. The scarce isotopic data (Tonarini et al., 1996) for unaltered Iblean xenoliths (bulk rocks and cpx separates) cluster around the HIMU component (Figure 13) ($^{87}Sr/^{86}Sr$ from 0.70271 to 0.70330, and $^{143}Nd/^{144}Nd$ from 0.51291 to 0.51325), thus suggesting that the latter represents the geochemical signature of the metasomatizing agents, as already observed for VVP xenoliths.

Sardinia Volcanic Province

Mantle xenoliths, included in Pliocene-Quaternary alkaline basic lavas from the type localities of eastern (Dorgali) and western (Scano) Sardinia, are typically characterized by anhydrous four-phase mineral assemblages with olivine, orthopyroxene, clinopyroxene and spinel; they represent the uppermost lithospheric mantle for a maximum depth of ca. 40 km, according to CO_2 fluid inclusion data and thermobarometric estimates (Beccaluva et al., 1989; Siena and Coltorti, 1993). Lherzolite compositions with cpx varying from 6% up to 16% characterize xenoliths from Dorgali, whereas at Scano harzburgites with subordinate cpx-poor lherzolites predominate (Figure 9). Primary textures are mostly protogranular, with the four main mineral phases occurring as large (> 0.2 mm) unzoned-equilibrated crystals. Secondary microdomains with pyrometamorphic textures have been observed in some samples, consisting of spongy pyroxenes and microcrysts of olivine, clinopyroxene, spinel, feldspar and rare phlogopite, as well as small interstitial glassy blebs and/or veinlets.

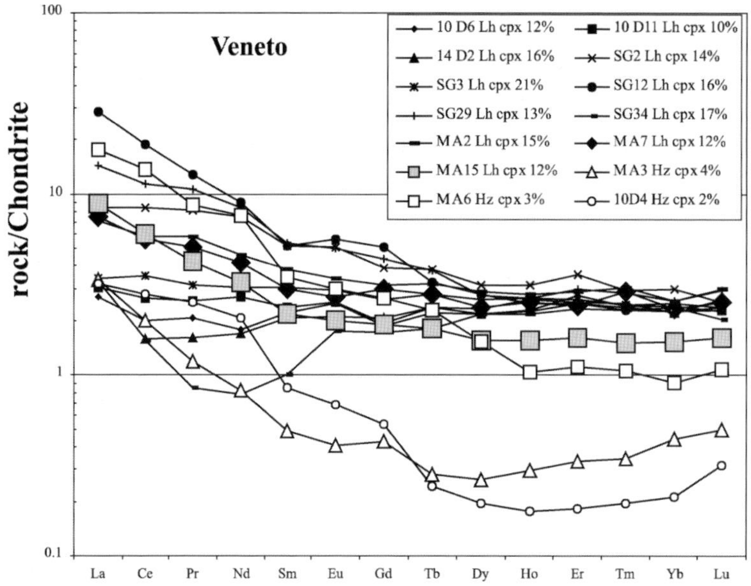
Veneto
rock/Chondrite
100
10
1
0.1
10 D6 Lh cpx 12%
10 D11 Lh cpx 10%
14 D2 Lh cpx 16%
SG2 Lh cpx 14%
SG3 Lh cpx 21%
SG12 Lh cpx 16%
SG29 Lh cpx 13%
SG34 Lh cpx 17%
MA2 Lh cpx 15%
MA7 Lh cpx 12%
MA15 Lh cpx 12%
MA3 Hz cpx 4%
MA6 Hz cpx 3%
10D4 Hz cpx 2%
La Ce Pr Nd Sm Eu Gd Tb Dy Ho Er Tm Yb Lu

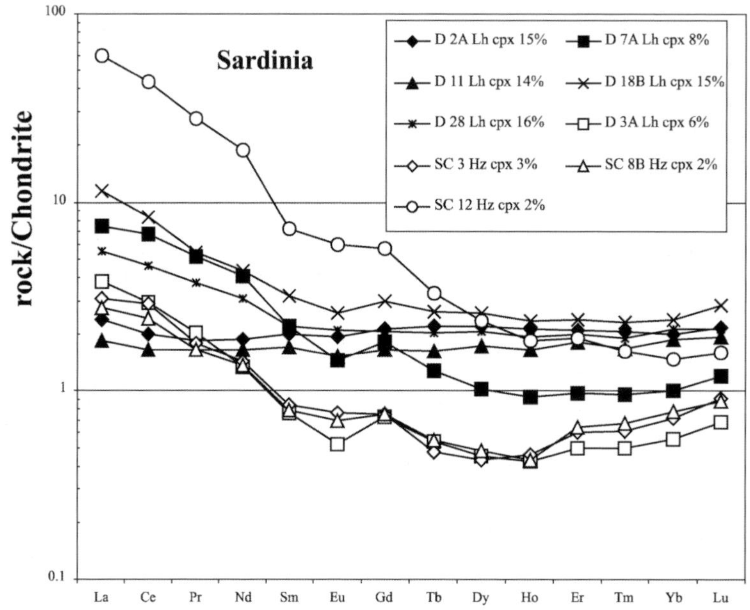
Sardinia
rock/Chondrite
100
10
1
0.1
D 2A Lh cpx 15%
D 7A Lh cpx 8%
D 11 Lh cpx 14%
D 18B Lh cpx 15%
D 28 Lh cpx 16%
D 3A Lh cpx 6%
SC 3 Hz cpx 3%
SC 8B Hz cpx 2%
SC 12 Hz cpx 2%
La Ce Pr Nd Sm Eu Gd Tb Dy Ho Er Tm Yb Lu

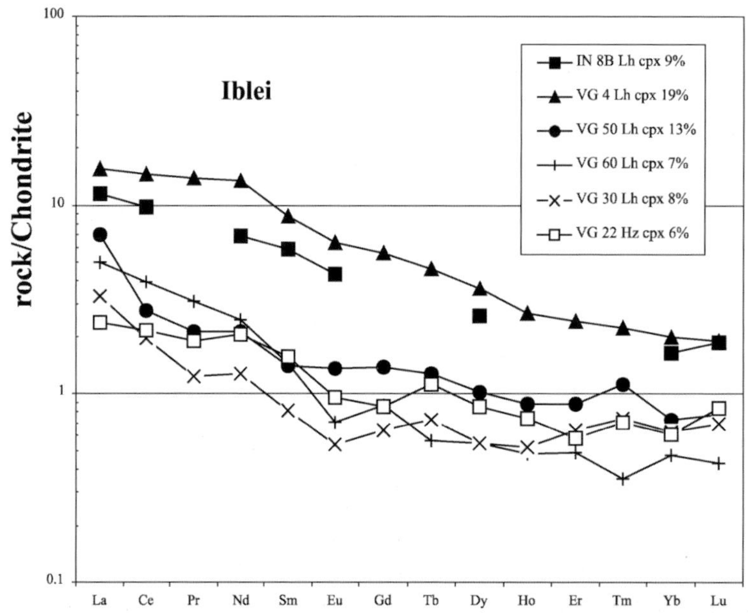
Iblei
rock/Chondrite
100
10
1
0.1
IN 8B Lh cpx 9%
VG 4 Lh cpx 19%
VG 50 Lh cpx 13%
VG 60 Lh cpx 7%
VG 30 Lh cpx 8%
VG 22 Hz cpx 6%
La Ce Pr Nd Sm Eu Gd Tb Dy Ho Er Tm Yb Lu

Figure 11: Chondrite-normalized REE patterns for whole rock mantle xenoliths entrained by alkaline lavas from Veneto, Iblei and Sardinia. Abbreviations: Lh – lherzolite; Hz – harzburgite. Modal % of clinopyroxene are also indicated in the legend. Normalizing factors from Sun and McDonough (1989).

On the whole, the xenolith populations define depletion trends from nearly undepleted lherzolites to depleted harzburgites, as observed for the VVP and Iblean mantle rocks (Figure 10). For the most fertile Dorgali lherzolites, Beccaluva et al. (2001b) reported variable REE distributions from nearly flat whole-rock REE patterns (1.7–2.4*chondrite) with LREE-depleted cpx (HREE = 11.5–12.7*chondrite; La_N/Yb_N down to 0.2) to LREE-enriched patterns in both whole rock (HREE = 2.0–2.5*chondrite; La_N/Yb_N = 2.5–4.8) and clinopyroxene (HREE = 8.2–13.9*chondrite; La_N/Yb_N = 1.5–4.9) (Figures 11 and 12); cpx-poor lherzolites from the same locality display comparatively lower HREE abundances (bulk-rock = 0.5–1.1*chondrite, cpx = 3.6–5.8*chondrite) and subparallel LREE-enriched patterns for both whole-rock (La_N/Yb_N up to 7.5) and cpx (La_N/Yb_N up to 8.5). Similar patterns are observed in the most refractory Scano harzburgites, which are characterized by low HREE contents (bulk-rock = 0.5–0.6*chondrite, cpx = 3.5–7.0*chondrite) and LREE-enrichment in both whole rock (La_N/Yb_N = 3.5–4.5 up to 40.7) and clinopyroxenes (La_N/Yb_N = 7.8–17.5 up to 45.0) (Figures 11 and 12). As reported by Beccaluva et al. (2001b), these REE data could be accounted for by (1) partial melting events and extraction of basic magmas leading to progressive HREE depletion, and (2) subsequent enrichment by metasomatizing agents resulting in variable increase in the most incompatible elements (e.g. LREE).

The Sr–Nd isotope compositions of whole rock and cpx separates (Figure 13) display a large range of variation, extending from DM throughout HIMU to EMI end-members. In particular, clinopyroxenes from Dorgali lherzolites show the wider compositional variation ($^{87}Sr/^{86}Sr$ 0.70262–0.70461 and $^{143}Nd/^{144}Nd$ 0.51323–0.51254), while those from Scano harzburgites cover a more restricted range ($^{87}Sr/^{86}Sr$ 0.70418–0.70432 and $^{143}Nd/^{144}Nd$ 0.51259–0.51252) between HIMU and EMI. The observed elemental and isotopic characteristics could be accounted for on the whole by alkali silicate metasomatizing agents with EMI and HIMU affinity which enriched a previously depleted lithospheric mantle (DM) (Beccaluva et al., 2001b).

4. DISCUSSION AND COMPARISON BETWEEN THE EUROPEAN AND AFRICAN LITHOSPHERIC MANTLE

As shown in the previous sections, the compositional characteristics of the parental basic magmas for the three investigated provinces are compatible with their segregation from the lithospheric mantle between about 30 and 100 km depth. In fact, their P–T segregation conditions (Figure 14) show a reasonable agreement with phase equilibria constraints for shallow silica-saturated to deep strongly silica-undersaturated basic melts (Falloon and Green, 1988; Falloon et al., 1988; Hirose and Kushiro, 1993; Hirose and Kawamoto, 1995): tholeiitic basalts, 10–16 kb, 1150–1250 °C; alkali-basalts, 14–22 kb, 1200–1280 °C; basanites and nephelinites, $\geqslant$ 22 kb, 1250–1350 °C. An additional depth constraint for strongly alkaline carbonated magmas (most nephelinites) is provided by the carbonate-amphibole mantle stability field at pressure $\geqslant$ 20–22 kb (Olafsson and Eggler, 1983; White. and Wyllie, 1992), which limits generation of these melts to $\geqslant$ 65–70 km depth. The P–T segregation trend plots between the experimental dry and hydrated-carbonated mantle *solidi*, in agreement with petrological modelling which invariably requires significant amounts (5–10%) of volatile-bearing phases, such as amphibole and phlogopite, in all the modelled sources. This trend is also fairly close to the regional geotherm inferred for the central-western Mediterranean area (Figure 14), thus suggesting that partial melting processes could have easily been triggered by local decompression effects related to a limited intra-plate tensional regime.

Incompatible element distributions and isotopic characteristics suggest that parental magmas of the three volcanic provinces were produced from alkali silicate-metasomatized lithospheric mantle sources which were also enriched by carbonatitic components in the deeper portions where nephelinites were generated. In the Veneto and Iblean magmas, a prevalent HIMU signature (plus subordinate EMII) is recorded, whereas Sardinian magmas are characterized by both EMI and HIMU geochemical components (Figure 4). In each investigated province, the alkaline and deeper magmas show a more marked HIMU signature compared to the subalkaline basalts; this suggests a more intensive, and probably more recent enrichment of the deeper lithospheric mantle sources (> 60–70 km, i.e. in the Thermal Boundary Layer, TBL: Latin et al., 1990; Anderson, 1994) by the HIMU metasomatizing agents. On the other hand, the sub-alkaline magmas generated from shallow lithospheric mantle sources display a variable influence of EM components: Veneto tholeiites extend from HIMU toward EMII, whereas Sardinian sub-alkaline

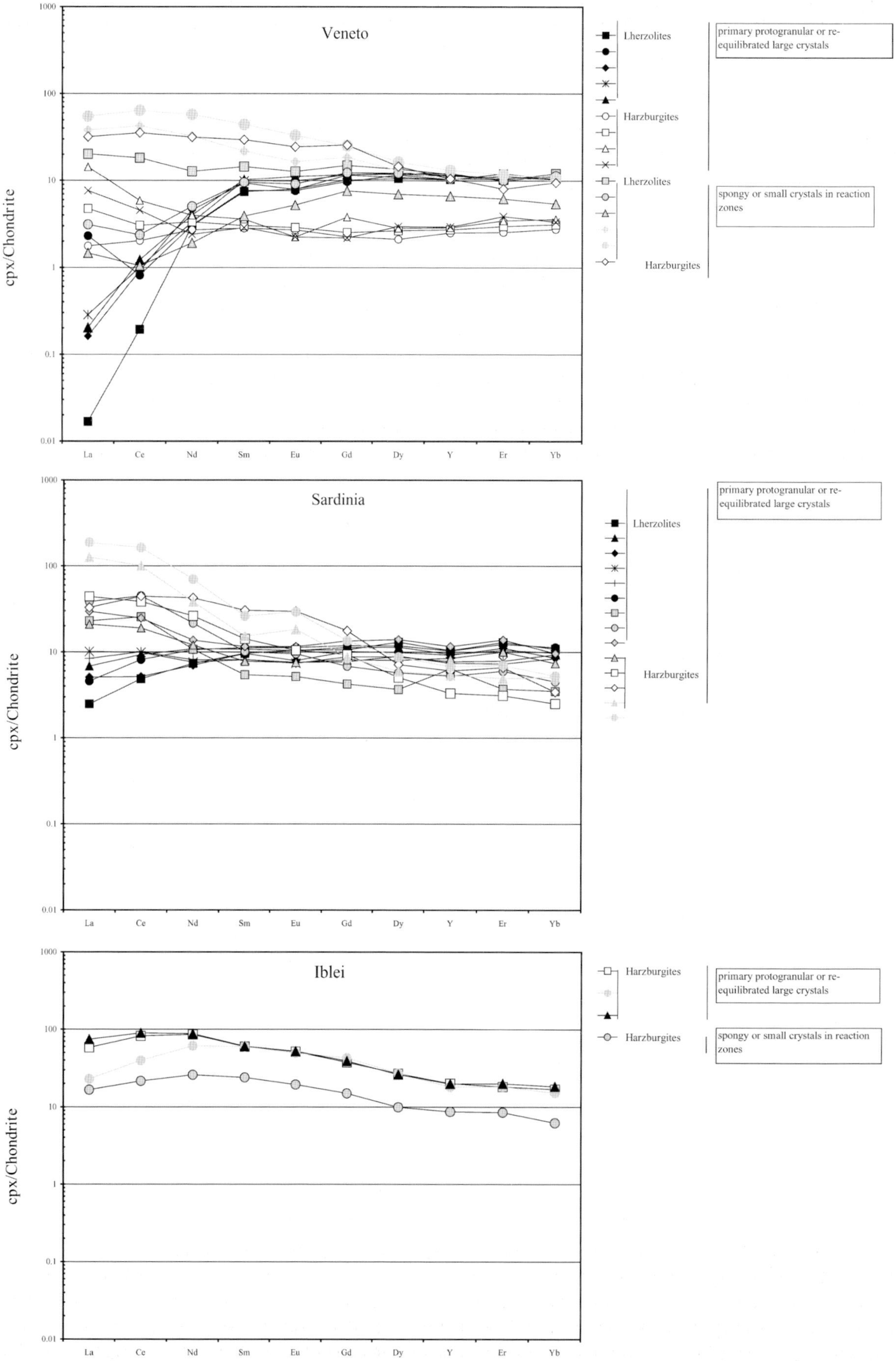
Veneto
cpx/Chondrite
1000
100
10
1
0.1
0.01
La Ce Nd Sm Eu Gd Dy Y Er Yb
Lherzolites
Harzburgites
primary protogranular or re-equilibrated large crystals
Lherzolites
spongy or small crystals in reaction zones
Harzburgites
Sardinia
cpx/Chondrite
1000
100
10
1
0.1
0.01
La Ce Nd Sm Eu Gd Dy Y Er Yb
primary protogranular or re-equilibrated large crystals
Lherzolites
Harzburgites
Iblei
cpx/Chondrite
1000
100
10
1
0.1
0.01
La Ce Nd Sm Eu Gd Dy Y Er Yb
Harzburgites
primary protogranular or re-equilibrated large crystals
Harzburgites
spongy or small crystals in reaction zones

Figure 12: Chondrite-normalized REE patterns for clinopyroxenes from mantle peridotite xenoliths entrained by alkaline lavas from Veneto, Iblei and Sardinia. Normalizing factors from Sun and McDonough (1989).

lavas are the most displaced toward EMI (Figure 4). Therefore, the shallower lithospheric mantle sources (< 60–70 km, i.e. in the Mechanical Boundary Layer, MBL: Latin et al., 1990; Anderson, 1994) were presumably less affected by the HIMU signature, and could thus have better preserved the geochemical imprint of the older EMI and EMII components, which metasomatized the previously depleted lithospheric mantle (DM).

Mantle xenoliths associated with alkaline lavas of the three volcanic provinces generally represent shallow portions of the lithospheric mantle column (< 40–50 km depth, i.e. in the MBL), in accordance with thermobarometric estimates and rheologic characteristics (Verde, 1996). Petrological data invariably indicate that these mantle xenoliths underwent a complex compositional evolution, characterized by at least two different types of processes: (1) depletion(s) by extraction of basic magmas, mostly occurring in the pre-Palaeozoic, as shown by both major element variations and progressive HREE depletion in whole rocks and constituent clinopyroxenes; (2) reaction(s) between sub-lithospheric metasomatizing agents and previously depleted lithospheric mantle (DM), resulting in variable enrichments of the most incompatible elements (e.g. LREE, LFSE, etc.). This is reflected in the compositional variations of primary minerals, particularly pyroxenes, and the variable occurrence of pyrometamorphic textures

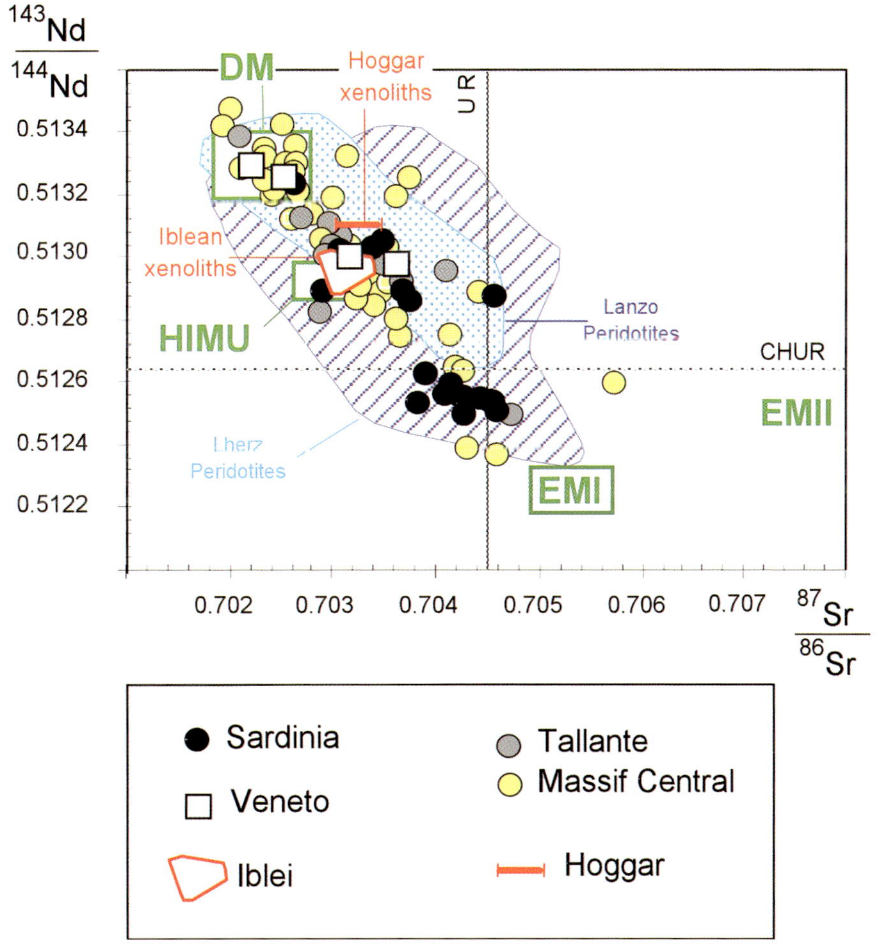

Figure 13: Sr–Nd isotope systematics for whole rocks and cpx separates of mantle peridotite xenoliths entrained by alkaline lavas from Veneto, Iblei and Sardinia. The compositional fields of peridotite xenoliths from Hoggar (Southern Algeria, Dautria et al., 1987), Tallante (Southern Spain, Beccaluva et al., 1999b), the Massif Central (Downes and Dupuy, 1987; Zangana et al., 1997) and Alpine peridotite massifs (Lherz, Downes et al., 1991; Lanzo, Bodinier et al., 1991) are also reported for comparison. Mantle end member components DM, HIMU, EMI, and EMII after Zindler and Hart (1986).

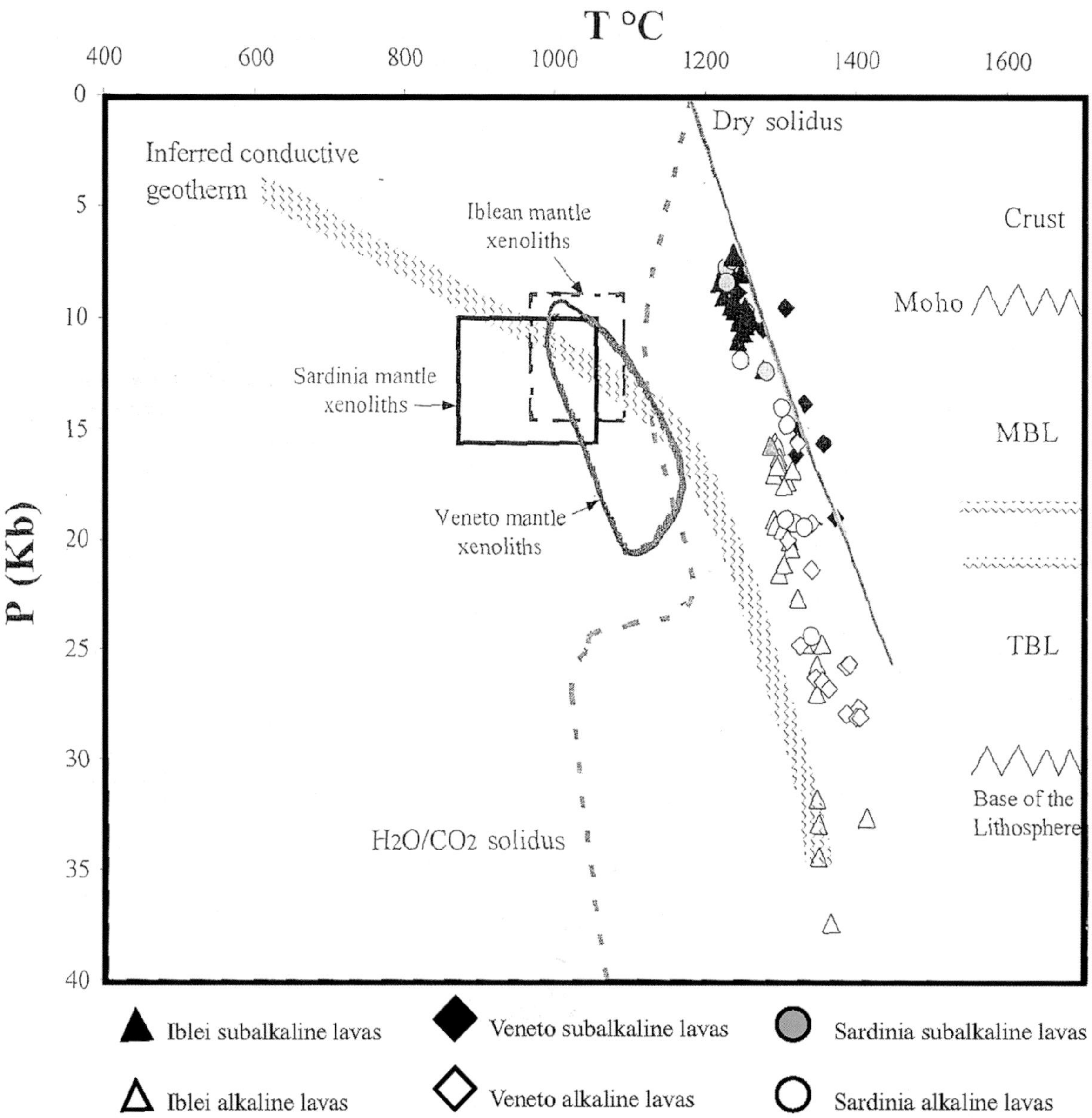

Figure 14: P–T segregation conditions of basic magmas from Veneto, Iblei and Sardinia, calculated according to Albarède (1992). The experimental mantle peridotite *solidi* for anhydrous and hydrated-carbonated conditions are reported after Green et al. (1987) and Wyllie (1987). The inferred regional conductive geotherm for the central-western Mediterranean area is also shown, based on P–T equilibration of associated selected mantle xenoliths from Sardinia, Veneto, and Iblei, using the geothermometer of Brey and Koehler (1990) and the geobarometer of Koehler and Brey (1990). MBL, TBL, Moho depth, and lithosphere thickness are based on the geophysical data quoted in the text.

and new phases such as amphibole, phlogopite, and glass. The inferred metasomatic agents may be attributed to small volumes (a few %) of strongly alkaline basic melts. The resulting isotopic signatures conform well to those of their host magmas, being dependent on the variable contributions of the DM, HIMU, and EM components: a prevailing HIMU imprint is recorded for both Veneto and Iblean mantle materials, whereas a predominant EMI component, in addition to HIMU, is observed for the Sardinian mantle.

Comparison on a circum-Mediterranean scale highlights the remarkable compositional similarities between Sardinian and European lithospheric mantle on the whole, whereas the Veneto and Iblean districts are closely comparable with the rest of the stable North-African domain. In fact, Sardinian mantle xenoliths share geochemical characteristics with many other European xenolith suites, such as those from the Massif Central (Zangana et al., 1997), Eifel (Witt-Eickschen et al., 2003), and Tallante (Southern Spain; Beccaluva et al., 1999b). The

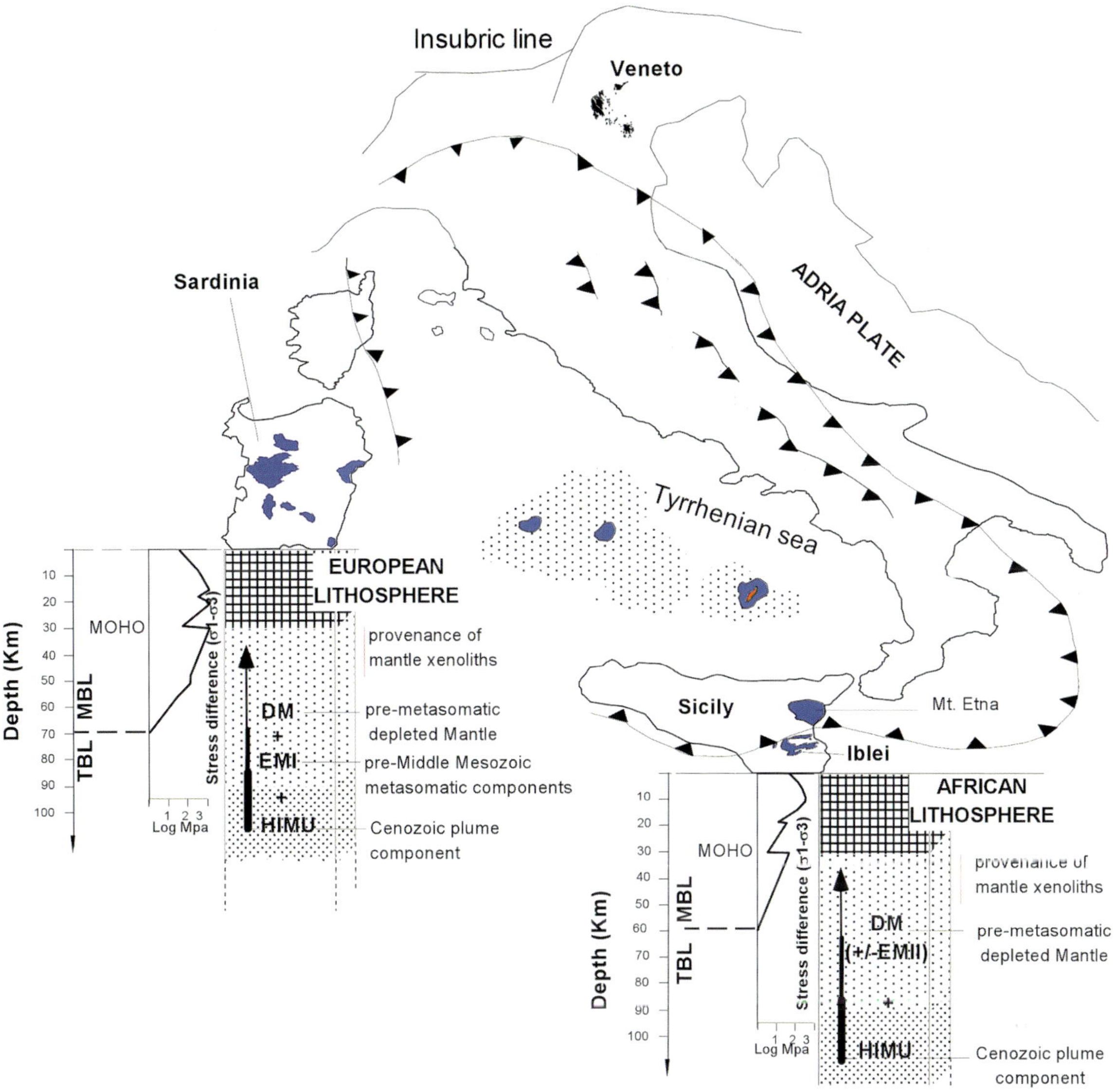

Figure 15: Schematic model for the compositional evolution of the European (Sardinia) and Adriatic–African (Veneto and Iblei) lithospheric mantle. The EMI metasomatic component is recorded mainly in the mantle sections underlying the European Variscan basement, where it could have been effective since the pre-middle Mesozoic. In both the African and European lithospheres, the effect of the HIMU component decreases (see arrow thickness) from the thermal to the mechanical boundary layer, and could be related to the Cenozoic plume region extending from the eastern Atlantic to Central Europe and western Mediterranean (see text for further explanation).

same geochemical signatures (i.e. predominant EMI in addition to HIMU) have been recognized in Cenozoic anorogenic magmas located within the Variscan basement of western-central Europe (Wilson and Downes, 1991; Cebrià and Wilson, 1995), with the most striking analogies observed between the Spanish (Cebrià et al., 2000) and Sardinian volcanic provinces. Moreover, Alpine-type peridotite massifs from the European lithosphere – including the Lherz, Lanzo and Ronda massifs – show enrichment trends ranging from DM to EMI components (Downes et al., 1991; Bodinier et al., 1991; Reisberg et al., 1989; Downes, 2001, and references therein) which could thus have been effective at least since the pre-middle Mesozoic. By contrast, the Veneto and Iblean magmatic associations lack notable EMI signature, and record a prevalent HIMU metasomatic imprint. They therefore compare favourably with other provinces of the stable North African domain – i.e. external to the Apennine–Maghrebian orogenic belt – such as Pietre Nere, SE Italy (Hawkesworth and Vollmer, 1979), the Sicily Channel (Civetta et al., 1998, and references therein), Hoggar (Allegre et al., 1981; Dautria et al., 1987), Canary Islands (Siena et al., 1991), Jebel Marra, Sudan (Davidson and Wilson, 1989), and the Cameroon Volcanic Line (Lee et al., 1996, and references therein).

From the above, a distinct compositional evolution of the European (Sardinia) and African (Iblei and Veneto) lithospheric mantle is proposed in Figure 15.

The enrichment of prevalent EMI component, at least since the pre-middle Mesozoic, appears an exclusive characteristic of the lithospheric mantle underlying the European Variscan basement. On the other hand, the addition of HIMU components seems to have been effective in both European and North-African lithospheres since the late Cretaceous (Wilson and Bianchini, 1999), as recorded by the Cenozoic volcanic provinces of these domains. Seismic tomography suggests that these components, referred to as the European Asthenospheric Reservoir (EAR: Cebrià and Wilson, 1995) or Low Velocity Component (LVC: Hoernle et al., 1995), could be related to a common sheet-like (Hoernle et al., 1995) or diapir-like (Granet et al., 1995) plume region, assumed to be present over an area extending from the eastern Atlantic to Central Europe and the western Mediterranean. It should be remarked that the geochemical characteristics discussed above do not correspond to marked geophysical differences in the structure of the African and European lithospheres, at least as far as Sardinia and Sicily are concerned (Figure 15). In fact, the transition between the MBL and TBL is found at ca. 60 km beneath Sicily and ca. 70 km beneath Sardinia, according to the strength profiles ($\sigma 1$–$\sigma 3$ vs. depth) calculated by Viti et al. (1997, and unpublished data). This stress difference is consistent with that estimated on the basis of the grain size of both Sardinian and Iblean mantle xenoliths (Verde, 1996), indicating their common provenance from the MBL. The mechanical boundary layer is thus to be considered a more appropriate long-term reservoir for preserving the older metasomatic components (e.g. EMI and EMII) which affected the previously depleted lithosphere. Conversely, the thermal boundary layer may represent an effective transient reservoir for more recent metasomatic agents (e.g. HIMU) rising directly from the convecting asthenospheric mantle (Wilson et al., 1995b; Beccaluva et al., 1998).

REFERENCES

Albarède, F., 1992. How deep do common basaltic magmas form and differentiate? J. Geophys. Res. 97, 10997–11009.

Allegre, C.J., Dupré, B., Lambert, B., Richard, P., 1981. The subcontinental versus suboceanic debate. Lead–neodynium–strontium isotopes in primary alkali basalts from a shield area: the Ahaggar volcanic suite. Earth Planet. Sci. Lett. 52, 85–92.

Anderson, D., 1994. The sublithospheric mantle as the source of continental flood basalts; the case against the continental lithosphere and plume head reservoirs. Earth Planet. Sci. Lett. 123, 269–280.

Barberi, F., Santacroce, R., Varet, J., 1982. Chemical aspects of rift magmatism. In: Continental and Oceanic Rifts. Geodynamic Series, vol. 8, pp. 223–258.

Beccaluva, L., Deriu, M., Macciotta, G., Savelli, C., Venturelli, G., 1977. Geochronology and magmatic character of the Pliocene-Pleistocene volcanism in Sardinia (Italy). Bull. Volcanol. 40 (3), 1–16.

Beccaluva, L., Colantoni, P., Di Girolamo, P., Savelli, C., 1981. Upper-Miocene submarine volcanism in the Strait of Sicily (Banco senza nome). Bull. Volcanol. 44, 573–581.

Beccaluva, L., Rossi, P.L., Serri, G., 1982. Neogene to Recent volcanism of the Southern Tyrrhenian-Sicilian area: implication for the geodynamic evolution of the Calabrian Arc. Earth Evol. Sci. 3, 222–238.

Beccaluva, L., Campredon, R., Feraud, G., Macciotta, G., 1983. Étude des relations entre volcanisme plio-quaternaire et tectonique en Sardaigne à l'aide de l'analyse structurale des dykes. Bull. Volcanol. 43, 365–379.

Beccaluva, L., Macciotta, G., Manetti, P., Peccerillo, A., Poli, G., 1984. Pliocene-Quaternary alkaline to subalkaline volcanism in Sardinia: Magma Genesis and evolution. Internal Report, Università di Firenze, pp. 1–50.

Beccaluva, L., Gabbianelli, G., Lucchini, F., Rossi, P.L., Savelli, C., 1985. Petrology and K/Ar ages of volcanics dredged from the Eolian seamounts: implication for geodynamic evolution of the southern Tyrrhenian basin. Earth Planet. Sci. Lett. 74, 187–208.

Beccaluva, L., Brotzu, P., Macciotta, G., Morbidelli, L., Serri, G., Traversa, G., 1987. Cenozoic tectono-magmatic evolution and inferred mantle in the Sardo-Tyrrhenian area. In: Advances in Earth Sciences Research. Accad. Naz. Lincei 80, 229–248.

Beccaluva, L., Macciotta, G., Siena, F., Zeda, O., 1989. Harzburgite to lherzolite xenoliths and clinopyroxene megacrysts of alkaline basic lavas from Sardinia (Italy). Chem. Geol. 77, 331–345.

Beccaluva, L., Di Grande, A., Lo Giudice, A., Macciotta, G., Siena, F., 1993. Geopetrographic map of northern-central Iblean area. S.E.L.C.A., Firenze.

Beccaluva, L., Siena, F., Coltorti, M., Di Grande, A., Lo Giudice, A., Macciotta, G., Tassinari, R., Vaccaro, C., 1998. Nephelinitic to tholeiitic magma generation in a transtensional tectonic setting: an integrated model for the Iblean volcanism, Sicily. J. Petrol. 39, 1547–1576.

Beccaluva, L., Coltorti, M., Milani, L., Salvini, L., Siena, F., 1999a. Petrogenesis of the Tertiary nephelinite to tholeiite volcanism in the Southern Alps. Plinius (Eur. J. Mineral., Italian suppl.) 22, 43–44.
Beccaluva, L., Bianchini, G., Coltorti, M., Del Moro, A., Siena, F., Vaccaro, C., 1999b. A case of multistage mantle metasomatism: peridotite xenoliths from Tallante (southern Spain). Plinius (Eur. J. Mineral., Italian suppl.) 22, 41–43.
Beccaluva, L., Coltorti, M., Milani, L., Salvini, L., Siena, F., Tassinari, R., 2000. Tertiary nephelinite to tholeite magma generation in the Veneto Volcanic Province, Southern Alps. In: Goldschmidt Conf. Abst. Vol.
Beccaluva, L., Bonadiman, C., Coltorti, M., Salvini, L., Siena, F., 2001a. Depletion events, natue of metasomatizing agents and timing of enrichment processes in lithospheric mantle xenoliths from the Veneto Volcanic Province. J. Petrol. 42, 173–187.
Beccaluva, L., Bianchini, G., Coltorti, M., Perkins, W.T., Siena, F., Vaccaro, C., Wilson, M., 2001b. Multistage evolution of the European lithospheric mantle: new evidence from Sardinian peridotite xenoliths. Contrib. Mineral. Petrol. 142, 284–297.
Bergamini, M.L., 1992. Petrologia degli xenoliti ultrafemici inclusi in vulcaniti alcaline dell'altopiano Ibleo (Sicila Sud-Orientale). Degree Thesis, Università di Ferrara.
Bianchini, G., Clocchiatti, R., Coltorti, M., Joron, J.L., Vaccaro, C., 1998. Petrogenesis of mafic lavas from the northernmost sector of the Iblean District (Sicily). Eur. J. Mineral. 10, 301–315.
Bianchini, G., Bell, K., Vaccaro, C., 1999. Mantle sources of the Cenozoic Iblean volcanism (SE Sicily–Italy): Sr–Nd–Pb isotopic constraints. Mineral. Petrol. 67, 213–221.
Bodinier, J.L., Menzies, M.A., Thirlwall, M.F., 1991. Continental to oceanic mantle transition-REE and Sr–Nd isotopic geochemistry of the Lanzo lherzolite massif. In: Orogenic Lherzolites and Mantle Processes. J. Petrol. spec. vol., 191–210.
Bonadiman, C., Coltorti, M., Milani, L., Salvini, L., Siena, F., Tassinari, R., 2001. Metasomatism in the lithosferic mantle and its relationships to magmatism in the Veneto Volcanic Province, Italy. Per. Mineral. 70, 333–357.
Brey, G.P., Koehler, T.P., 1990. Geothermobarometry in four phases lherzolites II. New thermobarometers and practical assessment of existing thermobarometers. J. Petrol. 31, 1353–1378.
Briot, D., Cantagrel, J.M., Dupuy, C., Harmon, R.S., 1991. Geochemical evolution in crustal magma reservoirs: trace element and Sr–Nd–O isotopic variations into continental intra-plate series at Monts Doré, Massifs Central, France. Chem. Geol. 89, 281–303.
Calanchi, N., Colantoni, P., Rossi, P.L., Saitta, M., Serri, G., 1989. The Strait of Sicily continental rift system: physiography and petrochemistry of the submarine volcanic centres. Marine Geology 87, 55–83.
Carter, S.R., Civetta, L., 1977. Genetic implications of the isotope and trace element variations in the eastern Sicilian volcanics. Earth Planet. Sci. Lett. 36, 168–180.
Cebrià, J.M., Lòpez-Ruiz, J., 1995. Alkali basalts and leucitites in an extensional intracontinental plate setting: the Late Cenozoic Calatrava Volcanic Province (Central Spain). Lithos 35, 27–46.
Cebrià, J.M., Wilson, M., 1995. Cenozoic mafic magmatism in Western-Central Europe: a common European asthenospheric reservoir? Terra Nova 7, 162.
Cebrià, J.M., Lòpez-Ruiz, J., Doblas, M., Oyarzun, R., Hertogen, J., Benito, R., 2000. Geochemistry of Quaternary alkali basalts of Garrotaxa (NE volcanic province, Spain): a case of double enrichment of the mantle lithosphere. J. Volc. Geotherm. Res. 102, 217–235.
Chauvel, C., Jahn, B.M., 1984. Nd–Sr isotope and REE geochemistry of alkali basalts from the Massif Central France. Geochim. Cosmochim. Acta 48, 93–110.
Civetta, L., D'Antonio, M., Orsi, G., Tilton, G.R., 1998. The geochemistry of volcanic rocks from Pantelleria island, Sicily Channel: petrogenesis and characteristics of the mantle source region. J. Petrol. 39, 1453–1492.
Clocchiatti, R., Joron, J., Treuil, M., 1988. The role of selective alkali contamination in the evolution of recent historic lavas of Mount Etna. J. Volc. Geotherm. Res. 34, 241–249.
Coltorti, M., Beccaluva, L., Bonadiman, C., Salvini, L., Siena, F., 2000. Glasses in mantle xenoliths as indicators of metasomatic agents. Earth Planet. Sci. Lett. 183, 303–320.
Condomines, M., Tanguy, J.C., Kieffer, G., Allegre, C.J., 1982. Magmatic evolution of a volcanic studied by $^{230}Th/^{238}U$ disequilibrium and trace elements systematics: the Etna case. Geochim. Cosmochim. Acta 46, 1397–1416.
Dautria, J.M., Liotard, J.M., Cabanes, M., Girod, M., Briqueu, L., 1987. Amphibole rich xenoliths and host alkali basalts: petrogenetic constraints and implications on the recent evolution of the upper mantle beneath the Ahaggar (Central Sahara, Southern Algeria). Contrib. Mineral. Petrol. 95, 133–144.
Davidson, J.P., Wilson, I.R., 1989. Evolution of an alkali basalt trachyte suite from Jebel Marra volcano, Sudan, through assimilation and fractional crystallisation. Earth Planet. Sci. Lett. 95, 141–160.
De Vecchi, G.P., Sedea, R., 1995. The Paleogene basalts of the Veneto Region (NE Italy). Mem. Sci. Geol. 47, 253–374.

Di Grande, A., Mazzoleni, P., Lo Giudice, A., Beccaluva, L., Macciotta, G., Siena, F., 2002. Subaerial Plio-pleistocene volcanism in the geo-petrographic and structural context of the north/central Iblean region (Sicily). Per. Mineral. 71, 159–189.

Downes, H., Dupuy, C., 1987. Textural, isotopic and REE variations in spinel peridotite xenoliths, Massif Central, France. Earth Planet. Sci. Lett. 82, 121–135.

Downes, H., 2001. Formation and modification of the shallow sub-lithospheric mantle: a review of geochemical evidence from ultramafic xenolith suites and tectonically emplaced massifs of Western and Central Europe. J. Petrol. 42, 233–250.

Downes, H., Bodinier, J.-L., Thirlwall, M.F., Lorand, J.-P., Fabries, J., 1991. REE and Sr–Nd isotopic geochemistry of Eastern Pyrenean Peridotite Massifs: sub-continental lithospheric mantle modified by continental magmatism. In: Orogenic lherzolites and Mantle Processes. J. Petrol. spec. vol., 97–115.

Esperança, S., Crisci, G.M., 1995. The island of Pantelleria: a case for the development of DMM-HIMU isotopic compositions in a long-lived extensional setting. Earth Planet. Sci. Lett. 136, 167–182.

Falloon, T.J., Green, D.H., 1988. Anhydrous partial melting of peridotite from 8 to 35 kb and petrogenesis of MORB. J. Petrol., Special Lithosphere Issue, 279–414.

Falloon, T.J., Green, D.H., Hatton, C.J., Harris, K.L., 1988. Anhydrous partial melting of a fertile and depleted peridotite from 2 to 30 kb and application to basalt petrogenesis. J. Petrol. 29, 1257–1282.

Gasperini, D., Blichert-Toft, J., Bosh, D., Del Moro, A., Macera, P., Télouk, P., Albarede, F., 2000. Evidence from Sardinian basalt geochemistry for recycling of plume heads into the Earth's mantle. Nature 408, 701–704.

Gillot, P.Y., Kieffer, G., Romano, R., 1994. The evolution of Mount Etna in the light of potassium-argon dating. Acta Vulcanol. 5, 81–87.

Granet, M., Wilson, M., Achauer, U., 1995. Imaging a mantle plume beneath the French Massif Central. Earth Planet. Sci. Lett. 136, 281–296.

Green, D.H., Falloon, T.J., Taylor, W.R., 1987. Mantle derived magmas – roles of variable source peridotite and C–H–O fluid composition. In: Mysen, B.O. (Ed.), Magmatic Processes: Physicochemical Principles. Geochem. Soc. Spec. Publ. 1, 139–154.

Hawkesworth, C.J., Vollmer, R., 1979. Crustal contamination versus enriched mantle: $^{143}Nd/^{144}Nd$ and $^{87}Sr/^{86}Sr$ evidence from the Italian volcanics. Contrib. Mineral. Petrol. 69, 151–165.

Hirose, K., Kawamoto, T., 1995. Hydrous partial melting of lherzolite at 1 GPa: the effect of H_2O on the genesis of basaltic magmas. Earth Planet. Sci. Lett. 133, 463–473.

Hirose, K., Kushiro, I., 1993. Partial melting of dry peridotites at high pressures: determination of compositions of melts segregated from peridotite using aggregates of diamond. Earth Planet. Sci. Lett. 114, 477–489.

Hoernle, K., Zhang, Y.-S., Graham, D., 1995. Seismic and geochemical evidence for large-scale mantle upwelling beneath the eastern Atlantic and western and central Europe. Nature 374, 34–39.

Hofmann, A.W., 1997. Mantle geochemistry: the message from oceanic volcanism. Nature 385, 219–229.

Irvine, T.M., Baragar, W.R.A., 1971. A guide to the chemical classification of the common rocks. Can. J. Earth Sci. 8, 523–548.

Koehler, T.P., Brey, G.P., 1990. Calcium exchange between olivine and clinopyroxene calibrated as a geothermobarometer for natural peridotites from 2 to 60 kb with applications. Geochim. Cosmochim. Acta 54, 2375–2388.

Kramm, U., Wedepohl, K.H., 1990. Tertiary basalts and peridotite xenoliths from the Hessian Depression (NW Germany), reflecting mantle compositions low in radiogenic Nd and Sr. Contrib. Mineral. Petrol. 106, 1–8.

Latin, D.M., Dixon, J.E., Fitton, J.G., White, N., 1990. Rift-related magmatism in the North Sea Basin: implications for stretching mechanism. In: Hardman, R., Brooks, M. (Eds.), Tectonic Events Responsible for Britain's Oil and Gas Reserves. Geol. Soc. London Spec. Publ. 55, 207–227.

Le Bas, M.J., Le Maitre, R.W., Woolley, A.R., 1992. The construction of the Total Alkali-Silica chemical classification of volcanic rocks. Mineral. Petrol. 46, 1–22.

Lee, D.C., Halliday, A.N., Davies, G.R., Essene, E.J., Fitton, J.G., Temdjim, R., 1996. Melt enrichment, of shallow depleted mantle: a detailed petrological, trace element, and isotopic study of mantle-derived xenoliths and megacrysts from the Cameroon line. J. Petrol. 37, 415–441.

Lustrino, M., Melluso, L., Morra, V., 2000. The role of continental crust and lithospheric mantle in the genesis of Plio-Pleistocene volcanic rock from Sardinia (Italy). Earth Planet. Sci. Lett. 180, 259–270.

Lustrino, M., Melluso, L., Morra, V., 2002. The transition from alkaline to tholeiitic magmas: a case study from the Orosei-Dorgali Pliocene volcanic district (NE Sardinia, Italy). Lithos 63, 83–113.

Macciotta, G., Venturelli, G., Beccaluva, L., 1978. Geochemistry of mafic Cainozoic volcanic rocks from Sardinia (Western Mediterranean). Bull. Volcanol. 41, 56–78.

Milani, L., Beccaluva, L., Coltorti, M., 1999. Petrogenesis and evolution of the Euganean Magmatic Complex, Veneto Region, North-East Italy. Eur. J. Mineral. 11, 379–399.

Montanini, A., Barbieri, M., Castorina, F., 1994. The role of fractional crystallization, crustal melting and magma mixing in the petrogenesis of rhyolites and mafic inclusions-bearing dacites from the Monte Arci volcanic complex (Sardinia, Italy). J. Volc. Geotherm. Res. 61, 95–120.
Nimis, P., 1995. A clinopyroxene geobarometer for basaltic systems based on crystal structure modelling. Contrib. Mineral. Petrol. 121, 115–125.
Niu, Y., 1997. Mantle melting and melt extraction processes beneath ocean ridges: evidence from abyssal peridotites. J. Petrol. 38, 1047–1074.
Olafsson, M., Eggler, D.H., 1983. Phase relations of amphibole-carbonate and phlogopite-carbonate peridotite: petrologic constraints on the asthenosphere. Earth Planet. Sci. Lett. 64, 305–315.
Panza, G.F., Suhadolc, P., 1990. Properties of the lithosphere in collisional belts in the Mediterranean – a review. Tectonophysics 182, 39–46.
Reisberg, L., Zindler, A., Jagoutz, E., 1989. Further Sr and Nd isotopic results from peridotites of the Ronda ultramafic complex. Earth Planet. Sci. Lett. 96, 161–180.
Romano, R., 1983. Succession of the volcanic activity in the Etnean area. Mem. Soc. Geol. It. 23, 27–28.
Schiano, P., Clocchiatti, R., Ottolini, L., Busà, T., 2001. Transition of Mount Etna lavas from a mantle-plume to an island-arc magmatic source. Nature 412, 900–904.
Siena, F., Coltorti, M., 1989. Lithospheric mantle evolution: evidences from ultramafic xenoliths in the Lessinian volcanics (northern Italy). Chem. Geol. 77, 47–364.
Siena, F., Coltorti, M., 1993. Thermobarometric evolution and metasomatic processes of upper mantle in different tectonic settings: evidence from spinel peridotite xenoliths. Eur. J. Mineral. 5, 1073–1090.
Siena, F., Beccaluva, L., Coltorti, M., Marchesi, M., Morra, V., 1991. Ridge to hot-spot evolution of the Atlantic lithospheric mantle: evidence from Lanzarote peridotite xenoliths (Canary Islands). In: Orogenic Lherzolites and Mantle Processes. J. Petrol. spec. vol., 271–290.
Sun, S.-S., McDonough, W.F., 1989. Chemical and isotopic systematics of oceanic basalts: implications for mantle composition and process. In: Saunders, A.D., Norry, M.J. (Eds.), Magmatism in the Oceanic Basins. Geol. Soc. Spec. Publ. 42, 313–346.
Tonarini, S., D'Orazio, M., Armienti, P., Innocenti, F., Scribano, V., 1996. Geochemical features of eastern Sicily lithosphere as probed by Hyblean xenoliths lavas. Eur. J. Mineral. 8, 1153–1173.
Tonarini, S., Armienti, P., D'Orazio, M., Innocenti, F., 2001. Subduction-like fluids in the genesis of Mt. Etna magmas: evidence from boron isotopes and fluid mobile elements. Earth Planet. Sci. Lett. 192, 471–483.
Trua, T., Esperanca, S., Mazzuoli, R., 1998. The evolution of the lithospheric mantle along the N. African plate: geochemical and isotopic evidence from the tholeiitic and alkaline volcanic rocks of the Hyblean Plateau, Italy. Contr. Mineral. Petrol. 131, 307–322.
Verde, M., 1996. Aspetti petrologici e reologici del mantello superiore nell'area mediterranea mediante lo studio di xenoliti di mantello inclusi in vulcanite basiche alcaline. Ph.D. Thesis, Università di Ferrara.
Viti, M., Albarello, D., Mantovani, E., 1997. Rheological profiles in the central/eastern Mediterranean. Ann. Geofis. 40, 849–864.
Weaver, B.L., 1991. The origin of ocean island basalt end-member compositions: trace element and isotopic constraints. Earth Planet. Sci. Lett. 104, 381–397.
Wedepohl, K.H., Gohn, E., Hartmann, G., 1994. Cenozoic alkali basaltic magmas of western Germany and their products of differentiation. Contrib. Mineral. Petrol. 115, 253–278.
White, B.S., Wyllie, P.J., 1992. Solidus reactions in synthetic lherzolite–H_2O–CO_2 from 20–30 kbar, with applications to melting and metasomatism. J. Volc. Geotherm. Res. 50, 117–130.
Wilson, M., Downes, H., 1991. Tertiary-Quaternary extensional related alkaline magmatism western and central Europe. J. Petrol. 32, 811–849.
Wilson, M., Bianchini, G., 1999. Tertiary-Quaternary magmatism within the Mediterranean and surrounding regions. In: Durand, et al. (Eds.), The Mediterranean Basin: Tertiary Extension within the Alpine Orogen. Geol. Soc. Spec. Publ. 156, 141–168.
Wilson, M., Downes, H., Cebrià, J.M., 1995a. Contrasting fractionation trends in coexisting continental alkaline magma series; Cantal, Massif Central, France. J. Petrol. 119, 181–196.
Wilson, M., Rosenbaum, J.M., Dunworth, E.A., 1995b. Melilitites: partial melts of the thermal boundary layer? Contrib. Mineral. Petrol. 119, 181–196.
Witt-Eickschen, G., Seck, H.A., Mezger, K., Eggins, S.M., Altherr, R., 2003. Lithospheric mantle evolution beneath the Eifel (Germany): Constraints from Sr–Nd–Pb isotopes and trace element abundances in spinel peridotite and pyroxenite xenoliths. J. Petrol. 44, 1077–1096.
Worner, G., Zindler, A., Staudingel, H., Schmincke, H.U., 1986. Sr, Nd and Pb isotope geochemistry of Tertiary and Quaternary volcanics from West Germany. Earth Planet. Sci. Lett. 79, 107–119.
Wyllie, P.J., 1987. Discussion of recent papers on carbonated peridotites, bearing on mantle metasomatism and magmatism. Earth Planet. Sci. Lett. 82, 391–397.

Zangana, N.A., Downes, H., Thirlwall, M.F., Hegner, E., 1997. Relationship between deformation, equilibration temperatures, REE and radiogenic isotopes in mantle xenoliths (Ray Pic, French Massif Central): an example of plume-lithosphere interaction? Contrib. Mineral. Petrol. 127, 187–203.

Zindler, A., Hart, S.R., 1986. Chemical Geodynamics. Earth Planet. Sci. Annual Review 14, 493–571.

CROP PROJECT: Deep Seismic Exploration of the Central Mediterranean and Italy
Edited by I.R. Finetti

CHAPTER 31

A Reappraisal of the Betic Ophiolitic Association: The Westernmost Relic of the Alpine Tethys

Encarnación Puga[1]

ABSTRACT

Available petrological, geochemical and geochronological data indicate that the Betic Ophiolitic Association (BOA) is formed by fragments of a Jurassic oceanic lithosphere overlain by Cretaceous sediments. The primary rock assemblage includes: mantle ultramafics mostly of spinel-lherzolite composition, containing numerous boudinated dolerite dykes; cumulus troctolites and gabbros, cross-cut by dolerite and basaltic dykes; olivine-basalt and basalt with well-preserved pillow and flow structures, cross-cut by basaltic and more differentiated dykes, occasionally of plagiogranite composition. All these basic rocks have a tholeiitic and T- to E-MORB character. The covering meta-sediments include a well-developed calc-schist unit containing foraminifera remains of probable Cretaceous age. Petrogenetic and age similarities between the BOA and the Alpine–Apennine ophiolites, as well as palaeogeographic reconstructions of the Mesozoic Tethyan Basin, suggest that the Betic ophiolites may have originated in the southernmost part of the Western Tethys, along an oceanic junction between the Alpine Tethys and the central Atlantic Ocean. Slices of different rock types from the BOA were tectonically imbricated with the crustal units of the Mulhacén Complex, during and after their eo-Alpine subduction and exhumation processes and, finally, incorporated into the continental margin, where they were partly covered by the intra-orogenic volcano-sedimentary Soportújar Formation. These slices of the Betic oceanic lithosphere preserve chemical and mineralogical evidence of an ocean-floor metamorphic process, responsible for the first serpentinization stage of the primary lherzolites, and the concomitant rodingitization of the associated dolerite dykes. This ocean-floor process also gave rise to high-gradient amphibolite assemblages that, locally, have been very well preserved in both the plutonic and volcanic sections of the BOA despite the superimposed Alpine metamorphism. In all the BOA sequences the orogenic metamorphism evolved from eclogite facies (developed during the Late Cretaceous eo-Alpine subduction process and partly retrograded to blueschist facies during a subsequent exhumation stage) to the Ab-Ep amphibolite facies that characterizes the Oligocene meso-Alpine event and was locally retrograded to greenschist facies during the Early Miocene neo-Alpine event.

Keywords: Betic ophiolites, metamorphism, geochemistry, Radiometric dating

1. INTRODUCTION

The basic and ultramafic rocks in the Mulhacén Complex of the Betic Cordilleras were previously interpreted as the metamorphic products of igneous materials originating from an early geosynclinal stage of the Alpine orogeny (Nijhuis, 1964; Voet, 1967; Puga, 1971). More detailed petrological and geochemical investigation of these rocks led us to reinterpret them as relics of a dismembered and metamorphosed ophiolitic suite (Puga, 1977; Burgos et al., 1980; Díaz De Federico, 1980; Bodinier et al., 1987; Puga et al., 1989a, 1989b), defined by Puga (1990) as the Betic Ophiolitic Association (BOA). Subsequently, palaeontological dating of scarce foraminifer remains preserved in the sedimentary unit covering the igneous rocks (Tendero et al., 1993), accompanied by advances in knowledge of trace-element, isotopic geochemistry and radiometric dating of the basic and ultramafic lithotypes (Puga et al.,

[1]Instituto Andaluz de Ciencias de la Tierra, CSIC-UGRA, Facultad de Ciencias, Avda. Fuente Nueva s/n, 18002 Granada, Spain. Fax: +34 958 243384; e-mail: epuga@goliat.ugr.es

1995, 1996, 1999a, 2002a, 2005), have confirmed our hypothesis on the ophiolitic character of this association. This interpretation differs, however, from that of Gomez-Pugnaire et al. (2000), who consider the metabasites of this association as subvolcanic intrusions into Triassic sediments of basaltic magmas generated, like the associated ultramafic rocks, in a continental rift setting.

The present work sets out the petrological, geochemical and geochronological characteristics of the BOA (made up of metamorphic rocks of basic, ultramafic and sedimentary origin) and argues its provenance from an oceanic lithosphere which during the Jurassic would have connected the Alpine Tethys to the central Atlantic. Most of the data presented in this chapter are unpublished, although some figures are reported from previous publications.

2. GEOLOGICAL SETTING

The Betic Cordilleras of SE Spain make up part of the peri-Mediterranean Alpine orogenic system and crop out in a relatively continuous band, delimited to the south by the Mediterranean Sea (Figure 1). The Internal Zone of these cordilleras has generally been subdivided into three nappe complexes called, from bottom to top: the Nevado-Filábride, Alpujárride and Maláguide. They are sometimes respectively correlated with the Penninic, lower Austroalpine and upper Austroalpine nappes of the Alps (Egeler and Simon, 1969; Michard et al., 1991). The Nevado-Filábride Complex was further subdivided into the Veleta and the Mulhacén Complexes, in accordance with the arguments presented in Puga et al. (2002a), making a total of four complexes.

The Veleta Complex is the deepest of the Internal Zones of the Betic Cordilleras (Figures 1 and 2). It was overthrust by the Mulhacén Complex (MC) after the first Alpine metamorphic event, which developed under an HP/LT (high-pressure/low temperature) metamorphic gradient only in the overlying MC. Both complexes were overthrust by

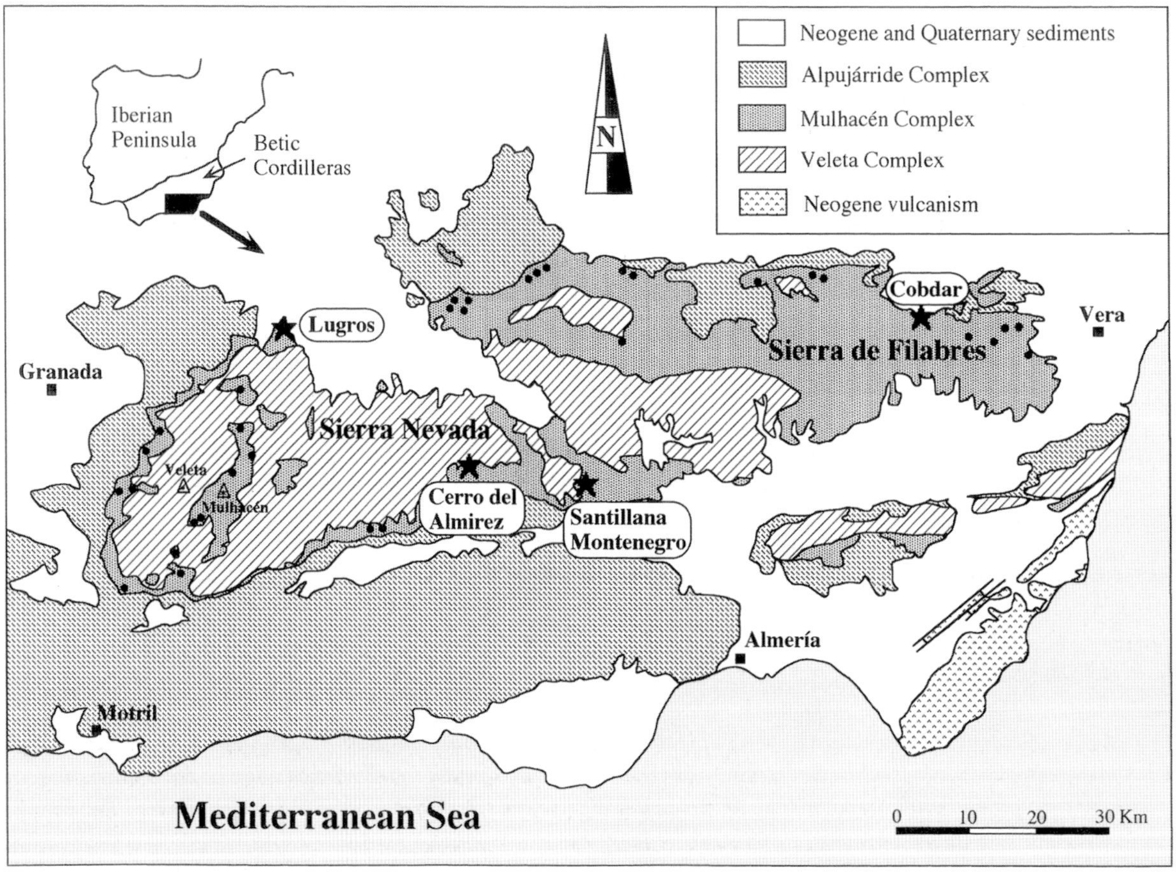

Figure 1: Geological sketch of the central-eastern sector of the Betic Cordilleras, showing the relationships between the Veleta, Mulhacén and Alpujárride metamorphic complexes and the post-orogenic volcanic and sedimentary formations (from Puga et al., 2000). Solid dots on the MC represent the location of the main outcrops of the BOA exposed in Sierra Nevada and Sierra de Filabres; stars indicate the location of the ophiolitic outcrops mentioned throughout the text.

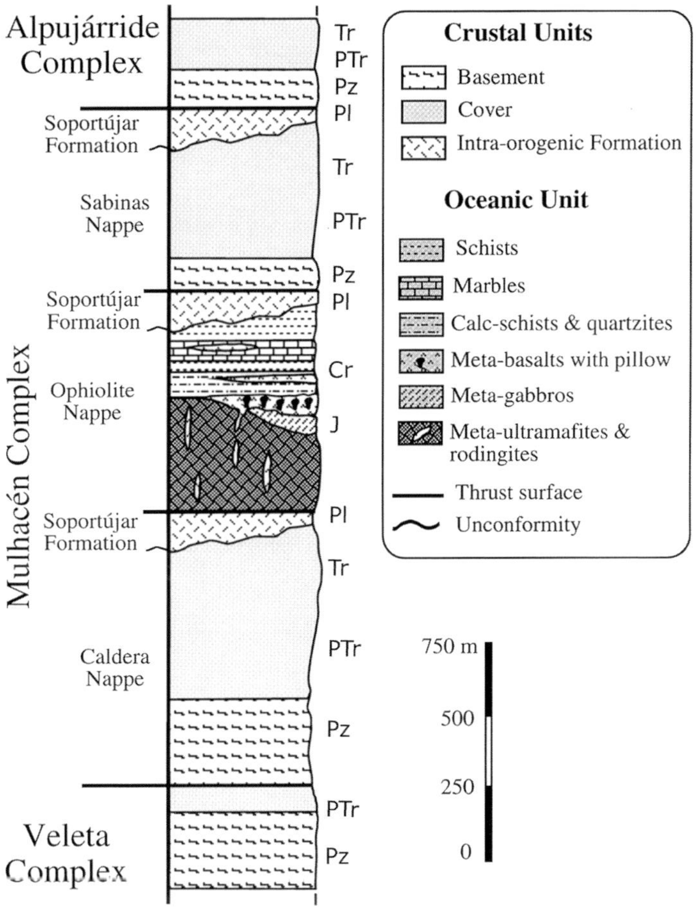

Figure 2: Simplified lithostratigraphic column of the metamorphic complexes in Figure 1, showing the main rock-types forming the BOA and a reconstruction of their relative positions during the primitive oceanic stage. The most probable ages, inferred from radiometric and palaeontological data, for the different formations making up this column are indicated as follows: Pz = Palaeozoic; PTr = Permo-Triassic; Tr = Triassic; J = Jurassic; C = Cretaceous; Pl = Paleocene.

the Alpujárride and Maláguide Complexes later than the second metamorphic event, as shown by their mutual cartographic contacts and the contrasting metamorphic evolutions in each of these complexes (Torres-Roldán, 1979; Díaz De Federico et al., 1990; De Jong, 1991, 1993).

The BOA is part of the MC, which crops out only in the central and eastern sectors of the Betic Cordillera overlying the Veleta Complex (Figure 1). These two complexes form a series of tectonic windows overthrust by the Alpujárride Complex (Figures 1 and 2). The MC is composed of two thrust units of crustal origin, the Caldera below and the Sabinas above, between which the oceanic BOA unit is tectonically intercalated (Figure 2). The Soportújar Formation is tectonically sandwiched between the different units of the MC (Figure 2).

The crustal units of the MC are made up of a Palaeozoic basement, consisting primarily of graphite-bearing micaschists containing metagranite slides and eclogitized skarn rocks, and a Mesozoic cover, mainly composed of marble, micaschists and rhyolitic orthogneisses. The BOA comprises numerous metre- to kilometre-sized lenses of metamorphic rocks, derived from mafic and/or ultramafic and sedimentary rock types, with mylonitic fabrics near the contacts with other Mulhacén nappes. This ophiolite unit is dismembered and metamorphosed in the same way as its counterparts in the Pennine nappes of the Alps (Dietrich, 1980; Desmons, 1989; Desmons et al., 1999), and it is discontinuously exposed along about 250 km in the MC (black dots in Figure 1). The Soportújar Formation is mainly formed by discontinuous levels of continental and evaporitic meta-sediments and meta-tuffites, which were deposited upon different formations of the MC during a intra-orogenic relaxation stage of probable Paleocene age (Figure 2 and Puga et al., 1996).

3. GEOCHRONOLOGICAL CONSTRAINTS

The metabasic rocks of the BOA have been radiometrically dated by Rb/Sr, K/Ar and ^{40}Ar/^{39}Ar as having originated between the Triassic-Jurassic boundary and the Late Jurassic (Table 1). Recent U/Pb dating of zircons from the BOA meta-gabbros gave an Early Jurassic age for the beginning of this magmatism, as well as a Late Jurassic age, interpreted as corresponding to the ocean-floor recrystallization stage (Puga et al., 2005). Moreover, ^{40}Ar/^{39}Ar laser probe dating of brown amphiboles filling millimetric veins in Cóbdar ophiolitic meta-basalts gave a Late Jurassic age for the oceanic metamorphism, which would be coeval or slightly younger than the Late Jurassic volcanic activity (Puga et al., 1991, 1995, 2002b). Radiometric age determinations of metamorphic mineral associations that developed during the Alpine orogeny in the MC are scarce, and their interpretation is in some cases ambiguous, mainly due to the possible existence of excess Ar in K-bearing minerals, and to the Sm/Nd isotopic disequilibrium observed in eclogitic assemblages (Nieto et al., 1997a). Two main events of Alpine metamorphic blastesis can, nevertheless, be envisaged on the basis of the available data, using K/Ar, ^{40}Ar/^{39}Ar, Sm/Nd and U/Pb on whole rocks and mineral separates from the different tectonic units of the MC (Puga, 1971; Portugal Ferreira et al., 1988; Andriessen et al., 1991; De Jong, 1991; Monié et al., 1991; Puga et al., 1989b, 1991, 1995, 2002a, 2005; Nieto et al., 1997b): an eo-Alpine Late Cretaceous-Paleocene event, and a meso-Alpine Late Eocene-Oligocene event (Tables 1 and 2). Some meta-andesites of the Soportújar Formation preserving igneous relics were dated by K/Ar as Paleocene, and their metamorphic minerals corresponding to the meso-Alpine event as Oligocene (Puga et al., 1996). Numerous radiometric age determinations, obtained by different methods, provide a Miocene age in the Mulhacén and other Betic complexes, and this has lead to

Table 1
Geochronological data on metabasic and associated rocks of the Betic Ophiolitic Association

Process	Type of sample	Locality	Method	Closing age (M.a.)	Reference
Jurassic magmatism	Intercumulus biotite in gabbro	Cóbdar	Ar/Ar	213 ± 2.5	Puga et al. (1991)
	Zircon in eclogite	Lugros	SHRIMP U–Pb	180–190	Puga et al. (2005)
	Gabbro (WR)	Cóbdar	K/Ar	174 ± 4	Portugal Ferreira et al. (1988)
	Relict igneous plagioclase in dolerite	Cóbdar	K/Ar	164 ± 4	Portugal Ferreira et al. (1988)
	Ocean-floor metamorphic amphibole in basalt	Cóbdar	Ar/Ar laser probe	158 ± 4	Puga et al. (1991)
	Zircon in eclogite	Lugros	SHRIMP U–Pb	150–155	Puga et al. (2005)
	Olivine dolerite mineral isochrone	Lubrin	Rb/Sr	146 ± 3	Hebeda et al. (1980)
Eo-Alpine event	Zircon in eclogite	Lugros	SHRIMP U–Pb	70–90	Puga et al. (2005)
	Glaucophane	Cóbdar	K/Ar	72 ± 1	Portugal Ferreira et al. (1988)
Meso-Alpine event	Barroisitic amphibole	Sª Baza	Ar/Ar	48.4 ± 2.2	Monié et al. (1991)
	Barroisite	Cóbdar	Ar/Ar	41.4 ± 2.3	Puga et al. (2002a)
	Zircon in eclogite	Lugros	SHRIMP U–Pb	30.9 ± 0.6	Puga et al. (2005)
	White mica	Cauchiles	Ar/Ar	28.5 ± 2.5	Puga et al. (2002a)
	Phengitic mica	Cóbdar	K/Ar	27 ± 1	Portugal Ferreira et al. (1988)
	Magnesiohornblende	Sª Baza	Ar/Ar	24.6 ± 3.6	Monié et al. (1991)
Neo-Alpine event	Zircon in pyroxenite	Almirez	SHRIMP U–Pb	15.0 ± 0.6	López Sanchez-Vizcaino et al. (2001)
	White mica	Cauchiles	Ar/Ar	13.0 ± 1.5	Puga et al. (2002a)
	Paragonite	Camarate	Ar/Ar	12.4 ± 1.1	Puga et al. (2002a)
	Actinolite	Almirez	K/Ar	11 ± 3	Portugal Ferreira et al. (1988)
	Actinolite	Camarate	Ar/Ar	11 ± 0.9	Puga et al. (2002a)

the naming of the neo-Alpine event (Puga et al., 2002a). The Miocene radiometric ages have been interpreted as due to partial reopening of the isotopic systems, most probably related to late extensional tectonic stages (Monié et al., 1991) and/or to the volcanic activity that generated the Cabo de Gata volcanic province, as well as the associated hydrothermal alteration and mineralization in the nearby metamorphic rocks (Bellon et al., 1983; De Jong et al, 1992).

The Alpine metamorphism in the three mentioned tectonic units making up the MC developed under eclogite facies conditions in the eo-Alpine event, and under Ab-Ep amphibolite facies in the meso-Alpine event (Puga and Díaz de Federico, 1978; Vissers, 1981; Gomez Pugnaire and Fernandez Soler, 1987; Bakker et al., 1989; Puga et al., 1989a, 1989b, 1999a, 2000). The only exception to this metamorphic evolution is represented by the Soportújar Formation, in which only Ab-Ep amphibolites, corresponding to the meso-Alpine metamorphic event, were developed (Puga et al., 1996). The neo-Alpine event locally gave rise to greenschist facies paragenesis by retrogression of the eo-Alpine and meso-Alpine parageneses (Puga and Díaz de Federico, 1978; Puga et al., 2000, 2002a).

4. PRIMARY LITHOLOGY AND FIELD OCCURRENCE

Figure 2 shows a hypothetical reconstruction of the BOA which, in terms of primary rocks, is formed from bottom to top by: (1) an ultramafic sequence, consisting of serpentinized lherzolites, containing numerous boudinated dolerite dykes; (2) a layered plutonic sequence made up of cumulus troctolites and gabbros cross-cut by dolerite dykes; (3) a volcanic sequence composed of basalts with pillow structures, lying unconformably over the plutonic sequence levels; (4) a meta-sedimentary sequence made up mainly of quartzite and ankerite-bearing calc-schists with some micaschist and marble layers, directly overlying the ultramafic or basic sequences. Each of the last three sequences reaches about a hundred metres in thickness, whereas the ultramafic sequence may be up to several hundred metres thick (Figure 2).

In order to illustrate the lithology, metamorphic evolution and chemical composition of the BOA, we have chosen the Lugros and Cóbdar outcrops, located in the western zone of Sierra Nevada, and the eastern zone of Sierra de Filabres, respectively (Figure 1), where the volcanic and plutonic sequences are well represented. The main difference between these two outcrops is the lesser degree of pervasiveness of the HP/LT (high pressure/low temperature) eo-Alpine metamorphism in the Cóbdar rocks. This has made possible the local preservation of igneous and oceanic parageneses in many lithotypes from Cóbdar, which are never present in the Lugros outcrop, as well as the superior preservation of oceanic assemblages in the Cóbdar outcrop. Comparable differences can also be observed in other ophiolitic outcrops in the same geographical areas as the Cóbdar and Lugros outcrops (Morten et al., 1987; Puga et al., 1989b, Cámara, 1995); these have been explained by Puga et al. (2000) as a polarity in the field metamorphic gradient, increasing from east to west in the ophiolite unit, as a consequence of the generation of HP/LT metamorphism at different depths in the same subduction zone.

For a description of the metabasic and meta-ultramafic lithotypes making up the ultramafic sequence of the BOA, the Almirez outcrop in the eastern Sierra Nevada has been singled out as the most representative (Figure 1). This outcrop shows very well developed ocean-floor metasomatism followed by HP/LT eo-Alpine metamorphism (Puga et al., 1999a). The Santillana and Montenegro outcrops (Figure 1) are very similar to the Almirez outcrop although they present a less pervasive ocean-floor metasomatism (Puga et al., 1999a).

The sedimentary sequence of the BOA is well represented in the Cóbdar and Almirez outcrops, whereas it is lacking in Lugros. In this sequence, some thin-bedded and fine-grained quartzite levels at the base could derive from recrystallized radiolarites, whereas an intermediate level of graphite- and garnet-bearing micaschists may be related to similar levels present in the Western Tethyan ophiolites that originated in oceanic anoxic episodes during the Middle to Late Cretaceous (Jenkyns, 1980). Moreover, remains of Cretaceous foraminifera similar to those identified in the sedimentary sequences of the Alpine–Apennine ophiolites (Lemoine et al., 1984) have been found in the calc-schists of the Cóbdar outcrop (Tendero et al., 1993).

The abbreviations for minerals suggested by Kretz (1983) have been used in the text, tables and figures.

In the **Cóbdar** area (Figure 1), only one sequence of metabasic rocks several kilometres in length and about 200 metres thick, representing the plutonic and volcanic protoliths, and 70 metres of metasediments are preserved (Puga et al., 1989a). Scapolite-bearing evaporitic phyllites and carbonate rocks of the Soportújar Formation, affected only

Figure 3: Cóbdar outcrop. 3-1: Flow structures with ropy surfaces in amphibolitized metabasalts from the volcanic sequence. 3-2: Deformed pillow structure, preserving small vacuoles and a primitive radial disjunction in meta-basalts. 3-3: Porphyritic pillow basalt formed by plagioclase microphenocrysts in a very fine-grained interstitial matrix containing clinopyroxene, ore, chlorite and green biotite. 3-4: Micro-porphyric meta-basalt with ocean-floor brown titanian pargasite filling fissures and replacing the igneous clinopyroxene in the matrix. 3-5: Olivine-bearing pillow basalt with well-preserved olivine phenocrysts in an interstitial matrix formed by calcic plagioclase, clinopyroxene and occasional chromite. 3-6: Olivine phenocrysts replaced by phlogopite and talc, surrounded by a matrix formed by albite and ocean-floor brown amphibole in meta-basalt. 3-7: Cumulate meta-gabbro formed by well-preserved olivine

by meso-Alpine metamorphism, overlie the highly tectonized quartzites, micaschists, marbles and calc-schists of the sedimentary sequence of the ophiolites, or its volcanic sequence directly, since both sequences are affected by the eo-Alpine and meso-Alpine metamorphic events. A few metabasalt outcrops, directly overlain by the Soportújar Formation, also present scapolite as late poikiloblasts and filling veins. The ophiolitic association, with or without the overlying Soportújar Formation rocks, is overthrust by marbles and micaschists of the Sabinas unit of the MC (Figure 2). The volcanic sequence is composed of aphanitic and amygdaloidal metabasalts, some of which preserve flow structures (Figure 3-1). The basalts, locally cut by several decimetre-thick dolerite dykes, gradually merge into pillow lavas ranging from a few decimetres to more than one metre in diameter (Figure 3-2 and Puga et al., 1989a). In the upper basaltic level, some leucocratic veins several centimetres in thickness, or boudinated dykes several decimetres thick, of plagiogranite *s.l.* have been found recently. Basaltic layers unconformably overlie different levels of the plutonic sequence, which is formed by alternating horizons of troctolitic cumulates and gabbros cut across by dolerite dykes several centimetres to decimetres thick.

The Cóbdar metabasalts are generally greenish, very fine-grained, and contain numerous mm-sized vesicles which were filled with carbonates in the oceanic environment and partially replaced by biotite, chlorite, amphiboles and scapolite during subsequent metamorphic stages (Puga et al., 1989a). The troctolitic cumulates, very common in the Cóbdar suite, are dark grey in colour, medium-grained, and commonly show an onion-shaped disjunction; they alternate with greenish metagabbros with a doleritic texture and medium to coarse grain-size.

In the **Lugros** area (Figure 1) the ophiolitic association consists of a slice of eclogites, less than 100 metres thick, derived from volcanic and plutonic protoliths corresponding to three tectonic klippen several hundred metres to 1 km in length, thrust over garnet-bearing micaschists of the Caldera unit and covered by Quaternary deposits. Plutonic lithotypes, cut across by scarce decimetric dykes of dolerite, predominate in this suite. They consist mainly of Ol-gabbros and gabbros, with medium- to coarse-grained doleritic and locally pegmatitic textures (Cámara, 1995). The volcanic sequence is mainly restricted to isolated blocks within an outcrop several hundred metres in extension. Most of these blocks correspond to multiple, irregularly folded layers of basalt, several centimetres in thickness, formed by a quenched outer part, an aphyric black outer zone and a hypocrystalline central zone. Other blocks in the Lugros suite are formed by piles of molded minipillows, several centimetres in diameter, alternating with other drained pillow structures of a maximum of several decimetres in length. All these structures are the result of submarine eruptions, involving recurrent small volumes of highly fluid magma, giving rise to predominantly small pillow lavas and flow structures, which are locally very well preserved despite having experienced the Alpine metamorphism (Figures 4-1, 4-2 and Puga et al., 1995).

The **Almirez** is the largest outcrop in the BOA, predominantly consisting of ultramafic rocks. These occur in the Cerro del Almirez, and form four klippen, the largest of which is about 3 km in length and several hundred metres thick. These klippen tectonically overlie graphite- and garnet-bearing micaschists of the basement of the Caldera unit (Figure 2). The ultramafic rocks are mainly serpentinites containing irregular centimetric to metric pods of modal harzburgites with spinifex-like textures (Puga et al., 1997a, 1999a). Both meta-ultramafic rock-types are traversed by numerous decimetre- to metre-thick rodingitized dolerite dykes, which are strongly boudinated and elongated following the main schistosity surfaces that developed during the meso-Alpine metamorphic event. The ultramafic sequence is directly overlain by ophicarbonates and/or quartzite and micaschist layers, with some calc-schists and marble lenses containing hydrothermal Fe–Mn mineralizations.

5. POLYMETAMORPHIC EVOLUTION OF THE BOA

The various igneous lithotypes comprising the Betic Ophiolitic Association, illustrated in Figures 3 to 5, preserve chemical and mineralogical evidence of a metasomatic and metamorphic stage of blastesis prior to the development of the Alpine orogenic metamorphism (Table 2). During this oceanic stage, both the gabbros and basalts of this association were the site of high-gradient metamorphic parageneses in amphibolite and greenschist facies (Puga et al., 1989a, 1999b, 2002b), of which only relics have been preserved. Likewise, the ultramafic BOA rocks, and the dykes intruded in them, underwent an initial stage of serpentinization and rodingitization in an oceanic

phenocrysts surrounded by calcic plagioclase and intercumulus clinopyroxene, scarcely transformed during post-magmatic processes. 3-8: Coronitic eclogite deriving from a olivine gabbro in which olivine has been replaced by almandine plus barroisite; igneous clinopyroxene by omphacite, rutile and barroisite; and plagioclase phenocrysts by clinozoisite, albite and white mica.

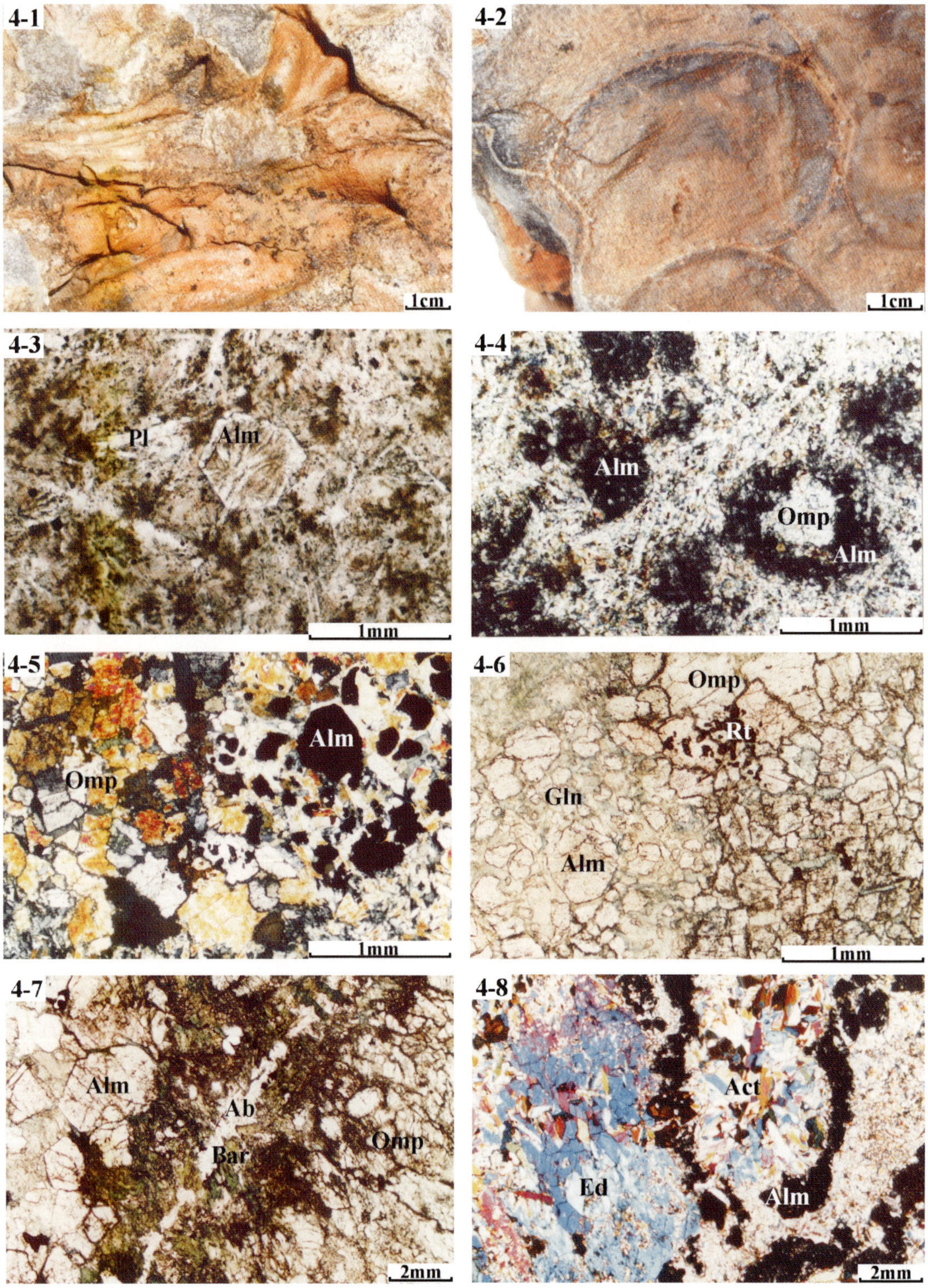

Figure 4: Lugros outcrop. 4-1: Drapes and folds on a lava flow surface in an eclogitized basalt. 4-2: Accumulation of eclogitized minipillows with ancient glassy rims, coated by a clearer millimetric outer rim, probably corresponding to pre-existing palagonite. 4-3: Eclogitized pillow-basalt showing an exceptionally well-preserved variolitic texture, formed by radial grouping of plagioclase microliths, due to its inclusion in an almandinic garnet poikiloblast forming part of the eclogite paragenesis. 4-4: Coronitic eclogite deriving from an olivine-bearing pillow lava in which the olivine phenocrysts have been transformed into omphacite, in the core, surrounded by a corona of almandine. 4-5: Eclogitized meta-gabbro formed by tectonized omphacite, replacing igneous clinopyroxene, and coronitic association of almandine and omphacite replacing olivine. 4-6: Same thin section as in previous photo, showing glaucophane surrounding garnet and omphacite, and albite plus sodic-calcic amphibole simplectitic rims replacing omphacite. 4-7: Eclogite partly

Table 2
Sequence of petrogenetic processes generating the different metamorphic lithotypes composing the BOA

Metamorphic event and probable time stage	Process and metamorphic facies	Rock types			
Jurassic ocean-floor accretion		Lherzolitic ultramafites and clinopyroxenites		Dolerite dykes, basalts and gabbros	
		UM-1	UM-2	B-1	B-2
		↓	↓	↓	↓
Late Jurassic ocean-floor metamorphism and metasomatism	Serpentinization Chloritization Rodingitization Amphibolitization	Slightly serpentinized lherzolites	Cl-bearing serpentine and chlorite rocks with scarce Ol relics	Metabasic rocks	Rodingites
		↓	↓	↓	↓
Late Cretaceous-Paleocene eo-Alpine event	Eclogite facies	TiChum-Di bearing meta-lherzolites	Spinifex-like secondary harzburgites	Eclogites	Meta-rodingites
		↓	↓	↓	↓
Eocene-Late-Oligocene meso-Alpine event	Ab-Ep amphibolite facies. Continental Serpentinization	Cpx-bearing serpentinites with Chl and Tr	Atg-T-Chl-bearing secondary harzburgites	Partly amphibolitized eclogites	Meta-rodingites

environment prior to the Alpine orogenic metamorphism, which mainly changed the chemical composition of the primary protoliths (Puga et al., 1999a). The saline hydrous fluids active during this oceanic stage could explain the high chlorine content present as NaCl submicroscopic inclusions in some metasomatized igneous minerals, such as brown olivine (Puga et al., 1999b), as well as in the mineral phases newly formed during this stage, such as brown amphibole and several phyllosilicates (Puga et al., 1999a, 2002b; Ruiz Cruz et al., 1999).

Most of the oceanic parageneses were transformed during the climax of the eo-Alpine metamorphic event into eclogitic ones and later, during the meso-Alpine event, into Ab-Ep amphibolites (Table 2). The eo-Alpine event was followed by a retrogression stage producing glaucophane-schist facies assemblages, whereas the meso-Alpine event climax was followed by greenschist-facies conditions (Puga et al., 1989a, 1989b, 2000). Some minerals that formed during the orogenic metamorphism, replacing oceanic minerals, also present a high chlorine content inherited from the oceanic minerals (Ruiz Cruz et al., 1999; Puga et al., 2002b).

In the **Cóbdar outcrop** the primary igneous paragenesis, composed of labradorite $\pm$ forsterite $\pm$ augite phenocrysts in a plagioclase + clinopyroxene + ore microcrystalline matrix, may be locally preserved in some metabasalts and metagabbros, which also preserve the igneous fabric (Figures 3). Nevertheless, igneous plagioclase phenocrysts are very often albitized (Figures 3-3, 3-4), and olivine phenocrysts commonly present thin kelyphitic rims of amphibole and a brown colour, due to exsolution of magnetite during the oceanic and orogenic superimposed metamorphism (Puga et al., 1999b). In the matrix, the igneous minerals are frequently replaced by metamorphic minerals (Figure 3-5).

The ocean-floor metamorphic process gave rise to some high-T hydrous phases such as brown calcic amphiboles, phlogopite, talc and sheridanite, which fill submicroscopic to millimetric veins and vesicles in basaltic rocks, and partly replace and rim igneous mineral in doleritic and gabbroic lithotypes (Figures 3-4, 3-6 and Puga et al., 1989a). Within the paragenesis of ocean-floor metamorphism the most characteristic mineral is brown amphibole, which is found dispersed throughout the matrix of the gabbros and basalts, as well as filling veinlets, or carpeting vesicles

amphibolitized following fissures, along which albite, barroisite and epidote replace the eclogitic paragenesis. 4-8: Amphibolitized eclogite deriving from an olivine-clinopyroxene gabbro, in which almandine coronas of eclogitic paragenesis are preserved, while omphacite has been transformed mainly into sodic-calcic amphiboles.

Figure 5: Almirez outcrop. 5-1: Clinopyroxene-bearing serpentinite mainly formed by an antigorite matrix on which diopside neoblasts, overgrown by tremolite, developed during the successive Alpine metamorphic events. 5-2: Mantle-derived porphyroclastic relics of clinopyroxene (Di1), with numerous inclusions of magnetite, isolated by diopside neoblasts (Di2) and surrounded by antigorite, in a clinopyroxenite. 5-3: Spinifex-like textured secondary harzburgite, formed by acicular olivine with subparallel elongation in a matrix of enstatite partly transformed into talc; Cr-magnetite crystals surrounded by Cr-chlorite aggregates probably replace pre-existing chromite crystals. 5-4: Grouping of thin prismatic enstatite crystals, coexisting with spinifex-like olivine and Ti-clinohumite, in a secondary harzburgite rich in Cr-chlorite aggregates. 5-5: Decimetre-thick dyke of rodingitized dolerite intruded into a secondary harzburgite with spinifex-like

in the basalts. It normally appears as relics more or less transformed into green amphiboles and other paragenetic minerals during the Alpine orogenic metamorphism. In the veins where brown amphibole dominates, this mineral, whose composition generally varies between kersutite and titanian pargasite, forms a network with sodic-calcic plagioclase predominating in the interstices (Puga et al., 2002b).

The influx of saline hydrous fluids during this stage also affected olivine phenocrysts, in which submicroscopic inclusions of NaCl have been identified by EMPA and TEM-AEM (Puga et al., 1999b). These hydrous fluids facilitated the exsolution of submicroscopic amphiboles and pyroxene crystals, together with magnetite, in the brownish altered zone of the igneous olivines, during a stage which began prior to the development of the orogenic minerals (Puga et al., 1999b).

The eo-Alpine metamorphic event developed eclogite or glaucophane schist assemblages only in some metabasic rocks of the Cóbdar region, whereas the meso-Alpine amphibolitization extensively affected both igneous relics and eo-Alpine assemblages. The lack of pervasiveness of eo-Alpine metamorphism in the Cóbdar ophiolites is suggested by the local preservation not only of igneous relics but also of the successive mineral phases forming the eo-Alpine coronitic textures. These developed in the plutonic and volcanic sequences by the successive reactions that took place between olivine and surrounding igneous plagioclase through oceanic and eo-Alpine metamorphic conditions (Figure 3-7, and Puga et al., 1989a, 1999b, 2000).

In some gabbro and dolerite dykes, the eclogitic assemblage omphacite + almandine + rutile ± glaucophane ± clinozoisite formed locally with blastoporphyric and coronitic textures (Figure 3-8). Other metamorphic lithotypes, such as omphacitites and glaucophanites, which also developed during the eo-Alpine event, are rarely present in the Cóbdar outcrop (Puga et al., 1989a).

The most common meso-Alpine Ab-Ep amphibolite assemblage is formed by the following minerals: sodic-calcic and calcic amphiboles, which replace omphacite and almandine, paragonite plus clinozoisite replacing plagioclase phenocrysts-, and epidote plus titanite, which overgrow clinozoisite and rutile, respectively (Figure 3-8). Chlorite and micas are also commonly present in this paragenesis. The most common textures in the meso-Alpine event are nematoblastic, although the microfolded schistose texture is also common in the phyllosilicates-rich amphibolite layers.

In the **Lugros outcrop** the igneous fabrics of the basic rocks are preserved (Figure 4), but the igneous parageneses, unlike those from Cóbdar, are entirely transformed into kyanite-bearing eclogites, and less commonly into co-facial garnet-glaucophanites (Puga et al., 1995). In metabasalts deriving from thin lava flow units or pillow lavas (Figures 4-1, 4-2) the variolitic or porphyritic textures are well preserved (Figure 4-3), although the igneous phenocrysts and primary glassy matrix have been completely replaced by metamorphic minerals. Relics of brown amphibole, similar to that formed during oceanic metamorphism in the Cóbdar ophiolites, are present in some samples. This mineral contains abundant tiny exsolved rutile crystals and shows a gradual transition to green sodic-calcic amphibole, which occurred during subsequent orogenic stages.

In Lugros basic rocks, the igneous olivine surrounded by calcic plagioclase was replaced during the eo-Alpine event by omphacite aggregates rimmed by almandine garnets (Figures 4-4, 4-8). In the same event, augite was transformed into omphacite + rutile (Figures 4-5, 4-6), and calcic plagioclase was replaced, during the prograde stage, first by clinozoisite-paragonite intergrowths and later by omphacite + kyanite aggregates (Figures 4-3, 4-4). Veins filled with kyanite ± omphacite may be found in these eclogites (Puga et al., 1995, 2000).

During the meso-Alpine event, eo-Alpine parageneses were only slightly overprinted by Ab-Ep amphibolite assemblages. Omphacite, previously developed from different igneous microdomains, was partially replaced first by amphibole-albite simplectitic intergrowths (Figures 4-5, 4-6) and later by sodic-calcic amphiboles, mainly barroisite, katophorite, taramite and edenite. These amphiboles may fill veins together with albite, partly replacing

texture formed by brown acicular olivine in a grey coloured matrix of enstatite. 5-6: Meta-rodingite made of recrystallized grossularitic garnet, coexisting with diopside, clinozoisite and pargasite, in a matrix of Mg-chlorite. 5-7: Diopside and tremolite-bearing chlorite-schist, with clinochlore elongated following the contact surfaces between the rodingite dyke and the enclosing ultramafic rocks. 5-8: Amphibolitized eclogite deriving from the inner part of a several-decimetres-thick dolerite dyke, formed by almandine, omphacite and rutile of the eo-Alpine paragenesis, partly replaced by meso-Alpine albite, katophorite and epidote.

the eclogitic paragenesis (Figure 4-7). Actinolite is the most common amphibole formed directly by olivine replacement (Figure 4-8). This fact suggests that, even in this outcrop, some igneous mineral relics could have been preserved until the meso-Alpine event.

Metamorphic structures, mainly meso-Alpine mylonitization bands, are superimposed over the igneous and eo-Alpine types and locally obliterate them, with the exception of localized low-strain zones. In these less-deformed zones primary structures predominate within decimetre-thick boudins, despite the complete replacement of igneous parageneses by eclogitic ones.

In the **Almirez outcrop** and throughout the entire ultramafic sequence of the BOA, the primary parageneses are almost completely replaced by metamorphic minerals during the subsequent oceanic and orogenic processes. The oceanic metasomatism in the BOA ultramafic sequence was much more pervasive than in the plutonic and volcanic sections. This process conditioned the development of two types of rock characterizing this sequence: the secondary harzburgites and meta-rodingites, which originated as a consequence of intensive chemical interchange of elements enhanced by hydrous saline fluids under high-T ocean-floor metamorphic conditions.

The meta-ultramafic rocks in this outcrop are mainly Cpx-bearing serpentinites (Figure 5-1), which locally contain centimetric layers of clinopyroxenite (Figure 5-2), and variably serpentinized secondary harzburgites (Figures 5-3, 5-4). Both types of ultramafic rocks host numerous boudinated rodingite and meta-dolerite dykes.

Cpx-bearing serpentinites mainly consist of an antigorite matrix surrounding porphyroclastic and mesh-serpentinized olivine relics, together with scarce olivine neoblasts probably formed through antigorite breakdown during the prograde eo-Alpine stage. Some olivine crystals were partly replaced by Ti-clinohumite during the HP eo-Alpine climax. Moreover, Cr-chlorite and Cr-magnetite aggregates, probably replacing mantel-derived chromite, and diopside relics and neoblasts, locally overgrown by tremolite, were also formed in this HP event (Figure 5-1). Cpx-bearing serpentinites, probably derive, according to their mineralogy and chemical composition, from a spinel lherzolite protolith that was only slightly serpentinized during the oceanic stage (UM-1 in Table 2). During the meso-Alpine event the eo-Alpine paragenesis was extensively replaced by antigorite with minor chlorite, talc and tremolite.

Clinopyroxenite is present in the serpentinized lherzolites forming groups of subparallel layers, several centimetres in thickness, containing mantle-derived porphyroclastic relics overgrown by eo-Alpine diopside neoblasts in a matrix of antigorite and chlorite (Figure 5-2).

Secondary harzburgites crop out as decimetre- to metre-wide irregular bodies that often develop a spinifex-like texture, and contain numerous millimetric-to-centimetric nodules formed by porphyroclastic mantle-derived olivine and magnetite. The spinifex-like texture consists of randomly or subparallel oriented blades, or branching and dendritic needles, of brown olivine, several millimetres to centimetres in length, and less than one to several millimetres in thickness, in a light-grey matrix consisting mainly of subparallel or radiating enstatite aggregates (Figures 5-3, 5-4). Brown spinifex-like olivine contains numerous opaque ores and saline inclusions deriving from the chlorine–bearing serpentine and chlorite (Ruiz Cruz et al., 1999). Cr-magnetite is surrounded by Cr-chlorite aggregates similar to those found in Cpx-bearing serpentinites (Figure 5-3). Clinohumite exceptionally occurs in paragenesis with brown spinifex olivine formed during the metamorphic eo-Alpine event (Figure 5-4). The spinifex-like texture development was enhanced by the very abundant fluids that arose from dehydration of the phyllosilicates (mainly serpentine) formed in the oceanic stage from the ultramafic lithotypes (Bodinier et al., 1993; Puga et al., 1997a, 1999a; Trommsdorff et al., 1998). During the meso-Alpine event, talc, antigorite, chlorite and tremolite formed in secondary harzburgites from their anhydrous eo-Alpine minerals (Table 1).

Secondary harzburgites seem to derive from spinel lherzolite, similar to the mantel-derived protolith from which the Cpx-bearing serpentinites derive, but strongly depleted in CaO (Table 3C and Puga et al., 1999a). The Ca-depletion must have been produced by Cpx-breakdown during the oceanic metasomatic process which would locally have transformed the lherzolite into chlorine-bearing serpentine and chlorite (UM-2 in Table 2). This process would be coetaneous and complementary with the rodingitization of their hosted dolerite dykes, which would have incorporated the CaO leached from the serpentinite (Puga et al., 1999a). This metasomatic change during the oceanic stage would explain the fact that metamorphic eo-Alpine paragenesis in the secondary harzburgites is clearly different from that forming the Cpx-bearing serpentinites, which would represent the less-metasomatized lherzolite lithotypes (Table 2 and Figures 5-1 to 5-4).

In the ultramafic sequence of the BOA, dolerite dykes commonly crop out as decimetre-thick boudins, consisting of a Ca-rich silicate association, which are surrounded by centimetre-thick layers of chlorite-schist, and enclosed in foliated Cpx-bearing serpentinites and in secondary harzburgites (Figure 5-5). The largest rodingite bodies form dykes up to one metre thick and several metres long which sometimes preserve remnants of igneous textures, indicating a doleritic or basaltic protolith.

The mineral composition of the dykes and boudins seems to depend on the grade of pre-Alpine metasomatism during the rodingitization process that affected the igneous protoliths, which is generally more intense towards the contacts with the host serpentinized peridotites (Puga et al., 1997a, 1999a). The small boudins and the exterior zone of the large dykes, normally more enriched in CaO, show a meta-rodingitic association (B-2 in Table 2), consisting of grossularite-rich garnet, diopside, Ca-rich amphibole, chlorite, titanite and epidote, with granoblastic or porphyroblastic textures (Figure 5-6), while the chlorite-schist borders normally comprise clinochlore, diopside, tremolite and apatite (Figure 5-7). Dykes and boudins with low Na_2O and intermediate CaO contents present transitional parageneses from rodingites to eclogites, with diopsidic pyroxene and pyralspite-rich garnets, named eclorodingite by Puga et al. (1999a). Moreover, the inner part of some dykes, which occasionally have a lower CaO and higher Na_2O content, consists of an eclogitic paragenesis, partly transformed into an Ab-Ep amphibolite (Figure 5-8) similar to those formed during the orogenic metamorphism in gabbros and basalts from the plutonic and volcanic sequences (B-1 in Table 2).

The described metamorphic evolution of the different lithotypes forming the Cóbdar, Lugros and Almirez outcrops, which are among the most representative of the BOA, together with the geochronological constraints for igneous and metamorphic events referred to in Section 3, have allowed us to establish the sequence of petrogenetic processes for the ophiolite suite shown in Table 2. In this table, UM-1 and B-1 represent petrologic evolutions followed by ultramafic and basic lithotypes, respectively, which were slightly affected by the oceanic metasomatism, whereas the UM-2 and B-2 evolution lines correspond to decimetric to metric portions of the same lithotypes, but having been pervasively affected by this metasomatism.

6. OCEAN-FLOOR AND OROGENIC METHAMORPHIC CONDITIONS

Figure 6 summarizes the most probable physical conditions estimated for the climax of the successive metamorphic events, during which the basic and ultramafic primary lithotypes of the BOA were transformed into polymetamorphic rocks. The reactions taken into account in order to deduce these metamorphic conditions are represented in this figure, and the provenance of each reaction is indicated in the corresponding figure caption.

The position of the arrows labelled "oceanic metamorphism" in Figure 6 indicates the range of LP/HT (low pressure /high temperature) conditions deduced for this process by applying the geothermometer of Holland and Blundy (1994) to brown amphibole and sodic-calcic plagioclase filling veins in Cóbdar meta-basalts (Puga et al., 2002b), and taking into account the presence of phlogopite, actinolite and chlorite replacing olivine in gabbros. A range of T between 700 °C and 880 °C has been obtained by applying the Blundy and Holland geothermometer to brown amphibole-plagioclase veins at the very low P-conditions of the submarine volcanic setting in which this paragenesis developed. The calculated T-range for the paragenesis filling veins can also be considered the most probable for the first amphibolitization process in their host basalts, in which the same type of brown amphibole was developed (Figure 3-4). Oceanic brown amphibole in the Cóbdar meta-basic rocks was partly replaced and overgrown, during orogenic metamorphism, by sodic-calcic amphiboles paragenetic with albite, for which T conditions lower than 550 °C have been calculated by applying the Holland and Blundy geothermometer.

In most of other oceanic and ophiolitic metabasic rocks, the upper T limit calculated for the ocean-floor metamorphism is around 700 °C (Mevel, 1988; Vanko and Stakes, 1991; Gillis and Thompson, 1993; Cortesogno et al., 1994). Nevertheless, Gaggero and Cortesogno (1997) have calculated a T of 800° to 900 °C for the first episode of ocean-floor metamorphism in gabbros recovered near the intersection of the Kane Fracture Zone and the Mid-Atlantic Ridge, by applying the geothermometer of Holland & Blundy to brown amphibole–plagioclase pairs from these rocks.

The inclined hatched areas in Figure 6, labelled "Cóbdar eo-Alpine" and "Lugros eo-Alpine", represent the highest P-T condition fields deduced for eclogitic paragenesis from each of these outcrops (Puga et al., 2000). The Cóbdar eo-Alpine field is delimited by applying the garnet-omphacite geothermometer of Ellis and Green (1979), the jadeite content isopleths for omphacite, and the lack of lawsonite in these eclogites. The metamorphic climax

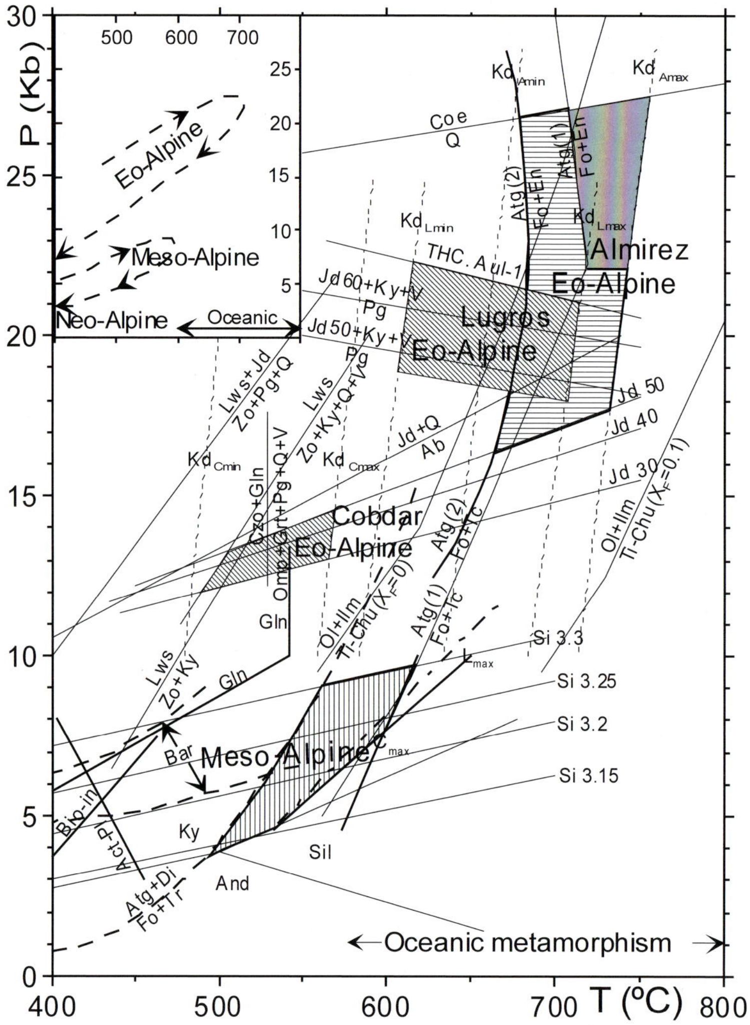

Figure 6: P-T-t diagram showing the estimated climax conditions for oceanic, eo-Alpine and meso-Alpine metamorphic events in the basic and ultramafic rocks of the BOA. Reactions plotted in this figure are the following: Q = Coe (Bohlen and Boetcher, 1982); Atg(1) = Fo + En (Ulmer and Trommsdorff, 1995); Atg(2) = Fo + En (Wunder and Schreyer, 1997); Atg(1) = Fo + T (Ulmer and Trommsdorff, 1995); Atg(2) = Fo + T (Wunder and Schreyer, 1997); Ol + Ilm = Ti-Chu, for $X_F = 0$ and $X_F = 0.1$ (Engi and Lindsley, 1980); Jd + Ky + V = Pg (Holland, 1979a); Lws + Jd = Zo + Pg + Q (Holland, 1979b); Lws = Zo + Ky + Q + V (Newton and Kennedy, 1963); Czo + Gln = Omp + Grt + Pg + Q + V (Ridley, 1984); isopleths of Jd content in omphacite (Gasparik and Lindsley, 1980); Si a.p.f.u. content in phengite for 3.15 to 3.3 values (Massone and Schreyer, 1987); Atg + Di = Fo + Tr (Berman et al., 1985); lower stability limit for glaucophane (Maresch, 1977); Amph-Pl pairs according to the Holland and Blundy (1994) geothermometer; stability Al-silicates fields according to Holdaway (1971); Kd_{Cmin} and Kd $_{Cmax}$ = minimum and maximum P-T curves respectively, obtained by applying the Gt-Cpx geothermometer of Ellis and Green (1979) to eclogites from Cóbdar outcrop; Kd_{Lmin} and Kd_{Lmax} = similar curves obtained for Lugros eclogites; Kd_{Amin} and Kd_{Amax} = similar curves for Almirez eclogitized dykes. Dark grey field = eo-Alpine conditions for Almirez outcrop according to Ulmer and Trommsdorff (1995) experimental work; horizontal hatched field together with the dark grey field = eo-Alpine conditions for Almirez outcrop according to Wunder and Schreyer (1997) experimental work; upper and lower diagonally hatched fields = eo-Alpine conditions for Lugros and Cóbdar outcrops, respectively; vertical hatched field = meso-Alpine conditions for the three outcrops. Superior inset = simplified P-T-t path for the successive metamorphic events registered in the BOA rocks for the following order in time: oceanic, eo-Alpine, meso-Alpine and neo-Alpine.

conditions in this event correspond to about 14 kbar and 570 °C for the Cóbdar metabasic rocks, i.e., at the limit between glaucophane schists and eclogites. These eo-Alpine metamorphic conditions are similar to those in other ophiolitic outcrops located in the eastern Sierra de Filabres, such as Lubrín (Morten et al., 1987) (Figure 1). The Lugros eo-Alpine field, in Figure 6, has been delimited by applying the jadeite content in equilibrium with kyanite and paragonite, the garnet-omphacite geothermometer of Ellis and Green (1979), and some THERMOCALC (Powell and Holland, 1988) determinations represented by the curve THC.Aul-1. The metamorphic climax conditions deduced for the eo-Alpine event at the Lugros outcrop correspond to about 21–22 kbar and 675 °C (Puga et al., 2000). Other outcrops located in the central and western zones of Sierra Nevada present similar eclogitic parageneses.

The eo-Alpine retrograde stage developed, under glaucophane schists conditions, during the exhumation of the oceanic floor towards the surface (see inset in Figure 6, and Puga et al., 2000). Very scarce relics of this retrograde assemblage have been preserved in eclogites from Cóbdar and Lugros after the pervasive blastesis and deformational stages of the meso-Alpine event.

The estimated P-T stability field for the eo-Alpine conditions in the ultramafics and their hosted dolerite dykes is represented by the horizontal hatched and dark grey areas labelled "Almirez Eo-Alpine" in Figure 6. The minimum P-T metamorphic conditions in the secondary harzburgites are constrained by antigorite breakdown, giving olivine, enstatite plus water, which would have occurred above 22 kbar at T lower than 720 °C, according to Ulmer and Trommsdorff (1995), but only above 15 kbar and 650 °C according to Wunder and Schreyer (1997) (Figure 6). The existence of F-bearing Ti-clinohumite, in Cpx-bearing serpentinites and in some secondary harzburgites, indicates a wide P-T stability range falling between the Ol + Ilm = TiChu curves for $X_F = 0$ and $X_F = 0.1$. The most probable and restricted range of T equilibria, shown in the hatched field, has also been delimited using Ellis and Green (1979) Grt-Cpx geothermometer applied to eclogites and meta-rodingites, whose upper limit is given by the Kd_{Amax} curve. The jadeite content isopleth Jd_{50} would indicate minimum-pressure conditions for the eclogitized dolerite dykes and their ultramafic host rocks, while the upper P limit is given by the absence of coesite in eclogites and in the oceanic metasediments overlying this ophiolitic association. Estimated P-T conditions for the high-pressure assemblages in ultramafic and dolerite dykes range from 650–750 °C and 16–25 kb (Puga et al., 1997a, 1999b).

The meso-Alpine assemblages developed within the field of Ab-Ep amphibolite facies. The physical conditions of metamorphism during this event, although poorly constrained in comparison with the eo-Alpine event, are nevertheless shown by the vertical hatched area, labelled "Meso-Alpine", in Figure 6. The lower-T limit of this field is given in ultramafic rocks by the curve corresponding to blastesis conditions of tremolite from antigorite and diopside breakdown, according to Berman et al. (1985). The upper-T limit is given by the corresponding curves resulting from the application of the Holland and Blundy (1994) geothermometer to albite and sodic-calcic amphibole pairs from Cóbdar (C_{max}) and Lugros (L_{max}) metabasic rocks. The upper P-limit of this field is given by the Si p.f.u. isopleth = 3.3 (from Massone and Schreyer, 1987), which corresponds to the highest Si content in meso-Alpine phengite from micaschists of the BOA sedimentary sequence; whereas the lower P-limit coincides with the lowest-P conditions of the kyanite stability field (according to Holdaway, 1971), which is the only Al-silicate present in the BOA rocks.

The BOA metamorphic evolution is very similar to that described for the Zermatt-Saas, northeast of the Corsica and Tauern ophiolitic eclogites (Holland, 1979a, 1979b; Oberhänsli, 1982; Barnicoat and Fry, 1986; Desmons, 1989; Pfeifer et al., 1989).

7. GEOCHEMISTRY OF MAJOR, TRACE AND REE ELEMENTS

Analytical Methods

Eighty whole-rock analyses, corresponding to different BOA lithotypes from the three selected outcrops (Cóbdar, Lugros and Almirez), have been plotted in Figures 7 to 18. Twenty-five of these analyses, representing the different lithotypes forming this ophiolite, are shown in Tables 3A to 3C. Major elements were analyzed by X-ray fluorescence, and trace elements by ICP-MS, at X-RAL (Ontario) and Granada University Laboratories, following standard procedures described in Baedecker (1987) and Bea (1996). Moreover, the $^{143}Nd/^{144}Nd$ vs $^{87}Sr/^{86}Sr$ results of 15 analyses of basic and ultramafic rocks from the above-mentioned outcrops have been represented in Figure 19. Sr and Nd isotopic analyses were performed at the Max Planck Institute of Chemistry (Mainz) and

Table 3A
Representative major and trace element analyses of the volcanic sequence

	C-1	C-2	C-3	C-4	C-5	L-1	L-2	L-3	L-4
SiO_2	44.37	48.30	47.96	47.81	64.03	47.20	46.06	47.50	47.50
TiO_2	1.56	1.91	2.24	2.01	1.18	1.55	1.12	1.44	1.46
Al_2O_3	14.66	17.30	16.73	17.00	15.80	16.90	14.00	17.30	17.40
Fe_2O_3	10.47	7.41	7.95	9.10	2.48	8.83	15.14	8.79	11.80
MnO	0.17	0.08	0.08	0.14	0.03	0.10	0.28	0.10	0.22
MgO	11.90	7.47	6.19	6.20	2.38	7.56	11.85	7.26	7.28
CaO	10.40	11.00	8.89	10.30	4.10	9.97	8.07	10.10	9.71
Na_2O	2.40	4.42	6.05	3.62	7.72	4.56	2.19	4.71	4.00
K_2O	0.42	0.31	0.35	0.88	0.17	0.20	0.20	0.16	0.19
P_2O_5	0.20	0.13	0.46	0.39	0.33	0.11	0.08	0.19	0.13
PC	2.97	1.39	2.42	1.89	1.70	0.70	0.91	1.00	
Total	99.51	99.72	99.32	99.34	99.92	97.68	99.90	98.55	99.69
#Mg	0.72	0.69	0.64	0.61	0.68	0.66	0.64	0.60	0.58
Li	18.4	36.0	31.3	7.6	7.5	20.0	10.5	22.0	17.0
Rb	4.7	<2	3.0	7.4	4.0	7.0	1.7	4.0	<2
Sr	464.7	576.0	519.6	432.7	273.7	29.0	37.6	37.0	15.0
Ba	302.8	62.0	22.8	74.1	96.3	71.0	136.9	81.0	86.0
Sc	32.8	25.4	32.9	32.0	9.0	31.0	27.7	29.0	27.4
V	199.2	140.0	273.5	256.4	82.7	232.0	117.1	218.0	231.0
Cr	565.1	204.0	115.8	224.8	58.3	590.0	267.2	650.0	480.0
Co	56.7	19.0	31.2	48.8	32.3	27.0	55.5	30.0	36.0
Ni	317.0	88.0	66.1	119.0	25.3	111.0	40.3	96.0	94.0
Cu	90.0	5.6	8.5	51.1	3.0	4.0	6.3	3.8	12.6
Zn	135.4	28.9	44.5	90.8	5.1	17.2	64.1	18.9	25.0
Y	22.3	25.0	30.0	29.1	45.7	23.0	15.4	21.0	40.0
Nb	7.2	24.0	19.4	18.9	26.3	9.0	2.9	9.0	4.0
Ta	0.5	1.0	4.2	1.3	2.0	0.4	0.2	0.4	<1
Zr	98.4	162.0	244.3	216.5	275.3	114.0	48.0	96.0	122.0
Hf	1.2	4.1	1.0	3.6	6.9	2.9	1.0	2.4	2.0
Th	0.3	1.2	1.1	1.1	5.8	0.8	0.3	0.5	<0.5
La	5.5	8.0	16.8	15.7	14.5	8.6	1.5	7.5	12.0
Ce	15.0	26.5	40.2	37.4	30.9	20.8	3.9	17.5	27.3
Pr	2.2	4.0	5.1	4.9	4.2	2.9	0.6	2.5	3.0
Nd	10.1	19.7	21.8	20.3	19.8	14.1	2.9	12.1	18.7
Sm	3.0	4.0	5.2	5.0	5.6	4.1	1.1	3.6	6.0
Eu	1.1	1.4	1.9	1.7	1.3	1.4	0.4	1.2	2.0
Gd	3.3	5.0	5.6	5.4	6.3	4.5	1.8	3.9	6.0
Tb	0.6	0.9	0.9	0.9	1.0	0.7	0.4	0.6	0.8
Dy	3.8	5.0	5.6	5.1	6.9	4.4	2.7	3.9	5.0
Ho	0.8	1.1	1.2	1.1	1.4	0.9	0.6	0.8	1.1
Er	2.4	3.0	3.1	3.0	3.6	2.6	1.6	2.5	2.0
Tm	0.4	0.4	0.4	0.4	0.7	0.3	0.2	0.3	0.4
Yb	2.0	2.0	2.1	2.5	2.8	2.4	1.4	2.4	2.0
Lu	0.3	0.4	0.3	0.4	0.6	0.3	0.2	0.3	0.4
Al_2O_3/TiO_2	9.4	9.1	7.5	8.5	13.3	10.9	12.5	12.0	11.9
TiO_2/P_2O_5	8.0	14.7	4.9	5.2	3.6	14.1	14.0	7.6	11.2
Zr/Y	4.4	6.5	8.2	7.4	6.0	5.0	3.1	4.6	3.1
Zr/Nb	13.7	6.8	12.6	11.5	10.5	12.7	16.4	10.7	30.5
Ce_N/Yb_N	1.9	3.4	4.9	3.9	2.9	2.2	0.7	1.9	3.5

C = Cóbdar metabasites; L = Lugros eclogites. Igneous lithotypes: C-1 = olivine basalt; C-2 = basalt; C-3 = pillow lava; C-4 = basaltic dyke; C-5 = plagiogranite; L-1 = olivine-bearing pillow lava; L-2 = olivine basalt; L-3 = pillow lava; L-4 = basalt.

Table 3B
Representative major and trace element analyses of the plutonic sequence

	C-6	C-7	C-8	C-9	L-5	L-6	L-7	L-8
SiO_2	43.92	45.20	47.29	45.30	49.16	48.97	48.00	48.40
TiO_2	0.85	1.79	0.90	2.08	0.77	1.57	1.57	1.93
Al_2O_3	9.48	13.78	17.78	16.10	17.30	9.80	15.20	14.20
Fe_2O_3	13.40	12.40	10.17	9.35	9.45	13.61	12.00	12.90
MnO	0.17	0.17	0.15	0.16	0.12	0.18	0.17	0.19
MgO	24.75	12.24	8.61	5.46	9.87	13.88	6.34	5.34
CaO	5.25	8.78	10.16	9.96	10.16	8.60	9.06	8.92
Na_2O	0.77	2.85	2.76	3.49	2.32	1.63	4.91	5.54
K_2O	0.28	0.71	0.33	0.89	0.12	0.11	0.28	0.19
P_2O_5	0.13	0.40	0.12	0.30	0.06	0.13	0.00	0.18
PC	0.00	0.70	1.27	3.93	0.88	1.31	1.00	0.85
Total	99.00	99.02	99.53	97.02	100.21	99.79	98.53	98.64
#Mg	0.81	0.69	0.66	0.57	0.70	0.70	0.54	0.48
Li	4.2	17.2	10.7	8.0	9.3	21.3	38.0	41.0
Rb	3.0	6.9	3.6	13.0	1.1	0.8	4.0	6.0
Sr	178.5	308.5	299.9	538.0	150.6	29.2	85.0	108.0
Ba	24.2	67.7	58.1	4850.0	60.5	5.9	326.0	183.0
Sc	16.3	26.0	24.5	32.4	33.5	33.6	31.0	33.6
V	109.1	191.9	157.3	260.0	141.4	208.1	265.0	207.0
Cr	1324.2	611.2	338.1	226.0	614.2	232.9	260.0	319.0
Co	99.1	63.9	63.2	38.0	62.7	46.9	29.0	27.0
Ni	942.1	336.7	196.4	104.0	134.5	145.1	42.0	23.0
Cu	100.9	68.0	79.2	60.3	16.0	27.1	11.8	4.0
Zn	84.7	95.9	85.8	50.8	79.9	109.8	46.2	48.1
Y	10.3	21.3	13.7	15.0	10.1	15.0	19.0	20.0
Nb	6.1	11.9	5.5	26.0	2.3	4.1	15.0	20.0
Ta	0.4	0.8	0.4	1.0	0.2	0.3	1.0	<1
Zr	65.0	110.5	63.3	171.0	32.0	75.0	76.0	119.0
Hf	1.4	2.3	0.8	3.0	–	1.0	1.0	2.0
Th	0.4	0.8	0.3	0.8	0.1	0.0	<0.5	<0.5
La	4.7	8.9	4.6	7.0	2.3	0.6	2.0	1.0
Ce	11.1	21.0	11.0	17.7	5.7	1.8	5.0	2.0
Pr	1.4	2.8	1.4	2.0	0.8	0.3	1.0	0.7
Nd	6.3	12.2	6.4	12.2	4.3	2.2	4.0	1.0
Sm	1.7	3.4	1.8	3.0	1.5	1.4	1.0	0.5
Eu	0.6	1.2	0.8	0.8	0.7	0.6	0.4	0.2
Gd	1.9	3.7	2.3	3.0	1.9	2.9	2.0	1.0
Tb	0.3	0.6	0.4	0.6	0.3	0.5	0.6	0.5
Dy	1.8	3.8	2.4	3.0	1.9	2.9	4.0	3.0
Ho	0.4	0.8	0.5	0.6	0.4	0.6	1.1	1.1
Er	1.1	2.2	1.4	2.0	1.0	1.5	2.0	3.0
Tm	0.2	0.3	0.2	0.3	0.1	0.2	0.3	0.5
Yb	1.0	1.8	1.2	1.0	0.8	1.3	2.0	2.0
Lu	0.2	0.3	0.2	0.2	0.1	0.2	0.2	0.3
Al_2O_3/TiO_2	11.2	7.7	19.8	7.7	22.5	6.2	9.7	7.4
TiO_2/P_2O_5	6.5	4.5	7.6	6.9	12.8	12.1	–	10.7
Zr/Y	6.3	5.2	4.6	11.4	3.2	5.0	4.0	6.0
Zr/Nb	10.7	9.3	11.6	6.6	14.0	18.1	5.1	6.0
Ce_N/Yb_N	3.0	3.0	2.4	4.6	1.8	0.4	0.6	0.3

C = Cóbdar metabasites, L = Lugros eclogites. Igneous lithotypes: C-6 = Cumulus troctolite; C-7 = olivine gabbro; C-8 = gabbro; C-9 = dolerite dyke; L-5 = cumulus olivine gabbro; L-6 = olivine gabbro; L-7 = gabbro; L-8 = dolerite dyke.

Table 3C
Representative major and trace element analyses of the ultramafic sequence

	A-1	A-2	A-3	A-4	A-5	A-6	A-7	A-8
SiO_2	40.12	40.36	44.04	39.44	46.41	49.19	35.95	42.21
TiO_2	0.07	0.08	0.12	0.18	0.15	2.15	3.15	1.89
Al_2O_3	2.29	2.70	2.88	4.40	1.65	15.49	14.85	13.82
Fe_2O_3	7.11	7.80	8.13	8.73	10.07	5.27	13.44	10.25
MnO	0.10	0.13	0.10	0.15	0.06	0.06	0.41	0.29
MgO	36.92	39.55	39.97	33.62	19.95	7.57	12.68	8.96
CaO	0.57	0.30	0.09	2.61	18.63	13.58	14.05	19.80
Na_2O	0.04	<0.01	<0.01	<0.01	0.10	5.05	1.30	0.99
K_2O	0.01	<0.01	<0.01	<0.01	0.00	0.12	0.40	0.20
P_2O_5	0.01	0.02	0.02	0.01	0.02	0.27	0.50	0.28
PC	11.68	7.90	3.67	9.95	2.16	1.00	2.12	0.57
Total	98.92	98.84	99.03	99.09	99.20	99.76	98.85	99.26
#Mg	0.92	0.92	0.92	0.90	0.82	0.76	0.68	0.66
Li	0.9	4.3	3.1	0.8	6.8	175.9	22.7	41.1
Rb	0.1	0.0	0.1	0.1	0.5	0.6	2.1	1.3
Sr	1.7	1.9	6.0	5.5	14.3	731.7	131.3	259.5
Ba	2.6	1.0	10.0	2.8	4.2	55.7	10.8	11.0
Sc	12.9	10.6	14.5	15.5	40.5	46.0	44.2	36.0
V	46.5	32.1	70.1	84.5	122.2	230.4	423.7	283.2
Cr	2258.7	1922.1	2549.0	2284.0	2749.5	75.3	206.5	92.2
Co	92.5	92.6	108.2	81.6	40.4	24.2	46.2	37.5
Ni	1919.1	2177.3	2119.3	1605.4	512.4	240.0	386.4	69.8
Cu	11.9	1.7	175.9	19.6	2.2	13.3	18.7	13.5
Zn	42.4	49.7	47.4	39.7	26.8	93.2	108.4	75.1
Y	1.7	0.4	1.0	2.9	6.2	30.1	47.0	33.6
Nb	0.0	0.6	0.3	0.0	0.7	9.5	20.9	10.3
Ta	0.0	0.2	0.0	0.0	3.7	0.7	1.6	0.7
Zr	5.8	7.4	5.5	9.2	6.4	164.0	289.0	176.0
Th	0.0	0.1	0.3	0.0	0.3	0.6	1.1	0.6
La	0.2	0.1	0.1	0.2	1.8	7.2	25.3	9.5
Ce	0.7	0.2	0.2	0.6	5.0	16.9	50.8	25.2
Pr	0.1	0.0	0.0	0.1	0.7	2.4	6.2	3.6
Nd	0.5	0.1	0.1	0.7	3.6	11.5	26.5	17.3
Sm	0.2	0.0	0.0	0.3	1.1	3.5	6.8	4.9
Eu	0.1	0.0	0.0	0.1	0.4	1.1	4.4	2.1
Gd	0.2	0.0	0.1	0.4	1.3	4.4	8.1	5.6
Tb	0.0	0.0	0.0	0.1	0.2	0.8	1.4	1.0
Dy	0.3	0.1	0.1	0.5	1.2	5.1	8.4	5.9
Ho	0.1	0.0	0.0	0.1	0.3	1.1	1.7	1.3
Er	0.2	0.0	0.2	0.3	0.6	3.0	4.7	3.4
Tm	0.0	0.0	0.0	0.1	0.1	0.5	0.7	0.5
Yb	0.2	0.1	0.2	0.3	0.6	2.6	3.9	3.0
Lu	0.0	0.0	0.0	0.0	0.1	0.4	0.6	0.5
Al_2O_3/TiO_2	32.7	35.1	23.2	24.9	11.0	7.2	4.7	7.3
TiO_2/P_2O_5	7.0	4.1	5.2	13.6	7.5	7.9	6.3	6.8
Zr/Y	3.5	16.5	5.5	3.1	1.0	5.5	6.2	5.2
Zr/Nb	134.9	13.1	19.5	340.7	9.3	17.2	13.9	17.0
Ce_N/Yb_N	1.0	0.6	0.3	0.5	2.3	1.7	3.3	2.1

A = Almirez ultramafites and hosted rodingitized or eclogitized dykes. Igneous lithotypes: A-1 lherzolite; A-2 and A-3 = Ca-depleted lherzolite; A-4 = lherzolite; A-5 = clinopyroxenite; A-6, A-7 and A-8 = dolerite dyke.

Granada University Laboratories, following the analytical procedures described in Thöny and Jagoutz (1992) and Montero and Bea (1998).

Chemical Characterization of the Different BOA Lithotypes

Figure 7, showing the Fe + Ti : Al : Mg cationic relationships, has been chosen to represent the different metamorphosed lithotypes forming the BOA: ultramafic, basic and differentiated leucocratic veins of plagiogranite. The latter, and most of the metabasic rocks, plot in the tholeiitic field. The minor group of rocks plotting in the Mg-rich tholeiite field are troctolite cumulates from the Lugros and Cóbdar outcrops, and meta-rodingites with Mg-enrichment and Al-depletion due to metasomatic interaction of their primary lithotypes with the ultramafic host rocks during the rodingitization process. The tholeiitic character of the BOA basic rocks is also confirmed by their FeO* vs. FeO*/MgO and TiO_2 vs. FeO*/MgO plots, represented in Figures 8 and 9, which mainly follow the trend of abyssal tholeiites. In addition, the lowest FeO*/MgO values correspond to troctolite cumulates and some meta-rodingites, whereas the highest FeO* and TiO_2 contents correspond to differentiated ferrobasaltic magmas, enriched in Ti-magnetite and ilmenite, which are present in minor amounts in the different BOA outcrops (Puga et al., 1989a, 1989b, 1995).

The tholeiitic character of the BOA metabasites is also confirmed by the relationships between certain trace elements, such as Ti, Zr, Nb, Y and Yb, which are less mobile during the post-magmatic events. Based on their Zr/TiO_2 *vs.* Nb/Y ratios, the magmas from which these metabasites derived were tholeiitic basalts, although some of them, mainly in the Cóbdar outcrop, are transitional to alkaline basalts (Puga et al., 1989a, 1989b).

The Harker diagrams (Figure 10) were made using Zr as a differentiation index, also due to its low mobility in the metasomatic and metamorphic processes that affected these rocks. These diagrams reveal the similar differentiation trends followed for major and trace elements by the metabasic rocks from the Lugros and Cóbdar outcrops, whereas the dolerites from the Almirez outcrop follow these trends for trace elements, but their major-element contents present differences which increase with the degree of rodingitization (Puga et al., 1999a). The most notable change during this process is the CaO increase, up to 30% in some samples, which is concomitant with the corresponding decrease in the other major elements. The scattering in Na_2O is partly due to metasomatic enrichment, mainly of the volcanic rocks with oceanic seawater. The highest V values are present in basic rocks from the three outcrops deriving from differentiated ferrobasaltic magmas. The less-differentiated magmas, with Zr contents lower than about 80 ppm, produced some olivine-rich gabbros with MgO contents of up to 25% (Table 3B), in which olivine and plagioclase are often cumulus phases. The scattering in Al_2O_3 contents shown by the less-differentiated rocks in the Harker diagram is due to the different proportions between olivine and calcic-plagioclase present in some of these cumulates.

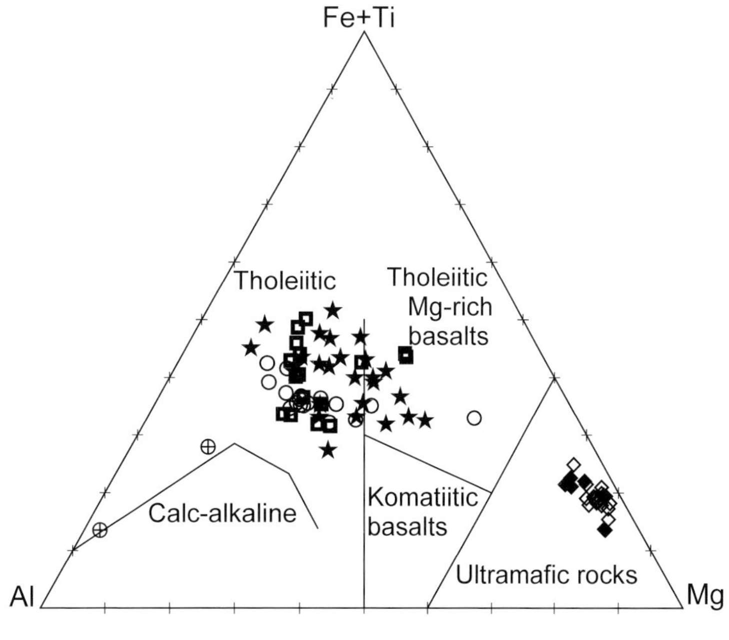

Figure 7: Fe+Ti : Al : Mg diagram (Jensen, 1976) for basic and ultramafic rocks of the BOA. Key to symbols: ○ = Metabasic rocks from Cóbdar; ◘ = eclogites from Lugros; ★ = meta-rodingites from Almirez; ◆ = Cpx-bearing serpentinites from Almirez; ◇ = secondary harzburgites from Almirez.

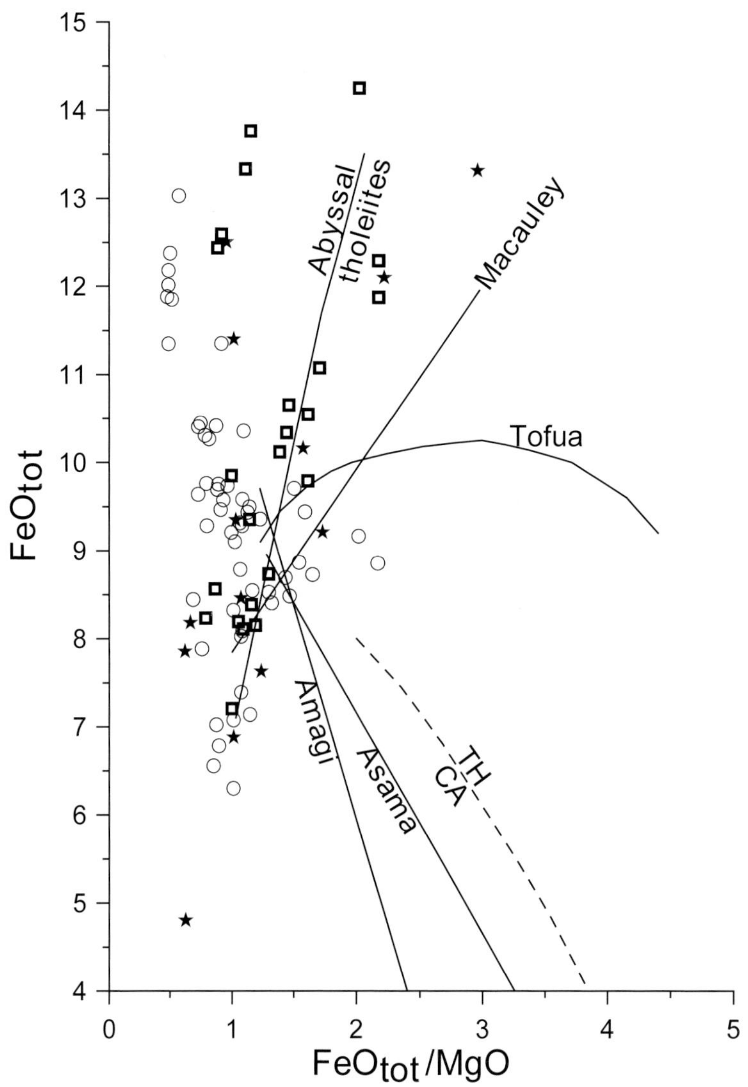

Figure 8: FeO* *vs.* FeO*/MgO diagram for the BOA basic rocks compared with the fractionation trends of abyssal tholeiites, island-arc tholeiites (Macauley and Tofua) and calc-alkaline series (Asama and Amagi) in Miyashiro (1975). Key for symbols as in Figure 7.

Plagiogranites display high SiO_2, Na_2O, Zr and Y values and low K_2O, CaO, MgO, FeO*, Rb, V, Cr and Ni values, similar to comparable rocks in oceanic (Thompson, 1973; Aldis, 1981; Pearce et al., 1984) and ophiolitic outcrops (Coleman and Peterman, 1975; Pognante et al., 1985; Bertrand et al., 1987; Borsi et al., 1996).

The major- and trace-element evolution during the magmatic differentiation of basaltic magmas (Figures 8 to 10 and Tables 3A to 3C) corresponds to a tholeiitic fractionation trend that produces an enrichment in Fe, Ti, Zr, Y and V, and a decrease in Mg, Cr and Ni (Miyashiro, 1975; Beccaluva et al., 1983).

The main types of meta-ultramafic rocks present in the BOA ultramafic sequence, Cpx-bearing serpentinites and secondary harzburgites, cannot be clearly distinguished in the Harker diagrams due to their similar Zr content. This element did not change during the oceanic metasomatic process which affected the primary lithotype, causing the main chemical differences between the two meta-ultramafic rock types (Puga et al., 1999a). Nevertheless, these differences can be clearly observed in their CaO vs. Al_2O_3 plot (Figure 11) and their REE patterns normalized to chondrites (Figure 12).

Cpx-bearing serpentinites follow the trend of the Ronda peridotites, which have been plotted in Figure 11 for comparison. The CaO vs. Al_2O_3 trend of these serpentinites suggests a lherzolitic lithotype, which has not been strongly affected by metasomatic processes. Moreover, the Al_2O_3/TiO_2 ratio of 20 to 35, their average TiO_2 contents of about 0.12%, and Cr contents of around 2800 ppm, together with their REE patterns

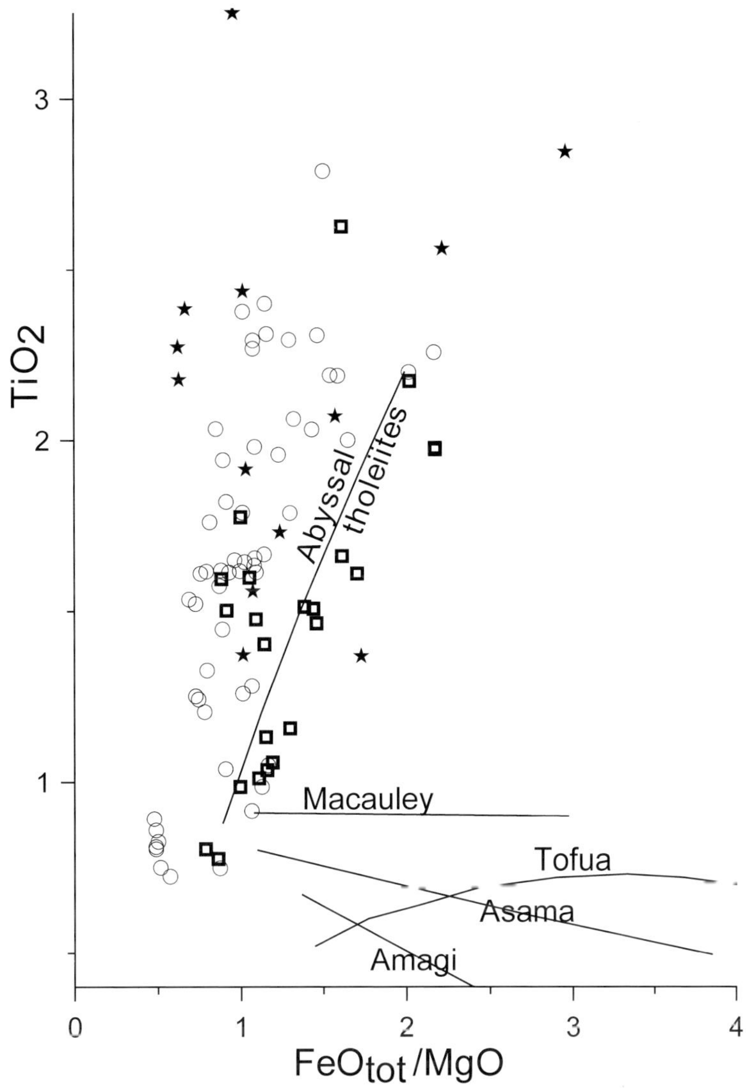

Figure 9: TiO2 *vs.* FeO*/MgO diagram for the BOA basic rocks compared with the different tholeiites and calc-alkaline series according to Miyashiro (1975). Key for symbols as in Figure 7.

(Figure 12, Table 3C and Puga et al., 1999a), all correspond to ophiolitic lherzolites (Beccaluva et al., 1983; Pearce et al., 1984). Some Cpx-bearing serpentinites, however, show notable CaO-enrichment (Figure 11). These high CaO values can be considered as due to a secondary increase in this element during the orogenic metamorphism, taking into account that these rocks contain abundant calcite, tremolite and/or diopside and talc, locally concentrated and filling millimetric veins that cut across the meso-Alpine schistosity.

Secondary harzburgites show CaO contents sharply lower than both Cpx-bearing serpentinites in the same outcrop and the different Ronda peridotites represented in Figure 11. Nevertheless, the Al_2O_3 range of these rocks is higher than 1.5%, their Al_2O_3/TiO_2 ratio averages 20 to 35%, and their TiO_2 contents vary from 0.07 to 0.2%, which allows us to classify them as deriving from a lherzolitic protolith, similar to the Cpx-bearing serpentinites but depleted in CaO. Moreover, Figure 12, showing the REE representative patterns of the different BOA lithotypes, presents similar HREE contents for both meta-ultramafic rock types, but much lower MREE and LREE contents in the secondary harzburgite (A-3 type, Table 3C) with respect to the Cpx-bearing serpentinites (A-4 type, Table 3C).

The peculiar REE pattern of the secondary harzburgites, together with their extremely low CaO contents, has been explained by Bodinier et al. (1993) and Puga et al. (1997a, 1999a) by a pervasive clinopyroxene breakdown process that presumably took place during the oceanic serpentinization, which would be coeval and complementary with the rodingitization process that occurred in the basaltic dykes hosted by these ultramafites. This hypothesis is confirmed not only by the progressive CaO enrichment experienced by the doleritic dykes during their transformation to rodingites, but also by the comparison of their REE patterns (Figure 12). In fact, the most rodingitized dykes,

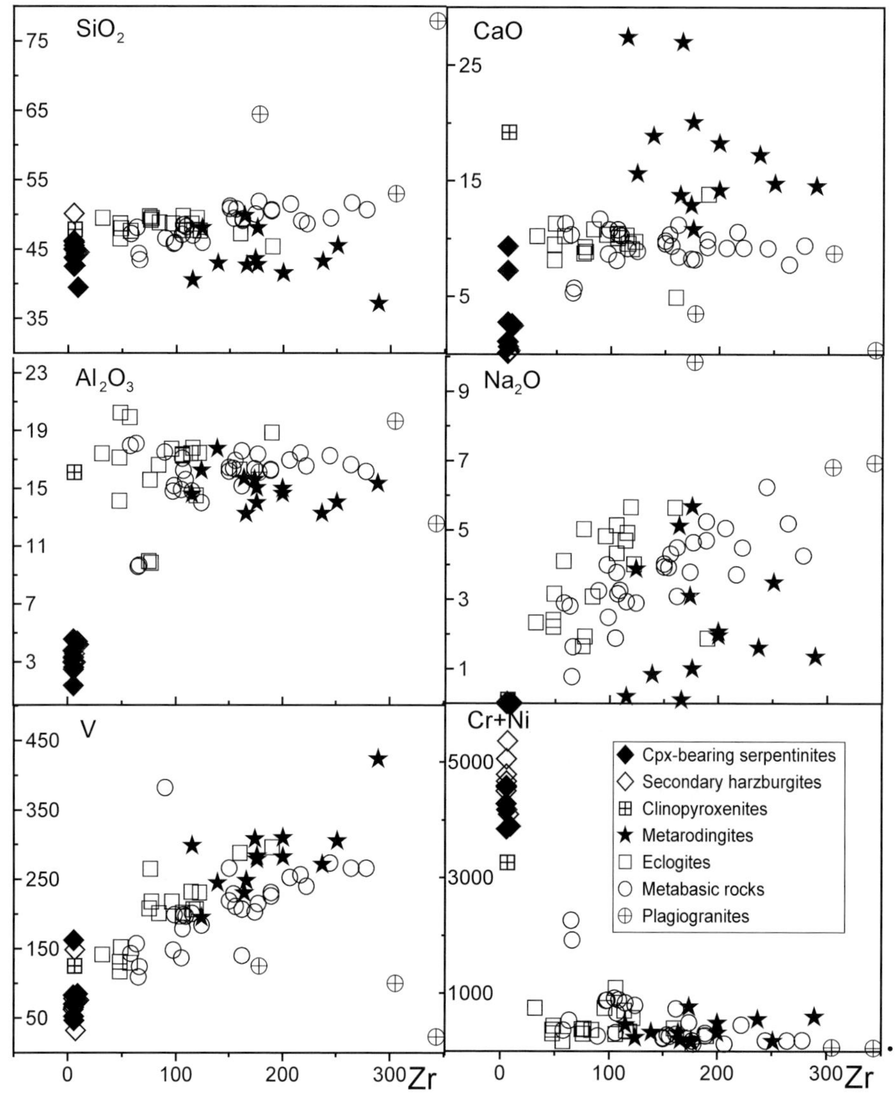

Figure 10: Harker diagrams of some major and trace elements vs. Zr (used as differentiation parameter) for the different lithotypes forming the BOA, which are indicated in the lower-right corner inset. The provenance of the different rock-types is similar to that in Figure 7 for the same symbols, and plagiogranites come from the Cóbdar outcrop.

with a composition similar to A-7 (Table 3C and Figure 12), show increasing enrichment from MREE to LREE, as compared to the REE patterns of the less-rodingitized samples, such as the A-6 type (Table 3C and Figure 12). The REE-enrichment of some rodingitized dykes might be complementary to the REE depletion experienced by the secondary harzburgites during their oceanic serpentinization. Even the positive Eu anomaly presented by the rodingite REE patterns, and the negative Eu anomaly, which is characteristic of the secondary harzburgite patterns, seem to be a post-magmatic, complementary chemical change produced during the oceanic metasomatism (Bodinier et al., 1993; Puga et al., 1997a, 1997b, 1999a). Finally, secondary harzburgites also show lower Y contents with respect to the Cpx-bearing serpentinites, which could also be compensated by a higher content of this element in rodingites than in basic rocks coming from the volcanic and plutonic sequence of the BOA (Tables 3A to 3C).

Figure 12 also shows the REE pattern of the less-rodingitized dolerite dykes (A-6 type) which falls within the variation range of the metabasic rocks from the Lugros and Cóbdar outcrops. L-1 and C-4 patterns correspond to volcanic rocks from Lugros and Cóbdar outcrops, respectively. The REE patterns C-6 and L-5 correspond, respectively, to troctolite from Cóbdar and olivine gabbro from Lugros outcrops; both types of rocks have olivine and calcic plagioclase as cumulus phases, the Lugros gabbro being more enriched in plagioclase, as indicated by its more positive Eu anomaly. The range of HREE values greatly overlap for the basic rocks from the different BOA outcrops, suggesting a similarly deep mantle source. Nevertheless, the slope of the Cóbdar magmas is somewhat

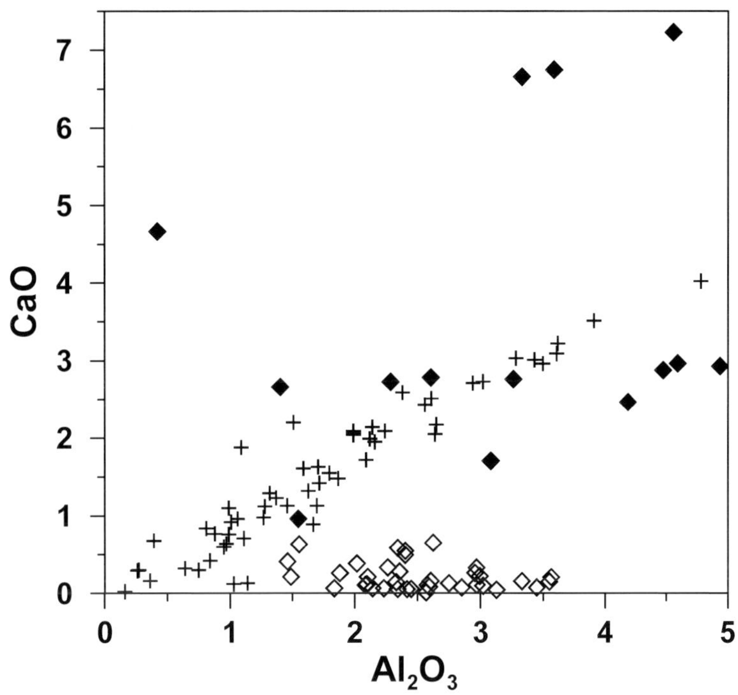

Figure 11: Al_2O_3 vs. CaO diagram showing CaO-depletion in the BOA secondary harzburgites, with spinifex-like textures from Almirez (◇), compared to Cpx-bearing serpentinites from the same outcrop (◆) and to Ronda peridotites (✚). (Slightly modified from Puga et al., 1999a).

higher than that for Lugros, with average values for Ce_N/Yb_N of about 3.0 and 1.5, respectively, whereas this ratio in the less rodingitized dolerite dykes is about 2.5 (Tables 3A to 3C).

Other minor lithotypes from the BOA whose REE patterns are represented in Figure 12 are clinopyroxenite (A-5) and plagiogranite (C-5). Clinopyroxenite presents a regular REE pattern with HREE values within the range of the BOA ultramafic rocks, but an increasing content from HREE to LREE, attaining LREE values similar to the most depleted basic magmas from the Lugros outcrop. The plagiogranite pattern, represented in Figure 12, is the average value of three leucocratic veins, which show high variability in their LREE/HREE ratios, but whose patterns all present the negative Eu anomaly indicative of a differentiated magma resulting from calcic-plagioclase segregation.

8. TECTONIC SETTING OF THE BOA MAGMATISM

The tholeiitic trend observed for major and trace elements in the BOA intrusive and effusive basic magmas (Figures 8, 9 and 10), together with the crystallization order of their igneous relics, consisting of spinel → olivine → plagioclase → clinopyroxene → apatite → Fe–Ti oxides, resembles the differentiation trend presented by oceanic magmas, as well as the basic rocks in the high-Ti ophiolite type which is characteristic of the Western Mediterranean Ophiolites (Pearce, 1980; Venturelli et al., 1981; Beccaluva et al., 1983).

Ti and V in the various BOA outcrops increase steadily towards high Zr and low #Mg values (Figures 9–10 and Tables AAABC3,3,3), with a Ti/V ratio ranging from 30 to 60. These values match those of tholeiitic and transitional suites, and significantly differ from those recorded in orogenic and alkaline volcanics (Shervais, 1982). The same affinities are shown by the relationships between some incompatible and compatible elements, such as Th/Yb vs Ta/Yb (Pearce, 1982) and Th : Tbx3 : Tax2 (Figure 13), both also indicating the oceanic character of this magmatism and its chemical similarity to the E-MORB and T-MORB types.

Other comparative trace-element correlations used to deduce the tectonic settings of the BOA magmatism were the following: Ti vs. Zr, Cr vs. Y, Ti/Cr vs. Ni and La/Yb vs. Zr/Y (Bodinier et al., 1987; Puga et al., 1989a, 1989b, 1995). According to all these parameters, the basic magmas of the BOA are similar to T-MORB and E-MORB types presently forming in various zones along the oceanic ridges (Wood et al., 1979; Sun et al., 1979; Hoffman, 1997).

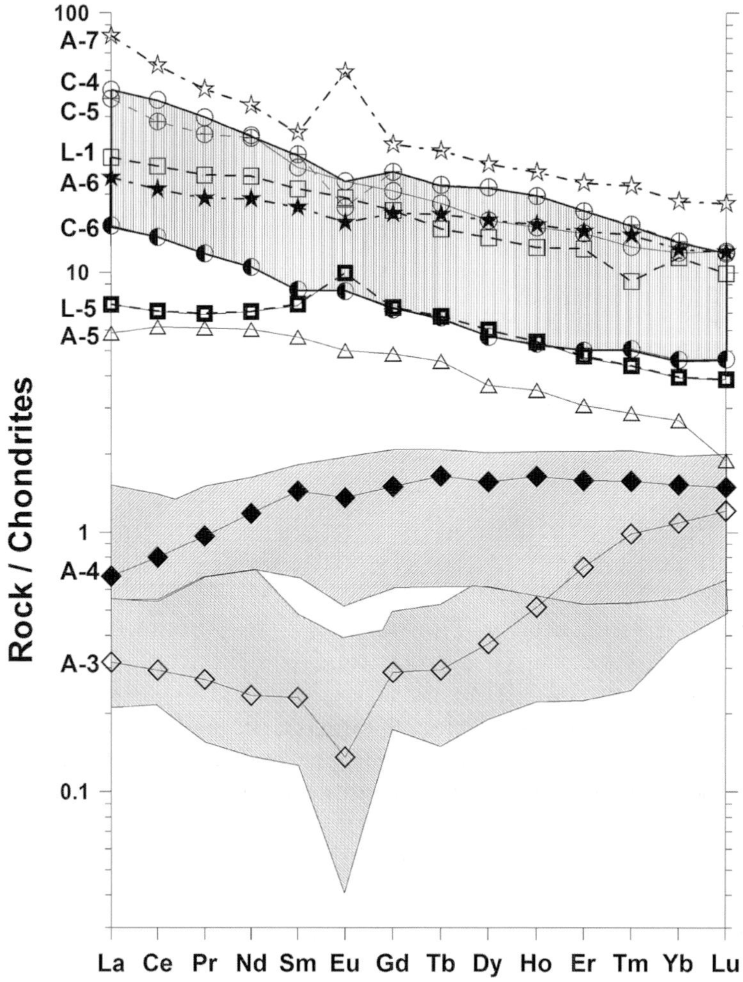

Figure 12: Chondrite-normalized REE plot for representative basic and ultramafic rocks of the BOA, using normalizing values from Boynton (1984). Key to symbols, grouped by outcrops, is as follows: (a) Cóbdar outcrop: ⊕ = plagiogranite average value, ○ = amphibolitized pillow lava, ◐ = meta-troctolite; (b) Lugros outcrop: □ = eclogitized pillow lava, ◘ = eclogitized troctolite; (c) Almirez outcrop: ◆ = Cpx-bearing serpentinite, ◇= secondary harzburgite, △ = clinopyroxenite, ★ = eclogitized dolerite, ☆ = rodingite. The analyses represented in this diagram are shown in Tables 3A, 3B and 3C using the same labels as indicated for each rock together with its La_N value.

Figure 14 shows the spidergram of a series of elements ranging from incompatible to compatible in character, normalized to the N-MORB average values of Pearce (1980), for three representative magma types of the Cóbdar (C-4), Lugros (L-1) and Almirez (A-6) outcrops. These three samples have also been represented in Figure 12 with the same labels. The BOA spidergrams are compared with those representative of the tholeiitic basaltic magmas from different tectonic settings, such as Continental Lithosphere (CT), Oceanic Islands (OIB), LREE-enriched Oceanic Ridges (E-MORB) and Island Arcs (IAT). The basaltic magmas that originated from a subcontinental mantle source (CT and IAT) always have a negative Nb–Ta anomaly in this spidergram. This anomaly is due to Th and K enrichment in these magmas during their ascent to the surface traversing the continental crust. This anomaly is, therefore, not presented by the basaltic magmas generated from a suboceanic mantle such as the OIB and MORB. Moreover, the greatest differences between these four magma types, which are characteristic of magmatism generated in different tectonic settings, can be seen in their respective contents of the more incompatible elements, which are located on the left-hand side of the spidergram, from Sr to Rb. The three samples of BOA basic rocks do not present the Nb–Ta negative anomaly, which is characteristic of magmatic crustal contamination, thereby suggesting that they most probably originated from a sub-oceanic mantle source; furthermore, they fit better with the E-MORB patterns, particularly for the more discriminant incompatible elements, than with the OIB pattern. All the BOA samples present a clear similarity with basic magmas enriched in incompatible elements currently forming in mid-oceanic ridges, although they show a higher inter-elemental variability due to differential mobility of elements

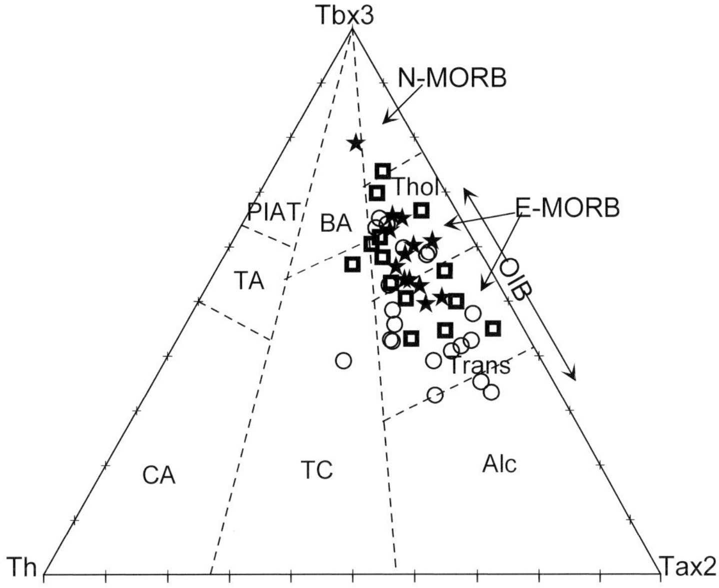

Figure 13: Th : Tbx3 : Tax2 plot of the metabasic rocks of the BOA in the Cabanis and Thiéblemont (1988) discriminative diagram, showing the fit of these rocks with the field corresponding to the tholeiitic and transitional magmas generated along the mid-oceanic ridges in the location characterized by LREE-enriched basaltic magmatism (E-MORB). Key for symbols as in Figure 7: ○ = Metabasic rocks from Cóbdar; ◘ = eclogites from Lugros; ★ = meta-rodingites from Almirez. CA = calc-alkaline rocks; TA = arc-tholeiites; PIAT = primitive island arc tholeiites; BA = fore-arc and immature back-arc basin basalts; TC = continental tholeiites; N-MORB = mid-oceanic-ridge basalts; E-MORB = mid-oceanic-ridge basalts enriched in incompatible elements; OIB = oceanic island basalts; Thol = tholeiitic; Trans = transitional; Alc = alkaline.

during the metasomatic and metamorphic processes undergone (Wood et al., 1979; Sun and McDonough, 1989; Hoffman, 1997).

Ce_N/Yb_N ratios higher than one, alongside the Zr/Y, Zr/Nb, TiO_2/P_2O_5 and Al_2O_3/TiO_2 ratios (shown in Tables 3A to 3C), allow us to classify the basic magmas forming the plutonic and volcanic sequences of the BOA, as well as those that intruded its ultramafic sequence, forming dykes, as similar to T- and E-MORB (Kay and Hubbard, 1978; Sun et al., 1979; Saunders et al., 1980).

Figure 15 shows the Zr/Y vs. Zr plot of representative average values of basic rocks from Lugros, Cóbdar and Almirez in the spreading-rate discriminating diagram of Pearce (1980), compared with the average values of some Tethyan ophiolitic metabasalts, and the average value of E-MORB basalts dredged at 45°N and 63°N on the Mid-Atlantic Ridge. According to this diagram, the BOA magmas from Cóbdar and Lugros were generated under similar conditions to those of the magmatism presently occurring along slow-spreading oceanic ridges, such as in the Atlantic Ocean, with a spreading rate less than 2 $cm\,y^{-1}$, and to those generating several Alpine–Apennine ophiolites (according to Figure 10 in Pearce, 1980). The higher spreading rate value indicated by the plot of the Almirez dykes in this figure probably does not correspond to reality due to the Zr increase and Y decrease that these rocks underwent during their rodingitization process (Tables 3A to 3C).

Finally, Figure 16 depicts the Ti vs. Y plot of the different meta-ultramafic rocks of the BOA in the discriminating diagram of Pearce and Peate (1995) for the tectonic setting of peridotites in which Cpx-bearing serpentinites, chemically more similar to the lherzolitic primary lithotype, fit better with peridotites coming from normal ridge segments with a low degree of melting. In contrast, secondary harzburgites are outside this field, probably due to a Y-depletion resulting from the oceanic serpentinization of these lithotypes.

9. GENETIC RELATIONSHIPS BETWEEN BASIC AND ULTRAMAFIC ROCKS

Figure 17 represents the plot of Al_2O_3/TiO_2 vs. TiO_2 for some representative basaltic and ultramafic rocks from the BOA on the Beccaluva et al. (1983) discriminant diagram for ophiolite types. According to this diagram, the BOA basic rocks and their plagiogranite differentiates correspond to a high-Ti ophiolite type, and they could be

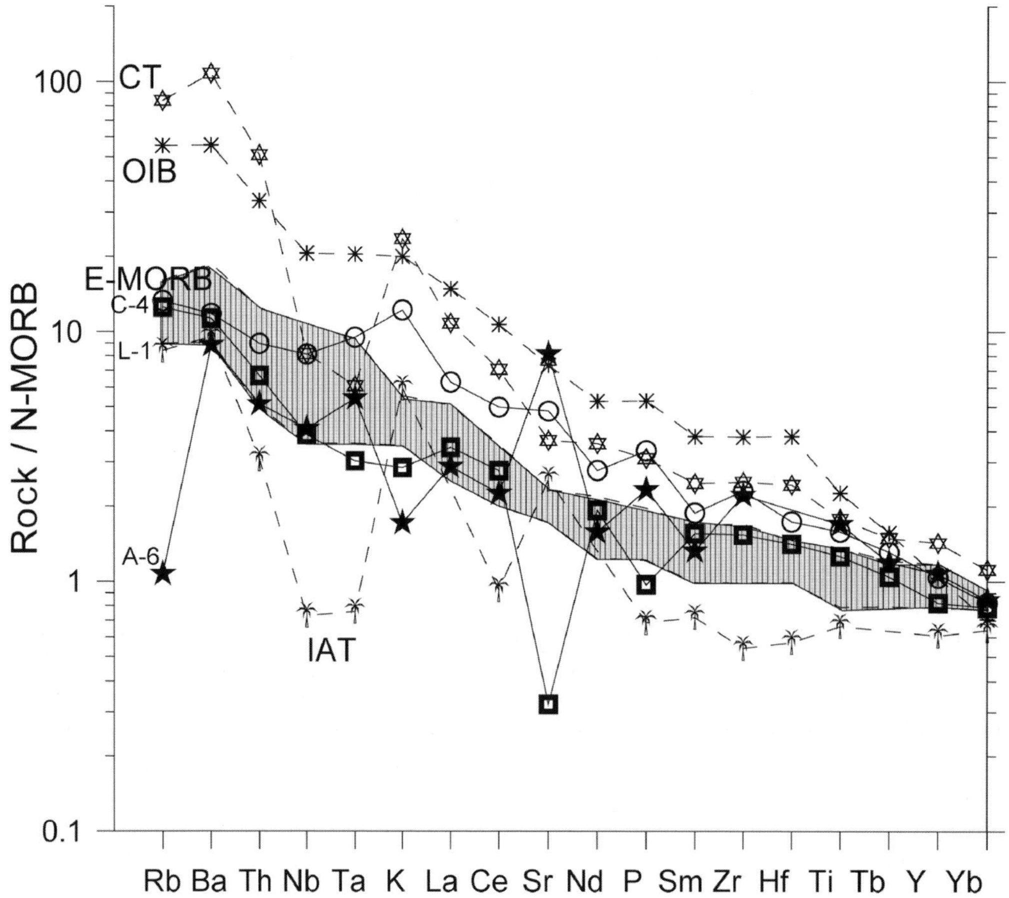

Figure 14: Spidergram of incompatible to compatible elements normalized to N-MORB (Sun and McDonough, 1989) for representative metabasic rocks of the BOA, coming from Cóbdar (C-4), Lugros (L-1) and Almirez (A-6), respectively. CT = continental tholeiites from Columbia River; OIB = average values for oceanic island basalts (Sun and McDonough, 1989); E-MORB shaded field has as lower limit the average values for LREE-enriched basalts from oceanic ridges (Sun and McDonough, 1989), and as upper limit the average values calculated for the LREE-enriched basalts from 36°N to 63°N along the Mid-Atlantic ridge (Wood et al., 1979); IAT = average values for the island arc tholeiites according to Pearce (1982).

genetically related to the BOA primary lherzolites by a melting degree lower than 15%, similarly to other Alpine–Apennine ophiolites. A similar degree of melting and the existence of probable genetic relationships between the BOA basic magmas and ultramafic rocks may be deduced from the Ni:Co:Sc diagram (Figure 18), taken from Ottonello et al. (1983). In this diagram, numerous BOA basic rocks plot along the dashed line corresponding to the olivine fractional crystallization curve, calculated by Ottonello et al. (1983) for Ligurian ophiolites. This fact confirms the marked influence of olivine segregation in the differentiation of the BOA magmas. A melting degree of less than 15% can also be deduced, as necessary to the origin of the BOA basic magmas, from the La/Yb vs. Zr/Y plot, and from the Cr vs. Y plot of the BOA metabasites (Bodinier et al., 1987; Puga et al., 1989a, 1989b). Moreover, clinopyroxenite veins plot along the fractional melting curve in Figure 18, suggesting a probable origin of these veins in a low degree of partial melting of their host lherzolites during a stage previous to the intrusion of these rocks by the dolerite dykes.

Despite the small size of the BOA meta-ultramafic outcrops and their high degree of metasomatic and metamorphic transformation, a probable oceanic origin has been suggested for these rocks by Puga et al. (1999a) based on the chemical composition of the scarce mantle-derived mineral relics preserved in them. In fact, the residual clinopyroxenes present in serpentinites and clinopyroxenite layers (Table 2 and Figures 8 and 9 in Puga et al., 1999a) are similar to diopsides of abyssal peridotites from fracture zones along the America–Antarctica and Southwest Indian Ridges, particularly to those from zones with a low degree of partial melting (Johnson et al., 1990). Moreover, the chemical composition of the residual olivine, prevalently Fo_{90-91}, and the #Cr value of the spinel, about 0.5–0.6 (Tables 3, 4 and Figure 9 in Puga et al., 1999a), indicate a probable MORB-type suboceanic environment for the BOA

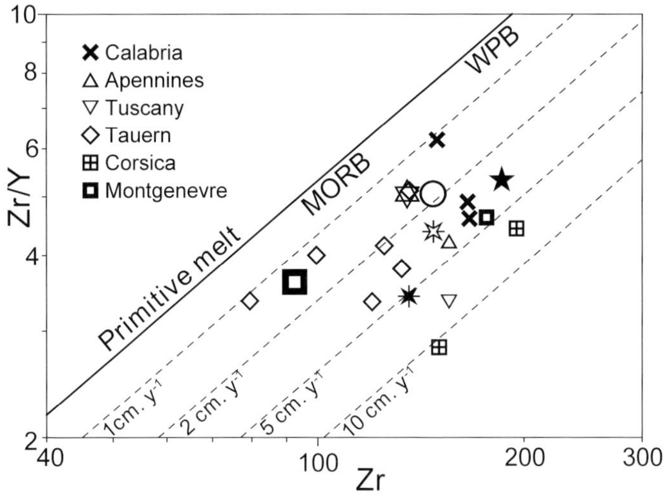

Figure 15: Zr/Y vs. Zr plot for average values of the BOA basic rocks in the spreading-rate discriminating diagram of Pearce (1980) based on average analyses from different oceanic ridge segments. Key to symbols is as follows: large square = Lugros eclogitized basalts; large circle = Cóbdar metabasic rocks; large full star = Almirez rodingites. The small symbols in the inset correspond to average values of Tethyan ophiolitic metabasalts (data from Pearce, 1980). Other meta-basalts plotted for comparison: large six-pointed star = average value of present E-MORB from the Mid-Atlantic Ridge at 45°N and 63°N (data from Wood et al., 1979); ☆ and ✱ = average values of Antrona and Zermatt-Saas ophiolitic meta-basalts respectively (data from Beccaluva et al., 1984; Pfeifer et al., 1989).

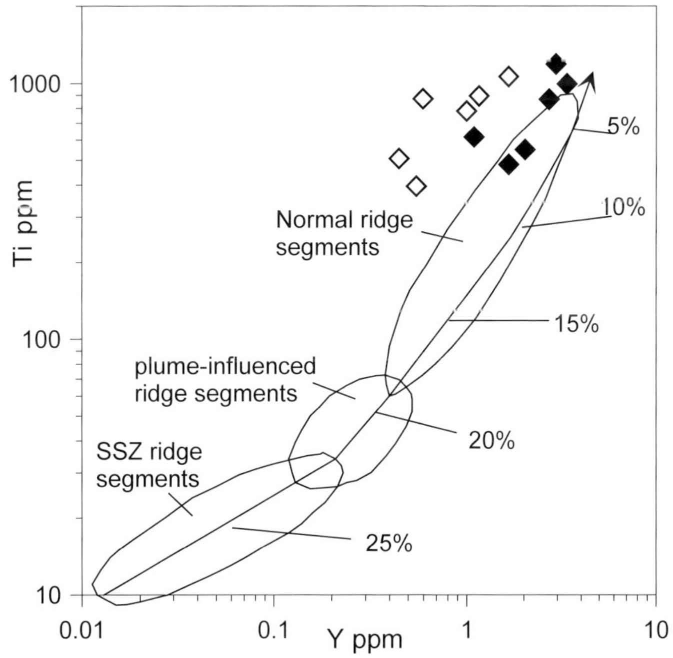

Figure 16: Ti : Y plot of the BOA ultramafic rocks in the discriminating diagram of Pearce and Peate (1995) for peridotites from different tectonic settings that underwent variable degrees of melting. Key to symbols: ◆ = Cpx-bearing serpentinites; ◇ = secondary harzburgites.

lherzolites, in agreement with the review of Arai (1994) on the characterization of spinel peridotites. Furthermore, the bulk-rock chemical compositions of the BOA lherzolites, particularly some discriminant element contents such as their TiO_2 and Cr_2O_3 ranges (Table 3C and Puga et al., 1999a) and their Ti vs. Y plot (Figure 16), are typical of ultramafic rocks with a MOR origin, according to Pearce et al. (1984) and Pearce and Peate (1995), respectively. The temperature of the decompressional trajectory followed by the ultramafic slices towards their oceanic emplacement has been used by several authors to hypothesize a subcontinental mantle provenance of some ultramafic

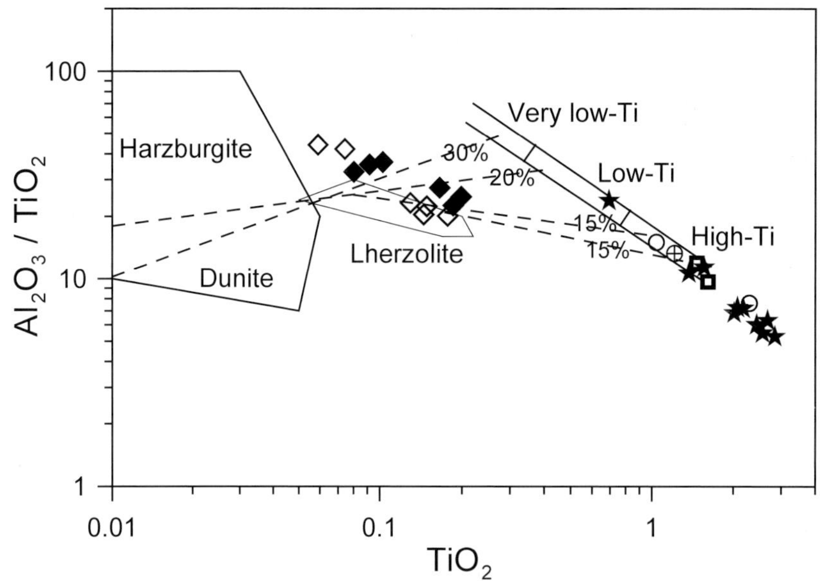

Figure 17: Al_2O_3/TiO_2 : TiO_2 plots of the BOA basaltic and ultramafic rocks in the discriminative diagram of Beccaluva et al. (1983) for high-Ti, low-Ti and very low-Ti ophiolites, showing dashed lines that join basaltic parental melts and peridotitic residua produced by different degrees of non-modal equilibrium partial fusion. Key to symbols: ○ = amphibolitized basalts from Cóbdar; ⊕ = plagiogranite from Cóbdar; ◘ = eclogitized basalts from Lugros; ★ = meta-rodingites from Almirez; ◆ = Cpx-bearing serpentinites from Almirez; ◇ = secondary harzburgites from the same locality.

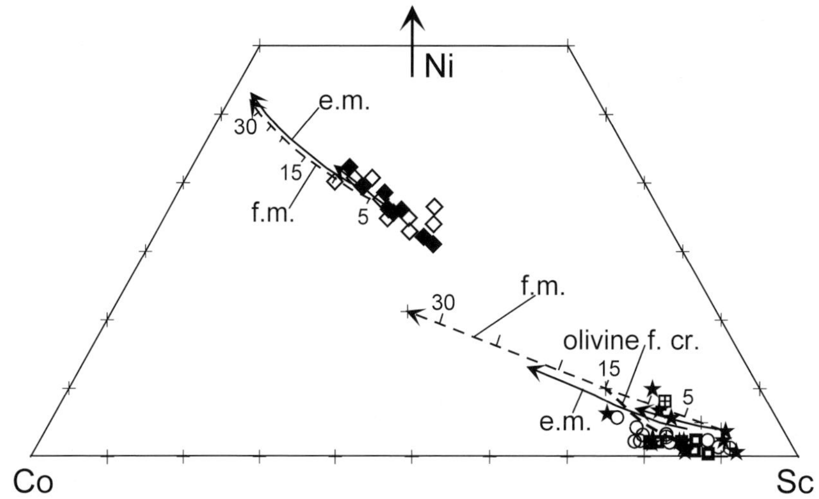

Figure 18: Ni: Co : Sc plot of the BOA basaltic and ultramafic rocks in the diagram of Ottonello et al. (1983), showing computed distributions of these elements, in fractional melting (f.m.) and equilibrium melting (e.m.), between liquids and residua for different percentage degrees of melting. Compositions have been normalized to the mantle estimate of Jagoutz et al. (1979). Key to symbols as in previous figure, and ⊞ = clinopyroxenite.

rocks forming part of the Alpine–Apennine ophiolites (Pognante et al., 1985, 1986; Trommsdorff et al., 1993; Rampone et al., 1995). Nevertheless, the lack of preservation of mantle-derived orthopyroxene relics, together with clinopyroxene, in the BOA ultramafic rocks unfortunately does not allow us to calculate this trajectory.

Figure 19 shows the similar range in $^{143}Nd/^{144}Nd$ ratios in the basic and ultramafic rocks of the BOA, and their $^{87}Sr/^{86}Sr$ isotopic differences (data from Puga et al., 2002a, 2004). The similarity in the $^{143}Nd/^{144}Nd$ values of Almirez peridotites and rodingites with those of the basic rocks from Cóbdar and Lugros outcrops is a significant argument in favour of a common asthenospheric origin for the basic and ultramafic BOA lithotypes. BOA basic rocks, especially those forming the plutonic sequence, present $^{143}Nd/^{144}Nd$ and $^{87}Sr/^{86}Sr$ isotopic ratios similar to mag-

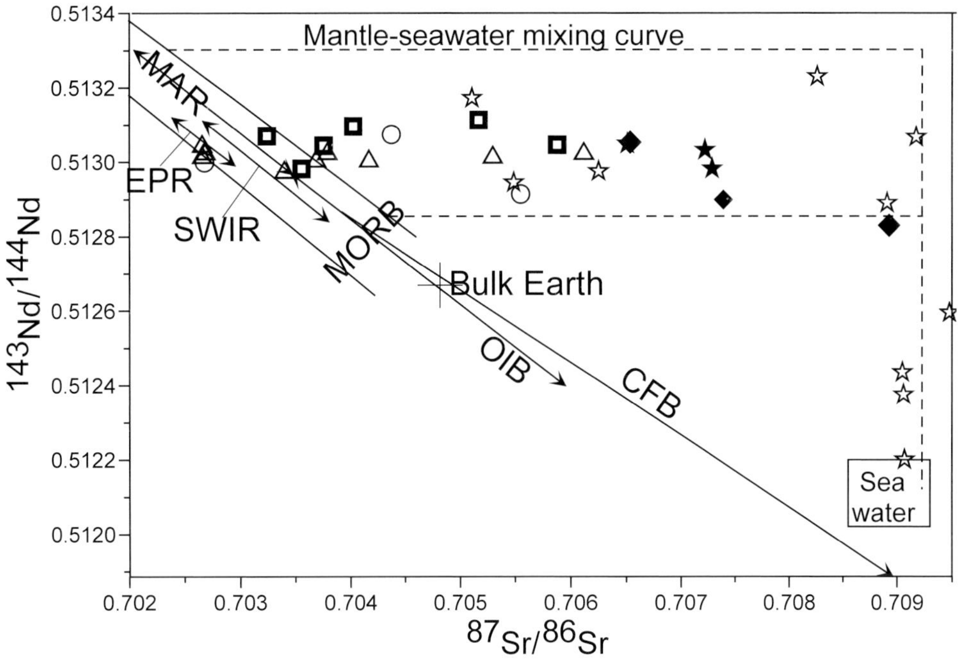

Figure 19: Present day $^{143}Nd/^{144}Nd$ *vs.* $^{87}Sr/^{86}Sr$ in the different BOA lithotypes compared with abyssal peridotites (empty stars) occurring in several mid-ocean ridge fracture zones (data from (Snow et al., 1994)). Broken lines show mantle-seawater mixing curves calculated by these authors for the abyssal peridotites represented in this figure. Key to symbols: ○ = metabasic rocks from Cóbdar; ◘ = metabasic rocks from Lugros; ★ = meta-rodingites from Almirez; ◆ = meta-ultramafic rocks from Almirez; △ = meta-basic rocks from Cóbdar (in Gomez-Pugnaire et al., 2000). MORB values for different ridges: EPR (eastern Pacific Rise), MAR (Mid Atlantic Ridge) and SWIR (south western Indian Ridge) from Hoffman (1997); OIB values (oceanic island basalts) from Hoffman (1997); CFB values (continental flood basalts) from Marzoli et al. (1999).

mas occurring in the various mid-oceanic ridges. However, the samples coming from the volcanic sequence, and particularly the rodingites and meta-ultramafites from the ultramafic sequence, present higher $^{87}Sr/^{86}Sr$ isotopic values. The abyssal peridotites from various mid-oceanic ridge fracture zones, which have been plotted in Figure 19 for comparison, present similarly high $^{87}Sr/^{86}Sr$ isotopic values according to Snow et al. (1994). These authors interpret the Nd–Sr array of abyssal peridotites, which follows a mantle-seawater mixing curve (see Figure 19), as due to contamination of peridotites by present-day seawater with a $^{87}Sr/^{86}Sr$ isotopic composition of 0.7092. A similar contamination, mainly by Jurassic seawater, with a $^{87}Sr/^{86}Sr$ isotopic ratio of 0.7072–0.7075 (Burke et al., 1992; Jones et al., 1994), intervening in the ocean-floor metasomatism that affected the BOA rocks, and more pervasively in those forming the ultramafic sequence, could explain the progressive enrichment in the $^{87}Sr/^{86}Sr$ isotopic ratio in these samples. In fact, the ocean-floor metasomatism giving rise to the concomitant serpentinization and rodingitization that occurred in the BOA, as well as in other Alpine–Apennine ophiolites, seems to have begun close to the spreading centres where the oceanic lithosphere formed, due to seawater circulation in the cooling lithosphere during the ocean-floor metamorphism (Alt et al., 1986; Nicolas, 1989; Rösli et al., 1991).

10. ORIGIN AND GEODYNAMIC EVOLUTION OF THE BOA

Figure 20 (slightly modified from Puga et al., 2002a) summarizes the origin and evolution of the BOA in the context of the deepest Betic metamorphic complexes in which the dismembered relics of this metamorphosed ophiolite crop out (Figure 1). According to our U/Pb dating on zircons from the Lugros outcrop (Puga et al., 2005) and other radiometric data for the Lugros and Almirez outcrops (see Section 3), together with the petrological and geochemical data given above, the oceanic floor from which the basic and ultramafic rocks of the BOA derive began its accretion in the Early Jurassic, between the crustal domains from which the Caldera and Sabinas of the MC derive, and continued to develop up to the Late Jurassic. The oceanic BOA magmatism was preceded by acidic to intermediate intraplate magmatism, present in the crustal units of the MC as thin pyroclastic layers of Permian-Triassic to Triassic rhyolites s.l., formed from mantle-derived magmas that traversed and assimilated an increasingly thinned

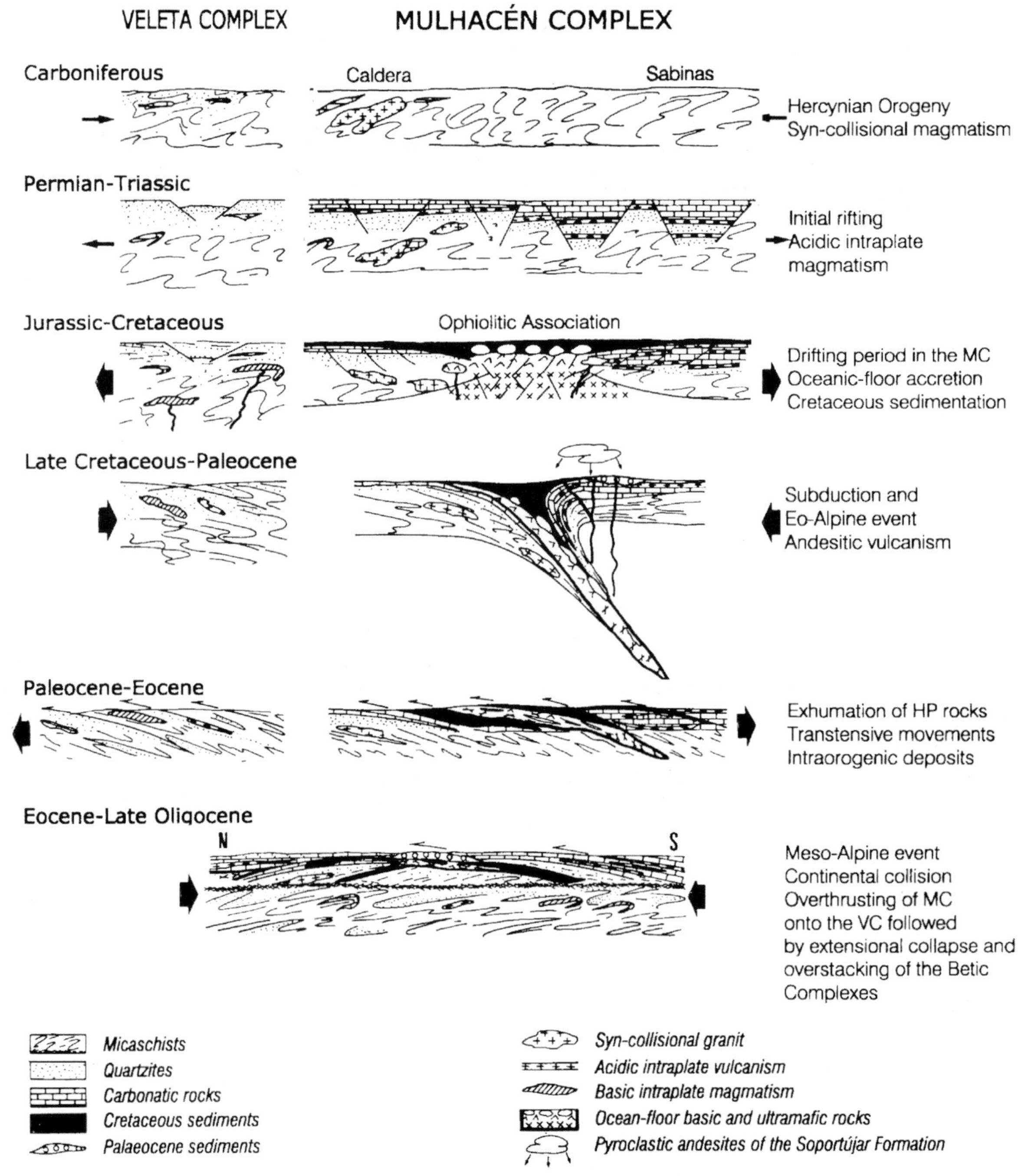

Figure 20: Tentative sketch of the evolution of the Veleta and Mulhacén Complex from the Carboniferous to Late Oligocene (slightly modified from Puga et al., 2002a), showing the Jurassic-Cretaceous accretion of the ocean-floor from which the BOA derives and its later partial disappearance due to the eo-Alpine subduction, which was followed by successive deformational and metamorphic stages during the meso-Alpine and neo-Alpine events.

continental crust, indicating a geodynamic evolution from a rifting toward a drifting environment (Nieto et al., 1997b, 1999; Puga et al., 2002a).

The Jurassic ocean-floor accretion was followed by a stage of ocean-floor metasomatism and metamorphism, coeval with volcanism, which has been dated as Late Jurassic (Tables 1 and 2). Subsequently, a period of sedimentation generated the sedimentary sequence of the BOA (Figures 2 and 20), which is similar to that covering the Alpine–Apennine ophiolites (Lagabrielle et al., 1984). These sediments may be overlying the basic rocks, as in the Cóbdar outcrop, or directly covering the ultramafic sequence, as in the Almirez outcrop. This fact has been interpreted by

Puga et al. (1999a) as due to tectonic uplift of ultramafic rocks throughout the oceanic crust, following transform or normal faults which are very common along the slow-spreading ridges (Bonatti, 1976; Bonatti et al., 1979; Francis, 1981; Coleman, 1984; Lagabrielle et al., 1984; Nicolas, 1989; Lagabrielle and Cannat, 1990; Cannat et al., 1992). In fact, as a result of these intra-oceanic tectonics, serpentinites crop out on the oceanic floors, making up (*f.i.*) about 20% of the current Atlantic ocean-floor (Cannat et al., 1995).

In the Late Cretaceous, a period of convergence between the Iberian and African plates caused the subduction of the ocean floor generated during the Jurassic and its overlying sediments, followed by the subduction of the Caldera unit (Figure 20 and Puga et al., 2000). The HP eo-Alpine parageneses developed under eclogite facies conditions in the BOA rocks during this compressive period, which was followed by a Palaeocene relaxation stage as a result of the transcurrent movements between the Iberian and African plates (Dewey et al., 1989). During this extensional stage, rapid exhumation of a small part of the previously subducted materials must have occurred, while scarce calc-alkaline magmas generated by subduction of the oceanic crust and their overlying sediments attained the surface producing pyroclastic levels and subvolcanic bodies of basaltic andesites, which were dated by K/Ar as Paleocene (Puga et al., 1996). These orthoderived rocks, together with scarce continental and lagunar sediments, form the Soportújar Formation, which was deposited onto the eclogitized and exhumed ophiolites and over other units of the MC also affected by the HP eo-Alpine metamorphism (Puga et al., 1996, 2002a).

A new period of convergence between the Iberian and African plates beginning at the Eocene-Oligocene boundary (Dewey et al., 1989) gave way to the meso-Alpine metamorphic event, probably during an intra-continental subduction stage (Puga and Díaz de Federico, 1978) which had been produced in a continental collision setting. In the course of this metamorphic event, Ab-Ep amphibolite facies developed in the intra-orogenic Soportújar Formation and in the other units of the MC (Puga et al., 2000). During a subsequent exhumation stage, the MC overthrust the Veleta Complex, after which both complexes were in turn overthrust by the Alpujárride and Maláguide complexes as a consequence of the extensional collapse affecting the whole Betic orogenic edifice (Platt and Vissers, 1989; García Dueñas et al., 1992; Crespo Blanc et al., 1994; Vissers et al., 1995).

11. OCEANIC PROVENANCE AND CONTINENTAL EMPLACEMENT OF THE BETIC OPHIOLITES

Similarities in age, as well as in petrological and geochemical characteristics, between the BOA and Alpine–Apennine ophiolites (Pearce, 1980; Beccaluva et al., 1983, 1984; Pfeifer et al., 1989; Lagabrielle and Cannat, 1990), together with palaeogeographic reconstructions of the Alpine Tethys for the Mesozoic (Beccaluva et al., 1984; Ricou et al., 1986; Lemoine et al., 1987; Trümpy, 1988; Andrieux et al., 1988; Favre and Stampfli, 1992; Guerrera et al., 1993) all suggest that the Betic ophiolites may have originated along a southwestern prolongation of the Piemontes Ocean. This oceanic branch of the Alpine Tethys, which Guerrera et al. (1993) refer to as the Nevado-Filábride Ocean, would have been connected to the central Atlantic through the Maghrebian Ocean (Figure 21). The ocean floor originated along the Nevado-Filábride Ocean, was subducted below the Mesomediterranean Terrane (Guerrera et al., 1993) during the Late Cretaceous eo-Alpine event and then partly exhumed onto the Iberian margin as eclogitized ophiolites, forming the BOA, or ophiolite unit, of the MC (Figure 2).

Figure 22 depicts the probable location of the BOA in the palinspastic sketch of Favre and Stampfli (1992) for the Ligurian Tethys at the Jurassic-Cretaceous boundary. In this paleogeographic reconstruction, similarly to that in Figure 21, the BOA would have derived from an ocean-floor segment located to the west of the Alboran plate, which constitutes the southwestern end of the Mesomediterranean Terrane. The lithological log in Figure 22 represents, according to Favre and Stampfli (1992), the relationship between the continental margin of the Alboran microplate and the oceanic branch bordering it to the west, from which the Betic ophiolites derive. It also shows the provenance of the Ronda and Beni Boussera ultramafic massifs from the upper mantle of this microplate denuded during the Jurassic.

The tentative locations of Cóbdar, Lugros and Almirez outcrops in the palinspastic sketches in Figures 21 and 22, have been reconstructed taking into account some of their petrological and geochemical peculiarities, together with their relative present-day geographical settings. Thus, the somewhat greater enrichment in LREE and incompatible elements of the Cóbdar magmas compared to those of Lugros (Figures 12 to 15), as well as the presence of Al-silicate relics in some cumulus meta-gabbros from Cóbdar, suggesting a very limited crustal assimilation (Puga et al., 1989a; Gomez-Pugnaire and Muñoz, 1990) is indicated by a location for the Cóbdar ophiolites in these figures derived from a segment of the ocean-floor lying closer to the continental margin than that of Lugros. Moreover,

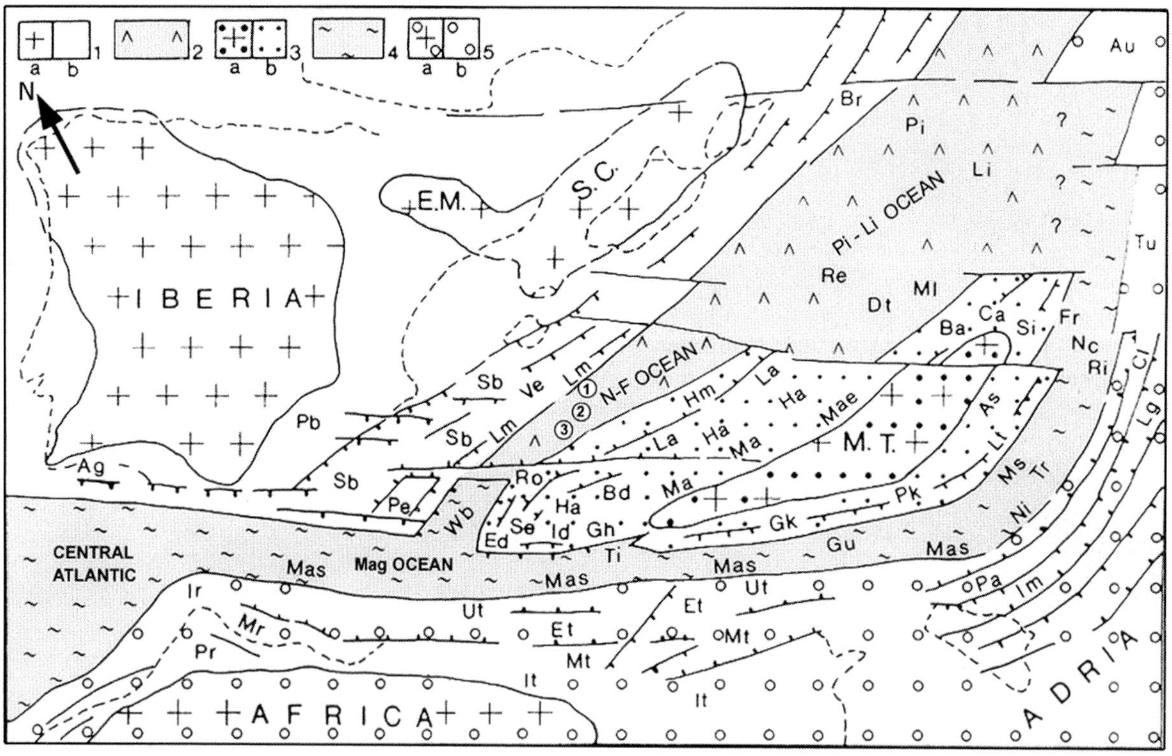

Figure 21: Early Cretaceous palaeogeography of the Central-Western Mediterranean region and distribution of the oceanic basins around the Mesomediterranean Terrane (slightly modified from Guerrera et al., 1993), showing the provenance of the Cóbdar (1), Almirez (2) and Lugros (3) ophiolitic outcrops located tentatively along the Nevado-Filábride (NF) Ocean, which connected the Piemontese-Liguride (Pi-Li) and the Maghrebian (Mag) oceans. Key: 1 = Europa–Iberia plate: (a) Emerged areas (E.M. = Ebro Massif; S.C. = Sardina–Corsica; (b) Margins (Ag = Algarve; Pb = Prebetic; Sb = Subbetic; Pe = Penibetic; Ve = Veleta; Lm = Lower Mulhacén or Caldera Unit; Br = Brianzonese). 2 = Northern Oceanic Basin (Nevado-Filábride ocean = Ophiolite Unit of the Mulhacén Complex; Re = Monte Reventino; Dt = Diamante-Terranova; Ml = Malvito; Pi = Piemontese; Li = Liguride). 3 = Mesomediterranean Terrane: (a) Emerged areas; (b) Margins (Ed = Rifian External "Dorsale"; Ro = Rondaides; Hm = Higher Mulhacén = Sabinas Unit; Se = Sébtides; La = Lower Alpujarrides; Ha = Higher Alpujarrides; Id = Rifian Internal "Dorsale"; Bd = Betic Internal "Dorsale"; Gh = Ghomarides; Ma = Malaguides; Mae = Malaguides-Espuña; Ba = Bagni; Ca = Castagna; Si = Sila; Gk = Grande Kabylie; Pk = Petite Kabylie; As = Aspromonte-Stilo; Lt = Longi-Taormina). 4 = Southern Oceanic Basin (Wb = Western Betic; Ti = Tisirène; Gu = Guerrouch; Ms = Monte Soro; Fr = Frido; Tr = Troina; Nc = North Calabrian; Mas = Massylian; Ni = Nicosia; Ri = Rocca Imperiale). 5 = Africa-Adria Plate: (a) Emerged areas; (b) Margins (Ir = Intrarif; Mr = Mesorif; Pr = Prerif; Ut = Ultratellian; Et = Epitellian; Mt = Mesotellian; It = Infratellian; Pa = Panormide; Im = Imerese; Cl = Campania-Lucania; Lg = Lagonegrese; Tu = Tuscanides; Au = Austroalpine).

the location of the Almirez outcrop along a ridge/transform intersection fracture zone represents the most feasible tectonic setting for the ultramafic sequence to have welled up to the bottom of the ocean floor, where it was directly overlain by the oceanic sediments.

The microporphyric texture of the basic dykes intruded into the ultramafic rocks, together with their chemical similarity to the basic magmas from the volcanic sequence, suggests a similar mantle source. Given their textures, these magmas, some of which are ferrobasaltic in composition (Figure 8) must have intruded the ultramafic host at shallow depth, but before the oceanic serpentinization occurred. Similar intrusive relationships between extensively fractionated gabbros and mantle peridotites have been observed in samples from the Mid-Atlantic Ridge axial valley south of the Kane transform, in dredged samples from southwestern Indian Ocean fracture zones, and in the ophiolites of the western Alps. In these outcrops the intrusion of basic magmas in the peridotites took place after their ductile deformation, which originated during their tectonic uplift through the oceanic lithosphere and before their serpentinization (Cannat et al., 1992 and references herein).

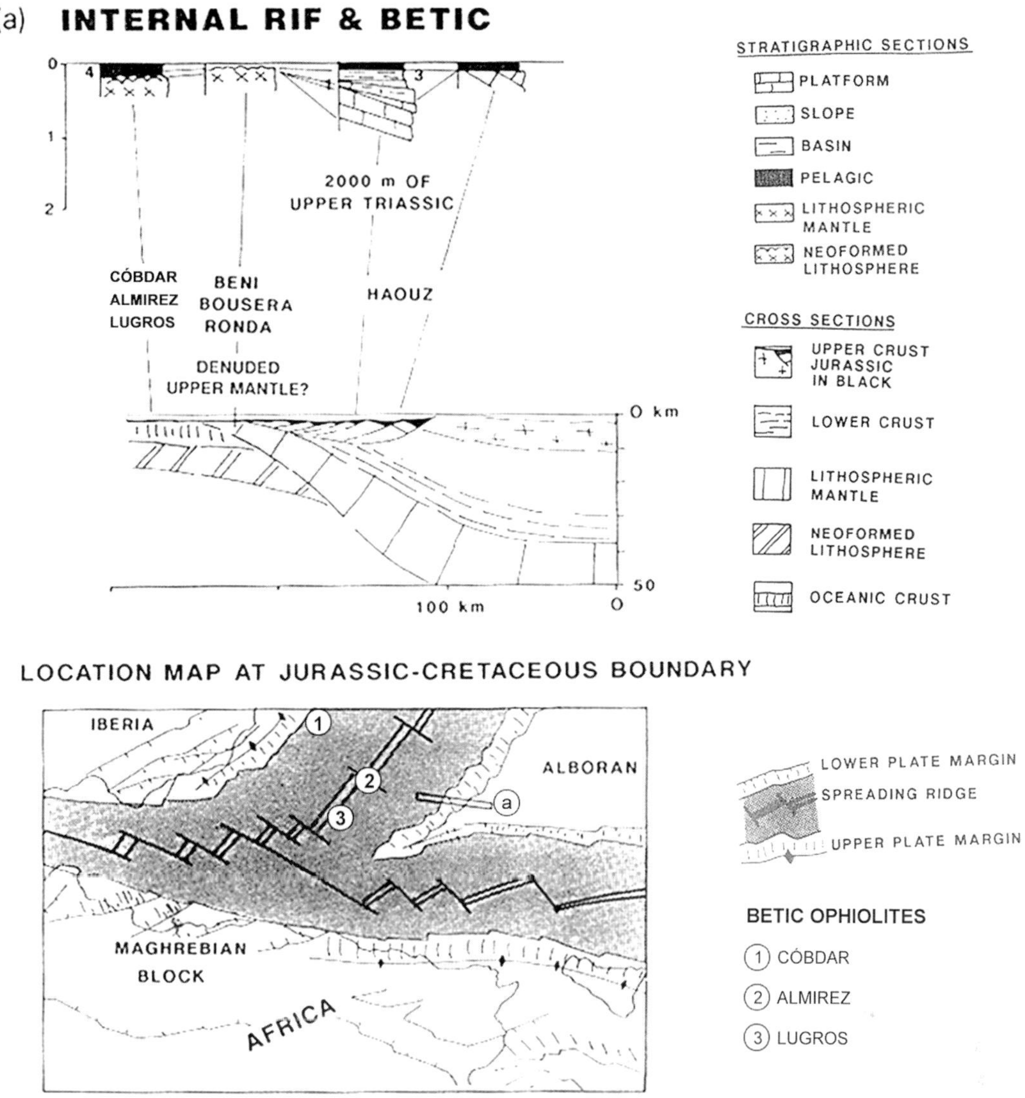

Figure 22: Palaeogeographic reconstruction of the Alpine Tethys and cross-section of the Alboran margin at the Jurassic-Cretaceous boundary, slightly modified from Favre and Stampfli (1992), showing the relative locations along the Nevado-filábride Ocean of the Cóbdar, Almirez and Lugros ophiolitic outcrops. The lithologic log is based on the following papers: Haouz units, Wildi (1983), El Hatimi et al. (1991); Beni Boussera and Ronda, Reuber et al. (1982); Cóbdar region, Puga et al. (1989b).

The relics of the different sequences composing the BOA were incorporated into the continental margin most probably after their subduction, high-pressure metamorphism and exhumation. During these processes, the ophiolitic slices were tectonically imbricated with the other units of the MC, of crustal origin, and all of them followed a common orogenic evolution from this post-eo-Alpine stage onward.

ACKNOWLEDGEMENTS

I would like to express my gratitude to all the members of the "Gruppo di Lavoro sulle Ofioliti del Mediterraneo", whose fruitful comments during our joint visit to the Betic Ophiolites motivated me to begin this work, as well as to my indispensable collaborator Antonio Díaz de Federico, without whose constant help I would not have been able to carry it out successfully. I am very grateful to L. Beccaluva, G.M. Stampfli and an anonymous reviewer for their fruitful critical reviews, which greatly improved the manuscript, and to my colleagues A. Martín Algarra and J.M. Nieto for very stimulating discussions on the Betic geology. I am also indebted to Christine Laurine for polishing the English version of the text, and to A. Díaz Puga for his valuable help with the figures. Funding for this project was provided by Project PB98-1334 of the Spanish Ministry of Science and Technology, and the Research Groups RNM 187 and 333 of the Junta de Andalucía.

REFERENCES

Aldis, D.T., 1981. Plagiogranites from the ocean crust and ophiolites. Nature 289, 577–578.

Alt, J.C., Muehlenbachs, K., Honnorez, J., 1986. An oxigen isotopic profile through the upper kilometer of the oceanic crust, Deep Sea Drilling Project Hole 504 B. Earth Planet. Sci. Lett. 80, 217–229.

Andriessen, P.A.M., Hebeda, E.H., Simon, O.J., Verschure, R.H., 1991. Tourmaline K–Ar ages compared to other radiometric dating systems in Alpine anatectic leucosomes and metamorphic rocks (Cyclades and southern Spain). Chem. Geol. 91, 33–48.

Andrieux, J., Frizon de Lamotte, D., Braud, J., 1988. A structural scheme for the Western Mediterranean area in Jurassic and Early Cretaceous times. Geodinamica Acta 3, 5–15.

Arai, S., 1994. Characterization of spinel peridotites by olivine-spinel compositional relationship: review and interpretation. Chem. Geol. 113, 191–204.

Baedecker, P.A., 1987. Methods for geochemical analysis. US Geol. Surv. Bull. 1770, 1–183.

Bakker, H.E., De Jong, K., Helmers, H., Biermann, C., 1989. The geodynamic evolution of the Internal Zone of the Betic Cordilleras (south-east Spain): a model based on structural analysis and geothermobarometry. J. Metamorphic Geol. 7, 359–381.

Barnicoat, A.C., Fry, N., 1986. High-pressure metamorphism of the Zermatt-Saas ophiolite zone, Switzerland. J. Geol. Soc. London 143, 607–618.

Bea, F., 1996. Residence of REE, Y, Th and U in granites and crustal protoliths: Implications for the chemistry of crustal melts. J. Petrol. 37, 521–552.

Beccaluva, L., Dostal, J., Macciota, G., Zerbi, M., 1983. Trace element geochemistry of some ophiolites from Calabria (Southern Italy). Ofioliti 8, 325–332.

Beccaluva, L., Dal Piaz, G.V., Macciota, G., 1984. Transitional to normal MORB affinities in ophiolitic metabasites from the Zermatt-Saas, Combin and Antrona units, Western Alps: implications for the paleogeographic evolution of the Western Tethyan Basin. Geol. Mijnb. 63, 165–177.

Bellon, H., Bordet, P., Montenat, C., 1983. Chronologie du magmatisme néogène des Cordillères bétiques (Espagne méridionale). Bull. Soc. Géol. France 25, 205–217.

Berman, R.G., Brown, T.H., Greenwood, H.J., 1985. An internally consistent thermodynamic data base for minerals in the system Na_2O–K_2O–CaO–MgO–FeO–Fe_2O_3–Al_2O_3–SiO_2–TiO_2–H_2O–CO_2. Atomic Energy Commission of Canada Technical Report 377, 62.

Bertrand, J., Dietrich, V., Nievergelt, P., Vuagnat, M., 1987. Comparative major and trace element geochemistry of gabbroic and volcanic rock sequences, Montgenévre ophiolite, Western Alps, Schweiz. Mineral. Petrog. Mitt. 67, 147–169.

Bodinier, J.L., Morten, L., Puga, E., Diaz de Federico, A., 1987. Geochemistry of metabasites from the Nevado-Filabride Complex Betic Cordilleras Spain: relicts of a dismembered ophiolitic sequence. Lithos 20, 235–245.

Bodinier, J.L., Puga, E., Diaz de Federico, A., Leblanc, M., Morten, L., 1993. Secondary harzburgites with spinifex-like textures in the Betic Ophiolitic Association (SE Spain). Terra Nova Abstract 4, 3.

Bohlen, S.R., Boetcher, A.L., 1982. The quartz-coesite transformation: a precise determination and the effects of other components. J. Geophys. Res. 87, 7073–7078.

Bonatti, E., 1976. Serpentinite protusions in the oceanic crust. Earth Planet. Sci. Lett. 32, 107–113.

Bonatti, E., Chermak, A., Honnorez, J., 1979. Tectonic and igneous emplacement of crust in oceanic transform zones. Amer. Geophys. Union 2, 239–248.

Borsi, L., Schärer, U., Gaggero, L., Crispín, L., 1996. Age, origin and geodynamic significance of plagiogranites in lherzolites and gabbros of the Piedmont–Ligurian ocean basin. Earth Planet. Sci. Lett. 140, 227–241.

Boynton, W.V., 1984. Geochemistry of rare earth elements: meteorite studies. In: Henderson, P. (Ed.), Rare Earth Element Geochemistry. Elsevier, Amsterdam, pp. 63–114.

Burgos, J., Díaz de Federico, A., Morten, L., Puga, E., 1980. The ultramafic rocks of the Cerro del Almirez, Sierra Nevada Complex, Betic Cordilleras, Spain. Preliminary report. Cuad. Geol. 11, 157–165.

Burke, W.H., Denison, R.E., Hetherington, E.A., Koepnick, R.B., Nelson, H.F., Otto, J.B., 1992. Variation of seawater $^{87}Sr/^{86}Sr$ throughout Phanerozoic time. Geology 10, 516–519.

Cabanis, B., Thiéblemont, D., 1988. La discrimination des tholéiites continentales et des basaltes arrière-arc. Proposition d'un nouveau diagramme le triangle Th-3xTb-2xTa. Bull. Soc. Géol. France, Ser. 8 IV (6), 927–935.

Cámara, F., 1995. Estudio cristalquímico de minerales metamórficos en rocas básicas del complejo Nevado-Filábride (Cordilleras Béticas). Tesis Doctoral Univ. Granada, 484 p.

Cannat, M., Bideau, D., Bougault, H., 1992. Serpentinized peridotites and gabbros in the Mid-Atlantic Ridge axial valley at 15°37′N and 16°52′N. Earth Planet. Sci. Lett. 109, 87–106.

Cannat, M., Mevel, C., Mala, M., Deplus, C., Durand, C., Gente, P., Agrinier, P., Belarouchi, A., Dubuisson, G., Humler, E., Reynolds, J., 1995. Thin crust, ultramafic exposures, and rugged faulting patterns at the Mid-Atlantic Ridge (22°–24°N). Geology 23, 49–52.

Coleman, R.G., Peterman, Z.E., 1975. Oceanic plagiogranite. J. Geophys. Res. 80 (8), 1099–1108.
Coleman, R.G., 1984. Pre-accretion tectonics and metamorphism of ophiolites. Ofioliti 9, 205–222.
Cortesogno, L., Gaggero, L., Molli, G., 1994. Ocean floor tectono-metamorphic evolution in the Piedmont–Ligurian Jurassic basin: a review. Mem. Soc. Geol. Ital. 48, 151–163.
Crespo Blanc, A., Orozco, M., García Dueñas, V., 1994. Extension versus compression during the Miocene tectonic evolution of the Betic Chain. Late folding of normal faults systems. Tectonics 13, 78–88.
De Jong, K., 1991. Tectono-metamorphic studies and radiometric dating in the Betic Cordilleras (SE Spain). Ph.D. Thesis, Vrije Universiteit Amsterdam, 203 p.
De Jong, K., 1993. The tectono-metamorphic evolution of the Veleta Complex and the development of the contact with the Mulhacen Complex (Betic Zone SE Spain). Geol. Mijnb. 71, 227–237.
De Jong, K., Wijbrans, J.R., Feraud, G., 1992. Repeated thermal resetting of phengites in the Mulhacén Complex (Betic Zone, southeastern Spain), shown by $^{40}Ar/^{39}Ar$ step heating and single grain laser probe dating. Earth Planet. Sci. Lett. 110, 173–191.
Desmons, J., 1989. Different metamorphic evolutions in the Alpine-Apenninic Ophiolites (France–Italy–Switzerland–Austria). Chem. Geol. 77, 229–250.
Desmons, J., Compagnoni, R., Cortesogno, L., 1999. Alpine metamorphism of the Western Alps: II High P/T and related pre-greenschist metamorphism. Mineral. Petrogr. Mitt. 79, 11–134.
Dewey, J.F., Helman, M.L., Turco, E., Hutton, D.H.W., Knott, S.D., 1989. Kinematics of the western Mediterranean. In: Coward, N.P., Dietrich, D., Park, R.G., (Eds.), Alpine Tectonics. Geological Society Special Publication, pp. 265–283.
Díaz De Federico, A., 1980. Estudio geológico del Complejo de Sierra Nevada en la transversal del Puerto de la Ragua. Tesis Doctoral Univ. Granada, 597 p.
Díaz De Federico, A., Torres Roldan R.L., Puga E., 1990. The rock series of the Betic Substratum. In: Les Bassins Néogènes du domaine Bétique Oriental, Documents et Travaux IGAL Paris, vol. 12–13, pp. 19–29.
Dietrich, V., 1980. The distribution of ophiolites in the Alps. Ofioliti, Spec. Issue 1, 7–51.
Egeler, C.G., Simon, O.J., 1969. Sur la tectonique de la zone Betique (Cordillères Bétiques Espagne). Ver. Kon. Ned. Akd. Wet. 25, 1–90.
El Hatimi, N., Duée, G., Hervouet, Y., 1991. La dorsale calcaire du Haouz : ancienne marge continentale passive téthysienne (Rif, Maroc). Bull. Soc. Géol. France 162 (1), 79–90.
Ellis, D.J., Green, D.H., 1979. An experimental study of the effect of Ca upon garnet-Cpx Fe–Mg exchange equilibria. Contrib. Mineral. Petrol. 71, 13–22.
Engi, M., Lindsley, D.H., 1980. Stability of titanian clinohumite: experimental and thermodynamic analysis. Contrib. Mineral. Petrol. 72, 415–424.
Favre, P., Stampfli, G.M., 1992. From rifting to passive margin: the examples of the Red Sea, Central Atlantic and Alpine Tethys. Tectonophysics 215, 69–97.
Francis, T.J.G., 1981. Serpentinization faults and their role in the tectonic of slow-spreading ridges. J. Geophys. Res. 86, 11616–11622.
Gaggero, L., Cortesogno, L., 1997. Metamorphic evolution of oceanic gabbros: recrystallization from subsolidus to hydrothermal conditions in the MARK area (ODP leg 153). Lithos 40, 105–1311.
Gasparik, T., Lindsley, D.H., 1980. Experimental study of pyroxenes in the system $CaMgSi_2O_6$–$CaAl_2Si_2O_6$–$CaO_5AlSi_2O_6$. EOS 61, 402–403.
Gillis, K.M., Thompson, G., 1993. Metabasalts from the Mid-Atlantic Ridge: new insights into hydrothermal systems in slow-spreading crust. Contrib. Mineral. Petrol. 113, 502–523.
García Dueñas, V., Balanyá, J.C., Martinez Martinez, J.M., 1992. Miocene extensional detachments in the outcropping basement of the Northern Alboran Basin (Betics) and their tectonic implications. Geo-Mar. Lett. 12, 88–95.
Gomez Pugnaire, M.T., Fernandez Soler, J.M., 1987. High-pressure metamorphism in metabasites from the Betic Cordilleras (SE Spain) and its evolution during the Alpine orogeny. Contrib. Mineral. Petrol. 95, 231–244.
Gomez-Pugnaire, M.T., Muñoz, M., 1990. Xenolitos de metapelitas de alto grado en metabasitas: evidencia del emplazamiento en corteza continental del magmatismo básico Nevado-Filábride (Cordilleras Béticas). Geogaceta 7, 13–15.
Gomez-Pugnaire, M.T., Ulmer, P., Lopez-Sanchez-Vizcaino, V., 2000. Petrogenesis of the mafic igneous rocks of the Betic Cordilleras: a field, petrological and geochemical study. Contrib. Mineral. Petrol. 139, 436–457.
Guerrera, F., Martin-Algarra, A., Perrone, V., 1993. Late Oligocene-Miocene syn-/-late-orogenic successions in Western and Central Mediterranean Chains from the Betic Cordillera to the Southern Apennines. Terra Nova 5, 525–544.
Hebeda, E.H., Boelrijk, N.A.I.M., Priem, H.N.A., Verdurmen, E.A.Th., Verschur, R.H., Simon, O.J., 1980. Excess radiogenic Ar and undisturbed Rb–Sr systems in basic intrusives subjected to Alpine metamorphism in southeastern Spain. Earth Planet. Sci. Lett. 47, 81–90.

Hoffman, A.V., 1997. Mantle geochemistry: The message from oceanic volcanism. Nature 385, 219–229.

Holdaway, M.J., 1971. Stability of andalusite and the aluminum silicate phase diagram. Am. J. Sci. 271, 97–131.

Holland, T.J.B., 1979a. Experimental determination of the reaction paragonite = jadeite + kyanite + H_2O, and internally consistent thermodynamic data for part of the system Na_2O–Al_2O_3–SiO_2–H_2O, with applications to eclogites and blueschists. Contrib. Mineral. Petrol. 68, 293–301.

Holland, T.J.B., 1979b. High water activities in the generation of high pressure kyanite eclogites of the Tauern window, Austria. J. Geol. 87, 1–27.

Holland, T.J.B., Blundy, J., 1994. Non-ideal interaction in calcic amphiboles and their bearing on amphibole-plagioclase thermobarometry. Contrib. Mineral. Petrol. 116, 433–447.

Jagoutz, L., Palme, H., Badenhausen, H., Blum, K., Cendales, M., Dreibus, G., Spettel, B., Lorenz, V., Wanke, H., 1979. The abundances of major, minor and trace elements in the earth's mantle as derived from primitive ultramafic nodules. In: Proc. 10th Lunar Planet. Sci. Conf., pp. 2031–2050.

Jenkyns, H.C., 1980. Cretaceous anoxic events: from continents to oceans. J. Geol. Soc. London 137, 171–188.

Jensen, L.S., 1976. A new cation plot for classifying subalkalic volcanic rocks. Ont. Div. Mines. Misc. Pap. 66, 1–22.

Johnson, K.T.M., Dick, H.J.B., Shimizu, N., 1990. Melting in the oceanic upper mantle: An ion microprobe study of diopsides in abyssal peridotites. J. Geophys. Res. 95, 2661–2678.

Jones, Ch.E., Jenkyns, H.C., Coe, A.L., Hesselbo, S.P., 1994. Strontium isotopic variations in Jurassic and Cretaceous seawater. Geochim. Cosmochim. Acta 58 (14), 3061–3074.

Kay, R.W., Hubbard, N.J., 1978. Trace elements in ocean ridge basalts. Earth Planet. Sci. Lett. 38, 95–116.

Kretz, R., 1983. Symbol for rock-forming minerals. Am. Mineral. 68, 277–279.

Lagabrielle, Y., Polino, R., Auzende, J.M., Blanchet, R., Caby, R., Fudral, S., Lemoine, M., Mevel, C., Ohnenstetter, M., Robert, D., Tricart, P., 1984. Les témoins d'une tectonique intraocéanique dans le domaine téthysien : analyse des rapports entre les ophiolites et leurs couvertures métasédimentaires dans la Zone Piémontaise des Alpes franco-italiennes. Ofioliti 9 (1), 67–88.

Lagabrielle, Y., Cannat, M., 1990. Alpine Jurassic ophiolites resemble the modern central Atlantic basement. Geology 18, 319–322.

Lemoine, M., Boillot, G., Tricart, P., 1987. Ultramafic and gabbroic ocean floor of the Ligurian Tethys (Alps, Corsica, Apennines): in search of a genetic model. Geology 15, 622–625.

Lemoine, M., Marthaler, M., Caron, M., Sartori, M., Amaudric Du Chaffaut, S., Dumont, T., Escher, A., Masson, H., Polino, R., Tricart, P., 1984. Découverte de foraminifères planctoniques du Crétacé supérieur dans les Schistes lustrés du Queyras (Alpes occidentales). Conséquences paléogéographiques et tectoniques. C. R. Acad. Sci. Paris 299, 727–732.

López Sanchez-Vizcaino, V., Rubatto, D., Gomez Pugnaire, M.T., Trommsdorff, V., Müntener, O., 2001. Middle Miocene high-pressure metamorphism and fast exhumation of the Nevado-Filábride Complex, SE Spain. Terra Nova 13, 327–332.

Maresch, W.V., 1977. Experimental studies on glaucophane: an analysis of present knowledge. Tectonophysics 43, 109–125.

Marzoli, A., Renne, P.R., Piccirillo, E.M., Ernesto, M., Bellieni, G., De Min, A., 1999. Extensive 200-Million-year-old continental flood basalts of the Central Atlantic Magmatic Province. Science 284, 616–618.

Massone, H.J., Schreyer, W., 1987. Phengite geobarometry based on the limiting assemblage with K-feldspar, phlogopite, and quartz. Contrib. Mineral. Petrol. 96, 212–224.

Mevel, C., 1988. Metamorphism in oceanic layer 3, Gorringe Bank, Eastern Atlantic. Contrib. Mineral. Petrol. 100, 496–509.

Michard, A., Goffé, B., Chalouan, A., Saddiqui, O., 1991. Les corrélations entre les Chaînes bético-rifaines et les Alpes et leurs conséquences. Bull. Soc. Géol. France 162, 1151–1160.

Miyashiro, A., 1975. Classification, characteristics and origin of ophiolites. J. Geology 83, 249–281.

Monié, P., Galindo-Zaldívar, J., González-Lodeiro, F., Goffé, B., Jabaloy, A., 1991. $^{40}Ar/^{39}Ar$ geochronology of Alpine tectonism in the Betic Cordilleras (southern Spain). J. Geol. Soc. London 148, 289–297.

Montero, P., Bea, F., 1998. Accurate determination of $^{87}Rb/^{86}Sr$ and $^{143}Sm/^{144}Nd$ ratios by inductively-coupled-plasma mass spectrometry in isotope geoscience: an alternative to isotope dilution analysis. Anal. Chim. Acta 358, 227–233.

Morten, L., Bargossi, G.M., Martinez, J.M., Puga, E., Diaz de Federico, A., 1987. Metagabbro and associated eclogites in the Lubrín area, Nevado-Filábride Complex, Spain. J. Metamorphic Geol. 5, 155–174.

Newton, R.C., Kennedy, G.C., 1963. Some equilibrium reactions in the join $CaAl_2Si_2O_8$–H_2O. J. Geophys. Res. 68, 2967–2984.

Nicolas, A., 1989. Structures of Ophiolites and Dynamics of Oceanic Lithosphere. Kluwer Academic, Dordrecht. 366 p.

Nieto, J.M., Jagoutz, E., Puga, E., Díaz De Federico, A., 1997a. Sm–Nd isotopic disequilibrium in eclogites from the Mulhacén Complex (Betic Cordillera, Spain). Terra Nova Abstract Suppl. 1, 22.

Nieto, J.M., Puga, E., Monié, P., Díaz De Federico, A., Jagoutz, E., 1997b. High-pressure metamorphism in meta-granites and orthogneisses from the Mulhacén Complex (Betic Cordillera, Spain). Terra Nova Abstract Suppl. 1, 22–23.

Nieto, J.M., Puga, E., Diaz De Federico, A., 1999. Late Variscan pyroclastic rocks from the Mulhacén Complex (Betic Cordillera, Spain). In: Leyrit, H., Montenat, C. (Eds.), Volcaniclastic Rocks, from Magmas to Sediments. Harwood Academic Press, pp. 215–233.

Nijhuis, H.J., 1964. Plurifacial alpine metamorphism in the Southeastern Sierra de los Filabres South Lubrin SE Spain. Ph.D. Thesis, University of Amsterdam, 151 p.

Oberhänsli, R., 1982. The P-T history of some pillow lavas from Zermatt. Ofioliti 2 (3), 431–436.

Ottonello, G., Joron, J.L., Piccardo, G.B., 1983. Rare Earth and 3d transition element geochemistry of peridotitic rocks: II. Ligurian peridotites and associated basalts. J. Petrol. 25 (2), 373–393.

Pearce, J.A., 1980. Geochemical evidence for the genesis and eruptive setting of lavas from Tethyan ophiolites. In: A. Panayiotou, (Ed.), Cyprus. Proc. Int. Ophiolite Symp. Cyprus, 1980, pp. 261–272.

Pearce, J.A., 1982. Trace element characteristics of lavas from destructive plate boundaries. In: Thorpe, R.S. (Ed.), Orogenic Andesites and Related Rocks, pp. 525–548.

Pearce, J.A., Harris, N.B.W., Tindle, A.G., 1984. Trace element discrimination diagrams for the tectonic interpretation of granitic rocks. J. Petrol. 25, 956–983.

Pearce, J.A., Peate, D.W., 1995. Tectonic implications of the composition of volcanic arc magmas. Annu. Rev. Earth Planet. Sci. 23, 251–285.

Pfeifer, H.R., Colombi, A., Ganguin, J., 1989. Zermat-Saas and Antrona Zone: a petrographic and geochemical comparison of polyphase metamorphic ophiolites in the West-Central Alps. Schweiz. Mineral. Petrogr. Mitt. 69, 217–236.

Platt, J.P., Vissers, R.L.M., 1989. Existensional collapse of thickened continental lithosphere: a working hypothesis for the Alboran Sea and the Gibraltar arc. Geology 17, 540–543.

Pognante, U., Rösli, U., Toscani, L., 1985. Petrology of ultramafic and mafic rocks from the Lanzo peridotite body (Western Alps). Lithos 18, 201–214.

Pognante, U., Perotto, A., Salino, C., Toscani, L., 1986. The ophiolitic peridotites of the western Alps: record of the evolution of a small oceanic-type basin in the Mesozoic Tethys. Tschermaks Mineral. Petrogr. Mitt. 35, 47–65.

Portugal Ferreira, M., Ferreira, J.D., Puga, E., Diaz De Federico, A., 1988. Geochronological contribution to the petrogenetic picture of the Betic Chain (SE Spain). In: II Congr. Geol. Esp. Com., vol. 2, pp. 55–58.

Powell, R., Holland, T.J.B., 1988. An internally consistent data set with uncertainties and correlations. III. Applications methods, worked examples and a computer program. J. Metamorphic Geol. 6, 173–204.

Puga, E., 1971/1976. Investigaciones petrológicas en Sierra Nevada Occidental. Tesis Doctoral, Univ. de Granada, 269 p.

Puga, E., 1977. Sur l'existence dans le complexe de la Sierra Nevada (Cordillère Bétique, Espagne du sud) d'éclogites et sur leur origine probable à partir d'une croûte océanique mésozoique. C. R. Acad. Sci. Paris 285, 1379–1382.

Puga, E., 1990. The Betic Ophiolitic Association (Southeastern Spain). Ofioliti 15, 97–117.

Puga, E., Díaz de Federico, A., 1978. Metamorfismo polifásico y deformaciones alpinas en el Complejo de Sierra Nevada (Cordillera Bética): Implicaciones geodinámicas. In: Proc. Reun. Geodinámica de la Cordillera Bética y Mar de Alborán, Univ. Granada, pp. 79–111.

Puga, E., Díaz de Federico, A., Bargossi, G.M., Morten, L., 1989a. The Nevado-Filabride metaophiolitic association in the Cobdar region (Betic Cordillera SE Spain): preservation of pillow structures and development of coronitic eclogites. Geodinamica Acta 3, 17–36.

Puga, E., Diaz de Federico, A., Fediukova, E., Bondi, M., Morten, L., 1989b. Petrology geochemistry and metamorphic evolution of the ophiolitic eclogites and related rocks from the Sierra Nevada Betic Cordilleras southeastern Spain. Schweiz. Mineral. Petrogr. Mitt. 69, 435–455.

Puga, E., Diaz de Federico, A., Bodinier, J.L., Monié, P., Morten, L., 1991. The Betic ophiolitic eclogites (Nevado-Filabride Complex SE Spain). In: Terra Nova Abstract Second Eclogite Field Symposium, vol. 6, pp. 9–10.

Puga, E., Díaz de Federico, A., Demant, A., 1995. The eclogitized pillows of the Betic Ophiolitic Association: relics of the Tethys Ocean floor incorporated in the Alpine chain after subduction. Terra Nova 7, 31–43.

Puga, E., Nieto, J.M., Diaz de Federico, A., Portugal, M., Reyes, E., 1996. The intra-orogenic Soportújar Formation of the Mulhacén Complex: evidence for the polycyclic character of the Alpine orogeny in the Betic Cordilleras. Eclog. Geol. Helv. 89 (1), 129–162.

Puga, E., Bodinier, J.L., Diaz de Federico, A., Morten, L., Nieto, J.M., 1997a. Pseudo-spinifex meta-ultramafic rocks containing eclogitized rodingite dykes in the Betic Ophiolitic Association (SE Spain): evidence of Alpine subduction following an ocean-floor metasomatic process. Quaderni di Geodinamica Alpina e Quaternaria 4, 98–99.

Puga, E., Diaz de Federico, A., Bodinier, J.L., Morten, L., Nieto, J.M., 1997b. High-pressure metamorphism in the Betic Ophiolitic Association (Betic Cordilleras, SE Spain). Terra Nova Abstract 9, 26–27.

Puga, E., Nieto, J.M., Díaz de Federico, A., Bodinier, J.L., Morten, L., 1999a. Petrology and metamorphic evolution of ultramafic rocks and dolerite dykes of the Betic Ophiolitic Association (Mulhacén Complex SE Spain): evidence of eo-Alpine subduction following an ocean-floor metasomatic process. Lithos 49, 23–56.

Puga, E., Ruiz Cruz, M.D., Diaz de Federico, A., 1999b. Magnetite-silicate inclusions in olivine of ophiolitic metagabbros from the Mulhacén Complex, Betic Cordillera, southeastern Spain. Canad. Miner. 37, 1191–1209.

Puga, E., Nieto, J.M., Diaz de Federico, A., 2000. Constraining P-T path in eclogites of the Betic Ophiolitic Association, Mulhacén Complex, Southeastern Spain. Canad. Miner. 38, 1137–1161.

Puga, E., Diaz De Federico, A., Nieto, J.M., 2002a. Tectono-stratigraphic subdivision and petrological characterisation of the deepest complexes of the Betic Zone: a review. Geodinamica Acta 15, 23–43.

Puga, E., Ruiz Cruz, M.D., Díaz de Federico, A., 2002b. Polymetamorphic amphibole veins in metabasalts from the Betic Ophiolitic Association at Cóbdar southeastern Spain: Relics of ocean-floor metamorphism preserved through the Alpine orogeny. Canad. Miner. 40, 67–83.

Puga, E., Díaz de Federico, A., Nieto, J.M., Diaz Puga, M.A., Rodriguez Martinez-Conde, J.A., Manteca Martínez, J.I., 2004. Argumentos petrológicos y geoquímicos para la subdivisión del Complejo Nevado-Filábride en los Complejos del Veleta y del Mulhacén. In: VI Congreso Geológico de España, Zaragoza (España). Geo-Temas, vol. 6–1, pp. 101–104.

Puga, E., Fanning, C.M., Nieto, J.M., Díaz de Federico, A., 2005. New recrystallisation textures in zircons generated by ocean-floor and eclogite facies metamorphism: a cathodoluminescence and U–Pb SHRIMP study with constraints from REE elements. Canad. Miner. 42, 1349–1368.

Rampone, E., Hofmann, A.W., Piccardo, G.B., Vannucci, R., Bottazzi, P., Ottolini, L., 1995. Petrology, Mineral and Isotope Geochemistry of the External Liguride Peridotites (Northern Apennines, Italy). J. Petrol. 36, 81–105.

Reuber, I., Michard, A., Chalouan, A., Juteau, T., Jermoumi, B., 1982. Structure and emplacement of the alpine-type peridotite from Beni-Bousera, Rif, Morocco: a polyphase tectonic interpretation. Tectonophysics 82, 231–251.

Ricou, L.E., Dercourt, J., Geyssant, J., Grandjacquet, C., Lepvrier, C., Biju-Duval, B., 1986. Geological constrains on the alpine evolution of the Mediterranean Tethys. Tectonophysics 123, 83–122.

Ridley, J., 1984. Evidence of a temperature-dependent "Blueschist" to "Eclogite" transformation in high-pressure metamorphism of metabasic rocks. J. Petrology 25 (4), 852–870.

Rösli, U., Hoernes, S., Köppel, V., 1991. Isotope data of metarodingites and associated rocks from the Lanzo and the Bracco ophiolitic massifs: indication on the evolution of the Alpino-type ultramafic-mafic complexes. Schweiz. Mineral. Petrogr. Mitt. 71, 125–141.

Ruiz Cruz, M.D., Puga, E., Nieto, J.M., 1999. Silicate and ore-mineral exsolutions in pseudo-spinifex olivine of the metaultramafic rocks from the Betic Ophiolitc Association: a TEM study. Amer. Miner. 84, 1915–1924.

Saunders, A.D., Tarney, J., Marsh, N.G., Wood, D.A., 1980. Ophiolites as ocean crust or as marginal basin crust: A geochemical approach. In: Panayiotou, A. (Ed.) Ophiolites, Proc. Int. Ophiolite Symp., Cyprus, vol. 1979, pp. 193–204.

Shervais, J.N., 1982. Ti–V plots and the petrogenesis of modern and ophiolitic lavas. Earth Planet. Sci. Lett. 59, 101–118.

Snow, J.E., Hart, S.R., Dick, H.J.B., 1994. Nd and Sr isotope evidence linking mid-ocean-ridge basalts and abyssal peridotites. Nature 371, 57–60.

Sun, S., McDonough, W.F., 1989. Chemical and isotopic systematics of oceanic basalts: implications for mantle composition and processes. In: Saunders, A.D., Norry, M.J. (Eds.), Magmatism in the Ocean Basins. Geol. Soc. Spec. Publ. 42, 313–345.

Sun, S.S., Nesbitt, R., Sharaskin, A.Y., 1979. Geochemical characteristic of midocean ridge basalts. Earth Planet. Sci. Lett. 44, 119–138.

Tendero, J.A., Martin Algarra, A., Puga, E., Diaz De Federico, A., 1993. Lithostratigraphie des métasédiments de l'association ophiolitique Nevado-Filabride (SE Espagne) et mise en évidence d'objets ankéritiques évoquant des foraminiféres planctoniques du Crétacé : conséquences paléogéographiques. C. R. Acad. Sci. Paris 316, 1115–1122.

Thompson, G., 1973. Trace-element distribution in fractionated oceanic rocks. 2. Gabbros and related rocks. Chem. Geol. 12, 99–111.

Thöny, M., Jagoutz, E., 1992. Some new aspects of dating eclogites in orogenics belts: Sm–Nd, Rb–Sr, and Pb–Pb isotopic results from the Austroalpine Saualpe and Koralpe type-locality (Carnthia/Styria, southeastern Austria). Geochim. Cosmochim. Acta 53, 347–368.

Torres-Roldán, R.L., 1979. The tectonic subdivision of the Betic Zone (Betic Cordilleras Southern Spain): its significance and one possible geotectonic scenario for the westernmost alpine belt. Amer. J. Sci. 279, 19–51.

Trommsdorff, V., Piccardo, G.B., Montrasio, A., 1993. From magmatism through metamorphism to sea floor emplacement of subcontinental Adria lithosphere during pre-Alpine rifting (Malenco (Italy). Schweiz. Miner. Petrog. Mitt. 73, 191–203.

Trommsdorff, V., Lopez Sanchez Vizcaino, V., Gomez Pugnaire, M.T., Müntener, O., 1998. High pressure breakdown of antigorite to spinifex-textured olivine and orthopyroxene, SE Spain. Contrib. Mineral. Petrol. 132, 139–148.

Trümpy, R., 1988. A possible Jurassic-Cretaceous transform system in the Alps and the Carpathians. Geol. Soc. Am. Spec. Paper 218, 93–109.

Ulmer, P., Trommsdorff, V., 1995. Serpentine stability to mantle depths and subduction-related magmatism. Science 268, 858–861.

Vanko, D.A., Stakes, D.S., 1991. Fluids in oceanic layer 3: Evidence from veined rocks, hole 735B, Southwest Indian Ridge. Proc. O. D. P., Scientific Results 118, 181–215.

Venturelli, G., Thorpe, R.S., Potts, P.J., 1981. Rare earth and trace element characteristics of ophiolitic metabasalts from the Alpine–Apennine Belt. Earth Planet. Sci. Lett. 53, 109–123.

Vissers, R.L.M., 1981. A structural study of the central Sierra de los Filabres (Betic Zone SE Spain) with emphasis on the formational processes and their relation to the alpine metamorphism. Ph.D. Thesis, University of Amsterdam, GUA Papers of Geology, 154 p.

Vissers, R.L.M., Platt, J.P., Van Der Wal, D., 1995. Late orogenic extension of the Betic Cordillera and the Alboran Domain: a lithospheric view. Tectonics 14, 786–803.

Voet, H.V., 1967. Geological investigations in the northern Sierra de los Filabres around Macael and Cóbdar, SE Spain. Thesis Univ. Amsterdam, 122 p.

Wildi, W., 1983. La chaîne tello-rifaine (Algérie, Maroc, Tunisie); structure, stratigraphie et evolution du Trias au Miocène. Rev. Géol. Dyn. Géogr. Phys. 24 (3), 201–297.

Wood, D.A., Tarney, J., Varet, J., Saunders, A.D., Bougault, H., Joron, J.L., Treuil, M., Cann, J.R., 1979. Geochemistry of basalts drilled in the North Atlantic by IPOD Leg 49: implications for mantle heterogeneity. Earth Planet. Sci. Lett. 42, 77–97.

Wunder, B., Schreyer, W., 1997. Antigorite: high-pressure stability in the system $MgO–SiO_2–H_2O$ (MSH). Lithos 41, 213–227.

CROP PROJECT: Deep Seismic Exploration of the Central Mediterranean and Italy
Edited by I.R. Finetti

CHAPTER 32

Evolutionary Reconstruction of the Mediterranean Region: Extrusion Tectonics Driven by the Convergence of Africa Arabia and Eurasia

Enzo Mantovani[1]

ABSTRACT

The post-collisional structural/tectonic evolution of the Mediterranean region, from the Oligocene to the present, has been mainly determined by the development of a number of trench-arc-back arc systems, i.e. migrating accretionary belts (arcs) associated with roll back of subducted lithosphere in front of the advancing arc and crustal stretching (back arc extension) in the wake of the arc. This kind of tectonic process has led to strong deformation and lateral migration, for hundreds of km, of various segments of the Alpine belt and to opening of major back-arc basins, such as the Balearic, Tyrrhenian, Aegean and Pannonian. In this work, the generation of these systems is interpreted as an effect of extrusion processes, driven by the convergence of the confining plates, Africa, Eurasia and Arabia. In particular, it is argued that trench-arc-back arc systems developed where buoyant orogenic zones, facing low buoyancy oceanic domains, were stressed by belt-parallel compression. This kind of context causes the lateral escape of crustal wedges at the expense of the adjacent oceanic domains, while extensional tectonics occurs in the wake of the migrating wedges. No significant contribution to crustal deformation is expected from subduction-related forces. This conviction is supported by previous quantifications of the effects of subduction processes, which suggest that the tensional stress induced by slab-pull in the overriding plate is largely weaker than the mechanical resistance of the lithosphere, and by the difficulty of reconciling the implications of subduction-related driving mechanisms with the observed deformation throughout the Mediterranean area. The geodynamic interpretation and evolutionary reconstruction here proposed, based on the minimum work principle, allow a considerable improvement, with respect to the alternative models so far proposed, in simplicity, internal coherence and physical plausibility of the explanations provided for the complex time–space distribution of deformations observed in the Mediterranean area since the Oligocene. In each evolutionary phase, the nature, location and timing of major observed tectonic events, are consistent with the implications of the minimum work principle in the related geodynamic context, tentatively deduced by the kinematic boundary conditions and the distribution of buoyant orogenic belts and low buoyancy oceanic lithosphere.
It is argued that the difficulties so far encountered in identifying a widely accepted geodynamic interpretation of the Mediterranean deformation pattern are mainly due to the fact that most previous attempts started from two misleading assumptions about major aspects of the geodynamic context. The first is that the kinematic boundary conditions in the study area are only constituted by the convergence of Africa and Eurasia, which neglects the important role played by the westward displacement of the Anatolian–Aegean–Balkan system. The second is that the convergence trend of Africa and Eurasia in the last 9 My has been SE–NW. This choice, which implies a very peculiar hook in the trajectory of Africa, has been considerably influenced by the implications of the Africa–Eurasia kinematics predicted by present-day global kinematic models. However, such kinematics is crucially connected with the hypothesis that Eurasia is a unique rigid block, which contrasts with the occurrence of seismotectonic activity in several zones of Western Europe, and cannot account for major features of the recent tectonic setting in the Mediterranean area. The difficulties generated by the above assumptions have led many authors to emphasize the role of subduction-related driving forces.

[1]Department of Earth Sciences, University of Siena, Siena, Italy

Keywords: Mediterranean evolution, trench-arc-back arc systems, extrusion processes, minimum-work principle

1. INTRODUCTION

The structural/tectonic setting and morphological configuration of the Mediterranean area has undergone a profound change in the last 30–40 My, which has mainly developed through the lateral migration and strong distortion of orogenic belts and the opening of new basins (e.g., Biju-Duval et al., 1977; Dewey and Sengor, 1979; Dercourt et al., 1986). This kind of deformation is connected with a tectonic process, usually defined as "arc-trench-back arc migration", which has occurred in some zones of the Mediterranean area, determining the formation of the Balearic, Tyrrhenian, Aegean and Pannonian back arc basins (Figures 1 and 2). These extensional events have occurred in the wake of various segments of the Cretaceous-Eocene Alpine chain, which have experienced considerable lateral migration, even for several hundreds of km. Such displacement of orogenic belts has been accommodated by the consumption of relatively large oceanic domains, which were left in the Mediterranean area after the continental collision of the Adriatic promontory with Eurasia.

The structural features and the evolutionary history of these migrating systems are well documented by a large amount of evidence from the various branches of Earth Sciences (e.g., Dercourt et al., 1986; Royden 1993, 1993; Mantovani et al., 1997, 2002, and references therein), whereas the driving mechanism of such tectonic process is still under debate. The general tendency of geodynamic interpretations so far proposed in literature is that of invoking driving forces internal to the system, since the external forces, generally attributed to the Africa–Eurasia convergence, seem to be inadequate to explain the relatively fast deformation observed in the study area, concerning the rates of trench retreat, crustal extension and arc migration. In the recent literature, the most popular of these interpretations is the slab-pull model (e.g., Malinverno and Ryan, 1986; Royden, 1993, 1993; Faccenna et al., 2001), that explains the roll back of the subducted lithosphere, the migration of the arc and the consequent formation of the back arc basin as an effect of the gravitational sinking and retreat of the slab. However, this dynamic interpretation has to face several difficulties, concerning its physical plausibility and compatibility with the observed deformation pattern, as argued by various authors (Uyeda and Kanamori, 1979; Taylor and Karner, 1983; Mantovani et al., 1997, 2001a, 2001b, 2002) and discussed later in this work.

Here, it is suggested that the observed deformation pattern throughout the Mediterranean region may simply and coherently be explained without invoking subduction-related forces. The only driving forces considered are those induced by the convergence of the confining plates, i.e. Africa, Arabia and Eurasia. In the proposed evolutionary reconstruction the relative motions of plates and microplates are compatible with the complex time–space distribution of tensional and compressional tectonic events in the study area and are also coherently connected with the interpretational scheme described in the next section.

2. MAIN TECTONOPHYSICAL CONCEPTS

In order to enable the reader to check the proposed interpretation more easily, this section describes the main concepts that are presumed to control the connection between causes and effects in each phase of the evolutionary reconstruction.

The convergence of Africa, Arabia and Eurasia in the Mediterranean region has been accommodated by various kinds of shortening processes, such as subduction of lithosphere, crustal thickening and lateral escape of buoyant crustal wedges at the expense of low buoyancy lithosphere lying in the surrounding zones. The time–space distribution of these processes is controlled by the minimum-work principle, that is the need for minimizing the total work of horizontal, kinematically induced forces against any kind of resistance (e.g., buoyancy forces, frictional forces, viscous resistance of the mantle, etc.). The plausibility of this principle has been proven for simple elastic systems (e.g., Saada, 1974), while such a demonstration is much harder for more complex systems. However, the qualitative and quantitative arguments provided by some authors (e.g., Sleep et al., 1979; Kleinrock and Morgan, 1988; Molnar and Lyon-Caen, 1988; Jamison, 1993; Masek and Duncan, 1998; Mantovani et al., 1997, 2001a) suggest that this principle controls the nature and distribution of tectonic processes.

The principal shortening process that accommodates plate convergence is the subduction of lithosphere. In this process, the amount of work of driving forces mostly depends on the structural/rheological properties of the subducting lithosphere, which control the volume of buoyant crustal material that must be scraped off the descending lithosphere and accumulated in the vicinity of the trench zone. The sinking of the non-buoyant part of the lithosphere does not require any (or negligible) work from horizontal driving forces, since it occurs at the expense of the potential gravitational energy stored in the system.

The above considerations imply that the most convenient subduction process is the one that involves the consumption of cold dense oceanic lithosphere not including intracrustal asthenospheric layers (Ranalli and Murphy, 1987; Lobkovski and Kerchmann, 1991). In fact, this kind of lithosphere subducts as a whole (without any delamination) and only leaves very limited amounts of accreted material at the surface, mostly deriving from the sedimentary layer. When, instead, weak ductile layers are present within the lithosphere, generally in the continental lower crust, they may behave as decoupling zones between the non-buoyant mantle part of the lithosphere and the buoyant crust. In this case, the amount of material non participating in subduction increases considerably, with the consequent increase in the work needed for the fracturation and embrication of brittle crustal slivers, and for the flow of ductile lower crustal material into or under the crust of the overriding plate (e.g., Van den Beukel, 1992; Meissner and Mooney, 1998).

The tentative reconstruction of rheological profiles for the various structural provinces of the Mediterranean area (Viti et al., 1997), based on structural information and heat flow data (e.g., Fadaie and Ranalli, 1990; Kohlstedt et al., 1995), indicates that lithosphere with no ductile decoupling layers only exists in the Ionian and Levantine old oceanic domains, while a ductile lower crust is present in most Mediterranean zones. A very thin lithosphere (less then 30 km) characterizes the zones which have undergone recent crustal thinning, as for example the Tyrrhenian and southern Aegean basins. The above information suggests that the consumption of the low buoyancy oceanic domains which were left in the Mediterranean region after the Eocenic continental collision, was the preferred shortening process for accommodating the convergence of the confining plates.

Once a minimum work configuration of shortening processes is reached throughout the system, it lasts until a significant change in the dynamic conditions at one or more consuming zones makes it convenient stopping the ongoing subduction processes and activating new shortening processes in the surrounding regions, where more favourable conditions are present. This change may happen, for instance, when the shortening work increases significantly at a consuming boundary, due to the arrival of continental lithosphere and to the consequent accumulation of a large amount of buoyant material in the vicinity of the trench zone. In this case, the minimum work condition may imply a halt to that consuming process and the activation of new subduction zones in the surrounding regions where less buoyant lithosphere is available. The transmission of the most intense compressional stresses from the suturing trench to the weakest constraints in the surrounding area is generally achieved by the lateral escape of crustal wedges, as occurred, for instance, at the Arabia–Eurasia collisional border by the westward extrusion of Anatolia (e.g., McKenzie, 1972; Dewey and Sengor, 1979).

Depending on the degree of confinement, the mechanical properties of the lithosphere, the rate of indentation and the presence of pre-existing (or incipient) major discontinuities, lateral escape may be preceded by a more or less long and pronounced phase of crustal thickening, and consequent uplift, of the accretionary belt lying in front of the indenter. This occurred, for instance, in the eastern Anatolian zone, where an uplift of about 2 km took place in response to the indentation of the Arabian promontory (Dewey et al., 1986; Barka and Reilinger, 1997).

Thickening starts developing when a lithospheric domain with an intermediate crustal thickness enters the trench zone. During this phase, the amount of ductile buoyant crustal material, which is scraped off the descending lithosphere and accumulated in the trench zone and below the upper crust of the overriding plate, increases significantly, causing crustal thickening and, consequently, isostatic readjustments by uplift (e.g., Zhao and Morgan, 1985; Westaway, 1995; Meissner and Mooney, 1998). This process encounters greater and greater resistance from gravitational forces as thicker and thicker crust enters the trench zone. Contemporaneously, shear stresses increase in the zones surrounding the indenter (e.g., Tapponnier and Molnar, 1976). The end of this process, with the suture of the consuming boundary, may be determined, for instance, by the formation of one or more major decoupling strike slip faults, which favour the lateral escape of wedges towards weak lateral boundaries and the consequent release of stresses in front of the indenter (Hubert-Ferrari et al., 2003). Gravitational collapse from uplifted zones may contribute to the lateral escape of crustal wedges towards thinner adjacent structures (e.g., Dewey et al., 1986; Ratschbacher et al., 1991; Sengor, 1993; Meissner and Mooney, 1998).

In the Arabia–Eurasia collision zone, the direction of the indenter was roughly perpendicular to the belt and no major extension occurred in the wake of the extruding wedge (Anatolia). In other cases, however, where an oblique indentation took place and a weak lateral boundary was present, the extrusion process caused the formation of a

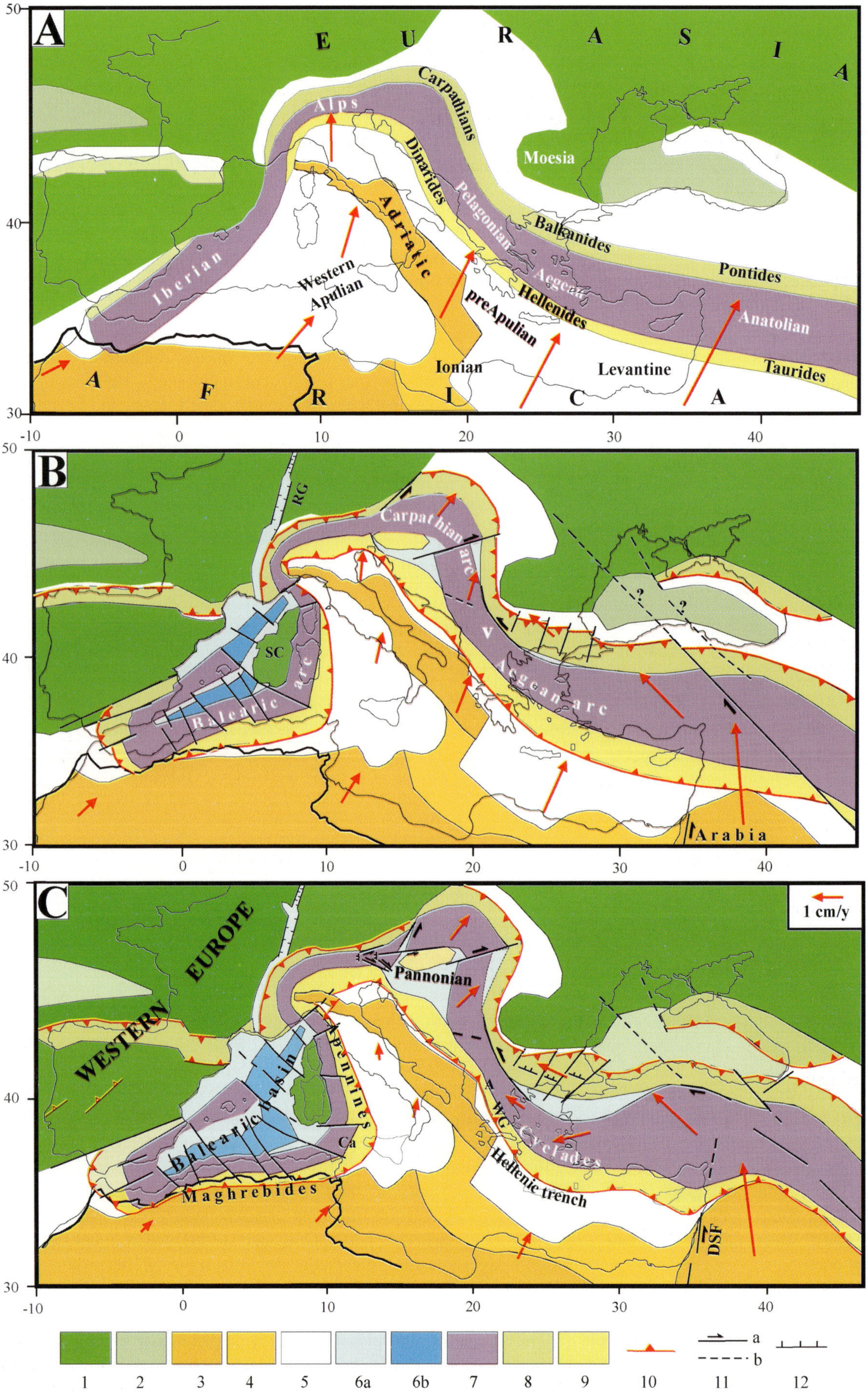
A
E U R A S I A
Carpathians
Alps
Moesia
Dinarides
Adriatic
Pelagonian
Balkanides
Pontides
Iberian
Western Apulian
Aegean
Hellenides
preApulian
Anatolian
Ionian
Levantine
Taurides
A F R I C A
B
RG
Carpathian arc
SC
V
Balearic arc
Aegean arc
?
Arabia
C
1 cm/y
WESTERN EUROPE
Pannonian
Apennines
Balearic basin
Ca
A
WG
Cyclades
Hellenic trench
Maghrebides
DSF
50
40
30
-10
0
10
20
30
40
a
b
1
2
3
4
5
6a
6b
7
8
9
10
11
12

Figure 1: Tentative reconstruction of the Mediterranean evolution for the period Oligocene-Middle Miocene. This evolutionary phase was characterized by a profound tectonic reorganization of both the western and eastern Mediterranean structural/tectonic settings, with the lateral migration and considerable distortion of various sectors of the Cretaceous-Eocene orogenic belts and the opening of the Balearic and Pannonian basins. (1)–(4) Eurasian and African–Apulian domains reported with the present size, the one shown in Figure 2(c): (1) and (2), respectively identify the continental and thinned parts of the Eurasian domain; (3) and (4) the continental and thinned parts of the African/Apulian domain. (5) Parts of the Eurasian and African domains which have been consumed during the evolution. (6) Zones affected by moderate (a) or intense (b) crustal thinning. (7) Orogenic belts built by the closure of the Neotethys Ocean, formed from oceanic remnants, metamorphic bodies and crystalline massifs. (8), (9) Accretionary belts formed from units of the European and African domains, respectively. (10) Compressional features. (11) Transcurrent faults, active (a) and inactive (b). (12) Normal faults. (A) Oligocene paleogeographic setting. (B) Lower Miocene: RG = Rhine Graben system, SC = Sardinia–Corsica microplate, V = Vardar zone. (C) Middle Miocene: A = Albanides, Ca = Calabria, DSF = Dead Sea fault, WG = western Greece. The kinematic patterns of Africa and Arabia adopted in the evolutionary reconstruction are discussed in Section 3. Motion rates in the maps are only indicative. Present geographical contours (thin black lines) and the paleoposition of the African coastline (thick black line) are reported for reference in each map.

back arc basin. This kind of mechanism, sketched in Figure 3, is expected to occur where a segment of the accretionary belt, under a belt-parallel compressional regime, undergoes lateral extrusion/bending and partly separates from the overriding plate. Crustal stretching in the back arc zone accommodates this separation. Simultaneously, the adjacent foreland zone, loaded by the extruding buoyant wedges, undergoes gravitational sinking, which causes the progressive retreat of the trench zone. The tectonic context which produces such deformation of the arc may be quite variable from case to case. Most often, this phenomenon occurs when a buoyant rigid structure enters a sector of the consuming border with a direction of motion not perpendicular to the trench (Figure 3). The belt-parallel compression induced by this oblique indentation is accommodated by lateral extrusion of wedges at the expense of the adjacent lithosphere.

This mechanism can occur only when the buoyancy of the stressed belt is significantly higher than that of the lithospheric domain lying in front of it. Thus, for instance, the lateral extrusion of the arc is strongly favoured when it faces old oceanic lithosphere, which is presumably characterized by very low or even negative buoyancy (Cloos, 1993). Another basic requisite for the occurrence of the proposed mechanism, is a brittle behaviour of the belt, which results in the formation of a few relatively large crustal wedges sliding and rotating each other, as shown in Figure 3. The extrusion of a plastic or greatly fragmented material, for instance, would tend to occupy the entire space available (e.g., Sornette et al., 1993), preventing separation of the arc from the overriding plate and consequent back arc extension.

The physical plausibility of extrusion processes and their possible importance in the generation of back arc basins have already been quantitatively demonstrated by a number of authors (e.g., Tapponnier et al., 1982, 1986; Peltzer and Tapponnier, 1988; Ratschbacher et al., 1991; Lavé et al., 1996; Faccenna et al., 1996; Keep, 2000; Mantovani et al., 2000a, 2000b, 2001a; Hubert-Ferrari et al., 2003; Seyferth and Henk, 2004). In the following, it is argued that the above mechanism is responsible for the development of the trench-arc-back arc systems recognized in the Mediterranean region.

Since the relative motions of the confining plates play an important role in the dynamic interpretation here proposed, an exhaustive explanation of the evidence and arguments supporting the adopted plate kinematics for Africa and Arabia with respect to Eurasia is reported in the next section.

3. KINEMATIC BOUNDARY CONDITIONS

3.1. Post Eocene Africa–Eurasia Relative Motion

It is widely agreed, on the basis of North Atlantic spreading rates, that Africa and Eurasia have undergone a roughly SW–NE to SSW–NNE convergence from the Oligocene to the Upper Miocene (e.g., Savostin et al., 1986; Dercourt et al., 1986; Dewey et al., 1989; Mazzoli and Helman, 1994; Rosenbaum et al., 2002), whereas the more

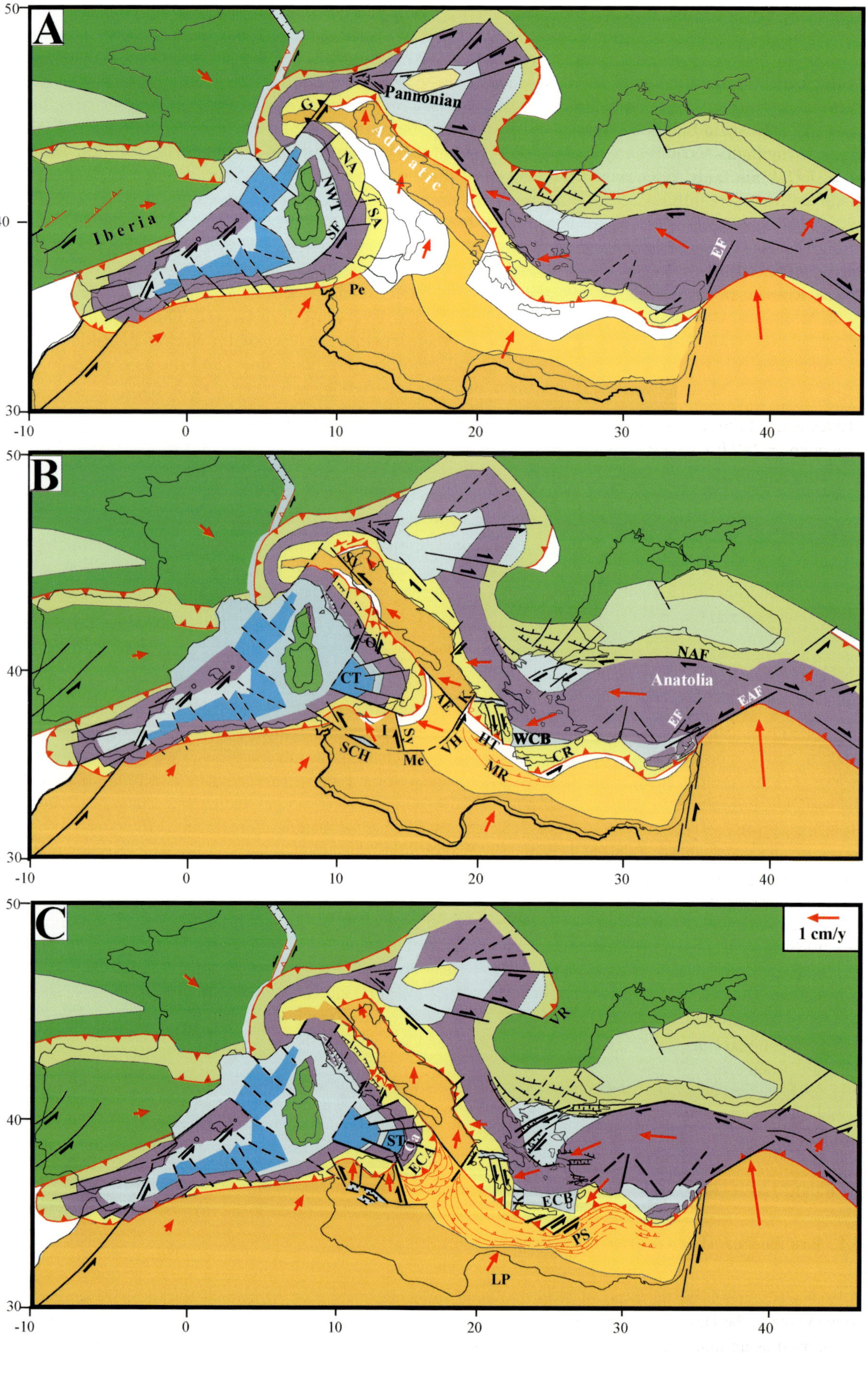
A
Pannonian
G
Adriatic
NA
NWT
SF
SA
Iberia
EF
Pe
B
SV
O
CT
NAF
Anatolia
AE
K
I
Sy
VH
HT
WCB
EF
EAF
SCH
Me
MR
CR
C
1 cm/y
VR
ST
Ca
EC
ECB
PS
LP
50
40
30
-10
0
10
20
30
40

Figure 2: Tentative reconstruction of the Mediterranean evolution since the Upper Miocene. This phase was characterized by a profound reorganization of the central and eastern regions, leading to the formation of the Tyrrhenian and Aegean basins. Symbols as in Figure 1. (A) Upper Miocene: EF = Ecemis fault, G = Giudicarie transpressional fault system, NA, SA = Northern and Southern Apennines, NWT = North-western Tyrrhenian basin, Pe = Pelagian zone, SF = Selli fault. (B) Late Pliocene: A = Ancona-Anzio fault, AE = Apulian escarpment, CR = Crete–Rhodes arc, CT = Central Tyrrhenian basin (Magnaghi–Vavilov), EAF = East Anatolian fault, EF = Ecemis fault, HT = Hellenic trench, K = Kefallinia fault, I = Iblean–Ventura microplate, Me = Medina fault, MR = Mediterranean ridge, NAF = North Anatolian fault system, O = Ortona-Roccamonfina fault, SCH = Sicily Channel fault system, Sy = Syracuse escarpment, SV = Schio–Vicenza fault, VH = Victor Hensen fault, WCB = western Cretan basin. (C) Present: Ca = Calabrian wedge, ECA = external Calabrian arc, ECB = eastern Cretan basin, KI = Kithira trough, LP = Libyan promontory, PS = Pliny and Strabo trenches, ST = Southern Tyrrhenian basin (Marsili), VR = Vrancea zone. See text and caption of Figure 1 for description of plate kinematics.

recent relative motion of these two plates is still object of debate. For this last time interval, some authors (e.g., Dewey et al., 1989) postulate a roughly SE–NW convergence trend in the Mediterranean area, while other authors (Tapponnier, 1977; Dercourt et al., 1986; Mantovani et al., 1997, 2002) propose a roughly N to NE ward Africa–Eurasia convergence.

The drastic change in convergence trend implied by the first hypothesis around the upper Miocene (e.g., Dewey et al., 1989) is not much credible, since it would have involved a significant variation of deformation styles in the zones surrounding Africa and Eurasia and, above all, in the Mediterranean region, which is not supported by any clear evidence. Furthermore, the geodynamic justification of such an important change in motion trend of large plates as Africa and Eurasia is rather obscure.

The choice of assuming a SE–NW trend for the recent Africa–Eurasia convergence (Dewey et al., 1989) has most probably been influenced by the fact that such trend was provided by present-day global kinematic models (e.g., Minster and Jordan, 1978; Chase, 1978) which are representative of the last 2–3 My plate kinematics. In fact, such change was not recognized (e.g., Dewey et al., 1973) before the publication of global kinematic models. Thus, it is crucially important evaluating the reliability of the Africa–Eurasia kinematics provided by global models, with particular regard to the plate configuration they have assumed (DeMets et al., 1990). In this regard, Albarello et al. (1995) have argued that the Africa–Eurasia relative motion predicted by the above models, and by the ones that have been successively proposed (e.g., Argus et al., 1989; DeMets et al., 1990), might be unreliable since the underlying approach is based on the assumption that Eurasia moves as a unique rigid plate, which cannot account for the significant seismotectonic activity recognized in several zones of Western Europe and can hardly be reconciled with the observed deformation pattern in the Mediterranean region (Mantovani et al., 1997, 2000a, 2001a, 2002). Since this appears to be a crucial problem for the choice of the relative motion between Africa and Eurasia in the last 9 My, a detailed description of the evidence which can constrain the recent/present Africa–Eurasia relative motion is given in Section 3.3.

The kinematic history of Africa that has been finally adopted on the basis of the above considerations is shown in the evolutionary maps (Figures 1 and 2). It has been assumed that since the continental collision between the Adriatic promontory and Eurasia, the eastward component of Africa–Eurasia relative motion in the Mediterranean area has increased significantly, whereas its northward component has progressively decreased. This choice is based on the hypothesis that since the above collision the African plate has encountered higher and higher resistance to northward movement, as also suggested by Jolivet and Faccenna (2000). In particular, it may be supposed that this resistance was higher than that encountered in the shortening of the regions lying NE and east of the Adriatic promontory, from the Eastern Alps to the Carpatho–Pannonian–Balkan area.

3.2. Arabia–Eurasia Relative Motion

It has been tentatively assumed, on the basis of the time evolution of spreading in the Red Sea and of transcurrent movements in the Dead Sea fault system (e.g., Girdler, 1985; Hempton, 1987 and references therein), that Arabia has moved as an independent block since roughly the Lower Miocene. In the first map of Figure 1 (Oligocene),

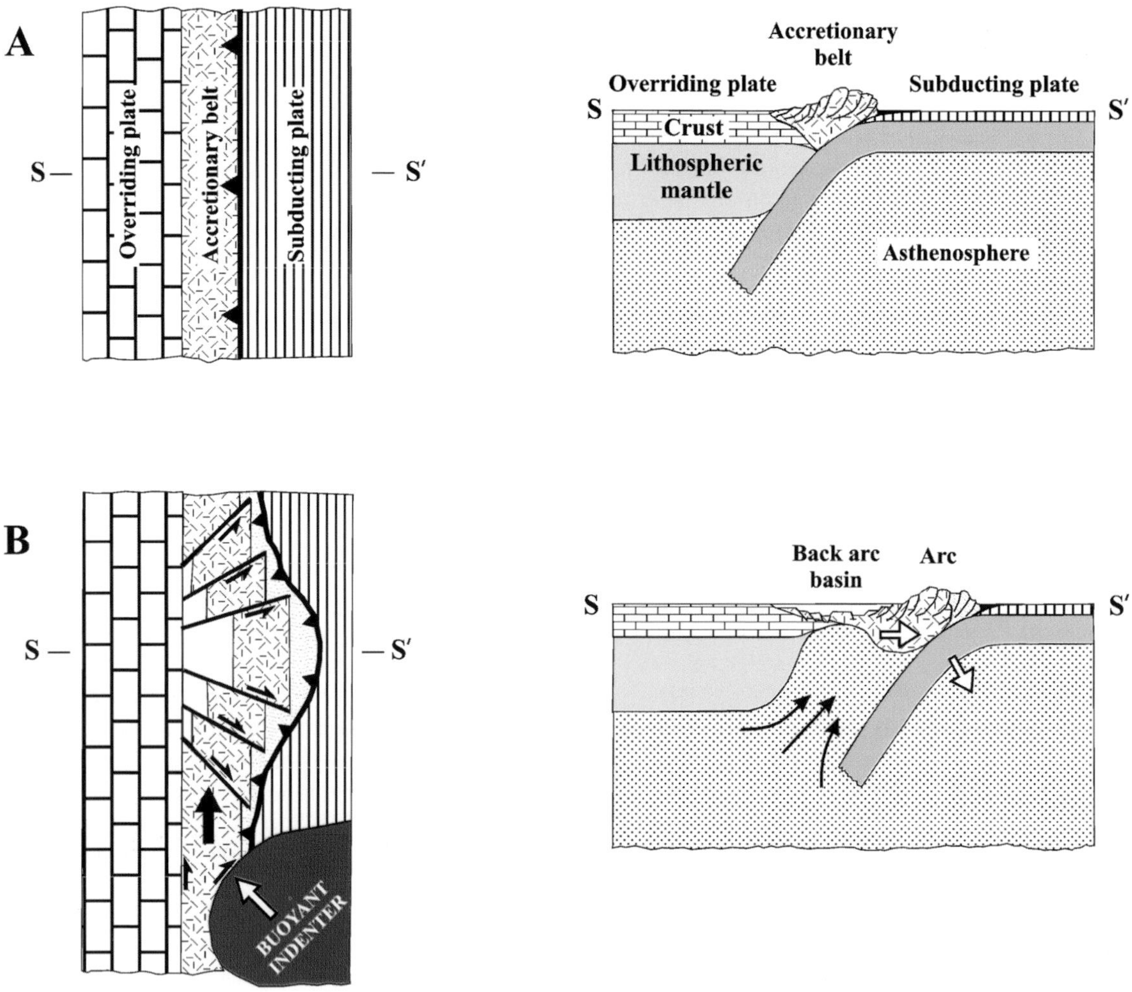

Figure 3: Proposed driving mechanism of a trench-arc-back arc system. (A) Structural/tectonic conditions which may precede the opening of a back arc basin: an accretionary belt has been formed by a subduction process; the belt is flanked by a low buoyancy oceanic zone. (B) A belt-parallel push (black arrow), induced for instance by the oblique indentation of a buoyant block (empty arrow), is accommodated by the lateral expulsion of crustal wedges, which results in an outward bending of the arc, at the expense of the adjacent low buoyancy lithosphere. The divergence between the arc and the overriding plate causes crustal stretching in a limited sector of the boundary, the back arc zone (white area), while outward extrusion of wedges causes accretionary activity on the front of the arc. The cross section shows the sinking and consequent roll back of the slab, caused by the load of the advancing arc. Empty arrows indicate motions. Slab roll back attracts asthenospheric material from the surrounding mantle (thin black arrows).

Arabia moves in close connection with Africa. From the Lower to Middle Miocene (Figure 1(b) and (c)), Arabia moves faster than Africa. From the Middle-Upper Miocene to the late Miocene-early Pliocene (Figure 2(a)), the scarce, or at least reduced, spreading activity in the Red Sea, and transcurrent movements in the Dead Sea fault system (e.g., Hempton, 1987) suggest that Arabia moved slightly faster than Africa. Since the late Miocene–early Pliocene (Figure 2(b) and (c)), Arabia has accelerated again (e.g., Hempton, 1987). The motion trend of Arabia shown in the maps is compatible with the Arabia–Eurasia kinematics proposed by most authors (e.g., DeMets et al., 1990; Jackson et al., 1992; Albarello et al., 1995).

3.3. Present-Day Africa–Eurasia Kinematics from North Atlantic and Mediterranean Evidence

In spite of the fact that the Mediterranean region is the main boundary zone between Africa and Eurasia, the relative motion between these two plates is generally derived from evidence lying outside this region (e.g., Argus et al., 1989; DeMets et al., 1990). This choice is generally justified by the consideration that no clear kinematic constraint on the Africa–Eurasia relative motion can be recognized at such a complex boundary, where deformation is distributed over a large zone. Even though I agree about the complexity of the Mediterranean deformation pattern, I am convinced that some significant constraints on the Africa–Eurasia kinematics can be identified. Evidence and arguments in support of this hypothesis are described in the following.

Plate motions mentioned in the text are referred to a fixed Eurasian frame, unless otherwise specified.

Kinematics of the Adriatic Promontory. Neotectonic, seismological and geodetic data clearly indicate that the Adriatic domain (Figure 4) is presently undergoing a counter clockwise rotation, that involves a roughly NE to NNE ward motion in the southernmost Adriatic zone (e.g., Anderson and Jackson, 1987; Westaway, 1990; Mantovani et al., 2001a; Nocquet and Calais, 2003). In my opinion, this evidence provides an important constraint on the motion of Africa in the central Mediterranean, since the Adriatic moves as a part of Africa. This last condition is suggested by the fact that no significant evidence of a recent (Quaternary) relative motion between these two domains can be recognized. Seismic activity inside the Adriatic is very weak (Figure 5). Few earthquakes of magnitude greater than 4 have occurred in this zone (Console et al., 1993; Renner and Slejko, 1994; Working Group Parametric Catalog of Italian Earthquakes, 1999, available at http://emidius.itim.mi.cnr.it/CPTI). Geological and geophysical evidence (e.g., Finetti and Del Ben, 1986; Argnani et al., 2001; De Alteriis, 1995) indicates that no major decoupling fracture exists in the Adriatic area. Tectonic and seismic activity is recognized in the Gargano zone, belonging to the Adriatic foreland (Figure 4), but such deformation has a limited seaward extension.

In spite of the above evidence, the fact that the kinematics of the Adriatic is not compatible with the motion of Africa predicted by global kinematic models has led a number of authors to look for a decoupling zone between Africa and Adriatic. Anderson and Jackson (1987) tentatively suggested that such decoupling zone could be located in the southern Adriatic, between southernmost Italy and Albania. However, this hypothesis is not supported by clear evidence, either in terms of seismicity or tectonic features, as argued by some authors (e.g., Westaway, 1990; Argnani et al., 2001).

Meletti et al. (2000) suggested that the Syracuse escarpment (Figure 2(b)) is the Africa-Adriatic boundary zone. However, this hypothesis is not plausible, since such discontinuity does not interrupt the structural continuity between the Adriatic domain and the Ionian zone, which is widely recognized as an integral part of the African plate (e.g., Dercourt et al., 1986; Finetti and Del Ben, 1986; Catalano et al., 2001).

Other authors have proposed various decoupling zones located in the central Adriatic area. A SW–NE oriented fault system running from the northern Gargano zone to the central Dinarides is suggested by Westaway (1990). Favali et al. (1993) and Console et al. (1993) suggest a roughly E–W fault system running from the southern Gargano zone to Albania. A more complex plate boundary between Africa and Adriatic, involving a transversal fault in the Middle Adriatic Sea, is proposed by Oldow et al. (2002). It must be pointed out that these last three hypotheses postulate that the southern part of the Adriatic moves in close connection with Africa. However, this condition, associated with the fact that the southern Adriatic moves NE to NNE ward, as suggested by geological, geophysical and geodetic data, would imply that Africa at central Mediterranean longitudes moves in that direction too which contrasts with the predictions of the NOVEL-1 model. The importance of this evidence is emphasized by the fact that the geodetic data set in this region includes the velocity vector of Matera (Figure 4), that can be considered as the most stable and reliable geodetic constraint in the central Mediterranean area (e.g., Devoti et al., 2002), due to the long period of observation (more than ten years) and to the coherence of the results obtained by different geodetic techniques (VLBI, SLR, GPS).

Morphology of the Hellenic and Cyprus Arcs. In the eastern Mediterranean zone, the northern oceanic margin of the African plate, the Ionian–Levantine domain, subducts under the Anatolian–Aegean wedge, which extrudes westward in response to the indentation of the Arabian promontory (e.g., McKenzie, 1978; Dewey and Sengor, 1979). The consuming boundary between these converging systems, the Hellenic and Cyprus arcs, is constituted by thrust fronts oriented SE-NW and left lateral transcurrent fault systems oriented SW–NE (Figure 4). These morphological features clearly indicate a SW–NE direction of plate convergence. To understand the implications of this evidence on the motion of Africa one must know the recent/present kinematics of the Anatolian–Aegean system. It is widely recognized that Anatolia moves roughly westward and the Aegean roughly SW ward (e.g., Barka, 1992; Sengor, 1993; McClusky et al., 2000). However, the short and long-term motion rates of this system, respectively deduced by geodetic and geological data, are very different. Geodetic observations suggest a velocity of 15–25 mm/y for Anatolia and of 30–40 mm/y for the Aegean zone (McClusky et al., 2000). Geological estimates are based on several measurements of fault offsets along the North Anatolian fault (NAF) since the early Pliocene (Barka, 1992, and references therein; Hubert-Ferrari et al., 2002) which indicate a fairly regular distribution of

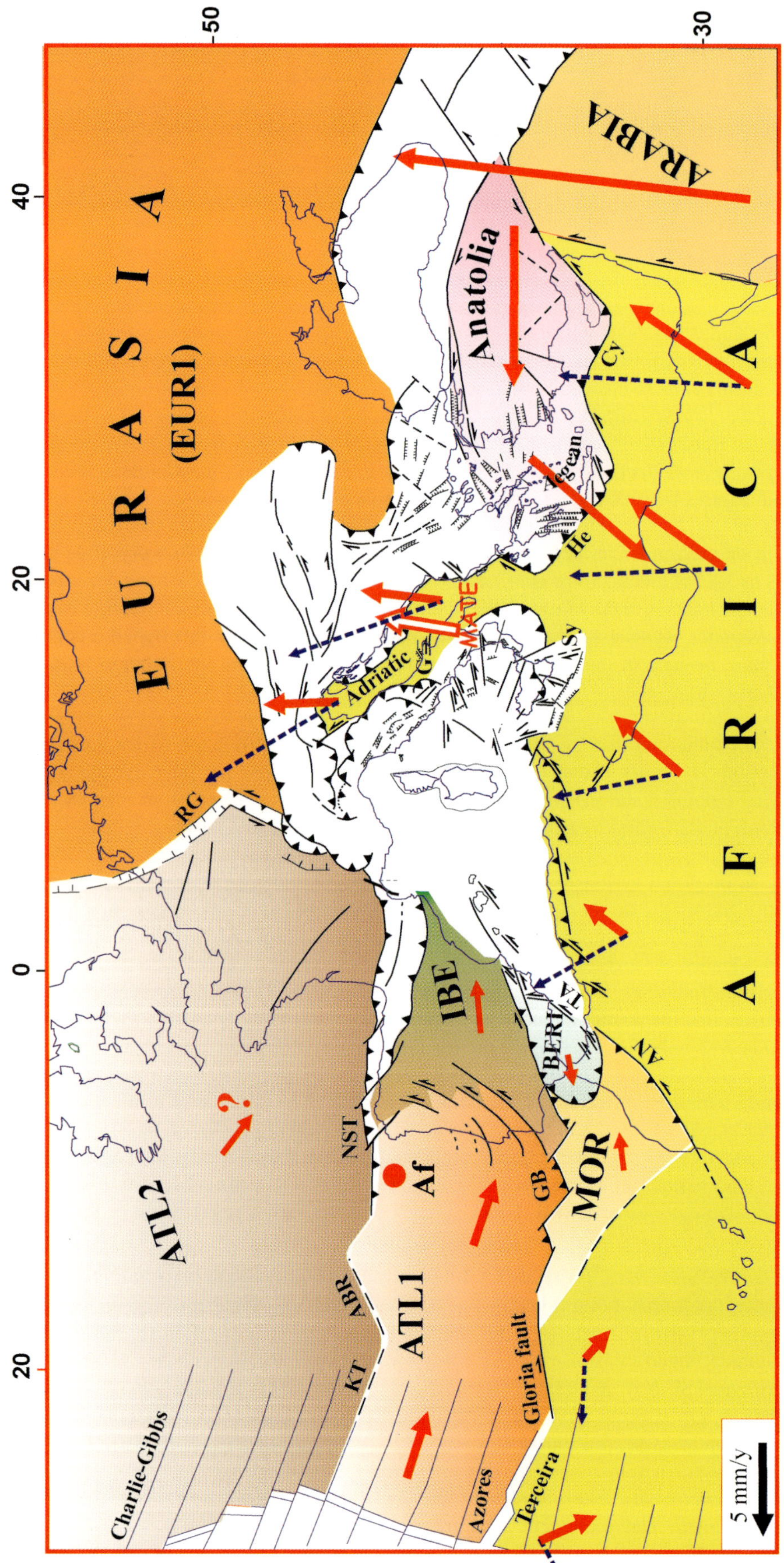
EURASIA
(EUR1)
AFRICA
ARABIA
Anatolia
Aegean
He
Cy
Sy
Adriatic
G
MATE
RG
IBE
BERI
AN
MOR
GB
Af
NST
ATL1
ATL2
ABR
KT
Gloria fault
Azores
Terceira
Charlie-Gibbs
5 mm/y
50
40
30
20
0
20

Figure 4: Proposed plate configuration and related recent/present (last 2–3 My) kinematics (red arrows) with respect to stable Eurasia (EUR1) in the Mediterranean and Atlantic regions. The motions of the Africa/Adriatic plate, the southern North Atlantic microplate (ATL1) and Arabia are compatible with the Eulerian poles given by Albarello et al. (1995): the Africa-EUR1 pole (Af) is located at 41.6°N, 11.8°W, with an angular velocity $\omega = 0.12°/\mathrm{My}$, the ATL1-EUR1 pole at 49.9°N, 129.6°E, $\omega = 0.04°/\mathrm{My}$ and the Arabia-EUR1 pole is located at 28.4°N, 17.0°E, $\omega = 0.51°/\mathrm{My}$. Blue dashed arrows show the Africa–Adriatic kinematics predicted by the NUVEL-1 model. The proposed motions of ATL2, IBE, MOR and BERI are discussed in the text. The empty arrow in southern Italy indicates the geodetic velocity at the Matera station (Anzidei et al., 2001). Tectonic symbols as in Figures 1 and 2. AN = Agadir–Nekor fault system, ABR = Azores–Biscay rise, ATL2 = northern North Atlantic microplate, BERI = Betic-Rif crustal wedge, Cy = Cyprus arc, G = Gargano zone, GB = Goringe bank, He = Hellenic arc, KT = King's trough, IBE = Iberian microplate, MATE = GPS station of Matera, MOR = Morocco microplate, NST = North Spain trench, RG = Rhine Graben system, Sy = Syracuse escarpment, TA = Trans Alboran fault system.

---

offsets from about 40 km, at the NAF's easternmost edge, to 25 km, at the Marmara zone. In terms of slip rate, the above evidence implies values ranging between 5 and 10 mm/y. Comparable values of slip rate have been obtained by the analysis of the long-term features of seismic activity in the NAF (Barka, 1992). Similar estimates at the Eastern Anatolian fault system (Cetin et al., 2003) suggest a slip rate of 11 mm/y in the last 2.5 My. The analysis of seismic strain rate at the Hellenic trench suggests a convergence rate of 4–15 mm/y (e.g., Jackson et al., 1992).

A possible explanation of this remarkable difference between short and middle long-term kinematics of the Anatolian–Aegean system has been provided by the study of post seismic relaxation induced by the well known sequence of strong earthquakes along the NAF started in 1939 (Mantovani et al., 2001c; Cenni et al., 2002).

Considering that the morphology of the Hellenic arc is a long-term feature, it must be related with the long-term motion rate of the Anatolian–Aegean system. Since this motion rate is comparable to that of Africa the orientation of trench zones and strike-slip faults at the Hellenic boundary implies a motion trend of Africa not much different from NE to NNE ward.

An even stronger argument in this direction is given by the morphology of the Cyprus arc. Since the recent deformation and mobility of this arc seems to be very limited (e.g., Pinar and Kalafat, 1999; Papazachos and Papaioannou, 1999), the orientation of the trench zone and of strike-slip faults implies a roughly NE to NNE ward motion of Africa.

North Atlantic Kinematic Constraints and New Plate Configuration in Western Europe. The Africa–Eurasia Eulerian poles provided by global kinematic models (Argus et al., 1989; DeMets et al., 1990), derived from North Atlantic kinematic indicators, predict a NW to NNW ward motion of Africa in the central-eastern Mediterranean region, which is significantly different from the motion trend suggested by the above Mediterranean constraints. However, this kinematics has been determined under the assumption that Eurasia moves as a unique rigid plate from the North Atlantic ridges to the Pacific trenches, which contrasts with the presence of seismotectonic activity in several zones of Western Europe, from Portugal to the Rhine Graben system (Ahorner, 1975; Buforn et al., 1988; Nicolas et al., 1990; Viti et al., 2001). This last evidence suggests that some parts of Western Europe do not move in close connection with the stable part of Eurasia, the one lying east of the Rhine Graben system (e.g., Biju-Duval et al., 1977; Albarello et al., 1995; Mantovani et al., 1997; Mantovani et al., 2002), hereafter identified as EUR1.

The distribution of recent deformation suggests that at least four microplates, the southern North Atlantic (ATL1), northern North Atlantic (ATL2), Iberia (IBE) and Morocco (MOR), and one extruding orogenic crustal wedge, the Betic-Rif (BERI), may be recognized in this zone (Figure 4). If such plate configuration is adopted, the North Atlantic kinematic indicators (DeMets et al., 1990) lying in the sector of the Mid-Atlantic ridge facing Western Europe and those located along the Azores-Gibraltar belt would not be representative of the North America-EUR1 and Africa-EUR1 relative motions respectively. In order to provide insights into the implications of the above plate mosaic, Albarello et al. (1995) have identified the Africa-EUR1 Eulerian poles that are compatible with the data set used by DeMets et al. (1990) deprived from the above kinematic indicators. This analysis has shown that a number of these poles (for instance the one located at 41.6°N and 11.8°W, with a rotation rate of 0.12°/My) predict a NE to NNE ward motion of Africa in the Mediterranean region (Figure 4). This result is very important, since implies

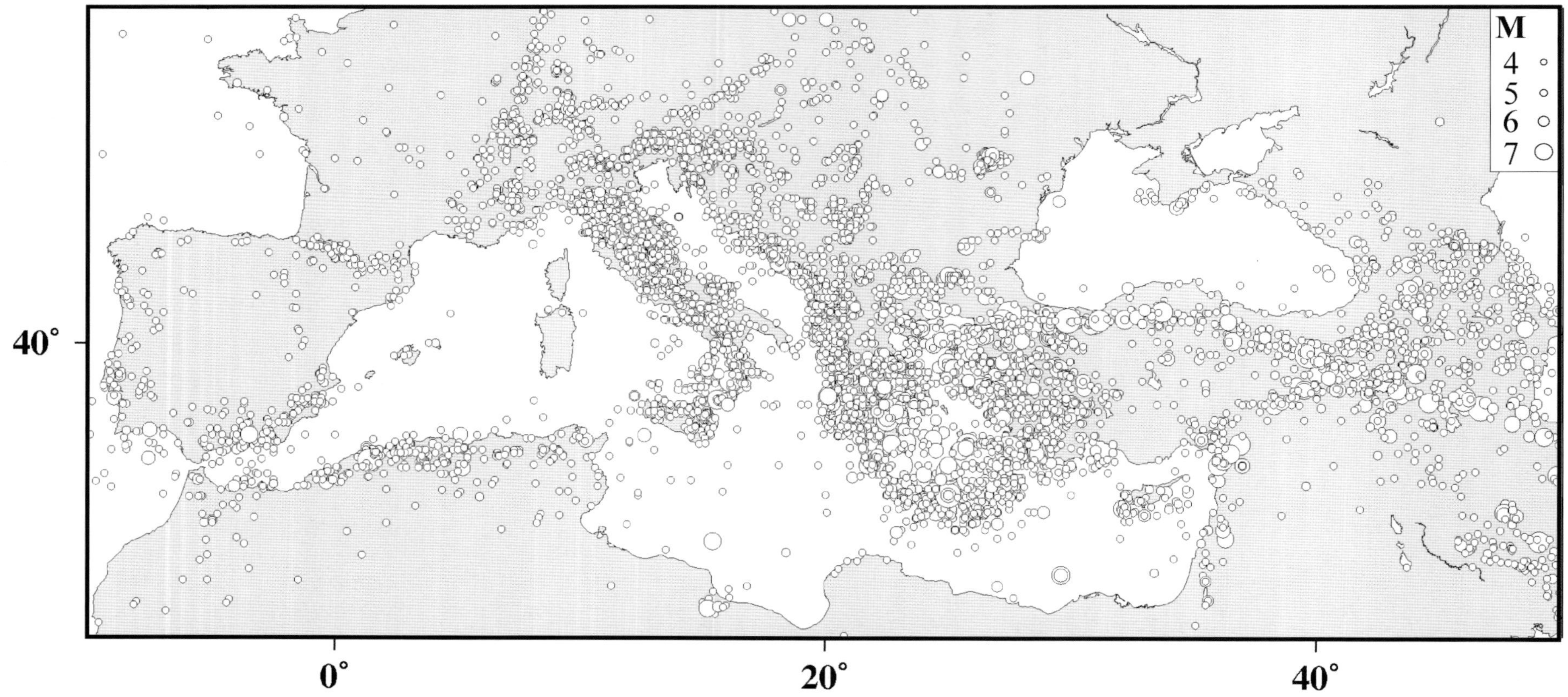

Figure 5: Major earthquakes ($M > 4$) occurred in the Mediterranean area since 1600. Data taken from Karnik (1969), Papazachos and Comninakis (1982), Mezcua and Martinez Solares (1983), Comninakis and Papazachos (1986), Postpischl (1985), D'Amico and Albarello (1999) and references therein, Working Group Parametric Catalog of Italian Earthquakes (1999).

that a linear trajectory of Africa from the Oligocene to the present, not involving the unlikely hook suggested by previous models at 9 My (e.g., Dewey et al., 1989), does not necessarily contrast with the North Atlantic kinematic constraints considered by DeMets et al. (1990). Furthermore, the motion of Africa predicted by the above pole in the central-eastern Mediterranean is consistent with the kinematics of southern Adriatic and with the morphology of the Hellenic and Cyprus arcs.

Albarello et al. (1995) have also determined the relative motion between the Atlantic microplate (ATL1) and EUR1 (Figure 4) compatible with the North Atlantic data facing this block and with those located along the Azores-Gibraltar belt, and that can also satisfy the closure condition of all three-plate circuits involved: North America-Africa-ATL1, Africa-EUR1-ATL1 and North America-EUR1-ATL1. The ATL1-EUR1 Eulerian pole provided by this analysis is located at 49.9°N and 129.6°E, with a rotation rate of 0.04°/My.

The plate motions predicted by the above Africa-EUR1 and ATL1-EUR1 poles are shown in Figure 4. The same figure also proposes a kinematics of the MOR, IBE, BERI and ATL2 blocks compatible with the strain styles recognized at the related plate boundaries. The proposed motion of IBE is slightly different from that of ATL1. The decoupling between these two blocks is accommodated by the system of sinistral transpressional faults recognized in western Iberia (e.g., Philip, 1987; Ribeiro et al., 1990, 1996; Jabaloy et al., 2002). The convergence between IBE and ATL2 is accommodated by seismotectonic activity in the Pyrenean belt (e.g., Nicolas et al., 1990; Grellet et al., 1993; Souriau et al., 2001) and the Bay of Biscay (Alvarez-Marron et al., 1997; Andeweg et al., 1999). The oblique convergence between IBE and Africa is accommodated by the lateral escape of the BERI wedge and by the deformation recognized in the Alboran basin, characterized by NW–SE shortening and left lateral shear (e.g., Philip, 1987; Andeweg and Cloetingh, 2001). The lateral decoupling between IBE and BERI is accommodated by dextral transpression at the Guadalquivir and Cadiz-Alicante fault system (e.g., Weijermars, 1987; Rebai et al., 1992). The decoupling between BERI and Africa is accommodated by sinistral transpression in the Alboran sea (e.g., Philip, 1987; Jacobshagen, 1992; Andeweg and Cloetingh, 2001) and southern Spain (e.g., Jabaloy et al., 2002; Faulkner et al., 2003).

The sinistral transpression observed at the Agadir-Nekor (AN) fault system (e.g., Brede et al., 1992; Mezcua et al., 1992; Hatzfeld et al., 1993; Ait Brahim et al., 2002) accommodates the motion of Africa with respect to MOR (Figure 4).

The convergence between ATL1 and MOR accounts for the thrusting observed at the boundary between these two microplates, mainly evidenced by recent deformation and strong seismic activity in the zone of the Goringe Bank (e.g., Grimison and Chen, 1986; Morel and Meghraoui, 1996).

At the Gloria fault, the relative motion of ATL1 and Africa is obviously compatible with the observed strike-slip movements (e.g., Buforn et al., 1988; DeMets et al., 1990; Viti et al., 2001), since this evidence has been used for the determination of the ATL1 kinematics.

The divergent relative motion between ATL1 and Africa at the westernmost sector of the boundary between these blocks, the Terceira Rift, is compatible with the NE–SW extensional strain observed in this zone (Buforn et al., 1988; Vogt and Jung, 2004).

The occurrence of moderate seismic activity at most of the northern paleo-boundary of the Iberian block, the King's Trough–Azores Biscay Rise–Northern Spain Trench (KT-ABR-NST) (Grimaud et al., 1982; Buforn et al., 1988; Andeweg et al., 1999; Lopez-Casado et al., 2001) suggests that a minor relative motion may occur between ATL1 and ATL2. However, the orientation and amplitude of this possible motion is very poorly defined. In any case, it seems unlikely that such relative motion is larger than 1–2 mm/y. This implies that ATL2 moves at a rate of at least 2–3 mm/y, which involves a relative motion of 2–3 mm/y between ATL2 and EUR1. The interaction between these two blocks involves a large region, extending from the Rhine graben system to the Biscay Bay. The occurrence of intraplate deformation in this zone is indicated by neotectonic and seismotectonic activity (e.g., Nicolas et al., 1990; Rebai et al., 1992; Grellet et al., 1993). In my opinion, to reliably reconstruct the deformation pattern of this zone it would be necessary taking also into account the possible effects of the roughly N to NW ward indentation of the Adriatic promontory against central Europe. The NW ward push of this indenter could help explaining the complex and heterogeneous deformation pattern observed in the Rhine Graben system and central Europe (e.g., Laubscher, 2001; Michon et al., 2003) and Western Alps (e.g., Eva et al., 1997; Sue et al., 1999).

The present kinematic pattern of the western Mediterranea region (Figure 4) started developing in the Miocene, as an effect of the oblique collision between the ATL1-IBE microplate, when it still was a unique block, and Africa.

The convergence of these two blocks has been accommodated by the deformation of their colliding edges, that is Iberia and Morocco. Under the ENE ward push of the westernmost African edge (Morocco), IBE decoupled from ATL1 through the system of sinistral transpressional faults in Portugal, assuming a more eastward motion trend with respect to ATL1 (Figure 4). Contemporaneously, the collision of the Moroccan edge of Africa with the continental Iberian block caused a considerable slowdown of MOR and its consequent decoupling from mainland Africa.

The Africa–Eurasia convergence predicted by the NUVEL-1 model (Figure 4) in the western Mediterranean region is compatible with the SE–NW shortening trend observed at the IBE-Africa collision zone, but cannot easily account for the observed left lateral shear, since most of the fault systems identified in this zone have a strike more or less perpendicular to the direction of such plate convergence. In the framework of the kinematic scheme predicted by the NUVEL-1 model the sinistral transpressional activity at the AN fault is generally interpreted as a lateral guide for the SW ward extrusion of the Morocco–Gibraltar crustal wedge (Rebai et al., 1992; Royden, 1993; Andeweg and Cloetingh, 2001). However, if this wedge were really extruding SW ward compressional deformation would mark its outer front (wherever it may be located in the Atlantic area), which, instead, is not recognized. Most likely, the above interpretations have not recognized the different kinematics of MOR and BERI.

Geodetic Constraints. Independent information on the kinematics of Africa could be derived by the presently available space geodetic measurements in this plate (Table 1 and Figure 6). Some attempts in this direction have been made by a number of authors (Sella et al., 2002; Nocquet and Calais, 2003; McClusky et al., 2003). However, the uncertainty that might affect the results of these preliminary evaluations can hardly be estimated. The main problem is that few permanent stations are currently operating in Africa (e.g., Altamimi et al., 2002), and that most of them are located near the active deforming boundaries of this plate (Figure 6), as recognized by Sella et al. (2002). The station of Lampedusa (LAMP) is located in the Sicily Channel tectonic belt where geological, geophysical and volcanological evidence clearly indicates the occurrence of recent/present trans-tensional activity (Finetti and Del Ben, 1986; Argnani, 1993; Reuther, 1987). The station of Mas Palomas (MASP) is located in the Canary Islands, where recent/present deformation is suggested by the occurrence of volcanic and seismic activity (e.g., Mezcua et al., 1992; Anguita and Hernàn, 2000). The station of Malindi (MALI) could belong to the Somalia plate rather than to Africa (e.g., Sella et al., 2002). The station of the Gough Island (GOUG) lies very close to the south Atlantic spreading ridge. The stations lying in the western (NKLG, YKRO) and southern (SUTH, HART) parts of Africa have been excluded from the data set used for the determination of the ITRF2000 global solutions since they do not satisfy the quality criteria adopted for site selection (Altamimi et al., 2002). Moreover, tectonic, seismic and volcanic activity is recognized near the Gulf of Guinea (Foster and Jackson, 1998; Meyers et al., 1998), where NKLG is located.

Even if one assumes that the problems cited above can be ignored, the use of geodetic data for constraining the motion of Africa is discouraged by the fact that the properties of the Africa–Eurasia poles that can be derived

Table 1

North (u) and east (v) components of GPS velocities for 8 African sites (ITRF2000 solution available at http://lareg.ensg.ign.fr/) and MATE (southern Italy). See text for explanation of abbreviations. The related standard deviations are in brackets. The values reported are residual velocities with respect to the Eurasia absolute rotation vector (57.96°N, 99.37°W, $\omega = 0.26$ °/My) given by Altamimi et al. (2002)

Station			ITRF2000	
Name	Lat (°)	Lon (°)	u (mm/y)	v (mm/y)
MATE	40.65	16.70	4.31 (0.36)	0.68 (0.13)
LAMP	35.50	12.60	2.64 (1.46)	−2.17 (0.94)
MASP	27.76	−15.63	1.43 (0.54)	−4.28 (0.34)
YKRO	6.83	−5.24	5.12 (1.10)	0.35 (1.10)
NKLG	0.35	9.67	3.23 (1.30)	−9.66 (2.39)
MALI	−2.97	40.19	7.19 (0.40)	−2.83 (1.34)
HART	−25.88	27.70	5.63 (0.41)	0.03 (0.35)
SUTH	−32.38	20.81	5.66 (1.69)	−0.23 (1.36)
GOUG	−40.34	−9.88	3.18 (3.59)	1.41 (1.89)

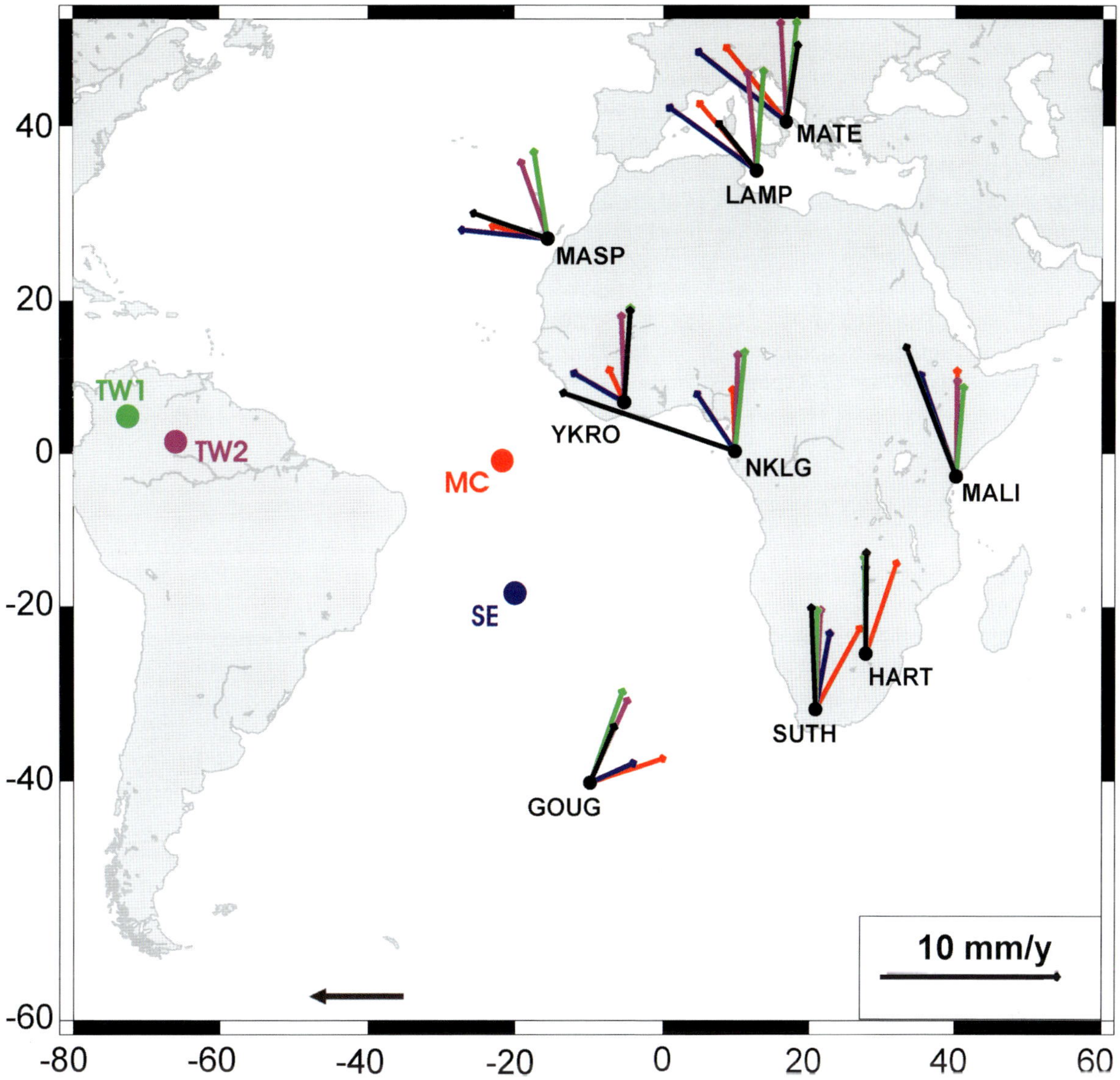

Figure 6: GPS permanent stations (black abbreviations, see text for full names) presently operating in the African plate. Coloured dots with abbreviations (MC, SE, TW1, TW2) identify geodetic Africa–Eurasia Eulerian poles. Information on poles and the respective sets of stations considered is given in Table 2. Coloured arrows in station sites indicate the velocities (see scale) predicted by the respective poles. Black arrows indicate the residual ITRF2000 geodetic velocities (Table 1) with respect to the absolute Eurasia rotation vector (Altamimi et al., 2002). See caption of Tables 1 and 2 for explanations.

Table 2

Africa–Eurasia Eulerian poles (latitude, longitude and angular velocity) taken from literature (SE, MC) and determined in this work (TW1, TW2). The pole of McClusky et al. (2003) has been also constrained by three non-permanent GPS stations in Egypt. For each pole, the stations considered and the misfit value per degree of freedom (χ^2/D) are reported, see Sella et al. (2002) for explanations. The location of poles is shown in Figure 6

Geodetic stations		Rotation vector				
		Lat (°)	Lon (°)	ω (°/My)	D	χ^2/D
SE	(Sella et al., 2002) (GOUG, SUTH, HART, MASP)	−18.2	−20.0	0.062	7	0.8
MC	(McClusky et al., 2003) (GOUG, SUTH, HART, MASP) + Egypt	−0.9	−21.8	0.060	11	0.3
TW1	This work (GOUG, SUTH, HART)	1.6	−66.0	0.050	3	0.2
TW2	This work (GOUG, SUTH, HART, YKRO)	5	−72.5	0.051	5	0.1

from these data (Table 2 and Figure 6) strongly depend on the set of stations used for computations. For instance, if the set of stations considered by Sella et al. (2002) is slightly modified by excluding MASP or substituting it with YKRO, the properties of the related rotation vector (poles TW1 and TW2 in Table 2 and Figure 6) undergo a considerable change, with a significant decrease of the misfit between model predictions and observations, here defined by the parameter χ^2/D (Table 2). The term χ^2 is the total weighted error (e.g., DeMets et al., 1990), which depends on the differences between predicted and observed velocities and on the uncertainty affecting geodetic data. D is the number of degrees of freedom of the kinematic model that, for a two-plate problem, such as the one here considered, is $D = 2N - 3$, N being the number of geodetic stations used to constrain the Eulerian pole (e.g., Nocquet et al., 2001).

It could be worth noting that the vectors predicted by the poles TW1 and TW2 in the southern Adriatic are compatible with the observed geodetic velocity of MATE, which would support the hypothesis that relative motion between Africa and Adriatic is negligible. However, as also recognized by Altamimi et al. (2002), I believe that any indication on the Africa kinematics provided by the presently available geodetic data should be taken with great caution.

4. EVOLUTIONARY RECONSTRUCTION

The proposed evolution is illustrated by a series of paleogeographic maps from the early Oligocene to the present (Figures 1 and 2). With respect to previous attempts (Mantovani et al., 1996, 1997), evolutionary reconstruction has been extended to the western Mediterranean region and to a longer time interval. Clearer explanations are also tentatively provided on how the proposed locations in space and succession in time of major tectonic events may be reconciled with the tectonophysical concepts described in Section 2.

The proposed scheme only aims at recognizing the large and regional scale evolutionary features. This may provide a basis for trying a reconstruction of local deformation patterns.

4.1. Initial Tectonic Setting (Early Oligocene)

Figure 1(a) shows a tentative reconstruction of the tectonic setting in the early Oligocene, mainly based on previous paleogeographic reconstructions of this evolutionary stage (Burchfiel, 1980; Cohen, 1980; Rehault et al., 1984; Burtman, 1986; Dercourt et al., 1986; Royden and Baldi, 1988; Finetti et al., 1988, 2001).

The African and Eurasian continental plates were separated by a rather heterogeneous domain, encompassing an inner continental core (the Adriatic plate), relatively large oceanic domains with low buoyancy lithosphere, such as the western Apulian, the Ionian and Levantine zones (e.g., Finetti et al., 1996, 2001), and a long orogenic system built up by the consumption of a large part of the Neotethys ocean and adjacent plate margins from the Cretaceous to the Oligocene (e.g., Dercourt et al., 1986). This last system, hereafter cited as the Tethyan belt, was particularly well developed in the eastern Mediterranean area, where it was formed by three parallel chains: a northern one with European affinity (Carpathians, Balkanides and Pontides), an inner one, consisting of oceanic remnants, metamorphic zones and crystalline massifs (Pelagonian–Aegean–Anatolian), and a southern one with African affinity (Dinarides, Hellenides and Taurides), as suggested by a number of authors (e.g., Brunn, 1976; Biju-Duval et al., 1977; Boccaletti and Dainelli, 1982; Aubouin, 1984; Burtman, 1986; Royden and Burchfiel, 1989). The external accretionary chains were most probably characterized by mechanical/rheological properties different from those of the inner metamorphic belt, due to different lithological compositions and thermobaric histories. In particular, it is reasonable to suppose that the accretionary belts, being colder than the inner massifs, were characterized by higher rigidity. This hypothesis could explain the different responses to oroclinal deformation that the two kinds of structure presented during their evolutionary history (e.g., Sengor, 1993).

A low buoyancy zone was also present in the Eurasian domain (e.g., Royden and Baldi, 1988), north of the Tethyan belt (Figure 1(a)).

4.2. From the Oligocene to the Middle Miocene

Western Mediterranean. Most reconstructions of the post-Eocene evolution of the western Mediterranean region (e.g., Rehault et al., 1984, 1987; Dercourt et al., 1986; Vigliotti and Laugenheim, 1995; Doglioni et al., 1999; Finetti et al., 2001; Lopez-Casado et al., 2001) agree that, around the late Oligocene, the central sector of the

Iberian belt, along with a fragment of the European foreland, the Sardinia–Corsica microplate, detached from western Europe (Figure 1(b)) and then underwent considerable E to SE ward migration and bending to reach the configuration shown in Figure 1(c). In between the migrating arc (Balearic arc in Figure 1(b)) and the European margin extensional tectonics occurred, with the formation of the Balearic basin.

Here, it is argued that the development of this trench-arc-back arc system was an effect of the convergence between Africa and Europe. This process started when the African continental margin collided with the southernmost edge of the Iberian belt (Figure 1 (a) and (b)), inducing a longitudinal, roughly SW–NE, compressional regime in this belt (Le Pichon et al., 1988; Sengor, 1993; Gelabert et al., 2002). Due to the high buoyancy of that accretionary body, the most convenient shortening pattern was presumably the lateral expulsion of crustal wedges at the expense of the adjacent low buoyancy oceanic domain, the western Apulian zone. Quantitative arguments in support of this presumed convenience, based on energy budget estimates in analogous tectonic contexts, are reported by Mantovani et al. (2000a). This extrusion mechanism led to the progressive outward migration and bending of the arc, with the consequent development of extensional tectonics in the Balearic back arc basin.

The halt to this process was imposed by the collision of the southern segment of the migrating arc, oriented roughly E–W, with the northern continental margin of Africa (Algeria–Tunisia zone). The opening of the Balearic basin came to its definitive end around the Middle Miocene, when the Sardinia–Corsica microplate ceased to migrate. At that final stage, the northern segment of the Balearic arc, comprising the Sardinia–Corsica fragment, remnants of the Alpine belt and the Oligo-Miocene Apenninic accretionary chain, was oriented roughly S–N (Figure 1(c)).

The proposed geodynamic interpretation provides plausible explanations for the major aspects of the observed deformation pattern in the western Mediterranean area.

- The structural/tectonic setting of the Balearic zone, reconstructed from geophysical investigations (e.g., Rehault et al., 1984), is characterized by a number of wedges decoupled by strike-slip faults (Figure 1 (b) and (c)), with a general geometry very similar to that predicted by an extrusion mechanism (Figure 3).
- The strong bending that the Balearic arc underwent during its migration, changing from a more or less straight configuration (Figure 1(a)) to its final shape, characterized by two almost perpendicular sectors (Figure 1(c)), is consistent with a SW–NE oriented driving force, like the one induced by the convergence between Africa and Europe during this phase (e.g., Dercourt et al., 1986; Gelabert et al., 2002).
- A sinistral transpressional deformation is recognized in the Iberian belt during the collisional phase which preceded its detachment from western Europe (Marroni and Treves, 1998; Finetti et al., 2001). This kind of strain regime is consistent with the oblique Africa–Europe convergence. The agreement between predictions and observed features is also confirmed by the fact that the successive extensional phase which affected this plate border was associated with a sinistral transcurrency (Finetti et al., 2001).
- Another significant feature which fits the proposed interpretation is the strong curvature that the arc shows in the sector running from the Western Alps to Corsica (Figure 1(c)). To explain such a distortion, one must presume a roughly north to NE ward displacement of the N–S segment of this arc, which is compatible with the evidence mentioned in the previous point and with the push of Africa implied by the proposed reconstruction.
- On the basis of geological evidence, a number of authors have suggested that the consuming process which built up the Cretaceous-Eocene Iberian belt involved the southward subduction of the Neotethys oceanic lithosphere under the Apulian oceanic domain (Boccaletti et al., 1980; Elter and Marroni, 1991; Doglioni et al., 1999; Finetti et al., 2001). This interpretation implies that the commencement of extension in the Balearic basin, coinciding with an inversion of the subduction direction at that plate border, can hardly be attributed to subduction-related forces, since the new embrional northward subducting lithosphere was not yet developed enough for inducing an appreciable slab-pull force.
- The westernmost sector of the Iberian belt, the Betic-Rif system, being squeezed between Africa and Iberia underwent, since the early Miocene, a westward escape at the expense of the adjacent Atlantic oceanic domain. In the wake of the extruding wedge extensional deformation occurred leading to the opening of the Alboran basin (e.g., Dercourt et al., 1986; Royden, 1993, 1993; Lopez-Casado et al., 2001).

The push of Africa, after its collision with the Iberian belt in the early Oligocene, could be the driving force for the generation of a major transtensional fracture, the Rhine graben system, inside the European foreland. This Oligocenic event, clearly documented by geological evidence (e.g., Le Pichon et al., 1988; Sissingh, 1998; Laubscher, 2001; Michon et al., 2003), could be an effect of the proposed mechanism, as sketched in Figure 7. After its collision with Africa, the Iberian belt transmitted a roughly NE ward push to central Europe. Simultaneously, the incipient outward bending/extrusion that this belt was undergoing under the belt-parallel compression exerted a roughly SE ward traction on Western Europe (France and Iberia). In this con-

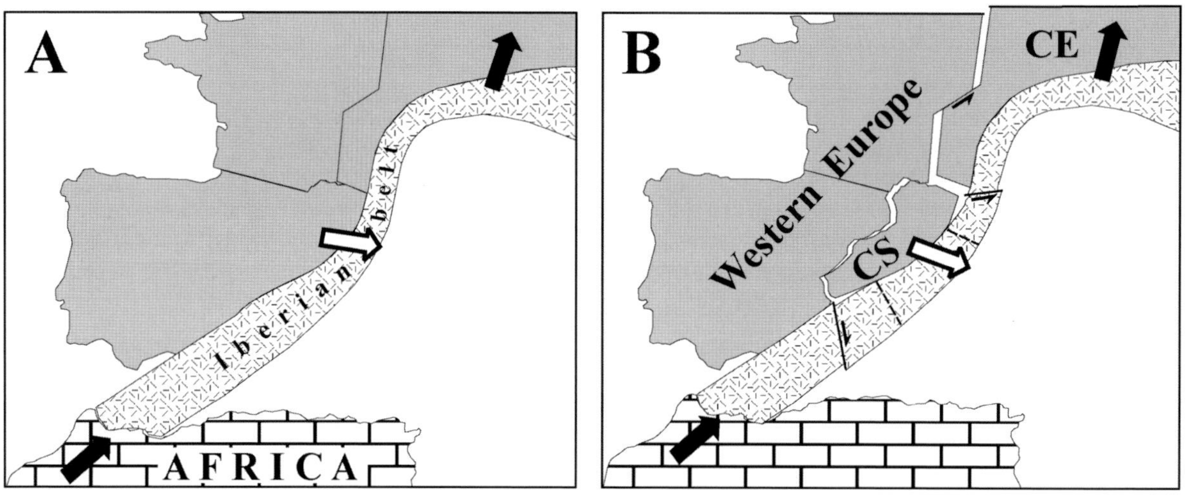

Figure 7: Sketch of the proposed genetic mechanism for the Rhine Graben system. (A) The NE ward push of Africa (black arrow) is transmitted to Central Europe (CE) by the Iberian belt. Contemporaneously, the tendency of this belt to extrude outward (under a belt-parallel compression) exerts a roughly east ward pull (empty arrow) on Western Europe. (B) The dynamic context sketched in (A) causes a sinistral transtensional tectonics in the Rhine graben system and then the detachment of the Corsica–Sardinia fragment (CS) from Western Europe. See text for comments.

text, central Europe and Western Europe, respectively pushed roughly NE ward and pulled roughly SE ward by the Iberian belt, started diverging in a oblique way, inducing the sinistral transtensional tectonics evidenced by geological data (e.g., Schreiber and Rotsch, 1998). One can reasonably expect that the effects of this driving mechanism have considerably attenuated after the detachment of the Iberian arc (in particular of the Sardinia–Corsica fragment) from Western Europe, since this event implied the end of the SE ward traction of the arc on Western Europe. Another reason for this driving mechanism to attenuate is that the progressive outward bending of the Iberian arc made the transmission of the Africa's push against Central Europe less and less efficient. This interpretation can explain why transtensional tectonics in the Rhine graben system has lasted for a relatively short geological time interval, up to the Upper Oligocene (e.g., Le Pichon et al., 1988; Laubscher, 2001).

Eastern Mediterranean. The evolutionary reconstructions so far proposed for the Carpatho–Balkan–Pannonian system (e.g., Burchfiel, 1980; Burtman, 1986; Royden and Burchfiel, 1989; Fodor et al., 1998) provide a progressive outward migration and bowing of the Carpathian arc, at the expense of a low buoyancy sector of the European foreland, through a deformation pattern similar to that shown in Figure 3. In the internal part of this migrating arc, transtensional tectonics occurred, generating the Pannonian back arc basin. During the Oligocene–early Miocene phase, activation of major strike-slip fault systems allowed the E to NE ward displacement of orogenic wedges in the northern part of the Pannonian area, while in the successive evolution, this kind of tectonic activity mainly occurred in the southern Pannonian region (e.g., Balla, 1986; Royden and Burchfiel, 1989; Royden, 1993, 1993; Fodor et al., 1998; Hippolyte et al., 1999).

During this phase, extensional activity is also recognized in the central-northern Aegean and northwestern Anatolian regions, as clearly indicated by geological and volcanological evidence, simultaneously with accretionary activity along the Hellenic trench (e.g., Fytikas et al., 1984; Mercier et al., 1989; Papadopoulos, 1989).

The migration and distortion of the Anatolian–Aegean–Balkan–Pannonian–Carpathian system is here interpreted as a side effect of the indentation of the Arabian promontory against the orogenic system which lay in front of it. It is largely accepted that this indentation caused the lateral escape of Anatolia (e.g., McKenzie, 1972; Dewey and Sengor, 1979). However, this extrusion process did not simply involve the Anatolian wedge, it caused displacement and lateral bending of the sector of the Tethyan belt running from eastern Anatolia to the Carpathians, as tentatively reconstructed in Figure 1 (b) and (c). This hypothesis is suggested by the fact that, in spite of its considerable deformation, the inner metamorphic core of this belt has maintained a fairly good continuity up to the present (Figure 2(c)). The overall deformation pattern of the Tethyan belt was strongly influenced by the location and lateral extension of the low buoyancy lithospheric domains in the surrounding zones, which "attracted" the extruding buoyant wedges, and by the different horizontal flexures of the inner massifs and the external accretionary chains, which led to their separation in the sectors of strongest curvature.

In the first phase of extrusion (Figure 2(b)), the lateral escape of eastern Anatolia was oriented roughly NW ward, guided by a system of dextral decoupling faults (Dercourt et al., 1986; Hempton, 1987; Finetti et al., 1988). This displacement was accommodated by the roughly NE ward lateral escape of crustal wedges in the Carpatho–Pannonian region, at the expense of a low buoyancy sector of the European foreland, and simultaneously by the SW ward bending/extrusion of the Aegean arc, at the expense of the Ionian–Levantine low buoyancy domain. In the wake of the extruding/bending Carpathian arc, transtensional tectonics occurred in the Pannonian area, as effect of pull-apart mechanisms or relative rotation of blocks (e.g., Hippolyte et al., 1999; Royden and Baldi, 1988).

The proposed deformation pattern in the Balkan area (Figure 1 (b) and (c)) involves a higher NW ward migration rate for the external belt with respect to the inner massifs, implying a left lateral decoupling between the Balkanides and the Pelagonian massifs (Figure 1 (b)). The imprints of this transcurrent decoupling might be represented by the sinistral shear recognized in the Vardar zone for the period involved (e.g., Brunn, 1960; Burtman, 1986; Zeilinga de Boer, 1989).

The progressive collision of various sectors of the Carpathian arc with the continental European margin caused the slowdown/end of trench-arc migration in this region and, consequently, of back arc extension in the northern Pannonian basin.

In the Aegean area, the hypothesized SW ward bending of the Aegean arc and the adjacent Hellenides belt, at the expense of the Ionian and Levantine oceanic zones, is consistent with the occurrence of accretionary activity at the Hellenic trench (Mercier et al., 1989) and of extensional tectonics in the northern Aegean and north-western Anatolian zones. This last deformation is interpreted as an effect of the local divergence between the Aegean massifs and the Balkanides belt (Figure 1 (b) and (c)).

4.3. From the Middle to the Late Miocene

Central Mediterranean. Around the Middle-Upper Miocene, an important tectonic event, the activation of the Giudicarie transpressional fault system (e.g. Semenza, 1974; Castellarin et al., 1992), caused a significant change in the tectonic setting of the central Mediterranean area. This event allowed the decoupling of the Adriatic promontory from its north western Padanian protuberance, which, after a long history of thrusting, was stuck into the Eurasian domain in the Western Alps. After this decoupling, the Adriatic underwent roughly NE ward accelerated motion, releasing the deformation accumulated during the long collisional phase of this promontory with the European plate. The occurrence of the Giudicarie decoupling might have been favoured by the previous eastward extrusion of buoyant crustal wedges (Figure 1(c)) from the Eastern Alps towards the northernmost Carpathian region (e.g. Ratschbacher et al., 1991).

The Giudicarie decoupling can account for the deformation pattern which occurred in the peri-Adriatic zones after that event. It explains why thrusting ceased or slowed down considerably in the southern Western Alps and only continued in the Alpine sector lying east of the decoupling Giudicarie fault system, i.e. the Eastern Alps (e.g., Castellarin and Vai, 1986). It also accounts for the location and timing of the extensional tectonics occurred in the northern Tyrrhenian zone (Figure 2(a)) from the Middle to the late Miocene (e.g., Sartori, 1990; Finetti and Del Ben, 1986; Sartori and Capozzi, 1998; Finetti et al., 2001). In fact, the post-Giudicarie kinematics of the Adriatic predicts crustal stretching in the Tyrrhenian zone just in the above time interval, during which the Adriatic promontory diverged from the Corsica–Sardinia microplate (fixed since the Middle Miocene), starting from the Giudicarie decoupling and ending with the collision of the Adriatic continental domain against the Aegean–Balkan system in the late Messinian (as discussed in the next section). As regards the location of crustal extension, the proposed interpretation predicts that such kind of deformation must only occur in the zone lying in between the migrating Adriatic and the Sardinia–Corsica microplate, that is the Tyrrhenian region located north of the Selli fault system, where extension has effectively occurred (Figure 2(a)).

The above reconstruction also explains why accretionary activity in the Northern Apennines has experienced a period of relative quiescence from the Middle-Upper to the late Miocene (e.g., Sartori, 1990). In fact, during the period of divergence between the Adriatic and the Corsica–Sardinia microplate no lithosphere consumption has occurred beneath the Northern Apennines and thus no accretionary activity is expected along that trench zone.

Eastern Mediterranean. Outward extrusion of crustal wedges still develops in the southern Carpathian arc (Figure 2(a)), as evidenced by geological and volcanological observations in the external compressional front and in the internal Pannonian back arc zone (e.g., Royden and Burchfiel, 1989). This deformation is interpreted as an effect of the NW ward migration of the Anatolian–Aegean–Pelagonian system. Arc-trench migration is no longer active in the northern Carpathians, due to the collision of this belt with the European continental margin.

In front of the Arabian indenter, the Taurides belt, along with some metamorphic fragments, extrudes faster than eastern Anatolia. This mechanism causes the bending of the Cyprus arc and its consequent detachment from the Anatolian block (Figure 2 (a) and (b)). This interpretation can explain the timing of thrusting activity at the outer front of the arc and of extensional deformation in the internal zone (e.g., Kempler and Ben Avraham, 1987; Kempler and Garfunkel, 1994).

During this phase, the Africa–Adriatic plate and the Anatolian–Aegean–Balkan system converged at the expense of the last remnant of low buoyancy oceanic lithosphere, the pre-Apulian zone. The map in Figure 2(a) shows the tectonic/kinematic setting at the end of this evolutionary phase, just before the late Miocene suture of the pre-Apulian consuming boundary (Mercier et al., 1989), the consequences of which are illustrated in Figure 2(b) and discussed in the next section.

4.4. From the Late Miocene to the Late Pliocene

Around the late Miocene, a profound reorganization of the tectonic setting took place in the central and eastern Mediterranean areas, as indicated by the simultaneous occurrence of several tectonic events (Figure 2(b)). A renewal of orogenic activity is recognized throughout the Apenninic belt, with particular in the southern arc (e.g., Di Nocera et al., 1976; Ortolani, 1979; Castellarin and Vai, 1986; Patacca and Scandone, 1989). Extensional deformation ceased in the northwestern Tyrrhenian and crustal stretching began in the central Tyrrhenian, with the formation of the Magnaghi–Vavilov basin (e.g., Sartori, 1990; Sartori and Capozzi, 1998). Transtensional tectonics occurred in the Sicily channel, with the formation of some troughs (e.g., Finetti and Del Ben, 1986; Reuther, 1987; Argnani, 1993). Thrusting took place in the Maghrebian belt lying north of the Iblean–Ventura block (Grasso et al., 1992; Catalano et al., 1994; Finetti, 2004). Since the late Messinian, a major transtensional fault system (the Medina and Victor Hensen faults, Figure 2(b)) developed in the central Ionian area (Ryan, 1978; Della Vedova and Pellis, 1989; Hieke and Wanninger, 1985; Hieke and Dehghani, 1999). Around the late Messinian, a major left lateral shear zone, the Schio-Vicenza fault, formed in the northern Adriatic foreland and the orientation of shortening in the eastern Southern Alps passed from roughly NE ward to NW ward (e.g., Castellarin and Vai, 1986; Cantelli and Castellarin, 1994). A reactivation of vertical movements occurred since the late Miocene at the Syracuse and Apulian escarpments, the two transition zones from the Ionian oceanic lithosphere to the continental lithosphere of the Iblean plateau, on the western side, and of the Adriatic plate, on the eastern side (Carbone et al., 1982; Finetti, 1982; Auroux et al., 1984; Finetti and Del Ben, 1986).

In the eastern Mediterranean, dextral transcurrent activity developed in the western segment of the North Anatolian fault system (e.g., Hempton, 1987; Barka, 1992; Westaway, 1994). Roughly N–S extensional tectonics occurred in western Anatolia and northern Aeged (e.g., Eyidogan, 1988; Zanchi et al., 1990). E–W compressional deformation affected the central Aegean zone, i.e. the Cyclades massifs (Mercier et al., 1987; Avigad et al., 2001). Accretionary activity accelerated at the Hellenic trench zone and extensional tectonics occurred in the southwestern Aegean zone, with the formation of the western Cretan basin (Dewey and Sengor, 1979; Mercier et al., 1989; Yilmaz, 1989; Taymaz et al., 1991; Meulenkamp et al., 1994).

In the following, it is argued that the above tectonic events, in both the central and eastern Mediterranean regions, can be coherently explained as effects of a strong increase in resisting forces at the pre-Apulian subduction zone, due to the complete consumption of the low buoyancy oceanic lithosphere which lay between the Adriatic and the Aegean continental domains. Geological data suggest that this border (western Greece) almost sutured, around the late Miocene, after considerable crustal thickening and uplift in that collision zone (e.g., Stiros, 1988; Mercier et al., 1989; Underhill, 1989). Since the occurrence of this suture, the convergence of the confining plates, which was previously accommodated by the consumption of the pre-Apulian domain, was taken up by new tectonic processes or by an acceleration of already active shortening mechanisms, both in the central and eastern Mediterranean areas, as discussed in the following.

Central Mediterranean. When the continental Adriatic domain collided with the Tethyan belt (Figure 2(a)), a relatively extended low buoyancy lithospheric domain was left west of the Adriatic continental domain, the western Apulian and Ionian zones (e.g., Serri et al., 1993; Beccaluva et al., 1994; Francalanci and Manetti, 1994; Finetti et al., 1996, 2001; Babbucci et al., 1997, and references therein). In this context, the minimum-work principle implies that the consumption of those zones should be the main focus of the kinematic/tectonic reorganization that happened in response to the suture of the Adriatic-Aegean border. This consumption actually occurred, as testified by the large amount of accretionary material that accumulated at the Apenninic and Calabrian consuming boundaries during the Pliocene (e.g., Patacca et al., 1990; Ortolani et al., 1992). The key tectonic event which determined such consumption was the clockwise rotation of the Adriatic (Figure 2(b)), driven by the roughly westward push of the Tethyan belt. This rotation began around the late Messinian, when the generation of major fractures allowed the decoupling of the Adriatic block from its northwestern (Padanian) protuberance and from Africa. The decoupling of the northern part of the Adriatic plate from its Padanian protuberance was accommodated by the Schio-Vicenza fault system (Figure 2(b)). This hypothesis is suggested by the occurrence of a sinistral shear along this fault since the late Miocene (Castellarin and Vai, 1986; Cantelli and Castellarin, 1994) and, coherently, by the change from NE ward to NW ward in the orientation of the compressional axis in the eastern Southern Alps (Castellarin et al., 1992; Cantelli and Castellarin, 1994).

Decoupling of the Adriatic plate from Africa was accommodated by sinistral shear at a major fracture zone, the Victor Hensen, Medina and Sicily channel fault system (Figure 2(b)). This hypothesis is suggested by the fact that sinistral transtensional activity is recognized at the above fault sectors starting from the late Messinian (Ryan, 1978; Finetti, 1982; Hieke and Wanninger, 1985; Finetti and Del Ben, 1986; Della Vedova and Pellis, 1989; Reuther, 1987; Argnani, 1993; Hieke and Dehghani, 1999).

The clockwise rotation of the Adriatic plate, connected with the northern Ionian domain, caused a constrictional regime in the central Mediterranean area, which was accommodated by the consumption of the remnant low buoyancy lithosphere in that region. However, to achieve a satisfactory explanation of the time–space distribution of deformation in the Tyrrhenian–Apennines region and surroundings it is necessary to invoke the occurrence of extrusion processes in the central Mediterranean area.

The roughly E–W convergence between the Adriatic–Ionian plate and the African continental domain (Tunisia) was accommodated by the lateral (NW ward) expulsion of an African fragment, the Iblean–Ventura microplate, guided by two lateral strike-slip faults (Figure 2(b)). Along the southern of these guides, the Sicily Channel, characterized by a leaky transform geometry, pull-apart extensional tectonics developed, with the formation of some troughs (e.g., Finetti, 1982; Finetti and Del Ben, 1986; Reuther, 1987; Argnani, 1993).

In its turn, the roughly NW ward indentation of the extruding Iblean–Ventura wedge into the Alpine–Apenninic orogenic body lying south of the Selli fault, caused the eastward escape of crustal wedges, at the expense of the remnant low buoyancy lithosphere of the western Apulian and Ionian zones (Figure 2(b)). This latter consuming process was responsible for the renewal of orogenic activity in the Apenninic belt, with particular regard to the Southern Apennines and Calabrian arc, after the late Miocene (e.g., Patacca et al., 1990; Ortolani et al., 1992). In the wake of the above eastward extruding wedges, crustal stretching occurred in the central Tyrrhenian area, with the formation of the Magnaghi–Vavilov basin. This interpretation accounts for the E–W trend of the above extensional phase, for its timing (from early Pliocene to early Pleistocene, as discussed in the next section) and for the peculiar shape of the stretched zone.

Under the load of the extruding crustal wedges, the northern Ionian lithosphere underwent a downward flexure. This deformation accounts for the late Miocene–early Pliocene reactivation of relative vertical movements along the two transition zones (the Syracuse and Apulian escarpments, Figure 2(b)) between the Ionian oceanic domain and the adjacent continental domains.

A more detailed tentative reconstruction of the configuration of extruding crustal wedges and of the resulting back arc extension in the central-southern Tyrrhenian since the late Miocene is illustrated in Figure 8. The position and geometry of the lateral transcurrent guides of the extruding wedges has been deduced from the fault systems identified by deep seismic soundings (Finetti and Del Ben, 1986). The proposed dynamic boundary conditions (Figures 2(c) and 8) imply that the Pliocenic orogenic phase in the southern Apenninic arc was connected to a consuming process driven by both the westward drift of the Adriatic plate (pushed by the Anatolian–Aegean–Balkan system) and by the eastward extrusion of Alpine-Apenninic crustal wedges (stressed by the NW ward indentation of the Iblean microplate). This double driving mechanism can explain the relatively high rate of trench

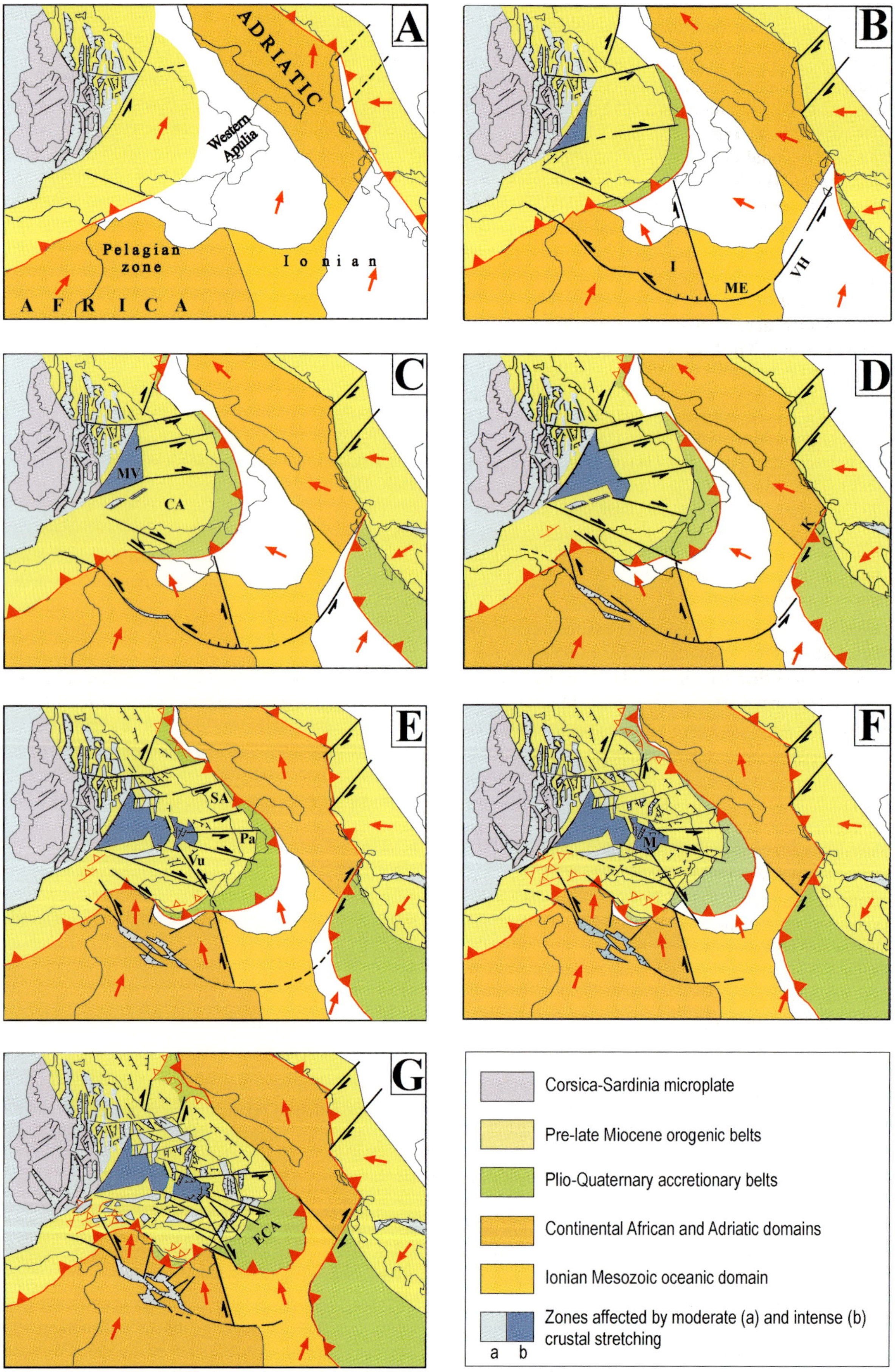
A
ADRIATIC
Western Apulia
Pelagian zone
Ionian
AFRICA
B
I
ME
VH
C
MV
CA
D
K
E
SA
Pa
Vu
F
M
G
ECA
Corsica-Sardinia microplate
Pre-late Miocene orogenic belts
Plio-Quaternary accretionary belts
Continental African and Adriatic domains
Ionian Mesozoic oceanic domain
Zones affected by moderate (a) and intense (b) crustal stretching
a b

Figure 8: Tentative reconstruction of the configuration of extruding wedges in the Tyrrhenian-Apennines trench-arc-back arc system since the late Miocene, taking into account the information on major faults provided by deep seismic soundings (Finetti and Del Ben, 1986). Symbols as in Figures 1 and 2. (A) Upper Miocene tectonic setting, preceding the Adriatic-Aegea continental collision (see Figure 2(a)). (B)–(D) Late Miocene to Upper Pliocene: the Adriatic plate, pushed by the Anatolian–Aegean system, undergoes a clockwise rotation (see Figure 2(b)). The decoupling of the Adriatic–Ionian plate from Africa, is accommodated by transtensional deformation at the Victor–Hensen (VH) and Medina (ME) faults, in the central Ionian area, and by the NW ward expulsion of an African fragment, the Iblean–Ventura (I) block. The indentation of this continental fragment into the orogenic body lying in front causes the eastward escape of crustal wedges at the expense of the western Apulian and Ionian low buoyancy zones. This extrusion triggers a renewal of orogenic activity in the Apenninic belt and produces crustal stretching in the Magnaghi–Vavilov (MV) basin. (E)–(G) Late Pliocene to Present: the collision between the Southern Apennines (SA) and the Adriatic continental domain causes a halt to the escape of crustal wedges in the central Tyrrhenian and, consequently, accretionary activity in the Southern Apennines and extensional tectonics in the MV basin come to an end. The lateral escape of the Calabrian arc (CA), guided by the Vulcano (Vu) and Palinuro (Pa) transpressional faults, accelerates. Thrusting along the external front of this wedge leads to the formation of the external Calabrian Arc (ECA), while the Marsili basin (M) opens in the internal zone. Under the SW–NE compression between the converging African/Iblean and Adriatic domains, the Calabrian arc undergoes a considerable fracturation, with the formation of troughs and sphenocasms, and fast uplift. Red arrows only indicate the motion trends of plates and microplates.

---

retreat (greater than 5 cm/y) observed in the Southern Apennines and Calabrian arc since the late Miocene (e.g., Patacca and Scandone, 1989).

The geodynamic framework shown in Figure 2(b) implies a constrictional regime in the northern Apenninic arc, even if weaker than the one which affected the southern arc. This regime was induced by the roughly N–S push of the Iblean indenter, partly transmitted by the southern Apenninic arc, and by the roughly NW ward drifting of the Adriatic plate. This belt-parallel compression caused the lateral extrusion/bending of the northern Apenninic arc, a deformation mechanism which can account for the occurrence, since the late Miocene, of thrusting along the external front of the arc and of extensional tectonics in the internal (Tuscan) area (e.g., Elter et al., 1975; Castellarin and Vai, 1986; Finetti et al., 2001). The plausibility of this driving mechanism is quantitatively demonstrated by numerical modelling (Viti et al., 2004). In the general deformation pattern of the Apenninic belt, the Latium–Abruzzi platform (central Apennines) played the role of a hinge zone. This hypothesis is suggested by the fact that the above sector has undergone a much lower outward migration with respect to the northern and southern Apenninic arcs (e.g., Sartori and Capozzi, 1998 and references therein). The decoupling of the Latium–Abruzzi platform from the northern Apenninic arc was accommodated by a dextral transpressional shear zone, including some major faults, such as the Ancona–Anzio and Valnerina (e.g., Castellarin et al., 1978; Decandia et al., 2002), while the decoupling from the Southern Apennines was taken up by the sinistral Ortona–Roccamonfina fault system. The importance of this last fault as a possible border between the oceanic-like and continental-like character of the Adriatic subducting margin is suggested by geological and volcanological evidence (Royden et al., 1987; Patacca et al., 1990; Serri et al., 1993; Francalanci and Manetti, 1994).

The fact that at the Kefallinia fault the sense of shear (dextral) is opposite to that observed in the Victor-Hensen and Medina faults does not contradict the above interpretation, since deformation along this fault is not related to the relative motion between Africa and the Adriatic, but is rather an effect of the dextral relative motion between the Adriatic domain and the SW ward extruding western Hellenic Arc.

Eastern Mediterranean. After the continental collision between the Adriatic and the Aegean/Balkan system in western Greece, the southward bowing of the Hellenic arc at the expense of the Levantine and Ionian low buoyancy domains accelerated. This process was also favoured by the late Miocene–early Pliocene activation of the western sector of the North Anatolian fault system, which marked the beginning of the westward extrusion of the whole Anatolian block (e.g., Barka, 1992; Westaway, 1994). The above interpretation provides plausible explanations for the major Pliocenic deformation recognized in the Aegean region (Figure 2(b)):

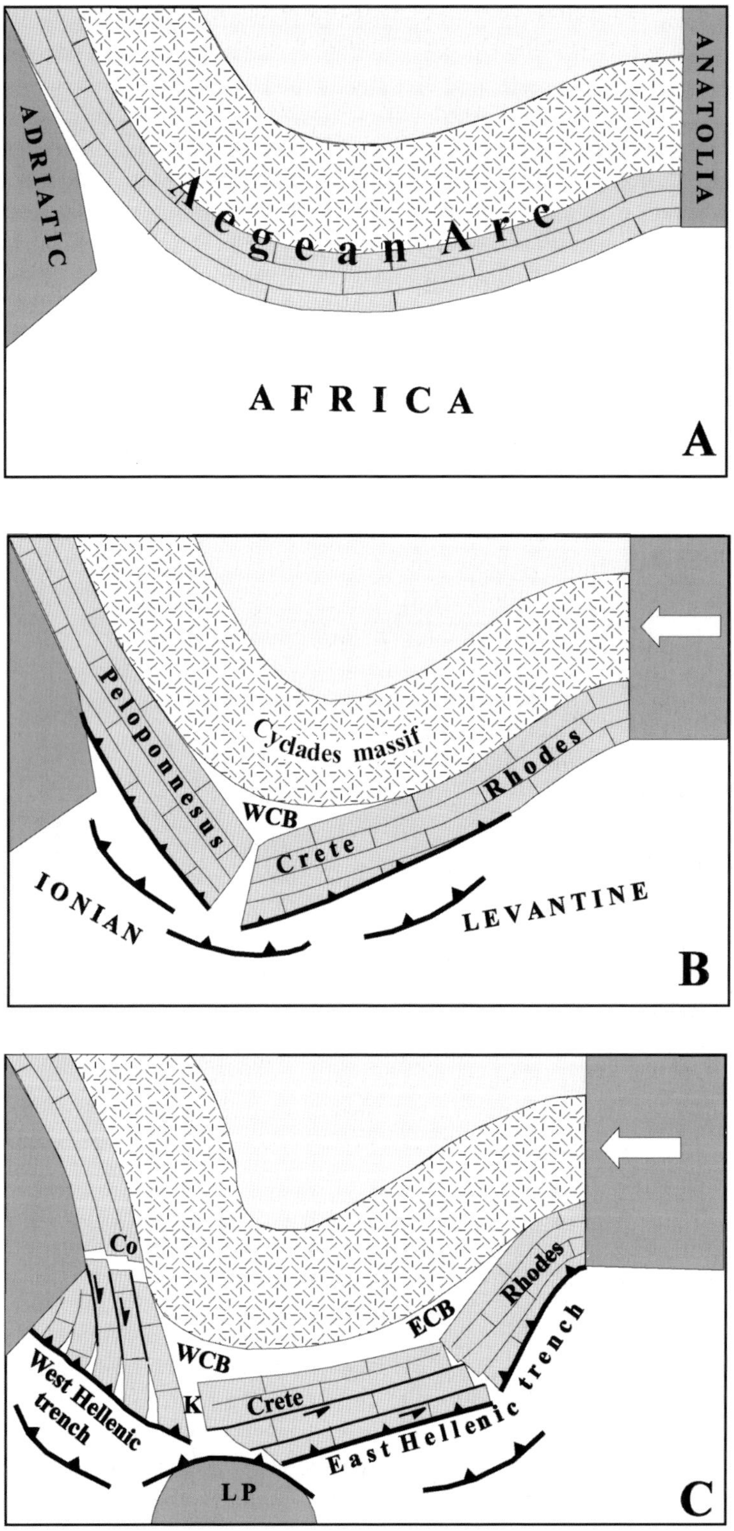

Figure 9: Proposed genetic mechanism for the western (WCB) and eastern (ECB) Cretan basins. (A) Late Miocene: in response to the E–W convergence of the Anatolian and Adriatic blocks, the Aegean arc bends southward. (B) Pliocene: the external belt (Hellenides) breaks in two sectors: the Peloponnesus and the Crete–Rhodes. The angular divergence between the Cyclades massifs and these two sectors produces crustal extension in the WCB. (C) Early Pleistocene: due to the incipient collision of the Libyan promontory (LP) with Crete, the Crete–Rhodes sector bends SE ward at the expense of the Levantine low buoyancy zone. The divergence between this arc and the Cyclades massifs causes crustal extension in the ECB. The clockwise

- The most intense accretionary activity along the external front of the Aegean arc occurred during the Pliocene, leading to the formation of the so-called Mediterranean ridge (e.g., Finetti, 1976; Underhill, 1989; Mascle and Chaumillon, 1997).
- Around the late Miocene–early Pliocene, transtensional tectonics accelerated in the North Aegean and western Anatolian regions, leading to the formation of a system of SW–NE dextral strike slip faults (Sengor et al., 1985; Hempton, 1987; Yilmaz, 1989; Mercier et al., 1989; Zanchi et al., 1990; Taymaz et al., 1991; Barka, 1992). This kind of deformation is expected in a dynamic context simultaneously characterized by E–W compression, as that induced by the convergence between eastern Anatolia and the Adriatic, and by N–S extension, as that induced by the divergence between the south to SW ward migrating Aegean arc and the Balkanides belt.
- Crustal thinning, with a dominant S–N extensional trend, occurred in the western Cretan basin (WCB) up to the late Pliocene, when this basin almost reached its present morphological configuration (Buttner and Kowalczyk, 1978; Angelier et al., 1982; Mercier et al., 1989; Meulenkamp et al., 1994). The fact that crustal stretching only occurred in such a limited zone, with an almost triangular shape (Figure 2(b)), and only developed during the Pliocene, imposes important constraints on the driving mechanism of this extensional event. A possible interpretation of these peculiar features is illustrated in Figure 9. The proposed mechanism suggests that the opening of the WCB was generated by the divergence between the inner arc (Cyclades massif) and the external Hellenides belt. This divergence was produced by the fact that the outer belt, presumably due to its greater bending and/or higher fragility, broke up in two sectors, the Peloponnesus and Crete–Rhodes (Figure 9(b)), which both separated from the Cyclades belt. This interpretation accounts for the timing of extensional tectonics in the WCB and for the peculiar location and shape of that basin.
- In the late Messinian–early Pliocene, the Cyclades zone was affected by intense compressional deformation and uplift, which generated pluri-kilometric sized faults and the first deposition of continental facies, after the Miocene marine sedimentation (Buttner and Kowalczyk, 1978; Durr et al., 1978; Mercier et al., 1987, 1989; Avigad et al., 2001). This deformation is interpreted as an effect of the E–W compressional stress field which affected the Cyclades belt during this phase.
- Around the late Miocene–early Pliocene, the outer front of the eastern Hellenic Arc (Crete–Rhodes segment) changed from a convergent to a sinistral transpressional boundary (Jacobshagen et al., 1978; Dewey and Sengor, 1979). This deformation is coherent with the proposed relative motion between the Hellenic Arc, migrating SW to southward, and the Levantine foreland, moving roughly NE as a part of Africa (Figure 2(b)).

After the collision of the Carpathian arc with the Eurasian continental domain, the resistance to any further northward displacement of the Tethyan belt has increased considerably. Thus, the extrusion path of the Tethian belt came to be controlled principally by the consumption of the Ionian and Levantine oceanic domains, which implied a sort of decoupling of the southern, more mobile part of the belt (Aegean–Pelagonian) from the northern, much less mobile segment (Carpathian). The zone where this decoupling mainly developed corresponds to the Albanides, where a dextral lateral shift of the Alpine and younger accretionary chains is clearly evidenced by morphological features and by geological and paleomagnetic observations (e.g., Kissel et al., 1985, 1995).

4.5. From the Late Pliocene–Early Pleistocene to the Present

A new kinematic/tectonic reorganization occurred in the central and eastern Mediterranean regions around the late Pliocene–early Pleistocene (Figure 2(c)). Extensional activity ceased in the central Tyrrhenian basin and began, with a NW-SE trend, in the southernmost Tyrrhenian, the Marsili basin. In the Apenninic belt, accretionary activity came to an end in the Southern Apennines and accelerated in the Calabrian Arc and Northern Apennines (e.g., Vai, 1987; Patacca and Scandone, 1989; Sartori, 1989; Bigi et al., 1989; Bartole, 1995). Renewal of accretionary activity occurred in the Outer Hellenides (western Greece). Extensional tectonics stopped in the WCB and began more to the east with a NW–SE trend, leading to the formation of the eastern Cretan basin (ECB). The eastern sector of the Hellenic Arc (Crete–Rhodes) underwent a considerable disruption which led to the land interruption between Crete and Rhodes.

As discussed in the following, the above major tectonic events can coherently be explained as effects of a new minimum-work reorganization in the central and eastern Mediterranean, in response to the incipient continental

---

rotation and internal deformation of the Peloponnesus crustal wedge, due to its oblique collision with the southernmost edge of the Adriatic continental plate, causes the formation of the Corinthian trough (Co). After the Libya–Crete collision, the divergence between the Peloponnesus, still moving SW ward, and Crete produces the formation of N–S trending extensional features, such as the Kithira trough (K).

collision at two consuming boundaries: the Southern Apennines (e.g., Patacca and Scandone, 1989) in the central Mediterranean and the Crete sector (Ryan et al., 1982; Lyon-Caen et al., 1988; Armijo et al., 1992; Mascle and Chaumillon, 1997) in the eastern Mediterranean.

Central Mediterranean. The arrival of the Adriatic continental crust in the Southern Apennines consuming boundary (e.g., Casnedi et al., 1982; Patacca and Scandone, 1989) caused a strong increase of resisting buoyancy forces. The progressive growth of this resistance made it convenient, around the late Pliocene, stopping the consuming process under the Southern Apennines and activating alternative shortening processes in the surrounding zones. As in previous cases, these processes were initially represented by crustal thickening, with consequent uplift, in the collision zone. Then, they turned to the lateral escape of crustal wedges towards the weakest lateral boundaries, as soon as major strike-slip lateral guides of extruding wedges were activated. The most evident effect in this sense was the roughly SE ward escape of the Calabrian wedge towards the Ionian low buoyancy domain (Figures 2(c) and 8(e)–(g)). This extrusion, guided by two major lateral transcurrent zones, the Vulcano and Palinuro fault systems (e.g., Finetti and Del Ben, 1986; Finetti et al., 1996), is coherently indicated by the simultaneous occurrence of several tectonic processes in the Calabrian Arc and surrounding zones:

- The accretionary activity along the external Calabrian Arc, at the border with the Ionian area, developed mainly in the Quaternary (e.g., Rossi and Sartori, 1981; Barone et al., 1982; Sartori, 1989), after the suture of the southern Apenninic consuming boundary.
- Extensional activity in the Tyrrhenian area underwent a drastic change around the late Pliocene–early Pleistocene: crustal stretching ceased in the central Tyrrhenian basin, in response to the suture of the southern Apennines boundary, and began in the southernmost Tyrrhenian (Marsili basin), with a SE ward trend (e.g., Kastens, 1988; Sartori, 1990), in the wake of the escaping Calabrian wedge (Figures 2(c) and 8(e)–(g)).
- Tectonic activity in the Calabrian wedge accelerated around the late Pliocene, as indicated by the formation, or reactivation, of a number of major transverse discontinuities and longitudinal troughs (e.g., Selli et al., 1978; Ghisetti, 1979; Finetti and Del Ben, 1986; Van Dijk and Okkes, 1991; Del Ben, 1993). The above evidence and the relative block rotations indicated by paleomagnetic data (Sagnotti, 1992; Scheepers et al., 1994; Meloni et al., 1997) suggest that the outward migration of the Calabrian Arc was accompanied by a significant bowing and strong fracturation. This deformation pattern is coherent with the fact that the lateral escape of Calabria occurred under the SW–NE compression induced by the convergence between Africa (Iblean wedge) and the Adriatic plate.
- Acceleration of uplift (1–2 mm/y) occurred around the late Pliocene-early Pleistocene in the whole zone directly involved in the Africa–Adriatic convergence, i.e. the Calabrian Arc and Southern Apennines (e.g., Ciaranfi et al., 1983; Westaway, 1993; Bordoni and Valensise, 1998).

The pieces of evidence described above and, in particular, the acceleration of uplift in the Calabrian Arc and Southern Apennines have tentatively been explained as effects of slab detachment under this zone (e.g., Westaway,

---

Figure 10: Reconstruction of the shortening pattern along a cross section from Corsica–Sardinia (CS) to the Adriatic foreland since the Oligocene, based on the interpretation of deep seismic reflection data (Finetti et al., 2001). (1) Oceanic lithosphere (Neotethys). (2) Alpine belt. (3)–(5) Upper and lower crust and pre-Pliocenic sedimentary cover of the Adriatic and Corsica–Sardinia continental domains respectively. (6) Lithospheric mantle (7) Asthenosphere. (a) Restored structural setting in the late Oligocene (modified after Finetti et al., 2001, Figure 8). (b) Upper Miocene: shortening of crustal units (71 km, Finetti et al., 2001) occurs in the western margin of the Adriatic foreland. This deformation is interpreted as an effect of the eastward migration of the Corsica–Sardinia block and the Alpine belt (Balearic Arc), driven by the Africa–Eurasia convergence (Figure 1). (c) Late Miocene: extensional tectonics affects the zone lying to the east of Corsica, with the formation of the northern Tyrrhenian basin and the reactivation, with tensional style, of the main thrust faults generated in the previous compressional phase (Finetti et al., 2001, Figure 9). This tensional event is interpreted as an effect of the divergence (12 km, Finetti et al., 2001) between the Adriatic promontory and the Corsica–Sardinia microplate (Figure 2(a)). (d) Present structural setting (modified after Finetti et al., 2001, Figure 8 and Plate 1). During the Pliocene, the previously undeformed Adriatic foreland underwent a shortening of 85 km (Finetti et al., 2001), accommodated by a number of major thrusts. This compressional phase is interpreted as an effect of the clockwise rotation of the Adriatic plate, under the push of the Anatolian–Aegean system (Figure 2(b)). See comments in the text.

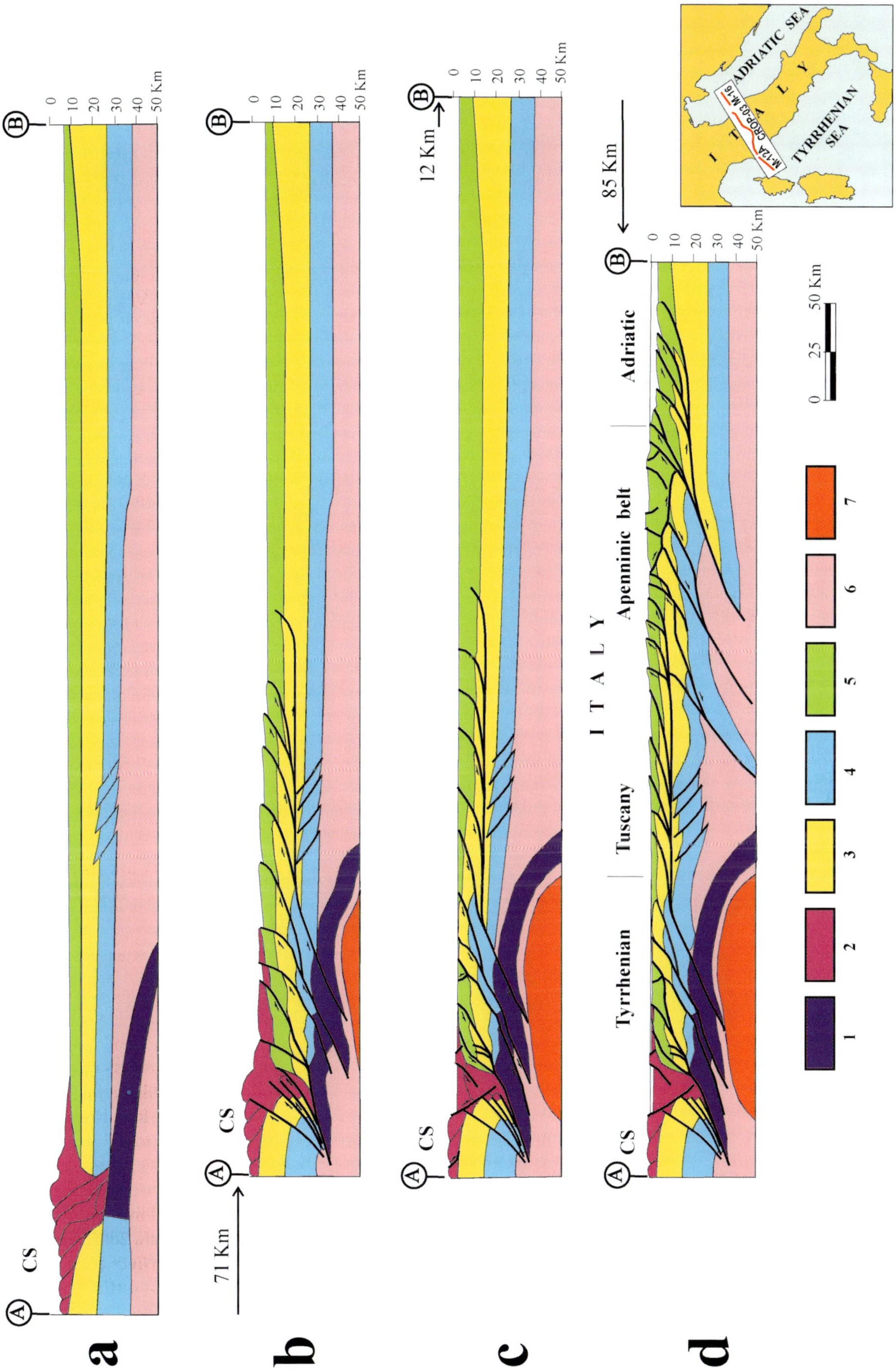
a
b
c
d
A
B
CS
0
10
20
30
40
50 Km
71 Km
12 Km
85 Km
Tyrrhenian
Tuscany
I T A L Y
Apenninic belt
Adriatic
1
2
3
4
5
6
7
0 25 50 Km
ADRIATIC SEA
TYRRHENIAN SEA
I T A L Y
CROP-03 M-16
M-12A

1993). However, this hypothesis does not take into account the results of the study of high-frequency seismic propagation from deep Tyrrhenian earthquakes to stations in the Calabrian Arc and the results of deep seismic soundings (Finetti, 2004; Mele, 1998), which seem to rule out any important break in the slab beneath the Calabrian Arc.

The redistribution of stresses in the central Mediterranean area that followed the suture of the Southern Apennines consuming boundary was mainly accommodated by the lateral expulsion of the Calabrian wedge, as argued above, but it also involved a belt-parallel compression in the Central-Northern Apennines. This stress regime emphasized the outward extrusion/bending of that arc, as indicated by the simultaneous occurrence of thrusting on the external part of the arc and extensional tectonics in the internal area, accompanied by uplift in the whole belt (e.g., Castellarin and Vai, 1986; Vai, 1987; Bartole, 1995; Piccardi et al., 1997; Boccaletti et al., 1997, 1999). Bowing of the northern Apenninic arc during this phase is also suggested by paleomagnetic evidence (Meloni et al., 1997).

The recent accurate reconstruction, from deep seismic reflection data (Finetti et al., 2001), of the crust-upper mantle structural setting along a cross section from the northern Tyrrhenian to the Adriatic domain (Figure 10) provides important insights into the deep implications of the extrusion processes invoked in this work. During the opening of the Balearic basin, from the Oligocene (Figure 10(a)) to the Middle-Upper Miocene (Figure 10(b)), the consumption of the Apulian foreland involved the embrication of lower crustal slivers under the Alpine belt and of upper crustal units, more to the east, in the Apenninic region. The geodynamic interpretation here proposed (Figure 1(b)) provides that this shortening process was caused by the forced eastward migration of the Balearic Arc driven by the SW–NE convergence between Africa and Europe. The fact that shortening of the Adriatic foreland affected both the upper and lower crust (Figure 10(b)) implies that the advancing block was characterized by strong mechanical resistance at least down to the Moho.

In the successive phase, from the Upper Tortonian (Figure 10(b)) to the late Messinian (Figure 10(c)), the Alpine–Apenninic belt was affected by tensional tectonics, with the formation of the northern Tyrrhenian basin (e.g., Sartori, 1990) and the tensional reactivation of thrusting faults generated by the previous shortening phase (Finetti et al., 2001). As argued earlier (Section 4.3), this tensional event is interpreted as an effect of the divergence between the Adriatic continental domain and the Corsica–Sardinia microplate, that followed the Giudicarie decoupling.

The last cross section (Figure 10(d)) shows that from the late Miocene to the present the Adriatic foreland has undergone a shortening of 85 km, accommodated by the underthrusting of a number of lithospheric slivers. As argued in Section 4.4, this reactivation of shortening is interpreted as an effect of the clockwise rotation of the Adriatic plate (Figure 2(b)) and of the simultaneous outward bending (with eastward migration) of the northern Apenninic Arc (Figure 2(b)).

A renewal of accretionary activity occurred along the southeastern Adriatic border (western Greece) around the early Pleistocene (e.g., Mercier et al., 1989). This reactivation could provide further basic insights into how the minimum-work principle controls tectonic processes. In particular, the above reactivation could indicate that lithosphere consumption at a plate boundary may stop in a given tectonic-dynamic context, when it no longer represents the most convenient shortening process, but can again become a convenient process in a successive stage, when low buoyancy lithosphere in the surrounding zone is considerably reduced or even absent. Indeed, the suture of the pre-Apulian consuming boundary occurred in the late Miocene (after the complete consumption of the pre-Apulian low buoyancy domain), when the most consumable lithosphere (the western Apulian zone) lay on the other side of the Adriatic plate (Figure 2(a)). Then in the late Pliocene, after the complete consumption of the western Apulian low buoyancy lithosphere (Figure 2(b)), the underthrusting of the Adriatic domain beneath the outer Hellenides could have again become the most convenient shortening process, in spite of the continental-like nature of the Adriatic lithosphere.

Discussion on the compatibility between the recent evolutionary pattern here proposed and the major features of seismicity and of recent volcanic activity in the central Mediterranean area are reported by Mantovani et al. (1997) and Tamburelli et al. (2000).

Eastern Mediterranean. The minimum-work principle may also help understanding the pattern by which tectonic processes reorganized in this region around the late Pliocene–early Pleistocene. The key event which triggered such reorganization was, once more, the slowdown/cessation of a consuming process at a continental collision zone. This occurred in the central Hellenic trench, between Crete and the Libyan promontory, as suggested by geological and morphological evidence (Ryan et al., 1982; Lyon-Caen et al., 1988; Armijo et al., 1992; Mascle and Chaumillon, 1997). Due to this incipient continental collision, the southward motion of the central Hellenic Arc (Crete) slowed down, causing an increase in longitudinal compression on the eastern Hellenic Arc, i.e. the Crete–Rhodes sector (Figures 2(c) and 9). This sector, being squeezed between western Anatolia and the Libyan promontory, underwent SE ward bending/extrusion, at the expense of the adjacent Levantine zone. This mechanism involved a considerable fracturation and microblock rotation of the stressed arc, which caused land interruption between Crete and Rhodes (Buttner and Kowalczyk, 1978; Lyon-Caen et al., 1988; Armijo et al., 1992). Contemporaneously, extensional tectonics occurred in the wake of the bending arc, leading to the formation of the ECB.

After the Crete–Libya collision, the zone lying between the Peloponnesus and Crete was affected by E–W extension, as indicated by the formation of the Kithira trough (Figures 2(c) and 9) and of other minor N–S tensional features (e.g., Armijo et al., 1992). This kind of deformation could be interpreted as a consequence of the fact that Crete, being slowed by the collision with the Libyan promontory, diverged from the Peloponnesus, which was still moving SW ward (Figure 2(c)).

5. BACK ARC EXTENSION IN THE WORLD

In the previous sections, it has been argued that generation of back basins in the Mediterranean area can plausibly be explained as an effect of extrusion tectonics. However, since the development of trench-arc-back arc systems is not confined to the study area, it may be useful to discuss on how the proposed interpretation and the most popular alternative model (slab-pull) can account for the major features of trench-arc-back arc systems in the world:

(a) Back arc extension is always associated with a subduction process, but conversely subduction does not always produce back arc extension. This implies that subduction is not a sufficient condition for the occurrence of back arc opening (e.g., Uyeda and Kanamori, 1979; Uyeda, 1986).
(b) Opening of back arc basins has been systematically accompanied by a considerable bending of the related arc (e.g., Uyeda, 1986; Mantovani et al., 2001b)
(c) Back arc opening does not develop along the entire length of the consuming boundary (e.g., Uyeda, 1986).
(d) No significant correlation can be found between any parameter of subduction processes (such as age of the subducting lithosphere, slab dip, slab length, arc length, subduction rate or velocity of the overriding plate) and the occurrence of back arc extension (Mantovani et al., 2001b).
(c) Consuming boundaries associated with trench-arc-back arc systems, such as the Marianatype, are generally characterized by much weaker seismicity and higher slab dips than consuming boundaries not associated with back arc basins, such as the Chileantype (e.g., Uyeda and Kanamori, 1979; Taylor and Karner, 1983; Scholz and Campos, 1995).

5.1. Extrusion Model

This interpretation postulates that a trench-arc-back arc system only develops where an orogenic chain is stressed by belt-parallel compression and low buoyancy lithosphere is present aside the stressed belt (Figure 3). This can explain why such systems have only developed in a number of subduction zones. The recognition of the required dynamic and tectonic conditions in those zones is described in this chapter for the Mediterranean area, and by Mantovani et al. (2001b) for circum-Pacific zones.

The evidence that some areas of former back arc extension are now inactive even though they remain adjacent to active subduction zones, for instance the Japan and Kurile basins, is explained as an effect of a change in boundary conditions, which no longer favour the occurrence of the extrusion mechanism. In several cases, this change was caused by the arrival of continental buoyant lithosphere in the sector of the trench zone facing the migrating arc and, in some cases, by the fact that the deforming arc has reached such a configuration, with respect to the dynamic boundary conditions, to inhibit any further deformation (Mantovani et al., 2001b).

The evidence that in trench-arc-back arc systems the migrating belt has systematically undergone a considerable curvature and that back arc extension has only occurred in a limited sector of the consuming boundary is fairly compatible with the deformation pattern predicted by the extrusion model (Figure 3). More generally, the geometry of a trench-arc-back arc system is controlled by the direction and size of the indenter, by the initial configuration of the accretionary belt and by the size of the low buoyancy domain lying in front of the extruding arc. This last factor, for instance, had a crucial influence on the deformation pattern of the Calabrian Arc–south Tyrrhenian and Carpatho–Balkan systems (e.g., Mantovani et al., 1997, 2000a).

The extrusion model may explain why back arc basins are systematically associated with subduction without invoking a causal relationship between the two processes. In fact, this model implies that lithosphere consumption and back arc extension are both side effects of a third process, the outward migration of the arc induced by belt-parallel compression (Figure 3). Trench retreat and slab sinking occurs under the load of the advancing arc, while back arc extension is produced by the divergence between the extruding crustal wedges and the overriding plate.

To explain the very different dip angles of slabs beneath subduction zones associated with back arc opening (e.g., the Mariana type) and not associated (e.g., the Chilean type), one should consider the rather different rheological properties of the mantle through which the slab must penetrate in these two cases. Geophysical investigations have shown that the asthenospheric layer is much more developed beneath oceanic domains than under continents (e.g., Pollack and Chapman, 1977; Artemieva and Mooney, 2001 and references therein). This implies that viscous resistance to slab penetration is much higher under a continental block (like the south America plate) than under a trench-arc-back arc system, like the Mariana one, where a "softer" mantle is present. Consequently, the steepening of the slab, driven by its negative buoyancy, encounters much less resistance beneath a Marianatype than beneath a Chileantype arc. This could explain why the Mariana slab has already reached an almost vertical configuration after a relatively short lifetime, whereas the Chilean slab is still dipping at a very low angle after a much longer lifetime.

Insights into this problem may also be gained by considering the features of the Calabrian and Hellenic subduction zones. The slabs under those arcs present quite different dip angles (65° and 30° respectively, see Mantovani et al., 2001b, and references therein), although the subducting lithosphere (the Ionian–Levantine oceanic domain) is the same in the two zones (e.g., Dercourt et al., 1986; Finetti and Del Ben, 1986). Such a difference in dip angles might be explained, for instance, by the different rheology of the mantle through which the respective slabs have penetrated. Beneath the Calabrian arc and the southern Tyrrhenian basin one can reasonably expect a soft mantle, due to the relatively large size of the stretched zone, whose formation has certainly induced an abundant flow of asthenospheric material from the surrounding mantle. This hypothesis is strongly supported by geophysical investigations in the Tyrrhenian area, which indicate a well developed asthenospheric mantle (Mele, 1998; Morelli and Piromallo, 2000; Martinez et al., 2000). Beneath the Aegean zone, instead, crustal stretching is very limited, being mostly confined to the small Cretan sea, where a crustal thickness of roughly 20–25 km is estimated. In the remaining Aegean area, the continental character of the lithosphere is widely recognized (e.g., Makris, 1978; Meissner et al., 1987; Papazachos and Nolet, 1997). Thus, the resistance to slab steepening beneath this zone, where the flow of asthenospheric material from the surrounding zones has not been as abundant as in the Tyrrhenian area, is expected to be much higher than in the southern Tyrrhenian.

The implications of the extrusion model could also provide the grounds for explaining another basic feature of trench-arc-back arc systems, the weak shallow seismicity observed at this kind of trench zone, like the Mariana one, compared with the very strong earthquakes recorded at Chileantype subduction zones (e.g., Uyeda and Kanamori, 1979; Scholz and Campos, 1995).

In this regard, one must consider that the dimensions of the subduction fault (controlling the magnitude of decoupling earthquakes) implied by the extrusion mechanism are expected to be considerably smaller than those of Chileantype subduction zones. In fact, in migrating arcs the overriding structure is formed from relatively thin crustal wedges (Figure 3) and thus the frictional interface with the descending lithosphere cannot exceed a few tens of km. In Chileantype boundaries, instead, the subduction fault may involve the entire lithosphere, and thus the magnitudes of the decoupling earthquakes may be much larger (e.g., Wang, 2000). Another factor which could help to mitigate shallow seismic activity at trench-arc-back arc systems with respect to Chileantype subduction zones is the reduced compression that one can expect between the subducting and the overriding lithosphere at the subduction fault in a context where the sinking and steepening of the slab is much less resisted by mantle forces (e.g., Uyeda and Kanamori, 1979; Carlowicz, 1995; Scholz and Campos, 1995).

5.2. Slab-Pull Model

The main difficulty of this interpretation is that the extensional stress induced in the overriding plate by the slab-pull force is too weak for producing the separation of the arc from the overriding plate. In fact, the results of numerical

(Becker et al., 1999; Pacanovsky et al., 1999) and analogue (Shemenda, 1993) modelling suggest that the above tensional stress may range between 5 and 40 MPa, which is significantly lower than the average tensional strength of an intact lithosphere (100–250 MPa, e.g., Dunbar and Sawyer, 1989; Fadaie and Ranalli, 1990; Mueller et al., 1996; Okaya et al., 1996). This problem could be overcome only in case the lithosphere of the overriding plate is considerably weakened by some tectonic process. For instance, Shemenda (1993) has tentatively suggested that such weakening might be caused by uprising of magma. However, this possibility is not much corroborated by the observed features, since in most trench-arc-back arc systems the location of the magmatic arc is clearly distinct from that of the back arc basin. In this regard, it must also be pointed out that widespread volcanic activity occurs in a number of subduction zones where back arc opening is no longer active, such as the Japan Sea, or never occurred, such as the Andes.

Another difficulty of the slab-pull model is that extensional stresses in the overriding plate can only occur when the friction coefficient at the subduction fault is lower than 0.2 (e.g., Shemenda, 1993; Hassani et al., 1997; Chemenda et al., 2000), which is out of the range of values (0.4–0.6) generally recognized for this parameter at subduction zones (e.g., Townend and Zoback, 2000).

The above evidence and arguments suggest that a slab-pull force can hardly produce deformation in the overriding plate. On the other hand, if the arc cannot detach from the overriding plate, the roll-back of the slab is inhibited and steepening of the slab becomes the only feasible effect of slab-pull (Shemenda, 1993). Other arguments in support of the hypothesis that slab steepening is more likely than slab roll-back, under the action of gravitational forces, are derived by considering the energy balances of these two kinds of deformation (Mantovani et al., 2001b). The evidence that most slabs are not vertical and that no simple relation exists between slab dip and age of subducted lithosphere (see, e.g., Taylor and Karner, 1983; Mantovani et al., 2001b) suggests that other forces act on the slab, in addition to its negative buoyancy. These forces may be, for instance, resistance to bending (Conrad and Hager, 1999), suction between the overriding and underthrusting lithospheres along their frictional interface (Jischke, 1975), uplift of the subducted slab due to the flow induced in the asthenosphere (Tovish et al., 1978; Turcotte and Schubert, 1982) and viscous resistance to slab migration through the mantle (Davies, 1980; Scholz and Campos, 1995). What the resultant effect of all these (and possibly other) forces could be, is not yet clear.

Analysis of the major features of subduction zones, both associated and not associated with back arc extension, has not provided any significant evidence which could help to explain why the slab-pull force causes back arc extension in some subduction zones and not in others (e.g., Mantovani et al., 2001b). Any interpretation of the above evidence should also explain why subduction zones associated and not associated with back arc extension are characterized by very different shallow seismicity and slab dip (e.g., Scholz and Campos, 1995).

To explain the curved shape of arcs as an effect of slab-pull, one must face a number of difficult problems. For instance, one should explain why slab sinking rates are higher in the central sectors of the arc with respect to the lateral ones. This behaviour could be due, for instance, to different densities in the various parts of the subducting lithosphere. However, it seems rather unlikely that such peculiar density distribution was systematically present in all trench-arc migrating systems of the world. Some authors have argued that arc buckling might be due to the earth's sphericity (e.g., Bayly, 1982; Yamaoka, 1988), but this hypothesis can hardly account for the extremely variable curvature of arcs at consuming boundaries.

Another difficulty of the slab-pull model is explaining why slab-pull only produces back arc opening in a limited sector of the consuming boundary. Since the above mechanism does not imply any significant lateral variation of the gravitational forces involved, one should find local peculiar structural rheological conditions capable of explaining such a laterally heterogeneous behaviour for each context where back arc extension is observed.

A discussion about the major difficulties encountered in interpreting the space–time distribution of tectonic events in the Mediterranean region as an effect of the slab-pull model is given by Mantovani et al. (1997, 2000a, 2002).

6. CONCLUSIONS AND DISCUSSION

The post-Eocene tectonic evolution of the Mediterranean area has been mainly determined by the convergence of the confining plates, i.e. Africa, Arabia and Eurasia. The presence in such a constricted zone of a considerable amount of buoyant accretionary belts and of large sectors of low buoyancy oceanic lithosphere has created the conditions for the occurrence of extrusion processes. This kind of process took place every time that the most intense stresses, induced by plate convergence, affected buoyant orogenic belts. In this context, the lateral escape of buoyant crustal wedges allowed to transmit compressional stresses to the zones where low buoyancy lithosphere was present. In particular, the belt-parallel compression that stressed various segments of the Cretaceous-Eocene Alpine belt determined their lateral bending/extrusion, with the development of trench-arc-back arc migrating systems in the Balearic, Tyrrhenian, Aegean and Pannonian regions. Such lateral displacement of accretionary belts was accommodated by the consumption of a large part of the low buoyancy oceanic domains originally present in the study area. Opening of the back arc basins occurred in the wake of the migrating and bending arcs.

The proposed evolutionary reconstruction provides plausible and coherent explanations for the location, nature and period of activity of the major observed tectonic processes. It must be pointed out that such a result is achieved by adopting a unique driving mechanism (the convergence of the confining plates) and a unifying conceptual framework, based on the minimum-work principle. The applicability of this last principle to the tectonic evolution of the study area cannot quantitatively be proved yet, due to the very high complexity of the tectonic context and to insufficient knowledge of the structural/rheological features. In this first attempt, the above approach only relies on the assumption that the location of the densest lithospheric domains has crucially influenced the direction of wedges extruded from the most constricted collisional zones. Although apparently oversimplified, this hypothesis can coherently account for the time–space distribution of tectonic events which have occurred in the study area in response to the respective kinematic boundary conditions.

The main objection advanced against the applicability of the extrusion model to the Mediterranean back arc basins, with particular regard to the Tyrrhenian (Gueguen et al., 1997; Jolivet et al., 1998) is that the migration rate of arcs was higher than the convergence rate between Africa and Eurasia. However, this hypothesis arbitrarily assumes that this plate convergence was the unique driving mechanism of the Mediterranean tectonics, neglecting that the main responsible for the Plio-Quaternary deformation in the central Mediterranean region was the westward push of the Anatolian–Aegean–Balkan system.

Other arguments against the extrusion model have been proposed for the Aegean-Hellenic system, on the basis of space geodetic data (e.g., McClusky et al., 2000). The velocity field delineated by these data suggests that the Aegean zone is moving faster (roughly 30 mm/y) than the Anatolian wedge (24 mm/y), which would contrast with the hypothesis that the deformation pattern of the Aegean area is driven by the westward push of Anatolia. However, quantification of post-seismic relaxation in the Anatolian–Aegean system induced by the sequence of very strong earthquakes which have occurred along the North Anatolian fault since 1939 (Mantovani et al., 2001c; Cenni et al., 2002) has shown that the geodetic velocity field in the above zones can be interpreted as a transient effect of the above phenomenon. This result, along with other arguments and evidence (e.g., Anderson, 1975; Rydelek and Sacks, 1990; Barka, 1992), could imply that the short-term kinematic behaviour of the study area may be different from the long-term (geological) one. This is also suggested by the fact that the Plio-Quaternary slip rate (less than 10 mm/y) along the North Anatolian fault, estimated from an analysis of fault offsets and the long-term seismicity pattern (Barka, 1992) is significantly lower than the average geodetic velocity of Anatolia.

The most popular alternative interpretation of trench-arc-back arc systems in the Mediterranean area, the slab-pull model, cannot easily be reconciled with several major features of the observed deformation pattern. In addition, the physical plausibility of this model is far to be demonstrated since the results of various modelling experiments indicate that the tensional stress induced in the overriding plate by subduction-related forces is significantly lower than the lithospheric strength.

Further support to the hypothesis that the extrusion mechanism can better account for the observed features of trench-arc migrating systems than subduction-related models may be obtained by extending the comparative analysis to the whole earth (Mantovani et al., 2001b).

The geodynamic interpretation here proposed may be useful, at least as a working hypothesis, for the planning of further observations and for the choice of the most suitable modelling experiments to be carried out. Some preliminary attempts in this direction are reported by Mantovani et al. (2000a, 2001a), who tried a quantitative estimate of the work required by different geodynamic models, and a numerical simulation of the recent/present strain field observed in the central-eastern Mediterranean area. The first of the above papers points out, by quantitative arguments, that the most convenient shortening pattern in the constrictional tectonic contexts presumably responsible for trench-arc-back arc systems (Figure 3), is the lateral expulsion of crustal wedges. The second paper shows that the highly heterogeneous strain field observed in the central-eastern Mediterranean area may be satisfactorily reproduced by assuming, as driving mechanism, the kinematic boundary conditions adopted in the last stage (Quaternary) of the evolutionary reconstruction here proposed.

REFERENCES

Ahorner, L., 1975. Present-day stress field and seismotectonic block movements along major fault zones in central Europe. Tectonophysics 29, 233–249.

Ait Brahim, L., Chotin, P., Hinaj, S., Abdelouafi, A., El Adraoui, A., Nakcha, C., Dhont, D., Charroud, M., Sossey Alaoui, F., Amrhar, M., Bouaza, A., Tabayaoui, H., Chaouni, A., 2002. Paleostress evolution in the Moroccan African margin from Triassic to Present. Tectonophysics 357, 187–205.

Albarello, D., Mantovani, E., Babbucci, D., Tamburelli, C., 1995. Africa–Eurasia kinematics: main constraints and uncertainties. Tectonophysics 243, 25–36.

Altamimi, Z., Sillard, P., Boucher, C., 2002. ITRF2000: a new release of the International Terrestrial Reference Frame for earth science applications. J. Geophys. Res. 107. ETG 2.

Alvarez-Marron, J., Rubio, E., Torne, M., 1997. Subduction-related structures in the North Iberian Margin. J. Geophys. Res. 102, 22497–22511.

Anderson, D.L., 1975. Accelerated plate tectonics. Science 167, 1077–1079.

Anderson, H., Jackson, J., 1987. The deep seismicity of the Tyrrhenian Sea. Geophys. J. R. Astr. Soc. 91, 613–637.

Andeweg, B., De Vicente, G., Cloetingh, S., Giner, J., Munoz Martin, A., 1999. Local stress fields and intraplate deformation of Iberia: variations in spatial and temporal interplay of regional stress sources. Tectonophysics 305, 153–164.

Andeweg, B., Cloetingh, S., 2001. Evidence for an active sinistral shear zone in the western Alboran region. Terra Nova 13, 44–50.

Anguita, F., Hernàn, F., 2000. The Canary Islands origin: a unifying model. J. Volc. Geoth. Res. 103, 1–26.

Angelier, J., Lyberis, N., Le Pichon, X., Barrier, E., Huchon, P., 1982. The tectonic development of the Hellenic arc and the Sea of Crete: a synthesis. Tectonophysics 86, 139–196.

Anzidei, M., Baldi, P., Casula, G., Galvani, A., Mantovani, E., Pesci, A., Riguzzi, F., Serpelloni, E., 2001. Insights on the present day crustal motion in the central Mediterranean area from GPS surveys. Geophys. J. Int. 146, 98–110.

Argnani, A., 1993. Neogene basins in the Strait of Sicily (Central Mediterranean): tectonic settings and geodynamic implications. In: Boschi, E., Mantovani, E., Morelli, A. (Eds.), Recent Evolution and Seismicity of the Mediterranean Region. Kluwer Academic, Dordrecht, pp. 173–187.

Argnani, A., Frugoni, F., Cosi, R., Ligi, M., Favali, P., 2001. Tectonics and seismicity of the Apulian ridge south of Salento peninsula (Southern Italy). Ann. Geofis. 44, 527–540.

Argus, D.F., Gordon, R.G., DeMets, C., Stein, S., 1989. Closure of the Africa–Eurasia–North America plate circuit and tectonics of the Gloria fault. J. Geophys. Res. 94, 5585–5602.

Armijo, R., Lyon-Caen, H., Papanastassiou, D., 1992. East-west extension and Holocene normal-fault scarps in the Hellenic arc. Geology 20, 491–494.

Artemieva, I.M., Mooney, W.D., 2001. Thermal thickness and evolution of Precambrian lithosphere: a global study. J. Geophys. Res. 106, 16387–16414.

Aubouin, J., 1984. Mediterraneenne (Aire). In: Encyclopedia Universalis, vol. 11, pp. 1023–1030.

Auroux, C., Mascle, J., Rossi, S., 1984. Geologia del margine ionico dalle isole Strofadi a Corfù (estremità settentrionale dell'Arco Ellenico). Mem. Soc. Geol. It. 27, 267–286.

Avigad, D., Ziv, A., Garfunkel, Z., 2001. Ductile and brittle shortening, extension-parallel folds and maintenance of crustal thickness in the central Aegean (Cyclades, Greece). Tectonics 20, 277–287.

Babbucci, D., Tamburelli, C., Mantovani, E., Albarello, D., 1997. Tentative list of major deformation events in the central-eastern Mediterranean region since the middle Miocene. Ann. Geofis. 40, 645–670.

Balla, Z., 1986. Palaeotectonic reconstruction of the central Alpine–Mediterranean belt for the Neogene. Tectonophysics 12, 213–243.

Barone, A., Fabbri, A., Rossi, S., Sartori, R., 1982. Geological structure and Evolution of the Marine Areas Adjacent to the Calabrian Arc. In: Mantovani, E., Sartori, R. (Eds.), Structure, Evolution and Present Dynamics of the Calabrian Arc. Earth Evolut. Sci. 3, 207–221.

Barka, A.A., 1992. The North Anatolian fault zone. Ann. Tecton. 6, 164–195.

Barka, A.A., Reilinger, R., 1997. Active tectonics of the eastern Mediterranean: deduced from GPS, neotectonic and seismicity data. Ann. Geofis. 40, 587–610.

Bartole, R., 1995. The North Tyrrhenian – Northern Apennines post collisional system: constraints for a geodynamical model. Terra Nova 7, 7–30.

Bayly, B., 1982. Geometry of subducted plates and island arcs viewed as a buckling problem. Geology 10, 629–632.

Beccaluva, L., Coltorti, M., Galassi, R., Macciotta, G., Siena, F., 1994. The cainozoic calcalkaline magmatism of the western Mediterranean and its geodynamic significance. Boll. Geof. Teor. Appl. 141–144, 293–308.

Becker, T.W., Faccenna, C., O'Connel, R.J., 1999. The development of slabs in the upper mantle: insights from numerical and laboratori experiments. J. Geophys. Res. 104, 207–226.
Bigi, G., Castellarin, A., Catalano, R., Coli, M., Casentino, D., Dal Piaz, G.V., Lentini, F., Parlotto, M., Patacca, E., Praturlon, A., Salvini, F., Sartori, R., Scandone, P., Vai, G.B., 1989. Synthetic structural-kinematic map of Italy, Scale 1:2.000.000. CNR, Roma.
Biju-Duval, B., Dercourt, J., Le Pichon, X., 1977. From the Tethys ocean to the Mediterranean sea: a plate tectonic model of the evolution of the western Alpine system. In: Biju-Duval, B., Montadert, L. (Eds.), The Structural Hystory of the Mediterranean Basin. Technip, Paris, pp. 143–164.
Boccaletti, M., Coli, M., Decandia, A., Giannini, G., Lazzarotto, A., 1980. Evoluzione dell'Appennino settentrionale secondo un nuovo modello strutturale. Mem. Soc. Geol. It. 21, 359–373.
Boccaletti, M., Dainelli, P., 1982. Il sistema regmatico Neogenico-Quaternario nell'area Mediterranea: esempio di deformazione plastico-rigida post-collisionale. Mem. Soc. Geol. It. 24, 465–482.
Boccaletti, M., Granelli, G., Sani, F., 1997. Tectonic regime, granite emplacement and crustal structure in the inner zone of the northern Apennines (Tuscan, Italy): a new hypothesis. Tectonophysics 270, 127–143.
Boccaletti, M., Bovini, M., Moratti, G., Sani, F., 1999. Compressive Neogene-Quaternary tectonics in the hinterland area of the Northern Apennines. In: Boccaletti, M., Dahmani, M. (Eds.), Neogene Sedimentation and Tectonics in the Western Mediterranean. J. Petr. Geol. 22, 37–60.
Bordoni, P., Valensise, G., 1998. Deformation of the 125 ka marine terrace in Italy: tectonic implications. In: Stewart, I., Vita-Finzi, C. (Eds.), Late Quaternary Coastal Tectonics. Geol. Soc. London Spec. Publ. 146, 71–110.
Brede, R., Hauptmann, M., Herbig, H.–G., 1992. Plate tectonics and interaction mountain ranges in Morocco – The Mesozoic-Cenozoic development of the Central High Atlas and the Middle Atlas. Geol. Rundsch. 81, 127–141.
Brunn, J.H., 1960. Les zones helléniques internes et leur extension. Bull. Soc. Géol. France 7, 470–486.
Brunn, J.H., 1976. L'arc concave zagro-taurique et les arcs convexes taurique et egeen : collision et arcs induits. Bull. Soc. Géol. France 18, 553–567.
Buforn, E., Udias, A., Colombas, M.A., 1988. Seismicity, source mechanism and tectonics of the Azores–Gibraltar plate boundary. Tectonophysics 152, 89–118.
Burchfiel, B.C., 1980. Eastern European Alpine system and the Carpathian orocline as an example of collision tectonics. Tectonophysics 63, 31–61.
Burtman, V.S., 1986. Origin of structural arcs of the Carpathian–Balcan region. Tectonophysics 127, 245–260.
Buttner, D., Kowalczyk, G., 1978. Late Cenozoic stratigraphy and paleogeography of Greece. A review. In: Closs, E., Roeden, D., Schmidt, K. (Eds.), Alps, Apennines, Hellenides. E. Schweizerbart'sche Verlagsbuchhandlung Nagele u. Obermille, Stuttgart, pp. 494–499.
Cantelli, L., Castellarin, A., 1994. Analisi e inquadramento strutturale del sistema "Schio-Vicenza". Atti Tic. Sc. Terra 1 (special issue), 231–245.
Carbone, S., Cosentino, M., Grasso, M., Lentini, F., Lombardo, G., Patané, G., 1982. Elementi per una prima valutazione dei caratteri sismotettonici dell'avampaese ibleo (Sicilia sud-orientale). Mem. Soc. Geol. It. 24, 507–520.
Carlowicz, M., 1995. Anchor-like force proposed for subduction zones. In: EOS 76, pp. 497–498.
Casnedi, R., Crescenti, V., Tonna, M., 1982. Evoluzione dell'avanfossa adriatica meridionale nel Plio-Pleistocene, sulla base di dati di sottosuolo. Mem. Soc. Geol. It. 24, 243–260.
Castellarin, A., Colacicchi, R., Praturlon, A., 1978. Fasi distensive, trascorrenze e sovrascorrimenti lungo la "Linea Ancona-Anzio", dal Lias medio al Pliocene. Geol. Romania 17, 161–189.
Castellarin, A., Vai, G.B., 1986. Southalpine versus Po Plain Apenninic Arcs. In: Wezel, F.C. (Ed.), The Origin of Arcs, vol. 19. Elsevier, Amsterdam, pp. 253–280.
Castellarin, A., Cantelli, L., Fesce, A.M., Mercier, J.L., Ricotti, V., Pini, G.A., Prosser, G., Selli, L., 1992. Alpine compressional tectonics in the Southern Alps. Relationships with the N-Apennines. Ann. Tecton. 6, 62–94.
Catalano, R., Infuso, S., Sulli, A., 1994. The submerged alpidic chain from southern Sardinia shelf to the Pelagian rifting: tectonic history. Boll. Geof. Teor. Appl. 36, 139–158.
Catalano, R., Doglioni, C., Merlini, S., 2001. On the Mesozoic Ionian basin. Geophys. J. Int. 144, 49–64.
Cenni, N., D'Onza, F., Viti, M., Mantovani, E., Albarello, D., Babbucci, D., 2002. Post seismic relaxation processes in the Aegean–Anatolian system: insights from space geodetic data (GPS) and geological/geophysical evidence. Boll. Geof. Teor. Appl. 43, 23–36.
Cetin, H., Guneyli, H., Mayer, L., 2003. Paleoseismology of the Palu-Lake Hazar segment of the east Anatolian fault zone, Turkey. Tectonophysics 374, 163–197.
Chase, C.G., 1978. Plate kinematics: the Americas, East Africa, and the rest of the world. Earth Planet. Sci. Lett. 37, 355–368.

Chemenda, A., Lallemand, S., Bokun, A., 2000. Strain partitioning and interplate friction in oblique subduction zones: constraints provided by experimental modeling. J. Geophys. Res. 105, 5567–5581.
Ciaranfi, N., Guida, M., Iaccarino, G., Pescatore, T., Pieri, P., Rapisardi, L., Ricchetti, G., Sgrosso, I., Torre, M., Tortorici, L., Turco, E., Scarpa, R., Cuscito, M., Guerra, I., Iannaccone, G., Panza, G.F., Scandone, P., 1983. Elementi sismotettonici dell'Appennino meridionale. Boll. Soc. Geol. It. 102, 201–222.
Cloos, M., 1993. Lithospheric buoyancy and collisional orogenesis: subduction of oceanic plateaus, continental margins, island arcs, spreading ridges, and seamounts. Geol. Soc. Am. Bull. 105, 715–737.
Cohen, C.R., 1980. Plate tectonic model for the Oligo-Miocene evolution of the Western Mediterranean. Tectonophysics 68, 283–311.
Comninakis, P., Papazachos, B., 1986. Catalogue of earthquakes in Greece and surrounding area for the period 1901–1985. Publ. Geophys. Lab., vol. 1, University of Thessaloniki, 167 p.
Conrad, C.P., Hager, B.H., 1999. Effects of plate bending and fault strength at subduction zones on plate dynamics. J. Geophys. Res. 104, 551–571.
Console, R., Di Giovambattista, R., Favali, P., Presgrave, B.W., Smriglio, G., 1993. Seismicity of the Adriatic microplate. Tectonophysics 218, 343–354.
D'Amico, V., Albarello, D., 1999. The problem of regionalization in the study of magnitude-intensity correspondences in the Mediterranean area: a distribution-free approach. Pure Appl. Geophys. 156, 651–665.
Davies, G.F., 1980. Mechanics of subducted lithosphere. J. Geophys. Res. 85, 6304–6318.
De Alteriis, B., 1995. Different foreland basins in Italy: examples from the central and southern Adriatic Sea. Tectonophysics 252, 349–373.
Del Ben, A., 1993. Calabrian Arc tectonics from seismic exploration. Boll. Geof. Teor. Appl. 139, 339–347.
Decandia, F.A., Tavarnelli, E., Alberti, M., 2002. The double reactivation of the Valnerina line: implications for the tectonic evolution of the Umbria–Marche Apennines, Italy. Studi Geologici Camerti 1, 77–87.
Della Vedova, B., Pellis, G., 1989. New heat flow density measurements in the Ionian sea. In: Proceedings 8° Meeting GNGTS, Roma, November 7–9, pp. 1133–1146.
DeMets, C., Gordon, R.G., Argus, D.F., Stein, S., 1990. Current Plate Motions. Geophys. J. 101, 425–478.
Dercourt, J., Zonenshain, L.P., Ricou, L.E., Kazmin, V.G., Le Pichon, X., Knipper, A.L., Grandjacquet, C., Sbortshikov, I.M., Geyssant, J., Lepvrirer, C., Pechersky, D.H., Boulin, J., Sibuet, J.C., Savostin, L.A., Sorokhtin, O., Westphal, M., Bazchenov, M.L., Lauer, J.P., Biju-Duval, B., 1986. Geological evolution of the Tethys belt from Atlantic to the Pamirs since the Lias. In: Aubouin, J., Le Pichon, X., Monin, A.S. (Eds.), Evolution of the Tethys. Tectonophysics 123, 241–315.
Devoti, R., Ferraro, C., Gueguen, E., Lanotte, R., Luceri, V., Nardi, A., Pacione, R., Rutigliano, P., Sciarretta, C., Vespe, F., 2002. Geodetic control on recent tectonic movements in the central Mediterranean area. Tectonophysics 346, 151–167.
Dewey, J.F., Pitman, W.C., Ryan, W.B.F., Bonnin, J., 1973. Plate tectonic and the evolution of the Alpine system. Geol. Soc. Am. Bull. 84, 3137–3180.
Dewey, J.F., Sengor, A.M.C., 1979. Aegean and surrounding regions: complex multiplate and continuum tectonics in a convergent zone. Geol. Soc. Am. Bull. 90, 89–92.
Dewey, J.F., Hempton, M.R., Kidd, W.S.F., Saroglu, F., Sengor, A.M.C., 1986. Shortening of continental lithosphere: the neotectonics of eastern Anatolia a young collision zone. In: Coward, M.P., Ries, A.C. (Eds.), Collision Tectonic. Geol. Soc. London Spec. Publ. 19, 3–36.
Dewey, J.F., Helman, M.L., Turco, E., Hutton, D.H.W., Knott, S.D., 1989. Kinematics of the western Mediterranean. In: Coward, M.P., Dietrich, D., Park, R.G. (Eds.), Alpine Tectonics. Geol. Soc. London Spec. Publ. 45, 265–283.
Di Nocera, S., Ortolani, F., Torre, M., 1976. La tettonica messiniana nell'evoluzione della catena Appenninica. In: Meeting on: Il significato geodinamico della crisi di salinità del Miocene Terminale del Mediterraneo. CRN-PFG, Firenze, pp. 29–47.
Doglioni, C., Gueguen, E., Harabaglia, P., Monelli, F., 1999. On the origin of west-directed subduction zones and applications to the western Mediterranean. In: Durand, B., Jolivet, L., Horvath, F., Seranne, M. (Eds.), The Mediterranean Basins: Tertiary Extension within the Alpine Orogen. Geol. Soc. London Spec. Publ. 156, 541–561.
Dunbar, J.A., Sawyer, D.S., 1989. How preexisting weaknesses control the style of continental breakup. J. Geophys. Res. 94, 7278–7292.
Durr, St., Altherr, R., Keller, J., Okrusch, M., Seidel, E., 1978. The Median Aegean crystalline belt: stratigraphy, structure, metamorphism, magmatism. In: Closs, H., Roeder, D., Schmidt, K. (Eds.), Alps Apennines Hellenides. E. Schweizerbart'sche Verlagsbuchhandlung Nagele u. Obermille, Stuttgart, pp. 455–477.
Elter, P., Giglia, G., Tongiorgi, M., Trevisan, L., 1975. Tensional and compressional areas in the recent (Tortonian to Present) evolution of the Northern Apennine. Boll. Geof. Teor. Appl. 65, 3–18.

Elter, P., Marroni, M., 1991. Le Unità Liguri dell'Appennino Settentrionale: sintesi dei dati e nuove interpretazioni. Mem. Descr. Carta Geol. d'Italia 46, 121–138.

Eva, E., Solarino, S., Eva, C., Neri, G., 1997. Stress tensor orientation derived from fault plane solution in the southwestern Alps. J. Geophys. Res. 102, 8171–8185.

Eyidogan, H., 1988. Rates of crustal deformation in western Turkey as deduced from major earthquakes. Tectonophysics 148, 83–92.

Faccenna, C., Davy, P., Brun, J.P., Funiciello, R., Giardini, D., Mattei, M., Nalpas, T., 1996. The dynamics of back-arc extension: an experimental approach to the opening of the Tyrrhenian Sea. Geophys. J. Int. 126, 781–795.

Faccenna, C., Becker, T.W., Lucente, F.P., Jolivet, L., Rossetti, F., 2001. History of subduction and back arc extension in the central Mediterranean. Geophys. J. Int. 145, 809–820.

Fadaie, K., Ranalli, G., 1990. Rheology of the lithosphere in the East African Rift System. Geophys. J. Int. 102, 445–453.

Faulkner, D.R., Lewis, A.C., Rutter, E.H., 2003. On the internal structure and mechanics of large strike-slip fault zones: field observations of the Carboneras fault in southeastern Spain. Tectonophysics 367, 235–251.

Favali, P., Funiciello, R., Mattietti, G., Mele, G., Salvini, F., 1993. An active margin across the Adriatic sea (central Mediterranean sea). Tectonophysics 219, 109–117.

Finetti, I., 1976. Mediterranean ridge: a young submerged chain associated with the Hellenic Arc. Boll. Geof. Teor. Appl. 19, 31–65.

Finetti, I., 1982. Structure, stratigraphy and evolution of the central Mediterranean. Boll. Geof. Teor. Appl. 24, 247–300.

Finetti, I., Del Ben, A., 1986. Geophysical study of the Tyrrhenian opening. Boll. Geof. Teor. Appl. 28 (110), 75–156.

Finetti, I., Bricchi, G., Del Ben, A., Pipan, M., Xuan, Z., 1988. Geophysical study of the Black Sea. Boll. Geof. Teor. Appl. 117–118, 197–324.

Finetti, I., Lentini, F., Carbone, S., Catalano, S., Del Ben, A., 1996. Il sistema Appenninico Meridionale–Arco Calabro–Sicilia nel Mediterraneo Centrale: studio geofisico-geologico. Boll. Soc. Geol. It. 115, 529–559.

Finetti, I., Boccaletti, M., Bovini, M., Del Ben, A., Geletti, R., Pipan, M., Sani, F., 2001. Crustal section based on CROP seismic data across the North Tyrrhenian–Northern Apennines–Adriatic Sea. Tectonophysics 343, 135–163.

Finetti, I.R., 2004. Innovative CROP seismic highlights on the Mediterranean region. In: Special volume of the Italian Geol. Society for the IGC 32, Florence, pp. 131–140.

Fodor, L., Jelen, B., Marton, E., Skaberne, D., Car, J., Vrabec, M., 1998. Miocene-Pliocene tectonic evolution of the Slovenia periadriatic fault: implications for Alpine-Carpathian extrusion models. Tectonics 17, 690–709.

Foster, A.N., Jackson, J.A., 1998. Source parameters of large African earthquakes: implications for crustal rheology and regional kinematics. Geophys. J. Int. 134, 422–448.

Francalanci, L., Manetti, P., 1994. Geodynamic models of the southern Tyrrhenian region: constraints from the petrology and geochemistry of the Aeolian volcanic rocks. Boll. Geof. Teor. Appl. 141–144, 283–293.

Fytikas, M., Innocenti, F., Manetti, P., Mazzuoli, R., Peccerillo, A., Villari, L., 1984. Tertiary to Quaternary evolution of volcanism in the Aegean region. In: Dixon, J.E., Robertson, A.H.F. (Eds.), The Geological Evolution of the Eastern Mediterranean. Geol. Soc. London Spec. Publ. 17, 687–699.

Gelabert, B., Sàbat, F., Rodriguez-Perea, A., 2002. A new proposal for the late Cenozoic geodynamic evolution of the western Mediterranean. Terra Nova 14, 93–100.

Ghisetti, F., 1979. Evoluzione neotettonica dei principali sistemi di faglie della Calabria centrale. Boll. Geof. Teor. Appl. 98, 387–430.

Girdler, R.W., 1985. Problems concerning the evolution of oceanic lithosphere in the northern Red Sea. Tectonophysics 116, 109–122.

Grasso, M., Reuther, C.D., Tortorici, L., 1992. Neotectonic deformations in SE Sicily: the Ispica fault, evidence of late Miocene-Pleistocene decoupled wrenching within the central Mediterranean stress regime. J. Geodynamics 16, 135–146.

Grellet, B., Combes, Ph., Granier, Th., Philip, H., 1993. Sismotectonique de la France Métropolitaine. Mem. Soc. Géol. France 164, 76.

Grimaud, S., Boillot, G., Collette, B.J., Mauffret, A., Miles, P., Roberts, D.G., 1982. Western extension of the Iberian–European plate boundary during the early Cenozoic (Pyrenean) convergence: a new model. Marine Geol. 45, 63–77.

Grimison, N.L., Chen, W.P., 1986. The Azores–Gibraltar plate boundary: focal mechanisms, depths of earthquakes, and their tectonic implications. J. Geophys. Res. 91, 2029–2047.

Gueguen, E., Dogliosi, C., Fernandez, M., 1997. Lithospheric boudinage in the Western Mediterranean back-arc basin. Terra Nova 9, 184–187.

Hassani, R., Jongmans, D., Chery, J., 1997. Study of plate deformation and stress in subduction processes using two-dimensional numerical models. J. Geophys. Res. 102, 17951–17965.

Hatzfeld, D., Caillot, V., Cherkaoui, T.-E., Jebli, H., Medina, F., 1993. Microearthquakes seismicity and fault plane solutions around the Nékor strike-slip fault, Morocco. Earth Planet. Sci. Lett. 120, 31–41.

Hempton, M.R., 1987. Constraints on Arabian plate motion and extensional history of the Red sea. Tectonics 6, 687–705.

Hieke, W., Wanninger, A., 1985. The Victor Hensen Seahill (central Ionian sea). Morphology and structural aspects. Mar. Geol. 64, 343–350.

Hieke, W., Dehghani, G.A., 1999. The Victor Hensen structure in the central Ionian Sea and its relation to the Medina Ridge (Eastern Mediterranean). Z. Geol. Ges. 149, 487–505.

Hippolyte, J.C., Badescu, D., Constantin, P., 1999. Evolution of the transport direction of the Carpathian belt during its collision with the east European Platform. Tectonics 18, 1120–1138.

Hubert-Ferrari, A., Armijo, R., King, G., Meyer, B., Barka, A., 2002. Morphology, displacement, and slip rates along the North Anatolian Fault, Turkey. J. Geophys. Res. 107. ETG 9.

Hubert-Ferrari, A., King, G., Manighetti, I., Armijo, R., Meyer, B., Tapponnier, P., 2003. Long-term elasticity in the continental lithosphere; modelling the Aden Ridge propagation and the Anatolian extrusion process. Geophys. J. Int. 153, 111–132.

Jabaloy, A., Galindo-Zaldìvar, J., Gonzàlez-Lodeiro, F., 2002. Palaeostress evolution of the Iberian Peninsula (Late Carboniferous to present-day). Tectonophysics 357, 159–186.

Jackson, J.A., Haines, A.J., Holt, W.E., 1992. Determination of the horizontal velocity field in the deforming Aegean Sea region from the moment tensors of earthquakes. J. Geophys. Res. 97, 17657–17684.

Jacobshagen, V., 1992. Major fracture zones of Morocco: the South Atlas and the Transalboran fault systems. Geol. Rundsch. 81, 185–197.

Jacobshagen, V., Durr, S., Kockel, F., Kopp, K.O., Kowalczyk, G., Berckhemer, H., Buttner, D., 1978. Structure and geodynamic evolution of the Aegean region. In: Closs, H., Roeder, R., Schmidt, K. (Eds.), Alps, Apennines, Hellenides. E. Schweizerbart'sche Verlagsbuchhandlung Nagele u. Obermille, Stuttgart, pp. 537–564.

Jamison, W.R., 1993. Mechanical stability of the triangle zone: the backthrust wedge. J. Geophys. Res. 98, 20015–20030.

Jischke, M.C., 1975. On the dynamics of descending lithospheric plates and slip zones. J. Geophys. Res. 80, 4809–4813.

Jolivet, L., Faccenna, C., 2000. Mediterranean extension and the Africa–Eurasia collision. Tectonics 19, 1095–1106.

Jolivet, L., Faccenna, C., Goffé, B., Mattei, M., Rossetti, F., Brunet, C., Storti, F., Funiciello, R., Cadet, J.P., d'Agostino, N., Parra, T., 1998. Midcrustal shear zones in postorogenic extension: example from the northern Tyrrhenian Sea. J. Geophys. Res. 103, 12123–12160.

Karnik, V., 1969. Seismicity of the European Area, Part I. Reidel, Dordrecht. 364 p.

Kastens, K.A., et al., 1988. ODP Leg 107 in the Tyrrhenian Sea: insights into passive margin and back-arc basin evolution. Geol. Soc. Am. Bull. 100, 1140–1156.

Keep, M., 2000. Models of lithospheric-scale deformation during plate collision: effects of indentor shape and lithospheric thickness. Tectonophysics 326, 203–216.

Kempler, D., Ben Avraham, Z., 1987. The tectonic evolution of the Cyprean Arc. Ann. Tecton. 1, 58–71.

Kempler, D., Garfunkel, Z., 1994. Structures and kinematics in the northeastern Mediterranean: a study of an irregular plate boundary. Tectonophysics 234, 19–32.

Kissel, C., Laj, C., Muller, C., 1985. Tertiary geodynamic evolution of Northwestern Greece: paleomagnetic results. Earth Planet. Sci. Lett. 72, 190–204.

Kissel, C., Speranza, F., Milicevic, V., 1995. Paleomagnetism of external Southern and Central Dinarides and Northern Albanides: implications for the cenozoic activity of the Scutari-Pec transverse zone. J. Geophys. Res. 100, 14999–15007.

Kleinrock, M.C., Morgan, J.P., 1988. Triple junction reorganization. J. Geophys. Res. 93, 2981–2996.

Kohlstedt, D.L., Evans, B., Mackwell, S.J., 1995. Strength of the lithosphere: constraints imposed by laboratory experiments. J. Geophys. Res. 100, 17587–17602.

Laubscher, H., 2001. Plate interaction at the southern end of the Rhine graben. Tectonophysics 343, 1–19.

Lavé, J., Avouac, J.P., Lacassin, R., Tapponnier, P., Montagner, J.P., 1996. Seismic anisotropy beneath Tibet-evidence for eastward extrusion of the Tibetan lithosphere. Earth Planet. Sci. Lett. 140, 83–96.

Le Pichon, X., Bergerat, F., Roulet, M.J., 1988. Plate kinematics and tectonics leading to the Alpine belt formation; a new analysis. Geol. Soc. Am. 218, 111–131.

Lobkovski, L.I., Kerchmann, V.I., 1991. A two-level concept of plate tectonics: application to geodynamics. Tectonophysics 199, 343–374.

Lopez-Casado, C., Sanz de Caldano, C., Molina Palacios, S., Henares Romero, J., 2001. The structure of the Alboran Sea: an interpretation from seismological and geological data. Tectonophysics 338, 79–95.
Lyon-Caen, H., Armijo, R., Drakopoulos, J., Baskoutass, J., Delibassis, N., Gaulon, R., Kouskouna, V., Latoussakis, J., Makropoulos, K., Papadimitriou, O., Papanastassiou, D., Pedotto, G., 1988. The 1986 Kalamata (South Peloponnesus) earthquake: detailed study of a normal fault, evidences for East-West extension in the Hellenic arc. J. Geophys. Res. 93, 14967–15000.
Makris, J., 1978. A geophysical study of Greece based on: deep seismic sounding, gravity and magnetics. In: Closs, H., Roeder, D., Schmidt, K. (Eds.), Alps, Apennines, Hellenides. E. Schweizerbart'sche Verlagsbuchhandlung Nagele u. Obermille, Stuttgart, pp. 393–401.
Malinverno, A., Ryan, W.B.F., 1986. Extension in the Tyrrhenian sea and shortening in the Apennines as result of arc migration driven by sinking of the lithosphere. Tectonics 5, 227–245.
Mantovani, E., Albarello, D., Tamburelli, C., Babbucci, D., 1996. Evolution of the Tyrrhenian basin and surrounding regions as a result of the Africa–Eurasia convergence. J. Geodynamics 21, 35–72.
Mantovani, E., Albarello, D., Tamburelli, C., Babbucci, D., Viti, M., 1997. Plate convergence, crustal delamination, extrusion tectonics and minimization of shortening work as main controlling factors of the recent Mediterranean deformation pattern. Ann. Geofis. 40, 611–643.
Mantovani, E., Albarello, D., Babbucci, D., Tamburelli, C., 1997. Recent/present tectonic processes in the Italian region and their relation with seismic and volcanic activity. Ann. Tecton. 11, 27–57.
Mantovani, E., Albarello, D., Babbucci, D., Tamburelli, C., Viti, M., 2000a. Genetic mechanism of back-arc opening: insights from the Mediterranean deformation pattern. In: Boschi, E., Ekstrom, E., Morelli, A. (Eds.), Problems in Geophysics for the New Millennium. Editrice Compositori, Bologna, pp. 151–178.
Mantovani, E., Viti, M., Albarello, D., Tamburelli, C., Babbucci, D., Cenni, N., 2000b. Role of kinematically induced horizontal forces in Mediterranean tectonics: insights from numerical modeling. J. Geodynamics 30, 287–320.
Mantovani, E., Cenni, N., Albarello, D., Viti, M., Babbucci, D., Tamburelli, C., D'Onza, F., 2001a. Numerical simulation of the observed strain field in the central-eastern Mediterranean region. J. Geodynamics 3, 519–556.
Mantovani, E., Viti, M., Babbucci, D., Tamburelli, C., Albarello, D., 2001b. Back arc extension: which driving mechanism? J. Virtual Explorer 3, 17–44.
Mantovani, E., Viti, M., Cenni, N., Albarello, D., Babbucci, D., 2001c. Short and long-term deformation patterns in the Aegean–Anatolian systems: insights from space geodetic data (GPS). Geophys. Res. Lett. 28, 2325–2328.
Mantovani, E., Albarello, D., Babbucci, D., Tamburelli, C., Viti, M., 2002. Trench-Arc–BackArc systems in the Mediterranean area: examples of extrusion tectonics. J. Virtual Explorer 8, 125–141.
Marroni, M., Treves, B., 1998. Hidden terranes in the Northern Apennines, Italy: a record of late Cretaceous-Oligocene transpressional tectonics. J. Geol. 106, 149–162.
Martinez, M.D., Lana, X., Canas, J.A., Badal, J., Pujades, L., 2000. Shear-wave velocity tomography of the lithosphere-asthenosphere system beneath the Mediterranean area. Phys. Earth Planet. Int. 122, 33–54.
Mascle, J., Chaumillon, P., 1997. Pre-collisional geodynamics of the Mediterranean Ridge and the Tyrrhenian Sea. Ann. Geofis. 40, 569–586.
Masek, J.G., Duncan, C.C., 1998. Minimum-work mountain building. J. Geophys. Res. 103, 907–917.
Mazzoli, S., Helman, M., 1994. Neogene patterns of relative plate motion for Africa–Europe: some implications for recent central Mediterranean tectonics. Geol. Rundsch. 83, 464–468.
McClusky, S., Balassanian, S., Barka, A., Demir, C., Ergintav, S., Georgiev, I., Gurkan, O., Hamburger, M., Hurst, K., Khale, H., Kastens, K., Kekelidze, G., King, R., Kotzev, V., Lenk, O., Mahmoud, S., Mishin, A., Nadariya, M., Ouzounis, A., Paradissis, D., Peter, Y., Prilepin, M., Reilinger, R., Sanli, I., Seeger, H., Tealeb, A., Toksöz, M.N., Veis, G., 2000. Global positioning system constraints on plate kinematics and dynamics in the eastern Mediterranean and Caucasus. J. Geophys. Res. 105, 5695–5719.
McClusky, S., Reilinger, R., Mahmoud, S., Ben Sari, D., Tealeb, A., 2003. GPS constraints on Africa (Nubia) and Arabia plate motion. Geophys. J. Int. 155, 126–138.
McKenzie, D., 1972. Active tectonics of the Mediterranean region. Geophys. J. R. Astr. Soc. 30, 109–185.
McKenzie, D., 1978. Active tectonics of the Alpine–Himalayan belt: the Aegean Sea and surrounding region. Geophys. J. R. Astr. Soc. 55, 217–254.
Meissner, R., Wever, Th., Flüh, E.R., 1987. The Moho in Europe – implications for crustal development. Ann. Geophys. 5, 357–364.
Meissner, R., Mooney, W., 1998. Weakness of lower continental crust: a condition for delamination, uplift and escape. Tectonophysics 296, 47–60.
Mele, G., 1998. High-frequency wave propagation from mantle earthquakes in the Tyrrhenian Sea: new constraints for the geometry of the south Tyrrhenian subduction zone. Geophys. Res. Lett. 25, 2877–2880.

Meletti, C., Patacca, E., Scandone, P., 2000. Construction of a seismotectonic model: the case of Italy. Pure Appl. Geophys. 157, 11–35.

Meloni, A., Alfonsi, L., Florindo, F., Cagnotti, L., Speranza, F., Winkler, A., 1997. Neogene and Quaternary geodynamic evolution of the Italian peninsula: the contribution of Paleomagnetic data. Ann. Geofis. 40, 705–727.

Mercier, J., Sorel, D., Simeakis, K., 1987. Changes in the state of stress in the overriding plate of a subduction zone: the Aegean Arc from the Pliocene to the Present. Ann. Tecton. 1, 20–39.

Mercier, J.L., Simeakis, K., Sorel, D., Vergely, P., 1989. Extensional tectonic regimes in the Aegean basins during the Cenozoic. Basin Res. 2, 49–71.

Meulenkamp, J.E., Van Der Zwan, G.J., Van Wamel, W.A., 1994. On late Miocene to recent vertical motions in the Cretan segment of the Hellenic arc. Tectonophysics 234, 53–72.

Mezcua, J., Buforn, E., Udias, A., Rueda, J., 1992. Seismotectonics of the Canary Islands. Tectonophysics 203, 447–452.

Mezcua, J., Martinez Solares, J.M., 1983. Sismicidad del area Ibero-Moghrebi. Publicacion 203, Instituto Geografico Nacional, 302 p.

Meyers, J.B., Rosendahl, B.R., Harrison, C.G.A., Ding, Z.-D., 1998. Deep-imaging seismic and gravity results from the offshore Cameroon Volcanic Line, and speculation of African hotlines. Tectonophysics 284, 31–63.

Michon, L., Van Balen, R.T., Merle, O., Pagnier, H., 2003. The Cenozoic evolution of the Roer Valley Rift System integrated at a European scale. Tectonophysics 367, 101–126.

Minster, B.J., Jordan, T.H., 1978. Present day plate motions. J. Geophys. Res. 83, 5331–5354.

Molnar, P., Lyon-Caen, H., 1988. Some simple physical aspects of the support, structure and evolution of mountain belts. In: Clark, S.P. (Ed.), Processes in Continental and Lithospheric Deformation. Geol. Soc. Am. Spec. Papers 218, 179–207.

Morel, J.L., Meghraoui, M., 1996. Goringe–Alboran–Tell tectonic zone: a transpression system along the Africa–Eurasia plate boundary. Geology 8, 755–758.

Morelli, A., Piromallo, C., 2000. The late stage of retreating subduction in the Alpine-Mediterranean region: constraints from travel time seismic tomography. In: Boschi, E., Ekstrom, G., Morelli, A. (Eds.), Problems in Geophysics for the New Millennium. Editrice Compositori, Bologna, pp. 179–216.

Mueller, S., Choy, G.L., Spence, W., 1996. Inelastic models of lithospheric stress-I. Theory and application to outer-rise plate deformation. Geophys. J. Int. 125, 39–53.

Nicolas, M., Santoire, J.P., Delpech, P.Y., 1990. Intraplate seismicity: new seismotectonic data in Western Europe. Tectonophysics 179, 27–53.

Nocquet, J.M., Calais, E., Altamimi, Z., Sillard, P., Boucher, C., 2001. Intraplate deformation in western Europe deduced from an analysis of the International Terrestrial Reference Frame 1997 (ITRF97) velocity field. J. Geophys. Res. 106, 239–257.

Nocquet, J.M., Calais, E., 2003. Crustal velocity field of western Europe from permanent GPS array solutions, 1996–2001. Geophys. J. Int. 154, 72–88.

Okaya, N., Cloetingh, S., Mueller, St., 1996. A lithospheric cross-section through the Swiss Alps – II. Constraints on the mechanical structure of a continent-continent collision zone. Geophys. J. Int. 127, 399–414.

Oldow, J.S., Ferranti, L., Lewis, D.S., Campbell, J.K., D'Argenio, B., Catalano, R., Pappone, G., Carmignani, L., Conti, P., Aiken, C.L.V., 2002. Active fragmentation of Adria, the north African promontory, central Mediterranean orogen. Geol. Soc. Am. 9, 779–782.

Ortolani, F., 1979. Alcune considerazioni sulle fasi tettoniche mioceniche e plioceniche dell'Appennino meridionale. Boll. Soc. Geol. It. 97, 609–616.

Ortolani, F., Pagliuca, S., Pepe, E., Schiattarella, M., Toccaceli, R.M., 1992. Active tectonics in the southern Apennines: relationships between cover geometries and basement structure a hypothesis for a geodynamic model. Newsletter 5, 413–419.

Pacanovsky, K.M., Davis, D.M., Richardson, R.M., Coblentz, D.D., 1999. Intraplate stresses and plate-driving forces in the Philippine Sea plate. J. Geophys. Res. 104, 1095–1110.

Papadopoulos, G.A., 1989. Cenozoic magmatism, deep tectonics, and crustal deformation in the Aegean sea. In: Kissel, C., Laj, C. (Eds.), Paleomagnetic Rotations and Continental Deformation. Kluwer Academic, Dordrecht, pp. 93–113.

Papazachos, B.C., Comninakis, P.E., 1982. A catalogue of historical earthquakes in Greece in and surrounding area. In: Publ. Geophys. Lab., University of Thessaloniki, vol. 5, 24 p.

Papazachos, B.C., Nolet, G., 1997. P and S deep velocity structure of the Hellenic area obtained by robust nonlinear inversion of travel times. J. Geophys. Res. 102, 8349–8367.

Papazachos, B.C., Papaioannou, Ch.A., 1999. Lithospheric boundaries and plate motions in the Cyprus area. Tectonophysics 308, 193–204.

Patacca, E., Scandone, P., 1989. Post-Tortonian mountain building in the Apennines. The role of the passive sinking of a relict lithospheric slab. In: Boriani, A., Bonafede, M., Piccardo, G.B., Vai, G.B. (Eds.), The Lithosphere in Italy. In: Advances in Earth Science Research, vol. 80. Accademia Naz. Lincei, Roma, pp. 157–176.
Patacca, E., Sartori, R., Scandone, P., 1990. Tyrrhenian basin and Apenninic arcs: kinematic relations since Late Tortonian times. Mem. Soc. Geol. It. 45, 425–451.
Peltzer, G., Tapponnier, P., 1988. Formation and evolution of strike-slip faults, rifts and basins during the India–Asia collision: an experimental approach. J. Geophys. Res. 93, 15085–15117.
Philip, H., 1987. Plio-Quaternary evolution of the stress field in Mediterranean zones of subduction and collision. Ann. Geophys. 5, 301–320.
Piccardi, L., Sani, F., Bovini, M., Boccaletti, M., Moratti, G., Gualtierotti, A., 1997. Deformazioni quaternarie nell'Appennino centro-settentrionale: evidenze ed implicazioni. Il Quaternario 10, 273–280.
Pinar, A., Kalafat, D., 1999. Source processes and seismotectonic implications of the 1995 and 1996 Cyprus, Eastern Mediterranean region, earthquakes. Tectonophysics 301, 217–230.
Pollack, H.-N., Chapman, D.S., 1977. On the regional variation of heat flow, geotherms, and lithospheric thickness. Tectonophysics 38, 279–296.
Postpischl, D., 1985. Catalogo dei terremoti italiani dall'anno 1000 al 1980. CNR–PFG, Roma. 242 p.
Ranalli, G., Murphy, D.G., 1987. Rheological stratification of the lithosphere. Tectonophysics 132, 281–295.
Ratschbacher, L., Merle, O., Davy, P., Cobbold, P., 1991. Lateral extrusion in the eastern Alps. Part I: Boundary conditions and experiments scaled for gravity. Tectonics 10, 245–256.
Rebai, S., Philip, H., Taboada, A., 1992. Modern tectonic stress field in the Mediterranean region: evidence for variation in stress directions at different scale. Geophys. J. Int. 110, 106–141.
Rehault, J.P., Boillot, G., Mauffret, A., 1984. The western Mediterranean basin geological evolution. Mar. Geol. 55, 447–477.
Rehault, J.P., Moussat, E., Fabbri, A., 1987. Structural evolution of the Tyrrhenian back-arc basin. Mar. Geol. 74, 123–150.
Renner, G., Slejko, D., 1994. Some comments on the seismicity of the Adriatic region. Boll. Geof. Teor. Appl. 36, 381–398.
Reuther, C.D., 1987. Extensional tectonic within central Mediterranean segment of the Afro-European zone of convergence. Mem. Soc. Geol. It. 38, 69–80.
Ribeiro, A., Kullberg, M.C., Kullberg, J.C., Manupella, G., Phipps, S., 1990. A review of Alpine tectonics in Portugal: Foreland detachment in basement and cover rocks. Tectonophysics 184, 357–366.
Ribeiro, A., Cabral, J., Baptista, R., Matias, L., 1996. Stress pattern in Portugal mainland and the adjacent Atlantic region, West Iberia. Tectonics 15, 641–659.
Rosenbaum, G., Lister, G.S., Duboz, C., 2002. Relative motions of Africa, Iberia and Europe during Alpine orogeny. Tectonophysics 359, 117–129.
Rossi, S., Sartori, R., 1981. A seismic reflection study of the External Calabrian Arc in the Northern Ionian Sea (Eastern Mediterranean). Mar. Geophys. Res. 4, 403–426.
Royden, L.H., 1993. The tectonic expression of slab pull at continental convergent boundaries. Tectonics 12, 303–325.
Royden, L.H., 1993. Evolution of retreating subduction boundaries formed during continental collision. Tectonics 12, 629–638.
Royden, L.H., Patacca, E., Scandone, P., 1987. Segmentation and configuration of subducted lithosphere in Italy: an important control on thrust belt and foredeep-basin evolution. Geology 15, 714–717.
Royden, L.H., Baldi, T., 1988. Early Cenozoic tectonics and paleogeography of the Pannonian and surrounding regions. In: Royden, L.H., Horvath, F. (Eds.), The Pannonian Basin: A Study in Basin Evolution. AAPG Memoir 45, 1–16.
Royden, L.H., Burchfiel, B.C., 1989. Are systematic variations in thrust belt style related to plate boundary processes? (The western Alps versus the Carpathians). Tectonics 8, 51–61.
Ryan, W.B.F., 1978. Messinian badlands on the southeastern margin of the Mediterranean sea. Mar. Geol. 27, 349–363.
Ryan, W.B.F., Kastens, K.A., Cita, M.B., 1982. Geological evidence concerning compressional tectonics in the eastern Mediterranean. Tectonophysics 86, 213–242.
Rydelek, P.A., Sacks, I.S., 1990. Asthenospheric viscosity and stress diffusion: a mechanism to explain correlated earthquakes and surface deformation in NE Japan. Geophys. J. Int. 100, 39–58.
Saada, A.S., 1974. Elasticity Theory and Applications. Pergamon, New York. 643 p.
Sagnotti, L., 1992. Paleomagnetic evidence for a Pleistocene counterclockwise rotation of the Sant'Arcangelo basin, Southern Italy. Geophys. Res. Lett. 19, 135–138.
Sartori, R., 1989. Evoluzione neogenico-recente del bacino tirrenico ed i suoi rapporti con la geologia delle aree circostanti. Giorn. Geol., Ser. 3 51 (2), 1–39.

Sartori, R., 1990. The main results of ODP LEG 107 in the frame of neogene to recent geology of perityrrhenian areas. In: Kastens, K.A., Mascle, J., et al. (Eds.), Proc. of the Ocean Drilling Program, Scientific Results, vol. 107, pp. 715–731.
Sartori, R., Capozzi, R., 1998. Patterns of Neogene to Recent rift-related subsidence in the Tyrrhenian domain. In: Cloeting, S., Ranalli, G., Ricci, C.A. (Eds.), Sedimentary Basins: Models and Constraints. Tipografia Senese, Siena, pp. 147–158.
Savostin, L.A., Sibuet, J.C., Zonenshain, L.P., Le Pichon, X., Roulet, M.R., 1986. Kinematic evolution of the Tethys belt from the Atlantic Oceanic to the Pamirs since the Triassic. Tectonophysics 123, 1–35.
Scheepers, P.J.J., Langereis, C.G., Zijderveld, J.D.A., Hilgen, F.J., 1994. Paleomagnetic evidence for a Pleistocene clockwise rotation of the Calabro-Peloritan block (southern Italy). Tectonophysics 230, 19–48.
Scholz, C.H., Campos, J., 1995. On the mechanism of seismic decoupling and back arc spreading at subduction zones. J. Geophys. Res. 100, 22103–22115.
Schreiber, U., Rotsch, S., 1998. Cenozoic block rotation according to a coniugate shear system in central Europe-indications from paleomagnetic measurements. Tectonophysics 299, 111–142.
Sella, G.F., Dixon, T.H., Mao, A., 2002. REVEL: a model for recent plate velocities from space geodesy. J. Geophys. Res. 107. ETG, pp. 1–32.
Selli, R., Colantoni, P., Fabbri, A., Rossi, S., Borsetti, A.M., Gallignani, P., 1978. Marine geological investigation on the Messina Strait and its approaches. Giorn. Geol. 42, 1–70.
Semenza, E., 1974. La fase Giudicariense, nel quadro di una nuova ipotesi sull'Orogenesi Alpina nell'area Italo-Dinarica. Mem. Soc. Geol. It. 13, 187–226.
Sengor, A.M.C., 1993. Some current problems on the tectonic evolution of the Mediterranean during the Cainozoic. In: Boschi, E., Mantovani, E., Morelli, A. (Eds.), Recent Evolution and Seismicity of the Mediterranean Region. Kluwer Academic, Dordrecht, pp. 1–51.
Sengor, A.M.C., Gorer, N., Saroglu, F., 1985. Strike-slip faulting and related basin formation in zones of tectonic escape: Turkey as a case study. Soc. Econ. Paleontol. Mineral., spec. pub. 37, 227–264.
Serri, G., Innocenti, F., Manetti, P., 1993. Geochemical and petrological evidence of subduction of delamination Adriatic continental lithosphere in the genesis of the Neogene-Quaternary magmatism of Central Italy. Tectonophysics 223, 117–147.
Seyferth, M., Henk, A., 2004. Syn-convergent exhumation and lateral extrusion in continental collision zones-insights from three-dimensional numerical models. Tectonophysics 382, 1–29.
Shemenda, A.I., 1993. Subduction of the lithosphere and back-arc dynamics: insights from physical modeling. J. Geophys. Res. 98, 167–185.
Sissingh, W., 1998. Comparative Tertiary stratigraphy of the Rhine Graben, Bresse Graben and Molasse Basin: correlation of the Alpine foreland events. Tectonophysics 300, 249–284.
Sleep, N.H., Stein, S., Geller, R.J., Gordon, R.G., 1979. Comment on "The use of the minimum-dissipation principle in tectonophysics", by P. Bird and D.A. Yuen. Earth Planet. Sci. Lett. 45, 218–220.
Sornette, A., Davy, P., Sornette, D., 1993. Fault growth in brittle-ductile experiments and the mechanics of continental collisions. J. Geophys. Res. 98, 12111–12139.
Souriau, A., Sylvander, M., Rigo, A., Fels, J.-F., Douchain, J.-M., Ponsolles, C., 2001. Sismotectonique des Pyrénées: principales contraintes sismologiques. Bull. Soc. Géol. France 172, 25–39.
Stiros, S.C., 1988. Model for the N. Peloponnesian (C. Greece) uplift. J. Geodynamics 9, 199–214.
Sue, C., Thouvenot, F., Fréchet, J., 1999. Widespread extension in the core of the western Alps revealed by earthquake analysis. J. Geophys. Res. 104, 611–622.
Tamburelli, C., Babbucci, D., Mantovani, E., 2000. Geodynamic implications of subduction related magmatism: insights from the Tyrrhenian–Apennines region. J. Volc. Geoth. Res. 104, 33–43.
Tapponnier, P., 1977. Evolution tectonique du système alpin en Méditerranée: poinçonnement et écrasement rigide-plastique. Bull. Soc. Géol. France 19, 437–460.
Tapponnier, P., Molnar, P., 1976. Slip-line field theory and large scale continental tectonics. Nature 264, 319–324.
Tapponnier, P., Peltzer, G., Le Dayn, A.Y., Armijo, R., Cobbold, P., 1982. Propagating extension tectonics in Asia: new insights from simple experiments with plasticine. Geology 110, 611–616.
Tapponnier, P., Peltzer, G., Armijo, R., 1986. On the mechanics of the collision between India and Asia. In: Coward, M.P., Ries, A.C. (Eds.), Collision Tectonics. Geol. Soc. London Spec. Publ. 19, 115–157.
Taylor, B., Karner, G.D., 1983. On the evolution of marginal basins. Rev. Geophys. 21, 1727–1741.
Taymaz, T., Jackson, J., McKenzie, D., 1991. Active tectonics of the North and Central Aegean Sea. Geophys. J. Int. 106, 433–490.
Tovish, A., Schubert, G., Luyendyk, B.P., 1978. Mantle flow pressure and the angle of subduction: non-Newtonian corner flows. J. Geophys. Res. 83, 5892–5898.
Townend, J., Zoback, M.D., 2000. How faulting keeps the crust strong. Geology 28, 399–402.

Turcotte, D.L., Schubert, G., 1982. Geodynamics – Applications of Continuum Physics to Geological Problems. Wiley, New York. 450 p.
Underhill, J.R., 1989. Late Cenozoic deformation of the Hellenides foreland, western Greece. Geol. Soc. Am. Bull. 101, 613–634.
Uyeda, S., 1986. Facts, ideas and open problems on trench-arc-backarc systems. In: Wezel, F.C. (Ed.), The Origin of Arc. Elsevier, Amsterdam, pp. 435–460.
Uyeda, S., Kanamori, H., 1979. Back-arc opening and the mode of subduction. J. Geophys. Res. 84, 1049–1061.
Vai, G.B., 1987. Migrazione complessa del sistema fronte deformativo-avanfossa-Cercine periferico: il caso dell'Appennino settentrionale. Mem. Soc. Geol. It. 38, 95–105.
Van den Beukel, J., 1992. Some thermomechanical aspects of the subduction of continental lithosphere. Tectonics 11, 316–329.
Van Dijk, J.P., Okkes, M., 1991. Neogene tectonostratigraphy and kinematics of Calabrian basins; implications for the geodynamics of the Central Mediterranean. Tectonophysics 196, 23–60.
Vigliotti, L., Laugenheim, V.E., 1995. When did Sardinia stop rotating? New paleomagnetic results. Terra Nova 7, 424–435.
Viti, M., Albarello, D., Mantovani, E., 1997. Rheological profiles in the central-eastern Mediterranean. Ann. Geofis. 40, 849–864.
Viti, M., Albarello, D., Mantovani, E., 2001. Classification of seismic strain estimates in the Mediterranean region from a 'bootstrap' approach. Geophys. J. Int. 146, 399–415.
Viti, M., De Luca, J., Babbucci, D., Mantovani, E., Albarello, D., D'Onza, F., 2004. Driving mechanism of tectonic activity in the northern Apennines: quantitative insights from numerical modeling. Tectonics 23. TC4005, 1–16.
Vogt, P.R., Jung, W.Y., 2004. The Terceira Rift as hyper-slow, hotspot-dominated oblique spreading axis: a comparison with other slow-spreading plate boundaries. Earth Planet. Sci. Lett. 218, 77–90.
Wang, K., 2000. Stress-strain 'paradox', plate coupling, and forearc seismicity at the Cascadia and Nankai subduction zones. Tectonophysics 319, 321–338.
Weijermars, R., 1987. A revision of the Eurasian–African plate boundary in the western Mediterranean. Geol. Rundsch. 76, 667–676.
Westaway, R., 1990. Present-day kinematics of the plate boundary zone between Africa and Europe, from the Azores to the Aegean. Earth Planet. Sci. Lett. 96, 393–406.
Westaway, R., 1993. Quaternary uplift of southern Italy. J. Geophys. Res. 98, 741–772.
Westaway, R., 1994. Present-day kinematics of the Middle East and eastern Mediterranean. J. Geophys. Res. 99, 12071–12090.
Westaway, R., 1995. Crustal volume balance during the India-Eurasia collision and altitude of the Tibetan plateau: a working hypothesis. J. Geophys. Res. 100, 173–192.
Working Group Parametric Catalog of Italian Earthquakes, 1999. Catalogo Parametrico dei terremoti italiani. ING, GNDT, SGA, SSN. Editrice Compositori, Bologna. 92 p.
Yamaoka, K., 1988. Spherical shell tectonics: on the buckling of the lithosphere at subduction zones. Tectonophysics 147, 179–191.
Yilmaz, Y., 1989. An approach to the origin of young volcanic rocks of western Turkey. In: Sengor, A.M.C. (Ed.), Tectonic Evolution of the Tethyan Region. Kluwer Academic, Dordrecht, pp. 159–189.
Zanchi, A., Kissel, C., Tapirdamaz, C., 1990. Deformazione continentale Neogenico-Quaternaria in Anatolia occidentale: nuovi dati strutturali. Mem. Soc. Geol. It. 45, 481–488.
Zeilinga de Boer, J., 1989. The Greek enigma: is development of the Aegean Orogene dominated by forces related to subduction or obduction? Mar. Geol. 87, 31–54.
Zhao, W.L., Morgan, J., 1985. Uplift of Tibetan plateau. Tectonics 4, 359–369.

CROP PROJECT: Deep Seismic Exploration of the Central Mediterranean and Italy
Edited by I.R. Finetti

CHAPTER 33

Plate Tectonics of the Apulia–Adria Microcontinents

Gérard M. Stampfli[1]

ABSTRACT

New data for the East Mediterranean domain allow us to assign this oceanic basin to the larger Neotethyan oceanic system which opened in Permian times. This interpretation is supported by a large array of geological and geophysical data, including the CROP deep seismic data. The same data set shows that a Greater Apulia plate existed in Mesozoic times, including the autochthonous Greece and SW Turkey plate. This plate also included a united Adria and Apulia microplate from Early Jurassic times. This key information indicates the need to create a new post-Variscan continental fit for the western Tethyan area, where the relationships between the Adrian, Apulian and Iberic plates can be defined with greater confidence.

Keywords: Apulia–Adria, plate tectonic context

1. INTRODUCTION

In the light of the large amount of new data provided by the CROP Atlas, we are reviewing here the geodynamic evolution in space and time of the Adria and Apulia micro-continents. This evolution has so far been described in many ways, rich with controversies. A wealth of data has also been accumulated through nearly two centuries of generally excellent field work by Italian geologists, but recent approaches have shown that more could be added, particularly in the field of plate tectonics, supported by new geophysical datasets. Detailed plate tectonic schemes and geophysical datasets are presented in the present volume, but we focus here on the larger framework of plate tectonics in the Mediterranean area, particularly on some major issues regarding its Late Palaeozoic and Mesozoic evolution.

One of the main issues is the age of the East Mediterranean–Ionian Sea basin, and the nature of the seafloor in this area. This, in turn, influences the way that continental re-assembly for the Mesozoic can be done. These major problems were approached by the editor and the author from different points of view, but an agreement rapidly emerged about the Permo-Triassic age of the E-Mediterranean–Ionian seafloor, while disagreement occurred over the proposal of separate Adria and Apulia micro-continents in Jurassic and Cretaceous times.

A new continental fit was built up in an attempt at reconciling plate tectonics with geophysical data (Finetti et al., 2001) and is presented here: it differs significantly from previous proposed models (e.g. Stampfli and Borel, 2002b; Stampfli et al., 2001b), particularly regarding the position of Adria and Iberia.

Before presenting the reconstructions and geodynamic scheme, we first review the problem of the age of opening of the East Mediterranean basin and its link with the Neotethys oceanic realm. Then, we discuss the necessity of having a close to the present position of Adria and Apulia already from Late Triassic times. These two items can then be used to support our present new plate tectonic scenario.

2. THE EAST MEDITERRANEAN NEOTETHYS CONNECTION

This problem has already been reviewed in some detail in previous publications (Stampfli, 2000; Stampfli et al., 2001b); depending on the authors, the East Mediterranean–Ionian Sea basin is regarded as opening already in the

[1]Institute of Geology and Palaeontology, University of Lausanne, BFSH2, CH 1015, Lausanne, Switzerland
E-mail: gerard.stampfli@unil.ch

Late Paleozoic (Vai, 1994) or as late as the Cretaceous (e.g. Dercourt et al., 1993, 1985). Most people would regard this ocean as opening in Late Triassic or Early Jurassic (e.g. Garfunkel and Derin, 1984; Robertson et al., 1996; Sengör et al., 1984) and therefore possibly related to the Alpine Tethys–Central Atlantic opening (Stampfli et al., 1998).

A new interpretation showing that the East Mediterranean domain has corresponded to an oceanic basin since the Late Paleozoic was proposed by Stampfli (1989; IGCP 276 conference). Subsequently, new plate tectonic reconstructions considering this basin as part of the Neotethyan oceanic system have been developed (Stampfli et al., 1991, 2001b; Stampfli and Pillevuit, 1993), supported by a large array of geological and geophysical data, such as:

- the geophysical characteristics of the Ionian Sea and East Mediterranean basin (isostatic equilibrium, seismic velocities, elastic thickness) excluding an age of the sea-floor younger than Early Jurassic.
- synrift sequences of the Neotethys northern margin have been found in the Talea Ori, in Crete. There the Early Permian sequence (König and Kuss, 1980) presents a typical synrift evolution: rapid subsidence, important clastic input, rapid flooding with pelagic facies, followed by the progradation of a Middle to Upper Permian platform (Kock, 2003).
- subsidence patterns in areas such as the Sinai margin, the Tunisian Jeffara rift, Sicily and Apulia s.str. (Stampfli, 2000; Stampfli et al., 2001b) confirming a Late Permian onset of thermal subsidence for the East Mediterranean and Ionian Sea basins and the absence of any younger thermal events.
- Late Permian Hallstatt-type pelagic limestones, similar to those found in Oman where they sometimes rest directly on MORB (Niko et al., 1996; Pillevuit, 1993), reported from the Sosio complex in Sicily (Kozur, 1995). This Late Permian Sicanian pelagic macrofauna, and microfauna (Kozur, 1990, 1991b, 1991c) presents affinities with both Oman and Timor, and implies a Middle Permian direct deepwater connection between the westernmost East-Mediterranean Basin and the Neotethys.
- Triassic MORB found in Cyprus in the Mamonia complex (Malpas et al., 1993), and derived from the East Mediterranean seafloor
- Recent seismostratigraphy of the Ionian Sea (Finetti and Del Ben, this volume, Chapter 19; Stampfli et al., 2001b) have confirmed a Permo-Triassic age of the Ionian seafloor, as well as its oceanic nature.

In conclusion, we propose a Middle to Late Permian onset of seafloor spreading in the Eastern Mediterranean basin. This opening would be concomitant with the opening of the Neotethys eastward and the northward drifting of the Cimmerian continents since late Early Permian. This model implies also a late closure of the Paleotethys (Middle to Late Triassic) on a Mediterranean transect of the Tethyan realm, accompanied by the opening of back-arc basins on the active Eurasian margin (Meliata, Maliac and Pindos back-arc basins) (Stampfli, 1996; Stampfli et al., 2003; Ziegler and Stampfli, 2001).

The original position of Apulia with respect to Africa can be quite well determined by closing the Ionian sea with a 40° rotation of Apulia around a point located north of Tunisia (Finetti and Del Ben, this volume, Chapter 19). This represents the total rotation of Apulia from Late Carboniferous to Middle/Late Triassic, but a more complicated scheme should be envisaged when considering that Apulia belonged for a while to the drifting Cimmerian continent, whose rotation point changed in time. Therefore a slightly different placing and rotation of Apulia is shown in the following reconstructions.

3. THE APULIA ADRIA PROBLEM

Paleomagnetic data show that the Apulian plate s.l. (Italy) has suffered relatively little rotation with respect to Africa since the Triassic (e.g. Channell, 1992, 1996). Also, the continuity between the active subduction zone under Greece and the outer Dinarides (de Jonge et al., 1994; Wortel and Spakman, 1992) shows that there is a possible plate limit between Apulia s.l. and the autochthon of Greece. However, we regard this feature as recent and having no bearing on the fact that a Greater Apulia plate existed in Mesozoic times, in which all the Greek autochthon is included, as well as the Bey-Daglari massif of SW Turkey. The apparent present plate limit comes from the fact that the Hellenic orogenic/accretionary wedge is oblique with regard to former paleogeographic domains, still colliding with Apulia on an Albanides transect and already subducting the East-Mediterranean seafloor on a Greek transect.

An initial reassembling of the western Tethyan micro-plates in a pre-break-up position led us to consider that the Apulian plate s.l. is most likely cut into two pieces, an Apulian plate s.str. to the south and an Adriatic plate s.str. to the north (Stampfli et al., 2001a; Stampfli and Mosar, 1999). The CROP seismic lines through the Adriatic domain (Finetti and Del Ben, this volume, Chapter 23) clearly show that there has been no major tectonic event cutting Italy into two units, at least since the Jurassic, and that the previous model had to be reassessed.

Our basic hypothesis is still that the Apulian part of Italy is definitely an African promontory from the Middle Triassic to Recent times without much displacement with regard to Africa. It represents the easternmost Cimmerian element detached from Gondwana in Middle Permian times during the opening of the East Mediterranean–Neotethys basin. We also consider that the Adriatic and Apulian micro-plates were welded in an Eocimmerian collision phase during the Late Permian–Early/Middle Triassic, and both units then became part of the African plate, giving birth to the present Italian configuration, called indifferently Apulia s.l. or Adria s.l.

The need for cutting Italy into two micro-plates came from the fact that in a Triassic Pangean fit, as used in our former model (Stampfli and Borel, 2002a; Stampfli and Mosar, 1999), there was very little room to insert the present form of Italy in its proper place, knowing that the Apulian promontory had reached its present location already in the Middle Triassic. Thus, we had to reconsider the fit of Iberia with Europe, as well as the size and position of the Alboran micro-plates. We made a much tighter fit of these elements with Europe, following similar older proposals (Srivastava et al., 1990; Srivastava and Tapscott, 1986). Still, there was not enough room to put the entire present length of Adria s.l. in its proper position. The conclusion was that Adria s.l. was shorter in Triassic times that it finally is now (or already in the Late Jurassic). We came to the conclusion that a few hundred kilometres had been gained through major phases of rifting affecting Greater Apulia since the Triassic and related to the break-up of Pangea and opening of the Alpine Tethys. We had already shown through subsidence patterns that the Lombardian rift was an aborted branch of the Central Atlantic rifting extension towards the Alpine region (Stampfli, 2000). The CROP seismic data testify to pervasive Early to Middle Jurassic rifting affecting many areas around Italy, and extending up to Greece eastward and Pelagian domain southward, where it is accompanied by important volcanism (Finetti, 1985). The reason for this widespread Jurassic rifting is that for the Pangean break-up to succeed, the Atlantic rift system had to join an active plate limit located in the Neotethyan domain. All possible ways to break through the Alpine-Mediterranean lithosphere were tried and many resulted in aborted rifts. This pervasive Jurassic rifting finally gave birth, in the Alps, to very large passive margins, flanked by rim basins (e.g. the Subbriançonnais, Helvetic-Dauphinois domains, Lombardian basin, Subbetic basin), and a narrow oceanic strip dominated by mantle denudation (Stampfli and Marchant, 1997). This can be regarded as forced rifting through an already thinned lithosphere, and effectively all this rifting took place under water, with little isostatic rebound of rift shoulders.

The fact that the Alpine Tethys ocean floor is dominated by denuded continental mantle of mainly Permian age (Rampone and Piccardo, 2000) proves that the ocean was very narrow, and a large amount of extension was distributed elsewhere in Greater Apulia. Thus, the present length and geometry of the Adriatic plate would have been achieved by Middle to Late Jurassic times. However, deformation of this plate interior during Alpine times is most likely, but only on a small scale, despite the fact that most of its border was involved in subduction and/or crustal shortening.

4. THE GEODYNAMIC EVOLUTION OF GREATER APULIA AND SURROUNDING REGIONS

The reconstructions shown in Figures 1 to 8 are based on a tight pre-Pangea break-up, Permian fit as well as magnetic anomalies from the Central Atlantic. Plate tectonic concepts have been systematically applied to our palinspatic models of the western Tethys, moving away from pure continental drift models, not constrained by plate limits, to produce a model which finally is progressively more self-constrained. Most other constraints and data bases used for the reconstructions can be found in our previous publications (Stampfli and Borel, 2000; Stampfli et al., 2001a; Stampfli and Borel, 2002b; Stampfli et al., 2001b).

We shall review here some of the major steps of this peri-Apulia geodynamic evolution, starting with the Paleotethys ocean; then we shall move to the Eocimmerian event, and finally to the Alpine cycle proper.

(a) Paleotethys Evolution. The opening of Paleotethys is related to the drifting from Gondwana of ribbon-type micro-continents, grouped under the label of the Hun superterrane (von Raumer et al., 2002). Amongst these terranes we find the Adria s.str. part of Italy, which together with other European Hunic terranes was accreted to Laurussia in Devonian times (Stampfli et al., 2002b). Following this accretion, Paleotethys started subducting northward, creating the Variscan cordillera system. The Paleotethys was a circum-Gondwana ocean whose history

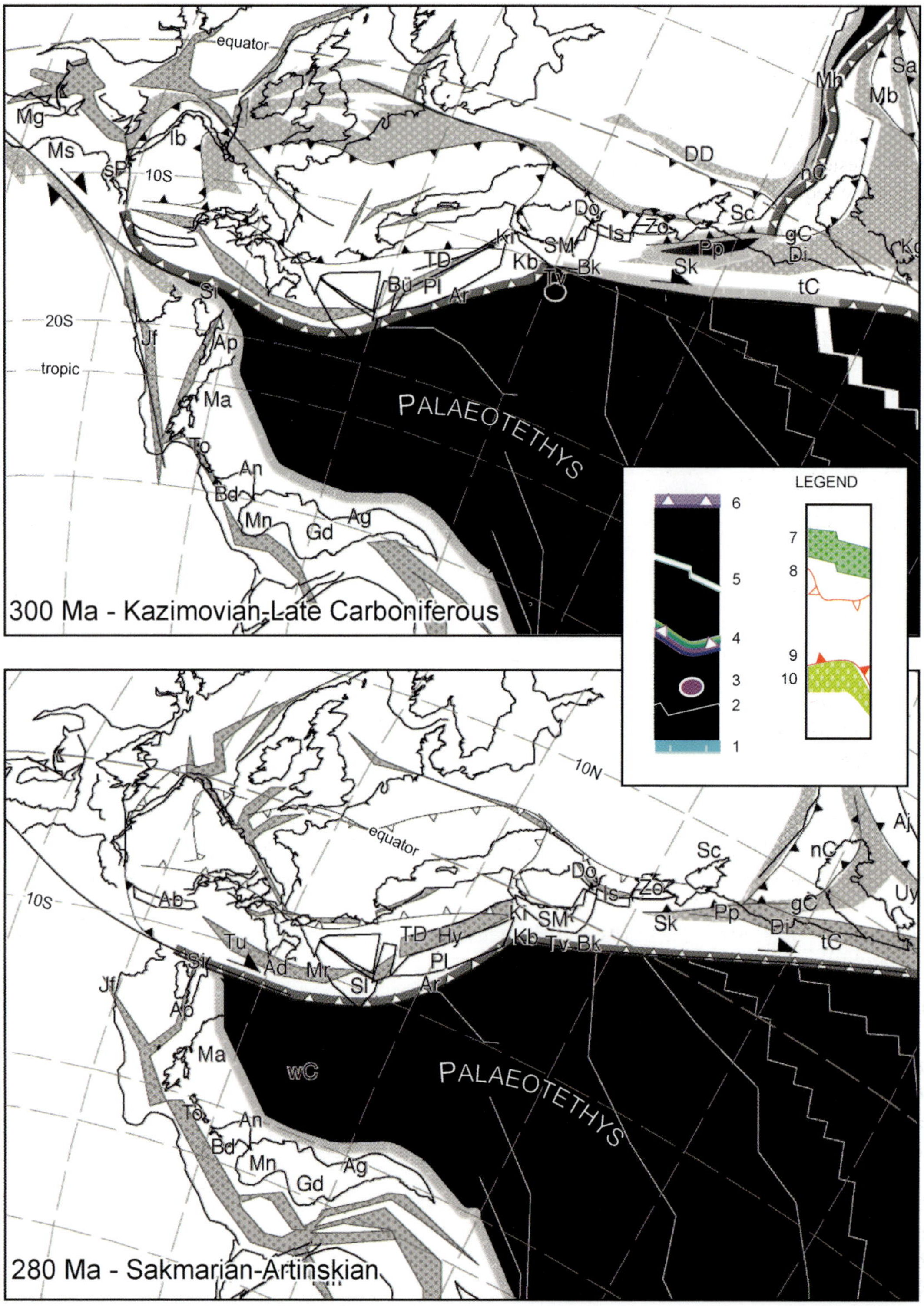

Figure 1: Western Tethys reconstructions for the Late Carboniferous and Early Permian, modified from Stampfli and Borel (2002a, 2002b). 1 – passive margin; 2 – magnetic anomalies or synthetic anomalies; 3 – seamount; 4 – intraoceanic subduction/arc complex; 5 – spreading ridges; 6 – subduction zone; 7 – rifts; 8 – sutures; 9 – active thrusts; 10 – foreland basins. Abbreviations of key localities: AA, Austro-Alpine; Ab, Alboran; Ad, Adria s.str.; Ag, Aladag; Aj, Ajat; Al, Alborz; An, Antalya; Ap, Apulia s.str.; AP, Aspromonte, Peloritani; Ar, Arna accretionary complex; As, Apuseni-south, ophiolites; At, Attika; Au, Asterousia; Av, Arvi; Ba, Balkanides, external; Bc, Biscay, Gascogne; Bd, Beydaghlari; Be, Betic; BH, Baer Basit-Hatay ophiolites; Bh, Bihar; Bk, Bolkardag; Bn, Bernina; Br, Briançonnais; Bs, Bisitoun seamount; BS, Bator–Szarvasko ophiolites; Bu, Bucovinian; Bü, Bükk; Bv, Budva; By, Beyshehir; Bz, Beykoz basin; Ca, Calabria autochthon; cB, central Bosnia; cD, central Dinarides ophiolites; Ce, Cetic; Ci, Ciotat flysch;

is quite well constrained on an Iranian transect (Alborz range, North Iran; Stampfli, 1978; Stampfli et al., 1991, 2001b) which represents the southern Gondwanan margin of the eastern branch of the ocean.

Toward the west, along the African domain, there are fewer constraints concerning the opening of the western branch of Paleotethys and the detachment of the European Hunic Terranes from Gondwana. In the High Atlas of Morocco (Destombes, 1971), the Silurian is unconformable on the Ordovician and presents locally, at its top, a conglomeratic sequence; the overlying Emsian-Eifelian sequence is locally very condensed and is represented by open marine carbonates. This starvation event represents the onset of important thermal subsidence, which can be related to the drifting of either Armorica or the Meseta fragments from Africa.

The northern margin of the Paleotethys ocean is well represented in the middle part of the European Hunic terranes in the Carnic Alps (Schönlaub and Histon, 1999), in Tuscany and Sardinia and the Alboran fragments (cf. Stampfli, 1996). It is also characterized by a Late Ordovician–Early Silurian clastic and often volcanic synrift sequence (Silurian flood basalts are also known in Sardinia and the Rif (Piqué, 1989)). Rift-related thermal uplift, erosion and tilting took place in Silurian times and is often wrongly related to the Taconic event. Open marine conditions started in the Silurian, being represented by a graptolite facies; a more general flooding took place in the Early Devonian and marked the onset of widespread thermal subsidence related to sea-floor spreading. On the northern margin, the Visean usually marks the onset of widespread flysch deposition, often accompanied by volcanic activity, corresponding to the change from a passive to an active margin.

Potential Paleotethyan accretionary sequences are located in the southern part of the Variscan orogen, and in all cases metamorphosed and intruded by subsequent Late Carboniferous granites (Vavassis et al., 2000) and usually involved in Eocimmerian and Alpine deformation. They are mainly found in the Chios-Karaburun region and in Crete (Stampfli et al., 2003), but certainly extend throughout most areas of Turkey comprised between the Pontides and Taurides domains (e.g. Kozur et al., 1998), and should also be present under Apulia, to reappear westwards in southern Spain (Stampfli et al., 2002b). The youngest fore-arc-type sequences related to the northwards subduction of Paleotethys are found in Crete, Turkey, Iran and Afghanistan, and usually extend up to the Carnian (Stampfli

---

CL, Campania-Lucania; Co, Codru; Cn, Carnic-Julian; CP, Calabria-Peloritani; cR, circum-Rhodope; Cv, Canavese; Da, Dacides; Db, Dent Blanche; DD, Dniepr–Donetz rift; Dg, Denizgören ophiolite (IP suture); Di, Dizi accretionary complex; Do, Dobrogea; Dr, Drina-Ivanjica; Ds, Drimos, Samothace ophiolites; Du, Durmitor; eA, east Albanian ophiolites; El, Elazig, Guleman ophiolites-arc; eP, east Pontides; Er, Eratosthen seamount; Fa, Fatric; Fc, Flamish cap; GB, Grand Banks; gC, great Caucasus; Gd, Geydag-Anamas-Akseki; Ge, Gemeric; GT, Gavrovo-Tripolitza; Gt, Getic; Gü, Gümüshane-Kelkit; hA, high-Atlas; Ha, Hadim; He, Helvetic rim basin; Hg, Huglu-Boyalitepe; hK, high karst; Hr, Hronicum; Hy, Hydra; Ib, Iberia, NW allochthon; Ig, Igal trough; Io, Ionian; Is, Istanbul; Ja, Jadar; Jf, Jeffara rift; Jv, Juvavic; Ka, Kalnic; Kb, Karaburun; Ke, Kotel flysch.; Kg, Karabogaz Gol; Ki, Kirshehir; Kk, Karakaya forearc; Ko, Korab; KS, Kotel-Stranja rift; Ku, Kura; Kü, Küre ocean; Ky, Kabylies; La, Lagonegro; lA, lower Austroalpine; Lb, Longobucco; Le, Lesbos ophiolites; Li, Ligurian; Lo, Lombardian; Ls, Lusitanian; Lu, Lut; Ly, Lycian ophiolitic complex; mA, middle Atlas; Ma, Mani; Mb, Magnitogorsk back-arc; Mc, Maliac rift/ocean; MD, Moldanubian; Me, Meliata rift/ocean; Mf, Misfah seamount; Mg, Meguma; Mh, Mugodzhar ocean; Mi, Mirdita autochthon; Mk, Mangyshlak rift; Ml, Meglenitsa ophiolite; Mm, Mamonia accretionary complex; Mn, Menderes; Mo, Moesia; MP, Mersin, Pozanti ophiolites; Mr, Mrzlevodice fore-arc; Ms, Meseta Morocco; MS, Margna-Sella; Mt, Monte Amiata fore-arc; Mz, Munzur dag-Keban; nC, north Caspian; Ni, Nilüfer seamount; Ns, Niesen flysch; nT, north Tibet; Nt, Nish-Troyan trough; OM, Ossa–Morena; Ot, Othrys-Evia-Argolis ophiolites; Oz, Otztal–Silvretta; Pa, Panormides; Pd, Pindos rift/ocean; Pe, Penninic; Pi, Piemontais; Pk, Paikon intra oceanic arc; Pl, Pelagonian; Pm, Palmyra rift; Pn, Pienniny rift; Pp, Paphlagonian ocean; Px, Paxi; Py, Pyrenean rift; Rf, Rif, external; Rh, Rhodope; Ri, Rif, internal; Sa, Salum; sA, south Alpine; sB, sub-Betic rim basin; Sc, Scythian platform; sC, south Caspian basin; Sd, Srednogorie rift-arc; Se, Sesia; Si, Sicanian; Sj, Strandja; Sk, Sakarya; sK, south-Karawanken fore-arc; Sl, Slavonia; Sm, Silicicum; SM, Serbo-Macedonian; Sn, Sevan ophiolites; sP, south Portuguese; Sr, Severin ophiolites; SS, Sanandaj–Sirjan; St, Sitia; Su, Sumeini; Sv, Svanetia; Tb, Tabas; TB, Tirolic–Bavaric; tC, Transcaucasus; TD, Trans-Danubian; Th, Thrace basin; Tk, Tuarkyr; To, Talea Ori; Tp, Troodos ophiolite; Tt, Tatric; Tu, Tuscan; Tv, Tavas + Tavas seamount; Ty Tyros fore arc; Tz, Tizia; uJ, upper Juvavic; UM, Umbria-Marches; Uy, Ust-Yurt; Va, Valais trough; Ve, Veporic; Vo, Vourinos (Pindos)-Mirdita ophiolites; wC, west-Crete (Phyl-Qrtz) accr. complex; Zl, Zlatibar ophiolites; Zo, Zonguldak.

et al., 2003). Then, they are involved in the Eocimmerian tectonic event corresponding to the final closure of Paleotethys in the western Tethyan region.

(b) Eocimmerian Event and the Marginal Oceans. An apparent lack of major tectonic events during the Permian and Triassic in SW Europe, or in the Appalachian domain, documents the welding of Gondwana with Laurasia to form the Permian Pangea. We already discussed this diachronous closure of the large Paleotethys ocean, insisting on the likely development of back-arc oceans or basins within the Permo-Triassic Eurasian margin (Stampfli, 1996; Ziegler and Stampfli, 2001). From the paleo-Apulian promontory eastward, it is quite clear that an Eocimmerian domain of deformation is found just south of a relatively undeformed Variscan, domain which is represented by a Late Carboniferous to Early Permian arc and clastic sedimentation of Verrucano type, mainly affected by extension. In contrast, the Cimmerian deformations are accompanied by Triassic flysch, mélanges and volcanics (Stampfli et al., 2003) and even intrusive events (Reischmann, 1998) of collisional type, marking the closing of either Paleotethys and/or the European marginal oceans.

As shown by the plate tectonic reconstructions (Figures 1 and 2), the Middle Permian margin of SE Europe is of transform type, and little subduction took place along that margin at that time due to the welding of Gondwana to Laurasia. However, the slab roll-back of Paleotethys rapidly induced "back-arc" rifting along the whole margin from Late Permian times. East of the paleo-Apulian promontory, this back-arc rifting graded into the seafloor spreading of a series of marginal oceans. Due to accelerated roll-back of the Paleotethys, ridge jumps took place from the Hallstatt-Meliata rift to the Maliac rift in the Early Ladinian, and again from the Maliac to the Pindos rift in the Early Carnian. Therefore the Late Carboniferous Pelagonia arc terrane got stranded between the Maliac and Pindos oceans and never collided with Greater Apulia, the limit between the two plates being more of a transform type. South of the Pindos, the small ribbon Sitia Variscan terrane and its Triassic fore-arc collided with Greater Apulia along a Greek transect during the Late Carnian.

Along the Cimmerian orogen, development of carbonate platform resumed in the Norian or Liassic, marking the end of this orogenic cycle. This Middle-Late Triassic tectonic pulse is well established in the Dolomites where it is also accompanied by the emplacement of diapirs (Castellarin et al., 1996) and preceded by important volcanism, first in the Permian, then in the Middle Triassic (Pietra Verde event), showing that northern Italy was deeply involved in the geodynamic processes described here (Beccaluva et al., this volume, Chapter 28).

(c) Other Areas. Besides these Hellenic Paleotethyan remnants, continental to marine Upper Carboniferous to Lower Permian sequences are found around the arc of the Eurasian active margin (Figures 2 and 3):

- in the Tuscan Apennines (e.g. Gattiglio et al., 1989, and references therein, and Engelbrecht, 1997; Engelbrecht et al., 1989);
- in the deep-water Kungurian to Roadian flysch (Kozur and Mostler, 1992; Kozur, 1999) of the clastic Trogkofel beds (Ramovs, 1968) found just south of the Periadriatic line (Mr:Mrzlevodice fore-arc).
- the latter extend to similar Permo-Carboniferous deep marine clastic sequences of the Carnic and Karawanken Alps eastward, and the deep water Permian of Sicily westward, with the Tuscan Monte Amiata Permian in between.
- Early to Late Permian pelagic sediments are also presented in the Sicannian basin in Sicily (Catalano et al., 1995, 1988) whose fauna presents clear Paleotethyan affinities (Kozur, 1990).

Most of these listed outcrops belong to late Variscan fore-arc-type basins usually located just south of a cordillera characterized by Late Carboniferous to Early Permian calc-alkaline plutonism (Stampfli, 1996; Vavassis et al., 2000). The fore-arc basins were slightly deformed or uplifted during Permian times and covered unconformably by Verrucano s.l. sequences, deposited in rift basins. In Tuscany (Aldinucci et al., 2001), a clear distinction can be made between rift-related upper Permian–lower Triassic deposits with marine incursions covering unconformably the Carboniferous fore-arc sequences, and younger Verrucano s.str. clastics of Middle to Late Triassic age (Cirilli et al., 2002). The oldest deposits correspond to the rifting phase responsible for the opening eastward of the Meliata–Maliac back-arc system. The continental Verrucano s.str. deposits of Late Triassic age are related to the Eocimmerian event.

(d) The Alpine Tethys, the Central Atlantic and the Vardar. Field work in the Canary Islands and Morocco (Favre and Stampfli, 1992) allowed dating the onset of sea-floor spreading to Toarcian in the northern part of the central Atlantic. Similar subsidence patterns between this region and the Lombardian basin (Stampfli, 2000) led us

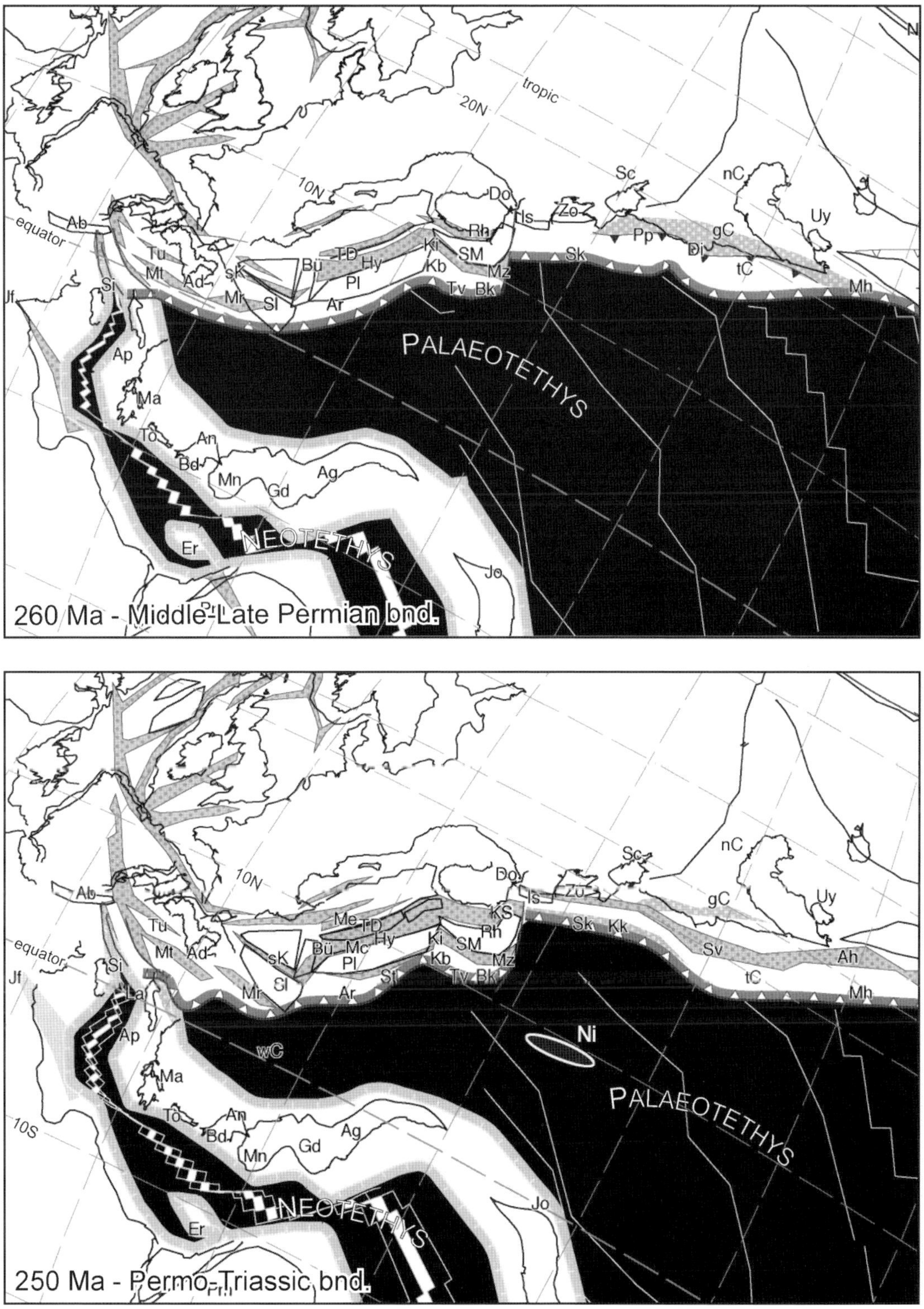

Figure 2: Western Tethys reconstructions for the Middle and latest Permian, modified from Stampfli and Borel (2002a, 2002b). For legend see Figure 1.

to propose a direct connection between these two areas (Figures 4 and 5). The Lombardian basin aborted (Bertotti et al., 1993) as it could not link up with the Meliata–Maliac–Pindos oceanic domains whose already cold oceanic lithosphere was rheologically unbreakable compared to surrounding continental areas. Therefore the Alpine Tethys rift opened north of the Meliata northern margin separating Adria and the Austro-Carpathian domain from Europe. Thermal subsidence of the Alpine area started in Aalenian times in the west (Briançonnais margin: Stampfli and Marchant, 1997; Stampfli et al., 1998) and Bajocian times eastward (Helvetic and Austroalpine margin: Bill et

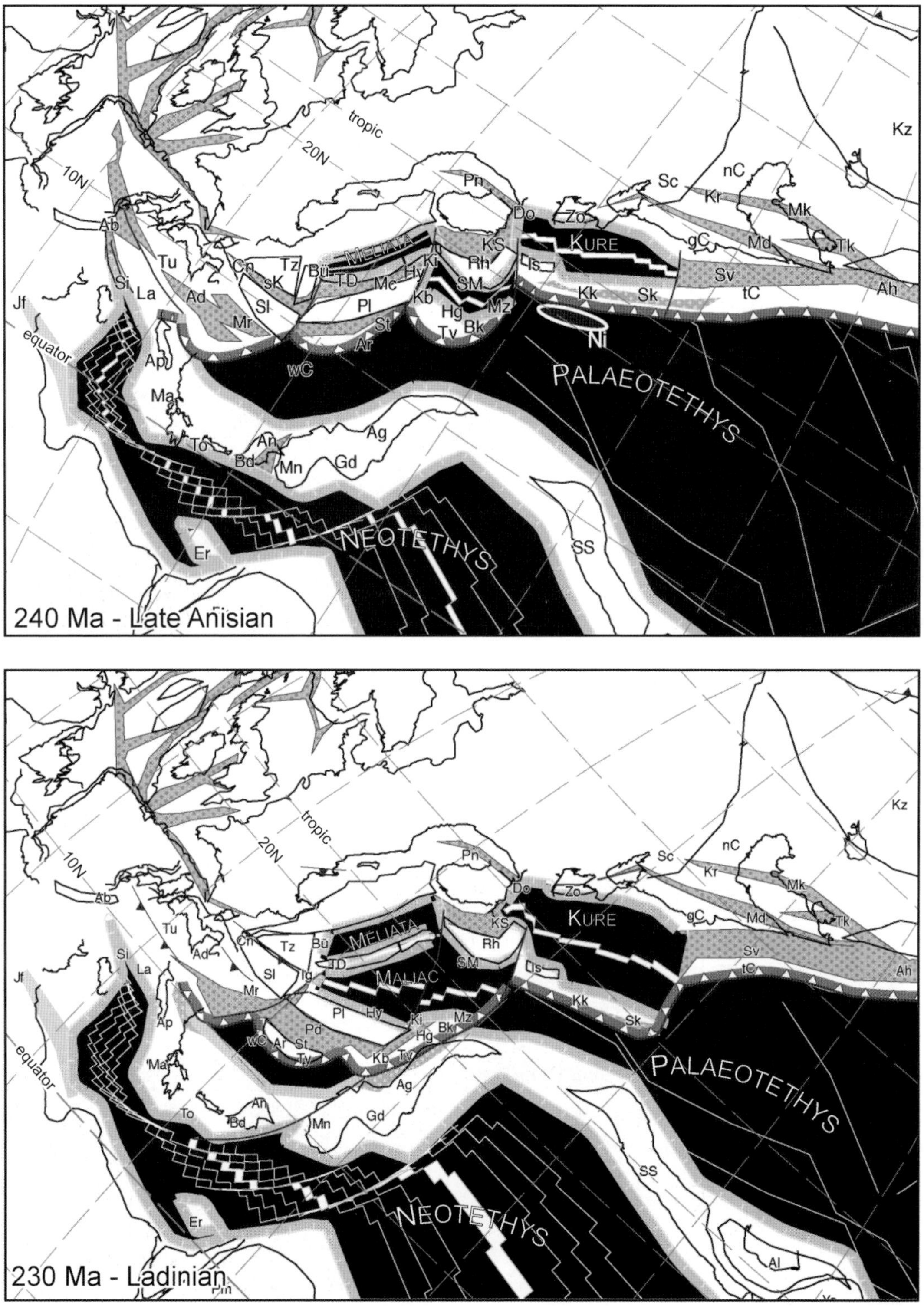

Figure 3: Western Tethys reconstructions for the Early and Middle Triassic, modified from Stampfli and Borel (2002a, 2002b). For legend see Figure 1.

al., 1997; Froitzheim and Manatschal, 1996). As discussed above, the Alpine Tethys spreading was considerably delayed, and spreading never gave birth to a real oceanic crust, the oceanic area being dominated by mantle denudation. The Briançonnais margin was certainly part of a transform segment of this ocean, reactivated in Late Jurassic times during the opening of the Pyrenean rift. A larger transform Maghrebide ocean linked up the central Atlantic and the Alpine Tethys, and is also characterized by local delay in thermal subsidence (Stampfli, 2000).

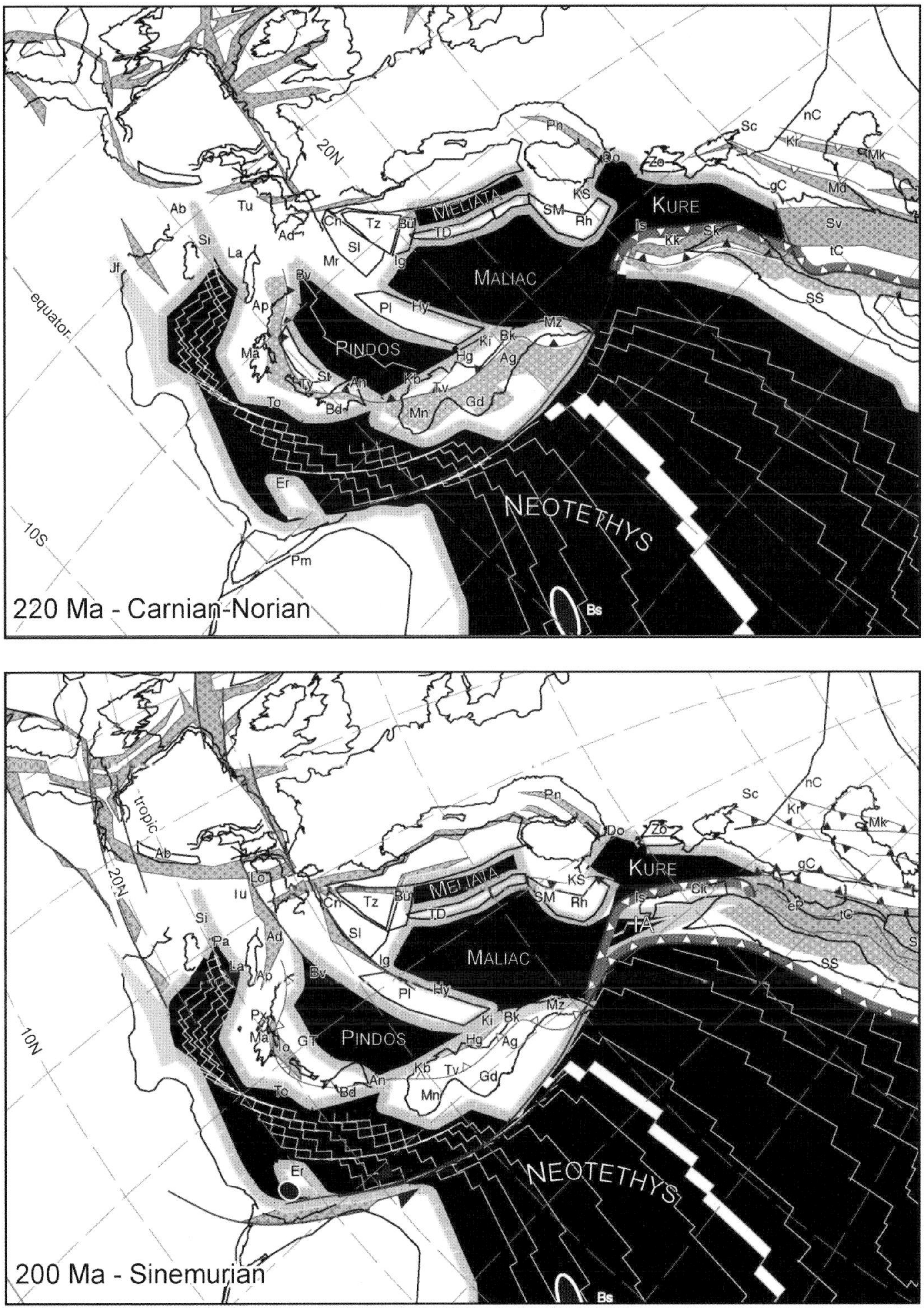

Figure 4: Western Tethys reconstructions for the Late Triassic and Early Jurassic, modified from Stampfli and Borel (2002a, 2002b). For legend see Figure 1.

The rotation of Africa with regard to Europe since the Late Triassic, and the southward subduction of the Küre back-arc, induced the eastward subduction of the Meliata–Maliac ocean under the young Neotethys oceanic crust and related margin. This subduction gave birth to the Vardar ocean, which totally replaced the Meliata–Maliac ocean by the end of the Jurassic. The Vardar ocean obducted southward onto the Pelagonian margin in the Late Jurassic (Figures 5 and 6), then subducted northward under Moesia, producing the large Late Cretaceous Srednogorie arc of Bulgaria and the opening of the Black Sea, representing the third generation of back-arcs in that region.

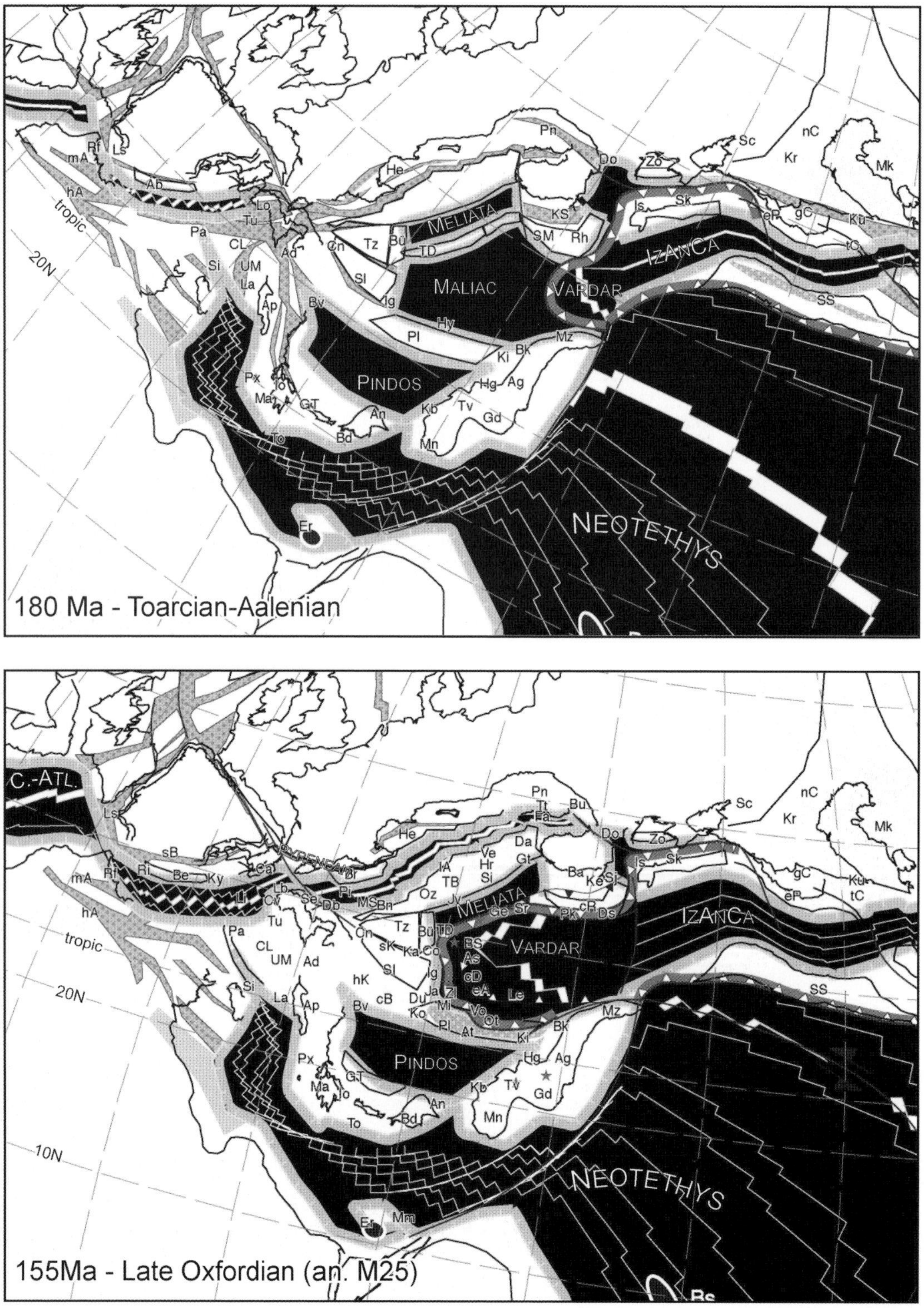

Figure 5: Western Tethys reconstructions for the Middle and Late Jurassic, modified from Stampfli and Borel (2002a, 2002b). For legend see Figure 1.

(e) Intra Apulia Jurassic Rifting. The word Ionian is used for two quite different geological entities, and we have already spoken about the Neotethyan origin of the Ionian sea floor, the western portion of the East Mediterranean basin. There is also a Ionian tectonic unit, defined in Greece (Aubouin et al., 1970), which is part of the Apulian para-autochthonous sequence of Greece and Albania. It consists of a thick carbonate platform starting in Norian times, and sealing the Paleotethys suture zone. Gypsum is sometimes found at its base (Papanikolaou et al., 1988). Then a shallow carbonate platform was locally replaced by a more pelagic sequence during the Jurassic. Pa-

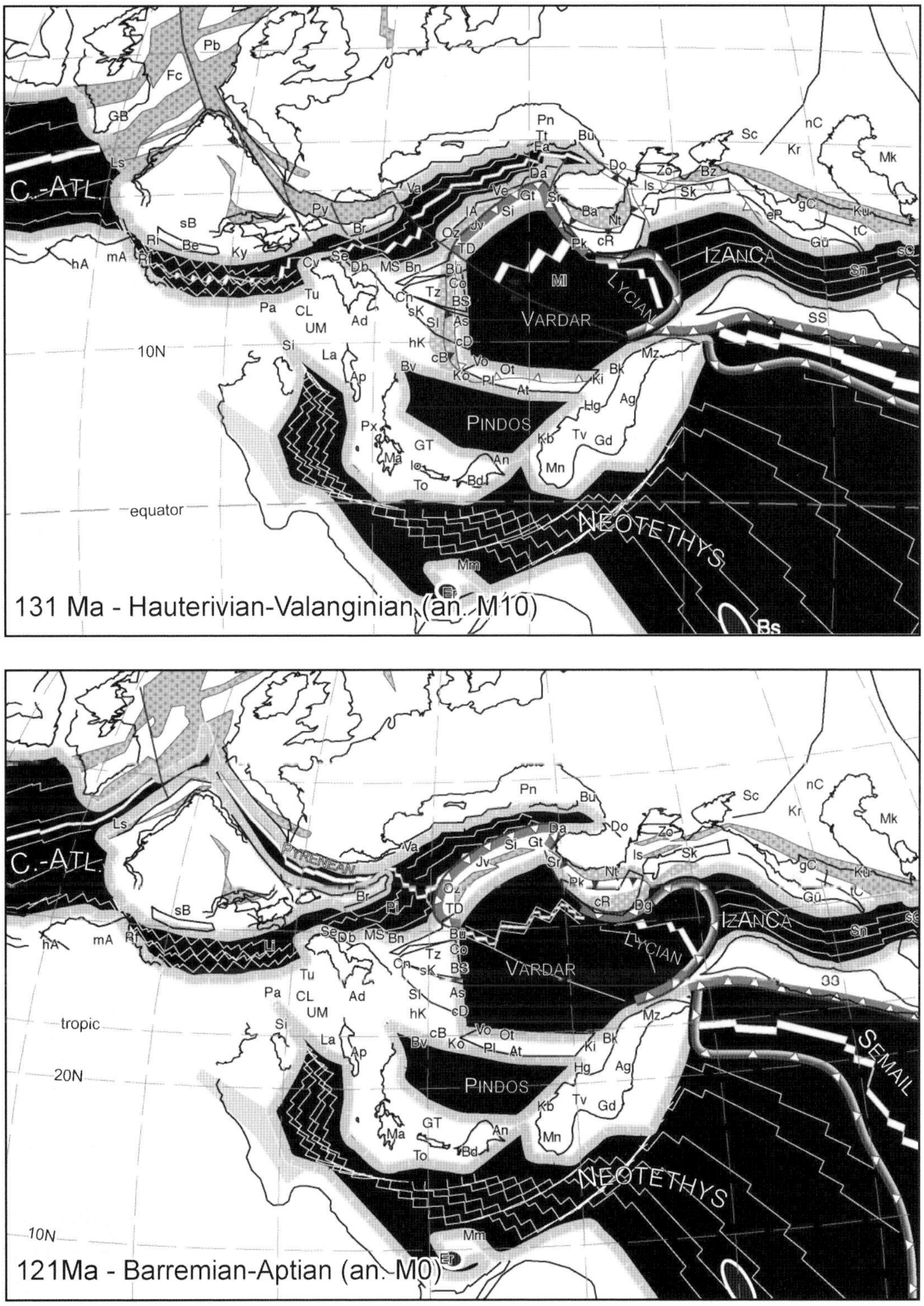

Figure 6: Western Tethys reconstructions for the Early Cretaceous, modified from Stampfli and Borel (2002a, 2002b). For legend see Figure 1.

leofaults have been observed in many places of the Ionian zone of Greece and Albania (Baudin and Lachkar, 1990; Dodona et al., 1994; Karakitsios, 1991; Karakitsios and Dermitzakis, 1997) and extend into the more autochthonous sequences found along the Adriatic coast and in the Adriatic sea, as seen on the seismic (Finetti and Del Ben, this volume, Chapter 23), and directly on land in Italy. Obviously, during the main phase of Pangea break-up, older structures such as the Paleotethys suture zone were reactivated, creating localized rift zones within the pre-existing large carbonate platform of Greater Apulia. These rifts aborted and gave place to deeper-water basins surrounded by carbonate platform locally active until the Miocene (e.g. Bernoulli et al., 1996; Vecsei et al., 1998). This platform and basin morphology was progressively covered by flysch deposits and two flexural basins developed around the Apulian carbonate promontory in Tertiary times, the Apenninic foreland to the west and the Dinaric-Hellenic foreland to the east (Carruba et al., this volume, Chapter 25).

Jurassic rifting also affected other areas of Italy. The major rift zone was located in the Lombard basin, as discussed above, but rifting accompanied by volcanism is also found in the Pelagian domain, south of Sicily. There too, it was emplaced within a former large Triassic carbonate platform which developed around the rift shoulder area of the Permo-Triassic Ionian ocean (Finetti and Del Ben, this volume, Chapter 19).

(f) The North Atlantic Ocean and the Pyrenean Domain and Cretaceous Tectonic. After the Early Cretaceous (Figure 6), there were few possibilities for the Atlantic mid-oceanic ridge to link up eastward with another ocean, due to the subduction of the Neotethys mid-oceanic ridge under Iran, and the shortening affecting the Vardar region. A last attempt was made through the opening of the Pyrenean and Biscay oceans, with a rifting phase starting in the Oxfordian, and the onset of spreading in the Aptian for the Portuguese–Galician ocean and Pyrenean area, and later in the Albian for Biscay (Stampfli et al., 2002a). By the end of the Santonian the break-up between North-America and Greenland took place, then in Campanian times the Biscay ocean aborted. Closing of the Pyrenean domain had already taken place during the opening of the Gulf of Biscay due to the accelerated rotation of the Iberic plate together with Africa. Therefore, it is not sure if the Pyrenean ocean progressed any further than mantle denudation, as expressed by the outcrop of the Lherzolites of Lherz (Fabries et al., 1998). This rotation also placed the Briançonnais peninsula in front of the Helvetic margin, creating a duplication of the European margin in the western Alps domain. The space in between the two marginal segments was called the Valais ocean or domain, and actually is a part of the Piemont ocean, trapped by the eastward displacement of the Briançonnais. The Valais domain extends eastward into the north Penninic domain, and both areas have been affected by HP/LT metamorphism (Bousquet et al., 1998; Goffé and Oberhänsli, 1992). The closing of this Valais domain had already taken place in Late Cretaceous times with the deposition of the Niesen flysch in Maastrichtian times, then the Meilleret flysch in Middle Eocene times. These Pyrenean tectonic inversion phases can be followed from the western Alps to the Pyrenees through Provence (Stampfli et al., 1998).

Iberia was rotating with Africa and Apulia–Adria, without any NS shortening, at least from the M0 magnetic anomaly (Aptian) up to the Maastrichtian. However, the southern margin of the Piemont ocean (the Austroalpine domain of the western Alps) was affected by tectonic movements from the Coniacian-Santonian, marked by the onset of flysch deposits (Gets and Dranse flysch, Caron et al., 1989). It is clear from the reconstruction (Figures 6 and 7) that large scale strike-slip movements affected the boundary between Adria and the Alpine domain, due the large-scale EW shortening taking place in the Vardarian area between Albian and Campanian times. These very large-scale lateral movements are well known (Trümpy, 1988, 1992) and finally placed Adria–Tizia behind the Austroalpine accretionary prism. During this process, the Piemont oceanic lithosphere got progressively disconnected from the northern margin of Adria and slowly started to subduct, resulting in a westward escape of the Austroalpine prism, which most likely collided locally with the northern margin already in Cretaceous times, to the north in the Tauern window and to the west with Calabria (Austroalpine Longobucco unit). Pieces of Adria margin were dragged into the subduction zone to give the Early Paleocene HP/LT metamorphism recorded in the Sesia domain (Oberhänsli et al., 1985; Rubatto, 1998) and Calabria (Colonna and Piccarreta, 1975) and affecting Austroalpine units.

(g) The Alpine Cycle. There is a fundamental difference between the Alpine cycle s.str. and Tethyan cycle s.l., or between the Alpine orogen (Alps and Carpathes) and the Tethysides (Dinarides-Hellenides, the Middle east mountain belts and the Himalayas s.l.). The Neotethys ocean, whose closure was responsible for the formation of the Tethysides orogen, actually does not directly interfere with Alpine geology s.str., except in Italy (Sicily). The Alpine Tethys should be regarded more as an extension of the central Atlantic ocean in the Tethyan realm rather than a branch of the larger and older Neotethys ocean (Stampfli, 2000). In that sense, the onset of the Alpine cycle could be placed in the Carnian, a period corresponding to the final closure of Paleotethys in the Mediterranean and Middle east regions, and to the onset of rifting in the central Atlantic-Alpine domain.

After several phases of rifting, as described above, and a more mature passive margin stage developing during the Cretaceous, the Alpine region entered a phase of convergence between the African plate and Europe, which closed the Alpine cycle. The tectonic evolution of the western Alps started with the formation of an accretionary prism related to the closure of the Alpine Tethys, where different geological objects, corresponding to different stages of accretion, can be recognized (see also Finetti, this volume, Chapter 7):

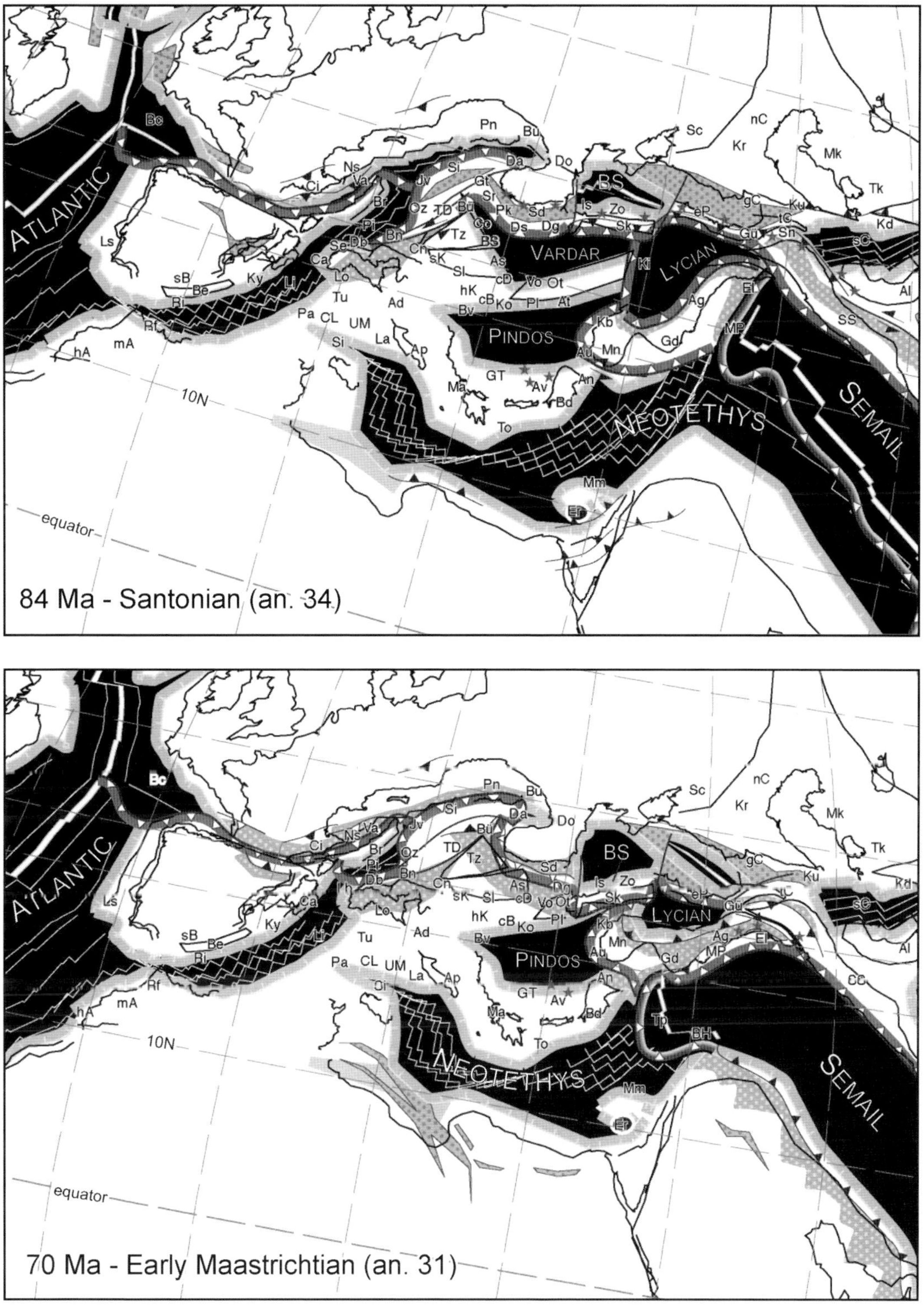

Figure 7: Western Tethys reconstructions for the Late Cretaceous, modified from Stampfli and Borel (2002a, 2002b). For legend see Figure 1.

- the Adriatic back-stop, comprising an aborted Jurassic rifted basin (Lombardian basin),
- the oceanic accretionary prism of the Piemont ocean (the western Alps portion of the Alpine Tethys), including crustal elements from the former toe of the southern passive margin (lower Austroalpine elements),
- accreted material of the Briançonnais terrain derived from the Iberic plate,
- accreted material of the Valais domain, representing the toe of the European (Helvetic s.l.) passive margin,
- accreted material of the former European continental margin and rim basin (Helvetic s.str. domain).

In time, one passes from the oceanic accretionary prism to the formation of the orogenic wedge that we place after the detachment or delamination of the subducting slab in the Early Oligocene (e.g. Stampfli and Marchant, 1995). The resulting heat flux allowed some more units to be detached from the European continental slabs and resulted in large-scale subduction of continental material (Marchant and Stampfli, 1997) and obduction of the more external units, such as:

- the external Variscan massifs and their cover,
- the molassic basin,
- the Jura mountains.

These were accompanied by retro-wedge thrusting of the southern Alps onto the Po Plain, which is a depression located above the former Lombard basin.

To these accretionary events one has to add other tectonic processes such as the "Pyrenean" inversion phase (Late Cretaceous–Middle Eocene) that affected the Helvetic margin, accompanied and followed by the Paleogene flexure event of the lower European plate (Paleocene to Miocene) in front of the advancing tectonic wedge.

If the Alpine chain is considered to go from Nice to the Carpathian domain, there is a fundamental difference between the western Alps sector that we have just described and the Austroalpine–Carpathian sector. The end of the story is quite similar in both areas – an orogenic wedge is finally emplaced onto the European margin, but the history of the Austroalpine Carpathian wedge is quite different in Jurassic and Cretaceous times (Faupl and Wagreich, 1999). Effectively these regions were first involved in the collision of the Vardar domain with the Triassic Meliata–Maliac margin (Bernoulli, 1981; Haas et al., 1995; Kozur, 1991a), giving birth to the upper Austroalpine nappes, represented mainly by the Northern Calcareous Alps. This means that the Austroalpine wedge has two sutures in its midst, the Meliata suture and the Penninic suture; HP/LT rocks developed in both (Thöni, 1999).

This Cretaceous collision can be followed from the Austrian Alps up to Bulgaria, where the Balkan orogen (Georgiev et al., 2001; Tari et al., 1997) represents the same Early Cretaceous collision between an intra-oceanic-arc domain (circum-Rhodope units) and a passive margin (Rhodope). The Cretaceous reconstructions presented in Figures 6 and 7, show that this collisional event extends even more eastward in Crimea and the Caucasus. The remnant Vardar and Izmir–Ankara oceans did not survive very long after this event: we have seen that the Vardar started to subduct under the newly formed Balkanic orogen to give the Srednogorie arc; eastward, the Izmir Ankara Jurassic ocean was replaced by a younger ocean, whose remnants are found in many parts of Turkey, forming upper structural Cretaceous ophiolitic units, as found in the Lycian nappes of western Turkey (De Graciansky, 1993; Gutnic et al., 1979; Okay et al., 1996).

As we have seen above, the southward subduction of the Alpine Tethys ocean is related to the history of the Meliata–Maliac and Vardar domain, and was inherited in the Western Alps from the pre-existing northward-vergence of the Austroalpine accretionary wedge. This northward-vergence is quite unique in the whole Alpine and Tethyan domain, where most orogens are south-vergent. The change of vergence of the Alpine Tethys subduction is found at the connection region between the Alps and the northern Apennines (Figures 7 and 8). It must be emphasized here that the Penninic Alpine prism is older (Late Cretaceous–Eocene) than the Apenninic (Oligocene–Pliocene): actually one started when the other one stopped. If there is confusion between the two it is because the Apenninic prism is re-mobilizing parts of the Penninic prism as exotic elements (e.g. the Bracco ophiolitic ridge, Elter et al., 1966; Hoogenduijn Strating, 1991).

In this framework, where large-scale constraints are taken into consideration, ophiolites of the Apenninic prism are regarded as mainly derived from the former Alpine prism which collided with the Iberic plate in Corsica and Calabria in Eocene times (Figure 8). These oceanic elements were re-mobilized when the remnant Alpine Tethys oceanic domain (Ligurian basin) south of the Iberic plate started to subduct northward in an ongoing process of shortening between Europe and Africa. The deep seismic data from the CROP project clearly show that the southward subducting Piemont ocean underplated the Adriatic indenter, and can still be recognized on the seismic sections (Finetti et al., 2001). The underplated material was thereafter displaced northward with Adria, whereas the not-yet-subducted Ligurian part of the Alpine Tethys started to subduct northward under the Iberic plate, generating HP/LT rocks (Puga, this volume, Chapter 31). On the Eocene reconstruction, nearly all of this remnant ocean is already subducted, creating an important roll-back which led to the opening of the Algero–Provençal back-arc basin in Oligocene times (Figure 8). On the Oligocene reconstruction, the toe of the Tuscan–Campano–Lucanian

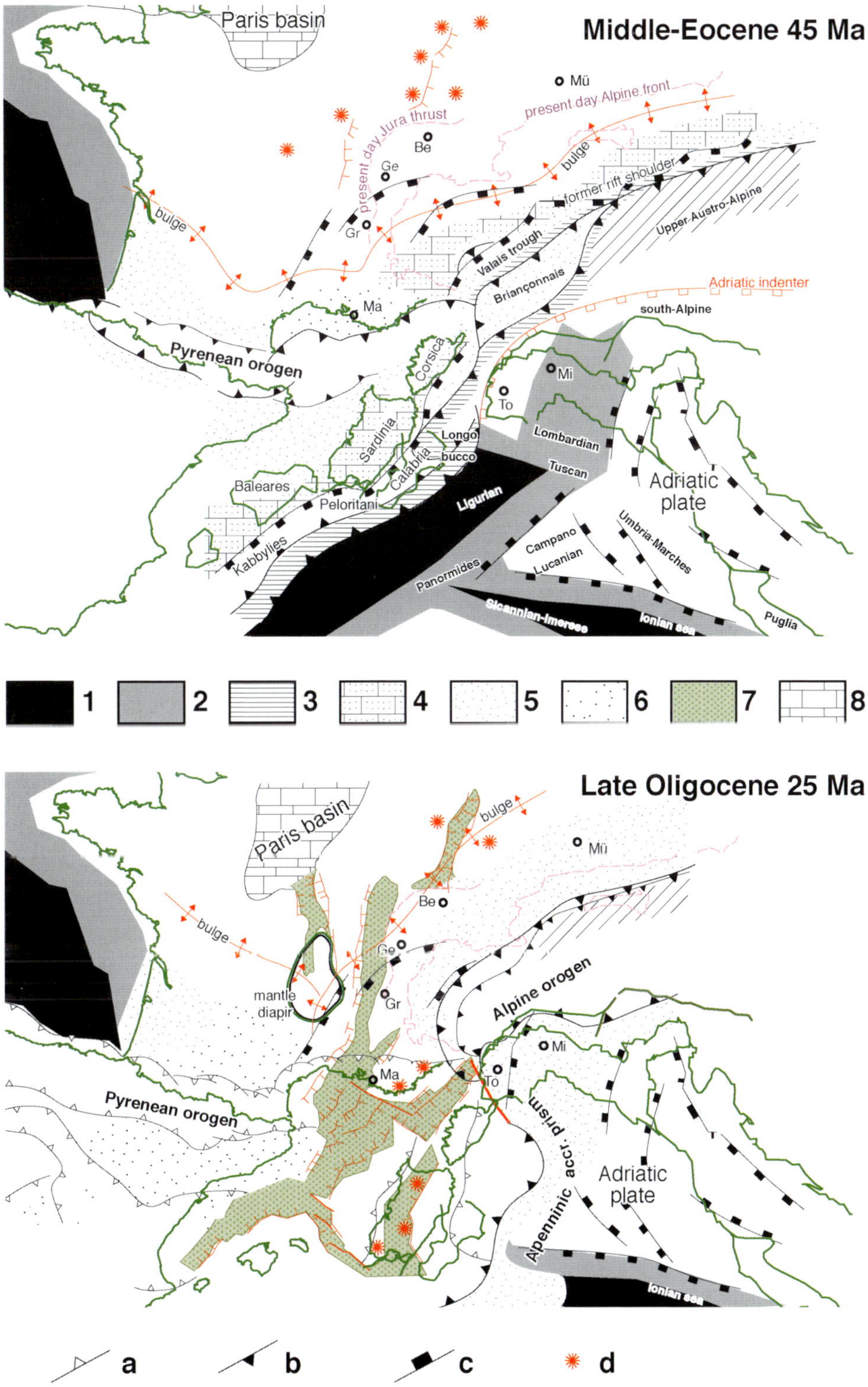

Figure 8: Western Tethys reconstructions for the Paleogene, modified from Stampfli et al. (2002a). 1 – ocean; 2 – passive margin; 3 – active margin; 4 – nummulitic platform; 5 – marine foreland basin; 6 – continental foreland basin; 7 – rifts; 8 – epicontinental basin; a, inactive thrusts; b, active thrusts; c, Jurassic normal faults; d, active volcanic centers.

margin would be located just under Calabria–Corsica. The Oligocene volcanism of Sardinia can therefore be regarded as subduction related, although the onset of the oceanic slab detachment could also be the source of this volcanism. The Alpine accretionary prism stranded on the border of Corsica and Calabria started to collapse backward, following the build up of the new Apenninic accretionary prism. These Ligurian units are now found in an upper structural position, whereas they were in a lower one during the Alpine collision.

The Lucano-Campanian promontory certainly formed an indenter separating the future northern and southern Apenninic prisms. Both prisms collapsed into pre-existing depressions, the Lombardian rift in the north and the Ionian Sea oceanic corridor in the south. These accretionary processes are still in movement today, and are largely covered in Chapter 12 by Finetti et al. (this volume), and in the preceding article by Mantovani (this volume, Chapter 32).

5. CONCLUSIONS

The input of the CROP data set to the plate tectonic model developed for the western Tethys area was of great importance. Two main factors – the Permo-Triassic age of the Ionian, East Mediterranean seas, and the unbroken Adriatic Late Mesozoic platform, give very important constraints on the post-Variscan continental fit of these regions. At that time the Adria s.str. and Apulia s.str plates were separated. The two plates were then welded together by an Eocimmerian event corresponding to the final closure of Paleotethys in late Permian times in these regions. This fit requires a tight position of the Iberic Alboran plates with respect to North America and Europe, and also implies a long-lasting rifting phase in the Atlantic regions now separating these domains. The final position of the Adria plate and its geometry close to the present day one, was reached through Jurassic rifting events which finally gave birth to the narrow Alpine Tethys.

REFERENCES

Aldinucci, M., Cirilli, S., Costantini, A., Decandia, F.A., et al., 2001. Stratigraphic-structural evolution of the Late paleozoic–Triassic sequences of southern Tuscany (northern Apennines). Dipartimento di Scienze della Terra, Siena University, Siena, 72 p.

Aubouin, J., et al., 1970. Contribution à la géologie des Hellénides : le Gavrovo, le Pinde et la zone ophiolitique subpélagonienne. Ann. Soc. Géol. Nord 90 (4), 277–306.

Baudin, F., Lachkar, G., 1990. Geochimie organique et palynologie du Lias superieur en zone ionienne (Grece) exemple d'une sedimentation anoxique conservee dans une paleo-marge en distension. Bull. Soc. Géol. France, Huitieme Serie 6 (1), 123–132.

Beccaluva, L., Coltorti, M., Saccani, E., Siena, F., Zeda, O., this volume. Triassic magmatism and Jurassic Ophilotes at the margins of the Adria Plate. Chapter 28.

Bernoulli, D., 1981. Ancient continental margins of the Tethyan Ocean. In: Bally, A.W., et al. (Eds.), Geology of Passive Continental Margins; History, Structure and Sedimentologic Record (with Special Emphasis on the Atlantic Margin). In: Education Course Note Series, vol. 19, no. 5. American Association of Petroleum Geologists, pp. 1–36.

Bernoulli, D., Anselmetti, F.S., Eberli, G.P., Mutti, M., Pignatti, J.S., Sanders, D.G.K., Vecsei, A., 1996. Montagna della Maiella; the sedimentary and sequential evolution of a Bahamian-type carbonate platform of the South-Tethyan continental margin. In: Anonymous (Ed.), Geologia delle aree di avampaese. In: Memorie della Societa Geologica Italiana, vol. 51, no. 1. Societa Geologica Italiana, pp. 7–12.

Bertotti, G., Picotti, V., Bernoulli, D., Castellarin, A., 1993. From rifting to drifting: tectonic evolution of the South-Alpine upper-crust from the Triassic to the Early Cretaceous. Sedimentary Geology 86, 53–76.

Bill, M., Bussy, F., Cosca, M., Masson, H., Hunziker, J.C., 1997. High precision U–Pb and $^{40}Ar/^{39}Ar$ dating of an Alpine ophiolite (Gets nappe, French Alps). Eclog. Geol. Helv. 90, 43–54.

Bousquet, R., Oberhansli, R., Goffe, B., Jolivet, L., Vidal, O., 1998. High-pressure-low-temperature metamorphism and deformation in the Bundnerschiefer of the Engadine Window; implications for the regional evolution of the eastern Central Alps. J. Metamorphic Geology 16 (5), 657–674.

Caron, C., Homewood, P., Wildi, W., 1989. The original Swiss Flysch: a reappraisal of the type deposits in the Swiss Prealps. Earth-Science Rev. 26, 1–45.

Carruba, I., Casnedi, R., Perotti, C., this volume. Mechanism and timing of the Pliocene-Pleistocene foredeep migration in the Abruzzo offshore area (Central Adriatic). Chapter 25.

Castellarin, A., Selli, L., Picotti, V., Cantelli, L., 1996. La tettonica Medio Triassica e il diapirismo nelle Dolomiti occidentali. In: Geologia delle dolomiti. Societa Geologica Italiana, San Cassiamo, pp. 25–46.

Catalano, R., Di Stefano, P., Kozur, H., 1988. New results in the Permian and Triassic stratigraphy of western Sicily with special reference to the section at Torrente San Calogero, SW of Pietra di Salome (Sosio valley). In: Soc. Geol. It., Atti del 74 Congresso Nazionale, A, pp. 119–125.

Catalano, R., Di Stefano, P., Vitale, F.P., 1995. Structural trends and paleogeography of the central and western Sicily belts: new insight. Terra Nova 7, 189–199.

Channell, J.E.T., 1992. Paleomagnetic data from Umbria (Italy): implications for the rotation of Adria and Mesozoic apparent polar wander paths. Tectonophysics 216, 365–378.

Channell, J.E.T. (Ed.), 1996. Paleomagnetism and Paleogeography of Adria. Paleomagnetism and tectonics of the Mediterranean region, vol. 105. Geol. Soc. Spec. Publ., pp. 119–132.

Cirilli, S., Decandia, F.A., Lazzarotto, A., Pandelli, E., Rettori, R.F.S., Spina, A., 2002. Stratigraphy and depositional environment of the Mt Argentario sandstones Fm. (southern Tuscany, Italy). Boll. Soc. Geol. It. spec. vol. 1, 489–498.

Colonna, V., Piccarreta, G., 1975. Metamorfismo di alta pressione/bassa temperatura nei micascisti di Zangarona–Ievoli–Monte Dondolo (Sila Piccolo, Calabria). Boll. Soc. Geol. Ital. 94 (1–2), 17–25.

De Graciansky, P.C., 1993. Basement-cover relationship in the Western Alps. Constraints for the pre-Triassic reconstructions. In: von Raumer, J., Neubauer, F. (Eds.), The Pre-Mesozoic Geology in the Alps. Springer-Verlag, pp. 7–28.

de Jonge, M.R., Wortel, M.J.R., Spakman, W., 1994. Regional scale tectonic evolution and the seismic velocity structure of the lithosphere and upper mantle: the Mediterranean region. J. Geophys. Res. 99 (B6), 12091–12108.

Dercourt, J., Ricou, L.E., Vrielinck, B., 1993. Atlas Tethys, Paleoenvironmental Maps. Gauthier-Villars, Paris. 307 p.

Dercourt, J., et al., 1985. Présentation des 9 cartes paléogéographiques au 1/20 000 000 s'étendant de l'Atlantique au Pamir pour la période du Lias à l'Actuel. Bull. Soc. Géol. France 8 (T I, No 5), 637–652.

Destombes, J., 1971. L'Ordovicien du Maroc. Essai de synthèse stratigraphique. Mémoires du Bureau de Recherches Géologiques et Minières Orléans 73, 237–263.

Dodona, E., Farinacci, A., Kanani, J., Nicosia, U., Tonielli, R., 1994. Mid Jurassic events in the Ionian Zone (SW Albania). Paleopelagos 4, 73–85.

Elter, G.P.E., Sturani, C., Weidmann, M., 1966. Sur la prolongation du domaine ligure de l'Apennin dans le Montferrat et les Alpes et sur l'origine de la nappe de la Simme s.l. des Préalpes romandes et chablaisiennes. Archives des Sciences, Genève 19 (3), 279–377.

Engelbrecht, H., 1997. From Upper paleozoic extensional basin fill to late Alpine low grade metamorphic core complex: preliminary note on the sedimentary and tectonic development of the Monticiano-Roccastrada-Zone (MRZ, southern Tuscany, Italy). Z. Deutsch. Geol. Gesell. 148 (3–4), 523–546.

Engelbrecht, H., Klemm, D.D., Pasini, M., 1989. Preliminary notes on the tectonics and lithotypes of the "ver rucono SL." in the Monticiano area (Southern Tuscany, Italy) and the finding of Fusulinids within the M. te Quoio FM. (Verrucano group). Riv. Ital. Paleontologia Stratigrafia 94 (3), 361–382.

Fabries, J., Lorand, J.P., Bodinier, J.L., 1998. Petrogenetic evolution of orogenic lherzolite massifs in the central and western Pyrenees. Tectonophysics 292 (1–2), 145–167.

Faupl, P., Wagreich, M., 1999. Late Jurasic to Eocene Palaeogeography and geodynamic evolution of the Eastern Alps. Mitteil. Österreich. Geolog. Gesell. 92, 79–94.

Favre, P., Stampfli, G.M., 1992. From rifting to passive margin: the example of the Red Sea, Central Atlantic and Alpine Tethys. Tectonophysics 215, 69–97.

Finetti, I., 1985. Structure and evolution of the Central Mediterranean (Pelagian and Ionian Seas). In: Stanley, D.J., Wezel, F.C. (Eds.), Geological Evolution of the Mediterranean Basin. Springer-Verlag, pp. 215–230.

Finetti, I.R., this volume. Crustal tectono-stratigraphic sections across the Western and Eastern Alps from ECORS-CROP and Transalp seismic data. Chapter 7.

Finetti, I.R., Bocaletti, M., Bonini, M., Del Ben, A., Geletti, R., Pipani, M., Sani, F., 2001. Crustal section based on CROP seismic data across the North Tyrrhenian–Northen Apennines–Adriatic sea. Tectonophysics 343, 135–163.

Finetti, I.R., Lentini, F., Carbone, S., Del Ben, A., Di Stefano, A., Guarneri, P., Pipan, M., Prizzon, A., this volume. Crustal tectono-stratigraphy and geodynamics of the Southern Apennines, from CROP and other integrated geophysical-geological data. Chapter 12.

Finetti, I.R., Del Ben, A., this volume. Crustal tectono-stratigraphy of the Ionian Sea from integrated new CROP seismic data. Chapter 19.

Finetti, I.R., Del Ben A., this volume. Crustal tectono-stratigraphic setting of the Adriatic Sea from new CROP seismic data. Chapter 23.

Froitzheim, N., Manatschal, G., 1996. Kinematics of Jurassic rifting, mantle exhumation, and passive-margin formation in the Austroalpine and Penninic nappes (Eastern Switzerland). Geol. Soc. Am. Bull. 108 (9), 1120–1133.

Garfunkel, Z., Derin, B., 1984. Permian–early Mesozoic tectonism and continental margin formation in Israel and its implications for the history of the Eastern Mediterranean. In: Dixon, J.E., Robertson, A.H.F. (Eds.), The Geological Evolution of the Eastern Mediterranean. Geol. Soc. Spec. Publ. 17, 187–201.

Gattiglio, M., Meccheri, M., Tongiorgi, M., 1989. Stratigraphic correlation forms of the Tuscan Paleozoic basement. In: Sassi, F.P., Zanferrari, A. (Eds.), Pre-Variscan and Variscan Events in the Alpine–Mediterranean belts, Stratigraphic correlation forms. Rend. Soc. Geol. It. 12 (2), 245–257.

Georgiev, G., Dabovski, C., Stanisheva-Vassileva, G., 2001. East Srednogorie-Balkan rift zone. In: Ziegler, P.A., Cavazza, W., Robertson, A.H.F., Crasquin-Soleau, S. (Eds.), PeriTethys Memoir 6: Peritethyan Rift/Wrench Basins and Passive Margins, IGCP 369. Mém. Museum Nat. Hist. Nat. 186, 259–294.

Goffé, B., Oberhänsli, R., 1992. Ferro- and magnesiocarpholite in the "Bündnerschiefer" of the eastern Central Alps (Grisons and Engadine window). Eur. J. Mineral. 4, 835–838.

Gutnic, M., Monod, O., Poisson, A., Dumont, J.F., 1979. Géologie des Taurides occidentales (Turquie). Mém. Soc. Géol. France 137 (LVIII), 109.

Haas, J., Kovacs, S., Krystyn, L., Lein, R., 1995. Significance of Late Permian-Triassic facies zones in terrane reconstructions in the Alpine–North Pannonian domain. Tectonophysics 242, 19–40.

Hoogenduijn Strating, E.H., 1991. The evolution of the Piemonte-Ligurian ocean, a structural study of ophiolite complexes in Liguria (NW Italy). Geologica ultraiectina 74, 127.

Karakitsios, V., 1991. Study of Jurassic paleofaults in the Ionian Zone of Epirus. In: D. Mountrakis (Ed.), 5th Congress of the Geological Society of Greece; proceedings. Deltio tes Ellenikes Geologikes Etaireias = Bulletin of the Geological Society of Greece. Ellenike Geologike Etaireia, Athens, Greece, pp. 307–318.

Karakitsios, V., Dermitzakis, M.D., 1997. Lacunes sedimentaires, discordances et phenomenes paleokarstiques (pliensbachien-tithonique) dans la zone ionienne (Epire, Grece nord-occidentale). In: Anonymous (Ed.), A la memoire du Professeur G.M. Paraskevopoulos. Annales Geologiques des Pays Helleniques. Laboratoire de Geologie de l'Universite, vol. 37, pp. 847–864.

Kock, S., 2003. Nouvelles donnnées stratigraphiques, métamorphiques et géochimiques sur la série autochtone des Talea Ori (Ctère centrale). Msc Thesis, University of Lausanne, 60 p.

König, H., Kuss, S., 1980. Neue Daten zur Biostratigraphie des permotriadischen Autochtonous der Insel Kreta (Grieschland). Neues Jahrbuch für Geologie und Paläontologie, Abhandlungen 9, 525–540.

Kozur, H., 1990. Deep-Water permian in Sicily and its possible connection with the Himalaya–Tibet region. In: 5th Himalaya–Tibet–Karakorum Workshop, Milano.

Kozur, H., 1991a. The evolution of the Meliata-Hallstatt ocean and its significance for the early evolution of the Eastern Alps and western Carpathians. In: Channell, J.E.T., Winterer, E.L., Jansa, L.F. (Eds.), Paleogeography and Paleoceanography of Tethys. Palaeogeogr. Palaeoclimatol. Palaeoecol. 87, 109–135.

Kozur, H., 1991b. Permian deep-water ostracods from Sicily (Italy), Part 1: Taxonomy. Geologisch Paläontlogische Mitteilungen Innsbruck, Sonderband 3, 1–24.

Kozur, H., 1991c. Permian deep-water ostracods from Sicily (Italy), Part 2: Biofacial evaluation and remarks to the Silurian to Triassic paleopsychrospheric ostracods. Geologisch Paläontlogische Mitteilungen Innsbruck, Sonderband 3, 25–38.

Kozur, H., 1995. First evidence of Middle Permian Ammonitico Rosso and further new stratigraphic results in the Permian and Triassic of the Sosio Valley area, Western Sicily. In: 1st Croatian Geological Congress. Zbornik Radova Proceedings, Opatija, pp. 307–310.

Kozur, H.W., 1999. Permian development in the western Tethys, In: Shallow Tethys, vol. 5. Dept. Geol. Sc. Chiang Mai Univ., Chiang Mai, pp. 101–135.

Kozur, H., Mostler, H., 1992. Erster paläontologischer Nachweis von Meliaticum und Süd-Rudabànyaicum in den nördlischen Kalkalpen (Oesterreich) und ihre Beziehungen zu den Abfolgen in den Westkarpaten. Geologische Paläontologische Mitteilungen Innsbruck 18, 87–129.

Kozur, H.W., Senel, M., Tekin, K., 1998. First evidence of Hercynian Lower carboniferous deep-water sediments in the Lycian nappes, SW Turkey. Geol. Croatica 51 (1), 15–22.

Malpas, J., Calon, T., Squires, G., 1993. The development of a late Cretaceous microplate suture zone in SW Cyprus. In: Prichard, H.M., Alabaster, T., Harris, N.B.W., Neary, C.R. (Eds.), Magamatic Processes and Plate Tectonics. Geol. Soc. London Spec. Publ. 76, 177–195.

Mantovani, E., this volume. Evolutionary reconstruction of the Mediterranean region: extrusion tectonics driven by plate convergence of Africa Arabia and Eurasia. Chapter 32.

Marchant, R.H., Stampfli, G.M., 1997. Subduction of continental crust in the Western Alps. Tectonophysics 269 (3–4), 217–235.

Niko, S., Pillevuit, A., Nishida, T., 1996. Early Late Permian (Wordian) non-ammonoid cephalopods from the Hamrat Duru Group, central Oman Mountains. Trans. Proc. Palaeontol. Soc. Japan, N.S. 183, 522–527.

Oberhänsli, R., Hunziker, J.C., Martinotti, G., Sten, W.B., 1985. Geochemistry, geochronology and petrology of Monte Mucrone: an example of eo-alpine eclogitisation of Permian granitoïds in the Sesia Lanzo zone, western Alps, Italy. Chem. Geology 52, 165–184.

Okay, A.I., Satir, M., Maluski, H., Siyako, M., Monie, P., Metzger, R., Akyüz, S., 1996. Paleo- and Neo-Tethyan events in northwestern Turkey: geologic and geochronologic constraints. In: Yin, A., Harrison, T.M. (Eds.), The Tectonic Evolution of Asia. Cambridge University Press, pp. 420–441.

Papanikolaou, D., Skarpelis, N., Alexopoulos, A., Markopoulos, T., 1988. First field meeting Crete, Greece, Guide Book, IGCP Project No 276. Techn. Univ. Crete, Chania, p. 36.

Pillevuit, A., 1993. Les blocs exotiques du Sultanat d'Oman, évolution paléogéographique d'une marge passive flexurale. Mémoires de Géologie (Lausanne) 17, 249.

Puga, E., this volume. A reappraisal of the betic ophiolitic association: the western most relic of the Alpine Tethys. Chapter 31.

Piqué, A., 1989. Variscan terranes in Morocco. In: Dallmeyer, D. (Ed.), Terranes in the Circum-Atlantic Paleozoic Orogens. Geol. Soc. Am. Spec. Paper 230, 115–129.

Ramovs, A., 1968. Biostratigraphy des klastichen Entwicklung der Trogofel stufe in den Karawanken und Nachbar gebieten. Neues Jahrbuch für Geologie und Paläontologie, Abhandlungen 131 (1), 72–77.

Rampone, E., Piccardo, G., 2000. The ophiolite-oceanic lithosphere analogue: new insights from the northern Apennines (Italy). In: Dilek, Y., Moores, E.M., Elthon, D., Nicolas, A. (Eds.), Ophiolites and Oceanic Crust: New Insights from Field Studies and the Ocean Drilling Project. Geol. Soc. Am. Spec. Paper 349, 21–34.

Reischmann, T., 1998. Pre-Alpine origin of tectonic units from the metamorphic complex of Naxos, Greece, identified by single Zircon Pb/Pb dating. In: 8th Intern. Congr. of the Geol. Soc. of Greece, Patras, Greece, pp. 101–111.

Robertson, A.H.F., et al., 1996. Alternative tectonic models for the Late Palaeozoic-Early Tertiary development of Tethys in the Eastern Mediterranean region. In: Morris, A., Tarling, D.H. (Eds.), Palaeomagnetism and Tectonics of the Eastern Mediterranean Region. Geol. Soc. London Spec. Publ. 105, 239–263.

Rubatto, D., 1998. Dating of pre Alpine magmatism, Jurassic ophiolites and Alpine subductions in the Western Alps. ETHZ, Zürich. 173 p.

Schönlaub, H.P., Histon, K., 1999. The Paleozoic Evolution of the southern Alps. Mitteil. Österreich. Geol. Gesell. 92, 15–34.

Sengör, A.M.C., Yılmaz, Y., Sungurlu, O., 1984. Tectonics of the Mediterranean Cimmerides: nature and evolution of the western termination of Paleo-Tethys. In: Dixon, J.E., Robertson, A.H.F. (Eds.), The Geological Evolution of the Eastern Mediterranean. Geol. Soc. Spec. Publ. 17, 77–112.

Srivastava, S.P., Roest, W.R., Kovacs, L.C., Oakay, G., Lévesque, S., Verhoef, J., Macnab, R., 1990. Motion of Iberia since the Late Jurassic: results from detailed aeromagnetic measurements in the Newfoundland Basin. Tectonophysics 184, 229–260.

Srivastava, S.P., Tapscott, C.R., 1986. Plate kinematics of the North Atlantic. In: Vogt, P.R., Tucholke, B.E. (Eds.), The Western North Atlantic Region. In: The Geology of North America, vol. M. Geol. Soc. Am., pp. 379–404.

Stampfli, G., 1978. Etude géologique generale de l'Elbourz oriental au sud de Gonbad-e-Qabus (Iran NE). Ph.D. 1868 Thesis, Univ. Genève, 329 p.

Stampfli, G.M., 1989. Late Paleozoic evolution of the eastern Mediterranean region. In: IGCP 276 Paleozoic geodynamic domains and their alpidic evolution in the Tethys. Short Course I–II, Abstr., Lausanne, Switzerland.

Stampfli, G.M., 1996. The Intra-Alpine terrain: a Paleotethyan remnant in the Alpine Variscides. Eclog. Geol. Helv. 89 (1), 13–42.

Stampfli, G.M., 2000. Tethyan oceans. In: Bozkurt, E., Winchester, J.A., Piper, J.D.A. (Eds.), Tectonics and Magmatism in Turkey and Surrounding Area. Geol. Soc. London Spec. Publ. 173, 1–23.

Stampfli, G.M., Borel, G., 2000. West-Tethys oceans. In: Geology 2000, Geol. Ver. 90th Meet. Terra Nostra Schr. Alf. Weg. St., Vienna, 113 p.

Stampfli, G.M., Borel, G., 2002a. A plate tectonic model for the paleozoic and Mesozoic of the Tethyan Domain constrained by dynamic plate boundaries and restored synthetic oceanic isochrone. In: 1st Int. Symposium of the Faculty of Mines (ITü) on Earth Science and Engineering. Istanbul Technical Univ., Faculty of Mines, Istanbul, 77 p.

Stampfli, G.M., Borel, G.D., 2002b. A plate tectonic model for the Paleozoic and Mesozoic constrained by dynamic plate boundaries and restored synthetic oceanic isochrons. Earth Planet. Sci. Lett. 196, 17–33.

Stampfli, G.M., Marchant, R.H., 1995. Plate configuration and kinematics in the Alpine region. In: Accademia Nazionale delle Scienze, Scritti e Documenti, 14 (Atti del congresso "Rapporti tra Alpi e Appennino"), pp. 147–166.

Stampfli, G.M., Marchant, R.H., 1997. Geodynamic evolution of the Tethyan margins of the Western Alps. In: Pfiffner, O.A., Lehner, P., Heitzman, P.Z., Mueller, S., Steck, A. (Eds.), Deep Structure of the Swiss Alps – Results from NRP 20. Birkhaüser, pp. 223–239.

Stampfli, G.M., Mosar, J., 1999. The making and becoming of Apulia. In: 3rd Workshop Alpine Geology. Mem. Sci. Geol. (Univ. Padova), spec. vol. 51 (1), 141–154.

Stampfli, G.M., Pillevuit, A., 1993. An alternative Permo-Triassic reconstruction of the kinematics of the Tethyan realm. In: Dercourt, J., Ricou, L.-E., Vrielinck, B. (Eds.), Atlas Tethys Palaeoenvironmental Maps. Explanatory Notes. Gauthier-Villars, pp. 55–62.

Stampfli, G.M., Borel, G., Cavazza, W., Mosar, J., Ziegler, P.A. (Eds.), 2001a. The paleotectonic atlas of the Peritethyan domain. European Geophysical Society, CD ROM.

Stampfli, G.M., Mosar, J., Favre, P., Pillevuit, A., Vannay, J.-C., 2001b. Permo-Mesozoic evolution of the western Tethyan realm: the Neotethys/East-Mediterranean connection. In: Ziegler, P.A., Cavazza, W., Robertson, A.H.F., Crasquin-Soleau, S. (Eds.), PeriTethys Memoir 6: Peritethyan Rift/Wrench Basins and Passive Margins, IGCP 369. Mém. Museum Nat. Hist. Nat. 186, 51–108.

Stampfli, G.M., Borel, G., Marchant, R., Mosar, J., 2002a. Western Alps geological constraints on western Tethyan reconstructions. J. Virtual Explorer 8, 77–106.

Stampfli, G.M., von Raumer, J., Borel, G.D., 2002b. The Palaeozoic evolution of pre-Variscan terranes: From peri-Gondwana to the Variscan collision. In: Martinez-Catalan, J.R., Hatcher, R.D., Arenas, R., Diaz Garcia, F. (Eds.), Variscan Appalachian Dynamics: The Building of the Upper Paleozoic Basement. Geol. Soc. Am. Spec. Paper 364, 263–280.

Stampfli, G.M., Marcoux, J., Baud, A., 1991. Tethyan margins in space and time. In: Channell, J.E.T., Winterer, E.L., Jansa, L.F. (Eds.), Paleogeography and Paleoceanography of Tethys. Palaeogeogr. Palaeoclimatol. Palaeoecol. 87, 373–410.

Stampfli, G.M., Mosar, J., Marchant, R., Marquer, D., Baudin, T., Borel, G., 1998. Subduction and obduction processes in the western Alps. In: Vauchez, A., Meissner, R. (Eds.), Continents and Their Mantle Roots. Tectonophysics 296 (1–2), 159–204.

Stampfli, G.M., Vavassis, I., De Bono, A., Rosselet, F., Matti, B., Bellini, M., 2003. Remnants of the Paleotethys oceanic suture-zone in the western Tethyan area. In: Cassinis, G., Decandia, F.A. (Eds.), Stratigraphic and Structural Evolution on the Late Carboniferous to Triassic Continental and Marine Successions in Tuscany (Italy): Regional Reports and General Correlation. Boll. Soc. Geol. It. spec. vol. 2, 1–24.

Tari, G., Dicea, O., Faulferson, J., Georgiev, G., Byrne, P., Mountney, N.P., 1997. Cimmerian and Alpine stratigraphy and structural evolution of the Moesian platform. In: Robinson, A.G. (Ed.), Regional and Petroleum Geology of the Black-Sea and Surrounding Region. American Association of Petroleum Geologists Memoir 68, 63–90.

Thöni, M., 1999. A review of geochronological data from the Eastern Alps. Schweiz. Miner. Petrog. Mitt. 79, 209–230.

Trümpy, R., 1988. A possible Jurassic-Cretaceous transform system in the Alps and the Carpathians. Geol. Soc. Am. Spec. Paper 218, 93–109.

Trümpy, R., 1992. Ostalpen und Westalpen-Verbindendes und Trennendes. Jb. Geol. B.-A. 135 (4), 875–882.

Vai, G.B., 1994. Crustal evolution and basement elements in the Italian area: paleogeography and characterization. Boll. Geofis. Theor. Appl. 36 (141–144), 411–434.

Vavassis, I., De Bono, A., Stampfli, G.M., Giorgis, D., Valloton, A., Amelin, Y., 2000. U–Pb and Ar–Ar geochronological data from the Pelagonian basement in Evia (Greece): geodynamic implications for the evolution of Paleotethys. Schweiz. Miner. Petrog. Mitt. 80, 21–43.

Vecsei, A., Sanders, D.G.K., Bernoulli, D., Eberli, G.P., Pignatti, J.S., 1998. Cretaceous to Miocene sequence stratigraphy and evolution of the Maiella carbonate platform margin, Italy. In: de Graciansky Pierre, C., Hardenbol, J., Jacquin, T., Vail Peter, R. (Eds.), Mesozoic and Cenozoic Sequence Stratigraphy of European Basins. In: Special Publication – Society for Sedimentary Geology, vol. 60. Society for Sedimentary Geology (SEPM), pp. 53–74.

von Raumer, J., Stampfli, G.M., Borel, G.D., Bussy, F., 2002. The organisation of pre-Variscan basement areas at the north-Gondwanan margin. Internat. J. Earth Sci. 91, 35–52.

Wortel, M.J.R., Spakman, W., 1992. Structure and dynamics of subducted lithosphere in the Mediterranean region. Proceedings of the koninklijke nederlandse Akademie van Wetenschappen. Series B: Palaeontology, Geology, Physics, Chemistry, Anthropology 95 (3), 325–347.

Ziegler, P.A., Stampfli, G.M., 2001. Late Paleozoic Early Mesozoic plate boundary reorganisation: collapse of the Variscan orogen and opening of Neotethys. In: Cassinis, R. (Ed.), The Continental Permian of the Southern Alps and Sardinia (Italy) Regional Reports and General Correlations. Annali Museo Civico Science Naturali, Brescia 25, 17–34.

CROP PROJECT: Deep Seismic Exploration of the Central Mediterranean and Italy
Edited by I.R. Finetti

CHAPTER 34

Geodynamic Evolution of the Mediterranean Region from the Permo-Triassic Ionian Opening to the Present, Constrained by New Lithospheric CROP Seismic Data

I.R. Finetti[1,*]

ABSTRACT

From the interpretation of the CROP seismic dataset a stepwide increase in knowledge of the Mediterranean Region lithosphere has been obtained. Significant data constitute now important new constraints for the reconstruction of a much better controlled geodynamic evolution. Beneath complex tectonic stacks of Tyrrhenian margin allochthonous units, the boundaries of the subducted Ionian Sea have been clearly imaged, showing that the Adria plate detached from Africa in the Permian–Early Triassic, opening the interposed Ionian Neotethys. Segmented margins of Adria and Africa, in the Ionian Tethys, can be reconnected by rotating Adria back about 40°. Seismic imagings of E-dipping slab subducted in the Eo-Alpine stage and W-dipping slab subducted in the Balearic stage of the Alpine Tethys have also been interpreted respectively beneath the NW Adria margin and the Corsica basin zone. In the Balearic stage (Oligocene–Lower Miocene), driven by W-dipping subduction of the Alpine Tethys remnant, the rotation of the Corso-Sardinian block, the opening of the back-arc homonymous basin and the generation of a single large arc which included the Northern Apennines, the Southern Apennines, the Calabrian Arc, the Sicilian Apennines and the Maghrebian chain took place. Corsica collided with NW Adria (Etrurian Promontory) and SE Sardinia collided with the Panormid Promontory. In the Tyrrhenian stage (Middle Miocene to Present), driven by WNW-dipping subduction of the Ionian lithosphere, in continuity to the end of previous stage, the orogeny of the Southern Apennines–Calabrian Arc–Sicilian Apennines, and the Tyrrhenian back-arc basin opening took place.

Keywords: Mediterranean Evolution, Permo-Triassic to Present, from CROP

1. INTRODUCTION

The geodynamics of the Mediterranean Sea have been discussed for many years. Various proposals have been made on the evolution of different sectors and stages of the Eurasia–Africa–interposed Tethyan oceans. It was soon perceived that a key point for understanding the complex geodynamics of the Mediterranean region was represented by time-space description of the Paleo and Neo-Tethyan seas evolution involved. Initially, there was no clear interpretation about what Tethys existed at different times (Sengör, 1985). Progressively there emerged an increasing agreement regarding the idea of a Paleozoic ocean (Paleotethys) north of the Cimmerian continent (Stöcklin, 1974; Sengör, 1979). To the south of this continent a *Neotethys* of Late Paleozoic-Mesozoic opening was located (Stöcklin, 1974). Based on large-scale seismic exploration data for the whole Mediterranean (Finetti et al., this volume, Chapter 1) it was found that the Ionian Sea–East Mediterranean basin is floored by an *old (Mesozoic) ocean crust* (Finetti, 1982, 1985). Successively, the Neotethyan ocean was well defined kinematically in the Tethys realm (Stampfli and Pillevuit, 1993). A third Tethyan ocean of Middle Jurassic opening (the *Alpine Tethys*), considered as an extension of the central Atlantic (Favre and Stampfli, 1992), composed the pre-Alpine Tethyan framework.

[1]Department of Geological, Environmental and Marine Sciences, Exploration Geophysics Group, University of Trieste, Via Weiss 1, 34127 Trieste, Italy
*Tel.: +39 40 5582276; e-mail: finetti@units.it (I.R. Finetti)

A well argued geodynamic reconstruction based on extrusion driven by Africa-Arabia-Europe convergence is presented by Mantovani (this volume, Chapter 32). New data in support of the opening of the Eastern Mediterranean in Permian times and reconstruction of the Mediterranean geodynamics are presented by Stampfli (this volume, Chapter 33).

Between Europe and Africa, from the Permo-Triassic Ionian opening, the Adria plate entered into the paleogeographic context and played a major role. Both Tethyan and Alpine geoscientists have dedicated much interest to this small continent. Correct understanding of the Mediterranean tectono-dynamics requires a correct regional reconstruction of its pre-orogenic paleogeography. The CROP deep seismic project provides significant new data that has helped reconstruct a much better constrained paleogeography of the pre-Alpine continent-ocean framework of the Mediterranean region (central-western sector). It is quite clearly evident from CROP data that the continental-oceanic setting established at the end of the Early Cretaceous played a dominant role in the successive Eo-Alpine, Balearic and Tyrrhenian geodynamic stages. In this chapter, some of the new and most relevant lithospheric results obtained from interpretation of the CROP seismic dataset, fully described in various chapters of this volume, are taken up and some paleogeographic schemes based on this new CROP data are proposed.

2. PERMO-TRIASSIC IONIAN OPENING

New CROP geophysical data, integrated with other geophysical-geological information, converge toward the conclusion that the Ionian–East Mediterranean originated in the Permian–Lower Triassic, when the Adria plate and Cimmerian continental blocks detached from Africa creating this interposed Neotethyan ocean basin. Regarding the Ionian opening, new CROP data show not only its boundary on the unsubducted old remnant ocean, but also where the deep buried boundaries lie beneath complex tectonic stacks of the Eastern Tyrrhenian and Northern Sicilian margins (Figure 1). On the CROP-controlled reconstruction of the Ionian boundaries (contact between the upper portion of the subducted Ionian slab and the Apulian or Sicilian continental slope, well seismically imaged) it is observed that, rotating the Adria plate back (clockwise) about 40° (Figure 1), the angled boundary A′B′C′ of Adria reconnects with the ABC angled boundary of the corresponding African sector. Clear seismic images of the Ionian boundary can be seen in different chapters of this volume (i.e. segment AB: Chapter 15 (Finetti et al., this volume); segment BC: Chapter 19 (Finetti and Del Ben, this volume); segment A′B′: Chapter 21 (Finetti and Del Ben, this volume); and segment B′C′: Chapters 12 and 21). The segment A′B′, at the end of the Eo-Alpine stage, constituted the northern boundary of the Ionian Sea, along with the 41° parallel. To define the time of opening of the Ionian Sea, careful stratigraphic studies of the *Ionian cover units tectonically transported in the Southern Apennines* (Finetti et al., this volume, Chapter 12; Ciarapica and Passeri, this volume, Chapter 11) *and in Sicily* (Finetti et al., this volume, Chapter 15) have been conducted. Significant data from these specific investigations show that the Ionian cover units which detached from the subducted Ionian slab and thrust over the Northern Apennines (known in literature as the *Lagonegrese*) are equivalent to the Ionian-derived units thrust over the Sicilian Apennines (and known in the literature as the *Imerese* and *Sicanian* units). Both allochthonous Ionian units exhibit basal depositions of Permian age, followed by Triassic to Oligocene deep water sediments. All Ionian-derived deep water basin (and slope) units of the Southern Apennines and Sicily have been named with the unifying term of Ionides (Finetti, this volume, Chapter 10).

3. ALPINE AND IONIAN NEOTETHYS PALEOGEOGRAPHY AT THE END OF THE LOWER CRETACEOUS

For the Alpine Tethys configuration (Middle Jurassic opening) the CROP seismic network does not furnish as many direct seismic images on preserved paleo-boundaries as for the Ionian Sea, but supplies various important information from which to derive indirect but realistic constraints, and some major points concerning its relationship with the Ionian Tethys. Significant new information obtained from CROP data is constituted by the identification and structural setting of the Alpine Tethys crust in the Corsica Basin–Western Tuscan archipelago area. Here the Alpine Tethys oceanic crust and asthenosphere forms a huge compressive bulge (*Etrurian Asthenospheric Bulge*) which subducts under Adria to the east and Corsica to the west (Figure 5). The CROP network, the crustal sections across the Alps (Finetti, this volume, Chapter 7), other MS lines data, and the above mentioned Tethyan bulge allow us to reconstruct an Alpine Tethys ocean basin which was fully communicating with the Ionian Tethys (Figure 2). Two important promontories delimited the passage from the Alpine Tethys and the Ionian Tethys: the *Etrurian Promontory* on NW Adria and the *Panormid Promontory* on the African plate. These two angular plate boundaries, as will be seen later on, constitute important elements for understanding the allochthonous units transported into tectonic stacks during the contractional geodynamics.

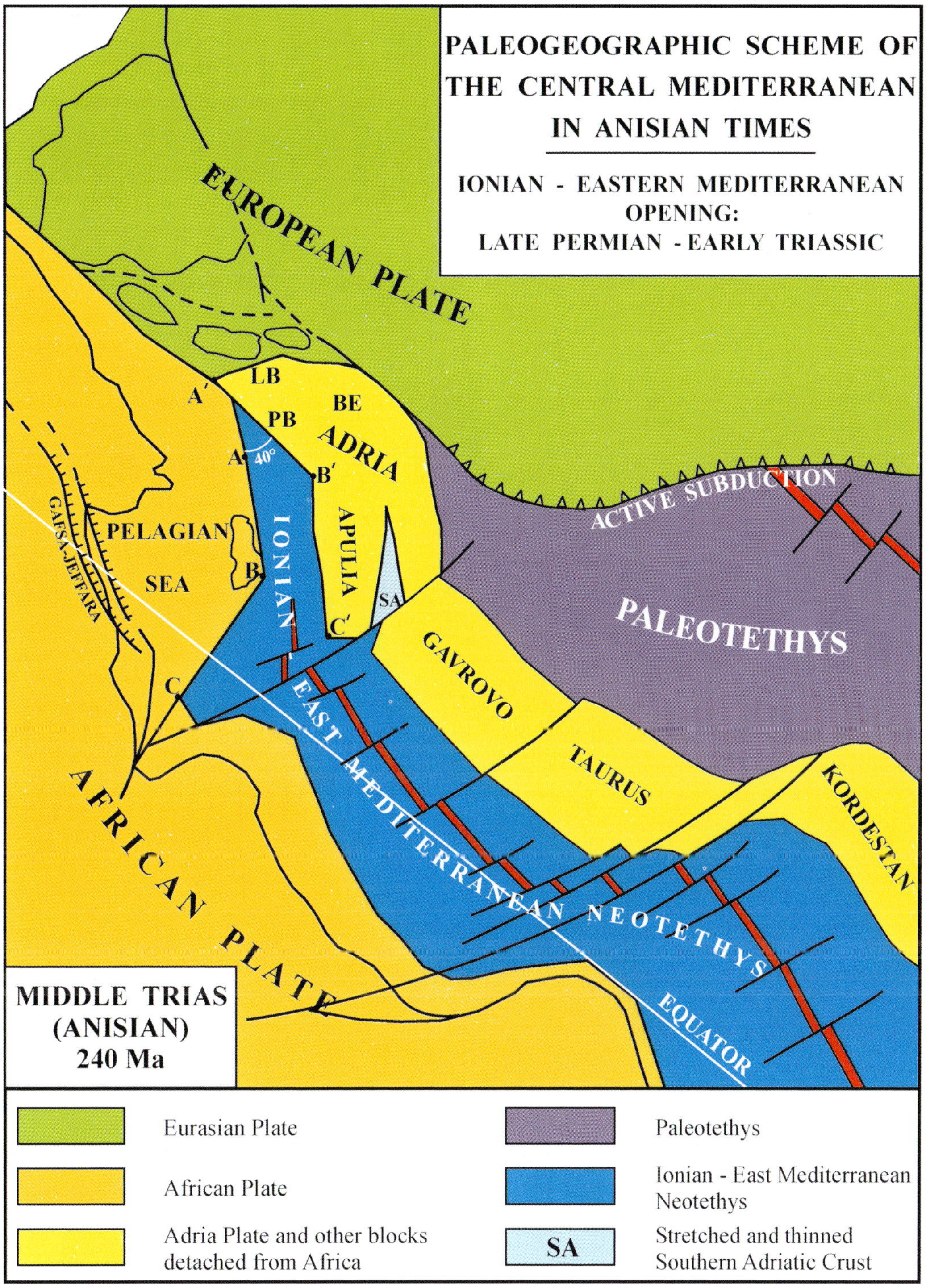

Figure 1: CROP-data-constrained reconstruction of the Ionian opening and Adria plate detachment within a schematic paleogeographic context of the central Mediterranean, referred to the Anisian (240 Ma). There is good CROP evidence that the angular points A, B and C of the Africa plate–Ionian boundary correspond to the angular points A′, B′ and C′ of the rotated Adria plate–Ionian boundary. The Adria plate is rotated anticlockwise about 40° with respect to Africa. Opening of the Ionian–Eastern Mediterranean took place from the Late Permian to Lower Triassic. The observed Permo-Triassic rift tectonics of Northern Adria (PB = Punta Bianca, LB = Lombard basin, BE = Bellunese basin) were generated by transtensive left-transcurrent faulting during Adria detachment and Ionian opening.

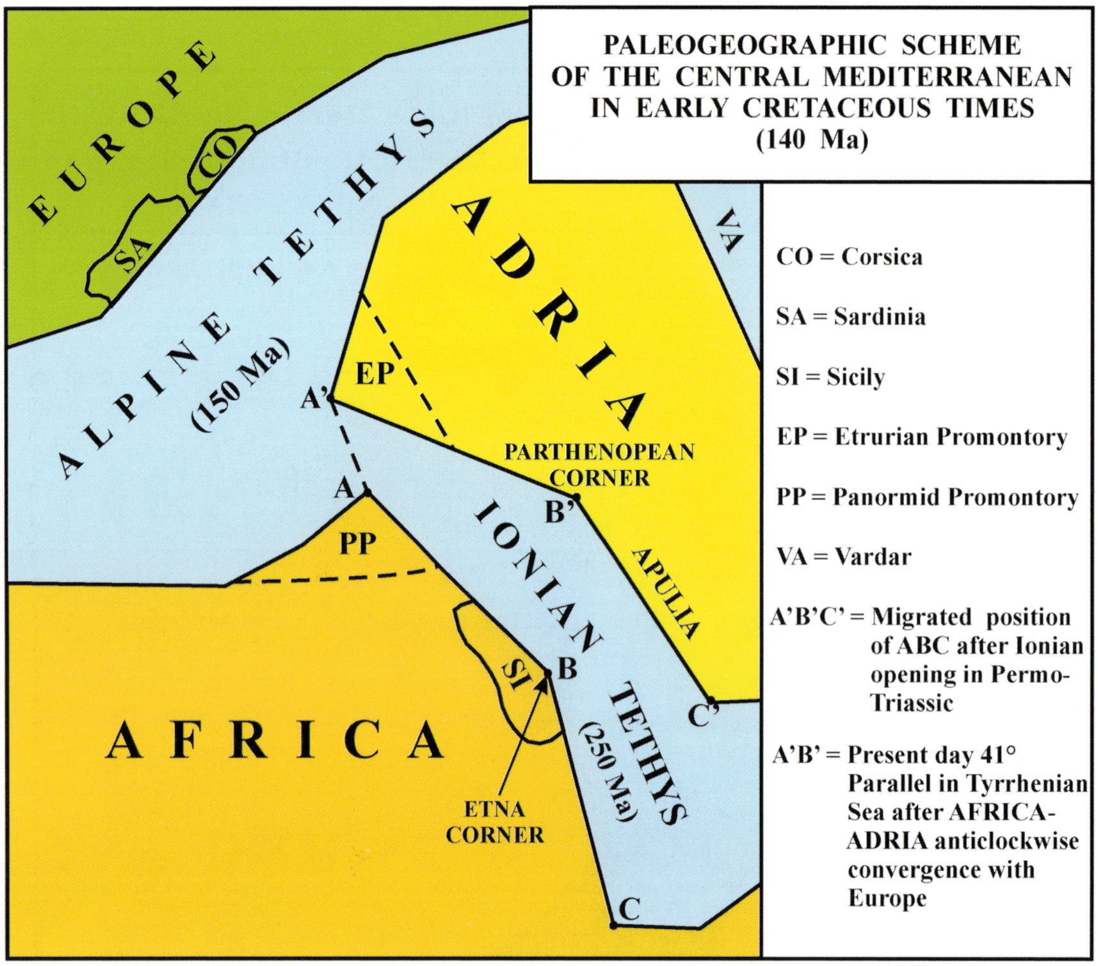

Figure 2: At the end of the Early Cretaceous, three plates (Africa, Adria and Europe), two Neotethyan ocean basins fully intercommunicating (Ionian Tethys of Permian–Early Triassic opening and Alpine Tethys of Middle Jurassic opening) composed the paleogeographic setting of the central Mediterranean region. Two promontories (Etrurian and Panormid) of significant geodynamic importance indicated the passage from the Ionian to the Alpine Tethys.

4. OROGENIC STAGES OF THE MEDITERRANEAN REGION

There is a consensus that the central-Western Mediterranean evolutive geodynamics can be subdivided into three main stages (*Eo-Alpine*, *Balearic* and *Tyrrhenian*). The Eo-Alpine stage (Upper Cretaceous–Lower Eocene) determined the Adria-Europe collision (Figure 3), the second stage (Balearic) gave rise to the rotation of the Corso-Sardinian block, with opening of the Balearic back arc basin, and formation of a wide arc (Figure 4); the third one (the Tyrrhenian) determined the subduction of large part of the Ionian Lithosphere, with formation of the Southern Apennines–Calabrian Arc–Sicilian Apennines orogenic system (Figure 6).

4.1. Eo-Alpine Stage

CROP data indicate that Adria, Africa and the interposed Ionian Tethys converged with Europe as a single block. So, in the Eo-Alpine stage, the convergence of Africa–Ionian–Adria with Europe determined the closure of the Alpine Tethys in the sector of the Alps, and restricted, but did not close, the Western Alpine Tethys, where subduction, very likely, occurred as an intra-ocean process (close to the European margin). Obduction of Alpine Tethydes units (including ophiolite-bearing units), from the Betic zone to the Alps, is Europe-verging along the whole Alpine Tethydes wedge. At the end of the Eo-Alpine stage, a wide West Mediterranean Alpine Tethys remnant was preserved (Figure 3). This explains the presence of the unsubducted Alpine Tethys ocean crust beneath thrust units in the Corsica Basin and western edge of the Tuscan Archipelago (Etrurian Bulge; Figure 5).

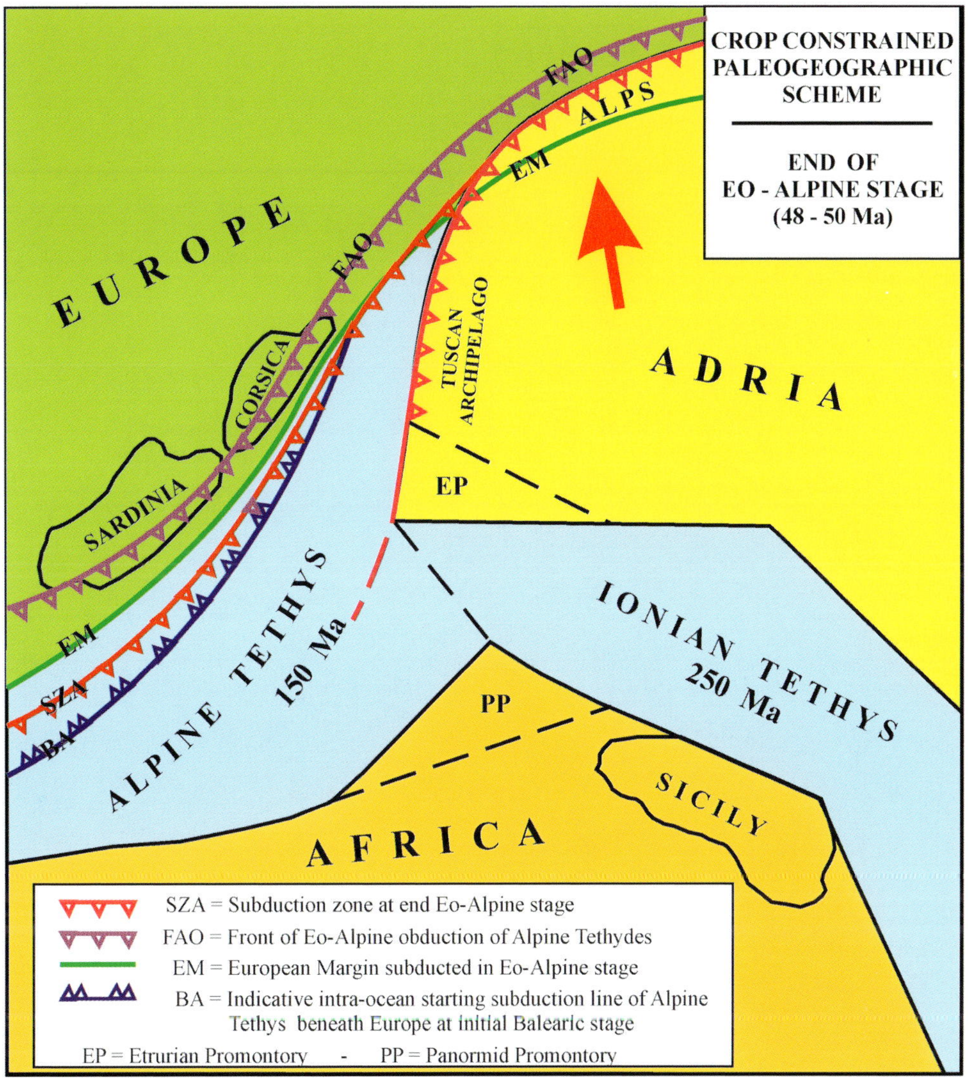

Figure 3: Paleogeographic scheme of the central Mediterranean at the end of the Eo-Alpine geodynamic stage. After closure of the Tethyan ocean in the Alps sector with Adria–Europe collision, west of the Adria plate, a wide Alpine Tethys, communicating with the Ionian, remained unsubducted. There is CROP evidence that in the Eo-Alpine stage (Upper Cretaceous–Lower Eocene) and successively, in the Balearic stage (Upper Oligocene–Lower Miocene), Africa–Adria, and the interposed Ionian Tethys converged as a unique block. The Africa–Europe convergence created an intraocean, Europe-verging, subduction on the northwestern zone of the unsubducted Alpine Tethys, which gave rise to detachment and obduction of Alpine Tethydes (including the Ophiolite) on the southern European margin (i.e. Betic segment of ophiolites). At the end of the Lower Oligocene (about 30 Ma), probably triggered by an intra-ocean Eo-Alpine backthrust, the Balearic geodynamic stage (BA) started to develop, driven by the southeastward slab retreat of the Alpine Tethys and consequent detachment-rotation of the Corsica–Sardinia block and detachment of fragments of the European margin, followed by drifting-thrusting on the Africa margin (Kabilides). In front of a southeastward migrating arc, accreted wedges of Alpine Tethydes in the Balearic stage and of Ionides in Tyrrhenian stage and back-arc basins, were formed.

4.2. Balearic Stage

Probably triggered by an intra-ocean Eo-Alpine backthrust and Africa–Europe convergence dynamics, in the Late Oligocene (about 30 Ma) the Balearic geodynamic stage (Oligocene–Lower Miocene) started to develop, driven by W-dipping subduction of the Western Alpine Tethys remnant. *At the end of the Balearic stage, a large orogenic arc (Maghrebian–Apenninic arc) including the Northern Apennines, the Southern Apennines, the Calabrian Arc, the Sicilian Apennines and the Maghrebian chain was formed* (Figure 4). Corsica was colliding with NW Adria (Tuscan Archipelago margin; Etrurian Promontory EP) and SE Sardinia entered in collision with the Panormid

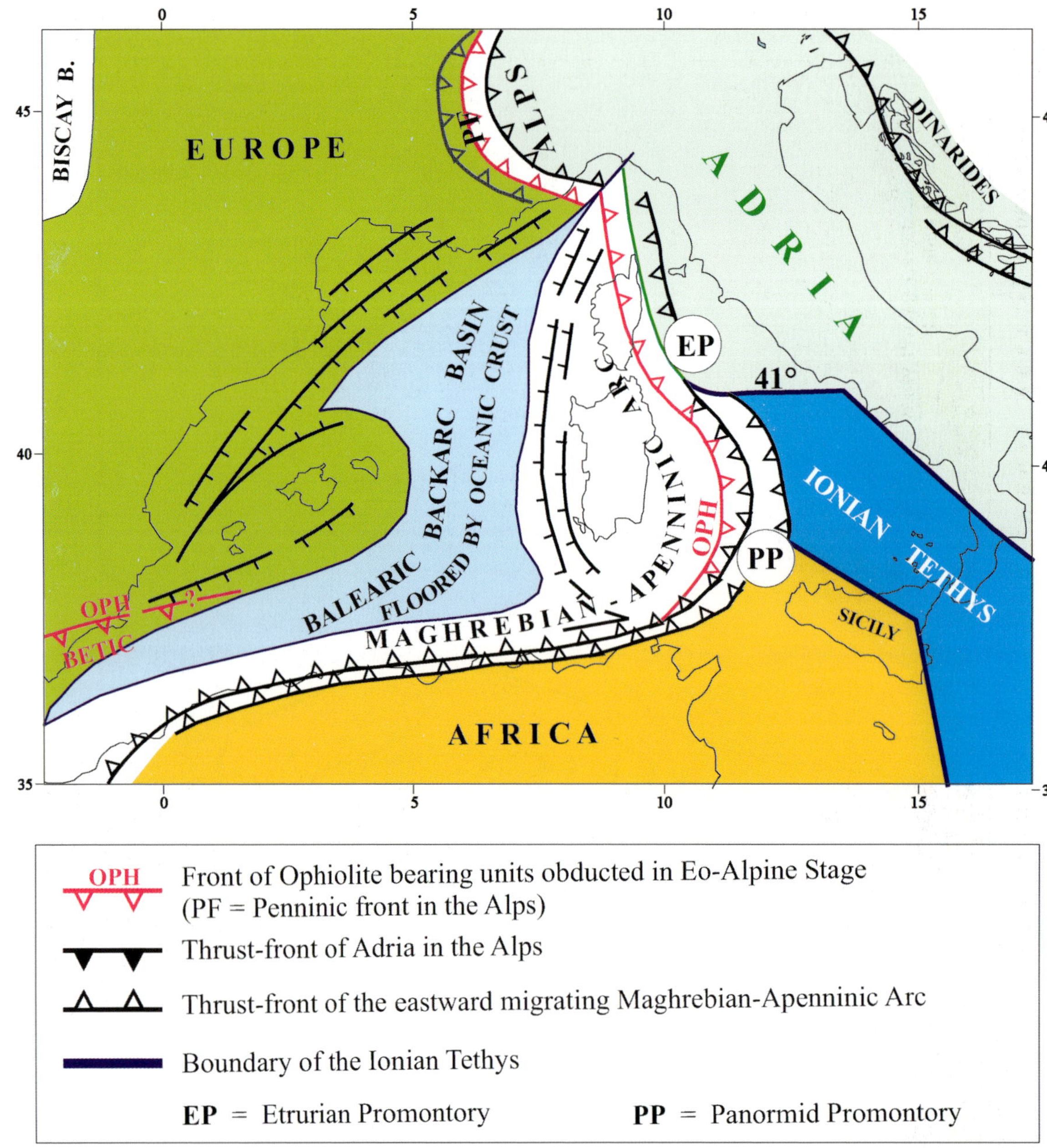

Figure 4: Paleogeographic scheme of the central-Western Mediterranean after the Balearic stage. A single wide arc included the Northern Apennines, the Southern Apennines, the Calabrian Arc, the Sicilian Apennines and the Maghrebian chain. Corsica collided with NW Adria (Etrurian Promontory) and SE Sardinian collided with the Panormid Promontory. Slab roll-back continued without stop into the Ionian Tethys (Tyrrhenian stage).

Promontory (PP) (Figure 4). Back to this arc, the Balearic basin opened, as described in other part of the Volume (Finetti et al., this volume, Chapter 20). *The CROP sections M10 and M11 image the W-dipping subduction of the Alpine Tethys remnant* (Finetti et al., this volume, Chapter 8). On the western Tuscan Archipelago, a large asthenospheric eastward-migrating compressive bulge originated (*Etrurian Asthenospheric Bulge*; Figure 5). This bulge is confined between two slabs: one E-dipping, subducted laterally beneath Adria in the Eo-Alpine stage, and the other W-dipping, subducted beneath Corsica in the Balearic stage (Finetti et al., this volume, Chapter 8). This proposed geodynamic evolution of the Balearic stage and its driving force is well supported by the CROP-based reconstruction of the asthenospheric bulge (Figure 5). After the Corsica–NW Adria collision in the Corsica basin area, and initial deformation of the internal Apennines, from the Middle Miocene to the Present, the eastward moving mantle flux related to this bulge constituted the main driving force of the continuing contractional Northern Apennines deformation in the Tyrrhenian stage.

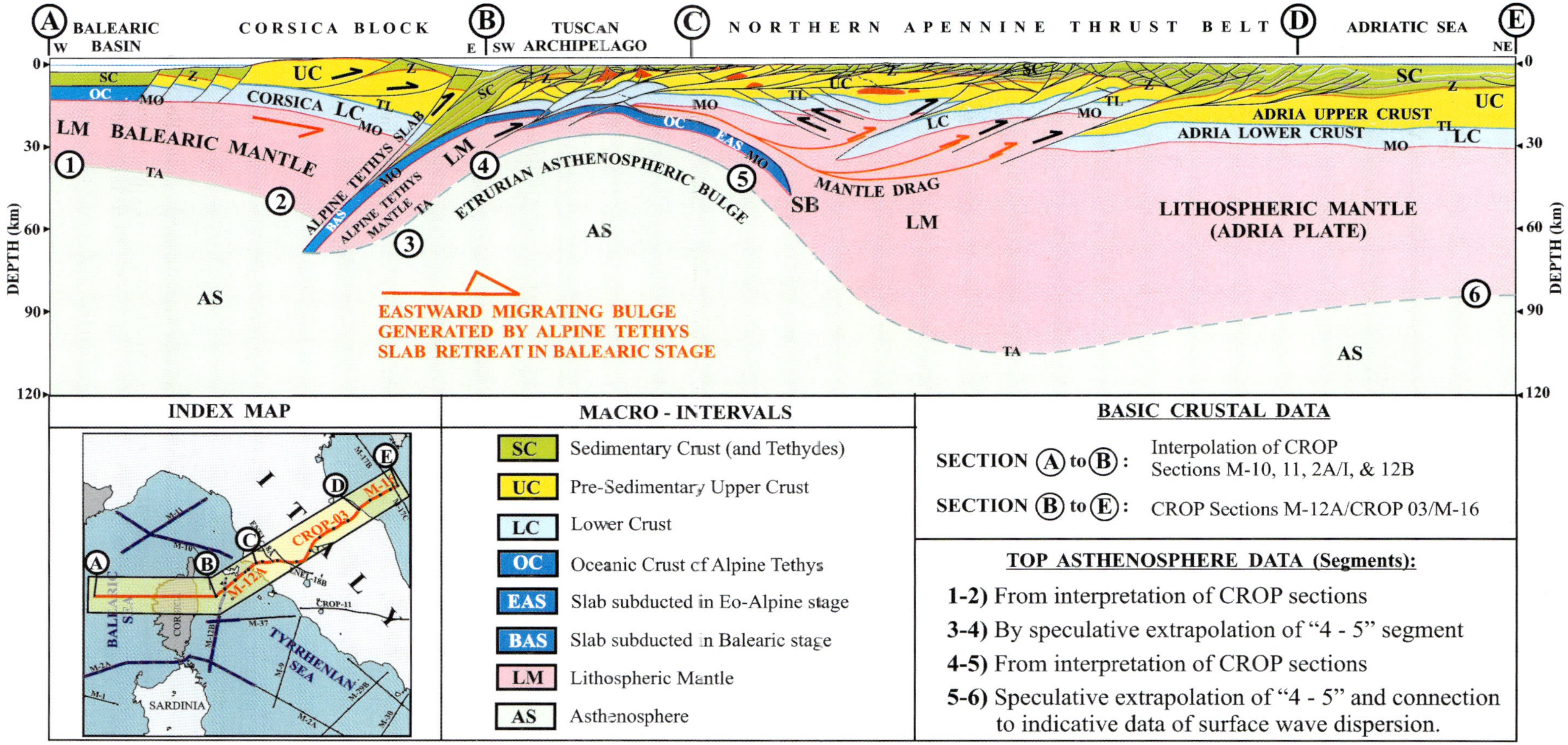

Figure 5: Basic lithospheric section from the Balearic Sea to the Adriatic foreland, across the Corsica block–Northern Apennines Thrust Belt. This reconstruction, mostly based on interpretation of new CROP seismic data, is very helpful for understanding the complex geodynamics of the Eo-Alpine, Balearic and Early Tyrrhenian stages and their driving forces. The oceanic basement crust of the Alpine Tethys gave rise to two slabs: the "EAS" to the east that subducted under the NW Adria margin in the Eo-Alpine stage, and the "BAS" to the west which subducted under the European margin in the Balearic stage. Between the two slabs, the eastward Balearic roll-back originated by compression a prominent asthenospheric bulge, which we name the "*Etrurian Asthenospheric Bulge*". The thicker blue CROP sections, or parts of sections of the index map represent the CROP network used for the lithospheric reconstruction of the interpolated segment "AB". The EAS extremity "SB" is considered the subducted southeastern boundary of the Alpine Tethys.

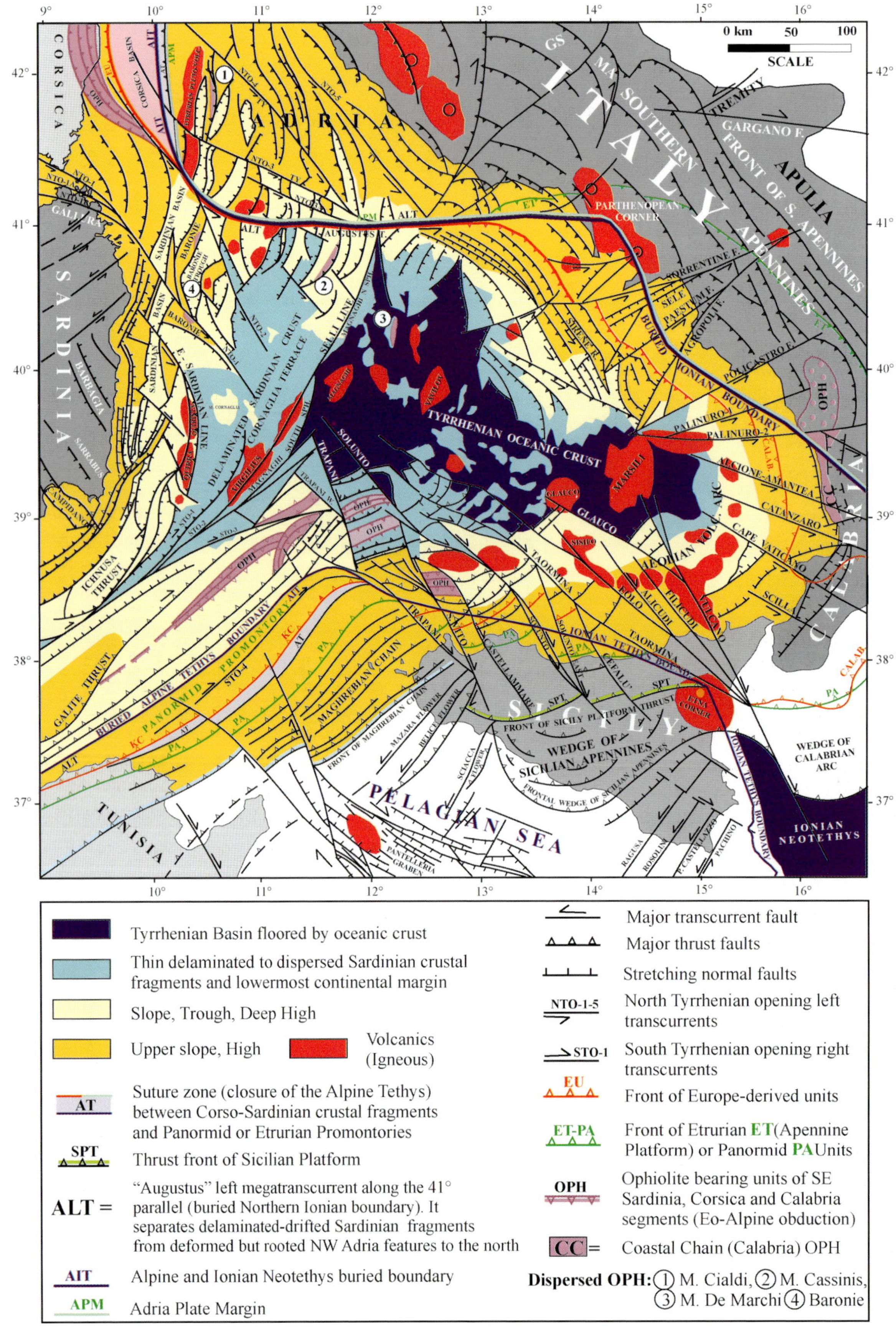

Figure 6: Present tectonic setting of the central Mediterranean region (Tyrrhenian and surrounding geological provinces).

4.3. Tyrrhenian Stage

At the end of the Balearic stage, E-migrating roll-back dynamics completed the subduction of the Alpine Tethys crust, but the slab-retreating process continued without stop into the Ionian Tethys (Tyrrhenian stage: Middle Miocene to Present). The Etrurian Promontory dissected the Northern Apennines from the Southern Apennines, the Panormid Promontory dissected the Maghrebian chain from the Sicilian Apennines and a new arc, driven by the Ionian slab roll-back, gave rise to the Southern Apennines–Calabrian Arc–Sicilian Apennines thrust belt.

Complex tectonic stacks of crustal units coming from different realms thrust over the Tyrrhenian margins of Adria and Sicily and in the Calabrian arc. In various chapters of this volume, the stacking history of these tectonic units in different orogenic provinces is described. The subduction-related Balearic and Tyrrhenian geodynamic stages were fast processes that propagated at speeds up to 10 cm/y. Simultaneously, the Africa–Europe convergence operated at an average speed of about 4 to 5 mm/y. The present tectonic setting of the central Mediterranean Region (Tyrrhenian Sea and surrounding provinces) is outlined in Figure 6. ***This geodynamic description is deliberately left schematic, thus leaving the reader to gather the much relevant new CROP data exhibited and commented on in the previous chapters of the volume for his own evolutive interpretation***.

ACKNOWLEDGEMENTS

This work was supported by MIUR (COFIN 1998, 2000 and 2002) with grants to I. R. Finetti, University of Trieste. The Author thanks all collaborating Geoscientists, Universities and Institutions that have contributed to the content of this CROP volume.

REFERENCES

Ciarapica, G., Passeri, L., this volume. Ionian Tethydes in the Southern Apennines. Chapter 11.

Favre, P., Stampfli, G., 1992. From rifting to passive margin: the Red Sea, the central Atlantic and the Alpine Tethys as examples. Tectonophysics 215, 69–97.

Finetti, I.R., 1982. Structure, stratigraphy and evolution of Central Mediterranean. Boll. Geof. Teor. Appl. 24, 247–426.

Finetti, I.R., 1985. Structure and evolution of the Central Mediterranean (Pelagian and Ionian Seas). In: Stanley, D.J., Wezel, F.C. (Eds.), Geological Evolution of the Mediterranean Basin. Springer-Verlag, pp. 215–230.

Finetti, I.R., this volume. Crustal tectono-stratigraphic sections across the Western and Eastern Alps from ECORS-CROP and Transalp seismic data. Chapter 7.

Finetti, I.R., this volume. Understanding the Ionides and their geodynamics. Chapter 10.

Finetti, I.R., Del Ben, A., this volume. Crustal tectono-stratigraphy of the Ionian Sea from new integrated CROP seismic data. Chapter 19.

Finetti, I.R., Del Ben, A., this volume. Ionian Tethys lithosphere roll-back sinking and back-arc Tyrrhenian opening from new CROP seismic data. Chapter 21.

Finetti, I.R., Del Ben, A., Forlin, E., Prizzon, A., this volume. Review of the main results of pioneering MS seismic exploration programme in the deep water of the Mediterranean Sea (OGS, 1969–1980). Chapter 1.

Finetti, I.R., Boccaletti, M., Bonini, M., Del Ben, A., Pipan, M., Prizzon, A., Sani, F., this volume. Lithospheric tectono-stratigraphic setting of the Ligurian Sea–Northern Apennines–Adriatic foreland from integrated CROP seismic data. Chapter 8.

Finetti, I.R., Lentini, F., Carbone, S., Del Ben, A., Di Stefano, A., Guarnieri, P., Pipan, M., Prizzon, A., this volume. Crustal tectonostratigraphy and geodynamics of the Southern Apennines from CROP and other integrating geophysical-geological data. Chapter 12.

Finetti, I.R., Lentini, F., Carbone, S., Del Ben, A., Di Stefano, A., Forlin, E., Guarnieri, P., Pipan, M., Prizzon, A., this volume. Geological outline of Sicily and lithospheric tectono-dynamics of its Tyrrhenian margin from new CROP seismic data. Chapter 15.

Mantovani, E., this volume. Evolutionary reconstruction of the Mediterranean Region: extrusion tectonics driven by the convergence of Africa Arabia and Eurasia. Chapter 32.

Sengör, A.M.C., 1979. Mid-Mesozoic closure of Permo-Triassic Tethys and its implications. Nature 279, 590–593.

Sengör, A.M.C., 1985. The story of Tethys: how many wives did Okeanos have. Episodes 8, 3–12.

Stampfli, G.M., Pillevuit, A., 1993. An alternative Permo-Triassic reconstruction of the kinematics of the Tethyan realm. In: Dercourt, J., Ricou, L.-E., Vrielinck, B. (Eds.), Atlas Tethys Palaeoenvironmental Maps. Explanatory Notes. Gauthier-Villars, pp. 55–62.

Stampfli, G.M., this volume. Plate tectonics of the Apulia–Adria microcontinents. Chapter 33.
Stöcklin, J., 1974. Possible ancient continental margin in Iran. In: Burk, C.A., Drake, C.L. (Eds.), The Geology of Continental Margins. Springer, pp. 873–887.

Subject Index

Abruzzo foredeep migration, 563
Adriatic CROP interpretation, 521
Adriatic crustal tectono-stratigraphy CROP controlled, 519
Adriatic Sea, 119, 519
Adriatic seismic stratigraphy, 521
Alboran Sea, 6
Alpine Neotethys opening, 103
Alps Moho, 600
Amanda 1 bis borehole, 524
Apennine platform units, 239
Apulian platform, 230, 540
Augustus line, 491, 495

Balearic Basin, 471
Balearic Basin Moho, 479
Balearic Basin CROP-ECORS data interpretation, 472
Bedrock geology, 69
Betic ophiolites, 665

Calabrian Arc, 393
Calabrian Arc seismicity, 408
Calabrian CROP interpretation, 396
Calabrian Wedge, 17, 407, 450
Capo d'Orlando Flysch, 345
Cenozoic tectono-magmatic evolution, 623
Central Apennines, 159
Central Apennines geodynamic evolution, 191
Central Apennines tectono-stratigraphy, 163
Central Apennines transcrustal section, 181
Central Italy evolutive geodynamics, 159
Central Mediterranean region grav-mag, 57
Central Sicily setting, 351
Central-Southern Adriatic CROP sections, 531
Central-Western Mediterranean, 623
Collision Corsica-NW Adria, 125, 126
Corsica basin, 128
Corso-Sardinian Block, 413
Corso-Sardinian crustal tectonics, 413
CROP constrained geodynamic evolution, 767
CROP exploration Ionian Sea, 447
CROP processing, 81
CROP seismic acquisition, 81
CROP seismic modelling in Adriatic, 549
CROP Sicily interpretation, 353
CROP-ECORS Western Alps, 109
CROP-ECORS-TRANSALP tectono-stratigraphic setting, 109
Crotone structure, 544
Crustal seismic modelling, 549
Crustal Seismogeological section Central Italy, 159

Etrurian asthenospheric bulge, 125, 126
Extrusion processes, 705

Geochemistry, 665
Geodynamc significance of Tyrrhenian morphology, 505
Gran Sasso crustal thrust, 167
Gravity field, 57
Gulf of Syrte, 25

Hellenic arc, 18, 20, 21, 22
Hellenic wedge, 17, 19, 21, 462
Hyblean Plateau, 323, 583

Igneous rocks, 69
Interpretative CROP reprocessing, 81
Ionian and Alpine Tethys opening, 103
Ionian cover seismic stratigraphy, 199
Ionian geodynamic evolution, 462
Ionian neotethys opening, 103, 198
Ionian oceanic crust, 16, 19
Ionian Sea, 447
Ionian seismic stratigraphy, 447, 450
Ionian slab, 393, 406, 409
Ionian subduction, 483
Ionian tectonic map, 411
Ionian Tethys, 209
Ionides in Southern Apennines, 209
Ionides tectonic distribution, 201
Ionides: from Ionian to South Apennines-Calabrian wedge-Sicily, 197

Jurassic Ophilites, 607

Lagonegro sequence, 200, 209, 212
Lampedusa, 586
Lampedusa high, 584
Larderello and M.Amiata plutons, 142, 143, 144
Levantine Sea, 23
Liguiran Sea, 119
Linosa graben, 24, 586
Lithospheric mantle evolution, 641
Lithospheric tectono-stratigraphy, 119

M. Cetona thrust, 139
Magnetic field, 57
Magnetotellurics Southern Apennnines, 263
Maiella thrust, 169, 172
Main pioneering results, 1
Malta graben, 24, 586
Marsili volcano, 13
Mediterranean Cenozoic volcanism, 641
Mediterranean evolution, 705, 747, 767
Mediterranean ridge, 21, 22
Messinian salt domes, 7
Metamorphism, 665
Minimum-work principle, 705

Moho depth contour map Central Mediterranean CROP controlled, 567
Morrone thrust, 169170, 171
MS seismic lines, 1
MT data acquisition-processing, 266
MT data modelling, 269
Multibeam Tyrrhenian bathymetry, 505
Multiscale derivative grav-mag analysis in Southern Apennines, 281

Northern Apennines, 119
Northern Apennines geodynamic evolution, 152
Northern Apennines geodynamics, 119, 225
Northern Apennines regional setting, 122
Northern Apennines transverse geothermal structures, 142
Northern Adriatic CROP sections, 531
Northern Calabrian Arc, 404
Numidian Flysch, 350
NV-WA CROP modelling, 377

Palinuro volcano, 14
Panormide units, 339
Pantelleria rift basin, 584
Paola basin, 15
Paxos Isle structure, 545
Pelagian CROP sections interpretation, 582
Pelagian foreland, 581
Pelagian Sea, 24, 581
Pelagian Sea Moho, 593
Pelagian Sea tectonics, 594
Permo-Triassic to Present evolution, 767
Plate tectonic context, 747
Pliocene evolution of Abruzzo foredeep, 570
Processing, 81

Radiometric dating, 665

Southern Apennines geodynamic evolution, 255
Southern Apennines seismogeological secion, 247
Sardinian rift system, 423
Sardinia-related CROP interpretation, 433
Sciacca platform, 325
Sciacca strike-slip fault, 583 590
Seismic modelling, 549
Seismic modelling technique, 552
Seismogeological section Southern Apennines, 225
Sicanian units, 332
Sicilian Alpine Tethydes, 340
Sicilian Calabride Chain, 343
Sicilian Ionides, 332
Sicily, 319
Sicily Channel rift zone, 24, 582
Sicily foreland, 321
Sicily geologic outline, 319
Sicily lithospheric tectono-dynamics, 319
Sicily orogenic domain, 327
Siena basin, 137, 138
South Apennines chain units, 233
Southeast Apulia CROP sections, 540
Southern Adriatic basin, 26
Southern Adriatic structure, 547
Southern Apennines, 225, 263
Southern Apennines aeromagnetics, 303
Southern Apennines gravity data, 285
Southern Calabrian Arc, 397
Southern Europe Moho, 600

Tectono-dynamics of Pelagian Sea, 589
Tectono-dynamics of Tyrrhenian Sea, 483
Tectonostratigraphic units of Sicily, 312
Thrust propagation, 225
TRANSALP Eastern Alps, 109
Trench-arc-back arc systems, 705
Triassic magmatism, 607
Triassic-Jurassic, 209
Tyrrhenian margin of Sicily, 319
Tyrrhenian morphology, 505
Tyrrhenian opening, 483
Tyrrhenian opening geodynamics, 501
Tyrrhenian Sea, 27, 69, 483
Tyrrhenian Sea CROP interpretation, 485
Tyrrhenian Sea tectonics, 500
Tyrrhenian seafloor multibeam, 505
Tyrrhenian sonobuoy, 8, 9, 10

Umbria-Marche seismicity, 149

Vavilov basin, 11
Vavilov volcano, 12
Vertical/wide-angle seismics, 377

WA acquisition-processing, 378
Western-Eastern Alps, 109
Wide-angle reflection-refraction, 31
Within-plate Cenozoic volcanism, 641

Author Index

Beccaluva L., 607, 623, 641
Bernardelli P., 57
Bianchini G., 623, 641
Boccaletti M., 119
Bonadiman C., 641
Bonini A., 119

Calamita F., 159
Carbone S., 225, 319
Carcione J.M., 549
Carruba S., 563
Casnedi R., 563
Cassinis R., 31
Cavalli C., 57
Cella F., 281
Ciarapica G., 209
Coltorti M., 607, 623, 641
Crescenti U., 159

Del Ben A., 1, 119, 159, 225, 319, 377, 413, 447, 483, 519, 581
De Luca L., 377
Di Fiore B., 263
Di Stefano A., 225, 319

Fais S., 413
Fedi M., 281
Finetti I.R., 1, 81, 103, 109, 119, 159, 197, 225, 319, 377, 393, 413, 447, 471, 483, 519, 549, 581, 595, 767
Florio G., 281
Forlin E., 1, 81, 159, 319, 377, 413, 471

Gei D., 549
Giori I., 57
Guarnieri P., 225, 319

Improta L., 263

Kingelé E., 413

Lecca L., 413
Lentini F., 225, 319
Longoni R., 57
Lozej A., 31
Luzio D., 377

Maciotta G., 641
Mantovani E., 705
Marani M., 505

Passeri L., 209
Patella D., 263
Perotti C.R., 563
Petrillo Z., 263
Pipan M., 81, 119, 159, 225, 319, 377, 413, 471
Prizzon A., 1, 119, 159, 225, 319, 377, 413
Puga E., 665

Rapolla A., 281
Rusciadelli G., 159

Saccani E., 607
Sani F., 119
Sartori R., 69
Scarascia S., 31
Scisciani V., 159
Siena F., 607, 623, 641
Siniscalchi A., 263
Stampfli G.M., 747

Vaccaro C., 641
Verde M., 623
Vitale M., 377

Zeda O., 607